PHYSICAL GEOLOGY TODAY

PHYSICAL GEOLOGY TODAY

Damian Nance
Ohio University

Brendan Murphy
St. Francis Xavier University

NEW YORK OXFORD
OXFORD UNIVERSITY PRESS

Oxford University Press is a department of the University of Oxford.
It furthers the University's objective of excellence in research,
scholarship, and education by publishing worldwide.

Oxford   New York
Auckland   Cape Town   Dar es Salaam   Hong Kong   Karachi
Kuala Lumpur   Madrid   Melbourne   Mexico City   Nairobi
New Delhi   Shanghai   Taipei   Toronto

With offices in
Argentina   Austria   Brazil   Chile   Czech Republic   France   Greece
Guatemala   Hungary   Italy   Japan   Poland   Portugal   Singapore
South Korea   Switzerland   Thailand   Turkey   Ukraine   Vietnam

For titles covered by Section 112 of the US Higher Education Opportunity Act,
please visit www.oup.com/us/he for the latest information about
pricing and alternate formats.

Published by Oxford University Press
198 Madison Avenue, New York, NY 10016
http://www.oup.com

Oxford is a registered trademark of Oxford University Press.

Library of Congress Cataloging-in-Publication Data
Nance, Damian.
Physical geology today / Damian Nance, Ohio University,
    Brendan Murphy, St. Francis Xavier University.
       pages cm
    ISBN 978-0-19-996555-7
    1. Physical geology—Textbooks. I. Murphy, Brendan E. II. Title.
    QE28.2.N36 2016
    551—dc23
                                                      2015016494

Printing number: 9 8 7 6 5 4 3 2 1

Printed in the United States of America
on acid-free paper

Brief Table of Contents

PART **I**   **The Solid Earth** 2

Chapter 1   **Foundations of Modern Geology** 4

Chapter 2   **Plate Tectonics** 38

Chapter 3   **Minerals** 68

Chapter 4   **Origin and Evolution of Igneous Rocks** 98

Chapter 5   **Weathering and Soils** 136

Chapter 6   **Sedimentation and Sedimentary Rocks** 162

Chapter 7   **Metamorphism and Metamorphic Rocks** 194

PART **II**   **Evolution of the Solid Earth** 222

Chapter 8   **Geologic Time** 224

Chapter 9   **Plates and Plate Boundaries** 254

Chapter 10   **Deformation and Mountain Building** 296

Chapter 11   **Earthquakes and Earth's Interior** 334

PART **III**   **Sculpting the Solid Earth** 368

Chapter 12   **Mass Wasting** 370

Chapter 13   **Running Water** 400

Chapter 14   **Groundwater** 436

Chapter 15   **Glaciers and Glaciation** 464

Chapter 16   **Deserts and Winds** 500

Chapter 17   **Coastlines and Coastal Processes** 528

PART **IV**   **Earth's Resources** 564

Chapter 18   **Mineral Deposits and Industrial Materials** 566

Chapter 19   **Energy Resources** 604

PART **V**   **Earth and Beyond** 642

Chapter 20   **Physical Geology and Climate Change** 644

Chapter 21   **The Planets** 678

Contents

Preface xiii

 **PART I The Solid Earth**

Chapter 1 Foundations of Modern Geology 4

1.1 Geology and the Scientific Method 6
1.2 The Birth of Geology 9
1.3 Geologic Time 10
1.4 Earth Materials: Minerals, Rocks, and Fossils 13
1.5 The Rock Cycle 14
1.6 Origin of Earth and the Solar System 19
1.7 Earth's Internal Heat 27
1.8 Earth as a System 30
1.9 What Does a Geologist Do? 34

Key Terms 36
Key Concepts 36
Study Questions 37

SCIENCE REFRESHER: Some Basic Concepts
in Chemistry and Physics 16

IN DEPTH: Origin of the Universe—The Big Bang
Theory 20

LIVING ON EARTH: Earth's Seasons—A 4.5-Billion-
Year Legacy? 22

SCIENCE REFRESHER: The Flow of Heat 28

Chapter 2 Plate Tectonics 38

2.1 Continental Drift 40
2.2 Paleomagnetism 46
2.3 Seafloor Spreading 50
2.4 Subduction 54
2.5 Moving Plates and Plate Boundaries 58
2.6 Hotspots: A Plate Tectonic Enigma 64

Key Terms 65
Key Concepts 66
Study Questions 66

LIVING ON EARTH: The Remarkable Journey
of the Green Turtle 53

IN DEPTH: Failed Rifts 62

Chapter 3 Minerals 68

3.1 Elements: Building Blocks of Minerals 70
3.2 Minerals: Orderly Expressions of Matter 77
3.3 Physical Properties of Minerals 82
3.4 Silicate Rock-Forming Minerals 87
3.5 Nonsilicate Rock-Forming Minerals 91
3.6 Minerals and Rocks 93
3.7 Minerals and People 93
3.8 Minerals and Plate Tectonics 95

Key Terms 96
Key Concepts 96
Study Questions 97

LIVING ON EARTH: Origin of the Chemical
Elements 74

SCIENCE REFRESHER: Some Basic Definitions
in Science 76

SCIENCE REFRESHER: Solids, Liquids,
and Gases 77

**Chapter 4 Origin and Evolution
of Igneous Rocks** 98

4.1 Igneous Rocks and the Rock Cycle 100
4.2 Magma Formation and Transport 101
4.3 Textures of Igneous Rocks 104
4.4 Classifying Igneous Rocks 113
4.5 Evolution of Igneous Rocks 116
4.6 Volcanic Eruptions 121
4.7 Igneous Rocks and Plate Tectonics 131

Key Terms 134
Key Concepts 134
Study Questions 135

SCIENCE REFRESHER: Heat, Temperature,
and Magma 104

LIVING ON EARTH: Volcanoes and the Air
We Breathe 108

IN DEPTH: Can Explosive Eruptions Be
Predicted? 130

Chapter 5 Weathering and Soils 136

5.1 Weathering 138
5.2 Mechanical Weathering 139
5.3 Chemical Weathering 143
5.4 Soils 149
5.5 Weathering, Soils, and Plate Tectonics 157

Key Terms 160
Key Concepts 160
Study Questions 160

SCIENCE REFRESHER: What Is an Acid? 144
IN DEPTH: The Role of Oxidation in Earth's
    History 148
LIVING ON EARTH: The Food Chain 152

**Chapter 6 Sedimentation and Sedimentary
              Rocks** 162

6.1 From Sediment to Sedimentary Rock 164
6.2 Types of Sedimentary Rocks 166
6.3 Classifying Sedimentary Rocks 174
6.4 Interpreting Sedimentary Rocks 179
6.5 Depositional Environments and Sedimentary
     Facies 185
6.6 Sedimentation, Sedimentary Rocks,
     and Plate Tectonics 190

Key Terms 192
Key Concepts 192
Study Questions 193

LIVING ON EARTH: Salt of the Earth 173
LIVING ON EARTH: King Coal 178
SCIENCE REFRESHER: Settling Velocities 182

**Chapter 7 Metamorphism and Metamorphic
              Rocks** 194

7.1 Metamorphic Rocks and the Rock Cycle 196
7.2 Factors That Control Metamorphism 198
7.3 Types of Metamorphism 204
7.4 Metamorphic Textures 207
7.5 Classifying Metamorphic Rocks 209
7.6 Metamorphic Zones and Facies 214
7.7 Metamorphism and Plate Tectonics 217

Key Terms 219
Key Concepts 220
Study Questions 221

SCIENCE REFRESHER: Pressure Units 200
IN DEPTH: Shock Metamorphism 206
IN DEPTH: Metamorphism and the Rock Cycle 215

PART II Evolution of the Solid Earth

Chapter 8 Geologic Time 224

8.1 Measuring Geologic Time: Relative
     and Absolute Ages 226
8.2 A Short History of Geologic Time 227
8.3 Relative Dating: Determining Chronological
     Order 230
8.4 Correlating Rock Strata 234
8.5 Relative Dating and the Geologic Time
     Scale 237
8.6 Absolute Dating: Finding a Geologic
     Clock 241
8.7 Absolute Dating and the Geologic Time
     Scale 248
8.8 A Sense of Time 250
8.9 Geologic Time and Plate Tectonics 251

Key Terms 251
Key Concepts 252
Study Questions 252

LIVING ON EARTH: Radioactivity and Radon Gas 242
SCIENCE REFRESHER: Types of Radioactive
    Decay 244
SCIENCE REFRESHER: Linear versus Exponential
    Relationships 247
IN DEPTH: The Quest for the Age of Earth 249

Chapter 9 Plates and Plate Boundaries 254

9.1 Plates on Earth's Surface 256
9.2 Plates and Isostasy 258
9.3 Plate Boundaries 261
9.4 Divergent Boundaries: Creating Oceans 263
9.5 Convergent Boundaries: Recycling Crust
     and Building Continents 269
9.6 Transform Boundaries: Fracturing
     the Crust 280
9.7 Hotspots: Tracking Plate Movements 286
9.8 Plate Tectonics and Plate-Driving
     Mechanisms 292

Key Terms 293
Key Concepts 294
Study Questions 294

LIVING ON EARTH: The Bizarre World of the
    Mid-Ocean Ridges 267
LIVING ON EARTH: Tsunami 272
IN DEPTH: Ophiolites—Clues to the Structure of
    Oceanic Crust 279

Chapter 10 Deformation and Mountain Building 296

10.1 Deformation 298
10.2 Orientation of Geologic Structures 302
10.3 Fractures 304
10.4 Folds 309
10.5 Unconformities Revisited 313
10.6 Plate Tectonics and Mountain Building 315
10.7 Mountain Building in the Geologic Past 327

Key Terms 332
Key Concepts 333
Study Questions 333

IN DEPTH: Partial Melting and the Making of Continental Crust—Nature's Ultimate Recycling Program 317
LIVING ON EARTH: In the Shadow of the Himalayas 328

Chapter 11 Earthquakes and Earth's Interior 334

11.1 Heat and Density: Clues to Earth's Interior 336
11.2 Earthquakes and Elastic Rebound 337
11.3 Seismic Waves 337
11.4 Journey to the Center of the Earth 347
11.5 Earth's Structure and Composition 352
11.6 Circulation in Earth's Interior 359
11.7 Plate Tectonics and the Fate of Subducted Slabs 365

Key Terms 366
Key Concepts 367
Study Questions 367

LIVING ON EARTH: Predicting Earthquakes—Some Lessons from California 340
SCIENCE REFRESHER: Density, Volume, and Seismic Velocity 344
SCIENCE REFRESHER: Wave Refraction 349
IN DEPTH: Earth's Magnetism and Its Dynamic Core 357

PART III Sculpting the Solid Earth

Chapter 12 Mass Wasting 370

12.1 Mass Wasting: Downslope Movement 372
12.2 Factors That Influence Mass Wasting 373
12.3 Mass Wasting Mechanisms 380
12.4 Minimizing Mass Wasting Hazards 395
12.5 Mass Wasting and Plate Tectonics 397

Key Terms 398
Key Concepts 399
Study Questions 399

SCIENCE REFRESHER: Surface Tension, Sand Castles, and the Goldilocks Solution 378
IN DEPTH: Mass Wasting on the Moon and Mars 386
LIVING ON EARTH: The Oso Landslide, Washington State 390

Chapter 13 Running Water 400

13.1 Running Water and the Hydrologic Cycle 402
13.2 Streamflow 404
13.3 Stream Erosion 409
13.4 Stream Transport 418
13.5 Stream Deposition 423
13.6 Floods and Flood Prevention 428
13.7 Running Water and Plate Tectonics 432
13.8 Running Water on Other Worlds 433

Key Terms 434
Key Concepts 435
Study Questions 435

LIVING ON EARTH: The Death of the Aral Sea— An Environmental Disaster 419
IN DEPTH: Dams and the Human Exploitation of Surface Water 430

Chapter 14 Groundwater 436

14.1 Groundwater: A Vital Resource 438
14.2 The Water Table 439
14.3 Aquifers and Groundwater Flow 442
14.4 The Dissolving Power of Groundwater 451
14.5 Exploitation of Groundwater 457
14.6 Groundwater and Plate Tectonics 462

Key Terms 462
Key Concepts 462
Study Questions 463

SCIENCE REFRESHER: Water Pressure at Depth 444
LIVING ON EARTH: The Historical Importance of Aquifers 447
LIVING ON EARTH: Water in the Sahara Desert 450
IN DEPTH: Exploring for Groundwater 456

Chapter 15 **Glaciers and Glaciation** 464

15.1 Glaciers and Glacier Ice 466
15.2 Glacial Erosion 471
15.3 Glacial Deposition 476
15.4 The Pleistocene Ice Age 485
15.5 Plate Tectonics and Glaciation 492
15.6 Ice on Other Worlds 496

Key Terms 497
Key Concepts 498
Study Questions 498

SCIENCE REFRESHER: Air Temperature in the Lower Atmosphere 468
LIVING ON EARTH: The Origin of the Great Lakes 477
IN DEPTH: Lake Missoula and the Channeled Scablands 490

Chapter 16 **Deserts and Winds** 500

16.1 The Origins of Deserts 502
16.2 Weathering and Erosion in Deserts 508
16.3 Wind in Deserts 511
16.4 Desertification: Natural and Human Induced 521
16.5 Deserts and Plate Tectonics 523
16.6 Deserts on Mars 524

Key Terms 525
Key Concepts 526
Study Questions 526

SCIENCE REFRESHER: Ocean Currents and the Coriolis Effect 506
IN DEPTH: Dust Devils 512
LIVING ON EARTH: The Dust Bowl 515

Chapter 17 **Coastlines and Coastal Processes** 528

17.1 Coastal Processes 530
17.2 Wind-Driven Waves 532
17.3 Tides 537
17.4 Coastal Erosion 540
17.5 Coastal Deposition 544
17.6 Coastlines 549
17.7 Coastal Management 551
17.8 Coastlines and Plate Tectonics 559

Key Terms 561
Key Concepts 561
Study Questions 562

SCIENCE REFRESHER: Wave Terminology 533
LIVING ON EARTH: Sea Level and Polar Ice 547
LIVING ON EARTH: Hurricanes and Storm Surges 553
IN DEPTH: Sea Level and Plate Tectonics 558

PART IV Earth's Resources

Chapter 18 **Mineral Deposits and Industrial Materials** 566

18.1 Consumption of Minerals and Industrial Materials 568
18.2 Metals and Metallic Minerals 569
18.3 Nonmetallic Minerals and Industrial Materials 575
18.4 Formation of Mineral Deposits 582
18.5 Plate Tectonics and Mineral Deposits 588
18.6 Mineral Exploration 592
18.7 Future Demands on Mineral Resources 596

Key Terms 602
Key Concepts 602
Study Questions 603

IN DEPTH: Mechanisms of Mineral Concentration 583
LIVING ON EARTH: The Environmental Impact of Mining 600

Chapter 19 **Energy Resources** 604

19.1 Energy Use 606
19.2 Petroleum—Oil and Natural Gas 608
19.3 Coal 622
19.4 Nuclear Power 625
19.5 Renewable Energy 629
19.6 Plate Tectonics and Energy Resources 637

Key Terms 641
Key Concepts 641
Study Questions 641

IN DEPTH: Correlation and the Search for Oil and Gas 614
LIVING ON EARTH: The Dilemma of Fracking 618
IN DEPTH: Is Ice That Burns the Fuel of the Future? 621

PART V Earth and Beyond

Chapter 20 Physical Geology and Climate Change 644

20.1 Taking Earth's Pulse 646

20.2 The Greenhouse Effect: A Matter of Balance 649

20.3 The Carbon Cycle 652

20.4 Climate Change: A Geologic Perspective 654

20.5 Earth-Sun Geometry 668

20.6 Feedback Systems 670

20.7 Plate Tectonics and Climate Change 673

20.8 Putting It All Together: The Greenhouse Effect and Global Warmth 676

Key Terms 676
Key Concepts 677
Study Questions 677

LIVING ON EARTH: Camels in the Arctic! 648

SCIENCE REFRESHER: The Physics of Global Warming 652

IN DEPTH: Melting of the Arctic Ice—An Omen of Global Warming? 666

Chapter 21 The Planets 678

21.1 Earth and Its Moon 680

21.2 The Terrestrial Planets and Their Moons 686

21.3 Plate Tectonics: A Planetary Perspective 696

21.4 The Jovian Planets, Their Moons, and the Kuiper Belt 698

21.5 Asteroids, Meteors, Meteorites, and Comets 705

21.6 Exoplanets and Other Solar Systems 711

Key Terms 712
Key Concepts 712
Study Questions 713

SCIENCE REFRESHER: Center of Mass 685

IN DEPTH: Martian Life on Earth? 695

LIVING ON EARTH: Visits from the Celestial Junkyard 707

Appendix A: SI and Customary Units and Their Conversions 714

Appendix B: Periodic Table of the Elements 715

Glossary 716

Credits 736

Index 739

Preface

We live in amazing times. In the past 20 years we have learned much about our planetary home. We can scarcely glance at a newspaper today without reading about new and exciting discoveries that relate to Earth and our position in the cosmos. At the same time, we are receiving mixed messages about the state of our local and global environment. The information is coming at us at such a bewildering pace that we struggle to assimilate it all.

As we navigate through the stormy waters of the twenty-first century, we need to remind ourselves that the world of 100 years ago was very different from the world of today. Our forebears could never have visualized the scientific and technological advances of the twentieth century. We now have a comprehensive understanding of the origin of mountains and oceans, and explanations for earthquakes and volcanic activity. We have devised ingenious methods that allow us to image Earth's interior, discover its composition, and understand some of the deep-seated processes that occur thousands of kilometers below the surface. Our forays into space have given us a new appreciation of the tiny portion of the universe we occupy. At the same time, we have come to realize that our planet is fragile and that our own actions may compromise our very existence.

The mission of this textbook is to convey the wonder and understanding of our planet. As authors, we are passionate about geology and hope that this comes across in this textbook. Each of you has a daily impact on the planet. Cumulatively, this impact may be enormous. One day, you will be making decisions, big and small, that will have an impact on the local, regional, or even global environment. We hope this book, in some small way, will have helped you make the right decisions.

Whether you choose a career in earth sciences or another field, some knowledge of Earth processes is essential in today's world, and a healthy respect and admiration for Mother Nature is a great calming influence when the waters of life get rough. Our challenge is to capture the excitement and wonder of our planet in these printed pages. So, welcome aboard, and if you are sitting comfortably, we'll begin.

Key Themes and Organization of the Text

The goal of this text is to provide you with an understanding and appreciation of planet Earth. A thorough grasp of any science requires some basic knowledge of facts and terminology and familiarity with the language of the science. Geology is about more than key terms; it is about understanding the processes that shape the planet. Understanding these processes will enable you to enjoy the science and appreciate what makes geology fascinating.

Our textbook presents concepts, processes, and principles in a straightforward manner that explains the linkages among them. We use the principles of plate tectonics, the fundamental unifying concept of geology. Plate tectonics lays the foundation for understanding our dynamic Earth. We first discuss plate tectonics in Chapter 2, and every subsequent chapter has a section demonstrating the relationship between the chapter content and plate tectonic processes. In this way, the content of each chapter is viewed in a plate tectonic context, so you can see how the various processes described relate to one another.

The book is divided into five parts:

Part I—The Solid Earth
Part II—Evolution of the Solid Earth
Part III—Sculpting the Solid Earth
Part IV—Earth's Resources
Part V—Earth and Beyond

This structure begins with foundational geology concepts and moves outward from Earth's core, investigates surface processes, resources, and environmental issues, then explores challenges such as the greenhouse effect, global warming, and climate change.

The book has many features to facilitate an understanding of geology.

- **Plate tectonic focus.** Plate tectonics, the fundamental unifying concept of geology, forms the foundation for much of our understanding of how Earth works. The book employs plate tectonics to demonstrate the relationships between the content of all chapters. After Chapter 2, where plate tectonics is first introduced, every chapter features a section demonstrating the relationship between the chapter content and plate tectonic processes.

- **Emphasis on conceptual understanding.** The book emphasizes concepts, processes, and principles over facts and excess terminology. Presenting concepts and processes in a straightforward manner and illustrating the connections between them allows for a deeper, more durable understanding and supports our plate tectonics theme.

- **Art.** The art program was meticulously developed to augment and support the book's plate tectonics, process, and concept themes. Furthermore, every chapter begins with a two-page overview graphic that preludes and summarizes the various topics and processes to be covered in the upcoming discussion. Portions of the chapter-opening overview graphic are then examined more closely throughout the chapter.

- **Boxed features.** The chapter discussion is broadened by a number of boxed features.

  - *Living on Earth* case studies highlight the many intersections between everyday life and the environment, and emphasize the reasons why the chapter content is important and relevant.

  - *Science Refresher* boxes review key scientific concepts that are critical to understanding the topic at hand.

  - *In Depth* features provide additional background and detail about special topics that are relevant to each chapter.

- **Chapter pedagogy.** Every chapter is supported by a variety of features designed to help students progress smoothly and effectively through the material:

  - *Check Your Understanding* questions throughout the text encourage students to review key points in context.

  - *Key Terms* in every chapter provide a useful review of important vocabulary.

  - Each chapter concludes with a summary of *Key Concepts* and *Study Questions* to facilitate review of the material.

Teaching and Learning Package

This book is supported by an extensive and carefully developed set of ancillary materials designed to support both professors' and students' efforts in the course.

- *Oxford University Press Animation Series.* Animation and visualization are very helpful when studying Physical Geology. Recognizing this, the authors have worked with leading animators to produce clear, dramatic, and illustrative animations and visualizations of some of the most important concepts in Physical Geology. Animations are available to adopting instructors at no charge.

- *Digital Files and PowerPoint Presentations.* Instructors will find all the animations—the Oxford University Press Animation Series, all of the images from the text, and some animations and visualizations from other sources—available to them, pre-inserted into PowerPoint. In addition, our ancillary author team has created

suggested lecture outlines, arranged by chapter, in PowerPoint. These materials are free to adopting professors.

- *Interactive Animation and Visualization Exercises.* Our animation and visualization exercises begin with the Oxford University Press Animation Series and incorporate interesting animations and visualizations from other sources. The exercises then guide the students through a series of activities based on the visualization at hand. Responses are automatically graded by the computer.

- *Review Questions for Students.* Carefully crafted to highlight the most important concepts in each chapter, these computer-graded review questions accompany each chapter in the textbook. Professors can assign them for homework or students can use them independently to check their understanding of the topics presented in the book. Responses are automatically graded by computer.

- *Test Questions and Testing Software.* Answerable directly from the text, these questions provide professors with a useful tool for creating and administering tests.

- *Dashboard.* A text-specific, integrated learning system designed with clear and consistent navigation. It delivers quality content and tools to track student progress in an intuitive, web-based learning environment. Dashboard features a streamlined interface that connects instructors and students with the functions they perform most, simplifying the learning experience to save time and put student progress first.

- *Course Cartridges.* Instructors may order selected digital supplements free of charge in ready-to-upload form for the most popular course management systems, including Blackboard, D2L, Moodle, Canvas, and Angel, by contacting their Oxford University Press representative.

Acknowledgments

An undertaking such as this is impossible to complete without the help and expertise of many people. The team at Oxford University Press encouraged and advised us every step of the way. Their expertise, diligence, and cheerfulness kept us motivated, on schedule (more or less), and made the production of this book a memorable experience. In particular, Dan Kaveney, Executive Editor, shared our initial enthusiasm for the project, expertly guided us through the revisions, deadlines, and revised deadlines, and made sure that all tasks were completed in a timely manner. Keith Faivre, Senior Production Editor, reduced our stress levels enormously by keeping all the balls juggling in the air, and guided the book through all the many stages involved in its production. Christine Mahon, Associate Editor, cheerfully kept the lines of communication open, provided timely words of advice and encouragement, kept track of the endless drafts, and helped enormously with the endless selection of photographs. Friday afternoon emails from Christine were a ritual call to arms! John Appeldorn, Editorial Assistant, was instrumental in commissioning manuscript reviews and keeping the administrative trains running on time. Stan Maddock, Andrew Recher, and their team of illustrators at Precision Graphics (Lachina) did an excellent job on the artwork. Stan made many valuable suggestions that considerably enhanced the line art. We'd like to offer a special word of thanks to David Fierstein, who provided invaluable assistance conceptualizing the chapter opening artworks.

We want to thank the Oxford development team of Thom Holmes (manager of development) and the impressive editing skills of Erin Mulligan. Erin's patient, diligent, and thorough editing resulted in substantial improvements to the manuscript while also keeping the process moving forward at all times. Development editor Jane Tufts deserves special recognition for helping us draft the glossary and development assistants Anna Langley and Kait Johnson for contributing additional content help. Thanks also go to David Jurman (Marketing) and Jolene Howard (Market Development) for their enthusiasm and expertise. While we were preoccupied with completing the book, Jolene and David were devising the strategies that would help the book reach its target readership. In production, we would also like to thank design manager Michele Laseau for the stunning look of the book, as well as copy editor Marcia Youngman and proofreader Linda Westerhoff for adding the final touches to the text.

But we owe our greatest debt to our wives, Caroline and Cindy, who allowed "the book" into the house, and gave us the time and space to finish it. Their love, patience, understanding, and assistance were a constant source of inspiration. Without them, this book would not have become a reality.

Manuscript Reviewers

We have greatly benefited from the perceptive comments and suggestions of the many talented scholars, instructors, and colleagues who reviewed the manuscript of *Physical Geology Today*. Their insight and suggestions contributed immensely to the published work.

A special word of appreciation goes to super-reviewer Cindy Murphy, who read virtually all of the text and caught many errors and omissions.

Academic reviewers of our text and figures included:

Katherine Amey, Kent State University
Joel Aquino, University of North Georgia
Deniz Ballero, Georgia Perimeter College–Newton
James Baxter, Harrisburg Area Community College–Wild
Raymond Beiersdorfer, Youngstown State University
Rob Benson, Adams State University
Christopher Berg, University of West Georgia

Paul Bierman, University of Vermont
Mark Boardman, Miami University–Oxford
Andy Bobyarchick, University of North Carolina–Charlotte
Howell Bosbyshell, West Chester University
Polly Bouker, Georgia Perimeter College–Clarkston
Michael Bourne, Wright State University–Dayton
Timothy Bralower, Penn State University–University Park
Scott Brame, Clemson University
Amy Brock-Hon, University of Tennessee–Chattanooga
Charles E. Brown, George Washington University
Samuel Castonguay, University of Wisconsin–Eau Claire
Winton Cornell, University of Tulsa
John W. Creasy, Bates College
Michael Dalman, Blinn College
Craig Dietsch, University of Cincinnati
Joachim Dorsch, St. Louis Community College–Meramec
Alberto Patiño Douce, University of Georgia
Carolyn Dowling, Ball State University
Mary E. Dowse, Western New Mexico University
Melissa M. Driskell, University of North Alabama
Anne Egger, Central Washington University
Lisa Emili, Penn State University–Altoona College
Stewart S. Farrar, Eastern Kentucky University
Mark Feigenson, Rutgers University–New Brunswick
Sara Fulton, Southwestern College
Alexandra Geddes, Lane Community College
Tracy Gregg, University at Buffalo
Kevin Hefferan, University of Wisconsin–Stevens Point
Marc Helman, CUNY Queens College
Daniel Hembree, Ohio University
Pennilyn Higgins, University of Rochester
Julie Hoover, Durham Technical Communiy College
Paul Hudak, University of North Texas–Denton
Solomon Isiorho, Indiana University–Purdue University Fort Wayne
Steven D. Kadel, Glendale Community College
David T. King, Auburn University
Heidi Lannon, Santa Fe College–NW
Arthur Lee, Roane State Community College
Denyse Lemaire, Rowan University
Stephen Dana Lewis, California State University–Fresno
Kyle Mayborn, Western Illinois University
Vali Memeti, University of Southern California
Margaret K. Menge, Delgado Community College
Laura Moore, University of North Carolina–Chapel Hill
David Mustart, San Francisco State University
Jay Muza, Broward College
Pamela Nelson, Glendale Community College–AZ
Klaus Neumann, Ball State University
Daria Nikitina, West Chester University

David Omole, Indiana University–Purdue University Fort Wayne

Kaustubh Patwardhan, Bucknell University

Edward Petuch, Florida Atlantic University–Boca Raton

Jason Polk, Western Kentucky University

Dawn Roberts-Semple, CUNY York College

Morgan F. Schaller, Rutgers University

Adam Schoonmaker, Utica College

Kimberly Schulte, Georgia Perimeter College–Dunwoody

Cecilia Scribner, University of Florida

Katherine Shaw, Green River Community College

Gregory Shellnutt, National Taiwan Normal University

Laura Sherrod, Kutztown University

Arif Sikder, Virginia Commonwealth University

Bruce Simonson, Oberlin College

Sheldon Skaggs, Bronx Community College

Jonathan E. Snow, University Of Houston–Main

Alycia Stigall, Ohio University

Gabrielle Tegeder, University of Nebraska at Omaha

Michael Vadman, California State Polytechnic University, Pomona

John Van Hoesen, Green Mountain College

Stacey Verardo, George Mason University

Cornelia Winguth, University of Texas–Arlington

PHYSICAL GEOLOGY TODAY

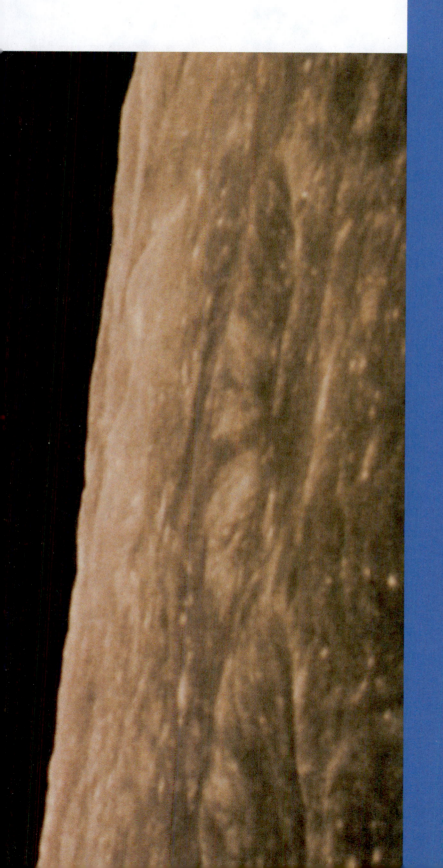

Chapter 1 | Foundations of Modern
            Geology   04

Chapter 2 | Plate Tectonics   38

Chapter 3 | Minerals   68

Chapter 4 | Origin and Evolution
            of Igneous Rocks   98

Chapter 5 | Weathering
            and Soils   136

Chapter 6 | Sedimentation and
            Sedimentary Rocks   162

Chapter 7 | Metamorphism and
            Metamorphic Rocks   194

1 Foundations of Modern Geology

1.1 Geology and the Scientific Method

1.2 The Birth of Geology

1.3 Geologic Time

1.4 Earth Materials: Minerals, Rocks, and Fossils

1.5 The Rock Cycle

1.6 Origin of Earth and the Solar System

1.7 Earth's Internal Heat

1.8 Earth as a System

1.9 What Does a Geologist Do?

▶ Key Terms

▶ Key Concepts

▶ Study Questions

SCIENCE REFRESHER: Some Basic Concepts in Chemistry and Physics 16

IN DEPTH: Origin of the Universe—The Big Bang Theory 20

LIVING ON EARTH: Earth's Seasons—A 4.5-Billion-Year Legacy? 22

SCIENCE REFRESHER: The Flow of Heat 28

Earth

Planets

Birth of solar system

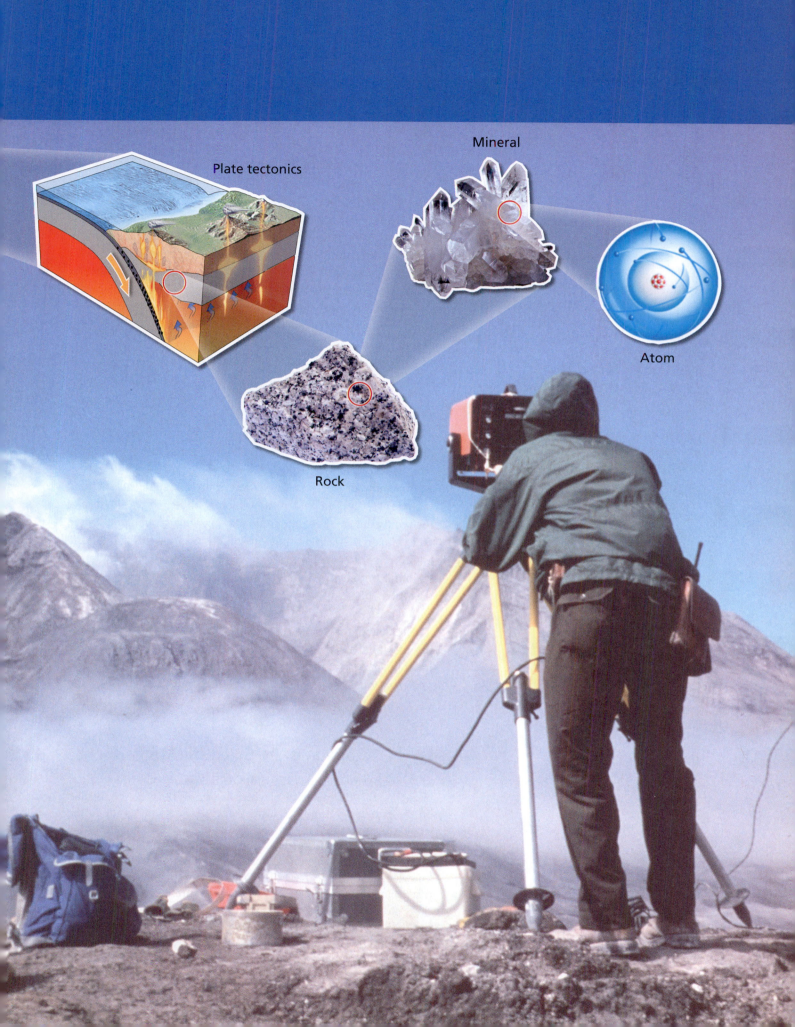

Plate tectonics

Mineral

Atom

Rock

1.1 List the goals of the study of geology and explain the scientific method as it applies to the study of Earth.

1.2 Explain the birth of geology as a science.

1.3 Discuss the enormity of geologic time and the unique insights geology offers on modern environmental issues.

1.4 Identify the materials of the solid Earth: minerals, rocks and fossils, and the three basic rock groups: igneous, sedimentary, and metamorphic.

1.5 Explain the relationship between the basic rock groups that the concept of the Rock Cycle portrays.

1.6 Summarize the processes that formed Earth and the planets of the Solar System.

1.7 Describe how Earth is cooling and how its evolution is related to its cooling history.

1.8 Identify Earth as an interactive system with four intimately linked components: the solid Earth; the hydrosphere; the atmosphere; and the biosphere.

1.9 Describe the pivotal role of geology and geologists in modern society.

Our planet is unique in the Solar System. It not only sustains a remarkable diversity of life forms but it also offers scenes of stunning beauty, from majestic mountains and luxuriant tropical islands to the stark splendor of desert landscapes.

The spectacular photograph on page 2 taken from *Apollo 8* dramatically reveals Earth's four fundamental surface components: rock, water, air, and life. These components interact with one another in a variety of intricate ways. Volcanoes, for example, not only produce lava that becomes new crust but they also vent gases and dust particles into our atmosphere that affect climate and, hence, the environments of living things, or *ecosystems*, that depend on climate.

Earth provides us with such plenty that for years humans have taken its resources for granted. The water in our morning coffee; the dozens of raw materials in our smart phones and the appliances in our homes; and the energy, metals, and plastics in the modes of transport we take to commute to work and school are just some of Earth's many resources we use before the workday even starts.

But many of these resources are finite because we consume them far faster than nature can replenish them. The rapid pace of technological change and our ever-growing global population have placed a strain on these resources and are threatening the environment in which we live. If we are to survive on this planet, we must endeavor to understand the processes that have governed Earth's evolution so we can learn how to live in harmony with nature.

The goal of this textbook is to provide you with an understanding of planet Earth, its minerals and rocks, the processes that have governed the planet's evolution throughout geologic time, and the processes that currently occur upon its surface.

1.1 Geology and the Scientific Method

The American Geological Institute defines **geology** as the study of planet Earth—the materials of which it is made; the processes that act on these materials; and the history of the planet and its life forms. Geologists investigate the physical forces that act on Earth, the chemistry of its constituent materials, and the biology of its past inhabitants as revealed by fossils. These investigations provide the basic knowledge for the discovery of valuable minerals and fuels, help identify geologically stable sites for major structures, and enable predictions of the dangers associated with the mobile forces of a dynamic Earth. **Physical geology**, the subject of this book, emphasizes the processes that interact at Earth's surface and have shaped its evolution throughout geologic time.

Figure 1.1 Geologist at Work

A geologist points a radar gun at molten lava to track its speed through an underground volcanic tunnel on Hawaii's Kilauea volcano.

A geologist's work (Fig. 1.1) involves the study, mapping, and interpretation of Earth and its evolution as revealed by rocks, minerals, and fossils, as well as the atmosphere and oceans. Geologists are also responsible for the exploration, development, and stewardship of Earth's valuable resources, and the evaluation of the environmental implications of these activities. These resources include the metals, hydrocarbons, and industrial materials upon which modern society depends, as well as soils, the cornerstone of agriculture, and our most precious resource, water. We explore the many things that geologists do in more detail in Section 1.9, "What Do Geologists Do?"

The physical forces that have acted on Earth, the processes that govern the chemistry of its minerals and rocks, and the biology of its ancient life as revealed by fossils are equally important features of Earth history. However, as we discuss in Section 1.2, many clues to Earth's early history have been lost. Just as a forensic scientist has difficulty evaluating the evidence at an old crime scene, so the geologist faces challenges in evaluating Earth's early history because much of the evidence has been erased by more recent geological events. To remedy this, geologists often rely on indirect and circumstantial evidence, much of which comes from other celestial bodies, including the Moon, the *terrestrial* (Earth-like) planets Mercury, Mars, and Venus, and *meteorites* (rocks that have fallen from space and landed on Earth) that are thought to represent the remnants of the early Solar System. The record for Earth's more recent history is more complete and understandable than that of its

ancient past but the evidence that a geologist depends upon is invariably incomplete.

To investigate evidence and test ideas based on interpretation of that evidence, the geologist, like all scientists, makes use of a methodical approach. This approach is known as the **scientific method** and is the fundamental doctrine upon which all modern science is based.

Scientists assume that the natural world behaves in a predictable way. To provide the knowledge upon which we base our understanding of the natural world, scientists employ the scientific method, which uses observation and theory to test scientific ideas (Fig. 1.2).

First, a scientist, or team of scientists, identifies a gap in our understanding of the natural world and sets about to learn more about the process and fill in the gap. The first stage in this task involves the collection of data. Scientists obtain data by measurement, experiment, and observation. Then they formulate a **hypothesis**, which is a tentative explanation of the data. A good hypothesis advances the simplest possible explanation of the available data. By definition, a hypothesis is unproven and must be rigorously tested. To be effective, a test must distinguish between rival hypotheses, and any legitimate test must be one that could disprove the proposed hypothesis. The most common test is that of *prediction*. If the hypothesis is true, then something the hypothesis predicts must also be true. The hypothesis can then be tested by seeing if the prediction is realized.

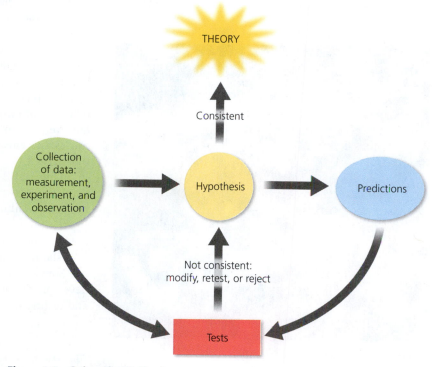

Figure 1.2 Scientific Method

A flow chart for the scientific method illustrates how observations lead to hypotheses. The predictions of hypotheses are then tested and theories are established.

The proposal that there may once have been life on Mars is an example of a hypothesis that is currently undergoing rigorous testing. This hypothesis has focused the attention of the scientific world and has heightened public awareness of the importance of the Pathfinder mission to Mars and the subsequent exploration of the planet's surface by the rovers *Spirit* and *Opportunity* (Fig. 1.3). Because liquid water is essential to life, the primary goal of the Mars Exploration Rovers is to search for evidence of past water activity in Martian rocks and soils by seeing if they contained minerals formed by processes involving water, and to determine whether the landscape was sculpted by flowing water. The evidence the rovers found is generally supportive. However, even the strongest proponents of the hypothesis do not claim it to have been proven; they only maintain that it is the simplest explanation of their current observations.

If a hypothesis is tested and results are not consistent with predictions, the hypothesis must be modified, retested or rejected. A hypothesis that passes a number of tests that rival hypotheses have failed becomes a **theory**. A hypothesis that becomes a theory has received a strong vote of confidence from the scientific community. But theories are not facts, and are still subject to further testing that may possibly disprove them. But as a theory passes more and more tests and the results of these tests are shown to be consistent with predictions, the scientific community regards the theory with ever-higher degrees of confidence. One such theory is the *theory of plate tectonics,* which provides a unified explanation for a wide variety of natural phenomena, such as the cause of volcanic eruptions and earthquakes, and the origin of mountains, continents, and

oceans. Throughout this book we explain the tenets of this theory and refer to its power to explain geological processes. Like the theories of evolution in biology and the big bang theory in astronomy, plate tectonics has become the cornerstone of physical geology and many of the concepts discussed in this book are linked to it.

It is very important that all hypotheses and theories are reexamined as new data become available. The results of these reexaminations are often communicated to the scientific community at conferences and in publications. The thousands of new scientific papers published annually in geology alone indicate that our knowledge of planet Earth is increasing. Most of these papers are subjected to peer review before publication, that is, they are scrutinized and criticized by colleagues who recommend acceptance or rejection of the paper to the journal editor. In most cases, the authors must respond to the comments and criticisms of colleagues before the paper is accepted for publication.

After many tests, numerous observations, measurements, and experiments, a group of scientific ideas is formulated into a **law**. A law is an expression or statement that accounts for a natural process or processes from which there is no known deviation. Although a law must provide an exact description of a process, it need not provide an explanation for it. For example, when Johannes Kepler proposed the laws of planetary motion in the early seventeenth century, he could not provide a satisfactory explanation for them. In fact, there was no explanation for planetary motion until the late seventeenth century, when Sir Isaac Newton proposed the law of universal gravitation.[1]

Although plate tectonics is now a well-established theory, geologists recognize that there is still much more to know and learn about it. Technological advances have helped us to understand processes occurring in the deep Earth that might be a driving mechanism for plate tectonics, and satellite global positioning systems (GPS) can measure the relative rates of movement of Earth's plates to a precision of millimeters per year. Despite these advances, however, we are still learning about and debating the causes and effects of plate tectonics and the mechanisms that drive it.

So much still remains to be done. Just as the first athlete to break the 4-minute mile extended the physical capabilities of humankind and led others to strive for new records, so science, in its relentless quest for knowledge, advances the frontiers of our understanding.

It should be noted that the scientific method has its critics. Some point out that its application slows down the development of science, and that the constant probing and testing the method requires is too pedestrian for the

Figure 1.3   Mars Rover *Opportunity*

The Mars rover *Opportunity* visited Endurance Crater, a Martian impact crater in 2004. Shown here is a simulated view of the rover.

[1]The term *scientific method* was not used in the time of Kepler and Newton. Since their deaths, however, their laws have been scrutinized by scientists who applied this method to their research.

rapidly changing modern world. Others point out that the scientific method has fundamentally failed to predict the consequences of certain scientific findings.

While acknowledging these criticisms, it is important to realize that modern science is founded on the application of the scientific method. This connection is apparent in several parts of this book. In addition, scientists need a consistent philosophical approach to science similar to the one provided by the scientific method in order to communicate effectively.

1.1 SUMMARY

- Geology is the study of Earth; geologists employ physical, chemical, biological, and mathematical methods to understand Earth, its materials, behavior, environment, and evolution.

- Scientists obtain information by measurement, experiment, and observation. They then formulate a tentative explanation of the data, called a hypothesis. A hypothesis is repeatedly tested and becomes a theory if it passes a number of tests and gains widespread acceptance.

- A group of scientific ideas may be formulated into a law.

1.2 The Birth of Geology

The Scotsman James Hutton is considered to be the founder of modern geology. In 1795, Hutton published a book entitled *The Theory of Earth,* in which he proposed that Earth experiences continual but gradual change, constantly building up and wearing down under the power of its own internal heat engine. He envisaged the history of Earth as "a succession of former worlds" with "no vestige of a beginning, no prospect of an end" in which mountains were uplifted and worn down in an endless cycle of creation and destruction.

Hutton found support for his idea of an ancient and gradually changing Earth at Hadrian's Wall, a defensive structure built close to the Scottish border by the Romans in AD 122 (Fig. 1.4). Hutton observed that the wall had changed very little, despite almost 1700 years of exposure to wind, rain, and ice. This observation led Hutton to reason that in order for entire mountains to be worn away, Earth must be unimaginably old.

Hutton's ideas ran counter to the prevailing view of his day, which held that Earth was unchanging and that all modifications to its surface were caused by catastrophic events, such as Noah's flood, brought about by divine intervention. The age of Earth was also thought to be a mere 6000 years, based on the Bible's record of births and deaths that placed the creation in the year 4004 BC. Hutton's vision of "worlds built upon worlds," of old landscapes being worn away and new ones being built, necessitated an ancient Earth with a surface that was continuously but gradually changed by processes that could be witnessed by

Figure 1.4 Hadrian's Wall

Hadrian's wall in northern England is a Roman fortification built in the reign of Emperor Hadrian in AD 122. James Hutton concluded that because the wall had suffered little deterioration in almost 17 centuries, the time required for the erosion of mountain belts must be vast.

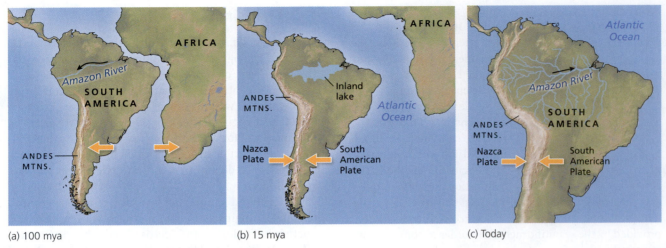

(a) 100 mya   (b) 15 mya   (c) Today

Figure 1.5   Mountain Building and the Amazon River

The ancestral Amazon River (a) flowed westward into the Pacific Ocean before the rise of the northern Andes Mountains, about 15 million years ago. When this occurred (b) an inland lake formed, which eventually evolved into the modern eastward drainage system that flows into the Atlantic Ocean.

his contemporaries. Later expressed in the phrase "the present is the key to the past," Hutton's insights have become the foundation of modern geology. For example, it is by examining the modern settings in which sediments are deposited, such as rivers or shallow seas, that the environments in which ancient sedimentary rocks were deposited can be recognized.

Since Hutton's time, major advances have allowed us to determine how old rocks are, and the duration of major geological events, such as *mountain building*. These analyses have confirmed Hutton's views. We now know that the oldest rocks on Earth are a little over 4.0 billion years old, and we have indirect evidence that Earth itself is 4.54 billion years old.

We also know that the creation and destruction of mountains takes hundreds of millions of years. Mountains rise in pulses of activity, as demonstrated by the history of the Amazon River in South America (Fig. 1.5). One hundred million years ago, before the rise of the Andes Mountains, the ancestral Amazon River flowed from east to west, and drained into the Pacific Ocean. About 15 million years ago, plate tectonic processes pushed up the Andes, and the resulting barrier blocked this drainage and created a large inland lake, eventually leading to the modern drainage basin of the Amazon River, which drains into the Atlantic Ocean.

> **CHECK YOUR UNDERSTANDING**
>
> ● How do modern environments help geologists interpret ancient rocks?

1.2 SUMMARY

• James Hutton's insights demonstrate that Earth is in continuous but gradual change, constantly building up and wearing down under the power of its internal heat engine.

• Using present-day settings of sediment deposition to determine the environments of ancient deposits is an application of the concept that the present is the key to the past.

1.3 Geologic Time

Hutton's ingenious insights made it clear that understanding geology is impossible without an appreciation of the enormity of geologic time. Most geologic processes operate so slowly that their influence is barely recognizable in the span of a single human lifetime. But over the course of geologic time, even the slowest of processes can bring about profound change. The lofty heights of the Grand Tetons (Fig. 1.6), for example, look much the same today as they did to early explorers some 150 years ago. But in the context of geologic time, mountains are short-lived features that have been uplifted and eroded away many times. We discuss geologic time in detail in Chapter 8: Geologic Time but we introduce this topic here in Chapter 1 because it allows us to put the study of geology in context.

The age of meteorites and rock samples brought back from the Moon by the Apollo program of lunar landings indicates that Earth is more than 4.5 billion years old. That is a staggering 4,500,000,000 years. With this length of time at its disposal, a garden snail crawling at a rate of just 3 meters (10 feet) per hour could travel 118 billion kilometers (73 billion miles). That's enough to get to the Moon and back 154 thousand times! This example illustrates the enormity of geologic time, and makes it clear that Earth history has been more than long enough for mountains to be uplifted and eroded away, for life to have risen and evolved, and for the positions and shapes of the continents and oceans to have changed beyond recognition.

Figure 1.6   Grand Tetons
In the context of geological time, even the impressive Grand Tetons of Wyoming are short-lived features on a geological time scale.

WHY DO WE STUDY GEOLOGIC TIME?

There are urgent and practical reasons why we need to understand Earth's evolution. If we are to practice responsible stewardship of our planet, it is crucial that we distinguish between the kinds of global change induced by human activity and changes that are part of natural cycles. For example, are industrial development and our consumption of energy responsible for global warming, ozone depletion, and the pollution of our waters by acid rain? Or are these human activities dwarfed by natural processes such as volcanic eruptions and changing patterns in ocean circulation, both of which result in dramatic changes in the environment? What kinds of human development can be sustained by our planet, and is sustainable development even possible?

In this context, the study of the record of Earth's evolution offers a unique and clear perspective. During this evolution, natural processes unaffected by the influences of the modern era are recorded in the rocks, sediments, and soils they produced. By understanding these processes, we can more clearly recognize the signals that result from human interference with the natural order and so formulate policies to deal with this interference.

Earth has its own pulse, its own natural rhythms, many of which profoundly influence phenomena as all-embracing as the motion and growth of continents, and the birth and destruction of oceans. These rhythms relate to the interaction between the solid Earth beneath our feet, its watery surface, its atmosphere, and the realm of life. These rhythms were in operation long before the modern era of industrialization, and may extend back to the very earliest stages of Earth's development. However, we must be able

to recognize them if human interference with the environment is to be distinguished from the natural order of things. An analogy can help us understand the value of studying the past. If a familiar song is cut off before it ends, we can usually continue to hum the tune because we recognize its rhythmic beat. In a similar way, if we are to gain an understanding of the modern Earth, we must first recognize the rhythms and beats of its geologic past.

By striving to understand Earth's evolution, geology offers a unique perspective in our efforts to address modern environmental issues. The accuracy of our predictions about the future depends on how well we have learned Earth's past rhythms. We seek distortions in the song, and these distortions are instantly recognizable just as they are in a recording of familiar music. If human activity truly represents a distortion in the evolution of our planet, it can be readily identified. However, this is possible if, and only if, we know enough about Earth's history to recognize its natural pulse.

> **CHECK YOUR UNDERSTANDING**
>
> ● How can learning about the geologic past help us understand modern environmental issues?

WHAT IS GEOLOGIC TIME?

Charting the course of geologic history is of fundamental importance because we need to understand the rate at which natural processes occur. To this end, geologists have developed a systematic two-pronged approach, much like the approach a forensic scientist uses in trying to reconstruct the scene of a crime. First, the forensic scientist tries to establish a sequence of events, placing these events in the order in which they occurred. Second, the forensic scientist tries to determine at what time the crime took place, using evidence such as a smashed wristwatch, the body temperature of the deceased, or the testimony of witnesses.

The "forensic arena" for the geologist is an **outcrop**, the place where consolidated rock, called **bedrock**, is exposed at Earth's surface. Like the forensic scientist, geologists have endeavored to place natural events in the sequence in which they occurred (known as *relative time*), and also establish the exact time when these events took place (known as *absolute time*). Using this two-pronged approach, geologists have painstakingly established a calendar known as the **geologic time scale** (Fig. 1.7), constructed from numerous observations and measurements in thousands of outcrops from all over the world. Just as we divide a calendar year into months, weeks, and days, so geologic time is divided into *eons, eras,* and *periods*. **Eons** are the longest subdivisions of the time scale, some having lengths of billions of years. Just four, the Hadean, Archean, Proterozoic, and

Phanerozoic eons, encompass all of Earth history. The Phanerozoic Eon, which takes in the last 541 million years, is subdivided into shorter time intervals lasting hundreds of millions of years called **eras**, of which there are three, the Paleozoic, Mesozoic, and Cenozoic. Each of these eras is then subdivided into still-shorter time intervals, called **periods**, lasting tens of millions of years. For example, the Jurassic is a period during the Mesozoic Era when dinosaurs ruled the world.

Even though the geologic time scale is a well-established concept, like all important concepts in science, it is subject to constant scrutiny, testing, and modification, as we discussed in Section 1.1. The calibrations get more precise as previous errors are identified and corrected. International committees constantly evaluate new data and discoveries to determine if any changes to the time scale are necessary. For example, when we, the authors of this book, were undergraduate students we were taught that the Phanerozoic Eon began 590 million years ago. By the time we became graduate students, that time estimate had been modified to 570 million years. The very latest data indicate that it comprises the last 541 million years.

One way of capturing the enormity of geologic time is to imagine all of Earth history compressed into a single calendar year with the formation of Earth taking place on January 1. In this way, we can think of geologic time in terms of months rather than billions of years. On this humanized time scale, the Hadean Eon would last until the end of February, the Archean Eon until mid-June, and the Proterozoic Eon until mid-November. The Paleozoic Era would not begin until November 18, while the Mesozoic Era would start on December 12, and the Cenozoic Era on December 26. The Jurassic Period would last a mere five days, from December 15 until December 20. On this condensed time scale, the rise of modern man would not take place until 11:37 p.m. on December 31, and the average span of a single human lifetime would last just half a second.

1.3 SUMMARY

- The record of Earth's evolution over geologic time provides a unique and clear perspective on modern environmental issues.

- Geologists strive to place natural events in a sequence of relative order (known as relative time) and to establish the exact time when these events occurred (known as absolute time).

- Geologic time is divided into eons, eras, and periods. Although a well-established concept, like all important concepts in science, geologic time is subject to constant scrutiny, testing, and modification.

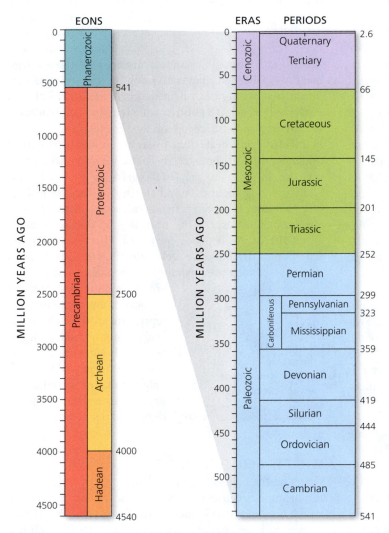

Figure 1.7   Geologic Time Scale
The major subdivisions of geologic time are eons, eras, and periods.

1.4 Earth Materials: Minerals, Rocks, and Fossils

Minerals, rocks, and fossils are the materials that record Earth's evolution. Minerals and rocks are the building blocks of the solid Earth. As we will discuss in greater detail in Chapter 3, a **mineral** is a naturally occurring substance that forms crystals and has a chemical formula. The common mineral *quartz*, for example, has the chemical formula SiO_2, and forms beautiful six-sided (hexagonal) crystals. Its external appearance reflects the orderly internal arrangement of the silicon (Si) and oxygen (O) atoms of which it is made up (Fig. 1.8). The science of minerals is known as *mineralogy*.

A **rock** is an aggregate of one or more minerals and does not have a specific chemical composition. The common rock-type *granite,* for example, contains the mineral quartz, but also contains other minerals, such as feldspar and mica (Fig. 1.9).

Figure 1.8   Mineral

This cluster of six-sided crystals of the mineral quartz (SiO_2) is from Brazil.

Figure 1.9   Rock

A sample of granite is made up of several minerals of different colors, including glassy gray quartz, plagioclase and ortho-clase feldspar (the white and pink minerals, respectively), and biotite and hornblende (the black minerals).

Figure 1.10   Fossil

This fossil fish was found in the Green River Formation, a 50-million-year-old lake deposit in Wyoming.

As we describe in detail in Chapters 4, 6, and 7, rocks fall into one of three groups: **igneous rocks** (from the Latin *ignis* meaning fire) form from the crystallization of molten rock, which is called **magma**. **Sedimentary rocks** (from the Latin *sedere* meaning to settle) are formed by the accumulation or precipitation of material on Earth's surface or in bodies of water. Before its consolidation into solid sedimentary rocks, this material is called *sediment*. **Metamorphic rocks** (from the Latin "to change form") are produced when preexisting rocks are transformed by heat and pressure. The relationship of the three rock types is discussed in Section 1.5.

A **fossil** (from the Latin *fossilis,* meaning "obtained by digging") is the *petrified remains* of ancient life or an *impression* of an ancient life form (Fig. 1.10). Petrified remains may be large or small and may include body parts, such as bones, teeth, skeletons, or shells. Impressions are signs of life, such as footprints or burrows, that the organism created while it was alive. The study of fossils is the science of *paleontology,* and the complete history of fossils through geologic time is known as the *fossil record.* The life forms preserved as fossils lived at or near Earth's surface, and with very few exceptions, fossils are found only in sedimentary rocks.

> ### CHECK YOUR UNDERSTANDING
> ◉ What is the difference between a mineral and a rock?
> ◉ What is the difference between sediment and sedimentary rock?
> ◉ What is a fossil?

1.4 SUMMARY

- Earth's evolution is recorded by minerals, rocks, and fossils.
- Rocks are made up of minerals and fall into one of three groups: igneous, sedimentary, and metamorphic.
- Fossils are the petrified remains or the impressions of ancient life forms.

1.5 The Rock Cycle

Hutton's insightful view that Earth is in constant but gradual change is vividly illustrated in the relationship between the three basic groups of rocks (introduced in Section 1.4) that make up Earth's continents and ocean floors.

Each of the three basic rock groups and the geologic processes that produce them are closely interrelated by the continual pattern of interchange we call the **rock cycle** (Fig. 1.11). The cycle illustrates how rock types transform into each other in an endless pattern of creation and destruction. The rock cycle is thus a vivid demonstration of a dynamic Earth undergoing gradual but continuous change, just as Hutton envisaged.

Igneous rock, the first of the three basic rock groups, forms when magma freezes. Igneous rocks can form at Earth's surface, as the product of volcanic eruptions, in which case they cool very rapidly and are immediately subject to *weathering* and *erosion,* processes that we describe in this section of the chapter. Basalt (Fig. 1.12a) is the most common igneous rock formed in this way. Alternatively, igneous rocks can solidify within Earth's crust, in which

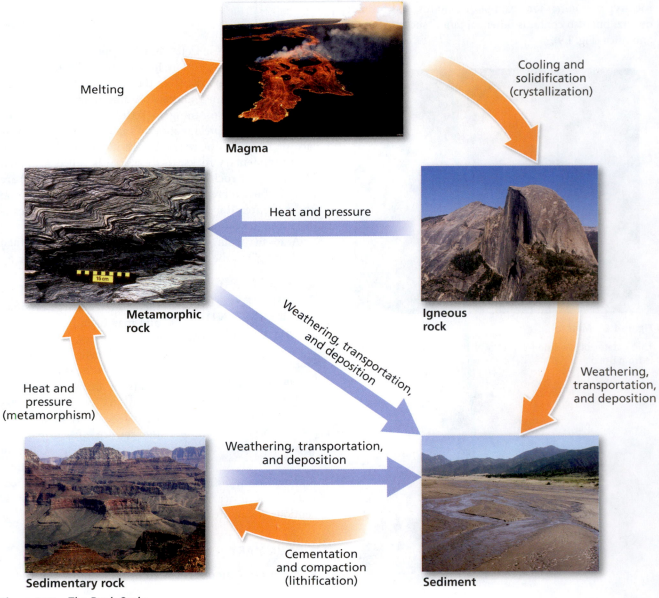

Figure 1.11   The Rock Cycle

The rock cycle shows the relationship between the three main groups of rocks: igneous, sedimentary, and metamorphic (orange arrows). Blue arrows within the circle show possible shortcuts.

case they cool slowly. Slow-cooled igneous rocks may later become exposed at the surface if the rocks that overlie them are uplifted by processes such as mountain building, and subsequently worn away. Granite (Fig. 1.12b) is an example of a slow-cooled igneous rock. Chapter 4: Origin and Evolution of Igneous Rocks is devoted to igneous rocks.

If they are exposed at Earth's surface, all rocks undergo **weathering**, during which they slowly break down through

Figure 1.12   Examples of Common Rocks

(a) Basalt and (b) granite are igneous rocks. Sedimentary rocks include (c) sandstone and (d) shale. (e) Slate and (f) marble are metamorphic rocks.

disintegration and decomposition as a result of their inter-action with water, ice, and air. Chapter 5: Weathering and Soils covers the weathering process in greater detail. **Erosion** is the process that loosens, dissolves, and trans-ports rocks and sediments. All weathered materials are eventually picked up and *transported,* typically by one of several *agents of erosion:* running water, wind, waves, or glaciers (large bodies of flowing ice). Eventually the eroded material is *deposited* as loose, *unconsolidated* sediment, such as gravel, sand, and mud (see Figure 1.11). Although some of this sediment is deposited in river valleys, lakes, or deserts, most is transported to the sea.

Once deposited, various processes combine to turn un-consolidated sediment into sedimentary rock, the second of the three basic rock groups. In Chapter 6: Sedimentation and Sedimentary Rocks we cover the process of sedimenta-tion in detail. The sediment is slowly buried as more sedi-ment is deposited, and it becomes compacted by the weight of the overlying material. It may also come into contact with ground water that moves between the grains of sedi-ment, depositing material that cements the grains together. This conversion of unconsolidated sediment into solid sed-imentary rock by compaction and cementation is called **lithification** (see Figure 1.11). Sandstone (Fig. 1.12c) and

shale (Fig. 1.12d) are examples of sedimentary rocks formed by this process. Uplift and erosion of sedimentary rock produce new unconsolidated sediment and the pro-cess repeats. In fact, the cycle of deposition, burial, and lithification may be repeated many times, each time result-ing in a new sedimentary rock.

If sedimentary rocks come into contact with magma, or are involved in large-scale crustal movements, such as mountain building, and become deeply buried as a result, they are subjected to intense pressure and heat (see Science Refresher: **Some Basic Concepts in Chemistry and Physics**). This is because the weight of the overlying rocks exerts pressure and because Earth's interior is hot. As the sedi-mentary rock responds to these new conditions, some of the minerals it contains may become unstable and react to form new minerals that are stable at higher pressures and temperatures. These reactions cause the original chemical and physical makeup of the rock to alter, producing the third basic rock group, known as metamorphic rock. Examples of metamorphic rocks include slate (Fig. 1.12e) and marble (Fig. 1.12f). Metamorphic rock is the subject of Chapter 7: Metamorphism and Metamorphic Rocks.

If metamorphic rocks continue to be buried, they may encounter high enough temperatures that they melt,

SCIENCE REFRESHER

Some Basic Concepts in Chemistry and Physics

Much of our knowledge of the natural world has been gained by observation of matter. *Matter* is actually a poorly defined term, but for our purposes we can employ the most common definition, which is that matter is some-thing that has both mass and volume (i.e., it occupies space). Solids, liquids, and gases have mass and volume and are therefore considered to be matter. An *atom* is a fundamental unit of matter, and is the smallest unit of a chemical element having all the characteristics of that ele-ment. An *element* is a substance that cannot be broken down using chemical methods.

Observations of matter typically involve measurements of physical quantities such as mass, length, and time. Many of the terms and concepts used in this textbook are some combination of these three physical quantities. Al-though many are also used in casual conversations, these concepts and terms have precise definitions in science.

Mass is a measurement of the quantity of matter in an object. In normal conversation, mass and *weight* are used interchangeably. But in science they have different mean-ings. Under most conditions, the mass of an object is con-stant because the amount of matter it contains does not vary. However, weight is the measure of the force of gravity acting on an object. The mass of an object would be the same if it were measured on Earth and on the Moon. But the Moon is much smaller than Earth (about 25 percent of

its size) and so its gravity is much less. So an object weigh-ing 80 kilograms (176 pounds) on Earth would weigh about 13.3 kilograms (29.3 pounds) on the Moon.

Length is measured in meters. *Area* is the size of a closed region in a plane. Area has a length and a width, and is therefore two-dimensional. The area is the length times the width (l $\times$ w), and as both are measured in meters, the unit of area is m². *Volume* has a third dimen-sion, height, also measured in meters, so the unit of volume is m³ (Fig. 1A).

Density is the mass of an object divided by its volume (m/V), or mass per unit volume. The units for density are kilograms per cubic meter (kg/m³). Density is a measure of how tightly packed the constituent atoms are of that object (Fig. 1B) and is a fundamental property of solids, liquids, and gases. For example, a large block of granite has a greater mass than a small piece of granite simply because it contains more matter. But the amount of mass per unit volume for each sample of granite is much the same, since density is independent of the size of the rock. It is therefore an important property common to all granites.

A rock's mass does not change with temperature and pressure, but its volume does. Increasing temperature causes rocks (and the minerals they contain) to expand. When this happens, the same mass occupies a larger volume and so has a lower density. Conversely, increasing

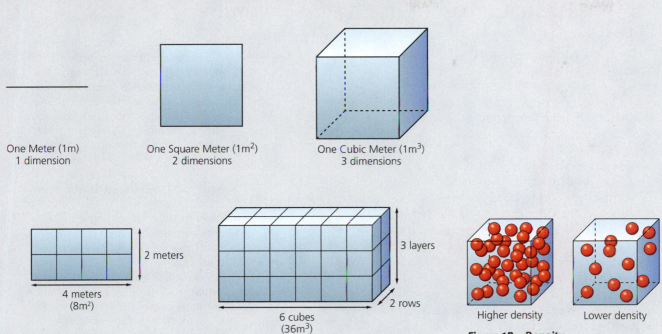

One Meter (1m)
1 dimension

One Square Meter (1m²)
2 dimensions

One Cubic Meter (1m³)
3 dimensions

2 meters

4 meters
(8m²)

3 layers

2 rows

6 cubes
(36m³)

Higher density   Lower density

Figure 1A   Length, Area, Volume

Length has one dimension, area two dimensions, and volume three dimensions. The area of an object can be determined from how many 1 m × 1 m squares fill the space. The volume of an object is determined from how many 1 m × 1 m × 1m cubes fill the volume occupied by the space.

Figure 1B   Density

Density is a measure of the amount of mass in a given volume. The object on the left has more mass within the same volume and therefore has a higher density.

pressure causes rocks and minerals to contract. When this happens, the same mass occupies a smaller volume and therefore has a higher density.

Throughout this book we will discuss how dense objects tend to sink relative to less dense (light) objects and this fundamental property has had a profound influence on the evolution of Earth. In fact, Earth is layered according to the density of its chemical elements. Dense elements, such as iron and nickel are concentrated in the center of Earth, whereas the relatively light elements such as nitrogen, oxygen, and argon are the dominant gases in Earth's atmosphere. Elements such as sodium, potassium, calcium, and magnesium that have densities between these extremes are concentrated in Earth's outermost rocky layer, the crust.

Velocity is speed in a specific direction and so the unit of velocity is meters per second (m/s). A change in velocity over time is called *acceleration*. In a dynamic Earth, there are many instances in which an object changes its velocity over time. In its simplest form acceleration can be determined from the change in velocity divided by time. According to Newton's First Law of Motion, objects continue to move with constant velocity unless acted upon by an external *force*. The concept of force relates to the cause of the acceleration, that is, the "push or pull" that changes its velocity. The force needed to achieve a particular acceleration

depends on the mass of the object, and if the force is doubled, so too is the acceleration. Force is therefore mass times acceleration, or F = ma. Pressure is a special case where the force per unit area is the same on all surfaces of the object (Fig. 1C). The pressure on an object increases with depth due to the increasing mass of rocks above it and around it.

Gravity is a fundamental property of nature by which all physical objects attract each other. The force of gravity between two objects is proportional to their masses but inversely proportional to the square of the distances between them. As a result of this attractive force, dispersed bodies in space are drawn ever closer until they unite, and once united are held together. The influence of gravity is therefore fundamental to understanding the origin of the Solar System (see Section 1.6). Every planet in the Solar System has its own gravitational field. Earth is the largest terrestrial planet and therefore has the strongest attraction for the gases around it. This is one of the main reasons why Earth has an atmosphere.

Energy is the capacity to do work. There are many forms of energy including kinetic energy, potential energy, gravitational energy, mechanical energy, radiant energy, and thermal energy. The transformation of energy from one form to another is the primary driver of Earth's evolution. A body has kinetic energy because of its motion, and

(Continued)

SCIENCE REFRESHER

Some Basic Concepts in Chemistry and Physics (Continued)

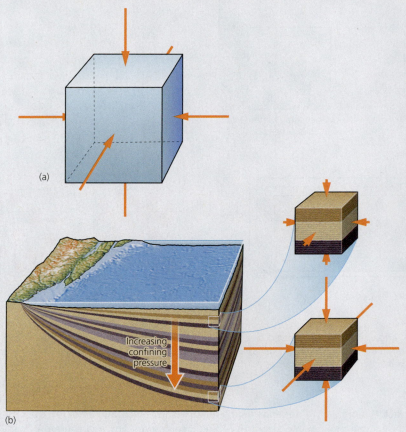

(a)

Increasing confining pressure

(b)

Figure 1C   Pressure

(a) In this example, the force per unit area is the same on all surfaces of the object. (b) The pressure on an object increases systematically with depth, due to the increasing mass of rocks above it and around it.

potential energy is stored in it because of its position. When the body is set in motion, its potential energy is converted into kinetic energy (Fig. 1D). As a mountain gets worn down over time, for example, potential energy is converted into kinetic energy as fragments of the mountain are transported by water, wind, or ice from higher to lower elevations. In physical geology, the conversion from potential to kinetic energy primarily occurs under the influence of gravity, and so for our purposes we can consider potential energy and gravitational energy as essentially the same concept.

Mechanical energy is the sum of potential and kinetic energy. Radiant energy is carried by sunlight and is a key component of fundamental surface processes such as the circulation of water and the sustenance of life. Thermal energy is responsible for the temperature of a body. *Temperature* is a measure of the average kinetic energy of all the particles in the body, whereas *heat* is a measure of the total kinetic energy of all the particles. Thus for two identical substances at the same temperature, the larger body has more heat simply because it is bigger.

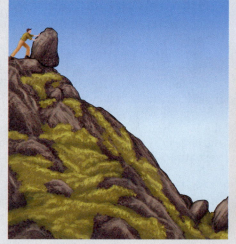
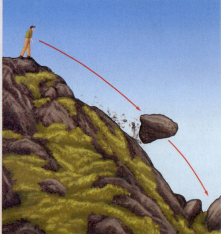

Figure 1D   Conversion of Potential Energy into Kinetic Energy

The boulder has potential energy (left) because of its elevation. Potential energy is converted into kinetic energy (right) once the boulder moves downhill under the influence of gravity.

creating magma, which is the raw material for new igneous rocks. In this way the rock cycle will have come full circle.

There are also shortcuts in the rock cycle. If an igneous rock is subjected to the great pressures and temperatures of mountain building, it too may undergo alteration and revert to a new metamorphic rock. Similarly, if a metamorphic or sedimentary rock is uplifted, it may ultimately become exposed at Earth's surface where it is subject to weathering, and the product of its erosion provides the raw material for new sedimentary rocks to form.

CHECK YOUR UNDERSTANDING

○ How do igneous rocks transform into sedimentary rocks?

○ How are metamorphic rocks transformed into sedimentary rocks?

○ How do metamorphic rocks transform into igneous rocks?

Although the rock cycle is a powerful concept that describes processes that transform one category of rocks to another, it does not adequately explain the natural forces that drive the cycle. The relentless quest to understand these forces led to the discovery of a unifying theory, known as plate tectonics, which has revolutionized geology as profoundly as the theories of evolution and relativity respectively revolutionized the sciences of biology and physics. Chapter 2: Plate Tectonics introduces the theory of plate tectonics and explains the crucial role it plays in the story of Earth and its evolution.

1.5 SUMMARY

* Hutton's view of a continuously changing Earth is expressed in the relationship between the igneous, sedimentary, and metamorphic rocks that make up Earth's continents and ocean floor.

* Each rock type is transformed into one of the others in a continuous process known as the rock cycle.

1.6 Origin of Earth and the Solar System

To understand how Earth evolved to become the modern life-sustaining planet we are familiar with, we must first learn how it formed. This is a challenge. Because Earth is such a dynamic planet with an ever-changing surface, no rocks older than 4.0 billion years are preserved, so there is no direct record of Earth's early history. The challenge is similar to the challenge a forensic scientist faces when trying to understand a crime in which the critical evidence has been repeatedly tampered with or erased. To address the challenge, we must put Earth's history in the context of the theories about the origin of the Solar System (see In Depth: **Origin of the Universe—The Big Bang Theory**).

THE SOLAR NEBULA THEORY

Earth shares its origin with that of the other members of the Solar System, the largest of which are the eight planets that orbit the Sun. The origin of Earth therefore bears on the history of the Solar System, and it is here that our introduction to our planet starts.

A basic understanding of the origin of the Solar System was first provided by French philosopher and mathematician René Descartes in 1644. These ideas were expanded on some 100 years later by the German philosopher Immanuel Kant and the French mathematician Pierre-Simon Laplace. Their collective thinking has evolved into the *solar nebula theory* (Fig. 1.13), which retains many essential concepts envisaged by Kant and Laplace more than 200 years ago.

Figure 1.13   Origin of the Solar System

(a) The solar nebula, a swirling cloud of gases and dust, forms. (b) Contraction flattens the nebula into a rotating disk. (c) Matter within the disk coalesces, forming the Sun and planetesimals. (d) The planetesimals coalesce in turn to form protoplanets and, eventually, the planets.

IN DEPTH

Origin of the Universe—The Big Bang Theory

The solar nebula theory accounts for the origin, some 4.6 to 4.5 billion years ago, of the Sun and all the planets (including Earth) that comprise the Solar System. The Sun is just one rather insignificant member of the *Milky Way Galaxy,* a systematic cluster of between 100 and 400 billion stars bound together by mutual gravitational attraction. Latest estimates indicate that there are between 150 and 200 billion galaxies in the universe. By definition, the *universe* contains everything, all matter and energy, all celestial objects, and all the spaces between them.

At one time, Albert Einstein believed that the universe was *static,* that it was neither expanding nor contracting. He later admitted that this was one of the most serious blunders in his career. Most astronomers are now convinced that the universe originated in an explosion of unimaginable proportions, an event known as the *Big Bang,* and has been expanding ever since. The most recent estimates of the age of this event place it about 13.8 billion years ago. From this perspective, our Solar System is a relatively recent development in the evolution of the universe.

But what is the evidence for the Big Bang, and how is the age of the universe estimated? In 1929, astronomer Edwin Hubble observed that the galaxies are moving away from each other, implying that the universe is expanding. If this is correct, they must have been closer together in the past. This has become a cornerstone of the Big Bang theory, which describes the universe as originating in a gigantic explosion and expanding ever since.

The expansion of the universe can be likened to the expansion of a loaf of raisin bread when it is baked. As the dough rises and the bread expands, the raisins become increasingly separated from each other (Fig. 1E). Although the bread expands uniformly, neighboring raisins move apart more slowly compared to raisins that are far apart. Similarly, although the universe is expanding at a uniform rate, neighboring galaxies move apart from one another more slowly than ones that are far apart.

Looking back into the past is like watching a DVD as it rewinds. According to the Big Bang theory, the age of the universe is the time it would take to "rewind the tape" such that all the galaxies in the universe, and all the mass within them, would return to a single region of dense hot gas.

It is tempting to envision that, at the time of the explosion, the universe originated from a single point. However, since at any one time the universe incorporates everything that exists, it has no edges and therefore no center. In the strictest sense, then, the universe at the time of the Big Bang should be thought of as an infinitely small, high-energy region with an infinite density. Immediately after the Big Bang, (just 10 millionths of a second later), the universe was filled with high-energy particles having a temperature of over 1 trillion (10^{12}) °K.

This region would have had so much heat that only radiation could exist. Space and time, in our conventional sense, would have had no meaning; they would have been infinitely warped around this region and nothing, not even empty space, would have existed beyond it.

According to Einstein's theories, space and time are linked. Without space, there is no time. So the question "what happened before the Big Bang?" cannot be asked because there was no "before." Time and space would have begun in the same instant as the explosion.

If the Big Bang had been a perfect explosion, all the components of the universe would have been homogeneously distributed. There would be no stars, and therefore no planets. But the explosion did not distribute matter homogeneously. The inhomogeneity allowed regions of concentrated matter to become centers of high gravity that rapidly accumulated mass to become large groupings of stars, or galaxies.

The original hypothesis of the Big Bang has passed so many tests that it was elevated to the status of a theory. In fact, in 1965, two astrophysicists detected its "echo." While measuring the brightness of the sky, they inadvertently recorded a peculiar noise with their antennae. At first, they attributed the noise to pigeons living inside the antennae. However, the noise persisted after the pigeons and their droppings were removed. Eventually, they concluded that the noise represented a weak cosmic microwave radiation that had been predicted as the remnant of the catastrophic explosion implicit in the Big Bang theory.

(a)   (b)

Figure 1E   *The Universe as a Loaf of Bread*

The uniform expansion of the universe is represented by the expansion of raisin bread when it is baked. As the bread expands, all raisins move apart at a rate that is proportional to their separation. Neighboring raisins move apart from one another more slowly than raisins that are far apart.

According to this theory, the Solar System began when a swirling cloud of dust and gas—called a *nebula*—flattened into a disk as it contracted under the influence of gravity.

What caused the initial contraction is uncertain, although the demise of a nearby star in a massive explosion known as a *supernova* may have provided the trigger. Whatever the cause, the contraction, once it had begun, initiated an irreversible chain of events. As the cloud shrank, its particles moved closer together, causing the gravitational attraction between them to increase (see Science Refresher: **Some Basic Concepts in Chemistry and Physics**). As a result, further contraction took place until 90 percent of the nebula had condensed to form the Sun. In the process, temperatures within the Sun soared until it ignited to become a star.

After the Sun formed, the remaining portions of the nebula slowly cooled and, as they did so, substances that require high temperatures to remain molten started to solidify. These substances, which included elements such as iron, nickel, silicon, calcium, sodium, and potassium, condensed and joined together to form small bodies called *planetesimals,* ranging in size from a few centimeters to several tens of kilometers. Owing to gravitational attraction and turbulence in the swirling nebula, collisions occurred between the planetesimals and they coalesced to form larger objects, called *protoplanets.* Protoplanets, which are bodies that are more than 100 kilometers (62 miles) in diameter, are large enough to become planets. One of these protoplanets eventually became Earth. Others became the terrestrial planets Mercury, Venus, and Mars. The lighter *volatile* elements (i.e., those that are gaseous at very low temperatures), escaped the gravitational clutches of the terrestrial planets. These volatile elements, including hydrogen and helium, were blown from the inner to the cooler outer part of the Solar System by the *solar wind,* the high energy stream of particles emanating from the Sun.

> **CHECK YOUR UNDERSTANDING**
> ● How does the solar nebula theory account for the origin of the Solar System?

THE FORMATION OF THE PLANETS

The planets in the Solar System that are closest to the Sun, Mercury, Venus, Earth, and Mars, have relatively high densities, indicating that, like Earth, they are essentially rocky spheres. Together they are termed the terrestrial planets after *terra,* the Latin word for land.

There are two popular models for the formation of the terrestrial planets by the process of accretion. *Accretion* is the process of growth by the gradual accumulation of additional layers or matter. According to the *homogeneous model* of planetary accretion, the protoplanets formed from planetesimals that had essentially the same composition. This model implies that the internal layering of the innermost planets must have been the result of internal processes that took place after accretion (Fig. 1.14a).

The *inhomogeneous model,* on the other hand, maintains that the nebula continued to cool as the planets formed, implying that the composition of the newly formed planetesimals may have changed with time (Figure 1.14b). The wide variety of meteorite compositions suggests that the raw materials for the innermost planets were indeed quite varied, which is consistent with the inhomogeneous model. According to this model, an iron-rich core accumulated first from the earliest particles. The planets then grew by the accretion of lighter more silicate-rich planetesimals.

According to both models, the formation of the innermost planets would have occurred so rapidly that the heat generated by the process of accretion would not have had time to dissipate. The intense heat in the primitive Earth may have caused the interior of the planet to melt. Dense chemical elements such as iron and nickel sank and concentrated in its core. Lighter, more volatile elements, by contrast, would have risen to the surface of the planet. The heat released as the innermost planets accreted may have boiled off the lightest gaseous elements. Elements with densities in between these extremes, such as silicon, potassium, sodium, and calcium, became concentrated in the outermost layers of Earth.

In the outer part of the Solar System, temperatures would have been low enough for volatile components to condense around any small rocky core, forming a layer of hydrogen ice containing smaller quantities of helium and minor amounts of silicate minerals. Like, the terrestrial planets, these outer planets also grew rapidly. When they attained a mass about 10–15 times that of Earth, they had sufficient gravitational attraction to capture gas directly from the solar nebula, until the gas in that region of the nebula was exhausted.

> **CHECK YOUR UNDERSTANDING**
> ● How does the solar nebula theory account for the rocky nature of the terrestrial planets?

These outer planets have densities that are significantly lower than those of the terrestrial planets, in spite of their greater size. They are essentially gaseous bodies with poorly defined surfaces.

ORIGIN OF THE EARTH-MOON SYSTEM

Recent models suggest that the Moon originated from the impact of a very large object (perhaps as large as Mars) with Earth's surface soon after Earth formed (see Living on Earth: **Earth's Seasons—A 4.5-Billion-Year Legacy?**). There has been some debate whether the Moon formed from the material ejected from Earth or from the remnants of the colliding object itself (Fig. 1.15). Chemical analyses of *lunar rocks* (rocks found on the Moon) show subtle chemical differences from those of Earth rocks, and at present the balance of the evidence suggests that the Moon may be largely made up of fragments of the colliding object. Irrespective of which model is correct, some of the debris from the impact was caught in Earth's gravitational field and cooled and condensed to form the Moon.

LIVING ON EARTH

Earth's Seasons—A 4.5-Billion-Year Legacy?

Earth is not an upright planet; it leans a little on its side and is tilted relative to the plane of its orbit around the Sun. The amount of tilt has varied throughout geologic time from about 21.5° to 24.5°. It is presently inclined at 23.5° (Fig. 1F) and its tilt is decreasing by a small amount annually.

But how did the tilt originate? The most widely accepted hypothesis is that the tilt was caused by the same massive impact that gave rise to our Moon. The tilt of Earth's axis is the reason we have seasons, so if this explanation is correct, the legacy of this impact that occurred more than 4.5 billion years ago is still felt by all of Earth's inhabitants.

Because Earth's tilt is 23.5°, on one day each year the Sun shines directly overhead at 23.5° N latitude. This day is June 21, and the latitude is the Tropic of Cancer. On this day, all localities in the Northern Hemisphere have maximum daylight (summer), while all localities in the Southern Hemisphere have their longest nights (winter). On December 21, the situation is reversed and the Sun shines directly overhead at 23.5° S latitude, which is the Tropic of Capricorn. In the six months between these two cardinal dates, the Sun shines most directly on differing positions between these latitudes, gradually shifting over a six-month period from one tropic to the other. From June 21 until December 21,

the length of daylight in the Northern Hemisphere progressively decreases to a minimum, while it increases to a maximum in the Southern Hemisphere (Fig. 1G).

This means that, in any year, there are only two occasions when the Sun shines directly on the equator. These two occasions occur on March 20 and September 22, the days that are known as the equinoxes. On these two days, all areas on Earth have 12 hours of daylight and 12 hours of darkness.

Earth is hottest in the tropics and coldest near the poles because the Sun's rays strike Earth most directly in the tropical region and least directly at the poles. In turn, this influences the amount of solar energy that reaches the surface. Solar energy reaching Earth's surface in higher latitudes is less intense than it is in lower latitude equatorial regions because the radiation travels through a thicker layer of atmospheric gases, which absorb, reflect, and scatter it (see Figure 1F).

If Earth's axis were not tilted, the amount of solar energy reaching its surface at any particular location would remain the same throughout the year and we would not experience seasons. The familiar cycle of spring, summer, fall, and winter, which we all take for granted, may therefore owe its origin to a chance impact very early in Earth's history.

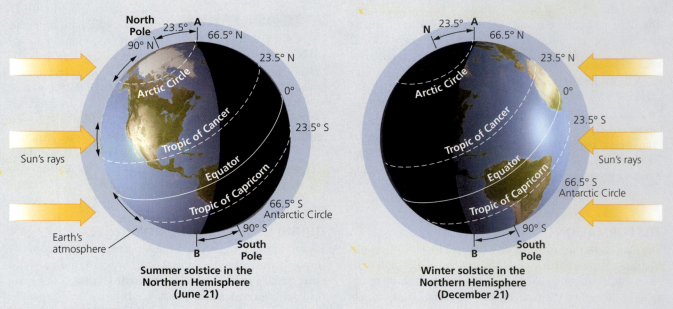

Summer solstice in the Northern Hemisphere (June 21)

Winter solstice in the Northern Hemisphere (December 21)

Figure 1F Earth's Tilt

The tilt of Earth's axis relative to the Sun is 23.5°, so the boundary between night and day (AB), is inclined at an angle of 23.5° to Earth's axis. The northern hemisphere is tilted toward the Sun in June, and experiences more direct and longer hours of sunlight. However, the sunlight that reaches Earth's surface in the far northern latitudes passes through a thicker layer of atmosphere and so is less intense than the sunlight shining on equatorial latitudes. On June 21 (left), the Sun shines directly on latitude 23.5° N, which defines the northern limit of the tropics (the Tropic of Cancer). In this orientation, areas south of the Antarctic Circle receive no sunlight. On December 21 (right), the Northern Hemisphere is tilted away from the Sun and experiences longer hours of darkness. Locations north of the Arctic Circle receive no sunlight at all.

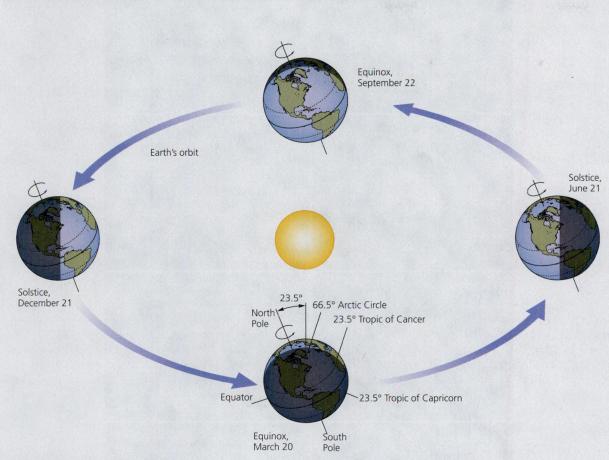

Figure 1G   Earth's Revolution

Earth revolves around the Sun with the Northern Hemisphere pointing toward the Sun during the Northern Hemisphere summer and away from it during the Northern Hemisphere winter. Because Earth's axis is tilted at 23.5°, the Sun shines directly on latitude 23.5° N (the Tropic of Cancer) on June 21. This date marks the summer solstice in the Northern Hemisphere when the day is longest. On December 21 (the winter solstice), the Sun shines directly on latitude 23.5° S (the Tropic of Capricorn) and the Northern Hemisphere day is shortest. On the equinoxes of March 20 and September 22, day and night are the same length (12 hours) everywhere on Earth. The diagrams are not to scale.

EARTH'S INTERNAL LAYERING

The solar nebula theory and the models portrayed in Figure 1.14 imply that the development of Earth's internal layering occurred very early in Earth history. Gravitational attraction caused planetesimals to grow into a protoplanet. This process released sufficient heat to cause large-scale melting of Earth's interior, allowing the densest elements, such as iron, to sink toward the center of the planet and lighter elements, such as silicon, to rise toward the surface. In this way, Earth became layered with the heaviest elements at its core, and the lightest materials, such as gases, escaping from the surface altogether to form Earth's gaseous envelope, which we refer to as Earth's *atmosphere*. As the planet cooled, gaseous water in the atmosphere cooled to form Earth's oceans. Hence, Earth is layered according to the "heaviness" or *density* of its component materials, and its layers vary in both *chemical composition* (the relative abundance of chemical elements) and *physical composition* (the relative abundance of solid, liquid, or gas) (Fig. 1.16).

Crust, Mantle, and Core

The principal divisions of Earth's interior based on variations in their chemical and physical properties are shown in Figure 1.17. The three main divisions, those of crust, mantle, and core, are based on differences in their chemical

> **CHECK YOUR UNDERSTANDING**
> ◉ Explain the role differences in density played in the formation of Earth's internal layers.

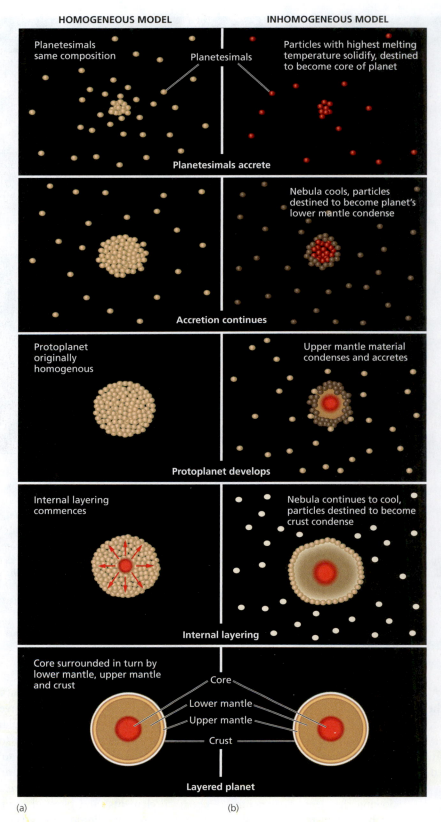

Figure 1.14   Two Models of Planetary Accretion

Two models shown here side-by-side in sequence have been proposed for the growth and early development of the terrestrial planets. (a) According to the homogeneous model, the planets grew by the accretion of planetesimals with uniform composition. Layering (core, red; lower mantle, dark brown; upper mantle, pale brown) developed from internal processes that took place after accretion. (b) In the inhomogeneous model, planetesimals of different composition were generated as the solar nebula cooled. The first particles to accumulate were heavy metals (red) followed by silicates (brown). The planet's internal heat further aided differentiation into layers.

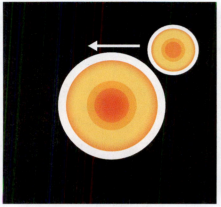

(a) A Mars-sized asteroid nears the planet

(b) The asteroid collides with Earth

(c) Many fragments are hurtled into space

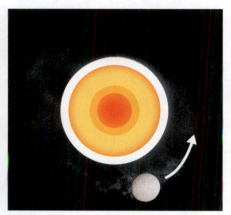

(d) Some particles coalesce to form the Moon

Figure 1.15   Origin of the Earth-Moon System

Origin of the Earth-Moon system from (a–b) the impact of a large Mars-sized body on Earth's surface. Note that the Earth's internal layers were formed at the time of impact (although they were probably not totally solid). Condensing debris from the impact of both bodies (c) formed an orbiting cloud around Earth (d) from which the Moon formed.

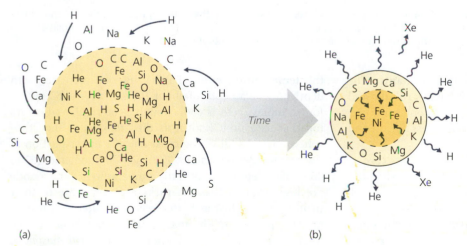

Figure 1.16   Earth's Internal Layers

This idealized model shows how a chemically layered Earth could evolve from a chemically uniform protoplanet (a) to a planet layered according to density (b). The elements in the figure are represented by their symbols from the periodic table (see Appendix B). The densest elements (iron, Fe, and nickel, Ni) sank to the core, whereas the lightest elements (hydrogen, H, and Helium, He) rose out of the Earth and escaped into space. Other elements (silicon, Si; aluminum, Al; oxygen, O; magnesium, Mg; calcium, Ca; potassium, K; and sodium, Na) became concentrated in Earth's mantle and crust.

composition. The outermost layer of Earth is a rigid shell called the **crust**, which is subdivided into two types: oceanic and continental. *Oceanic crust* is dominated by iron- and magnesium-rich igneous rocks called basalts. Oceanic crust is quite thin, ranging from 3 to 15 kilometers (2 to 9 miles) in thickness, and is generally less than 180 million years old. *Continental crust,* on the other hand, is dominated by igneous rocks that contain significantly more silicon and less iron and magnesium. Continental crust is lighter and thicker than oceanic crust, averaging 35 kilometers (22 miles) in thickness, and contains rocks up to 4 billion years old.

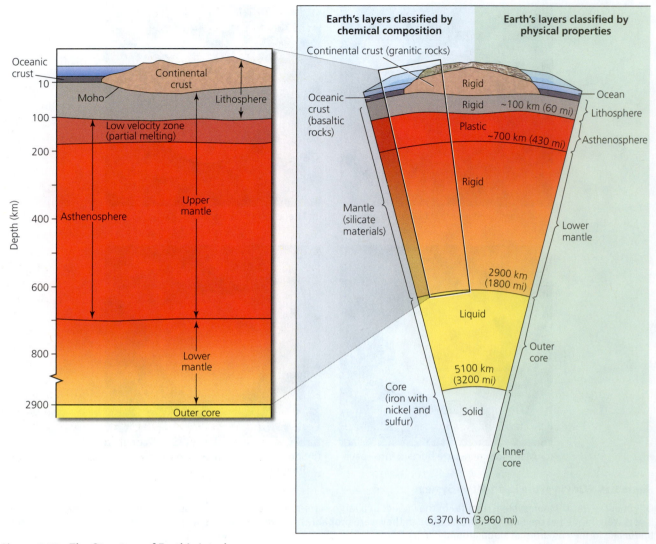

Figure 1.17   The Structure of Earth's Interior

The main figure shows the arrangement and composition of Earth's internal layers. Note the figure is not to scale and that thicknesses of the outer layers have been exaggerated for the purpose of clarity. The inset shows subdivision of Earth's outer layers into crust and mantle based on composition, and lithosphere and asthenosphere based on rigidity.

Beneath the crust is a solid layer called the **mantle**. The mantle comprises about 83 percent of Earth's volume and is compositionally distinct from the crust, containing significantly more iron and magnesium, and less silicon and aluminum.

Earth's center or **core** is dominated by iron, with subordinate amounts of nickel, sulfur, and other elements. It is subdivided into a liquid **outer core** and a solid **inner core**. Flow of the iron-rich liquid in the outer core is responsible for Earth's magnetic field.

> **CHECK YOUR UNDERSTANDING**
> ● What are the two types of crust at Earth's surface? What are the key differences between them?

Lithosphere and Asthenosphere

We can also divide the outer parts of Earth's interior strictly on the basis of changes in physical composition as the temperature within Earth's interior increases with depth. Above a depth of about 100 kilometers (62 miles), for example, Earth is cool enough to be rigid, and moves together as a unit, known as the *lithosphere*. The lithosphere includes all of the crust *and* the upper portion of the mantle. Between about 100 and 700 kilometers (62 to 455 miles) depth, however, the temperatures in the mantle are high enough that the rocks are very close to their melting points. Just as butter becomes softer as it warms up, this portion of the mantle is soft as a result of the high temperatures and is known as the *asthenosphere*.

As we shall see in Chapter 2, it is because the asthenosphere is soft, and so physically weak, that the rigid lithosphere is able to move above it, an essential feature of plate tectonics.

> **CHECK YOUR UNDERSTANDING**
> ● Compare and contrast Earth's inner layers.

1.6 SUMMARY

- The solar nebula theory accounts for the formation of Earth and the other planets in the Solar System.

- Earth's internal layers were formed very early in its history. Denser elements sank toward the core, whereas lighter elements rose toward the surface, resulting in layering.

- Earth's internal layers are subdivided on the basis of chemical differences into the core, which is surrounded by the mantle overlain by the crust.

- Earth's internal layers are also subdivided on the basis of physical differences. The hard outer shell of Earth is the lithosphere, which includes all the crust and part of the upper mantle. The lithosphere overlies a weak layer in the upper mantle known as the asthenosphere.

1.7 Earth's Internal Heat

It has long been observed that temperatures in mines increase with depth. As we will discuss in Chapter 8, in 1897 Lord Kelvin used this observation to argue that Earth is losing heat and that it must have been hotter in the past. In fact, if we had to sum up the driving force behind Earth's 4.54 billion years of evolution in a single sentence, it would be this: Earth is cooling down. Throughout its history, nearly all of the processes affecting Earth on a global scale—such as the formation of its rocks and minerals, the evolution of life, the origin and composition of the atmosphere, continents and oceans, and the development of our present-day environment—are related to this cooling.

Recent research has shown that the temperature within Earth's core is about 6000°C (10800°F)—some 1000°C (1800°F) hotter than previously thought. Experiments and theoretical calculations show that the temperature at the base of the mantle is about 4000°C (7200°F). The temperature at the top of the mantle (or the base of the crust) depends on a number of factors we will discuss later in this book, but in many places it is about 600°C (1080°F). As Lord Kelvin correctly deduced, Earth gets progressively hotter with depth. Much like heat flows out of the surface of a hot radiator to warm a cold room, Earth's heat flows from its relatively hot core toward its relatively cool surface. The dominant mechanism of heat flow within Earth is by **convection**, where material adjacent to a heat source is warmed, becomes less dense and rises, and cold material moves in to takes its place (see Science Refresher: **The Flow of Heat**).

Earth loses more heat from its surface than it absorbs from the Sun. This is achieved by the flow of heat (known as *heat flow*) from Earth's core to its surface, where the heat escapes.

The heat flow escaping into the atmosphere from Earth's surface can be measured at any location on Earth regardless of how geologically tranquil the locality may appear to be. The amount of heat flow varies from one location to another (Fig. 1.18) implying that heat escapes more efficiently from some parts of the surface than it does from others. The figure reveals an obvious correlation between

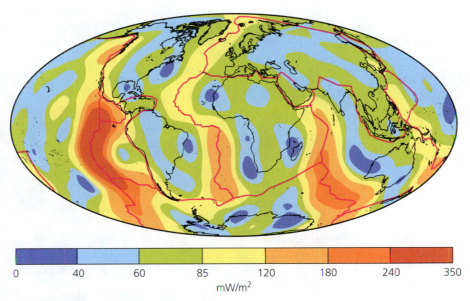

Figure 1.18   Heat Flow on Earth's Surface

Heat flow measurements on Earth's surface show the escape of heat from Earth's interior. Note how the amount of heat flow varies significantly from one location to another. The units are milliwatts (mW) per square meter (mW/m²). A watt is a unit of power, most familiar in its use in light bulbs.

SCIENCE REFRESHER

The Flow of Heat

The flow of heat from warmer to cooler regions has been one of the most important processes governing the evolution of Earth. Regions with higher temperature have more energy, and this energy flows toward cooler regions, which have less energy. This flow occurs by several mechanisms, the most important of which are conduction, convection, and radiation.

Conduction primarily occurs in solids (Fig. 1H). When a solid object is heated at one end, heat flow occurs as collisions between atoms within the substance. This transfers energy from the warm end to the cool end. Some substances (such as metals) are excellent conductors and heat is efficiently transferred from one end to the other. Other substances (such as wood) are poor conductors, which is why a hot pot with a wooden handle is safer to lift from a stove than one with a metal handle.

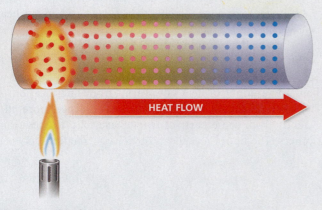

HEAT FLOW

Figure 1H   Conduction

When a solid body is heated at one end, energy is transferred toward the other end as collisions occur between atoms within the substance.

Convection is the process that takes place in liquids, gases, and solids that are close to their melting point. In these substances, hotter material moves upward toward cooler regions, while cooler material sinks to replace it. In this way, heat is transferred from the hotter region to the cooler one. We see convection in our everyday lives in the slow overturn of a simmering saucepan of soup, in the way heat circulates in a room, and in the upward movement of warm air in the atmosphere.

Convection actually involves the movement of a substance from the hot region to a cool region, as opposed to the conduction where the transfer occurs on a microscopic scale. In the case of a room (Fig. 1I), heat transferred from the radiator to the nearby air causes the air to expand so that its density decreases and it starts to rise through the surrounding colder, heavier air. As the hot air rises, it cools and becomes denser once again, while cold air from below moves in to replace it. However, this air in turn is heated by the radiator and so expands and rises. In this way, the air in the room circulates, with the warm air rising and the cooler air sinking in compensation.

Just such a process is believed to cause circulation in Earth's mantle where the rocks, although solid, are hot enough to flow like a slow-moving fluid. In the mantle, it is this circulation that is ultimately responsible for the movement of Earth's tectonic plates (which we discuss in detail in Chapter 2).

Radiation is the heat you feel near a stove or a campfire. It is also the way heat travels from the Sun to Earth (Fig. 1J), and exposure to this radiation causes sunburn. Unlike conduction and convection, radiation is a form of heat flow that does not operate through a substance.

In most instances, conduction, convection, and radiation combine to transfer energy from hot regions to cold regions (Fig. 1K), although one form may dominate over the others.

WARM AIR

HOT
AIR

Heat
source

COOL
AIR

COLD AIR

Figure 1I   Convection

Convection is illustrated by the circulation of heated air in a room.

Figure 1J Radiation

Radiation is the form of heat transfer that occurs from the Sun to Earth. All bodies radiate thermal energy in proportion to their temperature.

Figure 1K Conduction, Convection, and Radiation Working Together

Conduction, convection, and radiation all transfer energy from hot regions to cold regions. In this example, conduction transfers the energy through the metal handle of the cooking pot, convection transfers it through the water, and radiation through the air.

CHECK YOUR UNDERSTANDING

○ Why do measurements of heat flow at Earth's surface vary from one locality to another locality?

regions of high heat flow and Earth's crust beneath the oceans, indicating that Earth is unusually hot in these regions and that heat flow is highest where the surface is hottest. As we shall see in Chapter 2, these regions of high heat flow are areas of voluminous magmatism intimately linked to the process of plate tectonics.

1.7 SUMMARY

- Earth gets progressively hotter with depth and heat flows from Earth's hot core to its cooler surface.

- As a consequence of heat flow, Earth is cooling down.

1.8 Earth as a System

Until very recently, scientists obtained much of the evidence related to the evolution of Earth from the rocks and minerals that occur at or near Earth's surface. The term *geology* became synonymous with the study of rocks and minerals despite the wide-ranging scope of its definition. However, recent technological advances have allowed geologists to model the chemical, physical, and organic features that rocks and minerals preserve, and have provided insight into both ancient and modern atmospheres and oceans. Voyages into space have also allowed us to view Earth in the context of its neighbors within the Solar System.

As a result of these developments, geologists now know that events within the lithosphere, such as volcanic eruptions, have a profound effect on the composition of the atmosphere and oceans. Even a single volcanic eruption

(a)

Figure 1.19   1991 Eruption of Mount Pinatubo

(a) The eruption of Mount Pinatubo in the Philippines on June 12, 1991, was the second largest volcanic eruption of the twentieth century. It is estimated to have ejected more than 5 cubic kilometers (1.2 cubic miles) of material, five times more than the eruption of Mount St. Helens in 1980. Ash, gas, and aerosols were ejected into the atmosphere to heights of up to 30 kilometers (19 miles). (b) These maps illustrate the distribution of the aerosols before (above) and after (below) the eruption. Within 21 days the ejected aerosols had become distributed into a band that encircled Earth. The ejected gases included an estimated 18 million tons of sulfur dioxide, a gas that combines with rain droplets in the atmosphere to produce acid rain.

Aerosol distribution from Pinatubo Volcano

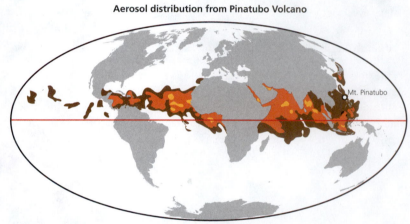

May 28 through June 6, 1991 (before eruption)

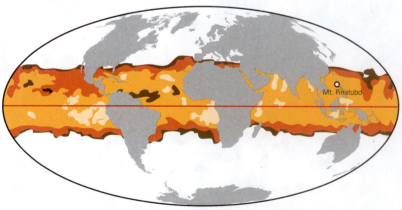

July 7 through July 10, 1991 (after circling the Earth)

| Low aerosol density | Medium aerosol density | High aerosol density | Very high aerosol density |

(b)

can have a dramatic effect. For example, in 1991, the eruption of Mount Pinatubo in the Philippines (Fig. 1.19) rocketed particles into the atmosphere to heights of up to 30 kilometers (19 miles). In only 21 days, the ejected *aerosols* (gas mixtures) encircled Earth. These aerosols included an estimated 18 million metric tons of sulfur dioxide, a gas that combines with rain droplets in the atmosphere to produce *acid rain,* which is harmful to plants and animals and is associated with the emissions from coal-fired power plants. The volume of aerosols ejected by Mount Pinatubo is comparable to the entire annual industrial output of sulfur dioxide in the United States.

Given that this is the effect of only a single volcanic eruption, imagine the profound influence that all the volcanic eruptions that have occurred throughout Earth's history have had on the composition of the atmosphere and the condition of life on Earth's surface.

The release of aerosols is not the only way that volcanoes create connections. For example, after magma erupts from volcanoes and reaches the surface, it is called *lava.* Lavas that erupt from volcanoes bring new chemicals from the interior of Earth to the surface. As lavas cool, they solidify to form part of the solid Earth. Because of this chemical enrichment, soils in the vicinity of volcanoes are often very fertile and promote plant life (Fig. 1.20). Exposure of the lavas and their soils to the air and rain causes the lava and soil to decompose and disintegrate, so that they are more readily removed and transported to the sea. This process brings vital chemical nutrients, such as calcium, to the sea, and helps ocean life to flourish. Recall that the same volcano also vents gases that, over the course of Earth history,

have profoundly influenced the composition of our atmosphere. In addition to the sulfur dioxide mentioned earlier, one of these gases is *water vapor* (water in its gaseous form), which cools and condenses to form rain and may ultimately drain to the sea. The circulation of water vapor in the atmosphere and the oceans is driven by heat from the Sun, that is, by solar energy.

Other than volcanoes, there are few other ways by which new chemicals are brought to the surface of Earth. Since our bones, soft tissue, and blood are made up of chemicals, and we live on the planet's surface, it follows that many of the chemicals in our bodies are likely to have originally come from Earth's interior by way of a volcano. When or where this happened we can never know. Indeed, even adjacent chemicals in our bodies are most unlikely to have arrived at Earth's surface at the same time, or even from the same volcano. So if you feel a little scrambled today, there's a good, natural reason for it!

A volcanic eruption is an example of how a *geological* event results in *chemicals* being transported (a *physical* process) from the heated interior of Earth to the exterior, and upward into the atmosphere or across Earth's surface to the oceans. As this happens, life—Earth's *biology*—flourishes. This example shows that the natural world does not subdivide scientific disciplines the way college courses do. Our ability to understand our planet and its environment, and our willingness to live in harmony with nature, may therefore depend on our ability to break down the artificial barriers between the traditional sciences. Then we can begin to understand Earth, and its complex evolution through geologic time.

Figure 1.20 Bali

Fields on the slopes of a volcano in Bali, Indonesia, are terraced to retard erosion and retain nutrients. Periodically replenished by volcanic ash rich in elements such as calcium and phosphorus that are essential to plant growth, Balinese soils produce some of the most productive croplands in the world.

These examples speak to the connections between different components of the planet, which rather than behaving independently, interact together as a *system* on, or close to, Earth's surface. The four main components of the Earth system—the solid Earth, oceans, the atmosphere, and life—are intimately linked (Fig. 1.21). These *surface components,* so-called because they interact at Earth's surface, are as follows:

- The **geosphere**: the solid Earth comprising the rocky outer layer—the lithosphere—and its hot interior.

- The **hydrosphere**: Earth's water, whether it is in oceans, lakes or rivers, frozen in ice and snow, or trapped underground in soils and rock fractures.

- The **atmosphere**: Earth's gaseous envelope and the air we breathe. It is mostly made up of nitrogen and

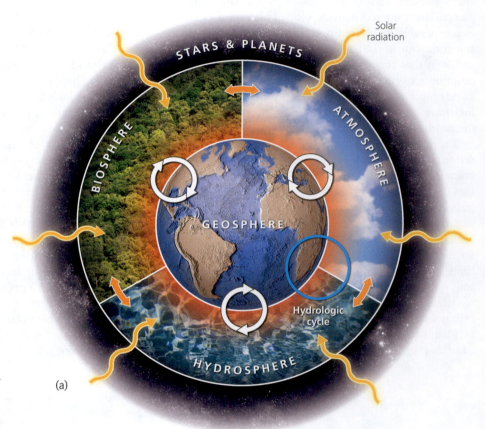

(a)

Figure 1.21   Earth as a System

(a) Earth's surface components interact and mutually influence one another. Processes in Earth's interior have led to the formation of its rocky outer layer (solid Earth), air (atmosphere), and water (hydrosphere). These components, in turn, provide the basic prerequisites for life (biosphere). The solid Earth is placed in the center because it has been the main source for the chemicals of the atmosphere and hydrosphere, and has provided nutrients for the biosphere. Arrows connecting components go in both directions, depicting the interactions between them. Interactions between the atmosphere, hydrosphere, and biosphere are also greatly influenced by solar radiation, shown in the outer part of the diagram. The position of the hydrologic cycle of Figure 1.22 is the circle in the bottom right of the diagram. (b) In this view of the French Polynesian island of Bora Bora in the southwestern Pacific Ocean, all four of Earth's surface components are visible: the solid Earth—the island itself; the hydrosphere—the ocean around the island; the atmosphere—the sky above; and the biosphere—the life the island supports.

(b)

oxygen, with smaller but important concentrations of other gases such as carbon dioxide, water vapor, and ozone.

- The **biosphere**: the realm of life, comprising plants and animals that are either living or in the process of decay.

Earth's four surface components, which are sometimes referred to more colloquially as *rock, water, air,* and *life,* act as *reservoirs* (storehouses) of the energy and the resources of our planet. These natural reservoirs are not only linked together at Earth's surface but also interact with the light and heat energy radiated from the Sun. One important example of this interaction is the **hydrologic cycle** (Fig. 1.22), which describes the motion of water at or near Earth's surface. Radiant energy from the Sun strikes the ocean surface and initiates the evaporation of ocean water into the atmosphere. Winds then blow this moisture onto land where it falls as rain or snow. A small amount of this moisture is absorbed by plant and animal life, or the ground. But most eventually drains into major river systems and is returned to the oceans to close the cycle.

At first glance, it may seem as if tracking the evolution of Earth's systems throughout geologic time is an impossible task. For example, how can we trace the evolution of the atmosphere or the hydrosphere in the past? Because the biosphere, hydrosphere, and atmosphere all interact with the solid Earth, much of that information is stored in the minerals and rocks of the solid Earth. To retrieve this information, we need only to crack the code. We can monitor the evolution of the biosphere, for example, by studying the fossil record. And the fundamental insight that "the present is the key to the past" provides us with the means to investigate the evolution of the hydrosphere and atmosphere.

For example, the existence of ancient oceans can be deduced by comparing ancient sedimentary deposits with those that occur along the flanks of modern oceans. Using this approach we can

CHECK YOUR UNDERSTANDING

- How is the hydrologic cycle an example of the interaction among Earth's components?

- How do volcanic eruptions bring new chemicals to Earth's surface?

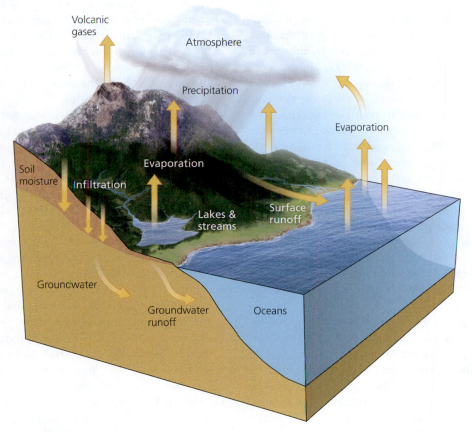

Figure 1.22   The Hydrologic Cycle

The hydrologic cycle describes the movement of water at or near Earth's surface. All of Earth's components participate in this cycle, which is driven by energy from the Sun. First, this solar energy causes evaporation of ocean water. The moisture in the atmosphere is blown onto land, where it falls as rain or snow. The moisture may also percolate into the soil and rocks, or pond in lakes, or be retained in plants and animals. However, the vast majority of the water drains into rivers, which return the water to the oceans, completing the cycle.

identify ocean basins as old as 3.5 billion years. Similarly, some rocks, such as coal, form under anomalously warm conditions, whereas others, such as glacial deposits, require conditions to be anomalously cool. Thus, the predominance of such rocks in particular intervals of geologic time provides insights into atmospheric composition, and the natural triggers of climate change.

1.8 SUMMARY

- Earth has four main surface components, or reservoirs, that interact: the geosphere (solid Earth), the hydrosphere (water), the atmosphere (air), and the biosphere (life).

- Earth's hydrologic cycle, the way water circulates at or near its surface, and volcanic eruptions, which bring new chemicals from Earth's heated interior to the surface, are examples of linkages between Earth's four surface components.

1.9 What Does a Geologist Do?

As the content in this chapter makes clear, geology is arguably the most diverse of all the sciences and geologists are employed in a wide variety of professions, many of which command high salaries. Geologists are also in demand, with jobs in geology expected to increase by more than 20 percent over the next decade. Geology is an "outdoor" science and geologists often work "in the field" as well as the office and so commonly have the opportunity to travel. For this reason, jobs in geology are routinely ranked among the best in North America, combining high salaries with an excellent quality of life, as gauged by such criteria as job satisfaction, opportunity, variety, and benefit to society. *Forbes Magazine*, for example, recently ranked geology as the seventh most valuable college major.

But what do geologists do? The variety of their occupations (Fig. 1.23) can be surprising. As custodians of Earth's resources and stewards of the environment, geologists are involved in solving some of society's biggest questions and some of its most challenging problems—from pondering the possibility of life on Mars to probing the causes and effects of global climate change.

Geologists are engaged in the search for the natural resources upon which society depends, such as water, fossil fuels, and metals, and in developing these resources in ways that safeguard the environment. Geologists are at the forefront of assessing natural hazards, such as landslides, storms, earthquakes, and volcanic eruptions, and in finding ways to reduce the suffering and property damage that such natural hazards cause.

From the siting of major construction projects to the placement of septic tanks, geologists are involved in monitoring the impact of human activity on the environment and, where the environment has been damaged by such activity, geologists are employed to assess the damage

caused and devise methods for its remediation. Geologists advise governments on issues such as climate change and energy initiatives, and help political authorities devise policies for environmental protection, resource management, and health and human safety. Geologists also form the core of our nation's geological surveys. They are involved in the exploration of the Solar System, in the study of the oceans and our atmosphere, and in agriculture, land use, and soil conservation efforts. And the friendly faces of the park rangers who bring alive the splendor of our national parks are often those of geologists.

Geologists are also employed in academia and work at universities and research institutes where much of the fundamental research is conducted and where many of the major new discoveries that move science forward are made. This research and these new discoveries provide the basis of our ever-evolving understanding of Earth.

Geology is a young science and was revolutionized by the unifying theory of plate tectonics just 50 years ago (see Chapters 2 and 9). As a result, geology is an exciting discipline with much to discover and many major advances yet to be made. What causes continents to move and why do they sometimes come together to form supercontinents? How has the movement of continents over geologic time affected the evolution of the planet? Why is our climate changing and what are the consequences of global warming? How did life first evolve on Earth and what caused the extinction of animals such as the dinosaurs? Is there life on other planets and how can we tell? Questions like these highlight just a few of the important problems geologists are considering and exploring.

The broad scope of geology is reflected in its many subdisciplines and specializations, including many topics which we cover in depth in this book. *Petrologists* and *mineralogists,* for example, specialize in the study of rocks and minerals. *Paleontologists* study fossils. *Geophysicists* specialize in the physics of Earth and phenomena such as magnetism, gravity, and earthquake waves. *Geochemists* study Earth's chemistry recorded in rocks and water on land, as well as in its atmosphere and oceans. *Planetary geologists* study planets and moons, and the evolution of the Solar System. *Petroleum geologists* specialize in the exploration and production of oil and natural gas, whereas *economic geologists* do the same for Earth's mineral resources.

Hydrologists and *hydrogeologists* respectively study the movement of water on Earth's surface and in the subsurface, while *oceanographers* study the world's oceans. *Environmental geologists* work to solve problems associated with natural hazards and the effects of human activity on the environment. *Volcanologists* specialize in the study of volcanoes and the prediction of volcanic eruptions. *Engineering geologists* apply geology to the construction industry and the building of bridges and dams. And the list goes on.

If you have natural curiosity or an interest in scientific detective work, if you are curious about our planet and the way it works, if you are fascinated by landscape or concerned for the environment, if you wish to become involved

Figure 1.23 What Geologists Do

(a) A geologist of Crew 125 EuroMoonMars B mission simulates living on Mars at the Mars Desert Research Station in Utah. (b) At Baffin Island in the Canadian Arctic, geologists use a coring system to sample mud below the surface in order to test for evidence of global warming. These cores enable evaluation of a more complete and continuous record of changing climate conditions. (c) Geologists core for coal in the Gobi Altai of Mongolia to try to determine the volume of coal that may be hidden below the surface. (d) Geologists analyze digital data to determine the location of subsurface oil and gas. (e) In Columbia, geologists monitor an active volcano. (f) A water trace is conducted by a geologist in Greenland using a fluorescent dye as part of a climate change study. The direction and rate of flow of meltwater from the glacier is determined by following the dye. (g) A geologist examines a core to assess the development of a quarry in New Zealand. (h) A geologist works as a ranger at Hawaii Volcanoes National Park. A ranger's duties typically include park conservation, natural resource management, and the development and operation of interpretive and recreational programs.

CHECK YOUR
UNDERSTANDING
○ Name and explain the
various subdisciplines and
specializations in geology.

in the management of our natural resources or the science of global change, if you wish to study the Solar System and the search for life on Mars or the billions of habitable *exoplanets* (planets outside our Solar System) that are suspected to exist, or if you are interested in any of geology's many subdisciplines, geology may be the career for you.

1.9 SUMMARY

- Geologists are employed in a wide variety of professions that reflect society's need for natural resources and stewardship of Earth.

Key Terms

atmosphere 32	fossil 13	law 8	physical geology 6
bedrock 12	geologic time scale 12	lithification 16	rock 13
biosphere 33	geology 6	magma 13	rock cycle 14
convection 27	geosphere 32	mantle 26	scientific method 7
core 26	hydrologic cycle 33	metamorphic rock 13	sedimentary rock 13
crust 25	hydrosphere 32	mineral 13	theory 8
eon 12	hypothesis 7	outcrop 12	weathering 15
era 12	igneous rock 13	outer core 26	
erosion 16	inner core 26	period 12	

Key Concepts

1.1 GEOLOGY AND THE SCIENTIFIC METHOD

- Geology employs science to understand Earth, its materials, behavior, environment, and evolution.
- Geologists, like other scientists, use the scientific method, whereby data and observation support scientific hypotheses, which are continually challenged and tested.

1.2 THE BIRTH OF GEOLOGY

- James Hutton proposed that Earth is in continuous but gradual change.

1.3 GEOLOGIC TIME

- An appreciation of the enormity of geologic time is necessary to understand geology.
- Geologists use a two-pronged approach to place natural events in their relative order and also establish the absolute time when these events took place.

1.4 EARTH MATERIALS: MINERALS, ROCKS, AND FOSSILS

- A mineral is a naturally occurring substance that forms crystals and has a chemical formula. A rock is an aggregate of one or more minerals.

- There are three rock groups: igneous, sedimentary, and metamorphic.
- Fossils are either the petrified remains or the impressions of ancient life forms.

1.5 THE ROCK CYCLE

- Earth's three groups of rocks, igneous, sedimentary, and metamorphic, transform into one another in a continuous process known as the rock cycle.

1.6 ORIGIN OF EARTH AND THE SOLAR SYSTEM

- Earth and other planets in the Solar System formed about 4.54 billion years ago when a nebula contracted and flattened into a disk under the influence of gravity.
- Earth's inner layers were formed very early in Earth's history as its elements either sank or rose according to their density.
- The crust, mantle, and core differ in chemical composition, whereas the lithosphere and asthenosphere have different physical properties.

1.7 EARTH'S INTERNAL HEAT

- Heat flows from Earth's hot core to its cooler surface and as a consequence, Earth is cooling down

1.8 EARTH AS A SYSTEM

- Earth has surface reservoirs: the solid Earth, hydrosphere, atmosphere, and biosphere

- Earth's surface components are linked together as a system and also interact with energy from the Sun.

1.9 WHAT DOES A GEOLOGIST DO?

- Geology is arguably the most diverse of all the sciences, and geologists are employed in a wide variety of professions including the search for natural resources, designing strategies to evaluate and minimize environmental impact, and assessing natural hazards.

Study Questions

1. What does the science of geology study and what tools does it employ?

2. Why must geologists depend on an incomplete record of evidence?

3. Do you think the scientific method is a useful way to investigate natural processes? If so, why? If not, why not?

4. According to the solar nebula theory, all the terrestrial planets formed at the same time. Explain.

5. Why are Earth's inner layers thought to have formed very early in Earth history?

6. Why is Earth's lithosphere defined as including all of the crust plus the uppermost mantle?

7. How does volcanic activity influence each of Earth's components (the geosphere, hydrosphere, atmosphere, and biosphere)?

8. Why did Hutton state that Earth had "no vestige of a beginning and no prospect of an end"?

9. How are each of the three rock groups transformed into each of the others?

10. Not all rocks follow the full rock cycle because shortcuts are possible. Citing examples, explain how shortcuts occur.

2 Plate Tectonics

2.1 Continental Drift

2.2 Paleomagnetism

2.3 Seafloor Spreading

2.4 Subduction

2.5 Moving Plates and Plate Boundaries

2.6 Hotspots: A Plate Tectonic Enigma

▶ Key Terms

▶ Key Concepts

▶ Study Questions

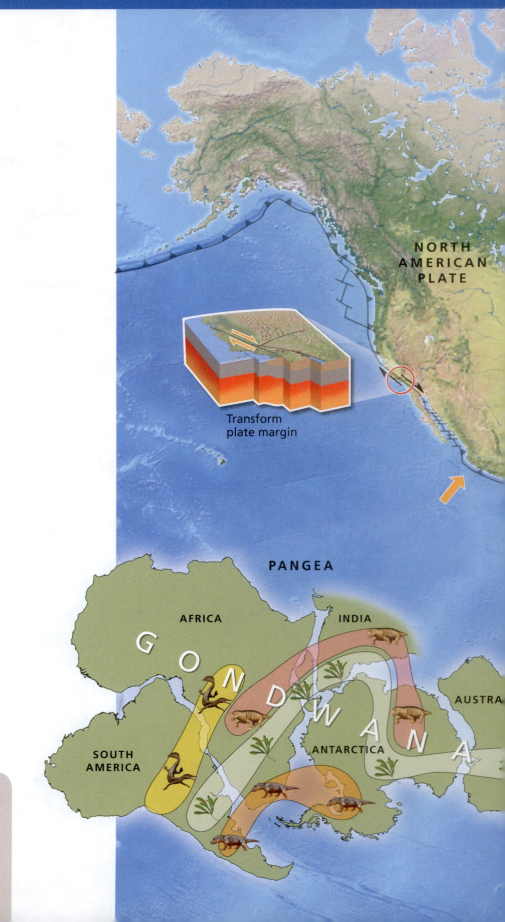

NORTH AMERICAN PLATE

Transform plate margin

PANGEA

AFRICA

INDIA

GONDWANA

SOUTH AMERICA

ANTARCTICA

AUSTRA

LIVING ON EARTH: The Remarkable Journey of the Green Turtle 53

IN DEPTH: Failed Rifts 62

Iceland

Magnetic stripes

CALEDONIAN MOUNTAINS

EURASIAN PLATE

VARISCAN BELT

Narrow ocean

Mid-Atlantic Ridge

APPALACHIAN MOUNTAINS

Island arc

AFRICAN PLATE

Oceanic transform fault

Plate motion

SOUTH AMERICAN PLATE

Continental rift

Hotspot

Continental arc

Wide ocean

2.1 Outline Wegener's idea of drifting continents and describe the evidence he cited to support his hypothesis as well as the primary objection to his claims.

2.2 Explain how paleomagnetism provides support for continental drift by providing evidence for the concepts of apparent polar wander and magnetic polarity reversal.

2.3 Summarize the process of seafloor spreading and the evidence that indicates that new ocean floor is continuously being created at mid-ocean ridges.

2.4 Summarize the evidence that old ocean floor is continuously being destroyed beneath deep ocean trenches by a process called subduction.

2.5 Explain how the theory of plate tectonics provides an explanation for continental drift and describe the three types of plate boundaries.

2.6 Describe how volcanic hotspots are attributed to plumes of hot material rising from deep within Earth's interior.

Although scholars have pondered the origin of the forces that shape our planet for centuries, it was not until the 1960s that scientists first developed the comprehensive theory called plate tectonics. According to this theory, Earth's rigid outer shell is broken into large, moving slabs or plates. Driven by heat from Earth's interior, these plates slowly move, carrying the continents as passengers along with them. Interaction between neighboring plates occurs along plate boundaries and accounts for the origins of the world's largest geologic features, including ocean basins, continents, and mountains. Plate tectonics has shown Earth to be a dynamic planet with an internal power source that continually shifts and reshapes the planet's surface. The evolution of this theory from its proposal and initial rejection to its universal acceptance is one of the most fascinating chapters in the history of science and forms a uniquely instructive lesson in the scientific method.

2.1 Continental Drift

Ever since maps of the Atlantic Ocean were first published in the mid-sixteenth century, scholars have pondered the remarkable jigsaw fit between the ocean's opposing coastlines (Fig. 2.1). Was this fit significant or was it just coincidence? The precision of the match became more apparent as maps of the Atlantic improved and, at the beginning of the twentieth century, this jigsaw fit was one of the many lines of evidence that led the German meteorologist **Alfred Wegener** to propose the hypothesis of **continental drift**. According to this hypothesis, the fit of the coastlines was no coincidence. Wegener proposed that the continents of Europe, Africa, and the Americas were once joined together and had subsequently broken up and moved enormous distances apart.

Wegener's concept of how Earth's geography had changed with time is dramatically illustrated by his own maps (Fig. 2.2). In these, he depicts the continents some 300 million years ago reassembled into a single landmass or supercontinent. He called this supercontinent **Pangea** (or *Pangaea*), meaning "all lands." Although Wegener presented his ideas in a book first published in 1915, almost half a century would pass before they gained much support. In fact, as recently as the late 1960s many introductory geology textbooks mentioned continental drift only as a curious, discredited idea.

A CONTROVERSY UNLEASHED

Although others had united the continents in much the same way before Wegener, the publication of his book sparked an international controversy that was to last almost a half century. In the early 1900s, most geologists believed that major features of Earth's surface were fixed and permanent, having been formed during the formation of the planet. Because it challenged established science, Wegener's hypothesis of moving continents was met with fierce

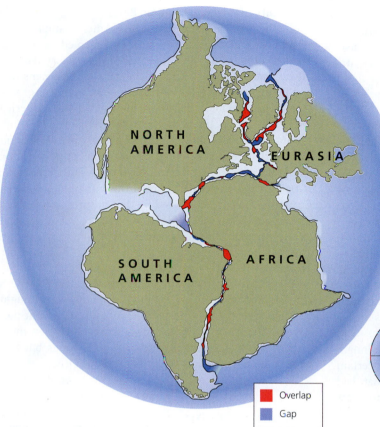

Figure 2.1 Jigsaw Fit of Continents

In 1965 Sir Edward Bullard and his colleagues at Cambridge University used a computer to obtain the best fit of the continents around the Atlantic Ocean. It was found to lie at a depth of about 1000 meters (3300 feet) below sea level, about halfway down the continental slope.

opposition from a shocked scientific community. Yet his ideas were not dismissed out of hand, because Wegener was a reputable scientist and had amassed an imposing collection of facts and opinion in support of his hypothesis. Although some of his evidence was convincing, much was speculative and provoked fierce debate. In the mid-1920s, at a conference in London, Wegener failed to convince the majority of his audience. A comment from a critic in attendance illustrates a contemporary view. "In examining ideas so novel as those of Wegener," he said, "it is not easy to avoid bias. A moving continent is as strange to us as a moving Earth was to our ancestors, and we may be as prejudiced as they were. On the other hand, if continents have moved, many former difficulties disappear, and we may be tempted to forget the difficulties of the theory itself and the imperfections of the evidence."

Criticism intensified when a professor of paleontology at Johns Hopkins University said that Wegener's method "is not scientific, but takes the familiar course of an initial idea, a selective search through the literature for corroborative evidence, ignoring most of the facts that are opposed to the idea, and ending in a state of auto-intoxication

in which the subjective idea comes to be considered as an objective fact." Eventually, following a decade during which continental drift was one of the most hotly debated issues in science, Wegener's hypothesis was discredited and fell into obscurity for more than 25 years.

Today we know that continents do indeed move, a realization that has revolutionized our picture of Earth. So what happened to change our understanding of the planet and why did it take so long? To answer these questions, let's first consider the evidence available in Wegener's day.

(a) About 300 million years ago

(b) About 50 million years ago

(c) About 1 million years ago

Figure 2.2 Pangea

(a) This map shows Wegener's reconstruction of the supercontinent Pangea. (b–c) These representations show the distribution of continents produced by the breakup of Pangea (light blue areas represent shallow seas). The approximate time assigned to each map was unknown to Wegener and has been added.

THE JIGSAW FIT OF CONTINENTS

It was easy for the skeptical world of science to dismiss any significance in the jigsaw fit of continents because coastlines were known to be ephemeral features of Earth's surface that could be changed by a single storm. How could their shape have any meaning after millions of years of coastal erosion?

In fact, the best fit between the Atlantic continents has since been shown by computer analysis to lie, not at their coastlines, but along lines drawn halfway down their continental slopes at a depth of 1000 meters (3300 feet) below sea level (see Fig. 2.1). Unlike coastlines, which mark only the edge of the land, the **continental slopes**, which separate areas of land and shallow seas from the deep ocean basins, mark the edge of the continents themselves and so are far more enduring features of Earth's surface. But the ocean floor had not been mapped in Wegener's day, so he had no way of knowing the shape of the continental slopes.

> ### CHECK YOUR UNDERSTANDING
>
> ○ Why was the jigsaw fit of continents so easily dismissed by scientists of Wegener's day?

Still, if the continents were fragments of what was once a much larger landmass, as Wegener proposed, they might be expected to share certain large-scale features that had formed before the landmass broke up. Like a correctly assembled jigsaw puzzle, not only would the shapes fit together but the pieces would recreate a meaningful picture. Wegener found evidence of such features in ancient mountain ranges and in major crustal fractures, each of which was brought into continuity when the continents were reassembled.

Continuity of Ancient Mountain Belts

Along the Atlantic seaboard of North America and Europe, several ancient mountain ranges that formed before the breakup of Wegener's Pangea are abruptly truncated or cut off by coastlines (Fig. 2.3a). Geologists of Wegener's day had no explanation as to why this should be so. For example, the Appalachian Mountains of eastern North America terminate along the Atlantic coast of Newfoundland. Mountains of very similar age and geology are also found in eastern Greenland, but the belt is abruptly truncated along the coast. On the other side of the Atlantic, the same is true of the ancient Caledonian Mountains of western Europe, which are truncated by the coastlines of western Britain and Scandinavia. Rocks of similar age and geology to the Appalachians also occur in the Variscan Belt, an ancient mountain range abruptly cut off by the coastlines of southern Europe and northwest Africa.

When the continents are reassembled into Wegener's single landmass, these ancient mountain ranges are brought together to form two continuous mountain chains. The patterns on the jigsaw pieces matched. In Wegener's view these ancient mountain ranges were once continuous but had been severed by the opening of the Atlantic Ocean.

> ### CHECK YOUR UNDERSTANDING
>
> ○ Which mountain belts in Europe did Wegener believe to be a continuation of the Appalachians?

In 1937, a similar argument was made for the southern Atlantic by the celebrated South African geologist Alexander du Toit. The Cape Fold Belt of South Africa, an ancient mountain range that reaches the coast at Cape Town, is closely comparable

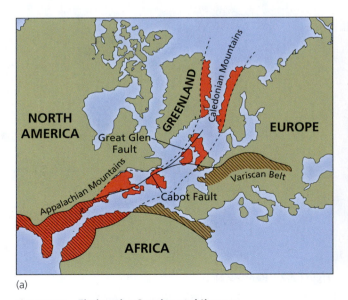

(a)

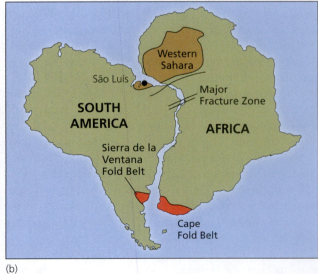

(b)

Figure 2.3 Fitting the Continental Jigsaw

(a) In the North Atlantic, ancient mountain ranges in North America and Europe come together to form continuous chains. Similarly, a major crustal fracture, or fault, in North America (Cabot Fault) lines up with one in Scotland and northern Ireland (Great Glen Fault). (b) In the South Atlantic, ancient rocks line up with similar rocks in South America, major crustal fractures in the bight of Africa line up with those in eastern Brazil, and an ancient mountain range in South Africa lines up with one in Argentina.

in age and geology to the Sierra de la Ventana Fold Belt of Argentina, an ancient range of mountains in South America located just south of Buenos Aires. When these two continents are brought together in the manner proposed by Wegener, the two mountain ranges connect up to form another perfect match (Fig. 2.3b).

Continuity of Major Faults

In addition to the continuity of mountain belts, Wegener's reassembly of the European and North American continents also brought together major crustal fractures or **faults**. In 1962, for example, the eminent Canadian geophysicist J. Tuzo Wilson proposed that the Great Glen Fault of Scotland was a continuation of the Cabot Fault of Newfoundland (see Fig. 2.3a). Both structures are major crustal fractures that formed before the assembly of Wegener's supercontinent. The continuity and alignment of these two faults when Europe and North America are reassembled is another pattern match on Wegener's continental jigsaw puzzle.

Likewise, major crustal fractures that reach the coast in the bight of Africa line up with similar fracture zones in eastern Brazil, and a distinctive region of rocks, now known to be 2 billion years old, in the western Sahara continues into the São Luis region of northeastern Brazil (see Fig. 2.3b).

> **CHECK YOUR UNDERSTANDING**
>
> ● How did the continuation of faults support Wegener's hypothesis of continental drift?

The Enigma of Ancient Climates

Other matching patterns lay in the rocks themselves and provided one of Wegener's most compelling lines of argument. On many of the southern continents, very distinctive glacial deposits had been found that were of the same age (now known to be about 300 million years), and had formed before the breakup of Pangea. Their widespread distribution across large areas of South America, South Africa, India, and Australia (Fig. 2.4) suggested that these continents had experienced a massive continental glaciation at this time, the scale of which was comparable to the continental glaciation the northern continents experienced during the last Ice Age. However, if the continents had fixed positions, as scientists of Wegener's day believed, then the distribution of these glacial deposits defied explanation. Rather than a single polar ice cap, the distribution of these deposits suggested several improbable ice caps at widely different latitudes. If Africa was stationary, for example, the presence of these deposits suggested that continental glaciation had reached the equator. Similarly, the distribution of these deposits suggested the wholesale glaciation of tropical India. The flow direction of the ice (arrows in Figure 2.4), as recorded in the grooves and scratch marks carved in the bedrock by the glaciers, also defied common sense. In both India and South Africa, the flow of ice was directed toward the poles from equatorial regions, exactly opposite to what one would expect. In South America and Australia, the direction of the glacier movement suggested that the ice had defied gravity, flowing inland and uphill from the sea!

All of these apparently contradictory observations were accounted for when Wegener reconstructed the southern continents into a portion of Pangea now known as **Gondwana** (Fig. 2.5). With the continents reassembled in a jigsaw fit, Gondwana could be home to a single ice cap centered on the South Pole, then located near the coast of Antarctica. The evidence of glacial movement could be accounted for by an outward flow of ice from the South Pole in a manner typical of a polar ice cap. With the intervening oceans removed, ice no longer appeared to have flowed inland (and uphill) from the sea.

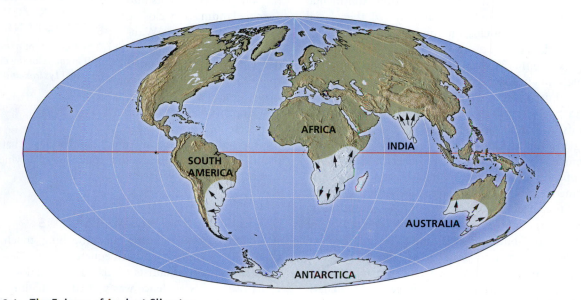

Figure 2.4   The Enigma of Ancient Climates
The distribution of ancient glacial deposits (in white) is difficult to explain in terms of separate continental glaciers. The directions of ice flow (shown by arrows) are also highly improbable.

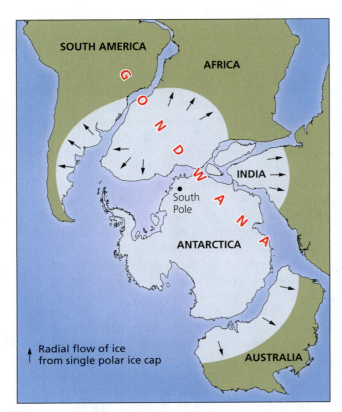

Figure 2.5 Gondwana

When the southern continents are reassembled into the southern landmass Wegener called Gondwana, we see that a single ice cap, centered on the South Pole and flowing outward (shown by arrows), provides an elegant explanation for the glacial evidence previously considered contradictory.

According to Wegener, the ancient continental glaciation did not take place at tropical latitudes. Instead, the southern continents at that time were assembled near the South Pole. Long after the ice age was over, continental drift dispersed these continents and moved the glacial deposits toward regions closer to the equator.

Other rock types that are indicators of past climate also suggested either movement of the continents or unaccountable shifts in Earth's major climatic zones. The presence of coal deposits in Antarctica and the Sahara Desert suggested that swampy subtropical conditions (under which coal forms) had once existed in areas of present-day continental glaciation and desert. Similarly, the distribution across much of northeastern North America and western Europe of desert sandstones and massive deposits of minerals formed by evaporation, such as rock salt and gypsum, suggested that arid conditions had occurred in the past in areas that today enjoy temperate climates. Remarkably, the occurrence of reef-building coral limestones in Greenland and northern Canada suggested that tropical seas had once existed north of the Arctic Circle. Because Earth's major climatic zones are controlled largely by their latitude, radical shifts in Earth's latitudes were implied. But if the continents were fixed, then radical shifts in Earth's latitudes could only be accomplished by equally radical

changes in Earth's spin axis, and this has not changed appreciably for billions of years. According to Wegener, all of these enigmas could be explained by the drift of continents through time.

THE PUZZLING FOSSIL RECORD

Wegener also found evidence for continental drift in the distribution of certain fossil species (Fig. 2.6). For example, the Permian reptile *Mesosaurus* had been found on both sides of the South Atlantic. Although it was a marine reptile, *Mesosaurus* was not thought to be a strong swimmer and, like the modern alligator, would not have been able to swim across an entire ocean. Similarly, fossils of the Triassic reptile *Cynognathus* had been found in South America, Africa, and Antarctica, while those of the Triassic reptile *Lystrosaurus* had been found in Antarctica, Africa, and India. Today, these regions are separated by major oceans. While it is conceivable that *Lystrosaurus*, a sheep-sized, land-dwelling reptile, could have walked from Africa to India, it certainly could not have crossed the polar ocean to Antarctica.

Wegener found similar evidence in the distribution of an unusual, apparently stunted assemblage of fossil ferns known as the *Glossopteris* flora. This unique flora had been found in South America, South Africa, India, Australia, and Antarctica, as well as in the Falkland Islands and Madagascar. Because the seeds of *Glossopteris* were several millimeters across, they could not have been carried great distances by the wind and are unlikely to have been able to float. As birds had not yet evolved to carry the seeds, how had the distribution of this flora been achieved?

The seemingly paradoxical distribution of fossil species such as *Mesosaurus*, *Cynognathus*, *Lystrosaurus*, and the *Glossopteris* flora is readily accounted for on Wegener's maps, which reassembled the southern continents into a single landmass by closing the intervening oceans. Without intervening oceans, the distribution of these and other species plotted on Wegener's maps posed no dilemmas. Those opposed to continental drift were forced to advocate the improbable existence of former "land bridges" that had at one time connected the present continents but which had now sunk below the surface of the oceans.

> ### CHECK YOUR UNDERSTANDING
> ◗ How does continental drift account for the distribution of *Lystrosaurus* fossils?

THE REJECTION OF WEGENER'S VIEWS

Although Wegener's ideas were compelling, much of his evidence was viewed to be inadequate, and his views were rejected by most scientists of his day. By analogy with a court of law, his evidence was viewed by the jury as circumstantial and insufficient to secure a conviction. While appealing to some, his ideas were dismissed by geophysicists who found the process of continental drift in defiance

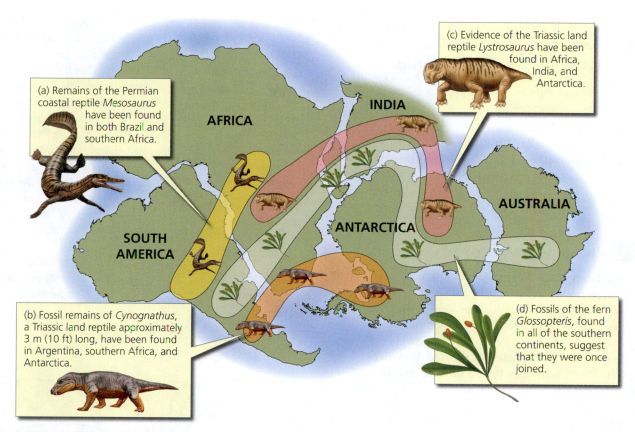

Figure 2.6 Puzzling Fossils

The distribution of fossil plants and animals provides further evidence of continental drift. (a) The 2-foot-long freshwater reptile *Mesosaurus* lived in Brazil and Africa. (b) Fossils of the land reptile *Cynognathus* are found in Argentina, southern Africa, and Antarctica. (c) The sheep-sized land reptile *Lystrosaurus* lived in Africa, India, and Antarctica. (d) The large-seeded *Glossopteris* flora is found on all five continents. The ranges of the reptiles (none of which were long-distance swimmers) and the large-seeded flora on continents now separated by vast oceans are readily accounted for on this modern reconstruction of Gondwana.

of the laws of physics. As was repeatedly pointed out by the eminent British geophysicist Harold Jeffreys, the chief weakness in the hypothesis of drifting continents was Wegener's inability to provide a viable mechanism, one that would allow the process to operate. What force, the geophysicists argued, could possibly cause the continents to plough their way across the ocean floors?

In an attempt to address these criticisms, Wegener described the drift of the southern continents away from the South Pole as *polflucht*, or "flight from the poles," which he attributed to the gravitational attraction of Earth's equatorial bulge. Such forces certainly exist, but as Jeffreys was able to demonstrate, they are far too weak to drag the southern continents northward. Wegener also suggested that the westward movement of the Americas had occurred as a result of tidal forces in Earth's crust produced by the gravitational attraction between the continents and the Moon. Again, such forces exist but are hopelessly inadequate for the task. Indeed, since tidal friction acts like a brake on the spinning Earth, tidal forces strong enough to move continents would long ago have brought Earth's eastward rotation to a halt, as Jeffreys was quick to point out.

So while continental drift was able to explain many enigmas: the surprising fit of continents, the enigma of ancient climates, the puzzling fossil record, and the distribution of today's animals and plants, Wegener was unable to provide a workable driving force. As a result, his hypothesis was rejected and his ideas fell into obscurity until strange discoveries on the floor of the ocean some 30 years later revived his challenging vision.

2.1 SUMMARY

- In 1915 German meteorologist Alfred Wegener proposed that all the continents were once joined together and had later broken up and moved apart. Using the jigsaw fit of their coastlines, Wegener reassembled the continents into a single landmass he called Pangea, meaning *all lands*. His idea of slowly moving continents, termed continental drift, unleashed an international controversy that lasted almost 50 years.

CHECK YOUR UNDERSTANDING

◯ What mechanisms did Wegener propose to account for continental drift?

◯ Why were Wegener's ideas rejected by the scientists of his day?

- Wegener found evidence of a jigsaw fit to continents in features such as ancient mountain belts, major fault lines, and certain distinctive rock formations. Wegener also found matching patterns among continents through evidence of ancient climate and from the distribution of fossils.

- Despite all the data that supported continental drift, Wegener's ideas found little acceptance because he was unable to provide a viable mechanism that would permit the process to take place.

2.2 Paleomagnetism

The vindication of Wegener's hypothesis of continental drift came from geophysics, the very field of science that had so effectively discredited the idea during Wegener's lifetime. The key lay in the phenomenon of magnetism as revealed by studies of Earth's magnetic field. The research that ultimately supported Wegener's hypothesis was not meant to be a study of continental drift, but an examination of past patterns in Earth's magnetic field. Yet the discoveries would serendipitously become the key to the acceptance of continental drift and the birth of the theory of plate tectonics.

EARTH'S MAGNETIC FIELD

If you have used a compass, you have witnessed Earth's magnetic field. It is the force of this field that makes a compass needle point north. Scientists believe that the origin of this magnetism is related to the motion of a liquid layer deep within Earth's interior, known as the *outer core*. The outer core of Earth is rich in iron and spins like a dynamo under the influence of Earth's rotation. The spinning dynamo, in turn, generates the planet's magnetic field (see In Depth: **Earth's Magnetism and Its Dynamic Core**, in Chapter 11). Because Earth's magnetic field is in part generated by its rotation about its axis, the position of Earth's magnetic poles broadly corresponds to that of its geographic poles. The field envelopes Earth and extends well out into space. A compass needle at Earth's surface detects only a tiny portion of Earth's vast magnetic field.

Earth's magnetic field behaves as if a simple bar magnet were aligned north-south at Earth's center, producing lines of magnetic force that emerge from the Southern Hemisphere and loop through space to return in the Northern Hemisphere (Fig. 2.7). Therefore, the lines of magnetic force are parallel to Earth's surface at the equator but plunge ever more steeply as the poles are approached, pointing up out of the ground in the Southern Hemisphere and down into the ground in the Northern Hemisphere.

CHECK YOUR UNDERSTANDING

- How does Earth's magnetic field control the orientation of a compass needle?
- What is magnetic inclination and how can it be used to determine latitude?

The orientation of a compass needle is controlled by these invisible lines of magnetic force. The lines cause the compass needle to point north but also cause the needle to tilt, the north end tilting downward in the Northern Hemisphere and upward in the Southern Hemisphere. So a compass reading really has two components—a horizontal component, or *swing*, used in navigation and pathfinding, and a vertical component, or *tilt*, known as **magnetic inclination**.

Because the amount of inclination progressively increases with distance from the equator, and the direction of tilt is different in the Northern and Southern hemispheres,

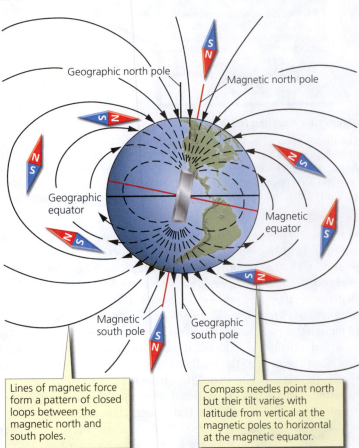

Lines of magnetic force form a pattern of closed loops between the magnetic north and south poles.

Compass needles point north but their tilt varies with latitude from vertical at the magnetic poles to horizontal at the magnetic equator.

Figure 2.7   Earth's Magnetic Field

Earth's magnetic field, which governs the orientation of compass needles, resembles that of a simple bar magnet. Lines of magnetic force form a pattern of closed loops that emerge from the planet in the Southern Hemisphere and reenter in the Northern Hemisphere. Because the magnetic field owes its origin to the spin of the Earth, the position of the magnetic poles broadly coincides with Earth's geographic poles. Compass needles align themselves with the magnetic lines of force and so are parallel to Earth's surface at the equator, but tilt at ever increasing angles as the magnetic poles are approached.

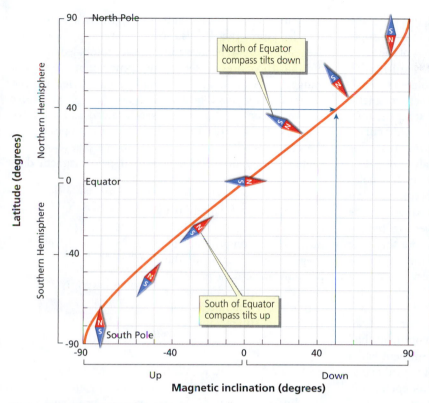

Figure 2.8   Magnetic Inclination

This figure shows the relationship between latitude and magnetic inclination (orange line), and its effect on the tilt of a compass needle at different locations on Earth's surface. For example, at a latitude of 40° N, the north end of a compass needle tilts downward 50° from the horizontal.

we can use the inclination of a compass needle to determine latitude (Fig. 2.8).

But how can we determine the patterns of Earth's magnetism in the deep past? As we shall see, certain rocks are able to record these two components of Earth's magnetic field at the time of their formation. The records from these rocks provided the first direct geophysical evidence for continental drift.

THE PALEOMAGNETIC RECORD

When basalt lava flows across Earth's surface and starts to cool, tiny magnetically susceptible grains of iron oxide (*magnetite*) start to crystallize within it. As they cool, these grains become magnetized by Earth's magnetic field. Once frozen into the rock, they function like thousands of tiny compass needles, creating a permanent record of the north-south direction and the magnetic inclination at the point on Earth's surface where the lava erupted at the time it solidified (Fig. 2.9). Although the resulting magnetic field is very

weak, we can measure its orientation with a highly sensitive instrument known as a *magnetometer*. In modern basalt lavas, the magnetic orientation points to the north magnetic pole, which lies close to the geographic North Pole.

The study and measurement of Earth's magnetic field from ancient rocks is called **paleomagnetism**. When the history of Earth's magnetic field was first examined in this way, the results were startling. Rather than pointing to the present-day magnetic poles as expected, the tiny magnetite compass needles in ancient basalt lavas were found to point in many different directions. These data indicate that either the poles or the continents (or both) moved through geologic time.

As we shall now see, yet more startling were results that suggested the north and south magnetic poles had in the past repeatedly swapped positions. Could this be true? The attempts to answer these puzzling questions led to the revival of the old ideas of continental drift.

CHECK YOUR UNDERSTANDING

◉ What is paleomagnetism and what instrument is used to study it?

APPARENT POLAR WANDER

The position of the magnetic poles is thought to be geographically fixed by Earth's axis of rotation. For example, when modern basalt lava flows, like those of Hawaii, are examined paleomagnetically, their tiny magnetite compass needles are found to point toward the present day

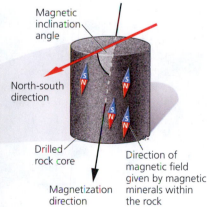

Figure 2.9   Paleomagnetism

Rock cores drilled from a basalt lava flow indicate the direction in which the magnetic field of the tiny magnetic minerals within the rock is aligned. From this, the north-south direction and magnetic inclination angle at the time the lava solidified can be determined.

geographic poles. But for ancient basalt lavas this is not the case. Scientists who analyzed basalts taken from Europe, for example, found that all magnetized samples of the same age pointed to the same location, but not toward Earth's poles. Moreover, samples of a different age pointed to a different location. When scientists connected these magnetic pole positions for successively older basalt samples, they found that the resulting line of locations resembled a curve.

Had Earth's magnetic poles moved with time? If so, this curve suggested a path of **polar wander**. Analysis of paleomagnetic rocks in North America also suggested polar wander. Amazingly to the scientists, the path of the European polar-wander curve differed from the North American one (Fig. 2.10a). Since the locations of the poles are tied to Earth's spin axis, surely there couldn't be more than one North Pole and one South Pole at a time! If the continents had fixed positions, these data would be very difficult to explain.

Supporters of continental drift reasoned that an alternative explanation for the polar-wander curve of an individual continent could be that the continent had moved relative to the poles. And as each continent had a different polar-wander curve, then the most logical explanation was that the continents had moved independently of each other. If so, the curves tracked the movement of the continents, not the poles, and recorded, not polar wander, but **apparent polar wander** misleadingly produced as a consequence of continental drift. This hypothesis was confirmed when Europe and North America were reassembled into the Pangea configuration proposed by Wegener. The two polar wander curves coincided, indicating that their present separation came about only when Pangea broke up and Europe and North America became separate continents (Fig. 2.10b).

The paleomagnetic record was heralded as conclusive evidence for continental drift. Wegener's proposal that the continents had moved relative to the poles was now supported by geophysical data. Indeed, paleomagnetic data from the southern continents proved to be consistent with the existence of Gondwana and placed the South Pole near the position proposed by Wegener. However, while the paleomagnetic data provided strong support for continental drift, the quest to find a viable mechanism still remained.

> **CHECK YOUR UNDERSTANDING**
>
> ◑ What is apparent polar wander and how has it been used to support continental drift?

PALEOMAGNETIC REVERSALS

A key piece of the puzzle was provided by an even more startling paleomagnetic discovery—evidence that Earth's magnetic field periodically reversed itself. Magnetized samples of basalt from the same locality were repeatedly found in which the North and South poles had apparently

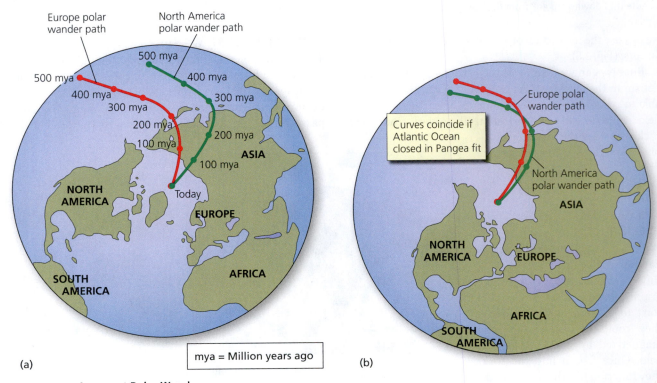

(a)

mya = Million years ago

(b)

Figure 2.10   Apparent Polar Wander

(a) Apparent polar-wander paths for North America (green) and Europe (red) for the past 500 million years define similar but separate curves indicating that the two continents moved relative to both the poles and each other during this time interval. (b) The separation of the two paths is attributed to the opening of the Atlantic Ocean; it is reconciled if the two continents are reassembled in their Pangea configuration.

been interchanged. Scientists noticed that the tiny compass needles produced in basalt lavas from certain eruptions pointed south rather than north. They quickly realized that there were only two plausible explanations for this. Did this interchange mean that rocks could somehow reverse their magnetization, or did it mean that Earth's magnetic field periodically reversed itself? As larger databases were compiled, all rocks of the same age were found to have the same polarity. This implied that the process was global in scale and that it was Earth's magnetic field that periodically reversed itself during an event known as a **magnetic polarity reversal**.

Although the reason why such magnetic reversals take place is still poorly understood, reversals are now known to have occurred quite frequently in the recent geologic past, with major reversals occurring every million years or so and shorter flips lasting a few thousand to a few tens of thousands of years. For example, for the past 700,000 years, Earth has experienced **normal magnetic polarity** during which the north ends of all compass needles point toward the north geographic pole. Prior to this, however, Earth experienced almost 2 million years of **reverse magnetic polarity** during which the north end of a compass needle would have pointed toward the south geographic pole. Some unknown motion of Earth's liquid outer core is believed to produce these reversals. But whatever their cause, this discovery would provide the ultimate confirmation of continental drift and would unlock the dramatic mechanism by which it was accomplished.

As more data on the magnetic orientation of dated basalt samples accumulated from all over the world, a precise time scale was constructed that charted the duration of each magnetic reversal. From this time scale, Earth's magnetic field was found to have reversed itself quite quickly many times, nineteen such reversals having occurred in the past four and a half million years (Fig. 2.11). With no obvious repetition in the duration of these events, the pattern of magnetic reversals resembles a bar code on produce in a supermarket. But as we shall now see, it was this pattern that unlocked the secret to a dynamic Earth history with the discovery of the very same magnetic bar code in rocks on the ocean floor.

CHECK YOUR UNDERSTANDING

 ⬥ What happens to Earth's magnetic field during a magnetic reversal?

2.2 SUMMARY

- Wegener's ideas were revitalized when geophysicists started studying Earth's magnetic field.

- When basalt lava cools, iron-rich minerals become magnetized and record the north-south direction and the magnetic inclination at the time the lava solidifies.

- The magnetic records frozen into ancient basalts were found to have widely varying magnetic directions and inclinations. Initially, this record seemed to indicate that the magnetic poles had been "wandering" for the past 500 million years. But when the continents were reassembled into the map of Pangea, the puzzling polar-wander paths coincided. This discovery indicated that instead of the poles having wandered, it was the continents that had drifted.

- Through studies of ancient basalts, scientists also discovered that Earth's magnetic field periodically reverses itself, interchanging the north and south magnetic poles.

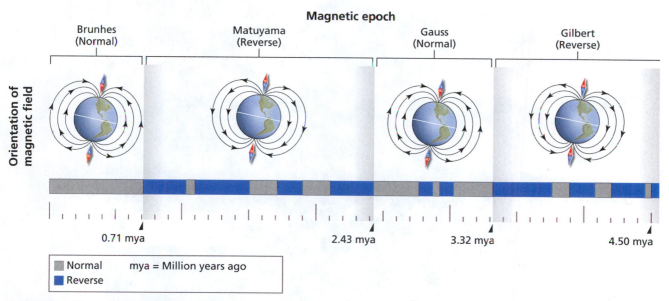

Figure 2.11 Magnetic Reversals

Times of normal and reverse polarity in Earth's magnetic field during the past 4.5 million years have been established from the study of magnetized samples from basalt lava flows of known age. In the normal configuration, the north magnetic pole is near the north geographic pole. In the reverse configuration the north magnetic pole is near the south geographic pole.

2.3 Seafloor Spreading

The clinching evidence for continental drift did not come from paleomagnetic studies on continents. Instead, it came from paleomagnetic studies of the ocean floors. These studies eventually led three geoscientists in the 1960s to propose the hypothesis of **seafloor spreading**. In 1963, the Canadian geophysicist Lawrence Morley and the British geophysicists Fred Vine and Drummond Matthews independently proposed that the seafloor moves symmetrically away from major ridges in the ocean, like two large conveyor belts, moving the ocean floor and continents along with it. But in a cruel twist of fate that reveals both the ongoing opposition to continental drift and a weakness in the scientific process of peer review, both of Morley's papers on the subject were rejected by scientific journals. So Vine and Matthews were the first to publish and, hence, the first to receive credit for the idea. Today, however, the explanation that these three geoscientists provided for the strange magnetic patterns found when the study of paleomagnetism was applied to the ocean floor is correctly known as the Vine-Matthews-Morley hypothesis.

With the scientific confirmation of seafloor spreading in the 1970s, Wegener's hypothesis of continental drift was finally provided with the one element it lacked: a mechanism. But what paleomagnetic discoveries were made on the ocean floor, and why did they lead Vine, Matthews, and Morley to propose the provocative hypothesis of seafloor spreading? To answer these questions and see how this new hypothesis emerged, we must look at curious features on the ocean floor that were first discovered in the 1950s and 1960s.

MAGNETIC REVERSALS ON THE SEAFLOOR

After the Second World War, the strategic importance of the oceans greatly increased. With the advent of nuclear submarines, a major effort was made to chart the ocean floor. Far from being flat, featureless plains, the ocean floor was found to contain vast **mid-ocean ridges**, which rose as much as 3 kilometers (1.9 miles) above the deep ocean floor and girdled Earth like the seam on a baseball. At the same time, efforts were being made to chart the magnetic properties of the ocean floor. So while the mystery of magnetic reversals was being unraveled on land, scientists were towing sensitive magnetometers across the world's oceans. These magnetometers revealed that long zebra-like stripes of high-intensity and low-intensity magnetism alternated across the floor of the oceans. But no theory existed that could account for these patterns.

One such remarkable survey charted magnetically imprinted rocks on the floor of the Atlantic Ocean southwest of Iceland. Here, the stripe-like departures from the expected magnetic patterns, called **magnetic anomalies**, were found to be centered on a segment of the Mid-Atlantic Ridge. The Mid-Atlantic Ridge is part of the system of mid-ocean ridges that had been found to traverse all of the world's oceans. Science would later show that the magnetic anomalies were actually long strips of rock with normal and reverse polarity that alternated in a symmetrical fashion on either side of the ridge axis (Fig. 2.12).

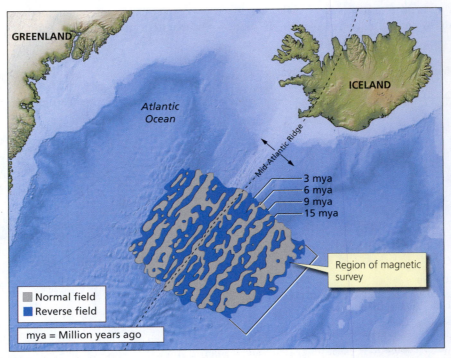

Figure 2.12 Magnetic Stripes

Evidence of magnetic reversals is visible in part of the Mid-Atlantic Ridge southwest of Iceland. Blue stripes identify ocean floor rocks showing reverse magnetization. Intervening gray stripes correspond to rocks with normal polarity.

A PATTERN EXPLAINED—SEAFLOOR SPREADING

The pattern of symmetrical magnetic reversals went unexplained until 1963, when Morley and Vine and Matthews independently showed that it could be accounted for by a hypothesis known as seafloor spreading. According to this hypothesis, partial melting of hot rocks beneath the mid-ocean ridges causes molten rock or **magma** to ooze up toward the seabed as the seafloor spreads away to either side as if it were carried on a giant conveyor belt. If, Morley and Vine and Matthews reasoned, lava flowed out along the center of mid-ocean ridges, then the iron minerals in the cooling lava would become magnetized in the prevailing direction of Earth's magnetic field. Because magma is continually being inserted into the ridge, previously crystallized rock is pushed progressively sideways (Fig. 2.13a). The process is similar to that of repeatedly inserting books into the middle of a bookshelf. The other books on the shelf are progressively pushed to the side (Fig. 2.13b). The magma is inserted from below rather than the side, but the overall effect is similar.

As the newly generated, magnetically imprinted rock is carried away from the ridge at a uniform rate in both directions, more lava erupts along the ridge and is, in turn, magnetized. As the process is continuous, any reversal in Earth's magnetic polarity is recorded as a new

strip of rock at the ridge crest displaying reverse magnetization (see Fig. 2.13).

Morley and Vine and Matthews realized that the symmetrical, stripe-like pattern of high and low intensity magnetism on either side of the ridge corresponds (respectively) to periods of normal and reverse magnetization. The stripes of high intensity magnetism are located where the magnetized rocks enhance the existing magnetic field of Earth because they have the same polarity. Conversely, stripes of low intensity magnetism are located where the magnetized rocks weaken the existing magnetic field of Earth because they have opposite polarity.

Because the process of seafloor spreading is symmetric with respect to the ridge axis, the seafloor patterns of magnetism on either side of a ridge mirror each other with matching stripes of normal and reverse polarity distributed symmetrically about the ridge axis. In this way, continuous injection of magma at the ridge axis over millions of years produces symmetrical stripes of normal and reverse polarity. Each stripe forms at the ridge axis only to be torn in two lengthwise as the seafloor spreads (see Fig. 2.13a). The pattern of magnetic stripes consequently provides a continuous record of Earth's history of magnetic reversals frozen into the basaltic ocean floor at the time of its formation. In conclusive confirmation of this

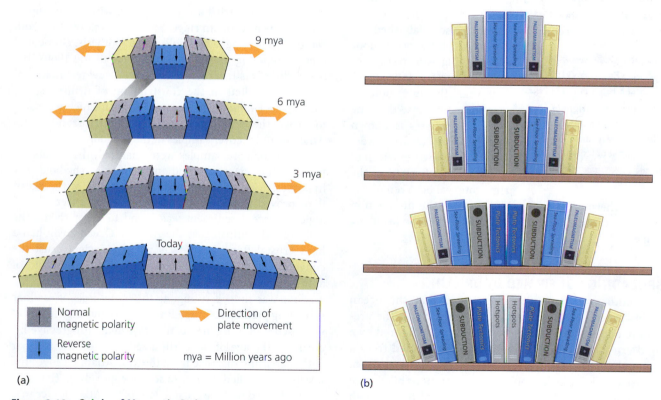

9 mya

6 mya

3 mya

Today

| ⬛ | Normal magnetic polarity | ➡ | Direction of plate movement |
| 🟦 | Reverse magnetic polarity | | mya = Million years ago |

(a)

(b)

Figure 2.13   Origin of Magnetic Stripes

(a) Basalt lava erupting along the crest of the ridge records Earth's prevailing magnetic polarity as it cools and solidifies. Seafloor spreading carries the rock away from the ridge in both directions as new lava erupts along the ridge axis. Occurring continuously, this process captures each reversal in Earth's magnetic field, leading to alternating strips of rocks with normal and reverse polarity distributed symmetrically about the ridge axis. (b) The process of seafloor spreading is similar to that of repeatedly inserting a pair of books into the middle of a bookshelf. Each time new books are inserted, the other books on the shelf are pushed sideways so that the distribution of books is symmetric about the center of the shelf.

explanation, the pattern of reversals on the seabed was found to precisely match the pattern of reversals derived from the continents (see Fig. 2.12). In other words, the bar codes are the same. This match provided the connecting link between the continental and sea-floor magnetic records. Because the timing of the magnetic reversals on the continents had been determined, this match also placed a timescale on the magnetic stripes of the ocean floor.

So Morley and Vine and Matthews were able to correctly predict that ocean floor rocks at the ridge crest are very young and have today's normal polarity. On either side of the ridge, ocean floor rocks become increasingly older with distance from the ridge crest, and have polarities that match Earth's recent history of magnetic reversals found on land. These relationships exist because the rocks at the center of the ridge are the most recent (like the books in the center of our bookshelf), whereas those at the edge of the ocean are the oldest.

The remarkable match, stripe for stripe, with the magnetic timescale developed on land established the hypothesis of seafloor spreading and showed the site of mid-ocean ridges to be the start of two huge conveyor belts that moved the ocean floor aside. It also established the rate at which the ocean floor was moving—an average of 2.5 centimeters (one inch) each year. This rate could be determined because the time between reversals is known and therefore scientists were able to measure the width of oceanic crust formed in the time interval between reversals. Furthermore, they could determine the direction of motion because it must be perpendicular to the magnetic stripes.

> **CHECK YOUR UNDERSTANDING**
>
> ● Explain the origin of magnetic anomalies on the floor of the Atlantic Ocean southwest of Iceland.

Figure 2.14 *Glomar Challenger*
Until her removal from active service in 1983, the research vessel *Glomar Challenger* was capable of drilling and retrieving sediment cores from the ocean floor several kilometers below the sea surface.

The concept of seafloor spreading offered a radically different view of the evolution of ocean basins, implying that the ocean basins are very young, with the youngest rocks at the mid-ocean ridge. So the conventional wisdom was being challenged by a bold new hypothesis. How could these competing ideas be tested? Prior to the 1960s, the ocean floor was inaccessible. However, new technological advances resulted in the construction of drilling ships, such as the *Glomar Challenger* (Fig. 2.14), which were able to probe the ocean depths between Africa and South America and recover samples of the seafloor.

When the core samples were analyzed, the results astounded the scientific world and confirmed the bold new hypothesis of seafloor spreading. The rocks of the ocean floor proved to be young and only a thin veneer of geologically recent sediment was found to cover them. The drilling did confirm that the Atlantic Ocean basin had a hard crustal floor, but it was not the ancient crust the conventional view had anticipated. Instead, the record of approximately 95 percent of geologic time was missing. The seafloor was indeed young, as the hypothesis of seafloor spreading predicted. The sediment also contained fossils of microscopic sea organisms that revealed a progressive increase in the age of the earliest sediment on the seafloor with increasing distance from the Mid-Atlantic Ridge. The fossils also verified the rate of seafloor spreading. The drilling confirmed that the crustal floor of the Atlantic Ocean is in motion, moving as if it were part of two huge conveyor belts that continually carry it away from the Mid-Atlantic Ridge in opposite directions (see Living on Earth: **The Remarkable Journey of the Green Turtle**).

SPREADING CONFIRMED BY DRILLING

For centuries, scientists had believed that the ocean basins were ancient features of Earth's surface. The prevailing consensus was that the oceans were floored by a rigid crust formed when the planet first cooled from a molten ball. The ocean floors were believed to consist of a crust as old as Earth itself, with a continuous blanket of sediment above it. It was expected that drilling through the layers of sediment into the crust below would yield samples that would read like an encyclopedia of geological time, each layer becoming older as the core got deeper.

LIVING ON EARTH

The Remarkable Journey of the Green Turtle

Every year, the tiny equatorial island of Ascension in the middle of the Atlantic Ocean witnesses a remarkable mystery of evolution. This barren volcanic peak on the Mid-Atlantic Ridge is the nesting ground of the green turtle (Fig. 2A). Each year, large numbers of green turtles, some weighing as much as 180 kilograms (400 pounds), crawl ashore on the island's few small beaches to lay their eggs in the sand. What makes this event so extraordinary is the enormous distance that each turtle must swim in order to reach this remote destination. From their home along the coast of Brazil the turtles embark on a hazardous, two-month voyage that takes them across half the width of the Atlantic Ocean, a journey of some 2000 kilometers (1250 miles). They make this journey without food and against a strong equatorial current. Why do they attempt such an arduous passage and why do they travel so far to hatch their young?

It is understandable that turtles might choose an island for their nesting site because there are far fewer predators on islands. Although hundreds of miles of sandy beaches lie along the coast of Brazil, the beaches are within reach of a variety of carnivores that would devour the eggs as well as the nesting adults. But what made the turtles choose such a distant island and how

Figure 2A   Green Turtle

The green turtle of South America makes its home along the coast of Brazil. But to lay its eggs, it swims all the way to the lonely island of Ascension in the mid-Atlantic, only to swim laboriously back again.

did they know that it was there? Continental drift and seafloor spreading provide an elegant explanation of just how such a situation might have slowly come about over millions of years.

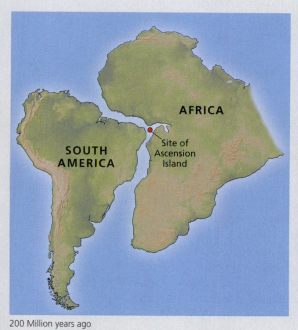

200 Million years ago

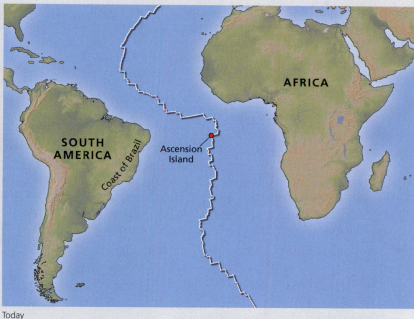

Today

Figure 2B   Odyssey of the Green Turtle

The green turtle may have acquired its habit of journeying to the distant island of Ascension to lay its eggs before the breakup of Pangea. Each generation traveled farther than their ancestors as the Atlantic Ocean widened, until today's marathon journey of some 2000 kilometers (1250 miles) was established.

(Continued)

The Remarkable Journey of the Green Turtle (Continued)

Green turtles first evolved about 200 million years ago when the supercontinent of Pangea was breaking up. At this time, the ocean separating Africa from South America was a narrow sea about as wide as a broad valley. Any island on the Mid-Atlantic Ridge would therefore have been both visible and readily accessible from the South American coastline. As the Atlantic Ocean slowly widened, the distance from South America to any island located on the Mid-Atlantic Ridge would have grown progressively longer (Fig. 2B). Unusually active volcanism at this location on the ridge would have ensured the continued existence of an island where Ascension Island now stands. Guided by instinct once the island could no longer be seen, each generation of turtles would have traveled a little further than their ancestors. As they did so, natural selection would have favored the sturdiest swimmers, slowly giving rise to the powerful turtles of today that are strong enough to swim halfway across an ocean to an island that only their instinct tells them is there.

If this is the explanation, then the very genetic code of the green turtle may have been programmed by continental drift. Shaped by the drifting continents, their remarkable behavior would provide living testimony to the process of seafloor spreading, which every year carries South America a little further from the Mid-Atlantic Ridge.

CHECK YOUR UNDERSTANDING

◔ What did the drilling vessel *Glomar Challenger* discover that astounded the scientific world?

In the 1970s studies near the crest of mid-ocean ridges also revealed lava flows and fissures. These studies gave evidence that not only is the ocean floor separating symmetrically along ridge axes, but new ocean floor is being created there as well. But what happens to all this crust once it is created? The answer to this question lay beneath another dramatic feature of the ocean floor and, with it, the final piece in our modern understanding of plate tectonics was revealed.

2.3 SUMMARY

* In surveys of the basalts of the ocean floor, scientists found that long stripes of high and low intensity magnetism alternated in a symmetrical fashion on either side of mid-ocean ridges. The pattern of these stripes exactly matched the pattern of Earth's magnetic reversals, suggesting conveyor-like movement of the crust of the oceans by the process of seafloor spreading.

* According to the hypothesis of seafloor spreading, the ocean floor continuously spreads away from the world's mid-ocean ridges in opposite directions, like two huge conveyor belts.

* The hypothesis of seafloor spreading was confirmed by analysis of rock samples that scientists recovered though ocean drilling.

2.4 Subduction

Scientists quickly realized that the same volume of ocean crust created at mid-ocean ridges must be destroyed elsewhere or Earth would swell up like a balloon. So where are the sites of this destruction? The solution lay in the origin of vast features on the ocean floor known as **deep ocean trenches**. Discovered by oceanographic surveys before the advent of the seafloor spreading hypothesis, these trenches are narrow, curved depressions in the ocean floor that can reach depths of up to 11.5 kilometers (7.2 miles) and are thousands of kilometers in length. Here, oceanic crust was found to lurch violently back into Earth's heated interior through a process known as *subduction*.

DEEP OCEAN TRENCHES

Evidence for ocean floor destruction came from mapping the location of earthquakes. Earthquakes are not distributed randomly, instead they define narrow zones (Fig. 2.15). Numerous shallow earthquakes trace the winding course of the world's mid-ocean ridges as they circle the globe on the ocean floor. Others occur beneath the world's great mountain belts. But most earthquakes define narrow zones that lie near another important feature of the ocean floor, the deep ocean trenches. The trenches occur in many regions but are especially characteristic of the margins of the Pacific Ocean.

It was the worldwide network of seismic stations set up to monitor underground nuclear testing during the Cold War that confirmed trenches as the sites of ocean floor consumption. These stations revealed that earthquakes bordering the Pacific always occurred on one side of a trench and not the other. In addition, the depth of the source of the earthquakes generally increased with increasing distance from the trench (Fig. 2.16). The point in Earth's interior where an earthquake is generated is the **earthquake focus**. The deep ocean trenches mark the start of inclined zones of earthquake foci that plunge into Earth's interior. These inclined zones of earthquake foci had been identified before the advent of plate tectonics, but

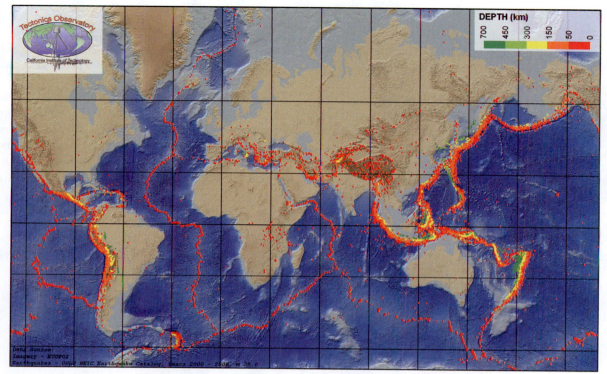

Figure 2.15   Earthquake Zones

The world's earthquake zones are defined by the distribution of earthquakes (magnitude >5) over a period of nine years between 2000 and 2008. The earthquakes are color-coded according to the depth at which the earthquakes occurred.

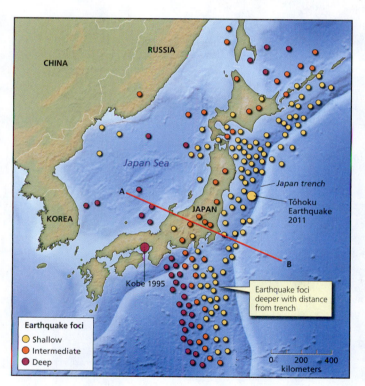

were now seen as reflecting the violent descent of oceanic crust back into Earth's heated interior.

SUBDUCTION ZONES

The process by which ocean floor is consumed is called **subduction** (Fig. 2.17). The inclined zone associated with this process is termed a **subduction zone**. The inclined zone of earthquake foci is generated along the upper boundary of the subducting slab of oceanic crust as it is jolted downward. The process of subduction also generates molten rock (magma). Such magmas might be expected since the downgoing slab is heated as it reaches greater depths. But only under extreme circumstances are the magmas derived from the subducting slab itself. Instead, the magma is generated in the rocks above the subduction zone by water vapor driven off the downgoing slab as it is heated. Water has a great capacity to carry heat and destabilize minerals. So as this water vapor moves upward, it promotes melting in the rocks above. The resulting magmas rise buoyantly toward the surface to fuel bow-shaped (arcuate) lines of

Figure 2.16   Destruction of Ocean Floor

This map and cross-section show the distribution of shallow, intermediate, and deep earthquakes for part of the Pacific Ring of Fire in the vicinity of the Japan trench. Note that the earthquakes occur only on one side of the trench and trace out an inclined zone of earthquake foci that plunges into Earth's interior. The earthquakes are color-coded according to the depth at which the earthquakes occurred.

volcanoes. The pronounced line of volcanoes bordering the trenches that encircle the Pacific is often called the *Pacific Ring of Fire* (Fig. 2.18).

The distribution of volcanoes around the rim of the Pacific shows that they occur in two quite different settings. In the Aleutian Islands, for example, the volcanoes form a curved line of volcanic islands, but in the Andes, they occur along the edge of a continent within a major mountain belt (see Fig. 2.18). This difference reveals the

existence of two types of subduction zones, one where subduction occurs beneath oceanic crust and the other where it occurs beneath the margins of continents.

The Aleutian Islands are representative of subduction beneath oceanic crust. In this setting, the magmas produced by subduction ultimately rise through the rocks of the ocean floor to fuel volcanoes that grow upward from the seabed, eventually rising above the level of the sea to form islands (Fig. 2.19). The resulting line of volcanic

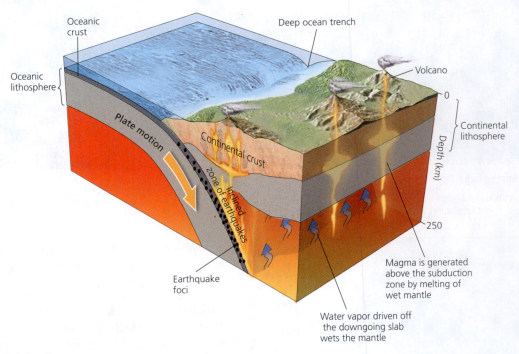

Figure 2.17   Subduction

The subduction of oceanic crust shows its relationship to the inclined zone of earthquakes bordering deep sea trenches, and its role in generating magmas.

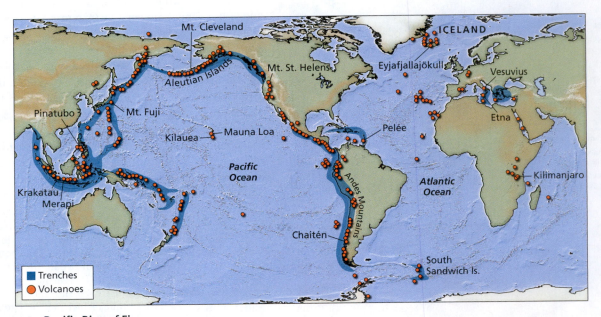

Figure 2.18   Pacific Ring of Fire

This map shows the global distribution of recent volcanoes. The almost continuous belt of volcanism around the Pacific Rim is the Ring of Fire, which lies adjacent to the deep trenches of the Pacific Ocean.

Figure 2.19   Aleutian Island Volcano
Mt. Cleveland volcano erupted on the Aleutian Island of Chuginadak on May 23, 2006.

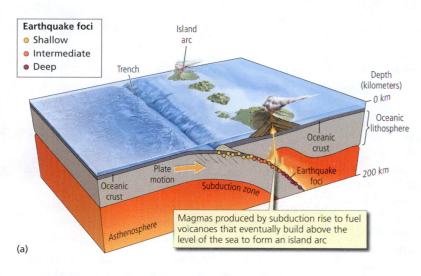

(a)

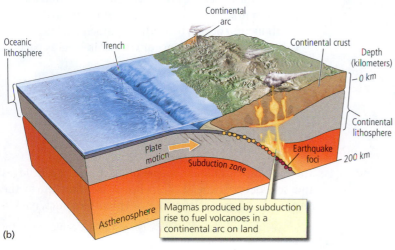

(b)

islands, which is distinctly curved and parallel to the adjacent ocean trench, is known as an **island arc** (Fig. 2.20a). Island arcs are curved due to the curvature of Earth. Because Earth is spherical, the ocean floor has a curved surface. The trench produced by the subduction of ocean floor is consequently curved, like a knife cut on the surface of an apple, or an indentation on the surface of a ping-pong ball. The curvature of the line of volcanoes above a subduction zone is inherited from the curved shape of the downgoing slab as it cuts into the crust.

The Andes, on the other hand, result from subduction beneath a continental margin (Fig. 2.20b). In this setting, the magmas produced by subduction eventually rise to fuel volcanoes on land (Fig. 2.21). En route to the surface, however, the rising magmas may pond beneath the continental crust. When this happens, the base of the continental crust may become sufficiently heated so that it, too, starts to melt. In addition, because continental crust is much thicker than oceanic crust, much of the magma never reaches the

Figure 2.20   Subduction Zones
Subduction zones beneath (a) oceanic and (b) continental crust show their relationship to the island and continental arcs, with which they are respectively associated.

Figure 2.21 Andes Mountains Volcano
The three vents of the Mt. Hudson volcano high in the Andes Mountains of southern Chile erupted on October 29, 2011.

surface but, instead, cools to form large bodies within the continental crust. This combination of surface volcanoes and deep-seated intrusive bodies forms what is known as a **continental arc**. Continental margins where such arcs develop are termed **active margins** because they coincide with active plate boundaries.

- Scientists have confirmed deep ocean trenches as the sites where the ocean floor is consumed. This violent descent of oceanic crust back into Earth's heated interior is called subduction.

- The zone along which oceanic crust is consumed is called a subduction zone. The subduction zone is characterized by an inclined zone of earthquake foci.

- If subduction occurs beneath an oceanic plate, curved lines of volcanoes are produced known as island arcs, the form of which reflects the curvature of Earth. If subduction occurs beneath a continental plate, volcanoes form within major mountain belts as part of a continental arc.

2.4 SUMMARY

- Deep ocean trenches are narrow, curved depressions that can reach depths of over 11 kilometers (7 miles) and extend for thousands of kilometers.

- Ocean trenches are associated with earthquakes and volcanic activity and are especially common at the margins of the Pacific Ocean.

2.5 Moving Plates and Plate Boundaries

The recognition of seafloor spreading at mid-ocean ridges and subduction beneath deep ocean trenches led to the revolution in Earth sciences we call **plate tectonics**, named from the realization that Earth's rigid outer shell is broken into huge moving plates and from the Greek *tecton*, which means "to build." Nearly 50 years after Wegener's provocative ideas were first published, plate tectonics provided the missing mechanism for continental drift, and in so doing, transformed geology as profoundly as evolution transformed biology and relativity transformed physics.

According to plate tectonics, the world's earthquake zones outline the boundaries between large slabs or **plates** that may be thousands of kilometers across but are only 50 to 150 kilometers (30 to 100 miles) thick (Fig. 2.22). They are approximately as thick in proportion to Earth as an eggshell is to an egg. These plates are in constant motion, and their incessant jostling, one against the other, causes frequent earthquakes. Plate boundaries therefore lie within Earth's most seismically active regions. So it is no surprise that the earthquake zones of Figure 2.15 match the plate boundaries of Figure 2.22.

So far, about a dozen large plates and numerous smaller ones have been identified. Some, such as the Pacific plate, are almost entirely oceanic. Others like the Eurasian plate, are mainly continental. However, most plates, like the North American, South American, and African, carry both continents and oceans so that as the ocean floors move, the continents embedded in the plates travel passively with them like bags on a conveyor belt. This shift in focus from continental drift to plate tectonics provided an explanation for the movement of continents and ushered in our modern view of a dynamic Earth.

Earth's plates define the **lithosphere**, or outer rocky layer of Earth, and "float" on a weak and partially molten portion of Earth's mantle below known as the **asthenosphere** (Fig. 2.23). The lithosphere includes Earth's *crust*, which is made up of rocks like those on Earth's surface, and the solid uppermost portion of Earth's *mantle*. The asthenosphere is a zone within the mantle, which is a broad region of denser rocks that separates the crust from Earth's iron-rich *core*. We will learn more about the lithosphere and asthenosphere when we examine Earth's internal structure in Chapter 11. Plates move away from the mid-ocean ridges and toward the deep ocean trenches, but each moves independently so that dramatic effects are produced where two plates meet. It is at these **plate boundaries**, where plates separate, converge, or slide past each other, that the effects of plate motion are most obvious.

At **divergent plate boundaries**, where plates move apart, continents may be torn in two with new oceans forming between them, much as Wegener envisaged with the breakup of Pangea and the formation of the Atlantic Ocean. The process starts with **continental rifting**, which is the extension and splitting of a continent as the lithosphere is stretched and thinned. The crust breaks and settles along faults to form a **rift valley** (Fig. 2.24a). At the same time, magma rising from the asthenosphere below

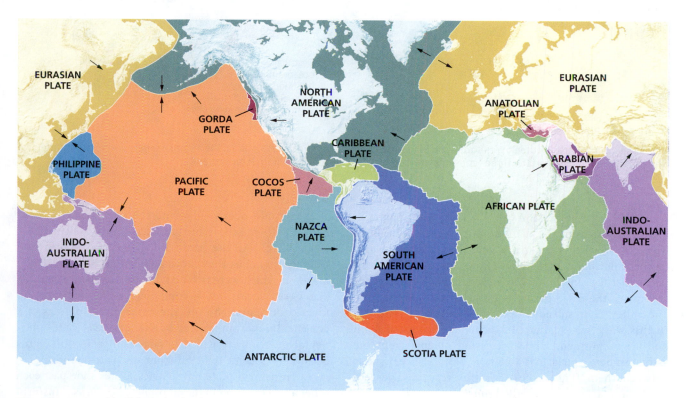

Figure 2.22   Earth's Lithospheric Plates

Earth's outer shell is broken into jagged, rigid fragments, much like a giant cracked eggshell. These fragments, or plates, are constantly in motion, as shown by arrows.

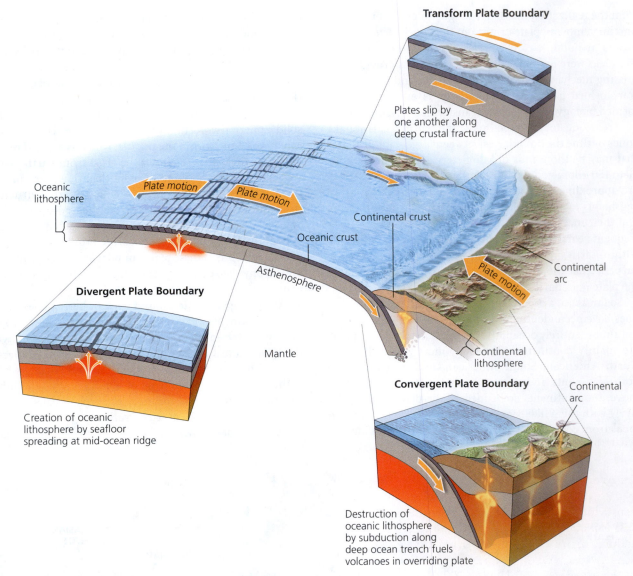

Figure 2.23   Plate Boundaries

A cross section of Earth's outer layers shows the lithosphere, and makes up Earth's plates, the asthenosphere within the mantle over which the plates move, and the three types of plate boundaries that result from their movement.

is injected into the crust. As we shall learn in Chapter 4, melts derived from the mantle have the composition of basalt, so when this magma reaches the surface, it erupts on the valley floor as basalt lava flows. Some rifts proceed no further (see In Depth: **Failed Rifts**). But if divergence continues, the rift eventually tears the continent in two. The valley floor subsides until it is flooded by the sea. In this way, two smaller continents are produced from one large continent while the rising magma from the asthenosphere starts to build new basaltic ocean floor between them (Fig. 2.24b). For this reason, divergent plate boundaries are also called *constructive plate boundaries*. At this stage, continental rifting gives way to continental drift and the opening of a new ocean. Oceanic lithosphere is created by seafloor spreading

at a new mid-ocean ridge and the lavas and the rocks beneath them cool and thicken into a rigid layer (Fig. 2.24c). At the same time, the two continental fragments cool as they move away from the ridge. Their margins, which were thinned during divergence, subside below sea level and become buried by sediment to form **continental shelves**, the name given to areas of a continental margin flooded by shallow seas. These continental margins are called **passive continental margins** because they come to lie far from any plate boundary.

At **transform plate boundaries**, plates slide past each other along fractures in the lithosphere known as **transform faults**. Transform faults are most common in the oceans where they offset the mid-ocean ridges (see Fig. 2.23), but some form great crustal fractures on land, such as the

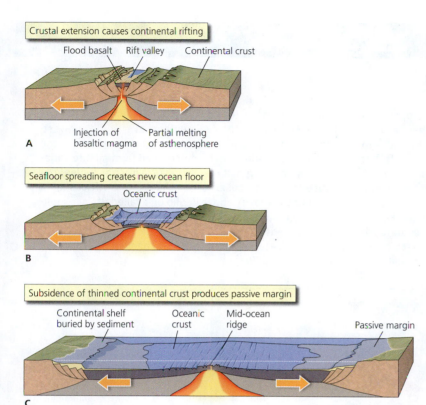

Figure 2.24   Divergent Plate Boundary

A divergent plate boundary forms through continental rifting and the opening of a new ocean basin.

San Andreas Fault in California (Fig. 2.25). Crust is neither created nor destroyed along transform faults, so transform plate boundaries are also called *conservative plate boundaries*. However along great faults like the San Andreas, destructive earthquakes are common as the two plates grind against each other.

At **convergent plate boundaries**, one plate bends downward and is subducted beneath the other. There are two possible scenarios and both of them involve subduction of the denser lithosphere. Either old oceanic lithosphere is consumed by subduction beneath younger ocean floor, or oceanic lithosphere is subducted beneath more buoyant (less dense) continental crust (see Fig. 2.23). This type of plate boundary—also called a *destructive plate boundary*—consists of a subducting *lower* plate and an overriding *upper* plate. Where the subducting plate bends before sinking, a great ocean trench up to 11.5 kilometers (7.2 miles) deep is formed. As the subducting slab angles down into Earth and becomes heated, water vapor is driven off, which, as we learned earlier, generates basaltic magmas in the overlying mantle that rise to fuel volcanoes in the upper plate. Where the upper plate lies within an ocean, the result is an island arc like that of the Aleutians. But where the upper plate is the margin of a continent, the subduction causes uplift and the result is a volcanic mountain range or continental arc like that of the Andes. In continental arcs, the basaltic magmas rising from the mantle may

cause additional melting at the base of the continental crust. This produces magmas of quite different composition, much closer to that of granite. As a result, the volcanoes associated with continental arcs differ from those of island arcs and, as we shall learn in Chapter 4, are often more explosive.

If, on the other hand, both converging plates carry continents, the continents may eventually collide. Neither plate can subduct completely because continental lithosphere is too buoyant (we discuss this in more detail in Chapter 9). Instead, one continent overrides the other and the collision forces the crust skyward to build great mountain ranges like those of the Himalayas. Volcanoes are not common in these continent–continent collision zones because the subduction that creates new magma grinds to a halt and those magmas that do form are unable to rise through the greatly overthickened crust.

Plate tectonics is a comprehensive theory about the forces that shape the planet. With

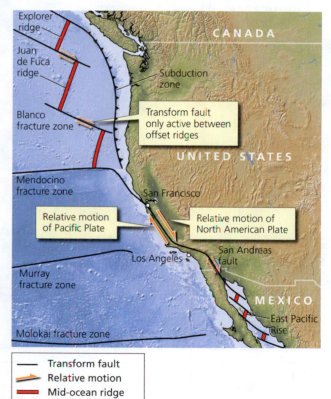

Figure 2.25   Transform Plate Boundaries

Transform plate boundaries are delineated by transform faults, along which one plate jostles past another. As illustrated by the west coast of North America, transform faults form major oceanic fracture zones that offset mid-ocean ridges, and major crustal fractures on land, such as the earthquake-prone San Andreas Fault in California.

IN DEPTH

Failed Rifts

Not all continental rift valleys successfully open into new oceans. Those that fail to do so, either because initial spreading is not sustained or because it shifts to some other location, are known as **failed rifts**. A rift that fails to open ultimately ceases to grow wider, and once the crust cools, it subsides and becomes buried by sediments. Even though they are not plate boundaries, these buried rifts may remain seismically active zones of weakness. Like wounds that never heal, these zones may continue to be the source of earthquakes triggered by stresses transmitted through the surrounding stable continent. Just such a failed rift system that is more than a billion years old can be traced beneath the sedimentary cover in the midwestern United States (Fig. 2C). Although invisible from the surface, its outline can be determined from minute variations in Earth's gravity known as **gravity anomalies**, because the dense basalt on the floor of the rift causes a slight rise in gravity above it. Slip on a failed rift is responsible for earthquake activity near New Madrid, Missouri, where a series of major earthquakes in 1811–12 caused church bells to ring in Boston and changed the course of the Mississippi River.

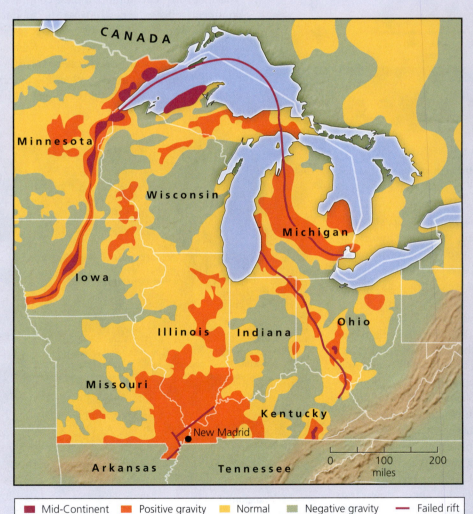

Figure 2C   Failed Rift

In this gravity anomaly map of the midwestern United States, red colors correspond to areas of fractionally higher gravity (positive gravity anomaly) and trace a feature known as the Mid-Continent Gravity High, a Precambrian failed rift system floored by dense basalt lava flows. Reactivation of the ancient faults bordering this and younger failed rifts is responsible for the present-day earthquakes of western and northern Ohio, and the New Madrid region of Missouri.

■ Mid-Continent Gravity High   ■ Positive gravity anomaly   ■ Normal gravity   ■ Negative gravity anomaly   — Failed rift

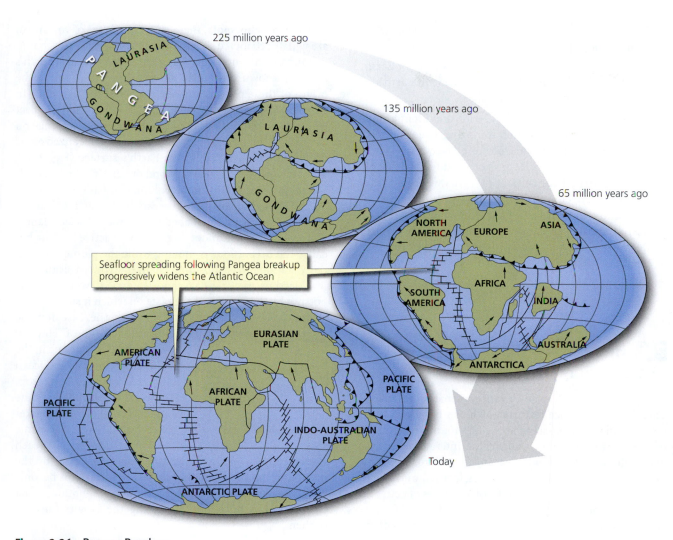

225 million years ago

135 million years ago

65 million years ago

Seafloor spreading following Pangea breakup progressively widens the Atlantic Ocean

Today

Figure 2.26   Pangea Breakup

These maps show stages in the breakup of Pangea from 225 million years ago to the present. Compare with Figure 2.2.

this revolution in scientific thought, the idea of moving continents gained widespread acceptance and it became universally accepted by geoscientists that the oceans are quite young geological features of Earth's surface, and the planet is an active one with a complex and dynamic history. In this new view of Earth, mountains, oceans, and continents lost their permanence and became just ephemeral expressions of a continual cycle of creation and destruction.

Plate tectonics showed that the supercontinent Pangea had indeed existed, just as hypothesized by Wegener. The southern continents were once united to form Gondwana, and Europe and Africa had separated from North and South America, just as their coastlines suggest (Fig. 2.26). But the continents did not have to plough their way through the solid rock of the ocean floors as Wegener had thought. Instead, the ocean floors were themselves moving, while the continents drifted passively in response to that movement, much like baggage on a conveyor belt at the airport.

The ways in which plates move, and the ways in which they interact with each other, are fundamental to our

CHECK YOUR UNDERSTANDING

● Why are plate boundaries associated with earthquakes?

● How does continental rifting lead to the formation of new oceans?

● What are the three types of plate boundary?

understanding of many of the major geologic processes affecting Earth's surface. In Chapter 9, we will examine plate boundaries and the processes associated with them in more detail. As we do, the theory of plate tectonics will further reveal itself to be a unifying principle, or **paradigm**, that has provided science with a powerful new way of looking at Earth's dynamic systems.

2.5 SUMMARY

- Continental drift, seafloor spreading, and subduction unite to form one coherent theory of plate tectonics.

- According to the theory of plate tectonics, Earth's outer shell consists of huge plates in constant motion. These plates move in response to the circulation of

Earth's heated interior, and carry the continents with them as they do so.

- Earth's plates comprise the outer rocky layer of Earth, or lithosphere, and move on a weak zone in Earth's mantle known as the asthenosphere.

- The effects of plate motion are most evident along plate boundaries, of which there are three types—divergent plate boundaries where plates move apart and new ocean floor is created following continental rifting, convergent margins where plates collide and oceanic crust is destroyed by subduction, and transform margins where plates slide past each other along great crustal fractures. Because plates constantly jostle each other for position, plate boundaries coincide with Earth's earthquake zones.

2.6 Hotspots: A Plate Tectonic Enigma

Plate tectonics provides an elegant explanation for many of Earth's major surface features, such as mid-ocean ridges, deep ocean trenches, island arcs, and mountain belts. Each of these features is the result of the interaction between plates along plate boundaries, and it is here, where two plates meet, that geologic activity is focused and we find earthquakes and volcanoes. Yet isolated areas of active volcanism also occur far from plate boundaries, and for these volcanoes plate tectonic theory has no clear explanation.

Known as **hotspots**, these enigmatic areas of volcanism are widely attributed, not to plate tectonic processes, but to columns of hot material rising from deep within Earth's interior, known as **mantle plumes**. We will examine the origin of these plumes and the volcanoes they produce in Chapter 9. For now, what is important to know is that where these plumes rise through the mantle to impinge on the base of the lithosphere, they give rise to localized areas of volcanism, or hotspots, on Earth's surface (Fig. 2.27). This volcanism may occur on land or beneath the sea, leading to the eruption of huge volumes of basalts. Where a plume occurs beneath the ocean, individual volcanoes build upward from the seabed to become volcanic islands. However, these volcanoes do not remain active for long because the ocean floor is moving while the plume is essentially stationary. A subsiding chain of extinct volcanoes is produced as each new volcano is carried away from the hotspot by the motion of the plate on which it sits.

Such is the case for the volcanic island of Hawaii, which sits atop a plume in the middle of the Pacific Ocean (Fig. 2.28) far removed from any plate boundary. The other Hawaiian islands likewise formed over the same stationary plume but each has since been moved away and become inactive by the motion of the Pacific plate. As a result, the islands become progressively older in the direction of plate motion, a fact that has been borne out by dating the basalts on each of them.

> ### CHECK YOUR UNDERSTANDING
> ○ What are hotspots and how do they form?

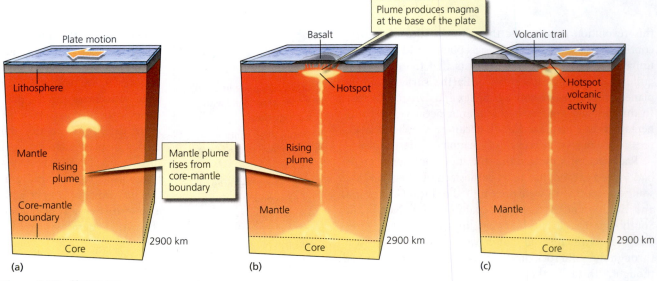

Figure 2.27 Hotspots

(a) A plume of hot material rises from deep within Earth's interior. (b) The plume impinges on the lithosphere, leading to the outpouring of basalts on the ocean floor. (c) Waning volcanic activity produces individual volcanoes that are carried away by plate motion to form a chain of extinct volcanoes.

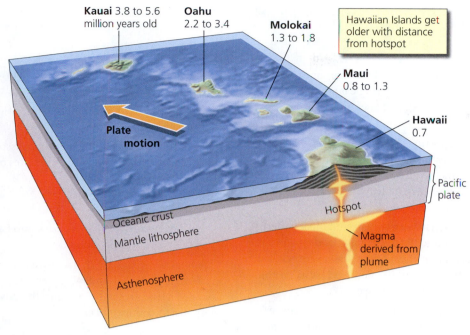

Figure 2.28 Hawaiian Islands

Movement of the Pacific plate over the fixed Hawaiian hotspot has led to the formation of the Hawaiian island chain. Each island formed as a volcano arose over the plume and then was carried away by plate motion. The basalt lavas on each island consequently get older as the island's distance from the plume increases (ages given in million years).

2.6 SUMMARY

- Hotspots are isolated areas of volcanism that can occur far from plate boundaries and are attributed, not to plate tectonic processes, but to near-stationary columns of hot material, known as mantle plumes, rising from deep within Earth's interior.

- Where hotspots occur beneath oceans, volcanoes grow upward from the seabed to become volcanic islands that are then carried away by the motion of the plate on which they sit to form an extinct volcanic chain.

Key Terms

active margin 58
Alfred Wegener 40
apparent polar wander 48
asthenosphere 59
continental arc 58
continental drift 40
continental rifting 59
continental shelf 60
continental slope 42
convergent plate boundary 61
deep ocean trench 54

divergent plate boundary 59
earthquake focus 54
failed rift 62
fault 43
Gondwana 43
gravity anomaly 62
hotspot 64
island arc 57
lithosphere 59
magma 51
magnetic anomaly 50

magnetic inclination 46
magnetic polarity reversal 49
mantle plume 64
mid-ocean ridge 50
normal magnetic polarity 49
paleomagnetism 47
Pangea 40
paradigm 63
passive continental margin 60
plate 59

plate boundary 59
plate tectonics 58
polar wander 48
reverse magnetic polarity 49
rift valley 59
seafloor spreading 50
subduction 55
subduction zone 55
transform fault 60
transform plate boundary 60

Key Concepts

2.1 CONTINENTAL DRIFT

- In 1915 Alfred Wegener proposed that the continents were once joined together into a single landmass called Pangea (meaning *all lands*), that later broke up and moved apart.

- Wegener found supporting evidence to the jigsaw fit of continents in features such as ancient mountain belts, major fault lines, and certain distinctive rock formations. If the continents were reassembled, these features could be seen as continuous.

- Wegener also found matching climate and fossil patterns, including the distribution of ancient glacial deposits, the flow direction of ice as it carved the ancient bedrock, and the presence of strikingly similar flora and fauna on continents now separated by major oceans.

- Despite the data supporting continental drift, Wegener's ideas found little acceptance among the scientists of his day because he was unable to provide a mechanism that would permit the process to take place.

2.2 PALEOMAGNETISM

- When basalt lava cools, iron-rich minerals become magnetized and record the north-south direction and the magnetic inclination at the time the lava solidified. The position of the north and south poles determined from ancient basalts suggest they have moved along apparent polar wander paths.

- Since Earth's poles are fixed, apparent polar wander paths cannot track movement of the poles and must, instead, reflect movement of the continents on which the lavas erupted.

- Paleomagnetism has shown that Earth's magnetic field periodically reverses itself, interchanging the north and south magnetic poles.

2.3 SEAFLOOR SPREADING

- Seafloor spreading describes the conveyor-like movement of ocean floor away from mid-ocean ridges.

- The hypothesis of seafloor spreading has been confirmed by ocean drilling.

- Deep ocean trenches, such as those associated with the Pacific Ring of Fire, mark the sites of ocean floor consumption.

2.4 SUBDUCTION

- The descent of oceanic crust back into Earth's heated interior is called subduction.

- Subduction zones are characterized by an inclined zone of earthquake foci, and by curved lines of volcanic islands called island arcs or by volcanically active mountains known as continental arcs.

2.5 MOVING PLATES AND PLATE BOUNDARIES

- According to the theory of plate tectonics, Earth's outer shell is broken into large moving plates that ride upon a yielding layer of Earth's interior. This comprehensive theory explains the origins of Earth's ocean basins, continents, and mountains.

- Plates comprise the rocky outer layer of Earth, known as the lithosphere, and move over a weak zone in Earth's mantle called the asthenosphere.

- There are three types of plate boundaries—divergent, along which plates move apart, convergent, where plates collide, and transform, along which plates slide past each other. Plate boundaries coincide with Earth's earthquake zones.

2.6 HOTSPOTS: A PLATE TECTONIC ENIGMA

- Hotspots are isolated areas of volcanic activity generated, not by plate tectonics, but by plumes of hot material rising from deep within the mantle.

Study Questions

1. Wegener did not live to see his hypothesis of continental drift vindicated. Why was this and what does it tell us about science and the scientific method?

2. Why did continental drift remain popular with paleontologists but not with geophysicists?

3. How is the polar wander curve of a continent established, and why does the curve show only apparent polar wander?

4. Explain how scientists deduced that Earth's magnetic field periodically reverses.

5. Explain how the pattern of magnetic reversals on land may be linked with the magnetic anomaly patterns across the Mid-Atlantic Ridge.

6. How does seafloor spreading account for the pattern of magnetic anomalies on the ocean floor?

7. Explain the statement: "Continental drift is a passive response to seafloor spreading."

8. What do oceanic trenches represent? Why are these trenches more abundant in the Pacific Ocean than in the Atlantic Ocean?

9. Using Figure 2.16, explain the relationship between earthquake foci and distance from the oceanic trench.

10. Account for the striking correlation between the distribution of earthquakes (see Fig. 2.15) and the shape of Earth's lithospheric plates (see Fig. 2.22).

11. Why are there only three types of plate boundaries?

12. What are hotspots and why are they not accounted for by plate tectonic theory?

3 Minerals

3.1 Elements: Building Blocks of Minerals

3.2 Minerals: Orderly Expressions of Matter

3.3 Physical Properties of Minerals

3.4 Silicate Rock-Forming Minerals

3.5 Nonsilicate Rock-Forming Minerals

3.6 Minerals and Rocks

3.7 Minerals and People

3.8 Minerals and Plate Tectonics

▶ Key Terms

▶ Key Concepts

▶ Study Questions

LIVING ON EARTH: Origin of the Chemical Elements 74

SCIENCE REFRESHER: Some Basic Definitions in Science 76

SCIENCE REFRESHER: Solids, Liquids, and Gases 77

Rock

Granite

Magma

Cooling magma produces igneous rock

Mineral

Crystal structure

Biotite

Feldspar

Quartz

3.1 Recognize that minerals are chemical compounds made up of atoms linked together by a variety of chemical bond types.

3.2 Explain what minerals are and explain how the characteristic physical properties of minerals are determined by the internal arrangement of their constituent atoms.

3.3 Describe the characteristic physical properties that we use to identify minerals, including crystal shape, color, luster, and hardness.

3.4 Compare and contrast the properties of the two primary categories of silicate minerals and cite examples of each.

3.5 Discuss and cite examples of the important properties and characteristics of the nonsilicate rock-forming minerals.

3.6 Compare and contrast rocks and minerals.

3.7 Cite examples of the role minerals play in society.

3.8 Explain what mineral formation can tell us about plate tectonics and the evolution of Earth.

As we learned in Chapter 1, there are three groups of rocks that make up Earth's continents and ocean floor: igneous, sedimentary and metamorphic. Most rocks are made up of mineral grains in much the same way that stained-glass windows are mosaics of pieces of colored glass. Minerals are the basic components of rocks and they are also the building blocks of the solid Earth, and so it is with minerals that our examination of Earth starts. This chapter provides a basic introduction to minerals, how they form, and why they are important. In the chapters that follow, we will discuss the major rock classifications and see how just a few minerals combine to form these rocks. From the study of minerals and rocks, we can learn a great deal about the environment in which rocks form, and about Earth's dynamic evolution.

3.1 Elements: Building Blocks of Minerals

The building blocks of all minerals, and indeed all chemical compounds, are the chemical elements. To understand minerals, it is crucial to know what elements are and how they bond to form compounds. **Elements** are the fundamental substances that make up all matter (see Living on Earth: **Origin of the Chemical Elements** on page 74). Each element has a name and a chemical symbol; for example, the chemical symbol for the element oxygen is O. More than 100 elements exist, but only 90 of them occur naturally. The rest are produced artificially in nuclear reactions. The elements are all listed on the periodic table in Appendix B. Some minerals, such as gold and sulfur, are made from a single element. Most minerals, however, are combinations of elements that form stable chemical compounds. A **compound** is a substance formed by bonding between two or more elements. For example, the mineral *pyrite*, or "fool's gold," is a compound made up of the elements iron and sulfur. To understand the relationship between minerals and elements, we must first look at atoms because atoms comprise the elements. (See Science Refresher: **Some Basic Definitions in Science** on page 76).

ATOMS

Each chemical element is made up of fundamental units known as atoms. An **atom** is the smallest unit of an element that displays all the physical and chemical properties of that element. The diameter of an atom is typically about 0.00000001 (10^{-8}) centimeters; approximately 10 billion atoms lined up side by side would form a chain about 1 meter (3 feet) in length. For convenience in expressing such small lengths, 10^{-8} centimeters is defined as 1 *angstrom* (or 1Å), after Swedish physicist Anders Jonas Ångström.

Each atom consists of a dense central unit, called the **nucleus**, where most of the mass of the atom is concentrated. Surrounding the nucleus is a cloud of orbiting particles, called **electrons**, which have almost no mass and carry a negative electrical charge. Electrons revolve around the central nucleus in a series of orbits, called *shells* that reflect differences in their energy levels. We will soon discuss how the number of electrons in the outermost shell of an atom is crucial in determining how an atom bonds with its neighbors. The atom's nucleus contains two types of particles: **protons**, which carry a positive electrical charge, and **neutrons**, which have an almost identical mass to protons but carry no charge.

To characterize atoms, we classify them on the basis of the number of protons they contain. Each atom of the same element has the same number of protons and this number is different from that of all other elements. The number of protons in an atom is its **atomic number**. For example, the simplest of the elements, hydrogen, has a single proton in its atomic nucleus (Fig. 3.1a) and is therefore assigned an atomic number of 1 in the periodic table of elements (see Appendix B). By definition, all hydrogen atoms have one proton. An oxygen atom, on the other hand, has 8 protons in

its nucleus (Fig. 3.1b), whereas an iron atom has 26. The atomic numbers for oxygen and iron are 8 and 26, respectively. Uranium, the heaviest of all the naturally occurring elements, has an atomic number of 92, indicating that each uranium atom contains 92 protons. The atomic number of an element determines many of its properties, including chemical bonding.

CHECK YOUR UNDERSTANDING

◉ What are atoms and what particles do they contain?

CHEMICAL BONDING

The nucleus of an atom is tiny, about 100,000 times smaller than the atom itself, but it is a relatively stable part of an atom and is not involved in chemical bonding. The orbiting electrons, on the other hand, are loosely held within the atomic structure and orbit in shells at discrete distances from the nucleus. The outermost electrons are particularly loosely held because the attraction between the protons in the nucleus and the orbiting electrons decreases with distance. The most stable arrangement for most elements is to have a full outer shell consisting of eight electrons. A few elements have this stable arrangement and, as a result, tend not to bond with other elements. However, most elements have less than the maximum number of electrons in their outer shells and tend to combine, or *bond*, with one or more neighboring atoms until they have a complete outer shell of eight electrons. This chemical "rule of thumb" is known as the *octet rule* and applies to the bonding arrangements of the most common elements in minerals. *Bonding* is an attractive force that holds two or more atoms together. Atoms bond in order to reach a more stable arrangement.

Some elements have a tendency to lose electrons, whereas others have a tendency to accept them. Still others share electrons with neighboring atoms. This mobility of electrons is the essence of bonding and allows elements to combine with their neighbors to form stable combinations of atoms called **molecules**. In this way, for example, one atom of carbon (C) combines with two atoms of oxygen (O) to form a single molecule of carbon dioxide (CO_2), as we can see in Figure 3.2. The bonding between carbon and oxygen is such that the outer shells of each atom share eight electrons.

CHECK YOUR UNDERSTANDING

◉ What is the octet rule?

Atoms combine to form molecules by bonding. There are three types of bonds: *ionic bonds*, *covalent bonds*, and *metallic bonds*. Each of these types of bonding is important in the formation of minerals. There are also weak attractive forces, known as *van der Waals forces*, which can occur between atoms or molecules.

Ionic Bonds

Ionic bonds are the simplest of the bond types and result from the *transfer* of an electron from one atom to another. Metals such as sodium, potassium, and calcium have only

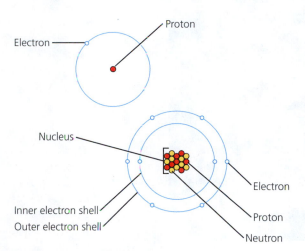

Figure 3.1   Structure of Atoms

(a) The atomic structure of hydrogen features an electron orbiting a single proton. (b) Oxygen has eight electrons orbiting a dense nucleus containing eight protons (red) and eight neutrons (yellow). Note that the first two electrons are in a lower orbit (shell) than the remaining six.

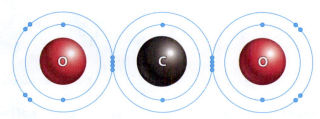

Figure 3.2   A Molecule of Carbon Dioxide

One atom of carbon (C) combines with two atoms of oxygen (O) to form a molecule of carbon dioxide. The outer shells of each atom conform to the octet rule.

a few electrons in the outer orbits of their atoms, and these are detached with relative ease from a metal atom, leaving the atoms' outermost shell with a stable arrangement of eight electrons. On the other hand, nonmetals such as chlorine and fluorine with several electrons in the outer orbits of their atoms can achieve a more stable configuration by accepting more electrons to fill their outermost shells. Consequently, if a metal with a tendency to lose an electron, such as sodium (Na), encounters a nonmetal with a tendency to accept one, such as chlorine (Cl), the two elements bond by the transfer of an electron from sodium to chlorine (Fig. 3.3a).

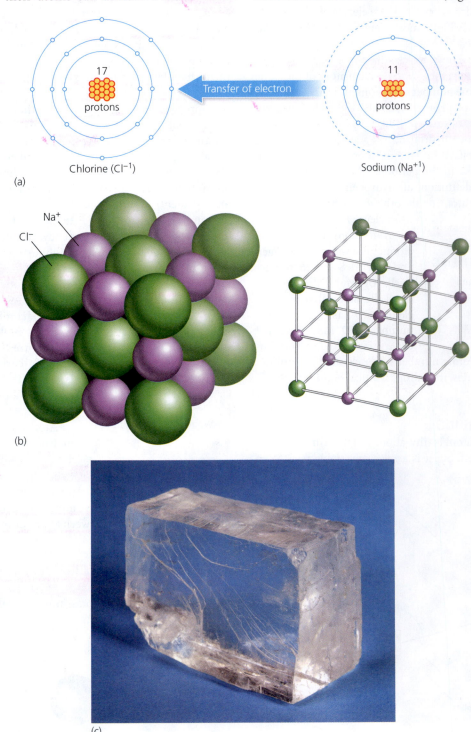

Figure 3.3   Structure of Halite (Table Salt): An Example of Ionic Bonding

(a) A sodium atom easily transfers its outermost electron to become a positively charged sodium cation. A chlorine atom readily accepts the electron from sodium to become a negatively charged chloride anion. The opposite charges of the sodium and chloride ions attract each other to form an ionic bond. (b) Halite (common salt) is held together by these ionic bonds. (c) A halite crystal's external appearance reflects its internal bonding structure.

With this transfer, the sodium and chlorine are not in their simple atomic states, but are charged atoms called **ions**. The sodium ion has a positive electrical charge because it has lost an electron, whereas the chloride ion has a negative charge because it has gained an electron. As a result of their opposing charges, the two ions attract each other and bond together. A positively charged ion, such as the sodium ion, is a **cation**, whereas a negatively charged ion, such as the chloride ion, is an **anion**.

The exchange of an electron from sodium to chlorine results in an ionically bonded molecule of the compound sodium chloride. Table salt (also called *halite*) is an orderly arrangement of alternating sodium and chloride ions, known as a **crystal structure** (Fig. 3.3b). As the sodium and chloride ions are bonded together in a 1:1 ratio, its chemical formula is Na^+Cl^-. The crystal structure is stable because it maximizes the attraction between oppositely charged ions and minimizes the repulsion between ions of the same charge. The orderly and stable arrangement means that the external appearance of the salt crystal reflects its internal bonding structure (Fig. 3.3c).

> **CHECK YOUR UNDERSTANDING**
>
> ● Why does sodium combine with chlorine to form halite (common salt)?

Covalent Bonds

Bonds created by *sharing* electrons are known as **covalent bonds**. Although each atom in a covalently bonded molecule donates electrons to the bond, these electrons are shared by the atoms that bond. The shared electrons provide the attraction that binds the molecule together (Fig. 3.4a).

Diamond, the crystalline form of carbon, is an example of a mineral held together by shared electrons (Fig. 3.4b). Each carbon atom has four nearest neighbors, so that four electrons are shared. This makes the covalent bonds between the carbon atoms very strong, and explains why diamonds are so hard and must be heated to such high temperatures (3500°C at Earth's surface) before they melt.

Metallic Bonds

In some metals, the outermost electrons are far enough from the nucleus that the nucleus cannot hold on to them. When atoms combine to form a piece of metal, each atom gives up an outer electron and becomes a positive ion. Collectively, the atoms form a layer of positive ions, with a common cloud of shared electrons swarming around them. These shared electrons roam throughout the material, holding together the metal cations (Fig. 3.5). This type of bonding is known as a **metallic bond** and the strength of metals is directly related to the strength of these bonds. The electrons within the cloud in bonded metals move freely and can carry energy from one end of a piece of metal to the other, which is why metals are generally excellent conductors of heat and electricity. These freely moving electrons also reflect solar radiation, which is why metals have shiny surfaces.

> **CHECK YOUR UNDERSTANDING**
>
> ● Compare and contrast the three types of chemical bonds that are common in minerals.

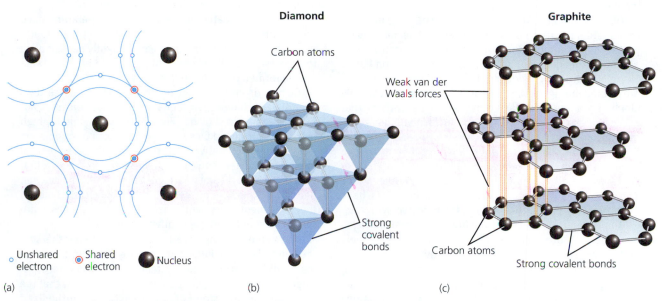

Diamond

Carbon atoms

Strong covalent bonds

Graphite

Weak van der Waals forces

Carbon atoms

Strong covalent bonds

○ Unshared electron   ◉ Shared electron   ● Nucleus

(a)   (b)   (c)

Figure 3.4 Atomic Structure of Diamond and Graphite

(a) Covalent bonds form when adjacent carbon atoms share electrons. (b) In diamond, covalently bonded carbon atoms form a strong three-dimensional structure. (c) In graphite, sheets of covalently bonded carbon atoms are linked by weaker van der Waals forces. The more tightly packed structure of diamond is related to its formation at depths greater than 150 km (93 mi) below Earth's surface, whereas the more open structure of graphite is consistent with its formation in the shallow crust.

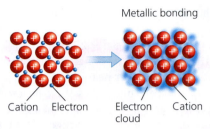

Metallic bonding

Cation   Electron   Electron   Cation
                           cloud

Figure 3.5   Metallic Bonding
Metallic bonding occurs when the outer electrons are shared across the entire metal. The atoms become positive ions, and the strength of the metallic bonds is related to the attraction between the positive layer of ions and the negative cloud of electrons.

Van der Waals Forces

In minerals, van der Waals forces are weak attractive forces that form between atoms or molecules. For example, the crystal structure of graphite is dominated by sheets of covalently bonded carbon. The sheets are held together by van der Waals forces (Fig. 3.4c), which are readily broken if a little stress is applied so that the sheets of graphite easily slide past one another. This characteristic property explains why graphite is such a soft mineral, and why it is used in writing and as a lubricant.

Most minerals that make up the rocks of the solid Earth form by a combination of ionic and covalent bonding. These minerals are relatively poor conductors (and so are good insulators) because the electrons are firmly bound to their adjacent cations. Because nonmetallic minerals are the dominant minerals of Earth's rocky outermost layer or *crust*, the crust as a whole is a poor conductor. On the other hand, rocks containing high concentrations of metals can be good conductors. Because of this, they can sometimes be detected by measuring how efficiently rocks below Earth's surface transmit an electric current. This technique is often used in mineral exploration to search for metal-rich rocks.

LIVING ON EARTH

Origin of the Chemical Elements

In this chapter, we discuss the various ways in which the chemical elements combine to form minerals, but this leads us to ask a broader question: How did these elements form in the first place? Naturally occurring chemical elements are numbered from 1 to 92, depending on the number of protons (or positive charges) in their atomic nuclei. The names we give these elements depend on this number because the number of protons determines an element's chemical properties.

Given that all protons are positively charged, how can their mutual repulsion be overcome so that as many as 92 protons can be packed into the nucleus of an atom? The answer to this question lies in the heavens, not on Earth. There is compelling evidence that the chemical elements are formed during the life cycle of stars. Because positively charged protons repel each other, it is not surprising that hydrogen, the simplest element, is by far the most abundant element in the universe, and the second simplest, helium, is the second most abundant. Together, these two elements comprise 98 percent of the chemistry of the universe and are believed to have been the only elements present early in its history. The heavier elements have been synthesized progressively as the universe evolved.

The normal tendency for positively charged particles to repel each other can only be overcome at extremely high temperatures where particles travel at very high velocities. For example, helium is formed by the fusion (joining) of two hydrogen protons, a process known as *thermonuclear fusion*. Thermonuclear fusion reactions occur within the Sun's core at temperatures of about 15 million °C (or 27 million °F) where hydrogen protons reach velocities of 500 km/sec (or over 1 million miles per hour). This process cannot be duplicated by natural means on Earth.

But how were the elements heavier than helium formed? In stars larger than the Sun, a succession of reactions may occur that can synthesize heavier elements. Helium atoms may fuse first to form carbon, then oxygen and, it is hypothesized, may ultimately form elements as heavy as iron, which has an atomic number of 26 (Fig. 3A). This process depends greatly on a star's core temperature since the production of heavier elements requires the temperature to be ever higher. Iron is the most tightly packed of all the elements, so no element higher in the periodic table can be formed by thermonuclear fusion.

So how do elements heavier than iron form? Large stars eventually exhaust their fuel and die in *supernova explosions* like the one shown in Fig. 3B. The unimaginably high pressures produced by such explosions can overcome the mutual repulsion between protons to synthesize chemical elements heavier than iron. Many scientists believe that supernovae are responsible for producing all of the heavier, naturally occurring elements in the universe. Because these elements are synthesized only during such catastrophic events, they are relatively rare.

The explosions of supernovae also disperse the synthesized elements into space, where they form the raw materials for future solar systems and become incorporated into other stars and planets. Thus it is likely that the chemical elements on Earth were actually synthesized during the life cycles of unknown and now extinct stars. In a very real way, we are all composed of stardust!

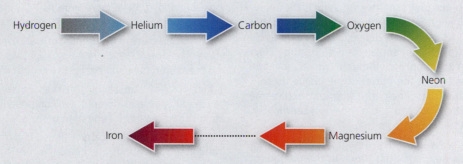

Figure 3A   Thermonuclear Fusion

Thermonuclear fusion, by which elements as heavy as iron (atomic number 26) are synthesized in the cores of massive stars, occurs in a succession of steps. Hydrogen atoms first fuse to form helium—the nuclear reaction that powers our own Sun. Helium atoms, in turn, fuse to form carbon, then oxygen and, depending on the star's core temperature, eventually elements as heavy as iron.

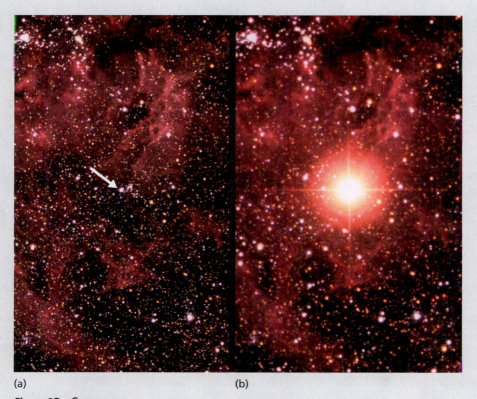

(a)   (b)

Figure 3B   Supernova

These two photos were taken before and after a supernova explosion. The first image (a) shows the location of the star before the explosion, and the second image (b) shows the flash of light emitted by the supernova explosion. This explosion was photographed in 1987, but actually took place about 170,000 years earlier (it took 170,000 years for the flash of light to reach Earth). Supernova explosions occur in large stars. The energy released by these explosions can synthesize elements with up to 92 protons.

Some Basic Definitions in Science

• **Atom:** A fundamental unit of matter. An atom is the smallest unit of an element having all the characteristics of that element. An atom consists of a dense, central, positively charged nucleus (pl. nuclei) surrounded by orbiting electrons. With the exception of hydrogen (which has one proton and no neutrons), the nuclei of the elements contain protons and neutrons. The entire atomic structure has an approximate diameter of 10^{-8} (0.00000001) centimeter, or more conveniently, 1 angstrom (1Å).

• **Atomic mass:** A measure of the amount of matter in an atom. The mass is approximately equal to the number of protons plus neutrons in the atom.

• **Atomic nucleus:** The center of an atom containing protons and (with the exception of hydrogen) neutrons and orbited by electrons. The nucleus contains more than 99 percent of the atom's mass but is about 100,000 times smaller than the atom.

• **Atomic number:** The number of protons in the atom of an element.

• **Bond:** An attractive force that binds two or more atoms together. Bonds may form by transfer of electrons (ionic), sharing of electrons (covalent), or by free movement of electrons (metallic).

• **Chemical reaction:** The transformation of one set of chemical substances to another set. Chemical reactions involve the breaking and formation of chemical bonds and can break molecules apart as well as form new molecules.

• **Compound:** A substance that contains two or more different elements that are joined by bonds. See and compare the definition of *molecule*.

• **Electron:** A negatively charged particle that is an essential component of an atom. Electrons surround and orbit the nucleus of the atom. An electron has a very small mass compared to a proton and a neutron.

• **Element:** An element is a substance that cannot be broken down using chemical methods. Each element consists of atoms, all of which must have the same number of protons (that is, the same atomic number).

• **Ion:** An atom that has an electrical charge because of an imbalance in the number of protons and electrons. Positive ions, or *cations,* have more protons than electrons. Negative ions, or *anions,* have more electrons than protons.

• **Mass:** A measure of the amount of matter in an object. The mass of an object is always the same, irrespective of where it is measured (see and compare the definition of *weight*).

• **Matter:** Any substance that has mass and occupies space. All physical objects are composed of matter. An atom is a fundamental unit of matter.

• **Molecule:** Two or more atoms linked together by chemical bonds. All compounds are molecules, but not all molecules are compounds. Water (H_2O) is a compound because the molecule links different elements together. H_2 is not a compound because its molecule links the same element.

• **Neutron:** A fundamental particle inside the atomic nucleus. A neutron has no net electrical charge and has approximately the same mass as a proton. In later chapters we will discuss how the number of neutrons in the atomic nucleus of an atom can vary, and how that number determines the *isotope* of the element.

• **Precipitate:** Solid that forms when ions in liquid solution combine and settle out of solution. Examples of precipitates include nonsilicate rock-forming minerals (such as halite or gypsum) produced by the evaporation of salt water from ancient seas.

• **Proton:** A positively charged particle inside the atomic nucleus. The number of protons in the nucleus determines the atomic number of an element, as shown in the periodic table of the elements.

• **Solution:** A homogeneous mixture of two or more substances. Typically, one substance (the *solute*) dissolves in the other (the *solvent*).

• **Substance:** Matter of definite composition and properties.

• **Weight:** Weight is the force that a given mass feels due to gravity. Unlike the mass of an object, which is the same no matter where it is measured, the weight of an object depends on the force of gravity, and this varies from location to location. For example, the weight of an object measured on the Moon is about one-sixth of its weight measured on Earth because the Moon's gravity is about one-sixth of that of Earth. The mass of the object, however, remains the same in both locations (compare definition of *mass*).

3.1 SUMMARY

• Minerals are the building blocks of rocks, and are important to our understanding of Earth's origin and dynamics.

• Elements are the building blocks of minerals, and are made up of atoms in which electrons orbit a central nucleus of protons and neutrons.

• The most stable configuration for most elements is to have a full outer shell consisting of eight electrons. Bonding of elements is the result of electron mobility and each atom's effort to achieve the most stable electron configuration.

• Ionic bonds involve the transfer of electrons from a metal to a nonmetal. Covalent bonds are formed by the sharing of electrons between elements. Metallic bonds form when atoms combine to form a piece of metal, and each atom gives up an electron to form a common cloud of shared electrons.

- Most minerals that make up the rocks of the solid Earth contain a combination of ionic and covalent bonding.

3.2 Minerals: Orderly Expressions of Matter

Minerals are the basic components of the rocks that make up the solid Earth. We define a **mineral** as a naturally occurring, inorganic crystalline solid with a specific chemical structure and formula. (See Science Refresher: **Solids, Liquids, and Gases**.) **Inorganic** compounds lack carbon derived from living matter. The term *crystalline* indicates that the mineral consists of atoms arranged in an orderly pattern; this orderly pattern gives the mineral a particular chemical composition.

NATURALLY OCCURRING AND INORGANIC

Because minerals are defined as having formed in the natural world, artificially manufactured inorganic solids are not usually regarded as minerals. For example, a synthetic diamond is not a mineral, although natural diamond is. Similarly, because minerals are defined as inorganic, **organic** chemicals, which are chemicals that contain carbon-hydrogen bonds, known as *hydrocarbons,* are excluded. For this reason, common table sugar (sucrose, $C_{12}H_{22}O_{11}$), as well as animal and vegetable matter, are not considered minerals. However, some materials produced by living things are not hydrocarbons and are therefore considered minerals. For example, some marine organisms secrete hard exterior shells composed of chemical compounds such as calcium carbonate ($CaCO_3$), which is commonly preserved as the mineral *calcite.*

Minerals form only under the appropriate physical conditions. For example, the minerals graphite (used as pencil lead) and diamond are both composed of a single essential element, carbon (C). However, their internal arrangement of carbon atoms is quite different (see Figure 3.4), and this difference affects the stability of the two minerals. In the diamond structure, the carbon atoms are very tightly packed because diamonds develop at very high pressures. Hence,

SCIENCE REFRESHER

Solids, Liquids, and Gases

In this text we deal with solids, such as minerals and rocks, liquids, such as magma or running water, and gases, such as the air we breathe. It is important that you understand some of the most basic properties of each of these *phases,* or *states,* of matter (Fig. 3C). Particles in a solid are closely packed. Although fixed in place, the particles are not motionless but vibrate in a limited fashion about fixed positions. As a result of these limited vibrations, a solid has a definite size, shape, and volume.

The particles in a liquid are about the same distance apart as they are in a solid, but they are more loosely bonded to one another and so are able to move relative to one another. As a result, liquids flow. For this reason, a liquid has a definite volume but has no size or shape and acquires the shape of its container. Most of the bonds in a liquid are temporary. They form, break, and reform as the liquid flows. The varying resistance of liquids to flow is a property known as *viscosity,* and reflects the strength of

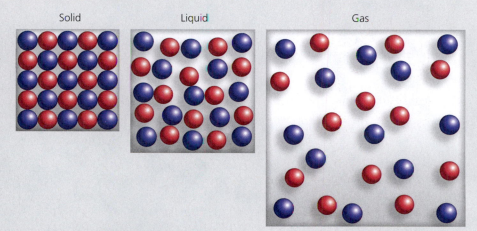

Figure 3C   Solids, Liquids, and Gases

Particles in a solid are fixed in place and packed closely to each other, those in a liquid can move around but stay close, and those in a gas are far from each other and move about freely.

each liquid's temporary bonds. In everyday usage, the viscosity of a liquid is a measure of its "stickiness"; for example, honey is more viscous than water.

In a gas, the particles move independently of each other. Therefore, a gas has neither a definite shape nor a volume and fills whatever container it is in.

diamonds can only form deep in Earth's interior where the pressures exerted by the weight of the overlying rocks are very high. Experiments show that this requires depths greater than 150 kilometers (93 miles). In contrast, graphite has a more open mineral structure and forms nearer Earth's surface, where the weight of overlying rocks is much less and the pressures are relatively low. Unlike diamond, some of the bonds in graphite are quite weak, and it is these bonds that break when you write with a pencil.

CRYSTALLINE SOLIDS

All minerals are crystalline solids, which means that their ions are arranged in regular three-dimensional frameworks. Solid materials such as glass are not considered minerals because they lack a regular atomic structure and are therefore not crystalline; these substances are **mineraloids**.

Left free to grow, minerals form **crystals**, which have perfect regular geometric shapes made up of flat **crystal faces** with straight edges and sharp corners. Crystals that have intrinsic value because of their beauty, durability, rarity or size are known as **gems**. In 1669, the Danish scientist Nicholas Steno deduced that the angle between the same two faces on different quartz crystals is identical (Fig. 3.6), and does not depend on the size of the crystal. He also proposed that crystals are composed of very small, identical building blocks and that the external appearance of the crystal—the *crystal form*—reflects how these small building blocks are put together, a basic principle that still applies today. We now

Figure 3.6   Crystal Faces
The angle between the equivalent faces of the same mineral is identical and does not depend on the size of the crystal. The angles between equivalent faces (labeled 1 to 4) of two crystals of quartz are identical.

know that the external form of a crystal (Fig. 3.7) is actually an expression of the orderly internal arrangement of a mineral's ions (see Figure 3.3b). This is a case where we can "judge a book by its cover," that is, the external appearance of a mineral provides important information about the internal geometry of its bonds.

Halite   Pyrite   Diamond   Quartz

Figure 3.7   Crystal Shape
Mineral crystals differ in shape for different minerals. Shown here are crystals of halite, pyrite, diamond, and quartz with their characteristic shapes.

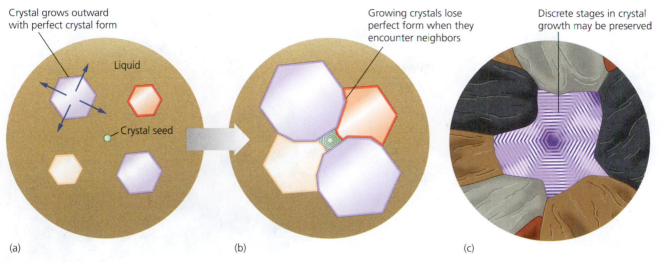

Crystal grows outward with perfect crystal form

Liquid

Crystal seed

Growing crystals lose perfect form when they encounter neighbors

Discrete stages in crystal growth may be preserved

(a)   (b)   (c)

Figure 3.8   Crystal Growth

(a) New crystals form as tiny seeds and grow outward as the ions in the liquid attach themselves to the crystal faces. (b) They maintain their external form until they encounter neighboring crystals and the resulting interference forces them to modify their shape. (c) In some instances, discrete stages in the crystal growth are preserved.

> **CHECK YOUR UNDERSTANDING**
>
> ● What is a crystalline solid? What is a mineraloid?

Crystals typically grow from a liquid in which the dissolved ions have become so concentrated that they are close enough to bond together (Fig. 3.8). This occurs when the liquid becomes *saturated,* meaning that it cannot hold any more ions in solution. The crystals that form are initially extremely small, but act as seeds and grow outward as other ions in the liquid attach themselves to the crystal faces. Growth continues until the local supply of atoms that can fit into the crystal structure is exhausted. As crystals grow, they maintain the same external shape, which reflects the internal structure of their bonds. In most situations, minerals eventually encounter obstacles as they grow, such as adjacent minerals. The minerals lose their characteristic external shape as they interfere with one another. However, the diagnostic internal structure of the mineral can still be deduced by X-ray studies, which penetrate the crystal interiors.

CHEMICAL COMPOSITION AND FORMULA

Because the ions of minerals have a regular arrangement, every mineral has a specific composition that can be expressed as a chemical formula. The **chemical formula** of any mineral lists its component ions according to the proportion in which they occur in the mineral. For example, the mineral halite (better known as table salt) has the chemical formula NaCl (sodium chloride). This means that it is made up of equal numbers of sodium (Na) and chloride (Cl) ions bonded together (see Figure 3.3b). The common mineral quartz has the chemical formula SiO_2. The subscript 2 after the oxygen symbol indicates that it

contains twice as many ions of the element oxygen (O) as it does of the element silicon (Si).

In reality, no natural mineral is absolutely pure. Modern analytical techniques provide chemical analyses of individual minerals, and it is clear from these that all minerals contain impurities. Quartz, for example, may contain small amounts of other elements. These impurities are evident in the various colors displayed by different forms of quartz (Fig. 3.9), from the deep purple of amethyst produced by trace amounts of iron (Fe), to the pale pink of rose quartz produced by traces of the metal titanium (Ti). These trace elements affect the appearance of various types of quartz but it is the availability of silicon and oxygen that is essential to the formation of quartz, because without these two elements the mineral simply could not form. A mineral formula is therefore an idealized expression of its composition. Just as chocolate is an essential ingredient in a chocolate cake, a mineral's chemical formula identifies the essential chemical elements needed for that mineral to form. The type of impurities and their abundance reveals important information about the physical conditions (such as pressure and temperature) under which the mineral grew.

> **CHECK YOUR UNDERSTANDING**
>
> ● What does the formula for orthoclase ($KAlSi_3O_8$) indicate about the composition of this mineral?

POLYMORPHS

Although each mineral has its own chemical formula, that formula is not necessarily unique to the mineral. In fact, two or more minerals can sometimes have the same chemical formula because they have the same chemical composition.

(a)

(b)

(c)

(d)

(e)

Figure 3.9 Varieties of Quartz

Several varieties of quartz are shown here including: (a) clear, (b) smoky, (c) amethyst, (d) agate, (e) rose. The characteristic crystal structure of quartz is evident in all examples, except for agate, which is *microcrystalline,* which means that its crystals are much too small to be seen with the naked eye.

Minerals with the same chemical composition but different crystal structures are known as **polymorphs**. As we have seen, the minerals diamond and graphite are made up of a single essential element, carbon, so that both have the same chemical formula (C) and are therefore polymorphs. However, because the internal arrangement of the carbon atoms within the crystal structure of the two minerals is quite different (see Figure 3.4), the appearance and properties of the two minerals differ. Which polymorph grows depends on the pressure and temperature at the time of growth.

> **CHECK YOUR UNDERSTANDING**
>
> ⦿ What is a polymorph?

ROCK-FORMING MINERALS

Although more than 4500 naturally occurring minerals have been documented, less than 1 percent are common in the rocks of Earth's crust. Minerals that are common in rocks are known as the **rock-forming minerals** (Table 3.1). But why are some minerals more common than others? Minerals are combinations of elements, and the most common minerals are composed of elements that are the most abundant in Earth's crust. Close to 84 percent of the crust is made up of oxygen and silicon atoms (Table 3.2). As a result, most rock-forming minerals are **silicate minerals**. Silicates are minerals in which silicon (Si) and oxygen (O) are chemically bonded, usually to a variety of metallic ions such as sodium (Na), potassium (K), calcium (Ca), aluminum (Al), iron (Fe), and magnesium (Mg). These six metals together make up about 16 percent of Earth's crust. Together with oxygen and silicon, these elements comprise more than 99 percent of the crust, and so the vast majority of minerals in crustal rocks are various combinations of them (Table 3.3). Some silicate minerals also contain water, not in its familiar liquid state, but incorporated as an integral part of the crystal structure. The presence of water in the mineral structure generally means that the mineral grew in the presence of water.

As we shall see, not all minerals are silicates. A small but important group of rock-forming minerals do not contain silicon, and these minerals are known as **nonsilicate minerals**. How do we distinguish between silicates and nonsilicates? As we shall see in Section 3.3, each mineral possesses a characteristic set of properties that help to determine its identity. These properties are an expression of the mineral's chemical formula and crystal structure.

3.2 SUMMARY

- Minerals are naturally occurring, inorganic, crystalline solids with specific chemical compositions and formulas.

- Left free to grow, minerals form perfect crystals with a regular geometrical form.

TABLE 3.1   SOME IMPORTANT ROCK-FORMING MINERALS

Mineral	Composition	Primary Occurrence
Ferromagnesian Silicates		
Olivine	$(Mg, Fe)_2SiO_4$	Igneous, metamorphic rocks
Pyroxene group		
Augite most common	Ca, Mg, Fe silicate	Igneous, metamorphic rocks
Amphibole group		
Hornblende most common	Hydrous* Na, Ca, Mg, Fe, Al silicate	Igneous, metamorphic rocks
Biotite	Hydrous K, Mg, Fe, Al silicate	All rock types
Non-ferromagnesian Silicates		
Quartz	SiO_2	All rock types
Potassium feldspar group		
Orthoclase, microcline	$KAlSi_3O_8$	All rock types
Plagioclase feldspar group	Varies from $CaAl_2Si_2O_8$ to $NaAlSi_3O_8$	All rock types
Muscovite	Hydrous K, Al silicate	All rock types
Clay mineral group	Varies	Soils and sedimentary rocks
Carbonates		
Calcite	$CaCO_3$	Sedimentary rocks
Dolomite	$CaMg(CO_3)_2$	Sedimentary rocks
Sulfates		
Anhydrite	$CaSO_4$	Sedimentary rocks
Gypsum	$CaSO_4 \cdot 2H_2O$	Sedimentary rocks
Halides		
Halite	NaCl	Sedimentary rocks

*Contains molecules of water

TABLE 3.2   COMMON ELEMENTS IN EARTH'S CRUST

| Element | Symbol | PERCENTAGE OF CRUST | |
		(By Weight)	(By Atoms)
Oxygen	O	46.6%	62.6%
Silicon	Si	27.7	21.2
Aluminum	Al	8.1	6.5
Iron	Fe	5.0	1.9
Calcium	Ca	3.6	1.9
Sodium	Na	2.8	2.6
Potassium	K	2.6	1.4
Magnesium	Mg	2.1	1.8
All others		1.5	0.1

TABLE 3.3   RELATIVE ABUNDANCE OF MINERALS IN EARTH'S CRUST

Mineral	Percentage
Plagioclase	39%
Quartz	12%
Orthoclase	12%
Pyroxenes	11%
Micas	5%
Amphiboles	5%
Clay minerals	5%
Other silicates	3%
Nonsilicates	8%

- Minerals which have the same chemical composition, but different crystal structures are known as polymorphs.

- Although there are more than 4500 known minerals, less than 1 percent are common in rocks. These are known as the rock-forming minerals. Most rock-forming minerals are silicates in which a variety of metallic ions are bonded with silicon and oxygen.

3.3 Physical Properties of Minerals

Mineralogy is the study of minerals, their physical properties, formation, occurrence, and composition. A *mineralogist* is someone who specializes in the study of minerals. To identify a mineral, a mineralogist performs a series of tests. In some instances, it is possible that identification may remain elusive and more sophisticated tests may be necessary. For example, microscopic studies can identify the existence of clay minerals in a sample, but cannot identify which of the many natural clays may be present. In forensic science, this distinction may be crucial if samples of soil taken from a shoe are to link a suspect to the scene of the crime. In these situations, the mineral may be subjected to high-energy X-rays or electron beams that can resolve crystal structure and composition on a very fine scale.

However, before all these more sophisticated tests are performed, an accurate identification of a mineral usually starts with an inspection of the sample, either with the naked eye or with a hand lens, in order to determine the physical properties that reflect the internal structure and composition of the mineral. Although current mineralogy employs sophisticated analytical tools to study minerals, it is remarkable how much information about a mineral's identity can be obtained by observation and a few simple tests. For example, a *gemologist* (someone who specializes in the study of gems) uses these tests to help distinguish between authentic and fake gems. Although we cannot determine the detailed chemistry of a mineral from this kind of examination, we can deduce its idealized composition and structure. Properties that we can examine in a hand specimen include a mineral's color and streak, luster, crystal form, hardness, cleavage, and specific gravity. To make a correct identification, all of these physical properties are considered together. In this section of the chapter, we discuss each of these properties in turn.

COLOR AND STREAK

The color of a mineral is one of the first things we are likely to notice. A mineral's **color** is often related to its overall composition. In the silicate minerals, for example, those that contain significant concentrations of iron and magnesium are dark and are called the *ferromagnesian silicates,* whereas those that are low in iron and magnesium but contain elements like sodium and potassium are generally lighter and are called the *non-ferromagnesian silicates* (Fig. 3.10). (We discuss the ferromagnesian silicates and non-ferromagnesian silicates in more detail in Section 3.4.)

Although useful, the color of a mineral is rarely diagnostic and, in some instances, can be quite misleading. For example, the semiprecious minerals amethyst and rose quartz have very different colors (see Figure 3.9), but to a mineralogist they are both merely varieties of quartz. Color in this case is not related to overall composition since the basic chemical formula (SiO_2) is the same for both minerals. The difference in color is entirely a consequence of minor impurities trapped within the crystals. Color, particularly in dark minerals, can also reflect grain size, which is not a fundamental property of a mineral. Many students learn this the hard way; a standard first mineralogy test often requires the identification of five samples of different colors, all of which turn out to be the same mineral! To alleviate this problem and facilitate identification, a mineral can be ground against an unglazed porcelain plate to produce a **streak**. Streak is the color of a mineral

CHECK YOUR UNDERSTANDING

- What is the difference between the mineral properties of color and streak?

Figure 3.10 Common Rock-Forming Silicates
Darker minerals (a–d) contain iron and magnesium and are ferromagnesian silicates. Lighter colored minerals (e–h) are rich in sodium or potassium and are non-ferromagnesian silicates. Pictured here are: (a) olivine; (b) pyroxene, (c) hornblende, (d) biotite, (e) quartz, (f) orthoclase feldspar, (g) plagioclase feldspar, and (h) muscovite.

silicate minerals, on the other hand, are held together by a variety of covalent and ionic bonds and are less reflective. The two main types of mineral luster are *metallic* and *non-metallic* (Fig. 3.12). Among the nonmetallic minerals, we see a variety of lusters, including *waxy, greasy, vitreous* (glassy), *brilliant* (diamond-like), and *dull*.

CRYSTAL FORM AND HABIT

If a mineral is left free to grow, it forms beautiful crystals with well-developed crystal faces. These faces are the external expression of the mineral's internal structure. Any two samples of the same mineral have the same **crystal form**, that is, the same geometric arrangement of crystal faces (Fig. 3.13). The crystals may differ in size, but every face on one crystal has an equivalent parallel face on every other crystal as we discussed in Section 3.2 (see Figure 3.6). This means that the angles between a mineral's crystal faces can be used to identify the mineral. However, mineral growth is usually hindered by the presence of neighboring minerals and this competition for space means that minerals are rarely given the opportunity to develop well-formed crystal faces. Fortunately, even in environments where their growth is restricted, minerals often have a characteristic outward appearance or **habit**. For example, some minerals tend to be *elongate* (i.e., long and narrow), others are typically *platy* (i.e., long and thin), and some minerals grow in fine elongated threads and are described as *fibrous* (Fig. 3.14).

> **CHECK YOUR UNDERSTANDING**
> ● How do neighboring minerals hinder crystal formation?

Figure 3.11   Streak

The color of a powdered mineral—its streak—is sometimes more diagnostic than the color of the mineral. Hematite, the mineral name for iron oxide or rust (Fe_2O_3), occurs in various forms. On a porcelain plate, however, its streak is always red.

in powdered form (Fig. 3.11). It is often a more consistent indicator than color, but not all minerals can produce a streak.

LUSTER

If we glance around a room, we can see that materials reflect light differently. A mirror is highly reflective whereas cushions and carpets are not. Likewise, different minerals reflect light differently. This property is known as **luster**. Recall from our earlier discussion of metallic bonds in Section 3.1 that metals have reflective, shiny surfaces. Most

(a)   (b)

Figure 3.12   Luster

Luster refers to the appearance of a mineral in reflected light. (a) The lead ore galena (PbS) has the appearance of a metal and is said to have a metallic luster. (b) Halite or rock salt (NaCl) has the appearance of glass and its luster is said to be nonmetallic and vitreous (glassy).

Figure 3.13 Crystal Form

Crystal form is the external expression of a mineral's orderly internal structure. These well-developed quartz crystals show a characteristic six-sided crystal form with pyramid-shaped ends.

Figure 3.14 Habit

Habit refers to the characteristic outward appearance of a mineral. Gypsum (shown here) sometimes grows in fibers and is said to have a fibrous habit.

HARDNESS

Recall from our discussion about bonding in graphite and diamond in Section 3.1 that a mineral's hardness relates to the strength of its chemical bonds. The strength of the covalent bonds in the mineral diamond, for example, makes it the hardest naturally occurring mineral. The mineral talc, on the other hand, has very weak bonds and is very soft. The contrasting hardness of these two minerals is reflected in their industrial uses; diamond is used as an abrasive, whereas talc is the principal ingredient of talcum powder.

Hardness is defined as the relative resistance of a mineral to being scratched. In 1812 Friedrich Mohs, an Austrian mineralogist, devised a scale of relative hardness. The scale is divided into 10 steps, each of which is defined by a common mineral (Table 3.4). Diamond, being the hardest mineral, is given a value of 10. In fact most gem minerals score high values on this scale. Ruby and sapphire, for example, are varieties of the mineral corundum, which has

TABLE 3.4 MOHS SCALE OF HARDNESS

Hardness	Mineral	Hardness of Some Common Objects
10	Diamond	
9	Corundum → rubies/sapphires	
8	Topaz	
7	Quartz	
		Steel file (6 ½)
6	Orthoclase	
		Glass (5 ½–6)
5	Apatite	Pocketknife (5)
4	Fluorite	
3	Calcite	Copper penny (3)
		Fingernail (2 ½)
2	Gypsum	
1	Talc	

Note that the scale is a relative one, and that diamond, the hardest mineral, is assigned a value of 10, and talc, the softest mineral is assigned a value of 1.

a hardness value of 9. Talc, the softest mineral, has a hardness of 1 on the Mohs scale. A mineral ranked higher on the scale will scratch one ranked below it on the scale. Your fingernail has a hardness between 2 and 3, and can therefore scratch gypsum (hardness 2) but not calcite (hardness 3). A pocketknife or a plate of glass has a hardness between 5 and 6, and so it can scratch the mineral apatite but not potassium feldspar (orthoclase). It is important to realize that the Mohs scale of hardness is not a linear one, but merely ranks minerals in terms of their hardness relative to ten common minerals.

> **CHECK YOUR UNDERSTANDING**
>
> ○ What is the Mohs scale of hardness and how is it used?

CLEAVAGE AND FRACTURE

Many minerals have a tendency to break (or "cleave") in the directions of their weakest bonds. This tendency is called **cleavage** and the preferred fracture surfaces are called *cleavage planes*. It is important to understand that cleavage planes are *not* crystal faces, because they are not expressions of the outward growth of a mineral. Instead, cleavages reflect systematic weaknesses in the internal structure of a mineral along which the mineral tends to break, in much the same way that logs tend to split along weaknesses in the wood. Gemologists exploit natural cleavage planes in gems to produce the majestic shapes of commercial gemstones.

There is a wide variety of cleavage patterns in silicate minerals. This variety reflects the wide variation in their mineral structure. However, all these cleavage patterns have one important feature in common. Because the bonds between silicon and oxygen are strong, cleavage develops along the weaker bonds *between* the silicate structures, rather than *within* them. Mica, for example, consists of long sheets of silicate structures. The bonds within an individual sheet are much stronger than the relatively weak bonds that occur between the sheets. As a result, micas are easily pried apart along cleavage planes that occur between these sheets (Fig. 3.15 a, b).

Minerals that break along irregular surfaces rather than cleavage planes are said to possess a **fracture**. Any mineral can be broken to produce irregular fractures, but some minerals break along fracture surfaces of characteristic shape. For example, the mineral quartz typically breaks like glass along smoothly curved surfaces like those of a conch shell and is therefore said to possess a *conchoidal* fracture (Fig. 3.15c).

> ### CHECK YOUR UNDERSTANDING
> ● How does a cleavage plane differ from a crystal face?
> ● Why does quartz have a conchoidal fracture rather than a cleavage?

SPECIFIC GRAVITY AND DENSITY

The **specific gravity** of a mineral is defined as the ratio of its mass relative to the mass of an equal volume of water at 4°C. For example, the common mineral quartz weighs 2.65 times the same volume of water and so has a specific gravity of 2.65. This value is similar to the mineral's **density**, which is calculated by dividing the mass of the mineral by its volume (given in units of g/cm³). The specific gravity and density of a mineral reflect both its chemical composition and mineral structure. For example, the contrasting densities of graphite and diamond (2.2 and 3.5 g/cm³, respectively) is directly related to the tighter bonding arrangement of diamond and the tighter packing of its carbon atoms (see Figure 3.4). As we have discussed, this difference in structure reflects the contrasting physical conditions under which these two minerals form, with the denser mineral (diamond) being more stable at the higher pressures found at deeper crustal levels.

> ### CHECK YOUR UNDERSTANDING
> ● Explain the difference in hardness between graphite and diamond.

With practice, you can estimate a mineral's density by simply holding it in your hand. When minerals are similar in other properties, how heavy the mineral "feels" can be an important clue in determining what the mineral is.

3.3 SUMMARY

- The physical properties of minerals reflect their internal structure and chemical composition, and aid in mineral identification.
- Color may yield some clues to a mineral's chemical composition, but the color of the mineral in powdered form, known as its streak, is often more reliable. Luster refers to the way in which minerals reflect light.

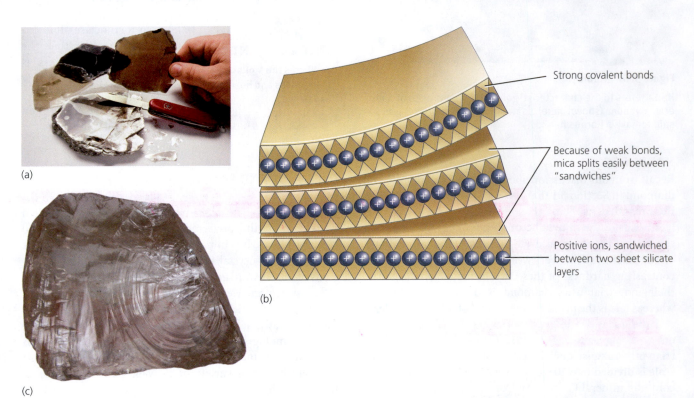

(a)

(b)

(c)

Strong covalent bonds

Because of weak bonds, mica splits easily between "sandwiches"

Positive ions, sandwiched between two sheet silicate layers

Figure 3.15 Cleavage and Fracture (a) Mica is pulled apart along its cleavage planes. (b) A simplified depiction of the relationship of cleavage to structure in mica is shown here. (c) Minerals like quartz, with networks of silica-oxygen bonds, exhibit no obvious cleavage, and instead display conchoidal fracture surfaces.

- The crystal form of a mineral reflects the arrangement of its crystal faces, and is an external expression of its internal bonding. The hardness of a mineral depends on the strength of its chemical bonds. Minerals are ranked in hardness relative to 10 well-known ones.

- Cleavage is the tendency of mineral to break along preferred directions and reflects systematic weaknesses in the internal structure of a mineral.

- The specific gravity of a mineral is the ratio of its weight relative to an equal volume of water, and reflects a combination of chemical composition and mineral structure. It is similar to a mineral's density, which is the mass of the mineral divided by its volume.

3.4 Silicate Rock-Forming Minerals

Recall from Section 3.2 that silicon and oxygen account for approximately 84 percent of all elements in the crust (see Table 3.2), so it is not surprising that silicates are the most common rock-forming minerals on Earth. The way in which silicon and oxygen combine is the fundamental building block of all silicate minerals. This building block is the basic structural unit of all rock-forming silicates and is known as the **silicate tetrahedron**. It is so-called because of its tetrahedral form, which is a pyramid-like shape with four triangular faces (Fig. 3.16). The silicate tetrahedron consists of a central silicon cation with a positive electrical charge bonded with four negatively charged oxygen anions positioned at the corners.

As Figure 3.16 illustrates, it is the relative sizes of the silicon and oxygen ions that governs the tetrahedron geometry. The smaller silicon ion has four positive charges, whereas the four larger oxygen ions each have two negative ones, so the tetrahedral building block has an overall negative charge of −4. As a result, silicate tetrahedra tend to bond with cations such as Na^+, K^+ and Ca^{2+}. Individual tetrahedra link together in a variety of forms that are characteristic of silicate minerals (Fig. 3.17). As we shall see, the vast number of naturally occurring silicate minerals is indicative of the many different ways in which this basic tetrahedral unit can combine with various metal ions to form stable crystal structures. The silicate tetrahedron is like a basic Lego building block and the vast array of silicate mineral structures depend on how these blocks are attached to one another.

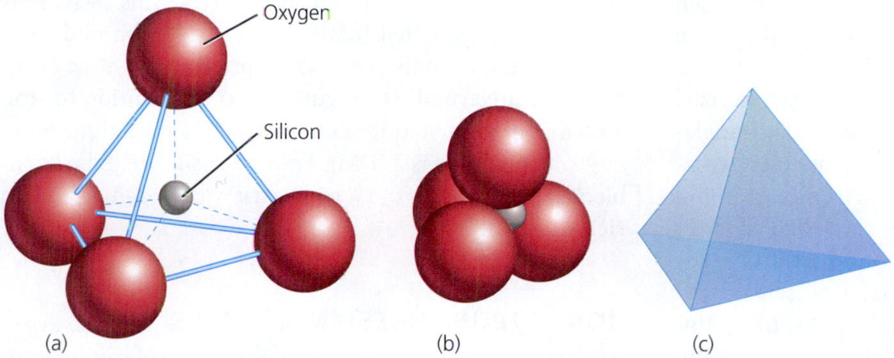

We divide the silicate minerals into two categories: the **ferromagnesian silicates**, which contain both iron (*ferrum* in Latin) and magnesium, and the **non-ferromagnesian silicates**, which do not contain either of these elements (see Fig. 3.10). This division is not merely a convenience; we will revisit this simple subdivision many times when we discuss the rocks in which these minerals occur.

> **CHECK YOUR UNDERSTANDING**
> - What is the difference between ferromagnesian and non-ferromagnesian silicate minerals?

FERROMAGNESIAN SILICATES

The presence of iron and magnesium in the crystal structure of silicate minerals has a considerable influence on the physical properties of the minerals. These minerals tend to be dark and have compact crystal structures. This is reflected in their densities, which typically range from 2.7 to 3.6 g/cm³. Ferromagnesian silicates dominate Earth's oceanic crust, which consequently has an average density of 2.9 g/cm³. Ferromagnesian silicates are subdivided into different mineral groups, each of which has a distinctive silicate structure.

The **olivine group** is characterized by isolated tetrahedra that are linked by iron and magnesium atoms (see Figure 3.17). The mineral *olivine* is typically dark to olive-green in color (see Figure 3.10a). You may be familiar with the gemstone variety of olivine known as peridot. Olivine is thought to be the most abundant mineral in the region of Earth's interior immediately below the crust, known as the mantle. It is also abundant in iron-magnesium rich igneous rocks, where it is commonly the first mineral to crystallize as the magma cools. Other minerals, such as *garnet*, have a similar silicate structure, but are different in composition and, as a result, occur in different types of rocks. The **garnet group** of minerals can occur in a wide variety of settings, but is most common in metamorphic rocks. Some clear varieties of garnet are used as gemstones.

The **pyroxene group** is a group of calcium-magnesium-iron (Ca-Mg-Fe) silicates that are usually green, brown, or black in color (see Figure 3.10b). In this group, the silicate

Figure 3.16 Silicate Tetrahedron
The silicate tetrahedron is the basic building block of silicate mineral groups. (a) An expanded view shows oxygen anions at the corners of the tetrahedron with a silicon cation at its center. (b) The second view shows the arrangement of the same ions as they actually occur. (c) The final view is a diagrammatic representation of the silicate tetrahedron.

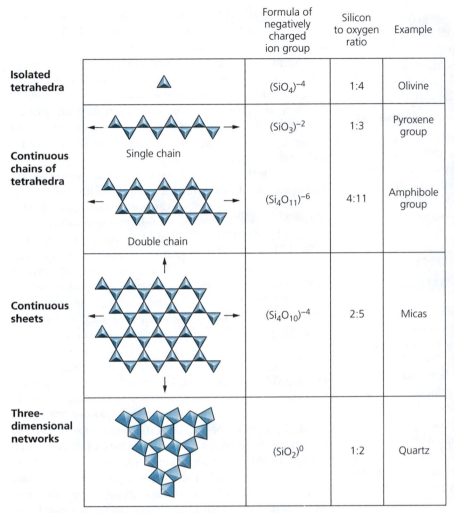

	Formula of negatively charged ion group	Silicon to oxygen ratio	Example
Isolated tetrahedra	$(SiO_4)^{-4}$	1:4	Olivine
Continuous chains of tetrahedra Single chain	$(SiO_3)^{-2}$	1:3	Pyroxene group
Double chain	$(Si_4O_{11})^{-6}$	4:11	Amphibole group
Continuous sheets	$(Si_4O_{10})^{-4}$	2:5	Micas
Three-dimensional networks	$(SiO_2)^0$	1:2	Quartz

Figure 3.17　Structure of Silicate Minerals
The arrangements of silica tetrahedra in some common rock-forming silicates are listed here in chart form.

tetrahedra are linked into a single chain by sharing an oxygen anion (see Figure 3.17). The chains are in turn linked to one another by calcium, magnesium, and iron cations. The silicon-oxygen bonds within a single chain are much stronger than the bonds that link the chains together. As a consequence, the pyroxenes tend to split between silicate chains to form a systematic cleavage pattern characterized by two cleavage directions that are almost at right angles (87° and 93°) to each other (Fig. 3.18). Like olivine, pyroxenes are abundant in Earth's mantle and in iron-magnesium rich igneous rocks where they typically crystallize at lower temperatures than olivine, between 1100 and 1300°C.

The **amphibole group** is a complex set of silicate minerals with highly variable chemical compositions. Most amphiboles are similar in composition to the pyroxenes but contain aluminum in addition to calcium, magnesium, and iron. In addition, water is also stored within the crystal structure (see Table 3.1).

The principal difference between amphiboles and pyroxenes lies in the silicate structure (Fig. 3.18). In amphiboles, the silicate tetrahedra are arranged in double chains

(see Figure 3.17). These chains are linked by bonds with cations such as calcium, magnesium, iron, or aluminum. This difference in structure imparts important property differences that allow us to distinguish between amphiboles and pyroxenes. Like those in the pyroxenes, the bonds linking the tetrahedral chains in the amphiboles are relatively weak, imparting a cleavage between the chains. However, the double chain geometry means that the angle between the two cleavages is close to 60° (56° and 124°). This contrasts with the pyroxenes where the two cleavages are almost perpendicular to one another. In addition, the long silicate double chains of the amphiboles are reflected in the elongate habit of the mineral (see Figure 3.18c), in contrast with the more stubby habit of pyroxene.

Dark-green *hornblende* is the most common member of the amphibole group and occurs in a wide variety of rocks (see Figure 3.10c). In iron-magnesium rich igneous rocks, hornblende typically crystallizes at lower temperatures than pyroxene, close to 1000°C. The trend in crystallization temperatures from olivine to pyroxene to amphibole shows that magmas crystallize more complex silicates at lower temperature, an issue we will explore in Chapter 4.

The **mica group** is a set of minerals made up of sheets of silicate tetrahedra connected to one another by cations. *Biotite* is a dark ferromagnesian mica in which the tetrahedral sheets are connected by cations such as potassium, iron, and magnesium. Because of the weak bonds between the sheets, all micas, including biotite, have an excellent cleavage parallel to the tetrahedral sheets. This cleavage is so well developed that biotite has a platy habit and individual biotite crystals can often be pried apart with a tweezers or a fingernail (see Figure 3.15a). In addition to iron and magnesium, biotite contains potassium, aluminum, and water, and so is a K-Mg-Fe-Al-OH silicate. Like hornblende, biotite occurs in a wide variety of iron-magnesium rich rocks.

NON-FERROMAGNESIAN SILICATES   Sodium + potassium

As the name implies, members of this group of silicate minerals are characterized by the *absence* of iron and magnesium in their crystal structures. As an example, silicates such as quartz and feldspar are dominated by relatively "light" elements and their specific gravities typically range from 2.6 to 2.9 g/cm³. Because these are the dominant minerals in Earth's continental crust, it is not surprising that the

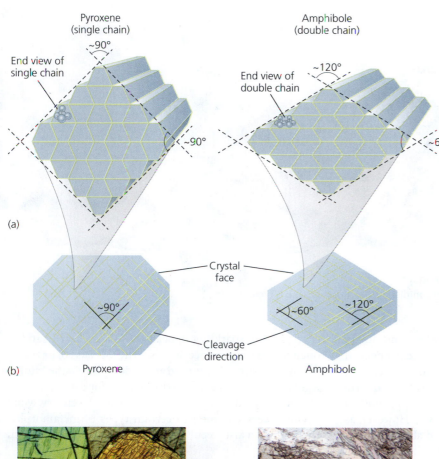

(c) Augite   Hornblende

Figure 3.18 Pyroxene and Amphibole
The crystal forms and cleavage orientations differ between pyroxene (on the left) and amphibole (on the right). (a) Pyroxene forms a single chain whereas amphibole forms double chains, implying the shape of their chains also differs. For both mineral groups, the bonds between the silicate chains are weak (green lines) and so control the orientation of their respective cleavages (dashed lines). Because of differences in the shape of their chains, the angle between their respective cleavages is different, about 90° for pyroxene, and 60° or 120° for amphibole. (b) The difference between the cleavage patterns of pyroxene and amphibole is most visible on their top faces. (c) Photos of pyroxene (augite) and amphibole (hornblende) illustrate the differences in cleavage patterns.

highly complex three-dimensional arrangement of silicate tetrahedra. In some of these tetrahedra, aluminum takes the place of silicon. The difference in charge between aluminum (+3) and silicon (+4) is balanced by the presence of other cations such as potassium (+1), sodium (+1) and calcium (+2) elsewhere in the crystal structure.

Because of the larger size of the potassium ion (1.33Å) compared to those of sodium (0.97Å) and calcium (0.99Å), two different families of feldspar occur. The *plagioclase feldspars* are characterized by varying abundances of sodium and calcium with negligible potassium. Conversely, the *potassium feldspars* contain abundant potassium but very minor sodium and calcium.

Plagioclase is the pale mineral you see reflected off the surface of the Moon, and is thought to be the most abundant mineral in the lower parts of Earth's crust (see Figure 3.10g). The potassium feldspars have three important polymorphs: sanidine, orthoclase, and microcline (Fig. 3.19). Sanidine crystallizes at relatively high temperature, orthoclase prefers intermediate temperatures, and microcline is the stable polymorph at relatively low temperature. Because lava flows cool quickly, they preserve the highest temperature minerals, so that sanidine is the most common potassium feldspar in lavas. However, magmas trapped beneath Earth's surface cool far more slowly, and so tend to form the lowest temperature polymorph, microcline.

The feldspar group of minerals occurs in a wide variety of rocks. However, their relatively weak bonds and the solubility of potassium, sodium, and calcium make the feldspar structure highly susceptible to weathering and dissolution during transport by water. (As we discuss in detail in Chapter 5, exposure at Earth's surface results in the disintegration and decomposition of rocks and minerals, a process known as weathering.) As a result, the presence of abundant feldspar in a sedimentary rock indicates proximity to its bedrock source.

Quartz is a silicate mineral made up almost entirely of silicon and oxygen. Like the feldspars, the mineral structure in quartz has a very complex three-dimensional geometry (see Figure 3.17) in which oxygen anions are shared between adjacent silicate tetrahedra. As a result of this sharing, there are, on average, two oxygen ions for every silicon ion. Hence quartz has the chemical formula SiO_2.

continental crust has an average density of 2.7 g/cm³. Non-ferromagnesian minerals also tend to be lighter in color and are less dense than the ferromagnesian silicates.

The most abundant non-ferromagnesian silicates in Earth's crust are members of the **feldspar group,** and are usually white or pink in color. Feldspars have a density of about 2.7 g/cm³, two cleavages with a 90° angle between them, and tend be rectangular in shape. They also have a

(a)   (b)   (c)

Figure 3.19   Polymorphs of Potassium Feldspar
Polymorphs of potassium feldspar include (a) sanidine, (b) orthoclase, and (c) microcline.

Because each oxygen ion carries a charge of -2 and silicon has a charge of $+4$, this arrangement is electrically neutral and, in contrast to most silicate minerals, no other cations are needed to balance the charge. In its purest form, quartz is clear and, if allowed to grow freely, will form beautiful six-sided (hexagonal) crystals (see Figure 3.6). However, as quartz grows, it commonly encounters and absorbs impurities that can result in a wide variety of colors including grey (smoky quartz), pink (rose quartz), and purple (amethyst) (see Figure 3.9).

Because its mineral structure is dominated by strong silicon-oxygen bonds, quartz is characterized by the absence of a cleavage. The spectacular six-sided shape of common quartz crystals (see Figure 3.6) reflects the regular internal bonding between silicon and oxygen, whereas the hardness of quartz (see Table 3.4) is due to the strength of the chemical bonds that binds these two elements together. These strong bonds cause quartz to be highly resistant to weathering, so that quartz-rich rocks tend to form local ridges and hills. Unlike feldspar, the quartz mineral structure survives transportation by water so quartz is common in sedimentary rocks. Quartz is also common in silicon-rich igneous and metamorphic rocks. Its strong silicon-oxygen bonds make quartz commercially useful as an abrasive in sandblasting and sandpaper.

Two important non-ferromagnesian silicates that, in addition to aluminum, incorporate water into their crystal structure (and so are Al-OH silicates), are *muscovite,* or "white mica," and the clay minerals, such as kaolinite or china clay. **Muscovite**, like biotite, belongs to the mica group and has a sheet-like arrangement of silicate tetrahedra and a strong cleavage. Although muscovite contains potassium, it does not contain iron or magnesium found in biotite and so tends to be light-colored (see Figure 3.10h). In fact, in thin sheets, muscovite is commonly highly reflective and transparent. On a sunny day at the beach, for example, the sparkling appearance of the beach sand is often due to the presence of muscovite. Because of its transparency, muscovite was used as window glass in the Middle Ages and its use as a heat resistant glass for ovens and fireplaces extended well past the mid-twentieth century. Muscovite is very common in certain types of metamorphic rocks but can also occur in silicon-rich igneous rocks such as granites.

Clay minerals have sheet-like structures similar to mica's structure, in which the silicate tetrahedra are bonded together in an open hexagonal structure (Fig. 3.20). There are many types of clay minerals, all of which tend to be so fine grained that it is difficult to distinguish between them even under the microscope. Sophisticated X-ray techniques are generally required to identify them. Most clay minerals form as a result of weathering. The breakdown of

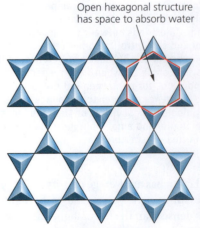

Open hexagonal structure
has space to absorb water

Figure 3.20   Structure of Clay Minerals
An idealized structure of clay minerals illustrates the way silicate tetrahedra are bonded in an open hexagonal array that allows them to absorb water.

the feldspar mineral structure during weathering and transport often results in the formation of clay minerals, and clay minerals are common in soils, unconsolidated sediments, and sedimentary rocks. However, as sedimentary rocks become buried and temperatures rise, clay minerals become unstable and commonly break down to form mica. As we shall see in Chapter 7, the identification of this breakdown reaction is important in the classification of rocks as either sedimentary or metamorphic.

CHECK YOUR UNDERSTANDING

- How do the appearance and characteristics of quartz reflect its crystal structure?
- Why is it difficult to identify the different types of clay minerals?

Clay minerals have several very important properties. Their open hexagonal structure allows them to absorb water and swell to several times their original size. The abundance of clay minerals in soils affects the quality of the land for agriculture. And one of the most common clay minerals, *kaolinite*, is used to make chinaware and gives the shine to high-gloss paper.

3.4 SUMMARY

- Because silicon and oxygen are the most common elements in Earth's crust, most minerals are silicates.
- All silicates have a basic tetrahedral form, known as the silicate tetrahedron, consisting of a central silicon cation bonded with four oxygen anions.
- Silicate minerals are subdivided into groups according to the arrangement of their tetrahedra, which can be isolated or can form single and double chains, continuous sheets, and three-dimensional networks.
- Silicates are classified as ferromagnesian or non-ferromagnesian, depending on the presence or absence of iron and magnesium in their crystal structures. Ferromagnesian silicates tend to be darker colored and denser than non-ferromagnesian silicates.

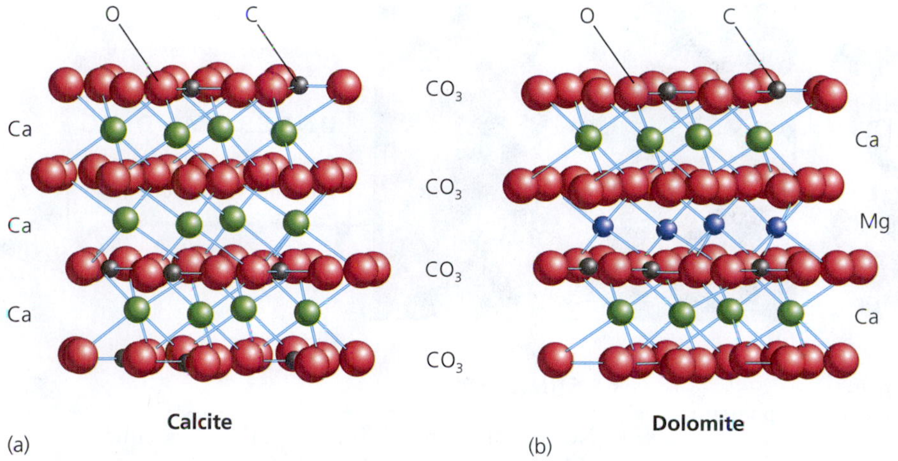

Figure 3.21 Structure of Carbonate Minerals
The structures of the carbonate minerals (a) calcite ($CaCO_3$) and (b) dolomite [$CaMg(CO_3)_2$]. These minerals share a similar structure and properties.

3.5 Nonsilicate Rock-Forming Minerals

A small but significant proportion of Earth's crust is made up of nonsilicate minerals (see Table 3.1). In nonsilicate minerals, ions of the common metals are bonded to elements other than a combination of silicon and oxygen. In general, the structures of these minerals tend to be simpler than those of the silicates and, as we shall see in Chapter 18, many nonsilicate minerals have very important properties that we take advantage of in our day-to-day lives.

CHECK YOUR UNDERSTANDING

- How do the structures of nonsilicates differ from the structure of silicates?

The most common nonsilicates are the **carbonate minerals** that make up rocks such as limestone and marble. In carbonate minerals, common cations such as calcium (Ca^{2+}) and magnesium (Mg^{2+}) are bonded to a combination of carbon (C) and oxygen (O) to produce minerals such as *calcite* ($CaCO_3$) and *dolomite* [$CaMg(CO_3)_2$] (Fig. 3.21). Because these minerals have similar chemistry and structure, they have very similar properties and are difficult to tell apart. However, calcite reacts vigorously with dilute hydrochloric acid whereas dolomite reacts much more slowly. The carbonate minerals in limestone are used in making cement and steelmaking.

Another important group of nonsilicate rock-forming minerals is produced by the evaporation of sea water from ancient seas (Fig. 3.22). These so-called **evaporite minerals** include *halite*, which is the mineral name for common table salt, sodium chloride (NaCl). Halite has a wide variety of uses, from food seasoning and preservation to highway deicing. *Gypsum* is another common evaporite mineral formed when calcium (Ca) is bonded to a combination of sulfur (S) and oxygen (O) to form a calcium sulfate ($CaSO_4{\cdot}2H_2O$). Gypsum is primarily used to make plaster, plasterboard, and other building materials.

There are many other nonsilicate minerals that are not sufficiently common to be considered rock-forming minerals. Among the most important are the **ore minerals**, which contain economically important

Figure 3.22   Common Rock-Forming Nonsilicate Minerals
(a) Calcite is a carbonate mineral. (b) Gypsum, and (c) halite are evaporite minerals.

Figure 3.23   Common Ore Minerals
Ore minerals include: (a) hematite, (b) galena, (c) sphalerite, and (d) gold.

metals (Fig. 3.23). Most ore minerals are relatively simple compounds in which a metal ion is either bonded to oxygen to form an *oxide,* as is the case for the iron ore *hematite* (Fe_2O_3), or is bonded to sulfur to form a *sulfide,* as is the case for the lead ore *galena* (PbS) and the zinc ore *sphalerite* (ZnS). Given sufficient quantities, sulfide minerals can form economically important metal deposits because the metal ions they contain can be extracted more easily than they can from more stable silicate mineral structures.

Ore minerals also include **native metals** (which are also referred to as uncombined metals). The most important native metals are gold (Au) and silver (Ag). More rarely, elements such as copper, tin, mercury, and nickel can occur as native metals, but they have a strong affinity for sulfur and more commonly occur as sulfides.

CHECK YOUR UNDERSTANDING

 Why are sulfide minerals economically important?

3.5 SUMMARY

- Nonsilicate minerals include carbonates, sulfates, chlorides, oxides, and sulfides; many have important industrial applications and economic value.

- Evaporite minerals include halite and gypsum. Evaporites are nonsilicate rock-forming minerals produced by the evaporation of sea water.

3.6 Minerals and Rocks

Just as the chemical elements are the building blocks of minerals, so minerals are the basic ingredients of rocks. Minerals, as we have learned, are inorganic solids with crystalline structures and characteristic physical properties that are determined by the regular internal arrangement of their constituent atoms. A rock, on the other hand, is an aggregate of one or more minerals or mineraloids. (Recall that mineraloids are solid materials that lack a regular atomic structure and are therefore not crystalline.) Like minerals, rocks are naturally occurring solids, but beyond this, rocks and minerals are quite different. Most rocks form from minerals that have either been naturally cemented together or have grown together in an interlocking mosaic. Rocks, therefore, do not possess a specific chemical structure and formula, and are not necessarily inorganic or crystalline. Volcanic glass, for example, is a rock dominated by mineraloids and it therefore has no crystal structure.

Most rocks contain several different minerals. The rock we call granite, for example, always contains interlocking minerals of quartz and feldspar, typically with variable amounts of muscovite and biotite (Fig. 3.24). Other rocks such as limestone are predominantly composed of only one mineral, in limestone's case calcite. Although similar in composition to calcite, limestone is not considered a mineral because it does not have a characteristic crystal structure. Instead, limestone is a rock predominantly made up of countless interlocking calcite grains. Likewise coal, which contains carbon and so is similar in chemical composition to both graphite and diamond, is a rock but not a mineral because its contents do not have a regular atomic framework and it is organic rather than inorganic in origin.

3.6 SUMMARY

* Minerals are the basic ingredients of rocks.
* A rock is an aggregate of one or more minerals or mineraloids that have been cemented or have grown together.

3.7 Minerals and People

Minerals have played a vital role in the development of civilizations. The Egyptians, Greeks, Romans, and many other ancient societies made use of ore minerals such as gold, copper, tin, and iron. Indeed, historians have named stages in the development of civilization—for example, the Bronze Age and Iron Age—after the materials humans first learned to master. Minerals have long been treasured as well for their beauty and their ability to form gemstones (Fig. 3.25).

Minerals also provide the source for valuable elements that can be harnessed in the name of progress. Elements such as copper, lead, and zinc can be extracted from sulfide minerals and are important in a whole range of commodities from industrial to medical. Cell phones, for example, contain a variety of elements such as lead, mercury, cadmium, gold, silver, and platinum that are either extracted from sulfides or occur as native metals. Other minerals are no less essential to modern living. For example, the *rare earth minerals,* which are those minerals that contain elements with atomic numbers between 57 (Lanthanum) and 71 (Lutetium), have become an essential source of the metals used in devices such as computer memories, DVDs, flat-panel TVs, cell phones, car catalytic converters, laser technology, magnets, and fluorescent lighting. To satisfy this demand, finding deposits of rare earth minerals is now a major goal of many mineral exploration companies. In 2013, rare earth elements extracted from these minerals totaled 111,000 metric tons (Fig. 3.26).

Correct mineral identification is also crucial to many professionals, from the forensic scientist who wants to link the minerals found in a suspect's shoes to the scene of a crime, to the gemologist who wants to distinguish between real and fake rubies. Distinguishing between minerals of the same family can even be important for your health. For example, the *serpentine* family of minerals consists of ferromagnesian silicates that are characteristically light and soft and have a waxy luster. These attributes make some members of this mineral family useful to sculptors and other artisans. However, one member of the family, the mineral *chrysotile,* is also known as "white asbestos" and is a considerable health hazard if it becomes airborne.

CHECK YOUR UNDERSTANDING
What is the difference between a mineral and a rock?

CHECK YOUR UNDERSTANDING
What role have minerals played in the telecommunication revolution of recent years?

Figure 3.24   Minerals and Rocks
The rock in the photo is granite; it is composed of the minerals quartz, feldspar, and biotite.

Granite (rock)
Quartz (mineral)
Biotite (mineral)
Feldspar (mineral)

Figure 3.25   Gemstones
Precious gemstones are commonly used in jewelry.

Rare earth elements

La	Ce	Pr	Nd	Pm	Sm	Eu	Gd	Tb	Dy	Ho	Er	Tm	Yb	Lu
57	58	59	60	61	62	63	64	65	66	67	68	69	70	71

(a)

Neodymium   Europium   Terbium   Dysprosium
(b)

Figure 3.26   Rare Earth Metals
(a) Rare earth metals belong to the lanthanide series, and have atomic numbers between 57 and 71. (b) Examples of rare earth metals include neodymium, europium, terbium, and dysprosium, which are shown here. (c) This graph summarizes production of rare earth mineral deposits by country in 2013.

World production

In 2013, world production of rare earth elements totaled 111,000 metric tons, with China producing 90% of the world's total.

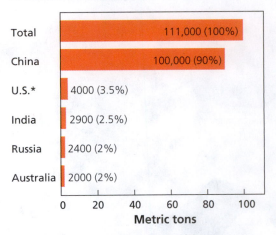

*The U.S. currently imports almost all of its rare earths from China, France, Japan, and Australia.

(c)

3.7 SUMMARY

- Minerals have played such an important role in human civilization that the stages in the development of civilization are named after the materials humans first learned to master.

- Minerals are resources that are essential to the maintenance of modern societies.

3.8 Minerals and Plate Tectonics

Because minerals are the basic building blocks of Earth's continents and ocean floor, being able to distinguish between them is fundamental to understanding the processes that have guided Earth's evolution through time. In Chapter 8, we will learn that the semiprecious mineral zircon contains radioactive elements, the decay of which can be used to deduce the mineral's age—that is, the time elapsed since it crystallized. This, in turn, allows us to determine the dates of important geologic events in Earth's history. Minerals also reflect the plate tectonic environment in which they were formed. This reality allows us to see the connections between mineral-scale and global-scale processes.

Although these connections apply to all minerals, in this section of the chapter we focus on polymorphs. When different minerals have identical chemistry, the polymorph that forms reflects the physical conditions (such as pressure, temperature, and the availability of fluids) at the time of crystallization. Each different polymorph reflects unique formation conditions. As we have seen, for example, distinguishing between the various polymorphs of potassium feldspar can help us understand if the mineral in question formed in rapidly cooling lava or in a slowly cooling magma chamber.

Let us consider what we can learn by revisiting the example of diamond and graphite. Repeated experiments show that diamond is much more likely to form than graphite when the lithosphere, which is normally about 100 kilometers (62 miles) thick, attains a thickness of more than 150 kilometers (93 miles). Such thickening occurs only in a convergent plate tectonic setting such as a continental collisional zone, where large slabs of continental crust are heaved on top of one another. In this way, diamonds provide important clues to the evolution of Earth by informing us about the location and movement of plates.

Another set of polymorphs includes the three silicate minerals kyanite, andalusite, and sillimanite, each of which has the chemical formula Al_2SiO_5. Because these minerals involve three of the most common elements in Earth's crust, they are far more common than graphite or diamond. Kyanite has the highest density (3.6 g/cm³), which means that it has the most tightly packed crystal structure. Not surprisingly, experiments show that kyanite is the most stable polymorph of the three at high pressures and relatively low temperatures. These are the conditions that favor tightly packed forms. In subduction zone settings, kyanite consequently forms deep in the crust where the pressure is high, in areas where the temperature is low (Fig. 3.27). In contrast, andalusite, with the lowest density (3.1 g/cm³) of the three polymorphs, has the most open crystal structure. Andalusite is therefore stable at lower pressures and higher temperatures than kyanite. Hence, in subduction zone settings, andalusite forms in shallow regions of the crust where the pressure is low, and where the temperature is high (especially adjacent to cooling magma chambers). These are the conditions that favor more open mineral structures. Sillimanite, with a density of (3.3 g/cm³), is stable at high pressures and high temperatures, and commonly forms where the deep crust is warmed by rising magma. In this way, the location of these three polymorphs paints a picture of tectonic activity and allows geologists to learn more about plate movement (Fig. 3.27).

CHECK YOUR UNDERSTANDING

- Show with the aid of a sketch how the formation of andalusite, sillimanite, and kyanite is linked to plate tectonics.

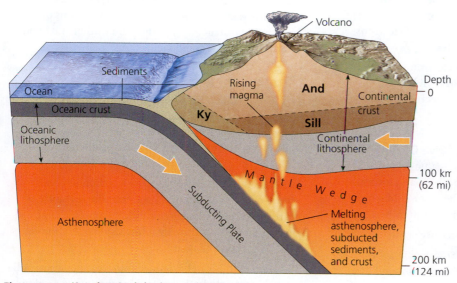

Figure 3.27 Kyanite, Andalusite, and Sillimanite

This schematic diagram shows the plate tectonic environments where the three polymorphs, kyanite, andalusite, and sillimanite, form. Kyanite (Ky) and sillimanite (Sill) form deep in the crust because they are both stable at high pressures. Sillimanite and andalusite (And) form in regions that are warmed by rising magmas because they are both stable at high temperatures, but andalusite forms at a relatively shallow depth because it is only stable at low pressures.

3.8 SUMMARY

- Minerals reflect the environment of their formation and are fundamental to understanding the processes that have governed Earth's evolution.

- High density minerals (such as diamond and kyanite) have tightly packed crystal structures and form at depth where the crust has been thickened by mountain building processes.

- Low density minerals (such as graphite and andalusite) have more open crystal structures and are more stable in the shallow crust.

Key Terms

amphibole group 88	crystal form 84	ion 73	nucleus 70
anion 73	crystal structure 73	ionic bond 71	olivine group 87
atom 70	density 86	luster 84	ore mineral 91
atomic number 71	electron 70	metallic bond 73	organic 77
carbonate mineral 91	element 70	mica group 88	polymorph 81
cation 73	evaporite mineral 91	mineral 77	proton 71
chemical formula 79	feldspar group 89	mineraloid 78	pyroxene group 87
clay mineral 90	ferromagnesian	molecule 71	quartz 89
cleavage 85	silicate 87	muscovite 90	rock-forming mineral 81
color 82	fracture 86	native metal 92	silicate mineral 81
compound 70	garnet group 87	neutron 71	silicate tetrahedron 87
covalent bond 73	gems 78	non-ferromagnesian	specific gravity 86
crystal 78	habit 84	silicate 87	streak 82
crystal face 78	inorganic 77	nonsilicate minerals 81	

Key Concepts

3.1 ELEMENTS: BUILDING BLOCKS OF MINERALS

- The fundamental unit of an element is the atom, which is made up of a central nucleus of protons and neutrons, and an orbiting cloud of electrons. The mobility of electrons allows atoms to bond to form molecules.

- Bonds between atoms are formed by electron transfer (ionic bonds), by electron sharing (covalent bonds), and by the development of a common electron cloud (metallic bonds).

3.2 MINERALS: ORDERLY EXPRESSIONS OF MATTER

- Minerals are naturally occurring, inorganic crystalline solids with a specific chemical composition and formula. The structure and formula express the manner in which the chemical elements bond together to form minerals.

- When left to grow freely, minerals produce crystals.

- Polymorphs are minerals with the same chemical composition but different crystal structures (e.g., diamond and graphite are both made up of carbon). Which polymorph grows depends on the physical conditions (such as pressure and temperature) at the time of growth.

3.3 PHYSICAL PROPERTIES OF MINERALS

- Most minerals have physical properties that we can use to identify them. These properties of minerals reflect their chemistry and the character of their bonding.

- Physical properties used to identify minerals, such as color, streak, luster, crystal form, habit, hardness, cleavage, density, and specific gravity, are expressions of a mineral's internal structure and the composition and bonding arrangement of its component atoms.

3.4 SILICATE ROCK-FORMING MINERALS

- Of the more than 4500 known minerals, less than 1 percent commonly occur in rocks. Most of these rock-forming minerals are silicates, in which metallic ions such as iron and magnesium or calcium, potassium, and sodium, are bonded to silicate tetrahedra linked in a variety of ways.

- Silicate tetrahedra either occur in isolation (olivine group) or are bonded to each other in single chains (pyroxene group), double chains (amphibole group), sheets (mica group and clay minerals) or 3D frameworks (feldspar group and quartz).

- Silicate minerals can be divided into ferromagnesian silicates that contain both iron and magnesium (e.g., olivine, pyroxene, amphibole) and the non-ferromagnesian silicates (e.g., feldspar, quartz), which do not contain these elements.

3.5 NONSILICATE ROCK-FORMING MINERALS

- A small but significant proportion of Earth's crust is made up of nonsilicate minerals. The most common nonsilicates are carbonate minerals. Evaporite minerals, which are produced by the evaporation of sea water, are another important group of nonsilicates.

- There are other important nonsilicate minerals that are not sufficiently common to be considered rock-forming minerals. Among these are the ore minerals, including the native metals.

3.6 MINERALS AND ROCKS

- Most rocks form from minerals that have either been cemented together or have grown together in an interlocking mosaic.

3.7 MINERALS AND PEOPLE

- Minerals provide humans with resources that have been harnessed by society for many reasons, including industrial, agricultural, technological, and medical purposes.

3.8 MINERALS AND PLATE TECTONICS

- Minerals form under characteristic chemical and physical conditions. As a result, they provide clues to the plate tectonic setting of the rocks in which they were formed.

Study Questions

1. How do ionic, covalent, and metallic bonds differ in their use of electron mobility?

2. Explain the relationship between metallic bonding and the characteristic properties of metals. How can these properties aid in mineral exploration?

3. Why do minerals have specific chemical formulas whereas rocks do not?

4. In what ways do crystal form, hardness, and cleavage reflect a mineral's internal structure?

5. Why are silicates the dominant rock-forming minerals and how are the characteristics of the members of the silicate groups related to the structural arrangement of the silicate tetrahedron?

6. Explain the differences in the cleavage patterns of pyroxenes, amphiboles, and micas.

7. Explain why quartz occurs in various colors.

8. Explain why there are two families of feldspar minerals.

9. How do halides and sulfates form? Explain.

10. Explain the relationship between graphite and diamond and the environments in which these minerals are formed.

4 Origin and Evolution of Igneous Rocks

4.1 Igneous Rocks and the Rock Cycle

4.2 Magma Formation and Transport

4.3 Textures of Igneous Rocks

4.4 Classifying Igneous Rocks

4.5 Evolution of Igneous Rocks

4.6 Volcanic Eruptions

4.7 Igneous Rocks and Plate Tectonics

► Key Terms

► Key Concepts

► Study Questions

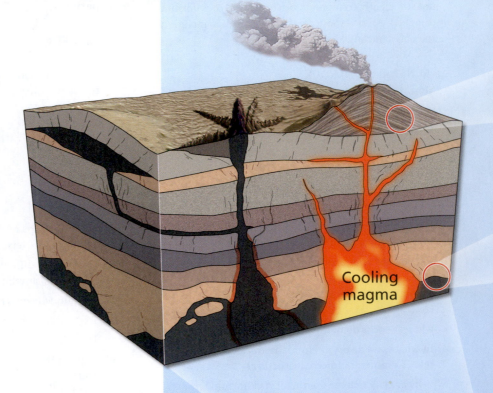

Cooling magma

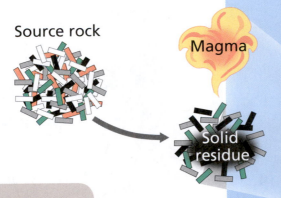

Source rock

Magma

Solid residue

SCIENCE REFRESHER: Heat, Temperature, and Magma 104

LIVING ON EARTH: Volcanoes and the Air We Breathe 108

IN DEPTH: Can Explosive Eruptions Be Predicted? 130

Extrusive rock

Intrusive rock

Subaerial lava flow

Submarine lava flow

4.1 Describe how igneous rocks relate to the two other rock groups (sedimentary and metamorphic).

4.2 Describe how magma forms and the factors that influence magma's ascent toward the surface and its cooling history.

4.3 Explain how magmas produce a variety of igneous rocks with textures that vary according to the environment of their formation.

4.4 Compare and contrast the different types of igneous rock and explain the basis of their classification.

4.5 Explain how the chemical composition and physical characteristics of magma evolve as magma cools.

4.6 Identify the processes that cause volcanic eruptions and the various types of volcanoes and volcanic rocks that eruptions produce.

4.7 Explain the role that plate tectonics plays in the formation and composition of magma and igneous rocks.

Throughout Earth's 4.54 billion year history, ongoing volcanic activity has brought chemistry from Earth's interior to its surface and has vented gases into Earth's atmosphere. Volcanoes are the surface expression of a complex underground labyrinth of chambers and conduits where magma—molten rock—resides. Igneous rocks form from cooling magma. By studying igneous rocks, we can discover much about the intense heat and pressure in Earth's dynamic interior, learn at what depth a magma formed, how it rose toward the surface, how fast it cooled, and how much it interacted with its surroundings.

Over geologic time, igneous activity has played a vital role in the formation of the planet on which we stand and the air we breathe. Volcanic eruptions provide dramatic evidence of the energy stored beneath Earth's surface, the unleashing of which can have devastating consequences. Indeed, this fact alone provides a strong motivation to understand the processes responsible so that more accurate predictions of eruptions can be made. But, while volcanic activity may have catastrophic consequences in the short-term, it is necessary in the longer term for the very survival of life on Earth. In order to understand igneous rocks, we must examine the conditions in Earth's interior where magmas form and the factors that influence their chemical composition.

4.1 Igneous Rocks and the Rock Cycle

Igneous rocks form from cooling magma. **Magma** is molten or partially molten material found beneath the surface of Earth. In addition to liquid, magma typically contains trapped gases, floating crystals, and fragments of variably digested rock. Magma that reaches the surface and flows from volcanoes is called **lava**.

Recall from Chapter 1 that igneous rocks are one of three basic rock groups that make up Earth's continents and ocean floors. The other two types are sedimentary rocks (Chapter 6) and metamorphic rocks (Chapter 7). Driven by Earth's internal heat engine, each of these rock groups may be transformed into one of the others, a concept elegantly portrayed in the **rock cycle** (see Fig. 1.11). The rock cycle is an idealized cycle of processes that describes the dynamic relationships among the three main rock types: *sedimentary*, *metamorphic*, and *igneous*.

The rock cycle clearly distinguishes between the three rock groups yet also demonstrates the relationships between them (Fig. 4.1). Igneous rocks (from the Latin word *ignis*, meaning fire) represent the products of cooling molten rock and form where magma cools and solidifies. Recall from Chapter 2 that most igneous rocks are the result of processes that occur near plate boundaries. The magma itself forms from the melting of preexisting rocks. Conversely, *sedimentary rocks* are the product of surface processes such as weathering, erosion, and deposition, and *metamorphic rocks* form within the crust where the pressures and temperatures are high. Within these three groups, rocks are further classified according to their appearance and their chemical composition.

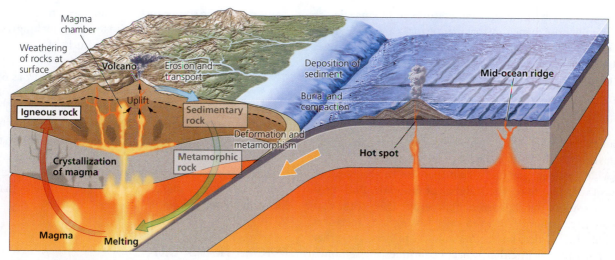

Figure 4.1   Igneous Rocks and the Rock Cycle

Processes that result in the generation, ascent, and crystallization of magma to form igneous rocks are emphasized in this diagram of the rock cycle.

We can see from Figure 4.1 that we need to address several issues in order to understand the origin and evolution of igneous rocks, especially why some magmas crystallize at depth in large caverns called **magma chambers**, whereas others rise and spew out on Earth's surface to form volcanoes. **Volcanoes** are openings in a planet's surface or crust, which allow hot magma, volcanic ash, and gases to escape from magma chambers below the surface.

Geologists who study igneous rocks try to determine three things. First, they want to learn about the magma's *formation* and the composition of the *source rock* that melted to produce the magma. Second, they investigate the *transportation* of the magma, from the site of melting toward the surface, by determining the processes that guided the magma's rise through the crust. Finally, they work to determine the extent to which the magma interacted with its surroundings as it rose and how it eventually cooled—what we refer to as the *cooling history* of the magma. Geologists use all this knowledge to understand why the melting began in the first place, a task that requires incorporating what we have learned from our discussion of plate tectonics in Chapter 2.

By careful examination of the chemical composition and appearance of igneous rocks, we gain insights into one of the most fundamental processes involved in Earth's evolution. In this chapter, we will find that the composition of the source rock profoundly influences the composition of the magma. In addition, a wide range of processes may change the chemical composition of the magma as it is transported away from its source and through the crust. As it rises, magma interacts with surrounding rocks. The extent of this interaction is governed by a variety of factors including the compositions of the magma and the adjacent rocks. We can obtain evidence of magma transport by careful field observations and measurements, and from

noting subtle changes to the magma chemistry as it interacts with its surroundings. Evidence of the final cooling history comes directly from the mineral content and the appearance of the rocks.

4.1 SUMMARY

* Igneous rocks, sedimentary rocks, and metamorphic rocks are the three main rock groups.

* Igneous rocks form when magma or lava cools; we classify igneous rocks according to their appearance and chemical composition.

* To learn more about igneous rocks, we study the nature of the source rock that melts to form magma, the magma's transport, and its cooling history.

4.2 Magma Formation and Transport

Volcanic eruptions are testament to the presence of molten magma in Earth's interior. However, although Earth's interior gets progressively hotter with depth, experiments show that most of Earth's mantle and crust is solid rock and magma only forms in specialized environments. To understand the origin and evolution of igneous rocks, we must start at the beginning, with the process of magma formation from the melting of solid rock.

SOURCE ROCK

Rocks are composed of minerals, and each mineral has a regular arrangement of ions in its crystal structure (see Chapter 3). For magma to form, sufficient energy must be supplied to weaken and break down the crystal structure of minerals in the **source rock** (see Science Refresher: **Heat, Temperature, and Magma** on page 104). Although the

---

CHECK YOUR UNDERSTANDING

○ What is lava?

○ What are the three main rock groups?

mantle is mostly solid, the temperatures in the mantle are sufficiently close to melting that mantle rocks are soft and can flow. However, for mantle rocks to melt, additional changes need to occur, such as a decrease in pressure, an increase in temperature, and/or an influx of volatiles. **Volatiles** are elements or compounds such as water, carbon dioxide, and nitrogen that are gases at relatively low temperatures. As we discuss in this section of the chapter, each of these changes can generate magma, especially (although not exclusively) when they occur near plate boundaries.

A decrease in pressure or an increase in temperature has similar effects on the stability of minerals. Both processes result in the expansion of minerals. This means that the average distance between neighboring ions increases so that the chemical bonds that hold the minerals together become strained. If either process continues, these bonds eventually break so that the solid rock begins to melt (Fig. 4.2a). Influx of volatile substances, such as water, also weakens mineral structures as water molecules tug at the bonds at the edges of the minerals. In addition, upward-migrating water is a very efficient carrier of Earth's internal heat and can supply the thermal energy needed to break these bonds.

Irrespective of the processes involved, field observations and laboratory experiments indicate that the amount of energy available is only sufficient to partially melt the source rock, a process known as **partial melting** (Fig. 4.2b). The most

stable part of the source rock does not melt and is known as **residue**. The temperatures required to initiate melting depend on the composition of the source rock. Rocks of the continental crust begin to melt at lower temperatures (as low as 650°C or 1200°F) than those of the mantle (as high as 1200°C or 2190°F). But in either case, melting preferentially takes place along the junctions between different minerals because these are the areas where the source rock is least stable. Hence, it is at these sites that the formation of magma generally begins (Fig. 4.2c).

CHECK YOUR UNDERSTANDING
● What is partial melting?

MAGMA TRANSPORT

Active volcanoes provide proof that magmas formed at depth in the crust or mantle can rise all the way to the surface. Magmas must therefore be able to intrude into the solid rock that lies above them. But how is this achieved? Although the exact processes involved are still controversial after more than a century of debate, there are some principles that provide a general understanding.

Magma is less dense than the surrounding source rock, so it has a tendency to rise because less dense material is more buoyant. When melting begins, the volume of magma is small and is easily held in the spaces between the mineral grains. But with continued melting, the proportion of magma increases until it reaches about 25–30 percent, at which point it gains sufficient buoyancy to begin its ascent through the overlying rocks. But how exactly does this happen? Does the magma shoulder these rocks aside as it ascends? Does it digest the rocks above it? Does it create and exploit fractures in the overlying crust? Or does it use a combination of these processes? Figure 4.3 shows a range of possibilities. Magma can escape through any available crack, ruthlessly exploiting fractures in the overlying rock to facilitate its ascent. As magma rises toward the surface, the resulting decrease in pressure causes the magma to expand like an inflated balloon, potentially producing new fractures in the surrounding rock, known as **wall rock**, which the magma can then use to rise even more rapidly. In addition, pieces of the wall rock can founder into the magma chamber. It is probable that all these mechanisms operate to some degree, but which are the most important is still vigorously debated.

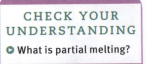

Figure 4.2   Processes of Partial Melting

(a) As the temperature increases or the pressure decreases, minerals expand. The average distance between neighboring ions in a solid increases until the bonds that hold the mineral together break and melting begins and eventually the solid becomes a liquid. (b) Partial melting of crustal or mantle rocks produces a magma and a solid residue. (c) A close-up look at partial melting shows how melting begins along the boundaries between mineral grains x, y, and z, and then migrates (red) along those boundaries.

CHECK YOUR UNDERSTANDING
● Why do magmas rise?

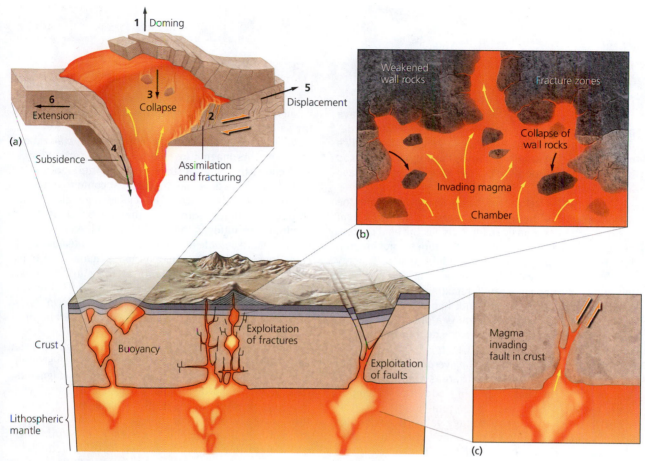

Figure 4.3   Mechanisms of Intrusion

Magma ascends because of buoyancy (i.e., it is less dense than the surrounding wall rock). Possible mechanisms for ascent include: exploitation of fractures; collapse of wall rock into the magma chamber; and exploitation of local faults. The insets illustrate examples of how mechanisms may work together. Inset (a) shows the rise of buoyant magma is accompanied by (1) doming of the roof; (2) fracturing of the wall rock as the magma expands; (3) collapse of wall rock into the magma chamber; (4) subsidence of wall rock; (5) lateral wall rock displacement by faulting; and (6) crustal extension. Insets (b) and (c) show that magma invading fracture zones encounters weakened and fractured wall rock, increasing the likelihood that chunks of the wall rock will collapse into the chamber.

MAGMA COOLING

Some magmas rise to Earth's surface where they fuel volcanoes. However, most magma does not reach the surface, but instead cools and solidifies at depth in magma chambers. This is because several factors work against the ascent of magma. The density contrast between the magma and the adjacent wall rock becomes less pronounced as the magma cools by losing heat to its surroundings. Digestion of wall rock by magma accelerates cooling because the magma expends energy in the process. The magma also becomes stickier as it cools, thereby increasing its resistance to flow. The cooling of magma results in the growth of minerals, a process known as **crystallization**, and the formation of solid igneous rock (see Science Refresher: **Heat, Temperature, and Magma** on page 104).

In this chapter, we will see that the wide variety of igneous rocks that occur in nature can be traced to the formation, evolution, and cooling history of magma. For example, magmas may cool at substantially different rates, and this profoundly affects the size of the crystals in the resulting igneous rocks and their appearance. Geologists use the appearance of a rock,

> **CHECK YOUR UNDERSTANDING**
>
> ● Why do some magmas fail to rise to Earth's surface?

as well as its mineral composition, to distinguish between various types of igneous rocks. In this way, the classification scheme for igneous rocks reflects the ways in which igneous rocks form.

4.2 SUMMARY

- A source rock melts to form magma when additional heat or reduced pressure causes minerals to become unstable. The energy available is usually only sufficient to partially melt the source rock.

- Magma is less dense than its source rock and tends to rise toward the surface when it attains sufficient buoyancy.

- If magma reaches the surface, it is called lava and is the fuel of volcanic eruptions.

- Most magma cools and crystallizes below the surface in magma chambers.

Heat, Temperature, and Magma

Heat and temperature are terms we use in casual conversation every day. In science, however, they have precise meanings. *Heat* is a form of energy, and is defined as the sum of the kinetic energy of all of a body's moving particles. Kinetic energy is the energy a body has because of its motion, and varies with the body's mass and velocity.

The higher the velocity of the particles, the greater the kinetic energy of the body and therefore the more heat the body contains. Consider, for example, the heat content of a crystal. The ions in a crystal structure are not motionless, but are actually vibrating in a limited fashion. If energy is added to the crystal from an external source, these vibrations become more rapid, resulting in an increase in kinetic energy of the crystal and, hence, an increase in its heat content. Therefore, there is a direct relationship between the energy supplied to a crystal from outside and the crystal's heat content.

Temperature is defined as the *average* kinetic energy of all the particles in a body. There is a direct relationship between the heat content of a body and its temperature because an increase in the total kinetic energy of a body also increases the average value. If two bodies of the same material have the same temperature, then the average kinetic energy of their molecules is the same. However, the larger body will contain more heat simply because it contains more moving molecules. Thus a pail of water contains more heat than a cup of water at the same temperature.

We can now apply these concepts to the crystallization of magma. As magma loses energy to its surroundings as it cools, the rapid, relatively disorganized movement of its ions slows down so that its heat content and temperature are both reduced. The ions also get closer and so begin to interact with each other and bond together, eventually resulting in the formation of minerals. Just as sugar is more soluble in hot tea than in cold tea, mineral components held in solution in the magma at high temperatures become less soluble as the magma cools. Eventually, the magma cools to a threshold level below which the components can no longer be held in solution, and crystallization begins as minerals start to form (see Fig. 4.5).

If a mineral is given sufficient time to organize its ions, and also has space to grow, it will form large, well-shaped crystals. This style of mineral growth is typical of igneous rocks that cool slowly beneath Earth's surface. However, in a volcanic eruption, the relatively disorganized ions in the lava have little time to organize themselves into more orderly arrangements. As a result, they form fine-grained crystals or, in the extreme case, structureless glass.

4.3 Textures of Igneous Rocks

When we examine specimens of igneous rocks, we are looking at the end result of melting, transport, and cooling of magma. We can learn a lot about these processes from a rock's overall appearance and from its mineral content. The appearance of the rock is determined by the **texture** of the rock, that is, the size, shape, and arrangement of its mineral grains. Texture can provide important information about the environment in which the rock formed and clues to the cooling history of the magma.

THE APPEARANCE AND TEXTURE OF IGNEOUS ROCKS

The size of a mineral in an igneous rock is a general indication of the rate at which the magma cooled, and therefore how far the magma rose toward the surface before it crystallized. Recall from Section 4.2 that magma that reaches the surface is called lava and it is the fuel of volcanic eruptions (Fig. 4.4a). Most lavas erupt at temperatures between 650°C and 1200°C (1200°F–2190°F), and cool rapidly as they lose heat to the atmosphere. Lava consists of relatively disorganized ions which have little time to organize and form minerals. As a result, any minerals that form at this temperature are small and the igneous rock is **fine-grained** (Fig. 4.4b). In fact, cooling is sometimes so rapid that minerals do not get a chance to form at all, and the product is a **volcanic glass** (Fig. 4.4c). This natural process is mimicked in the industrial procedure of manufacturing glass by rapidly chilling molten sand. Magma beneath Earth's surface cools more slowly and consequently has a larger-, coarser-grained texture, which we discuss in the material to come.

A fine-grained or glassy sample therefore implies rapid cooling, which usually indicates that the igneous rock is the product of a volcano (Fig. 4.5). However, the grain size of the rocks produced by volcanoes is not always uniform. A closer inspection of that same fine-grained or glassy sample (with the aid of a hand lens, for example) may reveal a small proportion of relatively large minerals surrounded by a much finer-grained material. This finer material is the **groundmass**, which consists of tiny interlocking crystals or glass. The relatively large minerals are **phenocrysts**. A rock with features like this is said to exhibit **porphyritic texture** (Fig. 4.4d). Phenocrysts crystallize in a magma chamber, but become incorporated into the magma during its ascent toward the surface. When phenocrysts are

(a)

(b)

(c)

(d)

(e)

Figure 4.4   Textures of Volcanic Rocks

(a) A volcanic eruption produces a flow of hot lava. (b) Fine-grained igneous rock and (c) volcanic glass are produced by the rapid crystallization of lava flows. (d) Rocks with porphyritic texture enclose coarser phenocrysts in fine-grained groundmass. (e) This example of flow texture shows the alignment of phenocrysts.

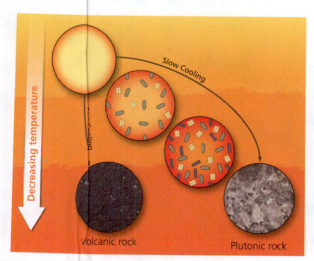

Figure 4.5   Cooling Rate and Crystal Growth

Lava cools rapidly at Earth's surface and crystallizes to form many small grains, and consequently has a fine-grained texture. Magma cools slowly beneath Earth's surface, and crystallizes to form relatively large grains, and consequently has a coarse-grained texture.

extruded with the molten lava at the surface, they sometimes become aligned like logs floating in a stream, a pattern known as **flow texture** (Fig. 4.4e).

A great many volcanic eruptions are explosive and produce little or no lava. Instead, fragments or **clasts** (from the Greek, *klastos*, meaning broken) are ejected from the volcano into the air (Fig. 4.6a), and eventually fall to earth to form a **pyroclastic deposit** (*pyro* is from the Greek word for fire) (Fig. 4.6b). Pyroclastic rocks—the rocks that make up pyroclastic deposits—are volcanic rocks that typically contain angular clasts surrounded by a very fine-grained groundmass. These fragments may vary in size. Coarser fragments are deposited closest to the site of the eruption, whereas the finest particles can be very widely distributed. This common pattern of particle distribution can tell us about past geologic activity. By systematically mapping the grain size distribution of pyroclastic deposits in the field, geologists can locate the **vents** of extinct volcanoes.

The volume of pyroclastic material associated with a single eruption can be substantial and its effects can be widespread. In April and May of 2010, pyroclastic

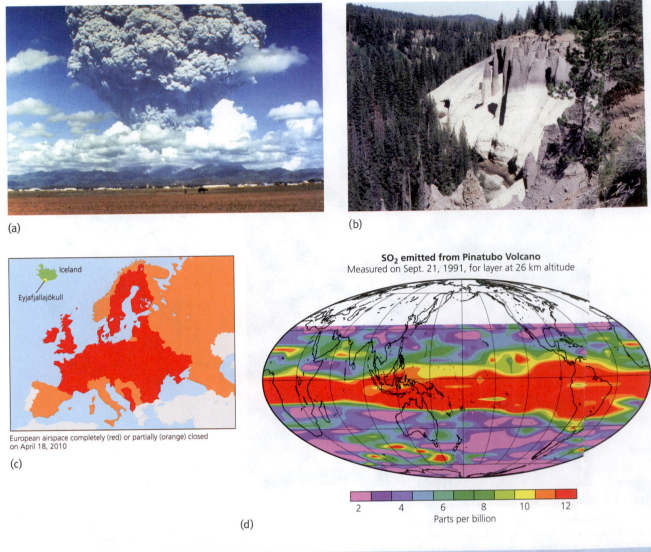

(a)

(b)

(c)

European airspace completely (red) or partially (orange) closed on April 18, 2010

SO₂ emitted from Pinatubo Volcano
Measured on Sept. 21, 1991, for layer at 26 km altitude

2   4   6   8   10   12
Parts per billion

(d)

(e)

Figure 4.6   Pyroclastic Deposits

(a) Eruption of Mount Pinatubo in the Philippines in 1991 jettisoned pyroclastic material into the atmosphere. (b) These ancient pyroclastic deposits are from Crater Lake National Park, Oregon. (c) The drift of the ash cloud associated with the eruption of the volcano Eyjafjallajökull in Iceland caused European airspace to completely (red) or partially (orange) shut down on April 18, 2010. (d) This map shows the distribution of sulfur dioxide at a height of 26 kilometers in the atmosphere about three months after the eruption of Mount Pinatubo. (e) Heavy ash fall from the Mount Pinatubo eruption damaged this DC-10 at a nearby air base.

eruptions of the volcano Eyjafjallajökull (pronounced "AY-uh-fyat-luh-YOE-kuutl-uh") in Iceland generated an ash cloud, estimated to have been between 0.1 and 0.2 cubic kilometers (0.02 and 005 cubic miles) in volume, that spread over northern and western Europe, causing enormous disruption to air traffic (Fig. 4.6c). In 1991, the eruption of Mount Pinatubo in the Philippines (see Fig. 4.6a) jettisoned an estimated 5 cubic kilometers (1.2 cubic miles) of debris into the atmosphere (see Living on Earth: **Volcanoes and the Air We Breathe** on page 108). This is five times more than the volume of debris associated with the 1980 eruption of Mount St. Helens in Washington State, but only a fraction of certain past eruptions, some of which are estimated to have ejected as much as 2500 cubic kilometers (600 cubic miles) of debris. Following the eruption of Pinatubo, the finest particles and ejected gases were caught up in atmospheric circulation patterns and distributed worldwide (Fig. 4.6d), and by blocking the flow of solar radiation to Earth, are thought to have been responsible for the relatively cool average global temperatures recorded in 1992. At least 20 commercial jet aircraft were affected by the eruption of Pinatubo, including two that lost engine power. About two dozen airplanes on the ground in the Philippines were also damaged by volcanic ash (Fig. 4.6e). Geologists have identified similar and even more violent eruptions in the geologic past by the widespread deposition of pyroclastic beds that are the fallout from large eruptions (see Fig. 4.6b).

Volcanoes are the end product of the ascent of hot, buoyant silicate magma and gas from the interior of Earth toward its surface. However, recall from Section 4.2 that much of this molten rock never reaches the surface and instead remains trapped in magma chambers in Earth's interior.

Magmas within these chambers cool very slowly so that their ions have far more time to become organized into crystal structures as they cool. As a result, igneous rocks produced deep within Earth's interior contain an interlocking mosaic of large minerals that are readily visible to the naked eye. Such deep-seated igneous rocks are described as **coarse-grained** and are termed **plutonic** (Fig. 4.5) after *Pluto*, the Greek god of the underworld.

Since plutonic rocks form beneath Earth's surface, you might wonder how we encounter them at all.

That these rocks are exposed at the surface is a testament to the rock cycle (see Fig. 4.1). We see them because the rocks that were above them at the time they crystallized have been removed by uplift and erosion.

EXTRUSIVE AND INTRUSIVE IGNEOUS ROCKS

Because magma cools and crystallizes below the surface as well as on the surface, there are two fundamentally distinct forms of igneous rocks; **extrusive rocks** that form on the surface and **intrusive rocks** that form below it. We distinguish between these two fundamentally different types of igneous rock based on their texture. As we shall see, this distinction is very important in the classification of igneous rocks.

All extrusive rocks are the product of volcanic eruptions and so are deposited on Earth's surface. They are produced either by cooling lava, in which case they are typically fine grained, or even glassy in appearance, or by explosive eruptions of volcanic material that falls to earth as pyroclastic deposits. All intrusive rocks, conversely, form below Earth's surface, where they cool and solidify. They are relatively coarse grained because they cool more slowly than extrusive rocks. They intrude into their surroundings and so commonly cut across any layering in the wall rocks around them.

Intrusive rocks occur in an intricate pattern of chambers connected by an array of pipes and fractures, much like the plumbing system in a house (Fig. 4.7). Intrusive rocks that are the crystallized remnants of large magma chambers cool to form large bodies known as **plutons**. Smaller plutons are *stocks* and larger ones are *batholiths*. Other types of intrusive rocks are thin sheet-like bodies that

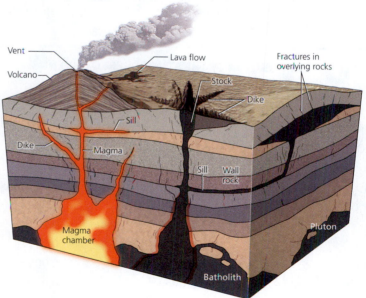

Figure 4.7   Types of Igneous Bodies

Magma cools slowly within magma chambers to form coarse-grained rocks that occupy large plutons (batholiths) or small plutons (stocks). Magma rising toward the surface may crystallize in narrow conduits, called dikes, which are oblique to bedding. Where magma intrudes along bedding planes in the surrounding rock, it cools and crystallizes to form sills. Plutons, dikes, and sills are intrusive igneous bodies. Magma that reaches the surface is called lava. It usually erupts from volcanic vents and cools rapidly to form fine-grained rocks.

LIVING ON EARTH

Volcanoes and the Air We Breathe

Earth's atmosphere consists of gases and trace amounts of suspended particles from volcanoes, nuclear explosions, and industrial pollutants. Although atmospheric gases can be detected up to 700 kilometers (435 miles) above Earth's surface, 99 percent of these gases typically occur within the first 32 kilometers (20 miles). Here their composition can be measurably influenced by individual volcanic eruptions. For example, the eruption of Mount Pinatubo in the Philippines in 1991 released 20 million tons of sulfur dioxide into Earth's atmosphere. Within three weeks of the eruption, a shroud of sulfur dioxide-enriched atmosphere had encircled Earth and become part of the atmosphere's global circulation pattern (see Fig. 4.6d).

If one volcano can have such a measurable effect, imagine the influence on the composition of the atmosphere of the 4.54 billion years of volcanic activity since Earth formed. If, as is generally believed, Earth has been cooling down since it first formed, then volcanic activity is likely to have been far more abundant early in Earth history. Most geoscientists maintain that our atmosphere was likely dominated by gases vented from volcanoes soon after Earth formed.

The venting of gases to create an atmosphere is the end product of the igneous processes in magma chambers. As magma chambers cool, the crystals that form first (olivine, pyroxene, and plagioclase) preferentially exclude volatile components such as water, carbon dioxide, and nitrogen from their crystal structures. As a result, the concentrations of the volatile components in the magma increase as the magma cools. Just as they do in a carbonated beverage, the volatiles rise to the top and are trapped at the roof of the chamber. Like a giant pressure cooker with a weak lid, the pressure exerted by the volatile gases steadily builds at the roof until the gases finally explode their way through the overlying rocks and vent into the atmosphere. These volatile components are important gases in our atmosphere.

The composition of Earth's atmosphere is unique in the Solar System. In contrast to its nearest planetary neighbors, Venus and Mars (the atmospheres of which are made up mainly of carbon dioxide—see Table 4A), Earth's atmosphere is dominated by nitrogen (78%) and oxygen (21%). The gas argon accounts for most of the remaining 1 percent. Carbon dioxide accounts for a mere 0.04 percent. These are average amounts, however, since the concentration of some gases (most notably water vapor) can be highly variable.

Volcanic activity releases the trapped volatile components of the inner Earth, and gravitational attraction prevents most of them from escaping into space. The larger the planet, the greater the gravitational hold on its atmosphere. The relatively small size of Mars provides an insufficient gravitational pull on its atmospheric gases. As a result, Mars' atmospheric gases have slowly leaked into space. Because volcanic activity on Mars virtually ceased

TABLE 4A COMPOSITION OF THE ATMOSPHERES OF VENUS, EARTH, AND MARS

VENUS		EARTH[a]		MARS[b]	
Gas	Percent Volume	Gas	Percent Volume	Gas	Percent Volume
CO_2	96.5	N_2 (nitrogen)	78.1	CO_2	95
N_2	3.5	O_2 (oxygen)	20.9	N_2	2.7
SO_2[c]	0.015	H_2O (water vapor)	0.05 to 4 (variable)	Ar	1.6
H_2O	0.01	Ar (argon)	0.9	CO	0.6
Ar	0.007	CO_2 (carbon dioxide)	0.04	O_2	0.15
CO	0.002	Ne (neon)	0.0018	H_2O	0.003
He	0.001	He (helium)	0.0005	Kr	Trace
O_2	≤ 0.002	CH_4 (methane)	0.0002	Xe	Trace
Ne	0.0007	Kr (krypton)	0.0001		
H_2S[c]	0.0003	H_2 (hydrogen)	0.00005		
C_2H_6[c]	0.0002	N_2O (nitrous oxide)	0.00005		
HCl[c]	0.00004	Xe (xenon)	0.000009		

[a]Compositions are for near-surface conditions, with terrestrial data other than H_2O tabulated for dry conditions. CO_2 on Earth is probably increasing by 2% to 3% of the listed amount each decade, because of fossil fuel consumption. This consumption may be modifying Earth's climate.
[b]Amounts of gases vary slightly with season and time of day. H_2O is especially variable. Some CO_2 condenses out of the atmosphere into the winter polar cap; changing polar cap sizes cause small changes in the total Martian atmospheric pressure.
[c]SO_2, sulfur dioxide; H_2S, hydrogen sulfide; C_2H_6, ethane; HCl, hydrogen chloride (hydrochloric acid).

about 1 billion years ago, the leaking of the Martian atmosphere into space is no longer compensated by the addition of volcanic gases. This explains why its atmospheric pressures are so much lower than those on Earth.

Although Earth's volcanic activity and gravitational attraction may account for the existence of its atmosphere, they do not explain its composition. There is an obvious mismatch between the composition of Earth's atmosphere (which is overwhelmingly dominated by nitrogen and oxygen) and the gases vented from modern volcanoes. Samples taken from active volcanoes show that water vapor typically comprises between 50 and 80 percent of the gases emitted. Volcanoes also emit smaller and highly variable amounts of carbon dioxide, nitrogen, sulfur dioxide, and hydrogen sulfide, and trace amounts of hydrogen, carbon monoxide, and chlorine. This composition is probably typical of the gases released from ancient volcanoes, because the magma was formed by melting similar source materials and would therefore have concentrated similar gases.

So why is there such a profound mismatch between the composition of the atmosphere and the composition of gases emitted from volcanoes? Water vapor emitted from a volcano becomes part of the hydrologic cycle. The vast majority of this water is extracted from the atmosphere as rain or snow and eventually ends up in the oceans. Carbon dioxide is highly soluble in water, and so is extracted in rainfall and absorbed by the oceans. Carbon dioxide is also extracted from the atmosphere by its interaction with life. In the presence of sunlight, organisms in the oceans (such as phytoplankton) and vegetation on land (like that of the tropical rain forests) not only extract carbon dioxide, but also introduce oxygen into the atmosphere in a process known as **photosynthesis**. But what about nitrogen, the most abundant atmospheric gas? Nitrogen is relatively unreactive, so that its concentration has built up throughout geologic time simply because it has not been extracted from the atmosphere in any significant way. So emissions from volcanoes may have given rise to Earth's atmospheric gases, but processes operating at or near Earth's surface have changed its composition, resulting in an atmosphere that is dominated by nitrogen and oxygen.

form when magma ascends from magma chambers toward the surface by exploiting weaknesses, such as fractures, in the overlying rock. These rocks also form in the passageways by which magma flows from one magma chamber to another. Magma trapped in these passageways cools to form planar, sheet-like intrusive bodies that are either dikes or sills, depending on their orientation relative to the layering in the surrounding wall rock. **Dikes** cut across existing layers at an angle and are therefore *discordant* with respect to the layering in the surrounding rock (Fig. 4.8a). **Sills**, on the other hand, intrude along the boundary between layers and are therefore *concordant* with respect to the layering (Fig. 4.8b).

The texture of dikes and sills varies markedly depending on their rate of cooling. Both tend to cool more rapidly than plutons because they are generally smaller bodies and

(a)

(b)

Figure 4.8   Dikes and Sills

(a) A discordant basalt dike cuts horizontal sedimentary rocks in the Grand Canyon. (b) A horizontal basalt sill (dark layer) intrudes between layers of sedimentary rock in Salt River Canyon, Arizona.

are closer to the surface. However, since they do not reach the surface, dikes and sills cool more slowly than volcanic rocks. As a result, they are commonly medium to fine grained. Porphyritic texture, characterized by phenocrysts surrounded by a fine-grained groundmass, is even more common in dikes and sills than it is in lava flows. This is because rising magma containing early-formed crystals may freeze in either a dike or a sill, so that the crystals become surrounded by a relatively fine-grained groundmass.

> ### CHECK YOUR UNDERSTANDING
> ⦿ How does the rate of cooling affect plutonic and volcanic rock texture?

Despite their different mode of occurrence and texture, neighboring intrusive and volcanic rocks at Earth's surface are often the product of the same event. Their varying expressions reflect the different levels of the crust at which they formed (see Fig. 4.7). Intrusive plutonic rocks are the crystallized remnants of magma chambers, intrusive dikes and sills represent magma that crystallized in sheet-like fractures as it was transported toward the surface or between magma chambers, and extrusive volcanic rocks are the product of magma that actually reached the surface. Because they are the product of the same magma, all three of these types of rocks typically have very similar chemical compositions. Their different rates of cooling, however, ensure quite different textures, varying from the coarse grain size of the slowly cooled plutonic rocks to the fine grain size of the more rapidly cooled volcanic rocks.

IGNEOUS ROCKS IN THE FIELD

Although textural features provide important clues about the origin of igneous rocks, they do not, on their own, provide proof of that origin. There are rare instances, for example, where extrusive rocks may be coarser grained, or intrusive rocks more fine grained, than one might expect. Also recall that a porphyritic igneous rock can form in either an extrusive or an intrusive environment, so the texture alone is not diagnostic of its origin. Proof of the origin of igneous rocks comes from observing **field relationships**. Field relationship is a term that describes how rocks relate to one another and to their neighbors. For igneous rocks, the most pertinent type of field relationship comes from observing the boundary, or **contact** between the igneous rock and adjacent rocks.

Contact Relationships

Examples of the distinction between extrusive and intrusive igneous rocks based on field observations are illustrated

in Figure 4.9 and Figure 4.10. Because extrusive rocks are either formed by cooling lava flows or the deposition of pyroclastic debris, they are laid down in discrete layers on Earth's surface (Fig. 4.9a). So a volcanic rock is deposited as a laterally extensive layer that is *younger* than the layer beneath it and *older* than the layer above it. The contact of a layer of volcanic rock with the layers beneath it may be somewhat irregular, reflecting irregularities in the topography at the time of eruption. The flow of hot lava on cold sediment simultaneously chills the lava and bakes the sediment. The result is the development adjacent to the contact of a **chilled margin**, where volcanic rock near the contact is fine grained or even glassy, as well as a **baked contact** where the heat from the lava locally transforms the sediment within a few centimeters of the contact into a metamorphic rock.

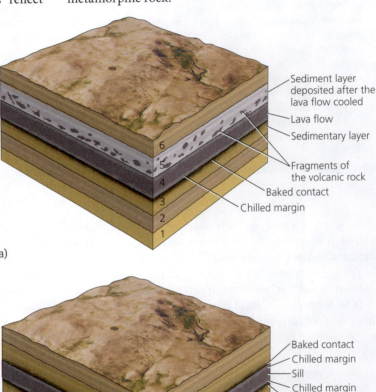

(a)

(b)

Figure 4.9   Field Relationships

These schematic diagrams illustrate the field criteria used to distinguish between extrusive and intrusive igneous rocks and to establish a sequence of events in relative order. (a) The sedimentary layer (layer 3) beneath the volcanic rock may show signs of baking. However, the layer above the volcanic rock (layer 5) does not (because it was deposited *after* the lava flow cooled) and instead may contain fragments of the volcanic rock. (b) A sill (layer 3) has chilled margins and exhibits baking along both contacts.

Figure 4.10   Contact and Thermal Effects

(a) Volcanic rocks in Central Washington State are found in the laterally extensive lava flows collectively known as the Columbia River Basalt. An individual layer of basalt is younger than the layers below it and older than the layers above. (b) This photo shows an intrusive contact where a dike (dark) intrudes a granite (white). The granite wall rock adjacent to the dike shows baking (pink discoloration). These relationships indicate that the dike is younger than the granite. (c) Closer inspection of the circled area in (b) shows that a portion of the baked granite has been ripped off from the wall rock by the dike to form a xenolith. (d) This is a typical xenolith in a granite pluton.

The upper contact of the volcanic rock typically displays very different characteristics from the lower one. This contact surface was exposed before the overlying sediment was deposited and so may show evidence of weathering, erosion, or even soil formation. Because the overlying sediment layer was deposited after the lava cooled, it shows no sign of baking, and may contain fragments of the underlying layer.

In contrast to extrusive rocks, intrusive rocks cut across or along the surrounding rock layers and are not part of the layered stratigraphy. *An intrusive rock is always younger than the rock it intrudes*, as we can see in Figure 4.9b and Figure 4.10b. This makes sense because only a younger rock can force itself into an older existing rock.

Because plutonic rocks are large bodies, geologists need to carefully map the distribution of the surrounding rocks to reveal the intrusive character. We distinguish plutons according to their size. Recall that smaller plutons are called **stocks** (see Fig. 4.7), whereas a pluton with a surface outcrop area greater than 100 square kilometers (38 square miles) is called a **batholith** (from the Greek for "deep rock"). Recently, it has become clear that many batholiths are *composite*; that is, they are the end product of many pulses of magma generation sometimes lasting over tens of millions of years. For example, the Sierra Nevada batholith in California and Nevada is the composite product of episodic magmatic pulses spanning an interval of some 120 million years between about 200 and 80 million years ago (Fig. 4.11).

The smaller size and sheet-like form of dikes and sills clearly distinguish them from plutonic bodies. Recall that dikes are also readily distinguishable from sills because dikes are discordant and cut across the layering in the wall rocks, whereas sills are concordant and are parallel to the layering. Although individual dikes are generally thin (with widths typically on the order of meters), they often occur in parallel or radiating patterns, or *swarms*, that may extend thousands of kilometers and are locally associated with major volcanic centers (Fig. 4.12). In fact, some of the magma in these dikes may reach the surface along a fracture or sets of fractures.

Thermal Effects

Because magma is generally much hotter than the wall rock that surrounds it, there is significant heat transfer across the contact between the two. As the magma cools, the wall rock is heated. As a result, minerals in the wall rock adjacent to a pluton may become unstable and react to produce new minerals that are more stable at elevated temperatures.

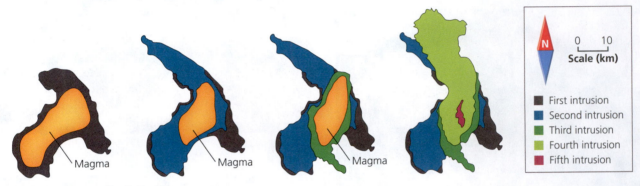

Figure 4.11   Composite Batholith

This series of diagrams shows the development of a composite batholith that is exposed in Yosemite National Park. This composite batholith was produced by five distinct episodes of intrusion over an 8 million year period, from about 93 to 85 million years ago.

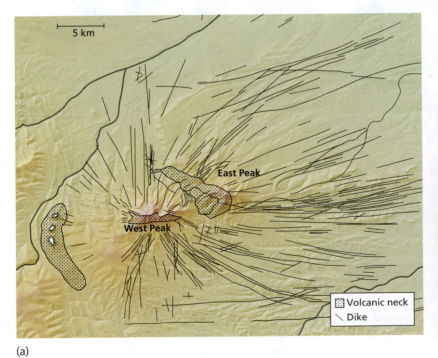

(a)

(b)

Figure 4.12   Dike Swarms

(a) This drawing shows the radial dike swarm around Spanish Peaks, Colorado. (b) The eroded remnant of a volcanic neck features radial dikes at Ship Rock, New Mexico.

Such a change in the mineral composition and texture of a rock is a form of metamorphism, a process that will be discussed further in Chapter 7. Hence wall rocks immediately adjacent to the contact with plutons commonly exhibit what is known as *contact metamorphism*.

Dikes and sills are typically thin so they cool more rapidly than plutons. However, the margins of dikes and sills cool more rapidly than their interiors and display chilled margins. At the same time, the wall rock on both sides of the contact warms up, forming baked contacts (Fig. 4.10c).

> **CHECK YOUR UNDERSTANDING**
>
> ○ Explain the origin of chilled margins and baked contacts.

As we already noted, identifying a sill is a little trickier than identifying a dike because sills are concordant with respect to layering in the overlying and underlying rocks. They may therefore be difficult to distinguish from volcanic rocks, which are also broadly concordant with respect to layering. However, since a sill is younger than both the underlying *and* the overlying rocks, it displays chilled margins and baked contacts along *both* its lower and upper contacts (see Fig. 4.9b). In contrast, a volcanic rock layer is deposited on Earth's surface, and so it is *older* than the rocks that overlie it (see Fig. 4.9a). These overlying rocks cannot exhibit baking along the contact, because they were not present at the time of the volcanic eruption. Furthermore, a volcanic rock is commonly weathered at its upper contact because its surface was exposed (see Fig. 4.8). In contrast, the upper surface of a sill is not weathered because sills are intruded at depth in the crust.

Inclusions

It is common for the force of intrusion to dislodge blocks of the neighboring wall rock into the magma. So the presence of fragments of wall rock in an igneous rock is evidence that the igneous rock has an intrusive origin. Some of these blocks are digested by the magma but others survive as unmelted remnants that we call **inclusions**. These inclusions either consist of rock fragments called **xenoliths** (Fig. 4.10c, d) or crystal fragments called **xenocrysts**. The word xenolith is from the Greek for "foreign rock" and xenocyrst means "foreign crystal." In either case, the inclu-

> **CHECK YOUR UNDERSTANDING**
>
> ○ On the basis of field relationships, how would you distinguish between a pluton and a dike?
>
> ○ On the basis of field relationships, how would you distinguish between a sill and a volcanic rock layer?

sions are derived from a rock that must be older than the age of the intrusive body. Recognizing inclusions also helps geologists deduce how much influence wall rock assimilation had on the magma evolution.

Field relationships provide important information on the nature and origin of igneous bodies. But igneous bodies also vary markedly in chemical composition. In Section 4.4, we discuss how we use the variations in chemical composition and textures to systematically classify igneous rocks.

4.3 SUMMARY

- Rock texture is defined by the size, shape, and arrangement of mineral grains.

- There are two fundamentally distinct types of igneous rock. Extrusive rocks are deposited at Earth's surface as lava flows or pyroclastic deposits. Intrusive rocks form below Earth's surface and cut across preexisting rock layers as they rise through the crust.

- When lava cools quickly, a fine-grained or glassy volcanic rock results that can display flow texture—an alignment of small crystals.

- Explosive volcanic eruptions produce pyroclastic deposits.

- Magma that crystallizes in magma chambers is coarse-grained; the resulting intrusive rocks are plutonic.

- Intrusive igneous rocks also form where magma crystallizes in narrow conduits to form dikes and sills.

4.4 Classifying Igneous Rocks

Geologists classify igneous rocks on the basis of their composition, textures, and field relationships. We describe the composition of an igneous rock in two ways. *Mineral composition* refers to the relative abundances of rock-forming minerals such as quartz, plagioclase, or pyroxene. *Chemical composition* describes the relative amounts of oxides such as silica (SiO_2), magnesium oxide (MgO), and iron oxide (FeO). Because these oxides are the building blocks of minerals, there is a direct relationship between mineral composition and chemical composition. The mineral content of igneous rocks varies with magma chemistry. For example, minerals such as olivine and pyroxene contain magnesium and iron, so a rock containing these minerals is rich in MgO and FeO. Alternatively, rocks that contain abundant quartz (SiO_2) and potassium feldspar ($KAlSi_3O_8$) are rich in SiO_2, Al_2O_3 and K_2O.

In coarse- or medium-grained rocks, we can determine mineral composition by examining a hand-sized sample, or *hand specimen*, often with the aid of a hand lens. For finer grained rocks, which may contain glass, mineral identification is more problematic and the rock's chemical composition, which is determined by laboratory analysis, may be the more useful method of classification.

For most igneous rocks, the total silica (SiO_2) content ranges between 30 percent and 80 percent by weight (Fig. 4.13). Using this variation, we subdivide igneous rocks into four categories: **felsic** (more than 66% SiO_2), **intermediate** (52 to 66% SiO_2), **mafic** (45 to 52% SiO_2), and **ultramafic** (less than 45% SiO_2). Each of these categories is further subdivided by the grain size of the rock, which in most situations depends

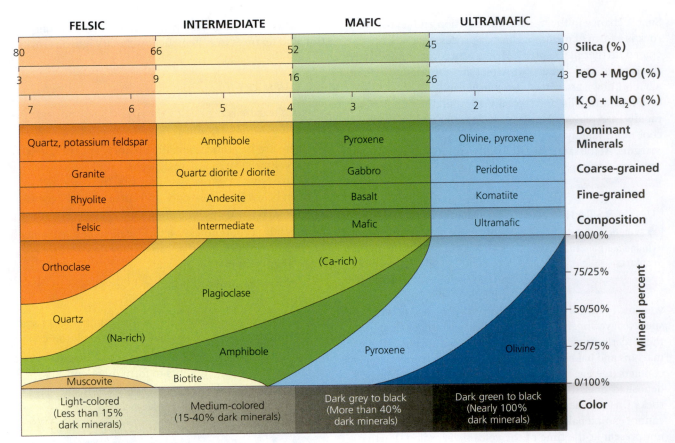

Figure 4.13 **Classification of Igneous Rocks**

This figure is a guide to the dominant minerals, chemical composition, color, and texture in each of the four main groups of igneous rocks.

on the rate at which the magma cooled. For example, a fine-grained felsic rock is called *rhyolite*, whereas a coarse-grained felsic rock is called *granite*.

Systematic variations also occur in important chemical components other than SiO_2. For example, felsic rocks with higher SiO_2 concentrations tend to contain more K_2O and Na_2O and much less FeO and MgO than mafic or ultramafic rocks (see Fig. 4.13). Chemical variations profoundly affect the mineralogy of igneous rocks so that identification of the minerals present aids in their classification. We shall now discuss each of the categories and subcategories of igneous rock in detail.

FELSIC IGNEOUS ROCKS

In any particular magma plumbing system, the composition of intrusive and extrusive rocks may be very similar. For example, any given intrusive rock may have a "volcanic equivalent." For igneous rocks rich in silica, the coarse-grained plutonic rock *granite* and its fine-grained volcanic equivalent *rhyolite*, may have crystallized from

the same magma but they formed in different parts of the plumbing system. Both rocks contain abundant quartz, K-feldspar and Na-rich plagioclase, and are consequently light in color. (Recall from Chapter 3 that plagioclase feldspar is the most common mineral in igneous rocks. It has a composition that varies between calcium-rich and sodium-rich varieties.) This mineralogy in granite reflects the relative abundance of SiO_2, K_2O, and Na_2O in the magma. Because of the abundance of <u>feldspar</u> and <u>silica</u>, the rocks are described as *felsic* (see Fig. 4.13). Felsic rocks also have low densities (around 2.7 g/cm³) because sodium and potassium are relatively light elements.

Granite is perhaps the most familiar igneous rock because of its use in public buildings and monuments. In addition to quartz, K-feldspar, and Na-rich plagioclase, granites may contain the shiny white mica known as muscovite (Fig. 4.14a). (Recall from Chapter 3 that the mica group is a set of minerals made up of sheets of silicate tetrahedra connected to one another by ions.) Together, these minerals typically account for 70 to 90 percent of the rock. The relatively low concentrations of FeO and MgO in granite are reflected in the low abundance (generally less than 15 percent) of iron- and magnesium-bearing minerals such as biotite and hornblende, which are the most common of these minerals.

Figure 4.14   Igneous Rock Types

Common types of igneous rocks include: (a) granite; (b) rhyolite; (c) diorite; (d) andesite; (e) gabbro; (f) basalt; (g) peridotite and (h) komatiite.

Individual mineral grains in coarse-grained granite are large (often about the size of a thumbnail). It is usually quite easy to distinguish the different minerals in granite on the basis of their appearance. The feldspars are generally rectangular and often the largest grains. They are pinkish in the case of K-feldspar, or white in the case of Na-rich plagioclase. Quartz, on the other hand, is a glassy grey mineral and tends to occupy the spaces between the feldspars. Biotite and hornblende are black and usually form smaller grains that are plate-like or rod-like, respectively. Like biotite, muscovite forms platy thin flaky grains but it lacks iron and magnesium, and so is silver rather than black. Although these last three minerals typically comprise less than 15 percent of a granitic rock by volume, the effect of this mineral combination is to produce a grey rock with a distinctive, speckled appearance (see Fig.4.14a).

Rhyolite has the same chemical composition as granite and is dominated by the same minerals. However, rhyolite is typically so fine grained that it is difficult to see individual crystals (Fig. 4.14b). Closer examination of rhyolite may reveal the presence of a small proportion of quartz or feldspar phenocrysts surrounded by a fine-grained groundmass, resulting in a porphyritic texture. Recall from Section 4.3 that phenocrysts usually reflect an earlier history of cooling and partial crystallization in a magma chamber prior to volcanic eruption.

INTERMEDIATE IGNEOUS ROCKS

Igneous rocks that lie between the compositional extremes of felsic and mafic rocks are classified as *intermediate*. Rocks with this composition are very common in modern mountain belts, and geologists cite their presence in the geologic record as evidence for the existence of ancient mountain belts. As a result of lower SiO_2, K_2O, and Na_2O contents, intermediate rocks typically contain much less quartz than granite. Similarly, K_2O-bearing minerals such as K-feldspar, muscovite, and biotite are rare or absent, and plagioclase feldspar contains less sodium and more calcium than the felsic rocks (see Fig. 4.13). Plagioclase in intermediate rocks is generally the dominant light-colored mineral.

Since intermediate magmas contain more FeO and MgO, iron- and magnesium-bearing minerals, such as hornblende and pyroxene, are more common in intermediate rocks than in felsic rocks. For the same reasons, the color, density, and mineral content of intermediate rocks lie between those of felsic and mafic rocks.

There are two important intermediate rocks, a plutonic rock named diorite, and a volcanic rock named andesite. **Diorite** (Fig. 4.14c) is a coarse-grained rock generally dominated by plagioclase and hornblende that can contain relatively minor amounts of quartz or pyroxene. The finer-grained **andesite** (Fig. 4.14d), named after its occurrence in the volcanic rocks of the Andes Mountains of South America, commonly contains phenocrysts of plagioclase and/or hornblende in a finer grained groundmass. Most of the phenocrysts are thought to have formed in magma chambers prior to eruption, whereas the fine-grained groundmass reflects chilling of the lava at the surface.

> **CHECK YOUR UNDERSTANDING**
> ◔ Where are intermediate igneous rocks commonly found?

MAFIC IGNEOUS ROCKS

Mafic igneous rocks are relatively poor in silica and are richer in the heavier elements magnesium (Mg) and iron (Fe). Because of their higher MgO and FeO contents, mafic rocks are dark in color and have quite high densities (around 3.0 g/cm³). The most common mafic igneous rocks are the coarse-grained plutonic rock known as gabbro and its fine-grained volcanic equivalent, basalt. In **gabbro** (Fig. 4.14e) the grains are usually large enough that we can readily deduce its mineral content and texture. Gabbro is typically composed of calcium-rich plagioclase and pyroxene, with or without hornblende and/or olivine.

Because both basalt and gabbro come from the same type of magma, **basalt** (Fig. 4.14f) contains the same minerals as gabbro, namely calcium-rich plagioclase and pyroxene, usually with some olivine or hornblende. When

> **CHECK YOUR UNDERSTANDING**
> ◔ How does the composition of felsic igneous rocks differ from that of mafic igneous rocks?

these minerals crystallize in the magma chamber prior to eruption, they occur as phenocrysts surrounded by a fine-grained or glassy groundmass. Microscopic examination of basalt reveals a groundmass typically consisting of very fine-grained plagioclase and pyroxene crystals that exhibit flow texture.

ULTRAMAFIC IGNEOUS ROCKS

As the name suggests, ultramafic rocks contain more magnesium and iron and less silicon and aluminum than most other igneous rocks (see Fig. 4.13). They are uncommon at Earth's surface but are thought to dominate Earth's interior in the region immediately below the crust. Ultramafic rocks are made up of iron- and magnesium-bearing minerals such as olivine and pyroxene and so have high densities (around 3.3 g/cm³). Ultramafic rocks are distinguished from mafic rocks by their relative lack of aluminum-bearing minerals such as plagioclase. Coarse-grained ultramafic rocks are known as **peridotite** (Fig. 4.14g). Ultramafic lavas, called **komatiites** (Fig. 4.14h), are rare in the geologic record; there are no modern volcanoes that produce lavas of this composition. However, there are important examples of such flows early in Earth's history, suggesting more extensive melting of the mantle, and therefore hotter conditions, at that time.

In this section, we have discussed the classification of igneous rocks based on their texture and composition. We have seen how composition affects important features such as color and density. We have also seen that the grain size and mineral content of an igneous rock depends on the rate at which the magma cools and on its chemical composition. In Section 4.5, we learn that magmas evolve by changing their chemistry as they cool. As a result, a single episode of magmatic activity can produce igneous rocks of variable compositions.

> **CHECK YOUR UNDERSTANDING**
> ◔ Why is plagioclase rare in ultramafic rocks?

4.4 SUMMARY

* On the basis of silica content, we divide igneous rocks into felsic, intermediate, mafic, and ultramafic categories.

* Felsic rocks are high in silica, and low in iron and magnesium. Mafic rocks are low in silica and high in iron and magnesium. The composition of intermediate rocks lies between those of felsic and mafic rocks. Ultramafic rocks are higher in iron and magnesium and lower in silica than other igneous rocks.

* The mineral content of igneous rocks varies according to magma chemistry.

4.5 Evolution of Igneous Rocks

In this section, we describe how the chemical composition and physical characteristics of a magma can change as the magma evolves. Some of these changes are internal processes that occur within the magma as it cools. Others are caused by external influences such as the interaction of the magma with the wall rock or mixing with magma from other magma chambers.

A cooling magma typically becomes richer in elements such as silicon and potassium and impoverished in elements like iron, magnesium, and calcium. These chemical changes occur because minerals that crystallize early in the cooling process are rich in iron, magnesium, and calcium. As a result, a mafic magma evolves toward a more felsic composition as it cools. At the same time, the physical properties of the magma also change since these properties depend on magma chemistry and temperature. For example, felsic magmas are cooler and more viscous (sticky) than mafic magmas, and so have a greater resistance to flow. We will learn more at the end of this section about how the physical property of viscosity has direct implications for the explosive character of volcanic eruptions, and consequently their impact on human life and the environment.

CHEMICAL EVOLUTION OF MAGMA

Both internal and external processes play a role in the chemical evolution of magma. In this section of the chapter we begin with the internal processes and then go on to discuss the external processes.

Internal Processes

In the early part of the twentieth century, the Canadian geoscientist Norman Bowen proposed that mafic magmas similar in composition to basalt or gabbro may evolve by cooling and crystallization to produce more silica-rich magmas like those represented by granite or rhyolite. Using laboratory experiments, Bowen monitored the cooling history of basaltic liquid and established a regular sequence in the order of crystallization of minerals.

Bowen's experiments showed that olivine is typically the first mineral to crystallize from the cooling mafic magma. Because it is the first to crystallize, we know that olivine is the most stable mineral in the mafic magma at high temperatures. However, as the magma continues to cool, it reaches a critical temperature at which pyroxene becomes the more stable mafic mineral. At that temperature, the olivine crystals react with the remaining magma to produce new crystals of pyroxene. Upon further cooling, a temperature is reached where amphibole becomes the more stable mineral. At this point, a reaction between the pyroxene crystals and the magma produces amphibole. At still lower temperatures, biotite is produced by reaction between magma and the amphibole crystals. This sequence, where one mineral becomes converted to another at a specific temperature in the cooling history, is known as **Bowen's Discontinuous Reaction Series** (Fig. 4.15).

At the same time that this sequence of mafic minerals is crystallizing, plagioclase feldspar is also crystallizing in the magma. Recall from Section 4.4 that plagioclase is the most common mineral in igneous rocks and has a wide range of compositions that varies between calcium-rich and sodium-rich varieties. As the magma cools, the most calcium-rich variety of plagioclase crystallizes first (Fig. 4.15). With continued cooling, however, progressively more sodium-rich varieties become more stable. Each new plagioclase crystal consequently reacts with the remaining liquid almost as soon as it forms, in a continuous cycle of crystallization and reaction. This process is known as **Bowen's Continuous Reaction Series**. With each reaction in the continuous series, progressively more sodium-rich plagioclase is produced.

It is very important to realize that both series of reactions—the discontinuous branch that produces mafic minerals and the continuous branch that produces increasingly sodium-rich plagioclase—operate simultaneously. However, the continuous reactions differ from the discontinuous reactions because they involve only one mineral (plagioclase) and the reactions occur continuously with falling temperatures. In contrast, the discontinuous reactions occur only at specific temperatures during the magma's cooling history and produce a sequence of mafic minerals.

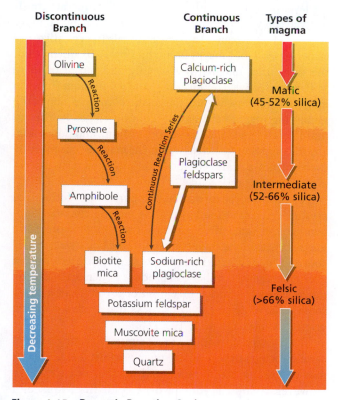

Figure 4.15 Bowen's Reaction Series
With decreasing temperature, early-forming olivine crystals become unstable and react with the melt to form pyroxene (left-hand side). Eventually, pyroxene becomes unstable and reacts with the magma to form amphibole, and so on. Each reaction takes place at a discrete temperature. At the same time, early-forming calcium-rich plagioclase is continuously replaced by more sodium-rich plagioclase as the magma cools (right-hand side).

Most mafic magmas initially crystallize olivine and calcium-rich plagioclase. If the magma completely solidifies at this stage, mafic rocks such as basalt and gabbro form. If magma is still present, however, these early forming minerals become unstable as they cool and react with the remaining magma to produce pyroxene and a slightly less calcium-rich plagioclase. With further cooling, if magma still remains, the pyroxene is replaced by amphibole and biotite, and the plagioclase acquires a composition intermediate between the calcium- and sodium-rich varieties.

Identification of these reaction series allowed Bowen to make one of the most important discoveries about igneous processes. Bowen proposed that the composition of magmas may change as the magma cools, a process known as **fractionation**. Fractionation occurs when early-forming minerals such as olivine and pyroxene are prevented from completely reacting with the magma. For example, because these minerals are denser than the magma, they may sink and accumulate as layers of gabbro or peridotite on the floor of the magma chamber. This process is known as **crystal settling** (Fig. 4.16), and was actually first described by Charles Darwin between 1831 and 1836.

During crystal settling, as each layer forms, it buries the layer beneath it. In this way, the mafic minerals are effectively prevented from reacting with the magma above (Fig. 4.16a–c). Because the first minerals to crystallize are richer in iron and magnesium and poorer in silicon than the remaining magma, the settling and burial of these minerals effectively removes iron and magnesium from the magma. As a consequence, the chemical composition of the remaining magma changes, becoming richer in the remaining elements such as silicon, sodium, and potassium, and in components such as water. The magma consequently evolves from a mafic composition toward a felsic one as crystal settling proceeds.

For example, the formation of gabbro or peridotite from crystals settling at the base of the magma chamber happens at the same time as the formation of more felsic magmas within the chamber itself. These liquids ultimately cool and crystallize to form diorite or granite. In this way a cooling magma chamber can produce a chemically diverse range of igneous rocks. Like a set of furniture in a living room, a set of diverse igneous rocks that share a common origin is known as a *suite*.

The concept of fractionation is similar to eating a box of candies, which have a variety of colors (Fig. 4.17). The box may originally contain an equal number of each color. However, if you eat your favorite color first, this color becomes progressively depleted. If you then eat your second- and third-favorite colors, the remaining population of candies is very different from (and much less appealing than) the one with which you started. Similarly, the settling of mafic minerals to the floor of the magma chamber removes iron and magnesium from the magma so that the remaining magma is more felsic.

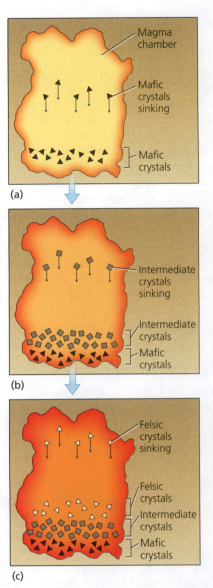

Figure 4.16 Fractionation by Crystal Settling
Early formed minerals have a higher density than the magma and so settle to the bottom of the magma chamber where they accumulate. Parts (a), (b), and (c) show progressive stages in crystal settling; the magma evolves from mafic to intermediate to felsic in composition.

Crystal settling is one example of an important igneous process known as **fractional crystallization**. During fractional crystallization early crystallizing minerals are isolated and prevented from further reaction with the magma. In crystal settling this happens when the crystals sink, but in other situations, crystals may float rather than sink because they are less dense than the magma. Plagioclase is an example of a mineral that commonly floats to the top of a mafic magma. Indeed the pale color of the

(a)　　　　　　　(b)　　　　　　　(c)

Figure 4.17　Fractionation

Changes in the relative proportion of colored candies as one color is preferentially removed is analogous to the process of fractionation as crystals are effectively removed from a magma by settling and the proportion of the remaining components is increased.

highlands on Earth's Moon has recently been attributed to the flotation of plagioclase to the top of ancient magma chambers that existed below the Moon's surface more than 4 billion years ago. Crystals can also become caught up in flow patterns of magma within large chambers or may adhere laterally to the cooler walls of the chamber. Just as crystal settling isolates the crystals that sink, the crystals in each of these situations are separated and prevented from reacting with the magma. As a result, the chemical composition of the magma changes.

External Processes

A magma chamber should not be viewed in isolation. The chamber is just part of a complicated interconnected plumbing system, not unlike the water system in a municipality that may have thousands of storage tanks interconnected by a labyrinth of pipes. In this context, magma may be pumped along fractures from one chamber into another as pressures rise and fall. This process of **magma mixing** produces magmas of intermediate composition if felsic magma is pumped into a chamber containing mafic magma (or vice versa).

There is strong evidence, for example, that mixing of mafic and felsic magmas occurred immediately prior to the 1991 eruption of Mt. Pinatubo. This volcano had been monitored for a long time, during which gases emanating from its vent (the opening of the volcano) were routinely analyzed. About two weeks before the Mt. Pinatubo eruption, however, the composition of these gases suddenly changed. From this, volcanologists immediately deduced that the chamber below the vent had been invaded by a magma of contrasting composition (Fig. 4.18). Like nature's equivalent of an upset stomach, this was viewed as a warning that the volcano was about to erupt.

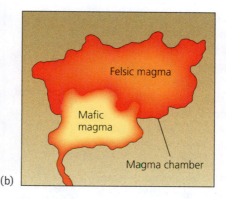

(a)　　　　　　　(b)　　　　　　　(c)

Figure 4.18　Magma Mixing

(a) A scientist monitors the composition of gases emanating from the vent of an active volcano. (b) Influx of mafic magma at the base of the chamber is detected by changes in the gas composition at a volcanic vent. The mafic magma mixes with a felsic magma to produce one of intermediate composition. (c) In this field example of magma mixing, signs of incomplete mixing of mafic (dark) and felsic (light) components are visible.

Magma mixing may also occur if magmas originating from different source regions exploit the same fractures and mix together as they rise toward the surface. In fact, geologists recognize the role of fractures in guiding the ascent of magma as one of increasing significance in magma mixing.

The chemical evolution of magma can also be influenced by its reaction with, and digestion of, the wall rocks that it encounters during its ascent. This process is known as **assimilation** and includes a combination of chemical (incorporation of soluble chemicals) and physical (incorporation and dismemberment of fragments) processes.

We have now learned that two contrasting processes can produce rocks of intermediate composition: fractional crystallization of mafic magma; and magma mixing between mafic and felsic liquids. It is not always possible to tell which of these two processes was the more important in an igneous body of intermediate composition. Magma mixing however rarely produces homogenous magma so there is often field evidence of incomplete mixing (Fig. 4.18c).

It is important to realize that the internal and external processes we have just described are not mutually exclusive. The chemical evolution of igneous rocks is a dynamic process that may involve one or all of the processes described above in any given area. For example, the energy expended by magma in digesting wall rock results in more rapid cooling that, in turn, promotes crystal fractionation. It is then the job of the geoscientist to decide which of these processes was the most dominant. If you take more advanced courses, you will learn that there are elegant ways to do this. In the material that follows, we explain that a close relationship also exists between the chemical and physical properties of magma—a relationship that explains the dramatic impact volcanic activity can have on human life.

> **CHECK YOUR UNDERSTANDING**
> ● What is assimilation?

PHYSICAL EVOLUTION OF MAGMA

The contrasting compositions and textures of igneous rocks reflect important variations in the physical properties of the magma. We have seen that in addition to liquid, magmas typically contain trapped gases, floating or sinking crystals and fragments of variably digested wall rock. Some magmas are very fluid and rise more rapidly than those that are sticky. This difference is expressed in the concept of **viscosity**, which, in everyday usage is a measure of a liquid's stickiness or resistance to flow. The less viscous a magma, the more easily it flows.

The viscosity of a magma is key to understanding the explosive nature of certain volcanic eruptions. Viscosity depends on the magma's temperature, chemical composition, and volatile content. Hotter magmas flow more readily than cooler magmas, just as hotter butter flows more readily than cooler butter. Magmas rich in silicon are more viscous

because silicon forms a network of bonds with oxygen in the magma and these bonds make the magma less fluid. Mafic magmas form at a higher temperature than felsic magmas and also have a lower silicon content. This combination of factors means that mafic magmas are far less viscous (and therefore flow more easily) than felsic magmas. Volatiles, on the other hand, have the opposite effect of silicon at high temperatures, and tend to break bonds apart, making magma more fluid.

We can find an explanation for the increasing viscosity of silica-rich magmas by comparing the sequence of crystallizing minerals in Bowen's Discontinuous Reaction Series with the arrangement of their respective silicate tetrahedra (Fig. 4.19). Notice the crystallization sequence; minerals with increasingly more complicated tetrahedral structures crystallize in sequence as magma cools. This is because the rapid movement of the ions in magma slows down as the magma cools, allowing the ions to adopt a more orderly arrangement. As a result, the magma becomes stickier and has more resistance to flow. Experiments show that *before* each mineral crystallizes, its tetrahedra are loosely linked together in the melt. This means that prior to

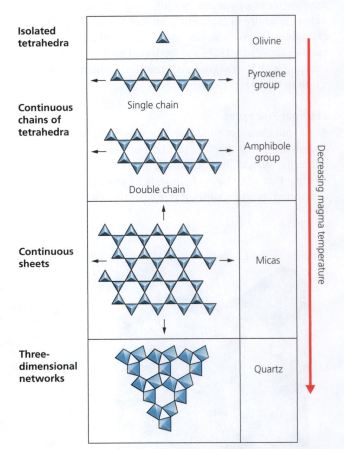

Figure 4.19 Bowen's Discontinuous Reaction Series and Crystal Structures

At high temperatures, olivine, which has the simplest crystal structure, is stable. As a magma cools it loses energy, allowing minerals with progressively more complex structures to become stable.

crystallizing minerals with complex silicate structures, felsic magmas contain tetrahedra that are loosely linked into relatively long chains. In contrast, mafic magmas crystallize less complex silicates, so we know that they contain relatively isolated tetrahedra. Because of their more extensive tetrahedral linkages, felsic melts are stickier—more viscous—than those of mafic composition.

In Section 4.6, we will see that differences in viscosity profoundly affect the explosiveness of volcanic eruptions, which range from the spectacular (but relatively harmless) lava fountains typical of Hawaii to the devastating pyroclastic eruptions of volcanoes like Mount St. Helens and Mt. Pinatubo.

4.5 SUMMARY

- Bowen's Reaction Series describes how mafic magmas change in composition during cooling as crystals react—discontinuously and continuously—with the melt.

- A wide variety of phenomena causes the composition of a magma to change as it cools and crystallizes, a process known as fractionation. For example, separation of early-formed minerals by crystal settling causes the composition of the remaining magma to become more felsic.

- A variety of processes produce intermediate to felsic rocks; they include fractional crystallization of mafic magma and magma mixing.

- The physical property of viscosity, which is key to understanding the explosive nature of certain volcanic eruptions, depends on a magma's temperature, chemical composition, and gas content.

4.6 Volcanic Eruptions

Volcanic eruptions testify to the awesome powers of nature and to the dynamic processes that occur in Earth's hidden interior. Throughout human history, volcanoes have been the source of many myths and legends. Over the past 100 years our understanding of volcanoes and the processes that cause them has increased dramatically, enabling geologists to predict eruptions and their potential impact on human life. In the short term volcanic eruptions are significant hazards, but over the longer term, we now realize that they are essential to the sustenance of life on Earth. They provide nutrients to soils and the life that depends on soils, and they replenish our atmosphere with their gaseous emissions.

There are two main styles of volcanic eruption, *fissure eruption* and *vent eruption*. We learned in Section 4.3 about the importance of dikes in transporting magma to the surface, and that dikes often occur in radiating swarms, some extending for thousands of kilometers from major volcanic centers (see Fig. 4.12). Some of the magma in these dikes may reach the surface along a fracture or sets of fractures, to fuel a **fissure eruption**. This style of eruption is common where the crust is under extension, and where the magma is relatively fluid. Fissure eruptions are the dominant style of eruption in regions, such as continental rifts, where vast volumes of basaltic lava are extruded to form deposits known as **flood basalts**, and at mid-oceanic ridges where oceanic crust is being pulled apart and magma is transported upward along steep fractures to the central axis of the ridge.

Vent eruptions occur where magma is extruded at a central vent. Figure 4.20 summarizes the various volcanic landforms—*shield, composite,* or *cinder cone* volcanoes—associated with this style of eruption. Vent eruptions stem from a central crater. Some volcanic landforms are tens of kilometers across (shield volcanoes) and are among the tallest mountains on Earth. Other landforms (cinder cones) are only 200–300 meters (650 to 1000 feet) high and less than 2 kilometers (1.2 miles) in diameter. In the material that follows we will see that the type of landform depends on several factors including the gas content and composition of the magma.

ERUPTION OF MAFIC MAGMA

Recall from Section 4.5 that mafic magmas are less viscous and more fluid (see Fig. 4.20) than intermediate or felsic magmas. Because mafic magmas are fluid, vent eruptions from basaltic volcanoes like those of Hawaii tend to flow rapidly, spreading out over wide areas to form relatively thin lava flows. As a result, the volcanic edifice is broad and its slopes are very gentle. Because their shape resembles that of a warrior's shield, they are known as **shield volcanoes** (Fig. 4.21). The vent is located at the summit and is typically a depressed area, up to 5 kilometers (3 miles) across, known as a **caldera**. Calderas form when magma in the chamber directly below the volcano's summit is emptied by an eruption on the volcano's flank. Lacking support, the summit collapses into the magma chamber. Subsequent eruptions may fill the caldera to form a **lava lake**. Recent studies show that this collapse is often followed by an influx of groundwater, which may result in short-lived but highly explosive eruptions.

Whether they emanate from fissures or from vents, basaltic lavas typically flow at rates between 10 to 300 meters (33 to 1000 feet) per hour, although flow rates of up to 30 kilometers per hour (19 miles per hour) have been measured. The exposed surface of the lava cools and congeals faster than its interior. Because of this, the otherwise smooth solid surface of the flow can develop wrinkles resembling irregular folds in cloth as the molten lava moves beneath it (Fig. 4.20c). Because of the resemblance to coiled rope, such flows are known as **ropy lava**. The native Hawaiian word for ropy lava is *pahoehoe* (pronounced pah-hoy-hoy). As the molten interior continues to advance, the surface may break up into very sharp, jagged blocks that resemble furnace clinkers. Such flows are referred to as **blocky lava**. The formation of blocky lava is exacerbated by escaping gases, which produce numerous voids and sharp spines. The Hawaiian term for jagged lava of this type is *aa*

Figure 4.20   Features of Mafic Lava Eruptions

Different features of mafic lava eruptions include: (a) a lava fountain (upper left) giving rise to fluid basaltic lava flows, Kilauea volcano, Hawaii; (b) a shield volcano (background) and a cinder cone (foreground) at Mauna Loa, Hawaii (note the broad gentle slopes of the shield volcano and the slight depression (left of center) at the site of the caldera); (c) contrasting wrinkled surfaces (foreground) typical of ropy (pahoehoe) lava flows and sharp, broken surfaces (background) typical of blocky (aa) flows; and (d) a lava tube at Kilauea volcano, Hawaii.

Figure 4.21   Volcanic Landforms

Volcanic landforms that develop around a single vent (all shown at the same scale). Note that shield volcanoes typically form much larger landforms than composite volcanoes or cinder cones.

(pronounced ah-ah). As the flow continues to solidify, the molten interior develops a complex labyrinth of passages, known as **lava tubes**, as the flow continues to move through its congealing surroundings. When the eruption ceases, the lava drains from these tubes, leaving them hollow (Fig. 4.20d).

Because of the low viscosity of mafic lava, gases trapped as bubbles in mafic lava can escape with relative ease, just as they do in a glass of soda pop. As a result, basaltic eruptions are rarely more violent than the familiar jets of incandescent lava we call **lava fountains** (Fig. 4.20a). Like the spray from a shaken bottle of champagne, these fountains result from rapid expansion and escape of the trapped gas bubbles, which occurs when the magma reaches Earth's surface and the pressure on the magma abruptly drops.

The eruption of mafic lava is the dominant volcanic process on the ocean floors. As these fissure eruptions occur underwater, they produce very distinctive bulbous structures known as **pillow lavas** (Fig. 4.22). Because the ocean floor

(a)

(b)

(c)

Figure 4.22   Pillow Lava

(a) Lava freezing on the cold seafloor produces pillows. (b) When the tips of pillows break off, the lava beneath escapes to form another pillow. (c) These pillow lava structures have been preserved in the geologic record.

covers 70 percent of Earth's surface, pillow lavas are the most common type of volcanic rock. However, pillow lavas are rarely observed on land. Observations by divers and from deep-sea submersibles show that when mafic lava is extruded onto the ocean floor, it rapidly freezes against the cold seawater, producing pillow-like bulges (Fig. 4.22a). But the interior of the pillow remains molten so that new pillows form when the hot lava from the interior oozes out through cracks in the pillow surface as the surface cools and shrinks. The pillow may also fall off the main body of lava, exposing hot lava to seawater to form yet another pillow (Fig. 4.22b). Identification of pillow structures in ancient mafic flows is important because it provides evidence of extrusion underwater, usually in an ancient marine environment (Fig. 4.22c).

ERUPTION OF INTERMEDIATE AND FELSIC MAGMAS

The higher viscosity of intermediate and felsic magmas results in more explosive styles of volcanism. Fissure eruptions are rare. In fact felsic magma is so viscous and moves so slowly that it is rarely extruded as lava at Earth's surface. Instead, it generally cools underground to form large plutons of granite.

Where felsic magma does extrude, its high gas content usually causes violent vent eruptions that produce steep-sided volcanoes and far more airborne fragments than surface lava. As a viscous felsic magma rises toward the surface, the reduced pressure results in the formation of gas bubbles, just as it does when you open a can of soda pop. In this case, however, the bubbles are initially held captive by the viscous melt. As the magma continues to rise, a threshold is eventually reached where the pressure from the gas bubbles explodes the magma into hot fragments. At first, the solid rock overlying the magma acts like a pressure cooker lid, temporarily sealing in the violent cauldron below. However, rocks are an imperfect seal, and when the lid breaks an explosive eruption is inevitable. Explosive eruptions of felsic magma vent mixtures of gases, or **aerosols**, into the atmosphere and jettison fragments of the hot magma and the surrounding rocks into the air. These eventually fall to Earth to form pyroclastic deposits.

Pyroclastic deposits are classified according to their particle size and whether or not they are consolidated (Fig. 4.23; Table 4.1). Unconsolidated pyroclastic deposits are called **tephra**. Recall from Section 4.3 that larger particles fall first so that coarser-grained deposits are found closest to the eruptive center. Particles greater than 64 millimeters (2.5 inches) in diameter are called **blocks** or **bombs**, depending on their shape. Blocks are solid angular fragments

(a)

(b)

(c)

Figure 4.23   Pyroclastic Deposits

Pyroclastic deposits are classified according to grain size. (a) A bomb is a large rounded object, often with a molten interior. These examples (glowing red) are 20 cm in length. (b) Lapilli and (c) ash are medium-grained to fine-grained material, respectively, ejected from a volcano. The darker fragments in the volcanic ash are blocks of older volcanic rocks incorporated during the eruption.

TABLE 4.1   CLASSIFICATION AND NOMENCLATURE OF PYROCLASTIC DEPOSITS BASED ON CLAST SIZE

| | | Pyroclastic Deposit | |
Clast Size in mm	Pyroclast	Mainly Unconsolidated (Tephra)	Mainly Consolidated (Pyroclastic Rock)
> 64	bomb, block	agglomerate bed of blocks or bombs, block tephra	agglomerate, volcanic breccia
64 to 2	lapillus	layer, bed of lapilli or lapilli tephra	lapilli tuff
2 to 1/16	coarse ash grain	coarse ash	coarse (ash) tuff
< 1/16	fine ash grain	fine ash (dust)	fine (ash) tuff

Source: After Schmid, 1981.

when they are ejected from the volcano and these angular fragments are found in many pyroclastic deposits. Bombs, on the other hand, are fluid when ejected, and their shapes are rounded during flight. Finer grained deposits, such as **lapilli** (2 to 64 mm or .07 to 2.5 inches in diameter) or *ash* (< 2 mm or .07 inches in diameter), and *volcanic dust* (about 0.001 mm or 3.9×10^{-5} inches in diameter) disperse more widely. Rocks that form from the deposition of blocks or bombs are **volcanic breccias** or **agglomerates**. The presence

of these rocks in the field can help geologists locate the sites of ancient volcanic centers. Ash is typically deposited in thin beds, known as **tuffs**, and may be dispersed hundreds of kilometers from its original source. The finest material may circulate the globe producing brilliant sunsets as it did following the 1883 eruption of Krakatau in Java, Indonesia.

The various products of explosive eruptions are summarized in Figure 4.24a. The most devastating of all violent eruptions is a ground-hugging avalanche of hot volcanic ash

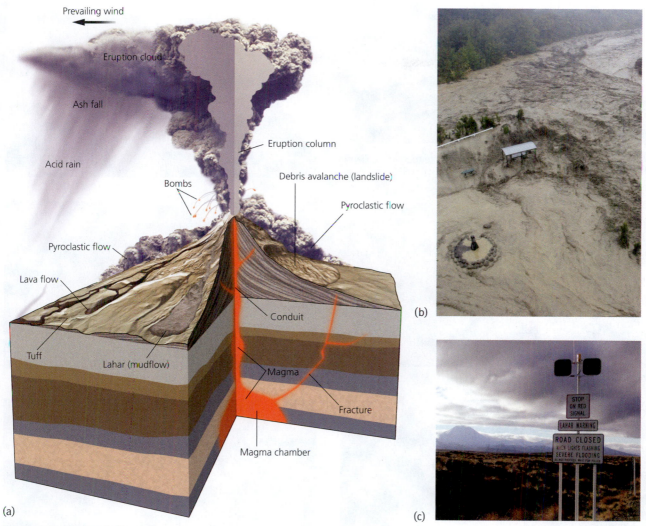

Figure 4.24 Volcanic Hazards

(a) This schematic diagram illustrates the typical products of explosive (pyroclastic) eruptions. Note the importance of wind direction on the eruption cloud. (b) The lahar in this photo was associated with the 2007 eruption of Mount Ruapehu, New Zealand. (c) A lahar warning posted near active volcanoes of the Taupo Zone, northern New Zealand, warns of road closure and flooding.

known as a **pyroclastic flow** (see In Depth: **Can Explosive Eruptions Be Predicted?** on page 130). In addition, the instabilities generated by pyroclastic eruptions often produce catastrophic mudflows known as **lahars** (Fig. 4.24b) when hot volcanic ash melts snow or glacial ice, or when rainfall that commonly accompanies volcanic eruptions renders the hot ash on the slopes unstable. These hot mudflows travel at speeds of up to 60 kilometers per hour (37 miles per hour) and sweep away everything in their path (Fig. 4.24c).

The explosion of Mount St. Helens in May of 1980 was a typical pyroclastic eruption (Fig. 4.25). The ascent of silica-rich magma to Earth's surface was temporarily impeded by the overlying solid rock. When the ascent of magma was renewed, the volcano vented steam and ash and there was increased low-intensity earthquake activity. In less than three minutes, a sequence of events cascaded like tumbling dominoes and triggered a series of violent eruptions. The ascending magma first caused the mountain to bulge, rendering its northern slope of old volcanic debris unstable.

An earthquake caused by the ascent of magma then triggered a landslide, which removed part of the unstable rock that was covering and trapping the magma. This abruptly reduced the pressure on the magma chamber and caused the gases held within the magma to rapidly expand, triggering a sideways blast. The blast, in turn, triggered another landslide that further reduced pressure, initiating an eruption from the volcano's vent. In less than three minutes, the ejected material had descended down the northern flank of the volcano and flowed into the forest below.

Because of their explosive character and high velocity, pyroclastic flows have claimed the lives of many inhabitants near active volcanoes. In 1902, for example, a pyroclastic flow erupting from Mt. Pelée destroyed the nearby port city of St. Pierre on the island of Martinique in the Caribbean, killing 29,000 people (Fig. 4.26). Although a pyroclastic flow resembles an avalanche as it thunders down the flanks of a volcano, it is in fact a glowing cloud containing a mixture of red hot rock fragments, molten

Figure 4.25 Mount St. Helens

The eruption of Mount St. Helens in the Cascade Ranges, May 18, 1980, is a classic example of a pyroclastic eruption. (a,b) An earthquake was generated by rising magma. Following the earthquake, the volcano's north face collapsed in a massive landslide. (c) The landslide released the pressure on the magma within the volcano, which exploded in a violent sideways blast of pyroclastic ash. (d) Photograph of landslide on the northern flank. (e) Subsequent central eruption from the volcanic vent ejected ash to a height of 19 kilometers (12 miles). (f) Resembling flattened straw, entire forests of tree were felled by the lateral blast.

lava, and explosive gases. Pyroclastic flows occur when this mixture becomes too dense to be vented upward, and so collapses under its own weight. Despite their felsic composition, these flows are highly mobile because the presence of gases significantly reduces their viscosity. They can attain velocities exceeding 300 kilometers per hour (190 miles per hour), and may spread as far as 100 kilometers (60 miles) from the site of the eruption.

Because it is full of gas bubbles, magma ejected during a pyroclastic eruption is typically a frothy and highly mobile mixture of liquid, solid, and gas. As the gas escapes into the

atmosphere, small cavities, known as **vesicles** (Fig. 4.27a), are left behind. These vesicles are stretched into tubular holes by the movement of the pyroclastic flow. When this frothy mixture cools, a rock called **pumice** is produced which consists of glass, rock fragments, and stretched vesicles (Fig. 4.27b). Pumice contains so many vesicles that it is often light enough to float on water. When this hot mixture comes to rest, it may retain enough heat that the plastic glass and rock fragments meld together and form a **welded tuff** (Fig. 4.27c). The presence of welded tuffs helps geologists recognize pyroclastic flows in the geological record.

Figure 4.26 Devastating Volcanic Eruptions

St. Pierre on the island of Martinique in the Caribbean was destroyed by a pyroclastic flow from Mt. Pelée in 1902; 29,000 inhabitants were killed.

(a)   (b)   (c)

Figure 4.27   Vesicles and Pumice

(a) When gas trapped within lava escapes into the atmosphere, vesicles may form. (b) As the lava continues to flow, these vesicles may become stretched into tubular holes. This frothy lava cools to form pumice. Pumice is light and can float on water. (c) If the pumice retains heat for a significant period of time, glass and fragments weld together to form a welded tuff.

In contrast to the relatively gentle slopes associated with mafic volcanoes, intermediate to felsic eruptions produce dramatic volcanic edifices. This is because loose pyroclastic material can remain stable on slopes of between 30 and 40 degrees. Around the vent, cinders ejected from the volcano rapidly build a relatively steep-sided cone-shaped edifice known as a **cinder cone** (see Fig. 4.20, Fig. 4.21). Over a two-year period, for example, the volcano Parícutin, near Mexico City, built a cinder cone up to 400 meters (1300 feet) in height (Fig. 4.28).

Earth's most picturesque volcanoes are **composite volcanoes** (also known as stratovolcanoes, Fig. 4.29). These consist of alternating layers of pyroclastic deposits from explosive eruptions and lava flows that erupted from either the central vent or along the flanks of the volcano. This alternation reflects variation in eruption types over time. Each pyroclastic deposit produces a steep slope that is protected from erosion by an overlying layer of lava. Over tens or even hundreds of thousands of years, successive eruptions produce large cone-shaped structures with slopes of between 10 and 25 degrees (Fig. 4.29).

Figure 4.28   Cinder Cone, Mexico

A cinder cone develops in Parícutin, Mexico, 1943–45.

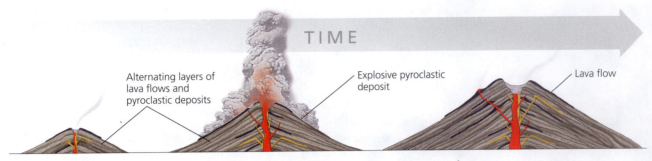

TIME

Alternating layers of
lava flows and
pyroclastic deposits

Explosive pyroclastic
deposit

Lava flow

Figure 4.29   Development of a Composite Volcano

These steep-sided volcanoes consist of alternating eruptions of intermediate to felsic lava flows and more explosive pyroclastic deposits. Over hundreds of thousands of years, alternating eruptions build a large cone with slopes of 10 to 25 degrees.

The majestic landscape of the Cascade Range, in the Pacific Northwest, owes its origins to composite volcanoes such as Mount Shasta, Mount St. Helens, and Mount Rainier (Fig. 4.30).

Gases may slowly build up in the magma chambers of composite volcanoes, eventually triggering violent eruptions in volcanoes that had been dormant for centuries. The devastating eruption of Vesuvius in AD 79, which

(a)

(c)

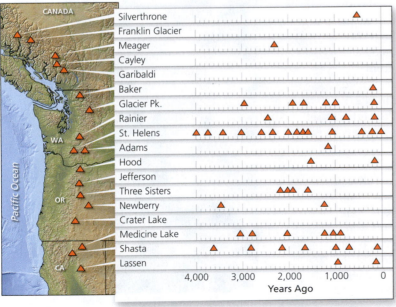

(b)

Figure 4.30   Composite Volcanoes

(a) Mount Rainier, Cascade Range is a composite volcano in the northwestern United States. (b) The Cascade Range stretches more than 1000 kilometers (600 miles) along the coastline of western North America from Mount Garibaldi in British Columbia to Mount Lassen in California. The range includes Crater Lake (see Fig. 4.31), Mount St. Helens, and Mount Rainier. (c) The ancient city of Pompeii was devastated in AD 79 by the eruption of Vesuvius, a composite volcano, in Italy. The incinerated bodies left their shapes as hollows in the pyroclastic tephra and are revealed here as plaster casts.

buried the ancient city of Pompeii, Italy, and killed most of its 20,000 inhabitants (Fig. 4.30c), was such an event.

In contrast to the relatively small calderas found at the summits of shield volcanoes, the calderas associated with violent pyroclastic eruptions can measure up to 100 kilometers (60 miles) across and have very steep walls. In fact, many calderas are so large that some escaped detection by field mapping and were only discovered when high quality aerial photographs and satellite images became widely available. Crater Lake in Oregon, is one of the most famous calderas in North America (Fig. 4.31). It occurs at the summit of Mount Mazama, an extinct composite volcano in the Cascade Range. Measuring 8–10 kilometers (5–6.5 miles) across and nearly 1200 meters (4000 feet)

deep, the caldera was formed about 7700 years ago when a violent eruption ejected between 50 and 70 cubic kilometers (12 and 17 cubic miles) of volcanic material. As a result, the volcanic edifice collapsed to create a caldera, which was subsequently filled with water. Later volcanic activity within the caldera built the small cinder cone known as Wizard Island, at the west end of Crater Lake.

In summary, the composition of magma profoundly affects its viscosity, which, in turn, strongly influences the style of volcanic eruptions.

> **CHECK YOUR UNDERSTANDING**
>
> ⊙ How does the viscosity of a magma influence its ascent to the surface?

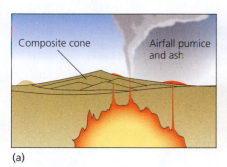
(a)

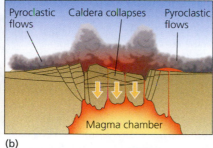
(b)

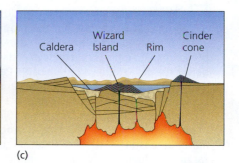
(c)

(d)

Figure 4.31 Origin of Crater Lake, Oregon

(a) Alternations of pyroclastic deposits and lava flows built a composite cone above the magma chamber. (b) About 7700 years ago, a violent eruption partly drained the magma chamber resulting in collapse and formation of the caldera. (c) Later eruptions produced a cinder cone called Wizard Island. (d) Crater Lake formed when the caldera, 8–10 kilometers (5–6 miles) across, subsequently filled with water.

Because of their lower viscosity, eruptions of mafic lavas are generally not explosive and produce broad volcanic structures with gentle slopes. Eruptions of relatively viscous intermediate and felsic magmas, on the other hand, are violent and produce steep-side volcanic edifices. In Section 4.7 we will discuss how these contrasting styles of volcanic activity are elegantly explained by plate tectonics.

4.6 SUMMARY

- Silica-poor (mafic) magmas are fluid because they have low viscosities and tend to erupt as fast-flowing lava flows.

- Silica-rich (felsic) magmas are more viscous, and therefore trap gases. As a result, felsic eruptions are explosive and produce pyroclastic deposits, which pose a much greater threat to society than eruptions that consist of lava alone.

IN DEPTH

Can Explosive Eruptions Be Predicted?

How can explosive volcanic eruptions be predicted? Like most geological phenomena, the difficulty in prediction is that volcanoes are the product of millions of years of geological activity, which does not lend itself to predictions on a timescale of days and weeks. A few days before the devastating eruption of Mt. Pelée, the residents of Martinique noticed that the volcano began emitting steam and fragments. They decided not to evacuate, a decision that had tragic consequences. Since that time, our greater understanding of volcanic processes has led to more successful predictions of volcanic eruptions because we can now identify and interpret the subtle warning signs.

Volcanic activity is a phenomenon that tends to recur, so a region that has not experienced any activity in the recent past will probably not experience any in the foreseeable future. Likewise, we can readily identify and monitor regions that are prone to eruptions in order to identify any unusual activity. Activities that commonly precede eruptions include inflation of the magma chamber beneath the volcano, increased earthquake activity, and subtle changes in the gas content emanating from the volcanic vent (Fig. 4A).

Active volcanoes are fed from an intricate plumbing system consisting of magma chambers. Magma usually ascends toward the surface by flowing from one magma chamber to another in response to the local buildup and release of pressure. For example, geophysical surveys

neat Mt. Vesuvius, Italy, reveal that a shallow magma chamber 4–5 kilometers (2.5–3 miles) below the surface is underlain by a much larger chamber 10–15 kilometers (6–9 miles) deep. As magma rises to the chamber immediately below a volcano, the chamber may become inflated. We can detect this inflation using sensitive tiltmeters and altimeters, tools which respectively measure the slope and elevation (relative to sea level) of Earth's surface. Satellites and global positioning systems (GPS) can also detect the changes in slope and elevation that accompany inflation of a magma chamber.

Earthquake activity of low to medium intensity often accompanies inflation of a magma chamber and the migration of magma from one chamber to another. Hence, an unusual amount of earthquake activity typically precedes an eruption. In addition, subtle changes in the gases emanating from the volcano may occur as new magma invades the magma chamber and mixes with the magma already present. For example, the 1991 eruption of Mount Pinatubo in the Philippines was preceded by small but measurable changes in gas emissions, which indicated the chamber beneath the volcano had been invaded. This, together with increased earthquake activity, was used to successfully predict the eruption, and tens of thousands of Filipinos were safely evacuated from the area closest to the volcano.

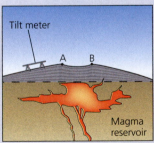
1. Inflation begins

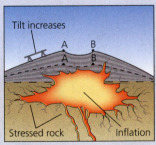
2. Peak inflation

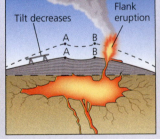
3. Volcano delfates

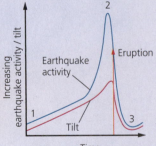

Figure 4A   Predicting Eruptions

Three stages are monitored in a typical Hawaiian eruption. 1. The volcano begins to inflate and earthquake activity increases as the magma chamber swells. 2. Inflation and earthquake activity peak. 3. The volcano erupts and abruptly deflates as earthquake activity diminishes. The graph shows the relationship between earthquake activity and the tilt of the volcano's flanks from stage 1 to stage 3.

4.7 Igneous Rocks and Plate Tectonics

We often think of volcanic eruptions as short-lived catastrophic events. Although this perception is correct, volcanoes themselves can have lifespans of hundreds of thousands or even millions of years, during which time they repeatedly erupt. Within this longer time frame, the theory of plate tectonics provides an explanation for the environment in which magma forms, the means by which it rises toward the surface, and the igneous rocks that are produced when it cools.

PLATE BOUNDARIES AND MAGMA COMPOSITION

Most magma is formed at or adjacent to plate boundaries, such as mid-ocean ridges and volcanic arcs (Fig. 4.32) where the heat necessary for magma to form is supplied.

Mid-Ocean Ridges

The world's mid-ocean ridges are regions of concentrated upwelling of mantle heat where temperatures are high enough to partially melt mantle rocks beneath the ridge crest. Even in these localities, however, there is only sufficient energy to partially melt the source rock—a process we described as partial melting in Section 4.3. A residue of unmelted solid material is left behind. Melting of mantle rocks in this fashion results in the generation of mafic magmas which are rich in iron and magnesium and low in silicon. Melting lasts as long as upwelling continues and

beneath some modern ridges it appears to have lasted for at least a hundred million years.

The magma at mid-ocean ridges is stored in magma chambers just a few kilometers below the ridge crest and is transported to the ridge axis along fractures, typically resulting in fissure eruptions. The extrusion of magma is related to seafloor spreading and the formation of oceanic crust (Fig. 4.33).

Seventy percent of Earth's crust lies beneath the oceans, so basaltic rocks produced by the extrusion of mafic magmas at mid-ocean ridges are the most voluminous igneous rocks on Earth's surface. These magmas undergo much less fractionation than those that ascend through continental crust. There are three main reasons for this. First, the magma chambers beneath the ridge crest are constantly being replenished by mafic magma generated in the mantle below, so that their temperature remains high, which inhibits cooling and fractionation. Second, the magmas rise through relatively thin crust, where they exploit fractures created by seafloor spreading. Because the magmas ascend rapidly to the surface, they have little time or opportunity to fractionate. Third, the magmas rise through oceanic crust, which is similar in composition to the magma. Hence, any contamination with the wall rocks produces relatively minor chemical change. As a result intermediate and felsic magmas at mid-ocean ridges are much more scarce than mafic magma.

> **CHECK YOUR UNDERSTANDING**
>
> ● Why is basalt the dominant igneous rock in oceanic crust?

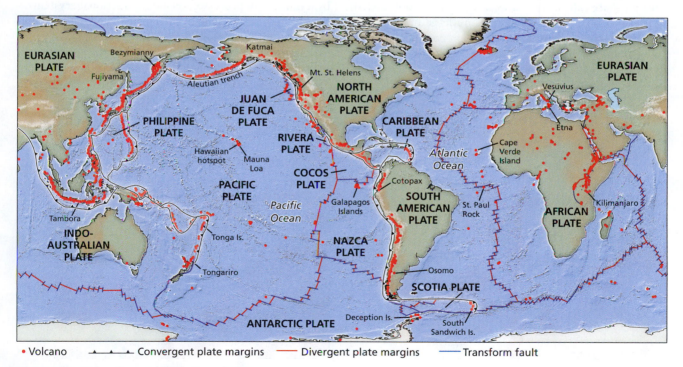

• Volcano   ━▲━▲━ Convergent plate margins   ━━━ Divergent plate margins   ━━━ Transform fault

Figure 4.32 Volcanism and Plate Boundaries

Most modern volcanoes are formed at or near plate boundaries and are located along mid-ocean ridges (divergent plate margins) or near subduction zones (convergent plate margins) like those around the Pacific Ocean.

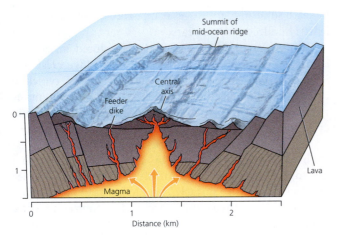

Figure 4.33   Mid-ocean Ridges

Melting beneath mid-ocean ridges results in mafic magma that rises through fractures in the oceanic crust to form fissure eruptions dominated by basalt.

Volcanic Arcs

Magmas are also created above subduction zones, forming **volcanic arcs** such as those of the western Pacific and in the Andes of South America (Fig. 4.34). In general, magma compositions in these settings are much more variable than those at mid-ocean ridges and include significant volumes of intermediate or felsic magma. In addition, there are important differences in the relative proportion of mafic to felsic magmas depending on the thickness and composition of the crust above the subduction zone. Volcanoes built on oceanic crust like those of the Marianas Islands in the western Pacific are called *island arcs*. These arcs are dominated by mafic to intermediate magmas and felsic magmas are subordinate. In contrast, arcs built upon continental crust (like the Andes) are called *continental arcs* and are dominated by intermediate to felsic magmas. Mafic magmas are subordinate.

In all arc environments, water is released from the subduction zone and rises into the overlying mantle as the oceanic slab descends. Water is an efficient transporter of heat and its presence weakens the bonds in minerals and lowers the melting temperature of silicate rocks (Fig. 4.35). As a result, the mantle above the subduction zone partially melts to produce mafic magmas that ascend toward the surface.

Magmas erupted from arc volcanoes, however, undergo much more fractionation than those at mid-ocean ridges, and so they produce volcanic rocks ranging in composition from basalt to rhyolite. In part, this is because the magmas ascend through much thicker crust and have time to cool and fractionate. In addition, the crust in most arc environments is under compression, rather than extension, so that there is less opportunity for magmas to exploit open fractures and to rise rapidly toward the surface. As a result, mafic magmas generated above subduction zones typically cool and evolve toward intermediate and felsic compositions before they reach the surface.

In continental arc environments, like that of the Andes, the ascent of mafic magmas is further impeded by a thick lid of continental crust. This means that an even smaller proportion of the mafic magma successfully reaches the surface without extensive fractionation. In addition, the ascending fluid-charged mafic magma is an efficient transporter of heat, and provides enough energy to partially melt the base of the continental crust. This process produces intermediate to felsic magmas, which may then mix with the mantle-derived mafic magma. Intermediate to felsic rocks in continental arc environments can therefore be produced by a number of processes, including fractional crystallization of a more mafic parent, partial melting of the continental crust, and mixing between mafic and felsic magmas.

Because of the importance of fluids in the generation of magmas above subduction zones, the volatile content of these magmas is significantly higher than that of mid-ocean

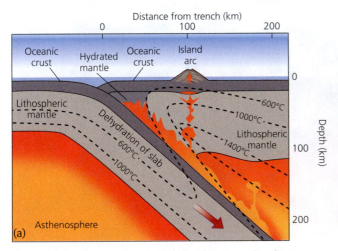
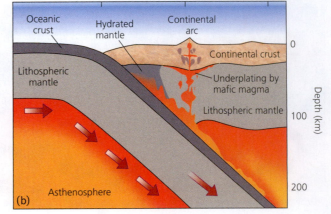

Figure 4.34   Magmatism in Subduction Zone Settings

(a) In oceanic subduction zone settings, magmas tend to be mafic in composition. (b) In continental subduction zone settings, magmas are more varied but intermediate and felsic compositions are dominant.

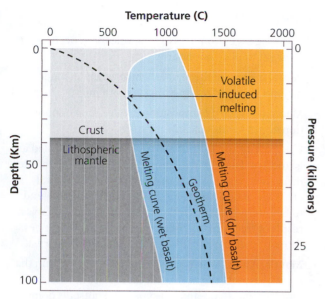

Figure 4.35 Influence of Water on Melting Temperatures

Water is an efficient transporter of heat, allowing magma to form at relatively low temperatures. Note how the infusion of water dramatically lowers the melting temperature required to melt rock of basaltic composition. For example at a 20-kilometer depth, the temperature required to melt dry basalt is about 1200°C, well above the temperature in the crust at that depth (about 700°C; given by the curve labeled "geotherm"). However, at the same depth, the temperature required to melt *wet* basalt is about the same as the temperature of the crust, implying that wet basaltic rocks will begin to melt at that depth while dry basalt will not.

CHECK YOUR UNDERSTANDING

○ Why are felsic rocks more common in volcanic arcs than at mid-ocean ridges?

ridge magmas. The volcanoes that tower above subduction zones, like Mount Pinatubo in the Philippines and Mount St. Helens in the western United States, are consequently much more explosive than those associated with mid-ocean ridges.

HOTSPOTS AND MAGMA COMPOSITION

Not all magmas are generated adjacent to plate boundaries. Some volcanoes, for example, occur in apparent isolation and many are more than a thousand kilometers from the nearest plate boundary (Fig. 4.36). These **hotspots** are the surface expression of thermal upwellings known as **mantle plumes** that originate beneath the plates, possibly at depths of up to 2900 kilometers (1800 miles) below Earth's surface. The material in the plume is solid, but it is soft enough to flow and rises in columns because it is less dense than the rocks of the surrounding mantle.

These hot upwellings induce melting in the mantle that produces mafic magmas. Plumes that occur beneath oceanic domains are overwhelmingly dominated by eruptions of basaltic lava. The well-known eruptions on the island of Hawaii are the end result of such a process. The island itself is the summit of an active shield volcano built up from the ocean floor. In contrast, when plumes occur beneath

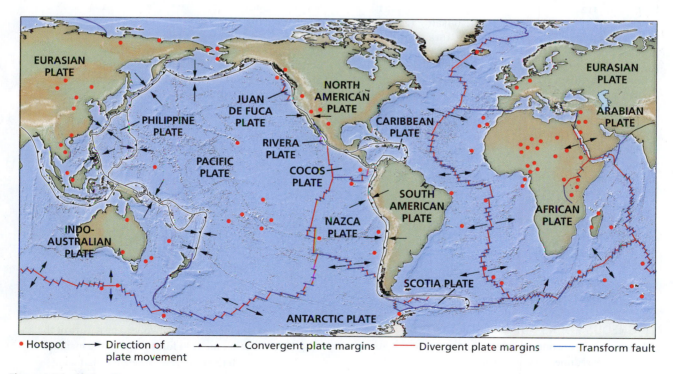

Figure 4.36 Hotspots

Hotspots are the surface expression of upwelling mantle plumes that may originate at the boundary between Earth's mantle and core, some 2900 kilometers (1800 miles) below the surface.

continental crust, as is the case at Yellowstone National Park, both mafic and felsic magmas are generated. The mafic magmas represent the successful penetration of melts from the mantle through the continental lid. The felsic magmas are generally attributed to partial melting of the continental crust caused both by the ascent of mafic magma and by rising heat and fluids above the mantle plume.

In summary, mafic igneous rocks are primarily produced by partial melting of the mantle, either in upwellings beneath mid-ocean ridges, or above subduction zones, or at hotspots where mantle plumes impinge on the lithosphere. Intermediate to felsic rocks are produced by a variety of processes including fractional crystallization of mafic magma, partial melting of the continental crust, and mixing of magmas of mafic and felsic compositions.

> **CHECK YOUR UNDERSTANDING**
>
> ○ What is the relationship between plumes and hotspots?
>
> ○ Compare and contrast the composition of magmas associated with oceanic and continental hotspots.

4.7 SUMMARY

- With the exception of hotspots, volcanoes occur at or adjacent to plate boundaries.

- Partial melting of the mantle, which generates mafic magma, is caused by upwelling beneath mid-ocean ridges.

- Magma of varied composition created above subduction zones forms volcanic arcs.

- If a volcanic arc is built on oceanic crust it is called an island arc. Island arc magmas tend to be mafic to intermediate in composition. Arcs built on continental crust are called continental arcs. Their magmas tend to be intermediate to felsic in composition, either as a result of fractionation of a more mafic parent, or because of melting of the base of the continental crust.

- Mantle plumes, which rise from great depths within the mantle, express themselves on the surface as hotspots. Plumes beneath oceanic crust produce hotspots dominated by mafic magmas; those beneath continental crust produce both mafic and felsic magmas.

Key Terms

aerosols 123
agglomerate 124
andesite 116
assimilation 120
baked contact 110
basalt 116
batholith 112
block 123
blocky lava 121
bomb 123
Bowen's Continuous Reaction Series 117
Bowen's Discontinuous Reaction Series 117
caldera 121
chilled margin 110
cinder cone 127
clast 105
coarse-grained 107
composite volcano 127
contact 110
crystal settling 118
crystallization 103

dike 109
diorite 116
extrusive rock 107
felsic 113
field relationships 110
fine-grained 104
fissure eruption 121
flood basalt 121
flow texture 105
fractional crystallization 118
fractionation 118
gabbro 116
granite 114
groundmass 104
hotspot 133
igneous rocks 100
inclusion 113
intermediate 113
intrusive rock 107
komatiite 116
lahar 125
lapilli 124

lava 100
lava fountain 122
lava lake 121
lava tube 122
mafic 113
magma 100
magma chamber 101
magma mixing 119
mantle plume 133
partial melting 102
peridotite 116
phenocryst 104
photosynthesis 109
pillow lava 123
pluton 107
plutonic 107
porphyritic texture 104
pumice 126
pyroclastic deposit 105
pyroclastic flow 125
residue 102
rhyolite 115
rock cycle 100

ropy lava 121
shield volcano 121
sill 109
source rock 101
stock 111
tephra 123
texture 104
tuff 124
ultramafic 113
vent 105
vent eruption 121
vesicle 126
viscosity 120
volatiles 102
volcanic arc 132
volcanic breccia 124
volcanic glass 104
volcano 101
wall rock 102
welded tuff 126
xenocryst 113
xenolith 113

Key Concepts

4.1 IGNEOUS ROCKS AND THE ROCK CYCLE
- Igneous rocks are one of three rock groups. The other groups are sedimentary and metamorphic rocks.

- Igneous rocks form by the crystallization of a cooling magma and are classified according to their texture, field relationships, and chemical composition.

- The study of igneous rocks involves the nature of the source rock that melts to form magma, the magma's ascent through the crust, and its evolution during crystallization.

4.2 MAGMA FORMATION AND TRANSPORT

- Source rock partially melts to form magma when sufficient heat is added to break the bonds of minerals.
- Magma tends to rise because it is less dense than its surroundings.
- Magma that reaches the surface is called lava and is the fuel of volcanic eruptions. But magma may also cool and crystallize below the surface in magma chambers.

4.3 TEXTURES OF IGNEOUS ROCKS

- The texture of an igneous rock is defined by the size, shape, and arrangement of its mineral grains.
- Lava cools quickly to produce a fine-grained or glassy volcanic rock. Explosive volcanic eruptions produce pyroclastic deposits of ejected material.
- Magmas that crystallize slowly below the surface in magma chambers produce intrusive coarse-grained plutonic rocks that cut across preexisting rock layers.
- Large bodies of intrusive rock are called plutons and may produce contact metamorphism in their wall rock. Tabular bodies form dikes and sills that often have chilled margins and produce baked contacts.
- Extrusive and intrusive igneous rocks are distinguished on the basis of their field relationships.

4.4 CLASSIFYING IGNEOUS ROCKS

- Igneous rocks can be subdivided into felsic, intermediate, mafic, and ultramafic on the basis of silica content.
- Associated extrusive and intrusive rocks often reflect different parts of the same magma plumbing system and so have similar chemical compositions.
- Felsic rocks are high in silica, and low in iron and magnesium. Mafic rocks are low in silica and high in iron and magnesium. Intermediate rocks have compositions between those of felsic and mafic rocks. Ultramafic rocks have the highest abundances of iron and magnesium and are lowest in silicon.

- The mineral content of igneous rocks varies with magma chemistry. Felsic rocks are dominated by quartz, K-feldspar, and Na-rich plagioclase, whereas mafic rocks typically contain pyroxene, olivine, or amphibole, and Ca-rich plagioclase.

4.5 EVOLUTION OF IGNEOUS ROCKS

- According to Bowen's Reaction Series, mafic magmas evolve during cooling as crystals react either discontinuously or continuously with the magma. Magma composition changes as it cools and crystallizes in a process known as fractionation.
- Intermediate to felsic rocks are produced by a variety of processes including fractional crystallization of a mafic parent or mixing of magmas of mafic and felsic compositions.

4.6 VOLCANIC ERUPTIONS

- Mafic magmas have low viscosities and tend to erupt as fast-flowing lava flows.
- Felsic magmas are more viscous and trap gases so felsic eruptions are explosive and produce pyroclastic deposits, which pose a much greater threat to society.
- Because of their different viscosities, mafic eruptions tend to produce broad volcanic structures with gentle slopes, whereas felsic magmas produce steep-side volcanic cones.

4.7 IGNEOUS ROCKS AND PLATE TECTONICS

- Most volcanoes occur at or adjacent to plate boundaries.
- Upwelling beneath mid-ocean ridges induces partial melting of the mantle, generating mafic magma.
- Magmas generated above subduction zones form volcanic arcs and tend to be mafic to intermediate if built on oceanic crust, and intermediate to felsic if built on continental crust.
- Hotspots are the surface expressions of mantle plumes. Plumes that rise beneath oceanic crust produce mafic magmas whereas those that rise beneath continental crust produce mafic and felsic magmas.

Study Questions

1. What is the relationship between the rock cycle and the origin and evolution of igneous rocks?
2. Why does cooling magma crystallize?
3. If a plutonic rock is exposed at the surface, what can you say about the history of the region following crystallization of the magma?
4. Explain the relationship between magma composition, viscosity, and the nature of volcanic eruptions.
5. Explain how crystal settling influences the composition of a cooling mafic magma.
6. Explain why caldera development associated with mafic eruptions differs from that associated with felsic eruptions.
7. If pyroclastic flows are associated with viscous felsic eruptions, why are these flows so mobile?
8. Why are basalt and granite more common than gabbro and rhyolite?
9. How does magma generate and exploit fractures?
10. Explain why magmas produced at subduction zones have more varied compositions than magmas produced at mid-ocean ridges.

5 Weathering and Soils

5.1 Weathering

5.2 Mechanical Weathering

5.3 Chemical Weathering

5.4 Soils

5.5 Weathering, Soils, and Plate Tectonics

► Key Terms

► Key Concepts

► Study Questions

Chemical Weathering

SCIENCE REFRESHER: What Is an Acid? 144

IN DEPTH: The Role of Oxidation in Earth's History 148

LIVING ON EARTH: The Food Chain 152

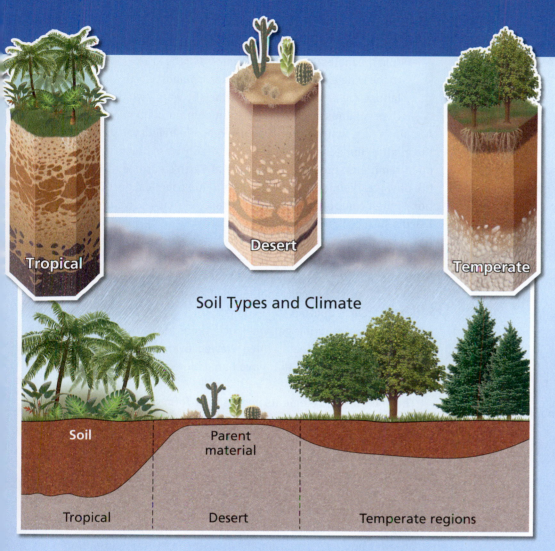

Soil Types and Climate

Tropical   Desert   Temperate

Soil

Parent material

Tropical   Desert   Temperate regions

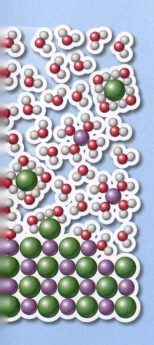

Mechanical Weathering

Learning Objectives

5.1 Recognize that weathering breaks down minerals and rocks and occurs as a result of both mechanical and chemical processes.

5.2 Explain the processes that cause mechanical weathering, which is responsible for rock disintegration.

5.3 Explain the reactions that cause chemical weathering, which is responsible for rock decomposition.

5.4 Describe how soils form and what factors control the development of soil profiles.

5.5 Discuss the role plate tectonic processes play in weathering, soil development, and global temperature variation.

From the majestic, rugged mountains of the Rocky Mountains to the wide expanses of the interior lowlands, North America's landscape is under constant attack by water, wind, and ice. Over millions of years, ancient mountain belts such as the Appalachians, which may have once reached heights that rivaled the mighty Himalayas, have been inexorably worn down into sub-dued remnants. Prolonged exposure at Earth's surface results in the physical disintegration of rocks and the chemical decomposition of their component minerals, a process known as *weathering*. Earth's surface is continually but gradually changing, just as James Hutton envisaged more than 200 years ago (see Section 1.2, "The Birth of Geology"). Weathering involves the breakdown of surface materials; it strips away Earth's surface layers, allowing deeper levels of the crust (typically composed of igneous and metamorphic rocks) to become uplifted and exposed.

Although weathering generally occurs far too slowly for us to easily observe its effects, when viewed over millions of years there are few processes that have had a more profound effect on Earth's evolution.

The minerals and elements necessary to form *soil*—that portion of Earth's surface sediment that can sustain life—are liberated by weathering of bedrock. Soil is one of our most valuable resources. The distribution of fertile soil has profoundly influenced human history since the dawn of civilization.

All rocks can transform into sediments, such as soil, and ultimately into sedimentary rocks by processes such as weathering that operate at or near Earth's surface. In this chapter, we study weathering and soil formation, the first step in the transformation process as a prelude to our examination of sediments and sedimentary rocks in Chapter 6.

5.1 Weathering

Weathering is the in-place (in situ) breakdown of surface materials into loose sediment consisting of broken rock and mineral fragments. It is the first step in the formation of sedimentary rock and involves the following two complementary processes:

- Mechanical weathering
- Chemical weathering

Mechanical weathering *disintegrates* a rock by breaking it into smaller fragments. Disintegration is a physical process and does not by itself cause changes in the rock's chemical composition. In contrast, **chemical weathering** involves chemical reactions that break down or *decompose* the unstable minerals in a rock and convert those minerals into stable products. As such, chemical weathering has the most impact, but in practice, mechanical and chemical weathering work hand in hand. For example, the physical disintegration of a rock into smaller pieces enhances chemical weathering by increasing the surface area where decomposition by chemical reactions can take place.

5.1 SUMMARY

- Weathering is the process by which rocks and minerals break down at Earth's surface and is the first step in the formation of sedimentary rock.

- Weathering takes place in two ways, mechanical and chemical; chemical weathering has a greater effect.

5.2 Mechanical Weathering

Mechanical weathering disintegrates rocks without causing any change in the chemical composition of the rocks (Fig. 5.1). This disintegration produces loose and disconnected fragments from what was originally coherent rock. The fragments are called **clasts** from the Greek "*klastos*," meaning broken. Several physical processes, acting individually or in concert, reduce rocks to small fragments. Human activities also play a role in mechanical weathering. In this section, we introduce and discuss these processes in detail.

JOINTS

Almost all rock outcrops contain regularly spaced cracks, collectively known as **joints**. Joints form in a wide variety of ways. For example, when basaltic lava is extruded onto

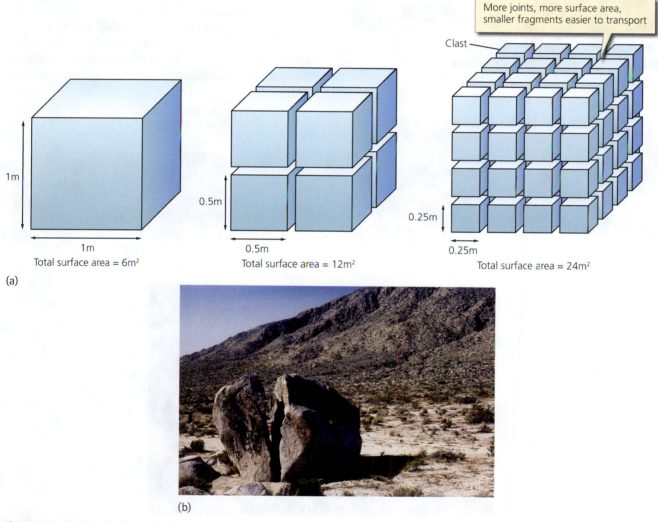

More joints, more surface area, smaller fragments easier to transport

Clast

1m

1m

Total surface area = 6m²

(a)

0.5m

0.5m

Total surface area = 12m²

0.25m

0.25m

Total surface area = 24m²

(b)

Figure 5.1   Mechanical Weathering

(a) Mechanical weathering reduces rock to smaller and smaller fragments, or clasts, which increases the total surface area of the rock and increases the effects of chemical weathering. (b) Mechanical weathering of a nearby granite outcrop has produced large fragments (foreground), which themselves have undergone mechanical weathering, breaking into smaller fragments and producing loose sand at the base of the boulder in this photograph.

Earth's surface it cools and solidifies very rapidly. As the temperature of the solid rock decreases, the rock shrinks and cracks appear at the surface that propagate downward to form joints, dividing the rock into very distinctive columns up to 30 meters in height, a feature known as **columnar joints** (Fig. 5.2).

Figure 5.2   Formation of Columnar Joints

The columnar joints at Giant's Causeway in Northern Ireland were formed by cooling of basalt lava.

The presence of joints greatly facilitates physical and chemical weathering processes.

As we will learn in Chapter 10, joints form when rocks experience stresses, which can result from a wide variety of causes. As noted above, common causes for joint formation include decreasing temperature as a rock cools. Granitic magma typically solidifies at temperatures between 650°C and 700°C, but the rock produced continues to cool until it attains the same temperature as the surrounding host rock. As it cools, the granite rock contracts and regularly spaced joints are the most common record of that cooling history.

Another common cause of joint formation is a decrease in pressure, which occurs when a rock is uplifted toward the surface. Miners will vouch for the fact that buried rocks are under pressure. As material is removed from a mine, the reduced support can result in reduction in pressure that causes rocks to burst away from the walls. In fact, rock bursts are one of the leading causes of mining accidents. Similarly, igneous and metamorphic rocks that crystallize well below Earth's surface are subject to the enormous pressure exerted by the weight of the overlying rocks. Buried igneous rocks, such as granite, expand as they are uplifted and the pressure is reduced as the overlying rocks are eroded. As a result, they fracture into onionskin-like joints, known as **exfoliation joints**, that are broadly parallel to Earth's surface (Fig. 5.3a).

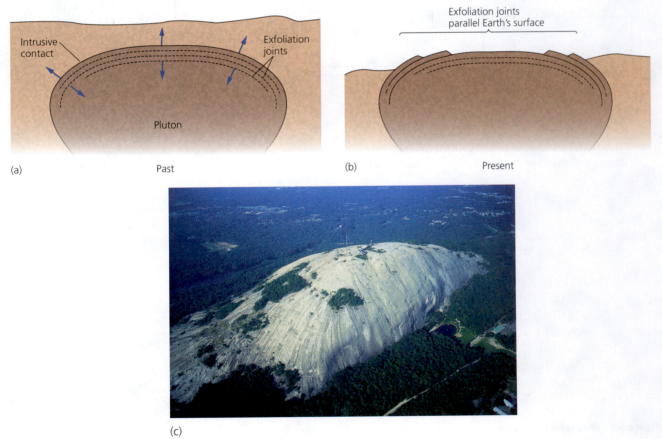

Figure 5.3   Exfoliation Joints

(a, b) Exfoliation joints in granite are roughly parallel to Earth's surface. The blue arrows denote expansion of rock caused by a decrease in pressure as the granite is uplifted toward the surface and the weight of the overlying rocks is removed by erosion. (c) Stone Mountain, Georgia, is a large granite dome that formed as slabs of jointed rock broke off.

These slabs of rock are unstable, and break off. Eventually smoothly rounded dome-shaped masses of rock form (Fig. 5.3b).

In summary, joints typically form from a reduction in temperature as rocks cool and from a reduction in pressure as rocks are uplifted. As we discuss in the following material, the presence of joints in nearly all surface outcrops facilitates a wide variety of physical and chemical weathering processes.

FROST WEDGING

When water freezes to form ice, it expands in volume by about 9 percent. This is why freezing water bursts pipes and why ice is less dense than liquid water and floats. This expansion can also cause rocks to mechanically disintegrate in a process known as **frost wedging**. In many regions, frost wedging is the most important agent of mechanical weathering. The process begins when water percolates into cavities, cracks, and joints in rocks (Fig. 5.4). If the air temperature then falls below the freezing point (0°C/32°F), the water starts to freeze and forms ice, first along the surface where it is in direct contact with the cold air and then below the surface as the cold penetrates downward. Because the ice first formed at the surface, the ice below cannot expand upward. As a result it expands outward, and the force it exerts enlarges the crack or cavity in the rock and widens the joint. When the air temperature rises and the ice thaws, these cracks may again fill with water, and widen still further with each new freeze.

CRYSTAL WEDGING

Water is a strong solvent; it has the ability to dissolve, or partly dissolve, many chemical compounds. Because of this, water that lodges in openings in rocks is not pure. It contains abundant dissolved salts. This situation is especially common in coastal environments, where rocks are bathed in sea spray that can invade cavities, cracks, and joints. If this water evaporates, salt crystals form and grow inward from the walls of the cavity or joint. If these crystals make contact either with the opposing wall, or with crystals that are attached to it, their growth can apply enough pressure to the sides of the cavity or joint to pry the walls apart (Fig. 5.5a).

Cleopatra's Needle, a granite obelisk that is now located in New York's Central Park, provides an impressive illustration of the effect of *crystal wedging* on weathering. The obelisk stood in Egypt for more than 3500 years, where its delicately engraved inscriptions remained well preserved in the warm, dry climate. Unfortunately, before it was moved to

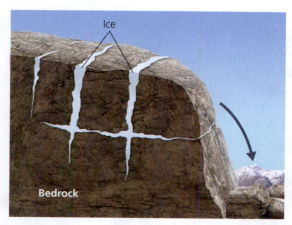

Figure 5.4 Frost Wedging

(a) Frost wedging occurs when water seeps into cracks, freezes, and expands to wedge the rock apart. (b) Loosened rocks eventually tumble downslope and accumulate at the base of the slope.

New York in 1879, it was stored at a site that let salty water invade the column, allowing salt crystals to grow. In New York's more humid climate, the salt crystals absorbed moisture and expanded, destroying the inscriptions in the process (Fig. 5.5b).

THERMAL EXPANSION AND CONTRACTION

Many field geologists believe that the shattered pebbles found in deserts are created by repeated *thermal expansion* and *contraction* associated with daily temperature fluctuations of up to 50°C (Fig. 5.6). Each mineral in a rock has a different susceptibility to expansion and contraction, depending on its bonding structure. For example, quartz expands three times more than feldspar for the same increase in temperature. When granite outcrops (which contain both feldspar and quartz) are heated,

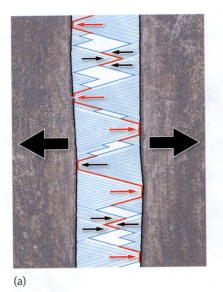

(a)   (b)   (c)

Figure 5.5   Crystal Wedging and Cleopatra's Needle

(a) When water that is lodged in cracks in rocks evaporates, the dissolved salts become more concentrated, eventually leading to the formation of crystals, which grow outward from the walls of the crack. Crystal wedging begins when these crystals come in contact with the opposite wall (red arrows) or start pushing against each other (smaller black arrows). The sum of all these small forces wedges the crack further apart. (b–c) In 1874, Cleopatra's Needle, a monument made of granite, was removed from Alexandria, Egypt (left), where it had stood for 3500 years. After 75 years in New York's Central Park (right), the inscriptions have suffered severe damage.

Expansion cracks form parallel
to the boulder surface due to
repeated expansion and contraction

Rock
fragments/clasts

(a)   (b)

Figure 5.6   Thermal Expansion and Contraction

(a) Extreme daily temperature fluctuations in a desert climate result in repeated expansion and contraction that can cause pebbles to shatter. (b) As the warmed surface rock of a boulder expands, it pushes away from the cooler rock's interior, causing the formation of cracks that are parallel to the boulder surface. With repeated heating and cooling, the surface layer crumbles to form clasts, and the newly exposed rock surface is subjected to repeated cycles of temperature fluctuation.

the quartz grains push against and may displace the neighboring feldspar grains. Because silicate rocks, like granites, are poor conductors of heat, this process takes place only within a few centimeters of the surface. As a result, the warmed surface rock may expand and push away from the cooler rock below, resulting in the formation of a crack that is parallel to the surface (Fig. 5.6b). With repeated heating and cooling, the surface layer may split apart from the main body of rock and crumble. The process continues as the newly exposed rock surface is subjected to the same conditions of repeated heating and cooling.

Initially, the interpretation of these field observations was not supported by simple laboratory experiments. Repeated heating and cooling of rocks, thought to simulate nearly 250 years of thermal expansion and contraction, failed to result in any measurable weathering. However, when the experiments were repeated with the addition of moisture during the cooling cycles (to simulate the

CHECK YOUR UNDERSTANDING

● How are the shattered pebbles found in desert regions thought to form?

presence of morning dew), the modified experiments did produce cracks near the rock surface, suggesting that moisture plays an important role in thermal expansion and contraction.

ORGANIC PROCESSES

Organic processes rank second to frost wedging as the most important agent of mechanical weathering (Fig. 5.7). As plants search for nutrients and moisture, their roots often invade cracks in rocks. As roots expand, they enlarge the cracks and exert a force powerful enough to disintegrate the rock. Burrowing activity by animals, particularly worms and gophers, is also effective in opening cracks, moving rock fragments, and overturning soil. Some regions have up to 50,000 earthworms per acre. In addition, experiments show that clays rich in microorganisms (i.e., organisms that can only be seen with the aid of a microscope) decompose twice as rapidly as sterile, lifeless clays.

> **CHECK YOUR UNDERSTANDING**
>
> ● In what ways do organic processes promote mechanical weathering?

HUMAN ACTIVITIES

Many forms of human activity, such as quarry excavation, road construction, and land development, also contribute to mechanical weathering. Activities such as digging or blasting uncover and fracture rocks that otherwise would be protected from weathering. While human-assisted weathering is less effective than that of burrowing organisms, it is one that has increased with economic development and our exponentially rising human population.

5.2 SUMMARY

* Mechanical weathering disintegrates rock by breaking it down into smaller fragments.
* The main processes of mechanical weathering are frost wedging, organic processes, crystal wedging, thermal expansion and contraction, and human activities.

5.3 Chemical Weathering

In contrast to mechanical weathering, which physically *disintegrates* rocks, chemical weathering *decomposes* surface rocks and minerals through chemical reactions. Owing to the extraordinary dissolving power of water, chemical weathering is by far the more important of the two processes. Although pure water is an excellent solvent, rainwater is an even better one because interactions with atmospheric gases and pollutants cause it to be mildly acidic (see Science Refresher: **What Is an Acid?** on page 144). For example, atmospheric carbon dioxide dissolves in water to produce carbonic acid according to the following reaction:

$$\underset{\text{water}}{H_2O} + \underset{\text{carbon dioxide}}{CO_2} \rightarrow \underset{\text{carbonic acid}}{H_2CO_3}$$

Carbonic acid dissociates (or splits) into hydrogen ions (H^+) and bicarbonate ions (HCO_3^-), and these ions are held in solution within the water. Rainwater containing high concentrations of hydrogen ions (H^+) is known as **acid rain**, and the higher the concentration of hydrogen ions, the greater the potency of acid rain as a solvent. Indeed, acid rain has attained notoriety for its potential to adversely affect a wide variety of things, from delicate ecosystems to public buildings and monuments.

Minerals have varying susceptibilities to chemical weathering and therefore do not dissolve at the same rate (Table 5.1). Some minerals completely dissolve, whereas others react with water or oxygen to form new minerals. In general, this variable chemical behavior reflects differences in the strength of the bonds within these

Joints widen as root grows

Soil

Decay of organic material releases acids that promote chemical weathering.

Figure 5.7   Organic Processes

(a) By enlarging cracks and fractures in rocks, the growth of this tree contributes to mechanical weathering. (b) In Ciudad Victoria, Mexico, a sidewalk is destroyed by growth of tree roots.

(a)

(b)

SCIENCE REFRESHER

What Is an Acid?

Acids are very common chemical substances that have a set of characteristic properties. Acids have a pungent odor and a sour taste, they corrode metals, and they turn some vegetable dyes from blue to red. We are most familiar with the acids in our stomach (hydrochloric acid), in vinegar (acetic acid), and the lactic acid that occurs in sour milk, yogurt, and cottage cheese.

Chemists use a number of definitions for acids, but the one most useful to us is that of Swedish chemist Svante Arrhenius. According to Arrhenius, an acid is a substance that can release a proton, or hydrogen ion (H^+). For example, when hydrogen chloride (HCl) dissolves in water it dissociates (or splits) into hydrogen ions (H^+) and chloride ions (Cl^-). Hydrogen ions are very reactive, so the strength of an acid depends on how readily it releases its hydrogen ions into solution. Strong acids include hydrochloric acid and sulfuric acid, which readily release hydrogen ions in water. In fact, it is the availability of highly reactive hydrogen ions that gives strong acids their pungent odor and their ability to attack metals. Other acids, such as acetic acid, are incompletely ionized, and are therefore weaker, because there is a smaller concentration of ionized hydrogen (H^+) available in the solution.

Although rainwater is naturally acidic from the interaction of water with atmospheric gases, evidence shows that in some locations it is excessively so, and this is generally attributed to human activities. For example, carbon dioxide produced by the burning of fossil fuels reacts with rainwater to form carbonic acid (H_2CO_3). Similarly, the nitrogen gases produced by internal combustion engines and the sulfur gases produced by industry react with rainwater to form nitric acid (HNO_3) and sulfuric acid (H_2SO_4), respectively. As a result, industrialization is often held responsible for an increase in the acidity of rainwater.

TABLE 5.1 SUSCEPTIBILITY OF MINERALS TO CHEMICAL WEATHERING

Halite	Least stable		Non-silicates
Gypsum-anhydrite			
Pyrite			
Calcite	Dissolution		
Dolomite			
Volcanic glass			
Olivine			
Ca-plagioclase			
Pyroxenes			
Ca-Na plagioclase		Increasing stability	Silicates
Amphiboles			
Na-plagioclase	Hydrolysis		
Biotite			
K-feldspar			
Muscovite			
Quartz			
Kaolinite (clay material)			
Hematite	Most stable		

Note: Minerals are listed in order of increasing resistance to weathering. (Exact positions for some minerals can change due to effects of grain size, climate, etc.)

minerals. In this section we discuss three important types of chemical weathering processes: dissolution, hydrolysis, and oxidation. We also discuss the processes of hydration and other factors that affect chemical weathering.

DISSOLUTION

Recall from Chapter 3 that nonsilicate minerals, especially the chlorides, carbonates, and sulfates, are dominated by weaker ionic bonds and are highly soluble, especially when the water is acidic and contains hydrogen ions. Because of this, minerals such as halite (NaCl), calcite ($CaCO_3$), dolomite ($CaMg(CO_3)_2$), and gypsum ($CaSO_5 \cdot 2H_2O$) dissolve readily, a process known as **dissolution**. Halite (which is also referred to as common salt) is the mineral most susceptible to chemical weathering and its crystal structure readily breaks down as the positive sodium ions are attracted to the negative ends of the water molecule, while the negative chloride ions are attracted to the positive end of the water molecule (Fig. 5.8). These ions are then released from the crystal structure and carried away in solution. As we will learn in Chapter 12, such charges exist in water molecules (H_2O) because their structure is V-shaped, with the negative oxygen ion at the base and a hydrogen ion at each tip. As a result, the electrical charges are not equally distributed, one side of the molecule being slightly negative and the other slightly positive. Such molecules are said to be *polar*, and the resulting attraction between their positive and negative sides is called a *hydrogen bond*.

Calcite, the dominant mineral in limestone, breaks down according to the following reaction:

$$CaCO_3 + H^+ + HCO_3^- \rightarrow Ca^{2+} + 2HCO_3^{2-}$$

calcite   hydrogen   bicarbonate   calcium   bicarbonate
          ions   ions   ions   ions

This reaction liberates calcium from the calcite mineral structure. There are no solid products, indicating that the mineral has completely dissolved. A limestone outcrop is therefore characterized by countless small grooves (Fig. 5.9a) that remain after calcite has dissolved. In addition, water

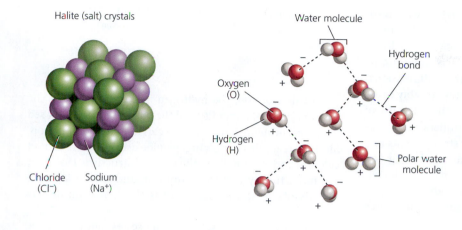

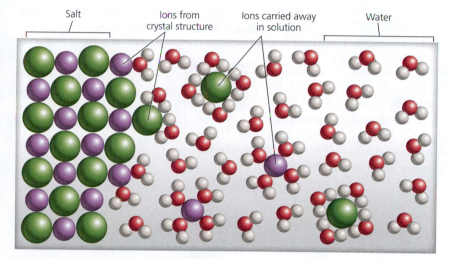

Figure 5.8   Dissolution of Halite (Salt)

Halite is the mineral most susceptible to chemical weathering; it readily dissolves because of the interaction between highly polarized water molecules, and the ionic crystal structure of salt. The salt structure breaks down as the positive sodium ions are attracted to the negative end of the water molecule, while the negative chloride ions are attracted to the positive end of the water molecule. As the halite dissolves, the ions are released from the crystal structure and carried away in solution.

seeping into joints in the limestone causes the joints to widen by dissolution into troughs and, if the water sinks into the subsurface, potholes and caves can form (see Section 14.4: The Dissolving Power of Groundwater).

Dissolution is especially important in regions where the surface or near-surface rocks are dominated by relatively soluble deposits of rock salt, gypsum, or limestone. This is the case in the central United States, Florida, and eastern Canada, where dissolution produces a variety of landforms including caves, and circular depressions in Earth's surface, known as *sinkholes*. In fact, gypsum and rock salt are so soluble that landforms such as these can form in a matter of days (Fig. 5.9b).

> ## CHECK YOUR UNDERSTANDING
>
> ❍ Name two distinctive landforms that are the result of chemical weathering.

HYDROLYSIS

As we mentioned in our discussion of dissolution and the breakdown of halite, the hydrogen ions present in water can extract ions from a mineral's crystal structure. This process is known as **hydrolysis**. In contrast to dissolution, which totally dissolves a mineral and removes its component ions in solution (see Figure 5.8), hydrolysis is a reaction that forms a new mineral and only liberates the more loosely bound ions from the mineral's

(a)

(b)

Figure 5.9   Dissolution

(a) Chemical weathering of limestone, which is dominated by the mineral calcite, is evident in the Burren, Ireland. (b) These sinkholes seen from the air near Hutchinson, Kansas, were produced by the sudden collapse of sediment and bedrock due to dissolution of salt in the subsurface. In this case, dissolution was caused by solution mining (see Chapter 18). Some of the collapse occurred beneath railway lines, which are visible in the photo.

structure. These liberated ions are then carried away in solution. But how does this happen? Most of Earth's crust is made up of silicate minerals, the structure of which is dominated by the silicate tetrahedron, either in isolation or arranged in chains, sheets, or complex three-dimensional forms. The tetrahedral structure itself is remarkably resistant to chemical weathering because of the strength of the silicate (Si-O) bond. Some silicate minerals, such as quartz and muscovite, are dominated by strong bonds and are therefore highly resistant to chemical weathering (see Table 5.1). However, for others, the bonds that link the tetrahedra to ions such as those of calcium, sodium, magnesium and potassium are much weaker and so are much more susceptible to attack.

The most important example of hydrolysis involves the feldspars, one of the most important families of silicate minerals. Although feldspars have a mineral structure dominated by silicate tetrahedra with strong internal bonds, these tetrahedra are only weakly bonded to soluble potassium, sodium, or calcium ions (Fig. 5.10a). Water from the atmosphere and stored in soil is generally acidic (see Science Refresher: **What Is an Acid?** on page 144) and preferentially attacks these weak layers, releasing ions into solution. As a result, the feldspar structure breaks down and clay minerals form. For example, potassium feldspar breaks down according to the reaction

$$2KAlSi_3O_8 + 2H^+ + 9H_2O = Al_2Si_2O_5(OH)_4$$

potassium   hydrogen   water   clay mineral
feldspar   ions

$$+ 4H_4SiO_4 + 2K^{2+}$$

silicic acid   potassium
ions

In this reaction, the hydrogen ions are present in the water because of its acidity and the potassium ions released from the mineral structure are carried away in solution. Similar reactions involving plagioclase feldspar extract calcium and sodium from the mineral structure. However, the more tightly bonded aluminum and silicon ions are not dissolved, so the end result is the formation of clay minerals (Fig. 5.10b).

Hydrolysis reactions, such as those responsible for the formation of clay minerals, have far-reaching implications. Feldspar is the most common mineral on the continents, so the ions carried off in solution as a result of the weathering of feldspar are a major component of streams that eventually find their way to the sea. Because of this, the process of chemical weathering is largely responsible for the salinity of the oceans and therefore has played a fundamental role in the evolution of life.

CHECK YOUR UNDERSTANDING

◯ Why do carbonate minerals undergo dissolution, whereas silicates undergo hydrolysis?

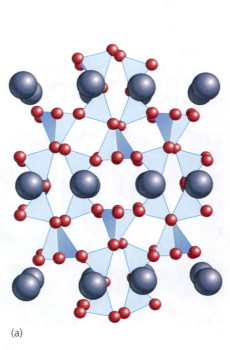

(a)

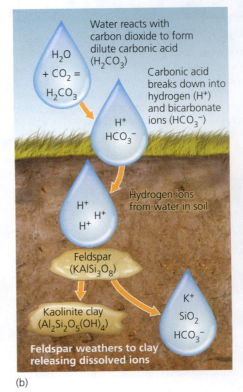

Water reacts with carbon dioxide to form dilute carbonic acid (H_2CO_3)

$H_2O + CO_2 = H_2CO_3$

Carbonic acid breaks down into hydrogen (H^+) and bicarbonate ions (HCO_3^-)

H^+ HCO_3^-

Hydrogen ions from water in soil

H^+ H^+ H^+

Feldspar ($KAlSi_3O_8$)

Kaolinite clay ($Al_2Si_2O_5(OH)_4$)

K^+ SiO_2 HCO_3^-

Feldspar weathers to clay releasing dissolved ions

(b)

Figure 5.10   Hydrolysis

(a) This schematic diagram shows the structure of a silicate mineral such as feldspar. The silicate tetrahedra are resistant to chemical attack. However, ions such as potassium, sodium, and calcium are located between these layers and are more readily attacked and removed in solution. (b) Feldspar in soil undergoes hydrolysis when attacked by acidic water and weathers to form kaolinite clay.

OXIDATION

Many chemical reactions between compounds involve the transfer of electrons. A compound is *oxidized* if it loses electrons during a chemical reaction, as occurs, for example, in the formation of oxides from reactions of cations with oxygen. When this occurs we say **oxidation** has taken place. Similarly, a compound is *reduced* if it gains electrons during a chemical reaction and we say **reduction** has occurred. A reaction that results in the loss of electrons by one compound and the simultaneous gain of electrons by another is known as an *oxidation-reduction* or *redox* reaction.

Minerals with abundant iron are especially susceptible to redox reactions because iron occurs naturally in both reduced and oxidized forms. In most rock-forming minerals, the reduced form of iron (Fe^{2+}) is more common than the oxidized form (Fe^{3+}), which contains one less electron. As a result, many rock-forming minerals readily react with oxygen in the atmosphere according to the reaction

$$\underset{\substack{\text{within}\\\text{mineral}}}{4Fe^{2+}} + \underset{\text{atmosphere}}{3O_2} \rightarrow \underset{\substack{\text{iron}\\\text{oxide (in } Fe^{3+}\text{ form)}}}{2Fe_2O_3} \qquad \text{(i)}$$

In sedimentary rocks, the effect of this oxidation reaction is easy to spot because the reaction results in iron oxide minerals such as hematite, which impart a characteristic red color to the rock (Fig. 5.11).

The oxidation of pyrite, an iron sulfide mineral that is present in small concentrations in many rocks, has recently received a lot of attention from environmental geologists because a reaction involving a combination of oxygen and

Figure 5.12   Acid Rock Drainage

Acid rock drainage often results in a severe water discoloration and occurs where sulfide minerals (such as pyrite) are oxidized. This redox reaction liberates sulfur from the crystal lattice. The sulfur then combines with water to form sulfuric acid, which commonly finds its way into streams. Acid rock drainage is especially prevalent in regions with abundant sulfide minerals, or near *mine tailings* (sites where waste rocks from a mine are dumped).

water liberates sulfur from the crystal lattice and forms sulfuric acid (H_2SO_4). Regions that have rocks with a lot of pyrite are therefore very prone to *acid rock drainage* (Fig. 5.12). A simplified version of the reaction is

$$\underset{\text{Pyrite}}{2FeS_2} + 7O_2 + 2H_2O = 2Fe^{2+} + \underset{\text{sulfate}}{4SO_4^{2-}} + 4H^+ \qquad \text{(ii)}$$

The Fe^{2+} ions liberated by redox reaction (ii) may then be oxidized according to reaction (i) shown previously. The extent of acid rock drainage is often amplified in current or former mining districts because many ore minerals are sulfides and are prone to a similar sequence of reactions.

The terms *oxidation* and *reduction* are used to describe *all* reactions in which the transfer of electrons occurs. Redox reaction (ii) is an oxidation reaction because the iron gives up an electron as it converts from Fe^{2+} to Fe^{3+}. Although strictly speaking oxygen need not be present for an oxidation reaction to occur, in practice, the abundance of oxygen in our atmosphere and oceans makes oxygen the most important agent of oxidation.

Oxidation occurs very slowly in dry environments. However, the rate of oxidation greatly increases in the presence of water. In water, for example, hydrolysis extracts reduced iron from silicate minerals such as olivine and pyroxene. The iron is then readily oxidized to produce hydrated iron oxide minerals, such as *limonite* ($FeO(OH)$), which is better known as *rust* (see Figure 5.11). Indeed, oxidation reactions are responsible for the rusty color of most outcrops and roadcuts adjacent to highways.

Figure 5.11   Oxidized Rock

A succession of red clastic rocks border the Murchison River in Western Australia. The color is due to the oxidation of iron rich minerals in the rock, which form red minerals such as hematite and limonite (a form of rust).

CHECK YOUR UNDERSTANDING

> What is oxidation and what role does it play in chemical weathering?

The Role of Oxidation in Earth's History

Oxidation reactions have played a significant role in Earth history. In the modern world, oxygen is primarily produced by organisms as a by-product of **photosynthesis**—the conversion of sunlight into the chemical energy used to fuel the activity of plants and other organisms. In the distant past there was less oxygen in Earth's atmosphere. Photosynthetic marine organisms probably existed more than 3.5 billion years ago, but the oxygen they produced was absorbed by soluble iron salts dissolved in ocean waters (Fig. 5A). The reaction produced insoluble hydrated iron oxides, such as ferric hydroxide, which sank to the seabed to form rocks called *banded iron formations* (Fig. 5B). Because it was consumed in these reactions with the salts in ocean water, oxygen was not released to the atmosphere. The atmosphere was consequently devoid of oxygen, and the evolution of life progressed in an oxygen-free (*anaerobic*) setting. About 2 billion years ago, however, the rate of oxygen production finally exceeded the ocean's capacity to absorb it and oxygen leaked into the atmosphere.

Evidence of the presence of oxygen in the atmosphere is recorded in 2 billion-year-old sedimentary rocks that were deposited in continental environments. Exposure to atmospheric oxygen results in oxidation reactions that produce minerals such as hematite and limonite. These minerals contain oxidized iron (Fe^{3+}), giving the rocks that contain them a rusty red color (see Fig. 5.11). These rocks are collectively called *continental red beds*. Continental red beds first appeared in the geologic record about 2 billion years ago. Their appearance signals the onset of oxygen in the atmosphere and broadly coincides with the disappearance of banded iron formations, which were no longer produced in the oceans once the oceans became fully oxygenated.

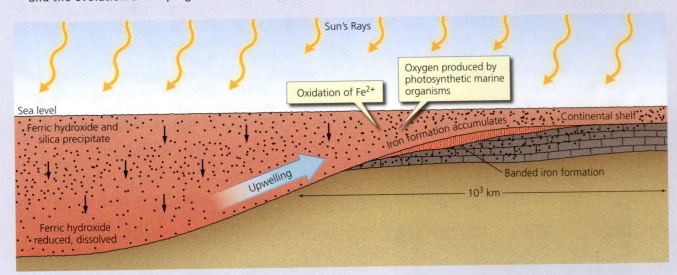

Figure 5A   Banded Iron Formations

Oxygen produced by photosynthetic organisms in the oceans oxidizes iron (Fe) from Fe^{2+} (ferrous) to Fe^{3+} (ferric). Fe^{2+} is soluble but Fe^{3+} is insoluble and forms compounds that precipitate. Dissolved Fe^{2+} is transported in upwelling currents toward the continental shelf, where it is oxidized to Fe^{3+} and precipitates to form banded iron formations.

Figure 5B   Banded Iron Formation

This banded iron formation in Western Australia formed before free oxygen first occurred in the atmosphere, about 2.0 billion years ago.

Oxidation reactions that involve water played a vital role in the evolution of life on Earth. For early life forms, free oxygen in the atmosphere and oceans was toxic. However, much of this oxygen was absorbed by Fe^{2+} ions, which were converted to minerals containing Fe^{3+} (see In Depth: **The Role of Oxidation in Earth's History** on page 148).

HYDRATION AND OTHER FACTORS THAT AFFECT CHEMICAL WEATHERING

If you have ever worked with ceramics, you know that clay minerals swell when you mix them with water. This is because clay minerals undergo **hydration**, that is, they absorb water into their crystal structure, resulting in expansion. Rocks that contain abundant clay minerals are weakened by this expansion.

The degree of chemical weathering is influenced by many other factors in addition to hydration. The most important of these are mechanical weathering, time, and climate. As noted earlier, because mechanical weathering disintegrates rock into smaller pieces, it provides a greater surface area on which chemical reactions can take place (see Figure 5.1). The extent of weathering also depends on the amount of time that the surface of bedrock or sediment is exposed. Outcrops along new highways tend to be less weathered than those along old highways, and these, in turn, are less weathered than nearby hillside outcrops of similar bedrock.

Chemical weathering is more intense in tropical and temperate regions, where there is abundant rainfall, than it is in desert and polar regions where net precipitation is limited. And because warm water is a stronger solvent than cold water, chemical weathering is especially pronounced in tropical regions. In fact, as we will learn in Chapter 18,

intense weathering of feldspar in tropical regions can lead to the formation of the aluminum ore, *bauxite* ($Al_2O_3 \cdot nH_2O$) (Fig. 5.13). Because aluminum is insoluble in water, it becomes concentrated in weathered sediment and soil when all other cations, including silicon, are carried away in solution.

Weathering is the necessary first step in providing solutions and loose material at Earth's surface. In Section 5.4, we will see how this loose material develops into the soils that nourish the ecosystems on land.

CHECK YOUR UNDERSTANDING

○ Why is time an important factor in influencing the extent of weathering in a given region?

5.3 SUMMARY

- Chemical reactions cause chemical weathering that decompose minerals in a rock.
- The main processes of chemical weathering are dissolution, hydrolysis, and oxidation. By dissolution, minerals with relatively weak bonds dissolve. In hydrolysis, soluble ions are removed from a mineral by interaction with water. In oxidation, cations in minerals combine with oxygen in air and water.
- The degree of chemical weathering rocks undergo is influenced by hydration, mechanical weathering, time, and climate.

5.4 Soils

Weathering produces a layer of mineral and rock fragments that overlies the bedrock and covers many parts of the land surface (Fig. 5.14). This loose layer of rock and mineral fragments is known as a **regolith** (from the Greek for "rock blanket"). **Soil** is the portion of this loose material that is capable of supporting plants and their root systems. The minerals and elements necessary to support life are liberated by weathering of bedrock. In this way, the processes occurring in soils form a vital bridge between the inorganic and organic worlds.

Two main types of soils exist:

- Residual soil
- Transported soil

Residual soil develops from the weathering of the bedrock directly beneath it. As a result, the weathered material in residual soil has not been subjected to significant transport. The mineral and rock material of **transported soil**, on the other hand, does not come from the underlying bedrock. Instead, this material has been transported from elsewhere by mobile agents such as running water, ice, or wind. In most instances, we

(a)

(b)

Figure 5.13 Bauxite
Intense chemical weathering of feldspar in tropical regions can lead to the formation of an aluminum ore called bauxite. Shown here are: (a) a typical outcrop of bauxite and (b) a sample of bauxite.

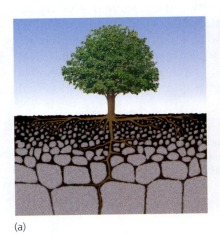

(a)   (b)

Figure 5.14   Regolith
A sketch (a) and field example (b) illustrate how weathering produces a loose layer of rock and mineral fragments, known as regolith, that overlies bedrock.

can distinguish residual soils from transported soils because residual soils contain fragments of the underlying bedrock.

Both residual and transported soils typically consist of disintegrated and decomposed rock, the decayed remains of animal and plant life, a variety of microorganisms that depend on plant life, and abundant minute holes, which are called *pore spaces* (up to 55% by volume), that contain trapped air and water (Fig. 5.15). The water in the pore spaces contains many dissolved nutrients and the circulation of water, together with that of air, nourishes the microorganisms living in the soil (Table 5.2). Nutrients such as carbon, oxygen, and hydrogen are primarily found in the air, whereas others such as nitrogen, phosphorus, potassium, calcium, magnesium, sulfur, iron, copper, manganese, boron, zinc, chlorine, and molybdenum are derived from the soil.

Processes that operate within soil enable it to sustain life and support agricultural activities (Fig. 5.16). A variety of factors influence the chemistry of soil, including the decomposition of organic matter, the disintegration of near-surface rocks, and the absorption of chemicals from water

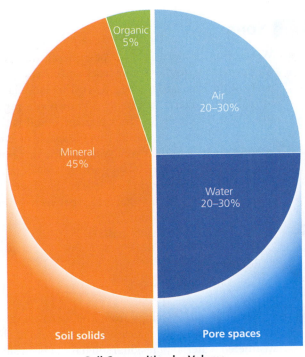

Soil Composition by Volume

Figure 5.15   Soil Composition
This pie chart shows the typical composition (by volume) of soil that can support plant growth.

TABLE 5.2   NUTRIENTS IN SOIL

Macronutrients from air and water
Carbon (C) C, O and H are essential to all organic tissue **Oxygen (O)** **Hydrogen (H)**

Macronutrients from soil
Nitrogen (N) growth of plant proteins **Phosphorus (P)** metabolism, cell membranes **Potassium (K)** cell permeability **Calcium (Ca)** plant cell wall structure **Magnesium (Mg)** production of chlorophyll **Sulfur (S)** synthesis of plant vitamins and proteins

Micronutrients from soil
Boron (B) **Copper (Cu)** **Chlorine (Cl)** **Iron (Fe)** **Manganese (Mn)** **Molybdenum (Mo)** **Zinc (Zn)**

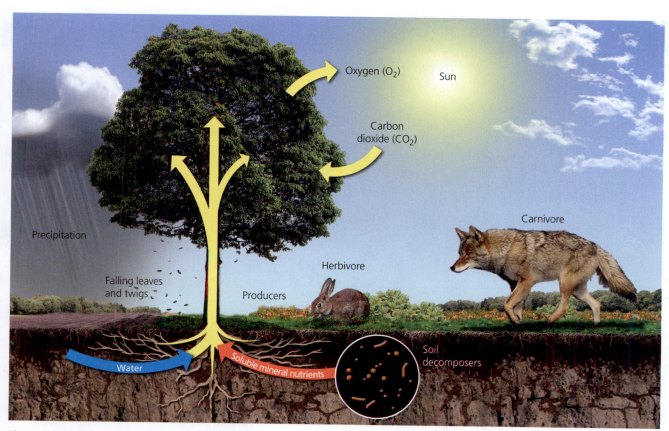

Figure 5.16   Soil Processes

Processes that operate within a soil support ecosystems. Trees and other rooted plants absorb carbon dioxide from the atmosphere and produce oxygen by photosynthesis and are consumed by herbivores (plant-eaters), who are in turn consumed by carnivores (meat-eaters).

and from the atmosphere. (See Section 3.2 Minerals: Orderly Expressions of Matter for a review of the terms organic and inorganic.)

Plants absorb inorganic chemicals from the soil and produce organic matter by way of photosynthesis, the conversion of sunlight into the chemical energy used to fuel the activity of plants and other organisms. Animals utilize this organic material, either by grazing on and digesting the plants directly (in the case of herbivores), or by eating other animals (in the case of carnivores). The decomposition of dead plants and animals eventually returns these chemicals to the soil. This process illustrates how vegetation is essentially a chemical extract from the soil, which it transforms from an inorganic form to an organic one.

Soil development underpins the *food chain* (see Living on Earth: **The Food Chain** on page 152) and has had a profound influence on the course of human development. In fact, many ancient civilizations were born in regions with fertile soil, and these regions are among the most densely populated today.

CONTROLS ON SOIL DEVELOPMENT

Other than water, there is probably no substance that influences our lives more than soil does. Agricultural regions rely on soil to provide food. But as any farmer or land developer knows, the quality of soil varies. Globally, the most fertile soils tend to occur in tropical and temperate climates rather than in arid regions.

Yet, even within a given region, the quality of soil varies considerably. So what factors control soil development? The nature of the underlying bedrock is a common control on the quality of residual soils. However, contrasting soils can occur on similar bedrock, suggesting that other factors are also involved. The most important factors affecting soil are the following:

- nature of the underlying bedrock
- extent of sediment transport

CHECK YOUR UNDERSTANDING

- What is the key difference between a residual soil and a transported soil?

- What is a regolith and how does it form?

- What essential nutrients for soil ecosystems are derived from the atmosphere?

- What essential nutrients for soil ecosystems are derived from soil itself?

LIVING ON EARTH

The Food Chain

By definition, soil must be capable of supporting life, and so soil is the foundation of how the food chain operates on land. The **food chain** describes the movement of energy and nutrients through the biosphere (see Chapter 1) from plants to herbivores and carnivores. Over the long term, plate tectonic processes play an important role in this movement because tectonic processes introduce inorganic chemicals to Earth's surface. At the surface, these chemicals are broken down by weathering and absorbed by plants and a variety of organisms. Solar energy stimulates photosynthesis, which stores energy in plants. The movement of energy and nutrients is further stimulated by the availability of moisture. But the food chain is only able to use a very small proportion of the Sun's radiant energy. About 50 percent of the solar radiation that enters Earth's atmosphere is either absorbed in the atmosphere or reflected back into space. We experience the effects of this reflection when the temperature drops as a cloud passes between us and the Sun. A further 50 percent of solar radiation is absorbed by Earth's surface. Altogether, less than 1 percent of the total radiant energy is utilized by plants and other organisms for photosynthesis.

The process of photosynthesis draws energy from the Sun's radiation to create carbohydrate sugar from carbon dioxide and water. Oxygen is produced as a by-product. The reaction is summarized as

$$6CO_2 + 6H_2O \rightarrow C_6H_{12}O_6 + 6O_2$$
Carbon   Water   Sugar   Oxygen
dioxide

The base of the food chain depends on photosynthesis and the presence of plants. The energy captured by photosynthesis is stored in the plant's sugars. That energy starts to be transmitted down the food chain when the plants are eaten by *herbivores*

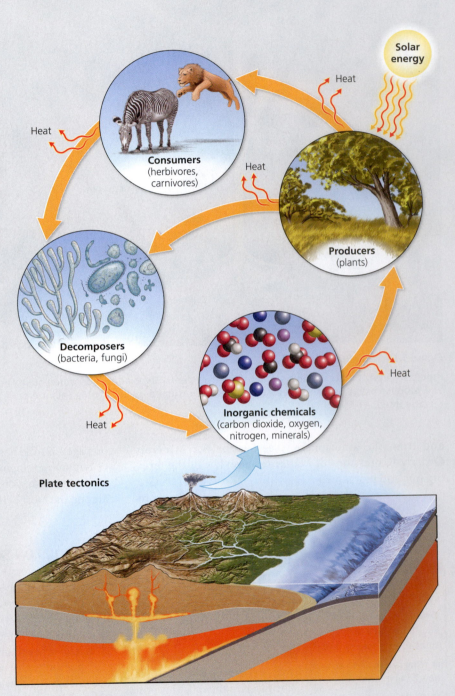

Figure 5C   The Food Chain

This simplified diagram illustrates how the transfer of energy and nutrients operates on land. Plate tectonics introduces inorganic chemicals to Earth's surface, which are broken down by weathering and absorbed by plants and a variety of organisms. Solar energy stimulates photosynthesis, which stores energy in plants. From there, the energy is transferred down the food chain by herbivores (plant-eaters) and carnivores (meat-eaters), until death and decomposition, which returns the energy temporarily into the inorganic world. These chemicals can then be reincorporated into the organic world by plants and transferred down the food chain again. Each step involves the loss of energy (heat) as plants and animals use energy in the process of living.

(plant-eaters). Herbivores may, in turn, be eaten by primary *carnivores* (herbivore eaters) or secondary carnivores (carnivore eaters) higher in the food chain or by *omnivores* (eaters of plants and animals) such as humans (Fig. 5C).

The transmission of energy and nutrients down the food chain is a very inefficient process. In fact, producer plants only transmit about 10 percent of the stored energy to herbivores, and only 10 percent of that is passed on to primary carnivores. Similarly, only 10 percent of this energy is passed on to secondary carnivores. The inefficiency of the food chain explains why there are fewer and fewer species as we follow its path from plants to secondary carnivores. There are vast numbers of edible plants and numerous insects and rabbits, but fewer foxes and even fewer lions.

One of the reasons for this inefficient transmission of energy is that, like humans, plants and animals use some of this energy and also burn off calories in the process of living. In Figure 5C this loss of energy is depicted by the arrows labeled "Heat." This heat ultimately returns to space by way of the atmosphere.

However, upon the death of a plant or animal, the remainder of its unspent energy may be stored if the environment is favorable. Members of the food chain that are not eaten eventually die and decompose. In unconsolidated sediments, decomposition of plant roots, bacteria, fungi, and animals occurs immediately below the surface and results in a nutrient-rich soil called **humus**. The energy stored in this layer is a vital part of the food chain and is used by the agricultural industry to generate much of the food we eat.

- amount of time weathering processes have been operating
- climate
- **topography** (the shape and arrangement of the physical features of an area, such as hills and valleys)
- activity of plants and animals

Soil development and quality involve a complex interplay of all these factors. In this section of the chapter we discuss each of these factors in turn.

Underlying Bedrock

The nature of the underlying bedrock accounts for much of the variation in the quality of residual soils within the same climate belt. For example, a granite regolith is likely to be a mixture of sand and clay. This is because granite is dominated by quartz, which is relatively resistant to weathering, and feldspar, which decomposes to form clay minerals. Basalt, on the other hand, contains virtually no quartz, so a basalt regolith contains only minor amounts of sand.

> ### CHECK YOUR UNDERSTANDING
>
> ◉ In what ways does the nature of the underlying bedrock account for variations in residual soil?

Sediment Deposition

In some areas, the underlying bedrock has no effect on how soil develops because the sediment overlying the bedrock has been transported to the site from another region. Transported soils develop in a wide variety of environments, and their characteristics depend on the transporting agent and other factors, including sediment supply, time since transport, and climate. Soil development on floodplains, for example, is independent of the underlying bedrock. This is because the sediment is typically transported great distances by streams and is deposited on the flood plain when the streams overflow their banks. Similarly, soils developed on glacial deposits in the prairies of southern Canada and the northern United States are unrelated to the bedrock because they were formed from glacially transported material that was dumped following the retreat of the last great continental ice sheets, which began some 11,000 years ago.

Time

The character of a soil changes with time. For example, a regolith that forms from the weathering of granite may initially be sandy because the feldspars have not had sufficient time to decompose. With time, however, the feldspars decompose to clay minerals and the regolith becomes a mixture of sand and clay. As weathering proceeds, soils tend to mature (that is, they become more clay-rich) and thicken. In volcanically active regions where soils develop on volcanic deposits between eruptions, lava flows that have been exposed to soil-forming processes for the longest time develop the thickest soils (Fig. 5.17). A soil must attain a certain thickness before it can support shrubs and trees, and sufficient time is required for this process to occur.

Climate

As soils mature, they develop increasingly in response to climate and are less influenced by the type of underlying

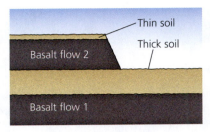

Figure 5.17   Increase of Soil Thickness with Time
Basalt flow 1 was exposed to soil forming processes for a longer time than basalt flow 2 and therefore a thicker layer of soil formed on top of it.

bedrock. In fact, on a global scale, climate is probably the most important control on soil development. Soils form most rapidly and richly in tropical regions and at mid-latitudes, where the climate is warm to temperate and there is abundant rainfall. As we learned earlier in the chapter, these are the conditions under which weathering is most intense and microorganisms are most likely to flourish (Fig. 5.18).

Topography

The physical landscape influences soil development. In steep, mountainous terrains, rain and melted snow often drain too quickly to penetrate far into the ground. These

environments inhibit chemical weathering and much of the soil that does develop is transported downslope. In relatively flat-lying areas, water has more opportunity to soak into the ground, so that chemical weathering is more extensive. Because of this, soils are generally thinner on steep slopes than they are in flat-lying regions.

The orientation of a slope also influences soil development. In the Northern Hemisphere, south-facing slopes receive more sunlight, which, in turn, influences soil temperature, moisture content, and the type of ecosystems that the soil can support. In the Southern Hemisphere, the same is true of north-facing slopes.

> **CHECK YOUR UNDERSTANDING**
> ● How does topography influence soil development?

Organic Activity

Plant and animal life play a role in soil formation, and their activity is an important indicator of the fertility of the soil. Organic activity accelerates the processes of chemical weathering and, hence, soil development. Plants roots and burrowing animals create passageways for air and water to migrate underground. When these plants and animals die, their decaying remains react with water to produce a complex mixture of acidic solutions, some of which are strong agents of chemical weathering. Decomposed

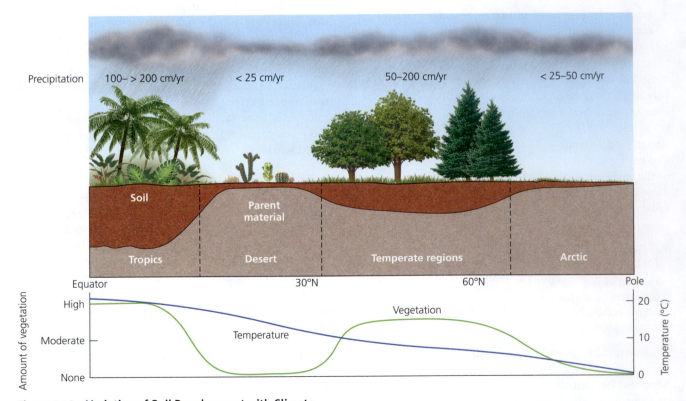

Figure 5.18   Variation of Soil Development with Climate
The thickest soil develops in tropical regions where there is abundant precipitation and solar radiation. In these climates, precipitation and warm temperatures facilitate weathering of bedrock, the development of vegetation, and flourishing microorganisms. Soil development is poor in desert regions due to a lack of sufficient rainfall and in polar regions because the cold climate restricts the development of vegetation.

organic matter also supplies nutrients to plants, and so forms a link in an excellent example of natural recycling. Microorganisms assist in the breakdown of plant and animal remains, making this recycling more efficient.

While each of these individual controls on soil formation may dominate in any one region, many soils owe their development to a combination of these factors. The rate of plant growth and decay, and hence the contribution of organic activity to soil development, is profoundly influenced by climate. In cold arctic regions, plant growth is slow and soils contain little organic matter. In contrast, tropical rain forests support a wonderfully diverse and abundant variety of plant and animal life. However, despite the range and complexity of these processes, well-developed soils show surprisingly simple and regular features, as illustrated by soil profiles, which we discuss next.

SOIL PROFILES

Processes operating within soil result in a layered structure, known as a **soil profile** (Fig. 5.19a). Soil profiles determine how soil is used for agricultural purposes. In mature soil (that is, soil in which soil-forming processes have been operating for a long time), a characteristic sequence of layers develops. The uppermost layer is the **O horizon**, so named because of its organic component. This layer is generally only a few centimeters thick and is rich in decaying organic matter. Beneath the O horizon is the **A horizon**, in which we find plant roots, microorganisms, and burrowing animals. Here, dead plant material is broken down by microorganisms, and vigorous organic activity results in the formation of an organic-rich humus. Depending on a variety of factors, it generally takes between 80 and 400 years to form a layer of humus one centimeter thick.

Together, the A and O horizons are generally referred to as **topsoil**. One gram of topsoil typically contains about 25 percent organic matter by weight, which includes over 1 billion bacteria, 300,000 fungi, 35,000 algae, 20,000 protozoa, and many larger organisms such as earthworms and insects.

In addition to organic material, the A horizon consists mostly of clays and stable minerals such as quartz. The A and O horizons together constitute a *zone of leaching*. In this zone, rainwater seeping down from the surface dissolves soluble ions such as calcium, magnesium, sodium, and potassium, and generally transports them downward. Commonly, the leached material accumulates immediately below the A horizon in a layer known as the **B horizon**, which is a *zone of accumulation* (Fig. 5.19b). In older, well-developed soils, however, the A and B horizons are separated by a zone, known as the **E horizon**, which has undergone substantial leaching of its mineral and organic content and is dominated by a pale layer of silicate minerals (Fig. 5.19c). Beneath the B horizon is the **C horizon** which consists mostly of fragmented and weathered bedrock. In residual soils, the C horizon forms a transition zone to the parent bedrock beneath. There are some important exceptions to the downward movement of water. For example, in arid climates, groundwater is drawn upward to the surface, where it evaporates. As a result, in arid regions, minerals precipitate at the surface instead of accumulating in the B horizon.

Because of the processes involved in soil development, the chemical composition of

> ### CHECK YOUR UNDERSTANDING
>
> ❍ Why are the A and O horizons together referred to as the *zone of leaching?*
>
> ❍ What is the relationship between the zone of leaching and the zone of accumulation in soils?

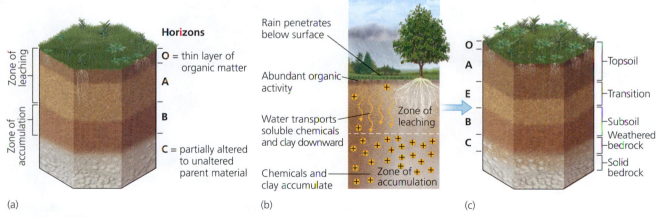

(a)

(b)

(c)

Figure 5.19 Soil Profiles

(a) A typical soil profile features a characteristic sequence of layers. (b) Processes result in the development of a zone of leaching and a zone of accumulation. (c) Sometimes a transition (E horizon) forms between the A and B horizons.

soils varies considerably from the top of the soil profile to the bottom. The chemistry of the uppermost layers (O and A horizons) is profoundly influenced by organic activity, whereas the chemistry of the lowermost layer (the C horizon) is controlled by the composition of the underlying bedrock. In the A horizon, the activity of microorganisms is especially important and this helps the soil to absorb nutrients essential to plant growth such as nitrogen and phosphorus from the atmosphere and from groundwater.

SOIL TYPES

Climate can result in important modifications to the typical soil profile. For example, in wet temperate climates, efficient leaching of soluble ions occurs in both the A and B horizons as water penetrates downward. The less soluble elements, such as silicon, aluminum, and iron, become concentrated in the B horizon, resulting in a soil type known as a **pedalfer**, a name derived from *pedon,* the Greek word for soil, *Al* for aluminum, and *Fe* for iron (Fig. 5.20a).

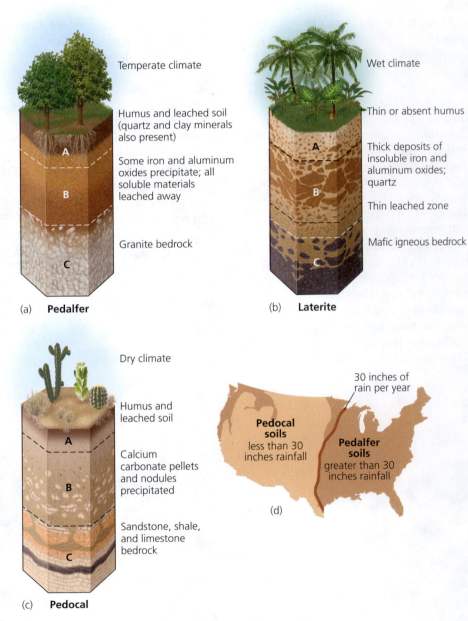

Figure 5.20   Soil Types

Climate can affect a typical soil profile. (a) Pedalfer soil develops in a temperate climate. (b) Laterite soil is common in wet, tropical climates. (c) Pedocal soil forms in arid climates. (d) The distribution of pedalfer and pedocal soil in the United States is influenced by the temperate climate in the East, which experiences greater than 75 centimeters (30 inches) of rainfall annually, and the arid climate in the West.

Pedalfer soils are typically rich in clay minerals and are red in color due to their high iron content.

In tropical regions with high rainfall and high temperatures, the leaching effect of water is so powerful that the soluble material may be completely removed from the soil, escaping with groundwater flow. Because only the most insoluble materials remain, they are especially rich in iron and aluminum, forming rust-colored soil known as **laterite** (Fig. 5.20b). Ironically, tropical rain forests, which are literally teaming with life, are commonly rooted in lateritic soils that are poor in nutrients. As a result, the clearing of rain forests does not yield fertile soils, many of which become barren after only a few years of agricultural activity.

In arid regions, the supply of moisture is limited to that provided by occasional cloudbursts. Initially, this water percolates downward through the O and A horizons, transporting soluble cations toward the B horizon. However, after the rainstorm, processes operating within the soil can reverse this flow direction by drawing the water upward toward the surface. For example, the root systems of plants draw water upward, as does the attraction between water and soil particles. As the water is drawn upward, the dissolved ions become saturated in the B horizon, and minerals such as calcite (calcium carbonate) precipitate. The resulting soil is known as a **pedocal**, after the Greek word for soil, and *cal,* for calcite (Fig. 5.20c). In some regions, calcite precipitation is so intense that a hard cement-like crust, known as a *caliche,* forms within the soil and severely affects soil quality. The development of caliche is an ongoing problem in relatively arid, agriculturally intense regions like the San Joaquin Valley of California.

The distribution of soils in North America is an example of the relationship between the amount of precipitation and the type of soil formed. In the rainier eastern North America pedalfer soils predominate, whereas the drier west is dominated by pedocal soils (Fig. 5.20d).

CHECK YOUR UNDERSTANDING

○ How does the soil profile differ between an arid climate and a temperate climate?

○ Why is the soil quality beneath tropical rain forests so poor?

○ What is the difference between pedalfer and pedocal soil and what is the origin of this difference?

5.4 SUMMARY

- Soil, which supports plant growth, is a product of weathering and is comprised of minerals, organic matter, water, and air.

- Residual soil derives from underlying bedrock whereas transported soil develops on sediment deposited from elsewhere by running water, wind, or ice.

- Soil quality depends on a complex interplay of several factors: the nature of the underlying bedrock, the origin of the sediment, time, climate, topography, and the activity of plants and animals.

- A mature soil has a well-developed soil profile consisting of uppermost organic-rich layers (O and A horizons), which together constitute a zone of leaching, a zone of accumulation (B horizon), and a lowermost zone of weathered bedrock (C horizon).

- Soil profiles are strongly influenced by climate.

5.5 Weathering, Soils, and Plate Tectonics

Plate tectonic processes influence weathering and soil development in a wide variety of ways, both regionally and locally. On a regional scale, plate tectonics is responsible for the formation of topographic features such as mountains and volcanoes that affect climate and soil production. We have already learned of the importance of frost wedging in elevated areas and of the challenges in developing soil on steeply sloping terrain. On a local scale, the stresses associated with plate tectonics result in features such as joints that accelerate physical and chemical weathering. But other effects of plate tectonics, although less obvious, may be equally profound. For example, some scientists believe there may be a relationship between plate tectonics and weathering that has affected global temperatures throughout Earth history.

JOINTS, ACID PRECIPITATION, AND TOPOGRAPHY

The formation of joints, which, as we discussed in Section 5.2, plays a critical role in mechanical weathering, often reflects the tectonic stresses present when rocks form below Earth's surface, or the subsequent removal of these stresses as the rocks are uplifted toward the surface during mountain building. Likewise, processes such as frost wedging are prevalent in mountainous areas, which preferentially occur near convergent plate margins.

Chemical weathering, on the other hand, is enhanced by the interaction of water with gases such as carbon dioxide and sulfur dioxide in the atmosphere, which results in precipitation that is acidic. Before the modern industrial era, the prime source of these gases was volcanoes, and volcanic activity, as we learned in Chapter 2, is directly connected to plate tectonic processes.

Volcanism associated with plate tectonic activity can also have a profound influence on soil quality. The two neighboring Indonesian islands of Java and Borneo, for example, share a similar climate, but Java lies adjacent to a

Figure 5.21   Java

The lush, fertile soils of Java are nourished by volcanic ash deposits that provide nutrients such as calcium and phosphorus. This picture shows the lush vegetation in the foreground and active volcanism (Mount Merapi, which erupted in October–November 2010) in the background. Java is situated above a subduction zone between Australia and Indonesia.

subduction zone and is volcanically active, whereas Borneo is not. The soils of Java are among the most fertile in the world, being nourished by ongoing volcanism that produces layers of volcanic ash rich in nutrients such as calcium and phosphorus (Figure 5.21). Borneo, on the other hand, lacks these ash deposits and its soil is depleted in essential nutrients.

Plate tectonic processes are also responsible for mountain building and therefore play

> **CHECK YOUR UNDERSTANDING**
>
> ◯ Give an example of how plate tectonics influences soil quality.

a role in the effect of topography on soil development. Recall from Section 5.4 that soils are generally thinner in mountainous regions than they are in flat-lying areas. Mountainous regions also tend to have a colder climate and, hence, less intense organic activity, which is important for soil fertility.

PLATE TECTONICS, WEATHERING, AND CLIMATE CHANGE

In addition to the role that plate tectonics plays in mountain building and soil development, many geoscientists advocate an even more profound long-term relationship between plate tectonics and weathering. This relationship may have acted like a thermostat, controlling global temperatures throughout much of Earth's history.

As we discussed in Section 5.3, the processes of dissolution and hydrolysis strip minerals of their most soluble components, which are transported to the oceans. Indeed, the high concentrations of elements such as sodium, magnesium, and calcium in ocean water (Fig. 5.22) are primarily controlled by weathering. Because these elements nourish marine organisms, which incorporate the elements as they absorb and expel ocean water, weathering has played an important role in the evolution of life on Earth.

Furthermore, by allowing elements such as calcium to be transported from continents to the oceans, weathering can act as a thermostat for climate (Fig. 5.23). The dissolved calcium transported to the oceans is absorbed by marine organisms that produce calcium carbonate shells by combining the calcium with carbon dioxide dissolved in ocean water. The oceans replace the carbon dioxide absorbed from ocean water by extracting it from the atmosphere.

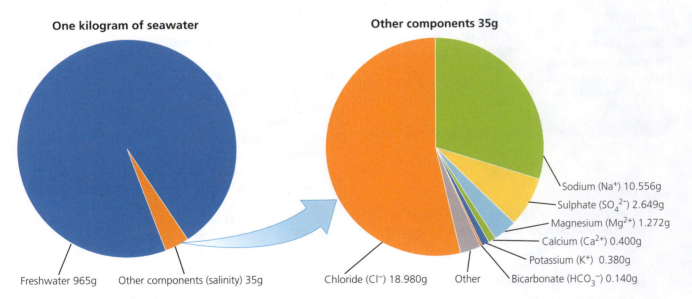

One kilogram of seawater

Freshwater 965g   Other components (salinity) 35g

Other components 35g

Chloride (Cl⁻) 18.980g   Other

Sodium (Na⁺) 10.556g
Sulphate (SO₄²⁻) 2.649g
Magnesium (Mg²⁺) 1.272g
Calcium (Ca²⁺) 0.400g
Potassium (K⁺) 0.380g
Bicarbonate (HCO₃⁻) 0.140g

Figure 5.22   Composition of Seawater

This chart shows the relative abundance of major constituents in a kilogram of ocean water.

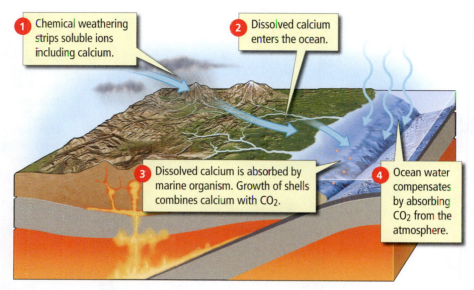

1 Chemical weathering strips soluble ions including calcium.

2 Dissolved calcium enters the ocean.

3 Dissolved calcium is absorbed by marine organism. Growth of shells combines calcium with CO_2.

4 Ocean water compensates by absorbing CO_2 from the atmosphere.

Figure 5.23 Plate Tectonics, Weathering, and Climate Change

The relationship between the plate tectonic environment, weathering, and climate change is illustrated in this schematic which shows how weathering can act as a thermostat for climate. Chemical weathering strips soluble ions including calcium from the crystal lattice of feldspars. The calcium is transported in solution to the ocean, where it is absorbed by marine organisms that combine it with carbon dioxide dissolved in ocean water to form hard calcium carbonate shells. The oceans respond to the loss of dissolved carbon dioxide by extracting more carbon dioxide from the atmosphere, thereby lowering the greenhouse gas content of the atmosphere and promoting global cooling.

Because carbon dioxide is an important greenhouse gas, this extraction lowers atmospheric greenhouse gas levels, thereby causing global cooling. The effectiveness of this process depends on the availability of calcium—the more calcium, the more carbon dioxide is extracted and the greater the cooling trend. The availability of calcium, in turn, depends on the rate of weathering, which increases as temperatures rise. So the more temperatures rise, the more weathering acts to lower temperatures and vice versa, thereby cooling a warming trend or warming a cooling one. In this way, weathering keeps global temperatures stable just as a thermostat acts to keep room temperature comfortable.

During times of intense mountain building, such as those that accompanied the numerous continental collisions associated with the formation of Pangea, the rate of continental weathering can be very high because of the abundance of elevated regions. In the case of Pangea, many of the mountain belts were additionally located in tropical regions where weathering rates are rapid. As a result, the rates of chemical weathering during the amalgamation of Pangea were likely to have been high, which would have led to a dramatic reduction in atmospheric carbon dioxide, and could have triggered continental glaciation, the widespread deposits of which Alfred Wegener used to help reconstruct Pangea (see Section 2.1).

CHECK YOUR UNDERSTANDING

◉ Give an example of how plate tectonics influences the rate of weathering.

In a human time frame, the rate of weathering is so slow that its effects seem to be inconsequential. Yet, from a plate tectonic perspective, few processes have played a more important role in steering Earth's evolution.

5.5 SUMMARY

- Plate tectonic processes promote weathering because they are often responsible for the formation of joints. Plate tectonics also leads to the formation of volcanoes, the volcanic gases of which react with water in the atmosphere to produce acid precipitation that affects soil development.

- Plate tectonic processes dictate topography that affects both weathering and soil development.

- Over the longer term, plate tectonics affects the rates of continental weathering, the transportation of soluble ions such as calcium to the sea, and the extraction of carbon dioxide from the atmosphere. These processes may, in turn, affect global temperatures.

Key Terms

A horizon   155
acid   144
acid rain   143
B horizon   155
C horizon   155
chemical weathering   138
clast   139
columnar joint   140
dissolution   144

E horizon   155
exfoliation joint   140
food chain   152
frost wedging   141
humus   153
hydration   149
hydrolysis   145
joint   139
laterite   157

mechanical
  weathering   138
O horizon   155
oxidation   147
pedalfer   156
pedocal   157
photosynthesis   148
reduction   147
regolith   149

residual soil   149
soil   149
soil profile   155
topography   153
topsoil   155
transported soil   149
weathering   138

Key Concepts

5.1 WEATHERING

- Weathering breaks down rocks and minerals at Earth's surface.

- Weathering occurs as a result of both mechanical and chemical processes.

5.2 MECHANICAL WEATHERING

- Mechanical weathering breaks rocks down into small fragments—a process called disintegration.

- Mechanical weathering can result from frost wedging, crystal wedging, thermal expansion and contraction, and organic processes. Human activities can also facilitate mechanical weathering.

5.3 CHEMICAL WEATHERING

- Chemical weathering causes rocks and minerals to decompose, principally through dissolution, hydrolysis, and oxidation.

- Chemical weathering is enhanced by mechanical weathering, which breaks rocks up into smaller pieces and increases their surface area.

5.4 SOILS

- Soil is a product of weathering that supports plant growth; soil can be residual or transported.

- Soil quality depends on the material being weathered, the duration of weathering, climate and topography, and the activity of organisms.

- A mature soil has a well-developed soil profile that can consist of O, A, B and C horizons and varies according to climate.

5.5 WEATHERING, SOILS, AND PLATE TECTONICS

- Plate tectonic processes promote weathering and soil formation through the formation of joints, production of volcanic gases, and the building of topography.

- Over the longer term, plate tectonics may act as a thermostat for the planet by controlling the rates of continental weathering and the transportation of soluble ions such as calcium to the sea.

Study Questions

1. Contrast the mechanisms of mechanical weathering you might expect to encounter in hot climates with those you would expect to find in cold ones.

2. How does mechanical weathering promote chemical weathering? Is the reverse true (does chemical weathering promote mechanical weathering)?

3. Why do minerals have different susceptibilities to chemical weathering? Why, for example, are nonsilicates generally more susceptible than silicates?

4. In hydrolysis reactions, why are only some cations removed?

5. Explain how chemical weathering influences the composition of ocean water.

6. Why are iron-rich minerals susceptible to oxidation reactions?

7. Explain how organic processes stimulate (a) mechanical weathering, (b) chemical weathering, and (c) soil development.

8. Explain the relationship between soil quality and the food chain.

9. In what way does weathering act as Earth's thermostat?

10. Continents were colonized by plants and animals about 450 million years ago. What types of mechanical and chemical weathering would have been less prevalent before that time?

6

Sedimentation and Sedimentary Rocks

6.1 From Sediment to Sedimentary Rock

6.2 Types of Sedimentary Rocks

6.3 Classifying Sedimentary Rocks

6.4 Interpreting Sedimentary Rocks

6.5 Depositional Environments and Sedimentary Facies

6.6 Sedimentation, Sedimentary Rocks, and Plate Tectonics

▶ Key Terms

▶ Key Concepts

▶ Study Questions

LIVING ON EARTH: Salt of the Earth 173

LIVING ON EARTH: King Coal 178

SCIENCE REFRESHER: Settling Velocities 182

Coarse

Fine

Deposition

Sandstone

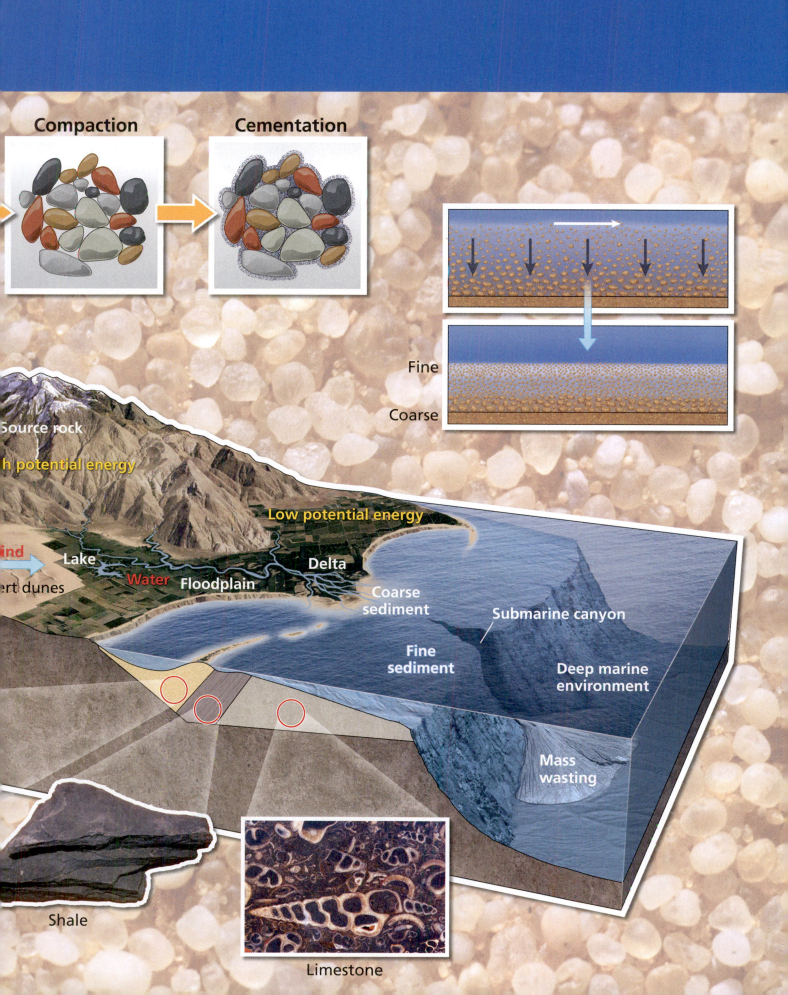

Compaction

Cementation

Fine

Coarse

Source rock

h potential energy

Low potential energy

ind

Lake

Desert dunes

Water

Floodplain

Delta

Coarse
sediment

Fine
sediment

Submarine canyon

Deep marine
environment

Mass
wasting

Shale

Limestone

Learning Objectives

6.1 Describe how sediment forms and consolidates to produce sedimentary rocks.

6.2 Compare and contrast the textures and compositions of sedimentary rocks and explain how sedimentary rocks vary according to the environment of their deposition.

6.3 Describe the different types of sedimentary rock and the basis of their classification.

6.4 Explain how the composition, fossil content, and presence of sedimentary structures allow us to interpret the origin of sedimentary rocks.

6.5 Identify the various sedimentary features that characterize deposition in continental, coastal, and marine settings.

6.6 Explain the role that plate tectonics plays in sedimentation and the formation of sedimentary basins.

In contrast to the igneous rocks we discussed in Chapter 4 that form by "hot" processes, sediments and sedimentary rocks are formed by "cold" processes that occur at or near Earth's surface. Sediment is the blanket of unconsolidated material that we see all around us. It occurs on land, in coastal regions, and at the bottom of our oceans. Eventually most of this sediment consolidates to form sedimentary rocks, which comprise about 75 percent of the exposed bedrock on continents and contain some of Earth's valuable natural resources, including most of our freshwater supply.

The formation of sedimentary rocks is one of nature's most important recycling projects. By converting old bedrock into new sedimentary rock, Earth's surface is continuously renewing itself, and in doing so, provides important new chemistry for sediments, soils, and the oceans. The characteristics of sedimentary rocks vary with the environment where they are deposited. As a result, we can learn much about the history of Earth by studying ancient sedimentary rocks.

The story of sediment formation began in Chapter 5; the process starts when surface bedrock is broken down by mechanical weathering and chemical weathering. Mechanical weathering produces loose debris that tumbles downslope and collects in valleys, where it is picked up and transported by streams, glaciers, or the wind. Eventually, this material is deposited to form sediment and consolidated into sedimentary rock. Chemical weathering plucks soluble ions from minerals and those ions are transported in solution. In some special environments, the ions become concentrated and crystallize to form sedimentary rocks.

The textures and compositions of sediments and sedimentary rocks preserve clues to the processes that formed them and the environments where they formed. In this way, sedimentary rocks serve as a window into the geologic past.

6.1 From Sediment to Sedimentary Rock

Sediment is naturally occurring unconsolidated material consisting of loose fragments of bedrock, minerals, shells, and/or crystals that precipitate directly from water. The physical settings in which sediments are deposited are referred to as their **depositional environments**. After deposition, a variety of processes cause sediment to bind together and consolidate into **sedimentary rock**. To understand the origin of sedimentary rock, we must understand how sediment is first formed by physical and chemical weathering, how both the solid and dissolved materials produced by weathering are transported and deposited, and finally, how loose sediment is converted into consolidated sedimentary rock.

FORMATION OF SEDIMENT

Weathering is the first step in the transformation of surface bedrock into sediment and, ultimately, into sedimentary rock. As we learned in Chapter 5, weathering involves the breakdown of bedrock, known as **source rock**, by both physical and chemical processes. *Mechanical weathering* physically disintegrates the rock into smaller fragments, whereas *chemical weathering* decomposes unstable minerals in the bedrock. These two types of weathering result in two fundamentally distinct types of sedimentary rocks, which we discuss in more detail in Section 6.2.

Chemical weathering strips bedrock minerals of their most soluble components. Soluble ions such as sodium and calcium are a component of streams and are transported in solution to lakes and oceans. Mechanical weathering, on the other hand, produces loose material, known as **detritus**, which is readily removed and transported to lower elevations (Fig. 6.1). Removal of material by gravity is known as **mass wasting** and we discuss that process in detail in Chapter 12. **Erosion** is the removal of sediment by mobile agents, such as wind, water, and ice. Wind is the dominant agent of erosion in hot desert environments, ice in cold polar regions and at high altitude, and water in climates between these extremes. We refer to these collectively as the *agents of erosion*. Debris that tumbles downslope by mass wasting typically collects in valleys where it is removed by a stream, a glacier, or by the wind.

Mass wasting and erosion combine to remove weathered material, thereby exposing fresh bedrock, which then becomes weathered in turn. The net effect is a conveyor belt of sediment that transports the products of weathering and erosion from elevated regions toward valleys and eventually to the sea. The detritus is deposited as unconsolidated sediment such as sand, gravel, or mud, either on **floodplains** (the area adjacent to a stream that experiences repeated episodes of flooding) or in the coastal zone where rivers enter the sea.

> ### CHECK YOUR UNDERSTANDING
> ◔ What are the primary agents of erosion and under which conditions does each agent dominate?

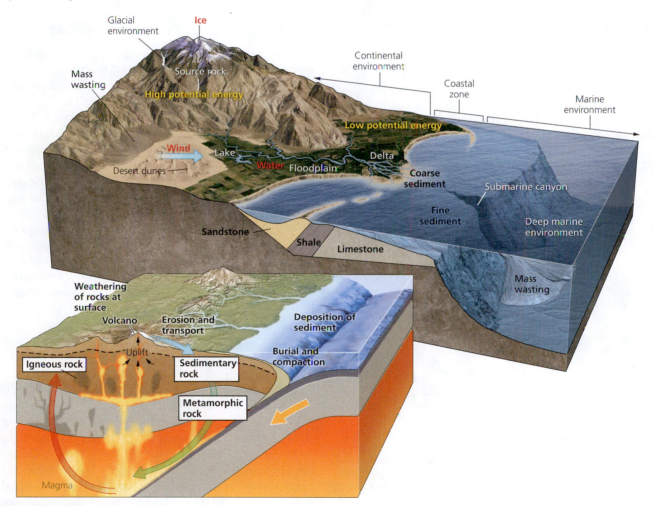

Figure 6.1   Sedimentation

This figure illustrates sediment production (by weathering and mass wasting), its transport by agents of erosion (water, wind, and ice), and its deposition. Sediments can be deposited in continental environments such as floodplains, lakes, and deserts, in shallow marine coastal zones, as well as deep marine environments. The inset shows the relationship of sedimentary processes to the rock cycle.

FORMATION OF SEDIMENTARY ROCK

After deposition, sediments typically undergo a variety of biological, physical, and chemical processes, collectively called **diagenesis**. During diagenesis, adjacent particles in loose unconsolidated sediment bind together to form a coherent sedimentary rock, a process called **lithification**. Spaces between the particles, known as **pores**, are filled with air or water, and the volume of the spaces relative to the volume of the sediment is known as **porosity**.

Biological processes include the activities of bacteria, of plants and their root systems, as well as of burrowing organisms. Physical processes include changes due to *compaction,* when the weight of the overlying sediment creates pressure that squeezes the broken mineral and rock fragments together. The fragments formed by compaction are known as **clasts** (after the Greek "klastos" meaning "broken"). Compaction also reduces the pore spaces between clasts, thereby reducing the porosity of the sediment.

Chemical changes result from water percolating through pore spaces. Because groundwater is a powerful solvent, this water often contains dissolved ions as a result of chemical weathering, and these ions crystallize minerals between the grains, reducing the porosity still further. These minerals also act as cement, binding the clasts together, in a process called *cementation.*

Groundwater commonly contains dissolved ions of calcium (Ca^{2+}) and bicarbonate (HCO_3^{-}) that result from the chemical weathering of the mineral calcite ($CaCO_3$). As the groundwater circulates though sediment, these dissolved ions may recombine to form calcite, according to the following reaction:

$$Ca^{2+} + 2HCO_3^{-} \rightarrow CaCO_3 + H_2 + CO_2$$
$$\text{Dissolved ions} \qquad \text{calcite}$$

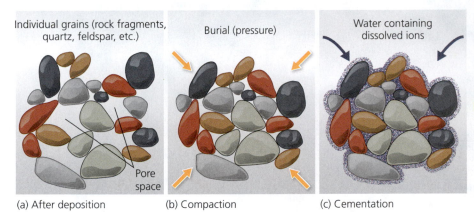

Individual grains (rock fragments, quartz, feldspar, etc.)

Burial (pressure)

Water containing dissolved ions

Pore space

(a) After deposition   (b) Compaction   (c) Cementation

Figure 6.2 Lithification

Lithification of sedimentary rocks occurs when cementation follows compaction. (a) Pore spaces between individual grains fill with air and water after deposition. (b) During compaction these pore spaces are squeezed as the sediment is buried by younger layers, thereby reducing the porosity of the rock. (c) Groundwater percolating between the grains contains dissolved ions that precipitate and bind adjacent grains together during cementation. Common cements include calcite, quartz, iron oxides, and clay minerals.

As the calcite crystals grow, they attach tightly to the sediment grains and bind them to one another. Because of this, calcite is one of the most common cements found in sedimentary rocks (see Chapter 3: Minerals). Depending on local conditions, quartz, iron oxide, and clay minerals also act as cements (see Fig. 6.2).

> **CHECK YOUR UNDERSTANDING**
>
> ⊙ How does sediment become sedimentary rock?

6.1 SUMMARY

- Sediment is formed by physical and chemical weathering and the transport of solid particles and dissolved ions.

- Following deposition, sediment is converted into sedimentary rock by compaction, which reduces pore spaces, and cementation, in which minerals precipitate from pore waters.

6.2 Types of Sedimentary Rocks

Sedimentary rocks fall into two main groups: *detrital sedimentary rocks* and *chemical* or *biochemical sedimentary rocks*. **Detrital sedimentary rocks** are the end product of mechanical weathering, which, as we established in Section 6.1, produces detritus that is transported, deposited, and consolidated. **Chemical** or **biochemical sedimentary rocks** are the end product of chemical weathering, which extracts soluble components from rock. Recall that under certain conditions, these chemicals become concentrated enough to precipitate from solution. This precipitation from solution may occur entirely as a result of inorganic processes. In that case, a chemical sedimentary rock forms. However, if precipitation is aided by the action of organisms, a biochemical sedimentary rock forms. For example, chalk, which is made up of the skeletal remains of microorganisms, is an example of a biochemical sedimentary rock.

Sedimentary rocks vary markedly in appearance, or *texture*. The textures of sedimentary rocks preserve important information about the processes responsible for their formation and the ancient environments in which they formed. For example, the composition, grain size, and shape of clasts in detrital sedimentary rocks differ between ancient continental and ancient marine environments. Similarly, chemical sedimentary rocks contain interlocking crystals, the result of precipitation from warm water where evaporation occurs. Biochemical sedimentary rocks commonly contain fossils of the organisms responsible for their formation. Because most lifeforms are sensitive to their environment, fossil content is an indication of the physical conditions in the region at that time.

DETRITAL SEDIMENTARY ROCKS

Detrital sedimentary rocks are the product of mechanical weathering, transport, and deposition of solid material. Because they contain clasts, they are described as having a **clastic texture** (Fig. 6.3). The mechanism of transport is either mass wasting or erosion. The finer grained material between clasts is called the **matrix**. All detrital rocks have a clastic texture; however, they vary considerably in other textural features such as clast composition, grain size, and grain shape. By interpreting these textures, geologists are able to deduce the environments where they were deposited.

> ### CHECK YOUR UNDERSTANDING
>
> ● What rock features are used to deduce the depositional environments of detrital rocks?

Clast Composition

Detritus generally consists of fragments or rocks, or individual minerals such as quartz, feldspar, mica, or clay minerals. By identifying rock fragments, geologists can deduce what source rocks were exposed in the region where erosion took place (Fig. 6.3a). These fragments can sometimes be matched to their original bedrock. Most sedimentary rocks contain different types of clasts indicating they were derived from a variety of source rocks. For this reason, geologists try to identify the **source area**, a geographic area from which the various clasts could have been derived. For example, some of the detritus that has accumulated at the mouth of the Mississippi River came from source areas in the continental interior of North America, as far away as Wyoming, Ohio, and Minnesota (Fig. 6.4a). Similar tracking

(a)   (c)

(b)   (d)

Figure 6.3 Clastic Textures

(a) The clasts in this rock include granite, which tells us that a granitic body was exposed in the source area. (b) The presence of feldspar clasts implies that this sediment traveled only a short distance from its source. (c) The dominance of quartz clasts in this sample indicates that the sediment traveled a long distance from its source. (d) The dominance of angular clasts in this rock implies that the sediment traveled only a short distance from its source.

Figure 6.4 Sediment Transport

(a) The Missouri-Mississippi drainage basin consists of thousands of streams and covers most of the central and southern United States. Some of the detritus deposited at the mouth of the Mississippi has been transported thousands of kilometers from its headwaters near the Canadian border. The inset shows the development of the Mississippi delta. (b) During the Ice Age, distinctive fragments of bedrock near Lake Superior and in South Dakota were transported hundreds of kilometers by an ice sheet and are found in glacial deposits in northeastern Kansas.

techniques show that continental ice sheets can transport giant boulders over great distances, only to drop them when the ice finally melts. For example, boulders deposited in northeastern Kansas consist of a distinctive collection of rocks (basalt, iron deposits, and quartzite) that were derived from bedrock hundreds of kilometers to the north (Fig. 6.4b). From this information, we can deduce that the ice sheets incorporated these boulders as they advanced southward from Canada to Kansas.

If the agent of erosion is running water, the various minerals such as quartz, feldspar, mica, and clay minerals that make up the detritus behave quite differently during transport. Recall from Chapter 3 that quartz and musco- vite (or white mica), for example, are chemically very stable. Quartz is also physically quite robust and so can survive many cycles of transport and deposition. Musco- vite, however, is physically quite fragile and is unlikely to survive either long transport or more than one cycle of transport and deposition. Because of this, the presence of muscovite in sedimentary rocks indicates nearby sources of muscovite-bearing bedrock.

Recall from Chapter 5 that feldspar is chemically un- stable and tends to break down by reacting with water to form clay minerals. Because of this instability, most feld- spar breaks down before it can form sedimentary rocks. This instability is reflected in the fact that feldspar typi- cally accounts for about 60 percent of the mineral content in source areas, but only 10–15 percent of the mineral content in average detrital sediment. Minerals such as olivine, pyroxene, and amphibole are also unstable and readily break down during weathering and transportation. When geologists find these unstable minerals in detrital sedimentary rocks they can assume that the sediment did not travel far (Fig. 6.3b). In contrast, quartz and clay min- erals are chemically stable and so are the dominant minerals in detrital sediments that are deposited far from their source area (Fig. 6.3c).

> **CHECK YOUR UNDERSTANDING**
>
> ◔ How does the relative stability of minerals affect the composition of sedimentary rock?

Grain Size
Sorting is the process by which particles of the same size, shape, or density are naturally selected and separated from dissimilar particles. The degree of sorting provides important clues to environmental conditions at the time of deposition.

For example, desert sand is typically made up of wind- blown grains. Because wind is a very effective sorter of de- tritus, the grains deposited from windblown sand are very similar in size. Such deposits of windblown sand are **well sorted** (Fig. 6.5a). In contrast, glacial ice is unselective and transports detritus that varies widely in size. When glaciers melt, debris ranging from boulders to mud is deposited to- gether. Such glacial deposits are **poorly sorted** (Fig. 6.5b).

Streams produce a sequence of deposits whose degree of sorting varies between these two extremes, depending on the energy level of the stream. In high-energy environments, such as a flooding stream, large boulders are transported along with fine particles, leading to poorly sorted sediments. In low-energy environments, the volume of water traveling downstream is much lower and water is a more effective se- lector of grains. The largest grains move only intermittently and for relatively short distances along the stream bed, while the finer grains are transported farther. This low-energy en- vironment produces well sorted deposits.

Grain Shape
Mechanical weathering typically produces angular frag- ments (Fig. 6.3d). However, sharp-edged shapes are usually modified during transport, a process called **rounding**. Larger grains which spend more time on the stream bed are particularly susceptible to rounding (see Figure 6.5a). Their sharp corners are chipped off by impacts with other grains or the bed of the stream. Silt and clay par- ticles, in contrast, are carried in suspension by the water, so that their original shape is less modified during trans- port. In general, however, grain size becomes smaller and fragments become more rounded as fragments move farther away from their source rock. Other more subtle changes to grain shape provide evidence about the jour- ney particles have taken. For example, during transport by wind, grains may become pitted and polished by mid- air collisions. Similarly, trans- port by ice can carve grooves known as *striations* on grain surfaces.

> **CHECK YOUR UNDERSTANDING**
>
> ◔ What can we learn from identifying the rock frag- ments in a clastic rock?

(a)

(b)

Figure 6.5   Textural Differences Related to Sorting
(a) Well-sorted, well-rounded quartz grains in clastic sedimentary rock are seen here through a microscope. (b) These poorly sorted glacial deposits feature clasts of granite ranging from 6 centimeters (2.4 inches) to less than 1 centimeter (0.4 inches) in diameter.

Biochemical + chemical both nonclastic

CHEMICAL AND BIOCHEMICAL SEDIMENTARY ROCKS

Less common than detrital sedimentary rocks are those that form through chemical and biochemical processes. The texture of both chemical and biochemical sedimentary rocks is described as **nonclastic** because they comprise interlocking crystals rather than broken fragments (Fig. 6.6).

Recall from Section 6.1 that chemical sedimentary rocks form from the precipitation of crystals from solutions. In hot, dry climates, water evaporates leaving dissolved ions behind. The concentration of these ions in the remaining water consequently increases, eventually reaching a maximum threshold, known as **saturation**. Crystals form from those dissolved ions and grow until the pore spaces between the grains are virtually eliminated, producing a texture consisting of interlocking crystals. **Evaporites** are a family of chemical sedimentary rocks that form by the precipitation of salts left behind by the evaporation of seawater. Evaporite deposits of salt and gypsum are examples of chemical sedimentary rocks (see Living on Earth: **Salt of the Earth** on page 173). Chemical sedimentary rocks are entirely inorganic in origin.

One of the best examples of evaporite deposits comes from the Mediterranean Sea, which dried up between 5 and 6 million years ago, when the Strait of Gibraltar, which connects the Mediterranean to the Atlantic Ocean, temporarily closed. This closure left the Mediterranean as an isolated sea in a warm climate. Because the amount of evaporation greatly exceeded the water entering the basin from rivers, sea level in the Mediterranean fell drastically, which concentrated the dissolved ions until they precipitated (Fig. 6.7a–b). The resulting evaporite deposits have been found in many localities at the bottom of the Mediterranean Sea (Fig. 6.7c), which refilled when the Strait of Gibraltar reopened. Today the Mediterranean contains a higher concentration of dissolved ions than the neighboring North Atlantic because the narrowness of the Strait of Gibraltar has left the sea nearly isolated. It has been calculated that if the strait closed again, the Mediterranean would dry up in about 1000 years.

A special type of chemical sedimentary rock, known as **travertine**, is in great demand as a decorative stone (Fig. 6.8). Travertine forms when hot underground water carrying dissolved calcium and carbonate ions suddenly chills when it reaches the surface or when it invades underground caves. The chilling causes the dissolved ions to precipitate.

Biochemical sedimentary rocks, in contrast to chemical sedimentary rocks, form by organic processes and commonly contain fossils of the organisms responsible for their formation. For example, marine organisms absorb ions such as calcium, magnesium, iron, silicon and phosphorus from seawater to form their hard shells and soft tissues. When these organisms die, they settle to the seabed and become compacted and cemented into biochemical sedimentary rock. Other biochemical sedimentary rocks, such as **coal**, which forms from the compaction of plant remains, are valuable natural resources and provide important clues for unraveling ancient environments and climate patterns.

(a)

(b)

(c)

Figure 6.6   Nonclastic Textures

(a) The nonclastic texture of interlocking calcite crystals is typical in inorganic limestone. (b) Oolitic limestone is a chemical limestone dominated by spherical grains of inorganically precipitated calcite, called ooids. (c) Biochemical limestone contains fossils.

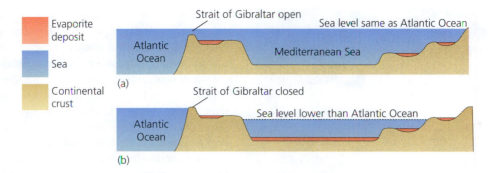

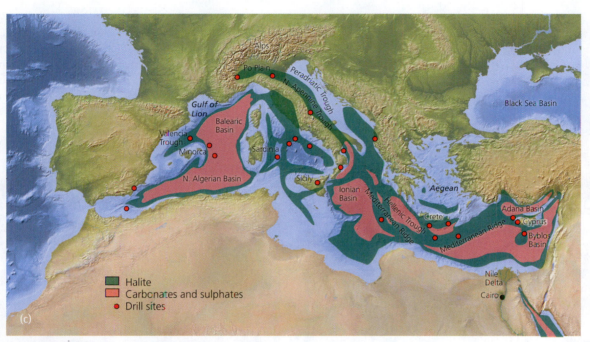

Figure 6.7   Evaporite Deposits in the Mediterranean Sea

(a) Connection between the Atlantic and Mediterranean led to localized evaporite deposition that was restricted to shallow water environments. (b) When the Strait of Gibraltar closed between 5 and 6 million years ago the Mediterranean Sea became isolated and its water evaporated, forming evaporite deposits along the seabed. (c) Evaporite deposits have been located in numerous places in the Mediterranean Sea.

Some rock types can be either chemical *or* biochemical in origin, depending on the environment of deposition. **Limestone**, for example, is a common sedimentary rock that is primarily composed of the mineral calcite ($CaCO_3$). Limestone can form in a variety of ways. In some environments, such as the warm, shallow seas of the Bahamas, the concentrations of calcium and carbonate ions in ocean water are so high that calcite precipitates as chemical sediment. Strong tidal currents may cause shells to roll back and forth along the seabed. Calcite crystals precipitate from the seawater and coat the rolling shell fragments. The end result is a spherical grain, called an **ooid**, consisting of concentric layers of inorganic calcite around a central fragment. Limestone that forms in this manner is **oolitic limestone** (Fig. 6.6b) and its origin is chemical rather than biochemical.

In other environments, however, calcium and carbonate ions are extracted from seawater by colonies of marine organisms to form a biochemical sediment. These organisms, such as corals and algae, secrete organic calcite to form their external skeletons (Fig. 6.6c). The life forms in these colonies are among the world's most diminutive homebuilders. Corals secrete organic calcite to form external skeletons, while the algae that live with them also secrete calcite and help cement and solidify the reef structure. The most famous modern **coral reef** is the Great Barrier Reef of Australia, a 2000-kilometer (1250-mile) network of coral ridges and islands (Fig. 6.9) similar in size to the state of Texas. Visible to orbiting astronauts, the Great Barrier Reef is the largest structure ever built by living organisms. When the organisms die, their descendants carry on reef building. Over millions of years, these reef-builders have constructed underwater platforms, ridges, cliffs, and canyons. The structure is made of **coral limestone**, which is biochemical in origin and consists of abundant fossils cemented by crystalline calcite. Remnants of reef structures similar to the Great Barrier Reef have been found in the sedimentary record and are thought to have formed in similar environments in the geologic past.

Figure 6.8   Travertine
The chemical sedimentary rock travertine is in great demand as a decorative stone.

Figure 6.9   Limestone Coral Reef
The Great Barrier Reef of eastern Australia as seen from the Space Shuttle, is a 2000 kilometer (1250 mile)-long network of limestone coral reefs and islands.

Because these types of limestone are made up of interlocking crystals of calcite, their textures are classified as nonclastic. In contrast, another type of limestone consisting of fragments of fossil shells and skeletal calcite forms as a result of wave action on reef communities (Fig. 6.10). The fragments in this type of limestone are cemented by inorganic calcite that crystallizes from pore water trapped between the fragments. This limestone has a clastic rather than a nonclastic texture. Such textures are described as *bioclastic*, to distinguish them from the clastic textures in detrital sediments.

Limestone can undergo diagenesis after it forms via chemical reaction with seawater. For example, magnesium, which is abundant in seawater, can replace some of the calcium in the rock, leading to the formation of a calcium-magnesium carbonate mineral called *dolomite*, which has the formula $CaMg(CO_3)_2$ (Fig. 6.11). A rock dominated by dolomite is a

Figure 6.10   Bioclastic Limestone
Bioclastic limestone consists of fragments of fossil shells and calcite formed by wave action on reef communities and cemented by inorganic calcite.

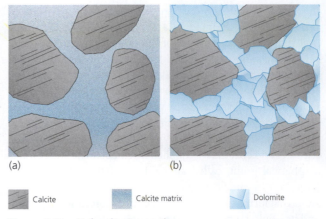

(a)   (b)

Calcite   Calcite matrix   Dolomite

Figure 6.11   Dolomite Formation
(a) Dolomite forms when large volumes of seawater percolating through the fine-grained matrix of limestone supply magnesium to replace calcium. (b) As a result, dolomite crystals form in the matrix.

LIVING ON EARTH

Salt of the Earth

Precious minerals such as diamond, gold, and silver are glamorous because their worth lies in their appearance as well as their distinctive properties. Many common minerals, however, also have a profound influence on our lives, although their roles go virtually unheralded. The common mineral halite is an excellent example of a vital but often overlooked resource. Halite (NaCl) has been produced naturally on several occasions in Earth history by the evaporation of great inland seas and is the dominant mineral in rock salt. For example, huge deposits of rock salt, 2000 meters (6600 feet) in thickness, formed when the Mediterranean Sea dried up some 6 million years ago.

More recently, the same process operating on a much smaller scale created Great Salt Lake in northern Utah. Starting about 30,000 years ago, during the last Ice Age, huge quantities of water ponded in northwestern Utah, forming a great freshwater lake that earth scientists named Lake Bonneville (Fig. 6A). Following the Ice Age, as the climate became warmer and drier, the evaporation of Lake Bonneville's fresh water caused it to shrink rapidly and become saltier. Eventually, large portions of the lake dried up, leaving behind deposits of rock salt, which is dominated by interlocking crystals of halite. Today, one such portion makes up the Bonneville Salt Flats, famed as a raceway for land-speed records. Only remnants of Lake Bonneville, such as Great Salt Lake and Utah Lake, now remain. However, the lake's former extent can still be seen in the prominent wave-cut terraces that were produced as its shoreline receded and which now lie high in the surrounding uplands (Fig. 6B).

Lake Bonneville was small in comparison to some of the great inland seas of the past. Much larger were the seas in North America that evaporated to produce the Louanne Salt, now buried beneath the Gulf Coast states, and the Salina Salt, now buried beneath the Great Lakes. Both of these deposits contain huge thicknesses of rock salt, which is mined in both Canada and the United States. Together, the United States and Canada produce over 50 million metric tons of salt each year, and the demand exceeds even this impressive production.

Exploitation of rock salt deposits, in one form or another, impacts each of us every day. Rock salt is used daily as table salt. The food industry in North America alone consumes well over two million metric tons of rock salt each year, but this use accounts for only about 3 percent of the total domestic demand. Most rock salt is used by the chemical industry in the production of chemicals such as chlorine, caustic soda,

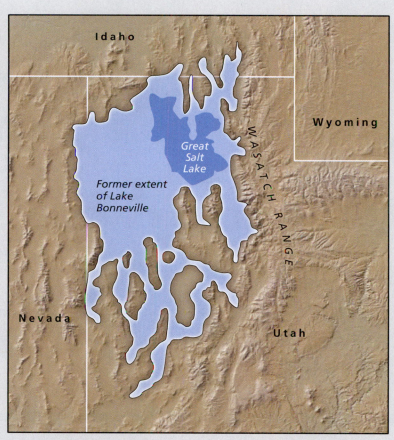

Figure 6A   Lake Bonneville

Figure 6B   Lake Bonneville Wave Cut Platform

Evidence for the former extent of Lake Bonneville includes the prominent wave-cut terraces in the surrounding highlands that were produced as its shoreline receded.

(Continued)

LIVING ON EARTH

Salt of the Earth (Continued)

and polyvinyl chloride. More familiar to us is its role in highway deicing, a use that accounts for some 20 million metric tons a year.

Rock salt is a highly mobile material and will flow under the weight of overlying sedimentary rocks just as glacier ice flows under the influence of gravity. It is also of low density and when buried by detrital sedimentary rocks of higher density it can become gravitationally unstable. Where this occurs, the salt tends to exploit weaknesses (such as fractures) in the overlying rock layers and pushes its way upward toward the surface in huge fingers called *salt domes* (Fig. 6C). As salt domes rise, the layers of sedimentary rock they pierce are arched upward to form structures in which petroleum can accumulate. Petroleum reservoirs formed in this way are important sources of oil and gas in the Gulf Coast region of the United States.

Rock salt is consequently a sedimentary rock of enormous economic importance, both because of the value of its component mineral, halite, and because of its ability to form reservoirs for oil and gas. It is a rock on which our lives depend. However, its use is not without consequences. Salt use has been criticized by both the environmental and medical communities. Highway deicing can damage roadside vegetation, contaminate groundwater supplies, and cause automobiles to rust. Although essential to the human diet, when consumed in excess, salt may

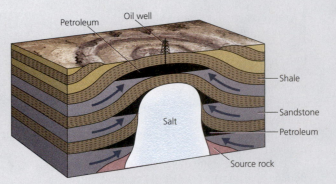

Figure 6C   Salt Domes

Because rock salt has a lower density than detrital sedimentary rocks, it tends to rise toward the surface to form salt domes. The overlying sedimentary layers arch upward to form structures in which petroleum can accumulate.

cause hypertension (high blood pressure). For this reason, significant efforts have been made to reduce the quantity of salt added to processed foods. There may also be health risks in using chlorine produced from salt for the treatment of municipal water supplies, and in using salt to produce chlorofluorocarbons (CFCs) that are known to contribute to ozone depletion in the upper atmosphere.

> **CHECK YOUR UNDERSTANDING**
>
> ○ Compare and contrast the formation of detrital and chemical and biochemical sedimentary rocks.

dolostone. Magnesium is a smaller cation than calcium (see Periodic Table, Appendix B), so this replacement is often accompanied by a volume loss. The volume loss results in voids in the rock. These voids sometimes become important reservoirs for petroleum and economic minerals.

6.2 SUMMARY

- Sedimentary rocks are of two types: detrital or chemical/biochemical.

- Most sedimentary rocks are detrital and are the product of mechanical weathering followed by the transport and deposition of rock waste. They contain mineral and rock fragments called clasts, and have a clastic texture.

- Minerals behave differently during transport depending on their stability. Feldspar and mafic minerals are particularly unstable and their presence in sediments indicates they were not transported long distances.

- Clastic rocks vary in the degree of sorting of their grains. In general, transport by running water rounds and sorts grains.

- Both chemical and biochemical sedimentary rocks are characterized by a nonclastic texture of interlocking crystals. Chemical sedimentary rocks are inorganic and form from the precipitation of crystals from solution. Biochemical sedimentary rocks are formed by organic processes and commonly contain fossils.

- Some rocks, such as limestone, can occur as either chemical or biochemical sediments, or a combination of both.

6.3 Classifying Sedimentary Rocks

Recall from Section 6.2 that the two groups of sedimentary rocks—detrital and chemical/biochemical—can be distinguished on the basis of texture. All detrital rocks have a clastic texture, whereas most chemical/biochemical rocks have a nonclastic texture. Although there are some important exceptions, we can use this textural distinction to classify the vast majority of sedimentary rocks.

In some instances, the texture of detrital rocks superficially resembles pyroclastic textures associated with explosive volcanic rocks (see Chapter 4: Origin and Evolution of Igneous Rocks). However, in explosive volcanic rocks, the predominant clasts may display several textural features indicating that they were hot when deposited. Similarly, the interlocking crystals that characterize the nonclastic textures of chemical and biochemical sedimentary rocks can be somewhat similar to the textures exhibited by plutonic igneous rocks. However, minerals such as halite, gypsum, and calcite are rare or absent in igneous rocks and the different mineral contents enable us to distinguish between these two types of rocks.

CLASSIFYING DETRITAL SEDIMENTARY ROCKS

To classify detrital sedimentary rocks, we need to name the particles the rocks contain. Because of the importance of sediment in our daily lives, many of the terms used for this purpose are already familiar. Whether it be in agriculture, land development, or as a source of water, sediment is one of our most valuable commodities, and the size and shape of the particles are important to resource potential.

Particles in detrital sedimentary rocks with diameters greater than 2 millimeters (0.08 inches), which is about the thickness of a nickel, are referred to by two collective terms; **gravel** if they are somewhat rounded, and **rubble** if they are more or less angular. A gravel particle that is less than 64 millimeters (2.5 inches) across, which is about the diameter of a tennis ball, is a **pebble**. A particle with a diameter

between 64 millimeters and 256 millimeters (10 inches), which is about the size of a soccer ball, is a **cobble**, and one with a diameter greater than 256 millimeters is a **boulder**.

A **sand** grain ranges in diameter from 0.067 millimeters (0.003 inches), which is about the thickness of your fingernail, to 2 millimeters. Farmers estimate the sand content of their soil by assessing the gritty feel of the sediment when they rub it between their fingers. Sand grains are just large enough to be visible to the naked eye and typically consist of minerals such as quartz or feldspar, or rock fragments. **Silt** grains range from 0.004 millimeters to 0.067 millimeters (0.0002 inches to 0.003 inches) and to see them you need to magnify them through a hand lens. Although silt is too fine to feel gritty between the fingers, it does feel gritty between your teeth. The finest particles, known as **clay**, are less than 0.004 millimeters across. Note the use of the term clay in this context is based on particle size. These particles are not necessarily clay minerals, which are defined on the basis of their composition and layered silicate structure (see Chapter 3: Minerals). Fortunately, most clay-sized particles are also clay minerals, but other minerals, such as quartz, may also be present.

We classify detrital sedimentary rocks according to the size of the grains they contain (Fig. 6.12). Detrital sedimentary rocks containing fragments greater than 2 millimeters (0.08 inches) in diameter, such as gravel, pebbles, and boulders, are **conglomerates** if the fragments are rounded or **breccias** if they are angular. Because of their coarse grain size, both conglomerates and breccias provide valuable evidence as to their origin, as we discussed in Section 6.2. The fragments themselves yield information on the source area where the sediment originated. The degree to which the fragments are rounded provides information on their mode of transport and deposition.

Sandstones are medium grained; they are made up of sand-sized grains with diameters between 0.067 millimeters and 2 millimeters (0.003 inches to 0.08 inches). Sandstones can be highly variable and this variety provides insights into the ancient sedimentary environments in which they were deposited. As a result, there is a classification scheme solely for sandstones. Classification depends on the composition of the clasts and the abundance of fine-grained material, known as the matrix, between those clasts. For example, a sandstone in which feldspar accounts for more than 25 percent of the grains is an **arkose**. Because feldspar is highly susceptible to chemical weathering, arkose forms in conditions that inhibit this process, such as a

Detrital Sedimentary Rocks

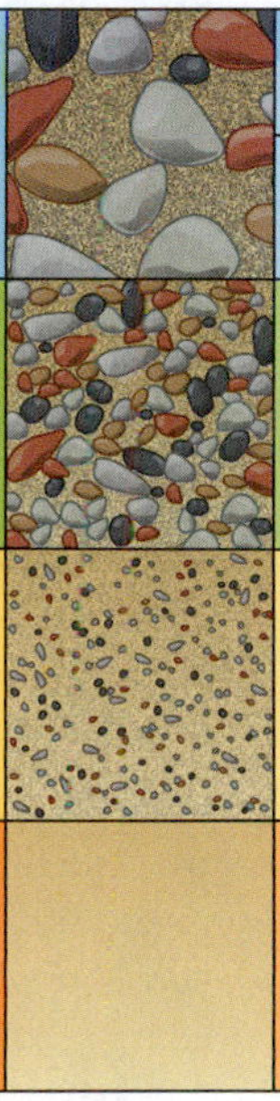

Texture (grain size)		Sediment Name	Rock Name
Coarse (over 2 mm)		Gravel (rounded fragments)	Conglomerate
		Rubble (angular fragments)	Breccia
Medium (1/16 to 2 mm)		Sand	Sandstone (if abundant feldspar is present the rock is called Arkose)
Fine (1/16 to 1/256 mm)		Silt	Siltstone
Very fine (less than 1/256 mm)		Mud	Mudstone or Shale

Figure 6.12 Classifying Clastic Sedimentary Rocks

[Handwritten margin notes: "inorganic carbonates - evaporites - calcite @ 50% - dolomite"; "Gypsum @ 80%"; "Halite @ 90%"]

dry climate or a period of relatively rapid erosion and limited transportation. In **quartz sandstone**, on the other hand, more than 90 percent of the mineral grains are made up of quartz, which is very resistant to chemical weathering. Hence, quartz sandstone forms in conditions that promote the weathering and removal of less stable grains, such as a humid environment, significant transportation, or long exposure to weathering.

Sandstones commonly contain a matrix of finer grained clay or silt between the sand-sized grains. Relatively pure quartz sandstones are typically light colored. In contrast the presence of a fine-grained matrix gives the rock a dirty appearance, and sandstones containing appreciable matrix are often dark in color. If more than 15 percent of the rock is composed of matrix, the sandstone is called a **graywacke**. Graywackes are predominantly deposited in marine environments by dense currents of sediment-rich water known as **turbidity currents**. Turbidity currents are generated when sediments slide off the edge of the **continental shelf**, which is the broad submerged region bordering a continent where the sea is quite shallow.

Mudstones and **siltstones** are very fine-grained varieties of detrital sedimentary rock dominated by either clay-sized or silt-sized particles, respectively. Mudstones that contain thin layers, and tend to split along those layers, are called **shales**.

> **CHECK YOUR UNDERSTANDING**
> ○ List the following rocks in order of increasing grain size: sandstone, conglomerate, siltstone, mudstone.

CLASSIFYING CHEMICAL AND BIOCHEMICAL SEDIMENTARY ROCKS

We classify chemical and biochemical sedimentary rocks according to their composition (Fig. 6.13). The oceans contain abundant dissolved salts thanks to the chemical weathering of minerals such as carbonates, sulfates, and feldspars. As we discussed in Section 6.2, when ocean water evaporates, pure water leaves the sea, and the dissolved salts are left behind. The concentration of the dissolved salts consequently rises and may increase to the point of saturation, which results in the precipitation and growth of crystals.

Recall from Section 6.2, that evaporites are a family of chemical sedimentary rocks that form by the precipitation of salts left behind by the evaporation of seawater and are entirely inorganic in origin. When seawater is heated in a laboratory, evaporite minerals first form when about 50 percent of the original water has evaporated. The first minerals to form are carbonates, typically calcite ($CaCO_3$) or dolomite ($CaMg(CO_3)_2$), which form deposits of limestone and dolostone, respectively. When about 80 percent of the original water has evaporated, the mineral gypsum ($CaSO_4 \cdot 2H_2O$) starts to crystallize, forming a rock also known as gypsum. At about 90 percent evaporation, halite crystallizes, forming deposits of rock salt. Anhydrite ($CaSO_4$) is another common evaporite mineral and forms in environments where gypsum is heated to about 40°C (104°F). Gypsum is used for plastering, and anhydrite is used as plaster of Paris in the casts that protect broken bones.

While evaporites form inorganically, organic processes are important in the formation of biochemical sedimentary rocks. Recall that biochemical sedimentary rocks are composed of the fossil remains of the organisms themselves or from the remains of their skeletons. In addition to limestone, which we discussed previously, other examples of biochemical sedimentary rocks are chalk and coal.

Chalk is a white, soft porous biochemical sedimentary rock made up of the skeletal remains of microorganisms no larger than a pinhead that drift near the sea surface. It is familiar to us as blackboard chalk and as the deposits that form the White Cliffs of Dover along the southeast coast of England (Fig. 6.14).

Coal is a black biochemical sedimentary rock comprised of highly compressed plant remains, such as tree trunks, roots, twigs, leaves, or moss. It is largely composed of carbon and is a major source of the energy that fuels modern society.

Coal forms almost exclusively in continental environments and

Chemical Sedimentary Rocks

Composition	Texture (grain size)	Rock Name	
Calcite $CaCO_3$	Fine to coarse crystalline	Crystalline limestone	
		Travertine	
	Shells and shell fragments cemented by calcite	Fossiliferous limestone	Biochemical limestone
	Microscopic shells and clay	Chalk	
Quartz SiO_2	Very fine crystalline	Chert (light colored)	
Gypsum $CaSO_4 \cdot 2H_2O$	Fine to coarse crystalline	Gypsum	
Halite $NaCl$	Fine to coarse crystalline	Rock salt	
Altered plant fragments	Fine-grained organic matter	Coal	

Figure 6.13 Classifying Nonclastic Sedimentary Rocks

Biochemical
are organic
coal

Figure 6.14   Chalk
Chalk is exposed along the White Cliffs of Dover.

provides evidence of a humid climate with lush vegetation (like that of modern tropical rain forests) at the time of its formation. Plant remains are not preserved if they are exposed to the atmosphere because they decompose in the presence of atmospheric oxygen. Accumulation of plant remains therefore only occurs in oxygen-deficient environments such as swamps, or in environments where they are rapidly buried by sediments. Coal is the end product of the progressive burial of plant remains in oxygen-starved conditions (Fig. 6.15). The first product is a water-saturated organic rich deposit called *peat*. With burial, compaction, and heat, the peat becomes more concentrated and turns into a brown coal called *lignite*. Continued burial increases the temperature and pressure still further, first forming *bituminous coal,* which is the most common form of coal in North America, and eventually a jet-black coal called *anthracite*. The formation process increases the carbon content (and so the fuel content) from about 60 percent in peat to about 90 percent–95 percent carbon in anthracite.

Most of the coal deposits in eastern North America and western Europe formed between 300 and 250 million years ago when both land masses resided in the tropics (Fig. 6.16). The matching distribution of these coal belts was one of the arguments used by Alfred Wegener (see Chapter 2) in proposing that North America and Europe had united to form part of the supercontinent Pangea at about this time (see Living on Earth: **King Coal** on page 178).

> **CHECK YOUR UNDERSTANDING**
>
> ● Which is *not* a chemical sedimentary rock? (a) rock salt, (b) gypsum, (c) coal, (d) anhydrite, (e) shale

SEDIMENTARY ROCKS OF BOTH CHEMICAL AND BIOCHEMICAL ORIGIN

As we noted in Section 6.2, some sedimentary rocks form in ways that can be either chemical or biochemical in origin.

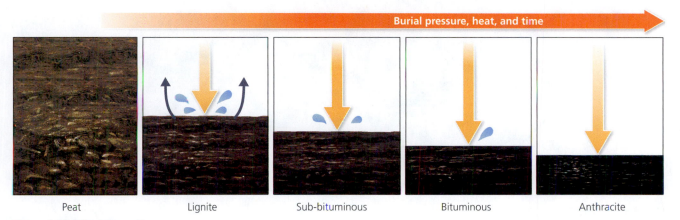

Burial pressure, heat, and time

Peat   Lignite   Sub-bituminous   Bituminous   Anthracite

Figure 6.15   Coal Formation
Coal is formed from highly compressed plant remains. Initially, peat is formed, but increasing compaction leads to the formation of lignite, bituminous coal, and anthracite.

LIVING ON EARTH

King Coal

The historical importance of coal cannot be overemphasized. The Industrial Revolution started in England in the late eighteenth century and by the early nineteenth century had spread to western Europe and North America. The change from wood to coal as a major energy source accompanied the introduction of steam power, which resulted in industrialization and the radical change from hand production to the use of heavy machinery.

The strategic importance of coal also led to conflict, and many wars were fought to secure a guaranteed supply of the resource. Prior to the outbreak of the Second World War, for example, Hitler invaded the French coal fields of the Ruhr valley expressly for the purpose of securing fuel for the Nazi war machine.

Because coal forms from the decomposition of vegetation and is a source of energy, it is classified as a *fossil fuel*. The energy it provides is that which was stored by the plants when they were alive. The proliferation of fossil-fuel dependent industry that started with the Industrial Revolution led to the first significant assault by humans on Earth's global ecosystems. When coal is burned to generate energy, *greenhouse gases,* which may contribute significantly to the problem of climate change, leak into the atmosphere.

The main centers of the Industrial Revolution (North America and Europe) owe their vital role in this relatively recent historic period to the way in which the coal deposits were distributed within the supercontinent Pangea some 300 million years ago (see Fig. 6.16). By geologic accident, when Pangea split up and the Atlantic Ocean formed, most of the coal deposits were left on the North American and European plates rather than in South America and Africa. Hundreds of millions of years later, this geological accident was key to the course of human history.

Best known of these are chert and limestone. **Chert** (see Fig. 6.13) is made up of microcrystalline quartz (SiO_2) and can form in two ways. It can be the product of inorganic chemical precipitation, in which case it is classified as a chemical sedimentary rock. But chert can also form from organisms that extract silica from seawater. Some organisms secrete silica rather than calcite, so when they die, their shells and other skeletal remains are deposited as silica-rich layers. In this case, the chert is a biochemical sedimentary rock. But distinguishing between these two modes of origin is often difficult.

> **CHECK YOUR UNDERSTANDING**
>
> ⊙ Which are clastic sedimentary rocks?
> (a) sandstone, (b) limestone, (c) mudstone, (d) rock salt

Limestone is a sedimentary rock dominated by the mineral calcite ($CaCO_3$) or calcium carbonate. As we noted in Section 6.2, limestone has several modes of origin (see Fig. 6.6). Limestone can form inorganically as chemical sedimentary rocks in shallow and warm open seas. However, other limestones, such as the

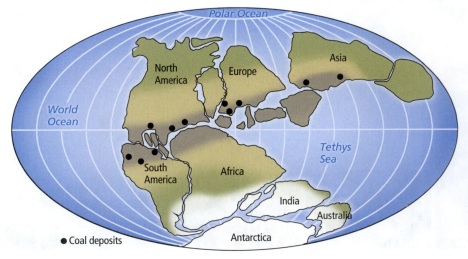

Figure 6.16   The Supercontinent Pangea

The distribution of 250- to 300-million-year-old coal deposits in North America and western Europe was one of several lines of evidence used by Alfred Wegener (see Chapter 2) to propose that these continents were part of one large supercontinent, Pangea, at the time the coal deposits were formed.

[handwritten notes at top: "reefs Chemical Sedimentary Rock = Limeston / Biochemical + chemical origins inorganic"]

coral limestone that makes up coral reefs as we discussed in Section 6.2, are biochemical in origin.

6.3 SUMMARY

- We classify sediment particles according to their size, and to some extent, their shape.

- We classify detrital sedimentary rocks according to the size of the grains they contain. Conglomerates, breccias, sandstones, mudstones, and siltstones are all detrital sedimentary rocks.

- In chemical sedimentary rocks, evaporation of sea water concentrates the dissolved salts until saturation causes the precipitation of evaporite deposits. Minerals and rocks formed by the evaporation of ocean water are inorganic.

- Coal is a biochemical sedimentary rock that is mostly carbon and is formed from highly compressed plant remains. Chalk is a biochemical sedimentary rock made up of the skeletal remains of microorganisms. Limestone has several modes of origin and can be either chemical or biochemical in origin or both.

6.4 Interpreting Sedimentary Rocks

Sedimentary rocks are the end product of a wide variety of processes. Fortunately, there are features within these rocks that allow us to determine the ancient environments in which they formed. For example, in Section 6.3 we discussed how evaporite deposits form in shallow, warm seas, whereas coal deposits form in tropical, humid environments. In this section, we concentrate on the methods used to interpret detrital sedimentary rocks.

SOURCE AREA

Sediments rarely come from just one place. For example, recall that the sediments deposited at the mouth of the Mississippi River contain clasts from innumerable localities within the vast Mississippi drainage system (see Fig. 6.4). In ancient sedimentary rocks, therefore, the identification of individual clasts is crucial to determining the source area.

Other features such as the mineral content, and the degree of rounding and sorting, indicate the proximity of a deposit to its source. For example, recall from Section 6.3 that we can deduce that a poorly sorted sandstone that contains angular fragments and unstable minerals (such as feldspar) was deposited closer to its source area than a well-rounded, well-sorted quartz sandstone. The combined effect of chemical and physical weathering on a granitic source rock is shown in Figure 6.17. Chemical weathering removes soluble ions (e.g., Ca, Na, Mg) throughout the process. Unstable minerals (amphibole, feldspar, biotite) are preferentially broken down. As material is transported, it is sorted and rounded. The end product of weathering and erosion is therefore a well-rounded, well-sorted deposit of quartz sand, which may consolidate into quartz sandstone.

FOSSIL CONTENT

Fossils are the preserved remains of ancient plant and animal life. Most fossils form when organisms are buried by layers of sediment and decompose so slowly that the sediment consolidates around them, thereby preserving their shape. Fossil identification provides information on the age of the rock and the environment in which the rock was deposited.

By studying the fossil content of sedimentary successions that span more than 3 billion years of Earth's history, paleontologists have produced a chronology documenting the evolution of life. We can compare the fossil content at

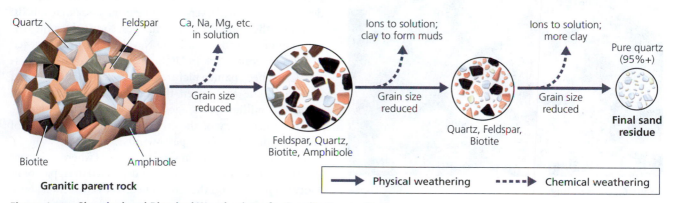

Figure 6.17 **Chemical and Physical Weathering of a Granite Source Rock**

At each stage, chemical weathering removes soluble ions (Ca, Na, Mg). Less stable minerals (amphibole, feldspar, biotite) are preferentially broken down and fine-grained clays produced by the weathering of feldspar are readily removed. The end result is a quartz-rich sand deposit.

any one locality with this chronology to deduce the depositional age of the sedimentary rock.

The fossil content of a rock also provides critical information about the nature of the organisms that lived at the time of sedimentation, and, therefore, the depositional environment. For example, coal deposits often preserve remnants of ancient vegetation whereas limestones may contain remnants of shelly organisms (see Fig. 6.10). From this information, we know that the coal formed inland and the limestone formed along an ancient seashore.

SEDIMENTARY STRUCTURES

Often, detrital sediments contain physical features, known as **sedimentary structures**, that reflect the conditions under which the sediment was deposited. There are many different types of sedimentary structures. Some occur within individual layers, some at the top of a layer, and others at the bottom.

Bedding

Spectacularly preserved in the Grand Canyon (Fig. 6.18), the most obvious and important sedimentary structure is **bedding**. Sediments are deposited in layers, or **beds**. Each bed has certain distinguishing characteristics such as the

Figure 6.18 Bedding
Sedimentary bedding is spectacularly exposed in the walls of the Grand Canyon, Arizona.

composition, shape, size, and sorting of its component particles. Subtle changes in depositional environment may produce a different set of characteristics in the overlying sedimentary layer. This results in a different bed that is separated from the underlying layer by a **bedding plane**. Differences between adjacent layers divide sedimentary rocks into a sequence of parallel beds, separated by bedding planes. Bedding planes represent temporary breaks in the depositional history of an area; they reflect the end of one depositional event and the beginning of another. Differences in sediment characteristics may affect the color of the adjacent layers and bedding planes can also be defined by changes in color. However, there are many processes that affect the color of rocks, so color differences are not the only criteria used to identify bedding planes.

How much time does the deposition of an individual bed represent? This is a very thorny problem, and in most regions the answer is unknown. In modern environments, sedimentation rates vary considerably. Large streams, like the Mississippi River, carry vast volumes of detritus, which they dump at the river mouth where the stream slows down as its waters empty into the ocean. If the ocean currents are not too strong, the sediment will accumulate at the river mouth to form a **delta**. Sedimentation rates in deltas can reach several meters per year and the beds that are formed in deltas are among the thickest of any depositional environment. In contrast, sediment transport into the deep oceans is very limited and sedimentation rates may be only millimeters per thousand years. Consequently, deep sea sediments commonly have very thin beds and typically consist of microscopic remains of organisms and clay minerals.

Sediment, whether transported by water, wind, or ice, is generally deposited in more or less horizontal beds. However, inclined layers, known as **cross beds**, sometimes occur. The particles that form typical horizontal beds settle through still water. But the particles that form cross beds are deposited by a current of moving air or water (Fig. 6.19a). For example, in desert environments, windblown sand is eroded from the upwind side of sand dunes and deposited on the downwind side. The steepest angle at which loose sand is stable is about 30°, so the beds on the downwind side are deposited as inclined layers, resulting in large-scale cross beds.

When the wind direction changes, so too does the orientation of the cross beds. The Navajo Sandstone exposed in Zion National Park in Utah, displays these characteristics. Cross beds paint a picture of the desert conditions that existed in the area when the sandstone was deposited, some 180 million years ago (Fig. 6.19b). We can deduce from the geometry of the cross-bedding the direction of ancient winds, known as **paleowinds**.

Cross beds also form in stream deposits where sediment transported by a moving current is deposited in ripples or in a local depression on the stream bed. The sediment is, once again, deposited at an inclined angle and so forms a cross bed that slopes in the downcurrent direction. As sedimentation

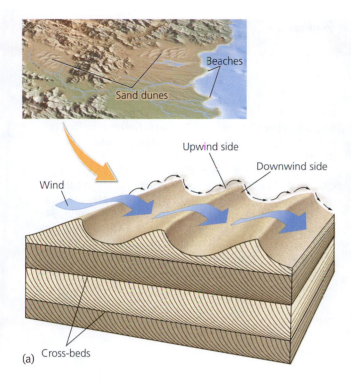

(b)

Figure 6.19 Cross-Bedding

(a) These cross beds were formed by wind. (b) Dune cross-bedding is seen in this outcrop of Navajo Sandstone in Zion National Park.

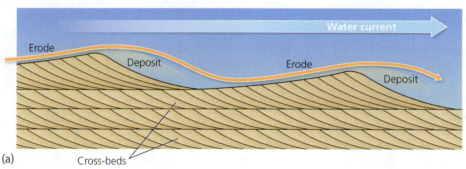

(a) Cross-beds

(b)

Figure 6.20 Formation of Cross Beds by Moving Water

These cross beds were formed by moving water. (a) Moving water causes grains to be eroded and transported downstream until they fall over the crest and are deposited at an inclined angle. As sedimentation proceeds, cross beds are covered by younger beds. (b) These cross beds are located at MacIsaac Point, Nova Scotia.

proceeds, these cross beds are covered by later cross beds, which are likewise deposited at an inclined angle in the down-current direction (Fig. 6.20). Cross beds can consequently provide important information concerning the flow direction of ancient currents, known as **paleocurrents**. Because streams constantly change their path as they meander back and forth across the floodplain, the direction of the current also changes, and this is recorded by the changing orientation of cross beds.

Cross beds also form in deltas because sediment is rapidly deposited when a stream's velocity is dramatically reduced upon entering a lake or the sea. Particles that drop immediately are deposited at an angle to form cross-beds that slope down current. As sedimentation proceeds the cross beds are, once again, covered by younger ones.

Because cross beds occur in a wide variety of environments, we cannot use them to identify the unique environment of deposition. When a single feature cannot uniquely distinguish between a range of possibilities, geologists rely on the overall *context* of the rock. For example, in the case of the Navajo Sandstone, the presence of dinosaur footprints, and well-sorted, polished sand grains indicates windblown deposition on land. Deltaic sediments, on the other hand, often include layers that contain marine or freshwater fossils. Geologists interpret information about sedimentary rocks and formulate the simplest explanation of all the available data.

Some beds display systematic changes in particle size. For example, the coarsest grains occur at the bottom of the bed and progressively give way to finer grains toward the top (see Science Refresher: **Settling Velocities** on page 182). This sedimentary structure is known as **graded bedding**.

Settling Velocities

Detrital sediments form when particles settle to the base of a body of water. It is a basic principle of physics that particles tend to move in the direction of an applied force. In a static body of water, particles tend to move vertically downward under the influence of the force of gravity.

If settling clasts are independent of one another, then there are two main forces to be considered: gravity which acts downward; and a drag force which resists the movement of the particle through the water column.

Initially, the velocity of a clast increases as it sinks. But eventually this downward force is balanced by the drag force and no further change in the clast's velocity occurs until the particle settles at the base of the water body. The velocity of the clast before it settles is the *terminal velocity* or the **settling velocity** of the clast and can be determined by experiment.

The settling velocity is affected by any property of the clast that affects the drag force through the water column. Such properties include the density of the clast, its size, and its shape. For spherical particles, settling velocity is predicted by *Stokes' Law* which states that the settling velocity increases with the density and size of the sinking clast. Stokes' Law offers an explanation of how graded bedding forms. The smaller particles move more slowly and therefore end up on top when they settle (Fig. 6.21e).

Graded bedding is a common feature of sediments deposited by turbidity currents. Recall from Section 6.3 that the rapid buildup of sediment eroded from continents can render localized areas of the continental shelf unstable, especially if deposition is accompanied by earthquake activity. The result is a turbidity current consisting of a dense current of sediment-laden water, acting like sludge pouring into a bathtub, which moves downslope into submarine canyons or off the edge of the continental shelf (Fig. 6.21a–c). Eventually, the turbidity current slows down and the sediment is deposited in a *submarine fan,* in which the coarsest particles settle to the bottom first, followed by progressively finer grained particles to form a graded bed of graywacke (Fig. 6.21d, e). A single graded bed may have gravel particles at its base and a clay-rich top.

> **CHECK YOUR UNDERSTANDING**
>
> ◔ What does bedding indicate about depositional environment?

Other Structures

Cross beds and graded bedding are examples of features that occur *within* a larger bed. Some sedimentary structures, such as *mudcracks* and *ripple marks,* occur on the surface of beds. Others structures, such as loading and fluting, occur at the base of a bed and are collectively known as *bottom structures.*

Mudcracks form when very fine-grained sediment is exposed to air, and shrinks as it dries out (Fig. 6.22). The shrinkage results in a polygonal pattern of cracks. Mudcracks may be preserved in the geological record if the cracked mud solidifies to form mudstone. When buried by the next layer of sediment, the cracks are filled, in which case they may be preserved as impressions, or *casts,* in the overlying layer.

Ripple marks are small curving ridges that form when wave motion or wind or water currents agitate the surface of a sediment layer. The back-and-forth motion of waves produces *symmetric* ripples with crests that are perpendicular to the wave motion (Fig. 6.23a, b). Where wind or water currents flow in a persistent direction, ripple crests also form perpendicular to the current direction but are *asymmetric,* with a relatively steep side that slopes downcurrent (Fig. 6.23c, d). Ripple marks consequently provide evidence of the sediment's environment of deposition and, in the case of asymmetric ripples, can be used when they appear in ancient rocks to determine the direction of the paleocurrent.

Whereas cross beds and graded bedding occur within a bed and mudcracks and ripple marks form on the bedding surface, **bottom structures**, such as *flute casts* and *load casts,* occur at the base of a bed. In marine settings, sediment deposition may fluctuate dramatically. Relatively tranquil conditions during which thin layers of mud are deposited, for example, are sometimes interrupted by turbidity currents that deposit coarse-grained, poorly sorted graywackes. When a turbidity current passes, it may initially scour spoon-shaped hollows in the mud layer, forming depressions known as *flutes.* The tail of the current then fills the flutes with sand to form **flute casts**. Flute casts are just one of a number of groove-like features that indicate the passage of a turbidity current. But while flute casts form as hollows or grooves in the mud layer, they are usually preserved at the base of the overlying graywacke bed (Fig. 6.24). Hence, they are bottom structures.

Another common bottom structure, known as **load casts**, forms in similar settings, but *after* the passage of the turbidity current. When graywacke sediment is deposited by the turbidity current, the coarser fragments are deposited first. This heavier sand sinks into the mud in the underlying bed to form bulbous protrusions while the mud

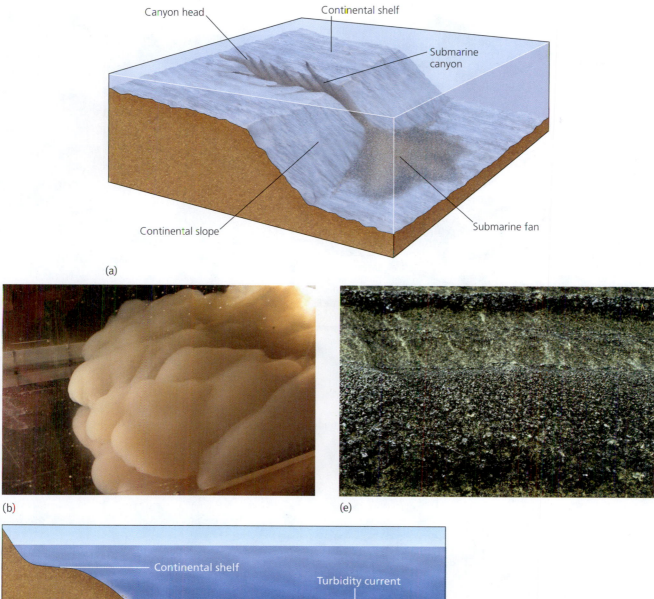

(a)

(b)

(e)

(c)

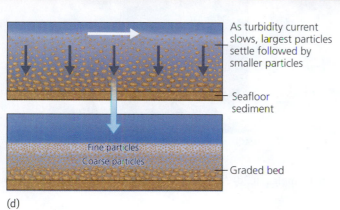

As turbidity current slows, largest particles settle followed by smaller particles

Seafloor sediment

Fine particles
Coarse particles

Graded bed

(d)

Figure 6.21 Graded Bedding

(a) Continental shelves are commonly dissected by submarine canyons. (b) Model showing how instabilities on continental shelves can cause submarine avalanches where sediment is carried into these submarine canyons. (c) These instabilities also carry sediment from the shelf in sediment-laden turbidity currents that move downslope along the seafloor. (d) Eventually, the turbidity current slows down and the sediment is deposited in a submarine fan, with the largest particles settling to the base of the bed. (e) Graded beds are deposited in order of decreasing grain size. This photo of an outcrop of turbidites in southern California shows grading from coarse to fine.

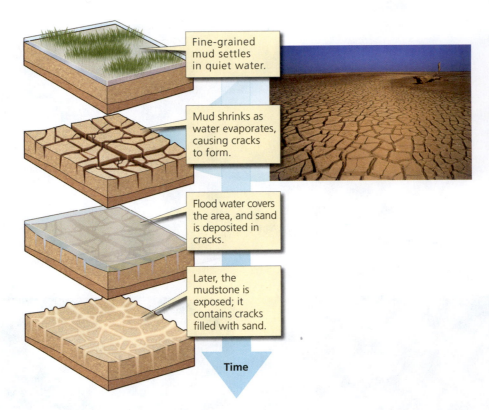

Fine-grained mud settles in quiet water.

Mud shrinks as water evaporates, causing cracks to form.

Flood water covers the area, and sand is deposited in cracks.

Later, the mudstone is exposed; it contains cracks filled with sand.

Time

Figure 6.22   Mudcracks

Mudcracks form where clay-rich sediments dry and shrink. The mudcracks in the photo are from a dry river bed in the Namib Desert, southern Africa.

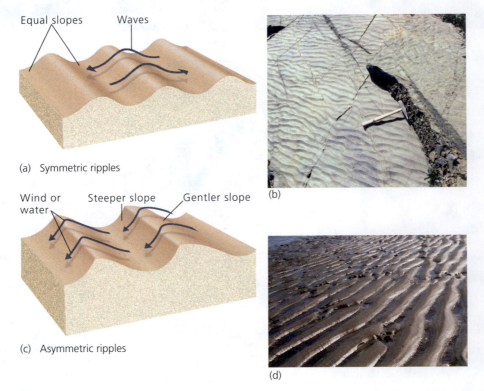

Equal slopes   Waves

(a)   Symmetric ripples

(b)

Wind or water   Steeper slope   Gentler slope

(c)   Asymmetric ripples

(d)

Figure 6.23   Ripple Marks

(a, b) The to-and-fro motion of waves produced symmetric ripples. (c, d) A current that flows in one dominant direction produces asymmetric ripples.

Figure 6.24   Flute Casts

Groove-like structures called flute casts form along the bottom of a bed when sand fills in the spoon-shaped hollows that are scoured out of the soft underlying layer.

Figure 6.25   Load Casts

Load casts form at the bottom of a bed when soft sand (yellow) sinks into an underlying layer of unconsolidated mud (dark brown).

rises into the sand in relatively narrow ridges (Fig. 6.25). In some situations, earlier structures such as ripple marks cause unequal loading of the mud so that their shape becomes modified as the mud starts to move.

We identify and interpret sedimentary structures to understand the environments in which sedimentary rocks are deposited. So far in this chapter, we have emphasized individual rock types and sedimentary structures. In the following sections, we apply these concepts to *sequences* of sedimentary rocks. The sequence of layers records the history of sedimentation and all changes in depositional environment. Because sedimentary layers are stacked one on top of the other, they can be read like the pages of a book. In Section 6.5, we will see how knowledge of depositional environments allows us to reconstruct ancient landscapes and track the rise and fall of sea level through time.

CHECK YOUR UNDERSTANDING

● What is the key difference between bottom and top structures in sedimentary rock?

6.4 SUMMARY

- To interpret sedimentary rocks we identify the source area, fossil content, and sedimentary structures.

- Bedding is the most important sedimentary structure. A bed is a layer of sediment with unique characteristics, such as the composition, shape, size, and sorting of its particles. A bedding plane represents a temporary break in the depositional history of an area.

- Other important sedimentary structures provide information about the environment of deposition and the direction of water or wind currents responsible for sediment transport. They include cross beds and graded bedding which occur within the bed, mud cracks or ripple marks which form at the top of a bed, and flute and load casts which occur at the base of the bed.

6.5 Depositional Environments and Sedimentary Facies

Much can be learned about the history of Earth by studying the ancient depositional environments preserved in the sedimentary rock record. For example, many of the features of limestone deposited over three billion years ago are similar to those of recent limestones. This observation tells us that depositional environments similar to those of today may have existed early in Earth's history.

In fact, most of our understanding of ancient depositional environments is derived from detailed studies of modern settings of sediment deposition. Geologists apply findings about modern depositional environments to ancient sedimentary rocks based on the supposition that the sedimentary characteristics of modern depositional environments are duplicated by those of the past. As we will learn in Chapter 8, this idea that "the present is the key to the past" is one of the basic principles of the science of geology.

In modern settings, sediments are deposited in continental or marine environments, or in the coastal zone lying between the two. Different sediments are deposited at the same time in each of these different depositional environments (Fig. 6.26). For example, in coastal zones, coarser clasts are typically deposited closer to shore than finer grained clasts. Likewise, finer grained clay particles become increasingly dominant oceanward, and finally, where the waters are relatively clear, limestone deposits may form. When these sediments consolidate, sandstone is located closer to the ancient shore than shale, while limestone forms farther out to sea. Should the continental shelf become unstable, a turbidity current may transport sediment into deeper ocean water, leading to the deposition of sediment that will later form graywacke. All of these sediments are deposited at about the same time, but in different parts of the same continental margin. This variability is expressed in the concept of *sedimentary facies*.

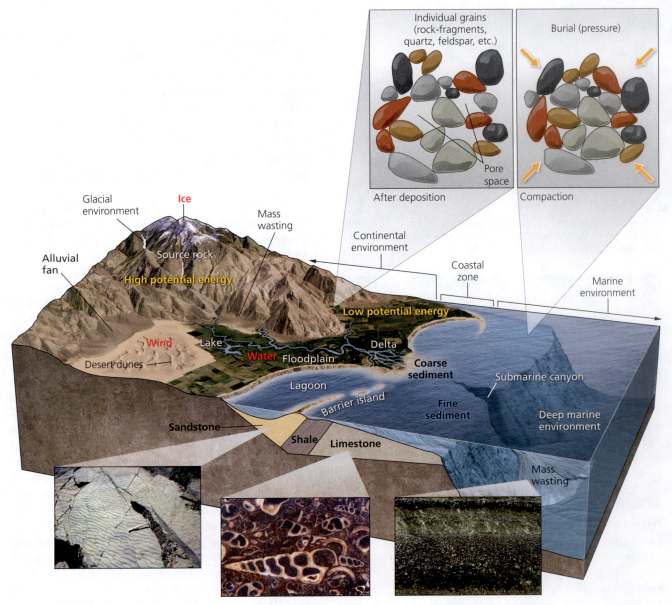

Figure 6.26 Major Depositional Environments
Sediment is deposited in continental environments near rivers and their floodplains, lakes, glaciers, or in deserts, in shallow or deep marine environments, or in the coastal zone that lies between the two and includes tidal flats, barrier islands, and lagoons. Photos show examples from the coastal zone (sandstone with symmetrical ripples), organic reefs (bioclastic limestone), and submarine fan (graywacke with graded bedding). Inset shows compaction and cementation of sediments in continental and shallow marine environments. The typical distribution of sandstone, shale, and limestone in coastal environments is shown in the foreground.

The term **sedimentary facies** is used to describe an association of sedimentary beds with differing characteristics, all of which were deposited at the same time within a particular sedimentary environment. For example, sandstone beds deposited close to an ancient shore may display a variety of features. Some beds may contain sedimentary structures such as wave-formed cross beds and ripple marks, whereas others may be interbedded with shelly or pebbly beds typical of beach settings. The assemblage as a whole is best described as a *sandstone facies*. In a coastal environment, there is usually a sandstone facies, a shale facies, and a limestone facies (Fig. 6.27). The boundary between these facies is typically gradual.

The dynamic nature of Earth's geologic past is reflected in the continual change in sedimentary environments and the resulting sedimentary facies developed upon its surface. In the rest of this section we discuss these different facies and what they can tell us about Earth's past.

> **CHECK YOUR UNDERSTANDING**
>
> ● In a typical coastal environment, why is sand deposited near shore?

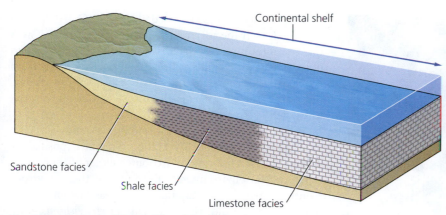

Figure 6.27   Coastal Depositional Environment
Sedimentary facies transition from sandstone to shale and, eventually, to limestone farther from the shore.

DEPOSITION IN CONTINENTAL ENVIRONMENTS

Continental environments are typically dominated by the action of streams, which dissect the landscape and carve out broad floodplains (see Fig. 6.26). In cold climates, however, ice may be the principal agent of erosion and deposition in continental environments, whereas in hot desert regions, winds have the dominant effect. Because each of these agents of erosion and deposition produces sediment of quite different character, the nature of the sedimentary rock formed in continental environments is strongly influenced by climate. Recall from our discussion of grain size in Section 6.2 that desert sands are often made up of windblown grains. Winds are effective sorters of detritus so these grains are very similar in size, and are well sorted. When glaciers melt, on the other hand, debris ranging from boulders to mud is deposited rapidly side by side and such deposits are poorly sorted.

This difference reflects the varying abilities of these agents of erosion to sort material as they transport it and is best displayed by flowing water, as we discuss in more detail in Chapter 13. Coarser clasts (sand and gravel sizes) roll, slide, or hop along the stream bed (a process called *saltation*), while finer clasts (silt and clay sizes) are held in suspension by the turbulent flow of the stream (Fig. 6.28). Channel deposits, therefore, tend to be dominated by coarser-grained gravel and sand. You may have observed that rivers become muddier following major storms. This is because when stream velocity is high the water has an enhanced ability to erode and transport detritus. Deposition occurs when the stream slows down, with the coarsest material being deposited first.

Stream velocity can drop suddenly in several environments. For example, steep, fast-flowing mountain streams abruptly slow down when they leave their narrow valleys and spread out onto the broad plain beyond. When this happens, a triangular-shaped lobe of sediment known as an *alluvial fan* is deposited where the mountain meets the valley (Fig. 6.29).

Similarly, during flooding, streams abruptly slow down when they overflow their banks and spread across the flat floodplain. When this happens, poorly sorted gravels and sands are deposited adjacent to the riverbank. These deposits form linear ridges, known as *natural levees* (Fig. 6.30). The finer-grained sediment, in contrast, is held in suspension and spreads across the floodplain to form deposits of mud and silt. Mudcracks (see Fig. 6.22) may form when floodplain deposits dry out between floods.

The sands and gravels deposited by streams may closely resemble those of an alluvial fan. However, their distribution distinguishes them. Deposits along the main course of a stream form levees and are found adjacent to finer-grained floodplain deposits. In contrast, we only find alluvial fan deposits in the areas that abut mountain fronts (see Fig. 6.26). By mapping rock distribution, we can

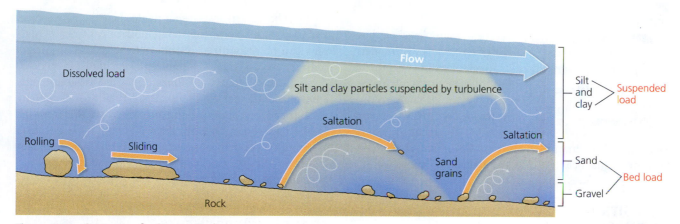

Figure 6.28   Contents of a Stream Bed
Material is sorted and transported in running water. Physical weathering produces clasts. Coarse clasts (sand and gravel) roll, slide, or hop along the stream bed. Finer particles (silt and clay) are held in suspension. The stream also carries dissolved material as a result of chemical weathering.

Figure 6.29   Alluvial Fan

Alluvial fan deposits like this one in Death Valley, California, form where fast-flowing mountain streams abruptly slow and deposit sediment when they reach the flat desert floor (see Figure 6.26).

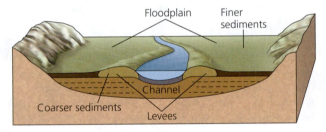

Figure 6.30   Floodplain Deposits

Sands and gravels form a natural levee adjacent to the stream channel; finer muds are distributed across the floodplain.

distinguish between the two environments in the geologic record.

Sediments are also deposited in slow-moving portions of the stream, or where the flow is slowed on entering a body of standing water such as a lake or the ocean. Initially, poorly sorted sediments are deposited. However these deposits may eventually be reworked by currents to form relatively well sorted deposits.

> **CHECK YOUR UNDERSTANDING**
>
> ● How do alluvial fans form?

These examples show how the characteristics of modern continental sediments provide the criteria for recognizing deposits in the geologic record. As we shall see in the following material, sedimentary deposits in coastal and marine environments also have characteristic features that help us to identify ancient coastlines.

DEPOSITION IN COASTAL AND MARINE ENVIRONMENTS

When streams meet the sea, the velocity of the stream diminishes dramatically and huge volumes of sediment are deposited. The Mississippi River, for example, deposits nearly one million metric tons of sediment daily into the Gulf of Mexico (see Fig. 6.4). If the ocean currents are not too strong, as is the case at the mouth of the Mississippi, the rapid deposition of sediments forms a delta. Because of the length of stream transport, well-rounded and sorted sand dominates deltaic sediments. These sediments also display sedimentary structures such as cross-bedding and ripple marks.

Where the ocean currents are strong, as is the case at the mouth of the Amazon River, the sediment deposited by streams is reworked and redistributed either offshore or to other coastal environments. Examples of these environments are *beaches,* muddy tidewater areas known as *tidal flats,* elevated sand ridges called *barrier islands,* and *lagoons,* which are long bodies of seawater isolated from the oceans by barrier islands or reefs (see Fig. 6.26). Each of these environments displays distinct sedimentary facies because they host sedimentary beds possessing characteristics that are distinct from those in adjacent environments. For example, we can distinguish a deltaic facies from a lagoonal facies by the contrasting characteristics of their respective deposits. Although the sediments deposited in each facies are about the same age, they show differences that reflect their different depositional environments.

Beach deposits tend to be composed of well sorted sands because movement of the water in the coastal zone preferentially removes silt and mud. Clay-sized particles are held in suspension longer than sand and so are deposited farther offshore, beyond the coastal zone. Accumulation of these clay-sized particles eventually forms layers of mudstone.

Where the supply of sediment is not overwhelming and the offshore environment is rich in nutrients, limestones will form. Such deposits typically contain abundant fossils. So in tracing a layer of sediment from nearshore to offshore, we may start in a sediment dominated by sand, pass through one dominated by mud, and end with one dominated by limestone, as the proportion of detritus decreases (see Fig. 6.27). Instabilities along the continental shelves can generate turbidity currents which carry sediment from the edge of the continental shelf into deeper water where they are deposited as graded beds (see Fig. 6.21).

The characteristics of sediments deposited in coastal and marine environments along modern continental margins can be recognized in the geologic record of ancient sedimentary rocks. For example, marine mudstones can be distinguished from mudstones deposited in continental environments, such as floodplains, by their fossil content, widespread distribution, and proximity to thick coastal sandstones. More generally, the vast thickness, widespread distribution, and the presence of marine fossils are key characteristics of rocks that allow us to understand where ancient continental margins were located.

COASTAL ENVIRONMENTS AND SEA LEVEL CHANGE

As sea level rises and falls, the position of the coastline continually changes. The distribution of sedimentary deposits in coastal regions is very susceptible to sea level changes, and sea level can fluctuate for a wide variety of reasons. For example, scientists are presently concerned that global warming may cause melting of the polar ice caps, resulting in a rise in sea level that could potentially swamp our coastal regions. Conversely, the spread of continental ice sheets at the onset of the Ice Age increased the volume of water on land, resulting in a dramatic drop in sea level that left the continental shelves exposed. As sea levels rise and fall, coastal regions migrate, as does the near-shore to offshore distribution of sandstone, mudstone, and limestone facies.

Rising sea level produces what is known as a **marine transgression**. During a marine transgression, the relative distribution of sandstone, shale, and limestone migrates inshore so that limestone is deposited on top of mudstone, and mudstone is deposited on top of sandstone (Fig. 6.31a).

Falling sea level, on the other hand, results in a **marine regression**, which produces the opposite pattern, with sandstone deposition on top of shale and shale deposition on limestone (Fig. 6.31b).

An example of a sedimentary sequence typical of deposition when a marine transgression is followed by a regression is shown in Figure 6.32. As sea level rises, the shore moves inland. As a result, the shale facies and the limestone facies migrate inshore so that the beach sandstone is overlain by shale, which is overlain by limestone. As sea level falls, the shore migrates in the opposite direction, and the limestone is overlain first by shale and then by sandstone.

Because the sequence of beds deposited during a marine transgression contrasts with that deposited during a marine regression, geologists can interpret the rise and

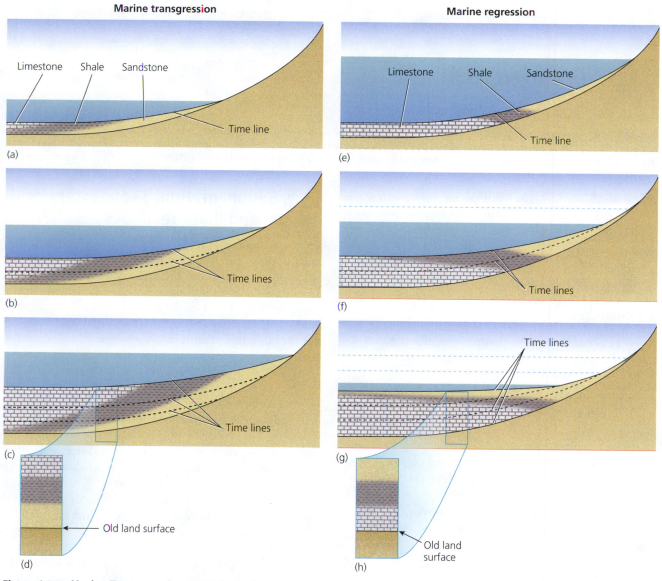

Figure 6.31   Marine Transgression and Regression
The illustrations in (a–c) show the distribution of facies during three stages of a marine transgression, and (d) shows the resulting sequence of sedimentary rocks. The illustrations in (e–g) illustrate three stages of a marine regression, and (h) shows resulting vertical sequence of sedimentary rocks.

Transgression followed by Regression

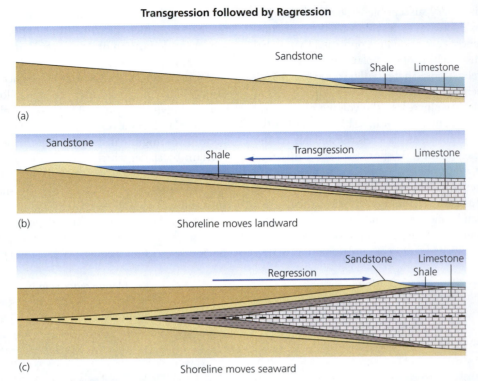

(a)

(b)   Shoreline moves landward

(c)   Shoreline moves seaward

Figure 6.32   Transgression Followed by Regression
This sequence of deposition of sedimentary rocks is recorded when a marine transgression is followed by a marine regression. As sea level rises (a to b), the shore moves landward and sandstone is overlain by shale, followed by limestone. When sea level drops (b to c), the shore moves seaward and the limestone is overlain first by shale and then by sandstone.

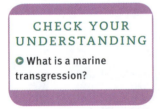

CHECK YOUR
UNDERSTANDING

○ What is a marine
transgression?

fall of sea level with time. These results have been compiled on a global scale, producing a fairly reliable record of sea level change for the past 600 million years.

6.5 SUMMARY

- We can characterize the environment of deposition of sedimentary rock by overall composition, sedimentary structures, and distribution.

- Sedimentary facies are assemblages of sedimentary rocks that possess a distinctive set of characteristics reflecting their deposition in a particular sedimentary environment.

- By understanding modern depositional environments (continental, marine, coastal), geologists can interpret the ancient sedimentary rock record.

- The relationship between erosion, transport, and deposition of sediment is controlled by stream velocity and particle size.

- Sandstone facies deposited in a variety of near-shore environments typically gives way to mudstone and limestone facies farther offshore.

- Rising sea level produces a marine transgression and falling sea level produces a marine regression.

6.6 Sedimentation, Sedimentary Rocks, and Plate Tectonics

In this chapter we have learned that sediment is transported downhill from topographically high regions to areas of lower elevation. The primary driver of such topographic differences is plate tectonics. Plate tectonics provides an elegant explanation for the origin of mountains, and mountains, in turn, provide the gravitational potential for the migration of sediment downslope.

Plate tectonics also explains why some regions have anomalously thick piles of sediment. Thick sediment is deposited in regions where depressions in the crust have formed as a result of *subsidence*. These regions, which we call **sedimentary basins**, trap sediment because of their relatively low elevation. In many cases, the weight of the sediments accentuates the crustal subsidence, providing further space for the deposition of additional sediment.

Sedimentary basins form in a variety of plate tectonic environments but all have a history of subsidence and sedimentation. Sedimentary basins primarily occur within continents or along continental margins, and they host some of the world's most important groundwater, oil, and natural gas resources. Three examples of sedimentary basins include *rift basins, thermal subsidence basins,* and *flexural basins,* all of which owe their origins to plate tectonic processes.

Rift basins are initiated when continents start to break apart (Fig. 6.33a). Rifting causes stretching and thinning of the continental crust, resulting in a *rift valley,* within which continental sediments start to accumulate as the floor of the rift valley subsides in response to faulting and crustal thinning. Because of their relatively low elevation, rift valleys typically have well-developed drainage systems and so are dominated by stream and floodplain deposits. Recall from Chapter 4 that thinning of the crust also leads to a reduction in pressure in the underlying mantle, initiating the rise of basaltic magma. Rift valleys therefore have thick piles of sediment, but may also feature layers of basalt.

Should rifting continue, the continent may eventually be split in two and a new ocean may open between the two continental fragments that steadily widens as a result of sea floor spreading at its mid-ocean ridge (Fig. 6.33b). The two continental fragments cool as they move away from the ridge. As they do so, their density increases, eventually causing their thinned margins to subside below sea level to form a continental shelf. Sediments derived from the adjacent continent progressively accumulate on the continental shelf as the shelf subsides. Because cooling triggers this subsidence, the continental shelf is an example of a **thermal subsidence basin.** Such basins can contain deposits up to 15 kilometers (9 miles) in thickness.

Rift and thermal subsidence basins form at divergent plate boundaries where continents are extended and may break apart. Basins can also form adjacent to convergent plate boundaries where the continental crust is squeezed horizontally so that one part rides up over the other. Here, the weight of the overriding crust causes the crust underneath to flex downward, producing a region of subsidence

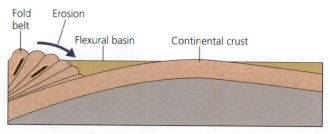

Figure 6.34   Sedimentary Basins in Convergent Plate Tectonic Settings

Flexural basins form in convergent settings, adjacent to regions where the crust has been thickened and uplifted as one part of the crust rides up over the other. The basin receives sediments produced by erosion of the uplifted area.

CHECK YOUR UNDERSTANDING

○ What role does plate tectonics play in sediment transport and deposition?

known as a **flexural basin** (Fig. 6.34). Flexural basins occur in the foothills of the Rocky Mountains where convergence responsible for the uplift of the Rockies has resulted in a region of subsidence. Sediments shed from the adjacent mountains are deposited in these flexural basins. The voluminous supply of this sediment means that organic remains are often rapidly buried, which is one of the most important factors in the formation of fossil fuel deposits. It is therefore no coincidence that flexural basins are host to some of the world's most important reserves of coal, oil, and natural gas.

6.6 SUMMARY

- Mountains, which are formed by plate tectonic processes, provide the gravitational potential for the migration of sediment.

- Sedimentary basins are characterized by thick deposits of sediments and sedimentary rocks; basins form where the crust has subsided.

- Rift basins are initiated when continents start to rift apart. Continental shelves are basins that form by thermal subsidence along the flanks of major oceans.

- Flexural basins form in convergent settings where one part of the continental crust is heaved on top of another and the additional weight causes the adjacent crust to flex downward.

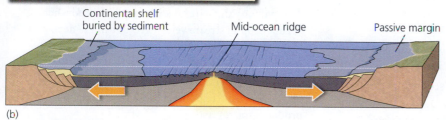

Figure 6.33   Sedimentary Basins in Divergent Plate Tectonic Settings

(a) Rift basins form as continents begin to rift apart. (b) If rifting proceeds and a new ocean opens, continental shelves form as thermal subsidence basins.

Key Terms

arkose 175
bed 180
bedding 180
bedding plane 180
biochemical sedimentary
  rock 166
bottom structures 182
boulder 175
breccia 175
chalk 166
chemical sedimentary
  rock 166
chert 178
clast 166
clastic texture 167
clay 175
coal 170
cobble 175
conglomerate 175
continental shelf 176
coral limestone 171

coral reef 171
cross bed 180
delta 180
depositional
  environment 164
detrital sedimentary
  rock 166
detritus 165
diagenesis 166
dolostone 174
erosion 165
evaporite 170
flexural basin 191
floodplain 165
flute cast 182
graded bedding 181
gravel 175
graywacke 176
limestone 171
lithification 166
load cast 182

marine regression 189
marine transgression 189
mass wasting 165
matrix 167
mudcrack 182
mudstone 176
nonclastic 170
ooid 171
oolitic limestone 171
paleocurrent 181
paleowind 180
pebble 175
poorly sorted 169
pore 166
porosity 166
quartz sandstone 176
rift basin 190
ripple mark 182
rounding 169
rubble 175
sand 175

sandstone 175
saturation 170
sediment 164
sedimentary basin 190
sedimentary facies 186
sedimentary rock 164
sedimentary
  structure 180
settling velocity 182
shale 176
silt 175
siltstone 176
sorting 169
source area 167
source rock 165
thermal subsidence
  basin 191
travertine 170
turbidity current 176
well sorted 169

Key Concepts

6.1 FROM SEDIMENT TO SEDIMENTARY ROCK

- Sediment is the product of physical and chemical weathering, and the transport of solid particles and dissolved ions.
- Sediment is converted into sedimentary rock by compaction and cementation.

6.2 TYPES OF SEDIMENTARY ROCKS

- Detrital sedimentary rocks contain clasts and are the product of mechanical weathering and transport.
- Clastic rocks vary in the degree of sorting and rounding of their grains.
- Chemical sedimentary rocks form from the precipitation of crystals from solutions.
- Biochemical sedimentary rocks are formed by organic processes and commonly contain fossils.

6.3 CLASSIFYING SEDIMENTARY ROCKS

- We classify detrital sedimentary rocks according to the size and shape of their constituent particles.
- We classify chemical and biochemical sedimentary rocks according to their composition.
- Some sedimentary rocks, like limestone, can have several modes of origin.

6.4 INTERPRETING SEDIMENTARY ROCKS

- Interpretation of sedimentary rocks requires knowledge of their source area, fossil content, and sedimentary structures.

- Sedimentary structures provide information about the environment of deposition. Bedding is defined by variations in the composition, shape, size, and sorting of particles. Bedding planes represent temporary breaks in the depositional history of an area.
- Cross beds and graded bedding occur within a bed, mudcracks and ripple marks form at the top of a bed, and flute and load casts occur at the base of the bed.

6.5 DEPOSITIONAL ENVIRONMENTS AND SEDIMENTARY FACIES

- A sedimentary facies is an assemblage of strata with distinctive characteristics that reflect the environment of deposition.
- Knowledge of modern depositional environments allows geologists to interpret the ancient sedimentary rock record.
- Alluvial fans and natural levees are produced when an abrupt drop in stream velocity results in the deposition of poorly sorted sands and gravels.
- Rising sea level produces a marine transgression where sandstone is overlain by mudstone and then limestone facies. A marine regression caused by falling sea level produces the reverse sequence.

6.6 SEDIMENTATION, SEDIMENTARY ROCKS, AND PLATE TECTONICS

- Plate tectonics plays a key role in the transport and deposition of sediment.
- Sedimentary basins form where the crust has subsided.

Study Questions

1. Sedimentary rocks account for about 75 percent of exposed bedrock, but only 5 percent of rocks in Earth's crust. Explain why this is so.

2. Why are sediments produced mainly at Earth's surface?

3. How does mass wasting differ from erosion?

4. Explain how weathering, mass wasting, and erosion act in concert to produce clastic sediments.

5. What characteristics of minerals influence their behavior during transport?

6. Explain why quartz and feldspar have contrasting resistance to weathering.

7. What features in clastic sediments are suggestive of a glacial origin?

8. Explain why limestone and chert may be either chemical or biochemical in origin.

9. Cross beds can form in a variety of environments. What features of the cross beds can you use to distinguish between those environments?

10. Explain how the distribution of sedimentary facies in coastal regions is sensitive to changes in sea level. How can you use the sequence of layer deposition to deduce whether ancient sea level has risen or fallen?

11. Compare the mechanisms responsible for sedimentary basin formation in divergent and convergent plate margins.

7 Metamorphism and Metamorphic Rocks

7.1 Metamorphic Rocks and the Rock Cycle

7.2 Factors That Control Metamorphism

7.3 Types of Metamorphism

7.4 Metamorphic Textures

7.5 Classifying Metamorphic Rocks

7.6 Metamorphic Zones and Facies

7.7 Metamorphism and Plate Tectonics

▶ Key Terms

▶ Key Concepts

▶ Study Questions

SCIENCE REFRESHER: Pressure Units 200

IN DEPTH: Shock Metamorphism 206

IN DEPTH: Metamorphism and the Rock Cycle 215

Regional metamorphism in subduction zones

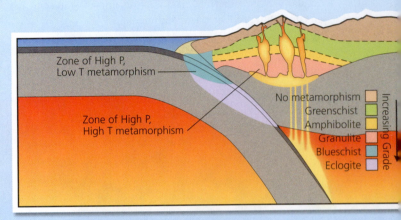

Zone of High P, Low T metamorphism

Zone of High P, High T metamorphism

No metamorphism
Greenschist
Amphibolite
Granulite
Blueschist
Eclogite

Increasing Grade

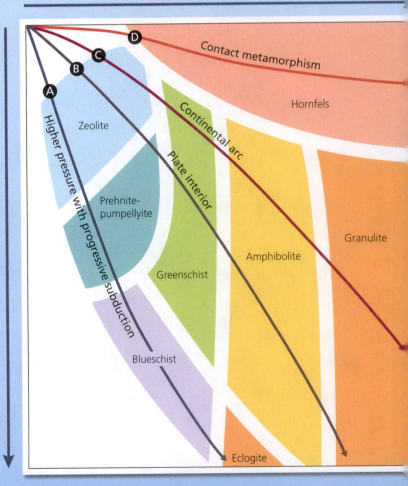

Temperature (°C)

Pressure (Kb)

Higher pressure with progressive subduction

Contact metamorphism

Hornfels

Continental arc

Plate interior

Zeolite

Prehnite-pumpellyite

Greenschist

Blueschist

Amphibolite

Granulite

Eclogite

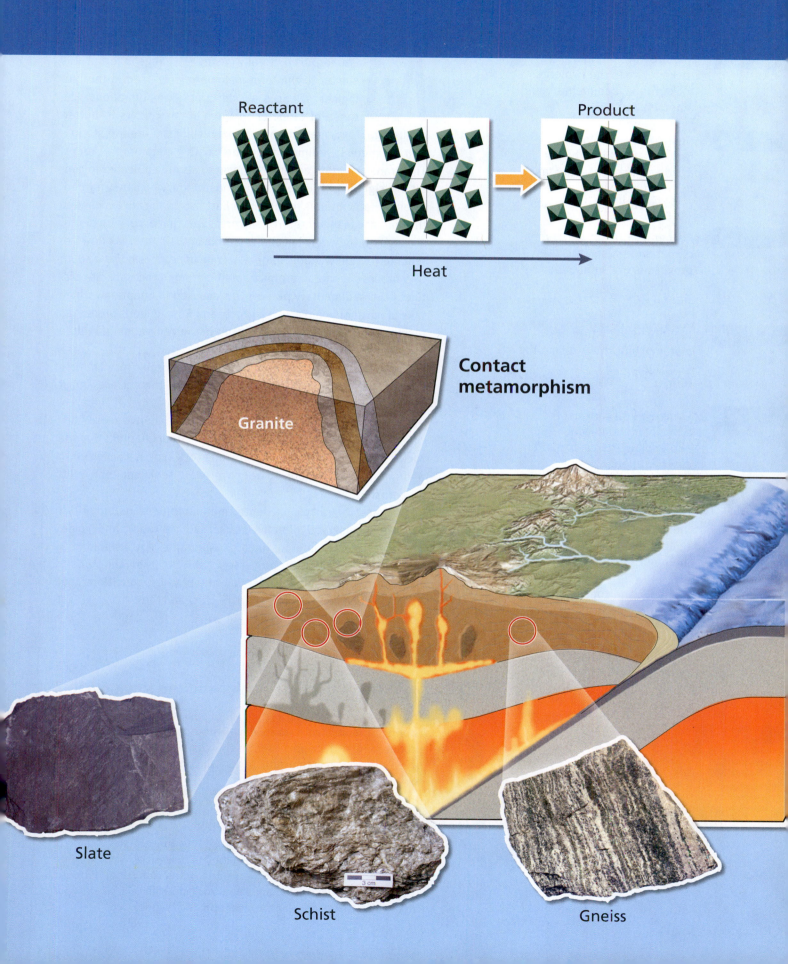

Reactant

Product

Heat

Contact
metamorphism

Granite

Slate

Schist

3 cm

Gneiss

Learning Objectives

7.1 Restate how metamorphic rocks relate to the two other rock groups (sedimentary and igneous).

7.2 Describe how metamorphic rocks are produced by the action of heat, pressure, and fluids on preexisting parent rocks over the course of time.

7.3 Recognize the different types of metamorphism, the processes that cause them, and the environments in which they occur.

7.4 Identify how the textures of metamorphic rocks reflect the environment of their formation.

7.5 Compare and contrast the types of metamorphic rock and explain the basis of their classification.

7.6 Explain how the mineral content of metamorphic rocks indicates the pressure and temperature conditions of metamorphism.

7.7 Compare and contrast the plate tectonic settings where metamorphism takes place and the variable conditions of metamorphism that the different settings produce.

Metamorphism literally means "a change in form." It is similar to the term *metamorphosis* used in the biological sciences to describe transformations like those of tadpoles to frogs or caterpillars to butterflies. In geology, metamorphic rocks result from the transformation of igneous, sedimentary, or preexisting metamorphic rocks. This transformation occurs whenever rocks are subjected to conditions significantly different from those under which they formed. Upon burial, minerals formed near Earth's surface are exposed to the increase in temperatures and pressures that occur with depth. As a result, they may become unstable and either break down or react to form new stable minerals *while the rock remains a solid*; this process is metamorphism. Metamorphic processes are distinct from sedimentary processes discussed in Chapter 6, which take place at or near Earth's surface, and from igneous processes discussed in Chapter 4, which result in the generation of molten rock, or magma.

Metamorphism occurs at depth and metamorphic rocks remain hidden from view until they are exhumed to the surface—most commonly by the uplift and erosion of overlying rocks. The presence of metamorphic rocks at Earth's surface therefore provides dramatic evidence of the rock cycle. Recall from Chapter 1 that the rock cycle is the process whereby one rock type is converted to another upon burial, only to be converted to sediment upon uplift and erosion.

Because they occur at depth, metamorphic transformations are rarely observed directly (unlike sedimentation and volcanism), so we use different lines of exploration and reasoning to understand metamorphic processes. As we shall see, the study of these rocks allows us to determine the physical conditions that exist at various levels within Earth's crust and the processes that occur there.

7.1 Metamorphic Rocks and the Rock Cycle

In previous chapters we learned that igneous rocks form from the cooling of molten rock, or magma, and that sedimentary rocks are produced by surface processes. In this chapter, we discuss the third group of rocks, metamorphic rocks, which result when preexisting rocks are altered in Earth's interior. A **metamorphic rock** is a rock that has been subjected to pressure, heat, and fluids so that its mineral composition and texture have been profoundly changed from their original state. This process of change is known as **metamorphism**. In contrast to sedimentary and igneous processes, which result in the formation of new rocks, metamorphic processes transform existing rocks into metamorphic ones (Fig. 7.1). This means that every metamorphic rock must have a *parent rock* from which that particular metamorphic rock was formed.

(a)

(b)

(c)

(d)

Figure 7.1   Metamorphism

This figure illustrates an example of progressive metamorphism that transforms a sedimentary shale (a) first to a fine-grained metamorphic rock, called slate, (b) then to a coarser-grained metamorphic rock called schist (c), and then to a gneiss (d).

When surface rocks are buried, they are subjected to the increased pressures and temperatures that exist at depth, and may interact with hot circulating fluids. As a result, the crystal structures of their minerals can become unstable, resulting in reactions that produce new minerals that are stable under the new physical conditions (see Section 3.8, "Minerals and Plate Tectonics"). The unstable minerals in the parent rock are the *reactants,* and the new stable minerals in the metamorphic rock are the *products.* In addition to changing the mineral content of rocks, metamorphism causes rocks to alter their appearance, or *texture,* when new conditions cause changes in their volume or shape. The combination of mineral and textural changes transforms the parent rock into a new kind of rock—metamorphic rock.

Metamorphism takes place at higher temperatures and pressures than sedimentary processes, such as weathering and cementation, but at lower temperatures than those required to generate magma. Although the temperatures involved in metamorphism are high enough to cause reactions between minerals, they are not sufficient to break all of a mineral's chemical bonds and generate molten rock. Metamorphic transformations occur *while the rock is still solid.* Under these conditions, a solid is typically *plastic,* like Play-Doh, and is therefore capable of bending and flowing under stress without breaking its bonds.

Metamorphic processes typically take place at depth, so we are usually unable to directly observe these processes. Unlike sedimentary and volcanic rocks, therefore, we rarely *see* metamorphic processes taking place, and we must *infer* what happened to the rocks based on limited observation backed up with theory and experiment.

But if metamorphism takes place only at depth, how can metamorphic rocks occur at Earth's surface? The answer is that metamorphic rocks are *exhumed,* that is, they are exposed by uplift and erosion as portrayed in the rock cycle (Fig. 7.2). Hence, our study of metamorphic rocks does not stop once we determine the conditions under which their metamorphism took place. It also includes investigation of the processes that exhume these rocks. The processes of uplift and erosion needed to expose metamorphic rocks are similar to those responsible for the exposure of plutonic igneous rocks at Earth's surface (see Chapter 4: Origin and Evolution of Igneous Rocks). Commonly, exhumation takes place during the later stages of mountain building and so provides clues to this important process.

CHECK YOUR UNDERSTANDING

◔ Explain how metamorphic transformations can occur while the rock is still a solid.

◔ Why are metamorphic processes hidden from view?

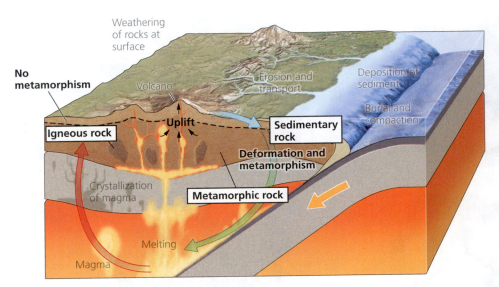

Figure 7.2   Metamorphism and the Rock Cycle
The rock cycle highlights the relationship between metamorphic, igneous, and sedimentary rocks.

To understand the evolution of metamorphic rocks, we must first gain insight into the factors that control metamorphic processes. As we shall see in Section 7.2, metamorphism results from a combination of factors.

7.1 SUMMARY

- Metamorphic processes occur below Earth's surface and involve changes that occur while rock is in the solid state.

- Metamorphism causes significant changes in the texture and mineral composition of preexisting rocks.

- Metamorphic processes differ from sedimentary processes, which take place at or near Earth's surface, and from igneous processes, which result in the generation of magma.

7.2 Factors That Control Metamorphism

When you prepare a meal, the end product depends on the starting ingredients, the oven setting, and the cooking time. In many respects, metamorphism is similar to cooking; the most important factors that control metamorphism are the "starting ingredients," or the chemistry of the parent rock, and the physical conditions, such as heat, pressure, and the circulation of fluids. Like cooking, sufficient time is required for mineral reactions and textural adjustments to proceed to completion.

Heat, pressure, and fluids are collectively known as the *agents of metamorphism.* They act together to help break down the bonds in unstable minerals, promote mineral

reactions, and aid in the growth of stable new mineral products. All agents act simultaneously but the relative contribution from each agent can vary markedly from one environment to another. Within Earth's crust, temperatures typically increase with depth by about 20°C to 30°C per kilometer (60°F to 100°F per mile), a rate known as the *geothermal gradient* (Fig. 7.3). As a result, sedimentary and volcanic rocks that form at Earth's surface become heated as they are buried. Depending on the plate tectonic environment, the geothermal gradient can be as low as 5°C per kilometer (14°F per mile) or as high as 60°C per kilometer (170°F per mile). As we shall soon see, this variation in environment is commonly reflected in the mineral content and the texture of the resulting metamorphic rock.

HEAT
Heat flows from bodies at high temperatures to those at lower temperatures. As discussed in Chapter 1, measurement of

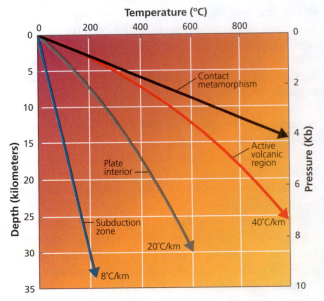

Figure 7.3   Geothermal Gradient
The rate at which temperature increases with depth is the geothermal gradient. Geothermal gradients are shown for three regions: an active volcanic region (about 40°C/km, 115°F/mi), the interior part of a tectonic plate (about 20°C/km, 58°F/mi), and a typical subduction zone (about 8°C/km, 23°F/mi).

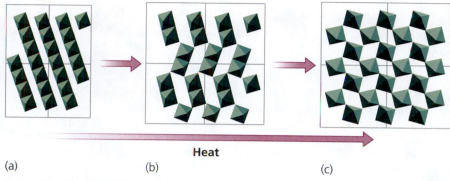

(a)   (b)   (c)

Figure 7.4   Heating a Mineral

This schematic illustrates the effect of heat on a mineral. (a) The compact crystal structure in the parent rock responds to heating by expanding (b), which increases the average length of the bonds and results in some distortion of the crystal structure. Continued heating causes the bonds to break, leading to (c) the formation of a new, less dense mineral with a more open crystal structure. Note that from (a) to (c), the same amount of mass occupies a greater volume, indicating that the crystal structure of (c) is less dense than (a).

heat flow at Earth's surface shows that heat flows from Earth's interior toward its surface and escapes into the atmosphere. In so doing, it provides the energy that fuels metamorphic transformations. Earth's internal heat is related to processes that formed Earth some 4.54 billion years ago, and to the ongoing decay of radioactive elements. As we will discuss in Chapter 8, radioactive elements are unstable and give off heat as they decay to more stable elements. Heat is also transported toward the surface by ascending magma. The outward flow of heat from Earth's interior toward its surface has been one of the most important influences on the evolution of our planet. Metamorphism is just one of several important consequences of this heat flow.

But how does heat affect the stability of minerals? Recall from Chapter 3 that each mineral has a regular arrangement of ions that vibrate in a limited fashion. As the temperature rises, the vibrations become more violent and collisions with neighboring ions may occur. The mineral also expands when heated so that the average distance between neighboring ions increases (Fig. 7.4a, b). This places a strain on the chemical bonds that hold the mineral together. Continued heat input may eventually cause the bonds to break, and new minerals are formed that have more open crystal structures (Fig. 7.4c). As a result, the new minerals are less dense.

If all the bonds in a rock break, the rock melts and becomes magma. Depending on the composition of the rock and other factors, such as the presence of fluid, the temperatures required for melting range from 700°C to 1200°C (1300°F to 2200°F). However, as we learned in Chapter 4, under most conditions, the heat is insufficient

to cause melting of the crust and the energy is absorbed in the formation of new stable mineral products. Even in extreme cases, where melting does occur it is usually only partial, and a solid residue of metamorphic rock remains.

Heat also plays an important role in accelerating the rate of metamorphic reactions (Fig 7.5). Rates of reaction rise dramatically with temperature. This means that metamorphic reactions in the warmer lower crust occur much more rapidly than reactions in the cooler upper crust. High temperatures increase the amount of energy at the site of the reaction and promote more rapid chemical breakdown and mineral growth. As a result, minerals tend to grow faster in the lower crust, and they are coarser grained. Although there are exceptions, in general the grain size of a metamorphic rock tends to increase with increasing temperature of metamorphism.

The difference between the parent rock and metamorphic rock depends on the **metamorphic grade**, which is a general measure of the intensity of metamorphism. For example, if the parent rock is a shale (a fine-grained sedimentary rock containing clay minerals), then its conversion to either a fine-grained metamorphic rock or a coarse-grained one would be called low grade and high grade metamorphism, respectively.

CHECK YOUR UNDERSTANDING

◔ What does metamorphic grade measure?

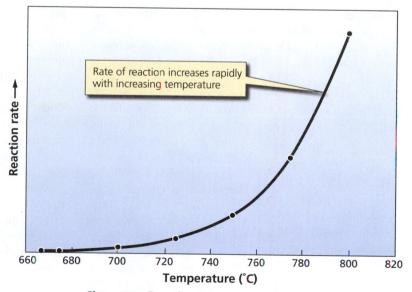

Figure 7.5   Reaction Rates

Chemical reaction rates increase exponentially when temperature increases.

SCIENCE REFRESHER

Pressure Units

There are various units of pressure; the ones most commonly used by geologists are bars and pascals. A bar is approximately equal to the pressure exerted by the atmosphere on Earth's surface at sea level. One bar is equal to 100,000 (10^5) pascals. Technically, a pascal is defined as a pressure of one newton per square meter, where a newton is the force required to give a mass of one kilogram an acceleration of one meter every second. Given the enormous pressures that exist below Earth's surface, the most convenient units for geologists are the **kilobar** (kbar), which equals 1000 bars, and the **gigapascal** (GPa), which equals a billion pascals. One kilobar is consequently equal to 10 gigapascals.

PRESSURE

We can think of pressure as a force pushing on a unit area of rock (see Science Refresher: **Pressure Units**). There are two important types of pressure that affect metamorphic rocks:

- **confining pressure**, pressure that acts on the rock equally in all directions, and
- **directed pressure**, pressure that pushes with more vigor in one direction than in others. *more common*

Rocks subjected to either pressure undergo a change in volume and/or shape, and are said to be under **stress**.

Confining Pressure

As you might expect, the vertical pressure exerted on a rock increases with depth as the effect of the weight of the overlying rock layers gets ever larger. Rocks buried to a depth of 3 kilometers, for example, have a pressure exerted on them of about 0.9 kilobar (or almost 900 times the pressure at Earth's surface), whereas rocks at 30 kilometers depth are subjected to pressures equal to 9 kilobars. But buried rocks are not only subject to pressure from overlying rock layers. They are also *confined* by neighboring rocks, which exert horizontal pressures on them. As a result, pressures acting in a horizontal direction also increase with depth.

The confining pressure that pushes on a rock is proportional to the weight of the overlying rocks. The effect is similar to that experienced by divers who feel the weight of the water above them. An increase in pressure causes a decrease in volume, requiring deep-sea divers to wear pressurized suits to resist these changes. Rocks, however, have no such protection and tend to contract (Fig. 7.6a). The crystal structures of minerals are consequently strained as the component ions of the crystal are forced closer together. Eventually, the crystal structures break down and metamorphic reactions take place that produce more stable, denser minerals. In addition, the pore spaces between the minerals close. A deeply buried rock is therefore compressed into a smaller volume and consists of relatively dense minerals. The effect of confining pressure consequently acts in opposition to that of heat, which causes minerals to expand, and produces more open crystal structures and, therefore, less dense minerals.

Directed Pressure

Although confining pressure causes rocks to change volume, it does not produce a change in shape, because the pressure acts equally in all directions. Confining pressure acts on all buried rocks but is most important in tectonically stable regions where other pressures are absent. In contrast, adjacent to plate boundaries or active faults, pressures exist that act in a horizontal direction. In these situations, rocks are subjected to directed pressure, which produces both a change of volume and a change in shape. In areas where pressures acting horizontally exceed the pressure acting vertically, the rocks are said to be under *compression*, and the rock layers will buckle and deform into folds (Fig. 7.6b). Compression occurs in places where portions of Earth's crust are being squeezed together, as is the case at the sites of *mountain building*. Where horizontal pressures are less than the vertical pressure, the rocks are said to be under *extension* and are pulled apart. Extensional stresses cause rock layers to lengthen and are most common where portions of Earth's crust are being torn apart, as for example at sites where continental rifting occurs (Fig. 7.6c).

A kind of deformation called **shear** occurs in fault zones where stresses cause rock masses to slide past one another. In zones of shear, slippage usually takes place on a large number of closely spaced planes, much like the slippage in a deck of cards

CHECK YOUR UNDERSTANDING

- Why does increased pressure favor the growth of more dense minerals?
- Why does increased temperature favor the growth of less dense minerals?

CHECK YOUR UNDERSTANDING

- What is the difference between confining pressure and directed pressure?

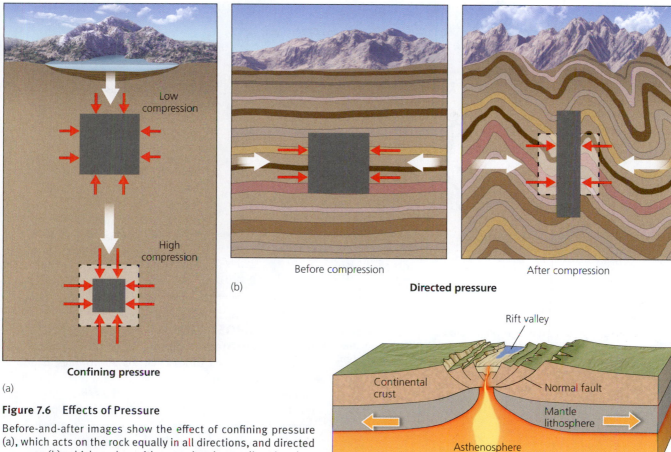

(b)

Before compression

Directed pressure

After compression

Confining pressure

(a)

(c)

Figure 7.6   Effects of Pressure

Before-and-after images show the effect of confining pressure (a), which acts on the rock equally in all directions, and directed pressure (b), which pushes with more vigor in one direction than another. Extension (c) thins the crust bringing the hot mantle closer to the surface, which steepens the geothermal gradient.

that you move in your hands. This process grinds and pulverizes the original grains so that their size is reduced.

Since directed pressure results in changes in shape, it not only affects the mineral content of a rock but also profoundly influences its texture, as we shall see in Section 7.4.

FLUIDS

Fluids are abundant in most metamorphic rocks and form an interconnected network between mineral grains. Experiments show that ions and atoms migrate much more rapidly in fluids than they do through solids. Therefore circulating fluids are the prime agents of chemical transport, aiding the breakdown of unstable minerals and the growth of new stable ones. For the mineral to grow, the ions needed for that growth must be transported to the site of the growing crystal and fluids are the most efficient transport system (Fig. 7.7). Water is the most common metamorphic fluid, but carbon dioxide is also important in the metamorphism of carbonate-rich rocks such as in the conversion of limestone to marble.

Fluids in metamorphic rocks have a variety of sources. They can be

- trapped in the rock during sedimentation;
- introduced by tectonic and igneous processes; or
- expelled from the crystal structures of unstable minerals when those minerals react.

We now discuss each of these sources in more detail.

For metamorphosed sedimentary rocks, much of the fluid may be water originally trapped between grains of sediment during deposition. Although lithification and subsequent burial may expel some of this fluid (see Chapter 6), that which remains may continue to bathe individual grains, resulting in what is known as a *pore water pressure*. Because the density of water is less than that of its host rock, this buoyancy is an upward force that acts against the weight of the overlying rock layers.

The tectonic environment also provides a potential water source. For example, oceanic crust absorbs a lot of fluid as it interacts with ocean water. Much of this water-rich fluid is released into the overlying mantle and lower crust when oceanic crust is subducted. On a more local scale, the cooling and crystallization of magmas may also be a source of water. Most early-forming igneous minerals

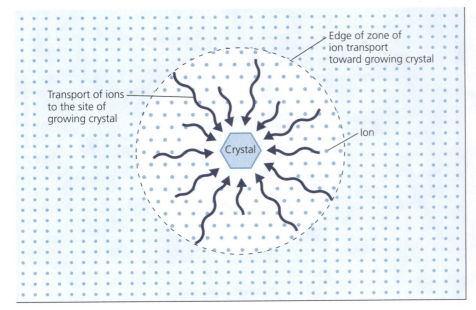

Edge of zone of
ion transport
toward growing crystal

Transport of ions
to the site of
growing crystal

Crystal

Ion

Figure 7.7   Crystal Growth and Ionic Transport
Ions must move toward the growing mineral crystal for the mineral to grow. The blue dots represent the distribution of the ions, which were evenly distributed before the mineral started to grow. Note how the supply of ions becomes depleted in a "halo" around the mineral. Eventually, this supply is exhausted and the mineral stops growing.

do not take water into their crystal structures. As a result, the concentration of water in a melt typically increases as crystallization proceeds. Eventually, the water separates from the melt and escapes into the surrounding host rock. Water has a tremendous capacity for storing heat, so fluids that ascend rapidly from subduction zones, or are expelled from cooling magmas, can provide the energy necessary for metamorphic reactions.

Fluids may also be derived from the metamorphic minerals themselves. Some mineral families, such as the micas and amphiboles, contain water in their crystal lattices. If these minerals become unstable, this water is liberated by metamorphic reactions to mix with the existing fluid in the rock.

ROCK COMPOSITION

If the appropriate amounts of kneaded flour, yeast, water, and salt are placed in an oven for 30 minutes at 300°C (570°F), the product is bread. However, a mixture of flour and egg placed in the same oven at the same temperature for the same amount of time yields a different product—pastry. In a similar manner, the chemical composition of the parent rock type provides the ingredients for the minerals formed during metamorphism. For example, under the same conditions, the minerals formed by the metamorphism of basalt are quite different from those formed by the metamorphism of mudstone. At 550°C (1020°F) and 6 kilobars pressure, basaltic rocks react to produce iron- and magnesium-rich minerals such as hornblende. In contrast, under the same conditions, mudstones produce aluminum-rich minerals such as kyanite (Fig. 7.8).

(a)   (b)

Figure 7.8   Influence of Rock Composition
Produced under the same conditions of temperature and pressure, these very dissimilar metamorphic rocks differ only in the composition of the parent rock. (a) Amphibolite is produced by the metamorphism of basalt, and (b) kyanite (blue crystals) is a mineral produced by the metamorphism of mudstone. The typical "salt and pepper" speckled appearance of the amphibolite reflects the dominance of hornblende (dark) and plagioclase (light).

TIME

Metamorphism is the path taken by a rock in order to reach a state of equilibrium with the new environment in which it finds itself. *Equilibrium* is an important concept in the study of metamorphic rocks and may be thought of as the "goalposts" toward which natural processes are moving. Given sufficient time, equilibrium, which is the most stable assemblage of minerals and textures possible for those conditions, will be obtained. As long as equilibrium persists, there is no overall change in the amount or the composition of the minerals present. However, if new conditions are imposed on a region, perhaps because of the movement of tectonic plates or the nearby ascent of magma, the equilibrium "goalposts" are moved and the minerals readjust toward this new position. Minerals react to these changes and one may form at the expense of another, or new minerals may form.

Heat and circulating fluids accelerate the rates of metamorphic reactions. However, recent data indicate that metamorphic mineral growth is a sluggish process even in the presence of hot fluids. Garnets, for example, may require as long as 5 million years to grow just 1 centimeter in diameter.

Because of the slow rate of mineral growth, it is possible that equilibrium may be approached but never attained if the conditions of metamorphism change. Where this occurs, metamorphic minerals record a very complex history indeed. Unraveling such a history may seem like an impossible task. However, careful analysis of minerals that grew during complex tectonic events shows that their compositions can act like a recording device, storing information that allows us to deduce the events.

In tectonically active regions, metamorphism may be constantly playing "catch-up" to changing tectonic environments. It is this process of catch-up that drives the formation of new minerals and new mineral growth. Using a technique known as *microprobe analysis,* geologists can determine the chemical compositions of minerals down to the micron scale (1 micron = 0.0001 or 10^{-4} centimeters). Thus a single garnet crystal can be analyzed many times from its core to its rim. Typically such garnets exhibit subtle but systematic variations in their chemistry as a result of their growth during changing metamorphic conditions (Fig. 7.9). Geologists can use these variations to deduce the tectonic history of the region during the time of garnet growth.

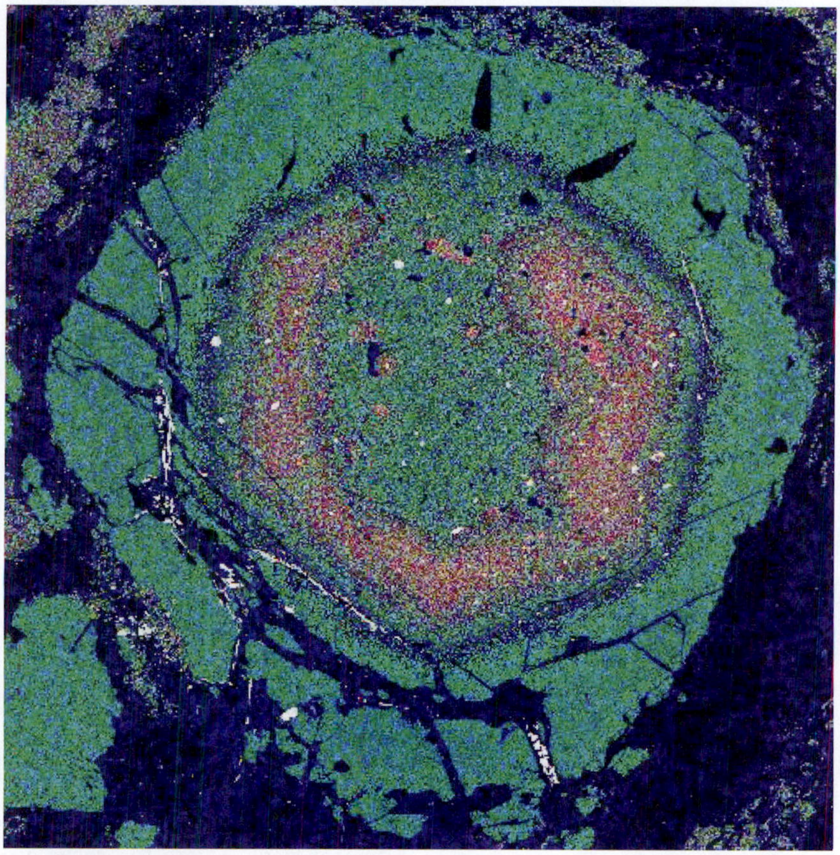

Figure 7.9   Zoned Garnet

A computer-generated color image of a garnet crystal, only a few millimeters across, shows its internal zoned structure. This image shows variations in the concentration of the element yttrium (Y). Subtle variations in the chemistry of metamorphic minerals like the ones seen in this garnet are often the result of growth under changing metamorphic conditions.

7.2 SUMMARY

- Heat, pressure, and fluids are the agents of metamorphism, but sufficient time is required to produce new stable minerals and textures.

- Temperatures in Earth's crust typically increase by about 20°C–30°C per kilometer. This increase is known as the geothermal gradient.

- Metamorphic reactions are sluggish in the upper crust but are more rapid and tend to go to completion in the lower crust, resulting in coarser-grained metamorphic rocks.

- Two types of pressure, directed pressure and confining pressure, influence the texture of metamorphic rocks. Confining pressure acts equally in all directions and directed pressure acts more in one direction than in others.

- Fluids facilitate metamorphism by transporting energy and ions to the site of mineral reactions.

- The chemistry of the parent rock profoundly influences the minerals and textures that form during metamorphism.

- Increasing temperature favors minerals that occupy greater volume, whereas increasing pressure favors more dense minerals.

7.3 Types of Metamorphism

Metamorphism takes place in a variety of environments. Each environment produces metamorphic rocks of different mineralogy, texture, and distribution. In this section we discuss the various types of metamorphism.

REGIONAL METAMORPHISM

When large regions of crustal rocks are subjected to elevated temperatures and pressures, the resulting changes in mineralogy and texture constitute **regional metamorphism**. Regional metamorphism produces extensive areas of metamorphic rock usually in association with mountain building. Because mountain building is typically followed by uplift and erosion, regional metamorphic rocks become exhumed and are exposed over wide areas of Earth's mountainous and former mountainous areas. The regional metamorphic rocks being produced today in the hidden interior of the Himalayas, for example, will become exposed when the Himalayas are sufficiently eroded to expose their roots. Ancient mountain chains like the Appalachians have undergone hundreds of millions of years of erosion so that regional metamorphic rocks are now widely exposed.

CONTACT METAMORPHISM

Metamorphic rocks formed by heating adjacent to a body of magma are the product of **contact metamorphism**. As we learned in Chapter 4, plutonic igneous rocks are formed as magma cools and crystallizes. This cooling reflects the transfer of heat from the hot igneous body to the cooler host rocks surrounding it. The heat transfer is especially marked in shallow crustal rocks where the temperature contrast between magma and adjacent host rocks may be several hundred degrees. The temperature contrast is not so high in the lower crust because rocks at these depths are relatively warm. Shallow host rocks adjacent to a cooling magma may consequently show progressive mineral changes close to the contact that are the result of *baking*.

The effects of contact metamorphism decrease with distance from the igneous body producing a halo or **aureole** of contact metamorphic rocks around the igneous body (Fig. 7.10). In general, larger igneous bodies have wider aureoles because they contain more heat. Thus, large granite batholiths may have aureoles several kilometers across, whereas the baked metamorphic rock margins adjacent to narrow dikes may be only a few centimeters in width.

Because contact metamorphism occurs only in the immediate vicinity of an igneous body, its effects are more localized than those of regional metamorphism. However, the high temperatures associated with mountain building invariably lead to the generation of magmas. As a result, metamorphism in mountain belts often involves a complex interplay of regional and contact metamorphic effects.

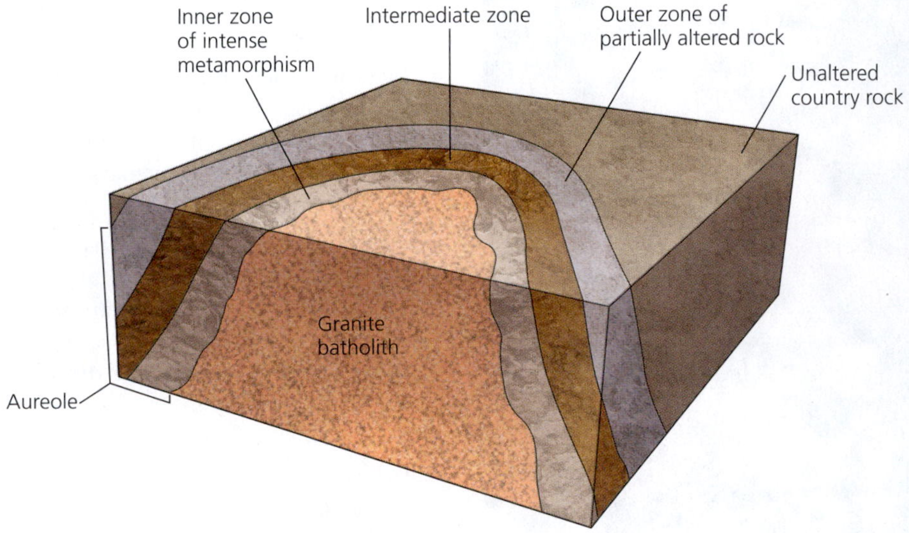

DYNAMIC METAMORPHISM

Metamorphism that occurs along faults in Earth's crust, that is, along fractures on which significant movement has taken place, is **dynamic metamorphism**. This type of metamorphism is generally limited to a zone within a few tens of meters of the fault. However, because faults rarely occur in isolation, but instead are part of larger fault systems, dynamic metamorphism can be quite widespread.

Dynamic metamorphic rocks form from either crushing or smearing along the zone of fracturing. Temperatures, pressure, and strain rate are important in determining the type of rock produced by dynamic metamorphism. **Strain rate** (which we discuss more in the material that follows) refers to the intensity of deformation.

Crushing occurs in near-surface environments where the rocks are cold and brittle, producing a rock called *fault breccia* that contains angular fragments (Fig. 7.11a). *Smearing*, on the other hand, requires higher temperatures and produces a distinctive streaky metamorphic rock called **mylonite** (Fig. 7.11b, c). Smearing results in a reduction of grain size, so that mylonites are typically fine-grained rocks. Their occurrence is the most important exception to the general

Figure 7.10   Contact Metamorphism
A zone of contact metamorphism, known as an aureole, typically surrounds igneous intrusions. The metamorphic aureole around this idealized granite pluton (in this case a granite batholith) contains three zones of metamorphic rock that reflect the decrease in temperature with distance from the pluton. A metamorphic aureole is most commonly developed adjacent to plutons in the shallow crust because of the high temperature contrast between the magma and the host rocks.

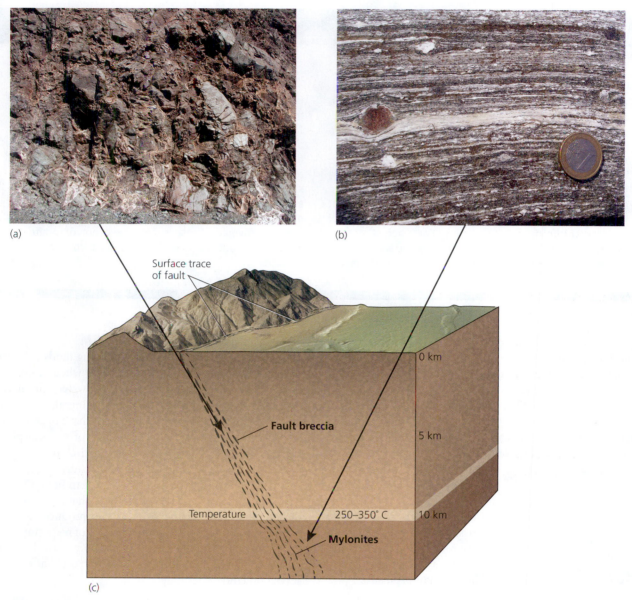

Figure 7.11   Dynamic Metamorphism

Crushing produces fault breccia (a); smearing produces mylonite (b). The schematic diagram (c) shows the locations of various rock types typically produced by dynamic metamorphism in a fault zone. Fault breccia at shallow depths gives way to mylonites at depth.

tendency for crustal rocks to become coarser grained with increasing depth (see Section 7.2).

> The strain rate adjacent to faults is a measure of the speed at which the crust moves along a fracture zone. The effect of strain rate can be illustrated with a simple analogy. If you take a stick of butter from the refrigerator and subject it to a high strain rate by abruptly bending it, the brittle stick will snap. If instead you subject it to a low strain rate by flexing it slowly, it is more compliant and will bend or smear without breaking. Rocks are similarly more prone to smear at low strain rates, whereas they tend to snap and pulverize when strain rates are high.

CHECK YOUR UNDERSTANDING

• Compare and contrast regional, contact, and dynamic metamorphism.

SHOCK METAMORPHISM

Metamorphism produced by the high velocity impact of an extraterrestrial object such as a meteorite or an asteroid on Earth's surface is known as **shock metamorphism**. Upon impact, the energy of the rapidly moving object is transformed into shock waves and heat. These shock waves cause extremely high pressures for a fraction of a second. As the shock waves pass, the minerals begin to warm up. Some minerals are vaporized and a vast amount of material is ejected. The effects of impact metamorphism are preserved

IN DEPTH

Shock Metamorphism

Shock metamorphism is relatively rare on Earth. However, it is likely to be the dominant type of metamorphism preserved on celestial bodies such as Mercury and the Moon. These bodies show visible evidence of abundant impacts, most of which occurred very early in their histories. This suggests that the early development of the Solar System was a tumultuous one, and that impacts on Earth were far more frequent at that time than they are today. Therefore, impact metamorphism may also have been the predominant type of metamorphism affecting the primitive Earth's crust.

Although the dynamic nature of Earth's surface has obliterated any evidence of these early events, more recent impacts are sometimes preserved as circular structures like that of Lake Manicouagan in Quebec, Canada (see Fig. 7.12b). The recent proposal that a meteorite impact led to the extinction of the dinosaurs 66 million years ago, has intensified the search for ancient impact sites. Geophysical evidence has now shown that the impact site of the dinosaur killer lies alongside the Yucatan peninsula of Mexico, and an even larger impact site may exist off the west coast of India where a 500-kilometer (310 mile) wide crater has been discovered. These events have not only affected Earth's crust but also may have played a pivotal role in the evolution of life.

in the intensely fractured and recrystallized rocks surrounding the impact site.

Some metamorphic minerals, such as *stishovite,* only form at impact sites. Stishovite is formed from quartz at extremely high pressures. Only at impact sites are such pressures generated. Stishovite has been found in the rocks surrounding Meteor Crater in Arizona and the Lake Manicouagan crater in Quebec, its discovery providing important evidence that both structures were formed by meteorite impact (Fig. 7.12a, b). We now know that Meteor Crater formed about 50,000 years ago, whereas the crater at Lake Manicouagan is 214 million years old (see In Depth: **Shock Metamorphism**).

CHECK YOUR UNDERSTANDING

◯ Name a mineral that forms only during shock metamorphism.

HYDROTHERMAL METAMORPHISM

The reaction of rocks with hot circulating fluids is **hydrothermal metamorphism**. For example, when basalts are extruded at mid-ocean ridges, hot water circulating through fracture systems promotes mineralogical reactions. Because mid-ocean ridge systems have been investigated by deep-sea submersibles, this form of metamorphism is one of the few to have been observed directly.

Reaction with seawater commonly results in important chemical changes in the rocks of the ocean floor. Oceanic basalts, for example, may absorb water that reacts with its high temperature igneous minerals to produce low temperature metamorphic grains. This style of metamorphism is commonly preserved in *ophiolite complexes,* the name given to fragments of ancient oceanic crust that are preserved on land.

(a)

(b)

Figure 7.12   Impact Craters

(a) Meteor Crater, near Flagstaff, Arizona, is about 1.2 kilometers (0.8 miles) in diameter and 200 meters (660 feet) deep. It formed about 50,000 years ago. (b) Lake Manicouagan in Quebec, Canada, is about 100 kilometers (62 miles) across and formed 214 million years ago.

BURIAL METAMORPHISM

Metamorphism of rocks caused by burial beneath a thick succession of overlying rock layers is **burial metamorphism**. Since temperatures within Earth's crust typically increase by about 20°–30°C per kilometer (60°F–100°F per mile), temperatures at the base of a pile of sediment and rock 10 kilometers (6 miles) thick may reach 200°C to 300°C (390°F to 570°F). Hence, burial and compaction of sediments can eventually cause some of the original minerals to break down to form new metamorphic grains. Burial metamorphism is occurring today, for example, at the base of the Mississippi delta, where the deposits reach a thickness of up to 11 kilometers (7 miles). Samples recovered from drill holes show that the original clay minerals have reacted to form very fine-grained metamorphic minerals such as muscovite and an iron-magnesium bearing sheet silicate called *chlorite*.

7.3 SUMMARY

- Regional metamorphism takes place over wide areas during mountain building.

- Contact metamorphism occurs when rock is heated by an adjacent igneous body.

- Dynamic metamorphism involves crushing or smearing adjacent to fault zones.

- Collision of extraterrestrial objects with Earth's surface produces shock metamorphism.

- Hydrothermal metamorphism occurs as the result of hot circulating fluids and is especially common in mid-ocean ridge settings.

- Burial metamorphism occurs beneath a thick succession of overlying rock layers.

7.4 Metamorphic Textures

The effects of metamorphism on a rock are not just recorded in mineral composition; they are also expressed in rock appearance, or *texture*. Textural analysis, like mineral composition, is a powerful tool for understanding metamorphic processes.

One of the most common textural descriptions of metamorphic rocks concerns the geometric arrangement of its grains, or **fabric**. This term is used in much the same way that the fabric of clothing describes the arrangement of its fibers. Most igneous and sedimentary rocks contain randomly oriented grains, and are said to have a *random fabric* (Fig. 7.13a). However, metamorphic rocks subjected to directed pressure may develop a nonrandom fabric, in which the mineral grains possess a *preferred orientation* (Fig. 7.13b). Mineral grains that have experienced only confining pressure, on the other hand, remain randomly oriented because the rocks do not change their shape. As a result, metamorphic rocks fit into one of two broad categories, *foliated* or *nonfoliated,* depending on whether or not their mineral grains possess a preferred orientation.

(a) Before metamorphism

(b) After metamorphism

Figure 7.13   Fabric and Foliation

(a) Igneous rocks, such as granites, typically display a random fabric. (b) If granites are metamorphosed under conditions of directed pressure, planar minerals, such as the micas, reorient so that the mineral grains possess a preferred orientation and their surfaces align approximately at right angles to the stress. The resulting mineral alignment gives the rock a planar texture, known as a foliation.

FOLIATED METAMORPHIC ROCKS

Foliation is a planar fabric produced by the alignment and segregation of minerals as they grow under conditions of directed pressure (Fig. 7.13b), such as those that accompany mountain building. Mountain building is associated with regional metamorphism and regional metamorphic rocks are typically foliated. Rocks that develop a closely spaced planar layering during metamorphism are **foliated metamorphic rocks**.

Studies of regionally metamorphosed rocks show that some minerals are more sensitive to directed pressures than others. *Platy* minerals, which occur in thin sheets, such as the micas or chlorite, for example, are particularly responsive to directed pressures, and rocks rich in these minerals are commonly strongly foliated. Because they lie flattened in the foliation plane, these minerals are said to define the foliation.

If a rock is under compression and undergoes shortening, any platy minerals in the rock become aligned approximately perpendicular to the directed pressure (see Fig. 7.13). Exactly how this is alignment is achieved is disputed, but most geologists agree that several processes contribute, including rotation, flattening, and ion migration (Fig. 7.14). For example, existing minerals may be *mechanically rotated* into an alignment that is roughly at right angles to the direction of shortening. Directed pressure may also *flatten* individual grains. In addition, the compression may cause some of the bonds between minerals to break. Unbonded ions may dissolve in the water trapped between these grains and *migrate* from areas of high pressure to areas of low pressure where they recrystallize. High-pressure sites occur parallel to the direction of shortening, whereas low-pressure sites are perpendicular to this direction. As a result, when new minerals precipitate from solution, they are oriented perpendicular to the directed pressure. The combined effect of all these processes is to form elongate aggregates of minerals with a preferred orientation approximately at right angles to the shortening direction.

Foliated metamorphic rocks can also be produced by dynamic metamorphism. A *mylonite,* for example, is a foliated rock produced by smearing of grains adjacent to faults (see Fig. 7.11b). This smearing process also involves rotation, flattening, and recrystallization. However, this type of foliation typically forms at about 45° to the shearing direction. To see how this happens, compare the effect of shearing stresses with those of directed stress. To visualize the random fabric in each rock before deformation, picture a circular area within the rock body (Fig. 7.15). Under directed stress, this circle becomes an ellipse whose long axis is perpendicular to the directed pressure and parallel to the foliation direction (Fig. 7.15a). Under shear, however, the long axis is oriented at 45° to the direction of shear and it is in this direction that the foliation forms (Fig. 7.15b).

The grain size of foliated minerals define various types of foliation. Since grain size tends to increase with the metamorphic grade, the type of foliation developed is an important clue to the metamorphic history (see Fig. 7.1). The finest-grained foliation is **slaty cleavage**, in which the aligned grains (typically chlorite and mica) are too small to be seen with the naked eye but are visible with the aid of a microscope. Slaty cleavage is produced by directed pressure during the metamorphism of clay-rich, or *pelitic*, rocks (mudstones and shales) at relatively low temperatures and pressures. Rocks with slaty cleavage form under low grade metamorphism and may preserve original sedimentary features such as bedding.

At higher grades of metamorphism, the platy minerals defining the foliation are visible. This type of foliation is **schistosity**. At still higher metamorphic grades, micas are much less abundant and **gneissic foliation** develops in which the new minerals are separated into distinct light and dark layers.

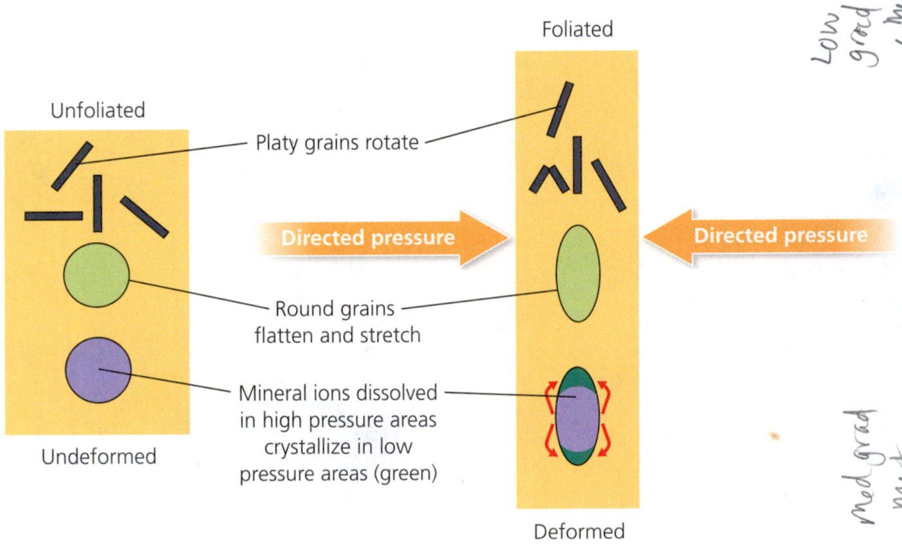

Figure 7.14   Origin of Foliation

Various processes give rise to foliation in rock. Grains respond to shortening by mechanical rotation (top), flattening (middle) or ion migration (bottom). Existing minerals are mechanically rotated. Flattening turns circular objects into ellipses. Ion migration occurs (see arrows) when portions of the minerals dissolve and ions *migrate* from high pressure areas and recrystallize in low pressure areas, which are perpendicular to the directed stress. These processes combined form aggregates of minerals that tend to be approximately at right angles to the directed stress.

> **CHECK YOUR UNDERSTANDING**
>
> ● Compare a metamorphic foliation formed by directed stress with that formed by shear.

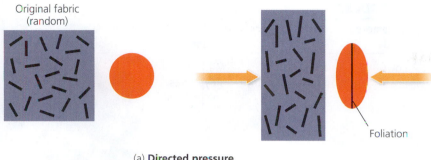

Original fabric (random)

(a) **Directed pressure**

Foliation

Original fabric (random)

(b) **Shear stress**

Foliation

Figure 7.15   Development of a Foliation
This schematic compares foliation development under (a) directed pressure and (b) shear stress in undeformed rocks with a random fabric. Under directed pressure, rocks undergo shortening in one direction and extension in another. A fabric develops perpendicular to the direction of shortening. Under shear stress, the average alignment of mineral grains is at about 45° to the direction of shear.

NONFOLIATED METAMORPHIC ROCKS

Other metamorphic rocks lack foliation, either because of the conditions under which they formed, or because of the minerals that they contain. These rocks have a random fabric and are called **nonfoliated metamorphic rocks**. Metamorphic rocks in which heat is the primary agent of metamorphism (see Fig. 7.10) are typically nonfoliated because they form without significant directed pressure. For example, contact metamorphism typically produces a nonfoliated metamorphic rock, known as *hornfels*, adjacent to cooling plutons. Similarly, burial metamorphism produces rocks that have a random fabric and are classified as nonfoliated. These rocks are nonfoliated because burial metamorphism is caused by confining pressure rather than directed pressure. Confining pressure is incapable of causing changes in shape.

Some metamorphic rocks are nonfoliated because of the minerals they contain. Coarsely crystalline rocks such as *marbles* (produced by the metamorphism of limestone), *quartzite* (produced by the metamorphism of quartz sandstone), and *amphibolite* (produced by the metamorphism of basalt) are common nonfoliated metamorphic rocks. All three of these rocks

> **CHECK YOUR UNDERSTANDING**
>
> ❂ Why are some metamorphic rocks foliated whereas others are nonfoliated?

are largely made up of just one mineral (calcite or dolomite in marble, quartz in quartzite, and hornblende in amphibolite) and each lacks platy minerals such as micas or chlorite. As a result, a foliation cannot form. Instead, all three rocks typically possess a random texture of interlocking crystals.

7.4 SUMMARY

- Metamorphic rocks are either foliated or nonfoliated and are respectively classified according to their grain size and composition.

- Foliated metamorphic rocks are the result of directed pressure and the preferred alignment of platy minerals perpendicular to those stresses.

- Nonfoliated metamorphic rocks form where rocks lack platy minerals or where they have been subjected to confining pressure rather than directed pressure.

7.5 Classifying Metamorphic Rocks

Recall from Section 7.4 that we divide metamorphic rocks into two main groups, foliated and nonfoliated, based on the presence or absence of a foliation (Table 7.1). Within these two groups, however, metamorphic rocks can be further classified according to their mineral content and grain size, both of which vary with the conditions of pressure and temperature during metamorphism.

CLASSIFYING FOLIATED ROCKS

There are four types of foliated rocks: slate, phyllite, schist, and gneiss. We divide these on the basis of grain size and the type of foliation (Fig. 7.16). The four foliated metamorphic rock types reflect an increase in the metamorphic grade from shallow to deeper levels in the crust. *Low grade metamorphic rocks,* such as slates and phyllites, retain many of their original characteristics, such as bedding. These rocks typically form at temperatures of less than 400°C (750°F) and at depths of less than 15 kilometers (9 miles). In *high grade metamorphic rocks,* such as schist and gneiss, on the other hand, the elevated temperatures (up to 1000°C, 1800°F) and greater depths (up to 100 km, 60 miles) have destroyed all of the original features. Bedding is rarely visible in schists and the layering in a gneiss is produced by the metamorphism and was not present when the parent rock was formed.

TABLE 7.1   CLASSIFICATION OF METAMORPHIC ROCKS

Texture	Metamorphic Rock	Typical Minerals	Metamorphic Grade	Characteristics of Rocks	Parent Rock	
Foliated	Slate	Clays, micas, chlorite	Low	Fine-grained, splits easily into flat pieces	Mudrocks, claystones, volcanic ash	
	Schist	Micas, chlorite, quartz, talc, hornblende, garnet, staurolite, graphite	Low to high	Distinct foliation, minerals visible	Mudrocks, carbonates, mafic igneous rocks	
	Gneiss	Quartz, feldspars, hornblende, micas	High	Segregated light and dark bands visible	Mudrocks, sandstones, felsic igneous rocks	
Nonfoliated	Marble	Calcite, dolomite	Low to high	Interlocking grains of calcite or dolomite, reacts with HCl	Limestone or dolostone	
	Quartzite	Quartz	Medium to high	Interlocking quartz grains	Quartz sandstone	
	Amphibolite	Hornblende, plagioclase	Medium to high	Dark-colored, weakly foliated	Mafic igneous rocks	
	Hornfels	Micas, garnets, andalusite, cordierite, quartz	Low to medium	Fine-grained, equidimensional grains, hard, dense	Mudrocks	

(a)

(b)

(c)

(d)

Figure 7.16   Foliated Metamorphic Rocks
The foliated metamorphic rocks types are, in order of increasing metamorphic grade: (a) slate, (b) phyllite, (c) schist, and (d) gneiss.

An increase in metamorphic grade affects rocks in two important ways. First, in high grade metamorphism, metamorphic reactions occur more rapidly because the transport of ions to the growing minerals is much more efficient at higher temperatures (see Fig. 7.5). This produces coarser grained minerals so that original features from the parent rock, such as bedding, become harder to recognize. Second, metamorphic reactions tend to release some of the fluid stored within the reacting crystals. As a result, low grade metamorphic rocks tend to contain many minerals with water in their crystal structures, whereas high grade metamorphic rocks contain a relatively small amount of these minerals.

These two trends are evident in the mineral content of foliated metamorphic rocks produced by the progressively increasing, or *prograde,* metamorphism of the parent rock shale. Shale is a very fine-grained sedimentary rock containing abundant clay minerals, which store significant quantities of water in their crystal lattices. The presence of water in a mineral is indicated in the mineral formula by the symbol [OH], and the relative amount of water is indicated by the number that follows the symbol. So the important

clay mineral *kaolinite* ($Al_2Si_2O_5[OH]_4$) contains more water in its structure than muscovite ($KAl_3Si_3O_{10}[OH]_2$). The prograde metamorphism of shale is consequently associated with an increase in grain size, a decrease in the water content of the metamorphic minerals formed, and the formation of metamorphic rocks with a progressively higher metamorphic grade. In the following section, we follow shale (the parent rock) as it undergoes prograde metamorphism to form the four types of foliated rocks: slate, phyllite, schist, and gneiss.

CHECK YOUR UNDERSTANDING

● How are metamorphic rocks classified?

Slate

As a shale undergoes metamorphism, its clay minerals become unstable and react to form very fine-grained platy minerals such as the white mica, muscovite ($KAl_3Si_3O_{10}[OH]_2$). As its formula indicates, muscovite contains less water in its mineral structure than the clay minerals from which it formed. Directed pressure aligns these platy minerals to form a slaty cleavage along which the metamorphic rock tends to split, much like wood splits along its grain. In this way, sedimentary shale is transformed into the fine-grained metamorphic rock we know as **slate** (Fig. 7.16a). The minerals in slate are too small to be seen with the naked eye. In fact, their identification may be difficult even under the microscope and a sample may have to be bombarded with X-rays in order to identify its mineral content.

Phyllite

With increasing metamorphism, the fine-grained metamorphic minerals in slate continually recrystallize and become coarser. At temperatures of approximately 300°C (570°F), a foliated metamorphic rock called **phyllite** (Fig. 7.16b) is produced. Although the crystals in phyllite are still not visible to the eye, they are large enough to reflect light off the foliated surface of a hand specimen, giving the rock a characteristic glossy appearance or *sheen* that aids in its identification. Surfaces that are oblique to the foliation, however, are relatively dull, indicating that it is the alignment of the grains on the foliation plane that gives the rock its reflective sheen.

Schist

At higher grades of metamorphism, the recrystallized micas are clearly visible to the naked eye and a schistosity is produced (Fig. 7.16c). This type of metamorphic rock is a **schist**. Slates, phyllites, and schists are all produced by the prograde metamorphism of shale, but schists occur in regions that experience the highest temperatures and pressures.

> ### CHECK YOUR UNDERSTANDING
> ● How would you distinguish between a schist and a slate in hand specimen?

Gneiss

At still higher metamorphic grades, the micas become unstable and break down to form minerals that do not give the rock a preferred direction of splitting. Instead, a gneissic foliation forms with alternating layers of different mineral composition. The coarse-grained rock produced this way is called **gneiss** (pronounced "nice") and is readily distinguished from schists because its foliation planes are not splitting surfaces and are typically spaced at intervals of several millimeters (Fig. 7.16d).

The precise mechanism that produces gneissic layering is uncertain, but it appears that light colored *felsic* minerals such as quartz and feldspar are segregated from darker *mafic* minerals such as biotite and hornblende to produce alternating layers of light and dark color. (See Chapter 4: Origin and Evolution of Igneous Rocks to review felsic and mafic minerals). Because gneisses form at high metamorphic grade, minerals such as biotite and muscovite are less abundant in gneisses than they are in schists, and minerals like garnet and sillimanite that contain little or no water in their crystal lattices (Fig. 7.17) tend to be more common. In addition, at these elevated temperatures, rocks readily flow so gneisses are characterized by intricate folds.

Migmatite

Crustal rocks eventually undergo partial melting if sufficient heat is applied. When gneisses reach temperatures between 600°C and 800°C, they start to melt. The felsic layers melt first because felsic minerals have lower melting points than mafic ones. The rock that forms consequently displays both an igneous component (felsic magma) and a metamorphic one (mafic residue). The resulting rock therefore has a mixed appearance and is called a **migmatite** (Fig. 7.18; see also Fig. 4.2).

We further classify most foliated metamorphic rocks with additional names that either reflect the minerals they contain or their parent rocks. In most cases, these names are self-explanatory. For example, a *mica schist* is a schist made of mica, and a *granite gneiss* is a gneiss formed from granite.

Chemical formula	Mineral		Geological environment
$Al_2Si_2O_5(OH)_4$	Clay		Sedimentary
$(MgAl)_3(SiAl)_2O_5(OH)_4$	Chlorite		
$KMg_3AlSi_3O_{10}(OH)_2$	Biotite	Increasing metamorphism	Greenschist facies
$Fe_3Al_2Si_3O_{12}$	Garnet		
$Fe_2Al_9O_6(SiO_4)_4(OH)_2$	Staurolite		
Al_2SiO_5	Kyanite		Amphibolite facies
Al_2SiO_5	Sillimanite		

Figure 7.17 Progressive Metamorphism of Shales
Minerals that develop from the progressive metamorphism of shales show a decrease in the amount of water (expressed by the symbol [OH]) in their mineral formulae as their metamorphic grades increase. Specific minerals and metamorphic facies are discussed in Section 7.6.

CLASSIFYING NONFOLIATED ROCKS

Some nonfoliated rocks are classified on the basis of their origin. For example, recall from Section 7.4 that hornfels refers to any rock that forms as a result of contact metamorphism from the heat of a nearby igneous intrusion. However, most nonfoliated rocks are classified according to their composition. *Marble*, for example, is a metamorphosed limestone or dolostone comprising largely calcite or dolomite, *quartzite* is a metamorphosed sandstone made largely of quartz, and *amphibolite* is a hornblende-rich rock often derived from metamorphism of basalt. In the following material, we discuss these types of nonfoliated metamorphic rocks in more detail.

Figure 7.18   Migmatite

Migmatite combines metamorphic and igneous features. Here folded gneissic banding is cut by granite veins produced when portions of the gneiss melted.

Hornfels

Hornfels is the product of recrystallization adjacent to a cooling igneous body and can form from any parent rock. Because recrystallization is primarily the result of heat and only rarely involves directed pressure, hornfels is typically nonfoliated. Adjacent to the pluton, the heat tends to drive off virtually all the fluids so only minerals that contain no water form. This paucity of fluids combined with the relatively rapid cooling of the pluton inhibits crystal growth so that hornfels is typically a fine-grained rock (Fig. 7.19).

Marble

Marble is a metamorphosed limestone or dolomite. On metamorphism, the original carbonate mineral grains (calcite or dolomite) recrystallize to form coarser interlocking crystals (Fig. 7.20). Since both minerals have well-developed cleavages, the individual crystals are highly

Figure 7.20   Marble

Marble is a typically nonfoliated rock produced by the metamorphism of limestone or dolomite.

reflective. This property, together with its white color and softness (hardness = 3), makes marble a popular decorative stone. In marine settings where deposition occurs, limestone and shale are often interbedded as sea level rises and falls. When this type of stratified rock is metamorphosed to high grades, the calcite grains slip along their cleavage planes in response to directed pressure, which enables the rock to flow. The result is a spectacular banded rock consisting of interbedded marble and schist (Fig. 7.21).

> ### CHECK YOUR UNDERSTANDING
>
> ● How does banded marble form?

Figure 7.19   Hornfels

Hornfels is a nonfoliated metamorphic rock produced by recrystallization adjacent to a cooling igneous body.

Figure 7.21   Banded Marble

When interbedded limestone and shale are metamorphosed to high grades, the calcite grains slip along their cleavage planes, which causes the rock to flow. The result is a spectacular banded rock consisting of interbedded marble and schist. (Photograph by Dr. S. A. Wellings)

Quartzite

Quartzite is metamorphosed quartz-rich sandstone. The quartz grains are welded together in an interlocking crystalline mosaic (Fig. 7.22). Quartzite, therefore, is a very hard rock. As field geologists will attest, it takes a hefty blow from a sledgehammer (wielded while wearing a pair of safety glasses) to safely dislodge a sample of quartzite from the outcrop. In contrast to most rocks, which split along grain boundaries when a sample is taken, the quartz grains of a quartzite are so tightly fused together that the fracture propagates across individual crystals. As a result, internal bonds within the crystal are broken so that sparks fly when a quartzite is sampled. Because quartz is generally colorless, quartzite is typically white. However, minor impurities such as iron may result in a red coloration.

> **CHECK YOUR UNDERSTANDING**
>
> ● Why do slates break very easily, whereas quartzites are very difficult to break?

Amphibolite

Amphibolite is metamorphosed basalt. When platy minerals such as chlorite that form at lower grades become unstable and react, they produce hornblende, the most common member of the *amphibole group* of minerals. Amphibolites also contain variable amounts of the feldspar mineral, *plagioclase*. The contrast in color between the black hornblende and pale feldspar gives amphibolite a speckled salt-and-pepper appearance (Fig. 7.23).

7.5 SUMMARY

- With increasing temperature, the grain size of foliated rocks increases from very-fine-grained slate, to fine grained phyllite, coarser schist, and even coarser grained gneiss. Under high grade metamorphic conditions, partial melting occurs, producing a mix of melt and solid residue called a migmatite.

- Nonfoliated metamorphic rocks form either in the absence of directed pressure or lack aligned platy minerals necessary to define a foliation.

- Some nonfoliated rocks are classified on the basis of their origin; others are classified according to their composition.

Figure 7.22   Quartzite
Quartzite is a nonfoliated rock produced by the metamorphism of quartz sandstone.

Figure 7.23   Amphibolite
Amphibolite is a nonfoliated rock produced by the metamorphism of basalt.

7.6 | Metamorphic Zones and Facies

In the late nineteenth and early twentieth centuries, the work of George Barrow in the southeastern Scottish Highlands revealed that systematic variations in the mineral content of metamorphosed shales were caused by differences in the conditions of metamorphism (see In Depth: **Metamorphism and the Rock Cycle**). Barrow's work showed that, as metamorphic grade intensifies, new metamorphic minerals grow at the expense of older minerals.

METAMORPHIC ZONES

Barrow divided the rocks of the southeastern Scottish Highlands into **metamorphic zones**. Each zone is characterized by the appearance of a new metamorphic mineral, which Barrow called an **index mineral** (Fig. 7.24). In all, Barrow identified six zones, which he defined, respectively, by the first appearance of the index minerals chlorite, biotite, garnet, staurolite, kyanite, and sillimanite. Note how these minerals correspond to the sequence of minerals that develop from progressive metamorphism of shale (see Section 7.5 and Figure 7.18). By drawing a line on a map marking the first appearance of each index mineral, Barrow mapped the distribution of these metamorphic zones with increasing

IN DEPTH

Metamorphism and the Rock Cycle

Because metamorphic rocks form at depth, their exposure at Earth's surface implies that they have been uplifted. Metamorphism is an integral part of mountain building, and the uplift and erosion of these mountains, as depicted by the rock cycle, are responsible for the exhumation of deeper crustal levels and the exposure of these deeper layers at the surface. During uplift, the pressure and temperature of the rocks decrease. Given this fact, one might expect high-grade metamorphic minerals to become unstable and react to form low-grade metamorphic minerals. So why don't minerals that have undergone metamorphism simply revert back to their parent minerals, which were stable under surface conditions, during uplift and erosion?

Although "reverse reactions" do indeed take place (a process known as *retrograde metamorphism*), in general these reactions are surprisingly rare. To understand why this is so, we need to recall several basic principles discussed earlier in the chapter, such as the rates of chemical reactions, the importance of time in achieving equilibrium, and the importance of the reactants themselves. As we have learned, the rate of a chemical reaction increases dramatically with temperature—the higher the temperature, the faster the reaction (see Figure 7.6). Chemical reactions are therefore rapid in the warm, deep crust but are much more sluggish in the cool, shallow crust. So as metamorphic rocks are uplifted and cooled, reaction rates slow down drastically. Retrograde reactions are consequently very slow, so that the metamorphic minerals formed at elevated temperatures are generally preserved.

Metamorphic minerals also require time to develop, and uplift is usually too rapid for this to occur. As a result, deeper crustal assemblages have insufficient time to convert to lower pressure minerals and are preserved.

In addition, many metamorphic reactions release fluids, which are driven out of the high-grade rocks and are therefore not present to participate in the retrograde reaction when crust is uplifted. A reaction cannot take place *unless all the ingredients are present,* so the absence of fluids allows high grade assemblages to remain stable even as the crust is exhumed.

Once at Earth's surface, however, metamorphic rocks are exposed to surface processes such as weathering and erosion, which ultimately break down the metamorphic minerals so that these minerals are rarely found in sedimentary rocks.

The continual cycling of crustal rocks from Earth's interior to the surface and back again, as depicted in the rock cycle, also cycles the minerals that the rocks contain. This cycling brings crustal rocks and their constituent minerals from Earth's inaccessible interior to the surface so that we can study the interior of the crust and the dynamic processes that occur within it. The preservation of metamorphic rocks during exhumation is an integral part of the rock cycle.

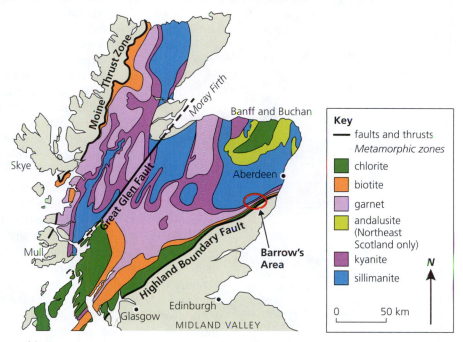

Figure 7.24   Metamorphic Zones

The metamorphic rocks in the eastern highlands of Scotland are divided into six zones on the basis of index minerals that occur in the rocks. With increasing metamorphic grade, they are the chlorite, biotite, garnet, staurolite, kyanite, and sillimanite zones. This sequence corresponds with the sequence of minerals produced by progressive metamorphism of shales (see Figure 7.17).

CHECK YOUR
UNDERSTANDING
◗ What is an index mineral?

metamorphic grade. These lines represent the location on Earth's surface of the reaction that produced each index mineral, and are known as **isograds**, which define lines of equal metamorphic grade.

METAMORPHIC FACIES

Building on the pioneering work of Barrow, the Finnish geologist Pentti Eskola proposed the concept of *metamorphic facies* and, in doing so, laid the foundation for our modern understanding of metamorphic rocks. Eskola compared metamorphic rocks from different areas and found that certain *mineral assemblages* (groups of minerals found together in a rock) were very common. He reasoned that metamorphic rocks with similar chemical compositions contain the same minerals if they form under similar conditions of pressure and temperature. Hence, the metamorphism of a collection of common rock types produces common minerals under the same metamorphic conditions. Our cooking analogy can illuminate this concept more fully. As any chef will vouch, given the same ingredients, oven settings, and time, the final dish should turn out the same. Likewise, rocks with the same chemistry, when subjected to the same physical conditions for the same amount of time, produce the same minerals. A **metamorphic facies** is an association of metamorphic rocks that were metamorphosed under similar conditions of temperature and pressure.

Eskola's reasoning enabled geologists to take the knowledge gained from the study of one metamorphic belt and apply it to another, thereby providing a framework for understanding the similarities and differences between different metamorphic belts. Eskola further reasoned that a common mineral assemblage must be stable over a specific range of pressures and temperatures. Hence, each metamorphic facies occupies a specific region on a *pressure-temperature diagram* (Fig. 7.25).

For rocks of similar chemical composition, *differences* in metamorphic mineral assemblages must therefore reflect differences in the temperature and pressure of metamorphism. These assemblages occupy different regions of the pressure-temperature diagram. Hence, different areas defined by different mineral assemblages represent different regions on the pressure-temperature diagram.

The conditions of metamorphism for a metamorphic rock must be within the range of pressures and temperatures over which all of its minerals are stable. This range can be determined experimentally and helps to define a given metamorphic facies.

The lowest grade of metamorphism produces minerals that belong to the **zeolite facies**. At these low metamorphic grades, most of the original features of the parent rock are preserved. The facies is named after a family of minerals called *zeolites,* which have very open crystal structures. As a result, they are relatively light minerals (densities between 2.0 and 2.4) and so are only stable at low pressures. Because of the low temperatures, metamorphic reactions are very sluggish and produce only fine-grained minerals.

The most common metamorphic facies is the **greenschist facies**. The metamorphism of many rock types at this grade

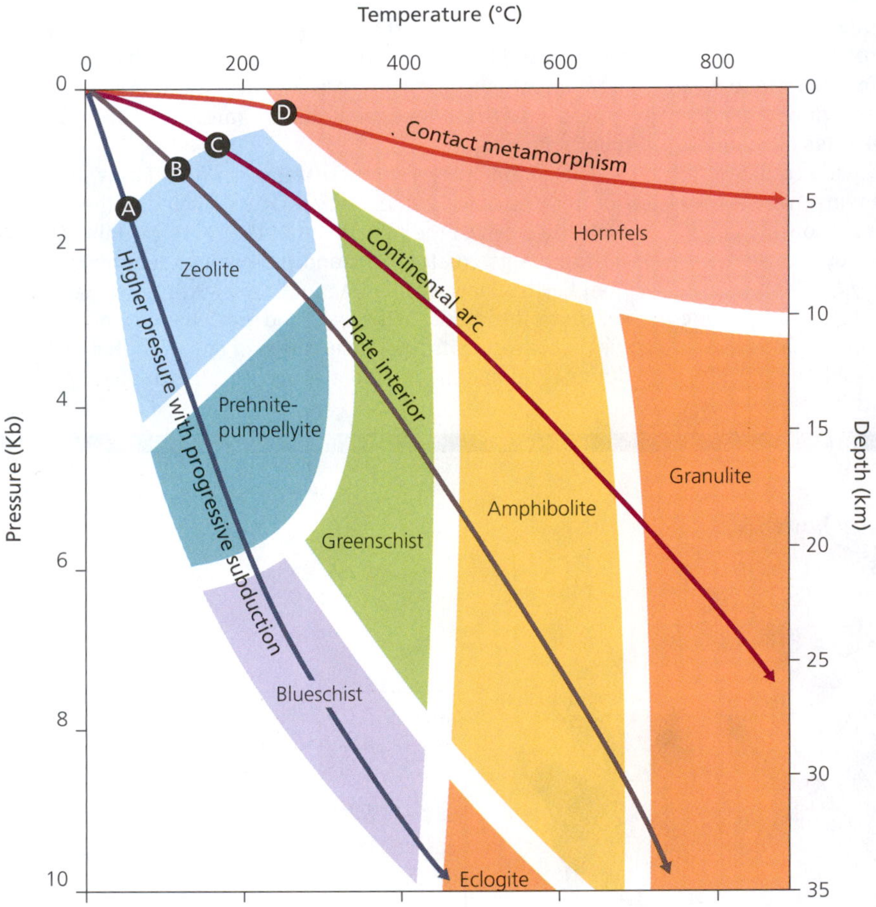

Figure 7.25   Pressure-Temperature Diagram

This simplified pressure-temperature diagram shows the temperature and pressure conditions of the various metamorphic facies and the average geothermal gradient near subduction zones (A), under continental arcs (B), in plate interiors (C), and adjacent to plutons (D). A metamorphic facies is an association of rocks that formed under similar conditions and each facies is characterized by particular minerals that form under similar pressure and temperature conditions. For example, the zeolite facies contains zeolites, the blueschist facies contains glaucophane, the eclogite facies contains omphacite, the greenschist facies contains chlorite, and the amphibolite facies contains hornblende.

produces a variety of green minerals, the most common of which is *chlorite*. Greenschist facies metamorphism is generally accompanied by directed pressures associated with regional metamorphism, so these rocks are usually foliated. In the lower grade part of the greenschist facies, metamorphic minerals are very fine grained and slates or phyllites are common. However, at higher grades within the facies, schists are typical.

At temperatures and pressures just above those of the greenschist facies, amphiboles such as hornblende become stable. At this point, metamorphism enters the **amphibolite facies**, characterized by coarse-grained schists and amphibolites. At still higher temperatures and pressures, such as those found in the lower crust, mineral assemblages of the **granulite facies** become stable. The granulite facies is so called because rocks metamorphosed at this grade are characterized by a granular texture. Platy minerals are rare or absent and the most common rock types are gneisses. Where temperatures exceed the melting point of a rock, we enter the realm of molten igneous rocks and reach the upper limits of metamorphism. The presence of migmatites is an important indicator of these extreme conditions.

In certain environments, very high pressures can develop at temperatures that are similar to those of the greenschist facies. The schists produced often contain a distinctive blue amphibole known as *glaucophane* (Fig. 7.26), which imparts its color to the rocks. These high-pressure, low-temperature conditions of metamorphism define the **blueschist facies**.

The highest grade of metamorphism, the rare **eclogite facies**, develops only under the extremes of crustal temperatures and pressures. Rocks containing garnet and the aluminous pyroxene, *omphacite,* characterize this facies. As we shall learn in Section 7.7, which focuses on metamorphism and plate tectonics, rocks of the blueschist and eclogite facies are usually associated with subduction zones, where tectonic plates

> **CHECK YOUR UNDERSTANDING**
>
> ◗ What is a metamorphic facies?

Figure 7.26 Glaucophane Blueschist
The blue amphibole, glaucophane, is characteristic of rocks of the blueschist facies. (Photograph by Dr. Cin-Ty A. Lee)

are consumed. Only through subduction do rocks encounter the necessary extremes of pressure and temperature to generate blueschist and eclogite facies rocks.

7.6 SUMMARY

- Metamorphic minerals vary with the pressure and temperature (or grade) of metamorphism and so provide clues to the conditions under which metamorphism took place.

- The distribution of metamorphic minerals on Earth's surface define metamorphic zones and reflect different intensities of metamorphism.

- An association of metamorphic rocks formed under similar conditions of pressure and temperature define a metamorphic facies.

- Differences in mineral content in rocks of similar composition reflect differing conditions of metamorphism.

7.7 Metamorphism and Plate Tectonics

Metamorphism is directly related to plate tectonics since plate tectonics controls the pressures and temperatures in Earth's crust. Metamorphic rocks form in the following three plate tectonic settings:

- at divergent margins near mid-ocean ridges where the seafloor spreads;

- in subduction zone environments where cold oceanic crust descends and is consumed; and

- in collisional zones where continents collide.

In this section we examine the conditions in each of these settings in detail.

DIVERGENT PLATE BOUNDARIES

At divergent plate boundaries, heat from the magma chamber below the mid-ocean ridge drives the circulation of hydrothermal fluids. This process results in hydrothermal metamorphism of the basaltic ocean crust soon after it forms. Oceanic ridges form an interconnected network 65,000 kilometers (40,000 miles) in length, so hydrothermal metamorphism is the dominant type of metamorphism on Earth. But this type of metamorphism is poorly preserved in the geologic record because most oceanic crust is destroyed by subduction. Under special circumstances, however, segments of oceanic crust escape subduction and are preserved on land.

> **CHECK YOUR UNDERSTANDING**
>
> ◗ What is the relationship between mid-ocean ridges and hydrothermal metamorphism?

These fragments of ocean floor are known as *ophiolite complexes,* which preserve evidence of mid-ocean ridge hydrothermal metamorphism.

SUBDUCTION ZONES

Recall from Section 7.2 that crustal temperatures typically increase with depth by about 20°C–30°C per kilometer (60°F to 100°F per mile). This typical geothermal gradient intersects the greenschist, amphibolite, and granulite facies (see Fig. 7.25), so it is not surprising that these metamorphic facies are the most common in the continental crust. In subduction zones, however, the descent of cold oceanic crust cools the surrounding rocks, resulting in anomalously low crustal temperatures at any given depth.

In fact the geothermal gradient in a subduction zone can be as low as 5°C per kilometer (14°F per mile), creating metamorphic conditions of high pressure and low temperature. This geothermal gradient intersects the blueschist and eclogite facies. When geologists find minerals such as glaucophane and omphacite that are characteristic of these facies, the minerals provide conclusive evidence of the existence of former subduction zones.

While subduction persists, rocks of the blueschist and eclogite facies continue to be generated. At the same time, fluids derived from the subducted slab promote melting in the overlying mantle. Recall from Chapter 4 that this results in the formation of large bodies of magma that give rise to continental arcs. As the fluids and magma rise to the surface, they transport heat to the crust resulting in anomalously warm temperatures. These environments consequently have a higher geothermal gradient than normal, sometimes as much as 60°C per kilometer (170°F per mile) (see Fig. 7.25). This geothermal gradient results in high-temperature, low-pressure metamorphism in the crust *above* subduction zones.

The contrast in geothermal gradient between a subduction zone environment and that of the continental arc is sustained as long as subduction persists, so that both types of metamorphism occur simultaneously (Fig. 7.27). This relationship produces *paired metamorphic belts* where high-pressure, low-temperature metamorphic rocks generated in subduction zones are juxtaposed against those produced by high-temperature, low-pressure metamorphism in continental arcs.

COLLISIONAL ZONES

In zones of continental collision, the continental crust of one plate is commonly thrust on top of that of the other. For example, in the Himalayan collision between India and Asia (Fig. 7.28a), the Asian crust has been thrust on top of the Indian crust, effectively doubling the crustal thickness. As a result, in collisional settings, the lower plate is deeply buried and experiences significant increases in pressure and temperature. Because silicate rocks are poor conductors of heat, the increase in pressure due to thrusting occurs much more rapidly than the increase in temperature. In fact, the thrusting of one continent over another can be considered almost instantaneous, so that its

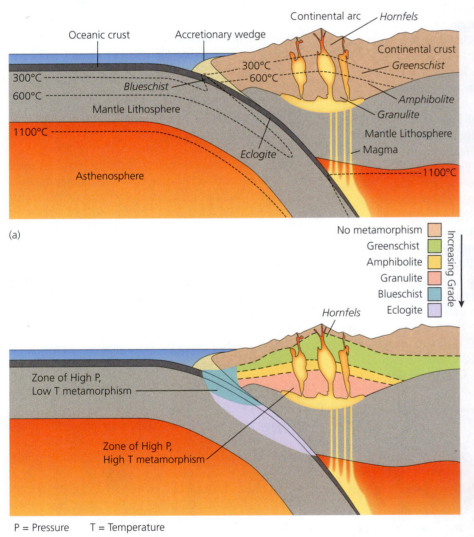

Figure 7.27   Metamorphism in Subduction Settings
These schematic diagrams show the typical distribution of (a) temperature and (b) metamorphic facies in subduction zone settings. As long as subduction is going on, the contrast in geothermal gradient between a subduction zone environment and that of the continental arc persists and metamorphism takes place in two contrasting settings. These contrasts create paired metamorphic belts.

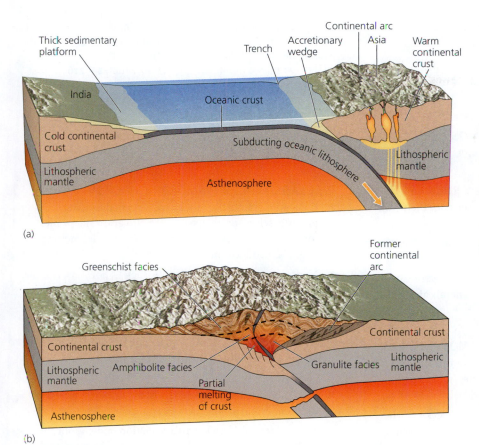

Figure 7.28   Metamorphism in a Collisional Setting

(a) Subduction beneath Asia results in a continental arc and a distribution of metamorphic zones similar to that shown in Fig. 7.27. The Asian crust is relatively warm because of the magmatism. The Indian crust, on the other hand is relatively cool because there is no subduction along its margin. (b) When collision occurs, warmer Asia is thrust over cooler India. The collision generates high grade (granulite facies) metamorphism and partial melting in the deep crust, and the rise of magma and fluids provide heat to cause metamorphism (amphibolite and greenschist facies) in the overlying crust.

geothermal gradient is effectively thrust with it (Fig. 7.28b). Over time, the crust eventually warms up (Fig. 7.28c). During this period, the crust is also uplifted due to erosion at the surface. Modeling of this dynamic process shows that

CHECK YOUR UNDERSTANDING

◉ Why do subduction zones have such a low geothermal gradient?

the lower crust undergoes high grade metamorphism and may even melt. As a result, granitic igneous rocks and migmatites form. Where sufficient volumes of melt are generated, magmas ascend and their heat causes metamorphism at shallower crustal levels.

7.7 SUMMARY

- Circulating hydrothermal fluids at mid-ocean ridges cause hydrothermal metamorphism of basaltic ocean crust.

- In subduction zones, the descent of cold oceanic lithosphere produces metamorphic rocks of the blueschist or eclogite facies that are stable at high pressures and relatively low temperatures.

- Above subduction zones, the ascent of fluids and magma heats the crust, producing metamorphic rocks that are stable at high temperatures and relatively low pressures.

- The contrast in geothermal gradient between a subduction zone and the adjacent continental arc produces paired metamorphic belts.

- In continental collision zones, the thickness of the continental crust may be doubled, promoting partial melting and the formation of migmatites.

Key Terms

amphibolite 214
amphibolite facies 217
aureole 204
blueschist facies 217
burial metamorphism 207
confining pressure 200
contact
  metamorphism 204

directed pressure 200
dynamic
  metamorphism 204
eclogite facies 217
fabric 207
foliated metamorphic
  rock 208
foliation 208

gigapascal 200
gneiss 212
gneissic foliation 208
granulite facies 217
greenschist facies 216
hornfels 213
hydrothermal
  metamorphism 206

index mineral 214
isograd 216
kilobar 200
marble 213
metamorphic facies 216
metamorphic grade 199
metamorphic rock 196
metamorphic zone 214

metamorphism   196
migmatite   212
mylonite   204
nonfoliated metamorphic
    rock   209

phyllite   212
quartzite   214
regional
    metamorphism   204
schist   212

schistosity   208
shear   200
shock metamorphism   205
slate   212

slaty cleavage   208
strain rate   204
stress   200
zeolite facies   216

Key Concepts

7.1 METAMORPHIC ROCKS AND THE ROCK CYCLE

- Metamorphism involves changes to rocks in their solid state that occur below Earth's surface.

- Metamorphism of preexisting rocks results in changes in their texture and mineral content.

- The agents of metamorphism are heat, pressure, and fluids, but time is required for new minerals and textures to form.

7.2 FACTORS THAT CONTROL METAMORPHISM

- Increasing temperature facilitates metamorphic reactions and favors minerals of greater volume. Increasing pressure favors minerals of greater density.

- Confining pressure acts equally in all directions, whereas directed pressure acts more strongly in one direction and produces a fabric.

- The mineral content of metamorphic rocks is profoundly influenced by the chemistry of the parent rock.

- Fluids facilitate metamorphism by transporting energy and ions to the site of mineral reactions.

7.3 TYPES OF METAMORPHISM

- Regional metamorphism occurs during mountain building.

- Contact metamorphism occurs adjacent to igneous bodies.

- Dynamic metamorphism occurs along fault zones.

- Shock metamorphism occurs during extraterrestrial impacts.

- Hydrothermal metamorphism is the result of hot circulating fluids.

- Burial metamorphism is caused by burial beneath a thick succession of rock layers.

7.4 METAMORPHIC TEXTURES

- Under directed stress, metamorphic rocks develop a planar fabric or foliation.

- Based on the presence or absence of a foliation, metamorphic rocks are said to be either foliated or nonfoliated.

7.5 CLASSIFYING METAMORPHIC ROCKS

- Foliated metamorphic rocks are classified as slate, phyllite, schist, or gneiss according to grain size, which increases with increasing temperature. Partial melting of gneiss produces migmatite.

- Nonfoliated metamorphic rocks, such as hornfels, marble, quartzite and amphibolite, are classified according to their origin or composition.

7.6 METAMORPHIC ZONES AND FACIES

- Metamorphic minerals vary with the pressure and temperature (or grade) of metamorphism.

- Metamorphic zones defined by index minerals reflect different intensities of metamorphism.

- A metamorphic facies is an association of metamorphic rocks formed under similar conditions of pressure and temperature.

7.7 METAMORPHISM AND PLATE TECTONICS

- Circulating fluids cause hydrothermal metamorphism at mid-ocean ridges.

- Subduction of cold oceanic lithosphere produces high-pressure, low-temperature blueschist and eclogite facies metamorphism in subduction zones.

- Rising magmas produce high-temperature, low-pressure greenschist and amphibolite facies metamorphism above subduction zones.

- Crustal thickening promotes partial melting and the formation of migmatites at continental collision zones.

Study Questions

1. Explain the distinction between metamorphic, igneous, and sedimentary processes.

2. How are the distinctions between metamorphic, igneous, and sedimentary processes reflected in rock textures?

3. Compare and contrast the varying roles of the agents of metamorphism in promoting metamorphic reactions and textural changes.

4. List and explain the varying sources of fluids in metamorphic rocks.

5. What is the relationship between time and equilibrium in the formation of metamorphic minerals?

6. Why do some marbles display a nonfoliated texture while others display intricate patterns of folds?

7. Why is metamorphism commonly accompanied by deformation?

8. Compare and contrast the application of the facies concept to metamorphic and sedimentary rocks.

9. Explain the origin of paired metamorphic belts.

10. What is a migmatite? Explain its characteristics and its relationship to continental collisions.

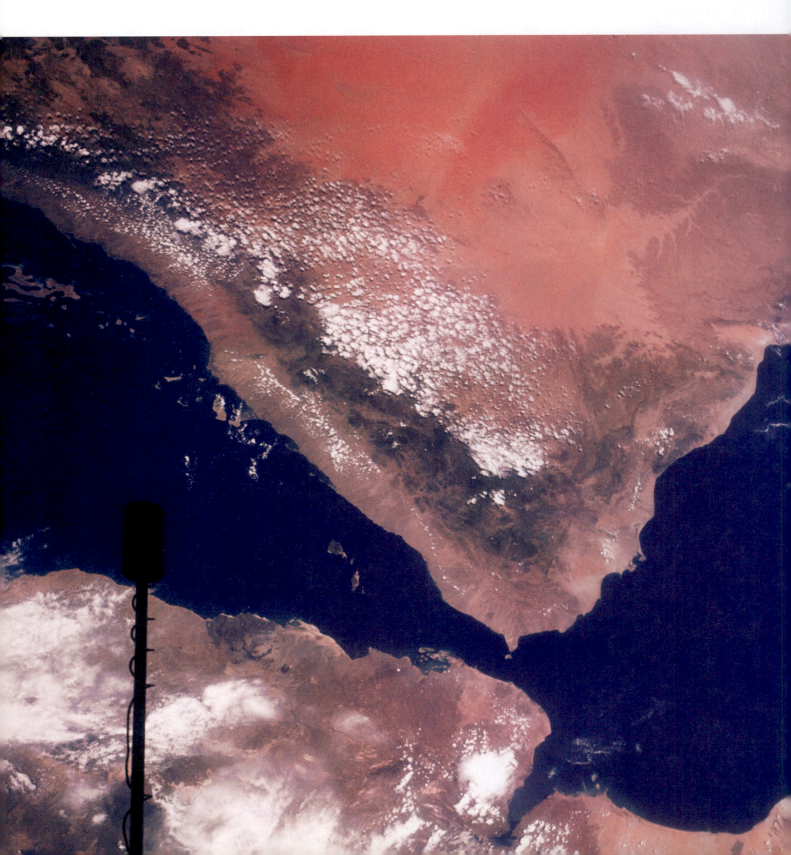

PART **II**
Evolution of the Solid Earth

Chapter 8 | Geologic Time   224

Chapter 9 | Plates and Plate
            Boundaries   254

Chapter 10 | Deformation
             and Mountain
             Building   296

Chapter 11 | Earthquakes
             and Earth's
             Interior   334

Geologic Time

8.1 Measuring Geologic Time: Relative and Absolute Ages

8.2 A Short History of Geologic Time

8.3 Relative Dating: Determining Chronological Order

8.4 Correlating Rock Strata

8.5 Relative Dating and the Geologic Time Scale

8.6 Absolute Dating: Finding a Geologic Clock

8.7 Absolute Dating and the Geologic Time Scale

8.8 A Sense of Time

8.9 Geologic Time and Plate Tectonics

▶ Key Terms

▶ Key Concepts

▶ Study Questions

LIVING ON EARTH: Radioactivity and Radon Gas 242

SCIENCE REFRESHER: Types of Radioactive Decay 244

SCIENCE REFRESHER: Linear versus Exponential Relationships 247

IN DEPTH: The Quest for the Age of Earth 249

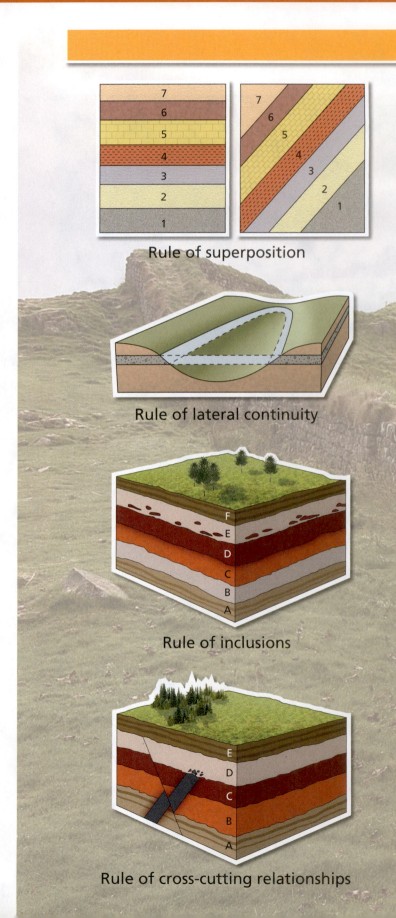

Rule of superposition

Rule of lateral continuity

Rule of inclusions

Rule of cross-cutting relationships

Relative Age: Rules, Correlation, and Faunal Succession

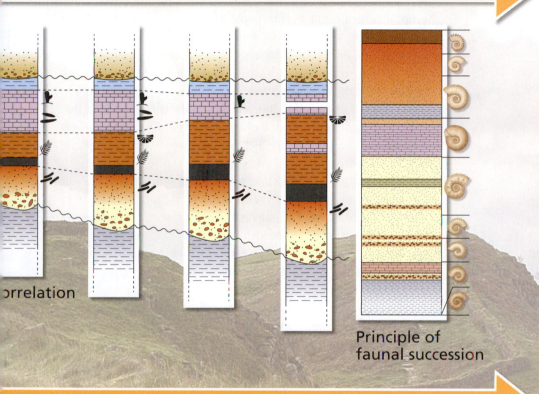

orrelation

Principle of faunal succession

Absolute Age: Radiometric Dating

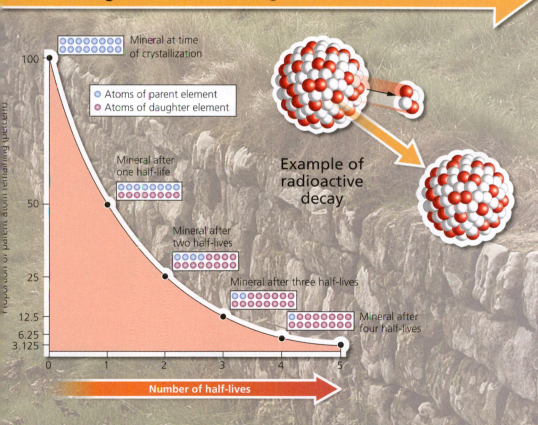

Mineral at time of crystallization

○ Atoms of parent element
● Atoms of daughter element

Mineral after one half-life

Mineral after two half-lives

Mineral after three half-lives

Mineral after four half-lives

Example of radioactive decay

Number of half-lives

Geologic Timeline

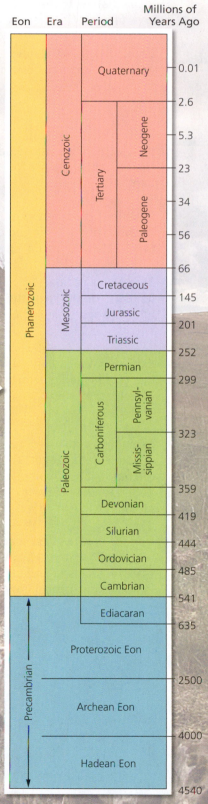

Eon	Era	Period		Millions of Years Ago
Phanerozoic	Cenozoic	Quaternary		0.01
		Tertiary	Neogene	2.6
				5.3
				23
			Paleogene	34
				56
				66
	Mesozoic	Cretaceous		145
		Jurassic		201
		Triassic		252
	Paleozoic	Permian		299
		Carboniferous	Pennsyl-vanian	323
			Missis-sippian	359
		Devonian		419
		Silurian		444
		Ordovician		485
		Cambrian		541
Precambrian		Ediacaran		635
		Proterozoic Eon		2500
		Archean Eon		4000
		Hadean Eon		4540

8.1 Define, compare, and contrast relative and absolute age dating.

8.2 Describe the development of ideas that led to the modern view of geologic time.

8.3 List and explain the rules that geologists apply to determine the relative age of geologic events.

8.4 Explain how correlation allows individual rock layers to be traced from one place to another.

8.5 Describe how relative dating methods allowed the geologic time scale to be established.

8.6 Summarize radioactivity and its use in determining the absolute age of geologic events.

8.7 Explain how absolute dating has established numerical ages for the geologic time scale.

8.8 Explain how relative and absolute dating together form the basis of our modern understanding of Earth history.

8.9 Discuss how the vast span of geologic time has allowed plate tectonics to profoundly influence on Earth history.

For most of us the immensity of geologic time is difficult to grasp. Yet comprehending the scope of geologic time is essential to any understanding of physical geology. Although some geologic activities, such as earthquakes and volcanic eruptions, are rapid events, most of the processes that shape Earth operate so slowly that it is only with the passage of vast stretches of time that their effects are realized. The profound changes these processes have wrought on Earth's surface are testament to the antiquity of the planet. However, during a human lifetime, their effects are often imperceptible. Our purpose in this chapter is to provide a sense of the vastness of geologic time by examining the methods by which geologists have established the age and history of Earth.

Geologic time and the slowness of natural processes are not just of academic interest. Many environmental concerns stem from the fact that humans consume resources far faster than nature can replenish them, and we produce pollutants far faster than nature can absorb them. In addition to helping us understand the impact of our resource consumption and waste production, understanding geologic time also influences our ability to predict catastrophic events such as earthquakes and volcanic eruptions. Even though these geologic events may only take seconds to have their impact, they are the culmination of natural processes that have built up over a very long time. Warning signals may therefore be evident only in the most imperceptible changes.

So how is geologic time measured and how has the history of Earth been established? Geologists use two main methods of measuring geologic time, one involving the determination of *relative age* and the other the determination of *absolute age*. Together, these two methods have been used to establish a geologic time scale, first by placing geologic events in sequence, and then by fixing the time at which they occurred. In this chapter, we discuss these two methods and how they work.

8.1 Measuring Geologic Time: Relative and Absolute Ages

It is often said that a geologist is like a detective searching for clues at the scene of a crime. When geologists examine an *outcrop* where rock formations appear at the surface of Earth, they try to deduce the sequence of events that produced the exposed rocks. In the outcrop shown in Figure 8.1, for example, the exposed layers of sedimentary rock have been bent into a U-shape. From this a simple history can be deduced in which deposition and burial to form sedimentary rock was followed by bending of the sedimentary layers.

At the outcrop, geologists can often place geologic events, like the deposition and bending of sedimentary layers, in sequence. That is,

Figure 8.1   Clues to the Past
An outcrop of folded sedimentary rocks on Interstate 68 at Sideling Hill in western Maryland reveals clues to the history of the Appalachian Mountains, where Sideling Hill is located.

their **relative age**, or the order in which the events occurred, can be determined. The actual time of deposition of the sedimentary layers can also be determined if fossils of known age are present in some of the folded rock layers. But fossils cannot, by themselves, reveal the duration of the events or tell us how long ago they occurred.

However, it is sometimes possible to determine directly just how old a rock really is. Igneous rocks, for example, can be dated by performing sophisticated laboratory measurements of the amount of *radioactive decay* in certain minerals present within them. The more decay, the older the mineral and, hence, the igneous rock. This approach provides a precise determination of the **absolute age** of the rock, that is, the amount of time that has elapsed since the rock formed.

But not all rocks can be dated in this way. As a result, final determination of the sequence of a particular set of geologic events is usually achieved by combining clues observed at the outcrop that reveal relative age with evidence provided by laboratory measurements of absolute age. To ensure correct diagnosis, it is important that the physical evidence visible at the outcrop is consistent with the laboratory data. Just as a detective who deduces the time of a crime from a smashed wristwatch needs to ensure that the wristwatch was not tampered with after the crime, so the geologist must ensure that samples used to determine absolute ages accurately reflect the sequence of events observed in the outcrop. In the outcrop shown in Figure 8.1, it is essential that the ages of deposition and bending, if these could be determined from samples collected at the scene, confirm the observation that the sedimentary layers were deposited before they were bent.

> **CHECK YOUR UNDERSTANDING**
> ○ What is the difference between relative age and absolute age?

8.1 SUMMARY

- Measurement of geologic time is based on the relative age of older versus younger events and the absolute age of minerals in years.

8.2 A Short History of Geologic Time

To develop a more complete appreciation for geologic time, it is useful to review major hypotheses made by previous generations of scientists as they endeavored to establish the age of Earth. Chief among these are the principles of catastrophism, gradualism, and uniformitarianism.

ARCHBISHOP USSHER AND THE PRINCIPLE OF CATASTROPHISM

In the mid-seventeenth century, the history of births and deaths in the Bible was used by Anglican Archbishop James Ussher to propose that Earth's creation had taken place in the year 4004 BC. According to Ussher, Earth was a mere 6000 years old. Other cultures, however, place different dates on the time of Earth's creation. In the Mayan calendar, for example, the deepest probings of eternity reach back 400 million years. In the Hindu calendar, the present world came into existence almost 2 billion years ago.

But in the Christian world of western Europe, Ussher's figure of 6000 years became dogma and was taken seriously for over 200 years by scholars who considered Earth to have formed as a molten ball that cooled and formed a lumpy crust with mountains, valleys, and oceans. According to this view, all geographic changes on Earth's surface were catastrophic in nature, such as Noah's Flood, which was viewed as one of a succession of destructions of Earth's surface by supernatural forces.

In support of this *principle of catastrophism*, eighteenth-century scholars pointed to the existence of terraces in river banks, and to valleys that were far too large to have been eroded by the streams they contained, as evidence of catastrophic flooding. Further evidence was seen in fossilized marine organisms in rocks exposed far from the sea and in the existence over much of northern Europe of great boulders scattered far from outcrops of similar rock (Fig. 8.2). These boulders were called *erratics,* because they had apparently moved from their original location. It is now known that erratics were deposited by glaciers during the last ice age (Fig. 8.3), but their existence was used by the catastrophists as evidence of sudden catastrophic flooding. Earthquakes and volcanic eruptions were also seen as sudden catastrophic events consistent with a 6000-year calendar for Earth history.

> **CHECK YOUR UNDERSTANDING**
> ○ What is the principle of catastrophism and how were erratic boulders used to support it?

(a)

(b)

Figure 8.2 Catastrophism

(a) A woodcut from Andrew Ramsay's book *Physical Geology and Geography of Great Britain,* first published in 1864, shows erratic boulders in the glaciated valley of Llanberis in North Wales. Lying far from outcrops of similar rock, boulders such as these were cited by the catastrophists as evidence of catastrophic flooding. (b) The boulder-strewn valley of Llanberis today.

Figure 8.3 Erratic Boulder

A glacial erratic in Yorkshire, England.

NEPTUNISTS VERSUS PLUTONISTS

Evidence that Earth might be older than Ussher's 6000 years started to accumulate when scholars turned their attention to rocks and the origin of rock layering. By the end of the eighteenth century, scholars who came to be known as the *neptunists* believed a great globe-engulfing sea laid down all of Earth's rocky layers. These layers were then gradually exposed as the water slowly receded. Crystalline rocks such as granite were thought to have formed from the chemical precipitation of crystals from seawater. Overlying sedimentary layers were interpreted to have been deposited as the ocean withdrew and the ancient crystalline rocks were exposed to erosion. Although incorrect in some of their interpretations, these scholars were right in thinking that Earth had to be ancient. Six thousand years was insufficient time for their explanation of how the layers formed.

After a debate that lasted many decades, the neptunists ultimately lost their case to scholars known as *plutonists,* who insisted that crystalline rocks such as granite were not chemical precipitates but had crystallized from a molten state. The plutonists were the first to argue for a dynamic and changing Earth, and in so doing were the harbingers of modern geological thinking.

JAMES HUTTON AND THE PRINCIPLE OF GRADUALISM

Foremost amongst the plutonists, and often considered the "father of modern geology," was the Scotsman James Hutton. In addition to his views on igneous rocks, Hutton observed the effects of weathering and erosion and became convinced that, given enough time, processes at work in the environment today could have produced all of the features previously attributed to catastrophic flooding. In his pivotal book *Theory of the Earth,* published in 1795, Hutton proposed that Earth, powered by its own internal heat, is in continuous but gradual change, and is constantly decaying, renewing, and repairing itself. This idea came to be known as *gradualism*. Rather than 6000 years of Earth history, gradualism required Earth to be truly ancient with, as Hutton put it, "no vestige of a beginning, no prospect of an end."

> ### CHECK YOUR UNDERSTANDING
> ● How did Hadrian's Wall convince James Hutton that Earth was ancient?

In defense of an ancient and slowly changing Earth, Hutton compared the level of erosion in old mountain belts with erosion observed in ancient monuments, which had changed little since they were built. Just south of the Scottish border, Hutton studied

Figure 8.4 Hadrian's Wall

Hadrian's Wall on the border of Scotland was built by the Romans in AD 122 and, in Hutton's day, was believed to be almost a third as old as Earth itself. Hutton used the preservation of its stonework to argue that an Earth capable of eroding away entire mountain belts must be truly ancient.

time needed to wear down mountains and carry sand and gravel to the sea must be enormous indeed.

CHARLES LYELL AND THE PRINCIPLE OF UNIFORMITARIANISM

In 1830, Charles Lyell, a British lawyer and geologist, elaborated on Hutton's idea of gradualism by proposing that geological processes at work on Earth today have acted with the same intensity from the earliest times to the present. This idea, which came to be known as the *principle of uniformitarianism,* is based on the view that only natural causes can explain natural events. Thus, past events can only be explained by analogy with modern ones. Furthermore, uniformity of process through time means a uniform rate of change, and therefore, necessitates a geologic time scale adequate for the slowness of natural processes.

As proof of the gradual nature of geologic processes, Lyell pointed to the Roman temple of Jupiter Serapis, near Naples in Italy (Fig. 8.5). Here, fragile columns remain standing despite evidence for considerable sea level change over the past 2000 years. The evidence takes the form of holes bored by marine organisms that are now set high in the columns. Surely, Lyell argued, only gradual processes

Hadrian's Wall. Built by the Romans between Scotland and England in AD 122, the wall had suffered little deterioration despite 16 centuries of erosion (Fig. 8.4). If, Hutton argued, 16 centuries could barely scar a 6-foot-tall wall, the

(a)

(b)

Figure 8.5 Gradualism

(a) A woodcut from Charles Lyell's book of 1830, illustrates the Temple of Jupiter Serapis near Naples, Italy. Lyell and other gradualists pointed to the holes bored by marine shells on the temple's columns as proof of the gradual nature of geologic processes. (b) The accompanying photo shows the temple as it looks now.

could have brought about such a significant change in sea level without toppling the columns.

The concept of uniformitarianism does not require all geologic processes to be gradual. Some processes, such as volcanic eruptions and earthquakes, are far from gradual as Lyell was well aware. Individual rocks layers also record quite short-lived events. For example, on shorelines subjected to occasional hurricanes, layers of shallow-marine sediments are frequently preserved only after severe storms. Each storm produces a single "storm deposit" layer after the hurricane abates. All sediment deposited between major storms is stirred up by the new storm and any layering it possessed is consequently destroyed. Therefore uniformitarianism, by which Lyell meant a uniformity of natural law, should not be taken to preclude catastrophic events. Instead, it means that "*the present is the key to the past.*"

The remarkable progression of ideas from those of Ussher to those of Lyell brought about a profound change in the way that scholars viewed the planet on which they live. To this day, the principles of Hutton and Lyell lie at the heart of our understanding of geologic processes and form the basis of our interpretation of the rock record.

CHECK YOUR UNDERSTANDING

❓ In your own words, explain the significance of the expression "the present is the key to the past."

8.2 SUMMARY

- As recently as 200 years ago Earth was widely believed to be a mere 6000 years old and all changes to its surface were attributed to supernatural disasters, such as Noah's Flood, a principle known as catastrophism.

- Scholars known as the neptunists incorrectly believed that all of Earth's rocky layers were deposited beneath a global ocean, but considered 6000 years to be insufficient time for their deposition.

- James Hutton considered Earth to be truly ancient and its surface to be in continuous but gradual change; a principle known as gradualism.

- Charles Lyell promoted the principle of uniformitarianism, which held that geologic processes act today as they have always done; a view that can be expressed as "the present is the key to the past."

8.3 Relative Dating: Determining Chronological Order

Geologists now follow established rules to determine the relative age of rock strata. The most commonly used of these are the rules of *superposition, inclusions,* and *cross-cutting relationships,* and a sedimentary relationship known as an *unconformity.*

STENO'S RULES

Our modern understanding of relative age begins with the observations of the Italian philosopher, Nicholas Steno. In the late seventeenth century, Steno proposed three fundamental rules that are based on three simple observations regarding sedimentary layers: (1) layers are deposited on top of each other in an ordered sequence of younger layer on top of older ones, (2) layers are flat-lying when they are deposited, and (3) layers are deposited over wide areas.

Imagine standing on the edge of a lake after a flood, with sediment being swept in by streams. This new sediment is deposited on the lakebed in a layer on top of the older sediment deposited previously. The rule of **superposition** states that if sedimentary *strata* (layers) are deposited in sequence, one on top of the other, then the oldest layers will occur at the base (Layer 1, Fig. 8.6a) and the youngest

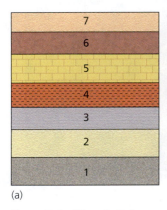
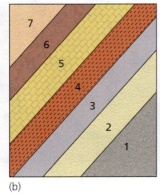
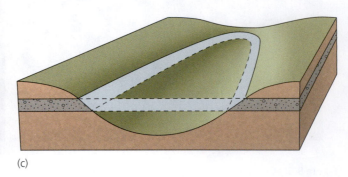

(a)   (b)   (c)

Figure 8.6 Steno's Rules

Steno's three rules concern the deposition of sedimentary strata. (a) A stacked sequence of horizontal layers (*strata*) is numbered 1 to 7 in the order in which the layers were deposited. (b) The same sequence is shown after tilting occurs. (c) The original continuity of a rock layer (shown here in grey) allows geologists to trace the layer from one side of a valley to the other.

at the top (Layer 7, Fig. 8.6a). This rule also applies to the fossils that the rock layers contain. In other words, fossil assemblages that occur at the base of the sequence are older than those that occur at the top.

The second rule, which holds that all sedimentary strata were deposited as flat-lying layers, is known as **original horizontality**. The rule implies that all inclined rock layers were once horizontal and were tilted *after* they were deposited (Fig. 8.6b).

Finally, if sedimentary strata are initially deposited over wide areas their layering must originally have been laterally continuous. The third rule is **original lateral continuity**. This rule asserts that otherwise similar layers separated across a valley must have been continuous when they were deposited. In the example illustrated in Figure 8.6c, the original continuity of strata was severed during the later development of the valley.

From Steno's rules, we know the rock layers were deposited as a laterally continuous horizontal sequence with the oldest layer at the bottom and the youngest layer at the top. Any departure from this original configuration requires that additional geological events happened in the region. Steno's rules may appear to be simple, even self-evident statements. But they provide an essential frame of reference that enables us to determine the relative age of geologic events. For example, where sedimentary layers are steeply inclined, as is the case for the folded strata in Figure 8.1, we know that some force must have tilted the strata after they were deposited.

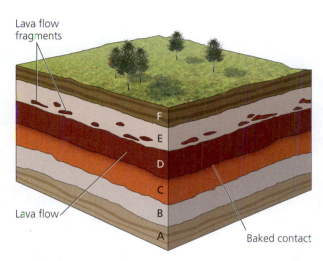

Figure 8.7 Relative Age

A determination of relative age can be made from inclusions of fragments from the lava flow (layer D) in sedimentary layer E.

was deposited after the lava had erupted. Note, also, that the bed beneath the lava flow (layer C) has been baked. Baking of this bed by the lava shows that eruption of the flow followed the deposition of layer C. Hence, the lava flow in layer D is younger than layer C and older than layer E.

RULE OF CROSS-CUTTING RELATIONSHIPS

Another rule of relative dating, known as the **rule of cross-cutting relationships**, holds that any feature that cuts across a rock must be younger than the rock it cuts. Accordingly, an intrusive body of igneous rock is always younger than the rocks it intrudes, and an offsetting fracture (or fault) is always younger than the rocks it offsets.

These relationships are illustrated in Figure 8.8. In this diagram an intrusive dike cuts across and bakes sedimentary

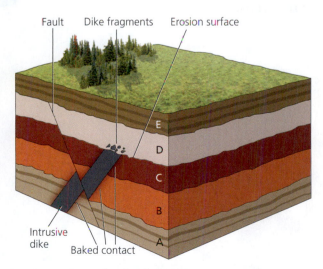

Figure 8.8 Cross-Cutting Relations

Relative age can also be determined by the cross-cutting relationships between sedimentary layers (A through E), an intrusion, and the offset of the layers and the intrusion along a fault.

CHECK YOUR UNDERSTANDING

● Summarize the rules of superposition, original horizontality, and original lateral continuity.

RULE OF INCLUSIONS

Recall from Chapter 6 that clastic sedimentary rocks contain fragments of older bedrock. Any rock containing fragments must be younger than the rock from which the fragments where derived because the fragments can only be pieces of *older* rock. The rule that holds that any rock must be younger than the fragments included within it is another principle of relative dating known as the **rule of inclusions**. Likewise, inclusions in igneous rocks, such as xenoliths and xenocrysts (see Chapter 4), must be older than the igneous rock in which they are found.

The sequence of sedimentary layers shown in Figure 8.7 illustrates some of these rules. According to the rule of superposition, the oldest unit is layer A while the youngest is layer F. Layer D, however, is a lava flow that was later buried by sedimentary layers E and F. Its relative age is shown by the presence of fragments of the lava in layer E. These fragments demonstrate that layer E

CHECK YOUR UNDERSTANDING

● How do geologists apply the rule of inclusions to determine the relative age of rock sequences?

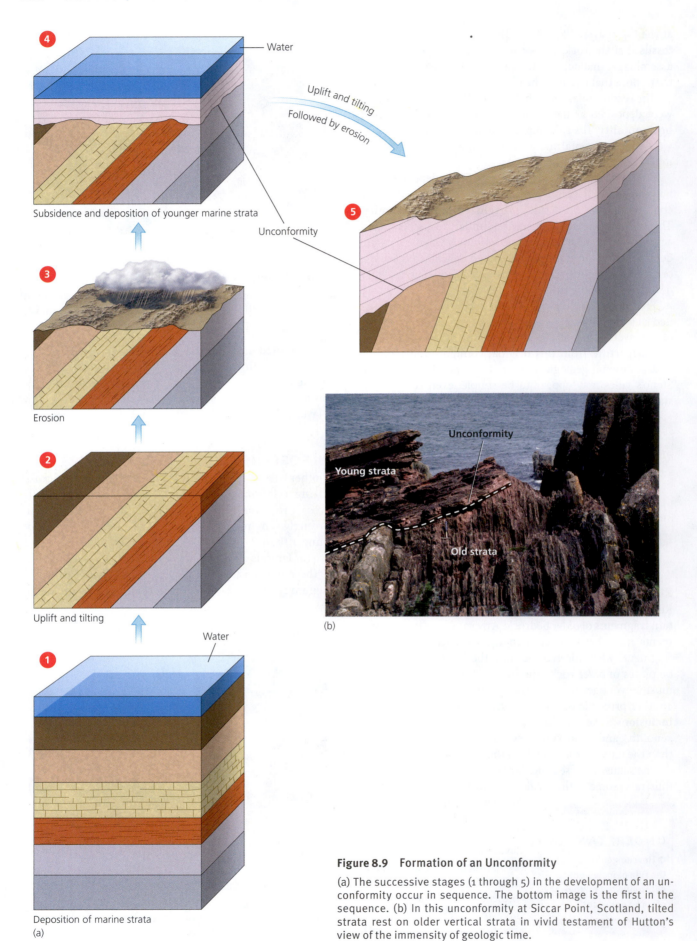

4 Water

Subsidence and deposition of younger marine strata

Uplift and tilting
Followed by erosion

3

Erosion

5

Unconformity

2

Uplift and tilting

Water

1

Deposition of marine strata

(a)

Young strata

Unconformity

Old strata

(b)

Figure 8.9   Formation of an Unconformity

(a) The successive stages (1 through 5) in the development of an unconformity occur in sequence. The bottom image is the first in the
sequence. (b) In this unconformity at Siccar Point, Scotland, tilted
strata rest on older vertical strata in vivid testament of Hutton's
view of the immensity of geologic time.

layers A, B, and C. Hence, according to the rule of cross-cutting relationships, its intrusion must have occurred *after* layer C was deposited. Fragments of the dike occur in layer D and so, according to the rule of inclusions, the fragments are older than layer D. Because the dike must have crystallized below Earth's surface, uplift and erosion are required to expose it, and so the top of layer C must have been eroded at Earth's surface before layer D was deposited. The upper surface of layer C is therefore an *erosion surface* that cuts across the dike. The fault, on the other hand, cuts and offsets the dike and all of the sedimentary layers except layer E. The fault is therefore younger than the dike and must have formed after the deposition of layer D but before that of layer E.

> **CHECK YOUR UNDERSTANDING**
>
> ● How does the rule of cross-cutting relationships establish the relative age of rock sequences?

UNCONFORMITIES

It was James Hutton in the late eighteenth century who first fully grasped the significance of Steno's rules. By recognizing the huge time intervals required to develop relative age relationships like those illustrated in Figure 8.9, Hutton was the first to realize the enormous length of geologic time. He found confirming evidence for an expanded view of geologic time at Siccar Point on the east coast of Scotland. Here, steeply inclined layers of sedimentary rock worn down by erosion were covered by gently tilted beds, a relationship known as an **unconformity** (Fig. 8.9a). Hutton recognized that both sets of rock layers contained marine fossils and so both were deposited below sea level. According to Steno's second rule, each set of rock layers had originally been laid down horizontally. Hutton's explanation of the history of Siccar Point therefore began with deposition of the lower strata as horizontal layers of marine sediment. Compressed into

sedimentary rock, the lower strata were tilted and uplifted above sea level by earth movements of unimaginable force. The exposed strata were then beveled off by erosion to form a relatively flat surface upon which, following its submergence below sea level, younger strata were deposited as horizontal layers. These layers, too, became compressed into sedimentary rock, after which the whole body of rock was once again tilted and uplifted. Erosion by the sea finally produced the shape of the rocks that Hutton could see, a shape that has remained virtually unchanged to this day.

The unconformity separating the two sets of rock layers is an erosion surface on which the upper layers were originally deposited horizontally. Because of this, the upper layers cut across the lower ones. The unconformity is an example of a cross-cutting relationship, in which a younger sequence of rock layers truncates an older sequence. Such relationships form fundamental criteria for determining the relative ages of geologic events because they help to establish the sequence of those events. Hutton realized that no single phenomenon could have produced the sequence of events represented in an unconformity like that at Siccar Point (Fig. 8.9b). Instead, it required time and enormous lengths of it. We now know that erosion of the upended layers alone took about 80 million years.

Unconformities consequently represent major time gaps in the geologic record during which no rock deposition is preserved. These gaps require sufficient time for the lower rock layers to be uplifted and eroded before being buried by deposition of younger rock layers.

Since Hutton's time many other unconformities have been identified and three types of unconformity have been recognized (Fig. 8.10). The unconformity that Hutton described, in which the strata above and below the unconformity surface lie at an angle to each other, is known as an *angular unconformity.* But in some cases, uplift and erosion of the lower strata occurred without tilting so that the rock layers above and below the unconformity surface remain

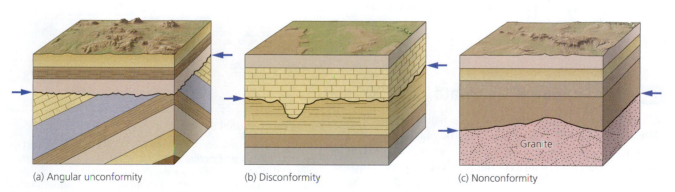

(a) Angular unconformity   (b) Disconformity   (c) Nonconformity

Figure 8.10   Types of Unconformity

(a) An angular unconformity is found between two sets of strata that make an angle with one another. (b) A disconformity is located between two sets of parallel strata, and (c) a nonconformity is found between sedimentary strata and older igneous or metamorphic rocks. Arrows point to unconformity surface.

parallel to each other. This type of unconformity is called a *disconformity*. In yet other cases, the rocks below unconformity surface are not sedimentary, but rather uplifted igneous or metamorphic rocks. This type of unconformity is called a *nonconformity*.

All three types of unconformity are now known to be common features of the rock record. Indeed, as we shall see later in this chapter, the sedimentary rocks exposed in the walls of the Grand Canyon in Arizona actually contain several unconformities, the largest representing a time gap of some 1.5 billion years. Unconformities are important geologic structures and we will return to examine them in more detail in Chapter 10: Deformation and Mountain Building.

> **CHECK YOUR UNDERSTANDING**
>
> ● What are the three types of unconformities and how do they form?

8.3 SUMMARY

- Steno's principles of superposition, original horizontality, and original lateral continuity hold that sedimentary strata are deposited in succession, as flat-lying continuous layers.

- According to the rule of inclusions, any rock layer is younger than the rock fragments it contains.

- According to the rule of cross-cutting relationships, any feature that cuts across a rock is younger than the rock it cuts.

- Unconformities are erosion surfaces separating two sets of rock layers, the upper layers having been originally deposited horizontally on the lower ones. First recognized by James Hutton, unconformities represent time gaps in the rock record during which no record of deposition is preserved.

- There are three types of unconformities: angular unconformities in which the upper layers lie at an angle to the lower ones; disconformities in which the two sets of layers are parallel; and nonconformities where sedimentary strata rest directly on igneous or metamorphic rocks.

8.4 Correlating Rock Strata

While Hutton was examining the Scottish coastline, farther south in England, the study of rock layers had also become the vocation of a canal engineer called William Smith. In the 1790s, while conducting a survey for a proposed canal near the city of Bath, Smith noticed that the sedimentary strata were consistently tilted to the east and

that each layer contained characteristic fossils that were distinct from those in the layers above and below. Smith was therefore able to identify the same layer in different parts of the country based on its fossil content. Developing on Steno's concept of original lateral continuity, "Strata Smith" (as he later became known) established the basis for the **principle of faunal succession** (Fig. 8.11). According to this rule, the fossils present in any set of sedimentary rock layers succeed each other in a specific order that can be identified over large areas. Hence, individual rock layers can be traced from one place to another on the basis of the fossils they contain, a process known as **correlation** (Fig. 8.12).

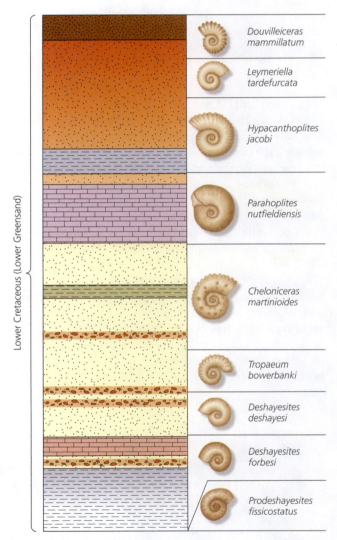

Figure 8.11   Faunal Succession

Rock layers in a cliff on the Isle of Wight off the south coast of England contain distinctive fossils that can be used to identify them. In this case the fossils are different species of *ammonites*, an extinct group of marine shellfish related to the modern octopus and squid.

Lower Cretaceous (Lower Greensand)

Douvilleiceras mammillatum

Leymeriella tardefurcata

Hypacanthoplites jacobi

Parahoplites nutfieldiensis

Cheloniceras martinioides

Tropaeum bowerbanki

Deshayesites deshayesi

Deshayesites forbesi

Prodeshayesites fissicostatus

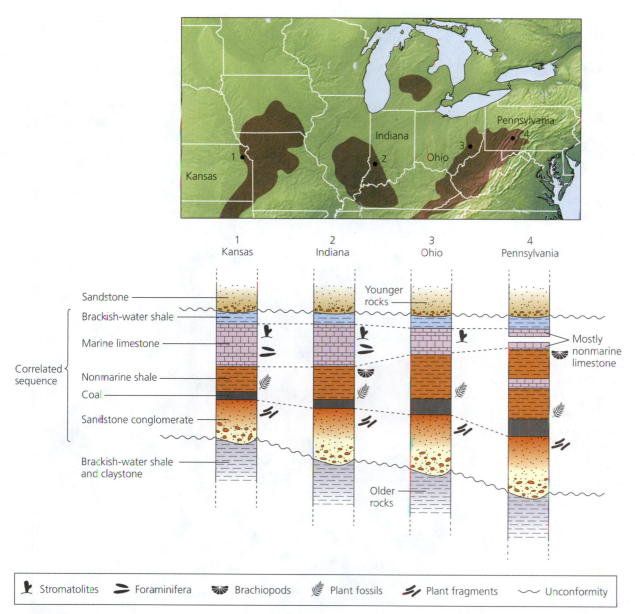

Figure 8.12   Correlation

Correlation of the sedimentary rock sequences and fossil assemblages above and below four coal fields in the U.S. Midwest indicates that the coal was deposited at or about the same time at all four sites. *Stromatolites, foraminifera,* and *brachiopods* are types of fossils. *Stromatolites* are produced by algae, *foraminifera* are microorganisms, and *brachiopods* are lamp shells.

By plotting the distribution of sedimentary strata containing the same fossils, Smith was able to produce the world's first large-scale geologic map (Fig. 8.13). Furthermore, by following Steno's principle of superposition, Smith was able to arrange each of the rock layers into a sequence of younger-upon-older strata known as a **stratigraphic column**. In so doing, Smith recognized that the collection of fossils in any one layer represented organisms that lived during the time that the layer was deposited. Because the same collection of fossils never occurred in earlier layers and never reappeared in later ones, the fossil content of a layer could be used to establish its position in the stratigraphic column. That is, the fossil content could be used to determine the layer's relative age. Smith's principle of faunal succession later became an important underpinning of Darwin's theory of evolution.

CHECK YOUR UNDERSTANDING

● What is the correlation of rock layers?

● How did William Smith use correlation to establish the stratigraphic column?

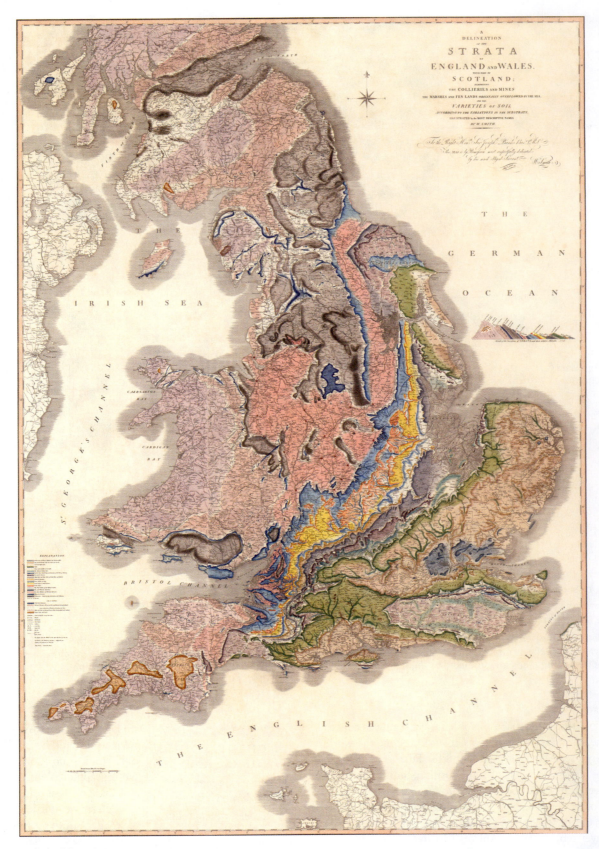

Figure 8.13　Smith's Map

William Smith's geologic map of England was made possible through correlation and the application of his principle of faunal succession. The map uses colors to show what the bedrock is at any given location. Green (lower right), for example, marks the start of the chalk plains where Stonehedge stands, whereas black (center) marks the distribution of coal fields. Published in 1815, this was the world's first large-scale geologic map of any country.

8.4 SUMMARY

- William Smith's principle of faunal succession asserts that individual rock layers can be identified on the basis of their fossil content.

- From fossil content, the relative ages of rock layers in separate successions can be determined, and layers that are from the same time period can be correlated with each other.

8.5 Relative Dating and the Geologic Time Scale

By applying the principles of relative dating, the geologic history of an area can be reconstructed. That is, each of the

geologic events that the area records can be correctly placed in the order in which it occurred. In this way, the principles of relative dating played a central role in the development of a geologic time scale.

RECONSTRUCTING GEOLOGIC HISTORY

In Figure 8.14 several sets of sedimentary strata have been variably offset by faulting and cut by intrusions, and from this we can decipher the sequence of events that occurred in the pictured area.

In the figure, the large image in the left corner shows the present-day distribution of strata. The small figures on the right show the sequence of events that created this final structure. Layers A through G form a continuous sequence of parallel sedimentary layers that were first deposited in succession (a), then later tilted and faulted (b), and finally

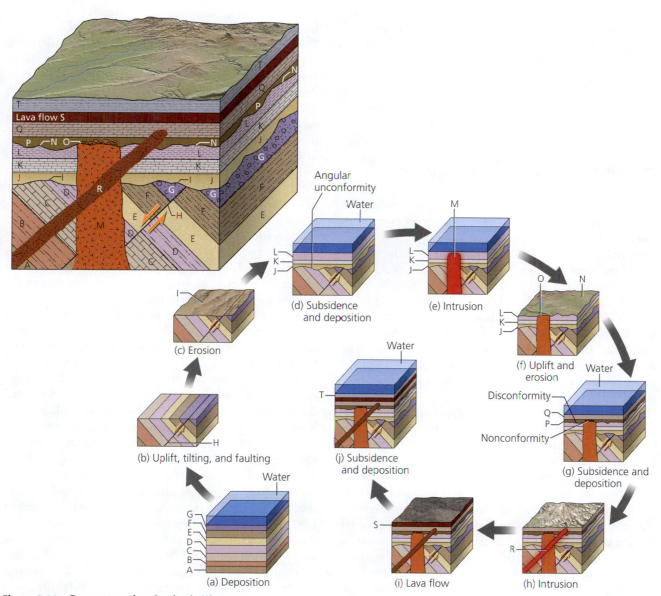

Figure 8.14 Reconstructing Geologic History

We can reconstruct the geologic history of an area through the application of the principles of relative dating.

uplifted and eroded (c). Following uplift and erosion, subsidence occurred and layers J through L were deposited above the erosion surface (d). The base of layer J cuts across layers A through G, producing an unconformity. It also cuts across a fault that offsets all the layers below the unconformity, indicating that faulting occurred after the deposition of layer G but before the deposition of layer J (rule of cross-cutting relationships). Using the same rule, the entire sequence was then intruded by unit M (e). The area was then uplifted and eroded (f), and, following subsidence, layers P and Q were deposited above the erosion surface to produce a second unconformity (g). This second unconformity is demonstrated by the presence of fragments of unit M in layer P. Next comes unit R, which, according to the rule of cross-cutting relationships, intruded after the deposition of layer Q (h). Lava flow S then flowed over and baked layer Q (i), after which layer T was deposited (j). Following final uplift, layer T is now being eroded by streams.

In this fashion, geologists can determine the relative sequence of events for any area. However, it is important to remember that the process only establishes the order of events, not their actual age. We cannot tell from relative age dating how long ago any one of these events took place. Nor can we determine the duration of an event or the time interval between events. Such information did not become available until the advent of absolute age dating.

STRATIGRAPHY AND THE STRATIGRAPHIC COLUMN

Throughout the nineteenth century, geologists applied the principles of relative age dating to compile an ever larger and more complete stratigraphic column of younger-upon-older strata. Clearly, no more than a fraction of the succession of sedimentary strata (or *stratigraphy*) can be observed at any one outcrop. But geologists realized that the information from each outcrop was like one piece of an enormous global jigsaw puzzle, and that the jigsaw could be put together by compiling and correlating the sedimentary and faunal successions from many outcrops. The end result of this compilation is a composite stratigraphic column in which all sedimentary layers are arranged in the order in which they had been deposited.

The construction of such a composite stratigraphic column is illustrated in Figure 8.15 using the geology of three national parks in the southwestern United States—Bryce, Zion, and the Grand Canyon.

The Grand Canyon in Arizona exposes one of Earth's most complete rock sequences, with a vertical thickness of almost 2 kilometers. From the law of superposition, we may conclude that the oldest unit is the Vishnu Schist (at the base) while the youngest is the Moenkopi Formation (at the top). Despite its great thickness, however, the exposed sequence represents only part of the complete stratigraphy of the region. Because they have the same fossil assemblage, the limestone at the top of the Grand Canyon sequence (known as the Kaibab Limestone) can be correlated with

the limestone near at the base of the succession at Zion National Park. Similarly, the sandstone at the top of the Zion succession (known as the Navajo Sandstone) can be correlated with the sandstone at the base of the sequence at Bryce Canyon. These correlations have allowed geologists to construct the complete stratigraphy of the region.

The sequence exposed at the Grand Canyon also contains some significant time gaps in the form of unconformities, in particular a disconformity between the Devonian Temple Butte Limestone (Ls) and the Cambrian Mauv Formation (Fm), and a major nonconformity between the Cambrian Tapeats Sandstone (Ss) and the Precambrian metamorphic Vishnu Schist (Fig. 8.15). Because the rocks representing these time gaps have either been removed by erosion or were never deposited, evidence of a rock record that might fill these missing intervals must be sought elsewhere.

COMPILING THE GEOLOGIC TIME SCALE

Throughout the nineteenth century, gaps in the rock record at one location were filled by those from another to produce an ever more complete stratigraphic column. Because the stratigraphic column arranged sedimentary strata according to their relative age, subdivisions of the column became subdivisions of geologic time. By the end of the century, this composite stratigraphic column had become the basis of a sophisticated geologic time scale (Fig. 8.16).

We can subdivide geologic time because strata contain fossils that change from layer to layer as new species emerge or existing species become extinct as we discussed in Section 8.3. Different parts of the stratigraphic column consequently contain different fossil assemblages and can be subdivided accordingly. In the mid-nineteenth century, this changing fossil record played a vital role in the development of Darwin's theory of evolution. The names given to the subdivisions remain with us today.

Many of the subdivisions of the geologic time scale are based on the emergence and extinction of fossil life forms. Long intervals of geologic time separated by major extinction events are **eras**. Each mass extinction event was global in scale and defines two major boundaries in the geologic time scale. So each event truly marks "the end of an era." But while mass extinctions could be used to subdivide geologic time, it was not known in the nineteenth century how many millions of years ago these events had occurred. Only with the advent of *radiometric dating* (see Section 8.6) has the absolute age of these events been determined.

The first of these mass extinctions defines the boundary that separates the era called the **Paleozoic**, meaning "ancient life," from the era called the **Mesozoic**, which means "middle life." The second separates the Mesozoic Era from the era called the **Cenozoic**, which means "recent life." Abundant shelly fossils first appear at the beginning of the Paleozoic Era. Indeed, the abrupt appearance of plentiful shelly fossils worldwide is used to separate the rocks in the Paleozoic Era from all rocks deposited beforehand. Although some of the

Grand Canyon National Park, Arizona   Zion National Park, Utah   Bryce Canyon National Park, Utah

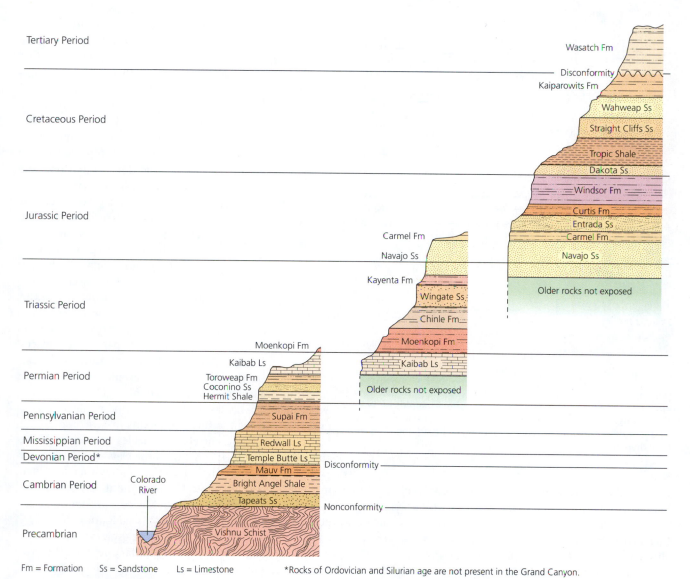

Fm = Formation   Ss = Sandstone   Ls = Limestone   *Rocks of Ordovician and Silurian age are not present in the Grand Canyon.

Figure 8.15   Grand Canyon

The Grand Canyon in northern Arizona affords one of Earth's most complete exposed rock sequences: almost 2 vertical kilometers (1.2 miles) of rock ranging up to 3 billion years in age. However, even here there are unconformities representing gaps in the local rock record, and correlations must be made with rocks exposed in other parts of the Colorado Plateau, such as Zion and Bryce Canyon, before the geological history of the entire region can be determined.

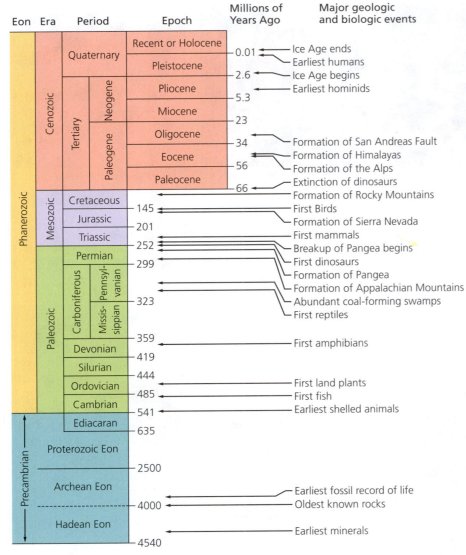

Figure 8.16 Geologic Time Scale
Geologic time is subdivided into eons, eras, periods, and epochs. These subdivisions are based on the principles of relative dating. Numbers (in millions of years ago) are determined by absolute dating and provide dates for the beginning of each time division.

during which mammals first become abundant.

The four largest intervals of geologic time, some with durations of billions of years, are **eons**. Eons represent fundamental stages in the planet's evolution. They include the **Archean**, meaning "primeval," which starts with Earth's oldest known rocks, and the **Proterozoic**, meaning "earlier life," which contains the first record of complex single-celled organisms. The Paleozoic, Mesozoic, and Cenozoic eras collectively belong to the **Phanerozoic Eon**, meaning "visible life." During this era fossils became abundant. Only during the Phanerozoic is the subdivision of geologic time based on plentiful fossils possible. The vast interval of geologic time that precedes the Phanerozoic Eon is often referred to simply as the **Precambrian**, a name that implies that it predates the Cambrian, which is the first subdivision of the Paleozoic Era.

Based on the advent and demise of literally thousands of different fossil life forms it is possible to subdivide each of the three eras of the Phanerozoic into shorter intervals called **periods**. These were named, for the most part, after places where classic exposures of rocks of that age are found. Thus, the Devonian Period of the Paleozoic Era is named after the county of Devon in southwest England, whereas the Jurassic Period of the Mesozoic Era is named for the Jura Mountains in Switzerland and France. Worldwide correlation enables the names to be applied worldwide. It was during the Devonian Period, for example, that the Temple Butte Limestone was deposited in what is now the Grand Canyon, and it was during the Jurassic Period that much of the Navajo Sandstone was deposited at Zion and Bryce Canyon (see Figure 8.15).

As we discussed in Chapter 1, the geologic timetable is continually being refined. The boundary between the Proterozoic and Phanerozoic eons has been looked at very closely in this regard. Although the Proterozoic Eon lacks fossils with body hard parts such as shells or skeletons, the very youngest rocks of this eon have been recognized worldwide to contain the impressions of soft-bodied multicellular organisms known as the *Ediacara fauna* after the hills in South Australia where the fossils were first found. In 2004, this led to the addition of a new period to the geologic timetable, the Ediacaran, immediately beneath the Cambrian (see Fig. 8.16).

fossils of the Paleozoic Era represent the ancestors of modern life forms, many have no living relatives left because they became extinct at the end of that era. The massive extinction that marks the end of the Paleozoic is estimated to have eliminated nearly 90 percent of all marine species.

The Mesozoic Era is dominated by fossils that do not occur in the Paleozoic. For example, dinosaur bones are not found in strata of the Paleozoic Era; they occur exclusively in the rocks of the Mesozoic Era. Fossil birds, mammals, and flowering plants are also seen for the first time in rocks of this era. The end of the Mesozoic is marked by another massive extinction, second only in scale to that which brought the Paleozoic to a close. This extinction, which claimed the dinosaurs and many other animal and plant species, heralds the start of the Cenozoic, the era

Figure 8.17   Index Fossils

This rock contains a well-preserved fossil of *Paradoxides*, a member of an extinct group of marine arthropods (animals such as crabs) called trilobites. *Paradoxides* lived during the mid-Cambrian Period of the Paleozoic Era and is an index fossil for that time interval.

Some fossils, known as *index fossils,* have proved particularly useful in subdividing geologic time (Fig. 8.17). This is because they are easily recognized, widely distributed around the world and, because they evolved rapidly, are restricted to a narrow interval of rock strata. Index fossils are sufficiently short lived that their occurrence accurately pinpoints the age of the strata in which they are found. Some can consequently be used to subdivide geologic periods into even shorter time intervals known as **epochs**. For example, the oldest *hominids,* the primate family that includes present-day humans, date back to the Early Pliocene Epoch at the end of the Tertiary Period.

> **CHECK YOUR UNDERSTANDING**
>
> ❍ What are the four eons of the geologic time scale?

Detailed studies of sedimentary strata and the fossils they contain enabled nineteenth century geologists to construct a detailed geologic timetable that allowed all rocks to be catalogued according to their relative age. By dividing geologic time in this fashion, they made it possible to examine the evolution of Earth and compare the geologic history of one part of the world with that of another.

8.5 SUMMARY

- We can reconstruct the geologic history of an area by applying the principles of relative dating. A stratigraphic column of strata forms the basis of the geologic time scale.

- The geologic time scale is divided into eons, eras, periods, and epochs. Most boundaries coincide with the emergence and extinction of fossil life forms.

8.6 Absolute Dating: Finding a Geologic Clock

Modern estimates of absolute time and modern determinations of the age of Earth did not become possible until 1896, when Henri Bequerel, together with Marie and Pierre Curie, discovered that the atoms of some chemical elements, such as uranium, are inherently unstable and break down in a predictable fashion. This property is known as **radioactivity**.

RADIOACTIVITY AND RADIOACTIVE DECAY

Radioactivity occurs where instabilities exist in the nucleus of an atom, as happens, for example, when the nucleus contains too many particles. All chemical elements are synthesized during the life cycles of stars, either within their thermonuclear cores or during the demise of large stars in supernova explosions (see Living on Earth: **Origin of the Chemical Elements** in Chapter 3). In these environments, atomic and subatomic particles are traveling at unimaginable speeds, constantly colliding into one another and sometimes synthesizing new elements as a result. It is hardly surprising that some of the particles formed under such extreme conditions are inherently unstable.

In Chapter 3, we learned that the nucleus of an atom contains two kinds of particles, protons and neutrons. Protons have a positive charge and their number in the nucleus (the *atomic number*) determines many of the properties of that atom. By definition, all atoms with the same number of protons in their nucleus belong to the same chemical element. Any atom with 8 protons, for example, is an atom of oxygen, and any atom with 92 protons is one of uranium. Neutrons, on the other hand, have no charge and their number in the nucleus can vary without affecting the atom's chemical properties. Neutrons have a mass that is virtually the same as that of a proton. So any variation in the number of neutrons causes the mass of an atom to vary. Atoms of the same element that have different masses are known as **isotopes** of that element. Isotopes of an element have the same number of protons but varying numbers of neutrons, and so have virtually the same chemical properties but differing atomic weights.

Isotopes are identified, not by their atomic number but by their *mass number,* which is the sum of the number of protons and neutrons in their atomic nuclei. Carbon, for example, has six protons and therefore has an atomic number of 6. However, carbon may have six, seven, or eight neutrons and so can have a mass number of 12, 13, or 14. It is therefore said to possess three isotopes, which are respectively termed carbon-12, carbon-13, and carbon-14 (Fig. 8.18).

Isotopes are very common in nature. In fact, most elements possess several isotopes, the vast majority of which are perfectly stable. However, isotopes that have too many particles in their nuclei are unstable and spontaneously break down to isotopes with more stable nuclei, releasing energy and particles in the process. Carbon-12 and carbon-13, for

LIVING ON EARTH

Radioactivity and Radon Gas

There is a general misconception that nature is entirely benevolent and that our environmental problems are all of our own making. This is not completely true. Radon gas is a natural constituent of our atmosphere, although it is present only in minute amounts. In the open air, radon is harmless. However, indoors, radon concentrations can build up to between 20 and 15,000 times that amount. At these concentrations, radon gas is deadly. It is estimated that inhalation of radon gas causes between 5000 and 25,000 cases of lung cancer every year in North America alone, a figure comparable with the number of deaths from automobile or home accidents. Radon gas inhalation is second only to smoking as the leading cause of lung cancer.

Radon is one of the links in the chain of reactions that converts radioactive uranium-238 (^{238}U) into stable lead (Fig. 8A). As uranium decays, it sets off a ripple effect like falling dominoes. Although radon is formed during this decay, it too is unstable and decays to polonium which itself decays and so on, until a stable variety of lead—lead-206 (^{206}Pb)—is ultimately produced. The very short half-life of radon decay (only 3.8 days) is a measure of radon's profound instability. Radon atoms decay by ejecting fundamental particles, known as alpha particles, from their nuclei. When radon gas is inhaled, the alpha particles damage biological tissue and cause lung cancer.

The danger with radon is not just its inherent instability. It is also extremely mobile and can readily escape from uranium-rich rocks and soils because it is a gas. It seeps into houses and buildings through any openings such as gaps between floors and walls, cracks in walls and foundations, and holes for plumbing (Fig. 8B). When radon seeps into a house or building, it circulates and accumulates as indoor air, especially in well-insulated, poorly ventilated houses. Since radon gas is heavy (its atomic mass is 222, see Figure 8A), its buildup is most concentrated in basements. If domestic water comes from a spring or well, it too may bring radon gas into the home. Well water can absorb radon from underground rocks and soils, only to release it, for example, during a shower. Buildings with uranium-rich building stone such as granite are also at risk. Modern buildings have air circulation systems that generally compound the problem. In the northern U.S. and Canada, accumulation of radon in domestic houses is most likely to occur in the wintertime. However, the problem can be greatly alleviated if basement and indoor air is continually exchanged with the air outside.

Because radon is radioactive, its presence can be readily detected at a minimal cost. However, not all locations are equally at risk. Some homes and buildings contain high concentrations of the gas; others have very low concentrations. There are two main factors that govern these variations: a building's ventilation and the local geologic environment. A poorly ventilated house built on uranium-rich rocks and soils is most at risk. However, once the possible radon threat has been identified, it is relatively easy and inexpensive to minimize exposure by adjusting ventilation systems.

Because radon is a product of uranium decay, radon emissions are highest near uranium-rich rocks and soils. Uranium occurs in very minor amounts in most rocks. Its average abundance in continental rocks is about 2.8 parts per million or 0.00028 percent. However, it is often concentrated by a factor of 10 to 100 in rocks such as granite, black shale, coal, and limestone. Recall that granite is an igneous rock, the product of the cooling and solidifying of magma well below Earth's surface. As the molten liquid cools, crystals form and the chemical elements separate and concentrate depending on their chemical and physical properties (see Chapter 4). These properties determine whether an element can fit into the crystal structures of minerals such as quartz and feldspar, which are the principal components of granite.

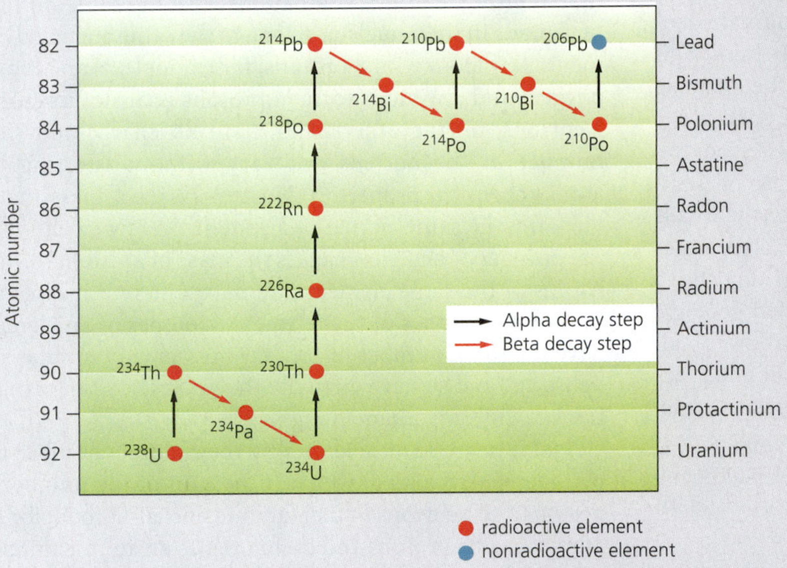

Figure 8A   Decay Path of Uranium

Radioactive uranium with an atomic mass of 238 (^{238}U) converts to nonradioactive lead-206 (^{206}Pb) in a number of steps. During these steps, radon-222 (^{222}Rn) is produced by radioactive decay of radium-226 (^{226}Ra). Radon then decays to polonium-218 (^{218}Po), with a half-life of 3.8 days. The alpha particle ejected from the radon nucleus destroys biological tissue, and is a leading cause of lung cancer.

Figure 8B Radon in the Home
Radon enters a house through cracks in the foundation, drains, joints, or between blocks in walls.

Uranium is an element that does not readily fit into these common minerals and, as a result, slowly accumulates and concentrates in the remaining molten liquid as crystals of these mineral forms. When the remaining liquid finally crystallizes, the uranium in these last crystals is very concentrated. After the solid granite forms, millions of years of erosion may strip off Earth's upper layers and eventually expose the solidified granite. If some of its uranium-rich minerals are exposed at the surface of the granite, or if soil forms on top of the granite, radon emissions may be significant.

Sedimentary rocks such as black shale, coal, and limestone also concentrate uranium. These rocks form from the consolidated remains of ancient life. Organic compounds typically concentrate uranium and as a result organic sediments form into uranium-rich rocks capable of generating radon. Soils and glacial deposits that overlie uranium-bearing rocks may also be uranium-rich. Radon can also escape from uranium-rich rocks that are deeply buried. Ancient faults and fractures that cut these rocks may leave scars that do not heal and these can concentrate and facilitate the escape of radon to the surface.

The potential hazard of radon demonstrates that not all natural processes are beneficial to humans. Some can be lethal. It also demonstrates the importance of linking scientists of different fields. In this case, we can only identify areas of potential risk if the medical and geoscience communities collaborate with their data and interpretations. There is a need for cross-referencing the regional distribution of lung cancer rates with the composition of the rocks and soils of a region. There is cause for hope, however, in evidence that once humans are removed from areas of excess radon, their bodies recover. If this is correct, identification of radon gas problems and remediation may not only halt future damage but also may reverse the damage already sustained.

example, are stable isotopes but carbon-14, the basis of *radiocarbon dating* used by archeologists, is radioactive and decays by the spontaneous breakdown of one neutron to a proton and an electron, which is ejected (see Science Refresher: **Types of Radioactive Decay** on page 244). The product is not carbon because it has seven protons rather than six. By definition, it is nitrogen, and the specific isotope, now with seven neutrons instead of eight, is nitrogen-14.

SCIENCE REFRESHER

Types of Radioactive Decay

The breakdown of radioactive isotopes occurs in one of three ways. The first type involves the spontaneous breakdown of a neutron to a proton with the ejection of a particle (known as a beta particle) with a negative charge. The total number of protons and neutrons (the mass number) stays the same. This form of decay is *beta particle emission* (Fig. 8C). By converting a neutron to a proton, the number of protons in the nucleus (the atomic number) increases by one, while the number of neutrons decreases by one. Carbon-14 is an example of this type of radioactive decay. Because a neutron is converted to a proton, the daughter element has 7 protons, and is therefore, by definition, Nitrogen-14. The negative charge ejected has negligible mass, so the atomic mass remains the same at 14.

Alpha particle emission is another form of decay that involves the ejection of a particle (known as an alpha particle) consisting of two protons and two neutrons from the nucleus of the unstable parent atom. As a result, the atomic number of the nuclei is decreased by two during alpha particle emission while the mass number is decreased by four.

A third form of decay, known as *electron capture,* involves the conversion of a proton to a neutron. The atomic number is consequently reduced by one while the mass number stays the same. But while some types of decay have no effect on the mass number, all three cause the atomic number to change. As a result, each type of decay produces a different element since it is the atomic number that determines the element to which an atom belongs.

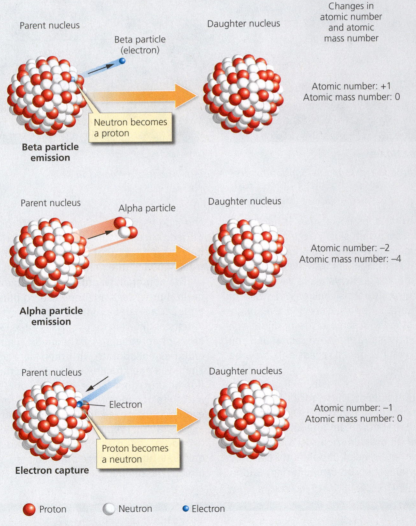

Figure 8C   Radioactive Decay

The three common types of radioactive decay each result in a change in the number of protons in the atomic nucleus and, therefore, a change in element.

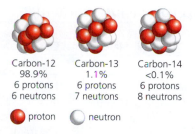

Figure 8.18   Carbon Isotopes

All carbon atoms have six protons inside their nuclei, but the number of neutrons can range from six to eight. As a result, carbon has three isotopes designated carbon-12, carbon-13, and carbon-14. Nearly all natural carbon (98.9%) is carbon-12; the rest is stable carbon-13 (1.1%) and radioactive carbon-14.

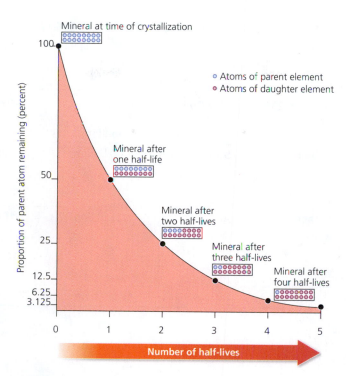

Figure 8.19   Half-Life

The half-life of a radioactive element is the time it takes for half the radioactive nuclei in a sample to undergo radioactive decay. After one half-life, half of the radioactive parent isotope will have decayed to the daughter isotope. After two half-lives, only a quarter of the parent isotope will remain, and so on.

If the new isotope produced by this *radioactive decay* is also radioactive, then several breakdown steps may separate the original radioactive **parent isotope** from the final stable nonradioactive or **daughter isotope**. For example, many steps are involved in the ultimate breakdown of uranium to lead, each step producing one radioactive isotope after another until finally a stable isotope of lead is formed (see Living on Earth: **Radioactivity and Radon Gas** on page 242). The energy released through each step of this process can be enormous; as we learned in Chapter 2, the decay of radioactive elements in Earth's mantle provides sufficient heat to power plate tectonics. But as we shall now see as we turn our attention to the absolute age of rocks, it is the pace of the process that is the key to its use in dating.

CHECK YOUR UNDERSTANDING

○ What is an isotope?

HALF-LIFE

For any radioactive element, the rate at which the parent isotope decays to its daughter is a fixed one that cannot be changed by any known means. Because of this, naturally occurring radioactive elements form the most reliable of clocks, decaying at a predictable rate regardless of any physical or chemical changes in their environment.

Ideally, the decay process starts with a set number of parent atoms and no daughter product. As the parent decays, the number of parent atoms decreases while the number of daughter atoms increases. At some point in time the amount of parent isotope will have decreased to exactly half its original quantity. The length of time it takes to reach this point is known as the isotope's **half-life** (Fig. 8.19). The length of an isotope's half-life is a measure of its rate of decay and is a fixed time interval that can be determined experimentally.

Radioactive decay converts a radioactive parent isotope to a stable daughter isotope. With the passage of each half-life, the number of atoms of the radioactive isotope is reduced by half, over the same time period the number of atoms of the daughter isotope increases by the same quantity.

Note that the total number of atoms (parent plus daughter) is constant. By measuring the ratio of the quantities of the parent to the daughter isotopes, geologists can determine the absolute ages of some rocks, providing the half-life of the parent isotope is known.

For example, the half-life for the decay of uranium-235 to lead-207 is 704 million years (Table 8.1). This means that 704 million years after the decay process starts, 50 percent of the uranium parent atoms will have decayed to atoms of lead, the daughter product. After another 704 million years, or a total of two half lives, half of the remaining parent atoms will have decayed to become atoms of the daughter product and the remaining number of atoms of uranium has shrunk to 25 percent of the original number. After yet another 704 million years, or three half-lives, the atoms of uranium will be just 12.5 percent of the original number, and so on.

Different parent isotopes decay at different rates (see Table 8.1). Compared to uranium-235, for example, uranium-238 decays (to lead-206) much more slowly, with a half-life of 4.5 billion years. On the other hand, carbon-14 decays (to nitrogen-14) much more rapidly, with a half-life of 5730 years. In fact, the half-life of carbon-14 is so short that it is not useful for dating rocks because the amount of parent remaining in anything older than about 60,000 years is too small to measure.

TABLE 8.1   THE SIX RADIOACTIVE ISOTOPE SYSTEMS MOST SUITABLE FOR RADIOMETRIC DATING

Isotopes		Half-Life of Parent (Years)	Effective Dating Range (Years)	Material That Can Be Dated
Parent	**Daughter**			
Uranium-238	Lead-206	4.5 billion	10 million to 4.6 billion	Zircon
Uranium-235	Lead-207	704 million		Uraninite
Thorium-232	Lead-208	14 billion		
Rubidium-87	Strontium-87	48.8 billion	10 million to 4.6 billion	Muscovite
				Biotite
				Orthoclase
				Whole metamorphic or igneous rock
Potassium-40	Argon-40	1.25 billion	100,000 to 4.6 billion	Glauconite
				Hornblende
				Muscovite
				Whole volcanic rock
				Biotite
Carbon-14	Nitrogen-14	5730 years	Less than 100,000	Shell, bones, and charcoal

The process by which the amount of the parent isotope is halved in each successive half-life is known as *exponential decay* (see Science Refresher: **Linear versus Exponential Relationships** on page 247). Because only one daughter is formed by the decay of a single parent atom, the total number of atoms does not change during the decay process. As a result, the number of atoms of the daughter product steadily increases, equaling the number of parent atoms after one half-life, reaching 75 percent of the total number of atoms after two, and 87.5 percent of the total number of atoms after three half-lives (see Fig. 8.19).

The half-life of any one radioactive isotope is fixed, but it varies markedly from one isotope to another. Obtaining an absolute age by radiometric means is therefore a matter of determining the parent-to-daughter ratio of a suitable radioactive element with a known half-life. For example, suppose in a uranium-bearing mineral the parent and daughter isotopes (uranium-235 and lead-207) were found to be present in the ratio 25 percent parent to 75 percent daughter. Reviewing Figure 8.19, we can see that, for 25 percent of the parent to be left, two half-lives must have elapsed. Given that the half-life of uranium-235 is 704 million years, the age of the mineral must therefore be twice this, or 1408 million years. Once the decay rates of the different radioactive isotopes had been determined, a means was finally available for measuring geologic time.

> **CHECK YOUR UNDERSTANDING**
>
> ● What is the half-life of a radioactive isotope?

RADIOMETRIC DATING

Of the many radioactive isotopes, only a few are suitable for dating purposes. To be useful, the parent isotope must occur in minerals commonly found in rocks, and both the parent and daughter isotope must be retained within the mineral being dated. The length of the parent's half-life must also be appropriate for measuring purposes. For example, although aluminum is an abundant element in rocks and minerals, the decay of radioactive aluminum-26 to its daughter isotope (magnesium-26) has a half-life of only 720,000 years. As a result, aluminum-26, while abundant in Earth's early history, rapidly decayed to immeasurably small quantities and is unsuitable for current dating purposes.

The parent-daughter isotope systems that have proved to be most appropriate for dating purposes, because they best meet the criteria listed above, are shown in Table 8.1. Of these, uranium-lead, rubidium-strontium, and potassium-argon are the isotope systems used most frequently. Because they are measured by radioactive decay, dates determined by these isotope systems are called **radiometric ages**.

But determining what a radiometric age means once it has been obtained requires interpretation. The answer, as the following descriptions of the uranium-lead and potassium-argon methods illustrate, often depends on the mineral selected for dating and the rock type that hosts the mineral.

Uranium-Lead Dating

The uranium-lead dating technique utilizes highly stable minerals, such as *zircon*, which contain trace amounts of uranium. Zircon crystals occur in very small quantities in many rocks and, when dated, the age they record is the age of their crystallization. In igneous rocks, zircons crystallize as the magma cools, so the age of the zircon crystals is the same as the age of the rock because it dates the time at which the rock solidified from a molten liquid. In this way, zircon allows us to accurately date the age of crystallization of the igneous rock from which it was obtained.

SCIENCE REFRESHER

Linear versus Exponential Relationships

Radioactive decay is an example of an *exponential relationship*, that is, a relationship in which the increase or decrease of something in a given time interval depends on its magnitude at the beginning of the time interval. This is quite different from a *linear relationship* in which something increases or decreases in a given time interval at a fixed rate. Most activities with which we are familiar involve linear relationships. At a fixed speed, for example, a car can be driven twice as far in two hours as it can be in one, or three times as far in three hours, and so on. Hence, the relationship between distance and time is a linear one, the distance traveled depending simply on the speed of the car.

But not all everyday activities are linear. For example, a familiar exponential process is that of population growth. If a family has two children, each of which grow to have two children of their own, and these children likewise have two children, and so on, then the size of the family will double with each generation. So rather than increasing in a linear fashion from 2 to 4 to 6 to 8 to 10, and so forth, the size of the family increases exponentially from 2 to 4 to 8 to 16 to 32, and so forth. Radioactive decay works in the same way except that instead of doubling with each generation, the amount of parent isotope halves with each half-life.

The zircon grains in sedimentary rocks, however, are derived from the erosion of preexisting rocks. So zircon is of no use in dating sedimentary rocks. A zircon grain taken from a beach in Florida, for example, will not provide the age of the beach. Rather, it will provide the crystallization age of the igneous rock from which the zircon grain was originally derived. Given that clastic sedimentary rocks may be derived from a variety of sources, a

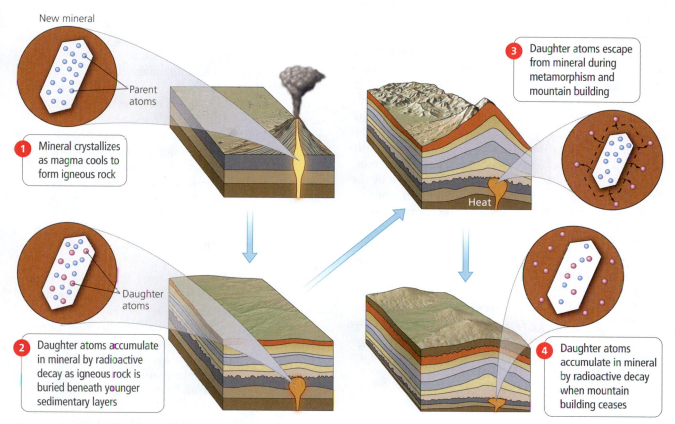

New mineral

Parent atoms

1 Mineral crystallizes as magma cools to form igneous rock

2 Daughter atoms accumulate in mineral by radioactive decay as igneous rock is buried beneath younger sedimentary layers

Daughter atoms

3 Daughter atoms escape from mineral during metamorphism and mountain building

Heat

4 Daughter atoms accumulate in mineral by radioactive decay when mountain building ceases

Figure 8.20   Resetting the Radiometric Clock

Daughter isotopes are lost from a rock heated during mountain building. (1–2) Following the original crystallization of an igneous rock, daughter atoms accumulate as atoms of the parent isotope decay. (3) Later, during mountain building, heat causes expansion of the mineral crystals, allowing the daughter atoms to escape and migrate elsewhere. Complete loss of daughter atoms during the heating event resets the radiometric clock to the time of mountain building. (4) The daughter isotope resumes accumulating after this heating episode ceases and the crystal contracts. The loss of daughter isotopes during the heating event resets the clock so that the age of the mineral records the time of mountain building, not the time of original crystallization of the mineral.

CHECK YOUR UNDERSTANDING

⦿ How is a radiometric age determined and what radioactive isotopes are most suitable for this purpose?

selection of zircons from a single sedimentary rock may yield a wide range of ages. However, the sedimentary rock must be younger than the youngest zircon it contains. The advantage of zircon in this case is its resistance to any form of alteration. (Note that in the form "cubic zirconia" zircon is used as a substitute for diamond, the most resistant mineral).

Potassium-Argon Dating

Radiometric techniques such as potassium-argon, on the other hand, are useful specifically *because* some of the minerals they are used to date are susceptible to alteration. If an area is reheated due to metamorphism, for example, the crystal structure expands, enabling the daughter isotope to escape from the crystal in which it was produced. In potassium-argon dating, the daughter product is a gas and is readily lost if the rock is reheated. As a result, the radiometric clock starts anew and is effectively reset to the time of reheating (Fig. 8.20). This poses a problem if the method is used to date the time the mineral originally crystallized but is an advantage in dating the metamorphic event.

The potassium-argon method is widely used for dating potassium-bearing micas such as muscovite and biotite which, like zircon, are common in some igneous rocks. If the igneous rock has not been affected by geologic events since its crystallization, the ages obtained from all three of these minerals is similar and dates the formation of the igneous rock. However, if the igneous rock was subsequently reheated as the result of a geologic event such as metamorphism during mountain building, the age obtained by the potassium-argon method is likely to date the most recent reheating. Hence, by dating zircon and mica in the same igneous rock, geologists can obtain the age of both its crystallization and subsequent metamorphism. Under these circumstances, however, the zircon age of the igneous rock differs from its mica age and care must be taken to ensure that the two ages are correctly interpreted.

CHECK YOUR UNDERSTANDING

⦿ How can the radiometric age of a mineral differ from the age of the rock that contains it?

8.6 SUMMARY

- Radioactivity is the phenomenon by which an unstable parent isotope spontaneously breaks down to a stable daughter isotope.

- The absolute age of a rock or mineral can be determined from the ratio of daughter isotope produced to that of the parent isotope remaining, providing the decay rate is known.

- Rate of radioactive decay is exponential and expressed in terms of the half-life of the parent isotope, that is, the time it takes for a given amount of the parent to decay to half that amount.

- Most radiometric ages are based on the parent-daughter systems uranium-lead, rubidium-strontium, and potassium-argon.

8.7 Absolute Dating and the Geologic Time Scale

As a result of the application of radiometric dating to rocks around the world, numerical ages have been determined for each of the major boundaries of the stratigraphic column (see Figure 8.16). We now know, for example, that the Paleozoic era started some 541 million years ago, that the Mesozoic era began 252 million years ago, and that the demise of the dinosaurs at the beginning of the Cenozoic era occurred 66 million years ago. Earth itself is thought to be 4540 million (or 4.54 billion) years old (see In Depth: **The Quest for the Age of Earth** on page 249).

However, assigning absolute ages to the geologic time scale is not as easy as it may seem at first. Major subdivisions are based on fossils, and the age of *fossiliferous* (fossil-bearing) sedimentary rocks are not easily determined by radiometric means. This is because, as we discussed in Section 8.5, most dating techniques measure when a mineral crystallized, not the time that it was incorporated into a sediment. So the relationship between the relative time scale (established on the basis of fossils) and absolute time (determined by radiometric dating) was constructed by carefully dating minerals of the same relative age as the fossiliferous sedimentary rocks.

An example of this approach is shown in Figure 8.21. In Section 1, dated lava flows are interlayered with fossiliferous sedimentary rocks. While the sedimentary rocks can be assigned to the Devonian Period on the basis of the fossils they contain, the actual age of the middle layer (layer 2) is bracketed by the absolute age of the lavas above and below. The lavas can be dated because they contain radioactive isotope-bearing minerals that crystallized when the lava cooled. In this case, the age of lavas A and B indicates that the fossils in the intervening sedimentary rocks of layer 2 are between 400 and 360 million years old.

We can also be sure that the rocks below lava A (layer 1) are more than 400 million years old, and that those above lava B (layer 3) are less than 360 million years old, but we do not know by how much. However, Section 2 exposes fossiliferous sedimentary rocks that correlate with layer 3, and an overlying sedimentary sequence (layer 4) that contains fossils belonging to the Carboniferous Period.

The Quest for the Age of Earth

In dealing with absolute age, it is important to distinguish accuracy from precision. Precision refers to the exactness of an estimate whereas accuracy refers to its correctness. For example, if when asked for the time you reply "it's just after twelve" when the correct time is 12:01 p.m., your response is accurate but imprecise. On the other hand, if you reply "16 minutes past twelve" because your watch is running 15 minutes fast, your response is precise but inaccurate.

To this day, the most precise determination of the age of Earth is that based on the seventeenth-century biblical chronology of Archbishop James Ussher. By carefully studying the Bible's continuous record of births and deaths, Ussher was able to place Earth's creation in the year 4004 BC. Other scholars subsequently duplicated Ussher's result and even refined it to the day and hour: 9:00 a.m. on October 23.

The determination is precise because it was given to the nearest minute, a resolution that modern science cannot hope to duplicate. Although it is precise, according to modern science it is wildly inaccurate. Modern estimates place the age of Earth at around 4540 million (or 4.54 billion) years. This determination is nowhere near as precise as that of Archbishop Ussher because it is known only to the nearest 10 million years or so. It is, however, more accurate.

The end of the nineteenth century witnessed several attempts to determine the absolute age of Earth by scientific methods that were neither accurate nor precise. In 1878, for example, the Irish geologist Samuel Haughton introduced the principle of estimating the duration of a particular geologic time interval from the total thickness of sediment deposited during that interval in a stratigraphic column. Unfortunately, much uncertainty existed about the rates at which the strata were deposited, and estimates of the amount of time that had elapsed since the beginning of the Paleozoic ranged from 18 to more than 700 million years.

Lord Kelvin, a noted Victorian physicist in England, used a different approach. It had long been observed that temperatures in mines increased with depth, implying a flow of heat from the interior of Earth to its surface. Therefore Earth, Lord Kelvin argued, is losing heat and must have been hotter in the past. In 1897, he attempted to determine its age by calculating how long it would take the planet to cool to its present state of heat loss, assuming it had begun as a ball of molten rock. Using this method, he determined that Earth's crust had first solidified 20 to 40 million years ago.

In 1898, the Irishman John Joly attempted to determine the age of the oceans from their saltiness, a suggestion first proposed by the celebrated astronomer Edmund Halley in 1715. Recognizing that the oceans owe their saltiness to the accumulation of salts supplied by rivers, he suggested that the age of the oceans might be determined from the total amount of salt in the sea. By determining the average salt content of river water and estimating the total flow of all the world's rivers, Joly calculated how much salt was delivered to the oceans in any given year. Dividing this figure into the total amount of salt in the sea gave Joly an estimate of how long it had taken to bring the world's ocean water to its present level of saltiness, assuming no loss of salt. Joly's answer was 80 to 90 million years.

Although Lord Kelvin's and Joly's estimates were of similar magnitude (both being in the order of the tens of millions of years), both wildly underestimated the age of Earth as modern methods were later to prove. This is because the underlying assumptions made in each case were later shown to be invalid. The reasons both estimates are erroneous are similar. Lord Kelvin assumed a steadily cooling planet, unaware that naturally occurring radioactive elements provide Earth with its own internal heat source. Just as an electric heater loses heat without cooling because electricity continues to supply it with the energy to maintain its temperature, Earth has an energy source that slows its rate of cooling. Therefore Kelvin's assumption of Earth's rate of cooling was flawed. Joly, on the other hand, assumed that the oceans were getting progressively saltier but the world's salt deposits testify to the removal of huge quantities of salt from the oceans by chemical precipitation. Therefore, Joly's answer does not give the age of the oceans; it is instead only a crude estimate of the average time that salt spends in the ocean after its removal from the land by rivers and before its removal from the ocean by precipitation or other processes.

The technology of radiometric dating estimates Earth to be about 4.54 billion years old. The oldest rocks yet discovered occur near Great Slave Lake in Canada and have a uranium-lead age of 4031 million years. Individual crystals of the mineral zircon have been found in sedimentary rocks in Australia with an age of 4404 million years. However, as we have learned, these zircons must have been derived from older rocks and so are older than the rocks in which they occur. Rocks as old as those near Great Slave Lake are extremely rare because Earth is a dynamic planet and has effectively recycled itself many times since the oldest rocks were formed.

In fact, the oldest rocks found on Earth are not from Earth itself. Instead, they are meteorites from space, the oldest of which date back 4567 million years. These meteorites date the formation of the Solar System and are thought to be representative of the primitive material from which Earth first formed. The oldest rocks brought back from the Moon by the Apollo missions date back about 4500 million years. Since the Moon is thought to be slightly younger than Earth, having formed following a massive impact between the young Earth and a Mars-sized body, the true age of Earth must lie between these two dates. At present, the best estimates suggest that Earth formed 4.54 billion years ago. A gap therefore exists between Earth's oldest rocks and the formation of the planet. Named after Hades, the Greek god of the underworld, this interval is called the **Hadean Eon** (see Figure 8.16).

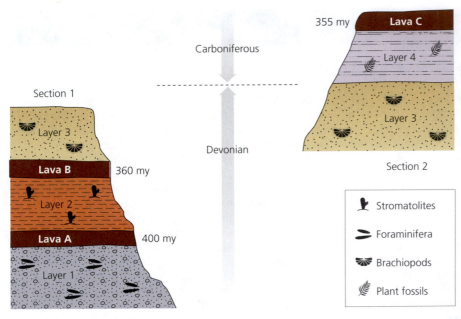

Figure 8.21   Dating Sedimentary Rocks
The absolute age of sedimentary rocks that contain fossils is determined by dating associated volcanic lavas. my = millions of years.

This sequence is capped by another dated lava (lava C) that is 355 million years old. From this we can tell that the deposition of layers 3 and 4 must have occurred between 360 and 355 million years ago. Because this must also be the age of the intervening fossils, the boundary between the Devonian and Carboniferous periods is similarly restricted to this 5-million-year time interval. It is by using this type of approach that the age of this boundary (now placed at 359 million years) has been determined.

An obvious drawback in this approach is the necessity for conveniently placed datable units such as lava flows. If no lavas were present in Sections 1 and 2 of Figure 8.21, it would not have been possible to date the boundary between the Devonian and Carboniferous periods to within 5 million years. Because of this, the ages assigned to the geologic time scale are under constant scrutiny and refinement. The recent revision of the age for the onset of the Paleozoic Era is an excellent example of this. Prior to the revision, an age of about 570 million years for this boundary was widely accepted. However, suspicions about the accuracy of this age arose with the dating of volcanic rocks in Newfoundland, Canada. These rocks yielded an age of 565 million years, yet they lay at least 4 kilometers beneath those containing the earliest Paleozoic fossils. This suggested a considerably younger age for the base of the Paleozoic. Consequently, attention was focused on similar sequences worldwide, as a result of which new dates have been

obtained that more accurately place the age of the boundary at about 541 million years.

8.7 SUMMARY

- By determining the radiometric ages of major boundaries of the stratigraphic column, geologists have established an absolute geologic time scale.

- Because only certain kinds of rocks can be dated, the geologic time scale is continually being refined as ongoing absolute age dating provides ever more precise estimates of the ages of the time scale's boundaries.

8.8 A Sense of Time

Radiometric dating provides the means for determining the absolute age of past geologic events and the rate at which they occurred. We can put our finger on the pulse of the planet. We can determine when events such as mass extinctions, ice ages, and major episodes of volcanism or mountain building occurred during Earth history, and so examine the pace and rhythms of natural activity. By using approaches similar to that shown in Figure 8.21, we can determine the age ranges of important fossil species (and hence the age of the sedimentary rocks that contain them).

We can also identify patterns of global change, such as the comings and goings of ice ages, and the time scale over which they occur. Our ability to examine Earth history in this way is of enormous importance. Today, for example, the possibility of climate change is a major environmental concern, yet we have limited understanding of either the impact of such a trend or the time it might take to become noticeable. With an absolute time scale we can identify natural patterns of climate change in the past and the rates at which they occurred. We can also test scientific ideas or computer models that predict events to occur at certain times or at certain rates.

Nevertheless, as we pointed out in Chapter 1, the geologic time scale should be viewed as a work in progress. In the last decade alone, several new versions of the time scale have been published in order to accommodate the increasing amount of information available from outcrops as well as the increasing precision of radiometric dating. Indeed, if you check an older physical geology textbook in the library, you may notice that the ages assigned to some of the geologic periods differ significantly from those in Figure 8.16. A vivid example of this progress is the recent introduction of a new geologic period, the Ediacaran, the first to be defined within the Precambrian.

CHECK YOUR UNDERSTANDING

- Why has it been difficult to assign absolute ages to the geologic time scale?

As geologists continue to calibrate the geologic time scale in ever greater detail, we should remember that its major subdivisions are still based on the principles established by Hutton and Smith over 200 years ago. Before any rock can be dated, its age relative to associated units must first be determined, and the area from which it was obtained must be mapped geologically. These techniques are as important today as they were in the past and still depend to a large extent on the principles of correlation and relative age dating developed by the pioneers of Physical Geology. Furthermore, the vast majority of rocks do not lend themselves to dating because they do not contain minerals appropriate for the purpose. And because we now know the age range of most fossil species, it is often easier to determine the age of a sedimentary rock on the basis of its fossil content than it is to date the rock itself. Consequently, the fossil-based subdivisions of the geologic time scale are still very much in use today.

> **CHECK YOUR UNDERSTANDING**
>
> ○ Why is the geologic time scale a work in progress?

8.8 SUMMARY

- By knowing the absolute age of past geologic events, we can determine the rates at which they occurred and so examine the pace and rhythms of natural activity.

- The geologic time scale is a work in progress and is constantly being modified as new data becomes available and radiometric age dates become increasingly precise.

8.9 Geologic Time and Plate Tectonics

As we learned in Chapter 2, the movement of Earth's tectonic plates is extremely slow. When the rate of seafloor spreading in the Atlantic Ocean was determined by matching the magnetic stripes found on the ocean floor to the record of magnetic reversals discovered on land, the average speed at which the ocean was opening was found to be a mere 2.5 centimeters (1 inch) per year. That is just 2 meters (6.6 feet) in an average human lifetime. So it is not surprising that, to the human eye, most of the effects of plate tectonics go unnoticed. Yet over the course of Earth history, plate tectonics has changed the face of the planet—opening and closing whole oceans, moving continents together and apart, and drastically changing Earth's climate and life. What has made this possible is the scope of geologic time, which has given plate tectonics a vast period over which to operate.

Between North America and West Africa, the floor of the Atlantic Ocean is some 4500 kilometers (2800 miles) wide. At a rate of seafloor spreading of 2.5 centimeters (1 inch) per year, the length of time it would have taken the Atlantic Ocean to reach this size is 180 million years. At the same rate, it would take a further 180 million years to close the ocean again. While these are vast time periods, they are small in relation to geologic time. (And as we shall learn in Chapter 9, many plates move much faster than 2.5 centimeters (1 inch) per year, with some mid-ocean ridges spreading at more than five times this rate.) With the age of Earth placed at 4.54 billion years, there is time enough to open and close an ocean the size of the Atlantic more than 12 times. Therefore from a geological perspective, Earth's modern geography is just one frame in a movie that tracks the ever-changing distribution of continents and oceans. Hence, the importance of plate tectonics to the history of Earth takes on a whole new significance, when viewed from the perspective of geologic time.

> **CHECK YOUR UNDERSTANDING**
>
> ○ How does geologic time impact plate tectonic processes?

8.9 SUMMARY

- It is the vast scope of geologic time that has allowed Earth's slow-moving plates to bring about such profound change to its geology, climate, and life.

Key Terms

absolute age   227
Archean   240
Cenozoic   238
correlation   234
daughter isotope   245
eon   240
epoch   241
era   238

Hadean Eon   249
half-life   245
isotope   241
Mesozoic   238
original horizontality   231
original lateral continuity   231
Paleozoic   238

parent isotope   245
period   240
Phanerozoic   240
Precambrian   240
principle of faunal succession   234
Proterozoic   240
radioactivity   241

radiometric age   246
relative age   227
rule of cross-cutting relationships   231
rule of inclusions   231
stratigraphic column   235
superposition   230
unconformity   233

Key Concepts

8.1 MEASURING GEOLOGIC TIME: RELATIVE AND ABSOLUTE AGES

- Relative age dating arranges geologic events in the order in which they occurred.
- Absolute ages measure the time elapsed since a geologic event took place.

8.2 A SHORT HISTORY OF GEOLOGIC TIME

- The principle of catastrophism holds that all changes to Earth's surface are the result of supernatural disasters and that Earth is only 6000 years old.
- The principle of gradualism advocated by James Hutton holds that Earth's surface is in continuous but gradual change and that Earth is ancient.
- The principle of uniformitarianism advocated by Charles Lyell can be expressed as "the present is the key to the past."

8.3 RELATIVE DATING: DETERMINING CHRONOLOGICAL ORDER

- Sedimentary strata are deposited in succession as continuous, horizontal layers such that younger strata lie above older ones.
- All rocks are younger than the inclusions they contain and older than the rocks that cut across them.
- Unconformities represent time gaps in the geologic record and serve to reveal the enormity of geologic time.

8.4 CORRELATING ROCK STRATA

- The principle of faunal succession permits individual sedimentary layers to be identified on the basis of their fossil content.
- Based on their fossil content, the age equivalence of rock layers in separate successions can be determined, such that age-equivalent layers can be matched or correlated with each other.

8.5 RELATIVE DATING AND THE GEOLOGIC TIME SCALE

- A stratigraphic column shows sedimentary layers arranged in vertical succession according to their age of deposition.

- Because the stratigraphic column arranges strata according to their relative age, subdivisions of the stratigraphic column became those of the geologic time scale.
- Geologic time is subdivided into eons, eras, periods, and epochs (as the level of refinement increases) based on the fossil record, with boundaries coinciding with the emergence and extinction of fossil life forms.
- Based on the principles of relative dating, the geologic time scale records the sequence of events but not their age in years.

8.6 ABSOLUTE DATING: FINDING A GEOLOGIC CLOCK

- The absolute age of a rock or mineral is determined from the ratio of the parent isotope to that of the daughter.
- The half-life of a parent isotope is the time it takes for a given amount of the parent to decay to half that amount.

8.7 ABSOLUTE DATING AND THE GEOLOGIC TIME SCALE

- Absolute ages have been determined for the geologic time scale through radiometric dating.
- Since many of the boundaries of the geologic time scale are based on fossils that cannot be directly dated by radiometric means, their age must be bracketed by dating suitable rocks immediately above and below.

8.8 A SENSE OF TIME

- Absolute ages confirm the immense span of geologic time and underline the profound changes that even the slowest of processes can produce.

8.9 GEOLOGIC TIME AND PLATE TECTONICS

- It is the immensity of geologic time that has made it possible for plate tectonics to influence Earth's geology, climate, and life so profoundly.

Study Questions

1. Describe the concepts of relative and absolute time and outline how both are used in the geologic time scale.
2. Describe the sequence of events that produces an outcrop in which a sequence of folded strata is unconformably overlain by a sequence of tilted strata.
3. What is the essential difference between the uniformitarian and catastrophic approach to geology? Which best explains features found on Earth's surface?
4. Use the principles of relative age determination to deduce the geologic history of the area shown in Figure 8.22.

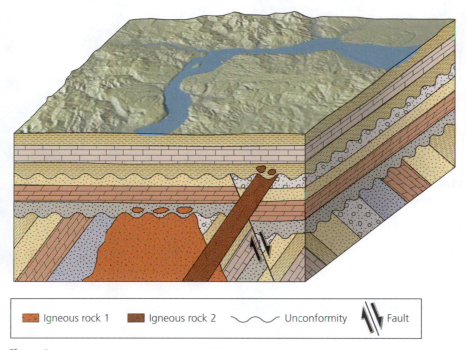

| Igneous rock 1 | Igneous rock 2 | ∿∿ Unconformity | ⫽ Fault |

Figure 8.22

5. Most of the major divisions of the geologic time scale coincide with the emergence and extinction of fossil life forms. Why do you think this is so?

6. Lord Kelvin calculated the age of Earth to be 20 to 40 million years based on its rate of heat loss, whereas John Joly calculated its age to be 80 to 90 million years based on the saltiness of the oceans. Why are both of these calculations seriously flawed?

7. What is the absolute age of Earth now thought to be and how has this age been determined?

8. What is radioactivity and how is the process of radioactive decay used in determining absolute age?

9. A feldspar crystal analyzed for potassium-argon dating is found to contain 25 percent of its original potassium-40. Refer to Fig. 8.19 to determine the age of the feldspar crystal, given that the half-life of potassium-40 is 1.25 billion years. What assumptions are being made in determining this age?

10. In a granite that had experienced a significant reheating event, how would you expect the age determined on the basis of the uranium-lead method to differ from that based on the potassium-argon technique?

11. A vertical sequence of Paleozoic sedimentary rocks contains two lava flows with ages of 415 and 420 million years (Fig. 8.23). The older flow is overlain by rocks containing Silurian fossils, whereas the younger flow overlies rocks with Devonian fossils. What are the ages of the rocks above, below, and between the two flows, and how can this be used to constrain the age of the boundary between the Silurian and Devonian periods?

Figure 8.23

9 Plates and Plate Boundaries

9.1 Plates on Earth's Surface

9.2 Plates and Isostasy

9.3 Plate Boundaries

9.4 Divergent Boundaries: Creating Oceans

9.5 Convergent Boundaries: Recycling Crust and Building Continents

9.6 Transform Boundaries: Fracturing the Crust

9.7 Hotspots: Tracking Plate Movements

9.8 Plate Tectonics and Plate-Driving Mechanisms

▶ Key Terms

▶ Key Concepts

▶ Study Questions

LIVING ON EARTH: The Bizarre World of the Mid-Ocean Ridges 267

LIVING ON EARTH: Tsunami 272

IN DEPTH: Ophiolites—Clues to the Structure of Oceanic Crust 279

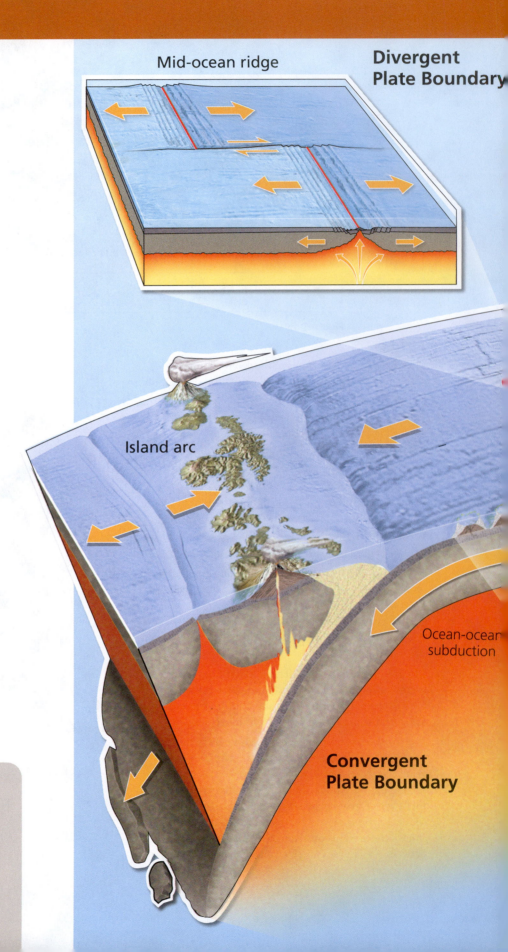

Mid-ocean ridge

Divergent Plate Boundary

Island arc

Ocean-ocean subduction

Convergent Plate Boundary

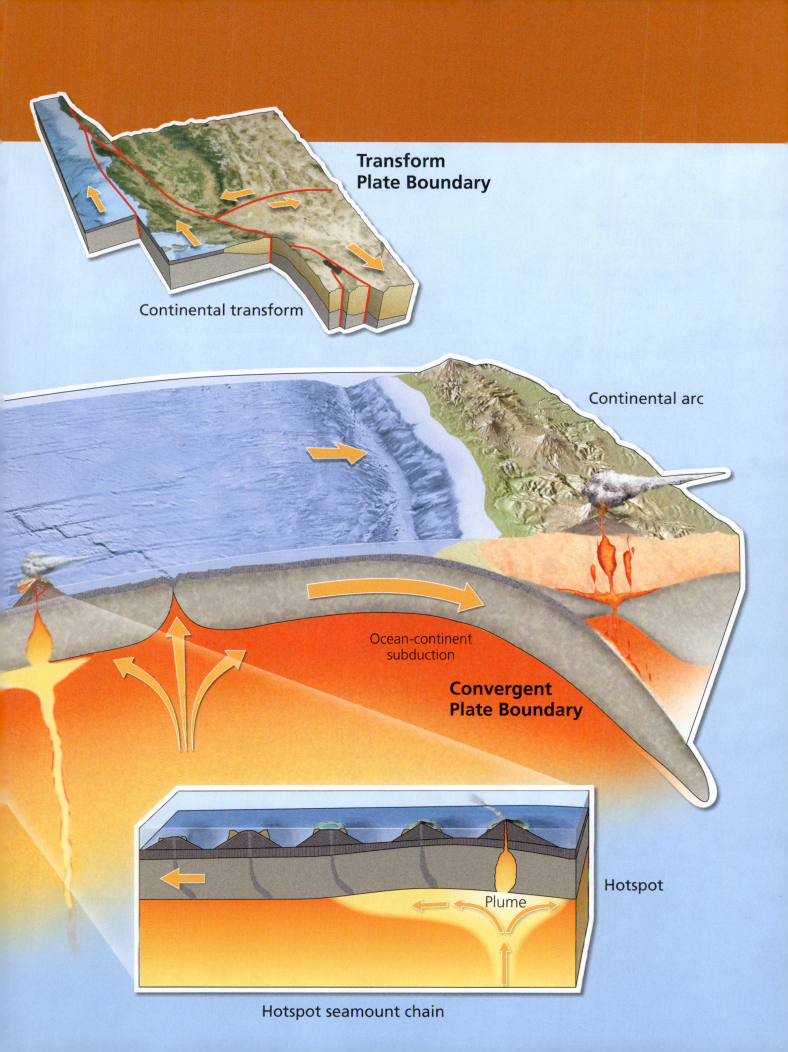

Transform Plate Boundary

Continental transform

Continental arc

Ocean-continent subduction

Convergent Plate Boundary

Hotspot

Plume

Hotspot seamount chain

9.1 Identify the physical and chemical divisions in Earth's outer layers.

9.2 Understand that the lithospheric plates are buoyant, and that this buoyancy controls the relationship between crustal elevation, crustal thickness, and crustal density.

9.3 Compare and contrast the three types of plate boundaries and describe the three main ways boundaries interact: spreading apart, coming together, and sliding past one another.

9.4 Describe the processes that occur at divergent boundaries and explain how new ocean floor is created.

9.5 Describe the processes that occur at convergent boundaries and explain how crust is recycled and continents are built.

9.6 Describe the motion along transform boundaries and compare and contrast the two principal types of transform faults.

9.7 Describe the enigmatic volcanic regions known as hotspots and explain how they can be used to track the movement of plates.

9.8 Compare and contrast the three types of force that may propel plates.

According to the theory of plate tectonics introduced in Chapter 2 Earth's rigid outer shell, the *lithosphere*, is divided into huge moving slabs called *plates*. One hundred kilometers (60 miles) or more in thickness, these lithospheric plates are in constant motion, repeatedly interacting along their boundaries as they jostle for position. At speeds of only centimeters per year, plates move slowly, not much faster than the rate at which your fingernails grow. Yet over millions of years, this movement can create and destroy entire ocean basins and carry continents across the face of Earth.

Plates interact in several ways along plate boundaries—splitting apart, colliding with one another, or sliding past each other—and it is at their margins that the geologic activity associated with plate movement is most intense. Today, we can actually measure this movement using GPS. By understanding how plates interact as they move, scientists can explain many of Earth's major geologic processes and features, among them the origin and distribution of earthquakes, volcanoes, and mountain belts. While many forces may be influencing plate motion, the main driving mechanism is clear: it is the flow of heat generated by radioactive decay from the interior of the planet to its surface. We can think of plate tectonics as nature's vast recycling plant and the underlying cause of almost every major feature on our planet's surface. In this chapter, we look at plate boundaries and examine the driving forces that we believe are responsible for plate movement.

9.1 Plates on Earth's Surface

Plates are rigid slab-like fragments that comprise Earth's outer rocky shell, known as the *lithosphere*. Our first step in understanding what plates are and how they move is to examine Earth's outer layers. In this section, we first discuss how heat influences the *physical* properties of these layers, allowing rigid plates to move above a softer, plastic layer called the *asthenosphere*. We then see how differences in the *chemical* composition of these layers cause plates to vary in their rock types and thickness.

EARTH'S OUTER LAYERS

Recall from Chapter 2 that Earth's outer layers (Fig. 9.1) can be subdivided based on the contrasting physical properties of their rocks:

- The **lithosphere**, from the Greek for *rock globe*, a rigid outermost layer broken into huge moving slabs, or plates

- The **asthenosphere**, from the Greek for *weak globe*, a broader region of soft rocks beneath the rigid lithosphere

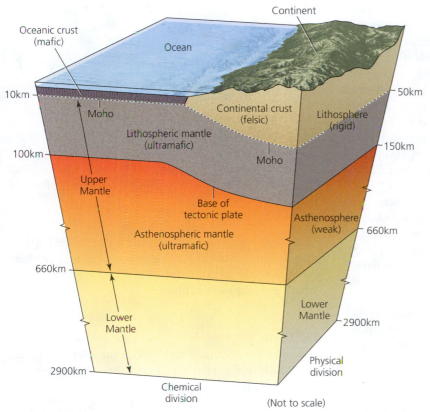

Figure 9.1   Earth's Outer Layers

Earth's outer layers are subdivided into crust and mantle based on rock type (chemical divisions), and lithosphere and asthenosphere based on rigidity (physical divisions).

The rigid rocky lithosphere is what makes up the plates, and the weak underlying asthenosphere allows the plates to move. The "floating" plates are free to move sideways if acted upon by a large enough force.

Away from plate boundaries, temperatures in the lithosphere—predominantly the upper 100 kilometers (60 miles) of Earth—are well below those required for melting. As part of the lithosphere, Earth's plates are therefore rigid and strong, and move as coherent slabs on the weak and partially molten asthenosphere beneath.

Much of our understanding about Earth's softer asthenosphere has come from the study of earthquakes. Earthquakes send out energy in all directions. Because of its potentially devastating consequences, we tend to focus on the energy directed upward toward Earth's surface. However, this energy is also propagated downward into Earth's interior. When measuring energy patterns released by earthquakes, scientists noticed that earthquake waves propagated into Earth's interior slowed down at depths of about 100 to 250 kilometers (60–150 miles) below Earth's surface. Because the speed at which energy propagates depends on the type of material it encounters along its path,

scientists realized that the rocks at this depth must be softer. Earthquake waves move more rapidly through rigid material, just as an athlete can run faster on a hard surface (such as a running track) than a soft one (such as sand).

The distinction between the asthenosphere and lithosphere is therefore based on the contrasting physical properties of the rocks in these regions rather than the chemical composition of the rocks themselves. In fact, the rocks at the base of the lithosphere are very similar in chemical composition to those in the asthenosphere, but they act in a rigid fashion because they are cooler. So scientists position the base of the lithosphere at a depth where temperatures first become high enough to make the rocks soft. Because this may occur over a broad zone, the boundary between the lithosphere and asthenosphere is gradual rather than sharp and its exact location is sometimes hard to detect.

LAYERS OF THE LITHOSPHERE

An important boundary occurs within the lithosphere that is based on contrasting *chemical* composition. The rigid lithosphere is divided into two main layers, the outermost, rocky *crust* and the denser rocks of the *lithospheric mantle* upon which the crust rests. There are two types of crust, *continental* and *oceanic*, which also differ in their chemistry and are distinguished by their contrasting rock types, density, and thickness (see Fig. 9.1).

In previous chapters, we learned that the continents are made up of a wide variety of rocks. In general, however, the **continental crust** is predominantly made up of broadly granitic rocks. The continental crust is the least dense of all Earth's solid layers and varies widely in thickness, averaging about 35 kilometers (22 miles) but increasing to as much as 80 kilometers (50 miles) below high mountain ranges such as the Himalayas. The **oceanic crust**, on the other hand, is overwhelmingly basaltic (or mafic). Oceanic crust is denser than continental crust because, as we learned in Chapter 4, basaltic rocks contain larger concentrations of dense elements such as iron and magnesium, whereas granitic rocks are dominated by relatively lightweight elements that typically produce less dense silicate minerals such as quartz and feldspar. Oceanic crust is also thinner than continental crust, typically averaging only 7 kilometers (4 miles).

The crust rests on the **upper mantle**, which consists predominantly of the even denser ultramafic rock called peridotite, which is dominated by olivine and pyroxene.

Located directly beneath the crust, the upper mantle is cool and rigid, and is part of the lithosphere. We therefore call it the **lithospheric mantle**. Experiments show that rocks of peridotite composition extend to a depth of at least 660 kilometers (410 miles), so from a strictly chemical standpoint, the upper mantle also includes the soft asthenosphere. However, the higher temperatures in the asthenosphere cause the peridotitic rocks it contains to become weak or plastic. It is important to understand that the rigid lithosphere therefore includes both the crust and part of the upper mantle, whereas the soft asthenosphere lies entirely within the upper mantle, as shown in Figure 9.1.

Although the boundary between the lithosphere and asthenosphere is gradual, that between the crust and upper mantle is sharp and readily detected from earthquake waves, which travel faster upon entering the denser peridotitic rocks of the upper mantle. This sharp boundary, named after its Croatian discoverer, is known as the *Mohorovičić Discontinuity*, or **Moho** for short.

But why do the crustal components of plates vary so much in thickness? Why, for example, is continental crust variably thick, whereas oceanic crust is consistently thin? And why does the continental crust mostly stand above sea level, whereas the oceanic crust is almost entirely submarine? Indeed, how do the plates "float" on the asthenosphere? The answers to these questions lead us to another dynamic property of Earth's interior, the phenomenon known as *isostasy*, which we discuss in Section 9.2.

> **CHECK YOUR UNDERSTANDING**
>
> ◯ How is it possible for mantle rocks to occur in both the asthenosphere and the lithosphere?

9.1 SUMMARY

- Earth's outer layers are divided into a rigid outermost layer, known as the lithosphere, and a broader region of soft rocks beneath it, known as the asthenosphere.

- The lithosphere—which is broken into tectonic plates—consists of the crust and the upper mantle, which the crust rests on.

- Earth's crust is of two types: continental and oceanic, and is separated from the upper mantle by the Moho.

9.2 Plates and Isostasy

We can obtain an image of Earth's interior analogous to that obtained by the passage of X-rays through a patient by systematically analyzing how earthquake waves move within Earth. These analyses consistently indicate that the crust is thin beneath the oceans and thickest below high mountain belts. In other words, the crust is thin where its surface is at a low elevation and thickest where the elevation is high. This means that the base of the crust (the

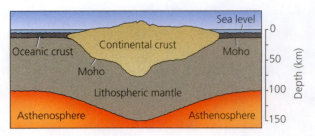

Figure 9.2   Crustal Roots
The shape of the Moho at the base of the crust mirrors Earth's surface. The crust thinnest beneath the oceans and thickest beneath high mountains, which are supported by deep crustal roots.

Moho) has a shape that mirrors in an exaggerated fashion the overlying shape of Earth's surface (Fig. 9.2). In short, the higher the crustal elevation, the deeper its crustal "root." But why should this be so?

BUOYANCY

The principle that governs this curious relationship between the Moho and Earth's topography is one of buoyancy. Floating container ships illustrate this principle rather well (Fig. 9.3). An empty container ship rides higher in the water than one laden with cargo because the water flows under the weight of the heavier ship. As a result, the laden container ship sinks to the depth where it displaces a volume of water equal in weight to that of the ship and its cargo.

We use the term **buoyancy** here to describe the way in which Earth's crust and the rigid lithospheric mantle beneath it "float" on the plastic asthenosphere below, which, like the water beneath the container ship, is capable of flow. Because the lithospheric mantle is a fairly uniform layer, this buoyancy depends to a large extent on two factors, the thickness of the crust and its density.

Thickness

Where the crust stands highest, it also extends downward. In other words, for the crust to stand high, it must have a deep "root" to support it. The continents stand highest where there are mountains. Therefore we know these high areas must be supported by crustal roots that project deep into the mantle (see Fig. 9.2). The crust beneath the Himalayas, for example, extends almost 80 kilometers (50 miles). The oceanic crust, on the other hand, is thin and has no thick roots.

> **CHECK YOUR UNDERSTANDING**
>
> ◯ Why does the Moho beneath the continents mirror their surface topography?

Density

Both continental and oceanic crust (and the lithospheric mantle below them) effectively float upon the denser asthenosphere below. But the granitic continental crust is lighter and so rides higher on Earth's

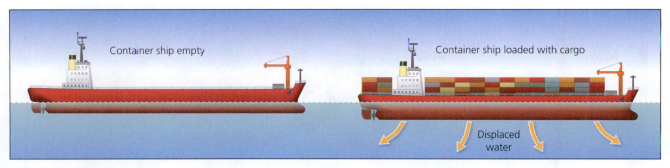

Figure 9.3   Buoyancy

Floating container ships illustrate the principle of buoyancy.

surface than the denser basaltic (or mafic) oceanic crust. Because of this, oceanic crust is confined to those areas where elevations are lowest. Water collects at these lowest elevations, so it is not surprising that most of Earth's oceanic crust lies beneath the water of the world's ocean basins.

Earth as a Split-Level Planet

Viewed from space, our planet shows a striking contrast in the elevation of the continents compared with that of the oceans. In fact, we live on a split-level planet, with much of Earth's surface comprising either continental crust at an elevation between 1000 meters (3300 feet) above sea level or oceanic crust between 4000 and 5000 meters (13,000 and 16,000 feet) below sea level. These two elevation ranges can be most easily seen on a graph known as the hypsometric curve. The *hypsometric curve*, named from the Greek word *hypsos*, or height, measures the proportion of Earth's surface that stands above a given elevation or depth (Fig. 9.4). For example the deepest trench plots at 100 percent because all of Earth's surface has a higher elevation, whereas Mt. Everest plots at 0 percent because there is no elevation higher than it. That portion of Earth's surface between sea level and 1000 meters forms the *continental platforms* and the surface between 3000 and 5000 meters (10,000 and 16,000 feet) below sea level forms the oceanic *abyssal plains* (Fig. 9.5).

Thus Earth's crust is in a state of buoyant equilibrium, the continents stand highest where they are thickest, and the oceans have formed in regions floored by denser basaltic crust. This relationship between crustal elevation, crustal thickness, and crustal density is described by the principle of isostasy.

CHECK YOUR UNDERSTANDING

◯ Why is the elevation of continents so much higher than that of the ocean floors?

ISOSTASY

Isostasy is the balance reached by Earth's crust as it (and the lithospheric mantle) floats upon the denser, more plastic asthenosphere. *Isostasy* comes from the Greek for "equal

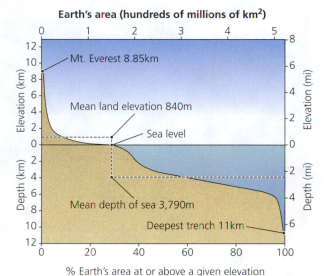

Earth's area (hundreds of millions of km²)

Figure 9.4   Hypsometric Curve

The hypsometric curve is a graph that illustrates the proportion of Earth's surface at or above a given elevation or depth. Note that less than a third of Earth's surface is above sea level and more than half lies below a depth of 3000 meters (10,000 feet). Mt. Everest is shown at 0 percent because none of Earth's crust is at a higher elevation than its peak. On the other hand, ocean trenches are shown at 100 percent because all elevations on Earth's surface are higher than them.

standing." Floating icebergs clearly demonstrate the principle of isostasy (Fig 9.6). According to this principle, the depth to which a floating object sinks depends upon its volume and its density. All icebergs, being of essentially the same density, sink in water such that about 90 percent of their volume is submerged. Therefore, the larger the iceberg, the greater this submerged volume will be; indeed, the common expression "tip of the iceberg" is derived from this concept. If ice above the water level melts, the iceberg rises to maintain a consistent proportion of ice above and below the water line. In Figure 9.6, the larger iceberg rides higher but also extends to greater depth than the smaller one.

Not only does the principle of isostasy explain why mountains stand high, but it also accounts for the subdivision of Earth's surface into continents and ocean basins

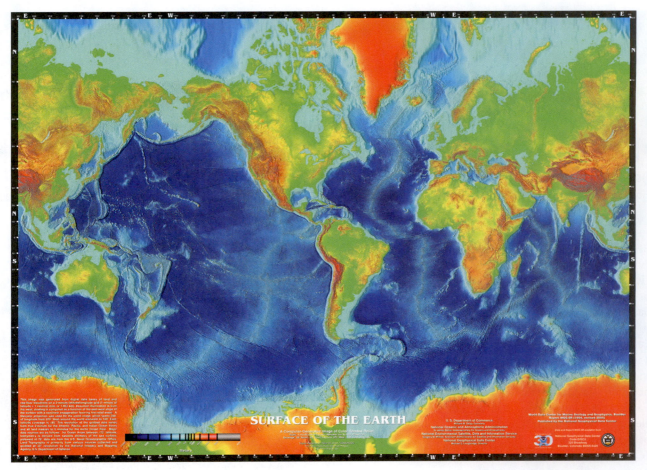

SURFACE OF THE EARTH

Figure 9.5   Earth's Surface

Satellite mapping of Earth's surface (color-coded according to elevation) shows that most of the planet's surface lies at the elevation of the continental platforms (green) or abyssal plains (deeper blue).

Figure 9.6   Isostasy

Floating icebergs demonstrate the concept of isostasy. These two icebergs float in such a way that the same proportion of each iceberg is above the water line.

(see Fig. 9.5). The edges of continents mark the transition between continental and oceanic crust. The crust therefore becomes thinner as it changes its composition from a lightweight, broadly granitic one to the dense basaltic composition that is typical of the ocean floors. As a result, the lighter continents ride high and the denser oceanic crust rides low, and so forms the floor of the deep ocean basins.

According to the principle of isostasy, Earth's major topographic variations are a function of crustal thickness and the density of crustal rocks. If either of these (thickness or density) changes, the crust responds by trying to establish a new "floating" balance or *isostatic equilibrium*. During the last Ice Age, for example, the crust beneath the advancing continental ice sheets was first depressed under the weight of the ice, only to rise again when the ice sheet retreated, a response known as *isostatic rebound* (Fig. 9.7).

In a similar fashion, as the summits of mountains are lowered by erosion, the crust rises in response to the reduced load and the crustal root supporting the mountains shrinks (Fig. 9.8). In this way uplift and erosion are processes that are inextricably linked. At the same time, the deposition of sediment produced by erosion causes the crust to sink. In this way, readjustments are constantly being made in an attempt to maintain isostatic equilibrium. These readjustments slowly reduce a mountain belt to normal crustal thicknesses. Indeed, it is this process of erosion compensated by crustal rebound that makes the rock cycle (see Chapter 1)

CHECK YOUR UNDERSTANDING

• What happens to crustal rocks and the upper mantle during the growth of continental ice sheets?

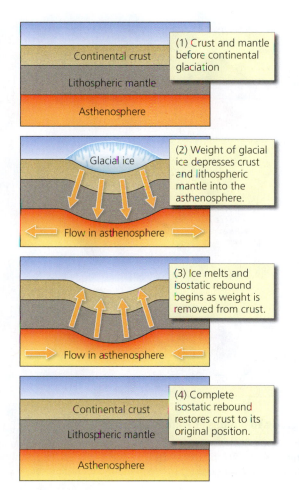

Figure 9.7   Isostasy and Ice Sheets

Continental ice sheets illustrate isostatic depression and rebound.

possible by bringing progressively deeper levels of the crust to the surface.

This combination of uplift and erosion has eliminated entire mountain belts that may have once rivaled the mighty Himalayas. In the process, the crust may rise as much as 50 kilometers (30 miles). Yet this distance is tiny compared to the horizontal distances moved by plates.

Because plates effectively float, they are also free to move sideways and can be carried in this way for thousands of kilometers in response to the flow of heat within Earth's heated interior. This is the motion that drives plate tectonics, and it is this movement and the plate boundaries along which it occurs that we discuss in Section 9.3.

9.2 SUMMARY

- The continental crust is relatively lightweight and, with the lithospheric mantle, floats on the pliable asthenosphere below. To stand high, continents must therefore have deep crustal roots to support their weight.

- Oceanic crust is denser and heavier than continental crust and so rides lower on the floors of the world's ocean basins. Oceanic crust is also of fairly uniform thickness.

- The relationship between crustal elevation, crustal thickness, and crustal density is called isostasy, which describes the balance reached by Earth's crust as it floats on the mantle beneath it.

9.3 Plate Boundaries

Recall from Section 2.5 that Earth's plates move relative to each other like rafts floating on a stream, and all plates interact with neighboring plates along their shared boundaries. In the simplified diagram in Figure 9.9, movement of Plate A to the left produces a gap behind, while an overlap is created in front. As a result, the margins of Plate A actually simultaneously experience three types of motion. At the rear, the motion relative to Plate B is one of separation and *extension*. At the front, it is one of convergence and *compression*, as one plate is forced beneath the other. The third type of motion occurs along its top and bottom margins where the plates simply slide past each other, a motion known as *shear*.

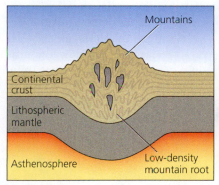

Before Erosion and Deposition

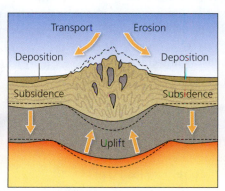

Erosion and Deposition

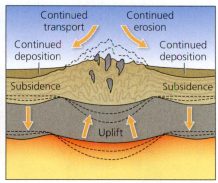

Continued Erosion and Deposition

Figure 9.8   Isostasy and Erosion

Erosion in mountainous regions causes uplift of the mountain root. Deposition of sediment, on the other hand, causes crustal subsidence.

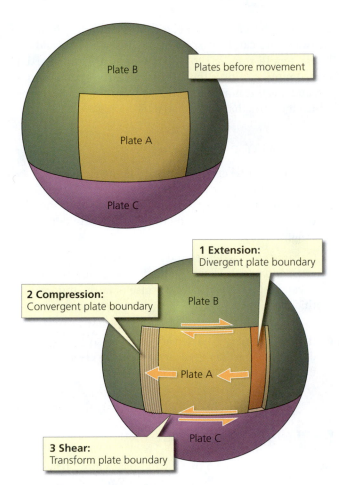

Figure 9.9   Plate Movement

As Plate A moves to the left, its margins simultaneously experience three types of motion: (1) extension, characteristic of divergent plate boundaries, (2) compression, characteristic of convergent plate boundaries, and (3) shear, characteristic of transform plate boundaries.

Actual plate boundaries behave in much the same way as the example in Figure 9.9. As the plates move away from mid-ocean ridges and toward deep ocean trenches, they jostle each other for position, interacting in the same three ways, separating, colliding, or sliding past each other. As we learned in Chapter 2, the three types of plate boundaries that result from these interactions are *divergent*, *convergent*, and *transform*, depending on their sense of movement (Fig. 9.10).

- **Divergent plate boundaries**, where two plates move apart, are marked by spreading. These boundaries are centered on a mid-ocean ridge and create new oceanic lithosphere. Because these boundaries form new lithosphere, we also refer to them as *constructive* plate margins.

- **Convergent plate boundaries**, where two plates collide, are marked by deep ocean trenches and subduction zones where one plate is forced beneath the other and consumed. Because these boundaries destroy

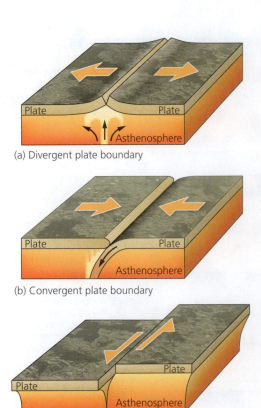

(a) Divergent plate boundary

(b) Convergent plate boundary

(c) Transform plate boundary

Figure 9.10   Types of Plate Boundaries

This figure illustrates the following three types of plate boundaries, which are distinguished on the basis of their relative motion: (a) divergent, (b) convergent, and (c) transform.

old lithosphere, we also refer to them as *destructive* plate margins.

- **Transform plate boundaries**, where two plates slide past each other, are marked by great crustal fractures called *transform faults*. Transform boundaries neither create nor destroy lithosphere but instead form links between one plate boundary and another. For this reason, we also refer to them as *conservative* plate margins.

Figure 9.11 illustrates the major lithospheric plates outlined by these three types of boundaries. Note how the boundaries are associated with earthquakes, which are a direct outcome of the jostling of plates along their margins. We now examine in more detail the nature of each of these plate boundaries and the surface features they create.

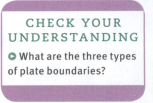

CHECK YOUR UNDERSTANDING

● What are the three types of plate boundaries?

9.3 SUMMARY

- Plate margins interact in three main ways—through extension, in which two plates move apart; through compression, in which two plates converge; and through shear, in which plates slide by each other.

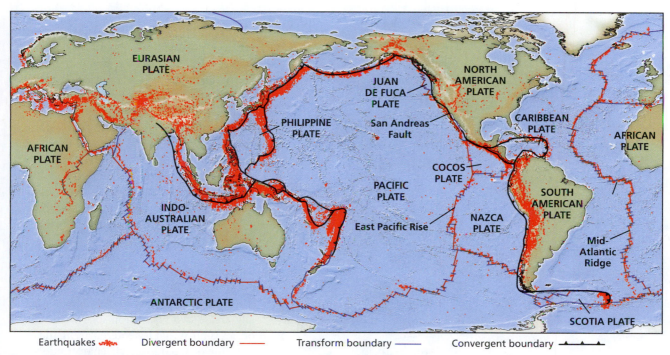

Earthquakes   Divergent boundary ———   Transform boundary ———   Convergent boundary ▲—▲—▲

Figure 9.11   Earth's Major Plates

This map highlights the boundaries of Earth's major lithospheric plates. Plate boundaries mark the sites of the main features on Earth: the mid-ocean ridges of divergent boundaries, the deep ocean trenches of convergent boundaries, and the great crustal faults of transform boundaries. Most earthquakes take place at plate boundaries.

- Three types of plate boundaries result from these interactions: extensional movement produces divergent boundaries, compressional movement produces convergent boundaries, and shear movement produces transform boundaries.

9.4 Divergent Boundaries: Creating Oceans

At divergent plate boundaries, plates move apart. Divergent boundaries are born when a large continent is torn in two. The process starts with continental rifting, during which progressive thinning of the continental crust occurs with the formation of a *continental rift valley*. Eventually, the continent separates into two halves and a new ocean forms between them. At this point, continental rifting becomes continental drift, and a new plate boundary (a *mid-ocean ridge*) is formed, along which new oceanic lithosphere is created by seafloor spreading.

CONTINENTAL RIFTING AND OCEAN FORMATION

The birth of oceans starts on dry land with the breakup of a continent. This process, which takes millions of years to complete, starts over an area of mantle upwelling that brings heat to the base of the plate. As the plate starts to separate, basaltic magma rises from the mantle below and

the overlying lithosphere bulges (Fig. 9.12a). As more magma wells up into the continental crust, the crust stretches and thins, which lowers the pressure in the mantle below, resulting in the formation of more basaltic magma. The rising batches of magma create a line of separation, and in the early stages of rifting, large blocks of continent settle, or *subside*, along fractures or **faults**. This settling of fault blocks forms a steep-walled **rift valley** (Fig. 9.12b).

This new valley commonly becomes a pathway for major rivers so that river and lake deposits accumulate on the floor of the valley in addition to flows of basaltic lava. The floor of the rift valley may eventually settle to an elevation that is below sea level but remains continental if access to the ocean is blocked. However, when the rift tears completely through the continent and intersects a coastline, seawater floods in and the first shallow-marine sediments are deposited on the rift floor. As the continental plates separate farther, more water flows in and the upwelling magma starts to build new basaltic ocean floor (Fig. 9.12c). In this way, continental rifting gives way to continental drift as a small but progressively widening ocean develops between the separating continents. The thinned margins of the continents slowly subside and settle as they move away from the heat source at the zone of divergence, and are progressively flooded and covered by marine sediments. That portion of the continental margin flooded by shallow seas is called the **continental shelf**. Continental margins formed in this way are known as **passive margins** and, as

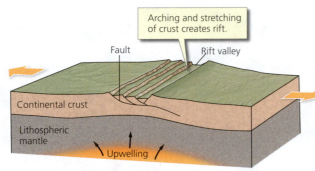

(a) Uplift and fracturing of crust above upwelling mantle

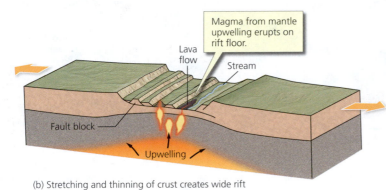

(b) Stretching and thinning of crust creates wide rift

(c) Crustal thinning continues until one continent becomes two that drift apart

Figure 9.12 Continental Rifting

A divergent plate boundary is the site of the breakup of a continent and new ocean formation.

Along this divergent plate boundary a slab of continental lithosphere known as the Somali plate may eventually split from Africa. As rifting takes place, magmas rise to fill the fissures and volcanoes are commonplace. The largest, Mt. Kilimanjaro in Tanzania, rises almost 5900 meters (19,350 feet) to form Africa's highest mountain. To the north, where the rift system is flooded by the waters of the Red Sea and the Gulf of Aden, separation has reached the next stage with the birth of a new ocean between Africa and the Arabian Peninsula. All three arms of the rift system meet in Ethiopia's Afar Triangle to form a Y-shaped plate junction known as a **triple point**. Given sufficient time, the Red Sea and Gulf of Aden may open into wide oceans just as the Atlantic Ocean has done over the past 180 million years.

Continental rifting, therefore, may eventually create two plates from one when a new ocean is born that is centered along the original line of separation. If separation occurs, the original site of continental rifting evolves to become a mid-ocean ridge once the continents drift apart.

MID-OCEAN RIDGES AND OCEAN OPENING

Once the two continents have separated and oceanic lithosphere has developed between them, the divergent plate boundary between the two plates is defined by a **mid-ocean ridge**, a submarine mountain chain buoyed by underlying mantle heat (See Fig. 9.12c). At the highest point of the ridge, seafloor spreading creates basaltic ocean crust. As the plates move apart, fractures open in the rift valley at the ridge crest and are continually filled by magma derived from the hot, partially molten mantle below. In this way, new oceanic crust continually arises at the summit of the ridge, and all previously formed ocean floor progressively slides farther and farther apart. As we learned in Chapter 2, the process is like inserting a series of books into the center of a bookshelf such that the newer books push the older ones aside. As older oceanic lithosphere is carried away from the ridge, it cools, contracts, and subsides. It also thickens because the asthenospheric mantle from which basalt is extracted at the ridge crest cools to become part of the lithospheric mantle as it too moves away from the ridge. If the plates separate at just a few centimeters per year, the ocean widens by tens of kilometers every million years. In this way, an ocean thousands

the ocean continues to open, come to lie far from the divergent plate boundary along which they first formed. The Atlantic Ocean, for example, is largely bordered by passive margins that first started to form as continental rifts during the breakup of Pangea, but now lie far from the Mid-Atlantic Ridge that marks the divergent plate boundary today.

Just such a progression of events is now taking place in East Africa (Fig. 9.13), where the Great Rift Valley cuts across Ethiopia southward to Mozambique, a distance of 4500 kilometers (2800 miles).

CHECK YOUR UNDERSTANDING

● What rock types might you expect to fill a continental rift as it opens into an ocean?

CHECK YOUR UNDERSTANDING

● What are mid-ocean ridges and why do they stand high?

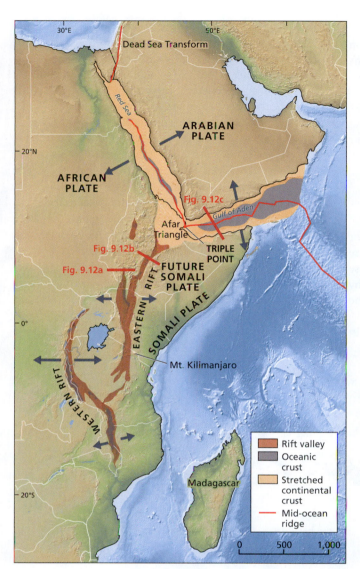

Figure 9.13   Great Rift Valley

The Great Rift Valley is a system of rifts along which eastern Africa (the future Somali plate) is separating from the rest of the continent (African plate).

of kilometers across may eventually open. The central Atlantic Ocean, for example, formed about 180 million years ago and is 6500 kilometers (4000 miles) wide.

A map of the age of oceanic crust (Fig. 9.14) is a confirmation of this process. The youngest crust occurs adjacent to modern spreading ridges, and the age of the crust increases with distance from them. Because the spreading ridge runs down the center of the Atlantic Ocean, the age distribution of its oceanic crust is symmetric. The youngest oceanic crust lies in the middle of the Atlantic Ocean, while the oldest occurs along its margins. In contrast, the spreading ridge in the Pacific Ocean is located closer to its eastern margin. Again, the crust is youngest adjacent to the ridge, but the age distribution is asymmetric and the oceanic crust in the western Pacific Ocean is significantly older than that in the eastern Pacific.

How fast do ridges spread? Direct measurements using GPS show that rates of divergence on modern mid-ocean ridges vary significantly, from 17 centimeters (7 inches) per year on the fast-spreading East Pacific Rise off South America, to a barely perceptible 0.1 centimeters (0.04 inches) per year in Africa's Great Rift Valley. Spreading rates along the Mid-Atlantic Ridge are considerably less than those on the East Pacific Rise, averaging about 3 centimeters (1.2 inches) per year. We can deduce the average spreading rate for any segment of the mid-ocean ridge for the past 160 million years or so by examining the width of each color band in Figure 9.14. From these data, it is clear that spreading rates in the Pacific Ocean have been greater than those in the Atlantic Ocean for much of this time interval.

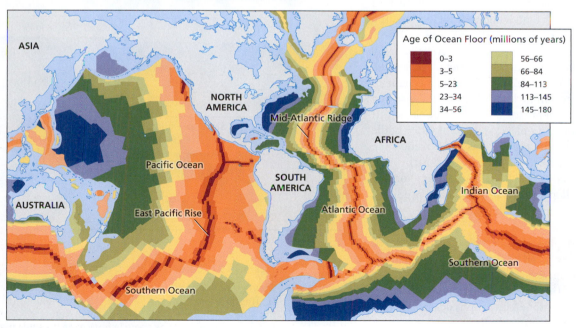

Figure 9.14   Mid-Ocean Ridge System

The age distribution of Earth's ocean floors reveals plate divergence at mid-ocean ridges.

DEVELOPMENT OF MID-OCEAN RIDGES

Mid-ocean ridges make up the greatest mountain range on Earth. They are found in all of the world's major oceans and form an interconnected system, 65,000 kilometers (40,000 miles) in length, that girdles the globe like a seam on a baseball. Although their slopes are gentle, they consistently rise 2 to 3 kilometers (1 to 2 miles) above the abyssal plains of the surrounding ocean floor. Only rarely do their summits stand above sea level, and only in Iceland is a significant portion of a mid-ocean ridge (the Mid-Atlantic Ridge) exposed.

Mid-ocean ridges are a product of extension and separation. The rift valley at the ridge crests, where basaltic lavas erupt to create new oceanic crust as the seafloor spreads, is evidence of this extension. Mid-ocean ridges are high, not because they are being compressed, but because they are swollen with heat, being thermally buoyed by the hot mantle beneath. As seafloor spreading carries the oceanic crust away from the ridge, the cooling seabed subsides and ultimately becomes part of the flat abyssal plain of the deep ocean.

Rift valleys continue to develop at the crests of mid-ocean ridges for the same reason that they form where continents are tearing apart. The influence of mantle heat not only uplifts the mid-ocean ridge, it also stretches the oceanic crust, causing it to settle at the crest of the ridge. As a result of this continual settling, mid-ocean ridges are heavily faulted regions that are prone to earthquakes (see Fig. 9.11). However, because seismic activity can occur only in brittle rocks and the rigid crust is thin at the summit of mid-ocean ridges, these earthquakes are relatively shallow, most being confined to depths of less than 5 kilometers (3 miles). The restriction of these earthquakes to a narrow band along the ridge crest shows that only here, where spreading is taking place, are the fault blocks actively settling. The mechanism of settling can be likened to the collapse of a row of unsupported books on a bookshelf. If an additional book is inserted into the middle of the row, and the row has no bookends, the books tip over and slide against each other as they fall (Fig. 9.15a). The pattern formed by the fallen books is similar to that produced by the collapse of fault blocks at mid-ocean ridges.

Although geologists have directly explored only a few hundred kilometers of the mid-ocean ridge system (see Living on Earth: **The Bizarre World of the Mid-Ocean Ridges** on page 267), it has been extensively probed by geophysical instruments. We therefore know it features high, jagged peaks in the Atlantic and smoother, broader ranges in the Pacific. These differences in ridge topography reflect variations in the rate of seafloor spreading (Fig. 9.15b). The elevation of oceanic crust depends on its temperature and the temperature is much the same everywhere along the ridge crest because it is governed by the heat from the mantle below. As a result, its elevation is quite constant. Away from the ridge (and the mantle heat source beneath it), the temperature of the crust decreases and the crust subsides to the level of the abyssal plain. However, the distance the crust moves away from the ridge before it subsides to that level depends on the spreading rate. Along slow-spreading ridges, the ocean floor does not move far before cooling so that subsidence to the level of the abyssal plain occurs close to the ridge. Along fast-spreading ridges, on the other hand, the ocean floor moves much farther away from the ridge before it cools and subsides to the level of the abyssal plain. Thus the slow-spreading Mid-Atlantic Ridge is narrow and steep, whereas the fast-spreading East Pacific Rise is broad and gentle. In both cases, the elevation of the seafloor corresponds to its age; only the spreading rates differ.

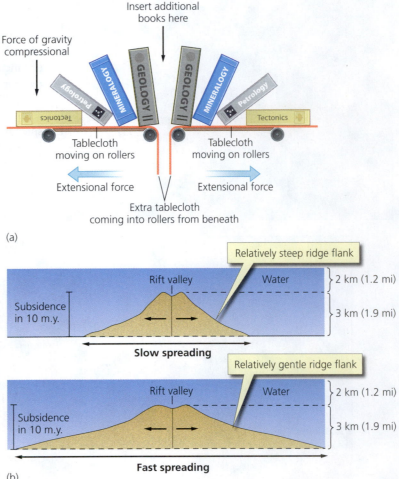

Figure 9.15   Features of Mid-Ocean Ridges

Characteristic features are associated with spreading ridges. (a) This bookshelf serves as a model for ridge faulting. (b) Spreading rate influences ridge topography. Note that the overall height of the ridge is the same in both examples because this is governed by the mantle temperature, which is the same in both cases.

LIVING ON EARTH

The Bizarre World of the Mid-Ocean Ridges

The mid-ocean ridge system is the largest geologic feature on Earth. It is populated by unique communities of animals, and is a natural laboratory for studying the creative forces of the planet at work. Yet humans had set foot on the Moon before scientists first explored Earth's mid-ocean ridges, just a little more than a mile below the surface of the oceans. When exploration of mid-ocean ridges finally got underway, the results were startling and provided insights into the nature of volcanic activity, the venting of superheated fluids, and the nature of the unique and delicate *ecosystems* (living communities) that depend on this activity. The surveys also provided fundamental information about the composition of the rocks of the ocean floor and its sediment cover, and confirmed scientists' suspicions about the source of chlorine in ocean water.

The first exploratory dives were carried out in the early 1970s. Using submersibles, scientists explored the ocean floor at a dive site chosen some 640 kilometers (400 miles) southwest of the Azores in an area typical of the Mid-Atlantic Ridge (Fig. 9A).

The site selected occurs along the divergent boundary between the African and North American plates, which are separating at about 2 centimeters (0.8 inches)/year. This separation is marked by a rift at the ridge crest, which measures 3 to 30 kilometers (2 to 20 miles) across and 1.5 kilometers (1 mile) deep. These dimensions are similar in scale to those of the Grand Canyon.

Scientists observed that the ocean floor at this depth is almost entirely volcanic, comprising basalt which drapes the valley slopes in bulbous flows called *pillow lavas*. Recall from Chapter 4 that these bolster-like forms are characteristic of lava extruded underwater, the freezing of its surface producing pillow-like bulges when it comes into contact with seawater. Because the interior of the pillow is still molten, the pressure eventually bursts open the pillows and stretches them into tubes or bulbous extrusions. But these eruptions and lava flows differ markedly from those on land. Instead of hot molten streams of lava, these pillow basalts displayed small red cracks that first gleamed and then winked out as they chilled against the near-freezing seawater. Unlike continental eruptions, their rapid solidification and the enormous pressures at such great depths prevented the escape of significant quantities of gases.

More dramatic discoveries accompanied a 1979 dive to the Galapagos Rift, northeast of the Galapagos Islands in the eastern Pacific. Although it was known that the mid-ocean ridge system was elevated because the rocks were hot, measurements taken at the dive sites found the rocks to be cooler than expected. Why would this be the case? Geologists at the time speculated that cold ocean water was penetrating the ocean crust through fractures, absorbing heat from the rocks, and returning to release that heat at the seafloor through some form of hot spring. It was not until the Galapagos Rift was explored, however, that such submarine hot springs were discovered. Analysis of the hydrothermal fluid (hydrothermal literally means "hot water") emanating from these hot springs, also called *hydrothermal vents*, indicated that they were rich in chlorine, and confirmed speculations for a submarine source of chlorine in ocean water.

The hot springs were surrounded by extraordinary life forms that make up one of Earth's few complex ecosystems not based on *photosynthesis* (see Fig. 9A). Photosynthesis is the process by which plants convert light energy from the Sun into the chemical energy they need to grow. But no light reaches the floors of the oceans. Instead, the bacteria that live near deep ocean hot springs harness energy liberated when the hydrogen sulfide from the hydrothermal vent combines with oxygen in the seawater (see Fig. 9A). They then use this energy to convert inorganic carbon dioxide in the seawater into organic compounds, a process known as *chemosynthesis*.

The vent bacteria form the base of a food chain involving animals that either live on the bacteria, live in collaboration with them (a relationship known as *symbiosis*), or live on the animals that eat the bacteria. Giant, foot-long albino clams and mussels live in the vent's warmth by filtering and digesting bacteria from the warm seawater, which contains as many as a billion bacteria per liter (see Fig. 9A). Red-plumed tube worms more than 3.6 meters (12 feet) in length also thrive near the vents where they live in symbiosis with the bacteria. The bacteria pack the tube worm's sack-like bodies and use oxygen filtered through the tube worm's plumes to metabolize hydrogen sulfide into food for their hosts. Higher in the food chain lie albino anemones, crabs, and vent fish that eat the clams, mussels, and tube worms.

Subsequent explorations found similar marine communities at many other locations suggesting that these seafloor oases are a common feature of mid-ocean ridges. Since the process is independent of the Sun and occurs in an environment shielded from ultraviolet radiation by ocean water, it may have a significant bearing on the origin of life. Perhaps it was at Earth's first mid-ocean ridges rather than in shallow seas that life first developed from lifeless molecules.

But what is the origin of the hot springs? At these submarine vents, seawater seeps down to hot rocks beneath the ridge and is superheated to temperatures as high as 350°C (660°F). This hot caustic brine becomes saturated with minerals leached from the surrounding basalt. As the superheated water spews back onto the ocean floor, it builds chimneys that form as the

(Continued)

LIVING ON EARTH

The Bizarre World of the Mid-Ocean Ridges (Continued)

minerals dissolved in the vent water crystallize on meeting the near-freezing seawater (see Fig. 9A). The crystalline precipitate forms a sooty soup of particles that belch from the chimneys, called *black smokers* or *white smokers* depending on the mineral mix. The minerals in the soup include iron and manganese oxides, zinc and copper sulfides, and a little silver. On precipitation, they form such commercially important ore minerals as chalcopyrite (copper-iron sulfide) and

sphalerite (zinc sulfide). But how exploitable these ore deposits are so far from land, and who they belong to, is uncertain. However, it is clear that many major ore deposits now on land may have originally formed in this way. As techniques to explore these resources become more advanced, exploitation of the seafloor with its abundant deposits and delicate ecosystems is likely to become one of the major environmental issues of the twenty-first century.

Figure 9A   Submarine Vents

Clockwise from upper left: Location of the Azores and Galapagos Islands; a black smoker; giant tube worms and albino clams and crabs thrive around a hot spring on the East Pacific Rise.

But where does all the oceanic lithosphere created at divergent plate boundaries go? To answer this question we must examine another type of plate boundary, one that is associated with plate collision. We now turn to convergent plate boundaries, where oceanic crust meets it fate.

9.4 SUMMARY

- Divergent plate boundaries are responsible for the rifting of continents, the opening of oceans, and the development of mid-ocean ridges and passive continental margins.

- Divergent plate boundaries are constructive plate margins, producing new ocean crust by seafloor spreading at mid-ocean ridges.

- As a result of seafloor spreading, the crust moves away from mid-ocean ridges so that the seabed cools and subsides to form the flat abyssal plain of the deep ocean.

9.5 Convergent Boundaries: Recycling Crust and Building Continents

Assuming that the size of Earth is constant, the creation of new lithosphere must be balanced by the destruction of old lithosphere. In contrast to the maximum age of continental crust (a little over 4 billion years), oceanic crust is less than 200 million years old (see Fig. 9.14). This contrast suggests that the oceanic lithosphere is being preferentially destroyed. As Figure 9.9 illustrates, the destruction of lithosphere is accomplished at convergent plate boundaries, where two plates overlap. Precisely what happens at such boundaries depends on whether the overlapping plates are continental or oceanic, that is, whether the lithosphere of the converging plates is capped by continental or oceanic crust. Three types of convergent plate boundary are possible: ocean-ocean, ocean-continent, and continent-continent. Although the first two types of convergent plate boundaries both consume oceanic lithosphere, they have quite different characteristics, and we examine each in turn. The third type of convergent plate boundary, that of

continent-continent collision, is summarized here as well, but we will examine it in more detail in Chapter 10. However, to understand how the overlap between two plates is achieved, we must first turn to the process responsible.

SUBDUCTION

Along convergent plate boundaries, two plates come together and one angles down beneath the other in a process known as **subduction** (Fig. 9.16). The inclined zone along which this happens is a **subduction zone**. In general, the denser plate is subducted beneath the lighter, more buoyant one. These two plates are known as the *lower plate* and *upper plate*, respectively.

When two oceanic plates converge (Fig. 9.16a), it is the denser of the two that is subducted. In this case, the outcome is determined by the relative age of the converging plates. Because oceanic lithosphere cools and becomes denser as it ages, ocean-ocean convergence tends to destroy the plate with the older, denser oceanic lithosphere. In this way, old oceanic lithosphere is preferentially recycled and the age of the ocean floor is kept young. This explains why the oldest oceanic lithosphere is less than 200 million years old—all ocean floor older than this has been preferentially destroyed by subduction (see In Depth: **Ophiolites—Clues to the Structure of Oceanic Crust** on page 279). Many examples of ocean-ocean subduction occur today in the western Pacific Ocean, where the oldest oceanic crust occurs (see Fig. 9.14).

When one of the colliding plates is capped by continental crust, the denser oceanic lithosphere angles downward

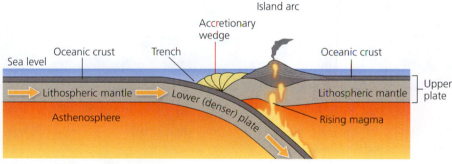

(a) Ocean-ocean convergence

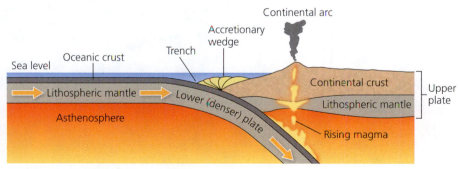

(b) Ocean-continent convergence

Figure 9.16 Subduction

Subduction occurs at convergent plate boundaries and can involve (a) two oceanic plates, or (b) an oceanic plate and a continental plate.

beneath it (Fig. 9.16b). The broadly granitic rocks of the continental crust are less dense and, hence, far more buoyant and more difficult to subduct than the dense basalts of the oceanic crust.

CHECK YOUR UNDERSTANDING

○ What is subduction?

A modern example of this process occurs along the western margin of South America, where the oceanic Nazca plate plunges beneath the continental South American plate. As we shall see in Chapter 10, this process results in the formation of mountains such as the Andes. Because of its contrast in density, oceanic lithosphere created at mid-ocean ridges tends to be destroyed and recycled, whereas buoyant continental crust is preserved.

When both converging plates carry continents, major collision ensues (Fig. 9.17). Such is the case when subduction consumes all of the oceanic lithosphere between two continents, so that eventual collision between them is inevitable. There are consequently two stages in the development of this type of plate boundary, one as the continents converge and the other when they meet. The first stage involves ocean-continent convergence and produces features like those described above (see Fig. 9.16b), such that Andean-style mountains may exist along the edge of one of the continents long before they meet. However, as the two continents continue to converge on each other, the oceanic crust between them is consumed (Fig. 9.17a), culminating in a second stage (Fig. 9.19b), that of continent-continent collision (Fig. 9.17b).

When the two continents collide, the one on the subducting plate is initially dragged down into the subduction zone. But continental lithosphere is too buoyant to be subducted completely and subduction eventually comes to a halt. Recent geophysical images show that the dense subducting slab of oceanic lithosphere may continue to sink and may eventually break off from the continent and fall into the mantle.

Continent-continent collision involves the destruction of an ocean and the eventual termination of plate boundary activity at the site of ocean closure. The continents and the plates to which they are attached become welded together and the mountains that are formed are stranded within a single unified plate. The effects of continental collision are dramatic, as we shall learn in Chapter 10 when we turn our attention to the process of mountain building.

Prior to the terminal development of collisional mountains, however, other major features are characteristic of the subduction process. All subduction zones, for example, are marked by earthquakes, deep ocean trenches, and volcanoes. But whereas some subduction zones create curved chains of volcanic islands, others produce explosive volcanoes along continental margins. We discuss these features and how they form in the following section.

CHARACTERISTICS OF SUBDUCTION ZONES

The downgoing oceanic plate in a subduction zone may penetrate deep into the mantle as a cold, rigid slab. Earthquakes only happen in cold brittle rock, so the earthquakes associated with subduction consequently occur at much greater depths than the earthquakes associated with mid-ocean ridges. (Recall that at a mid-ocean ridge, hot magma sits within a few kilometers of the surface, so the crust is rigid only at relatively shallow depths). In fact, earthquakes within a subducting slab may occur at depths as great as 650 kilometers (400 miles), in contrast to the very shallow depths—less than 5 kilometers (3 miles)—typical of earthquakes at mid-ocean ridges.

Subduction zone earthquakes have a variety of sources. Some subduction zone earthquakes are generated where the cold, and therefore brittle, oceanic plate bends before it subducts. Others occur if the downgoing slab breaks up before it sinks into the mantle. But most are produced by slippage along the subduction zone itself, where the two plates rub together. If these earthquakes displace the overlying seabed, they may also launch devastating seismic

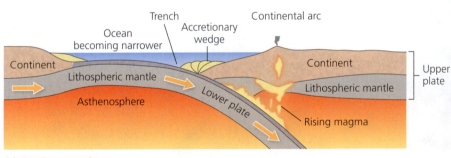

(a) Continent-continent convergence

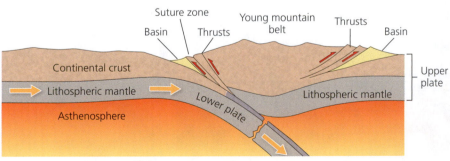

(b) Continent-continent collision

Figure 9.17 Collision

These figures illustrate the convergence and collision of continents. (a) When two continents converge on each other, the oceanic crust between them is first consumed. (b) In the second stage of convergence, continent-continent collision occurs.

sea waves called *tsunami* (see Living on Earth: **Tsunami** on page 272).

Because rocks absorb heat very slowly, it takes a long time for the subducting plate to heat up and become soft as it descends. Until this softening occurs, the downgoing slab maintains its rigidity and is capable of breaking along brittle, earthquake-producing fractures within it, in contrast to the far hotter and more plastic asthenosphere that surrounds it. Subduction zone earthquakes, like those of Japan (Fig. 9.18), are consequently confined to the subducting lithosphere, so their points of origin, or *foci*, become deeper in the direction of subduction. In fact, only in subduction zones can earthquakes originate from depths of more than 100 kilometers (60 miles).

Also recall from Chapter 2 that the site of subduction is marked by a deep ocean trench where the oceanic plate bends before sinking. Ocean trenches are produced by frictional drag between the colliding plates, which causes the subducting plate to pull the edge of the overriding plate down with it. This process creates some of the deepest points on the ocean floor. For example, at more than 11,000 meters (36,000 feet) below sea level, the floor of the Mariana Trench in the western Pacific is the deepest point on Earth's surface. This frictional drag also causes sediment to be scraped off the downgoing slab as it is subducted. Plastered against the overriding plate, this scraped-off sediment forms a triangular-shaped body known as an **accretionary wedge**, a feature we will discuss in more detail in Chapter 10. Subduction is also marked by curved lines of volcanoes known as **volcanic arcs**, which tower above the trench, either as volcanic islands (*island arcs*) or along continental margins (*continental arcs*), and are fueled by ascending magma that forms above the downgoing slab as it plunges into Earth's interior (see Fig. 9.16).

> Why is subduction associated with an inclined zone of earthquakes?

CHECK YOUR UNDERSTANDING

Some 75 percent of these volcanoes lie behind the deep ocean trenches of the Pacific Ocean, where they bear witness to the subduction of the Pacific beneath as many as six different plates (see Fig. 9.11). The volcanoes fueled by this subduction form the 48,000 kilometer (30,000 mile)-long chain, known as the *Pacific Ring of Fire*, that encircles this ocean.

Ocean-Ocean Convergence

Where the upper plate is oceanic, the volcanoes form a curved line of islands that run parallel to the adjacent trench (Fig. 9.19), like those of the Aleutian Islands (see Fig. 9.37) in the northern Pacific. These curved chains of volcanic islands are called **island arcs**. The curvature of island arcs reflects the spherical shape of Earth, and therefore the curvature of the surface of the subducting slab. As the subducting plate plunges into Earth's interior, it forces the trench to adopt a curved shape, just as a dent in a ping-pong ball adopts a circular outline.

Note that these volcanoes are not formed along the plate boundary itself, but instead occur in the upper plate. Experiments show that the subduction zone is too cold to form large volumes of magma. However, as the subducted lithosphere descends, it eventually becomes warm enough to release fluids such as water vapor. These fluids trigger melting in the overlying mantle and the melts so produced rise through the upper plate to fuel volcanoes at Earth's surface.

Where the subducting oceanic lithosphere is very old, the subduction zone is commonly very steep because the

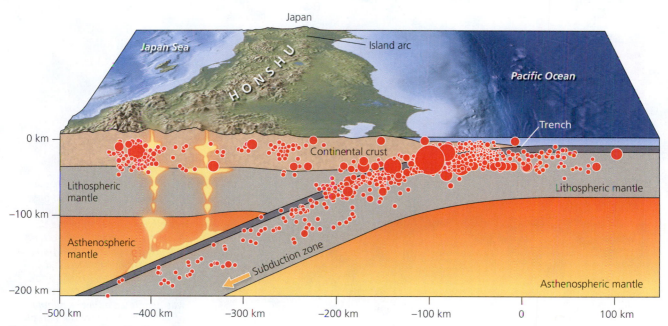

Figure 9.18 Japanese Earthquakes

This cross section of Japanese earthquakes shows the close correspondence of a subducting slab with the inclined zone of earthquake foci (red dots—size proportional to earthquake magnitude).

LIVING ON EARTH

Tsunami

One of the most devastating consequences of subduction is the generation of the massive ocean waves known as **tsunamis** (from the Japanese for "harbor wave"). Large subduction earthquakes, like the magnitude 9.1 earthquake that occurred off the island of Sumatra's northwest coast in December 2004, and the magnitude 9.0 Tōhoku earthquake that struck eastern Japan in March 2011, release vast amounts of energy and cause significant vertical movement of the seabed. When this occurs, a portion of the energy is transferred to the ocean water above the earthquake and generates powerful seismic sea waves capable of crossing an entire ocean basin. When these waves reach the coastline they unleash their energy and cause widespread devastation.

Tsunamis are commonly generated above locked (jammed together) segments of a subduction zone when stresses built up over a long period are suddenly released. If the two plates at a subduction zone lock together while the stress from plate motion continues, the front of the overriding plate is dragged downward, slowly storing

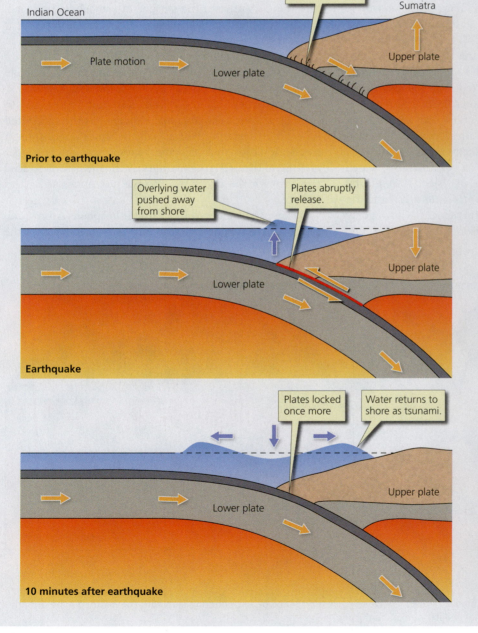

Figure 9B   Generation of Tsunami

Tsunamis are formed above subduction zones as a result of the drag and release of the upper plate (along red line) by the lower one. The overlying ocean water is first pushed away from the shore, only to return as a tsunami. In the case of the 2004 Sumatra earthquake, the tsunami was generated about 10 minutes after the earthquake, slamming into the coast of Sumatra just 15 minutes later.

energy like a stretched spring. Finally, it recoils upward, generating a powerful earthquake that releases the stored energy, lifting the seabed and the ocean water above (Fig. 9B).

So unlike a normal wave, which concentrates its energy at the surface of the sea, a tsunami carries its energy from the ocean bottom (where the displacement occurs) to the surface. Ships at sea scarcely notice a tsunami passing, for in the deep ocean a tsunami raises only a broad, gentle swell no more than a few feet high. But the shallower seafloor close to land compresses the wave and focuses its power into a series of major waves that can reach heights of up to 40 meters (130 feet).

Following the Sumatra earthquake, a killer tsunami sped across the Indian Ocean like a ripple across a pond but at speeds of about 800 kilometers (500 miles) per hour (Fig. 9C). Slamming first into Sumatra (Fig. 9D) and then Thailand, Sri Lanka, and India just two hours later, the waves killed over 230,000 people.

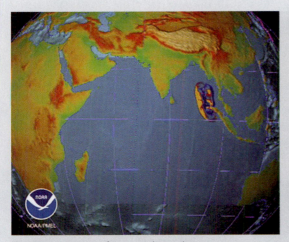

A 30 minutes after earthquake

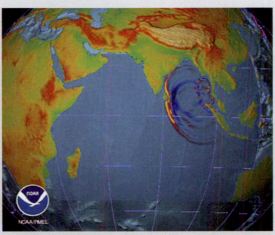

B 2 hours after earthquake

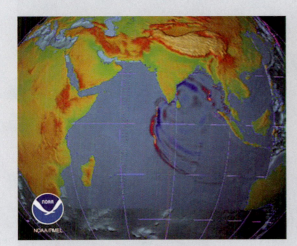

C 3.5 hours after earthquake

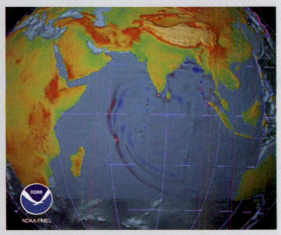

D 5.5 hours after earthquake

Figure 9C *Tsunami following Sumatra Earthquake*

This is a computer simulation of the tsunami generated by the 2004 Sumatra earthquake: (a) 30 minutes after earthquake, (b) after 2 hours, (c) after 3.5 hours, (d) after 5.5 hours. The tsunami crossed the entire Indian Ocean to reach South Africa 11 hours later.

(Continued)

LIVING ON EARTH

Tsunami (Continued)

Only seven years later, another massive subduction earthquake off the coast of Japan, the strongest ever recorded in that country, launched a devastating tsunami that slammed into Japan's east coast half an hour later (Fig. 9E). Locally reaching a height of 40 meters (130 feet), the waves obliterated coastal towns, caused meltdowns and radiation release at the Fukushima Daiichi nuclear power plant, and claimed the lives of almost 20,000 people.

Crossing the entire Pacific Ocean, the tsunami reached California and Oregon as a surge up to 2.4 meters (8 feet) high some ten hours later, damaging docks and harbors and causing over $10 million in damage. When it eventually reached the ice shelf of Antarctica, almost 18 hours after the earthquake, the tsunami broke off an iceberg the size of Manhattan Island.

Because it is almost entirely surrounded by subduction zones, it is the Pacific Ocean that is most at risk for tsunamis, and here an oceanwide tsunami warning system exists. Based in Hawaii, this detects tsunami-generating earthquakes and predicts the time of the tsunami's arrival. In this way, those areas at risk can be alerted before the tsunami waves arrive and the potential loss of life is greatly reduced. In 2011, this was the case for the Japanese tsunami for those living on the Pacific rim, but in Japan itself there was little time to warn the affected population because of its proximity to the earthquake site. In the Indian Ocean, where tsunamis are less frequent, no such early warning system exists. As a result, the death toll of the 2004 Sumatra tsunami was far higher.

Figure 9D   Tsunami Damage in Sumatra

A coastal village in Sumatra lies obliterated by the tsunami launched by the 2004 Sumatra earthquake. Traveling inland, the wave reached a height of 30 meters (100 feet).

Figure 9E   Tsunami Comes Ashore

This aerial view shows a tsunami wave coming ashore at Iwanuma on the east coast of Japan on March 11, 2011. The 4-meter (13-foot) tsunami swept boats, cars, buildings, and tons of debris miles inland.

downgoing slab is denser than the underlying asthenosphere. As a result, the slab sinks into the asthenosphere, rolling back oceanward as it does so (Fig. 9.20). You can mimic this "roll back" by holding a long horizontal sheet of thin cardboard at one end. Because cardboard is denser than air, its unsupported free end falls downward and the sheet rolls back toward you. The action of this roll back may tug the upper plate behind the arc, causing its oceanic crust to fracture. As molten rock in the mantle rises to fill the opening fracture, a small spreading zone is created behind the island arc, like a miniature version of a mid-ocean ridge (see Fig. 9.19). This process is **back-arc spreading**, and the

basin it opens behind the arc is a **back-arc basin** floored by oceanic crust. Much of the western Pacific Ocean is underlain by old, dense oceanic lithosphere (see Fig 9.14). As a result, this region has been particularly affected by subduction zone roll-back and back-arc spreading. This explains the presence of seas between the islands of the western Pacific and the coastline of Asia (see Fig. 9.18). The floors of the Sea of Japan and much of the Philippine Sea, for example, were formed by back-arc spreading.

As we saw in Figure 9.19, the volcanoes that make up an island arc are the product of subduction and occur on the upper plate. As the downgoing oceanic slab angles into

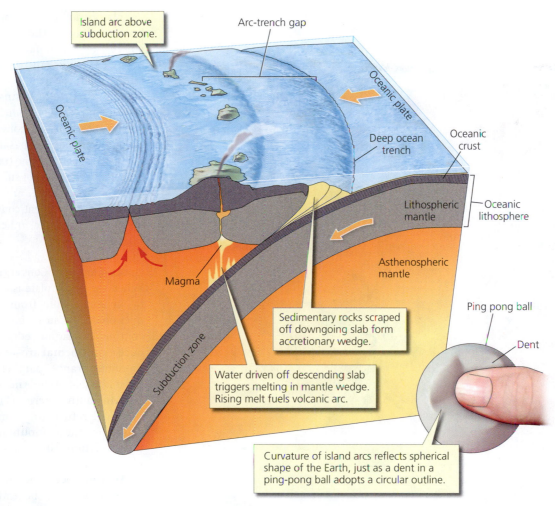

Island arc above subduction zone.

Arc-trench gap

Oceanic plate

Oceanic plate

Deep ocean trench

Oceanic crust

Lithospheric mantle

Oceanic lithosphere

Asthenospheric mantle

Magma

Ping pong ball

Dent

Sedimentary rocks scraped off downgoing slab form accretionary wedge.

Water driven off descending slab triggers melting in mantle wedge. Rising melt fuels volcanic arc.

Subduction zone

Curvature of island arcs reflects spherical shape of the Earth, just as a dent in a ping-pong ball adopts a circular outline.

Figure 9.19 Ocean-Ocean Convergence

Subduction beneath oceanic crust produces a curved island arc of predominantly basaltic volcanoes. The curvature of island arcs and the trench reflect the spherical shape of Earth just as a dent in a Ping-Pong ball adopts a circular outline. The volcanoes are not formed along the plate boundary; they occur in the upper plate.

Earth's interior it is heated and, when the temperatures become high enough, the subducted oceanic crust releases water and other gases formerly trapped in minerals and fractures. Experiments show that this occurs when the subducted lithosphere reaches a depth of about 100 kilometers (60 miles). The released fluids rise and invade the wedge-shaped area of mantle (or **mantle wedge**) above the subduction zone where they promote melting of the mantle rocks. They do so because the invading water molecules tug and weaken the bonds in the mantle minerals, thereby lowering the temperature required for melting. As a result, the rocks in the overlying mantle wedge begin to melt. The resulting magmas, like all melts from the mantle

(as we learned in Chapter 4), are basaltic in composition. Rising through the overlying oceanic plate, the magmas may reach the surface where eruptions build an arc-shaped line of largely basaltic volcanoes parallel to the trench.

The gap between the trench and the arcuate line of volcanic islands (the "arc-trench gap"; see Fig. 9.19) reflects the angle of subduction. Magmas cannot be produced until the downgoing slab becomes hot enough to release its trapped fluids, which requires the slab to reach depths of 100 kilometers (60 miles) or more. The volcanoes, which form above the region at which melting in the mantle first occurs, consequently lie behind the trench, and so (unlike earthquakes) are geographically displaced from the plate boundary.

Because steeply dipping subduction zones attain this depth closer to the trench than shallow-dipping ones, the arc produced is also closer to the trench, resulting in a

CHECK YOUR UNDERSTANDING

● Why are deep ocean trenches the sites of shallow earthquake activity but not volcanic activity?

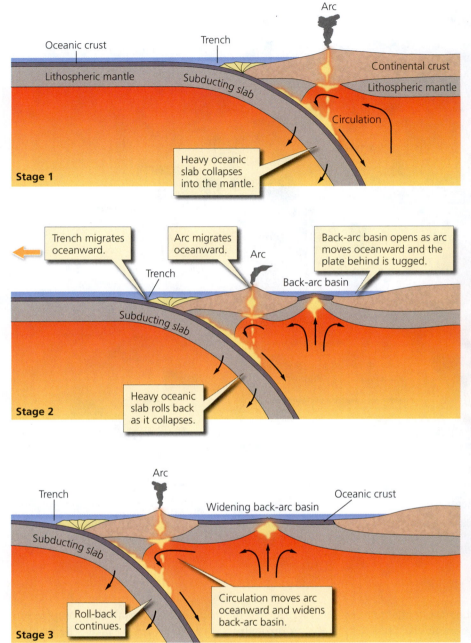

Figure 9.20   Slab Roll-Back

These figures illustrate the roll-back of a subduction zone as a subducting slab progressively collapses into the asthenosphere. The progression is shown in three stages.

arc always defines the upper plate because it is in the upper plate where arc magmatism always occurs. Deep ocean trenches, inclined zones of earthquakes, island arcs, and back-arc basins are all features typical of subduction involving ocean-ocean convergence. Volcanism is largely basaltic, and the distance between the trench and arc reflects the angle of subduction. Some of these features are also typical of ocean-continent convergence but, as we shall see, others differ markedly.

Ocean-Continent Convergence

When the upper plate is continental, magma rising from the subduction zone causes uplift of the continent's leading edge to produce a **continental arc**—a volcanic mountain range parallel to the coast like that of the Andes Mountains in South America (Fig. 9.21). (We will return to examine the origin of these mountains and their relationship to subduction in Chapter 10.)

As with ocean-ocean convergence, subduction beneath continents produces a deep ocean trench and an inclined zone of earthquakes. Roll-back and the development of back-arc basins may also accompany ocean-continent convergence, in which case a portion of the continental margin and its volcanic mountains may break away and move oceanward as a back-arc basin floored by oceanic crust opens behind it. Such is the case for the volcanic islands of Japan, which were once attached to Asia, but were separated with the opening of the Japan Sea (see Fig. 2.16).

Unlike ocean-ocean convergence, subduction beneath continents produces a volcanic arc amid mountains and results in volcanic eruptions that are often highly explosive. Such continental margins are therefore referred to as **active continental margins** since they coincide with a plate boundary and so are tectonically active, unlike the passive continental margins produced by ocean opening. Ocean-continent convergence along active continental margins shrinks the size of oceans, as dense basaltic ocean floor is subducted beneath more buoyant continental lithosphere.

comparatively narrow "arc-trench gap." In either case, however, the relative position of the arc and trench can be used to determine the direction or *polarity* of subduction, because the subduction zone always angles down from the trench toward and beneath the neighboring volcanic arc. Likewise, the position of the

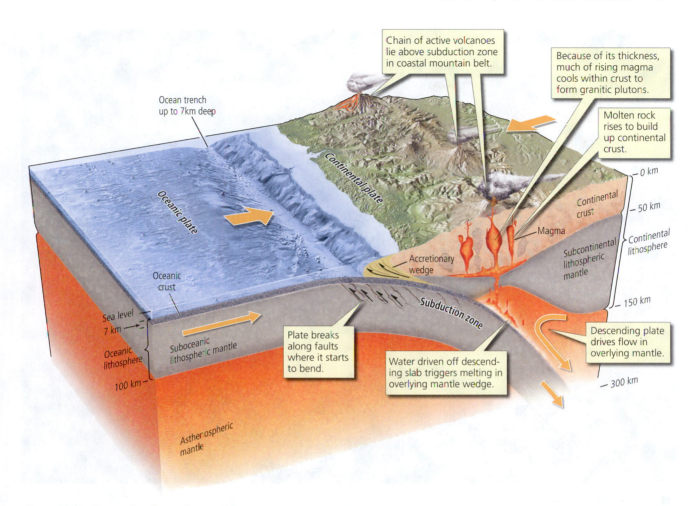

Figure 9.21   Ocean-Continent Convergence

Subduction beneath continental crust produces a continental arc comprising coastal mountain belts, such as the Andes, with andesitic or rhyolitic volcanoes.

As is the case for convergence between oceanic plates, the magmas produced above these subduction zones are mainly basaltic, and are formed by partial melting in the overlying mantle wedge where melting is triggered by the addition of water and other fluids from the subducted oceanic crust. Where the upper plate is continental, however, these fluid-charged mafic magmas must push their way through the overlying continental crust and often pond beneath it before doing so. The ascending fluid-charged magma is an efficient transporter of heat and promotes partial melting of the overlying continental crust to produce granitic melts. These coexisting but compositionally contrasting magmas may then mix.

Because of the thickness of continental crust, much of this rising granitic magma cools deep within the crust to form granitic plutons. But some reaches the surface where it becomes the fuel for active volcanoes. Unlike basalt magmas, which are very fluid and permit their dissolved gases to escape easily, continental crustal melts are more viscous because they are richer in silica, which tends to make melts sticky. The melts are generally intermediate to felsic (andesitic or rhyolitic) in composition (see Fig. 4.13), and are rich in liquefied gases under tremendous pressure, giving vent to explosive volcanic eruptions.

Recall from Chapter 4 that magma chambers can be like pressure cookers with lids of brittle crust. As the pressure within the chamber builds, cracks appear in the crustal lid, which are relentlessly exploited by the overpressured magma. As pods of melt rise to fill near-surface magma chambers, the gases trapped within them expand. This expansion fractures the overlying rock, enabling the magma to rise still farther. This, in turn, allows the magma to surge upward, culminating in violent explosions at the surface in which lava, gases, and solid particles are jettisoned out of the vent in the form of a pyroclastic eruption.

The devastating pyroclastic eruption of Mt. Pinatubo in the Philippines in June 1991 (Fig. 9.22), exemplifies the often explosive volcanism associated with ocean-continent convergence. This eruption, the largest in over 80 years, blasted 5 cubic kilometers (1.2 cubic miles) of ash and some 20 million tons of sulfur dioxide into the atmosphere in a

Figure 9.22   Mt. Pinatubo
This photo of the eruption of Mt. Pinatubo on June 15, 1991, was taken from Clarke Air Force Base in the Philippines.

matter of hours. By blocking out sunlight, this blanket of gas and dust cooled the planet by a degree or so for several years (see Fig. 1.19).

Subduction zone volcanism stands in sharp contrast to the generally passive or *quiescent* volcanism of mid-ocean ridges. Mid-ocean ridges do not form such serious "pressure cookers," because they lack a thick crustal lid and the basalt magmas they produce are both less viscous and less gas-charged than the magmas of subduction zones. Mid-ocean ridges are also highly fractured areas that facilitate the ascent of magma rather than allowing the buildup of pressure. Mid-ocean ridge volcanism is exemplified on land by the lava fountains and fast-flowing lava flows witnessed on the island of Iceland, which sits astride the Mid-Atlantic Ridge.

Continental arcs rather than island arcs are therefore typical of ocean-continent convergence. At the same time, volcanism tends to be explosive and is generally of a more silica-rich andesitic or rhyolitic composition rather than basaltic. Each of these characteristics reflects the presence of a continental plate above the subduction zone.

SUBDUCTION AND THE GROWTH OF CONTINENTS

Subduction of dense oceanic lithosphere is nature's vast recycling program. As dense oceanic lithosphere is destroyed by subduction, its destruction triggers melting and the ascent of more buoyant magmas which cool to form new crust in island arcs and the volcanic mountain belts of continental arcs. This new crust is more felsic, and hence more buoyant, than that of the ocean floors. Because of this buoyancy and the elevation of their crust, island arcs, once formed, generally resist being subducted. Instead, they tend, over time, to coalesce into larger composite bodies through collisions among themselves, or become welded to the leading edge of larger continents if they are swept into ocean-continent collision zones. In this way, they add to the continents so that continental crust can ultimately be viewed as a product of subduction and the destruction of oceanic lithosphere. Because ocean-continent subduction zones where these processes take place can form only at continental margins, subduction-zone magmas and colliding island arcs are typically added to the edge of continents. As a result, continents grow sideways with time and

IN DEPTH

Ophiolites—Clues to the Structure of Oceanic Crust

Most of the ocean floor is hidden from direct observation by a cover of ocean water. Rocks of the oceanic crust are also quite rare in the geologic record because of the efficient subduction of oceanic lithosphere by plate tectonics. In rare instances, however, sections of oceanic lithosphere may escape this process. This occurs most commonly near subduction zones. Fragments of ocean floor preserved in this way are called **ophiolites**, from the Greek *ophis* meaning "serpent." The name reflects the blotchy green, snakelike appearance of some of the rocks that ophiolites contain.

Some ophiolites, such as the Troodos Complex of Cyprus, preserve a complete cross section of seafloor that demonstrates how the oceanic lithosphere forms (Fig. 9F). At the top of the sequence are deep-ocean sediments laid down in the ancient sea. Directly beneath these sediments are bulbous *pillow lavas*, which formed when basalt magma was extruded onto the seabed and chilled against seawater (see Chapter 4). Beneath the pillow lavas is a layer dominated by dikes that represent mafic magma that solidified in conduits leading from the magma chamber to the surface. The repeated opening of fractures by seafloor spreading creates dike after dike, producing a complex of *sheeted dikes*. Below these lie coarse-grained mafic rocks called *gabbros*, which formed from the slow crystallization of magma that once filled the magma chamber beneath a spreading ridge. At the base of the sequence are dark, heavy, magnesium- and iron-rich ultramafic rocks called *peridotites*. Some of these peridotites are thought to have formed on the floor of the magma chamber that fed the overlying dikes and lavas. Here heavy crystals settling out of the magma chamber as it cooled would have slowly accumulated to produce concentrated layers of early crystallizing magnesium- and iron-rich minerals (see Chapter 4). Other peridotites, however, are upper mantle rocks left behind after partial melting had extracted the basaltic magma to form the overlying oceanic crust.

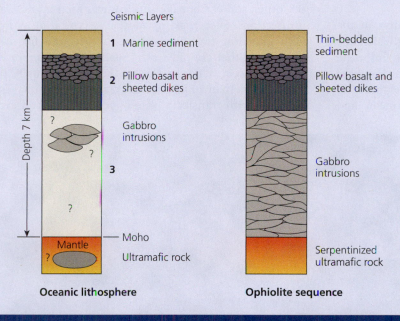

Figure 9F   Ophiolite Sequence

Compare the structure of oceanic crust determined from seismic studies and drilling, shown on the left, to the typical ophiolite sequence, like that of the Troodos Complex in Cyprus, found in mountain belts on land, shown on the right. Note that thicknesses are approximate because the sequence is usually faulted.

so often have a nucleus of ancient rocks surrounded by progressively younger ones. We will return to the dramatic story of collision and its role in the growth of continents when we examine mountain building in more detail in Chapter 10.

The material we have covered in Section 9.5 serves to emphasize the wide range of processes associated with the subduction of oceanic lithosphere at convergent plate boundaries. Subduction may occur beneath oceanic crust, in which case the product is an island arc, or it may be consumed below continental crust, in which case a continental arc of volcanic mountains is formed. Subduction causes magmatism, which may be largely plutonic or mostly volcanic, and fuels volcanoes that may be basaltic and quiescent, or intermediate to felsic and highly explosive. Subduction also leads to collision and, as we shall learn in Chapter 10, this may affect quite small regions, or may be climactic and involve entire continents. No other type of plate boundary exhibits such diversity and, as we shall see in Section 9.6, the complexity of convergent margins contrasts with the comparative simplicity of the third and last type of plate boundary—that of transform faults.

9.5 SUMMARY

- At convergent plate boundaries the denser plate angles down, or is subducted, beneath the more buoyant one. This boundary is characterized by earthquake-prone subduction zones, above which magmas form.

- Ocean-ocean convergence produces deep ocean trenches and island arcs such as the Aleutians.

- Ocean-continent convergence produces continental arcs such as the Andes that are home to explosive volcanoes.

9.6 Transform Boundaries: Fracturing the Crust

Transform plate boundaries are major fractures in the lithosphere, known as **transform faults**, along which one plate slides by another. Most transform faults link plate boundaries on the ocean floor so that the earthquakes produced by their jarring motion pose little threat to human populations. But some intersect continents and generate earthquakes that pose a serious threat to society. Like wounds that never properly heal, the fractures themselves remain zones of weakness for hundreds of millions of years after motion on the plate boundary has ceased.

Plate motion along transform faults is horizontal. The plate boundary is a "conservative" one, the transform fault simply being the link from one active plate boundary to another. Most transform faults link two segments of a mid-ocean ridge (see Fig. 9.11). But they may also be the link between an ocean ridge and a subduction zone, or the link between two subduction zones. So at either end of a transform fault, the movement on the fault abruptly ends and gives way to (or is "transformed" into) movement of another kind, such as spreading, subduction, or transform motion on another fault (Fig. 9.23).

The transform fault itself is rarely smooth. If the rocks along a transform fault lock while the stress from plate motion continues, strain builds up on either side of the fault so that the rocks slowly bend or deform, storing up

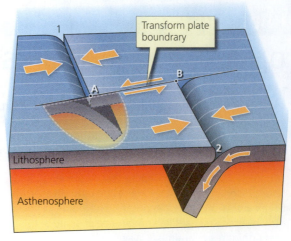

Figure 9.23   Transform Boundary

This transform plate boundary (A-B) links two subduction zones (1, 2). Note that motion on the transform fault abruptly ends at either end of the fault, where it gives way to, or is "transformed" into, convergent plate motion. In this example, the transform fault transfers the convergent motion from one subduction zone to the other. The hole cut into the lithosphere is for viewing purposes.

elastic energy like a spring. Finally, when the "spring" snaps, rocks on either side of the fault jerk violently past each other and earthquake shock waves are sent out in all directions. Such *stick-slip* motion is typical of many faults, including transform faults, and accounts for the earthquakes that are often associated with faults. Since the fault movement occurs close to the surface, usually at depths of less than 20 kilometers (12 miles), earthquakes associated with transform plate boundaries on land are

Figure 9.24   Haitian Earthquake

The Haitian National Palace in Port-au-Prince, Haiti (the second floor of which completely collapsed), was heavily damaged by the devastating earthquake of January 12, 2010. The magnitude 7 earthquake was generated by motion along the northern transform boundary of the Caribbean plate at a depth of just 13 kilometers (8 miles).

often more damaging than deeper ones of similar magnitude. Such was the case for the destructive earthquake in Haiti in January 2010, which devastated the country's capital city of Port-au-Prince (Fig. 9.24). Caused by abrupt movement along the transform fault that forms the northern boundary of the Caribbean plate (see Fig. 2.22), the focus of this magnitude 7 earthquake lay at a depth of only 13 kilometers (8 miles). As a result, the city felt the full force of the earthquake, which proved to be among the deadliest in history—one estimate placing the number of fatalities at 316,000.

Transform boundaries are of two principal types: *Oceanic transforms*, which link offset mid-ocean ridges and *continental transforms*, which separate two continental plates. In this section, we discuss each type in turn and then discuss some famous examples of each.

OCEANIC TRANSFORMS

Transform faults are most common in oceanic settings where they link two segments of a mid-ocean ridge that are offset from each other. They are perpendicular to the ridge crests, producing a rectilinear pattern of ridges and transforms that is characteristic of spreading centers (Fig. 9.25).

At first glance, it looks like the mid-ocean ridges were once continuous and were then offset by horizontal movements along the transform faults. But the direction of movement along oceanic transforms is precisely the opposite of that which seems to produce the ridge offsets. In Figure 9.26, for example, the ridge is offset to the left. However, the direction of movement is imposed by seafloor spreading at the ridge crests. Because of seafloor spreading, the crust to the

right of each ridge segment must move to the right. Only between the offset ridge crests is the seafloor moving in opposite directions, so only this segment is a transform fault separating different plates. Likewise, it is only this segment that is seismically active. Beyond the offset ridge crests, the seafloor moves in the same direction on either side, so the fracture is merely a seismically inactive scar and the lithosphere on both sides belongs to the same plate. Earthquakes on oceanic transforms are generally both frequent and shallow. However, they do not generate tsunami because the motion on a transform fault is largely horizontal, so that the seabed is not lifted and seawater is not displaced upward as it is with the tsunami-launching earthquakes of subduction zones.

It is thought that transform offsets in mid-ocean ridges result when two oceanic plates begin to diverge along a curved boundary (Fig. 9.27a). At that point, the original curves are forced to readjust into a series of right-angle segments. Offsets may also be inherited from faults that were active during the initial continental rifting stage of ocean creation (Fig. 9.27b) where preexisting lines of weakness in the continent produced rifts at an angle to the direction of spreading. As the continents begin to separate, these lines of weakness evolve into transform faults that offset the developing ridge segments. The geometry of the new ocean may be profoundly influenced by preexisting weaknesses in the flanking continents.

> **CHECK YOUR UNDERSTANDING**
>
> ● What is thought to cause transform offsets in mid-ocean ridges?

Other oceanic transform faults link mid-ocean ridges to subduction zones or link one subduction zone to another (see Figs. 9.9 and 9.23). In so doing, transform faults enable movement of oceanic lithosphere from the site of its creation to that of its destruction. In the northeastern Pacific, for example, the Mendocino transform fault permits oceanic crust of the Juan de Fuca plate to be transported to the Cascade subduction zone beneath the Pacific Northwest, whereas the Queen Charlotte transform fault off the coast of British Columbia transports ocean floor from the Juan de Fuca Ridge toward the Aleutian subduction zone beneath Alaska (Fig. 9.28).

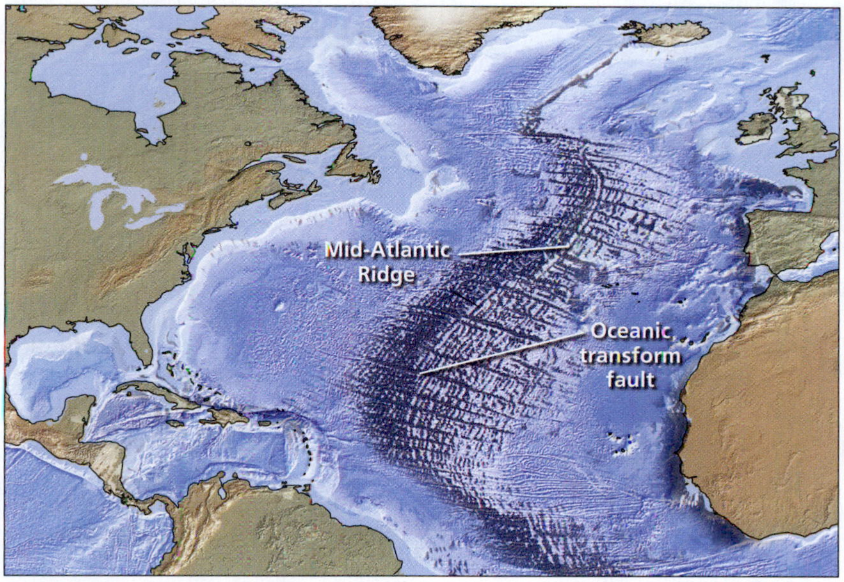

CONTINENTAL TRANSFORMS

Only a few transform faults intersect continents. Those that do are usually longer and more continuous than their oceanic counterparts. They also lack the simple rectilinear geometry of oceanic transforms. As they cut through the various materials of the continental crust, continental transform faults continually exploit any

Figure 9.25 Oceanic Transforms

The submarine topography of the central and northern Atlantic Ocean shows oceanic transform faults offsetting the Mid-Atlantic Ridge at more or less right angles.

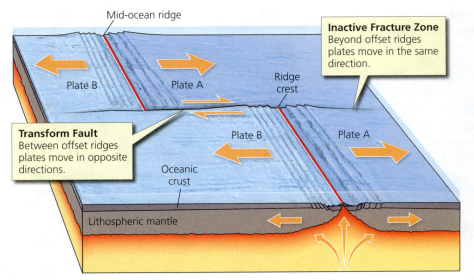

Figure 9.26 Offset Ridge
Look carefully at the arrows in this figure of a transform fault across a mid-ocean ridge. Between the two ridge crests, the crust is moving in opposite directions and the fault is active. But along the other segments, the crust is moving in the same direction and the fault is inactive.

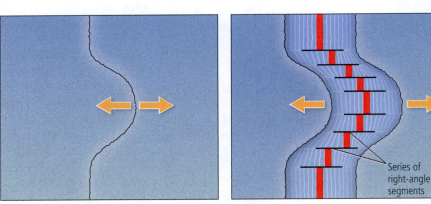

(a) Curved plate boundary

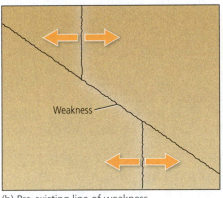

(b) Pre-existing line of weakness

Figure 9.27 Origin of Ridge Offsets
Transform offsets in mid-ocean ridges occur when plates diverge. They can result from: (a) a curved plate boundary between oceanic plates; or (b) a preexisting line of weakness between continental plates.

weakness they encounter. As they do so, they may bend and alter their path. These bends along the fault zone cause small basins to open and produce localized areas of uplift.

Continental transform faults serve as links between other plate boundaries. The resulting stick-slip motion threatens the surrounding region with earthquakes. When continental transform faults bend, the rocks on either side are either extended or compressed depending on the direction of the bend and the direction of movement of the fault (Fig. 9.29). As a simple demonstration of this, cut a piece of paper in two along a line with a kink in it like a Z, and then slide the two pieces of paper along the cut line. Depending on which way the paper is moved, the kink generates either a gap between the pieces or a zone of local compression where the edges stick and the paper folds. These two reactions mirror *releasing* and *restraining bend*s along a transform fault.

At *releasing* (separating) *bends* in a transform fault, the crust stretches until it breaks and collapses to create a long, narrow depression, called a **pull-apart basin**, which may collect sediment or become filled with water. The basin and its contents may superficially resemble that of a young continental rift. However, pull-apart basins are restricted to releasing bends whereas rifts are continuous linear troughs. Examples of pull-apart basins include the Dead Sea on the Dead Sea Transform between Jordan and the West Bank (Fig. 9.30), and the Salton Sea on the San Andreas Fault of southern California.

In contrast, at *restraining bends* where the transform fault curves in the other direction, plate motion jams the blocks of crust together creating folded mountains, called **transverse ridges**, across the line of the fault. California's Transverse Ranges, north of Los Angeles, occur at a restraining bend in the San Andreas Fault. In fact, it was movement on a buried or "blind" fault within this bend that was responsible for the devastating Northridge earthquake in January 1994, which

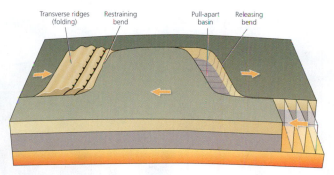

Figure 9.29   Origin of Pull-Apart Basins and Transverse Ridges

The model shows how local zones of compression and extension may occur along a transform fault leading to the development of a pull-apart basin and transverse ridges at bends the transform fault.

claimed 61 lives and resulted in damage estimated at 20 billion dollars.

Both pull-apart basins and transverse ridges vary in size but are often heavily faulted areas with minor volcanism and hot springs. As the San Andreas Fault system illustrates, both may develop simultaneously on the same continental transform depending on the direction of the bends in the fault trace.

> **CHECK YOUR UNDERSTANDING**
>
> ⦿ What happens when a continental transform bends?

The best known continental transform faults are the two we just introduced: the Dead Sea Transform, where the two plates (African and Arabian) move in the same direction but at differing speeds, and the San Andreas Fault, where the adjacent plates (North American and Pacific) move in opposite directions. We discuss these two faults in more detail in the following material.

The Dead Sea Transform

The Dead Sea Transform (see Fig. 9.30) is an extension of Africa's Great Rift system linking the spreading ridge in the Red Sea to part of the convergent zone of the Taurus-Zagros Mountains in Turkey (see Fig. 9.13). As the Arabian plate split away from the slower moving African plate, separating in a direction perpendicular to the Red Sea, horizontal motion along the Dead Sea Transform caused one side of the Jordan Valley to slip sideways past the other. Both sides are moving north, but the Arabian plate moves faster, causing displacement along this continental transform fault. Like two athletes running in the same direction but at different speeds, their relative motion creates an offset. The Dead Sea, at 430 meters (1400 feet) below sea level, is the lowest landlocked body of water on Earth. It formed in the pull-apart basin between two overlapping strands of the transform fault separating the two plates. Relative motion between the Arabian and African plates

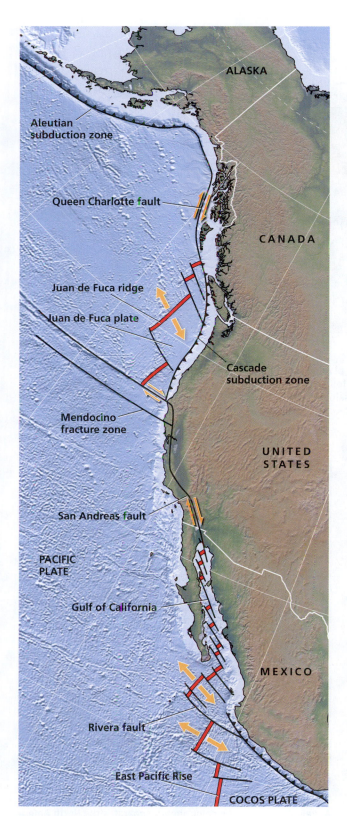

Figure 9.28   Western North America

This map of plate boundaries along the west coast of North America shows the Queen Charlotte, Mendocino, and San Andreas transform faults, the Juan de Fuca and Gulf of California ridges, and the trenches of the Aleutian and Cascade subduction zones.

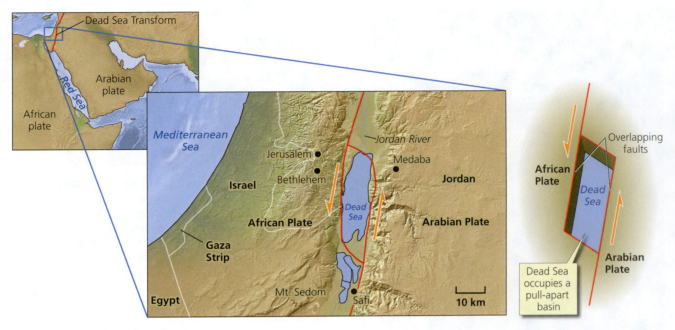

Figure 9.30   Dead Sea Transform

The Dead Sea Transform forms the eastern margin of the Arabian plate, which is separating from Africa with the opening of the Red Sea. The transform separates the northward moving Arabian plate (right) from the slower moving African plate (left). Overlapping faults between Jordan and the West Bank outline a crustal depression in which the Dead Sea resides.

began 15 million years ago and the total offset along the Dead Sea Transform has now reached almost 110 kilometers (70 miles).

The San Andreas Fault

The San Andreas Fault system is a family of faults that links the spreading center in the Gulf of California to the Cascade subduction zone (see Fig. 9.28). In California, the fault forms a 1600-kilometer (1000 mile)-long scar most clearly visible in the Carrizo Plain (Fig. 9.31), northwest of Los Angeles, where offset streams testify to active transform movement. Almost half of the over 20-meter (65-foot) offset on these streams occurred during the 1857 Fort Tejon earthquake (Fig. 9.32). In April 1906, up to 6.5 meters (21 feet) of abrupt movement on the San Francisco segment of the San Andreas Fault produced one of the worst earthquakes in U.S. history. At magnitude 8.3 on the Richter scale, the San Francisco earthquake rippled the ground in waves 1 meter (3 feet) high and 20 meters (65 feet) from crest to crest, tilting and destroying buildings at least 30 kilometers (20 miles) from the fault. However, most of the damage, and most of the estimated 3000 lives that were lost, occurred as a result of fires ignited by overturned kerosene lamps and wood stoves. Within a few days, the city of San Francisco sustained over half a billion dollars in damage, and almost 3000 acres of the city center lay in ruins.

The San Andreas Fault separates the North American and Pacific plates and, on average, moves at about 3.5 centimeters (1.5 inches) per year. Volcanic rocks cut by the fault at Pinnacles National Monument, southwest of Hollister (see Fig. 9.32), are found on the other side of the fault near

Figure 9.31   San Andreas Fault

The San Andreas Fault on the Carizzo Plain, northwest of Los Angeles (shown in the color photo), separates the North American (left) and Pacific (right) plates. The black and white photo shows destruction in the downtown area of San Francisco following the 1906 earthquake and fire.

Fort Tejon, northwest of Los Angeles. This offset of once uninterrupted volcanic rocks shows that 315 kilometers (195 miles) of movement has occurred on the fault in the 23 million years since the volcanic rocks erupted. In another

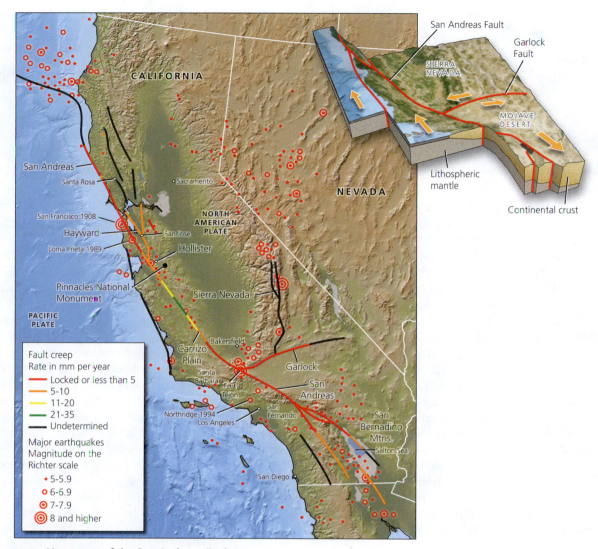

Figure 9.32   Movement of the San Andreas Fault

This map of the San Andreas Fault of California shows rates of fault creep and the distribution of major (magnitude greater than 5) earthquake epicenters.

30 million years, Los Angeles on the Pacific plate will have moved north of San Francisco on the North American plate.

However, segments on the San Andreas Fault are not moving at uniform rates. Some segments are not moving at all because they are presently locked, whereas other segments are moving slowly but continuously—a process known as *fault creep*. Those with rates of fault creep of over 2 centimeters (0.8 inches) per year are unlikely to experience severe earthquakes because continuous movement ensures that the stresses are relieved as fast as they build up. However, other segments, like the one that runs through San Francisco, are locked and therefore are building up strain that must one day be released in a major earthquake.

Because locked segments on the fault stay jammed, on average, 125 years before they snap forward, the segment in greatest danger is that north and east of Los Angeles, where such a jolt is now overdue since the Fort Tejon earthquake occurred well over 125 years ago. Figure 9.32 also shows that, while many earthquake epicenters lie along the fault itself,

numerous earthquakes occur on the California-Nevada border. This indicates that not all the plate motion is absorbed by the San Andreas Fault system and that the plate boundary is really a broad fragmented zone rather than a single major fault.

Ancient Inactive Transform Faults

Although both the Dead Sea and San Andreas transform faults are active plate boundaries, ancient inactive transform faults still scar mountain belts that developed along ancient plate boundaries. For example, Loch Ness on Scotland's Great Glen Fault is home not only to the fabled Loch Ness Monster but it also straddles part of an ancient transform fault that, in vivid testament to Wegener's theory of continental drift, was torn apart with the opening of the

> **CHECK YOUR UNDERSTANDING**
>
> ● Why are moving segments of a fault less likely to experience earthquakes than locked segments?

310 million years ago   Today

Figure 9.33   Great Glen and Cabot Faults

The Great Glen Fault of Scotland and the Cabot Fault of Newfoundland represent an ancient transform plate boundary that was active 310 million years ago but was severed when the Atlantic Ocean opened 180 million years ago.

Atlantic Ocean (Fig. 9.33). Some 310 million years ago, when Europe and North America were part of the same continent, a San Andreas-like transform fault cut across the future islands of Newfoundland, Ireland, and Great Britain. When the Atlantic Ocean opened about 200 million years ago, the old fault broke apart so that the continuation of the Great Glen Fault now lies more than 4800 kilometers (3000 miles) away in the Cabot Fault of Newfoundland.

These ancient scars never quite heal and remain as fundamental zones of weakness in the continental crust. In the future, when the Atlantic Ocean starts to close and subduction begins along its margins, these wounds will be among the first to reopen. In fact, modern stresses along the eastern seaboard of North America, although far smaller than those of active plate margins, generate hundreds of microseismic events each year, many of which are sited on the ancient transform faults of the Appalachian Mountains.

9.6 SUMMARY

- Transform plate boundaries are marked by major crustal fractures that occur either on the ocean floor, where they affect mid-ocean ridges, or between two continental plates, where they create earthquake-prone faults such as the San Andreas.

- At releasing bends in continental transform faults, pull-apart basins such as the Dead Sea are produced as the rocks on either side of the fault are extended. Conversely, at restraining bends, transverse ridges like those of California's Transverse Ranges are formed as the rocks on either side are compressed.

- Ancient inactive transform faults remain as zones of weakness in the continental crust.

9.7 Hotspots: Tracking Plate Movements

Although volcanoes are characteristically associated with plate boundaries, active volcanism also occurs in certain isolated areas that lie far from any plate boundary. Termed "hotspots," these puzzling features are not readily accounted for by plate tectonic theory, yet they are thought to play a fundamental role in the breakup of continents. Their existence has even been used to document plate motions. But how can such features be both related and unrelated to plate tectonics?

Although the existence of hotspots was first proposed by the eminent Canadian geophysicist J. Tuzo Wilson more than 50 years ago, their origin is still not fully understood. Recall from Chapter 2, that, as their name suggests, **hotspots** are small, isolated areas of higher than average heat associated with volcanoes. Hot material rising through the mantle from deep within Earth's interior impinges on the base of the lithosphere and gives rise to localized areas of volcanism—hotspots—on Earth's surface.

Although some hotspots lie on plate boundaries, most occur in the interior region of plates. Therefore, unlike most volcanic activity, hotspots cannot be directly tied to processes occurring at plate boundaries. Instead, they are widely attributed to giant plumes of heat, called **mantle plumes**, which well up from the deep mantle. When they reach the base of a plate, these mantle plumes, the existence of which was first proposed in 1971 by Jason Morgan at Princeton University, heat the overlying lithosphere, doming it like a blister. They also generate magmas that fuel volcanoes on the plate above, forming a "hotspot."

But the plate on which the volcano erupts is moving, whereas the mantle plume beneath it is essentially fixed. So just as a sewing machine generates a line of stitches as the needle repeatedly punctures the moving cloth, a plume repeatedly punctures the crust as the plate moves past. In this way, a line of extinct volcanoes is produced as the moving plate carries each hotspot-built volcano away while a new one is built over the plume. In the oceanic realm, hotspots form volcanic islands that slowly subside and become extinct. In tropical waters, the active volcanic island often develops a *fringing reef* of corals, while the line of extinct volcanic islands first develop *barrier reefs* as they subside and eventually become *atolls*. Finally, the island chain becomes a line of submerged *seamounts* as the ocean floor cools and subsides and the volcanoes are beveled by the sea (Fig. 9.34).

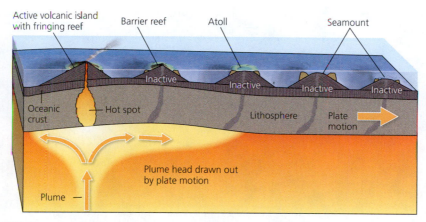

Figure 9.34   Seamount Chains

Linear chains of volcanic islands and seamounts are produced by movement of an oceanic plate over a stationary mantle plume. The age of the islands increases to the right. According to this model, the plume originates from a thermal disturbance deep within the mantle. The hotspot is the hot region within the plate directly above the plume.

ORIGIN OF HOTSPOTS

Although this model explains the pattern of volcanism, the origin of hotspots remains enigmatic because they often lie far from plate boundaries where most volcanoes are located. Given that hotspots remain in relatively fixed positions over very long periods of time, most geologists think that the plumes that underlie them must arise from a very deep source in the mantle, one that is much deeper than the base of the plates. The most commonly cited explanation attributes hotspots to columns of heat that rise from a thermally unstable layer, which exists deep within Earth's interior at the core-mantle boundary (Fig. 9.35). At this boundary, the solid lower mantle is in contact with the liquid outer core. What initially creates these rising columns is speculative but they are thought to rise in a manner analogous to the colored oils of a lava lamp. Other geologists, however, have

suggested that mantle plumes come from much shallower sources. The number of hotspots is also uncertain, with estimates ranging from 20 to 120. Figure 9.36 shows the distribution of some of the better-known hotspots.

We distinguish two types of hotspots, oceanic and continental, on the basis of the type of crust on which they occur. While the origin of the two types may be the same, there are major differences in the nature of their volcanism, which we discuss in the following material.

OCEANIC HOTSPOTS: EVIDENCE OF PLATE MOVEMENT

Oceanic hotspots provide dramatic evidence of the movement of Earth's lithospheric plates. Best known of all the world's hotspots is the Hawaiian hotspot.

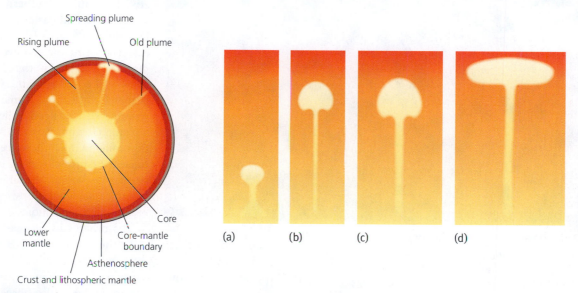

Figure 9.35   Mantle Plumes

Most geoscientists believe that mantle plumes ascend from the core-mantle boundary, about 2900 kilometers (1800 miles) below Earth's surface (left). The ascent of a mantle plume is modeled using a buoyant solution of glucose syrup (right). (a) The plume establishes and maintains a relatively narrow central feeder pipe known as a plume tail, and (b, c) balloons below the lithosphere to eventually form (d) a plume head some 400–1000 kilometers in diameter. The plume causes volcanism and uplift of the lithospheric plate above it. As the plume material cools, it may "underplate" (adhere beneath) the lithospheric plate.

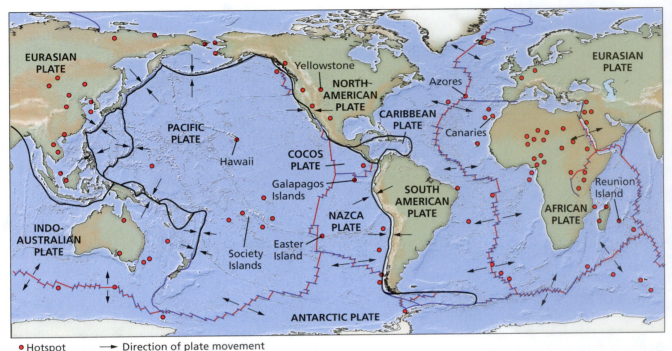

• Hotspot → Direction of plate movement

Figure 9.36 Global Distribution of Major Hotspots

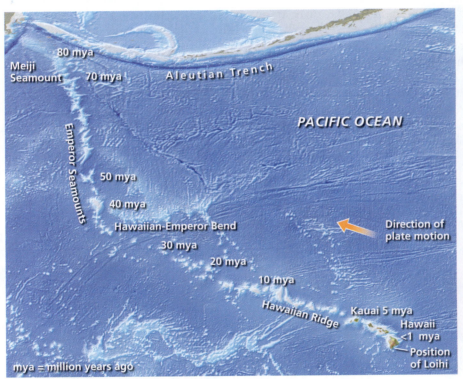

Figure 9.37 Hawaiian-Emperor Chain

Movement of the Pacific plate over the Hawaiian hotspot has carried away volcano after volcano to form a hotspot trail composed of the Emperor Seamounts and Hawaiian Ridge. The bend between the two reveals that the Pacific seafloor once followed a more northerly course. The usually hidden topography of the ocean floor in the northwest Pacific Ocean is shown in this view without the veneer of ocean water.

The Hawaiian-Emperor seamount chain (Fig. 9.37) marks the passage of the Pacific plate over a plume for the past 80 million years. The oldest seamounts, now eroded by time, stand at the line's northern end where the Pacific plate dives beneath the Aleutian Trench. Other hotspots pepper the Pacific. The Society Islands in the southern Pacific, for example, and Easter Island and Darwin's Galapagos Islands in the eastern Pacific formed over other mantle plumes. In the Atlantic Ocean, the Azores and Canary Islands formed above plumes. Likewise, the island of Reunion lies above a plume far from rifts and subduction zones in the Indian Ocean. Because the source of melting lies over plumes in the mantle, all of these volcanic islands are overwhelmingly basaltic in composition.

The progressive increase in the age of volcanic activity to the northwest of Hawaii is evident in the increasingly eroded—and hence older—appearance and lower elevation of the volcanoes in this direction. This apparent increase in age is also borne out by radiometric dating (Fig. 9.38). Present-day volcanic activity is observed on Mauna Loa and Kilauea, which lie over the Hawaiian plume on the southeast side of Hawaii. Further

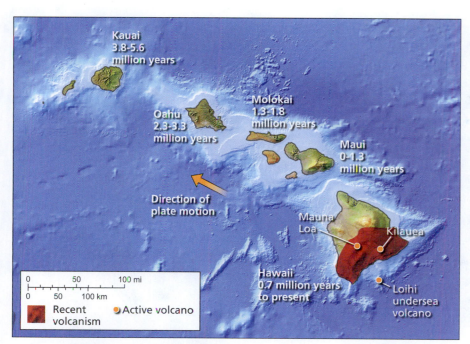

Figure 9.38   Age of Hawaiian Volcanism

Volcanism in the Hawaiian Islands progressively increases in age to the northwest, the direction in which the Pacific plate is moving.

thinking of how a line of stitches bends if there is a change in the direction of movement of the cloth. Because hotspots remain more or less stationary, seamount chains provide evidence of the direction and speed of the plate on which they sit. Geologists can reconstruct past plate configurations by sliding the plate back toward the present hotspot along the seamount track.

Thirty million years ago, for example, the island of Midway was located where Hawaii is today, directly above the plume. Since then, the Pacific Plate has moved 2600 kilometers (1600 miles), carrying Midway in a northwesterly direction away from the plume. Dividing distance by time, the average rate of plate movement during this time interval has been almost 9 centimeters (3.5 inches) per year. Hence the kink between the Hawaiian and Emperor chains suggested to Morgan that the presently northwest-moving Pacific plate was moving directly north prior to 40 million years ago. This idea has since been borne out by dating.

Morgan's proposal is further strengthened by the existence of similar kinks of identical age in nearby hotspot trails (Fig. 9.39). In a neat geochronological fit, the kink in each hotspot trail occurs in the vicinity of a 40-million-year-old seamount. The timing of this change in movement of the Pacific plate coincides with a global reorganization of plate motion thought to have been brought about by the collision of India with Asia.

> **CHECK YOUR UNDERSTANDING**
>
> ● What are kinked hotspot trails and how are they produced?

southeast, the submarine volcano Loihi, which is within 960 meters (3150 feet) of the sea's surface, is inching its way toward islandhood, and will one day become the newest member of the chain as the islands move northwest.

Northwest of Kauai, the Hawaiian island chain continues as a dotted line of atolls, sand islets, and seamounts that stretch all the way to Midway Island, a distance of 3500 kilometers (2200 miles). These little islands subsided as they were transported away from the plume and have been worn flat by the sea during their long journey. But they are, in fact, among the world's tallest mountains, built flow by flow from the ocean floor 6 kilometers (4 miles) beneath the sea to form a chain of more than fifty extinct volcanoes. On Hawaii, Mauna Kea, which has lain dormant for 3600 years, is the world's tallest mountain, rising 10,000 meters (33,000 feet) from the seafloor (Mt. Everest is the highest, but is not the tallest). Mauna Kea is also the world's largest single mountain, with a volume of 40,000 cubic kilometers (9600 cubic miles).

Beyond Midway Island, the hotspot track continues in the Emperor Seamount chain, which reaches as far as the Aleutian trench, a distance of 2500 kilometers (1550 miles). All of the extinct volcanoes in this chain have subsided below sea level as they cooled over time. In keeping with his idea of stationary mantle plumes, Jason Morgan attributed the kink between the Hawaiian and Emperor seamount chains to a change in the direction of movement of the Pacific plate at the age of the seamount at the kink, about 40 million years ago. This can be imagined by

CONTINENTAL HOTSPOTS: SOWING THE SEEDS OF CONTINENTAL BREAKUP

Continental hotspots may play an active role in determining the position of certain plate boundaries. Repeated experiments suggest that hotspots and their underlying plumes play a major role in the rifting of continents (Fig. 9.40), first doming and then cracking the continental crust into characteristic Y-shaped rifts, and then providing the magma that rises between the diverging continents. These rifts meet at a triple point like that of Ethiopia's Afar Triangle at the southern end of the Red Sea where the East African Rift meets the Gulf of Aden (see Fig. 9.13). Continental hotspots may consequently play a fundamental role in the initiation and geometry of continental breakup and the opening of ocean basins. The hotspots in these cases

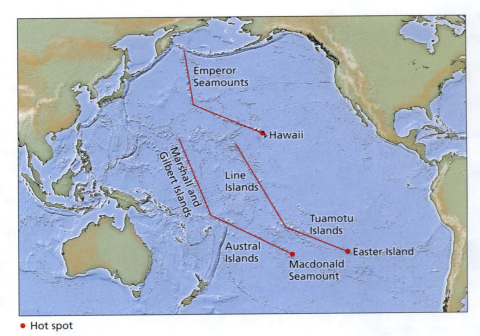

• Hot spot

Figure 9.39 Kinked Hotspot Trails
Kinked hotspot trajectories in the Pacific. The kink in each island chain occurred about 40 million years ago and is related to a change in the direction of motion of the Pacific plate.

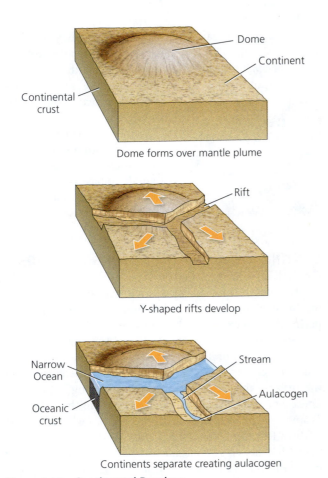

Dome forms over mantle plume

Y-shaped rifts develop

Continents separate creating aulacogen

Figure 9.40 Continental Breakup
Mantle plumes may create the wedges that drive continents apart.

would be located in the new ocean basin once the continents separated. So it is possible that some oceanic hotspots had earlier continental histories.

Typically two of the radial rifts that meet at a triple point widen into oceans while the third may either widen as well or becomes inactive to form a special type of *failed rift* called an **aulacogen** from the Greek *aulax*, meaning furrow. Aulacogens are depressions that channel major rivers into the new ocean and slowly become filled with sediment. An excellent example of an aulacogen occurs at the elbow of Africa's west coast where the Benue Trough brings the River Niger to the Atlantic Ocean (Fig. 9.41). Following continental breakup, the hotspot beneath such a triple point will ultimately lie near a mid-ocean ridge. This explains the large number of hotspots located at or close to the Mid-Atlantic Ridge as well as the mid-ocean ridges of the southern Indian Ocean (see Fig. 9.36). The hotspot that created the Benue Trough is thought to lie today beneath Ascension Island in the Central Atlantic (see Living on Earth: **The Remarkable Journey of the Green Turtle** in Chapter 2).

In contrast to oceanic hotspots like Hawaii, magmas produced by mantle plumes beneath continents must rise through many kilometers of granitic continental crust. As a result, the record of activity for continental hotspots suggests long volcanic silences punctuated by short cataclysmic explosions of pyroclastic material. Voluminous mantle-derived basaltic magma contains sufficient heat to melt the overlying granitic continental crust during its ascent. Melting of continental crust produces magma with the same composition as granite. This gas-rich magma is less dense but more viscous than basalt, so that it retains its

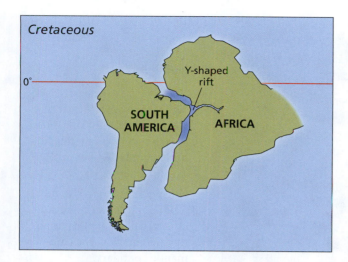

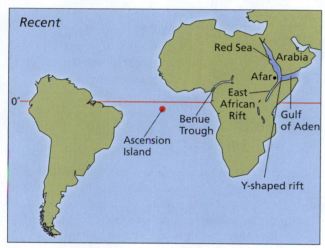

Figure 9.41   Benue Trough

The Benue Trough is thought to be an aulacogen that formed as a result of hotspot-facilitated continental rifting between Africa and South America during the Cretaceous period, about 140 million years ago. The hotspot is now located beneath Ascension Island in the central Atlantic. The East African Rift Valley may also be an aulacogen that formed when Africa and Arabia began separating 25 million years ago.

gases as it rises, allowing pressures to build explosively. Bulging beyond capacity, the ground above the magma chamber finally cracks, releasing the pressure and triggering a catastrophic eruption. Only after this eruption does the deeper basaltic magma flood out as lava flows (Fig. 9.42).

Such is the history of the Yellowstone hotspot in Wyoming, which is manifest at Earth's surface today by hot springs, geysers, and other hydrothermal activity. Although we do not associate Yellowstone National Park with volcanic eruptions, huge explosions rocked the region three times in the last 2 million years and are likely to do so again. Each has occurred on a fairly regular schedule of once every 700,000 years. The first and largest occurred about 2 million years ago, and blew 2500 cubic kilometers (600 cubic miles) of ash skyward in an explosion 15,000 times greater than the 1980 explosion of Mount St. Helens. Magma exploded

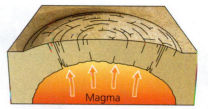

Swelling magma chamber arches and fractures overlying rocks.

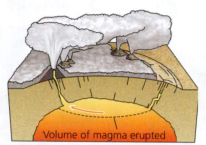

Fractures tap gas-charged rhyolite in magma chamber releasing pressure and triggering vast ash flows.

Crust collapses to form caldera.

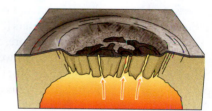

Later basalt lavas flood caldera floor.

Figure 9.42   Formation of Yellowstone Caldera

This figure illustrates four stages in the development of the Yellowstone caldera 600,000 years ago.

again 1.3 million years ago and again 600,000 years ago, both explosions strewing 1000 cubic kilometers (240 cubic miles) of ash across the western United States from Montana to Louisiana. In each case, collapse of the ground following the violent escape of huge volumes of ash and lava produced massive crater-like structures. Recall from Chapter 4 that large, steep-walled and broadly circular structures produced by the collapse of volcanoes following an eruption are called *calderas*.

The size of the Yellowstone caldera is so large that it was not discovered until detailed mapping was carried out in the 1950s and 1960s. Forests and younger volcanic flows now obscure much of its outline so that it is very difficult to see from ground level. But its size is readily apparent

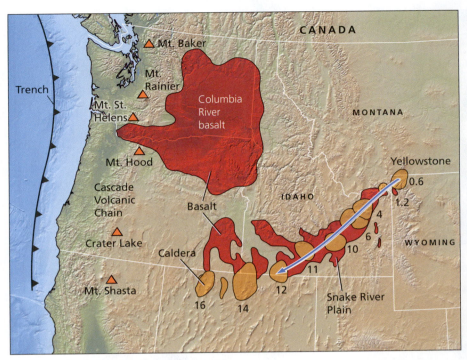

Figure 9.43 Yellowstone Hotspot Track
The track of the Yellowstone hotspot (arrow) across southern Idaho is revealed by the flood basalts (in red) of the Snake River Plain and the broadly circular calderas (in orange—numbers give age in millions of years) buried beneath them.

on satellite imagery and from the park's geology. Beneath the caldera, earthquake waves have detected a column of heated rock extending more than 400 kilometers (250 miles) down into Earth's interior, supporting the presence of a mantle plume below the national park. Southwest of the park, the Snake River Plain marks the passage of the North American plate over the plume for the past 20 million years (Fig. 9.43). Comprising flow upon flow of basaltic lavas, which have buried earlier calderas that become progressively older to the southwest, the length of the track indicates that the North American Plate has moved southwestward over the plume at a rate of about 4 centimeters (1.6 inches) per year.

CHECK YOUR UNDERSTANDING

• How can continental hotspots evolve to become oceanic hotspots?

• In what way might hotspots drive the wedge that splits continents apart?

Because all hotspots are essentially stationary, they provide dramatic evidence of plate tectonics, recording not only the direction but also the speed of plates that move over mantle plumes. They may also play an instrumental role in the breakup of continents and the opening of ocean basins. But hotspots do not tell us *why* the plates are moving. So what causes plate movement, what forces are involved, and to what do they owe their origin? To answer these questions, we turn in the next section to the potential driving mechanisms for plate tectonics.

9.7 SUMMARY

• Hotspots are isolated areas of higher than average heat that are associated with volcanoes. Most hotspots occur, not at plate boundaries, but in a plate's interior.

• Hotspots are thought to be produced by giant columns of heat, called mantle plumes, that well up from the core-mantle boundary.

• Oceanic hotspots provide dramatic evidence of the movement of Earth's tectonic plates. Continental hotspots may play a role in continental rifting.

9.8 Plate Tectonics and Plate-Driving Mechanisms

Although there may be many forces influencing plate motion, the main driving mechanism is clear. It is the flow of heat from the interior of the planet to its surface and the return flow into the interior as Earth cools by convection (see Science Refresher: **The Flow of Heat** in Chapter 1). The convective circulation of Earth's heated interior is powered by the energy released from the decay of naturally occurring radioactive elements in the mantle. This convective circulation locally causes plates to move apart or spread, to come together or collide, or to slide past each other in shear. How convection actually makes the plates move, however, is still not fully understood. Are they pulled from the front, pushed from behind, dragged from below, or is it some combination of these? Given these options, geologists have shown that plates could be propelled by three types of force: slab pull, ridge push, and mantle drag (Fig. 9.44).

• During **slab pull**, the weight of the subducting slab pulls the rest of the plate behind it toward a trench and down a subduction zone. This movement is much like a tablecloth, which slides off a table when enough of it is hanging over the edge. This gravity-driven force occurs because the subducting slab is denser and heavier than the asthenosphere through which it is sinking.

• During **ridge push**, the plate slides downslope under the influence of gravity toward a trench from the ridge crest. Here, the elevated position of the ridge allows the plate to simply slide downhill like a toboggan.

• During **mantle drag**, the plates are carried by convective circulation in the asthenosphere. In this case the plates move like rafts in a moving stream.

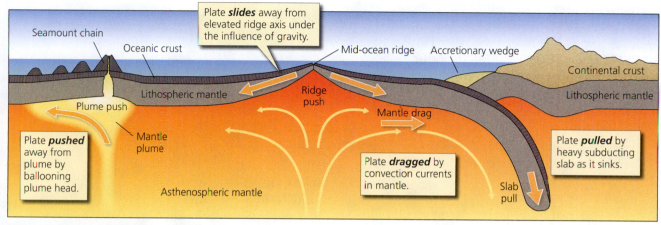

Figure 9.44   Forces That Move Plates

The forces that cause plates to move are thought to be created by gravity-driven slab-pull or ridge-push mechanisms. Mantle drag occurs between the upper mantle and the asthenosphere and may also contribute to lithospheric plate movement, or may slow it down.

In addition to these three forces, the ballooning heads of mantle plumes (see Fig. 9.35) may push plates away from hotspots. Such "plume push" forces may be important during continental rifting, given the role of mantle plumes in splitting continents (see Fig. 9.40).

The most important of these forces appears to be slab pull because the fastest moving plates, such as the Cocos and Nazca plates of the eastern Pacific, are those with the greatest lengths of subducting edges relative to their size. On the other hand, because nonsubducting plates like the North American plate also move, ridge push must also be important. The role of mantle drag is uncertain. If it was a major driving force, those plates with the largest surface areas over which the force could operate, like the Pacific plate, should be the fastest moving. The fact that they are not suggests that mantle drag is a frictional force that opposes slab pull and ridge push, and slows the plates rather than carries them. The presence of continental crust may also slow plates down. This is because the continental lithosphere has deep roots that may exert a drag as they plow through the asthenosphere, much like a sea anchor slows the movement of a boat. Clearly, there is much to be learned about the driving mechanisms behind plate motion.

9.8 SUMMARY

- The plates of Earth's outer shell collide, split apart, or slide past one another, driven by forces from Earth's heated interior.

- Plates may be propelled by three types of force: slab pull, ridge push, and mantle drag. "Plume push" forces may also be important during continental rifting.

CHECK YOUR UNDERSTANDING

○ Which plate-driving mechanism is the most important and how do we know?

Key Terms

accretionary wedge   271
active continental
    margin   276
asthenosphere   256
aulacogen   290
back-arc basin   274
back-arc
    spreading   274
buoyancy   258
continental arc   276
continental crust   257

continental shelf   263
convergent plate
    boundary   262
divergent plate
    boundary   262
fault   263
hotspot   286
island arc   271
isostasy   259
lithosphere   256
lithospheric mantle   258

mantle drag   292
mantle plume   286
mantle wedge   275
mid-ocean ridge   264
Moho   258
oceanic crust   257
ophiolite   279
passive margin   263
pull-apart basin   282
ridge push   292
rift valley   263

slab pull   292
subduction   269
subduction zone   269
transform fault   280
transform plate
    boundary   262
transverse ridge   282
triple point   264
tsunami   271
upper mantle   257
volcanic arc   271

Key Concepts

9.1 PLATES ON EARTH'S SURFACE

- Earth's outer layers comprise a rigid lithosphere with a weaker asthenosphere below.
- The lithosphere is broken into moving plates that comprise the crust and uppermost mantle.
- Earth's crust is either continental or oceanic, and is separated from the upper mantle by the Moho.

9.2 PLATES AND ISOSTASY

- Continental crust floats buoyantly on the mantle and, like an iceberg, stands highest where it is thickest.
- Dense oceanic crust is less buoyant and so forms the floors of the world's oceans.
- The balance between crustal elevation, crustal thickness, and crustal density is called isostasy.

9.3 PLATE BOUNDARIES

- Plate margins interact through extension, compression, and shear.
- Extensional movement produces divergent boundaries, compressional movement produces convergent boundaries, and shear movement produces transform boundaries.

9.4 DIVERGENT BOUNDARIES: CREATING OCEANS

- Divergent plate boundaries rift continents, open oceans, produce mid-ocean ridges, and create new oceanic crust.

9.5 CONVERGENT BOUNDARIES: RECYCLING CRUST AND BUILDING CONTINENTS

- At convergent plate boundaries the denser plate is subducted beneath the more buoyant plate.

- Subduction zones generate earthquakes and are the sites of volcanism.
- Ocean-ocean convergence produces deep ocean trenches and island arcs.
- Ocean-continent convergence produces the volcanic mountain ranges of continental arcs.

9.6 TRANSFORM BOUNDARIES: FRACTURING THE CRUST

- Transform plate boundaries are marked by major crustal fractures along which plates slide past each other.
- Oceanic transform faults offset mid-ocean ridges.
- Continental transforms create earthquake-prone faults like the San Andreas.

9.7 HOTSPOTS: TRACKING PLATE MOVEMENTS

- Hotspots are solitary areas of volcanism thought to result from mantle plumes.
- Most hotspots occur in a plate's interior rather than at plate margins.
- Oceanic hotspots provide a record of plate movement.
- Continental hotspots may play a role in continental rifting.

9.8 PLATE TECTONICS AND PLATE-DRIVING MECHANISMS

- Plates move in response to the flow of heat from Earth's interior.
- Three types of force may move plates—slab pull, ridge push, and mantle drag.

Study Questions

1. From the hypsometric curve (see Fig. 9.4), it is evident that there is a difference in elevation between continental and oceanic crust. Why is this and how does it support models of isostasy based on both crustal thickness and density?

2. Explain the marked difference in the depth of earthquakes at mid-ocean ridges and subduction zones.

3. What is back-arc spreading and how does it account for the presence of seas between the islands of the western Pacific?

4. Transform faults are so called because they permit one type of plate motion to be "transformed" into another. What is meant by this and how is it achieved?

5. Lines of volcanic islands such as those of the Hawaiian chain are not readily accounted for by plate tectonic

theory, yet can be used to deduce the direction of plate motion. Explain this paradox.

6. Deep ocean trenches are characterized by a deep topographic depression and abundant earthquakes. Explain how these features form.

7. Why do earthquake foci outline the upper boundary of a descending slab in a subduction zone?

8. Subduction zones predominantly rim the Pacific Ocean. However, the type of subduction varies significantly from the Andes of South America to the western Pacific. Compare and contrast these variable characteristics of subduction and provide an explanation for them.

9. Why does plate tectonic activity appear to be confined to Earth's lithosphere?

10. Explain the kinked track of the seamount chains in the Pacific Ocean.

10 Deformation and Mountain Building

10.1 Deformation

10.2 Orientation of Geologic Structures

10.3 Fractures

10.4 Folds

10.5 Unconformities Revisited

10.6 Plate Tectonics and Mountain Building

10.7 Mountain Building in the Geologic Past

▶ Key Terms

▶ Key Concepts

▶ Study Questions

IN DEPTH: Partial Melting and the Making of Continental Crust—Nature's Ultimate Recycling Program 317

LIVING ON EARTH: In the Shadow of the Himalayas 328

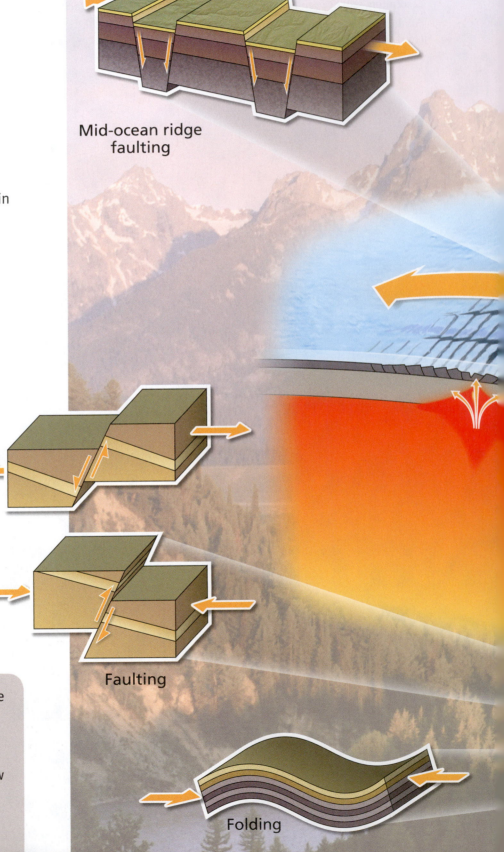

Mid-ocean ridge faulting

Faulting

Folding

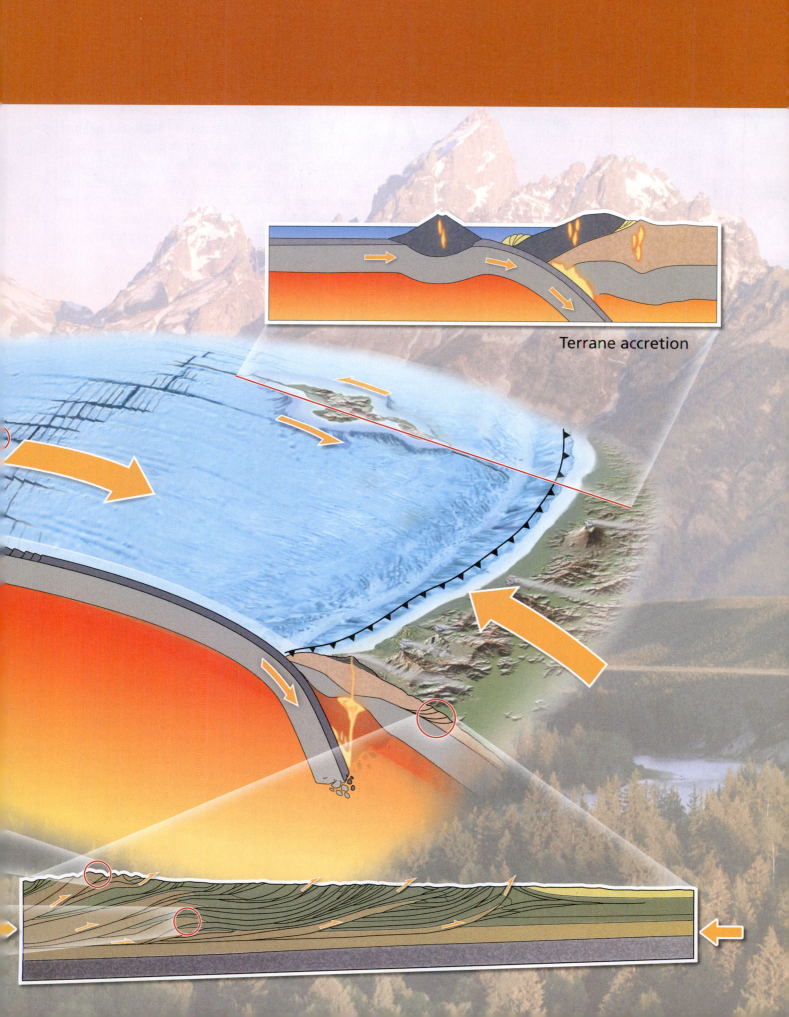

Terrane accretion

Learning Objectives

10.1 Describe the processes of rock deformation and compare and contrast ductile and brittle behavior in rocks.

10.2 Explain how strike and dip are used to measure the orientation of geologic structures.

10.3 Compare and contrast joints and faults and discuss how each type of fracture forms and the geologic structures produced as a result.

10.4 Identify types of fold structures and summarize how folds are described based on the orientation of their axial plane and fold hinge.

10.5 Compare and contrast different types of unconformities, and assess their relationship to deformation.

10.6 Discuss the plate tectonic causes of mountain building above subduction zones and at zones of continental collision.

10.7 Explain how the geologic record helps us to explore the role of plate tectonics in the evolution of ancient mountain belts

As Earth's great tectonic plates jostle against each other, the rocks along their boundaries are subjected to huge forces. As a result, the rocks break or bend and uplift or subside to form major mountain ranges and basins. The deformation of rocks creates dramatic structures that tell us much about the forces that produced these structures and the conditions under which those forces operated. Deformation of the crust also produces mountains, vivid testaments to plate tectonic theory.

Mountain building not only propels the rock cycle and fuels the growth of continents but also is responsible for some of Earth's most breathtaking scenery. Indeed, our desire to understand the world around us and the natural forces responsible for such scenery helped to pave the way to the recognition of plate tectonics as a fundamental geologic process.

Rocks deform to produce a variety of geologic structures. Near Earth's surface, where temperatures and pressures are low, rocks tend to be brittle and break to form *fractures*. Here, the abrupt movement of rocks along *faults* causes earthquakes. In contrast, at the elevated temperatures and pressures of deeper crustal levels, rocks are more pliable and bend to form *folds*. Both faults and folds are characteristic of mountains, which form by a variety of processes along plate boundaries and in plate interiors. The most important regions of crustal deformation occur at convergent plate boundaries where mountain building as a consequence of subduction and collision has continued intermittently throughout geologic time.

10.1 Deformation

In Chapter 8, we learned that rock layers deposited at Earth's surface are horizontal, laterally continuous, and arranged in a sequence with the oldest at the bottom and the youngest at the top. However, examination of outcrops shows that rock layers are often inclined and may abruptly terminate, and some sequences have older layers overlying younger ones (Fig. 10.1). Such complexities reflect the history of the rock layers following their deposition.

In Chapters 2 and 9 we learned about the enormous pressures generated by plate movement as the plates jostle against one another. These pressures cause the rocks along plate boundaries to change their original position, shape, and/or volume (Fig. 10.2). These changes are the result of **deformation**, and rocks so affected are said to be *deformed*.

Deformation of the crust is one of the most visible indications that Earth is a dynamic planet. Deformation is a major process in the building of mountains and is often closely associated with metamorphism. Like metamorphism, deformation reflects the history of the rocks after their formation, and the conditions under which rocks

Figure 10.1   Rock Layers in Outcrop

These folded rock layers on the north coast of Cornwall, UK, have been disturbed since their original deposition such that they are no longer horizontal or laterally continuous, and are locally arranged in a sequence with older layers overlying younger ones.

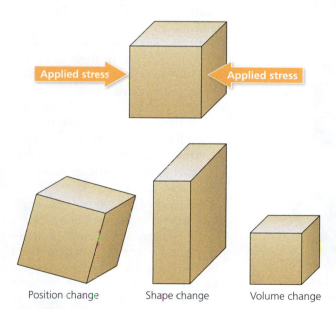

Position change   Shape change   Volume change

Figure 10.2   Deformation

Deformation causes rocks to change their original position, shape, and/or volume.

deform typically result in the growth of new metamorphic minerals as well. Deformation takes several forms depending on the physical conditions under which it occurs. To examine the effects of deformation, we must first return to a topic that we encountered during our examination of metamorphic rocks, the concepts of stress and strain.

STRESS AND STRAIN

In everyday conversation, we often use the terms *stress* and *strain* interchangeably. In geology, however, these two terms have very distinct meanings. **Stress** refers to the tectonic pressures responsible for deformation and is defined as the *force per unit area* acting on a rock body. **Strain** is the product of deformation and is a measure of the *change in position, shape or volume* that a rock body undergoes as a result of an applied stress. Strain, therefore, is the response of a rock body to an applied stress. Hence, stress and strain are respectively the cause and effect of deformation in much the same way that the stress of exams causes the strain of exam anxiety.

Recall from Chapter 7 that there are several forms of stress. A stress that acts uniformly in all directions, like the pressure of water on a submerged diver or a submarine, is *confining pressure*. This type of stress strains a rock body by changing its volume but not its shape. Confining pressures are produced at depth by the vertical stress generated by the weight of the overlying rocks. Just as divers experience increased water pressure the deeper they dive, so rocks are subjected to increased confining pressure the deeper they are buried.

Stresses that act unequally on a rock body and change its location and shape are *differential stresses*. Just as there are three kinds of plate boundaries, so differential stresses and the type of strain they produce can be of three types. **Tensional stresses** are outward acting; they pull rocks apart, causing them to strain by becoming longer and thinner (Fig. 10.3a). Such stresses are typical of divergent plate boundaries where Earth's crust is being ripped apart, as is the case in the East African Rift Valley. **Compressional stresses**, on the other hand, are inward acting; they push rocks together, causing them to strain by becoming shorter and thicker (Fig. 10.3b). Such stresses are characteristic of convergent plate boundaries where blocks of Earth's crust collide and mountain ranges such as the Andes and Himalayas are produced. **Shear stresses** act parallel to each other but in opposite directions and cause rocks to strain by slipping past each other sideways (Fig. 10.3c). Shear stresses are common along faults, including transform plate boundaries, such as the San Andreas Fault in California, where one crustal block slides past another.

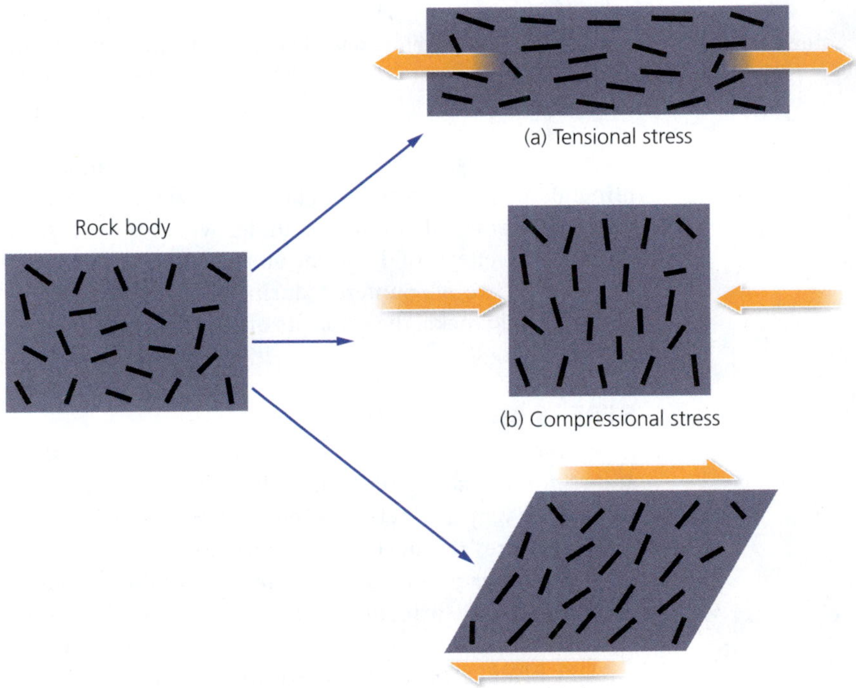

Regardless of the type of stress involved, if the rock body affected is displaced or undergoes a change in shape or volume, or any combination of these processes, it is deformed. Although the stress may eventually cease, the strain caused by the stress is a permanent record of that deformation in the rock. There are several ways in which deformation is accomplished, as we discuss in the material that follows. Just how a rock responds to stress depends on the type of deformation and the physical conditions under which it occurs.

> **CHECK YOUR UNDERSTANDING**
>
> ○ What is the difference between stress and strain?

TYPES OF DEFORMATION

When rocks are subjected to differential stress, whether tensional, compressional, or shear, they commonly bend or break. We can intuitively grasp why a rock body fractures under a high enough stress and why a small rock sample breaks if hit hard enough with a hammer. But what conditions permit a rock to bend without breaking? To answer this question, geoscientists have performed many experiments in which samples of rock are progressively stressed for various lengths of time under conditions of pressure and temperature that simulate the conditions within Earth's crust. These experiments have shown that rocks, like all materials, deform in two distinct ways.

Although different rocks may respond to stress quite differently, the general pattern of their behavior is similar. For low stresses, the change in shape and volume (or strain) is typically temporary so that the rock returns to its original form once the stress is removed. This behavior is described as **elastic deformation**. Beyond a critical level of stress, however, rocks no longer return to their original shape once the stress is removed. In these cases, the strain becomes permanent. This behavior is **plastic deformation**.

We can illustrate the difference between these two types of deformation by contrasting the behavior of an elastic band with that of a strip of modeling clay. When we stretch an elastic band, it remains extended only while it is being pulled and returns to its original length as soon as we let it go. This is *elastic* deformation. It is difficult to permanently stretch an elastic band because, when stretched too far, the band simply breaks. An elastic band consequently shows little in the way of plastic deformation before breaking. Materials that behave in this way are said to be **brittle**, like glass, and the

Figure 10.3 Differential Stresses
The three types of differential stress are: (a) tensional stress, (b) compressional stress, and (c) shear stress.

deformation produced when they break is known as **brittle failure**.

A strip of modeling clay, on the other hand, will not return to its original shape even when stretched only slightly. Instead, it deforms plastically and remains permanently extended. Stretched too far, the strip will break, but not before it is permanently extended many times its original length. Materials that behave in this fashion and flow when they are stressed are said to be **ductile**.

Rocks can deform in elastic brittle fashion or plastic ductile fashion. Brittle failure in rocks causes them to crack or fracture, whereas ductile behavior causes them to flow, smear, or fold (Fig. 10.4). But what controls which type of behavior a rock will exhibit? As we shall now discuss, several factors are involved.

> **CHECK YOUR UNDERSTANDING**
> ● Compare and contrast elastic and plastic deformation.

FACTORS THAT INFLUENCE DEFORMATION

Experiments show that the dominant factors influencing the way in which rocks deform are rock type, temperature and pressure, and time. We now describe each of these factors in more detail.

Rock type

Just as an elastic band and modeling clay respond quite differently when stretched because they are made of different materials, so rocks of different composition respond quite differently to stress. Some rocks, such as granite, deform in a brittle fashion except at very high stresses, when they become ductile. Others, such as rock salt, deform in a ductile manner even under quite low stresses. Such contrasts in behavior explain why identical stresses can cause some rocks to fracture while others bend. Thus, brittle and ductile deformation often occur side by side, the contrasting behavior reflecting the varying response of different rock types to the same applied stress.

Temperature and pressure

Experiments have shown that ductile behavior in rocks is enhanced if deformation occurs at higher temperature and pressure. Consequently, rocks tend to deform plastically at depth where temperatures and pressures are high, but elastically near Earth's surface. As a result, brittle failure in rocks is characteristic of shallow crustal levels whereas ductile folding and flow are a more common response at depth, where they typically accompany metamorphism. Because continental crust is predominantly granitic in composition, its behavior at depth is strongly dependent on that of its two dominant minerals, quartz and feldspar. Experiments have shown that, if water is present, quartz deforms plastically at about 300°C (570°F), which reduces the strength of the middle crust, whereas feldspar behaves plastically at 500°C (930°F), which reduces the strength of the lower crust.

Time

Experiments have also shown that rocks respond in a more ductile fashion when stresses are applied slowly over long periods of time. Hence, stresses that are insufficient to permanently deform a rock when they are first applied may deform the rock if they are maintained for a long enough period.

The role of time in rock deformation is usually described in terms of the *strain rate,* that is, the rate at which the rock changes shape. Experiments show that high strain rates favor brittle behavior, whereas low rates promote bending and smearing. The effect of strain rate on rocks is similar to the behavior of butter when it is taken from a refrigerator. If a cold stick of butter is subjected to a high strain rate by abruptly bending it, it will break in a brittle fashion. However, if the same stick is subjected to a low strain rate by flexing it slowly, it bends or smears in a ductile fashion. Ductile behavior in rocks is promoted by the extremely slow rates at which geologic changes take place.

> **CHECK YOUR UNDERSTANDING**
> ● What factors control whether a rock will behave in a brittle or ductile fashion?

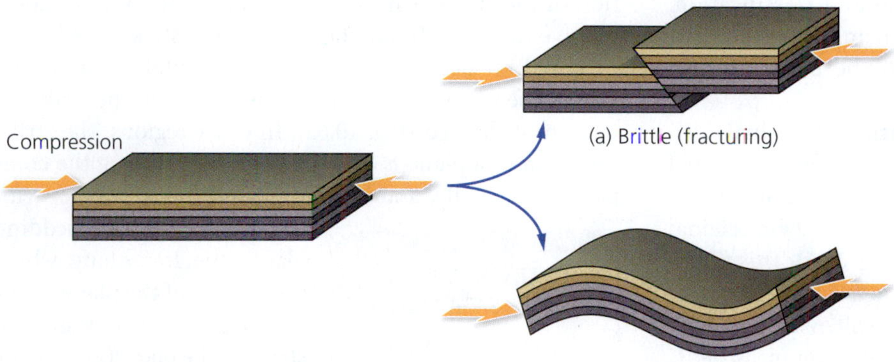

10.1 SUMMARY

- Deformation of Earth's crust is the result of differential stresses that displace and strain rock bodies by changing their shape and volume as a result of tension, compression, or shear.

- Rocks that deform elastically until failure are brittle and will fracture, whereas those that deform plastically are ductile and will smear or fold.

Figure 10.4 Brittle and Ductile Behavior in Rocks
Rocks respond to stress in either: (a) a brittle fashion and break to form a fracture, or (b) a ductile fashion and bend to form a fold.

Compression

(a) Brittle (fracturing)

(b) Ductile (folding)

- Elastic deformation is promoted by low temperatures and pressures and high strain rates, whereas plastic deformation is encouraged by high temperatures and pressures and low strain rates. As a result, rocks near Earth's surface tend to be brittle and break during deformation, whereas those at depth are more ductile and bend.

10.2 Orientation of Geologic Structures

Because of the various ways in which they deform, rocks exhibit a variety of geologic structures. Recall from Chapter 7, for example, that when deformation accompanies metamorphism, the growth of new minerals may occur in a preferred orientation resulting in a planar fabric known as a *foliation* (see Figure 7.16). In this instance, although the new minerals are the product of metamorphism, the foliation is the result of differential stress and is an example of a geologic structure.

In this chapter, we focus our attention on *fractures* and *folds,* two other types of common geologic structures that reflect the two ways (brittle and ductile) that rocks behave during deformation. We also revisit unconformities, which, although not the direct product of deformation, nevertheless can form as a consequence of it. Because fractures result from brittle behavior, they typically develop at the low temperatures and pressures and rapid stain rates characteristic of shallow crustal levels. Folds, on the other hand, are the product of ductile behavior and so they develop at some depth within the crust where temperatures and pressures are elevated and strain rates are usually slow. In contrast, unconformities are the result of uplift and erosion, as we learned in Chapter 8.

The type of geologic structure can tell us a lot about the conditions under which deformation took place, but it is from the orientation of structures such as fractures and folds that we can determine the directions of the stresses involved. The orientation of geologic structures also plays an important role in their description and classification. Before investigating the structures themselves, we must therefore examine the way in which the orientation of structures is measured in the field.

As we learned in Chapter 8, the principle of original horizontality holds that all sedimentary strata are deposited more or less horizontally. But as we have seen, strata in outcrops can be steeply inclined, implying that their original horizontal orientation has been tilted (Fig. 10.5). This tilting is attributed to tectonic stresses, and it is from the geometry of such strata that geologic structures are often identified. So how, then, is the orientation of a tilted rock layer measured?

Many geological structures (such as an individual bed, a foliation, or a fault) have discrete flat surfaces, or *planes,* and are said to have a planar geometry. It is a fundamental principle of geometry that any two lines can uniquely

Figure 10.5   Tilted Strata

These steeply tilted rock strata are located in Cape Breton Island, Canada.

define a plane. In geology, two specific lines are most commonly used for this purpose, the horizontal and the line of maximum slope. The horizontal line is known as the *strike,* and the line of maximum slope is the line of *dip* (Fig. 10.6). These two parameters are used in combination to describe the orientation of all planar structures. The **strike** of an individual surface, such as a bedding plane, is the orientation of an imaginary horizontal line on the plane. The orientation of the strike is measured with a compass as an angle relative to north. In coastal regions, the strike line is easy to visualize. Because sea level is horizontal, the intersection of the sea surface with a bed is parallel to the strike direction of the bed (Fig. 10.6a). In other regions, the strike direction of a plane is best visualized by holding the compass horizontally, and bringing its edge into contact with the surface of the bedding plane. The line along which the surface of the plane meets the edge of the compass is the strike of the plane (Fig. 10.6b).

Dip is a measurement of the maximum slope of a plane relative to the horizontal (Fig. 10.6). All inclined

CHECK YOUR UNDERSTANDING

○ What is meant by the strike and dip of a geologic structure and how is each parameter measured?

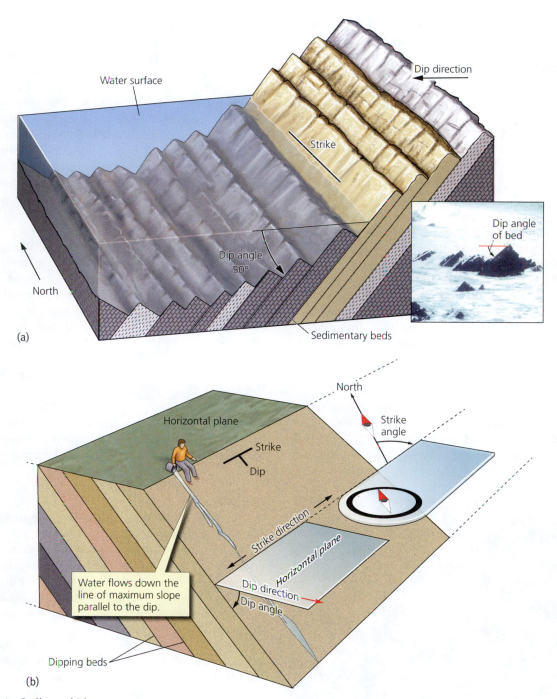

Figure 10.6   Strike and Dip

(a) The strike, dip, and direction of dip are indicated for a succession of tilted beds. The strike is parallel to the intersection of the sea surface with a bed (in this case north), whereas the dip is the bed's maximum angle of slope (in this case 50 degrees). The direction of dip is the direction downslope (in this case to the west). In the accompanying photo, the dip of sedimentary beds along the coast of North Devon in England is shown by the orientation of the beds with respect to the sea surface. (b) The strike of a bed is indicated by holding a compass horizontally and bringing its edge into contact with the surface of the bedding plane. The line along which the two meet is the strike of the bed.

planes have a line of maximum slope. If we sprinkle water on the plane, gravity would drive the water down this line. The angle the line of maximum slope makes with the horizontal is the dip. Horizontal beds have a dip of zero degrees; vertical beds have a dip of 90 degrees. To distinguish planes that dip at the same angle but in opposite directions, just as the two sides of a roof might slope 30° to the west on

one side of a house and 30° to the east on the other, we also note the *direction of dip* (either west or east in this case).

Used together, the parameters of strike and dip (and dip direction) allow us to fully describe the geometry of all planar geologic structures. The orientation of planar structures such as fractures is directly defined by the strike and dip of the fracture surface, whereas the geometry of nonplanar structures such as folds is described by the variations in the strike and dip of the folded surface.

10.2 SUMMARY

- Geologists use strike and dip (and direction of dip) of a planar geologic structure to measure its orientation.

10.3 Fractures

As we discussed in Section 10.1, because of the controls of temperature, pressure, and strain rate on rock behavior, rocks near Earth's surface typically behave in a brittle fashion and break during deformation. This results in the formation of **fractures**. Fractures are divided into two broad groups, known as *joints* and *faults,* depending on the amount of movement that has taken place along the fracture surface. The movement on joints is insignificant, whereas the movement on faults is appreciable.

JOINTS

Joints are rock fractures along which there has been no appreciable movement. Joints are one of the most common of all geologic structures and, as we learned in Chapter 5, play an important role in rock weathering. Most joints do not

occur in isolation but, instead, occur in broadly parallel, evenly spaced groups known as **joint sets** that are responsible for the blocky appearance of most outcrops (Fig. 10.7).

A variety of processes can produce joints. Some joints result from the contraction or expansion of a rock body and others result from deformation. **Columnar joints** are distinctive sets of contraction fractures that outline long polygonal columns (Fig. 10.8a). They typically form when bodies of fine-grained igneous rock, such as basalt lava flows, cool down and crack as they shrink. **Exfoliation joints** are gently curved expansion fractures that are broadly parallel to the land surface (Fig. 10.8b). These joints are most common in exposed granite bodies. Recall from Chapter 4 that granites crystallize at depth and are only exposed when the overlying rocks are removed by erosion. This removal of the weight of the overlying rocks is known as *erosional unloading* and greatly reduces the stress on the granite, causing it to expand and crack.

Most joints, however, are **tectonic joints**, which are produced by tectonic forces that cause rocks to break in a systematic fashion in response to regional stresses. Such tectonic joints often occur in steep, parallel sets (Fig. 10.8c) and may develop over wide areas that are being stretched, compressed, or sheared, by crustal stresses.

> **CHECK YOUR UNDERSTANDING**
>
> ● Compare and contrast the mode of origin for columnar joints and tectonic joints.

FAULTS

In contrast to joints, **faults** are fractures on which significant movement has occurred. Once formed, a fracture represents a plane of weakness, so it is not surprising that many fractures are also planes of movement. In modern settings, rapid displacements along faults are a direct cause of earthquakes, as we discuss in more detail in Chapter 11. Faults in the rock record provide evidence of ancient seismic activity.

In outcrops, faults are identified by the severing and displacement of rock layers that were once continuous (Fig. 10.9a). When the displacement is small, this movement usually takes place on a discrete surface known as the *fault plane*. In large-displacement faults, however, movement takes place on a network of fault surfaces that together define the *fault zone* (Fig. 10.9b). For example, literally thousands of individual faults occur within the San Andreas fault zone of California.

Faults also influence the landscape where they intersect Earth's surface. Because faults often move abruptly, rocks within a fault zone may be crushed to form *fault breccia* (Fig. 10.9b) or pulverized into finely powdered *fault gouge*. Gouge and breccia are easily eroded and as a result, the fault line may be marked by a depression in the land surface (see Fig. 9.31). Alternatively, the movement on a fault may bring together rocks whose resistance to erosion is quite different. One side of the fault consequently erodes

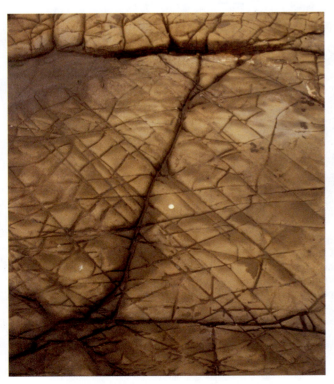

Figure 10.7 Joint Sets
The blocky appearance of this outcrop at Kimmeridge Bay in England is the result of the intersection of several sets of roughly parallel, evenly spaced joints.

Figure 10.8   Types of Joints

(a) Columnar joints are seen at Devil's Tower, Wyoming. (b) Sheet jointing is apparent in Yosemite National Park. (c) Regional stresses cause tectonic joints such as these in Arches National Park, Utah.

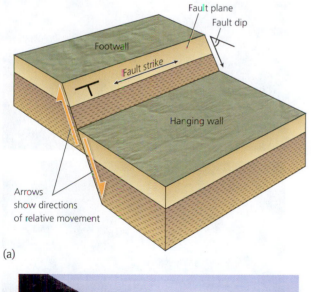

Figure 10.9   Parts of a Fault

(a) The different parts of a fault are illustrated in this schematic. (b) The occurrence of a fault breccia (broken rock) within a fault zone can be seen in a roadside outcrop of the Moab Fault in Utah. (c) Two fault scarps resulting from recent faulting are visible at Badwater, Death Valley.

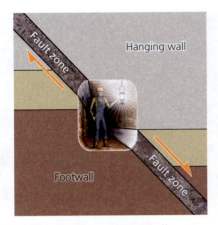

Figure 10.10   Hanging Wall and Footwall
The hanging wall is the name given to the rocks above a fault, whereas the footwall describes the rocks below. Both are mining terms. The hanging wall is named for the overhanging side of a mined opening; the footwall is the side on which a miner would stand.

faster than the other, and the line of the fault is marked by a cliff known as a *fault scarp* (Fig. 10.9c).

To assess fault movement, geologists compare rocks on either side of the fault. The rocks above and below the fault surface are respectively described as the *hanging wall* and

footwall. Both names are old mining terms. The hanging wall is named for the overhanging side of a mined opening (where a miner might hang a lantern) whereas the footwall is named for the side on which a miner would stand (Fig. 10.10). We distinguish between two main groups of faults: dip-slip faults and strike-slip faults, on the basis of the displacement of the hanging wall relative to the footwall. As we shall see, the distinction between these groups relates to the stresses responsible for the displacement.

CHECK YOUR UNDERSTANDING
❍ What is the difference between a joint and a fault?

Dip-slip faults

Faults in which the direction of movement—referred to as **slip**—is generally parallel to the dip of the fault plane are **dip-slip faults** (Fig. 10.11). There are two types of dip-slip faults. When the hanging wall moves *down* relative to the footwall the dip-slip fault is a **normal fault** (Fig. 10.11a). Conversely, when the hanging wall moves *up* relative to the footwall the dip-slip fault is a **reverse fault** (Fig. 10.11b).

As discussed earlier, faults rarely occur in isolation. Normal faults are the product of tensional stresses and occur in large numbers in areas where the crust is being

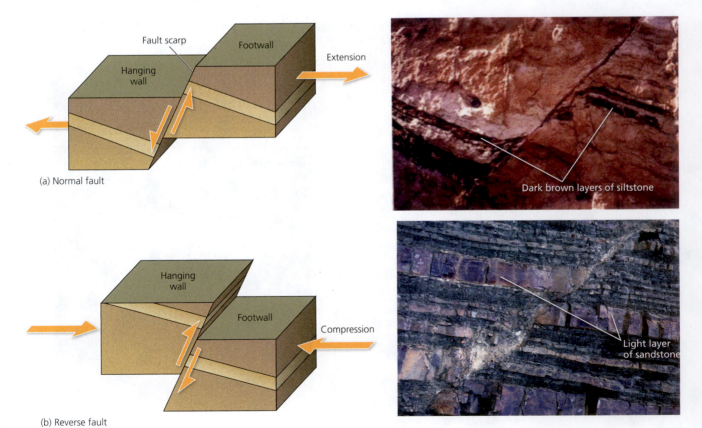

(a) Normal fault

(b) Reverse fault

Figure 10.11   Normal and Reverse Faults
(a) Tensional stresses produce normal faults. The rocks of the hanging wall move downward relative to those of the footwall. In the accompanying photo, dark brown layers of siltstone are truncated and offset by a normal fault. (b) Compressional stresses produce reverse faults. The rocks of the hanging wall are heaved upward relative to those of the footwall. In the photo, a light layer of sandstone is offset by a reverse fault.

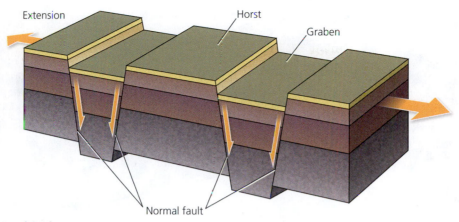

Figure 10.12   Horst and Graben

Horst and graben structures are characteristic of tensional tectonic settings. Horsts are crustal blocks between two outward dipping normal faults. Grabens are crustal blocks that lie between two inward dipping normal faults. Crustal blocks that define grabens have been displaced downward relative to the crustal blocks that define horsts.

extended. They are therefore characteristic of mid-ocean ridges and continental rifts at divergent plate margins along which plates separate (see Fig. 2.24 and Fig. 9.12). In continental rift settings, normal faults cause the crust to become thinner by breaking the crust into a number of down-dropped blocks. At the surface, the downward movement of these blocks can create a series of faulted ridges and valleys. Where this occurs, two German words are used to describe the features, **horst** (meaning "height") for the ridge and **graben** (meaning "grave") for the valley (Fig. 10.12). Grabens are crustal blocks that have been displaced downward between two inward dipping normal faults. Horsts are crustal blocks between two outward dipping normal faults that are high standing because the blocks on either side have dropped. The boundary between a horst and a graben is typically a fault scarp.

Reverse faults are the product of compressional stresses and occur preferentially in areas where the crust is being squeezed. They are characteristic of subduction zones and mountain belts at convergent plate margins where plates collide and the crust is thickened (see Fig. 9.17b). When viewed in outcrop, normal and reverse faults are typically steep, the dip of their fault planes lying between 45 and 90 degrees. However, on a crustal scale, both types of faults are typically curved surfaces that become more gently dipping with depth. Major normal faults often merge downward into low-angle extensional faults called **detachments** (Fig. 10.13a). The precise reason for this curvature is complex, but one of the main factors is that the vertical stress increases with depth because of the increasing weight of overlying rock. Detachment faults are often located near the boundary between the brittle upper crust and the ductile lower crust because the contrast in rigidity makes this boundary a fundamental zone of weakness that faults can exploit.

Detachment faulting is characteristic of a broad region of extension in the western United States and Mexico known as the Basin and Range Province, where countless normal faults divide the crust into parallel mountain ranges separated by valleys (Fig. 10.13b). In the Basin and Range Province, most of the normal faults dip to the west. The down-faulted blocks consequently have an asymmetric shape and are called **half-grabens** (see Fig. 10.13a), which give this region its distinctive topography.

Similarly, reverse faults often merge at depth with low-angle compressional faults known as **thrusts** (Fig. 10.14a). Thrust faulting is a very common feature of convergent plate margins and is clearly evident in all mountain ranges. One of the best-known examples occurs in the foothills of the Rocky Mountains in Alberta, Canada (Fig. 10.14b). Recall from Chapter 6 that the weight of the overriding thrust-thickened crust causes the overridden crust to flex downward, forming a flexural basin in which sediments shed from the upthrusted mountains are deposited.

> **CHECK YOUR UNDERSTANDING**
>
> ◔ Compare and contrast a normal fault and a reverse fault.
>
> ◔ Briefly describe each of the following geologic structures: detachment, thrust, horst, and graben.

Strike-slip faults

Faults where the slip is broadly parallel to the strike of the fault plane are **strike-slip faults** (Fig. 10.15). In contrast to dip-slip faults, in which movement is either up or down, the fault blocks on either side of a strike-slip fault move sideways. Many strike-slip faults are nearly vertical, in which case there is no hanging wall or footwall. Because a given fault block can either move left or right, we distinguish between two types of strike-slip faults. As we all know from giving directions to a destination, whether the turn is left or right depends on which way you are facing. We must therefore use a convention that avoids this problem. To do so, we observe the direction of movement of the

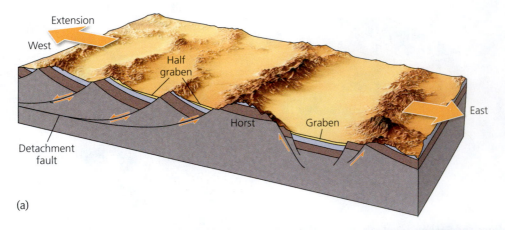

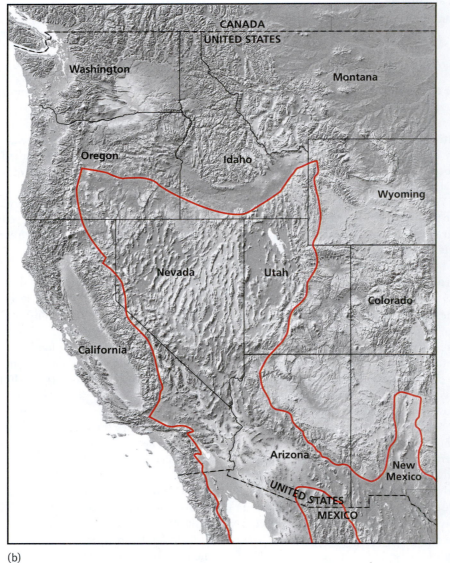

Figure 10.13   Detachment

(a) Major normal faults often flatten with depth into gently dipping extensional faults called detachments. (b) Detachment faulting is common in the Basin and Range Province of the western United States and Mexico (outlined in red), where it has resulted in the formation of the half grabens that are characteristic of the region's distinctive topography.

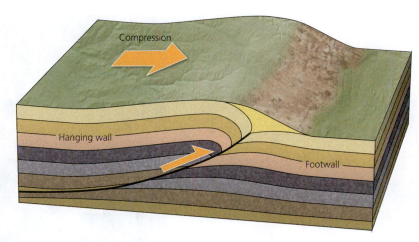

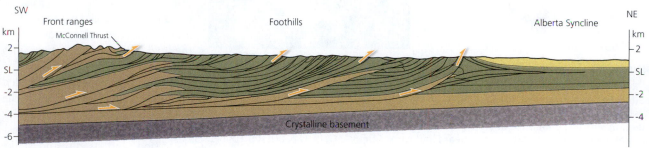

Figure 10.14   Thrust

(a) Major reverse faults often flatten with depth into gently dipping compressional faults called thrusts. (b) Thrusts are common in all mountain ranges, but are particularly well developed in the foothills of the Canadian Rockies in Alberta. Note: SL = sea level.

fault block on the *far* side of the fault line. In this way, the movement direction is the same for viewers on both sides of the fault. In Figure 10.15a, for example, the far side of the fault has moved to the right no matter which side you are standing on. This type of strike-slip fault is **right-lateral**. Conversely, in Figure 10.15b, the far side of the fault has moved left regardless of which side you stand on and the fault is **left-lateral**.

Strike-slip faults are the product of shear stresses and are characteristic of transform plate boundaries where plates slip past each other. A *transform fault* is a special type of strike-slip fault that is also a plate boundary (see Chapter 2). California's San Andreas transform fault (see Figure 9.31), for example, is a strike-slip fault zone along which right-lateral motion between the Pacific and North American plates has amounted to more than 550 kilometers (340 miles).

CHECK YOUR UNDERSTANDING

- How do we distinguish between right-lateral and left-lateral strike-slip faults?
- What is the difference between a joint and a fault?
- How does the movement of a dip-slip fault differ from that of a strike-slip fault?

10.3 SUMMARY

- Joints are fractures along which there has been no significant movement. Columnar joints are produced by contraction in cooling lava flows, and exfoliation joints are formed by expansion due to erosion of overlying rocks. Most joints are tectonic and record regional stresses caused by crustal movements.

- Faults are fractures along which significant movement has occurred. Movement is roughly parallel to the dip in dip-slip faults and parallel to the strike in strike-slip faults.

- Normal dip-slip faults are extensional and may form horst and graben structures. Low-angle normal faults are detachments. Reverse faults are compressional and are typical of convergent plate margins. Low-angle reverse faults are thrusts.

- Strike-slip faults, whether right-lateral or left-lateral, are the product of shear stresses and are characteristic of transform plate boundaries.

10.4 Folds

At deeper levels in Earth's crust, where temperatures and pressures are elevated and strain rates are low, rocks deform in a ductile fashion by bending, smearing, and flowing, rather than by breaking. Like the pages of a bent paperback book, rock layers at depth respond to deformation by flexing up and down to form wave-like buckles known as **folds**. In contrast to the brittle failure seen in joints and faults, folds are the product of plastic deformation in layered rocks.

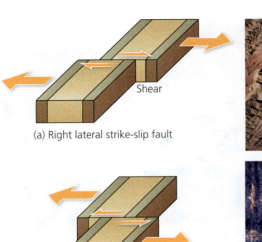

Shear

(a) Right lateral strike-slip fault

(b) Left lateral strike-slip fault

Figure 10.15   Strike-Slip Faults

(a) Right-lateral faults, in which the block on the far side of the fault moves right relative to the fault block on the near side, are illustrated by offset ridges in southern Nevada. (b) Left-lateral faults, in which the fault block on the far side moves left, are illustrated by offset gneissic banding near Charlevoix, Quebec.

Folds form when a rock layer is compressed and they can occur on a wide variety of scales, ranging from tiny folds that are barely visible to the naked eye to vast troughs and arches that are several kilometers across (Fig. 10.16).

Although folds have highly variable orientations, small-scale and large-scale folds in the same region often have similar orientations because they are the product of the same compressional event. Measurements taken on small folds in an outcrop can therefore provide valuable information about the geometry of regional folds.

Geologists measure the orientation of two important fold components, the *axial plane* and the *fold hinge* (Fig. 10.17), in the field. These orientations are the basis of classifying folds. The **axial plane** of a fold is an imaginary surface that divides the fold into two. The two sides of a fold are *limbs* and they meet at the *fold hinge*. The axial plane of a fold bisects the hinge zone of each folded layer. The orientation of the axial plane is perpendicular to the direction of compression, and its dip and strike provide the first of two measurements that define a fold's orientation. The second important measurement is the

CHECK YOUR UNDERSTANDING

○ What is the difference between the axial plane and the fold hinge of a fold?

(a)

15 cm

(b)

(c)

Figure 10.16   Folded Rocks at Various Scales

(a) Dime-sized crenulations are visible in deformed metamorphic rock in the Pieria Mountains in Greece. (b) Outcrop-sized folds deform sedimentary rocks in the Santa Monica Mountains in California. (c) A mountain-sized fold is visible in the Narcea Valley in northern Spain.

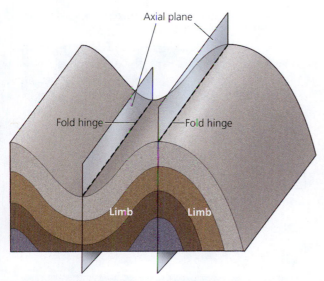

Figure 10.17   Axial Plane and Fold Hinge

The axial plane and fold hinge (dashed line) are called out in this diagram of a fold.

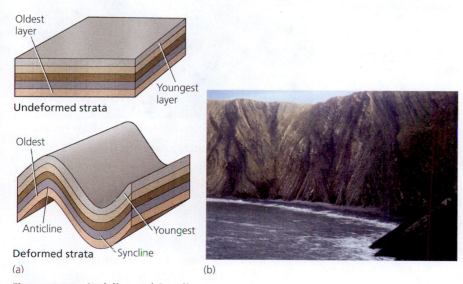

Figure 10.18   Anticline and Syncline

(a) Anticline and syncline form from undeformed strata. (b) Anticlines and synclines are visible in deformed sedimentary rocks on the north coast of Devon, England.

orientation of the **fold hinge**, which is the line that joins the points of greatest curvature in each folded layer, like the roofline of a house. The fold hinge is also the imaginary line along which the axial plane cuts across each folded layer and therefore is the line of intersection between the axial plane and the folded surface.

TYPES OF FOLD

We subdivide folds into two types—*anticlines* and *synclines*—that usually alternate with each other. An **anticline** forms when a sequence of rock strata in normal succession is arched upward, whereas a **syncline** forms when the strata sequence is flexed downward. As a result, in an anticline the oldest rocks occur in the center or *core* of the fold, whereas in a syncline the youngest rocks occupy the fold core (Fig. 10.18). Recall from Chapter 8 that the principle of superposition places the oldest strata at the base of such a succession. The sequence of strata does not change during folding, just as the sequence of pages in a paperback is unchanged when the book is bent. The folding of such a succession pushes the oldest rocks to the core of an anticline and the youngest ones to the core of a syncline.

We describe folds based on the geometry or orientation of the axial plane. Folds are *symmetric* if the limbs diverge from the axial plane at roughly the same angle, and *asymmetric* if they do not (Fig. 10.19). The axial plane bisects a symmetric fold, dividing it into two halves that are more or less mirror images of each other. The axial plane does not, however, bisect an asymmetric fold; instead it divides it into two halves of different shape. The folds in Figure 10.18b are symmetric whereas those in Figure 10.16a are asymmetric.

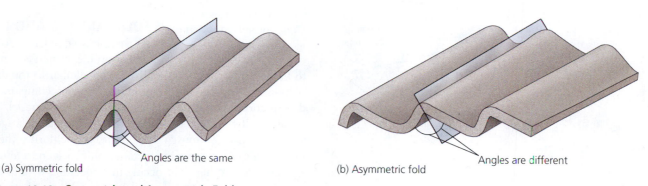

Figure 10.19   Symmetric and Asymmetric Folds

(a) Symmetric folds are divided into two equal halves by their axial planes. (b) Asymmetric folds are divided into two unequal halves by their axial planes.

We also describe folds based on the orientation of the fold hinge. Thus folds are either *horizontal* or *plunging* depending on the slope of their fold hinge (Fig. 10.20). Folds typically form at depth, but can subsequently be exposed by uplift and erosion. When plunging folds are exposed by erosion, as shown in Fig. 10.21a, their expression on a geological map has a distinctive zig-zag pattern in which the beds change direction at the fold hinges. The folded outcrop patterns in the Appalachian Valley and Ridge province of central Pennsylvania are an excellent example of the surface expression of folds (Fig. 10.21b). The anticlines point in the same direction that the folds plunge, whereas synclines point in the opposite direction to the plunge.

Some fold structures do not have a fold hinge at all, but, instead, plunge in many directions; these cannot be simply classified as anticlines or synclines. Such is the case for *domes* and *basins* (Fig. 10.22). Where rock layers are pushed upward from below, the result is a broadly circular bulge known as a **dome**. The surface of a dome slopes away in all directions from the summit. Conversely, when rocks are depressed into a bowl-shaped structure, a broadly circular **basin** forms where all rocks dip inward toward the center.

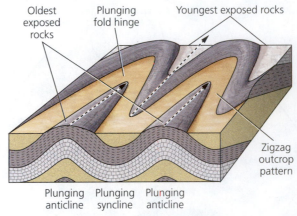

(a) Zigzag outcrop pattern of plunging folds

(b) Satellite view of the Appalachian Valley and Ridge Province

Figure 10.21   Zigzag Outcrop Patterns

(a) Zigzag outcrop patterns are produced by plunging folds. (b) A false color satellite view shows a characteristic zigzag pattern of folded sedimentary rocks in the Appalachian Valley and Ridge province of central Pennsylvania.

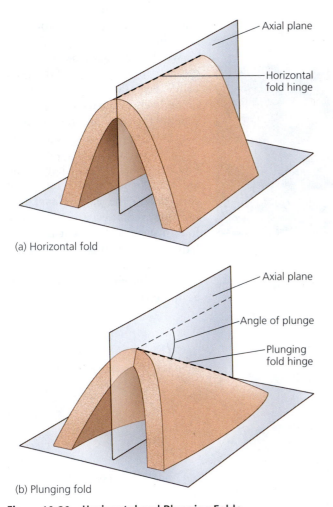

(a) Horizontal fold

(b) Plunging fold

Figure 10.20   Horizontal and Plunging Folds

(a) This fold has a horizontal fold hinge. (b) This fold has a plunging fold hinge.

CHECK YOUR UNDERSTANDING

○ Compare and contrast an anticline and a syncline.

○ What is a plunging fold and why does it generate a zigzag outcrop pattern?

Like the anticlines and synclines they resemble, domes expose the oldest rocks and basins expose the youngest rocks in their centers.

FOLDS AND FOLIATION

As we discussed in Chapter 7, metamorphic minerals tend to flatten, rotate, or recrystallize perpendicular to the direction of compression when metamorphism accompanies deformation. As a result, the minerals become parallel. If these parallel minerals possess a strong internal cleavage, as is the case for muscovite, biotite, and chlorite, they impart this property on the metamorphic rock and a *foliation* develops. Where this occurs in association with folding, the foliation develops parallel to the fold's axial plane, which, like the foliation, forms perpendicular to the direction of compression (Fig. 10.23). Such folds consequently

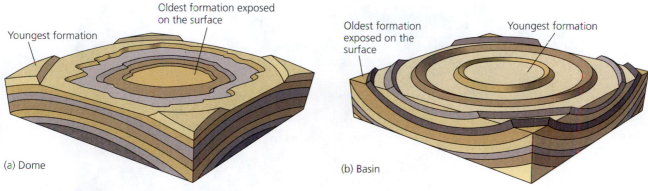

(a) Dome

(b) Basin

Figure 10.22   Domes and Basins

(a) In a dome, the oldest rocks are exposed in the core. (b) In a basin, the youngest rocks are in the core.

Figure 10.23   Folds and Foliation

In this example of folding and foliation, the bedding is broadly horizontal and defined by a difference in color (green), whereas the foliation is vertical. The foliation is parallel to the fold's axial plane and perpendicular to the direction of compression.

have two visible planar features: bedding that outlines the fold and is defined by differences in composition and color, and a foliation that parallels the axial plane and is defined by the preferred orientation of platy minerals.

10.4 SUMMARY

- Folds are produced when layered rocks bend. Folds possess an axial plane and a fold hinge. We use the orientations of these to measure the fold orientation and to describe fold types.

- Folded layers are alternatively arched upward and downward to form anticlines and synclines. Broadly circular uplifts and depressions are domes and basins.

- Where folding accompanies metamorphism, the folded rocks may develop a foliation parallel to the fold's axial plane and perpendicular to the direction of compression.

10.5 Unconformities Revisited

As we have seen, deformation is recorded in rocks in a brittle or ductile fashion, and deformation modifies the geometry of rock strata, creating geologic structures such as faults, folds, and foliations. Deformation also has other major consequences, one of the most important of which is crustal uplift and erosion. Uplift and erosion together exhume the guts of mountain chains, including deep-seated plutonic and metamorphic rocks. As we learned in Chapter 8, when new strata are deposited on top of these exhumed rocks, a fundamental surface known as an *unconformity* is produced. These unconformities are an important consequence of deformation, and their identification is a vital step in understanding the architecture and uplift history of mountain belts. In this section we discuss unconformities in more detail.

Recall from Chapter 8 how James Hutton deduced that the beds of strata beneath an unconformity at Siccar Point (see Fig. 8.9b) were tilted prior to erosion and the deposition of the upper layers. This tilting resulted from stresses applied to the lower layers some time after deposition. As Hutton correctly deduced, a significant time gap must have existed between the deposition of the upper and lower strata. During this time portions of the tilted lower strata were removed by erosion. As a result of the erosion, a segment of the geologic record is missing at that locality. The unconformity consequently records a gap in the history of deposition in the region.

At Siccar Point, we refer to the relationship as an **angular unconformity** because an angular discordance exists between the upper and lower strata. We are now in a position to place Hutton's ingenious insights into a plate tectonic context. For example, a spectacular angular unconformity is exposed along the shoreline of Nova Scotia, Canada (Fig. 10.24). The lower layers consist of steeply dipping

Figure 10.24   Angular Unconformity

In this angular unconformity in the cliffs of Rainy Cove, Nova Scotia, Canada, the steeply dipping beds (lower sequence) are continental clastic sedimentary rocks of early Carboniferous age that were upended by Appalachian continental collision during the formation of Pangea. The gently dipping beds (upper sequence) are also continental clastic rocks, but these rocks were deposited during the Triassic Period after the formation of Pangea and were derived from the erosion of the Appalachian Mountains as Pangea began to break up.

sedimentary rocks of Carboniferous age (about 350 million years old), whereas the upper layers consist of shallowly dipping Triassic rocks (about 220 million years old). Following their deposition, the Carboniferous rocks were deformed during continental collision associated with the amalgamation of the supercontinent Pangea. This collision generated mountains in which the Carboniferous rocks were uplifted and eroded. The Triassic rocks were then deposited on top of the steeply dipping Carboniferous strata as Pangea broke up, creating the angular unconformity we see today.

All unconformities reflect an interval of erosion that removed part of the geological history of the region, but not all are angular unconformities. In addition to angular unconformities, two other types of unconformity can occur. A **nonconformity** occurs when a plutonic igneous rock or a metamorphic rock is directly overlain by younger sedimentary strata (Fig. 10.25a). Because both plutonic and metamorphic rocks form beneath Earth's surface, they must be exhumed by uplift and erosion prior to deposition of younger strata. For example, in the modern Himalayas ongoing uplift and erosion have caused recent sand and gravel to be deposited on the exposed guts of the mountain chain. Because the overlying strata are significantly younger than the exposed metamorphic and plutonic igneous rocks, a significant portion of the geologic record is missing at the boundary between them.

The third type of unconformity—a **disconformity**—is more subtle (Fig. 10.25b). The basic definition of unconformities still applies, namely that the rock units above and below the unconformity surface have significantly different ages of deposition, implying a gap in the depositional record. However, in a disconformity, the lower units are not tilted, but merely uplifted and eroded prior to the deposition of the upper units. The lower units therefore remain essentially horizontal and parallel to the younger strata deposited on top of them. Because the older and younger strata are parallel, identification of a disconformity is more difficult than identification of an angular unconformity or a nonconformity. The two sequences in a disconformity must be precisely determined using either fossil assemblages or absolute dating of volcanic rocks above and below the unconformity surface. Dating determines if a significant portion of the geologic record is missing between the two sets of strata. Because the lower units in a disconformity are not folded, it is unlikely that they were directly involved in collisional mountain building. In fact, disconformities tend to occur in regions that are distant from mountain building. Disconformities can also develop as a result of the exposure and reflooding of continental shelves with the rise and fall of sea level.

Unconformities, like faults and folds, are important geologic structures. All are manifestations of deformation

> **CHECK YOUR UNDERSTANDING**
>
> ● How are angular unconformities distinguished from nonconformities?
>
> ● What are disconformities and why are they difficult to recognize in the geologic record?

(a)

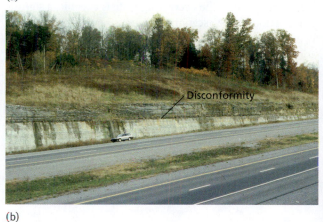

(b)

Figure 10.25   Nonconformity and Disconformity

(a) A nonconformity surface in the Black Hills of South Dakota separates bedded sedimentary rocks (above) from pink granite (below). Because the granite was formed in Earth's interior whereas the sedimentary rocks were deposited on the surface, the rocks do not represent a continuous sequence and a significant time gap separates them. (b) A disconformity in central Tennessee separates darker layers above from the lighter layers below. Although the two sets of layers are parallel, about a million years separates their deposition.

and so provide evidence that deformation has occurred. Together, they allow a region's history of deformation to be interpreted and enable us to assess the relationship of deformation to mountain building and plate tectonics.

10.5 SUMMARY

- Unconformities indicate a significant time gap in the geologic record.
- An angular unconformity shows an angular discordance between the upper and lower strata. In a nonconformity, sedimentary strata directly overlie plutonic or metamorphic rock. In a disconformity, the rock strata above and below the unconformity are parallel, but other evidence suggests a time gap between the deposition of the two sequences.

10.6 Plate Tectonics and Mountain Building

We now turn to the most dramatic of all processes associated with crustal deformation, the building of mountains. Mountains provide some of the most breathtaking scenery on Earth and are testament to the awesome powers of nature (Fig. 10.26). Indeed, the explanation of the process of mountain building or **orogeny** (from the Greek for "mountain origin") is one of the most impressive successes of the theory of plate tectonics.

Mountains form along every type of plate boundary, albeit by different processes. Mountains are also associated with the hotspots of plate interiors. Recall from Chapter 9 that Earth's most continuous system of mountains is delineated by its mid-ocean ridges, which are formed, not by collision, but by mantle upwelling at divergent plate margins as

Figure 10.26   The Grand Tetons

The majestic Tetons rise above the Snake River in northwestern Wyoming.

a consequence of seafloor spreading. As we learned in Chapter 9, mid-ocean ridges stand high because they are swollen with heat, and they rise to a consistent elevation because the temperature of the magmas beneath all mid-ocean ridges is the same. But since stresses are tensional, the mountains of mid-ocean ridges are associated with extensional structures such as normal faults and grabens. Although hidden from view by water, these mid-ocean ridges typically rise 2–3 kilometers (1–2 miles) above the surrounding ocean floor and form an interconnecting network of mountain chains 65,000 kilometers (40,000 miles) in length. As impressive as they are today, however, mountains such as these are rarely preserved in the geologic record because the oceanic crust of which they are built seldom survives the process of subduction.

Mountains also occur within oceanic plates where hotspots pierce the oceanic crust to form linear island chains. Recall from Chapter 9 that hotspots are thought to represent the surface expression of relatively fixed columns of hot upwelling mantle known as *mantle plumes* that rise from the core–mantle boundary. This upwelling provides sufficient buoyancy to uplift the lithosphere and create volcanic mountains. Uplift over the vigorous Hawaiian hotspot, for example, amounts to more than 6 kilometers (3.1 miles). The seamount chain produced as each volcano cools and subsides and is carried away from the hotspot by plate motion has created an underwater mountain chain over 6000 kilometers (3100 miles) in length (see Fig. 9.37). But as with mid-ocean ridges, such mountains only occasionally survive subduction to become part of the geologic record.

Mountains also form along transform plate margins, although these mountains are discontinuous. This may seem counterintuitive as crust is neither created nor destroyed at such margins. However, these margins have localized areas of compression where the crust is squeezed and uplifted to form localized mountains. Such areas occur where a transform fault curves and plate motion causes the crustal blocks on either side to jam together. Recall from Chapter 9 that such curves are *restraining bends* and the folded mountains that develop across the fault as a result of the convergence are *transverse ridges* (see Fig. 9.29).

The most important mountains preserved in the geologic record are the result of crustal deformation at convergent plate boundaries. At these plate margins, orogeny is a direct result of subduction or the product of plate collision. In the next section we first discuss the processes that form mountains during plate subduction and then we examine the mountains formed by plate collisions.

MOUNTAINS PRODUCED BY SUBDUCTION

When an oceanic plate is subducted, it is heated as it descends. Experiments show that at a depth of about 100 kilometers (62 miles), the subducted plate starts to release the water it had previously absorbed by its interaction with ocean water. Recall from Chapter 4 that the release of water vapor and other gases stimulates the generation of mafic magma in the overlying mantle wedge (see In Depth: **Partial Melting and the Making of Continental Crust—Nature's Ultimate Recycling Program** on page 317). The fate of the ascending magma, however, depends on the nature of the overriding plate. If the overriding plate is oceanic, a significant portion of the ascending magmas reach the surface to form a volcanic *island arc* of predominantly basaltic composition. But if the overriding plate is continental, the end result is the generation of a volcanically active mountain range such as that of the Andes (Fig. 10.27).

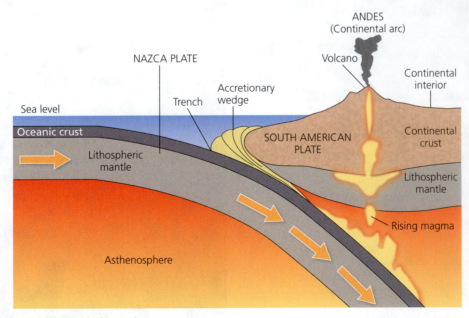

Figure 10.27   Subduction Beneath the Andes

As the Nazca plate is subducted beneath the South American plate, the descent of the slab initiates melting in the overlying mantle and continental crust. The ascent of these melts combined with compressional forces forms the Andes Mountains. Oceanic sediments scraped off the subducting slab mix with sediments derived from the Andes to form the accretionary wedge.

IN DEPTH

Partial Melting and the Making of Continental Crust—Nature's Ultimate Recycling Program

According to plate tectonics, new crust is created at mid-ocean ridges, is transported away from the ridge by seafloor spreading, and is ultimately consumed at a subduction zone. Collisions between buoyant continental crust and dense seafloor always result in the preferential subduction of oceanic crust. Continents may even grow laterally by scraping more buoyant crustal fragments off the descending oceanic plates. Hence, the interior regions of continents, such as the continental shield of Canada, contain ancient rocks and are typically surrounded by belts of progressively younger rock (Fig. 10A).

But why do mid-ocean ridges produce basaltic crust while subduction generates granitic rocks? The answer lies in the way magma is formed. When most substances begin to melt, there is always solid residue left behind

because the process of melting is incomplete. Recall from Chapter 4 that we refer to this process as *partial melting*. Some chemical components are preferentially incorporated into the melt, whereas others are selectively retained in the residue. As a result, the composition of the melt is different from that of the residue, and the compositions of both melt and residue differ from that of the original material. Consequently, the composition of a liquid produced by partial melting is rarely the same as the solid starting material. In general, lightweight components tend to concentrate in the melt that rises because these components are buoyant and separate from the relatively dense residue.

This general principle also holds for the melting processes that occur in the mantle. As the ocean floor spreads, partially molten mantle material wells up from beneath the

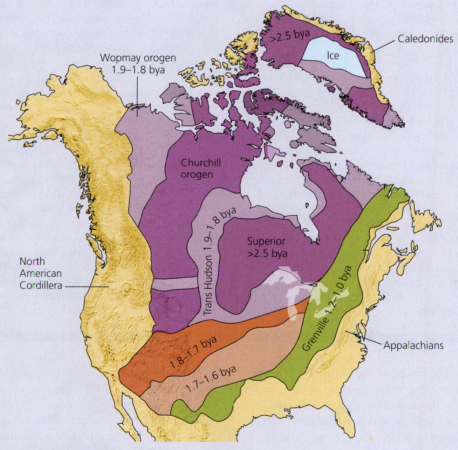

Figure 10A   Age Provinces of North America

In this map showing the major age provinces of North America, note how the continental core of ancient rocks (cratons) tends to be surrounded by progressively younger mountain belts, suggesting that the continent has grown through a series of collisions. (bya = billion years ago).

(Continued)

Partial Melting and the Making of Continental Crust— Nature's Ultimate Recycling Program (Continued)

mid-ocean ridges to form new ocean floor. The upper mantle is composed of the rock type known as peridotite, which consists of a variety of minerals dominated by the magnesium-rich silicates, olivine, and pyroxene (see Chapter 2). Mixed in with these minerals, however, are small amounts of lighter silicate minerals such as feldspar, which occurs in both basalt and granite. Partial melting of the mantle in the asthenosphere beneath the mid-ocean ridges selectively incorporates these lighter minerals much like distillation selectively concentrates lighter alcohol. The liquid produced by partially melting mantle rocks is consequently less dense than peridotite, and cools to produce the dark, moderately dense rock we call basalt. Thus, the magma that rises buoyantly to fill the gap produced as two plates separate is basalt because this is the composition of the magma produced by partial melting in the mantle. The residue left behind in the mantle is referred to as *depleted mantle* because it is mantle material from which basalt has already been extracted.

By the time basaltic crust reaches the subduction zone, however, it has been altered as a result of its chemical interaction with sea water. As a result, it contains many minerals that can store water in their crystal structures. It is also covered with a thick veneer of sediment, most of which is scraped off at the trench although some of it is subducted. The highly fractured oceanic crust also contains huge quantities of water, which is carried into the subduction zone. Because of this varied intake, the melting processes associated with subduction are far more complex than those occurring at mid-ocean ridges.

Heating of the downgoing slab as it subducts recycles the basaltic oceanic crust, its veneer of subducted sediment, and its content of ocean water. The water and other gases that are driven off the descending slab as it heats up rise into the mantle rocks above the subduction zone, where they promote partial melting in the mantle wedge. Because melting occurs in mantle material, basalt liquids are once again produced, although they differ subtly in composition from those generated at mid-ocean ridges because they are produced at different depths. In ocean–continent collision zones, these basalt liquids may, in turn, trigger partial melting of the continental crust as they rise toward the surface. This is possible because the melting temperature of basalt is above 1100°C (2000°F), whereas that of granite can be as low as 750°C (1380°F). Just as hot water poured on butter causes the butter to melt, so the ponding of basaltic liquids beneath continental crust may cause sufficient local heating to allow the granite to partially melt, further concentrating lighter minerals as a result. In this case, the ancient continental crust is recycled to produce magma, which ultimately cools and crystallizes to form new continental crust.

Each time partial melting occurs, lighter minerals are concentrated in the magma so that, with every episode of subduction, the continents become progressively richer in silicon and poorer in iron and magnesium. If this siliceous magma reaches the surface, the resulting volcanic rock is a rhyolite (recall from Chapter 4 that rhyolite is a felsic volcanic rock rich in lightweight elements). If the magma is trapped beneath the surface, it cools slowly to form the felsic plutonic rock granite.

The process of partial melting not only explains the compositional difference between oceanic crust and the mantle from which it is derived but also accounts for the composition of continental crust and its difference from that of the ocean floor. Partial melting selectively concentrates the lighter elements so that, with every episode of subduction, the continents have become progressively more granitic in composition. Because these compositional differences have a strong influence on buoyancy, partial melting is also responsible for the fact that plate tectonics preferentially recycles oceanic crust while preserving continental crust.

Where subduction occurs beneath continents, the heat associated with the ascending mafic magmas is sufficient to promote melting near the base of the continental crust. This produces felsic magma. Because of this, subduction zones beneath continental crust produce a much wider range of magma compositions than subduction zones beneath oceanic crust. Most of this magma never reaches the surface because its ascent is impeded by a thick barrier of continental crust. Instead, it becomes lodged within the continental crust, where the felsic magmas cool slowly to form great bodies of granitic rock. The magmas that do reach the surface to fuel volcanoes are predominantly intermediate to felsic.

The presence of abundant hot magma within the continental crust, coupled with the compression that accompanies plate convergence, causes the leading edge of the continental plate to crumple and rise into a mountain chain. Although subduction may be continuous, the rise of mountains appears to occur in discrete episodes. For example, 100 million years ago, the ancestral Amazon River flowed westward, draining into the Pacific Ocean (see Fig. 1.5). However, about 15 million years ago, the rise of the Andes was rapid enough to block this drainage pattern and form an inland lake. Eventually streams draining into the Atlantic Ocean eroded back until they tapped the lake water and the modern Amazon River was born. The rapid rise of the Andes some 15 million years ago may be due to an interval of more rapid westward movement of the South American plate, which increased compression on rocks already softened by heat from subduction-derived magmas.

Complex tectonic activity also occurs in the vicinity of the deep ocean trench that marks the initial descent of the subduction zone. Here, deep-water sediments and elevated portions of the ocean floor are scraped off the subducting slab and mixed with sediments derived from the erosion of the continent. This chaotic mixture is plastered onto the leading edge of the upper plate to form a wedge-shaped package of metamorphosed and highly deformed rocks, known as an **accretionary wedge** (see Fig. 10.27). In addition, friction at the edge of the continental plate may be sufficient to drag segments of the continental crust down the subduction zone, a process called *subduction erosion*. In the Andes, for example, the continental arc is very close to the trench, suggesting that large segments of continental crust in the vicinity of the trench have been subducted. Finally, small portions of the oceanic crust may break away and escape subduction by becoming incorporated into the mountains of the overriding plate as *ophiolites*. (Recall that ophiolites are fragments of preserved ocean crust; see In Depth: **Ophiolites—Clues to the Structure of Oceanic Crust** in Chapter 9.)

The mountains that result primarily from the subduction process are volcanic and lie above a subduction zone. However, where subduction removes oceanic lithosphere between small crustal blocks, collisions between these blocks occur, and the mountain building processes associated with these collisions are superimposed on earlier processes produced by subduction. As we shall now see, continental margins above subduction zones may collide with these small crustal blocks as subduction proceeds, thereby expanding the size of the continent. Eventually, the expanded continent may collide with another continent.

> **CHECK YOUR UNDERSTANDING**
>
> ○ Why are the mountains associated with seafloor spreading rarely preserved in the geologic record?

> **CHECK YOUR UNDERSTANDING**
>
> • Why does the fate of magmas rising above a subduction zone depend on the nature of the overriding plate?

MOUNTAINS PRODUCED BY COLLISION

The most dramatic mountains are produced by collision at convergent plate margins. Such collisional mountains can be the product of collision at two different scales. Some are caused by the collision between a continental margin and small crustal blocks known as *terranes*. Others are the result of the collision of two continents.

Terrane accretion

The process of subduction can ultimately lead to ocean closure and a collision between continents that once faced each other across an ocean. However, smaller collisions usually precede this climactic event because the subducting ocean floor itself is far from featureless. Instead, most oceans are dotted with island arcs like those of the western Pacific (see Fig. 9.5). Ocean floors are home to features such as mid-ocean ridges and seamount chains as well as high-standing areas called *oceanic plateaux*. In addition, oceanic plates carry small continental blocks, such as Madagascar and Japan, which are **microcontinents**. Like baggage riding passively atop a carousel at an airport, these features migrate passively until they encounter a subduction zone, where they are scraped off the descending plate and are added or *accreted* to the leading edge of the overriding plate as they collide with it (Fig. 10.28). This accretion jams the subduction zone, which consequently moves oceanward so that subduction can continue.

In this way, each such encounter produces a block of land known as an **accreted terrane**, or more simply, a **terrane**. Each accreted terrane has a specific set of geologic characteristics that differ from those of the overridden plate and from those of other terranes. Such accreted terranes may be plastered to the edge of a continent in great numbers along zones of ocean–continent convergence.

The western margin of North America has been built by a succession of terrane collisions (Fig. 10.29). It is thought that these terranes were once dispersed throughout the eastern Pacific Ocean. However, when North America moved westward following the breakup of Pangea, subduction of the eastern Pacific Ocean beneath the continent brought these terranes into collision with its advancing western margin.

We can roughly estimate how much oceanic crust was consumed by the westward motion of North America by looking at the position of the spreading ridge in the Pacific Ocean (see Fig. 9.14). Seafloor spreading is normally a symmetric process that creates new oceanic lithosphere in equal amounts on the two diverging plates. Spreading ridges, such as the one in the Atlantic Ocean, consequently tend to be located in the mid-ocean. In the Pacific Ocean, however, the spreading ridge occupies a highly asymmetric position near the ocean's eastern margin. This implies that much of the oceanic crust that once lay to the east of the ridge has been consumed. In fact, the North American plate has actually collided with the ridge along the Californian coastline. During the consumption of this oceanic crust, terranes that once lay embedded within the Pacific plate collided with the North American continental margin.

More than 50 such terranes have been identified along the continental margin of western North America between Mexico and Alaska. They have been found as far inland as Colorado and Utah. Over the past 200 million years, the piecemeal accretion of these terranes has added an average of 600 kilometers (310 miles) to the width of this margin.

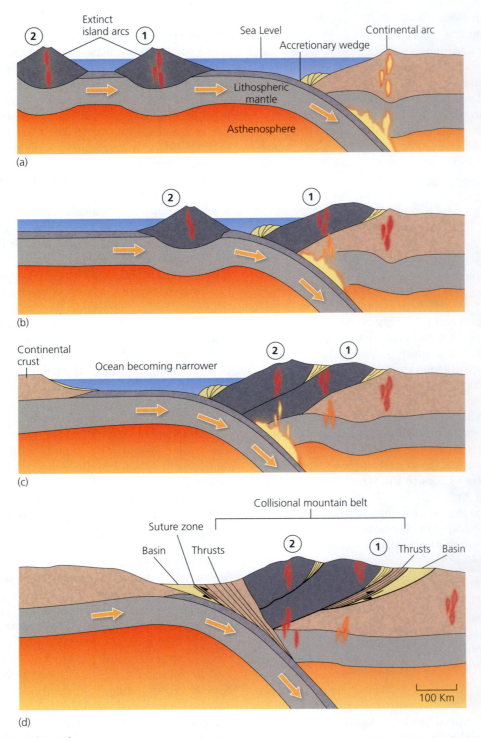

Figure 10.28   Terrane Accretion

This figure models the accretion of two extinct island arc complexes to a continental margin. (a) Two island arc complexes (1 and 2) sit far from a continental margin. (b) The nearest arc accretes when the intervening ocean floor is subducted and the subduction zone moves oceanward. (c) Eventually, the second island arc complex accretes and the subduction zone moves again. The net effect is the growth of the continental margin on the right oceanward. (d) If the ocean eventually closes, terrane accretion is followed by continental collision.

The addition amounts to about 20 percent of continental North America (see Fig. 10.29).

Terranes have two important stages to their history; the first stage reflects the setting in which they originated and the second stage reflects their history of convergence and collision with a continental margin. Evidence that the terranes of western North America originated within the Pacific Ocean is recorded in their geologic histories. Perhaps the best example of this history can be found in the rocks exposed around San Francisco Bay (Fig. 10.30). To

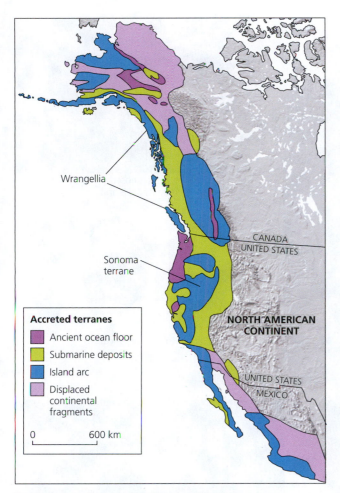

Figure 10.29   Terranes of Western North America
The terranes that have accreted to the western North American margin over the past 200 million years consist of remnants of island arcs, submarine sedimentary deposits, ophiolite suites uplifted from the ancient ocean floor, and continental fragments displaced by faults. Wrangellia, for example, is an island arc terrane that collided with North America some 100 million years ago. These terranes have since been dispersed along the continental margin by faults such as the present-day San Andreas.

the north of the Golden Gate Bridge, the sedimentary rocks of the Marin Headlands are typical of ocean floor sediments and contain microscopic oceanic organisms. This region is believed to be part of a terrane that lay within the Pacific Ocean for some 150 million years prior to its accretion to the North American margin. The geology of nearby Alcatraz Island, however, is quite different. Its fossils and sedimentary rocks are characteristic of near-shore deposits. To the north of Alcatraz, Angel Island contains deformed and metamorphosed rocks that may have been formed in a subduction zone or during the collision between two plates. These three terranes formed in very different environments at great distances from each other and at varying distances from the continental margin. Their present side-by-side location is attributed to their accretion to the continental margin of western North America.

Similar analysis of other terranes has revealed the scale and importance of this accretionary process to the evolution of continents. For example, paleomagnetic and fossil evidence indicates that much of British Columbia, the Yukon, and Alaska are made up of terranes that originated much nearer the equator. Other parts of Alaska may have drifted across the entire width of the Pacific Ocean (Fig. 10.31).

The Rocky Mountains comprise often far-traveled terranes that accreted to North America during the past 200 million years. Parts of Nevada, for example, originated as an island arc off the Pacific coast. In fact, when the dinosaurs roamed Earth, some 150 million years ago, the Rocky Mountains as we know them did not exist at all. At that time, the Pacific coast lay many hundreds of kilometers to the east of its present position and was characterized by a broad continental shelf similar to the current Atlantic seaboard.

Many microcontinental collisions are oblique sideswipes, or "fender-benders," rather than "head-on" collisions with the continental margin. As a result, many of the terranes of western North America were broken up by faults following their accretion and have since been dispersed by fault movement along the length of the continental margin (see Fig. 10.31). One hundred million years ago, for example, an island arc called *Wrangellia* (after the Wrangell Mountains in Alaska) that lay somewhere in the Pacific drifted north and collided with North America. Some blocks of Wrangellia were thrust onto the continent while others continued northward on faults similar to the present day San Andreas Fault. By 40 million years ago, fragments of Wrangellia had been dispersed along the continental margin from Idaho to Alaska.

Western North America has experienced at least 50 terrane collisions. Some of these terranes seemed to have formed at or near their present positions, but others, such as Wrangellia, clearly traveled greater distances. Because their original relationship with respect to North America is usually uncertain, fragments added to the continent in this way are often referred to as *exotic* or *suspect* terranes.

Many of the world's mountain belts include complex mosaics of accreted terranes. Some of these small crustal blocks may have been returned to their original locations. That is, they may have been broken off the continent, perhaps as a result of back-arc spreading (see Chapter 9), and then returned to that continent by closure of the back-arc basin (Fig. 10.32). Other crustal blocks, however, are clearly exotic and have traveled great distances. Regardless of their origin, it is clear that the wandering and accretion of these terranes play an important role in continental growth as a prelude to continent–continent collision.

Mountains produced by continental collision

The most dramatic plate collisions take place when two plates carrying continents collide. Recall that continents

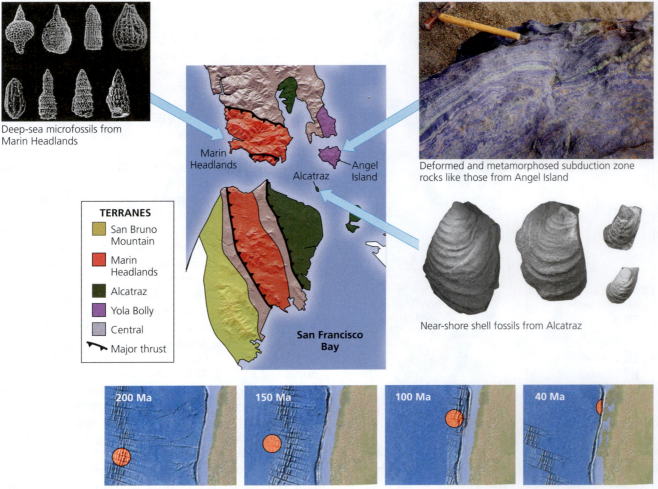

Deep-sea microfossils from Marin Headlands

Deformed and metamorphosed subduction zone rocks like those from Angel Island

Near-shore shell fossils from Alcatraz

TERRANES

- San Bruno Mountain
- Marin Headlands
- Alcatraz
- Yola Bolly
- Central
- Major thrust

San Francisco Bay

Transport history of Marin Headlands (orange circle) from the equatorial mid-Pacific to coastal North America

Figure 10.30   San Francisco Terranes

The rocks exposed around San Francisco Bay belong to several accreted terranes. Those of the Marin Headlands are typical of ocean floor sediments. They contain deep-sea microfossils and, over the past 175 million years, traveled to North America from the equatorial mid-Pacific by plate motion. The rocks of nearby Alcatraz Island, in contrast, are typical of shallow water deposits and contain fossils typical of shorelines, whereas those of Angel Island to the north were deformed and metamorphosed in a subduction zone.

ride passively on plates like baggage on a conveyor belt. However, subduction eventually destroys the oceanic crust between continents so that collisions between them are inevitable. Such collisions result in the formation of some of Earth's most spectacular scenery (Fig. 10.33) and some of its most complex geology, in part because continental collisions are inevitably superimposed on earlier mountain building processes related to subduction and/or terrane accretion (see Fig. 10.28).

Continental collisions are the consequence of the closure of ocean basins (see Fig. 9.17). If the rate of subduction in an ocean basin exceeds the rate at which new crust is generated at its mid-ocean ridge, the process of subduction ultimately leads to the complete closure of the ocean. Given that Earth has a constant size, any discrepancy between the rates of subduction and spreading in one ocean basin must be compensated for by enhanced subduction or spreading in another ocean basin. Plate reconstructions for the last

225 million years (Fig. 10.34), for example, show that the destruction of the ancient Tethys Ocean, which once separated Africa and India from southern Europe and Asia, was accompanied by the opening of the Atlantic Ocean.

The higher rate of subduction over spreading in a closing ocean has implications for the evolution of the ocean's subduction zones. Initially, the destruction of old oceanic crust is likely to produce subduction zones with back-arc basins, such as the ones found in the present-day western Pacific. Recall from Chapter 9 that when old lithosphere subducts, it also rolls backward to create back-arc basins. However, as the ocean basin shrinks, the subduction of younger oceanic crust gradually becomes dominant, producing volcanic mountains on the overriding continental plate that are parallel to the deep ocean trench. So before continental collision even begins, mountains similar to the modern Andes exist on the upper plate above the subduction zone. Subduction terminates when the continents,

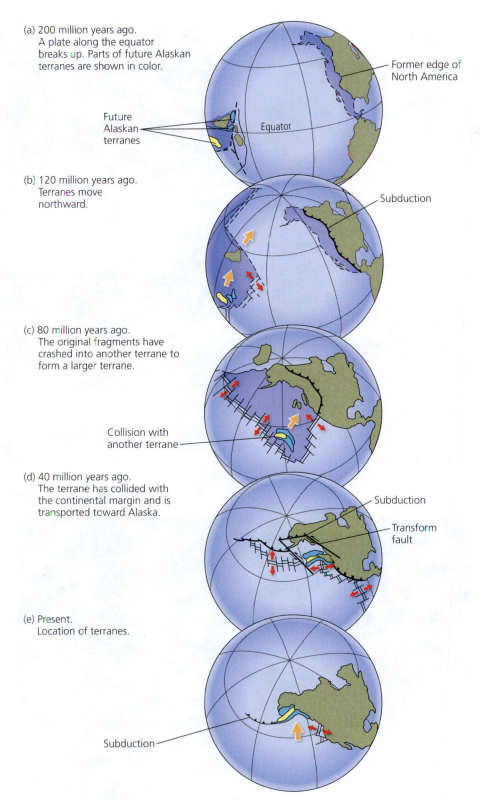

(a) 200 million years ago.
A plate along the equator breaks up. Parts of future Alaskan terranes are shown in color.

Former edge of North America

Future Alaskan terranes

Equator

(b) 120 million years ago.
Terranes move northward.

Subduction

(c) 80 million years ago.
The original fragments have crashed into another terrane to form a larger terrane.

Collision with another terrane

(d) 40 million years ago.
The terrane has collided with the continental margin and is transported toward Alaska.

Subduction

Transform fault

(e) Present.
Location of terranes.

Subduction

Figure 10.31 Trans-Pacific Journey of Alaskan Terranes
Crustal fragments from the southern Pacific may have accreted to the North American continental margin and been subsequently transported to Alaska.

riding like baggage atop a conveyor belt of mantle rocks, inevitably collide.

Like two colliding automobiles, the impact of the continents causes widespread crumpling, distortion, and dislocation of rock strata as the ocean finally closes. The opposing continental margins meet head on and meld together in massive mountain ranges now far from the sea. Leftover slices of ocean floor may also be pushed up to

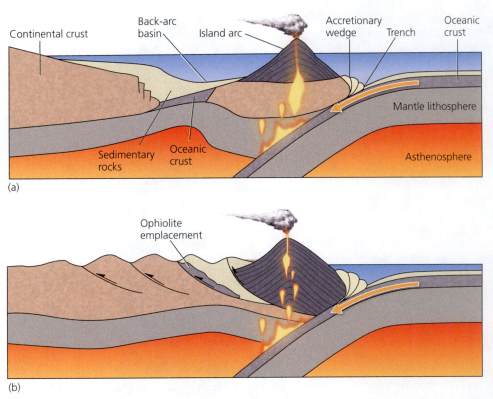

Figure 10.32   Terrane Accretion by Back-Arc Closure

Terranes can be accreted by consumption of a back-arc basin. (a) A volcanic island arc is initially separated from the continental margin by a back-arc basin. (b) Continued subduction results in closure and deformation of the back-arc basin, ophiolite emplacement and collision of the island arc with the continental margin.

Figure 10.33   Mt. Everest

Rising to a height of 8,848 meters (29,029 feet), Mt. Everest, the world's highest mountain, owes its splendor to the collision of India and Asia.

(a) 225 million years ago

(b) 135 million years ago

(c) 65 million years ago

Closure of Tethys Ocean accompanied by opening of the Atlantic

(d) Today

Figure 10.34   Plate Reconstructions

Continental reconstructions for 225 million years ago, 135 million years ago, 65 million years ago, and today illustrate the separation of India from Africa and the destruction of the Tethys Ocean as India moved rapidly northward and rammed into Asia. During the same period, the Atlantic Ocean progressively widened.

mark the **suture,** the boundary between the formerly separate continents (see Fig. 10.28d). The approaching continents usually have irregularly shaped margins that can profoundly influence the style of the resulting orogeny. Protruding parts of the margins, called *promontories,* for

example, tend to bear the brunt of the collision, while recesses called *embayments* in the margins are likely to experience less of an impact, and may even preserve vestiges of oceanic crust.

Because continental lithosphere is too buoyant to be subducted completely, plate convergence following continent–continent collision ultimately grinds to a halt. As one continent overrides the other, the ensuing collision and crustal thickening forces the crust skyward in an orogenic event that builds enormous mountain ranges. Huge faults develop in these ranges as the rocks in the impact zone are squeezed upward and sideways. Because cessation of subduction and the presence of overthickened continental crust stop the creation and ascent of magma to the surface, volcanoes are not common in continental collision zones.

Detailed studies of continental collision zones show them to be extremely complex regions that cause events to take place far from the former plate edges. Although the overall sense is one of compression and convergence, important regions of translational movement and even divergence may occur. Nowhere is this better illustrated than in eastern Asia, where the building of the Himalayas created a collisional logjam of enormous scale between the continents of India and Asia (Fig. 10.35).

Approximately 135 million years ago, the breakup of Pangea caused India to split from Africa's eastern flank (see Fig. 10.34), and by 65 million years ago India was drifting rapidly northward at about 15 centimeters (6 inches) per year. It continued to do so until about 40 million years ago, when it collided with the underbelly of the Eurasian plate to form the Himalayan Mountains and the Tibetan Plateau. Prior to collision, the Tethys Ocean, which separated the two continents, was being subducted beneath the Eurasian plate. India consequently sat on the lower plate as it approached the subduction zone and was overridden by the Eurasian plate when collision took place.

The crescent-shaped Himalayas stretch 2500 kilometers (1550 miles) and contain 30 of the world's highest peaks, including the 8848 meter- (29,029 foot-) high Mt. Everest (see Fig. 10.33). Behind the mountains lies the Tibetan Plateau where the average elevation is higher than most of the highest mountains of the continental United States. Beyond this stretches China whose earthquake-prone faults have developed thousands of kilometers from the point of collision as one continent pushes the other out of the way.

Forty million years after continental collision began, the building of the Himalayas is still in progress. Since the collision began, India has penetrated a farther 2000 kilometers (1240 miles) beneath the Eurasian plate, pushing up the Himalayas, and thickening the crust beneath the Tibetan Plateau. It is also breaking up China and Indochina, which are being squeezed sideways to give India room to move north, a process known as *tectonic escape.* The detailed explanation for the region's complex geology is controversial, but the effect has been modeled experimentally by pushing a rigid block, which represents India, into a slab of modeling clay, which represents Asia (Fig. 10.36).

Figure 10.35   Eastern Asia Continental Collision

The collisional logjam of eastern Asia started about 40 million years ago when India first made contact with Tibet. India has penetrated 2000 kilometers (1250 miles) beneath the Eurasian plate, pushing up the Himalayas and the high Tibetan Plateau. Pinned against stable Siberia, China and Indochina are being squeezed sideways toward the Pacific Ocean. Half arrows show motion along major strike-slip faults such as the Altyn Tagh and Red River; large arrows show movement of crustal blocks. As the crustal blocks jostle for position, one block far from the collision zone (red arrow) is pushed apart, opening up the Baikal Rift that harbors Siberia's Lake Baikal, which is over 1.6 kilometers deep.

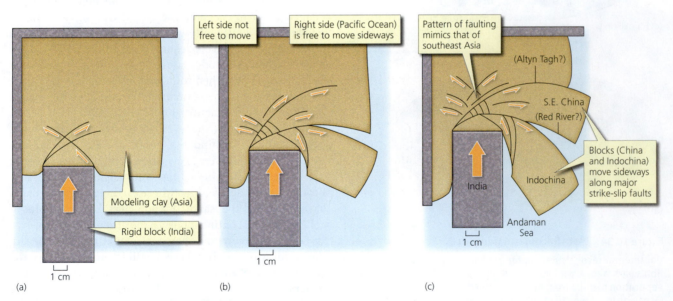

Figure 10.36   Model for Continental Collision in Eastern Asia

(a–c) Progressive collision of a rigid block (representing India) and modeling clay (representing Asia) explains Asian tectonics. Note the outward squeezing of the unconfined margin of the modeling clay (right).

The pattern of faulting produced mimics that of southern Asia when the modeling clay slab is free to move on one side (the Pacific Ocean) but not on the other (Siberia). According to the model, parts of China and central Asia (as far west as Iran and Armenia on the Caspian Sea) have been moved sideways along major strike-slip faults. As the fault blocks jostle each other, some far from the collision zone have been wedged apart, resulting in local extension. Such extension opened up the rift valley that today harbors Siberia's Lake Baikal (see Fig. 10.35). Formed by the forces of collision, Lake Baikal is the world's deepest lake, with a depth of over 1600 meters (5300 feet). It traps 20 percent of the world's freshwater, more than in all of the Great Lakes put together.

The Himalayan Mountains contain a complex mixture of rocks caught between the converging continents, including terranes that were accreted to the Eurasian plate prior to the arrival of India. Having originated in widely different environments and locations, these rocks are now welded together by collision. The force of collision also uplifted portions of oceanic crust from the subducting plate and squeezed them into the suture joining the two continents to form a band of ophiolites (see In Depth: **Ophiolites—Clues to the Structure of Oceanic Crust** in Chapter 9). Fossiliferous sedimentary rocks that were deposited in a tropical Tethys Ocean, now lie under a cover of snow and ice near the roof of the Himalayas. Parts of the uplifted crust of southern Asia have also been stripped of their sedimentary cover so that granites are exposed.

Today, the lowest elevation in the Tibetan Plateau is 3660 meters (12,000 feet), and more than 17,000 glaciers cover its surface (see Living on Earth: **In the Shadow of the Himalayas** on page 328). In fact, Tibet is so high that it tends to sag under its own weight, opening small, north-south rift valleys as it does so. As India continues to push northward, the Tibetan Plateau is squeezed against the crustal blocks that surround it. Stable Eurasia and the Siberian Shield constrain the plateau's movement to the north and west, so it moves east, cascading into southeast Asia and China as a major tectonic logjam (see Fig. 10.35). Hence, China is a tectonically active region despite its great distance from the site of the collision between India and Asia. Nowhere was this more tragically apparent than in Chengdu, the capital of China's Sichuan province, in 2008. The devastating, magnitude 7.9 earthquake that struck this city and the surrounding province killed over 69,000 people and may have owed its origin to the forces that push up the mighty Himalayas thousands of kilometers to the southwest.

> **CHECK YOUR UNDERSTANDING**
>
> • How do mountain belts produced by continental collision differ from those associated with subduction?
>
> • Why can faulting occur so far away from the site of the collision of India and Asia?

10.6 SUMMARY

- Mountains occur along every type of plate boundary and in plate interiors. They form at mid-ocean ridges by mantle upwelling, at hotspots over mantle plumes, on transform faults at convergent bends, and at convergent margins by subduction and collision.

- Mountains formed above subduction zones are related to compression and to the generation and ascent of magma to form a volcanic arc. Material scraped off the subducting plate forms an accretionary wedge.

- Elevated areas on the ocean floors, such as island arcs, seamounts, ocean ridges, and microcontinents are accreted to the continental margin at subduction zones as terranes. More than 50 such terranes, some of which traveled long distances, have accreted to western North America in the past 200 million years.

- Continental collision terminates subduction, causes crustal thickening and orogeny, and squeezes rocks in the impact zone both upward and sideways.

- The effects of continental collision extend far beyond the suture where collision occurs, and are often preceded by the accretion of terranes.

10.7 Mountain Building in the Geologic Past

As we have seen, a close relationship exists between plate tectonics and the origin of modern mountain belts. But what of ancient mountain belts? Does plate tectonics also provide the conceptual framework in which to interpret ancient episodes of mountain building? Most geologists believe the answer is yes, although just how far back the process of modern plate tectonics can be taken is a matter of considerable debate. The principal difficulty in extrapolating the process backward is that oceanic crust is destroyed by subduction and rarely survives for more than 200 million years. As a result, direct evidence of seafloor spreading in the ancient geologic record is difficult to find.

Today, most geologists believe that the essential elements of plate tectonic processes were in place by the beginning of the Proterozoic Era, some 2.5 billion years ago, and some believe that a modified form of plate tectonics may have commenced at the beginning of the Archean Eon, about 4.0 billion years ago. However, it took a provocative paper by the Canadian geophysicist, J. Tuzo Wilson, entitled "Did the Atlantic Ocean close and then re-open?" to provide the key to unlocking the mysteries of these ancient plate motions.

THE WILSON CYCLE

In 1966, J. Tuzo Wilson proposed that the Appalachian mountain belt of eastern North America and its continuation as the Caledonian belt in western Europe (see Chapter 2) formed by the destruction of an ocean that predated the Atlantic (Fig. 10.37). In Wilson's model (Fig. 10.38), the evolution of the Appalachian–Caledonian mountain belt began some 500 to 600 million years ago (in Late Precambrian to Cambrian time) with the deposition of thick sequences of shallow marine strata at the margins of the former ocean. The similarity between these deposits and those of modern continental shelves suggested that they were deposited in a similar environment. But two distinct

LIVING ON EARTH

In the Shadow of the Himalayas

The mountains of the Himalayas are the predominant influence on the environment in southern Asia, one of the world's most densely populated regions. Because the mountains are high enough to influence atmospheric circulation, they have a profound effect on the region's climate. The climate of India is dominated by the *monsoon*, which every summer brings rains from the Indian Ocean. As the moisture-laden clouds track northeast over India, they are forced up by the Himalayan foothills. As the air cools, the last of the monsoon's moisture falls as either rain or snow. Compaction of the snow into ice results in the thousands of glaciers that carve the mountains into rugged peaks. The rain and glacial meltwater feed the headwaters of some of the world's greatest rivers (Fig. 10B), including the Ganges, Indus, and Mekong. In addition, in the uplifted Tibetan Plateau to the north of the Himalayas, lie the headwaters of important easterly flowing rivers, such as the Yangtze and Huang He (the Yellow River).

The Himalayas also exert an influence far to the north. The air that travels northward over the mountains is dry because it has been stripped of its moisture. As a result, some of the world's largest deserts occur in western China.

Together, the action of glaciers and running water in the Himalayas ensure the continued cycle of uplift and erosion. The tremendous erosional power of the Ganges is indicated by the amount of sediment it carries to the Indian Ocean, as much as 1450 million metric tons annually. This is more than any other river in the world. By comparison, the Mississippi River carries a mere 145 million metric tons annually to the Gulf of Mexico.

As rivers flow to the sea, their paths are influenced by ongoing tectonic activity. Rivers exploit weaknesses in the bedrock, and fault zones are particularly weak because the rocks along them are highly fractured. As a result, many of the faults accounted for by the model illustrated in Figure 10.36 have major rivers flowing down their lengths. Some of the world's earliest civilizations began on the floodplains of these famous rivers, which provide both a water supply and a fertile soil for agriculture.

In densely populated Asia today, many communities still exist along these river valleys, relying on the river water for their survival. Unfortunately, this means that many of the places inhabited by these communities are prone to earthquakes, because the fault zones that the rivers follow are active ones. These rivers also overflow their banks each year, depositing fine-grained silt on the adjacent floodplain. The silt provides a vital replenishment of nutrients to the soil in these agriculturally intensive regions. However, the flooding also brings tragedy. The Huang He River is the deadliest in the world. In 1887, its flooding caused 900,000 fatalities, and in 1931, four million people died when flooding resulted in starvation and disease.

Figure 10B   Himalayan Range

In this composite satellite image of the snow-capped Himalayan Mountains, the high Tibetan Plateau is visible behind the mountains, and China's Taklamakan Desert (lighter area) is seen to the north (top). The drainage systems of the Himalayas feed tributaries of the Indus, Ganges, and Brahmaputra rivers, while those of the Tibetan Plateau feed the headwaters of the Mekong, Yangtze, and Huang He (Yellow) rivers.

In an effort to control the flooding of the Yangtze River (and harness its enormous energy for hydroelectric power), the Chinese constructed the controversial Three Gorges Dam (Fig. 10C). Completed in 2006, it is the most powerful dam ever built, and China's most ambitious project since the Great Wall was constructed 2000 years ago.

Despite the monsoon's erosional onslaught, the Himalayas remain the world's highest mountains because, with an uplift rate of 5 millimeters per year, they are still rising faster than they are being eroded. However, as erosion strips off the tops of the mountains ever deeper segments of the mountain belt are exposed at the surface as the crust rises to reestablish isostatic equilibrium (see Fig. 9.8). In this way the Himalayas are a vivid example of isostasy and the rock cycle (see Chapter 9). As the guts of the mountain belt are exposed, igneous and metamorphic rocks formed deep within the crust weather and erode, and the products of this erosion are transported downstream to be deposited as new sediment in vast submarine fans at the mouths of the Indus and Ganges rivers.

Figure 10C   Three Gorges Dam

The Three Gorges Dam on the Yangtze River in China is more than 2.3 kilometers (almost 1.5 miles) long and 110 meters (360 feet) high. Completed in 2006, the controversial hydroelectric dam is the world's largest power station and is intended to reduce the potential for floods downstream and increase the river's shipping capacity. But its reservoir flooded 1300 archeological sites and displaced some 1.3 million people.

fossil assemblages occurred in these rocks. One, known as the *Pacific realm,* occurred along the northwestern (or North American) side of the mountain belt, while the other, called the *Atlantic realm,* occurred along the southeastern (or European) side. The distinctiveness of these fossil assemblages suggested that the shallow marine sedimentary rocks in which they occurred must have been separated by an ocean at the time of their deposition (Fig. 10.38a). This ocean could not have been the modern Atlantic, because we know that the Atlantic Ocean only started to form some 200 million years ago with the breakup of Pangea. Instead, it must have been an older ocean that had been destroyed by the time Pangea formed.

The abundance of volcanic rocks of Ordovician age (444 to 485 million years old) with chemical compositions that resemble those of modern island arcs suggests that this former ocean began to subduct in the Ordovician Period (Fig. 10.38b). The extensive deformation that affected these

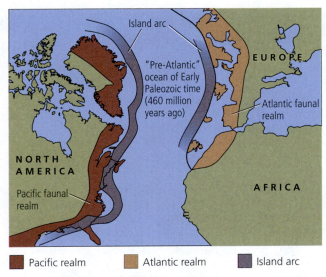

Pacific realm   Atlantic realm   Island arc

Figure 10.37   Pre-Atlantic Ocean

Tuzo Wilson's model for the Appalachian-Caledonian mountain belt was based on the recognition of different faunal realms and the inference that an ocean once lay between them. Wilson interpreted belts of volcanic rocks of Ordovician age (444–485 million years ago) to be remnants of ancient island arcs formed by the ocean's subduction.

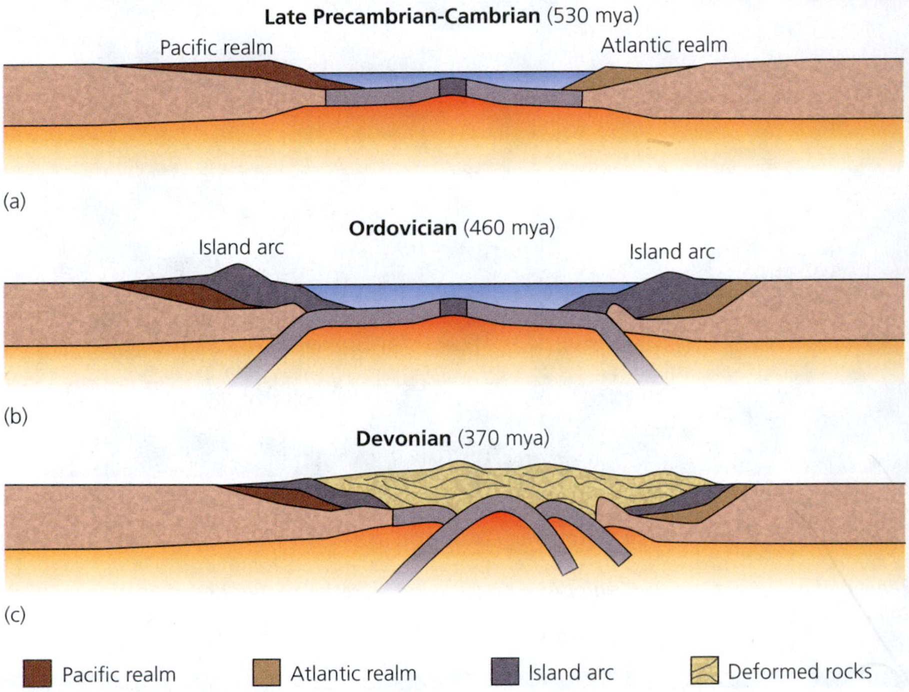

Late Precambrian-Cambrian (530 mya)

Pacific realm   Atlantic realm

(a)

Ordovician (460 mya)

Island arc   Island arc

(b)

Devonian (370 mya)

(c)

Pacific realm   Atlantic realm   Island arc   Deformed rocks

Figure 10.38   Wilson Cycle

(a) Continental rifting led to the formation of an ancient ocean flanked by continental shelf deposits in Late Precambrian-Cambrian time some 500 to 600 million years ago. (b) Subduction in this ocean commenced in the Ordovician Period, between 444 and 485 million years ago, and dismembered the continental shelves, resulting in the formation of island arcs. (c) Continued subduction led to closure of the ocean sometime during the Devonian Period, between 359 and 419 million years ago.

resulted in the formation of the Appalachian–Caledonian mountain belt. Subsequent field work in Newfoundland identified relics of this ancient ocean. Although the vast majority of its ocean floor was destroyed by subduction, slivers of oceanic crust trapped by the colliding continents have been preserved in Newfoundland as ophiolites (see In Depth: **Ophiolites—Clues to the Structure of Oceanic Crust** in Chapter 9).

The idea that oceans may open and then close ignited the concept of cycles of tectonic activity. In what came to be known as the **Wilson cycle**, continental rifting and breakup leads to the development of a continental margin on which thick sequences of sediments are deposited. Eventually, subduction commences leading to the development of island arcs. When the rate of subduction exceeds the rate of seafloor spreading, the ocean begins to close, culminating in continental collision. Many of the types of orogeny we examined in this chapter are part of this cycle. As the ocean begins to close, for example, mountains are produced by subduction. These mountains resemble those of island arcs or the modern Andes. As subduction proceeds and the ocean gets narrower, terranes become accreted to the continental margin. Eventually, when the oceanic crust is consumed, mountains are produced by continental collision. Thus the development of a mountain chain such as the Appalachian–Caledonian belt may involve several episodes of tectonic activity, beginning with subduction and terrane accretion, and culminating in continental collision.

> **CHECK YOUR UNDERSTANDING**
>
> ○ What is the Wilson cycle and how has it been applied to ancient mountain belts?

THE SUPERCONTINENT CYCLE

The Wilson cycle is one of ocean opening and closing. But what processes determine when oceans open and close? The answer to this question may lie in an even larger tectonic cycle known as the supercontinent cycle. Although subduction destroys most direct evidence for the existence of ancient oceans, Tuzo Wilson realized that the continental shelves of ancient oceans, as well as the magmatism associated with their opening and closure, and the deformation accompanying eventual collision, are likely to be preserved. It is essentially the same approach that we use today to deduce the existence of ancient

rocks is attributed to continent–continent collision during the Devonian Period and into the Carboniferous, between about 340 and 419 million years ago (Fig.10.38c). This collision brought the two fossil assemblages together and

oceans and mountain belts. Increased understanding of the processes involved in the Wilson cycle has led to an expansion of his ideas into what has become known as the **supercontinent cycle** (Fig. 10.39). This approach places mountain building activity in a global context, and in so doing, demonstrates that continental breakup in one part of the world necessitates mountain building activity in another.

To understand the driving mechanism of this cycle, we must first examine why Pangea or any other supercontinent would break up. Several theories that might account for the breakup of supercontinents have been proposed. The most popular theory centers on their insulating properties. Because continental crust is a poor conductor of heat compared to thinner ocean crust, supercontinents act like an insulating blanket that blocks the escape of heat from Earth's interior. Just as a book placed on an electric blanket gets hot along its base because it prevents heat from escaping, so a supercontinent causes heat to build up in the mantle beneath it. As the heat accumulates, magmas are generated in the mantle while the supercontinent bulges upward and cracks. The magma is injected into these cracks and other preexisting fractures and the supercontinent is progressively pried apart. Eventually, continental breakup takes place and new oceanic crust is created as the continental fragments slide off the thermal bulge and drift apart (see Fig. 10.39, stage 2). If this model is correct, supercontinents have in-built obsolescence; their very formation sowing the seeds of their own destruction. Subduction and the accretion of terranes along the western margin of North America may well be related to the opening of the Atlantic Ocean and the breakup of Pangea. This model consequently provides a framework for linking, on a global basis, the plate tectonic processes of continental rifting, seafloor spreading, subduction, and collision that we discussed in Chapters 2 and 9.

But the continents cannot continue to drift apart indefinitely. Recall from Chapter 9 that oceanic lithosphere becomes colder and denser as it ages, eventually becoming denser than the underlying asthenosphere so that its subduction is inevitable (see Fig. 10.39, stage 3). The geologic record suggests that

this occurs before the oceanic lithosphere attains an age of about 200 million years. The margins of the central Atlantic Ocean are floored by oceanic crust as much as 180 million years old. This part of the ocean is therefore approaching its maximum age and should soon begin to subduct. When that happens, the Atlantic may begin to shrink, and another supercontinent could form when its opposing continental margins collide. Alternatively, if the Atlantic Ocean continues to expand, the Pacific Ocean may eventually close and another supercontinent could form when the opposing margins of that ocean collide. These two possible future supercontinents have been named *Pangea Proxima* and *Amasia,* respectively, and are predicted to assemble sometime within the next 50 to 250 million years (Fig. 10.40).

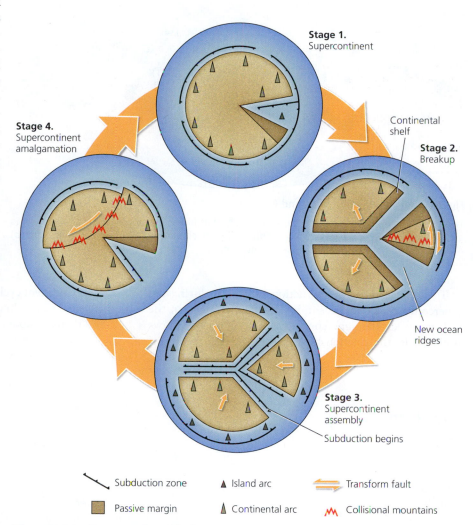

Figure 10.39 Supercontinent Cycle
The supercontinent (stage 1) is flanked by subduction zones which generate continental arcs. The supercontinent traps mantle heat beneath it resulting in breakup (stage 2). Breakup results in the formation of new oceans and mid-ocean ridges. The oceans are flanked by continental shelves, known as passive margins, which are characterized by thick layers of sedimentary rocks. Eventually, the aging oceans become gravitationally unstable and begin to subduct (stage 3) and a new supercontinent begins to reassemble. This results in the destruction of the continental shelves and the formation of island arcs. Ultimately, the oceanic lithosphere is destroyed by subduction, and a new supercontinent assembles (stage 4). This supercontinent traps mantle heat once again and the cycle starts anew.

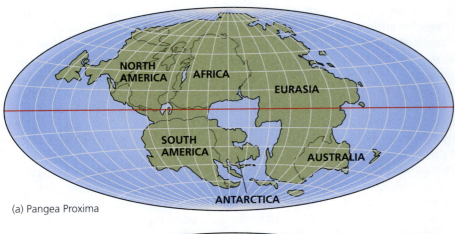

(a) Pangea Proxima

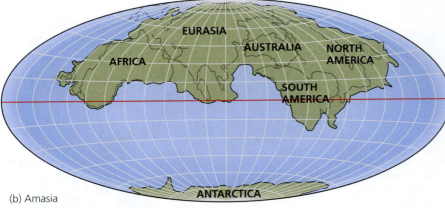

(b) Amasia

Figure 10.40   Pangea Proxima and Amasia

Two possible configurations of a supercontinent that might assemble within the next 50 to 250 million years: (a) Pangea Proxima, formed by the closure of the Atlantic Ocean, and (b) Amasia, formed by the closure of the Pacific Ocean.

The generation and destruction of oceans during the supercontinent cycle are similar to that envisaged in the Wilson cycle. The supercontinent cycle, however, draws attention to the relationship between the generation of

interior oceans, which are those that form when a supercontinent breaks up, and processes taking place in the exterior ocean, which surrounds a supercontinent following its assembly. In addition, the model offers a mechanism that may explain why supercontinents occur, and may account for the apparently episodic nature of mountain-building activity.

10.7 SUMMARY

- The Wilson cycle of tectonic activity commences with rifting and the development of an ocean followed by subduction and the formation of island arcs, and finally continent–continent collision.

- Intense episodes of mountain building and continental rifting are explained by a supercontinent cycle. Supercontinent amalgamation is followed by breakup and the generation of new oceans, and then by subduction that ultimately leads to reassembly of a supercontinent.

Key Terms

accreted terrane (terrane)   319
accretionary wedge   319
angular unconformity   313
anticline   311
axial plane   310
basin   312
brittle   300
brittle failure   301
columnar joint   304
compressional stress   300
deformation   298
detachment   307

dip   302
dip-slip fault   306
disconformity   314
dome   312
ductile   301
elastic deformation   300
exfoliation joint   304
faults   304
fold   309
fold hinge   311
fracture   304
graben   307

half-graben   307
horst   307
joint   304
joint sets   304
left-lateral fault   309
microcontinents   319
nonconformity   314
normal fault   306
orogeny   315
plastic deformation   300
reverse fault   306
right-lateral fault   309
shear stress   300

slip   306
strain   299
stress   299
strike   302
strike-slip fault   307
supercontinent cycle   331
suture   325
syncline   311
tectonic joint   304
tensional stress   300
thrust   307
Wilson cycle   330

Key Concepts

10.1 DEFORMATION

- Deformation is the result of stresses (compression, extension, or shear) that strain rocks by changing their shape or volume.
- Brittle rocks deform elastically and fracture, whereas ductile rocks deform plastically and fold.
- Low temperatures and pressures and high strain rates promote elastic deformation, whereas high temperatures and pressures and low strain rates promote plastic deformation. As a result, near-surface rocks tend to fracture, whereas those at depth will bend.

10.2 ORIENTATION OF GEOLOGIC STRUCTURES

- The orientation of a planar geologic structure is measured by its strike and dip.

10.3 FRACTURES

- Joints are fractures that show no appreciable movement.
- Faults are fractures showing significant movement and are grouped into dip-slip and strike-slip varieties.
- Normal faults and detachments are extensional and common in rift settings. Reverse faults and thrusts are compressional and typical of convergent plate margins.
- Strike-slip faults can be right-lateral or left-lateral and are typical of transform plate boundaries.

10.4 FOLDS

- Folds are bent rock layers and possess an axial plane and fold hinge that are used to describe the fold orientation and fold types.

- Folded layers are bent upward to form anticlines and downward to form synclines.
- When folding accompanies metamorphism, a foliation may form parallel to the fold's axial plane.

10.5 UNCONFORMITIES REVISITED

- Unconformities record time gaps in the geologic record and are of three types: angular unconformities, nonconformities, and disconformities.

10.6 PLATE TECTONICS AND MOUNTAIN BUILDING

- Mountains form at all plate boundaries—at mid-ocean ridges, at bends in transform faults, and by subduction. They also form at hotspots over mantle plumes.
- Mountains above subduction zones reflect compression and the ascent of magma.
- Elevated areas of ocean floor are accreted to continental margins at subduction zones as terranes.
- Continental collision causes significant crustal thickening and mountain building.
- Continental collision affects areas at great distances from the suture where collision occurs.

10.7 MOUNTAIN BUILDING IN THE GEOLOGIC PAST

- The Wilson cycle includes rifting, ocean opening, subduction, and continental collision.
- Episodic periods of continental rifting and collision suggest the existence of a supercontinent cycle.

Study Questions

1. What controls whether rocks respond to deformation in a brittle fashion or a ductile fashion?

2. How is the orientation of a planar geologic structure measured?

3. In what way is the type of fault related to the type of stress that caused the fault?

4. How are the orientations of the axial plane and fold hinge used in the description of folds?

5. How are the different types of unconformity recognized in the field and what series of events might lead to the formation of each type?

6. In what geologic settings are mountains produced and what processes are involved in their formation?

7. Explain how the accretion of terranes along the western margin of North America may be related to the breakup of Pangea.

8. How does the model for continental collision illustrated in Figure 10.34 help to explain the collision between India and Asia?

9. What is the relationship between the formation of oceans and mountain building?

10. In what way does the supercontinent cycle build on the concepts of the Wilson cycle?

11

Earthquakes and Earth's Interior

11.1 Heat and Density: Clues to Earth's Interior

11.2 Earthquakes and Elastic Rebound

11.3 Seismic Waves

11.4 Journey to the Center of the Earth

11.5 Earth's Structure and Composition

11.6 Circulation in Earth's Interior

11.7 Plate Tectonics and the Fate of Subducted Slabs

▶ Key Terms

▶ Key Concepts

▶ Study Questions

LIVING ON EARTH: Predicting Earthquakes—Some Lessons from California 340

SCIENCE REFRESHER: Density, Volume, and Seismic Velocity 344

SCIENCE REFRESHER: Wave Refraction 349

IN DEPTH: Earth's Magnetism and Its Dynamic Core 357

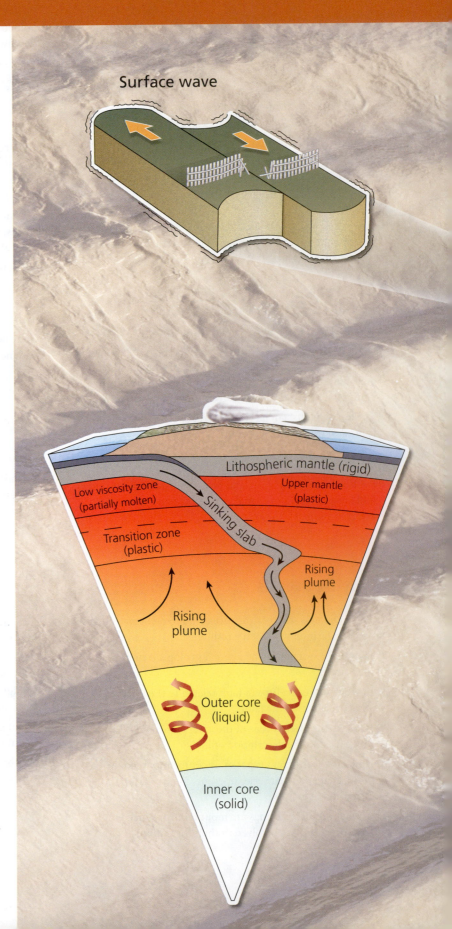

Surface wave

Lithospheric mantle (rigid)

Low viscosity zone (partially molten)

Upper mantle (plastic)

Sinking slab

Transition zone (plastic)

Rising plume

Rising plume

Outer core (liquid)

Inner core (solid)

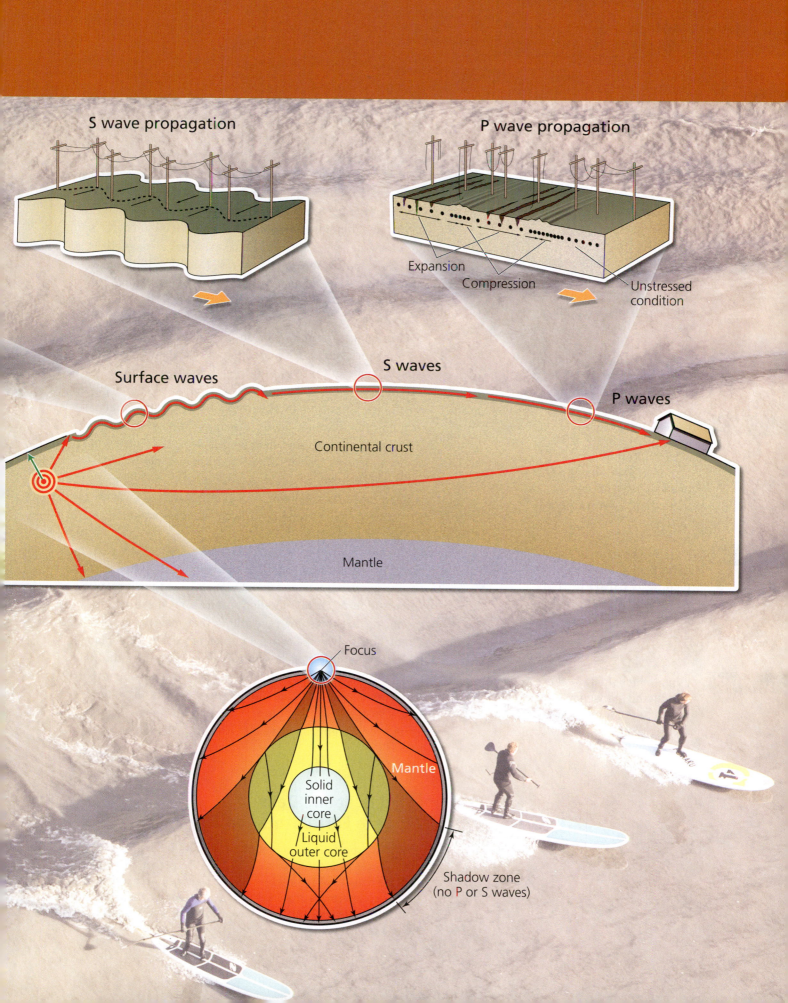

S wave propagation

P wave propagation

Expansion

Compression

Unstressed condition

Surface waves

S waves

P waves

Continental crust

Mantle

Focus

Mantle

Solid inner core

Liquid outer core

Shadow zone (no P or S waves)

11.1 Describe how geophysicists use geophysical methods and earthquake waves to probe Earth's interior.

11.2 Explain the elastic rebound theory, which provides an explanation for the origin of earthquakes.

11.3 Describe the characteristics of seismic waves and explain the difference between primary waves, secondary waves, and surface waves.

11.4 List the key observations that revealed Earth to possess a layered planetary interior.

11.5 Explain our current understanding of Earth's hidden interior using the evidence provided by Earth's internal heat, its density, and the path of earthquake waves through its interior.

11.6 Describe how it is now possible to produce three-dimensional pictures of Earth's interior and what these pictures indicate about the circulation of materials in the core and mantle.

11.7 Explain how the fate of subducted slabs may be linked to the formation of mantle plumes at the core–mantle boundary.

What is Earth's interior made of and how do we know? The only rocks available for study are those at the surface and the rocks we can recover by drilling. Our direct knowledge of Earth's interior is confined to the first few tens of kilometers immediately below the surface. As Earth has a radius of almost 6400 kilometers (4000 miles), this direct knowledge applies to just a tiny fraction of the planet.

We must therefore examine Earth's interior indirectly, by studying phenomena such as earthquakes, gravity, magnetism, and heat flow and by applying the laws of physics. With our knowledge of these phenomena, we can interpret images of Earth's interior, much like a physician interprets X-ray or CAT scan images of a patient.

Earthquakes occur when a locked fault suddenly lets go, sending out shock waves in all directions as the pent-up energy is released. Some of these earthquake waves travel through Earth's interior and, in doing so, reveal much about the materials through which they pass. Studies of earthquake waves show that Earth is divided into three main layers: the *crust,* the *mantle,* and the *core.* These studies have also identified a weak layer in the mantle, above which the tectonic plates can move. More recently, computer imaging of earthquake waves in Earth's mantle has shown in great detail the changing speeds at which these waves travel. This added knowledge tells us much about the variation and flow of heat within Earth.

Despite many challenges, the indirect application of these methods to the study of Earth's internal processes has led to surprising agreement among geophysicists about the structure and composition of Earth's interior. In addition, these methods are producing exciting new revelations that may finally provide a direct link between Earth's internal processes and plate tectonics, something that has remained elusive since the time of Wegener. Yet scientists still have much to learn about Earth's mysterious interior. In this chapter we explore Earth's interior—its layers and the circulation of heat through them—by studying Earth's movements as revealed through earthquake waves.

11.1 Heat and Density: Clues to Earth's Interior

The first important clues to understanding the nature of Earth beneath the crust come from studying the planet's internal heat and the density of its internal layers. *Density* is a measure of how closely compacted a material is, and it is one of the chief factors controlling the speed with which earthquake waves travel through Earth's interior.

It is a popular misconception that we are standing on solid ground that is anchored to the interior of Earth. We are not. As we learned in Chapter 9, the continents we are standing on are floating like rafts in a crowded stream, moving today at rates between 1 and 6 centimeters (0.5–2.5 inches) per year. Over millions of years this motion has created entire ocean basins, when moving continents diverge from each other, only to close these basins when continents converge and collide. These motions are powered by heat energy that has been escaping from the interior of Earth for more than 4.5 billion years. Volcanoes are the most obvious and dramatic manifestation of the escape of this internal heat. However, the flow of heat from the surface of Earth to the atmosphere, known as *heat flow,* can be measured everywhere on our planet, as we discussed in Chapter 1 (see Fig. 1.18). This heat flow indicates that our planet is cooling. But where is this heat coming from and how is it being transported from deep within Earth to the planet's surface? To answer these questions we must turn our attention to the structure and composition of Earth's interior.

Processes in the interior of our planet are hidden from view and can only be examined indirectly. So how different is Earth's interior from its surface? Laboratory experiments tell us that most of the minerals present in surface rocks would not be stable at depths any greater than about 40 kilometers (25 miles) below the surface. The average density of continental rocks is about 2.7 g/cm³, that is, an average continental rock weighs about 2.7 times as much as the same volume of water. However, the average density of Earth is 5.52 g/cm³. Calculations show that this difference in density cannot be due only to increasing pressure with depth, implying that the interior of Earth must be vastly different from its exterior. In the absence of direct observations, Earth scientists use *geophysical methods,* or the application of physical laws, to probe Earth's interior. These methods give us a good first approximation of the composition of Earth from core to surface.

We know for instance, that Earth is an internally layered planet, being stratified according to density from its upper atmosphere to its inner core. But how have we determined this? To answer this question we must first examine the nature of earthquakes and the shock waves they generate because it is these waves that have revealed the layered nature of our planet by providing us with an "X-ray" picture of Earth's interior.

> **CHECK YOUR UNDERSTANDING**
>
> ○ Why must we use indirect methods to study Earth's interior?
>
> ○ How does the density of Earth indicate that the planet's interior is vastly different from its surface?

11.1 SUMMARY

- In the absence of direct observation, geophysicists apply physical laws to probe Earth's interior.

- By measuring Earth's density and heat indirectly, geophysicists have learned how Earth's layers are segregated.

11.2 Earthquakes and Elastic Rebound

As we learned in Chapter 10, the boundary between two rock masses struggling to slide past each other forms a fracture in the crust known as a *fault.* If rocks along a fault lock together while stress continues, *strain* builds up in the rocks on either side of the fault. The sides slowly bend, storing up elastic energy (Fig. 11.1). This situation is analogous to bending a stick—you can feel the strain energy building in your arms as the stick is bent tighter. Finally, when the stick snaps, the tension in your arms is abruptly released. In the same way, when the strain stored in the rocks on either side of a fault reaches a threshold, the rocks abruptly rupture and jerk violently past each other. The energy released sends out shock waves, known as **seismic waves,** in all directions for periods that can last for several minutes. This is the *stick-slip motion* of faults we encountered in Chapter 9. This motion and its cause form an explanation for the origin of earthquakes known as the **elastic rebound theory,** which lies at the heart of ongoing research into earthquake prediction (see Living on Earth: **Predicting Earthquakes— Some Lessons from California** on page 340).

The vibrations of an earthquake begin at the **focus**—the point where a locked fault suddenly lets go. The focus of an earthquake generally lies at some depth below Earth's surface, whereas the **epicenter** is the point on Earth's surface directly above the focus. From the focus, seismic waves spread outward in all directions in much the same way that ripples radiate when a pebble is thrown into a pond. The waves soon ripple through Earth's interior, literally filling Earth in the same way that sound waves fill a ringing bell.

> **CHECK YOUR UNDERSTANDING**
>
> ○ What is the elastic rebound theory and how does it account for the occurrence of earthquakes?

11.2 SUMMARY

- According to the elastic rebound theory, earthquakes occur when stress builds up on locked faults, straining the rocks on either side until they abruptly rupture, sending out seismic waves.

11.3 Seismic Waves

There are several types of seismic waves, which we group into two categories: *surface waves* and *body waves* (Fig. 11.2). In this section of the chapter, we discuss the different kinds of waves and their characteristics.

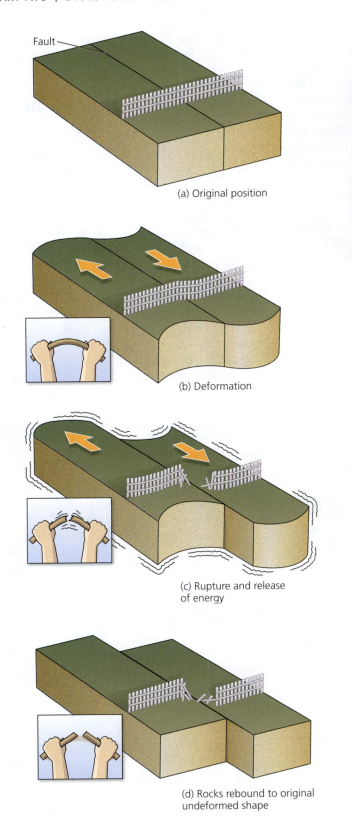

(a) Original position

(b) Deformation

(c) Rupture and release of energy

(d) Rocks rebound to original undeformed shape

(e) Movement along a fault

Figure 11.1   Elastic Rebound Theory

This sequence of figures illustrates the elastic rebound theory for the cause of earthquakes. (a–d) Stresses acting on rocks cause the rocks to bend (strain) until they break, releasing stored energy as they snap forward. (e) Following the 1906 San Francisco earthquake, a fence in Marin County, California, was offset by 2.5 meters (8 feet).

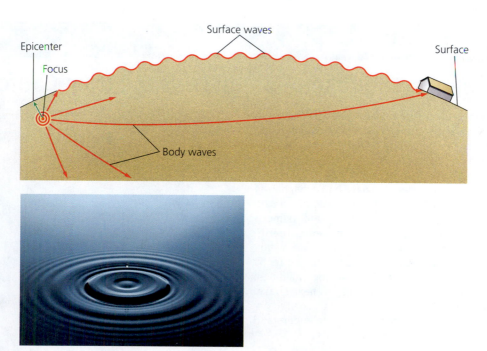

Figure 11.2   Seismic Waves
Seismic waves spread in all directions from the focus of an earthquake much like ripples on a pond.

SURFACE WAVES

The seismic waves that travel along Earth's surface are **surface waves**. Surface waves cause most of the damage associated with earthquakes, producing low-frequency vibrations that roll the ground like ocean waves or whip it sideways. Compared with other seismic waves, surface waves travel slowly and take longer to diminish. As a result, these waves can sway buildings located at great distances from the earthquake epicenter. Surface waves are of limited use in probing Earth because they only travel along the surface and do not penetrate the interior.

> **CHECK YOUR UNDERSTANDING**
>
> ● Why are surface waves of limited use in learning about Earth's interior?

BODY WAVES

> **CHECK YOUR UNDERSTANDING**
>
> ● How do surface waves differ from body waves?

In contrast to surface waves, **body waves** are not confined to Earth's surface, but instead travel through Earth's interior (see Fig. 11.2). We derive much of our knowledge of Earth's internal structure from these waves. There are two main kinds of body waves: *primary* and *secondary*.

Primary waves

Primary (P) waves are generated by sudden compression or extension of the ground at the site of an earthquake, like a push or pull on the end of a stretched spring (Fig. 11.3a).

Like sound waves, P waves temporarily affect the volume of the material they pass through by alternately compressing and expanding this material in the direction in which they travel. P waves travel through solids, liquids, and gases because all three materials react to the wave by resisting a change in volume when they are compressed. However, the exact speed of P waves depends on the nature of the material through which they pass. P waves travel most rapidly through dense, rigid rocks; they are slowed by hot rocks and by liquids and gases.

Secondary waves

Secondary (S) waves are generated by the shearing or sliding motion at an earthquake site. S waves are therefore shear waves and behave like oscillations in a rope (Fig. 11.3b). When a rope is shaken, a series of up-and-down or side-to-side oscillations travel along its length. In the same way, S waves cause the material in their path to heave vertically or sideways, perpendicular to the direction in which they are traveling. This motion affects the shape of the material through which the oscillation travels. As a result, S waves can only travel through solid materials because, in contrast to liquids and gases, only solids have a definite shape. S waves are blocked when they encounter liquids or gases.

TRANSMISSION OF SEISMIC WAVES

P waves temporarily cause changes in volume because, by alternately compressing and extending the materials through which they pass, P waves cause the materials to expand and shrink. In contrast, S waves cause materials to

LIVING ON EARTH

Predicting Earthquakes—Some Lessons from California

In 1868, after a major earthquake in the San Francisco region, the *San Francisco Daily Examiner* proclaimed that such earthquakes were too rare to be worried about. In 1906, more than 700 people perished when San Francisco was reduced to a pile of ashes by raging fires that followed a massive earthquake estimated to have had a magnitude of 8.3 on the Richter scale. In just a few seconds, a segment of the San Andreas Fault lurched forward as much as 6 meters (20 feet). Today the threat of serious earthquakes is an accepted fact of life that almost 40 million Californians live with, and billions of dollars are currently spent on renovating or retrofitting existing buildings, bridges, and roads to try to outmuscle the next quake. More than 100 years after these tragic events, plate tectonics has provided the conceptual framework to understand earthquakes.

Recall from Chapter 9 that the San Andreas Fault, which runs through San Francisco, is a continental transform fault along which earthquakes are inevitable. Modern estimates of an earthquake's magnitude are based on the total energy released by the earthquake, which is calculated, in part, by multiplying the area of the fault's rupture surface by the distance Earth moves along the rupture. In recent years, the two most important earthquakes were centered in Loma Prieta, about 100 kilometers (60 miles) south of San Francisco (in October 1989), and at Northridge, 40 kilometers (25 miles) northwest of Los Angeles (in January 1994). These earthquakes demonstrated how the complexity of earthquake patterns relates to the variations in the stress levels along the fault itself. This complexity means that earthquake prediction is even more difficult than had previously been anticipated.

The Loma Prieta earthquake, with a magnitude of 6.9, killed 63 people and caused extensive damage, estimated at six billion dollars, in the San Francisco area. The earthquake was caused by a rupture along the San Andreas Fault some 10 kilometers (6 miles) below the surface. A plot of seismic activity over the previous 20 years along the San Andreas Fault from San Francisco in the north to Monterey in the south (Fig. 11A) reveals that the Loma Prieta earthquake was located in a major *seismic gap*. Seismic gaps, such as the ones around San Francisco, are marked by few earthquakes (see Fig. 11A). These areas are locked segments of the fault and are therefore storing, rather than releasing, the energy associated with the fault motion. Loma Prieta snapped because the stored stress finally exceeded a critical threshold.

Figure 11B shows that the Loma Prieta earthquake and its aftershocks were preferentially concentrated in the Loma Prieta seismic gap, which, in a 1988 study, had been assigned a high probability of a major earthquake (magnitude >7) for the period 1988–2018 (Fig. 11C). The seismic gap revealed by Figure 11B near Parkfield, farther south on

Figure 11A   Bay Area Earthquakes

This map shows the epicenters on the San Andreas fault system in the San Francisco Bay area for the 20-year period prior to 1989. The main shock of the 1989 Loma Prieta earthquake is shown by the large symbol (the epicenters of the Loma Prieta earthquake and its aftershocks are in white; earlier epicenters are in yellow). Note the seismic gap revealed by the absence of yellow symbols in the San Francisco peninsula.

the San Andreas Fault, for which a very high probability of a major earthquake had also been assigned (Fig. 11C), was the site of a magnitude 6.0 earthquake in 2004. This analysis implies that the remaining seismic gap in the San Francisco peninsula is still at high risk. It is estimated that an earthquake between magnitudes 7.5 and 8 is required to alleviate the built-up strain in the San Francisco region.

The Loma Prieta earthquake, tragic as it was, is considered typical of seismic events along continental transform faults. In contrast, the Northridge earthquake of 1994 awakened our senses to the enigmatic character of seismic events. The fault that ruptured 18 kilometers (11 miles) beneath Northridge that January morning was a *blind fault*—one that is not exposed at the surface. Because of this, the pattern of stress buildup that led to its rupture was undetectable. The earthquake, with a magnitude of

Cross-Sections of Seismic Activity Along the San Andreas Fault, California

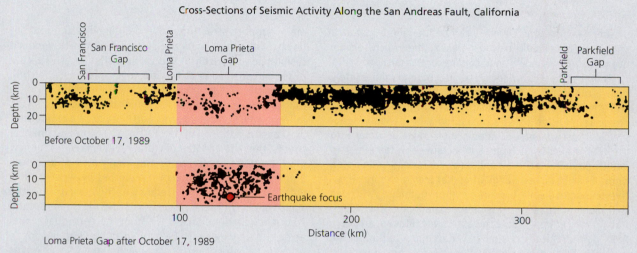

Figure 11B *Seismic Gaps*

The record of seismic activity on the San Andreas Fault in the 20-year period prior to the 1989 Loma Prieta earthquake shows the locations of the thousands of earthquakes that occurred during that time interval. The record also reveals three seismic gaps (marked by far fewer earthquakes) near San Francisco, Loma Prieta, and Parkfield, which lies farther south on the San Andreas Fault. The Loma Prieta seismic gap was filled (lower section) by the main shock (large circle) and aftershocks of the Loma Prieta earthquake.

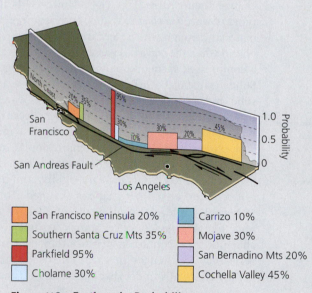

San Francisco Peninsula 20%
Southern Santa Cruz Mts 35%
Parkfield 95%
Cholame 30%
Carrizo 10%
Mojave 30%
San Bernadino Mts 20%
Cochella Valley 45%

Figure 11C *Earthquake Probability*

This probability map was published in 1988 for major earthquakes (magnitude >7.0) along the San Andreas Fault between San Francisco and San Bernardino for the period 1988–2018. Note the high probability (35%) assigned to the southern Santa Cruz Mountains where Loma Prieta is located, and the very high probability (95%) assigned to Parkfield, which experienced a magnitude 6.0 earthquake in 2004.

6.7, claimed 61 lives and resulted in damage estimated at 20 billion dollars, a figure for natural disasters exceeded in North America only by Hurricanes Andrew (1992), Katrina (2005), Ike (2008), and Sandy (2012).

The underlying problem with predicting earthquakes is that, in a matter of seconds, earthquakes can release the stored energy along a fault that may have taken hundreds or even thousands of years to accumulate. So while earthquake-prone areas can be readily identified from seismic risk maps, predicting exactly when an earthquake might occur is a very difficult task. For example, a 1985 study organized by the United States Geological Survey estimated that the San Andreas Fault at Parkfield had a 95 percent probability of generating a magnitude 6 earthquake before 1993. Although this should have encouraged local residents to take precautions such as storing emergency supplies and learning earthquake drills, predictions of this sort are too broad to be useful in preparations to minimize human exposure to a sudden catastrophe. As it turned out, a magnitude 6.0 earthquake did indeed occur, but not until 2004.

The main thrust of earthquake prediction is to identify *precursors*, or measurable properties whose patterns change immediately prior to the main quake. As with volcanic predictions, Global Positioning Systems using a network of satellites and receiving stations can measure movements of the crust on a centimeter scale. In this way, accelerated or retarded motions can be measured along the length of the fault and unusual activity, such as low intensity

(Continued)

LIVING ON EARTH

Predicting Earthquakes—Some Lessons from California

(Continued)

seismic (*microseismic*) events, can be identified. Just as a stick being bent creaks before it breaks, these microseismic events may foreshadow an oncoming earthquake.

Rocks undergo an increase in volume immediately before rupturing as a result of the creation of tiny cracks and the consequent influx of water to fill these cracks. These tendencies can be measured with sensitive geophysical instruments. Similarly, movement of water into the cracks results in changes in local well-water levels. In some instances, agitated animal behavior has been noted prior to earthquakes, suggesting that they are more sensitive to microseismic precursors than humans. All agree that much more research is needed, and given the potential for an unimaginable disaster, the stakes are very high indeed. But the challenge is daunting—to predict the instant a fault will succumb to stresses that may have built up for thousands of years.

In California, building codes require that a structure should be able to withstand a 25-second main shock. The main support structures of a building (its foundation, floors, walls, and roof) are all tied together to enhance resistance to horizontal and vertical shaking. The nature of the subsurface is also critical. For example, the 1985 earthquake that struck Mexico City and caused 9500 fatalities had its epicenter more than 200 kilometers (125 miles) away. So why was this earthquake so devastating? The reason is that all materials have a natural resonance, that is, they are capable of vibrating at a fixed frequency. Like the glass shattered by the high notes of an opera singer, resonance is energy transferred between bodies with the same natural frequency. In Mexico City, the buildings and the subsurface soil had the same natural frequency as the shock wave from the earthquake. Just as a child's swing has to be pushed at the right time for an efficient transfer of energy, the earthquake's shock waves were very efficiently transferred to the buildings with devastating consequences.

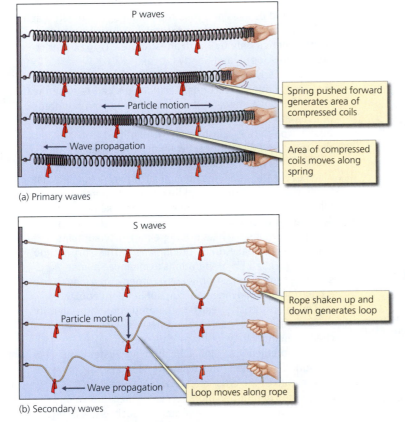

(a) Primary waves

(b) Secondary waves

Figure 11.3   Wave Propagation

(a) A P wave is like a sudden push on the end of a stretched spring. The particles vibrate parallel to the direction of wave propagation. (b) An S wave is analogous to shaking a rope. The particles vibrate perpendicular to the direction of wave propagation.

change their shape (Fig. 11.4). However, both types of wave are *elastic*. That is, both require the material through which they are traveling to rebound when the wave has passed through, just as a compressed spring rebounds when it is let go. Solids transmit both P and S waves because all solid materials are elastic. However, P waves travel through fluids (liquids and gases) because fluids behave elastically under compression and will expand again when the compression is removed. Fluids can therefore transmit P waves.

Unlike solids, fluids have no rigidity and do not rebound after their shape is changed. This is because fluids have no shape of their own and, like a glass of water, simply assume the shape of their container. Because they have no resistance to changes in shape, fluids do not behave elastically under shear and cannot transmit S waves.

For a simple analogy of the transmission of P and S waves through matter, imagine a line of people standing shoulder to shoulder (Fig. 11.5). A solid, because of the linkages between its atoms, is like a line of people with linked arms (Fig. 11.5a). If we mimic a P wave by pushing the first person toward the second, all the people in the line will eventually move in the same direction. Similarly, if we mimic an S wave, by moving the first person forward and backward all the people in the line will eventually be forced to

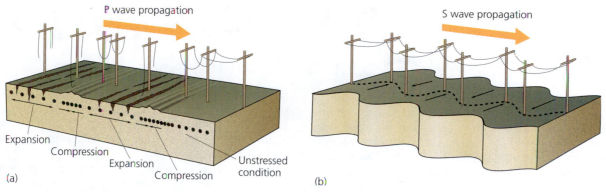

Figure 11.4   Ground Motion

(a) P waves compress and expand the ground. (b) S waves make the ground move from side to side or up and down.

move. Hence a solid, like a line of people whose arms are linked, can transmit both P and S waves.

In a liquid, the linkages between adjacent atoms are much looser than in a solid, as if the line of people, although still standing shoulder to shoulder, were holding their arms by their sides (Fig. 11.5b). If we push the first person into the second to mimic a P wave, all the people in the line will again be eventually affected. But if we now move the first person forward and backward to mimic an S wave, the second person will not follow because

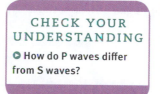
CHECK YOUR UNDERSTANDING
➲ How do P waves differ from S waves?

the two people are no longer connected so the movement is not transmitted. Hence a liquid, like a line of people whose arms are not linked, can transmit P waves but not S waves.

Solid materials respond fastest to body waves when the solids are dense and rigid (see Science Refresher: **Density, Volume, and Seismic Velocity** on page 344). The reason

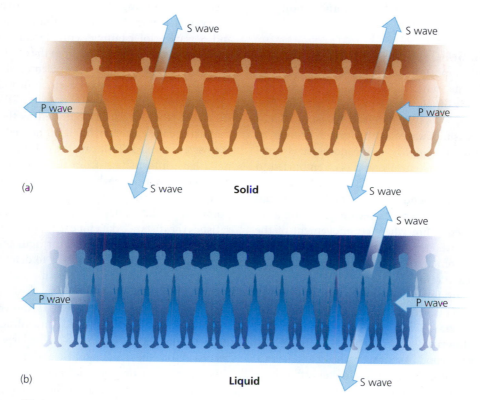

Figure 11.5   Human Waves

The response of solids and liquids to seismic waves is mimicked by a line of people. (a) Solids resemble a line of people standing side-by-side with their arms linked. If the first person in line is pushed into the second to mimic a P wave, everyone in the line will move in the same direction. Likewise, if the first person is moved backward and forward to mimic an S wave, the entire line will eventually move in the same fashion. (b) Liquids resemble a line of people standing shoulder-to-shoulder, with their arms at their sides. If the first person is pushed into the second to mimic a P wave, the second person will fall into the third and so on, so that everyone in the line will eventually move in the same direction. But if the first person is moved backward and forward to mimic an S wave, the rest of the line remains stationary. In the same way, both P and S waves travel through solids, but P waves can travel through liquids whereas S waves cannot.

SCIENCE REFRESHER

Density, Volume, and Seismic Velocity

The speed of seismic waves depends on the density of the material through which they are traveling. But why should this be so? **Density** is defined by the mass of an object divided by its volume. In essence, it is a measure of how tightly the constituent atoms in the object are packed.

Seismic waves are elastic waves and can only travel through materials that behave elastically. That is, the material must be able to rebound from an applied stress so that it vibrates as the seismic wave passes through. The stiffer the material, the easier it is for seismic waves to pass through and so the faster they travel. Materials with tightly packed atoms tend to be more rigid because the

bonds holding the atoms together are short, just as a roof supported by short girders is less flexible that one supported by long girders. Hence, denser materials tend to be more rigid and so transmit seismic waves at higher velocities.

All solid materials are rigid and so all types of earthquake waves can be transmitted through them. But liquids will only rebound from compression, not from sideways movement or shear. As a result, liquids can transmit P waves because these are compressional, but cannot transmit S waves, which depend on shear.

---

is that different types of solids react differently to compression and to shear. For example, a diving board and trampoline respond quite differently when jumped on, although both are solid materials. The speed of body waves depends on the material through which they are traveling; the speed increases with increased density and rigidity.

MEASURING SEISMIC WAVES

As we have noted, important information about Earth's interior, especially its density and rigidity, can be obtained from the study of seismic waves. To be used for this purpose the waves must first be carefully measured. Geophysicists detect seismic waves with devices known as **seismometers** (or *seismographs*). A seismometer is typically mounted directly on bedrock, and is often buried to avoid the noise of normal day-to-day activities. The instrument contains a weighted pendulum attached to a pen that produces a wiggly line (known as a **seismogram**) on a revolving paper-covered drum whenever movement of the ground shakes its frame. Depending on the way the instrument is set up, seismometers can be used to detect either horizontal or vertical motion (Fig. 11.6a, b).

P waves travel through crustal rocks at speeds up to 7 kilometers (4 miles) per second and are the first to arrive at seismographic stations. S waves are slower than P waves, traveling through the crustal rocks at speeds up to 4.5 kilometers (3 miles) per second. As a result, the characteristic signals of S waves are recorded by seismometers after those of P waves (Fig. 11.6c). The time interval between the arrival of the first P wave and the first S wave at a recording station depends on the distance between the recording station and the earthquake (11.6d).

> **CHECK YOUR UNDERSTANDING**
>
> ● How are seismic waves measured?

Like two runners in a foot race, one of whom runs faster, the greater the distance traveled by the waves, the greater the P wave's margin of victory and the greater the time gap between the arrival times.

Using Figure 11.6d, we can determine the distance between the recording station and the earthquake. A time interval of 8 minutes between the arrival of the first P and S waves, for example, corresponds to a distance of 6000 kilometers (3700 miles) from the earthquake's epicenter. From this data alone, the epicenter of the earthquake could lie at any locality on a circle of radius 6000 kilometers (3700 miles) centered on the station. So the data from one recording station can tell us the distance from the earthquake epicenter, but it cannot tell us the direction the waves traveled. To locate the epicenter (Fig. 11.6e), we must have information from three or more stations. We can then draw a circle whose radius corresponds to the distance to the earthquake epicenter from each station. The point where the circles intersect is the only location that satisfies the distance data for each station. It therefore is the location of the epicenter of the earthquake.

Seismograms can also be used to determine the size or *magnitude* of earthquakes. The size scale with which we are most familiar was developed by the American seismologist Charles Richter in 1935 and is known as the **Richter magnitude scale**. The Richter magnitude is calculated from the height or *amplitude* of the largest seismic wave recorded on the seismogram, and the distance of the seismic station from the epicenter as given by the time interval between the arrival of the first P and S waves. In Figure 11.7, for example, the amplitude of the largest seismic wave is 23 millimeters (1 inch) and the interval between the first P and S waves (P–S) is 24 seconds. This indicates a distance from the epicenter of 225 kilometers (140 miles). On Richter's chart, a straight line connecting 24 seconds on column A to 23 millimeters on column C

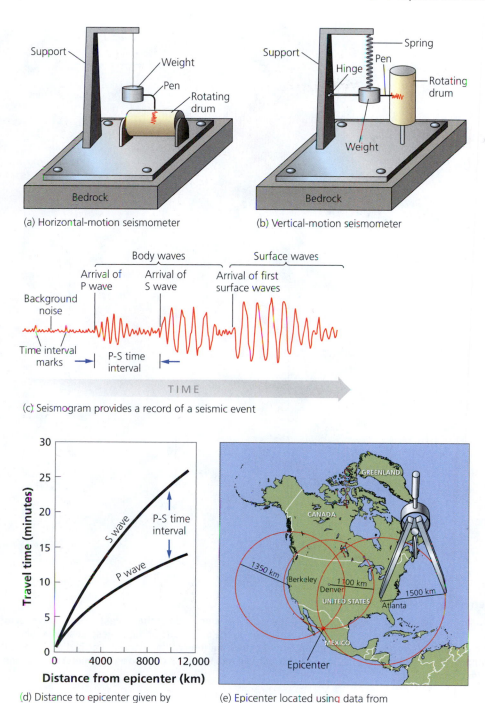

(a) Horizontal-motion seismometer

(b) Vertical-motion seismometer

(c) Seismogram provides a record of a seismic event

(d) Distance to epicenter given by P-S time interval

(e) Epicenter located using data from several seismic stations

Figure 11.6   Recording Earthquakes

Seismometers record (a) horizontal and (b) vertical motion in bedrock, producing (c) seismograms that (d) allow geoscientists to determine the distance to the earthquake epicenter from the time interval between arrival of P and S waves recorded on the seismogram. (e) To determine the location of the epicenter, data from several recording stations are needed.

determines the Richter magnitude on column B, in this case magnitude 5.0.

Unlike most of the scales with which we are familiar, the Richter scale is *logarithmic,* which means that each whole number increase measures a tenfold increase in ground motion. This means that ground motion during an earthquake of magnitude 8 is *10* times greater than an

earthquake of a magnitude 7 and *10,0000* times greater than one of magnitude 4. Earthquakes less than magnitude 3 are usually not felt even close to the epicenter. When it comes to earthquake energy, the multiplier is even larger. With each whole number increase on the Richter scale the amount of energy released by an earthquake increases by a factor of 32. This means that a magnitude 8 earthquake,

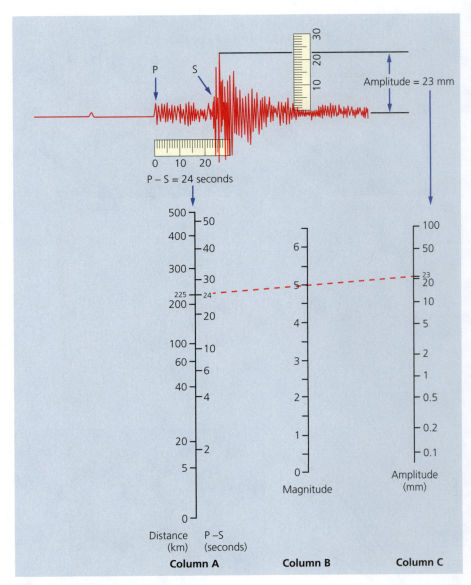

Figure 11.7   Determining Earthquake Magnitude

The Richter scale uses a chart that relates magnitude to the height or amplitude of the largest seismic wave recorded on a seismogram and the distance to the epicenter. In the example shown here, the amplitude of the largest seismic wave is 23 millimeters and the distance to the epicenter is 225 kilometers (given by P–S, the time interval between the arrival of the first P and S waves, of 24 seconds). By connecting 225 kilometers (24 seconds) on column A of Richter's chart to 23 millimeters on column C, a Richter magnitude of 5 is obtained from column B.

rather than being just twice as strong as one of magnitude 4, actually releases more than a million times the energy (32 × 32 × 32 × 32 = 1,048,576)—an amount equivalent to blowing up 6 million tons of TNT. Small differences in earthquake magnitude consequently imply large differences in their energy and destructiveness.

11.3 SUMMARY

- Seismic waves are divided into two broad categories: surface waves and body waves.

- Surface waves travel along Earth's surface. They are the main cause of destructive earthquake damage. Body waves travel through Earth's interior, radiating

out in all directions from the earthquake focus. They consist of primary (P) waves, which travel fast and are the first to arrive at a recording station, and secondary (S) waves, which travel about half as fast as P waves.

- P waves are compressional waves and temporarily cause changes in volume, by alternately compressing and extending the materials through which they pass. In contrast, S waves cause materials to change their shape. Like oscillations in a rope, they progress by shearing material one way and then the other. P waves behave like sound waves and travel through solid, liquid, and gas. S waves cannot travel through fluids (liquids or gases).

- Seismometers (seismographs) record vibrations that are transmitted through Earth. These records, called seismograms, reveal much about Earth's interior.

- Seismograms are used to locate the epicenters of earthquakes and to determine their magnitude.

11.4 Journey to the Center of the Earth

A journey to the center of Earth would reveal a layered planetary interior. But how do we know this? Our knowledge of Earth's interior is derived from several indirect sources. The most important of these is **seismology**, the study of earthquake waves. Because the speed with which seismic waves travel depends upon the properties of the material through which they pass, the pattern created by seismic waves as they travel through Earth's interior provides insight into the nature of the materials they encounter at depth. It was from this source that geophysicists first learned that Earth had a crust, a mantle, and a core, the outermost part of which is liquid. To understand how this knowledge was acquired, we must first examine closely the way seismic waves travel.

PROBING EARTH'S INTERIOR

Because Earth is a sphere, seismic waves radiating from an earthquake located close to Earth's surface encounter different regions of Earth's interior along their travel path. The patterns created by the interfering seismic waves are extremely complicated. To simplify matters, seismic waves are often described as **seismic rays** rather than waves, in the same way that the illumination from a light source is described as rays of light. We refer to the line of travel of an earthquake wave as a **ray path**. An analogy to the distinction between seismic waves and their ray path is illustrated in Figure 11.8, where the movement of the surf (or waves) can be described by the direction (or path) the surfers are traveling in.

Much of the seismic energy released by an earthquake eventually returns to Earth's surface where it can be analyzed at many recording stations. The deeper a wave penetrates, the longer is its ray path, so that seismic waves reaching a recording station located far from the earthquake have traveled farther into Earth's interior than those reaching a nearby station (Fig. 11.9).

Figure 11.8   Waves and Ray Paths

This photo of surfers illustrates how waves (in this case the surf) can be described by their path of travel (in this case the direction the surfers are moving).

The further a seismic ray travels from the earthquake, the deeper it penetrates into the Earth.

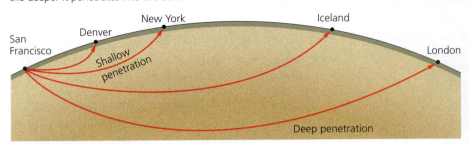

Figure 11.9   Seismic Rays

Seismic rays traveling in a concave path from an earthquake in San Francisco to recording stations in Denver, New York, Iceland, and London penetrate more deeply into Earth's interior the farther they travel from the earthquake focus.

As with any journey, the time it takes to travel depends on the distance involved and the conditions encountered along the route. Thus the time it takes for seismic waves to reach a recording station depends on the distance between the station and the focus of the earthquake, and the properties of the rocks encountered along the ray path. Because earthquake waves are sent out in all directions, the energy from any one earthquake is picked up by thousands of recording stations worldwide so that the travel times along thousands of different ray paths can be examined. Any variations in travel time that cannot be accounted for by differences in the distance traveled must reflect variations in the properties of the rocks encountered. Over the last 50 years, information from each of the hundreds of moderate earthquakes that occur annually in various parts of the world has been compiled and processed by powerful computers. The results have provided a surprisingly thorough understanding of the nature of Earth's interior (Fig. 11.10).

If Earth were perfectly homogeneous, the ray paths of seismic waves would spread out from the focus of an earthquake in all directions at constant speeds and along straight travel paths (Fig. 11.10a). In reality, the speed of earthquake waves increases with depth because the density of the materials they encounter increases as the pressure rises in Earth's deep interior. Because of this density increase, seismic waves move faster at depth than they do near the surface. This causes seismic waves to adopt curved, or *refracted* ray paths (Fig. 11.10b), bending up toward the surface as they travel downward into regions of faster velocity (see Science Refresher: **Wave Refraction** on page 349).

The effect is analogous to a car that drifts from the highway onto a soft shoulder. Because the passenger side wheels hit the soft shoulder and lose traction before those on the driver's side, the car slows on the passenger side and the faster-moving driver's side wheels cause the car to veer into the shoulder. In the same way, seismic waves veer toward regions of lower seismic velocity.

In addition to noting these gradual velocity changes, scientists recognized abrupt changes in the speed of seismic

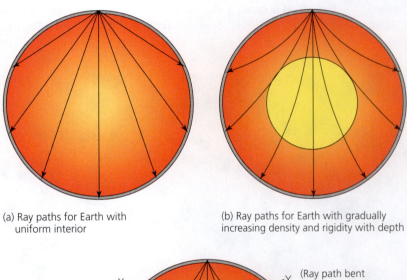

(a) Ray paths for Earth with uniform interior

(b) Ray paths for Earth with gradually increasing density and rigidity with depth

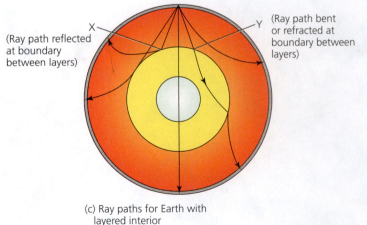

(Ray path reflected at boundary between layers)

X

Y

(Ray path bent or refracted at boundary between layers)

(c) Ray paths for Earth with layered interior

Figure 11.10   Ray Paths

These three figures contrast ray paths expected for Earth with (a) a homogeneous (uniform) interior, and (b, c) an inhomogeneous interior. (c) For an Earth with a layered interior some ray paths are reflected by the boundaries between layers (e.g., ray X), whereas others are bent, or refracted (ray Y).

SCIENCE REFRESHER

Wave Refraction

The change of direction waves experience when they pass from one material to another is known as **refraction** (Fig. 11D). When wave fronts encounter a different material, the ray path will generally meet the boundary at an angle. Because of this, one part of a wave front crosses into the new material before the other. When this occurs, one part of the wave front is traveling through the new material (segment AC in Layer 2), while the other remains in the old (segment AB in Layer 1). If the new material (Layer 2) is less rigid, the wave velocity of segment AC is lower than the velocity of segment AB and the spacing between wave fronts decreases. As a result, the wave front bends so that when the ray path emerges into the new medium, it is not parallel to the original ray path but is kinked or *refracted* and travels in a different direction.

To describe this refraction, a reference line is selected that is perpendicular to the boundary between the layers. In the case presented in Figure 11D, the ray path is refracted *toward* the reference line as it travels into the less rigid material. The opposite occurs when the wave travels from a less rigid material into a more rigid one. The ray path refracts *away* from the reference line.

We can better understand the effect by picturing a line of marchers walking arm in arm along a desert highway (Fig. 11E). The marchers maintain a constant speed when walking on the road, but abruptly slow down if they walk in the sand beside it. If the marchers head off the road at an angle, one by one they reach the sand while the rest are still marching on the road. As a result, their lines pivot and their paths bend because some are marching faster than others.

When seismic waves travel across a boundary between materials with sharply contrasting rigidity, wave refraction is pronounced. Even within essentially the same material, the degree of rigidity increases with density, and density increases with depth. (Denser materials are more rigid because their atoms are more closely packed and average bond lengths between atoms are consequently shorter.) Because densities increase with depth, seismic waves travel faster at depth and seismic ray paths progressively curve until they intersect Earth's surface. The result is a concave rather than linear path for the seismic ray as shown in Figure 11.9.

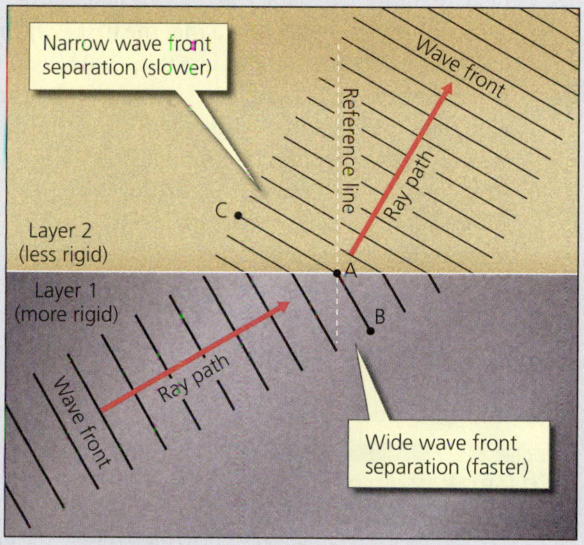

Figure 11D Refraction of Seismic Waves

A series of wave fronts emanating from an earthquake travel at the same speed (represented by their separation) in Layer 1 but at a slower speed (represented by their smaller separation) in the less rigid Layer 2. Because the ray path meets the boundary between the layers at an angle (for example, at A) one part of a wave front reaches Layer 2 before the other and starts to move at the slower speed sooner. Segment AB (in Layer 1) of the wave front therefore travels farther than segment AC (in Layer 2) in the same time interval and as a result, the wave front bends. The ray path perpendicular to the wave fronts is also bent or refracted on entering Layer 2, and the spacing between wave fronts becomes narrower.

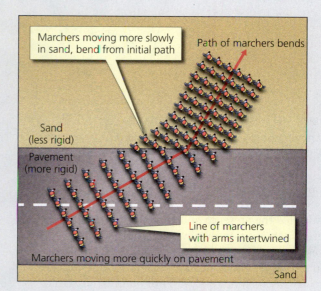

Figure 11E Desert Marchers

The path followed by ranks of marchers walking arm-in-arm on a desert highway is bent as they head off the road at an angle because those on the highway (more rigid surface) are marching faster than those on the sand (less rigid surface). Seismic waves bend in a similar fashion when they cross boundaries between materials of different rigidity in which they travel at different speeds.

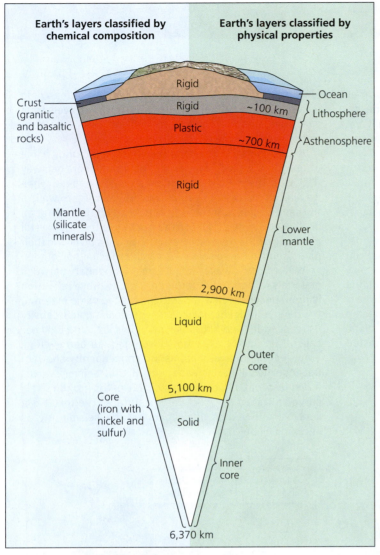

Earth's layers classified by chemical composition

- Crust (granitic and basaltic rocks)
- Mantle (silicate minerals)
- Core (iron with nickel and sulfur)

Earth's layers classified by physical properties

- Rigid
- Rigid — ~100 km
- Plastic
- ~700 km
- Rigid
- 2,900 km
- Liquid
- 5,100 km
- Solid
- 6,370 km

- Ocean
- Lithosphere
- Asthenosphere
- Lower mantle
- Outer core
- Inner core

Figure 11.11   Earth's Layered Interior
Earth's interior features and internal layers. Note that the thickness of layers in the figure is not to scale.

waves at particular depths within Earth's interior. Known as **seismic discontinuities**, these abrupt velocity changes indicated the presence of distinct layers of material of differing composition or physical properties. In this way, it was revealed that Earth is a layered planet with a very thin outer layer or **crust**, a massive metallic center or **core**, and a thick intervening rocky layer or **mantle** (Fig. 11.11).

A LAYERED PLANET REVEALED

Many seismic discontinuities are now known to exist within Earth's interior but the discovery of three discontinuities in particular revealed the planet's interior to be layered. The first, between the crust and the mantle, is the *Moho*. The second separates the mantle from the core and was detected with the discovery of a seismic shadow zone that ray paths failed to penetrate. The third occurs within

the core and was found when seismic signals were detected in the shadow zone.

The Moho

Imagine that you and a friend set out for the same destination at the same time, but along different paths. Your friend travels the more direct route, yet you arrive at the destination first. The only possible explanation for this is that you traveled at a faster speed. It was this sort of thinking that led Andrija Mohorovičić, a Croatian seismologist, to identify the existence of a compositionally distinct crustal layer in 1909. By analyzing seismic data recorded for nearby shallow earthquakes, he was able to show that P waves arriving at recording stations farther than 200 kilometers (125 miles) from the epicenter of an earthquake traveled significantly faster than P waves arriving at stations lying closer to the epicenter. He concluded that the P waves that followed the deeper ray paths to more distant stations traveled at greater speeds and so must have passed through rocks of higher density than the rocks encountered along the shallower ray paths to nearby stations (Fig. 11.12).

Mohorovičić determined that the first P waves to arrive at recording stations less than 200 kilometers from the epicenter were *direct waves* that had traveled the shortest route—directly through the continental crust. Beyond 200 kilometers, however, the first P waves to arrive at recording stations were *indirect* or *refracted waves* that had traveled the longer route at higher speed through the denser rocks at depth. The abrupt jump in speed observed at the more distant recording stations showed that the boundary separating the shallow crustal rocks from the denser rocks below was a sharp one, and that it occurred at a depth of 35 to 40 kilometers (22 to 25 miles). Since that time, similar studies have shown that the boundary between the oceanic crust and mantle typically occurs at a depth of about 5 to 10 kilometers (3 to 6 miles). The boundary, which separates rocks of the continental crust and oceanic crust from those of the underlying mantle, is formally named the *Mohorovičić Discontinuity*, in honor of its discoverer, but it is better known as simply the **Moho**.

> **CHECK YOUR UNDERSTANDING**
> ⬤ How did observations of P wave velocities lead to the discovery of the Moho?

The shadow zone

Shortly after the discovery of the Moho, German seismologist Beno Gutenberg discovered the second major boundary of Earth's interior, the one between the mantle and the core. This boundary was not detected by an abrupt change in the arrival times of seismic waves but instead

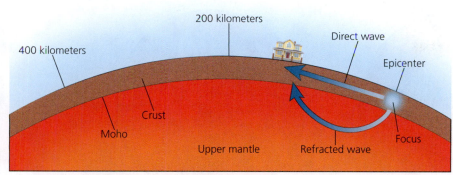

(a) Direct wave arrives first at stations within 200 kilometers of epicenter

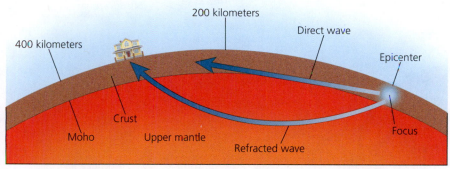

(b) Indirect (refracted) wave arrives first at stations greater than 200 kilometers from epicenter

Figure 11.12   The Crust-Mantle Boundary: Detecting the Moho

Ray paths of P waves traveling from the focus of a shallow earthquake to two recording stations reveal the existence of the Mohorovičić Discontinuity or Moho. (a) Within 200 kilometers (125 miles) of the epicenter, the direct wave arrives first at the recording station. (b)
Beyond 200 kilometers, the indirect (refracted) wave arrives first because the indirect
wave traveled at higher speed through the denser rocks of the upper mantle.

encountered everywhere at about
the same depth (2900 kilometers).
The refraction and late arrival of P
waves beyond the shadow zone
also supported the notion of a
liquid layer below the mantle, one
that slows and refracts the P waves
as they enter the core.

Since P waves reappear at latitudes of more than 143 degrees
from the epicenter (see Figure
11.13b), Gutenberg concluded that
the core must act on P waves in
much the same way that a lens focuses light by changing the path of
light rays. Because the liquid core is
less rigid than the lowermost
mantle, the P waves are bent inward
as they enter the core and bent
inward again as they exit the core.
In both cases, the refraction occurs
because seismic waves slow down
as they enter the liquid core and

**CHECK YOUR
UNDERSTANDING**

• What is the significance
of the shadow zone?

was revealed by the almost complete failure of the waves
to arrive. For any earthquake that is of sufficient magnitude to be detected at recording stations many thousands
of kilometers from the epicenter, there is always a belt encircling the globe in which neither P waves nor S waves
are detected (Fig. 11.13). This belt is known as the **shadow
zone**, the edge of which is located about 11,500 kilometers
(7000 miles) or 103 degrees around the globe from the
epicenter of an earthquake (Fig. 11.13a). At distances
beyond this, S waves cannot be detected. However, at recording stations located more than 16,000 kilometers
(10,000 miles) or 143 degrees of latitude from the epicenter, P waves reappear, but with much delayed arrival times
(Fig. 11.13b).

Thus, for any earthquake that scientists detected far
from its source, a ring-like shadow, free from direct P and
S waves, was produced between 103 and 143 degrees from
the epicenter. From the existence of this shadow zone,
Gutenberg concluded that Earth must have a core that affects the transmission of seismic waves. Because S waves
fail to reappear, at least a portion of this core is interpreted
to be liquid. All strong earthquakes, no matter where
their epicenters are located, produce a similar shadow
zone (Fig. 11.13c), which means that the liquid core is

accelerate as they exit it. The net result is that the P waves
are directed toward the area of the globe opposite the
earthquake epicenter and away from the 103 to 143 degree
sector. In this way, they are prevented from emerging
within the shadow zone. From the width of the shadow
zone, the radius of the core was determined to be 3470 kilometers (2155 miles).

A discontinuity within the core

It was later shown that the shadow zone is not totally dark,
suggesting that the core is not fluid throughout. In 1936 the
Danish seismologist Inge Lehmann proposed the existence
of an **inner core**, which she suspected to be solid, to explain the previously undetected presence of faint P waves
within the shadow zone that could only have been reflected
from a discontinuity within the core. The faster speed of
P waves within the inner core (consistent with its solid
nature) causes their ray paths to be bent into the shadow
zone because this time they
accelerate on entry and therefore bend outward. The diameter of this inner core is
now known to be 2540 kilometers (1580 miles), but its
exact size was not determined

**CHECK YOUR
UNDERSTANDING**

❯ Why is the inner core
thought to be solid?

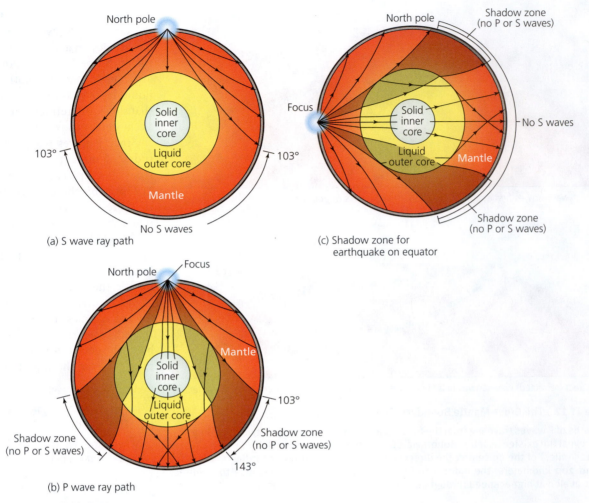

Figure 11.13   Ray Path Patterns
Traveling from an earthquake focus at the North Pole, (a) S waves are blocked by Earth's liquid outer core so that no S wave ray paths reach the surface beyond 103 degrees from the epicenter. (b) P waves are bent, or refracted, so that neither P wave nor S wave ray paths reach the surface in the shadow zone between 103 and 143 degrees from the epicenter. (c) Shadow zone produced by an earthquake epicenter located at the equator.

until underground nuclear tests performed in the early 1960s produced simulated earthquakes, the time and location of which were precisely known.

The liquid portion of the core is the **outer core**. The presence of the outer core entirely prevents S waves from penetrating the core–mantle boundary, thereby accounting for the absence of S waves within the shadow zone and beyond it (see Fig. 11.13b).

11.4 SUMMARY

- Seismic waves spread out in all directions, much like light rays from a light source. The time it takes rays to reach the surface depends on the distance between the earthquake focus and the recording station, and on the properties of the rocks the rays encounter.

- From seismic records, geophysicists deduced three seismic discontinuities where Earth's density changes abruptly. The Moho is the boundary between the crust

and the denser mantle beneath. The shadow zone is a region in which P waves are faint and S waves are absent because they have been refracted by a seismic discontinuity between the mantle and the core. Faint P waves beyond the shadow zone indicate that Earth has a solid inner core.

11.5 Earth's Structure and Composition

Earth is layered according to density, with a dense core whose outer part is liquid, a mantle, and a crust made up of relatively lightweight elements (Table 11.1). In fact, Earth's layering actually extends beyond the crust to the oceans and atmosphere. More recent studies of seismic wave velocities have shown that, like the core, Earth's crust and mantle can be subdivided (Fig. 11.14). In this section of the chapter, we examine each of Earth's major internal subdivisions in more detail.

TABLE 11.1   GENERAL CHARACTERISTICS OF EARTH'S INTERNAL LAYERS

Layer	Depth from Surface	General Composition	Average Temperature	Average Density	Proportion of the Earth's Total Mass	Proportion of the Earth's Total Volume
Crust			500°C		0.4%	<1%
Continental crust	From surface to 35 kilometers	Granite/diorite		2.7 g/cm³		
Oceanic crust	From sea level to 11 kilometers	Basalt		2.9 g/cm³		
Mantle	From 11 km to 2900 kilometers	Iron and magnesium silicates	2500°C	4.5 g/cm³	68.1%	83%
Core					31.5%	16%
Outer core	From 2900 km to 5100 kilometers	Liquid iron and sulfur	4600°C	11.8 g/cm³		
Inner core	From 5100 km to 6370 kilometers	Solid iron and nickel	5000°C	16.0 g/cm³		
Whole earth	6370 kilometers			5.5 g/cm³		

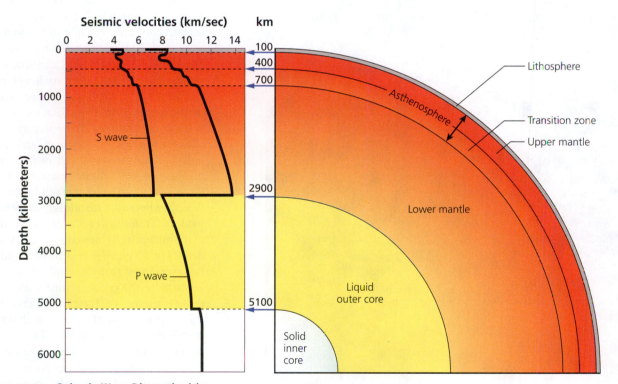

Figure 11.14   Seismic-Wave Discontinuities

Abrupt changes in the velocities of P and S waves define the boundaries between Earth's internal layers.

THE CRUST

The crust is not only the thinnest of Earth's many layers but it also shows the greatest variety in its composition. This variety results from the dynamic processes that occur at or near Earth's surface. On the continents, for example, the plants and animals of the biosphere sit atop a thin veneer of soil, sediment, and sedimentary rock. This veneer overlies a crust composed of metamorphic and igneous rocks of broadly granitic composition. Like all granitic rocks, the crust of the continents is thought to be rich in potassium, sodium, and silicon and to have a low density of about 2.7 g/cm³. But given that the average thickness of the continental crust is 35–40 kilometers (22–25 miles) and that the deepest borehole has yet to penetrate this crust to a depth of more than 12 kilometers (7.5 miles), little is known of the composition of continental crust at depth.

In 1925, a seismic discontinuity was detected within the continental crust. This boundary separates the upper crust from a slightly denser lower crust, where the velocity of seismic waves is somewhat greater. Based on the available evidence, the lower crust is thought to be broadly dioritic in composition and highly metamorphosed due to elevated temperatures and pressures. Recall from Chapter 4 that diorite is a coarse-grained igneous rock dominated by plagioclase and hornblende. Even faster seismic wave velocities occur in the crust beneath the oceans, demonstrating a compositional distinction between oceanic crust and continental crust. Despite being removed from direct observation by several kilometers of seawater, the broadly basaltic composition predicted for the oceanic crust on the basis of its seismic velocities has been confirmed by deep-sea drilling. The crust of the ocean floors is thought to be rich in calcium, magnesium, and iron and to have a density of about 2.9 g/cm^3.

As we discussed in Chapter 9, oceanic crust is also much thinner than continental crust (Fig. 11.15). The depth of the Moho (the base of the crust) below the ocean floor typically ranges from 5 to 10 kilometers (3 to 6 miles) but is commonly 6 to 7 kilometers (about 4 miles). In contrast, the Moho below the continental crust lies at a depth that is rarely less than 20 kilometers (12 miles) and may exceed 80 kilometers (50 miles) beneath major mountain ranges. In both cases, however, the Moho delineates an abrupt boundary separating lower crustal rocks of dioritic to basaltic composition from the ultramafic rocks of the uppermost mantle below.

> **CHECK YOUR UNDERSTANDING**
>
> ○ What distinguishes the upper continental crust from the lower continental crust?

THE MANTLE

Extending from the base of the crust to the core–mantle boundary at a depth of 2900 kilometers (1800 miles), the mantle is the largest of Earth's internal subdivisions and comprises 83 percent of Earth's volume. The mantle gets its name because it surrounds, or "mantles," the core. Mantle samples are rarely brought to the surface, so most of our knowledge about its composition is based on variation in seismic wave velocities and experimental studies of mineral behavior at very high pressures and temperatures. From these data, the mantle is generally assumed to be largely solid, chemically similar throughout, and made up of ultramafic rocks full of iron- and magnesium-rich silicate minerals. From the occasional samples brought to the surface, the uppermost mantle is thought to be made of *peridotite*, a rock composed mainly of the minerals olivine and pyroxene with a density of 3.3 g/cm^3. Experimental studies, however, show that these minerals could not survive the pressures that exist lower down in the mantle. Therefore we know that the mineral content of the rocks in the lower mantle must be different even if the overall composition of the rock is much the same.

As we learned in Chapter 9, the uppermost mantle and crust make up the outer rocky layer of Earth—the rigid *lithosphere* (see Fig. 11.15). The plastic *asthenosphere* beneath the lithosphere makes movement of the overlying lithospheric plates possible and is responsible for isostasy (see Chapter 9). The asthenosphere is thought to be inherently soft because the mantle at this depth interval is close to its melting point (Fig. 11.16).

The asthenosphere is thought to be a globally continuous layer, in contrast to

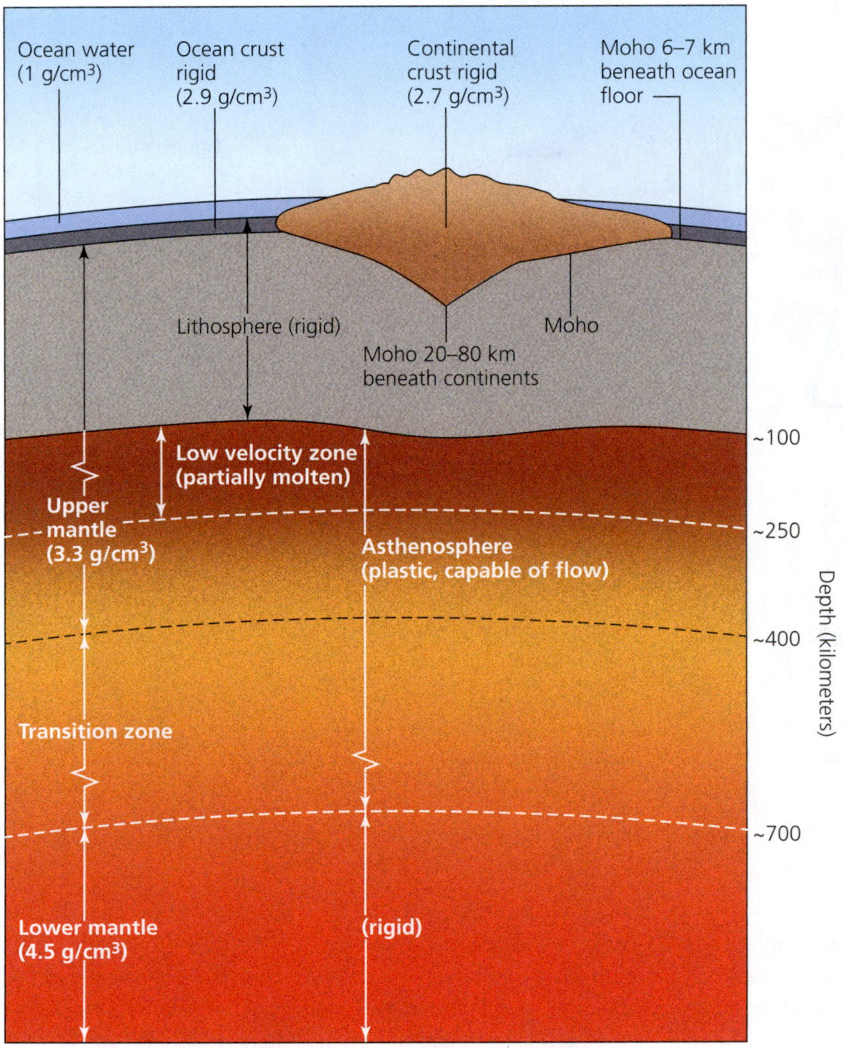

Figure 11.15 Crust and Mantle
Earth's crust and part of the mantle showing layering, density variations, and changes in rigidity with depth.

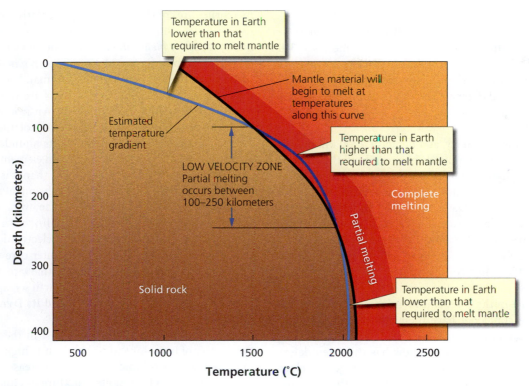

Figure 11.16   Low-velocity Zone

Melt is produced in the mantle where the mantle temperature (blue curve) exceeds that needed for melting to begin (black curve). The presence of melt in the mantle results in a zone of low seismic velocity within the asthenosphere, known as the low-velocity zone. The low-velocity zone is located between 100 to 250 kilometers (60 to 150 miles) beneath Earth's surface, where the temperature is sufficiently high to partially melt mantle rock (red zone). At shallower and deeper levels, the temperature in the mantle is insufficient to melt mantle rock (blue curve).

a discontinuous region within it that is known as the *low-velocity zone*. The **low-velocity zone** occurs within the upper mantle between the depths of 100 and 250 kilometers (60 to 150 miles) and is associated with a marked decrease in seismic wave velocities. The effects are more pronounced for S waves than for P waves and are best explained by the presence of a small amount of molten material in the zone. The presence of such melt is likely at this depth because it is in this region that the mantle comes closest to its melting point (see Fig. 11.16). Because we know that S waves penetrate the low-velocity zone, the amount of melt must be small. In fact as little as one percent liquid would produce the observed effects. Although detected beneath both the oceans and continents, the low-velocity zone is not entirely continuous. It is notably absent beneath areas of ancient continental crust, where the upper mantle has experienced many past cycles of subduction and so is less readily melted because melt has been previously extracted.

Other discontinuities discovered within Earth's mantle are thought to record the transformation of upper-mantle minerals to mineral structures that are stable at higher pressures. Increases in seismic wave velocities indicate that such transformations occur from 400 to 700 kilometers (250 to 435 miles) in depth (see Fig. 11.15). This interval, known as the **transition zone**, separates the **upper mantle** from the **lower mantle**. Unlike the abrupt compositional boundary at the Moho, the discontinuities of the transition zone are not

sharp. Instead, they range over a significant depth interval. For this reason, the discontinuities are generally attributed to changes in the crystal structure of the minerals present rather than to changes in the chemistry of the mantle rocks. Under the intense pressures of Earth's interior, minerals are forced to adopt new crystal structures in which their component atoms are more tightly packed, a process known as **phase change**. These new structures increase the density of the rocks concerned (and therefore the velocity of seismic waves passing through them) but do not affect their overall chemical composition. The transition zone is thought to be a region in the mantle where several different minerals, such as olivine and pyroxene, undergo such phase changes.

Below the transition zone, in the lower mantle, evidence of such phase changes is absent. Instead of the stepwise increases that occur in the transition zone, seismic velocities increase smoothly in response to the steady increase in density with depth. This trend abruptly ends at the base of the mantle, where a profound discontinuity marks the boundary of Earth's core.

> **CHECK YOUR UNDERSTANDING**
>
> ● What is the relationship between the low-velocity zone and the asthenosphere?
>
> ● What is the transition zone and what mineralogical changes are believed to take place within it?

THE CORE

Below the mantle lies Earth's massive metallic center or core, which, with a diameter of 6940 kilometers (4310 miles), is a little larger than the planet Mars. As we have learned, a seismic discontinuity exists within the core that separates the liquid *outer core* from a solid *inner core*.

The outer core

The core–mantle boundary, at a depth of 2900 kilometers (1800 miles), marks a profound change in both the composition and the physical state of the material of Earth's interior. When they encounter this boundary, P waves are dramatically slowed, S waves are halted entirely, and dramatic increases in density occur to values of nearly 12 g/cm^3. Because S waves stop, we know that the outer core must be liquid. At the same time, because density increases dramatically in the outer core, we know its composition is unlike that of the mantle. Even under the enormous pressures that exist in the core, silicate minerals such as those of the mantle could not have their atoms packed tightly enough to account for such a high density. So what is Earth's outer core made of and how has this been determined? Once again, geophysicists pieced together clues from a wide variety of sources. Among these, the most important clues come from estimates of Earth's density and the study of meteorites.

The total mass of Earth is 5.98 × 10^{24} kilograms (that is, 5.98 multiplied by 10, 24 times). This value has been determined from the gravitational attraction between the planets, from the time it takes Earth to revolve around the Sun (the length of an Earth year), and from the time it takes Earth to rotate about its own axis (the length of an Earth day). The volume of Earth is 1.08 trillion cubic kilometers (260 billion cubic miles). This figure can be determined because the shape of Earth is known. If Earth's mass is divided by its volume, an average density of 5.5 g/cm^3 is obtained. This means that an average piece of Earth weighs about five and a half times as much as the same volume of water.

Density is a measure of how tightly packed the internal atomic structure of a substance is and, as we have learned, is one of the main variables controlling the speed of earthquake waves. From the speed of such seismic waves, we can tell that most of the rocks within Earth's mantle and crust have densities well below 5.5 g/cm^3. This implies that the material of Earth's outer core must have a density well in excess of 5.5 g/cm^3. In fact, calculations show that core densities between 10 and 13 g/cm^3 are needed to achieve an average Earth density of 5.5 g/cm^3. For the outer core, the only liquids of high enough density are molten metals, and the only abundant metal in the Solar System with the appropriate density is *iron*.

Another line of evidence that supports an iron-rich core comes from the composition of *meteorites,* which is the name given to rocks that fall to Earth from space as we learned in Chapter 1. As members of the Solar System, meteorites are considered to be representative of the primitive material from which Earth originally formed. Those known as *stony meteorites* are made of rocky material that closely resembles the peridotite of Earth's mantle. Other meteorites, however, are metallic and mainly composed of iron and nickel. These *iron meteorites* are thought to resemble the material of Earth's core. An iron-rich core is also supported by the relative scarcity of this metal at Earth's surface when compared to its abundance in the Solar System. Although it is a common element, iron is far less abundant in Earth's crustal rocks than it is, for example, in meteorites. This suggests that the metal must be far more common in Earth's interior if the average composition of our planet is to be brought in line with that of the Solar System. Finally, iron is a good electrical conductor. The motion of an iron-rich molten outer core would help generate the electric current loops that are thought to power, in a dynamo-like fashion, Earth's magnetic field (see In Depth: **Earth's Magnetism and Its Dynamic Core** on page 357).

However, at the pressures that exist in the outer core, the density of iron alone would be slightly higher than that indicated by the seismic data. For this reason, the outer core is thought to be a molten mixture of either iron and sulfur or iron and oxygen. Both sulfur and oxygen are common light elements that mix with iron and facilitate melting. The addition of sulfur or oxygen to the mixture would consequently lower both the density of the outer core and the melting point of the material of which it is comprised. A liquid comprising 88 percent iron and 12 percent sulfur, for example, would not only have the correct density but would also produce the same P wave velocities as those in the outer core. However, other elements, including nickel, silicon and, possibly, hydrogen, are also likely to be present in small amounts.

> **CHECK YOUR UNDERSTANDING**
>
> ● What is the composition of the outer core and how has this been determined?

The inner core

At a depth of 5100 kilometers (3170 miles), a small increase occurs in P wave velocities (see Fig. 11.14) that is attributed to the presence of a solid inner core. The inner core's diameter, at 2540 kilometers (1580 miles), is similar to that of the dwarf planet Pluto. The bending of ray paths across the inner core boundary due to the velocity increase also accounts for the presence of faint P waves within the shadow zone. However, since temperatures within Earth's core are thought to rise from about 4000°C at the core–mantle boundary to values of about 5000°C at the inner core boundary, the existence of a solid inner core raises an obvious question. How can the inner core be solid when the outer core, which is not as hot, is molten? The answer is that the composition of the inner core is probably pure iron or a pure iron-nickel alloy. It

Earth's Magnetism and Its Dynamic Core

Albert Einstein stated that the origin of Earth's magnetic field was one of the five greatest unsolved problems in physics. A complete explanation remains elusive to this day. Earth's magnetic field is a region of magnetic force that surrounds and penetrates the planet. Within this region, invisible lines of magnetic force form a pattern of closed loops that emerge from the planet in the southern hemisphere and, after looping far out into space, return to Earth in the Northern Hemisphere. The surface expression of the magnetic force, used by humans for navigation and pathfinding, represents only a tiny fraction of the entire field.

The pattern of magnetic lines of force around Earth is similar to that displayed when iron filings are shaken onto a piece of paper placed over a simple bar magnet (Fig. 11F). Earth's magnetic field, like the bar magnet, can be said to have a north and a south pole. Within Earth's magnetic field, the north–south lines of magnetic force control the orientation of any smaller magnetized object providing the smaller object is free to move. As a result, a compass needle, which is a small magnet pivoted about its center of gravity, points toward the north and south magnetic poles. Because Earth's magnetic poles lie close to the geographic North Pole and South Pole, about which the planet rotates, the compass has proved to be a powerful tool in the exploration and navigation of our lands and seas.

The north magnetic pole is actually offset from the geographic North Pole by an angle of 11.5 degrees and wobbles slowly around it (see Figure 11F). Today, the north magnetic pole lies in the Arctic northwest of Ellesmere Island, whereas the south magnetic pole lies off the Antarctic coast in the Southern Ocean between Antarctica and Australia. Over the centuries, however, both magnetic poles have moved in a more-or-less circular fashion around their respective geographic poles.

Exactly how Earth's magnetic field is produced is not known with certainty. In 1600, the English physician William Gilbert attributed the field to the presence of permanently magnetized materials deep within Earth's interior. But this does not account for the field's variations with time and is precluded by the very high temperatures of Earth's interior. The magnetic field is now thought to be generated by electrical currents within Earth's outer core. Approximately 2200 kilometers (1370 miles) thick and 2900 kilometers (1800 miles) below Earth's surface, the outer core is a highly conductive metallic layer rich in iron.

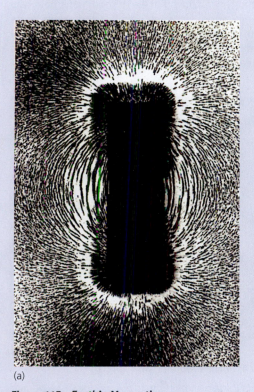

(a)

(b)

Figure 11F   Earth's Magnetism

(a) Magnetic field pattern of a simple bar magnet is revealed by iron filings aligning themselves along the lines of magnetic force. (b) Magnetic field pattern of Earth features lines of force that closely resemble those produced if a bar magnet was aligned north–south at Earth's center; the curved lines of magnetic force are parallel to Earth's surface at the equator and become more steeply inclined as the poles are approached.

(Continued)

IN DEPTH

Earth's Magnetism and Its Dynamic Core (Continued)

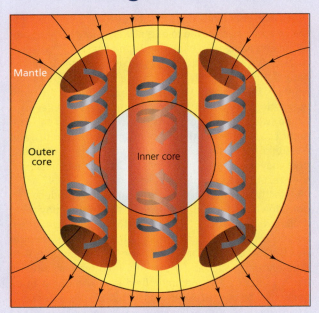

Figure 11G   Origin of Earth's Magnetism

Earth's magnetic field may be generated by a dynamo in Earth's outer core set up by the spiraling flow of electrically conducting metallic liquid that circulates under the influence of Earth's rotation.

At the very high temperatures (4600°C, 8310°F) that prevail in the outer core, a molten iron-rich liquid slowly circulates. This circulation is affected by Earth's rotation. The two forces of circulation and rotation combine to create columns of spiraling liquid that move parallel to Earth's rotational axis (Fig. 11G). As the columns interact with Earth's magnetic field, they generate electricity. The electric currents, in turn, produce magnetism that reinforces Earth's magnetic field. Thus Earth's outer core behaves in a self-sustaining fashion, generating electric currents, which enhance the magnetic field that produces them. For as long as circulation continues in its outer core, Earth's magnetic field will be sustained.

The direction of Earth's magnetic field at the surface is consequently imposed by motion of the liquid outer core, which is itself influenced by Earth's rotation. As a result, Earth's magnetic poles are closely aligned with its rotational axis and have been throughout Earth history.

As we discussed in Chapter 2, Earth's magnetic field is known to reverse its polarity, that is, the north magnetic pole becomes the south magnetic pole and vice versa. The cause of these reversals is not fully understood but, over millions of years, they are known to have occurred many times. Such reversals in Earth's magnetic field presumably reflect some chaotic aspect of the process that produces the field, perhaps due to disturbance or shifts in the pattern of flow in the outer core. Such changes in flow pattern are thought to cause the intensity of Earth's magnetic field to fluctuate with time.

Despite the uncertainties in the origin of Earth's magnetic field and the cause of its polarity reversals, scientists have replicated both phenomena in a supercomputer model using software that simulates the conditions in Earth's core, calculates the resulting magnetic field, and then runs the code for hundreds of thousands of simulated years to see what happens (Fig. 11H). Not only does the computer simulation generate a magnetic field closely resembling that of Earth but the field is also found to occasionally reverse itself. The model suggests that, during reversals, Earth's magnetic field weakens and becomes disorganized for several thousand years, and then slowly regenerates, but not necessarily with the same polarity. If it builds with the opposite polarity, a magnetic reversal is produced.

Interestingly, we know that Earth's magnetic field was much stronger 2000 years ago than it is today and, at its present rate of decay, the field should reach zero intensity within the next two millennia. So it is possible that Earth is currently heading toward another magnetic reversal.

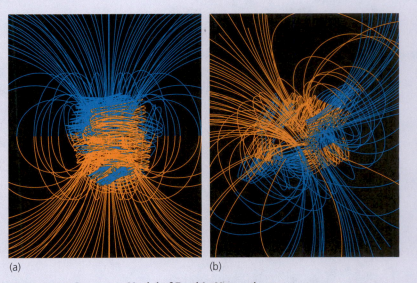

(a)   (b)

Figure 11H   Computer Model of Earth's Magnetism

(a) In this computer simulation of Earth's magnetic field during a period of normal magnetic polarity, the colored lines show the magnetic field (blue where north points downward, orange where north points upward). The tangled knot of field lines lies within Earth's core. (b) When the computer code in this model is run for hundreds of thousands of simulated years, the magnetic field is found to occasionally reverse itself.

therefore lacks the small quantities of lighter elements, like sulfur, that are thought to be present in the outer core and act to lower the melting point of the outer core material.

It is likely that Earth's solid inner core has evolved over time. That is, early in Earth's history the entire core may have been molten. But as Earth cooled, the inner and outer core slowly separated. The heavier, iron-rich components sank toward the inner core and solidified. The lighter components became concentrated in the outer core, where their presence ensures that the core material remains above its melting point. In this way the core developed an internal structure in which a circulating liquid shell surrounds a solid interior.

As Earth continues to cool, the inner core may continue to grow at the expense of the outer core. Indeed, billions of years from now, the core may conceivably become entirely solid, as is believed to be the case in the core of Mars. If this were to occur, Earth might lose its magnetic properties, perhaps losing at the same time the heat engine that drives plate tectonics.

> **CHECK YOUR UNDERSTANDING**
>
> ○ How can the inner core be solid when the outer core, which is not as hot, is molten?

11.5 SUMMARY

- Earth is layered according to its density, with a thin crust, a dense rocky mantle, and a metallic core.

- Seismic-wave velocities show that the continental crust, which is mostly granitic to dioritic, is thicker and less dense than the oceanic crust, which is largely basaltic.

- The mantle comprises 83 percent of Earth's volume and occupies the space below the crust and above the core. It is divided into the upper mantle and the lower mantle, with a transition zone between them.

- The core is the central part of Earth. It is divided into a liquid iron-rich outer core and a solid inner core that is probably pure iron.

11.6 Circulation in Earth's Interior

Although the bulk of Earth's interior is solid, much of it is thought to be slowly circulating (see **Science Refresher: Convection**, in Chapter 1). Some form of circulation is needed in the liquid outer core to generate Earth's magnetic field. Circulation within all or part of the solid mantle is also believed to be ultimately responsible for the motion of Earth's lithospheric plates. Two forms of circulation are possible, one driven by temperature contrasts and the other by differences in composition. Known respectively as *thermal convection* and *compositional convection* (Fig. 11.17), both are believed to operate within Earth's interior.

THERMAL CONVECTION

Circulation of Earth's interior is the principal means by which the planet loses its internal heat. Heat can travel through stationary materials by *conduction*. (Conduction is the process that makes a metal poker too hot to hold when its tip is held in a fire.) However, geologists believe circulation of Earth's mantle is driven by thermal convection (Fig. 11.17a). During **thermal convection**, warm, less dense material rises while cool, denser material sinks. This process is far more efficient than conduction as a means of transporting heat because the material itself moves. In this way, hot material is transferred from the interior toward the exterior of the planet, and the heat produced in the interior travels to the surface where it can be radiated away into space. Because the process also encourages the segregation of material according to their densities, it has also played a major role in developing Earth's internal layering.

Much of Earth's internal heat comes from the breakdown of naturally occurring radioactive elements. Radioactive elements are not highly concentrated in the mantle, but the mantle comprises about 83 percent of Earth's volume and radioactive decay of these elements provides a significant quantity of heat. Chief among these are the elements uranium (isotopes ^{238}U and ^{235}U), thorium (^{232}Th), and potassium (^{40}K). The decay of these three elements produces more than enough heat to cause the mantle to convect thermally.

COMPOSITIONAL CONVECTION

Although radioactive elements may occur in the core, they are unlikely to be present in sufficient quantities to allow the outer core to circulate by way of thermal convection. Instead, the mechanism for the circulation in the outer core, which powers the dynamo that generates Earth's magnetic field, is compositional convection (Fig. 11.17b).

> **CHECK YOUR UNDERSTANDING**
>
> ○ What is the difference between thermal convection and compositional convection?
>
> ○ In what areas of Earth's interior is each type of convection likely to be found?

Compositional convection relies, like thermal convection, on light material rising while heavy material sinks. In compositional convection, however, the contrast in density is due to differences in composition rather than differences in temperature. In the outer core, circulation is thought to be driven by the buoyancy of the less dense material left when iron solidifies and sinks to the surface of the inner core. This links convection in the outer core to the growth of the inner core.

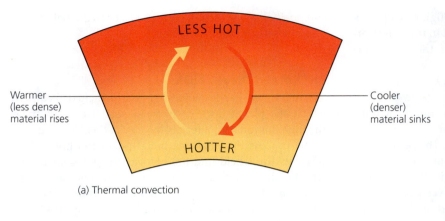

(a) Thermal convection

Warmer (less dense) material rises

Cooler (denser) material sinks

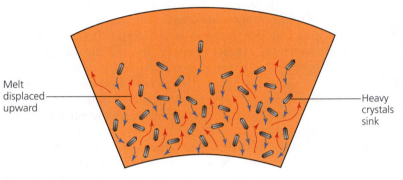

(b) Compositional convection

Melt displaced upward

Heavy crystals sink

Figure 11.17   Thermal and Compositional Convection
Convection within Earth's interior may be thermal or composition. (a) During thermal convection, warmer, less dense material rises from a heat source, while cooler, denser material sinks. (b) During compositional convection, dense crystals sink and displace lighter melt upward.

MECHANISM OF MANTLE CONVECTION

How can the mantle convect if it is solid? The answer is that materials do not have to be liquid in order to move. The slow downhill movement of valley glaciers provides an everyday example of such flow. Similarly, given enough time, a large ball of silly putty slowly flattens itself into a disk. In both cases, movement is slow and occurs while the material (ice or putty) is solid. This kind of movement is known as solid-state flow or **creep**. The ability of materials to creep is greatly increased at elevated temperatures, especially when temperatures near the melting point of the material involved. As we learned when examining the asthenosphere, materials become soft and yielding as they approach their melting points, and under such conditions are capable of flow. Although the mantle is closest to its melting point within the asthenosphere (and is thought to just exceed it in the low-velocity zone), nowhere is it very far from its melting point (Fig. 11.18). Therefore the entire mantle is likely to be capable of slow convective flow in the solid state.

> **CHECK YOUR UNDERSTANDING**
>
> ○ How is it possible for the mantle to convect if it is comprised of solid material?

MODELS OF MANTLE CONVECTION

If the mantle is convecting, it is tempting to link its circulatory motion with the movement of the lithospheric plates at Earth's surface. However, the pattern of mantle convection is not well understood and several different models of mantle convection have been proposed (Fig. 11.19): whole mantle convection, layered mantle convection, and mantle plumes.

Whole mantle convection

The simplest of these models is **whole mantle convection**. In whole mantle convection several large convection cells extend through the entire mantle from the core–mantle boundary to the base of the lithosphere, and include both the upper and lower mantle layers (Fig. 11.19a). To link this motion to plate movement, warm, buoyant mantle material rises beneath the mid-ocean ridges, then spreads laterally beneath the lithosphere, which it carries in a conveyor-belt fashion. Eventually, the material cools and, beneath sites of subduction, sinks back into the mantle and is reheated.

Layered mantle convection

Some geophysicists feel that the pattern of mantle flow is unlikely to be as simple as whole mantle convection. They argue that a major discontinuity, at a depth of about 660 kilometers (400 miles), separates the upper from the lower mantle and may profoundly influence the pattern of mantle circulation (Fig. 11.19b). A model called **layered mantle convection** proposes that more vigorous circulation occurs in the upper mantle because the mantle rocks are closest to their melting points. Proponents of this model argue that the upper mantle is probably compositionally distinct from the lower mantle because it is the source of mid-ocean ridge basalts. The upper mantle has therefore had basalt melt extracted from it, unlike the lower mantle, which is probably closer to its original composition. In this model, more vigorous convection in the upper mantle is thought to occur independently of the slower convection in the more rigid lower mantle so that the convective systems are decoupled across the transition zone. Convective circulation in the upper mantle is thought to have had a direct bearing on plate tectonics in that upwelling occurs beneath mid-ocean ridges. Downwelling, on the other hand, occurs where cold oceanic lithosphere descends in subduction zones.

Layered convection has to be the case if the boundary between the upper and lower mantle is a chemical one because whole mantle convection would mix the two layers and so remove any chemical distinction between them.

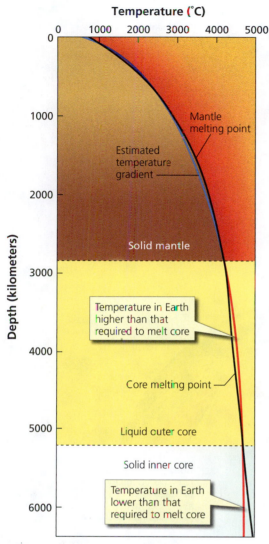

Figure 11.18   Earth's Internal Temperature

In this graph, temperatures within Earth, as revealed by seismic-wave velocities, are compared to the melting point of the materials of Earth's internal layers. The nature of the materials is determined from their composition and the pressure conditions that exist at depth. Melting occurs at depths where the temperature within Earth (blue curve in mantle; red curve in core) is higher than that required to melt the material present (black curve). Note that the material of the outer core is above its melting point, whereas that of the inner core is not.

Whole mantle convection is possible only if the boundary is produced by changes in the crystal structure (phase changes) of the minerals in the mantle, as suggested by experiments replicating conditions at this depth (see Section 11.5). Minerals carried across such a boundary by convection currents would simply change their internal structure to a more open one that is stable at lower pressure (in areas of upwelling) or to a more densely packed structure that is stable at higher pressure (in areas of downflow).

Mantle plumes

A third model for mantle convection (Fig. 11.19c) suggests that circulation takes the form of focused columns of mantle

upwelling and a much more diffuse system of return flow. These narrow zones of upwelling are the **mantle plumes** we examined in Chapter 9. The plumes are believed to arise from a thermally unstable layer at the core–mantle boundary and rise upward through the entire mantle until they reach the base of the lithosphere. Here they spread out sideways, carrying the lithospheric plates away from the sites of upwelling as they do so. The more diffuse return flow is thought, once again, to be focused at the sites of subduction, where cold, dense oceanic lithosphere descends into the mantle.

Which of these three models best describes mantle convection is not known. Indeed, in reality, mantle convection could involve several mechanisms, and is almost certain to be a more complex process than any of these models describe. Whatever the case, the answer may soon be revealed, thanks to the development of a powerful new technique for studying Earth's interior. This technique, known as *seismic tomography,* is capable of providing detailed three-dimensional maps of Earth's mantle. The technique has already revealed tantalizing clues to the pattern of temperature variation within the mantle, and shows the potential for revealing further details on the true nature of mantle convection.

> **CHECK YOUR UNDERSTANDING**
>
> ❍ Compare and contrast the three models of mantle convection discussed in the text.

SEISMIC TOMOGRAPHY: 3D MAPS OF THE MANTLE

Seismic tomography is a relatively new technique that promises to improve our knowledge of the structure of the mantle in a dramatic fashion. Although still being developed, it has already indicated a pattern of convective flow in the mantle that is far more complex than that suggested by any of the proposed models we have discussed. The principle of seismic tomography is similar to that of the *CAT (computerized axial tomography) scan* used in medical diagnosis. A CAT scan is an X-ray procedure in which numerous X-ray images of a patient are taken from various angles by rotating the X-ray beam about the patient's body. Because the behavior of X-rays depends on the internal structure of the patient, rotation of the X-ray beam allows the structure to be imaged in the third dimension. Computers are then used to combine the X-ray data so that a three-dimensional image of the patient's internal organs can be created.

Seismic tomography employs the same approach to create three-dimensional images of Earth's interior by combining seismic data from large numbers of earthquakes. As earthquakes occur in many places on Earth's surface, information can be synthesized from many sources, mimicking the CAT scan effect of the rotating X-ray beam. Because of its relatively uniform composition and structure, variations of seismic velocity in the mantle

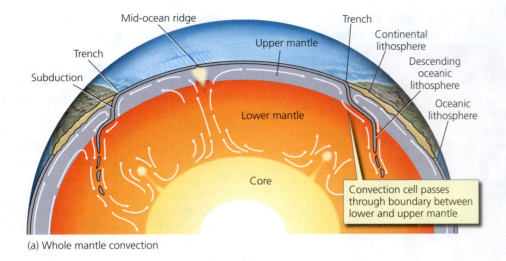

(a) Whole mantle convection

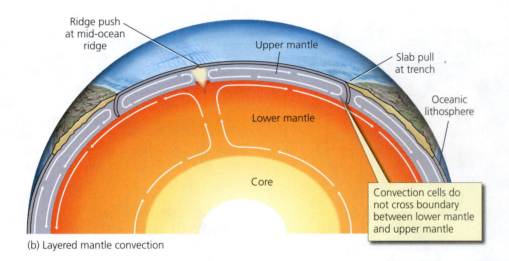

(b) Layered mantle convection

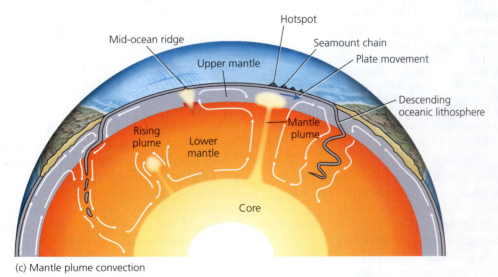

(c) Mantle plume convection

Figure 11.19 Mantle Convection Models

There are three commonly proposed models for mantle convection: (a) whole mantle convection, in which convection cells involving the entire mantle carry the lithosphere like a conveyor belt and mid-ocean ridges are the surface expression of deep mantle upwelling; (b) two-layer convection, in which faster circulation of the upper mantle, driven by slab pull at subduction zones and ridge push at mid-ocean ridges, is decoupled from the slower circulation of the lower mantle; and (c) mantle plumes, in which the upwelling of heat is confined to narrow plumes and mid-ocean ridges are a secondary effect of this upwelling. In all three models, downwelling occurs where subducting oceanic lithosphere plunges deep into the mantle. The core–mantle boundary may be the graveyard of such slabs.

are probably due to temperature differences that are, in turn, related to mantle convection. Hotter areas of the mantle, because they are less rigid, have slower seismic velocities than cooler areas at the same depth. Thus areas of unusually low seismic velocities are thought to indicate regions of hotter mantle, whereas areas of unusually high velocities are thought to depict cool regions. Seismic tomography's sophisticated computer analysis of many hundreds of slices through the mantle have made it possible for geophysicists to draw maps of seismic-wave velocity, and therefore mantle temperature, for various depths in the mantle.

The basic principle of seismic tomography is illustrated in Figure 11.20. In this diagram, the dark area represents a region of the mantle in which seismic velocities are anomalously slow. This, in turn, implies that the region is anomalously hot. The travel times for the rays (DD′, EE′, FF′, GG′) that pass through the shaded area are longer than the travel times of the rays that do not pass through this region, enabling the approximate location and shape of the anomalously hot region to be deduced. Given enough ray paths, it becomes possible to map out the anomalous region in three dimensions and, from their travel times, determine whether the area is anomalously hot or cold.

Although this technique works well in principle, there are some problems. For example, there are not enough seismic stations and there are too few earthquakes to produce the number of ray paths needed to map out the detailed structure of the mantle. In addition, less is known of the lower mantle because the resolution—that is, its ability to see detail—of the technique decreases with depth. Nevertheless, tomographic imaging of the mantle suggests a pattern of circulation that is far more complex than one of simple convection cells. For example, the images illustrated in Figure 11.21 show the distribution of unusually hot and cold areas in the upper mantle in both map view and cross section.

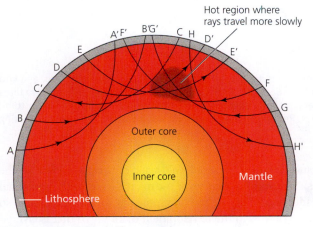

Figure 11.20   Seismic Tomography

The dark region has a lower seismic-wave velocity because it is hotter than the surrounding region. In this example, seismic travel times determined for rays that travel equal distances (for example, AA′, DD′, etc.) vary. The rays that travel through the shaded region take longer than those that do not. From this information, the approximate location and shape of the anomalously hot region can be deduced.

When portrayed in map view (Fig. 11.21a), tomographic results show a strong relationship between the velocity of rays in the mantle and the present distribution of lithospheric plates to a depth of 100 kilometers (60 miles). Thus hot areas where seismic-wave velocities are low (in red) generally coincide with spreading ridges, whereas cold areas of high seismic velocity (in blue) occur beneath the oldest parts of the continents. This implies that, at least in the upper 100 kilometers of the mantle, the pattern of convection is relatively straightforward.

At a depth of 300 kilometers (190 miles), however, the pattern is dramatically different, and the correlation with present-day plates significantly diminishes. The mantle beneath some spreading ridges, such as the one to the south of Australia, appears to be hot at 100 kilometers (60 miles) but is cold at depth. For others, like the mantle in the Red Sea, the reverse is true; that is, the mantle is hotter at depth. The pattern changes again at a depth of 500 kilometers (310 miles). According to geophysicists favoring a layered convection model, such changes argue against whole mantle convection. If, they argue, divergent plate boundaries are the direct expression of circulation rising from the core–mantle boundary, unusually hot conditions beneath the ridges should exist to great depths.

In contrast to the ridges, however, most of the continents maintain their velocity structure with depth. As you can see in Figure 11.21b, in which the patterns of hot and cold regions are shown in cross section to the base of the upper mantle at a depth of 660 kilometers (410 miles), the mantle beneath continents appears cold to depths as great as 500 kilometers (310 miles). This is more consistent with large-scale convection and suggests that continents have very deep, cold roots. At its present stage of development, the images produced by seismic tomography are not yet detailed enough to resolve these apparently conflicting results. But the resolution of the technique is being steadily improved and may soon achieve the level necessary to reveal the true nature of mantle convection. In fact, as we will learn in the next section, tomographic results have recently identified severed slabs of subducted oceanic lithosphere at the base of the mantle, which would argue against the layered mantle convection model.

Tomographic results for the entire mantle are still rudimentary. However, when viewed in three dimensions (Fig. 11.22), they suggest a very complex pattern of mantle circulation with irregular areas of hot (rising) and cold (descending) mantle that resemble the colored oils in a lava lamp. Although the nature of these areas is uncertain, they may be linked to the deep descent of subducting slabs and the compensating rise of mantle plumes from the core–mantle boundary.

Seismic tomography is raising important new questions about the nature of plates and their relationship to the lithosphere, the answers to which are certain to further our understanding of plate tectonic processes. The technique also promises to reveal many new details about Earth's interior as the quality of the pictures it provides improves. We may be on the verge of resolving the nature

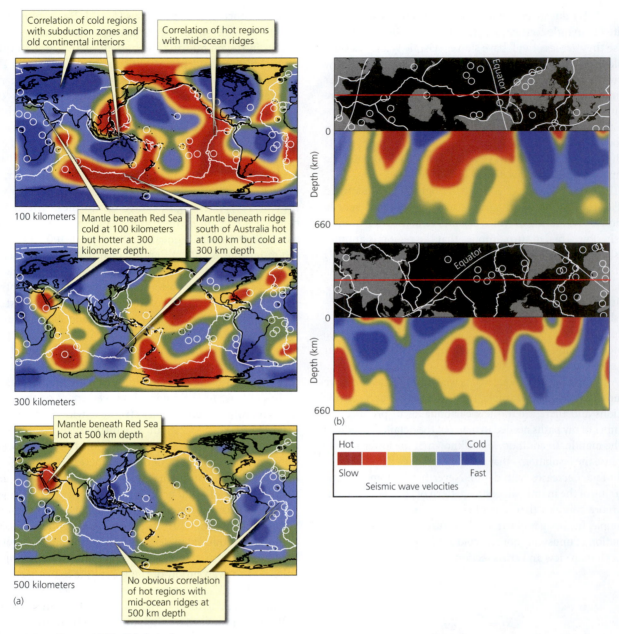

Figure 11.21   Scans of Earth's Interior

Global maps show tomographic scans of seismic-wave velocities in the upper mantle, (a) in map view at depths of 100, 300, and 500 kilometers (60, 190, and 310 miles), and (b) in cross section to a depth of 660 kilometers (410 miles) directly beneath the two lines indicated, one that girdles Earth from pole to pole (top) and the other running a diagonal course across the Pacific Ocean (bottom). Regions in red have low seismic-wave velocities indicating they are hot. Regions in blue have high seismic-wave velocities indicating they are cold.

of mantle convection and, in doing so, we may finally solve Wegener's problem with the mechanism of continental drift (see Chapter 2) by coming to understand how the movement of the plates is linked to that of the mantle.

The principal function of mantle convection is to bring the heat generated in Earth's interior to the surface, where it is eventually radiated into space. Plate tectonics is part of this process. Seafloor spreading brings hot new crust to the surface, where it cools, while subduction returns old, cold crust to Earth's interior. At the same time, circulation of the mantle ensures the continual addition of its lighter components to the crust, and so contributes to the development of an Earth that is layered according to its density.

11.6 SUMMARY

- Geophysicists believe that circulation of the mantle is driven by thermal convection and circulation in the liquid outer core is driven by compositional convection.

CHECK YOUR UNDERSTANDING

❂ How is seismic tomography analogous to a CAT scan?

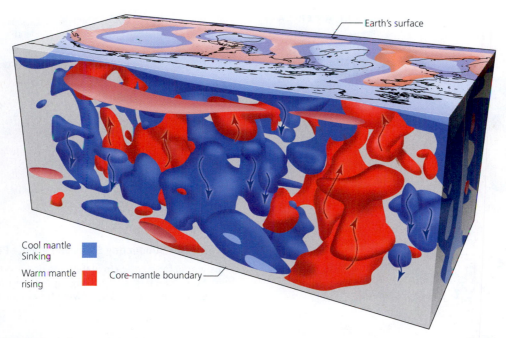

Earth's surface

Cool mantle Sinking

Warm mantle rising

Core-mantle boundary

Figure 11.22   Mantle Lava Lamp

A three-dimensional tomographic image of heat distribution in Earth's mantle resembles a lava lamp, with irregular areas of hot, rising mantle (in red) separated by cooler, descending regions (in blue).

- Three models have been proposed to describe the pattern of mantle circulation: whole mantle convection, layered mantle convection, and mantle plumes.

- Seismic tomography, a computer imaging technique, gives us an "X-ray" picture of Earth's interior and suggests that convective flow in the mantle is far more complex than any of the proposed models.

11.7 Plate Tectonics and the Fate of Subducted Slabs

Exciting results on the fate of subducted slabs have cast considerable doubt on the viability of the layered mantle convection model. Seismic tomographic results have identified anomalously cold bodies, thought to represent the remnants of severed subducted slabs. These cold bodies have penetrated the lower mantle as far as the core–mantle boundary (Fig. 11.23). Because the subducted slabs penetrate the transition between the upper and lower mantle, there is apparently no impenetrable barrier to mantle flow across this zone indicating that a simple layered convection model (portrayed in Fig. 11.19) is unlikely. In fact, it appears that the core–mantle boundary may be the graveyard of subducted slabs, potentially establishing a direct link between plate tectonics and the deep interior of Earth.

Even so, many mysteries must still be explained. First, the descent of such slabs demands a compensating return flow, like the water that rushes up when a diver enters the water. The pattern of this flow, however, remains ambiguous. The main candidate is that of mantle plumes, which many geophysicists believe to ascend from the core–mantle boundary. If so, a link may exist between the motion of plates at Earth's surface and the formation of mantle plumes at the core–mantle boundary (Fig. 11.24). However, velocity patterns beneath some subduction zones suggest that not all descending plates penetrate the transition zone between the upper and lower mantle. Some plates appear to be deflected by it instead. Perhaps only the oldest, coldest, and therefore densest oceanic plates are heavy enough to push their way through the transition zone, whereas younger, more buoyant subducting plates are not, and so are deflected by it. It is even possible that subducting slabs may collect above the transition zone until they become heavy enough to avalanche to the core–mantle boundary.

Some of the ambiguity inherent in seismic tomography is clearly related to its present level of resolution, and will doubtless become clearer in the future as the resolution of the technique improves with the gathering of more data. For example, there is currently no obvious evidence for deep mantle plumes, although their tails may simply be too narrow to detect at this level of resolution. Also, because mantle convection is such a slow process, the pattern of heat distribution and, therefore, the velocity structure of the mantle is likely to depend on past plate configurations as well as the present one. Thus the mantle may have a "better memory" for ancient tectonic events than the crust, which is involved in endless recycling and renewal. Much better correlations may therefore emerge when tomographic results of higher resolution are compared to the pattern of plate motions over very long

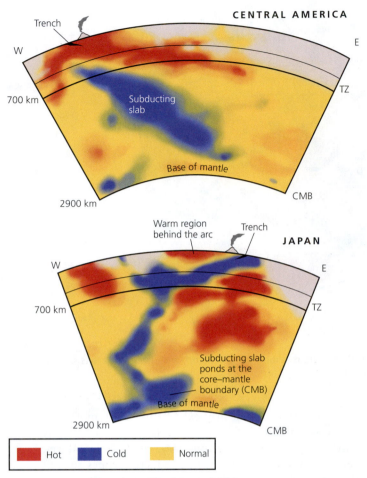

Figure 11.23   Fate of Subducted Slabs

Vertical slices through Earth's mantle beneath Central America and Japan show the distribution of colder (seismically fast) material (in blue) and hotter (seismically slow) material (in red) to the core–mantle boundary (CMB) based on seismic tomography. The distribution of colder material suggests that the subducting slabs beneath both areas have penetrated the transition zone (TZ) and extend to the base of the mantle.

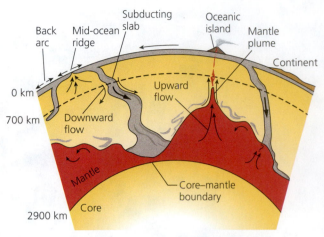

Figure 11.24   Subducting Slabs and Mantle Plumes

In this model for mantle circulation the downward flow brought about by subducting slabs plunging toward the core–mantle boundary is compensated by upward flow in the form of mantle plumes.

periods of time. Although the results of such studies are still some years away, seismic tomography clearly promises dramatic improvements in our understanding of mantle processes in the not-too-distant future.

11.7 SUMMARY

- The results of seismic tomography suggest that subducted slabs of oceanic lithosphere may ultimately reach the core–mantle boundary, compensated by return flow in the form of mantle plumes. A potential link may therefore exist between plate tectonics and the formation of mantle plumes in Earth's deep interior.

- Improvements in our understanding of mantle processes are likely to be dramatic as the resolution of seismic tomography improves.

Key Terms

body wave 339	inner core 351	primary (P) wave 339	seismology 347
compositional convection 359	layered mantle convection 360	ray path 347	seismometer (seismograph) 344
core 350	low-velocity zone 355	refraction 349	shadow zone 351
creep 360	lower mantle 355	Richter magnitude scale 344	surface wave 339
crust 350	mantle 350	secondary (S) wave 339	thermal convection 359
density 344	mantle plume 361	seismic discontinuity 350	transition zone 355
elastic rebound theory 337	Moho (Mohorovičić discontinuity) 350	seismic ray 347	upper mantle 355
epicenter 337	outer core 352	seismic tomography 361	whole mantle convection 360
focus 337	phase change 355	seismic wave 337	
		seismogram 344	

Key Concepts

11.1 HEAT AND DENSITY: CLUES TO EARTH'S INTERIOR

- Indirect measurement of Earth's density and heat shows how the planet's internal layers are segregated.

11.2 EARTHQUAKES AND ELASTIC REBOUND

- Earthquakes occur when a locked fault lets go and the rocks on either side elastically rebound.

11.3 SEISMIC WAVES

- Seismic waves are grouped into surface waves, which travel along Earth's surface, and body waves, which travel through Earth's interior.
- Body waves include primary (P) waves, which travel like sound waves, and slower secondary (S) waves, which move like oscillations in a rope and cannot travel through liquids.
- Seismometers produce a record of an earthquake's vibrations called a seismogram, from which much can be learned about Earth's interior.
- Seismic waves spread out in all directions from earthquake foci and travel at speeds that vary with the properties of the rocks through which the waves travel.

11.4 JOURNEY TO THE CENTER OF THE EARTH

- Through studying earthquake waves, geophysicists know that Earth is divided into three main layers: the crust, the mantle, and the core.
- A seismic discontinuity called the Moho separates Earth's crust from the mantle.
- Diffraction of earthquake waves by a seismic discontinuity that separates the mantle from the liquid outer core produces a shadow zone in which direct P and S waves are absent.
- The presence of faint P waves within the shadow zone reveals the existence of a solid inner core.

11.5 EARTH'S STRUCTURE AND COMPOSITION

- Earth is layered with a thin crust, a rocky mantle, and a metallic core.
- Continental crust is mostly granitic to dioritic, whereas the thinner and denser oceanic crust is largely basaltic.
- The mantle is divided into an upper and lower mantle, between which lies the transition zone.
- The core is divided into a liquid iron-rich outer core and a solid inner core of pure iron.

11.6 CIRCULATION IN EARTH'S INTERIOR

- Thermal convection drives mantle circulation but compositional convection drives circulation in the liquid outer core.
- The pattern of mantle circulation is variably attributed to whole mantle convection, layered mantle convection, or mantle plumes.
- Seismic tomography suggests that mantle flow is very complex.

11.7 PLATE TECTONICS AND THE FATE OF SUBDUCTED SLABS

- Subducted slabs may ultimately reach the core–mantle boundary with return flow accomplished by mantle plumes.

Study Questions

1. What are seismic waves and how are they distinguished?
2. Why do seismic waves travel with greater velocity at depth compared to their velocity near the surface of Earth?
3. Explain why S waves cannot be transmitted through liquid.
4. How is the boundary between the crust and mantle defined, and how does it differ from the boundary between the lithosphere and asthenosphere?
5. What do the variations of P and S wave velocities with depth (Fig. 11.14) tell us about the major internal divisions of Earth?
6. Explain the absence of S waves in the shadow zone.
7. Explain the presence of faint P waves within the shadow zone.
8. If an earthquake occurred at the equator, where would the shadow zone occur?
9. Why does the lithosphere include both the crust and the uppermost part of the mantle?
10. What do seismic tomographic scans of the upper mantle (see Fig. 11.20) reveal about mantle convection and the validity of mantle convection models (see Fig. 11.18)?
11. In what way might plate motion and the formation of mantle plumes be linked?

PART **III**
Sculpting the Solid Earth

Chapter 12 | Mass Wasting   370

Chapter 13 | Running Water   400

Chapter 14 | Groundwater   436

Chapter 15 | Glaciers and
Glaciation   464

Chapter 16 | Deserts and
Winds   500

Chapter 17 | Coastlines and Coastal
Processes   528

12 Mass Wasting

12.1 Mass Wasting: Downslope Movement

12.2 Factors That Influence Mass Wasting

12.3 Mass Wasting Mechanisms

12.4 Minimizing Mass Wasting Hazards

12.5 Mass Wasting and Plate Tectonics

▶ Key Terms

▶ Key Concepts

▶ Study Questions

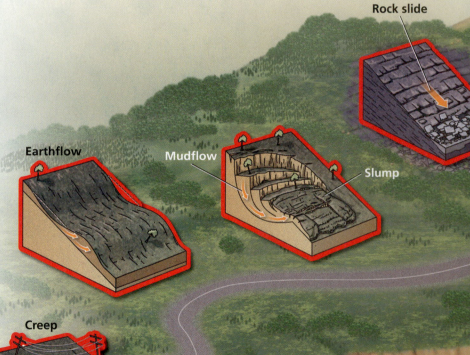

TYPES OF MASS WASTING

Rock slide

Talus

Earthflow

Mudflow

Slump

Creep

SCIENCE REFRESHER: Surface Tension, Sand Castles, and the Goldilocks Solution 378

IN DEPTH: Mass Wasting on the Moon and Mars 386

LIVING ON EARTH: The Oso Landslide, Washington State 390

Debris flow

Rock bolt

Rock netting

Grading

Rock fall

MASS WASTING MITIGATION

Drainage

Retaining wall

12.1 Define mass wasting and explain why it occurs.

12.2 Identify the various factors that govern mass wasting and the ways in which mass wasting is triggered.

12.3 Describe the mechanisms of mass wasting and the subdivision of their method of movement into fall, slide, and flow.

12.4 Identify methods that assess the threat of mass wasting and methods that reduce the hazards mass wasting poses.

12.5 Illustrate how mass wasting can be triggered by events linked to plate tectonics.

Although some surface processes are obvious enough or occur often enough to witness, others are more subtle or occur less frequently. The relentless pounding of our coastlines by the sea and the steady transport of sediment in the muddy waters of our streams are familiar examples of faster surface processes at work. In contrast, the breakdown of surface materials through weathering generally occurs far too slowly for us to observe. Earth materials can also be transported downslope *without* the aid of a transporting agent such as water, wind, or ice. Instead, they are moved by the pull of gravity. The downslope movement of material under these conditions is known as *mass wasting*. We can imagine mass wasting as a conveyor belt that brings sediment to the base of slopes, where it can be picked up and transported down valley by other agents of erosion, such as running water or flowing ice.

Earth is made up of slopes, which surface processes and tectonic activity can render unstable. Mass wasting is an inevitable and irreversible consequence of Earth's uneven surface. Mass wasting can be rapid, like landslides, or almost imperceptibly slow, like the downhill creep of soil. Mass wasting can also pose a serious threat to society and has been responsible for significant property damage and loss of life. The hazards mass wasting poses can be mitigated by a variety of innovative engineering techniques. Nevertheless, any careful evaluation of a property being considered for either domestic or commercial use includes an assessment of risk, and requires an understanding of the processes described in this chapter.

12.1 Mass Wasting: Downslope Movement

Mass wasting is the downslope movement of Earth materials under the direct influence of gravity and includes the phenomena popularly referred to as *landslides*. Following the breakdown of surface materials by weathering that we described in Chapter 5 (a process that is not considered a form of mass wasting because it does not involve downslope movement), mass wasting is often the first phenomenon to modify the landscape. But in contrast to the other surface processes that affect the landscape, mass wasting does not require a transporting medium such as water, ice, or wind.

Although the effects of mass wasting are confined to solid Earth, each of Earth's other surface reservoirs, which are described in Chapter 1—the hydrosphere, atmosphere, and biosphere—exert powerful influences on the style and effectiveness of mass wasting processes. The most familiar examples of mass wasting are those that are rapid and dramatic, such as landslides and *rockfalls* (both of which we discuss in detail later in the chapter). But mass wasting also encompasses less obvious processes, such as the continuous but almost imperceptible "creeping" downhill movement of soils.

Mass wasting occurs because Earth's surface is rarely flat. Instead, our planet is made up of sloping surfaces that under certain conditions can become unstable. This is especially true where slopes are steep or where other factors combine to reduce the resistance to downhill sliding. Such conditions are necessary for sliding to occur because Earth materials have an inherent natural strength, called **shear strength**, which resists sliding. Indeed, it is shear strength that prevents all slopes from simply flattening out under the influence of gravity like water on the surface of a lake. For mass wasting to occur, therefore, gravitational forces must first overcome the strength of surface materials. This requires the presence of a slope.

Although this chapter focuses on mass wasting, it is important to realize that mass wasting processes often work in concert with other forms of erosion. For example, material that travels downslope from higher elevations to lower ones by mass wasting can be picked up by other transporting agents, such as streams. As a result, mass wasting plays a vital role in the overall erosional process. Although the contribution of mass wasting may not be as obvious as those of other surface processes, it is responsible for shaping much of the landscape around us. Without mass wasting, for example, streams would simply cut downward to form steep-sided ravines, called *canyons*. As we shall see in this chapter, however, such steep slopes are inherently unstable. So instead of steep-walled canyons, we more often see the familiar "V" shape of stream valleys, which results from the influence of mass wasting of a valley's sides.

Although mass wasting occurs on all sloping surfaces, it takes a wide variety of forms. It also takes place at highly variable rates, and these rates do not depend simply on the steepness of the slope. So what governs slope stability, and what initiates mass wasting and controls its form and speed? To answer these questions we turn our attention in Section 12.2 to the factors that influence mass wasting.

12.1 SUMMARY

- Mass wasting is the downslope movement of surface materials under the influence of gravity.

- A slope is needed for mass wasting to occur because materials have a shear strength that resists downslope movement.

12.2 Factors That Influence Mass Wasting

Although mass wasting is driven primarily by gravity, other factors often play an important role in triggering downslope movement. The most obvious of these is the steepness of the slope and its resulting stability. Other factors relate not only to the local environment but also to regional climatic and tectonic influences. Among these influences are the water content of the slope material, the amount of vegetation on the slope, any natural or artificial controls on slope stability, and the occurrence of earthquakes and other triggering mechanisms.

SLOPE STABILITY AND ANGLE OF REPOSE

If we pour sugar onto a table, the mound of sugar develops a stable slope as it grows. If we continue the process, the mound grows but the stable slope remains unchanged. If we start again, a second mound develops with an identical slope to the first (Fig. 12.1). This principle also holds true for mounds of flour or salt, except that the slope of the mounds is slightly different in each case. This is because each substance develops its own

CHECK YOUR UNDERSTANDING

- What is mass wasting and why does it occur?

- Why is weathering not considered a form of mass wasting?

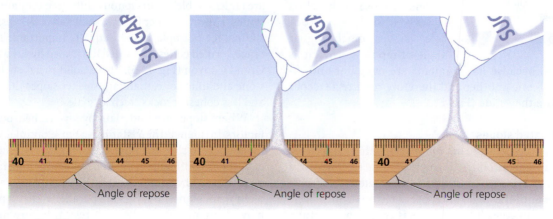

Figure 12.1 Angle of Repose

The slopes of sugar piles on a tabletop remain the same regardless of the size of the pile. This is because the maximum slope that the sugar can maintain is dictated by its angle of repose, which for granulated sugar is about 30 degrees.

(a)   (b)

Figure 12.2   Variations in Angle of Repose

(a) Slopes of sand dunes in the Sahara Desert of Mauritania, and (b) slopes of coarse angular talus near the summit of Mount Olympus, Greece, each have a different characteristic angle of repose.

characteristic stable slope known as the **angle of repose**. The steepness of a stable slope varies with the type of material: for most unconsolidated materials, the angle lies between 25 and 40 degrees. For example, sand dunes rarely have slopes that exceed 30 degrees. The reason is that the windblown sand grains that form sand dunes are smooth and rounded, and so readily slide past each other (Fig. 12.2a). On the other hand, coarse, angular material such as **talus**, the name given to accumulations of loose rock debris at the base of cliffs and rock faces, has a much higher angle of repose (as much as 45 degrees) because its fragments are irregular and interlock with one another and therefore require a steeper slope before they slide (Fig. 12.2b).

CHECK YOUR UNDERSTANDING

● What is an angle of repose and why does it differ for different substances?

Unconsolidated slopes

The angle of repose is the *maximum* slope that can be sustained by an unconsolidated material. If the steepness of a slope is increased beyond this angle, the material will slide downslope until the angle of repose is restored. Such *oversteepening* of unconsolidated slopes is common in nature, occurring, for example, whenever streams undercut their banks (Fig. 12.3) or when waves erode a coastline to form poorly consolidated cliffs. In each case, the loose material

above the area of undercutting becomes gravitationally unstable and so succumbs to mass wasting. Oversteepening of the stream bank or cliff face in these examples is the result of erosional processes. However, oversteepening of natural slopes can also occur as a result of human activities. New road cuts, for example, often have oversteepened slopes, as a result of which they are prone to collapse by mass wasting (Fig. 12.4).

Bedrock slopes

Where slopes are underlain by *solid* bedrock, steepness is only one factor that governs their stability. Dislodged rocks are far less stable on precipitous cliffs than they are on gentle slopes. However, other factors, such as a rock's resistance to sliding, for example, may be equally important. This resistance, termed **cohesion**, is well illustrated by the walls of the Grand Canyon (Fig. 12.5). There, solid rocks, such as sandstone and limestone, maintain steeper slopes than crumbly and less cohesive rocks, such as shale.

Where slopes are underlain by *layered* bedrock, several factors influence their stability. For example, rocks are more likely to slip when the slope and the layering (usually bedding in sedimentary rocks or foliation in metamorphic ones) dip in the same direction. Layers within a rock are often weak planes. If layers are parallel to the slope, sliding may occur more readily. As a result, failure may occur when road cuts or building-site excavations undermine slopes of this sort (Fig. 12.6a, b). In contrast, layered bedrock that dips in the opposite direction to that of the slope

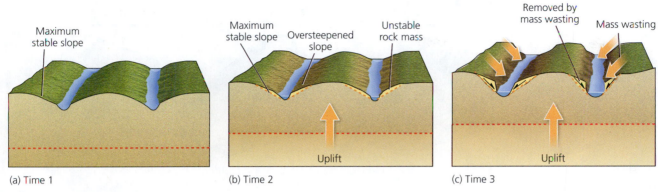

(a) Time 1   (b) Time 2   (c) Time 3

Figure 12.3   Oversteepening

(a) Stream erosion has produced banks with the maximum stable slope. (b) Uplift causes the stream to erode its bed, producing banks with oversteepened slopes. (c) Mass wasting removes the oversteepened material, recreating banks with the maximum stable slope.

Figure 12.4   Collapsed Road Cut

Following heavy rainfall and spring thaw, a road cut that was oversteepened during road construction collapsed on Route 33, near Haydenville, Ohio.

CHECK YOUR UNDERSTANDING

⊙ What is the cohesion of a rock and how does it influence slope stability?

often remains stable even when the slope is undermined (Fig. 12.6c, d). Because of relationships such as these, it is important that construction sites be surveyed geologically prior to their development.

WATER CONTENT

The angle of repose (and therefore mass wasting) in unconsolidated materials is strongly influenced by the presence of water, which fills the **pores** (the spaces between grains). The effect of water content, however, depends on how much water is present and how easily it can move from one pore to another. These factors hinge, respectively, on the **porosity** (pore volume) and **permeability** (pore connectivity) of the material. Water poured on sand, for example, quickly disappears because sand is both porous and permeable. But as any builder of sand castles knows, damp sand supports

Figure 12.5   Cohesion

The role cohesion plays in slope stability is illustrated in the walls of the Grand Canyon, where massive rock layers, such as sandstone and limestone, support far steeper slopes than less resistant rocks such as shale do.

much steeper slopes than dry sand. This is because water molecules are polarized and have opposite electrical charges at either end. The polarity of water molecules attracts other water molecules as well as the water-coated

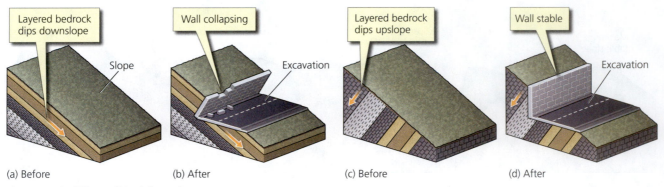

(a) Before   (b) After   (c) Before   (d) After

Figure 12.6   Effect of Rock Layering

(a) Slopes underlain by layered bedrock are more prone to slip when the rock layering dips downslope, (b) particularly when undercut by an excavation. (c) Slopes are more stable when the rock layering dips in the opposite direction to the slope, (d) even when undercut.

sand grains. This attractive force is known as **surface tension** and binds the sand together (see Science Refresher: **Surface Tension, Sand Castles, and the Goldilocks Solution** on page 378). The binding forces can be overcome if the sand contains too much water, and when the forces are overcome the walls of a sand castle collapse. This happens if the pores between grains become filled with water, a condition known as **saturation**. The grains become lubricated and the cohesion between them is destroyed, allowing them to slide past each other. For this reason, unconsolidated slopes, such as road cuts on the highway, are particularly prone to collapse following periods of heavy rainfall or during a prolonged spring thaw when the ground becomes water saturated (see Fig. 12.4).

As water invades the pore spaces in the slope material, it replaces air trapped between grains. As a result, water also adds weight to the slope material, and this alone may trigger downslope movement. Water can also act as a lubricant, causing clay-rich layers to become slippery, which encourages the sliding of any overlying material, as with a car on a muddy road. A close link also exists between water and two other important influences on mass wasting, those of climate and vegetation.

> **CHECK YOUR UNDERSTANDING**
> ● How is slope stability affected when materials become saturated with water?

CLIMATE AND VEGETATION

Mass wasting is also influenced by vegetation and climate. Plants have intricate root systems that bind soil together. Plants also absorb water and provide a cover that offers soil protection from erosion and water saturation. Mass wasting is therefore enhanced in arid climates that provide little vegetative cover and are subject to occasional heavy downpours during which water enters the soil, a process called **infiltration**. The resulting instabilities are especially pronounced where slopes are steep. In the winter of 1997–98, for example, numerous landslides were triggered on steep slopes in southern California when an El Niño brought heavy rains to the region's normally dry climate (Fig. 12.7). For the same reasons, mass wasting is enhanced whenever vegetation is removed from slopes, for example, by deforestation and forest fires. Without vegetation to protect and anchor it, the soil is only loosely held and commonly starts to migrate downhill.

> **CHECK YOUR UNDERSTANDING**
> ● In what ways do climate and vegetation influence the likelihood of mass wasting events?

EARTHQUAKES AND OTHER TRIGGERING MECHANISMS

Because the natural strength of materials resists downslope movement, a trigger is sometimes needed to overcome this resistance, even on unstable slopes. In addition to storms, mass wasting events may be triggered by tectonic events, such as earthquakes and volcanic eruptions. Indeed, it is the products of mass wasting triggered by such events that are commonly responsible for much of the destruction and loss of life earthquakes and volcanic eruptions cause.

Earthquakes shake the ground and commonly trigger the collapse of unstable slopes. The 1994 Northridge earthquake in the San Fernando Valley, California, provides a good example. This earthquake caused so many landslides in the Santa Susanna Mountains north of the epicenter that for a time the mountains were almost completely obscured by dust (Fig. 12.8). Not only did these landslides block roads and damage homes but the dust they produced also caused an outbreak of a respiratory disease known as valley fever, which claimed several lives.

Volcanic eruptions also shake the ground, and this shaking can cause unstable slopes to slide (see In Depth: **Can Explosive Eruptions Be Predicted?** in Chapter 4). In addition, meltwater released from the eruption of a snow-capped volcano can saturate the soil, triggering downslope movement on the volcano's flanks. Some eruptions also produce huge quantities of volcanic ash which, when mixed

Figure 12.7   Infiltration

Infiltration of El Niño rains is thought to have caused this massive landslide near Aromas, California, on April 22, 1998. The slide destroyed one house, blocked a road, and severed two natural-gas pipelines, cutting gas service to 60,000 customers in Santa Cruz and Monterey counties.

Figure 12.8   Earthquake-Triggered Landslides

Dust from thousands of landslides triggered by the Northridge earthquake on January 17, 1994, partially obscures the Santa Susanna Mountains north of the epicenter in the San Fernando Valley, California.

SCIENCE REFRESHER

Surface Tension, Sand Castles, and the Goldilocks Solution

As every kid who has played with sand knows, when it comes to building sand castles, the sand needs to be damp. Too dry, and the sand will collapse into a pile, too wet and it will run away like a fluid. But if it is just right, like Goldilocks' porridge, the sand will hold together, allowing turrets to be built bucket after bucket. What makes this possible is an attraction between water molecules called *surface tension*.

A molecule of water (H_2O) is made up of two kinds of elements, a single oxygen ion and a pair of hydrogen ions. The molecule is held together because each hydrogen has a positive electrical charge and is attracted to the oxygen, which has a pair of negative charges. The atoms bond in such a way that the molecule is V-shaped with the oxygen at the base and a hydrogen at each tip (Fig. 12A). As a result, the electrical charges are not equally distributed; the side of the molecule with the oxygen ion has a slight negative charge, whereas the hydrogen side is slightly positive. This charge difference is called a *dipole* and such molecules are called *polar molecules*.

Thanks to this charge difference, water molecules are attracted to each other because the slightly negative oxygen side of one molecule is attracted to the slightly positive hydrogen side of another (Fig. 12B). This attraction is called a *hydrogen bond* and is most obvious at a water surface where the attractive force causes the water to bead or form drops. This force is *surface tension*. Just as it is harder to break through a

row of people if their arms are locked together than if their arms are at their sides, so surface tension imparts an elastic strength to the surface of water. As a result, the surface can be pushed down without breaking. Surface tension is what allows insects such as water striders to skate across the surface of a pond.

It is also surface tension that allows us to form sand into sand castles. When dry sand is poured from a bucket it forms a pile (Fig. 12C), the slope of which is governed by the angle of repose (for dry sand this is about 30°). But if the sand is moistened so that the sand grains become coated with water, the surface tension of the water forms little elastic bridges between the water-coated grains of sand, which tends to hold the grains together (Fig. 12D). Too much water, however, and the sand becomes saturated. This causes the grains to become separated, eliminating the binding forces and allowing the mixture to flow like a fluid (Fig. 12E).

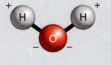

Figure 12A   Water Molecule

A molecule of water (H_2O) is V-shaped with two hydrogen ions (H), each with a positive electrical charge, bonded to a single oxygen ion (O) with two negative electrical charges.

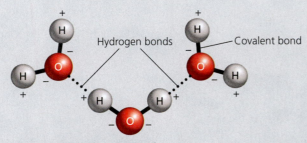

Figure 12B   Hydrogen Bonds

The positive side of one water molecule attracts the negative side of another. This attraction is known as a hydrogen bond and is responsible for surface tension.

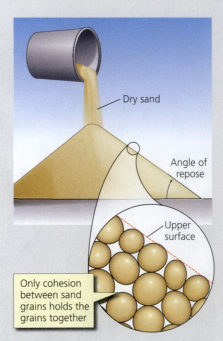

Figure 12C   Dry Sand

Dry sand forms a pile, and the slope (or angle of repose) of the pile depends on cohesion between the sand grains. For dry sand, the angle of repose is about 30°.

(Continued)

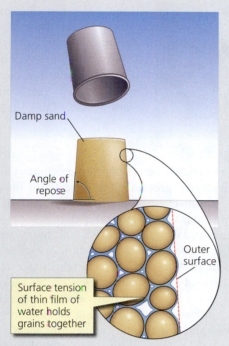

Figure 12D   Damp Sand

Damp sand can support steep slopes because surface tension on the film of water coating the sand grain holds the grains together.

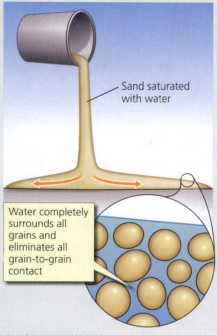

Figure 12E   Saturated Sand

In saturated sand, the sand grains become separated with water, which eliminates all grain-to-grain contact and allows the mixture to flow like a fluid.

with meltwater, form destructive rivers of mud with the consistency of wet concrete. During the May 18, 1980, eruption of Mount St. Helens, rivers of mud mobilized by melted ice swept down valleys west and east of the volcano at speeds up to 40 kilometers per hour (25 miles per hour), some traveling 90 kilometers (56 miles) to the Columbia River (Fig. 12.9).

Mass wasting can be triggered by a variety of human activities, particularly where these influence the stability of slopes. Oversteepening of slopes caused by road building and dumping, undercutting of slopes caused by the excavation of construction sites, and clearing of vegetation are just some of the activities that can lead to mass wasting.

It is clear that a variety of processes are involved in the downslope movement of surface materials and that the phenomena referred to as landslides have a variety of origins. So what exactly are these processes and what causes some forms of mass wasting to occur rapidly while others are slow? To answer these questions, we turn in Section 12.3 to the classification of mass wasting mechanisms. As we discuss these mechanisms, it is important to remember

CHECK YOUR UNDERSTANDING

⬥ What human activities trigger mass wasting events?

that while the classification scheme clearly distinguishes between different types of mass wasting, nature does not. In reality, one type of mass wasting will often trigger another type, or will trigger some other surface process in a devastating domino effect that can result in serious loss of life and property damage.

12.2 SUMMARY

- The strength of the slope material and the steepness of the slope both affect when, and how, mass wasting occurs.

- All consolidated material develops a stable slope known as the angle of repose. The steepness of the resulting slope varies with the type of material. Strong rocks support steeper slopes than weak ones, and layered rocks are more prone to sliding if the layering is parallel to the slope.

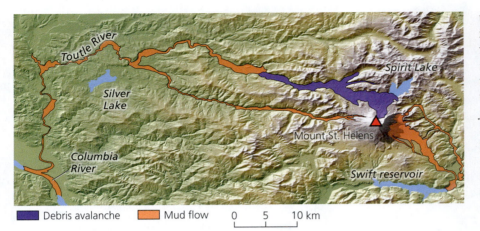

Debris avalanche   Mud flow   0   5   10 km

Figure 12.9 Volcanic Mudflows
Destructive rivers of mud triggered by the May 18, 1980, eruption of Mount St. Helens. Mud flowed down valleys west of the volcano as far as the Columbia River, a distance of more than 90 kilometers (56 miles).

- Slope stability is strongly influenced by water content. Once water saturates a slope, its material slides much more easily.
- Vegetation helps bind soil together, increasing its strength.
- Mass wasting can be triggered by storms, earthquakes, volcanic eruptions, and human activities such as road building and site excavation.

12.3 Mass Wasting Mechanisms

Rocks that fall from vertical cliffs move in a different manner (and at quite different speeds) than sediment that slowly migrates downhill. Rocks and sediment also consist of very different materials. Yet both falling rocks and migrating sediment are examples of mass wasting, because both the rocks and the sediment move under the influence of gravity. It is convenient to classify mass wasting processes according to *mechanism, material,* and *speed.* In this section, we identify and discuss three movement mechanisms— *fall, slide,* and *flow* (Fig. 12.10). Each mechanism is further qualified by the nature of the material involved (usually rock, debris, earth, or mud) and its rate of movement (slow or fast) (Table 12.1).

> **CHECK YOUR UNDERSTANDING**
>
> ● What are the three main mechanisms of mass wasting?

FALL

When material is detached from precipitous slopes and free-falls to the ground, either directly or in a series of bounds, the mechanism of mass wasting is **fall**. During a free fall each fragment moves separately. Because the material involved is usually bedrock, this mass wasting process is commonly called a **rockfall**. Rockfalls are often the result of *frost wedging,* whereby water trapped in cracks expands on freezing and wedges the rock apart. As we learned in Chapter 5, frost wedging is an important weathering process. By providing material for rockfalls, frost wedging can also be an agent for mass wasting in mountainous areas, as witnessed by the slope of angular fragments, called *talus,* found at the base of most rock faces (see Fig. 12.2a). Frost wedging shows how weathering, which does not move material, can contribute to mass wasting, which does. Other factors that contribute to rockfalls include the loosening of rocks by plant roots, the undercutting of rock faces by streams and wave action, and the oversteepening of bedrock slopes by human activities.

Rockfalls are the fastest mass wasting processes, happening so rapidly that they are rarely witnessed. Most are small falls that intermittently contribute angular fragments to talus slopes. Large rockfalls, however, can be very destructive and, on impact, may generate massive air blasts or trigger avalanches of rock capable of traveling great distances. Such was the case in May 1970, when a massive earthquake-triggered rockfall fell from the north face of Nevado Huascarán, the highest peak in the Peruvian Andes

TABLE 12.1 CATEGORIES OF MASS WASTING

Type of Movement	Description	Subcategory
Fall	Materials free-fall in air	Rockfall
Slide	Material moves as relatively coherent mass on well-defined basal surface	Rockslide Slump
Flow	Material moves as a viscous fluid and is commonly unconsolidated and water saturated	Mudflow Debris Flows Earthflow Solifluction Creep

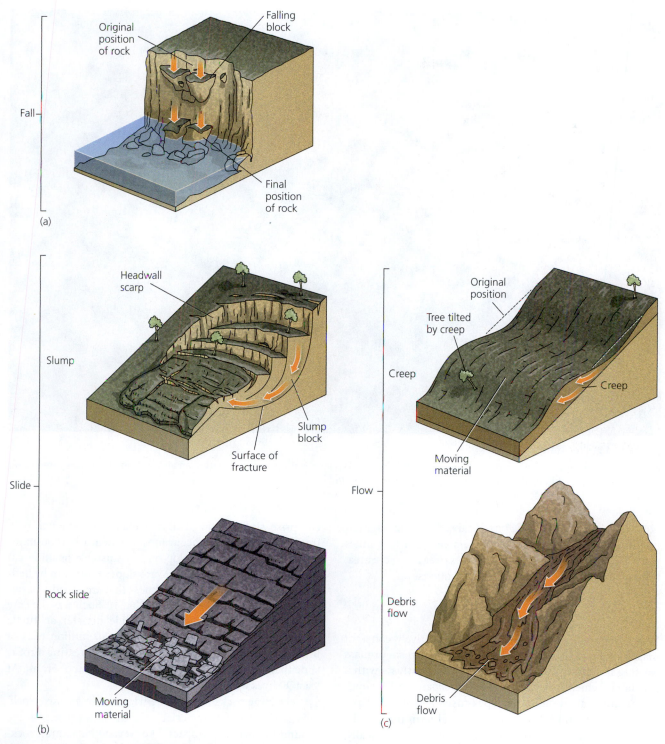

Figure 12.10 Categories of Mass Wasting

Mass wasting falls into three categories (a) fall, (b) slide, and (c) flow. These categories reflect the mechanism of movement.

(see Fig. 12.13), an event we discuss in more detail later in this section.

More recently, a much smaller rockfall caused extensive damage to part of Yosemite National Park (Fig.12.11). On July 10, 1996, a block of granite with an estimated volume of almost 60,000 cubic meters (79,000 cubic yards) broke off the rock face above the Happy Isles Nature Center. After

detaching, the rock mass slid down the steep rock face for 180 meters (590 feet) before free-falling 550 meters (1800 feet) to the talus slope below, impacting at a speed of 420 kilometers per hour

Figure 12.11   Rockfall

On July 10, 1996, a block of granite slid down this rock face in Yosemite National Park, creating an air blast that blew down trees over an area of 10 acres, killing one person and injuring several others.

(260 miles per hour). The fall itself caused little damage but the impact generated a 280 kilometers per hour (175 miles per hour) air blast that blew down trees over a 10-acre area, killing one person and injuring several others.

SLIDE

When material moves downslope in a relatively coherent mass along a well-defined surface, the mechanism of mass wasting is a **slide**. The cohesion of a slide contrasts with a rockfall in which each fragment moves separately. In popular literature, an event of this type is usually referred to as a landslide, but landslide is a nontechnical term that can be applied to almost any form of mass wasting. Technically, slides fall into one of two categories: *rockslides* and *slumps*.

Rockslide

Like rockfalls, **rockslides** involve the rapid downslope sliding of segments of bedrock. However, unlike rockfalls, rockslides occur along planes of weakness such as bedding planes, foliation surfaces, or fractures. The 1996 rockfall in Yosemite (see Fig. 12.11), for example, started as a rockslide when the detached mass of granite slid along a plane of weakness before falling. Since the detached rock mass usually breaks up as it moves downslope, most rockslides

comprise sliding and tumbling masses of angular rock fragments. Rockslides occur most frequently in mountainous regions but can develop on any unstable or undercut slope. Particularly susceptible are slopes underlain by bedrock containing planes of weakness that are broadly parallel to the slope surface (see Fig. 12.6a, b). Such was the case in 1903 for Canada's worst landslide disaster at Turtle Mountain, Alberta (Fig. 12.12), when 90 million tons of limestone broke away from the mountain, creating a rockslide that buried part of the coal mining town of Frank. At least 90 lives were lost.

Very large rockslides can be triggered by massive rockfalls if the rockfall material spreads outward across the ground following its impact. The Nevado Huascarán rockfall in Peru in May 1970, for example, generated a type of rockslide known as a **rock avalanche**, in which the rock fragments ride on a cushion of compressed air, much like an avalanche of snow. The trapped air reduces friction between the rock fragments and increases the buoyancy of the slide. A rock avalanche is a particularly destructive form of slide that can spread out over wide areas and reach speeds in excess of 350 kilometers per hour (220 miles per hour).

Estimated at close to 100 million cubic meters (130 million cubic yards), the rock avalanche generated by the 1970 Nevado Huascarán rockfall in Peru roared down a canyon

Figure 12.12   Frank Slide

The 1903 Turtle Mountain rockslide in Alberta, Canada, that created the scar seen here claimed the lives of at least 90 people. Thirty million cubic meters (a billion cubic feet) of rock broke away and, within two minutes, buried part of the town of Frank.

at speeds of 280–335 kilometers per hour (175–210 miles per hour) for some 10 kilometers (6 miles) before part of it surged over a 230 meter (750 foot) ridge into an adjacent valley. Here it fanned out and, led by a wave of rock debris some 80 meters (260 feet) high, obliterated the town of Yugay and killed all but 92 of its 18,000 inhabitants (Fig. 12.13). Continuing down the canyon, the main mass of the rock avalanche inundated a second town. It then surged across the Rio Santa and climbed 80 meters up the far side of the valley before falling back like a wave retreating from the shore.

CHECK YOUR UNDERSTANDING

◌ What is the difference between a rockslide and a rock avalanche?

Slump

A slide in which blocks of material move downslope along a curved surface is a **slump**. Because the surface on which they are moving is curved, slump blocks rotate as they slide so that their upper surfaces tilt and often face uphill (Fig. 12.14). As a result, trees rooted in these surfaces or any upright structures built upon the surfaces are tilted backward (i.e., uphill). Even if a slump is not observed, the tilted upper surface of blocks and the back-tilting of trees can provide evidence of slumping.

Slumping occurs most frequently when a slope is undermined and the material on the upper part of the slope is no longer supported by its base. Slopes can be oversteepened in this way as a result of natural processes, for example, where a stream undercuts its banks or where wave action undercuts a cliff. Oversteepened slopes can also result from human activities such as road building and the excavation of hillsides for building sites. In fact, road cuts are particularly susceptible to this form of mass wasting (see Fig. 12.4).

Because of the spoon-like shape of the surfaces on which they move, slumps have a distinctive amphitheater-like shape. Upslope, a crescent-shaped cliff, called a *headwall scarp*, marks the surface of rupture. Below this scarp, the body of the slump deposit is typically made up of several individual slump blocks, each with a curved headwall scarp. Downslope, the blocks break up and the slump becomes increasingly chaotic and may eventually give rise to more fluid mass wasting at the downslope end of the slump, or *slump toe* (Fig. 12.14), which can fan out to form an apron of hummocky ground. Interestingly, our knowledge of the distinctive geometry of slumps has also enabled us to

Figure 12.13   Rock Avalanche

Nevado Huascarán rock avalanche buried the Peruvian towns of Yungay and Ranrahirca in 1970. By the time it struck Yungay, the avalanche is estimated to have comprised over 100 million cubic meters (130 million cubic yards) of water, mud, and rocks.

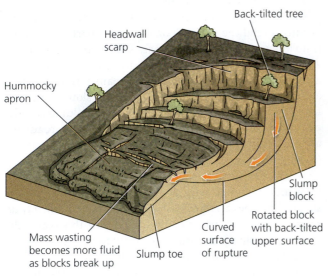

Figure 12.14   Slumping

During slumping, loosely consolidated material moves downslope along a curved surface, causing individual slump blocks to rotate so that their upper surfaces become tilted. Downslope, the blocks break up, giving rise to more fluid mass wasting at the slump toe.

recognize them on other terrestrial planets (see In Depth: **Mass Wasting on the Moon and Mars** on page 386).

Most slumps do not move rapidly, nor do they travel great distances. Yet they can be very destructive, particularly when they occur in developed areas where people live. Along the southern California coastline, for example, a variety of factors combine to make slumping a persistent problem. Much of this region is underlain by weakly consolidated materials, such as clay-bearing sands and gravels. The region is also tectonically active, with the result that steep slopes are widespread, fractures that water can seep down are numerous, and earth tremors that can trigger sliding are frequent. In addition, southern California experiences a climate that is dry but occasionally subject to heavy rains and stormy seas, especially during El Niño events. As with many types of mass wasting, slump movement is facilitated by heavy rainfall, which saturates unconsolidated materials, adding weight to sediments on slopes, and lubricating clay minerals, which causes them to become slippery. Preexisting rupture surfaces can also be reactivated by heavy rainfall because water infiltrating these ruptures reduces resistance to mass wasting and in this way promotes further sliding. For this reason, slumping is a process that tends to recur, so that slopes that have slumped in the past are prone to slump again.

In addition to heavy rainfall, El Niño events are associated with stormy seas and widespread coastal erosion. Sea cliffs are often undercut by wave action and slump as a result. In 1958, for example, undercutting of the steep cliffs at Pacific Palisades led to the development of a massive slump on the coastline west of Los Angeles (Fig. 12.15). Since then the easily eroded siltstone cliffs have experienced many slumps, repeatedly blocking the Pacific Coast Highway between Santa Monica and Malibu.

> **CHECK YOUR UNDERSTANDING**
> ◉ What features associated with slumping illustrate how slumps slide on curved surfaces?

FLOW

The term **flow** describes mass wasting in which unconsolidated and usually water-saturated materials move downslope like viscous fluids, often spreading out in lobate (toe-like) fans. Flows are therefore distinct from slides, which move downslope as coherent masses. Flows are often produced when other types of mass wasting, such as rockslides and slumps, move downslope because the fluids that these forms of mass wasting contain build up as they move down the slope. We subdivide flows into *mudflows, debris flows, earthflows, solifluction,* and *creep,* depending on the type of material transported and its rate of movement.

Mudflow

Mudflows are rapidly flowing masses of water-saturated material, more than half of which is silt sized or finer. Depending on the water content (which can exceed 50%), the

Slump blocks
Pacific Coast Highway

Pacific Palisades | Santa Monica

Los Angeles

Pacific
Ocean

Palos
Verdes

Long
Beach

0   10   20 km

Figure 12.15 Pacific Palisades
Undercutting of poorly consolidated cliffs at Pacific Palisades, near Los Angeles, led to the development of a massive slump, which completely blocked the Pacific Coast Highway in 1958.

consistency of mudflows varies from that of soup to that of wet cement, and their rates of movement range from 1 kilometer per hour to more than 150 kilometers per hour (0.5–95 miles per hour). They occur most frequently in mountainous, semiarid regions during occasional periods of heavy rainfall.

Under such conditions, the ground quickly becomes water saturated, and channels fill with coalescing muddy streams. Fanning out onto canyon floors, these short-lived streams swiftly evolve into highly fluid slurries of mud and boulders that can travel rapidly for significant distances, unleashing their fury on people and property farther down the valley.

Occasional heavy rains in an otherwise dry climate make mudflows, like slumping, a persistent problem in California, especially during El Niño events. The San Francisco Bay area, for example, has experienced many damaging mudflows and was particularly hard hit during the potent El Niño of 1997–98. Farther south, the coastal resort of Laguna Beach also witnessed the destructive power of mudflows when heavy El Niño rains fell in February 1998. The resort is located at the mouth of Laguna Canyon, which cuts through a sequence of alternating sandstones and weakly resistant mudstones that are prone to slide. In 1998, dozens of small slides developed in the water-saturated soil on the steep sides of the canyon (Fig. 12.16). These slides quickly liquefied and became transformed into mudflows that accelerated rapidly to the canyon floor. One mudflow actually cascaded over an 8-meter (26-foot) cliff before tearing a house, which lay directly in its path, in two.

Mudflows are by no means restricted to areas of semiarid climate like California. In fact, destructive mudflows can occur in almost any hilly area that is subject to prolonged heavy rainfall. For example, mudflow disasters are a threat in many parts of the subtropical Appalachian Mountains, and a period of unusually heavy rain preceded the fatal Washington mudslide of March 2014 in the temperate northwestern United States (see Living on Earth: **The Oso Landslide, Washington State,** on page 390).

Occasionally, mudflows are generated when clay-rich sediments, known as **quick clays**, become saturated with water and are disturbed by vibrations. This causes the quick clay to become highly fluid, a phenomenon known

IN DEPTH

Mass Wasting on the Moon and Mars

Mass wasting is driven by the pull of gravity, the natural phenomenon by which all physical bodies attract each other (see Science Refresher: **Basic Terminology in Physics** in Chapter 1). As a result, mass wasting is by no means unique to our own planet. Indeed, on rocky celestial bodies such as the Moon, which lack both water and an atmosphere, mass wasting is one of the few surface processes that modifies the impact-scarred landscape. But the Moon has maintained topography despite billions of years of mass wasting, which demonstrates that the wasting process is far less effective when it is acting alone, rather than in partnership with weathering and erosion as it does on Earth. Mass wasting on the Moon is best seen on crater walls, many of which show signs of landslides. Those landslides occurred because the initial slopes (created by the impacts that formed the craters) exceeded the angle of repose of the Moon's loose surface materials (Fig. 12F).

The most dramatic evidence of mass wasting on a planet other than our own occurs on Mars, where a landslide was photographed in action in 2008. Mass wasting on Mars takes the form of slumps, mudflows, and a variety of features attributed to frozen ground. In one Martian canyon, spectacular slump structures give way downslope to flow deposits. These structures are seen in the walls of the Valles Marineris (Fig. 12G), a massive canyon system up to 8 kilometers (5 miles) deep that runs east–west for more than 3,000 kilometers (1864 miles) near the planet's equator. Like slumps on Earth, the Martian slumps are clearly identified by their amphitheater-like geometry and jumbled slump blocks. The slumps have given rise to flows

Figure 12G   Slump on Mars

This vast amphitheater-shaped slump is on the south wall of the Valles Marineris canyon. Note the curved headwall scarp (which has partially removed a crater wall on the plateau surface), the backward-tilted slump blocks, and the apron of mudflow or debris flow deposits spreading across the canyon floor.

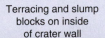

Figure 12F   Copernicus

This photograph of the crater Copernicus taken by the Apollo 17 mission in 1972 clearly shows terracing and slump bocks on the inside of the crater rim produced as a result of mass wasting. Copernicus is one of the youngest craters on the Moon and has a diameter of 93 kilometers (58 miles).

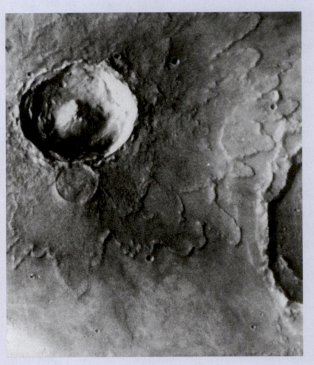

Figure 12H   Rampart Crater

Lobate (toe-like) deposits surrounding the Martian rampart crater, Yuty, are thought to be the result of mudflows produced by impact-melting of subsurface ice. The visibility of a preexisting crater outside (below) the rim of the 19 kilometer- (12 mile-) diameter rampart crater shows that the deposit is thin.

at the slump toe, some of which have traveled as much as 30 kilometers (19 miles) across the canyon floor.

Distinctive lobate (toe-like) deposits that closely resemble mudflows can also be seen surrounding certain types of Martian impact structures known as *rampart craters* (Fig. 12H). The similarity of these deposits to those that emanate from slump toes suggests that the rampart craters formed by flow. Such a mechanism would require at least the transient existence of water on the Martian surface. This water most likely originated as meltwater,

produced from ice in the Martian subsurface by the heat generated from meteorite impact. Water on Mars may therefore exist in the form of permanently frozen ground or *permafrost* not unlike the water of the polar regions on Earth, a prospect that has important implications for the possibility of life on Mars. Rather than representing material ejected on impact, the lobate deposits of rampart craters probably formed as ground-hugging flows produced when impact-generated meltwater mixed with the ejected material to form mud.

Figure 12.16 Mudflow

Two homes in Laguna Canyon, California, succumb to an El Niño–generated mudflow in 1998.

as **liquefaction**, because the vibrations cause the interlocking clay minerals to become aligned, greatly reducing their resistance to sliding (insert, Fig. 12.17). Liquefaction can be triggered by earthquakes, explosions, or even by heavy automobile traffic, and is particularly problematic in the St. Lawrence Lowlands of eastern Canada, an area that is underlain by sensitive marine clays that were deposited following glacial retreat at the end of the Ice Age. On May 4, 1971, for example, the newly built Saint-Jean-Vianney subdivision of La Baie, near Chicoutimi, Quebec, was devastated when 6.9 million cubic meters (240 million cubic feet) of sensitive clay abruptly liquefied and collapsed toward the Saguenay River at 25 kilometers per hour

(15 miles per hour). The flow carried away 40 homes and took the lives of 31 people (Fig. 12.17).

Better known than quick clay is **quicksand**, which forms when loose, water-saturated sand is suddenly agitated. As with quick clay, this process causes liquefaction of the sand if the water is unable to escape. During liquefaction, the sand becomes highly fluid and can no longer support a load. Stepping on quicksand consequently causes a person to sink. But not far, for quicksand is not the human-swallowing terror of the movies and is rarely more than a few feet deep.

A special type of mudflow known as a **lahar** occurs in association with volcanic eruptions, particularly volcanoes

Figure 12.17   Quick Clay

The quick clay destruction of Saint-Jean-Vianney, Quebec, took place in 1971 when a sensitive marine clay abruptly liquefied, destroying 40 homes and taking the lives of 31 people. The insert shows how shaking aligns the randomly oriented clay minerals, allowing the clay to slide.

that are capped by snow and ice or volcanoes that exist in wet climates. Lahars are chiefly composed of unconsolidated, hot ash deposits that are readily mobilized by meltwater or rainfall. Lahars rapidly accelerate down the flanks of volcanoes into local valleys, where they can travel great distances. Recall from Chapter 4, that meltwater-mobilized lahars were associated with the May 1980 eruption of Washington's Mount St. Helens. There, lahars invaded local valleys and traveled a farther 90 kilometers (56 miles) at speeds up to 40 kilometers per hour (25 miles per hour) (see Fig. 12.9). Lahars have also accompanied subsequent eruptions of Mount St. Helens (Fig. 12.18).

CHECK YOUR UNDERSTANDING

● What is quick clay and under what conditions does it generate mudflows?

● What distinguishes a lahar from a mudflow?

Debris flow

Debris flows are the downslope movement of a slurry of poorly sorted rock fragments, soil, and mud in which more than half the material is sand sized or coarser. Like mudflows, they are often initially confined to valleys but tend to spread out downslope to produce fan-shaped deposits, the upper surfaces of which are very irregular. Because they are coarser than mudflows and, hence, contain less muddy water, the rates of movement of debris flows are generally slower, ranging from about 1 meter (3 feet) per year to as much as 100 meters (328 feet) per hour.

Debris flows are especially common when heavy rains fall on areas that have been stripped of vegetation by

wildfires. Wildfires expose loose soil that is very susceptible to erosion during rainstorms. This combination of fire followed by rain—known as a *fire-flood sequence*—is particularly problematic in southern California. There, deadly debris flows have occurred on several occasions when heavy winter rains follow a summer of wildfires. On Christmas Day 2003, for example, a debris flow inundated Waterman Canyon in the San Bernardino Mountains, killing 13 people, when torrential rain fell on an area that had burned just two months before.

Like fire-flood sequences, heavy rainfall can also trigger debris flows in much the same way as it triggers mudflows. In the spring of 1983, for example, numerous debris flows occurred in the Wasatch Mountain in central Utah. These flows were generated by the buildup of water in the ground following unusually heavy rain. The buildup raised the level of the water table (see Chapter 14), saturating the ground and making slopes more prone to sliding. The largest and most destructive of these debris flows occurred during a two-week interval in April of 1983. Near the town of Thistle, southeast of Salt Lake City, a prehistoric slide was once again set into motion. Containing about four million cubic meters (over five million cubic yards) of rocks and mud, the debris flow slowly but relentlessly moved downslope. Advancing at 1.5 meters (5 feet) per hour and building in height at 0.6 meters (2 feet) per hour, the foot of the slide eventually dammed the Spanish Fork River despite all efforts to keep a channel open. In a deadly domino effect, floodwaters rising behind the dam produced a lake nearly 5 kilometers (over 3 miles) long and 60 meters (almost 200 feet) deep (Fig. 12.19). This lake inundated the town of Thistle, obliterated three major highways and the main line of a railway, and threatened downstream communities with flooding. The threat was eliminated only after the lake was drained by tunneling through one of the canyon walls. The total costs of the Thistle debris flow exceeded $400 million, making it one of the most costly landslides in US history. Heavy rain is also behind the persistent debris flow problem in southern British Columbia, Canada. There, many protective structures have been erected to protect property, roads, and railways.

Just as advancing debris flows can dam a river (and cause a domino effect like that of the flooding in Thistle, Utah), this sequence of events can also act in reverse. This happened in the early 1970s, near Vancouver, British Columbia. In the headwaters of Klattasine Creek, a massive debris flow was triggered by the collapse of a lake dammed by glacial deposits.

Disastrous debris flows are not, however, restricted to natural slopes and dams. They have also been caused by

Figure 12.18   Lahar
A volcanic mudflow or lahar triggered by melting snow and ice scarred the north flank of Mount St. Helens following an eruption in 1982.

Figure 12.19   Thistle Debris Flow
The 1983 debris flow near Thistle, Utah, became one of the costliest landslides in US history when it dammed the Spanish Fork River. The resulting lake inundated the town of Thistle and obliterated several highways and the main line of the Denver and Rio Grande Western Railway.

negligent human activities. One of the most tragic instances occurred in 1966 in the town of Aberfan, in South Wales (Fig. 12.20). For more than 50 years, the debris from underground coal mines had been dumped into large waste piles (called *tips*) along the slope above the town. On October 21, Tip 7 failed, and black sludge slid 800 meters (2600 feet) through cottages, across a canal, and over Pantglas Junior School. In just a few seconds 144 people perished, including 116 school children. The investigation into the tragedy showed that many of the

LIVING ON EARTH

The Oso Landslide, Washington State

On the morning of March 22, 2014, a portion of hillside on the North Fork of the Stillaguamish River east of Oso, Washington, collapsed. The resulting massive landslide sent mud, rock debris, and trees across the river, burying the neighborhood of Steelhead on the opposite bank (Fig. 12l) to a depth of 12 meters (40 feet). Almost 50 homes, many of which were occupied, were destroyed, and nearly a mile of State Route 530 was buried. Over 20 fatalities were attributed to the catastrophe immediately following the slide, and more than 30 residents were still unaccounted for days after the event. The landslide also dammed the Stillaguamish River, which backed up several miles eastward, raising concern of a downstream flash flood should the debris dam fail.

The 2.5-square-kilometer (1-square-mile) landslide, described by Steelhead residents as a "fast-moving wall of mud," occurred after more than six weeks of heavy rain, during which the area received twice its normal rainfall. The unstable, water-saturated hillslope north of the Stillaguamish River slumped, leaving a well-defined headwall scarp with rotated slump blocks at its foot (compare Fig. 12l with Fig. 12.14). But as the slump crossed the river, it liquefied and transformed into a mudflow, and it was this mudflow that overwhelmed Steelhead.

Sadly, the landslide came as no great surprise to local residents. Known to locals as "Slide Hill," the hillside had had a history of small landslips and was the site of a

Figure 12l   Oso Landslide

This aerial view of the Oso landslide shows features typical of a slump (see inset), including a crescent-shaped headwall scarp and rotated slump blocks to the right of the Stillaguamish River and, in the former neighborhood of Steelhead to the left of the river, an apron of hummocky ground at the slump toe.

significant landslide in 2006 (Fig. 12J). Millions of dollars were spent reinforcing the riverbank and the hillslope above it following the 2006 landslide, but slumping is a process that tends to reoccur. In this case it is likely that the river eroded the toe of the earlier slump, setting the stage for it to be reactivated when the ground became saturated again and water, infiltrating preexisting rupture surfaces, allowed these surfaces to slide farther.

Figure 12J   Before and After the Oso Landslide

These aerial views before (top) and after (bottom) the Oso landslide, clearly show the scar (with its young growth of trees) of an earlier landslide on the hillslope that collapsed.

Figure 12.20   Debris Flow

Tragedy struck Aberfan, South Wales, in 1966. Coal tips (background) were positioned above the town. Tip number 7 was located above a natural spring, which saturated the coal debris and turned it into sludge. Lubricated in this fashion, the debris moved downhill and buried part of the town.

tips, including Tip 7, were built on top of natural springs. The debris had become so saturated with water that the pressure of the pore water exceeded the friction between the grains. As a result, the sediment liquefied and flowed downhill like a fluid.

Earthflow

Earthflow is the downslope movement of tongue-shaped, fine-grained masses of clay-rich soil and weathered bedrock along a surface more or less parallel to the ground. Earthflows may evolve into mudflows, but are generally coarser and less fluid than mudflows, and hence move more slowly, and are covered with solid material that is carried along by the flow beneath (Fig. 12.21). They occur on grassy hillslopes and banks during times of heavy rainfall or spring thaws, but earthflows are most familiar in the road cuts of recently constructed highways. Most earthflows are small, with dimensions measured in meters, and most move slowly, typically from a few centimeters to a few meters per day over a period of several days or weeks.

> **CHECK YOUR UNDERSTANDING**
>
> ● What is the difference between a mudflow and a debris flow?

Figure 12.21   Earthflow
An earthflow scar mars a grassy hillslope in Mission Pass, California.

Solifluction

A downslope movement of waterlogged surface material that is even slower than an earthflow occurs on slopes underlain by frozen ground. The frozen ground acts as a downward barrier to percolation and rates of movement are typically on the order of a few centimeters per year. This phenomenon, called **solifluction**, is most common in regions where the ground is permanently frozen. Known as **permafrost**, this frozen ground underlies large areas of the northern hemisphere at high latitudes (Fig. 12.22a). In these areas, solifluction occurs during summer months when a thin zone at the surface of the permafrost, known as the *active layer*, thaws.

> **CHECK YOUR UNDERSTANDING**
>
> ● What is solifluction and where is it most common?

Because the water from the thawed layer is unable to percolate downward, the ground surface becomes completely sodden and slowly flows downslope in a characteristically lobate fashion (Fig. 12.22b).

Creep

Creep is the imperceptible but more or less continuous downslope movement of soil and weathered bedrock under the influence of gravity. As its name suggests, it is the slowest type of mass wasting. Yet in terms of the volume of material it moves in any given year, creep is the most important of all mass wasting processes. Rates of movement are usually measured in millimeters or centimeters per year. The rate of creep may be enhanced during periods of heavy rainfall simply because the soil becomes water saturated and loses its resistance to downslope movement. However, creep is most often caused by freeze-thaw or wet-dry cycles that cause the soil to alternately expand and contract. Wetting or freezing lifts any surface particles at right angles to the slope, and drying or thawing lowers them vertically, so in either type of cycle there is a net downhill component of movement (Fig. 12.23).

> **CHECK YOUR UNDERSTANDING**
>
> ● Compare and contrast earthflow and creep.

Unlike slumps, which rotate as they move downslope so that trees are tilted backward (see Fig.12.14), creep causes upright structures to slowly tilt forward. Structures tilt forward because a creeping mass moves faster at the surface, where there is less resistance, than it does at depth. Even though the process itself is far too slow to observe, evidence of creep can often be deduced from the orientation of trees on hillsides. The downslope tilt of utility poles and monuments, the downhill displacement of fence lines and retaining walls, tree trunks bent upward at their base, and the downhill bending of inclined beds at the top of road cuts and cliff faces (Fig. 12.24), all testify to the elusive downhill movement of surface materials known as creep.

Because mass wasting is an inevitable process that affects all parts of Earth's surface, the property damage and loss of life that these processes can cause might also

(a)

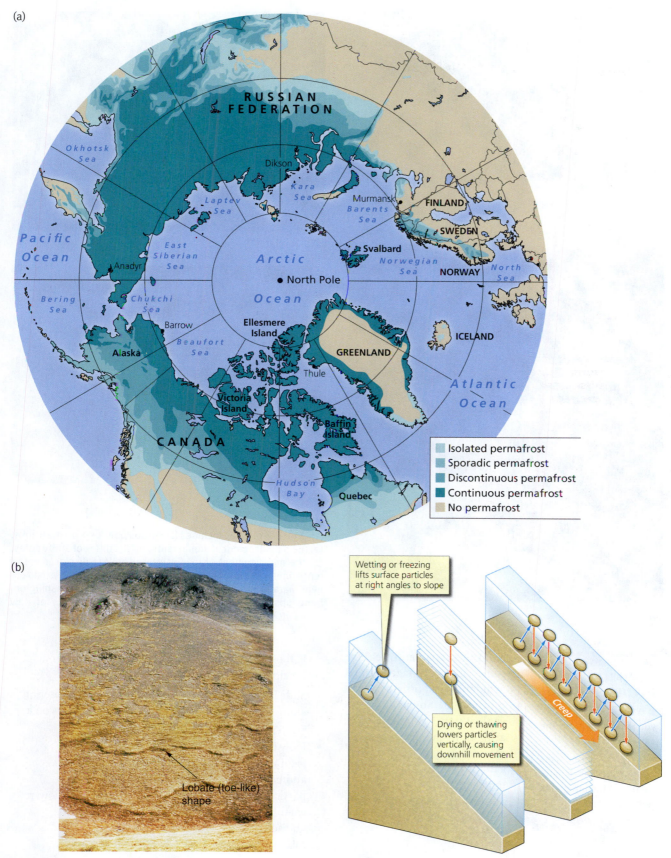

Isolated permafrost
Sporadic permafrost
Discontinuous permafrost
Continuous permafrost
No permafrost

Wetting or freezing lifts surface particles at right angles to slope

Drying or thawing lowers particles vertically, causing downhill movement

Creep

Lobate (toe-like) shape

Figure 12.22 Permafrost

(a) Zones of continuous, discontinuous, sporadic, and isolated permafrost cover the Northern Hemisphere. (b) This characteristically lobate, solifluction flow is located near Suslositna Creek, Alaska.

Figure 12.23 Creep

Surface particles respond to water by lifting during freezing or wetting and falling during thawing or drying. Occurring in cycles, this process causes creep by alternately lifting and then dropping particles fractionally downslope.

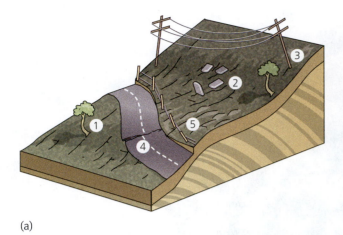

(a)

(b)

(c)

(d)

Figure 12.24   Evidence of Creep

(a) Evidence of creep includes (1) the upslope curve of tree trunks; (2) the displacement of monuments; (3) the tilting of utility poles; (4) the misalignment of roadways; and (5) the displacement of fence lines. (b) Dipping beds bend downhill at the top of a road cut near Marathon, Texas. (c) Tree trunks near Seattle, Washington, bend upslope. (d) A retaining wall at the University of New Brunswick, Canada, tilts as a result of creep.

---

seem inescapable. In reality, some of these hazards are avoidable, and others can be minimized. So how do we evaluate the potential hazards that the processes of mass wasting pose, and what actions can we take to counter them? In Section 12.4, we address the assessment and mitigation of mass wasting hazards.

12.3 SUMMARY

- Mass wasting is classified according to how the material moves, what it consists of, and how fast it travels.

- Fall is the free fall of material (such as rocks) from precipitous slopes.

- Slide is the movement of a coherent mass of material along a well-defined surface. A slide that moves on a curved surface is a called a slump.

- Flow is movement of material in a fluid-like fashion. Flow includes fast-moving mudflows and hot lahars, which consist of fine-grained particles; debris flows, which are coarser and slower; earthflows, which are less fluid than mudflows; and solifluction, which occurs when water-saturated soil moves slowly downslope, usually over permafrost.

- Creep, the slowest type of mass wasting, is the imperceptible downslope movement of soil under the influence of gravity.

12.4 Minimizing Mass Wasting Hazards

All forms of mass wasting are potentially very destructive processes. Yet the hazards they pose are fairly predictable and largely avoidable. Assessing and predicting potential slope failures begins with identifying and mapping geologic evidence of past events and features that might trigger mass wasting. Once this is done, laboratory testing of sediment and bedrock samples follows. Information obtained and compiled in this way can inform the decisions land-use planners and engineers make about land development. Coupled with knowledge of the weather outlook, this information can be used to forecast mass wasting events.

IDENTIFYING AND PREDICTING SLOPE FAILURE

Because most forms of mass wasting are recurrent or ongoing processes, earth scientists can use evidence of past events to identify areas prone to mass wasting. Slide and flow scars, hummocky toe terrain, tongue-shaped flow deposits, fissuring of the land surface, and tilting and displacement of formerly upright objects all identify past mass wasting events. Recurring forms of mass wasting, such as slides and flows, tend to follow the same paths, and an examination of past events can determine the potential magnitude and frequency of future events. Features that might trigger mass wasting, such as the location of faults in seismically active areas, can be mapped and their potential for movement assessed. In addition, geologists can collect samples of sediment and bedrock and test them to determine physical properties, such as their strength and cohesion, that effect susceptibility to mass wasting processes. They can also examine hillslopes to determine steepness relative to the angle of repose and to predict stability under various weather conditions.

The information gathered by such studies is assembled to create *slope-stability maps* (Fig. 12.25). These maps identify areas prone to mass wasting. Slopes are color-coded to show their relative stability, and features pertinent to hazard assessment are delineated. These features include fault lines and the geologic contacts between rocks of different stability. Slope-stability maps play a vital role in land-use planning because they allow planners and engineers to make informed decisions about land development. In this way, planners can evaluate the potential threat of slope failure, avoid areas most at risk, and implement plans to minimize the effects of a particular hazard before they develop a land area. Such maps also help governments make

responsible recommendations for land usage and decisions regarding building codes and zoning. Sites susceptible to slow mass wasting, for example, might prove suitable for recreational development but not for building purposes. Or susceptible sites might be rendered suitable for construction with appropriate building codes or the grading of particular slopes.

Because the likelihood of slope failure is greatly increased during times of heavy rainfall, geologic information is more useful when it is coupled with knowledge of weather patterns. For the powerful El Niño event of 1997–98, for example, the US Geological Survey prepared national landslide hazard outlook maps by combining forecast information for precipitation (from the National Oceanic and Atmospheric Administration) with slope-stability maps for the United States.

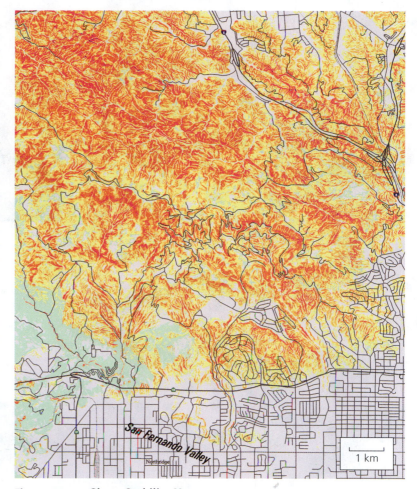

Figure 12.25   Slope-Stability Map

This slope-stability map of part of the Santa Suzanna Mountains north of Northridge, California, indicates areas where landslides are most likely to be triggered by earthquakes. The map is color-coded from green to orange to indicate increasing steepness of the mountain slopes and, hence, greater likelihood of landslides. The orange areas are steepest and green areas least steep. Flat areas where landslides are highly unlikely are grey.

PREVENTING SLOPE FAILURE

While it is not always possible to prevent mass wasting at a particular development site, it is sometimes possible to protect the site or reduce the hazard to an acceptable level by implementing methods to increase slope stability. Some remedies are as simple as planting trees or other deep-rooted plants. Other remedies require sophisticated engineering strategies. Whatever methods planners and engineers use, the methods usually serve one of three functions: to support the slope; to reduce the slope's water content; or to modify the slope's shape (Fig. 12.26).

Supporting a slope

A common way to support unstable slopes is through the use of retaining walls (Fig. 12.26a). Anchored to the bedrock or a solid foundation by reinforcing rods, retaining walls buttress (support) excavations and the base of slopes.

Alternatively, the slopes themselves can be strengthened with anchor rods, known as *rock bolts*, that are driven into the slope face. The face can also be sprayed with concrete or covered with wire mesh; these materials help hold the slope face in place (Fig. 12.26b).

Reducing water content

Because water content plays a key role in destabilizing slopes, one way to stabilize a potential slide mass is to remove the water from the mass. To decrease the water content of slope material, a two-pronged approach is often used. First, water is prevented from gaining access to the slope by creating drainage channels that either divert it at the top of the slope or intercept it and bring it to the base. Second, water within the slope material is removed by installing perforated drainage pipes into the slope face (Fig. 12.26c). Water can also be pumped from

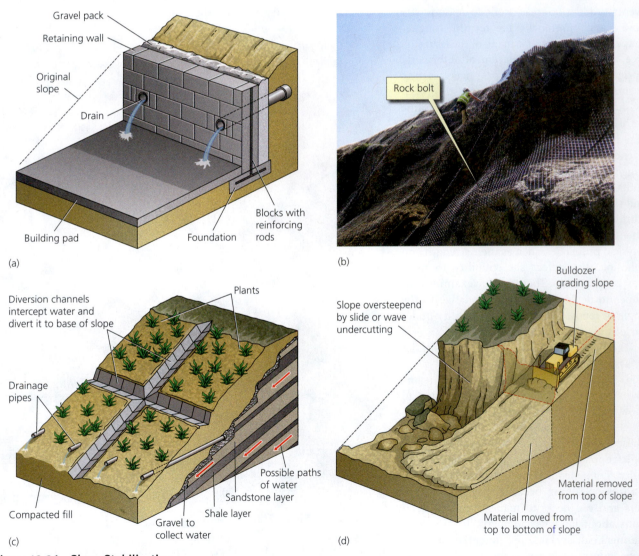

(a)

(b)

(c)

(d)

Figure 12.26 Slope Stabilization

Engineering techniques used to stabilize slopes include (a) retaining walls and (b) wire mesh and rock bolts that enhance slope stability; (c) diversion channels and drainage pipes that reduce the water content of slope materials; and (d) grading that modifies a slope's shape.

a slope to remove the threat to slope stability caused by its weight.

Modifying slope shape

A common means of stabilizing a slope by modifying its shape is through *grading,* a process that involves moving material from the top of the slope and adding it to the bottom (Fig. 12.26d). This lowers the slope angle, adding to its stability. One way of stabilizing oversteepened slopes, for example, is to grade them to a slope angle that is lower than the angle of repose of the slope material. Alternatively, the shape of an unstable slope can be modified by simply removing any potential slide mass altogether. In a similar fashion, artificial slopes such as road cuts are frequently stabilized by *benching,* a process that involves cutting steps into the road cut to reduce its overall slope.

> **CHECK YOUR UNDERSTANDING**
>
> ● Which engineering techniques can stabilize slopes?

Although potentially destructive, mass wasting is a reasonably predictable and often preventable geologic hazard. Mass wasting is simply the consequence of gravity acting on Earth's surface. However, on a larger scale it is the forces of plate tectonics that provide the slopes on which gravity acts. The important role of plate tectonics in mass wasting processes is the topic we address in Section 12.5.

12.4 SUMMARY

* Earth scientists can evaluate the potential for mass wasting events for any given area.

* Evidence of past events helps to identify areas prone to slope failure; geologic indications include slide and flow scars, flow deposits, fissuring, and tilting and displacement of objects.

* Identifying and predicting slope failure involves locating potential triggers such as faults and knowing the weather patterns in an area.

* Preventing or minimizing slope failure involves planting vegetation; stabilizing slopes with retaining walls, rock bolts, and sealing materials; reducing the water content by using diversion channels and drainage pipes; and grading.

12.5 Mass Wasting and Plate Tectonics

Mass wasting is caused by the action of gravity on slopes and, because slopes are the most common of all landforms, few areas of Earth's surface are unaffected by mass wasting processes. But gravity is only half the story. The other driving force is plate tectonics, which is responsible for the uplift and mountain building that creates and maintains the slopes upon which mass wasting processes operate. As a result, mass wasting is most common in tectonically active regions. Within the continental United States, for example, areas where mass wasting is most frequent (Fig. 12.27) correspond to the mountainous areas of the Appalachians, the Rocky Mountains, California, and the Cascades. These areas owe their origin to plate tectonics forces acting now or in the geologic past. As most of the world's major mountain belts are the product of plate collision, it should come as no surprise that the hazards of mass wasting are most acute at convergent plate boundaries.

Locally, plate tectonics can also be an immediate cause of mass wasting. For example, earthquakes caused by plate motion often trigger landslides on slopes rendered unstable by seismic shaking. The devastating Nevado Huascarán rockfall and rock avalanche in Peru in 1970 (see Figure 12.13) was triggered by a magnitude 7.9 earthquake located off the coast where the Nazca plate subducts beneath South America.

Earthquake waves can also cause mass wasting as a result of liquefaction. Recall from Section 12.3 that liquefaction occurs when unconsolidated water-rich sediments briefly lose their strength as a result of seismic shaking.

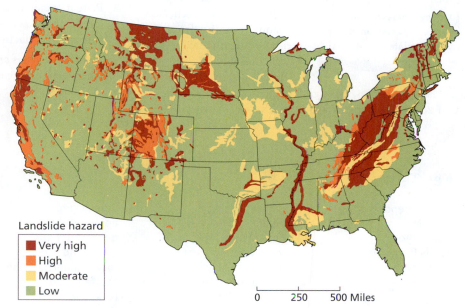

Landslide hazard

■ Very high
■ High
■ Moderate
■ Low

0   250   500 Miles

Figure 12.27   Landslide Hazard Map

This map highlights areas where landslides are most frequent and most likely to occur within the conterminous United States. Red areas have a very high likelihood, orange areas have a high likelihood, and yellow areas have a moderate likelihood. Landslides can and do occur in the green areas, but the likelihood is low.

Figure 12.28 Liquefaction
Apartment complexes in Nigata, Japan, were tilted, fell over, or settled into the ground as a result of liquefaction caused by an earthquake in 1964.

which can cause mass wasting in the form of mudflows and lahars (see Fig. 12.18). In addition, the slopes of volcanoes can be steep and, in response to the seismic shaking associated with eruptions, can collapse in massive landslides. Such was the case on Mount St. Helens in 1980, when a magnitude 5 earthquake caused the oversteepened north face of the volcano to collapse, triggering a massive eruption.

Driven by the combined forces of gravity and plate tectonics, mass wasting forms the first step in the slow but relentless process of landscape modification by surface processes. By carrying the products of erosion downslope, mass wasting often starts the process that slowly but surely transports surface materials toward the sea. In the chapters to come, we will examine the other surface processes, implemented by water, ice, and wind, that carry out this transportation.

12.5 SUMMARY

- Plate collision is responsible for mountain building that creates slopes upon which mass wasting can operate, so mass wasting is especially common along convergent plate boundaries.

- Plate movement causes earthquakes that can trigger landslides and cause sediment to lose its strength through liquefaction.

- Plate boundaries are often associated with volcanoes, where eruptions can produce mudflows, lahars, and landslides.

> **CHECK YOUR UNDERSTANDING**
>
> ● How does plate tectonics account for the distribution of landslides in Figure 12.27?
>
> ● How does liquefaction link plate tectonics to mass wasting?
>
> ● Why is mass wasting especially common at convergent plate boundaries?

When liquefaction occurs, sloping ground may slide, opening up large fissures, while buildings experience a sudden loss of support and settle into the ground or tilt (Fig. 12.28). Liquefaction, for example, was a primary cause of the extensive property damage in Christchurch, New Zealand, resulting from the series of earthquakes that struck that city in 2010 and 2011.

Plate boundaries, especially those associated with subduction zones, are also marked by volcanoes, the eruptions of

Key Terms

angle of repose 374	infiltration 376	pores 375	saturation 376
cohesion 374	lahar 387	porosity 375	shear strength 373
creep 392	liquefaction 387	quick clay 385	slide 382
debris flow 388	mass wasting 372	quicksand 387	slump 383
earthflow 392	mudflow 384	rock avalanche 382	solifluction 392
fall 380	permafrost 392	rockfall 380	surface tension 376
flow 384	permeability 375	rockslide 382	talus 374

Key Concepts

12.1 MASS WASTING: DOWNSLOPE MOVEMENT

- Mass wasting is the downhill movement of Earth materials under the pull of gravity.

- Mass wasting needs a slope because all materials have a shear strength that resists downslope movement.

12.2 FACTORS THAT INFLUENCE MASS WASTING

- Mass wasting is influenced by the steepness of the slope and the strength of the slope material.

- Unconsolidated materials develop a stable slope known as the angle of repose, the steepness of which varies with the material.

- Strong rocks support steeper slopes than weak rocks, and layered rocks are prone to sliding when the layering parallels the slope.

- If a slope is water saturated, its material will slide more easily.

- Vegetation binds soil and increases its strength.

- Mass wasting can be triggered by storms, earthquakes, eruptions, and human activities.

12.3 MASS WASTING MECHANISMS

- Mass wasting is classified according to the material and its method and speed of transport.

- Fall is the free fall of material from cliffs.

- Slide is the movement of coherent material along a surface. Most slides are slow but rockslides and rock avalanches are fast. A slide that moves on a curved surface is a slump.

- Flow is the fluid movement of material and includes fast-moving fine-grained mudflows and hot lahars, coarser and slower debris flows, less fluid earthflows, and solifluction or the downslope movement of water-saturated soil in areas of permafrost.

- Creep is the imperceptible downslope movement of soil under gravity. It moves the most volume of material of all mass wasting mechanisms.

12.4 MINIMIZING MASS WASTING HAZARDS

- Slide and flow scars, flow deposits, fissuring, and tilting or movement of objects help identify areas prone to slope failure.

- Locating faults, analyzing soil, and knowing weather patterns aid in the prediction of slope failure.

- Slopes can be stabilized with vegetation, retaining walls, rocks bolts, and sealing materials; diversion channels and drainage pipes; and grading and benching.

12.5 MASS WASTING AND PLATE TECTONICS

- Plate tectonics ultimately creates the slopes upon which mass wasting operates.

- Plate collision produces mountainous areas where mass wasting is most common.

- Plate boundaries are the location of earthquakes that can trigger landslides and cause liquefaction, and volcanoes, the violent eruption of which can lead to a variety of mass wasting processes.

Study Questions

1. What factors influence the stability of slopes and how do they do so?

2. Why does the influence of water on a material's angle of repose depend on how much water is present and how easily the water can move through the material?

3. What criteria are used to classify mass wasting mechanisms and how are they applied?

4. Mass wasting often involves a domino effect, one process triggering another or being triggered by it. Referring to examples other than those used in the chapter, discuss the ways in which this can happen.

5. In what ways might a powerful El Niño event influence mass wasting in southern California and what steps could be taken to minimize the threat to life and property?

6. What circumstances led to the tragic Aberfan disaster of 1966 and what steps could have been taken to avoid it?

7. What features of terrestrial mass wasting have been used to identify the mass wasting processes on the Moon and Mars?

8. How does plate tectonics influence mass wasting?

9. In an area prone to mass wasting, what steps can be taken to assess the threat to a land development project?

10. What factors are taken into account when a slope-stability map is produced?

11. Before starting construction in an area threatened by mass wasting, what engineering techniques could be used to reduce the hazard to an acceptable level?

13 Running Water

13.1 Running Water and the Hydrologic Cycle

13.2 Streamflow

13.3 Stream Erosion

13.4 Stream Transport

13.5 Stream Deposition

13.6 Floods and Flood Prevention

13.7 Running Water and Plate Tectonics

13.8 Running Water on Other Worlds

► Key Terms

► Key Concepts

► Study Questions

LIVING ON EARTH: The Death of the Aral Sea—An Environmental Disaster 419

IN DEPTH: Dams and the Human Exploitation of Surface Water 430

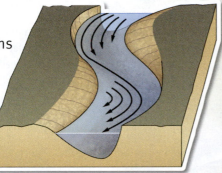

Meanders

Oxbow Lake

Floodplain

Waterfall

Precipitation

Runoff

Levee

Delta

Evaporation

13.1 Identify the role that running water plays in the hydrologic cycle.

13.2 Compare and contrast the two types of streamflow and explain the factors that effect a stream's velocity and gradient.

13.3 Describe the work of running water as an agent of erosion and the drainage patterns streams create.

13.4 Identify the three ways a stream transports its sediment load.

13.5 Describe the processes and products of stream deposition.

13.6 Discuss the causes and consequences of flooding.

13.7 Demonstrate how streams are closely linked to plate tectonics and how most of the world's great drainage basins owe their origin to the breakup of Pangea.

13.8 Explain that Earth is the only planet in the Solar System with running water on its surface, although there is compelling evidence for the former existence of running water on Mars.

No surface process plays a greater role in sculpting the landscape around us than that of running water. Running water, whether in a babbling brook or in one of the world's great rivers, dominates the continental drainage systems that transport water toward the oceans and is an essential element of the hydrologic cycle we discussed in Chapter 1. Collectively referred to as *streams*, running water in all its forms is the most effective agent of erosion and deposition, carving the valleys that dominate our landscape and acting as an endless conveyor of unconsolidated sediment from the headwaters to the sea. Streams occur in almost every corner of Earth's land surface and are responsible for the development of our most familiar landforms. Whether the local landscape is one of mountain ravines and waterfalls, rolling hills and valleys, or flat floodplains and lakes, streams have made an important contribution to its development. They also are a vital natural resource and have played a central role in the course of human history. Since the dawn of civilization, streams have irrigated our croplands, provided highways for transportation, and supplied us with energy and drinking water.

Indeed, running water has influenced the development of society more than any other medium. Stream water irrigated the crops that nourished the cradles of civilization, many of which developed along major river valleys, and running water has been used since antiquity as a source of power and as arteries of commercial transport. So it is no coincidence that many of today's great cities have grown up beside major rivers. But our dependence on streams for water, commerce, agriculture, and energy is not without its drawbacks. Rivers are prone to flood, and when they do so, they can spread disease and cause widespread property damage and loss of life. So humans live on floodplains at their peril.

Because of its influence on the landscape around us and the many ways it affects our lives, it is important that we understand the work of running water. In this chapter, we explore the role of running water as a surface process. We investigate the way in which streams flow and the work of streams as agents of erosion, transportation, and deposition. We explore the causes and consequences of flooding. Finally, we discuss how Earth is the only planet in the Solar System with running water on its surface, although there is compelling evidence for the former existence of running water on Mars.

13.1 Running Water and the Hydrologic Cycle

The **hydrologic cycle**, which we introduced in Chapter 1, describes the motion of water in Earth's atmosphere and on its surface. Heated by the Sun, water constantly evaporates from Earth's oceans and enters the atmosphere as moisture. In the atmosphere, water is carried by the wind until it condenses to form clouds and precipitates. Because Earth's surface is dominated by oceans most of this precipitation falls back into the sea. But some falls on land and, if it is not absorbed by the ground, is eventually returned to the oceans as running water that drains downhill as surface runoff. The drainage of this surface water is not uniform, but rather is concentrated in narrow ribbon-like depressions or **channels**. The flow of running water within these channels, whether it takes the form of a small babbling brook or a great river, is referred to in geology as a **stream**. The flow of water in the world's streams is part of the ceaseless circulation of water from the oceans to the continents and back again. As we

shall see, it is this return flow of water in streams that plays the greatest single role in the sculpting of our landscape.

The volume of water at Earth's surface is enormous, amounting to about 1.4 billion cubic kilometers (340 million cubic miles). However, fully 97 percent of this supply resides in the oceans, and of the remainder, most is stored on land in ice caps and glaciers (Fig. 13.1). Hence, only a minute proportion of Earth's total water budget takes part in the hydrologic cycle at any one time. Yet this cycle is a global process of gigantic scale. Each year, almost 500,000 cubic kilometers (120,000 cubic miles) of water evaporates from the world's oceans and land surface. On average, this moisture spends only about 11 days in the atmosphere before falling back to Earth's surface as rain or snow. As we just

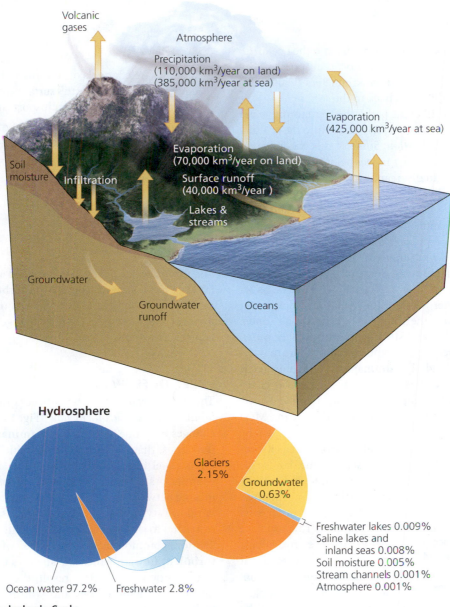

Figure 13.1   The Hydrologic Cycle

In the hydrologic cycle, water evaporated from the oceans and precipitated on land is returned to the sea. Numbers in parentheses show the amount of water involved (in cubic kilometers per year).

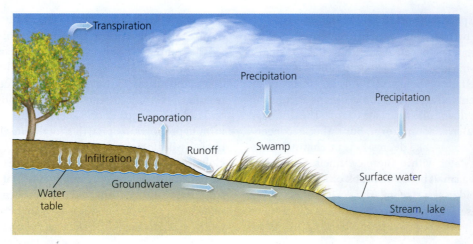

Figure 13.2   Movement of Water
Water takes a variety of paths at Earth's surface.

noted, the bulk of this precipitation falls into the ocean. Of the precipitation that falls on land, most is evaporated. But the remainder, some 40,000 cubic kilometers (9600 cubic miles), must eventually drain back to the sea.

There are several different paths this remaining water can take (Fig. 13.2). The vast majority seeps into the ground, a process known as *infiltration*. Some of this water evaporates again, but the remainder moves below Earth's surface as *groundwater,* eventually reappearing at the surface to form springs. The details of this underground flow of groundwater and the processes for which it is responsible are the subject of Chapter 14. A small amount of precipitation is absorbed by plant roots and eventually returned to the atmosphere by *transpiration,* the name given to the loss of moisture from plant leaves. Some of the water may also contribute to the growth of ice sheets and glaciers, as we will learn in Chapter 15. But the remainder flows over Earth's surface under the influence of gravity as *runoff,* which collects in lakes and streams to form *surface water.* Some runoff is subsequently evaporated and some may infiltrate into the ground. The drainage of surface water is not uniform, but instead, is concentrated in narrow channels and in lakes and swamps. As we noted previously, any form of channelized surface water is called a stream. Most water returns to the sea by way of streams, and as it does so, it sculpts the landscape and redistributes unconsolidated sediment in a manner unrivaled by any other surface process. This surface portion of the hydrologic cycle, which acts as an agent of erosion, transportation, and deposition, is the subject of this chapter. We begin by examining the flow of running water, and the way this flow enables streams to carry sediment.

13.1 SUMMARY

- Streams are channels of running water. They are a very common feature on Earth's surface and form Earth's most effective agent of erosion, transport, and deposition.

- Running water is a key component of the hydrologic cycle by which Earth's water balance is maintained.

- Streams return water from the land back to the oceans, where it originated.

13.2 Streamflow

The flow of water in streams is guided along channels and takes either a smooth course or a turbulent one. Streamflow depends on several factors, each of which is interdependent. So when one variable changes, the others must adjust accordingly.

TYPES OF FLOW

The flow of water in streams occurs in one of two ways, *laminar flow* or *turbulent flow* (Fig. 13.3), and commonly changes from one to another. In **laminar flow**, the stream's flow lines, that is, the lines along which the water moves, are parallel to each other and to the sides of the stream channel. As a result, the flow is smooth and there is no mixing between one part of the stream and another. Laminar flow is typical of slow-moving fluids and so is generally restricted to slow-moving streams with smooth channels.

More common in streams is the swirling movement of **turbulent flow**, in which the stream's flow lines crisscross in a chaotic fashion. As a result, the water is continually mixed, producing eddies and whirlpools. The degree of turbulence in streams depends on the velocity of the water and the irregularity of the surface, or *stream bed*, over which the stream flows. Most streams move fast enough

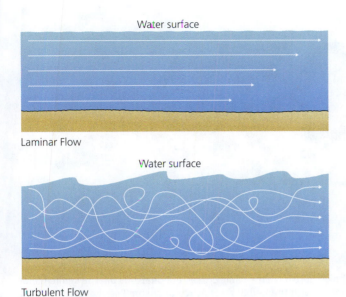

Figure 13.3 Laminar and Turbulent Flow

In laminar flow, the flow lines are parallel and the flow is smooth with no mixing. In turbulent flow, the flow lines cross and water continually swirls and mixes.

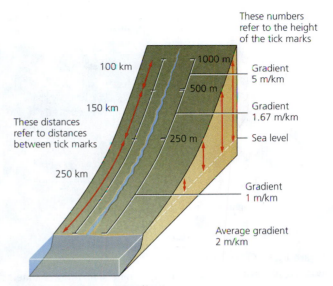

Figure 13.4 Stream Gradient

A stream's average gradient is the vertical drop between its headwaters and the sea, divided by the distance the stream travels. Note the gradient slackens from 5 m/km to 1 m/km from the headwaters to the sea.

> **CHECK YOUR UNDERSTANDING**
>
> ● What is the difference between laminar flow and turbulent flow?

that the streamflow is turbulent. As we shall see, this has important consequences for sediment transport. It is the turbulence of streamflow that keeps a stream's sediment load in suspension in much the same way that turbulent air keeps the lottery balls in motion before the winning number is called.

FLOW VELOCITY

Several factors affect the rate at which a stream flows:

- The slope, or gradient, of the stream
- The roughness of the stream bed
- The shape of the stream channel

For streams to flow, they must have a slope, or *gradient*. That is, the elevation of the stream valley must decrease from its source, referred to as its *headwaters,* to the sea. One of the most important influences on streamflow is the *average gradient,* the vertical drop between the headwaters of a stream and the sea, divided by the distance the stream travels. For example, if a stream's headwaters occur at an elevation of 1000 meters (3300 feet) above sea level, and it flows 500 kilometers (310 miles) to the sea, its average gradient is 2 meters per kilometer (10.6 feet per mile). However, the gradients of major streams are not uniform. Instead, like the curve of a playground slide, a stream's gradient changes along its length, typically slackening from its

elevated mountain source to the relatively gentle plains of the coastal region (Fig. 13.4).

The steepness of a stream's gradient is one of the factors controlling the rate at which it flows. A steep gradient promotes a more energetic streamflow, and all other things being equal, the steeper the gradient the faster the stream's velocity. The *velocity* of a stream is its rate of flow, measured in either meters per second or kilometers per hour. Most streams flow at velocities between about 1 kilometer and 35 kilometers per hour (0.6 to 22 miles per hour).

Other factors that influence a stream's velocity are the roughness of the stream bed and the shape of its channel (Fig. 13.5). Because frictional resistance between a stream and its bed causes the stream to slow down, fast-flowing streams tend to be those with smooth channels and broadly semicircular cross sections, because this shape minimizes the amount of water in contact with the stream bed at any one time. Stream velocity is at a maximum in midstream and decreases toward the stream bed and stream banks as a result of increasing frictional resistance. Maximum flow rates generally occur just below the surface in the center of the stream.

Viewed in cross section, the downslope path, or **longitudinal profile,** of a stream gradually decreases in gradient from the headwaters of the stream to its *mouth,* the location where the stream empties into a standing body of water such as the sea or a lake. One might expect a stream's velocity to decrease accordingly. In fact, the opposite is usually true. Contrary to our visual impression of fast-moving mountain streams and lazy, winding rivers, the *average velocity* of streams actually *increases* downstream.

There are several reasons for this. First, under the influence of gravity, water not only flows downslope but also accelerates as it does so. Second, the velocity of a stream is greatest where frictional resistance is least, and a change to

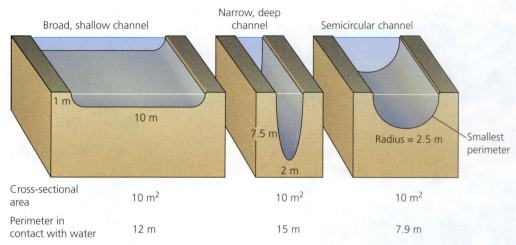

Broad, shallow channel   Narrow, deep channel   Semicircular channel

1 m
10 m
7.5 m
2 m
Radius = 2.5 m   Smallest perimeter

| Cross-sectional area | 10 m² | 10 m² | 10 m² |
| Perimeter in contact with water | 12 m | 15 m | 7.9 m |

Figure 13.5   Stream Channels
Streams flow fastest in semicircular channels. All three channels shown here have the same cross-sectional area but the semicircular channel has the smallest perimeter. As a result, less water is in contact with the bed of a semicircular channel so that streams with a cross section shaped this way experience less frictional resistance to flow than those occupying broad, shallow channels or narrow, deep channels.

smoother channel shapes generally accompanies the decrease in stream gradient. Headwater channels, while steep, tend to be boulder strewn and irregular, whereas channels nearer the stream mouth are smoother and more semicircular in shape. In fact, in the headwaters, nearly 95 percent of a stream's energy is dissipated by friction. As a result, the stream's average velocity is quite low (less than half of that at its mouth), although at any one time some of the water may be moving very rapidly. Farther downstream, the channel widens and less energy is lost to friction. Consequently, the average velocity is higher. Finally, the velocity of a stream tends to increase with the overall volume of water being carried, and this, too, increases downstream.

> **CHECK YOUR UNDERSTANDING**
> ○ What factors influence stream velocity?

STREAM DISCHARGE

The amount of water flowing in a stream at any one time and place is measured by its discharge. The **discharge** of a stream is the total volume of water that passes a given location within a given period of time. The discharge of a stream varies with time. For example, it decreases in times of drought and may rapidly increase after a heavy rainfall or spring thaw. Because any increase in the amount of precipitation or meltwater within an area of drainage requires streams to carry more water, increases in discharge are usually accompanied by proportional increases in the stream width, depth, and velocity. Discharge also increases progressively downstream as tributaries contribute ever more water to the main channel. As a result, discharge is at a maximum at the mouth of a stream.

We measure discharge by multiplying the stream's cross-sectional area by its velocity and we usually express discharge in cubic meters per second. By this measure, the world's greatest river is the Amazon in South America, which discharges water from its mouth at a staggering average rate of about 212,000 cubic meters (7.5 million cubic feet) per second, a volume that represents about 20 percent of Earth's total streamflow (Table 13.1). The Amazon River's discharge is more than twelve times greater than that of the Mississippi River, which discharges into the Gulf of Mexico at an average rate of about 17,300 cubic meters (611,000 cubic feet) per second. The Amazon River's huge discharge is partly due to its large area of drainage (see Fig. 1.5c) and partly due to the abundant supply of water from the wet tropical climate that dominates the continental interior.

> **CHECK YOUR UNDERSTANDING**
> ○ What is the discharge of a stream and how is it measured?

BASE LEVEL

The **base level** of a stream is the lowest level to which the stream can erode its channel. Because most streams flow into the sea, sea level is the *ultimate base level*. However, *local base levels* may occur along the path of a stream (Fig. 13.6). For example, where lakes or reservoirs occur in a valley, they form a barrier between the upstream and downstream segments of the stream. Because the upstream segment cannot erode below the level of the lake, the lake serves as a local base level. The same principle holds where one stream enters another. The nature of the underlying bedrock may also affect the local base level. Streams preferentially erode weaker rocks, so when a stream encounters

TABLE 13.1   THE WORLD'S LARGEST RIVERS RANKED BY DISCHARGE

			Drainage Area		Average Discharge	
Rank	River	Country	Square Kilometers	Square Miles	Cubic Meters Per Second	Cubic Feet Per Second
1	Amazon	Brazil	5,778,000	2,231,000	212,400	7,500,000
2	Congo	Zaire	4,014,500	1,550,000	39,650	1,400,000
3	Yangtze	China	1,942,500	750,000	21,800	770,000
4	Brahmaputra	Bangladesh	935,000	361,000	19,800	700,000
5	Ganges	India	1,059,300	409,000	18,700	660,000
6	Yenisei	Russia	2,590,000	1,000,000	17,400	614,000
7	Mississippi	United States	3,222,000	1,244,000	17,300	611,000
8	Orinoco	Venezuela	880,600	340,000	17,000	600,000
9	Lena	Russia	2,424,000	936,000	15,500	547,000
10	Parana	Argentina	2,305,000	890,000	14,900	526,000

resistant rock, a local base level for the upstream segment of the stream forms. In certain circumstances, this can lead to the generation of rapids or waterfalls, which develop close to the contact between weak and resistant rocks.

Waterfalls and *rapids* occur when a step develops in a stream profile, either because the stream crosses an outcrop of resistant rock (Fig. 13.7a), or because of uplift along a fault. In the case of a waterfall, the step is continually undermined by the cascading water, which excavates a deep *plunge pool* at the base of the falls. As a result, the step is progressively eroded away and the waterfall retreats upstream. Over the past 12,000 years, for example, the Niagara Falls have retreated 11 kilometers (7 miles) from their original starting point on the Niagara Escarpment, creating the Niagara Gorge in the process (Fig. 13.7b). As a waterfall retreats upstream, the step in the stream profile is gradually lowered and the falls degenerate into rapids before being removed altogether.

As local regions of steeper gradient are eroded, the entire stream eventually develops a smooth profile, the average gradient of which slowly decreases with time as the land over which it flows is worn down. All streams strive to reach a profile that produces a rate of flow sufficient to transport the stream's entire sediment load. Such a stream is said to have an *equilibrium profile* and is known as a **graded stream**.

Ideally, graded streams are those in which the gradient, velocity, channel qualities, discharge, and load are in balance with each other so that there is little erosion or deposition, and the stream acts simply as an agent of sediment transport. In reality, however, a stream's environment is dynamic. Streams must constantly respond to local changes, especially those caused by mass wasting (see Chapter 12). Suppose, for example, that collapse of a stream bank or mass wasting dumps a body of sediment into the stream channel (Fig. 13.8). Because the front of the sediment body is steeper than the stream bed, the gradient of the stream is locally increased. This causes the stream to flow faster so that the body of sediment is preferentially eroded. As a result, the sediment load is transported downstream and

the smooth gradient is restored. This is an example of how mass wasting and stream erosion work in concert to transport sediment downstream.

A stream is also affected by changes in base level—especially changes in sea level. Sea level changes can occur over thousands or millions of years, and are related to a variety of processes including the advance and retreat of continental ice sheets and tectonic uplift or subsidence. Because of this, the ultimate base level of an area of drainage may rise or fall, and the average gradient of a stream may change drastically. If sea level rises, for example, stream valleys are invaded by the sea, leading to the formation of drowned valleys. If sea level falls, the gradient and energy of streams increases, resulting in a new cycle of erosion, transport, and deposition as the streams endeavor to attain new equilibrium profiles—a process known as **rejuvenation**.

> **CHECK YOUR UNDERSTANDING**
>
> ○ How do local base levels differ from a stream's ultimate base level?

13.2 SUMMARY

- Water flows in streams two ways: smoothly, as in laminar flow, or in a more chaotic fashion, known as turbulent flow.

- How fast a stream flows varies according to its slope or gradient, the roughness of the stream bed, and the shape of the channel.

- Discharge, or the amount of water that flows through a channel in a given period, is an important gauge for quantifying streamflow.

- The lowest elevation to which a stream can erode its channel is the base level; base level is dictated by the point at which a stream enters an ocean, a lake, or another stream or meets at a waterfall or rapids.

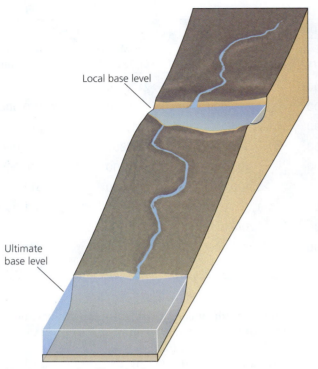

Local base level

Ultimate base level

(a) Lake forms a local base level

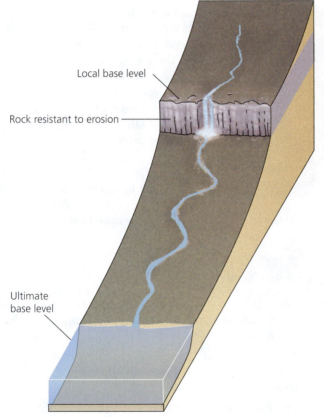

Local base level

Rock resistant to erosion

Ultimate base level

(b) Resistant rock forms a local base level

Figure 13.6   Local Base Levels

The upstream segment of a stream cannot erode below local base levels. (a) Where the stream flows into a lake, the lake forms a local base level, as shown here at Lake Louise in the Canadian Cordillera. The stream running through the valley (background) cannot erode below the base level represented by the water level in the lake. (b) Where a stream flows across a resistant rock layer to form a waterfall, the local base level is at the top of the waterfall. At Taughanock Falls, New York, a local base level occurs where a resistant sandstone at the top of the falls overlies a weaker layer of shale.

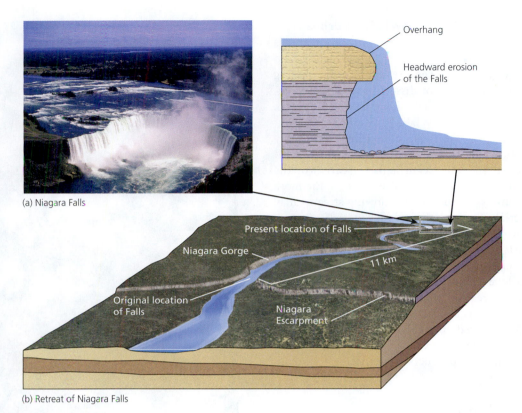

(a) Niagara Falls

(b) Retreat of Niagara Falls

Figure 13.7   Niagara Falls

(a) Niagara Falls is located on the border between New York State and Ontario, Canada. The local base level at the top of the falls is formed by the resistant Niagara Limestone, which overhangs less resistant shale. Headward erosion causes retreat of the waterfall. (b) Retreat of the Niagara Falls upstream from the outcrop of the Niagara Limestone on the Niagara escarpment has excavated the 11 kilometer- (7 mile-)long Niagara Gorge. The average rate of retreat is approximately 1 meter per year (3 feet per year).

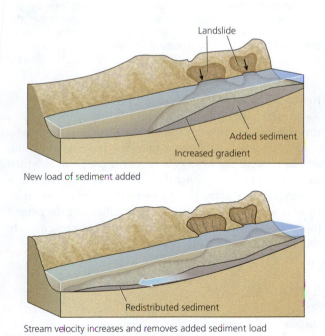

New load of sediment added

Stream velocity increases and removes added sediment load

Figure 13.8   Restoring Gradient

When sediment is dumped into a stream, the gradient of the stream increases, causing the water to flow faster. This results in removal of the sediment such that the gradient of the stream is restored. In this way, mass wasting and stream erosion work together to transport sediment downstream.

13.3 Stream Erosion

Running water is a particularly effective agent of erosion. Erosion removes rock and sediment, and creates channels that focus the flow of surface water into streams. The scouring action at the bottom of the stream channel and the undermining of the banks leads to the formation of stream valleys with a characteristic V-shaped profile. Although the collapse of the undermined stream banks occurs by mass wasting, the collapsed sediment is subsequently transported by the action of the stream. The V-shaped valley this process creates then acts as a guide to the flow of water to form a variety of drainage patterns. In this section of the chapter, we turn to the ways streams erode and to the valleys and drainage patterns they create in the process.

STREAM CHANNELS

A stream forms when water carves a clearly defined channel by removing the rocks and soil in its path. Streams erode their channels in three main ways:

- Removal of rock debris
- Downcutting by abrasion
- Headward erosion

Removal of rock debris

Streams erode their channels by lifting loose material off the stream bed or by wearing or dissolving away the material along the stream bed. Where streamflow is turbulent, the swirling water easily lifts loose particles off the stream bed or removes unconsolidated material from the stream bank, and carries off the debris downstream. The faster the stream flows, the more effective this process is. Indeed, fast-flowing streams may even pry material away from the stream bed itself.

In addition, streams receive a constant supply of unconsolidated debris as a result of physical weathering. Physical weathering continually produces loose material that is carried downslope by mass wasting processes and eventually washed into streams. Dissolved material produced by chemical weathering is also added to streams. Weathered material is either picked up and transported downstream by the stream's turbulent flow, or is carried downstream in solution. In this way, the products of weathering eventually find their way to the sea.

Downcutting by abrasion

Sediment-laden streams also erode their channels by scouring the stream bed, a process known as *abrasion*. As sand and gravel are swept downstream, their impact on the bed of the stream channel slowly wears away the bedrock surface, just as if the stream was a moving strip of coarse sandpaper. As a result, the stream cuts vertically downward and the stream valley is deepened. Steep-sided stream valleys or *canyons* in the desert states of the American West provide dramatic evidence of the power of this process of *downcutting* (Fig. 13.9). Canyons form where the rate of downcutting is rapid compared to the retreat of the valley walls as a result of weathering and mass wasting processes. Canyons are particularly common in arid areas, where weathering and mass wasting are slow but streamflow is occasionally very rapid due to intermittent flash flooding.

In some streams, abrasion produces distinctive circular depressions known as **potholes** in the bedrock of the channel floor (Fig. 13.10). Commonly several meters in diameter and several meters deep, potholes are produced by the drilling action of cobbles and boulders that are swirled around within the pothole by the stream current. The swirling motion abrades both the bedrock and the spinning cobbles, while the ongoing supply of cobbles allows the process to continue. The cobbles responsible for the most recent abrasion can usually be seen at the bottom of the pothole.

Figure 13.9   Downcutting

Carved through resistant rocks of Precambrian age by the scouring action of sand, gravel, and boulders carried in the swirling waters of the Wind River, the near-vertical walls of the Wind River Canyon, Wyoming, provide dramatic evidence of the power of abrasion.

Headward erosion

As streams deepen their channels and widen their valleys through erosion, they also tend to increase their lengths by eroding upslope. This process, known as **headward erosion**, occurs because the head of a stream is the point at which surface water moving downslope first becomes focused into a channel. Once focused in this way, the volume and velocity of flow abruptly increases, greatly enhancing the erosive power of the stream. The head of the stream consequently propagates upslope as erosion proceeds.

STREAM VALLEYS

Stream valleys are the most common landforms on Earth's surface and the most obvious consequence of stream erosion. Stream valleys are produced by the combined effects of downcutting of the stream channel by abrasion and mass wasting of the valley sides under the influence of gravity. Ongoing stream erosion and mass wasting widen, deepen, and lengthen the stream valley as the stream system ages,

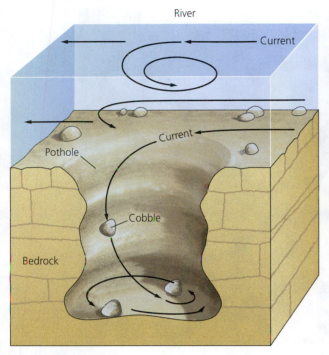

Figure 13.10   Potholes

Potholes in stream channels are drilled into the bedrock by trapped boulders and cobbles that are spun by the swirling motion of the stream current (see photo).

ultimately creating a broad, flat floodplain on the valley floor. Other factors can also influence stream valley development. The most important of these are changes in base level, which can lead to stream rejuvenation, and variations in the topography and bedrock structure of the stream course.

Valley profile

Where downcutting predominates, as is the case in arid regions, stream valleys take the form of narrow canyons with steep walls. In more humid environments, however, mass wasting is usually able to keep pace with downcutting, allowing the valley sides to retreat, and the valley to develop its typical V-shaped profile (Fig. 13.11).

Narrow, V-shaped valleys, often containing rapids and waterfalls, are characteristic of streams that are actively downcutting toward base level. But as the gradient of the stream decreases, downcutting becomes less important and the stream's energy is directed increasingly at widening the valley floor. This is because streams have a tendency to develop sinuous loops, called *meanders*, where the gradient of the valley floor is gentle. When the stream enters such a loop, frictional resistance is highest on the inner part of the bend, and the line of fastest flow shifts from the center of the channel toward the outside of the bend. As a result, the stream erodes first one bank and then the other, producing a flat valley floor, or **floodplain**, that progressively broadens with time (Fig. 13.12).

So called because they are inundated with water only during flooding, floodplains are characteristic of graded streams that are approaching an equilibrium profile.

Stream piracy

As headward erosion slowly extends a stream upslope, the barrier of land that separates it from an adjacent stream may be progressively eroded away. When this occurs, the stream's headward erosion may ultimately breach the land barrier between the two streams so that the stream on the far side is diverted into its headwaters. This is known as **stream piracy** (or *capture*) (Fig. 13.13).

Once capture has occurred, the discharge of the pirate stream increases, and with it, the stream's potential to erode and transport sediment. The discharge of the captured stream, on the other hand, is greatly reduced, because it is severed from its headwaters. Such severed streams are said to be *beheaded* and often look very small compared to the width of the valley. The portion of the former valley immediately below the point of capture may even be entirely dry. The dry region where a valley abandoned in this manner cuts through a ridge is known as a *wind gap*.

Stream terraces

As we have discussed, streams in valleys of gentle gradient tend to develop meandering loops that carve out broad, flat

Figure 13.11   V-Shaped Valley

Streams carve valleys into the landscape that are V-shaped in profile, such as the Grand Canyon of the Yellowstone River in Yellowstone National Park.

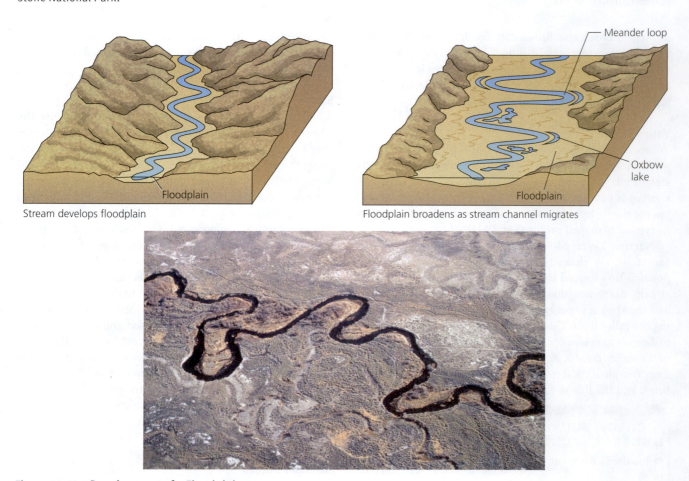

Figure 13.12   Development of a Floodplain

A floodplain develops and the valley floor widens as a result of migration of the stream channel. Photo shows meander loops and old river channels in the floodplain of the Owens River, California.

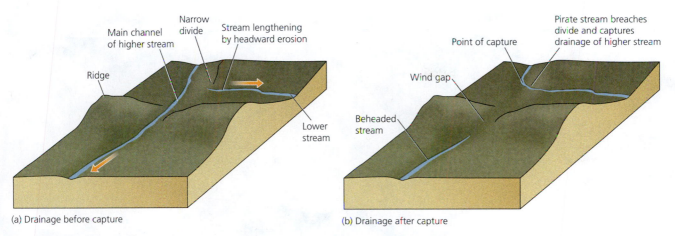

(a) Drainage before capture

Ridge

Main channel
of higher stream

Narrow
divide

Stream lengthening
by headward erosion

Lower
stream

(b) Drainage after capture

Point of capture

Pirate stream breaches
divide and captures
drainage of higher stream

Wind gap

Beheaded
stream

Figure 13.13 Stream Piracy

(a) Before capture, a narrow divide separates the headwaters of one stream from the main channel of another at higher elevation. (b) Headward erosion by the "pirate" stream breaches the divide and captures the drainage of the higher stream. As a result, the "beheaded" higher stream is severed from its headwaters and the valley through the ridge is deprived of its stream, creating a wind gap.

floodplains on the valley floor. In detail, the floodplains of many streams are not uniform surfaces. Instead, they step up toward the sides of the valley in a series of broad, flat benches known as **stream terraces** (Fig. 13.14). Stream terraces are produced when a stream repeatedly downcuts into its own floodplain, each time developing a new floodplain at a lower elevation. Preserved portions of the former floodplains remain as a series of elevated terraces (Fig. 13.15).

But why does a stream cut down into its own floodplain? The most common cause is a change in the base level toward which the stream is eroding, which, as we learned in Section 13.2, causes the stream to be rejuvenated so that it starts downcutting into its own deposits. Such

Figure 13.14 Stream Terraces
The flat steps in the stream deposits of this tributary of the San Juan River in western Argentina are stream terraces.

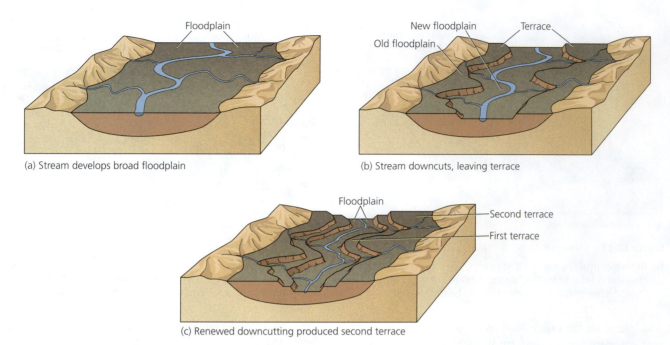

(a) Stream develops broad floodplain

(b) Stream downcuts, leaving terrace

(c) Renewed downcutting produced second terrace

Figure 13.15   Origin of Stream Terraces

Stream terraces form when a stream downcuts into its own floodplain. (a) A stream develops a broad floodplain. (b) When the stream downcuts and develops a new floodplain at a lower elevation, it leaves portions of the former floodplain as an elevated terrace. (c) Repetition of the process creates another terrace.

rejuvenation is typically a result of changes in sea level, but can also be linked to an increase in discharge or a reduction in the volume of sediment load.

Incised meanders

Rejuvenation of a meandering stream may also result in the formation of *incised meanders*. **Incised meanders** are meanders that retain their shape as they are deepened by downcutting, so that the stream comes to lie within a deep, meandering valley (Fig. 13.16a). Most incised meanders start as the normal floodplain meanders of a low-gradient stream. But following a significant change in base level, the stream cuts down through its floodplain deposits and into the bedrock below, while maintaining its pattern of meanders. Once in solid bedrock, the meanders are unable to migrate laterally as they would on a floodplain and the stream comes to occupy a deep meandering valley with no floodplain (Fig. 13.16b).

> **CHECK YOUR UNDERSTANDING**
>
> ● Explain why stream terraces and incised meanders are both products of stream rejuvenation.

Antecedent and superposed streams

When a prominent ridge lies across the path of a stream, the stream usually behaves as we would expect it to: it goes around the obstacle. However, rather than taking the easy route, some streams cut a steep-sided ravine directly through the ridge. In central New York State, for example, lie the headwaters of two great rivers, the Susquehanna, which drains into Chesapeake Bay near Baltimore, and

the Delaware, which empties into Delaware Bay south of Philadelphia. To take these routes both rivers have not just cut through ridges, they have cut across the entire width of the Appalachian Mountains. Why would a stream take such a counterintuitive course? There are two possible reasons. The stream may predate the uplift of the ridges, in which case it is known as an *antecedent stream*. Alternatively, the stream may have been established on a more uniform surface and cut down into the ridge structure from above, in which case it is called a *superposed stream*.

An **antecedent stream** cuts through a ridge because its course was established before the ridge was uplifted, and was maintained because downcutting kept pace with uplift as the ridge rose. In contrast, a **superposed stream** is one whose drainage was established on relatively flat-lying rocks but which subsequently cut down into tilted, folded, or faulted strata lying buried beneath (Fig. 13.17). The Delaware and Susquehanna rivers belong to this second category, having been developed on Mesozoic sedimentary rocks that once covered much of the Appalachian Mountains.

> **CHECK YOUR UNDERSTANDING**
>
> ● How do antecedent and superposed streams differ?

STREAM DRAINAGE

Streams do not occur in isolation. Instead, they form vast, elaborately branched networks that together define the *drainage system*. Depending on the topography and bedrock geology over which they flow, the network of stream

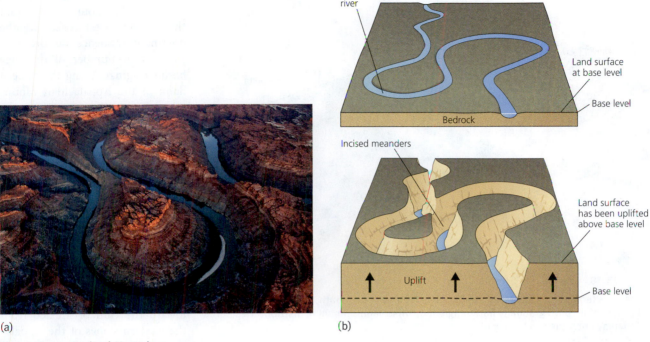

(a)   (b)

Figure 13.16   Incised Meanders

(a) "The Loop" in Canyonlands National Park, Utah, is a series of incised meanders on the Colorado River that formed when rapid uplift of the Colorado Plateau rejuvenated the river. (b) Incised meanders are formed when a meandering stream is rejuvenated as a result of uplift (or a fall in sea level), and cuts down into the bedrock while retaining its meander pattern.

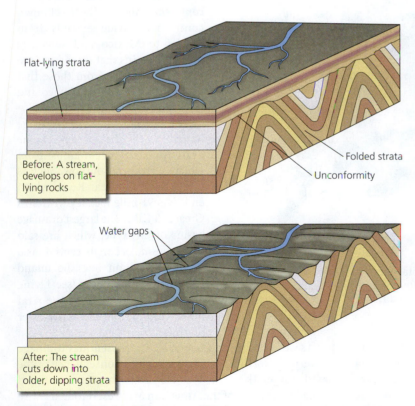

Figure 13.17   Origin of Superposed Drainage

The course of a superposed stream develops on relatively flat-lying rocks, before cutting down into buried dipping strata.

channels may adopt various configurations, giving rise to a variety of drainage patterns.

Drainage basins

As a stream flows from its headwaters to its mouth, the main *trunk stream* is fed by a complex set of side channels called **tributaries**, each of which may have tributaries of its own (Fig. 13.18). At its mouth, a stream may branch into a numbers of smaller channels, known as **distributaries**, which sometimes enter the sea by way of a triangular tract of land called a **delta**.

For example, the Mississippi River is fed by countless tributaries, the largest of which are the Missouri and Ohio rivers (Fig. 13.19). Each time a tributary enters the Mississippi, the amount of water in the channel of the trunk stream increases significantly. Thus the headwaters of the Mississippi River in Minnesota contain only a tiny fraction of the water that issues into the Gulf of Mexico from the river's mouth in Louisiana, where the trunk stream branches into several distributaries that feed the Mississippi delta.

The total land area that contributes water to a stream system is its **drainage basin**. The

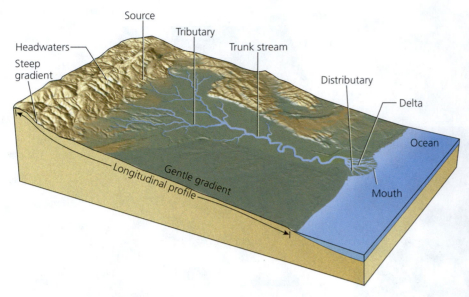

Figure 13.18   Course of a Stream

A stream is fed by tributaries as it flows from its source to its mouth, the slope or longitudinal profile of its path gradually flattening out downstream. Branching distributaries at the mouth may enter the sea by way of a delta.

Figure 13.19   Mississippi Drainage Basin

The drainage basin of the Mississippi River is the total area drained by the river and all its tributaries. The basin is 3.2 million square kilometers (1.2 million square miles) in area and extends from the Appalachians to the Rockies and from the Great Lakes to the Gulf Coast.

drainage basin of the Mississippi River, for example, covers 3.2 million square kilometers (1.2 million square miles) and is the largest in North America. Drainage basins are separated by narrow tracts of higher ground known as **divides**, and may be grouped together according to the major body of water into which the streams ultimately

discharge (Fig. 13.20). In North America, the mountain belts on the east and west coasts are the dominant influence on drainage patterns. A number of drainage basins originate along the eastern flank of the Appalachian Mountains and discharge into the Atlantic Ocean. Similarly, drainage basins on the western side of the Rocky Mountains discharge into the Pacific Ocean. In the continental interior, drainage is guided between these two mountain ranges. For example, the headwaters of the Missouri and Arkansas rivers drain from the eastern slopes of the Rockies toward the Gulf of Mexico, whereas those of the Ohio River drain toward the Gulf of Mexico from the western slopes of the Appalachians. To the north, in Canada, major drainage outlets occur in Hudson Bay and the Arctic Ocean. Boundaries between these groups of drainage basins separate streams flowing toward opposite sides of the continent and are known as *continental divides.* The best known continental divide extends from Alaska to Mexico and separates drainage basins that discharge into the Pacific Ocean from those that drain into the Gulf of Mexico and the Atlantic (see Fig. 13.19).

Not all of America's drainage basins reach the oceans. Some drain to lakes or swamps from which there is no outflow. Such is the case for the Bear River, which ends its 491-mile journey in Utah's Great Salt Lake. The largest drainage basins of this type, which are said to be *closed,* occur in central Asia where they drain into the inland Caspian and Aral Seas (see Living on Earth: **The Death of the Aral Sea—An Environmental Disaster** on page 419).

As we will learn in Chapter 15, the present-day drainage pattern of North America is actually a consequence of the last Ice Age. The enormous ice sheets that once occupied Canada and parts of the American Midwest rerouted entire drainage systems northward into the Arctic Ocean and southward into the Gulf of Mexico. The enormous weight

Figure 13.20   Major Drainage Basins of North America

All drainage basins are separated by areas of higher topography known as divides. For example, the Continental Divide of the Rocky Mountains separates streams that flow westward into the Pacific Ocean from those that flow eastward into the Gulf of Mexico. Black arrows indicate main drainage directions.

the pattern of drainage systems (Fig. 13.21). Under normal circumstances, the map pattern of trunk streams and their tributaries within a single drainage basin resembles the irregular branching habit of a tree and is termed **dendritic drainage** (Fig. 13.21a). Dendritic drainage patterns are produced when the drainage basin is underlain by broadly homogeneous material or by essentially flat-lying bedrock, and resistance to erosion is uniform. As a result, the underlying material has little influence on the pattern of streamflow, which is controlled, instead, by the regional slope.

Where the drainage basin is underlain by alternating bands of resistant and less resistant rock that outcrop in parallel belts, a rectangular pattern of drainage develops. In this pattern, the broadly parallel trunk streams are more or less perpendicular to their much shorter tributaries. This pattern is developed because the trunk streams tend to follow major valleys between ridges of resistant rock, while the short tributaries drain from the ridges themselves. Because the drainage pattern resembles the growth of vines on a trellis, it is termed **trellis drainage** (Fig. 13.21b). Trellis drainage is particularly common in the alternating valleys and ridges of the folded Appalachian Mountains.

Where the position of stream channels is strongly influenced by fractures in the bedrock, the pattern of drainage may be truly rectangular because prominent sets of bedrock fractures are commonly perpendicular to each other. As a result, in this drainage pattern, both the trunk streams and their tributaries show frequent right-angle bends and often join each other at right angles. The pattern is consequently termed **rectangular drainage** (Fig. 13.21c).

Finally, streams that radiate outward from a central high area like the spokes on a wheel, display **radial drainage**. This pattern of drainage is best developed on dome-shaped uplifts or on conical mountains such as volcanoes (Fig. 13.21d).

CHECK YOUR UNDERSTANDING

● Draw a labeled sketch of a drainage system and include headwaters, tributary, trunk stream, distributary, and delta.

● What is a drainage basin and how are the boundaries between drainage basins defined?

of these ice sheets also depressed the continental crust in the Canadian interior. This crust has been slowly rebounding since the ice sheets retreated, but has not yet fully recovered. As a result, major Canadian streams such as the Churchill River drain toward Hudson Bay, where the ice sheets were thickest and the crust was most depressed.

Drainage patterns

Topography and bedrock structure not only affect the course of individual streams but they can also influence

CHECK YOUR UNDERSTANDING

● What are the four main drainage patterns?

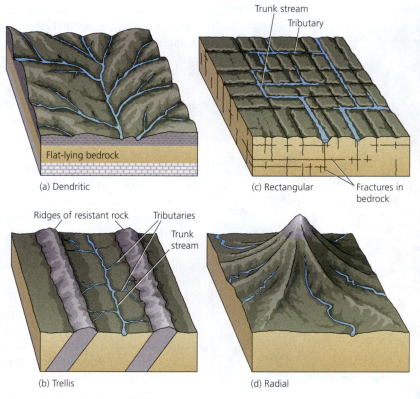

Figure 13.21 Drainage Patterns

Examples of drainage patterns include (a) dendritic drainage, (b) trellis drainage, (c) rectangular drainage, and (d) radial drainage.

- Topography and bedrock influence the pattern of drainage, which takes four main forms: dendritic drainage, trellis drainage, rectangular drainage, and radial drainage.

13.4 Stream Transport

The products of erosion, whether derived from the stream channel itself or supplied to the stream by weathering and mass wasting, are ultimately transported downstream in the form of a *sediment load* by the streamflow. A stream's sediment load has both a solid and dissolved component and is transported in one of three ways, depending on the size and solubility of the material:

- As a *suspended load,* if the particle size is small
- As a *bed load,* if the particles are large
- As a *dissolved load* if the material is soluble.

These three types of load are illustrated in Figure 13.22.

13.3 SUMMARY

- Streams erode their channels in three main ways: removing rock debris, downcutting by abrasion, and headwater erosion.

- As streams erode the land, they create V-shaped valleys.

- Where the gradient of the channel is gentle, streams develop sinuous loops or meanders that carve out a broad, flat floodplain on the valley floor.

- Occasionally, the stream of one drainage basin is diverted into the headwaters of another. This phenomenon is known as stream piracy.

- Stream downcutting in response to tectonic uplift or sea level fall can produce stream terraces—a series of elevated terraces that mark the former level of the floodplain—or incised meanders, which cut into bedrock and become deep meandering valleys with no floodplain. Both stream terraces and incised meanders are characteristic of rejuvenated streams.

- Streams that cut through ridges rather than going around them are antecedent streams if their path preceded ridge uplift, or superposed streams if they cut down into the ridge from a path established above.

SUSPENDED LOAD

Particles that are small enough to be lifted off the stream bed and held in suspension by the turbulent flow of the stream constitute the **suspended load**. It is this load that is most frequently visible, giving rise, particularly after heavy rainfall, to the cloudy or muddy appearance of

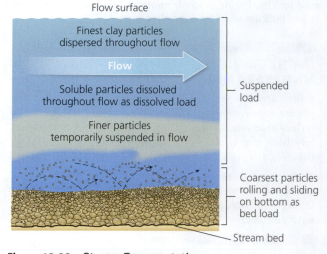

Figure 13.22 Stream Transportation

The sediment load of streams is transported in one of three ways: in suspension as suspended load; along the stream bed as bed load; or in solution as dissolved load.

LIVING ON EARTH

The Death of the Aral Sea—An Environmental Disaster

One of the world's largest closed drainage basins is that which brings central Asia's great Amu Darya and Syr Darya rivers to the Aral Sea of Kazakhstan and Uzbekistan (Fig. 13A). Fifty years ago, the Aral Sea was the world's fourth largest lake, covering 68,000 square kilometers (26,000 square miles). It was larger than all but the largest of the Great Lakes in North America and produced an annual harvest of about 45,000 tons of fish that supported 60,000 people. Since that time, the Amu Darya and Syr Darya rivers have been diverted to irrigate cotton, vegetable, fruit, and rice crops. The result has been an environmental catastrophe. The Aral Sea has shrunk by 90 percent and the rivers feeding it have been reduced to mere trickles. Only a ribbon of the main body of the lake (the South Aral Sea) remains along its western margin. Former fishing communities are now stranded hundreds of kilometers from the water and the fishing industry is dead because the fish were unable to survive in the increasingly salty water. Only the North Aral Sea, the water level of which was raised by a 13-kilometer (8-mile) dam built in 2005, retains some of its original water and is showing signs of recovery. But if nothing is done to reverse the decline of the South Aral Sea it will have essentially disappeared by 2020.

The region's climate has also been affected. Because the lake is no longer a moderating influence, the region has hotter summers and colder winters. The exposed sand is a salt desert—a toxic mixture of salt, pesticides, and defoliants, which are chemicals used on plants to cause their leaves to fall. Dust storms strip these chemicals from the soil, and the wind carries them for distances of up to 300 kilometers (almost 200 miles). As a result, wide regions of Uzbekistan and Kazakhstan have become contaminated. Now that there is less rain, crop yields are much lower. To combat this, fields are laced with herbicides, insecticides, and fertilizers, which have percolated downward to contaminate the groundwater.

Because of high concentrations of heavy metals and other toxic elements

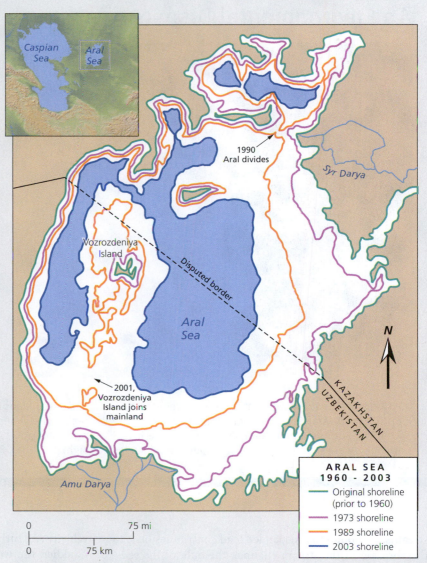

Figure 13A   Shrinking Aral Sea

The Aral Sea was once the fourth largest freshwater lake in the world. In 1960, the Syr Darya and Amu Darya rivers that flowed into it were diverted for irrigation purposes. The size of the Aral Sea has been shrinking ever since, with horrific environmental consequences. The satellite photographs of the Aral Sea on the next page were taken in 1989 and 2007.

in the drinking water, critical mineral deficiencies affect almost all of the region's population. Fully 99 percent of the population in northern Uzbekistan, for example, is anemic because their bodies are unable to absorb nutrients such as calcium and iron. Children eat chalk to satisfy their craving for calcium and have a high risk of developing brain damage. The rates of cancer, tuberculosis, typhoid, and liver and kidney diseases in children have also increased dramatically.

In 1992, the countries that border the Aral Sea signed an agreement on the management of its waters, and the dam built in 2005 has done much to restore

(Continued)

LIVING ON EARTH

The Death of the Aral Sea— An Environmental Disaster (Continued)

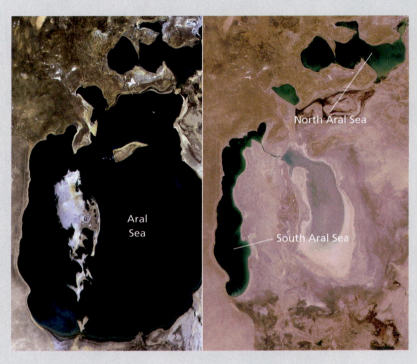

North Aral Sea

Aral Sea

South Aral Sea

Figure 13A   (Continued)

the North Aral Sea. But the dam has further reduced flow into the South Aral Sea. Environmentalists say that the South Aral Sea could be eventually restored if the irrigation canals were shut down, but that it would take decades and cost billions of dollars. Meanwhile, the economy of the region is still heavily dependent on revenue from agriculture, and no end to this environmental nightmare is in sight.

---

streams. Although the suspended load contributes little to the abrasion of the stream channel, the bulk of the sediment load is carried in this fashion. Indeed, almost three quarters of the sediment carried by the world's streams is transported in suspension.

Under normal conditions, only the finest material, mostly silt and clay, can be carried in suspension, but during floods, the particle size can be much larger. This is because the type of material that can be carried in suspension depends on the velocity of the stream and what is known as the *settling velocity* of the particle. Settling velocity is a measure of the rate at which a particle settles to the stream bed. This rate varies with the size, shape, and density of the particle, as well as the velocity and viscosity of the stream water. In other words, small, flat grains of light material, such as clays, are more easily suspended than larger, spherical grains of heavier material under the same conditions. During flooding, the amount of material carried in suspension dramatically increases with the rise in the velocity and turbulence of the stream. But as the stream gets muddier, the water that flows becomes thicker with suspended material. Hence, the *viscosity* of the stream also increases. For example, the suspended load of China's Huang He (or Yellow) River is so great during flooding (up to 70 percent by volume) that the river, the world's muddiest, acquires the consistency of molasses (Fig. 13.23). Because particles do not sink as fast in a more viscous fluid, their settling velocity decreases, making it easier to carry them in suspension. The increase in velocity and viscosity caused by flooding consequently combine to allow larger and denser grains to be carried in suspension. As the flood recedes, these grains are redeposited farther downstream.

BED LOAD

Particles that are too large or too heavy to be carried in suspension constitute the **bed load**. These particles move by either *traction* or *saltation* along the stream bed. Larger

Figure 13.23   Sediment Load

Hukou Waterfall on China's silt-laden Huang He (Yellow River), the world's muddiest river.

boulders and cobbles may hug the channel floor and move downstream in an intermittent sliding, dragging, or rolling motion known as **traction**. In the process, the sediment load particles are worn down and the channel floor is eroded by abrasion. On the other hand, smaller grains of sand and gravel may be briefly lifted off the channel floor by the stream's turbulent flow and so move downstream in a series of skips or hops. This intermittent skipping motion is known as **saltation** (Fig. 13.24). During saltation, grains are drawn up into the stream's flow by turbulent eddies, or are thrown upward by collisions with other saltating grains, and are carried a short distance downstream before falling back to the stream bed. The effectiveness of saltation as a mode of transportation depends on the size of the grains. Larger grains tend to take short hops close to the stream bed, while smaller grain are lifted higher and carried farther. In general, however, the bed load, whether moved by traction or saltation, accounts for only about 10 percent of a stream's total sediment load.

> **CHECK YOUR UNDERSTANDING**
>
> ● How does the suspended load of a stream differ from its bed load?

DISSOLVED LOAD

As much as a quarter of a stream's total sediment load is rarely seen because it is dissolved. The portion of the load that is carried in solution is the **dissolved load**. It represents the soluble products of chemical weathering and is largely made up of dissolved ions of calcium (Ca^{2+}) and bicarbonate (HCO_3^-). However, soluble ions of sodium (Na^+), magnesium (Mg^{2+}), potassium (K^+), chloride (Cl^-), and sulfate (SO_{42}^-) are also common, as are various amounts of organic acids, which sometimes color the water brown. Although some of these dissolved ions may precipitate as a result of evaporation or changes in the chemistry of the stream water, most eventually reach the sea and hence contribute to the chemistry of ocean waters. Indeed, almost 4 billion tons of dissolved ions are contributed to the sea in this fashion each year.

The abundance of elements such as sodium, calcium, and magnesium in ocean waters is consequently related to their abundance in streams and to the efficiency of chemical weathering (see Chapter 5). Depending on the nature of the bedrock and sediment in the stream valley, each stream has its own characteristic chemistry and so dispenses a distinctive chemical cocktail into the sea. However, the distinctive chemistry introduced by individual streams is very

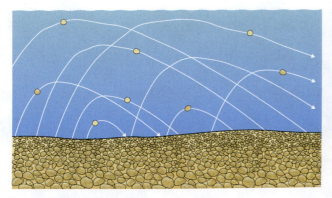

Figure 13.24   Saltation

Saltation is the intermittent downstream skipping motion of grains of sand and gravel that are small enough to be briefly lifted off the channel floor by a stream's turbulent flow.

efficiently dispersed and mixed by ocean currents so that most of the world's ocean water has a similar composition.

Most of a stream's dissolved load comes from the dissolution of rocks and the removal of soluble material from soils, and is largely supplied by groundwater. The ability of a stream to carry a dissolved load, in contrast to the solid component of the sediment load, is entirely independent of its velocity. This ability is because, once dissolved, soluble material remains in solution regardless of the rate at which the stream is flowing. However, the amount of the dissolved load varies widely, depending on such influences as climate, vegetation, and the local geology. It is usually measured as the number of dissolved ions for every million parts of water, or parts per million (ppm), with typical values ranging from 100 to as much as 1000 ppm.

> **CHECK YOUR UNDERSTANDING**
>
> ◉ In what way does a stream's dissolved load contribute to the chemistry of ocean waters?

CAPACITY AND COMPETENCE

The maximum amount of sediment that a stream can carry at any one time is its **capacity**, just as the capacity of an elevator is the maximum number of people it is authorized to carry. Not surprisingly, the capacity of a stream is proportional to the amount of water flowing within it, that is, its discharge. On a global basis, the combined capacity of streams is huge, with as much as 20 billion tons of sediment reaching the sea each year.

The maximum size of the material a given stream can carry is determined by the stream's velocity, and is known as the stream's **competence**. Not surprisingly, the faster a stream flows, the larger the particle size of the material it can transport. The competence of the stream is actually proportional to the square of the velocity. This means that when the velocity of a stream doubles, the maximum size of the material it can carry increases fourfold, and when it triples, the maximum size increases nine times. This

Figure 13.25   Boulders in a Mountain Stream

Mountain streams such as this one below Vernal Falls in Yosemite National Park often contain boulders that look far too large to be moved by the stream. However, the size of the material a stream can move increases with the *square* of the stream's velocity. As a result, even boulders as large as these can be rolled or dragged along the stream bed during floods, when the Merced River, here just a trickle of water, turns into a raging torrent.

> **CHECK YOUR UNDERSTANDING**
>
> ◉ What is the capacity of a stream and how does it differ from its competence?

relationship is most clearly seen in mountain streams, which often contain boulders that appear to be far larger than the streams are capable of moving (Fig. 13.25) because the largest of the boulders are moved only during occasional floods, when the streams become raging torrents. The discharge, and hence the capacity, of streams also increases during flooding. It is not surprising, therefore, that the greatest transportation of sediment occurs during floods when stream capacity and competence are at their highest.

13.4 SUMMARY

• The products of erosion are transported downstream in three ways: small particles, such as silt and clay, are carried in suspension; larger particles are carried as bed load, through rolling, traction, and saltation; and dissolved ions are carried in solution.

- Capacity and competence are two measures used to describe a stream's ability to carry solid particles. Capacity is the maximum amount of sediment a stream can carry at any one time. Competence is the maximum size of the material being carried. Faster streams can carry larger particles.

13.5 Stream Deposition

All solid sediment transported by streams is eventually deposited. Ultimately, the sediment load is deposited at the mouth of a stream, where the flow rate is slowed on entering standing water and the load settles to the sea bed. However, up to 75 percent of a stream's sediment load may be deposited in areas of slow streamflow, such as floodplains, along the stream course. All such stream-deposited sediment is collectively referred to as *alluvium*.

Stream deposition occurs whenever the velocity of a stream falls below that needed to keep the sediment load in motion. As a result, stream deposition is intermittent, occurring most frequently during flooding, when a stream overflows its banks and loses its velocity as it spreads out over the floodplain. The ability of a stream to either erode, transport, or deposit sedimentary particles is determined by the size of the particle and the velocity of the stream. The relationship between these variables is shown in Figure 13.26. For example, large gravel-sized particles require high stream velocities to be transported and are the first particles to be deposited as a stream slows down. On the other hand, fine-grained clays are readily transported even in the slowest of streams, and require the water to be almost stationary in order to settle. The varying mobility of sediment particles in streams leads to *sorting*, whereby particles become separated according to their size, as we learned in Chapter 6. Sorting is a fundamental characteristic of stream deposition, and gives rise to deposits that tend to be of similar grain-size at any given location.

The well-sorted alluvium deposited by streams produces a variety of distinctive landforms, some of which are produced in the stream channel itself, while others form on the floodplain or in deltas at the stream mouth.

CHANNEL DEPOSITS

Accumulations of sand and gravel that occur in midstream are **channel bars** (Fig. 13.27). Channel bars may be temporary landforms that come and go with changes in the stream's capacity and competence, or they may build up to become vegetated midstream islands. By diverting the stream's flow to one side or the other, channel bars serve to widen the stream by promoting undercutting of the stream banks. In streams that are heavily laden with sediment, many channel bars may be deposited, each one dividing and broadening the stream as they accumulate. This more or less continuous midstream deposition effectively chokes the trunk stream with sediment. In situations like this the stream develops an elaborate network of shallow, interlacing channels that repeatedly split and come together as

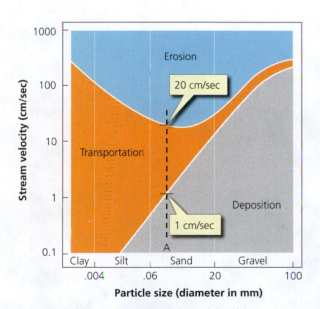

Figure 13.26   Stream Velocity and Particle Size

The velocity of the stream and the size of the particle control sediment erosion, transportation, and deposition. To pick up and remove a particle of size A, for example, the stream velocity must be in excess of 20 centimeters per second. The stream can transport this particle as long as the stream velocity is greater than 1 centimeter per second. When the stream flow falls below 1 cm/sec, the stream deposits the particle.

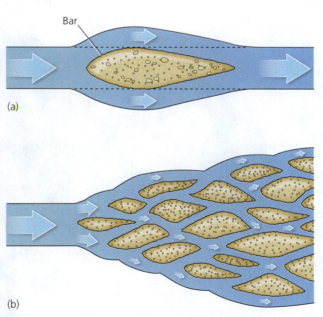

Figure 13.27   Channel Bar Development

(a) Deposition of a bar in mid-channel diverts the flow of water and widens the stream. (b) When this process is repeated, the flow of water is divided many times, the stream develops a braided pattern, and the channel is greatly widened.

Figure 13.28   A Braided Stream

This braided stream in southeast Iceland is choked with sandy sediment.

they weave a path between the bars. Because of the resemblance of their channel pattern to the strands of hair in a braid, streams of this sort are called **braided streams** (Fig. 13.28). Braided streams develop whenever more sediment is supplied to a stream than the stream can carry. Braided streams are particularly characteristic of the debris-laden streams produced by the melting of glacial ice.

Bars of a different sort accumulate in winding streams. Recall from Section 13.3 that in areas where the stream gradient is low, streams have a tendency to develop sinuous curves called meanders. Such streams are consequently called **meandering streams** (see Fig. 13.12). Because meandering streams flow fastest around the outside of the curve, the outer bank is progressively eroded and the stream channel develops an asymmetric shape, which is deeper toward this faster-flowing part of the bend. In contrast, on the inside of the curve, the stream slows down, thereby promoting sediment deposition (Fig. 13.29). The curved ridge of sand and gravel that forms as a result of this drop in flow rate occupies the "point" of the inside bend and is called a **point bar**.

With progressive erosion of their outer banks, meanders become increasingly sinuous, so that, eventually, only thin necks of land separate adjacent loops. With continued erosion, the intervening necks may be eroded away entirely, giving the stream a more direct route to the sea. The abandoned meander, once isolated from the main channel, is commonly left as a horseshoe-shaped lake called an **oxbow lake** because of its resemblance to the U-shaped harness around an ox's neck (Fig. 13.30).

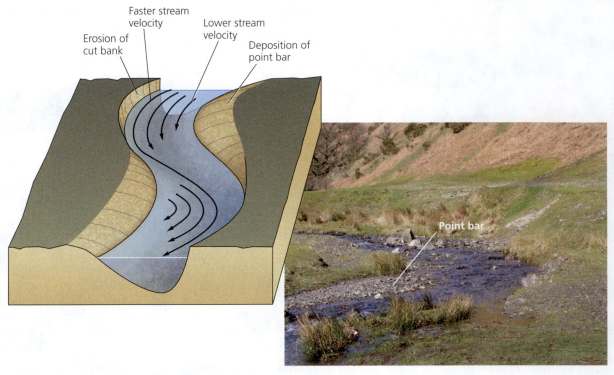

Figure 13.29   Erosion and Deposition in a Meander

In meandering channels, maximum flow rates are highest at the outside of each bend. The shift in the line of fastest flow from the center of the stream to the outside of the curve causes undercutting of the stream bank. At the same time, low stream velocities on the inside of the curve promote deposition of sediment to form point bars.

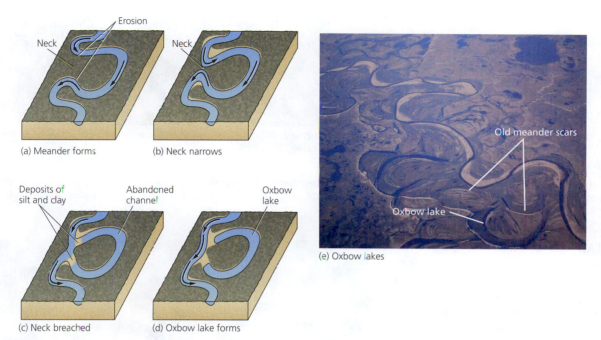

Figure 13.30   Formation of Oxbow Lakes

(a) Erosion occurs on the outside banks of a meandering stream. (b) This results in the formation of a narrow neck. (c) Eventually the stream cuts off the meander in order to form a straighter channel. (d) The abandoned channel becomes an oxbow lake. (e) In this aerial photo of oxbow lakes on the Yamal Peninsula in Russia, the old scars represent the migration of meanders.

The difference in the velocity of a meandering stream from one bank to the other means the position of the channel is constantly changing as erosion on one side is approximately balanced by deposition on the other. Stream deposits consequently migrate laterally as a result of the *lateral accretion* (sideways buildup) of a succession of point bars (Fig. 13.31), and so can become quite extensive. As we have discussed, over time, stream migration leads to a widening of the valley into one with a flat floor known as a floodplain.

> **CHECK YOUR UNDERSTANDING**
>
> ○ What is the difference between a braided stream and a meandering stream?

FLOODPLAIN DEPOSITS

As its name suggests, the floodplain is the area inundated whenever a stream overflows its banks during flooding. For most of the time, the floodplain lies above the level of a stream. It nevertheless forms an important area of stream deposition outside the stream channel. Not surprisingly, any water that escapes the fast-flowing channel abruptly slows down as it spreads out across the relatively flat floodplain. As the water overflows its banks, the finer-grained material held in suspension tends to be deposited on the floodplain, whereas the coarser-grained sediment is deposited adjacent to the riverbank. With each successive flood, this coarser material builds up to form a ridge, known as a **natural levee**, that becomes a protective barrier to further flooding. This levee is generally the highest point on the floodplain (Fig. 13.32). The levees on the lower reaches of the Mississippi River, for example, rise 6 meters (20 feet) above the level of the floodplain. A natural levee can act like a dam. As a result, the floodplain behind levees is often poorly drained and may develop into a marsh known as a *backswamp*.

During floods, stream water that breeches a levee deposits its load as a thin layer on the floodplain. Over time, therefore, floodplain deposits are built up both laterally, by the accretion of point bars, and vertically, by the deposition of sediment layers during episodes of flooding. As a result, floodplain areas typically comprise sand and gravel deposited on the shifting levees and point bars of migrating stream channels, interbedded with much finer silt and clay deposited on the floodplain itself.

> **CHECK YOUR UNDERSTANDING**
>
> ○ Describe the origin of a channel bar, a point bar, an oxbow lake, a natural levee, and a backswamp.

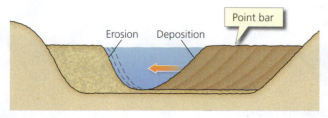

Figure 13.31   Lateral Accretion

Stream deposits migrate sideways due to the lateral accretion of point bars.

(a) Stream under normal conditions

(b) Floods deposit coarse material on river bank

(c) Natural levee is exposed when flood recedes

Figure 13.32   Natural Levees

Natural levees form along stream banks. (a) The stream is shown here under normal conditions. (b) As the stream begins to flood, water spreads across the floodplain and deposits coarser material near the banks. (c) When the flood recedes, a natural levee is exposed.

DELTAS

As a stream meets the sea, its velocity is dramatically reduced since the sea is essentially a standing body of water. As a result, the ability of the stream to carry its sediment load is abruptly lost and the stream dumps most of the load at the stream mouth. If this sediment is deposited faster than it can be carried away by coastal currents, it may *prograde* (build out) seaward into a low, fan-shaped deposit at the mouth of the stream (Fig. 13.33). Recall from Section 13.3 that this deposit is known as a delta. It was given this name because its shape often resembles the triangular symbol (Δ) for the Greek letter delta.

Within a simple prograding delta, a vertical succession of beds with distinctive sedimentary structures commonly develop. As Figure 13.34 illustrates, inclined *foreset beds* form the body of the delta, and represent deposits of the coarser sediment (mostly sand and silt) that is dumped at the mouth of the stream directly onto the sloping edge of the prograding delta. Finer material is carried farther out to sea, where it is deposited in horizontal layers to form *bottomset beds.* Later foreset beds are deposited on top of these bottomset beds as the delta builds out to sea. Finally, *topset beds* are deposited by the shifting streams that traverse the delta and spread broad horizontal beds over the delta surface. As we learned in Chapter 6, the result is the formation of cross beds.

The continual dumping of sediment on the delta surface impedes the streamflow. As a result, the main channel frequently divides into a network of smaller ones. Recall from Section 13.3 that these branches are known as distributaries because their channels act to distribute the streamflow and its sediment load across the delta. These channels may further divide as they, too, become choked with

Figure 13.33   Delta

A delta forms where a stream empties into the ocean. As the stream's flow stops, its sediment load is dumped and forms a triangular-shaped sedimentary deposit, in this case at the mouth of the Yukon River in Alaska.

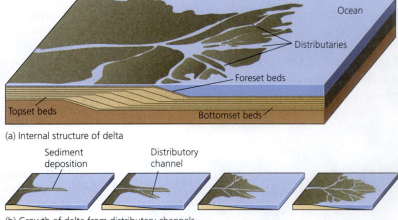

(a) Internal structure of delta

(b) Growth of delta from distributary channels

Figure 13.34   Growth of a Delta

(a) A simple prograding delta's internal structure consists of topset, foreset, and bottomset beds. (b) A delta grows as sediment deposition chokes the stream and distributary channels are successively developed and abandoned.

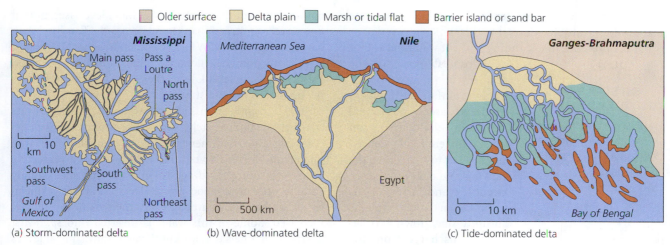

| Older surface | Delta plain | Marsh or tidal flat | Barrier island or sand bar |

(a) Storm-dominated delta   (b) Wave-dominated delta   (c) Tide-dominated delta

Figure 13.35   Types of Deltas

(a) Deltas like this stream-dominated delta of the Mississippi River are sometimes termed "bird's-foot" on account of their shape. (b) The delta of the River Nile in Egypt is a wave-dominated delta. (c) The Ganges-Brahmaputra in Bangladesh creates a tide-dominated delta.

sediment, or they may be abandoned because flooding breaches their levees and establishes a new channel course. As a result, the pattern of distributary channels frequently changes with time, such that the sediment load is deposited in a broad fan.

As the delta builds out into the sea, the shape of this fan depends on the relative influence of the distributary streams to that of waves and tides (Fig. 13.35). In *stream-dominated deltas,* like that of the Mississippi River, deposition at the mouths of the distributary channels cause fingers of sediment to build out seaward. Because of their shape, deltas of this type are sometimes called "bird's foot" deltas. In *wave-dominated deltas,* like that of the Nile River in Egypt, the sediment deposited by distributary channels is reworked and redistributed by wave action to form barrier bars at the seaward margin of the delta. As a result, the front of the delta comprises a series of arcuate barrier islands that may close off bodies of water to form lagoons. Finally, in *tide-dominated deltas,* like that of the Ganges-Brahmaputra in Bangladesh, much of the sediment is redistributed along the distributary channels and delta margin to form tidal sand bars aligned with the direction of tidal flow.

Deltas do not always develop at the mouths of major rivers. Neither the Columbia River of the Pacific Northwest, for example, nor the mighty Amazon and Parana rivers of South America, have deltas at their mouths because strong waves and ocean currents are able to disperse the sediment loads of these rivers before they have a chance to accumulate.

ALLUVIAL FANS

The triangular-shaped aprons of sediment, called **alluvial fans**, that form at the mouths of mountain canyons in desert regions like that of the southwestern United States (Fig. 13.36) owe their origin to an abrupt reduction in

Channels are dry most of the time

(a) Alluvial fan

(b) Alluvial fan, Death Valley

Figure 13.36   Alluvial Fans

(a) Delta-like alluvial fans form in deserts when occasional flash floods discharge sediment-laden water from mountain canyons onto the flatter desert floor. (b) This alluvial fan is located at the mouth of a canyon in Tucki Mountain, Death Valley, California.

stream velocity and competence. The drop in flow rate that produces fans, in contrast to that which forms deltas, is not the presence of a standing body of water but rather an abrupt decrease in the gradient of the stream. Although rainfall is infrequent in desert areas, occasional mountain cloudbursts can cause torrents of debris-laden water to rush down the canyons and out onto the adjacent lowland. Alluvial fans form when such streams abruptly slow down and deposit their sediment load as they discharge from the canyon and fan out onto the flatter desert floor.

CHECK YOUR UNDERSTANDING

○ What are the three types of beds that commonly develop as a delta progrades?

○ How does each type of bed in a delta form?

13.5 SUMMARY

- Stream deposition occurs whenever the velocity of a stream falls below that needed to keep the sediment in motion. Stream deposition occurs most frequently during flooding.

- Sand and gravel may accumulate midstream, as channel bars, and on the inside of meanders, as point bars. Channel bars are characteristic of sediment-laden braided streams with interlacing channels. Point bars develop in the sinuous channels of meandering streams.

- Stream deposits migrate laterally over time, creating a flat floodplain on the valley floor. Natural levees build up from coarser-grained sediment deposited next to the riverbank and act like dams during flooding.

- When a stream enters the sea, it deposits most of its sediment load at the stream mouth in a triangular shaped landform known as a delta. Sediment is distributed across the delta by distributary channels, which deposit the sediment load in a broad fan.

13.6 Floods and Flood Prevention

Floods occur when a stream's channel can no longer contain the water flowing through it. Like a traffic jam, which occurs when there are too many cars on the road, a flood occurs when more water is entering the drainage than is being discharged from it. The volume of water increases with time as the flood builds. This can happen for a variety of reasons. Most floods commence with a period of prolonged or heavy rainfall. Usually, the ground can absorb a portion of this rainfall. But if heavy rain continues, the ground becomes saturated and surface runoff is fed directly into the stream channel. This results in a temporary imbalance in the water budget of the stream, in which water input along the stream and its tributaries exceeds the

output at its mouth. Eventually the water level rises, until it flows over the stream banks and across the flat floodplain beyond (Fig. 13.37). When the water reaches its highest level at any point, the flood is said to *crest*. Once the flood crests, the water level falls as the crest of the flood follows the flood waters downstream.

Recall from Section 13.5, that when a stream leaves its channel, it immediately slows down and deposits its coarser-grained sediment load adjacent to the riverbank. The resulting ridge or natural levee provides a protective barrier to future flooding. To enhance this effect, many communities near flood-prone rivers construct artificial levees so that the volume of water the stream can hold is increased. Some of this water is diverted into *spillways*, overflow channels that are cut through the artificial levees to provide the controlled release of floodwater. However, to be effective, artificial levees must be built along the entire length of the stream. If they are only built along the lower part of a river, floods upstream invade the floodplain behind them. And if they are built only along the upper part, the concentrated fury of the flood is unleashed onto the floodplain downstream.

(a)

(b)

Figure 13.37   Flooding

Floods occur when there is too much water for the stream channel to carry. As a result, (a) streams overflow their banks and (b) wreak havoc on communities built on the floodplain.

In the summer of 1993, 60 centimeters (almost 24 inches) of rain fell in a four-day period in southern Minnesota. This brought about North America's worst flood since 1878, when records were first kept. Most of the flooding was concentrated along an 800-kilometer stretch of the upper Mississippi Valley, from St. Louis, Missouri, to St. Paul, Minnesota. Here, where the stream channels are relatively shallow and narrow, they could not hold the enormous volume of water. As a result, the levees were breached and floodwater inundated the floodplain (Fig. 13.38). The great Mississippi flood of 1993 claimed dozens of lives, displaced more than 50,000 people, and caused damage in excess of $10 billion.

Less than two decades later, in May 2011, a flood of similar proportions occurred when record rainfall from a series of major storms across northern Arkansas, southern Missouri, and parts of the Ohio River Valley combined with meltwater from thawing snow to inundate the lower Mississippi Valley from Illinois to Louisiana. So great was the threat of flooding in southern Louisiana that, for the first time since 1973, the Morganza Spillway was opened, deliberately flooding the Atchafalaya Basin in order to protect the cities of Baton Rouge and New Orleans. The 2011 flood displaced over 25,000 people and caused economic losses estimated at $3–4 billion.

In many parts of the world, ice jams are also a common cause of flooding. When frozen streams thaw in the spring, ice jams may form a natural dam which blocks the normal course of water flow and causes flooding up to the height of the dam in the valley upstream. In 1997, and again in 2009 and 2011, this was a cause of serious flooding on the Red River, which flows north along the North Dakota-Minnesota state line into Canada's Lake Winnipeg. The 1997 flood testifies to the importance of accurate flood forecasting. In parts of North Dakota, raised banks or *dikes* were built to withstand a forecasted flood of 15 meters (49 feet). Unfortunately, the river crested at 16.5 meters (54 feet) and so the floodwater spilled over the dikes. In Manitoba, up to 1400 cubic meters (almost 50,000 cubic feet) of water per second was diverted into artificial channels around the city of Winnipeg. As a result, the damage was not as severe as it was in North Dakota. These artificial channels offered the river a straighter route and, by increasing the stream gradient, the water moved more rapidly away from the zone of potential flooding. In the case of the 1997 flood, the artificial channels protected the city of Winnipeg from devastation; however, it is a dangerous strategy. A faster-flowing river has greater power to erode. Thus in the future, these artificial channels may offer Winnipeg only temporary protection from the ravages of the Red River.

As a protective measure, flood-control dams are often built upstream from densely populated areas. They are generally made of earth or concrete, and serve to block and divert water from its natural path. However, when they fail, they too can result in catastrophic flooding. One of the most infamous floods in North American history occurred in Johnstown, Pennsylvania, in 1889, when an earthen dam gave way on the Conemaugh River and 2200 inhabitants of the town perished.

Many scientists believe that floods are an inevitable consequence of surface runoff, and that it is only a matter of time before levees are breached. In their view, strategies may contain ordinary floods, but not extraordinary ones. A better strategy might be to bow to the forces of nature by controlling human use of flood-prone lands. This view was strengthened by the failure of levees during the massive flooding along the Missouri and Mississippi rivers in 1993.

Figure 13.38   Breached Levee
The great Mississippi flood of 1993 breaches a levee in Illinois.

Dams and the Human Exploitation of Surface Water

The exploitation of surface water is as ancient as civilization itself. Perhaps our most visible interference with the natural flow of surface waters is in the construction of dams, and the reservoirs of water that are ponded behind them. Dams are constructed for many purposes, including hydroelectric power (see Chapter 19), irrigation, municipal water supplies, recreation, and flood control. Although dam construction can bring many benefits, they are not without significant long-term disadvantages (Fig. 13B).

Large dams flood vast areas and destroy natural wildlife habitats. They also alter the natural pattern of stream sedimentation, with surprising results. The pattern of stream sedimentation changes because the reservoir that forms behind a dam is an artificial lake, the water level of which acts as a local base level (Fig. 13C). The average gradient behind the dam is therefore decreased, diminishing the stream's capacity to erode. In addition, the dam stops the flow of the stream when it reaches the reservoir, causing the stream to deposit its sediment load. As a result, the water released below the dam is essentially sediment-free. The stream consequently flows faster and its energy is used primarily to erode the underlying sediment and bedrock. In this way, the construction of a dam alters a stream's natural profile by making it so that its sediment load is prematurely deposited in the reservoir and its ability to erode downstream is enhanced.

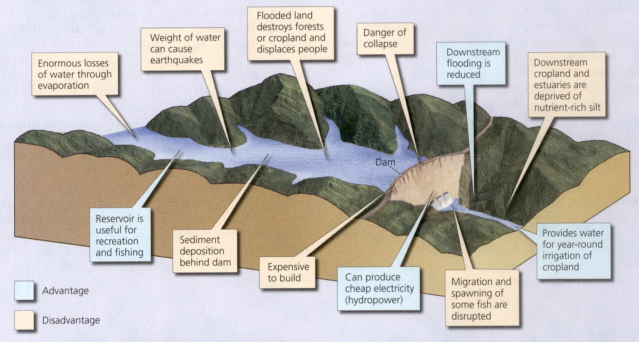

Figure 13B Advantages and Disadvantages of Dams
Large dams and reservoirs have positive and negative impacts.

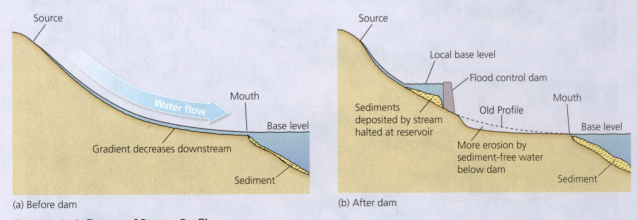

Figure 13C Influence of Stream Profile
These diagrams depict the profile of a stream (a) before and (b) after dam construction.

This change in the natural cycle of a stream can have many unforeseen consequences. Left to their own devices, streams deposit about 75 percent of their sediment load on their floodplains, where the sediment provides essential nutrients for soil development. Before the construction of the Aswan High Dam in Egypt, for example, annual flooding of the Nile River deposited from 6 to 15 centimeters (2.4 to 6 inches) of new soil per century. This natural fertilizer replenished the nutrients extracted from the soil by intense agricultural activity. Since construction of the dam was completed in 1970, the transport of sediment downstream has been blocked, with the result that artificial fertilizers must now be used. In addition, the downstream portion of the Nile now flows faster and the river has undermined bridges and smaller dams. The amount of sediment reaching the Mediterranean Sea has also been drastically reduced, and the Nile Delta is now subsiding faster than sediment can build it up. This sediment also provided nutrients for the marine ecology so that the fishing industry that once thrived in the region has been decimated, with the loss of 30,000 jobs.

Completed in 2006, the environmental impact of the world's largest dam, the Three Gorges Dam on the Yangtze River in China, has yet to be fully realized. However, there is increasing concern that it may be severe. The reservoir displaced a staggering 1.4 million people, flooding 13 cities, 140 towns and 1350 villages, and has been plagued by algae and pollution. There is also evidence that the dam may be triggering landslides and altering entire ecosystems, and there is concern that the weight of the reservoir's water may cause an increase in earthquake activity.

One of the world's most ambitious surface water projects was started in the early 1970s in northern Quebec, Canada (Fig. 13D). The original 50-year plan, devised to provide inexpensive hydroelectric power to Quebec and the northeastern United States, was to construct 600 dams that would alter or reverse the flow of 19 major rivers at a cost of $60 billion. By 1985, the first phase of the project on the La Grande River had been completed at a cost of $13.7 billion. The second phase of the project on the Eastmain River was begun in 1989, but was halted in 1994 because of objections by the indigenous Cree people and because environmental concerns led the New York State government to cancel contracts with Hydro Quebec. An agreement with the Cree people in 2002 paved the way for the completion of the second phase and the diversion of the Rupert River, but an ambitious additional project on the Great Whale River (James River II) was suspended indefinitely. Together, the two phases of the James River Project generate over 80 terawatt-hours of electricity a year (enough to meet the total demand of a small country like Belgium) but have flooded some 13,000 square kilometers (5000 square miles)—an area the size of Connecticut.

It is clear that humans still have much to learn about the consequences of our interference with the natural cycle of surface water. We depend on water for our survival, yet our methods of exploitation are often poorly conceived and the results may be unforeseen. Only through further study of our use of this priceless resource will we be able to leave future generations a more sustainable planet.

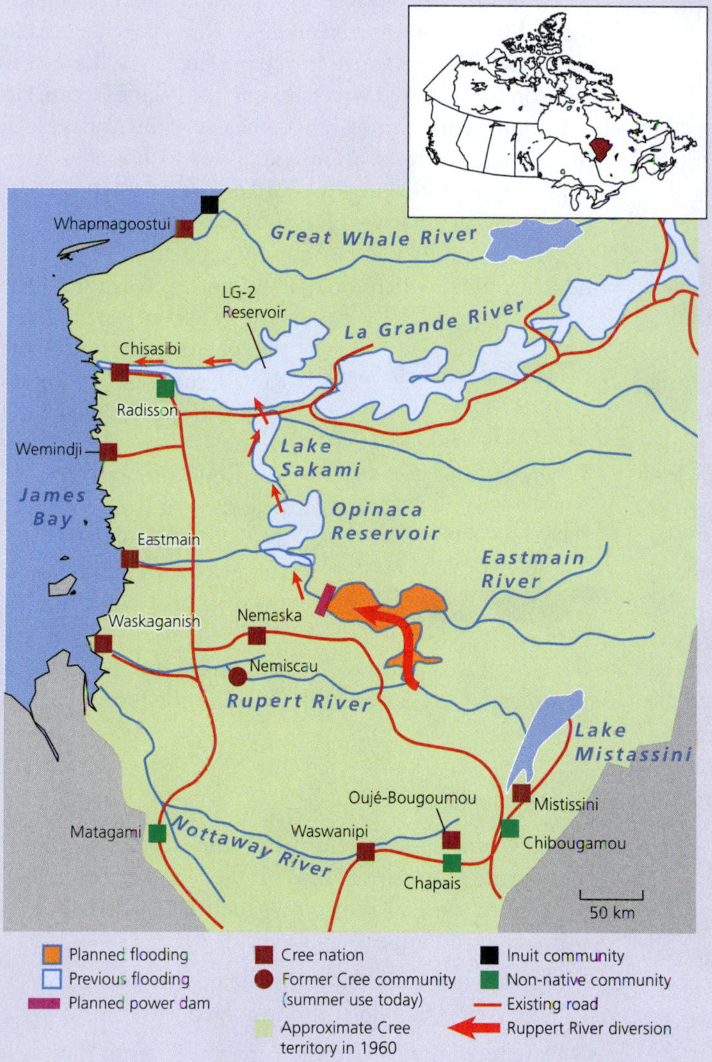

Figure 13D James Bay Project

The James Bay Project in northern Quebec is an ongoing program to construct a series of hydroelectric power stations on the La Grande River and its tributaries, and divert neighboring rivers into the La Grande watershed. If fully completed, the project would be the largest hydroelectric system in the world.

Map legend:
- Planned flooding
- Previous flooding
- Planned power dam
- Approximate Cree territory in 1960
- Cree nation
- Former Cree community (summer use today)
- Existing road
- Ruppert River diversion
- Inuit community
- Non-native community

50 km

431

CHECK YOUR UNDERSTANDING

○ What are artificial levees and what are the limits of their effectiveness?

In the aftermath of these floods, two diverging points of view emerged among the affected communities. Some suggested that the river should be allowed to reclaim its natural course and that development in flood-prone regions should be restricted. Others suggested that higher flood walls should be built to give them better protection (see In Depth: **Dams and the Human Exploitation of Surface Water** on page 430).

13.6 SUMMARY

* Floods occur when a stream can no longer contain the water flowing through it. When a stream leaves its channel, it slows down and deposits its coarser-grained material next to the river bank to form a natural levee.

* In some places, artificial levees are constructed to increase the effectiveness of a natural levee. In others, artificial channels are dug to divert water from a city. Finally, flood-control dams are sometimes built to block and divert water from its natural path.

13.7 Running Water and Plate Tectonics

The intimate link between running water and plate tectonics may not, at first, seem obvious. But at the largest of scales it is plate tectonics that have given us the world's high-standing continents, where all running water starts its surface journey, and the deep ocean basins, into which all running water ultimately flows. Like mass wasting, running water requires a slope, and it is uplift that occurs in response to the forces of plate tectonics, which ultimately provides this slope. The evolution of most of the world's rivers has been influenced either directly or indirectly by plate tectonics.

The influence of plate tectonics is evident in the world's great river systems; plate tectonic forces have determined the location, pattern, size, and longevity of these systems. Most of the world's major drainage basins owe their origin to the breakup of Pangea and the mountains produced by the resulting closure of the Tethys Ocean and the accretion of terranes along the Pacific margins of North and South America. Today, these mountains provide the source of many of the world's great rivers—the Amazon and Paraná rivers in the Andes; the Mackenzie and Yukon rivers in the Canadian Cordillera; and the Indus, Ganges-Brahmaputra, Irrawaddy, Salween, Mekong, Yangtze, and Huang He (Yellow) rivers in the Himalayas and Tibet Plateau. Even the Mississippi River, which rises in northern Minnesota far from mountains, drains the Rockies through the Missouri, Arkansas, and Red rivers and the Appalachians by way of the Ohio, Cumberland, and Tennessee rivers. Therefore, the drainage system of much of the US continental interior is hemmed in between two mountain belts generated by plate tectonics (see Fig. 13.20).

In addition, recall from Chapter 2 that the breakup of Pangea produced new oceans such as the Atlantic, which are bordered by passive continental margins. It is toward these margins, rather than the active margins of subducting oceans, that most of today's major rivers flow. Recall also from Chapter 9, that the breakup of Pangea also produced aulacogens, the name given to the third arm of a triple point rift that fails to open into an ocean (Fig. 13.39). Because aulacogens are depressions that extend into the continent, they frequently channel major rivers into the opening ocean. Thus, the Benue Trough, which sits at the elbow of Africa's west coast (see Fig. 9.41), is an aulacogen that brings the Niger River to the Atlantic Ocean. Likewise, both the Amazon and Mississippi rivers enter the sea by way of failed rifts, and Canada's Mackenzie River follows an aulacogen associated with the opening of the Arctic Ocean.

Plate tectonics also has a more direct influence on drainage by destroying old river systems and creating new ones. The history of the Amazon and Nile rivers provides excellent examples of this. As we discussed in Chapter 10, the ancestral Amazon River flowed westward into the Pacific Ocean. However, about 15 million years ago, the rise of the Andes blocked this drainage to form an inland lake, which was eventually tapped by rivers flowing eastward into the Atlantic Ocean to form the Amazon River of today (see Fig. 1.5).

Plate tectonics also profoundly influenced the course of the River Nile, which today flows northward from Africa's Lake Victoria to the Mediterranean Sea in Egypt. Prior to the opening of the Western Rift of the East African Rift Valley, however, most of the region's rivers flowed westward into the Atlantic Ocean by way of the Congo Basin. But as the Western Rift opened, some of these streams reversed their course with uplift of the rift rim and flowed backward to fill today's Lake Victoria. Once full, the lake water spilled into the Western Rift to drain northward as today's River Nile.

CHECK YOUR UNDERSTANDING

○ Why do so many of the world's major drainage basins owe their origin to the breakup of Pangea?

13.7 SUMMARY

* Plate tectonics has determined the location, pattern, size, and longevity of most of the world's great river systems, many of which owe their origin to the breakup of Pangea.

* Many major rivers entering the oceans formed by Pangea breakup do so along aulacogens.

* Plate tectonics caused the Amazon River to reverse its course with the uplift of the Andes and redirected the Nile River with the opening of Africa's Western Rift.

Figure 13.39   Aulacogens and the Breakup of Pangea

This map highlights the aulacogens (failed rifts) associated with the breakup of Pangea. Note that today several major rivers, including the Amazon, Mississippi, and Niger, reach the ocean by way of these failed rift valleys.

13.8 Running Water on Other Worlds

As far as we know, Earth is unique among the planets in the Solar System in having running water on its surface. Mercury and the Moon are both barren and waterless, and the surface temperature of Venus is so hot that any surface water would boil away. Conversely, planets beyond the orbit of Earth are generally so cold that any surface water can only exist as ice. But it may not always have been like this. Venus, for example, was once much cooler and so could have had running water on its surface in the distant past. Mars was once much warmer, and most certainly did have surface running water.

Liquid water does not exist on the surface of Mars today. However, there is widespread evidence in the Martian landscape that running water was at one time present in abundance. The evidence is based on the similarity of certain Martian landforms to the stream features we have discussed in this chapter. Satellite images have revealed networks of branched stream-like valleys (Fig. 13.40), dendritic drainage systems, braided channels, and terraced, meandering canyons that can only have been cut by running water.

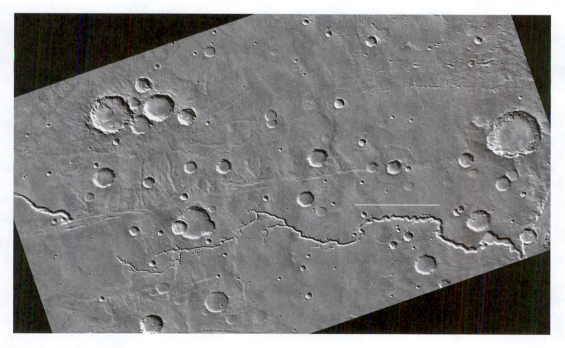

Figure 13.40   Nirgal Vallis

A Mars Global Surveyor image of Nirgal Vallis shows a long, winding valley with several tributaries.

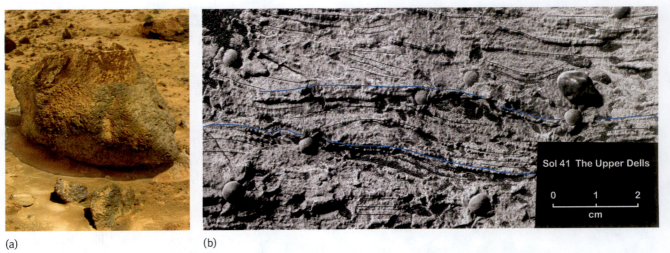

(a)   (b)

Figure 13.41   Clues to Past Water on Mars

(a) The rounded, water-worn appearance of "Yogi," a meter-size rock at the Mars Pathfinder site, suggests transport by running water. (b) A close-up view from the Mars rover Opportunity of a martian conglomerate called "Upper Dells" containing rounded pebbles and displaying sets (separated by blue lines) of inclined cross beds (black lines), which both suggest deposition by flowing water.

Close examination of the photos taken on the Mars Pathfinder mission also suggest that running water was once an important surface process. Although the scenes are those of bone-dry rocky desert, many of the rocks look water-worn and rounded, and some appear to be stream-deposited conglomerates (Fig. 13.41), the layering of which shows cross-bedding indicative of flowing water (see Chapter 6).

These features indicate that Mars once had abundant liquid water on its surface. So where did it all go? The Martian channels are ancient features, which formed about 3 to 4 billion years ago. Some of the water has doubtless leaked into space as Mars has lost much of its original atmosphere. However, most may still be present. Mars has polar ice caps and, beneath the planet's surface, there may be a frozen permafrost layer up to a kilometer (0.6 miles) thick. If so, this layer might have an important bearing on the existence of life on Mars today, and may prove crucial to the feasibility of the planet's eventual colonization.

13.8 SUMMARY

- Earth is unique in the Solar System in that it has running water at its surface.

- On Mars, the former existence of running water is recorded in dendritic drainage systems, in terraced, meandering, and braided channels, and in water-worn boulders and cross-bedded conglomerates. Much of this water may now lie frozen beneath the planet's surface.

> **CHECK YOUR UNDERSTANDING**
>
> • What is the evidence that Mars once had running water on its surface?

Key Terms

alluvial fan   427
antecedent stream   414
base level   406
bed load   420
braided stream   424
capacity   422
channel   403
channel bar   423
competence   422
delta   415
dendritic drainage   417

discharge   406
dissolved load   421
distributary   415
divide   416
drainage basin   415
floodplain   411
graded stream   407
headward erosion   410
hydrologic cycle   402
incised meander   414
laminar flow   404

longitudinal profile   405
meandering stream   424
natural levee   425
oxbow lake   424
point bar   424
pothole   410
radial drainage   417
rectangular drainage   417
rejuvenation   407
saltation   421
stream   403

stream piracy   411
stream terrace   413
superposed stream   414
suspended load   418
traction   421
trellis drainage   417
tributary   415
turbulent flow   404

Key Concepts

13.1 RUNNING WATER AND THE HYDROLOGIC CYCLE

- Running water is the most effective agent of erosion and deposition.
- By returning water from the land to the oceans, streams are key components of the hydrologic cycle.

13.2 STREAMFLOW

- Water flows either smoothly, in laminar flow, or chaotically, in turbulent flow.
- A stream's rate of flow depends on its gradient, the roughness of the stream bed, and the shape of its channel.
- Discharge is a means of quantifying streamflow by multiplying the cross-sectional area of the channel by the velocity of the stream.
- Base level is the lowest elevation to which a stream can erode its channel and occurs where a stream enters the sea, a lake or another stream, or encounters a resistant rock.

13.3 STREAM EROSION

- Streams erode their channels by removing debris, by abrasion, and by headward erosion.
- Stream erosion creates V-shaped valleys.
- Streams develop loops or meanders that carve out broad, flat floodplains where the gradient is gentle.
- Stream piracy occurs where the stream of one drainage system is diverted into the headwaters of another.
- Rejuvenated streams downcutting in response to uplift or sea level fall produce stream terraces and incised meanders.
- Streams that cut through ridges are antecedent if their path preceded uplift, or superposed if they cut down from a path established above.
- Topography and bedrock influence drainage patterns, which can be dendritic, trellis, rectangular, or radial.

13.4 STREAM TRANSPORT

- Streams transport small particles in suspension, larger particles as bed load, and dissolved ions in solution.
- Capacity is the maximum amount of sediment a stream can carry. Competence is the maximum size of the material carried.

13.5 STREAM DEPOSITION

- Deposition occurs when a stream slows down. Deposition within the stream channel produces bars.
- When a stream leaves its channel, it deposits coarse material on the bank and fine material on the floodplain.
- Deposition at the mouth of a stream is responsible for the formation of deltas.

13.6 FLOODS AND FLOOD PREVENTION

- Floods occur when a stream bursts its banks.
- Artificial levees, artificial channels, and flood-control dams are common methods of preventing flooding.

13.7 RUNNING WATER AND PLATE TECTONICS

- Plate tectonics has influenced most of the world's great drainage basins, many of which owe their origin to the breakup of Pangea.
- Many major rivers enter the ocean in aulacogens formed during the rifting of Pangea.
- Plate tectonics directly influences drainage by destroying old river systems and creating new ones.

13.8 RUNNING WATER ON OTHER WORLDS

- Earth is the only planet with running water at its surface.
- Evidence of the previous existence of running water on Mars includes dendritic drainage; terraced, meandering and braided channels; water-worn boulders; and cross-bedding.

Study Questions

1. Describe the hydrologic cycle and discuss how Earth's water supply is partitioned among its various reservoirs.
2. In what ways are the gradient, velocity, and discharge of a stream related?
3. How do streams attempt to acquire an equilibrium profile and become graded?
4. Describe the processes of stream erosion.
5. How does a floodplain form?
6. Describe the processes by which a stream transports its sediment load.
7. Why do braided streams form channel bars while meandering streams form point bars?
8. How are deltas influenced by streams, waves, and tides?
9. Why do rivers flood and what measures do communities take to prevent flood damage?
10. How have the forces of plate tectonics influenced major drainage basins?
11. Mars has no liquid water on its surface today, yet is believed to have done so in the past. How has this been determined and what is thought to have happened to the water?

14 Groundwater

14.1 Groundwater: A Vital Resource

14.2 The Water Table

14.3 Aquifers and Groundwater Flow

14.4 The Dissolving Power of Groundwater

14.5 Exploitation of Groundwater

14.6 Groundwater and Plate Tectonics

► Key Terms

► Key Concepts

► Study Questions

SCIENCE REFRESHER: Water Pressure at Depth 444

LIVING ON EARTH: The Historical Importance of Aquifers 447

LIVING ON EARTH: Water in the Sahara Desert 450

IN DEPTH: Exploring for Groundwater 456

Area of recharge

Flowing well

Non-flowing well

Shale   Sandstone

Confined aquifer

Water table

Fresh groundwater

Salty groundwater

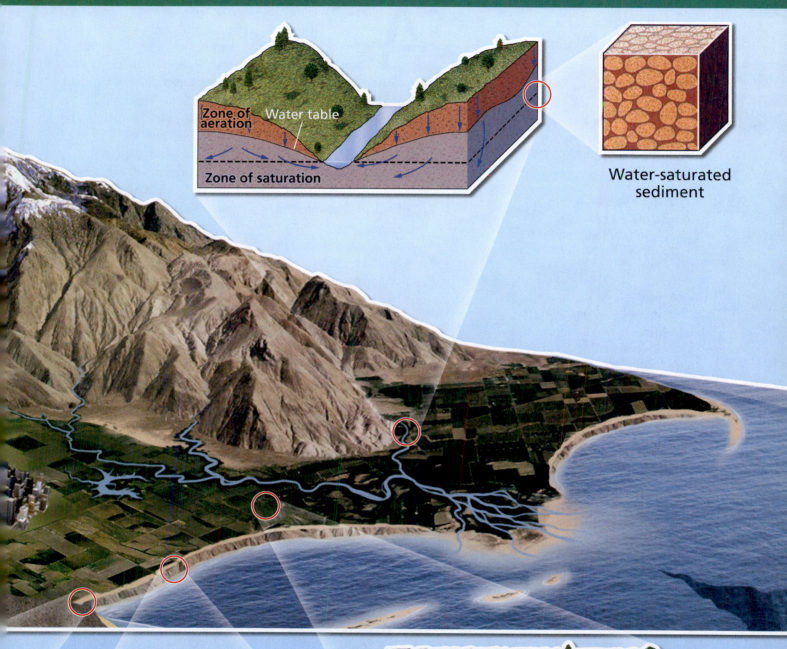

Zone of
aeration   Water table

Zone of saturation

Water-saturated
sediment

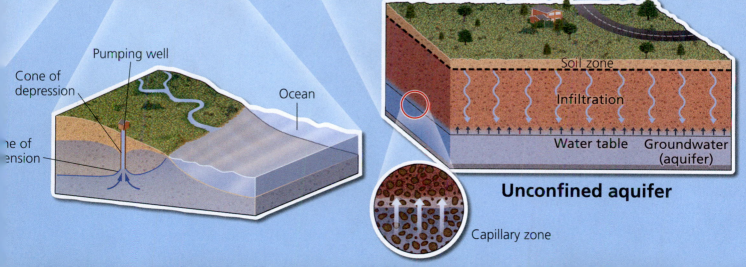

Pumping well

Cone of
depression

Ocean

...ne of
...ension

Soil zone

Infiltration

Water table   Groundwater
(aquifer)

Unconfined aquifer

Capillary zone

Learning Objectives

14.1 Recognize that groundwater is a vital source of accessible freshwater.

14.2 Describe how groundwater forms below the water table.

14.3 Explain the origin of aquifers, the controls of groundwater flow, and the origin of oases and hot springs.

14.4 Describe how the movement of groundwater through soluble rock creates caverns and a distinctive surface topography.

14.5 Explain how groundwater is exploited as a resource and how the unique qualities of groundwater render it susceptible to contamination by human activities.

14.6 Identify how plate tectonics governs the location and overall shape of some aquifers, as well as groundwater flow, by creating the mountains that influence topography and act as recharge areas for aquifers, as well as the faults that bring groundwater to the surface.

Although we are most familiar with the surface waters of streams, lakes, and swamps, Earth's main water resource is stored in the ground and is known as groundwater. Groundwater is the result of precipitation percolating downward from the surface to form subterranean *aquifers*, the name given to a layer of water-bearing sediment or rock. At a certain depth below the surface, water fills all fractures in the sediment or rock and all gaps between their constituent grains. The upper surface of this water-saturated zone is known as the *water table*, and its location is fundamental to the successful exploitation of this vital resource.

Because we consume groundwater faster than nature can replenish it, groundwater is a nonrenewable resource. Humans must therefore mine this resource, just like any other commodity. But to exploit groundwater we first must find it. Sophisticated exploration methods, such as electrical pictures of underground rocks and sediments, help identify aquifers and provide targets for the drilling of water wells.

In addition to being a very valuable resource, water is a very strong solvent and so is very susceptible to contamination both by natural processes and by human activities. Human activities can modify the composition of water and limit its use, thereby compromising our groundwater supply.

14.1 Groundwater: A Vital Resource

Humans can survive without food for weeks, but a person can only live for a few days without water. Our bodies require 4 liters (about 1 gallon) of water per day. Thus, freshwater is our most precious resource. The 350 million people in the United States and Canada require 1.5 billion liters of drinking water per day, and this is just the

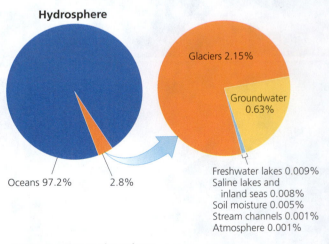

Figure 14.1 Earth's Hydrosphere
Most of Earth's water supply is in the oceans. The rest is in the atmosphere and on land.

TABLE 14.1   EARTH'S WATER INVENTORY IN CUBIC KILOMETERS

Location	Volume	Percent of Total	Percent Freshwater
Oceans	1,327,500,00	97.20	—
Glaciers	29,315,000	2.15	77
Groundwater	8,442,580	.63	22
Freshwater and saline lakes and inland seas	230,325	.016	0.6
Atmosphere at sea level	12,982	.001	Trace
Average in stream channels	1255	.0001	Trace

tip of the iceberg. We also use water for domestic purposes (in cooking, bathing, and cleaning), for agriculture (in irrigation and for livestock), and for industry (in manufacturing). In the United States and Canada, the world's most extravagant water users per capita, the total demand is a staggering 1.6 trillion liters per day.

Yet freshwater represents less than 3 percent of the water on Earth, and about three-quarters of this is presently inaccessible because it is locked up in ice sheets and mountain glaciers (Fig. 14.1). **Groundwater** constitutes about 22 percent of the world's freshwater (Table 14.1) and is stored in rocks and sediments below the ground, in what is termed the *subsurface*. Groundwater is by far the largest reservoir of *accessible* freshwater, supplying 25 percent of the total current demand in the United States and close to 30 percent of the demand in Canada. Groundwater is clearly an indispensable resource and its origin, exploitation, and management are of vital concern to society. We begin our examination of this essential commodity by describing how groundwater forms and the features that allow it to become a resource.

CHECK YOUR UNDERSTANDING

○ What forms of freshwater are not considered groundwater?

○ Why is groundwater a nonrenewable resource?

14.1 SUMMARY

* Groundwater is the largest reservoir of accessible water and represents 22 percent of the total freshwater.

* We consume groundwater much faster than nature can replenish it through precipitation so groundwater is a nonrenewable resource and is mined like any other resource.

14.2 The Water Table

When water evaporates from the oceans, it contains almost no dissolved salts (only a few milligrams per liter), and this relatively pure form of water is called "fresh" water. Most of the precipitation of freshwater that falls on land drains back to the sea under the influence of gravity as part of the hydrologic cycle (see Chapter 13). However, over thousands of years, water that has infiltrated into the subsurface has accumulated and formed substantial groundwater resources. This water stored in the ground nurtures life and facilitates our survival.

Sediment contains billions of tiny spaces and rocks contain abundant fractures. At a certain depth beneath the surface, all of these openings are filled with water. But how does the water get there? Groundwater forms as rain migrates downward into the subsurface through unconsolidated soil, sand, and gravel and into fractures and cavities in the rock below. The increasing pressures at depth tend to close fractures so that bedrock eventually forms an impenetrable barrier to this downward movement of water. But in the zone above this barrier, known as the **zone of saturation**, water fills all the fractures and pore spaces (the spaces between particles) in soil, sediment, or rock (Fig. 14.2). These underground water-rich layers contain the resource we call groundwater and the upper surface of this saturated zone is known as the **water table**.

If we stand outside in the pouring rain for long enough, we often say we are "saturated," by which we mean that we couldn't possibly get any wetter. The subsurface below the water table is also saturated, because water occupies all the spaces available to it, so that the sediments and rocks below the water table cannot get any wetter. We can determine the level of the water table by simply digging a hole. When we reach the water table, water will start filling the hole. When making sand castles on a beach, for example, water floods in as we dig below the water table because the sand below it is saturated with water.

Above the zone of saturation, the sediment is undersaturated and the pore spaces consist of a mixture of air and water. This region is the **zone of aeration**. The water table, therefore, is the surface separating these two zones. In a very thin zone immediately above the water table, a so-called *capillary fringe* (see Fig. 14.2) occurs as groundwater rises upward as a result of surface tension (which binds together the surface molecules of the water) and the tendency of water to wet solid surfaces. We can demonstrate the capillary effect if we place soda

CHECK YOUR UNDERSTANDING

○ Why is groundwater classified as "freshwater"?

○ How does the water table form?

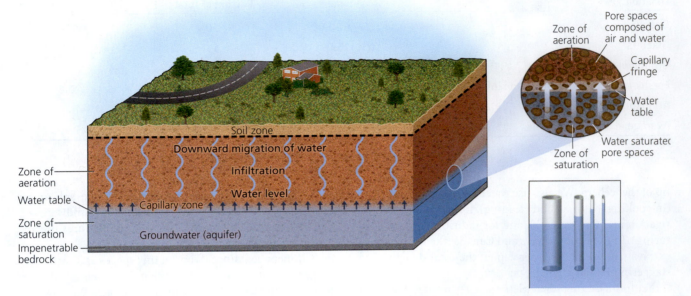

Figure 14.2 The Water Table

This figure illustrates the typical distribution of groundwater in the subsurface and the formation of the water table. The downward migration of water is eventually stopped by impenetrable bedrock. Above this barrier, a zone of saturation occurs. In the zone of saturation, water saturates any fractured bedrock or unconsolidated sediments by filling all the available pore spaces. Above this zone, pores in the zone of aeration are a mixture of air and water. The upper inset shows the capillary fringe where water rises a few centimeters upward into the zone of aeration because of surface tension. The lower inset shows the general principle of capillary rise related to adhesion of water onto surfaces. The smaller the diameter of the tube, the greater the capillary rise.

straws with different diameters into a glass of water. Because of surface tension, the capillary rise is greatest in the tube with the smallest diameter. Just as water rises in the soda straws, so the water climbs upward as bonds form between the water molecules and the solid surface of the sediment (see Science Refresher: **Surface Tension and Sand Castles** in Chapter 12).

POROSITY AND PERMEABILITY

Two factors—*porosity* and *permeability*—reflect how much water a subsurface sediment or rock can hold, and how easily groundwater can move through the sediment or rock. As water is absorbed into unconsolidated sediments, it can only occupy the spaces or pores between adjacent mineral grains. So the amount of water unconsolidated sediment can hold depends on the volume of space between the grains. As we discussed in Chapters 6 and 12, this volume of space is known as the **porosity**.

For example, where sediments are composed of rounded grains of roughly equal size (a condition known as *well-sorted*, as we discussed in Chapter 6), the grains are stacked together like billiard balls and large spaces exist between them (Fig. 14.3a). These sediments are said to have a high porosity. In contrast, sediments with grains of different sizes have severely reduced porosity because small grains may fill the pore spaces between the larger ones (Fig. 14.3b). Fine clays have a very high percentage of minute pores (high porosity), and can therefore hold large volumes of

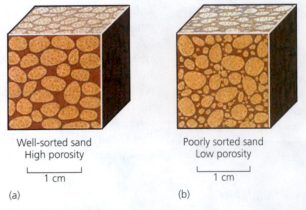

Well-sorted sand
High porosity

Poorly sorted sand
Low porosity

(a)   (b)

Figure 14.3 Porosity

A schematic diagram illustrates sediments with (a) high porosity and (b) low porosity. Well-rounded and well-sorted sand grains have high porosity, whereas poorly sorted sands have low porosity.

water when saturated. If clays are compacted to form shale rock, however, the porosity is greatly reduced (Table 14.2).

While porosity is a measure of the relative volume of pore spaces in a sedimentary material, recall from Chapters 6 and 12 that **permeability** is a measure of how well these pore spaces allow a fluid (in this case water) to flow through the material. Permeability is the crucial factor in determining the groundwater potential of a region because flow is essential if groundwater is to form a resource. Certainly many materials that are porous (such as well-sorted sandstone) are

TABLE 14.2   TYPICAL POROSITY VALUES FOR DIFFERENT MATERIALS

Material	Percentage Porosity
UNCONSOLIDATED SEDIMENT	
Soil	55
Gravel	20–40
Sand	25–50
Silt	35–50
Clay	50–70
ROCKS	
Sandstone	5–30
Shale	0–10
Solution activity in limestone, dolostone	10–30
Fractured basalt	5–40
Fractured granite	10

Source: US Geological Survey, Water Supply Paper 2220 (1983) and others.

also permeable, but there are also important exceptions. For good permeability, pores need to be a sufficient size and they need to be interconnected. Clays, for example, are highly porous, but they are not permeable because their pore sizes

> **CHECK YOUR UNDERSTANDING**
>
> ○ What type of material is porous but not permeable?
>
> ○ What type of material is permeable but not porous?

are too small. In contrast, limestone commonly contains large pores but can also be impermeable because the pores may not be connected.

There are also examples of permeable rocks with low porosity. Igneous rocks such as basalts or granites generally have very low porosity because their grains are interlocking. Yet, as we learned in Chapter 4, they are often systematically fractured (Fig. 14.4), which greatly increases permeability if the fractures are interconnected (see Table 14.2).

GROUNDWATER SUPPLY

Most groundwater occurs within 1 kilometer (4800 feet) of the surface. Surface water (streams, lakes, swamps, or springs) occurs where the water table intersects the surface. Groundwater supply is a delicate balance between **recharge** provided by precipitation and **discharge** onto the surface (Fig. 14.5a). **Springs** occur where the groundwater naturally discharges onto the surface. They are especially common in valleys within mountainous regions if the downward migration of groundwater is blocked by impermeable rock. In these situations, the

Figure 14.4   Fractured Igneous Rock
Systematically fractured igneous rock of high permeability but low porosity can be found at Land's End on the west coast of Cornwall, England.

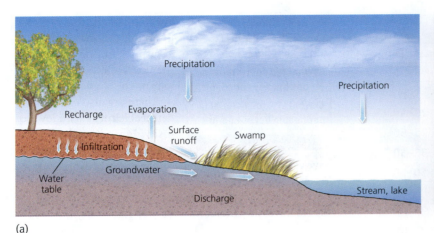

(a)

(b)

Figure 14.5   Water Pathways

(a) There are numerous pathways for water near Earth's surface. (b) A mountain spring flows in Co. Wicklow, Ireland.

groundwater flows laterally until it intersects the valley walls (Fig. 14.5b). Before the properties of the water table were understood, the origin of springs had no obvious explanation and they became the source of many myths and legends. Today we know that they are evidence of subsurface groundwater.

Changes in the balance between recharge and discharge are indicated by changes in the level of the water table. Because groundwater travels very slowly, a water table that is lowered by a period of extended drought or overuse may take years to return to its former level. For the groundwater to be recharged, precipitation must be able to penetrate the subsurface layers. This occurs most efficiently in regions with gentle slopes and light vegetation underlain by unconsolidated, porous and permeable sedimentary material or fractured bedrock. Too much recharge, however, may cause flooding. Once the subsurface becomes saturated, it cannot hold any more water so that any subsequent precipitation results in rapid surface runoff (excess discharge), swelling nearby streams.

Recharge is poor in regions where impermeable sediments or bedrock are exposed at the surface, where the topography is steep so that most precipitation becomes surface runoff, and where the vegetation is dense and impedes the fall of water, which may evaporate before it can reach the ground. Even regions with high rainfall may therefore have a rather limited supply of groundwater if they possess some or all of these characteristics.

> **CHECK YOUR UNDERSTANDING**
>
> ○ Why is the balance between discharge and recharge so important to the water table?

14.2 SUMMARY

- The downward passage of water is eventually blocked by a barrier of impermeable rock. Above this barrier, water accumulates to form the zone of saturation

where the pore spaces of the sediment are water filled. The water table is the upper surface of the zone of saturation. The zone of aeration occurs above the water table.

- Groundwater becomes a resource if the rock or sediment layer that contains it can store and transport water. In order to do so, the rock or sediment layer must be both porous (with a large volume of pore spaces between the sediment grains) and permeable (with pore spaces that are interconnected and large enough to allow water to flow).

- Groundwater supply depends on the balance between discharge and recharge. Groundwater becomes surface water where the water table intersects the surface and is discharged into streams, lakes, springs, or swamps. Groundwater is recharged when precipitation infiltrates the subsurface.

14.3 Aquifers and Groundwater Flow

A rock body or layer of unconsolidated sediment that can both store and transmit water below Earth's surface is an **aquifer**, from the Latin "to bear water." An aquifer is both porous and permeable, and is the most productive source of groundwater. Conversely, an **aquitard** is a rock body or sediment layer that does not readily transmit water but instead retards its motion. If an aquitard is impermeable, it is referred to as an **aquiclude**.

The flow of groundwater varies depending on the characteristics of the rocks, sediment, and soils through which the water passes. Because of frictional resistance, groundwater flow tends to be very slow, usually at rates of just a few meters per year. The highest flow rates measured in North America are less than 300 meters (1000 feet) per year and occur only in exceptionally permeable material. Because groundwater moves very slowly, groundwater forming today may take hundreds or even thousands of

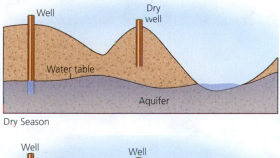

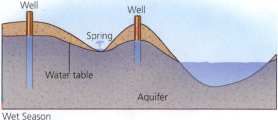

Figure 14.6   Wells and Water Table Levels

Wells drilled into an aquifer must penetrate below the water table. Note the lower level of the water table during the dry season. The water table feeds wells and springs, and changes in the level of the water table from the wet to the dry season affect the availability of water. The change shown here, for example, would cause the shallow well (on the right) to dry up during the dry season.

years to become surface water. Similarly, groundwater gushing from springs today may represent precipitation that occurred thousands of years ago. For example, the groundwater used to support the populations of the desert cities of the southwestern United States, such as Phoenix and Las Vegas, fell as rainfall in the neighboring mountains during the colder, wetter climate of the last Ice Age, more than 10,000 years ago. Eventually, however, most groundwater is gravity-fed into adjacent streams where it is transported to the sea.

In general, in order to extract groundwater, wells must be drilled into an aquifer. The wells must penetrate below the water table into the zone of saturation, so that changes in the level of the water table do not cause the well to dry up (Fig. 14.6). Ideally, wells should be deep enough to provide water even in times of drought. This is rarely the case, however, and long dry summers can cause havoc with well water supply.

There are two types of aquifers: *unconfined* and *confined*. The difference between them is a function of local geology and topography.

- **Unconfined aquifers** are those with upper surfaces that coincide with the water table (see Fig. 14.2).

- **Confined aquifers** occur where the water-bearing rocks or sediments are sandwiched between impermeable layers. This situation is common where well-sorted sandstones are interbedded between impermeable clay layers.

UNCONFINED AQUIFERS

Unconfined aquifers store vast quantities of water and are common in relatively flat regions. This is because many flat regions have long histories of meandering stream systems. Stream systems deposit well-sorted sands and gravels that are ideal for both storing and transmitting water. As a meandering stream develops, its channel migrates back and forth across the floodplain. The sand and gravel deposits therefore encompass the entire floodplain and may be very extensive (see Chapter 5).

Because Earth's surface is irregular, and because rocks and soils beneath the surface have varying capacities to absorb and store groundwater, the depth of the water table varies from place to place. In regions with appreciable rainfall, the water table crudely mirrors the surface topography (Fig. 14.7). Variations in the elevation of the water table create a **hydraulic head**, which reflects the difference in elevation between adjacent zones of high and low pressure. This difference in pressure results in the flow of groundwater "downhill" under the influence of gravity toward the valleys (see Science Refresher: **Water Pressure at Depth** on page 444). Where the water table intersects Earth's surface to form a spring, groundwater discharges into streams, lakes, and swamps and so becomes surface water. This natural flow of groundwater under the influence of a hydraulic head is an important link between groundwater and surface water.

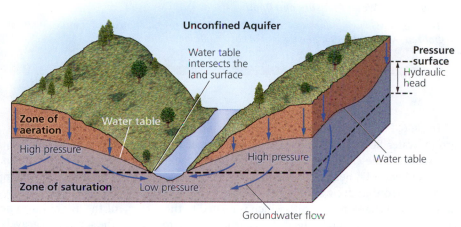

Figure 14.7   Groundwater Flow

Groundwater flow moves downward under the influence of gravity through the zone of aeration into the zone of saturation. Due to its hydraulic head, most water then migrates from high-pressure to low-pressure regions in the zone of saturation and may become surface water at springs, and drain into lakes, streams, or swamps, where the water table intersects the land surface.

SCIENCE REFRESHER

Water Pressure at Depth

Water exists on Earth's surfaces as a solid (in glaciers), a liquid (in oceans, streams and lakes) and as a gas (in the atmosphere). But how does water behave at depth in the crust? Recall from Chapter 2 that the pressure and temperature of the crust increase with depth. This greatly affects the composition and flow of groundwater.

As deep-sea divers will vouch, water pressure increases rapidly with depth. In fact, the pressure on water depends only on its depth. This means that the pressure on water at the bottom of each of the vessels in Fig. 14A is the same. A simple experiment demonstrates how water pressure increases with depth (Fig. 14B). Fill a Styrofoam cup with water, and cut holes in it with a nail at four different depths. Water will flow from each hole but notice how different each flow is. The water flows feebly from the top hole. However, with increasing depth, longer jets of water shoot out of the holes.

This simple experiment explains why the water pressure in the showers of university dormitories is much greater on the ground floor than it is on the top floor, and why, in any municipal water system, houses at the bottom of valleys have higher water pressure than those on the tops of hills. Taps or water outlets at higher elevations than the pressure surface must have water pumped to them. Many municipal water systems create an artificial hydraulic head by storing

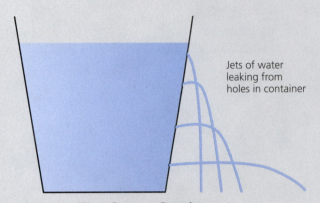

Jets of water leaking from holes in container

Figure 14B   Water Pressure Experiment
This simple experiment demonstrates how water pressure (and hence the strength of the jet of water) increases with depth.

the town water in elevated reservoirs or water towers so that water flows freely between the reservoir and the dwellings or commercial buildings of the town.

The temperature of water also increases with depth, as the water absorbs Earth's internal heat. As we discussed in Chapter 4, water has a tremendous capacity for absorbing heat so the water temperature increases relatively slowly with depth, about 2.5°C (4.50°F) for every 100 meters (330 feet) depth, although this rise can be considerably greater in volcanically active regions. The ability of water to store this heat is key to its use as a source of geothermal energy, which we discuss later in this section of the chapter and which we will address in greater detail in Chapter 19.

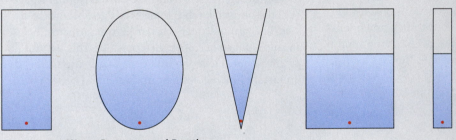

Figure 14A   Water Pressure and Depth
The pressure in water depends only on its depth. The pressure on a water molecule (symbolized by the red dot) at the bottom of each container is the same.

In 1856, French engineer Henri Darcy reasoned that the groundwater flow rate between two wells is related to the difference in the elevation of the water table in each well (i.e., the hydraulic head) and the distance between them. Darcy's reasoning drew an analogy between groundwater flow rate and the slope of a hill, which is related to the difference in elevation and to the distance from the top to the bottom. The shorter the distance between the top and bottom of a hill, the steeper the slope. Groundwater flow rates are generally greater in regions where significant variations in water table levels occur over short distances.

The High Plains (or Ogallala) aquifer extends from southern Wyoming, Nebraska, and South Dakota in the north to New Mexico and Texas in the south (Fig. 14.8) and is an important example of an unconfined aquifer. Consisting of sands, gravels, and clays, it provides water for much of the agricultural heartland of the central United States. The sands and gravels slope gently to the east away from the Rocky Mountains and the aquifer is recharged by precipitation falling in the mountains, which then flows eastward beneath the surface, within these permeable sediments. Precipitation in the region itself also makes a significant

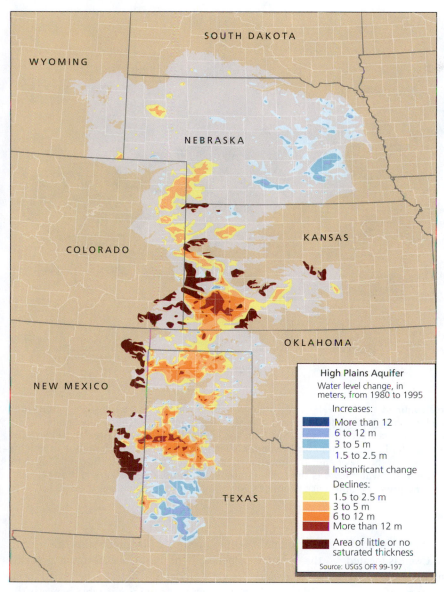

Figure 14.8   High Plains Aquifer

This map of the High Plains aquifer of the midcontinental United States shows changes in the level of the water table between 1980 and 1990. Although in some regions the water table rose during that time period, in most regions it dropped significantly. The Rocky Mountains are located to the west of the area shown in the map.

irrigation in the United States. The withdrawal rates are 10–50 times the rates of recharge. As a consequence, the thickness of the aquifer has changed. These changes are most easily seen as variations in the level of the water table in Figure 14.8. In some areas, this level has dropped by about 60 meters (200 feet) since 1980, and for most of the aquifer, the water table shows moderate to severe decline. Only a few areas show moderate to significant rise. Studies of the High Plains aquifer demonstrate the importance of monitoring in the long-range management of water resources.

CONFINED AQUIFERS

Confined aquifers (see Living on Earth: **The Historical Importance of Aquifers** on page 447) form when water flows into a permeable, porous layer of sediment that is sandwiched between layers of low permeability (Fig. 14.9). The water in confined aquifers is commonly supplied by precipitation in an adjacent mountainous region, which is known as the *recharge area*. If the layers are tilted away from the mountains, gravity drives the water flow downslope within the permeable layer and beneath adjacent valleys (see Science Refresher: **Water Pressure at Depth** on page 444).

If the confined aquifer is tapped by a well in the valley, the pressure difference causes the water to rise upward. If a large number of wells are drilled, the levels to which the water rises in each well collectively define an imaginary surface, known as a **pressure surface** (or *potentiometric surface*). This surface is broadly analogous to the horizontal water surface generated when water flows between two interconnected tanks (inset, Fig. 14.9). In a confined aquifer, however, the pressure surface is gently inclined away from the recharge area because friction causes the water to lose energy as it flows downhill between the sediment grains. This means that a well drilled near the recharge area intersects the pressure surface at a higher elevation than a well drilled farther away.

In most cases, the pressure surface in confined aquifers

contribution, typically ranging from 35 to 55 centimeters (14 to 22 inches) per year. However, in times of drought this contribution is severely reduced.

The depth to the water table, which forms the upper surface of the High Plains aquifer, varies locally from 30 meters to 120 meters (100 feet to 400 feet). The thickness of the aquifer itself is also highly variable, ranging from about 1 meter to more than 300 meters (1000 feet). The exploitation of this vast groundwater resource was triggered by the disastrous "Dust Bowl" droughts of the 1930s and late 1940s. Today it is mined by nearly 200,000 wells that together extract 30 percent of all groundwater used for

CHECK YOUR UNDERSTANDING

○ Compare and contrast a confined aquifer and an unconfined aquifer.

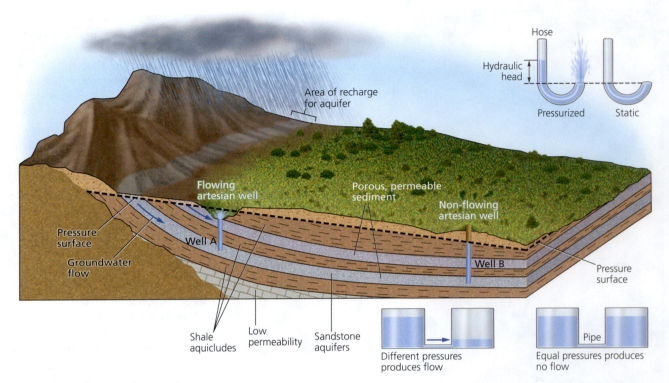

Figure 14.9   Confined Aquifer

Flowing and nonflowing wells are drilled into a confined aquifer. Water flows preferentially downslope in a permeable, porous rock or sediment that is confined above and below by a shale aquiclude. There must be sufficient rainfall to recharge the aquifer. Wells located at a lower elevation than that of the pressure surface (below dashed line, see well A) flow freely as artesian wells. Wells above the elevation are nonflowing (above the dashed line, see well B). The upper inset shows an analogous situation in which a "hydraulic head" in a hose causes water to flow out at the lower end. The lower inset shows examples of water flow influenced by pressure in two containers connected by a pipe. If the water level is the same in both containers, the pressure is the same and there is no flow. If the water level is different, the pressure difference produces flow until the water levels are the same in both containers.

is below the ground so water in a well drilled into the aquifer never reaches the surface. However, if a well is located on ground that is below the pressure surface, the water gushes out to form an **artesian well**. This type of well is named after the French village of Artois, which has had a supply of free-flowing water for centuries.

SPRINGS AND OASES: EVIDENCE OF SUBSURFACE GROUNDWATER

We have seen that springs occur where the water table intersects the surface, particularly in mountainous regions (see Fig. 14.5b). However, springs also occur where water at depth within the aquifer is forced upward to the surface along fractures. The water flows upward due to the hydraulic head in the aquifer, and reaches the surface if the surface is at a lower elevation than the pressure surface.

An **oasis** (Fig. 14.10) is the result of a spring that occurs in a desert region (see Living on Earth: **Water in the Sahara Desert** on page 450). In the Palm Springs region of southern California, for example, several oases are located along the edge of the Mojave Desert in a basin bounded on both sides by high mountains. A series of faults, which includes the San Andreas, carve their way through the basin.

Precipitation falling on the nearby mountains drains into the subsurface beneath the basin and forms a precious groundwater resource under the desert. The pronounced difference in the elevation of the water table beneath the mountains and its level below the desert lowlands forms a very strong hydraulic head. Because the desert surface is well below the pressure surface, the pressure difference forces the water up through fractures and faults beneath the desert to the surface. The resulting oases provide vital nourishment to the parched vegetation of the Mojave Desert.

Occasionally, the temperature of spring water is anomalously warm, as is the case in parts of the Palm Springs region. Where this occurs, the water is a potential source of geothermal energy. In the following section we discuss how groundwater becomes heated and examine several of the spectacular features that occur as a result.

HEATED GROUNDWATER: HOT SPRINGS, FUMAROLES, AND GEYSERS

At depths below about 20 meters (66 feet), water is generally unaffected by weather and climate at the surface. Instead the temperature of water is influenced by the heat inside Earth. As miners who work deep underground can

The Historical Importance of Aquifers

The London Basin (Fig. 14C) contains a famous example of a confined aquifer. This aquifer is nearly 100 km (62 miles) across and underlies much of southeastern England, including the city of London. The aquifer is in a layer of chalk, which is both permeable and porous, that is sandwiched between impermeable clay layers. These strata were deposited between 65 and 95 million years ago on top of older Jurassic and Lower Paleozoic rocks. In the Chiltern Hills to the north and west, and the North Downs to the south and east, the upper clay layer has been eroded away, exposing the chalk. Rainfall in these areas sinks into the chalk and migrates to the center of the basin.

The supply of water from this aquifer has been of great historical significance in the development of London as a major world city. The needs of the city, however, have taken their toll. The demand for water has exceeded the aquifer's replenishment, even in a region famous for its rain. When the fountains in Trafalgar Square were first constructed in the mid-nineteenth century, water tapped from the chalk aquifer gushed freely at the surface to form artesian wells. Today, the water level in the chalk has fallen by over 130 meters (450 feet). Because the chalk is no longer saturated, the aquifer is no longer artesian and water must now be pumped from depth if it is to reach the surface. Freshwater used to percolate from the chalk into London's Thames River. However, depletion of the aquifer has caused the flow to reverse itself. Today the Thames water seeps into the aquifer and contaminates it.

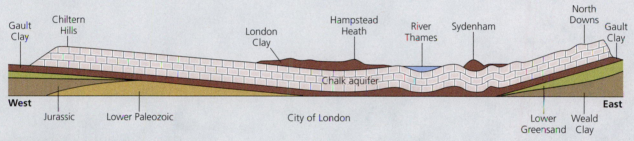

Figure 14C   London Basin

A cross section, about 100 kilometers (62 miles) across, showing the confined chalk aquifer beneath the London Basin. Precipitation in the Chiltern Hills and North Downs is absorbed by the layer of chalk and flows as groundwater beneath the London Basin. The aquifer is confined by impermeable clay layers.

verify, temperature increases with depth—a fact that reflects the flow of heat from the interior of Earth to its surface. On average, temperature increases about 2.5°C for every 100 meters (330 feet) depth, although this rise can be considerably greater in volcanically active regions. As we discussed in Chapter 4, water has a tremendous capacity for absorbing this heat.

Occasionally, it is possible for groundwater to penetrate the crust far below the regional water table. As water seeps down fractures or faults, it absorbs heat from the rocks around it (Fig. 14.11a). If this water is then able to rise rapidly to the surface, it may retain some of this heat and its temperature is significantly higher than that of surface water. As a result, water derived from deep wells is routinely warmer than ambient surface waters. If the water reaches the surface naturally, a **hot spring** will form. It is not surprising, therefore, that hot springs are abundant in tectonically active areas adjacent to deeply penetrating faults and fractures.

Hot springs are especially abundant in volcanically active regions, where magma or very hot rocks are close to the surface. Indeed the heated groundwater may be converted to steam, which can escape to the surface through fissures to form steam vents known as **fumaroles** (Fig. 14.11b,c). An area famous for its fumaroles is the Valley of Ten Thousand Smokes in Alaska.

In some situations, superheated groundwater may persist as liquid rather than turning to steam because of the confining pressure of the rocks. This situation resembles a pressure cooker in which the uppermost crust of Earth acts like the lid of the cooker. Just as steam escapes through the valve on the lid of a pressure cooker, superheated water endeavors to escape through any flaw it can exploit in the overlying crust. Such flaws are abundant because most crust is scarred by faults, caused by either current or ancient tectonic activity. Eventually the superheated water finds an avenue of escape and forces its way upward. As it does so, its pressure is reduced, allowing the water to turn into steam. This conversion results in rapid expansion because the steam occupies more space than the water did. The expansion creates a violent subsurface explosion that expels the water toward the surface where it forms a gushing spring

(a)

(b)

Figure 14.10   Oasis

(a) An oasis (center) exists in the desert of Morocco. Groundwater migrating underground from the mountainous areas (background) seeps up along fractures. The oasis occupies a narrow zone above these fractures. (b) Lush vegetation grows in an oasis in the Grand Staircase/Escalante National Monument, Utah.

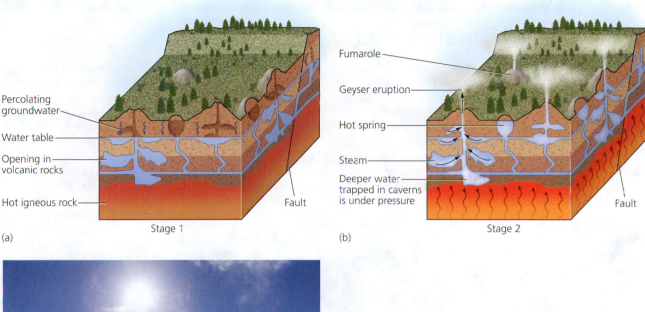

(a) Stage 1

Percolating groundwater

Water table

Opening in volcanic rocks

Hot igneous rock

Fault

(b) Stage 2

Fumarole

Geyser eruption

Hot spring

Steam

Deeper water trapped in caverns is under pressure

Fault

(c)

Figure 14.11   Heating and Pressurization of Groundwater

(a) Water penetrating into the deep subsurface (by way of faults or fractures) is warmed by the rocks beneath, especially in volcanically active areas where hot rocks lie close to the surface. Deeper water trapped in caverns is under pressure. (b) When the water temperature reaches its boiling point, a sudden drop in pressure causes the water to turn into steam, which rises to the surface to form geysers. (c) Steam may be continuously emitted from the region around a volcanic vent, as seen here in Volcanoes National Park, Hawaii.

of hot water and steam (see Fig. 14.11). After each eruption, the vacated space in the subsurface is filled with more groundwater and the process repeats itself in a cyclic fashion. The result is a **geyser** (from the Icelandic word for "gusher").

The most famous geyser in North America is Old Faithful in Yellowstone National Park, Wyoming, which produces a natural spectacle that attracts thousands of tourists each year (Fig. 14.12). The region where Old Faithful is located has the highest heat flow in continental North America, which is testament to its recent and prolonged volcanic history. This history is believed to reflect the presence of a hotspot directly beneath the Yellowstone region. Fortunately, its volcanic eruptions are far less frequent than those of its geysers. On average, the

Yellowstone hotspot produces major volcanic eruptions every 700,000 years. The time between eruptions of Old Faithful, on the other hand, varies from 35 to 80 minutes but the eruptions are typically about 66 minutes apart. The height of the geyser is usually around 45 meters (150 feet).

GEOTHERMAL ENERGY

Heated groundwater is a potential source of geothermal energy. In Iceland, where the Mid-Atlantic Ridge stands above sea level, more than 50 wells tap the geothermal energy produced from groundwater heated by local magmatism. These wells provide more than 50 percent of the island's energy needs and almost 90 percent of its heating and hot water requirements, including all-year-round heated outdoor public swimming pools! Geothermal energy stations near Vesuvius, the famous active volcano

CHECK YOUR UNDERSTANDING

● How does a geyser form?

LIVING ON EARTH

Water in the Sahara Desert

Aquifers can occur in some very surprising locations. Because underground water can travel hundreds of kilometers from its source, water supplies are sometimes located beneath the surface of very dry regions. In 1981, the space shuttle Columbia mapped the eastern Sahara desert using *radar*, which is a form of microwave radiation that can penetrate clouds and dry desert sands but cannot penetrate moist sediments. The radar map reveals an image of ancient topography beneath the desert sands characterized by wide river valleys flanked by rocky hills (Fig. 14D). This image is a powerful illustration of how climate changes and is thought to reflect conditions during the last ice age (approximately 10,000 to 70,000 years ago) when Earth's average temperature was cooler than today. Many regions that experience desert conditions today then had more hospitable climates with significantly higher rainfall.

The ancient drainage system was fed by precipitation, primarily in the adjacent highlands in Libya, and developed on top of a very porous and permeable type of rock known as the Nubian sandstone. This sandstone, which occurs beneath most of the Sahara Desert, is about 900 meters (3000 feet) thick and stores enormous quantities of groundwater, estimated to be in excess of 18,000 cubic kilometers (4300 cubic miles). The antiquity of this water has been confirmed by radiocarbon dating (see Chapter 8). These analyses show that the groundwater is about 35,000 years old, indicating that the water actually fell as precipitation during the last Ice Age.

Locally, the Nubian sandstone is folded upward into anticlines (see Chapter 10) and reaches the surface (Fig. 14E). Where this happens, the water seeps out at the surface to form an oasis, a local spring of artesian water in the desert. In some instances, the sandstone layer is offset by a fault. Like a leak in a plumbing system, water escapes from the sandstone layer and rises up the fault to form an oasis at the surface.

The Nubian aquifer is a major source of water in western Egypt and Libya. However as large as this resource is, it is no longer being renewed by modern rainfall. Its water is therefore a nonrenewable resource and care must be taken to ensure that it is well managed and used efficiently.

Figure 14D   Ancient Rivers in the Sahara

Satellite image (left) of the Safsaf Oasis in the eastern Sahara desert provides few clues as to the origin of the oasis. But a radar image (right), taken by the space shuttle Columbia in 1981, reveals the rock layer beneath the desert sand, unveiling an ancient topography with wide black channels cut by the meandering of an ancient river that once fed the oasis.

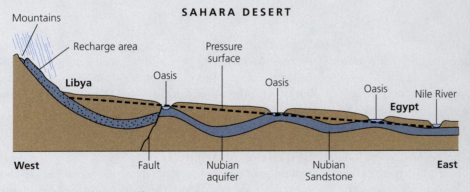

Figure 14E   Nubian Aquifer

A cross section beneath the Sahara Desert from the mountains of Libya in the west to the Nile River in the east unmasks the Nubian aquifer, which extends beneath the Sahara Desert.

Figure 14.12   Geyser

Old Faithful geyser in Yellowstone National Park, Wyoming, attests to the recent and prolonged volcanic history of the region where the park is located.

in Italy, provide energy equivalent to that produced by four coal-fired power plants.

In the southwestern United States, anomalously hot rocks occur within a few kilometers of the surface because of modern tectonic activity, and provide an important regional source of geothermal energy. But it is northern California, with its high regional heat flow, that boasts the largest area of geothermal development in the world. According to some estimates, this region has the potential to supply 25 percent of California's energy needs.

14.3 SUMMARY

- Unconfined aquifers are aquifers whose upper surface coincides with the water table. Confined aquifers occur where the water-bearing rocks or sediments are sandwiched between impermeable layers.

- Groundwater generally flows "downhill" toward valleys at a few meters per year under the influence of gravity from regions of high pressure to those of low pressure.

- Groundwater becomes surface water where the water table intersects the surface. The intersection of the

water table with the surface commonly generates springs. Springs also occur where water is forced up to the surface along fractures. An oasis is the result of a spring that occurs in desert regions.

- Water forced up to the surface is generally anomalously warm. Deep penetrating groundwater may absorb Earth's internal heat and so become a source of geothermal energy and hot springs. In volcanically active regions, groundwater can be converted to steam, which escapes to the surface to form fumaroles. Cyclic emissions of superheated groundwater are known as geysers.

14.4 The Dissolving Power of Groundwater

Water's property as a solvent enables groundwater to dissolve soluble material from rocks and unconsolidated sediments. This form of chemical weathering (see Chapter 5) is influenced by climate and is most extensive in humid, tropical regions, where there is abundant rainfall compared to polar regions, where permafrost hinders groundwater infiltration. The composition of groundwater is also strongly influenced by the subsurface materials it encounters and the ease with which the materials dissolve. Minerals have different susceptibilities to weathering and do not dissolve at the same rate (Table 14.3). Some minerals, such as quartz and muscovite, have strong bonds and are highly resistant to chemical weathering. Other silicates, such as plagioclase,

TABLE 14.3   MINERAL STABILITY

Resistance to weathering of common rock-forming minerals

Halite	
Gypsum-anhydrite	
Pyrite	Nonsilicates
Calcite	
Dolomite	
Volcanic glass	
Olivine	
Ca-plagioclase	
Pyroxene	Increasing stability
Ca-Na plagioclase	
Amphibole	Silicates
Na- plagioclase	
Biotite	
K-feldspar	
Muscovite	
Quartz	
Kaolinite (clay material)	
Hematite	Oxides

have relatively weak bonds and readily break down into clay minerals by releasing elements such as sodium and calcium into solution. Carbonate minerals, such as calcite and dolomite, which are typically rich in calcium, magnesium, and iron, are highly soluble and are almost entirely removed in solution. Not surprisingly, laboratory analyses show that groundwater tends to be enriched in these soluble elements (e.g., sodium, calcium, magnesium, and iron). Because most groundwater eventually finds its way to the sea, the process of chemical weathering has been a prime source of these elements in ocean water throughout most of geologic time.

Local bedrock has a profound influence on water chemistry. Much of the central United States and eastern Canada, for example, are underlain by a bedrock of limestone or gypsum, both rich in calcium. Water that contains large quantities of bedrock-derived dissolved ions (such as calcium, magnesium, and iron) is known as **hard water**. The presence of dissolved ions can pose problems for humans. In many domestic appliances and industrial machines, for example, dissolved calcium forms a scum residue as water that has been heated up cools. In regions underlain by other highly soluble materials, such as salt, the water is rarely suitable for drinking.

Elevated temperatures greatly enhance water's properties as a solvent. As a result, many hot springs have high concentrations of dissolved ions derived from soluble minerals. In many cultures, bathing in water with such elevated mineral contents is thought to be healthy, and regions where such waters occur commonly have health resorts known as *spas*. Hot springs have been used for bathing since the time of the Qin dynasty in China (third century BC), and in the first century AD, the Romans opened the hot springs of Bath, England, for public bathing and underground heating.

KARST TOPOGRAPHY

The dissolving power of water can produce caves and caverns in the subsurface and can have a profound impact on the landscape and the drainage of surface water. The features produced when significant amounts of bedrock are dissolved are collectively known as **karst** (from the Slavic, *kars,* meaning "bleak, waterless place"). The surface expression of a karst—**karst topography**—is characterized by a very irregular landscape that includes broad plains with tall *monoliths,* circular depressions called *sinkholes,* and disappearing streams. Beneath a karst topography, underground cavities, or caves, typically develop.

Karsts develop in regions with abundant precipitation that contain soluble bedrock. Gypsum, rock salt, and limestone, which are dominated by the nonsilicate minerals gypsum, halite, and calcite, respectively, are especially soluble because the water that passes through them dissolves these soluble minerals. In most cases, the dissolving agent is a combination of surface water and groundwater. But in rare instances it can be briny water trapped within the rock itself or hydrothermal water injected into these rocks from below. Because limestone is the more common rock type, karsts most commonly develop in areas where limestone is either exposed at the surface, or in the shallow subsurface.

A number of factors influence the rate at which limestone dissolves, including the degree of fracturing, climate, and topography. Fractures allow surface water to infiltrate the ground, permit groundwater to travel more efficiently, and provide a greater surface area on which reactions can take place. Climate controls the amount of precipitation and, hence, the amount of surface water and groundwater. Climate also influences the composition of the soil. Fertile organic-rich soils increase the acidity of groundwater and, hence, its dissolving power.

Karst topography primarily occurs in humid, tropical regions such as southeastern Asia and Central America where there is abundant rainfall and lush vegetation. But they can also occur in humid, temperate climates at mid-latitudes, such as the eastern United States and southern Europe (Fig. 14.13). Karst topography is developed on some 15 percent of the land surface in the United States, reflecting the country's abundance of limestone bedrock. It is most common in Alabama, Florida, Indiana, Iowa, Missouri, Texas, and Virginia. In regions of low rainfall, like those near the poles

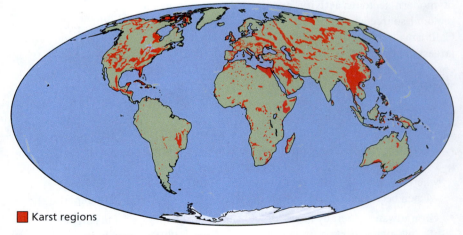

■ Karst regions

Figure 14.13   Global Distribution of Karst

The distribution of karst landforms is illustrated on this world map. Note the abundance of karst in tropical and temperate climates and the virtual absence of karst in polar and equatorial climates.

Figure 14.14   Karst Landforms

Typical landforms of karst topography include monoliths, which are composed of relatively insoluble rock and stand out from the surrounding landscape like these in Guilin, China.

scientists, tourists, and cave explorers ("spelunkers") for centuries. But how do these features form?

These subterranean caves form in a variety of ways depending on the source of the water. The most common form of cave develops when surface water infiltrates into subsurface limestone through its many fractures, and slowly dissolves the rock to make ever wider channels and passageways (Fig. 14.18). Wider passageways increase the amount of water that can infiltrate, thereby accelerating the process. Once water reaches the water table, it flows sideways as groundwater, expanding the openings still further as it continues to dissolve away the limestone. If the water table then drops, these passageways are left stranded above it and form an intricate system of interconnected caves and caverns. However, drops of groundwater continue to seep into the caves through fractures in the limestone roof. As the water evaporates from the ceilings, small deposits of calcite (from the dissolved calcium) form. With each drop, a thin layer of calcite is deposited around the preceding one. The layering creates a column of calcite known as a **stalactite** projecting downward from the roof (see Fig. 14.18). During this entire process, an opening is maintained in the center of the stalactite, such that the stalactite has a hollow tube in its center, known as a **soda straw**. As drops fall onto the floor of the cave, layers of calcite form stubby mounds that grow upward from the ground to form **stalagmites**. In caves, a downward growing stalactite may eventually connect with an upward growing stalagmite to form a **column**, an important structure that supports the roof of the cave.

One of the most famous caves of this type is Mammoth Cave in Kentucky, which consists of a labyrinth of passageways more than 630 kilometers (390 miles) in length. The limestone bedrock, deposited in shallow seas some 340 million years ago, has dissolved at various times since then. When surface streams incised through the limestone about 2 million years ago, the regional water table dropped and the labyrinth of passageways drained to leave empty caves and caverns.

Not all caves are formed by downward-migrating water. Some develop as soluble rock is dissolved by upward-migrating solutions. The best example of this type of cave in North America is the Carlsbad Caverns of New Mexico. The bedrock here is a limestone reef that was deposited about 250 million years ago in a shallow, semitropical inland sea. Because evaporation of the sea resulted in sulfur-rich water, the reefs contained hydrogen sulfide (H_2S) compounds. These compounds reacted with oxygen to produce sulfuric acid (H_2SO_4), which then migrated

or in deserts, limestone dissolution is rare and karst development is virtually absent.

Karst topography is characterized by a wide variety of landforms. **Monoliths** (Fig. 14.14), composed of relatively insoluble and isolated rock, stand out above broad plains underlain by soluble rocks. **Sinkholes** are circular depressions formed either directly by the dissolving process itself, or indirectly by collapse of the cave's roof under the weight of the overlying rocks (Fig. 14.15). A *solution sinkhole* forms when groundwater dissolves soluble bedrock at or near the surface. Because gypsum and salt are much more soluble than limestone, solution sinkholes in bedrock made up of these minerals can develop in a matter of days (Fig. 14.16a). A *collapse sinkhole* occurs when the roof of a cave can no longer support the weight of the rocks above it and collapses. If the cave expands or the water table is lowered the support for the overlying rock is reduced. These are the main causes of collapse sinkholes. A stream may encounter a sinkhole and plunge into it to form a **disappearing stream**, leaving a dry valley downstream (Fig. 14.16b). After flowing underground, the water may eventually return to the surface as a spring.

CHECK YOUR UNDERSTANDING

● Why do karsts most commonly develop in regions underlain by limestone?

● Explain the origin of monoliths.

CAVES

Subterranean limestone caves are perhaps the best known of all karst formations (Fig. 14.17). The bewildering labyrinths of caverns and passageways, and remarkable calcite columns known as *stalactites* and *stalagmites,* have attracted

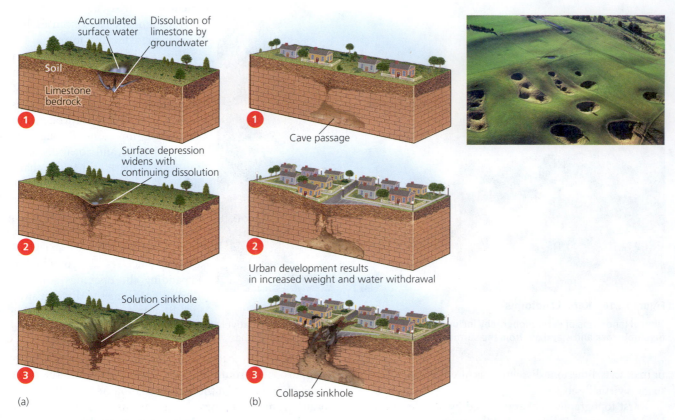

Figure 14.15   Sinkhole Formation

Sinkholes form by (a) dissolution of the soluble host rock (solution sinkhole), or (b) collapse of the roof of a cave (collapse sinkhole). Photo shows solution sinkholes, implying the existence of soluble rock in the subsurface.

Figure 14.16   A Sinkhole and a Disappearing Stream

(a) Solution sinkholes, such as this one in Florida, can form in a matter of days. (b) A disappearing stream occurs when a stream encounters, and plunges into, a sinkhole.

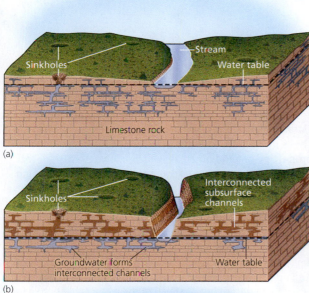

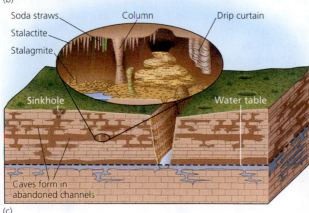

(a)

(b)

(c)

Figure 14.17   Cave Formation
(a) Groundwater migrating though the zone of aeration dissolves limestone rock. (b) Groundwater moves along the water table to form interconnected subsurface channels as dissolved rock is transported to streams on the surface. (c) As the stream continues to erode and the water table drops, caves form in the abandoned channels.

upward through the fractures to dissolve the limestone and initiate cave development. Over the past 2 million years, however, the region has periodically experienced wetter climates associated with the waxing and waning of polar ice sheets. During these wetter periods, the relatively abundant supply of surface water resulted in downward-migrating solutions that enhanced cave development.

We have now learned that groundwater is a vital source of freshwater. We have also looked at the factors that influence its flow, its ability to store heat and dissolve minerals, and some of the features that develop when groundwater interacts with soluble rocks. But our insatiable demand for

(a)

(b)

Figure 14.18   Stalactites, Stalagmites, and Sinkholes
(a) Stalactites project downward from the cave roof and stalagmites grow upward from the cave floor at Lehman Caves in Nevada. (b) A circular sinkhole was formed by the dissolving of gypsum bedrock in Eddy County, New Mexico.

freshwater creates a continuing need for this vital resource (see In Depth: **Exploring for Groundwater** on page 456), and it is to the exploitation of groundwater that we turn in Section 14.5.

IN DEPTH

Exploring for Groundwater

In the practice of dowsing, a "water witch" walks back and forth across an area holding a Y-shaped stick until the bottom part of the Y jerks downward. According to the water witch, underground water will be found directly below the location the witch's stick identifies. Although the technique has been heavily criticized by scientists for more than 200 years, there are over 25,000 practicing water witches in North America today. Scientists cannot find a valid reason for the behavior of the Y-shaped stick, and statistics show that the success rate of the method is no better than chance.

Most scientific exploration for groundwater focuses on finding the level of the water table. The first step is to examine topographic maps, which show the distribution of streams, lakes, hills, and valleys. Streams and lakes occur where the water table intersects the surface, so their positions provide important clues to groundwater distribution. But how do we use these clues to find water in the subsurface between locations?

Hydrogeologists concern themselves with the quantity and quality of water resources, and where water is located. As a first step, hydrogeologists determine the sedimentary sequence, also known as the *stratigraphy* of the region, to identify permeable and porous sediments. It is also important to locate impermeable sediments, such as clay, so that they can evaluate the nature of the potential aquifers (confined or unconfined). The region's stratigraphy may be exposed in road cuts, on hillsides, and along the banks of streams where erosion has cut down into older sediment layers. By piecing all this information together, hydrogeologists compile the stratigraphy of the region using the same methods of stratigraphic correlation discussed in Chapter 8.

In many regions, a series of exploratory *boreholes* are drilled in order to check the stratigraphy and accurately determine the level of the water table. Sensitive instruments are then lowered down these boreholes in order to measure critical physical properties (such as the electrical characteristics of the rocks and water) and identify the telltale signs of water saturation.

As the subsurface cannot be viewed directly, the most commonly used techniques attempt indirectly to create a picture, or an image, of the subsurface indirectly. Just as an X-ray image gives information on bone structure and body tissue, an electrical image of the subsurface can identify potential aquifers. Because pure freshwater is an extremely poor conductor of electricity, it is highly resistant to the passage of an electrical current. This property is analogous to that of bone, which blocks the passage of X-rays and enables physicians to view breaks in our bones on an X-ray image. In a similar way, the high resistance of freshwater makes its presence obvious in electrical images. These images are also sensitive to the differing electrical properties of sand and clay, which allows them to detect variations in porosity and permeability. They can also provide information on pore-water composition so that water quality may be assessed. For example, poor quality water that may contain a high concentration of salts is much more conductive (less resistant) than freshwater.

Borehole exploration, although useful, provides only limited information because it depends on the number of wells drilled, and the confidence with which the hydrogeologist can determine the level of the water table between well sites. Another method, known as *magnetotellurics*, has better success in pinpointing both the depth and extent of aquifers by providing a more complete electrical image of the subsurface. The magnetotelluric method utilizes natural electrical currents in the outer atmosphere (recall our discussions on Earth's magnetism in Chapter 2), which penetrate Earth's surface and interact with Earth materials. The currents induce electrical fields in Earth materials, which conduct electricity to varying degrees, depending on the nature of the material. Sensitive equipment can measure this interaction (Fig. 14F) and produce a regional electrical image that provides a picture of the subsurface.

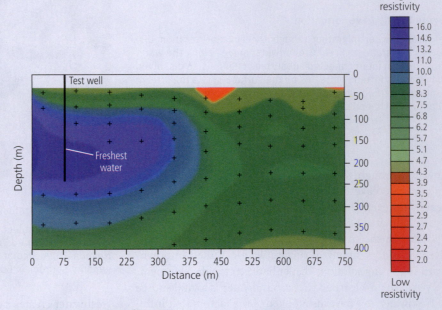

Figure 14F   Magnetotelluric Image

In this electrical image of the subsurface in Merced County, California, the freshest water occurs in highly resistive zones (blue) shown on the left of the image.

Magnetotelluric surveys have been successful in California, which, because of its dry climate and high population, has a particularly acute water problem. Figure 14F is an example of its application in Merced County, California, where the subsurface consists of sandstones and impermeable clay layers. A subsurface image produced by the magnetotelluric survey identifies potential water-bearing zones as regions of relatively low electrical conductivity (or high resistivity) (on the left in Fig. 14F). The image in Figure 14F shows that the water-rich sandstone occurs in a lozenge-shaped lens. Because of this shape, hydrogeologists interpret the sandstone to have been deposited in former stream channels. This particular image aided the selection of drill targets in Merced County that ultimately confirmed the predictions of the image. A test well drilled at a distance of 75 m obtained excellent freshwater from the highly resistant regions at depths between 100 and 150 meters (330 to 500 feet).

Given their incomplete nature, it is unlikely that borehole surveys could have detected these water-rich lenses. Many hydrogeological surveys now use magnetotellurics to constrain favorable target zones, and then follow the survey with borehole drilling directed at those targets.

14.4 SUMMARY

- Because water is a strong solvent, it is the primary agent of chemical weathering and the composition of groundwater is influenced by the extent of chemical weathering. This, in turn, is related to mineral stability. In general, minerals with strong bonds in their crystal structure (such as quartz) are relatively resistant, whereas those containing weak bonds (such as plagioclase) are more reactive.

- Karsts are features produced when groundwater dissolves bedrock. Gypsum, rock salt, and limestone are very soluble, and their dissolution produces surface features, collectively known as karst topography, that include monoliths, sinkholes and disappearing streams, and subsurface features such as caves.

- Caves consist of a labyrinth of subsurface passageways and are primarily formed when dissolution of soluble rock is followed by a drop in the water table.

- Columns of calcite produced by dripping water on the roof and floor of caves are stalactites and stalagmites, respectively.

14.5 Exploitation of Groundwater

When exploration locates a usable aquifer, a series of wells are sunk below the water table and the water is pumped out. When you suck water from a glass through a straw, a cone-shaped depression forms in the water surface around the straw. This depression only recovers when you stop sucking on the straw, by which time the water level in the glass is lower. In a similar way, a cone-shaped depression forms around a water well that levels out again only after pumping of the well stops. But if the amount of water withdrawn exceeds the recharge, the depression levels out at a lower elevation.

Communities almost always withdraw more water from the ground than nature is able to replenish. As a result, a **cone of depression** typically forms around each well (Fig. 14.19), which reflects the diminishing water supply. Eventually, the water table is lowered below the well and the well goes dry. Because groundwater flow is so slow, the cone of depression may persist for some time after the well is abandoned. Other complications can also occur as a result of extracting groundwater. When groundwater is extracted from unconsolidated sediment, there is a loss of volume in the subsurface, which may result in land settling or **subsidence**.

Between 1925 and 1977, large amounts of groundwater were pumped out of the ground beneath the agriculturally intensive San Joaquin Valley of California, a region that produces about one quarter of the table food in the United States on only 1 percent of its farmland. As a result, the land subsided by approximately 9 meters (30 feet) (Fig. 14.20a). In heavily populated regions, subsidence can severely damage a community's buildings, roads, water lines, and sewage schemes. The famous Leaning Tower of Pisa in Italy owes some of its tilt to subsidence associated with groundwater withdrawal from an aquifer (Fig. 14.20b). Mexico City, with the high demand for water posed by its population and climate, has subsided by more than 10 meters (33 feet) in the last 100 years, and some measurements suggest the eastern part of the city is subsiding by a staggering 37 cm (15 inches) each year.

In coastal communities, subsidence associated with the extraction of groundwater may lead to the invasion of seawater. Coastal Venice, in northern Italy, for example, has subsided 3 meters (10 feet) in the past 1500 years, so it is not surprising that the city has waterways for streets (Fig. 14.20c). The city provides a clear example of the potential problems of subsidence caused by the overtaxation of an aquifer.

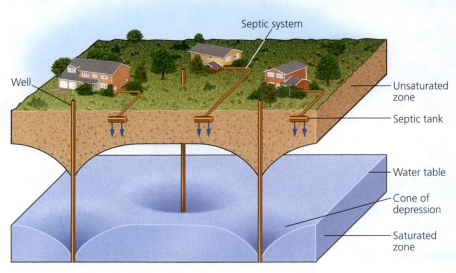

Septic system

Well

Unsaturated zone

Septic tank

Water table

Cone of depression

Saturated zone

Figure 14.19   Cones of Depression

If water is withdrawn from the ground faster than it is replenished, cones of depression form in the water table directly beneath wells. In areas where the wells are closely spaced, these cones of depression may overlap.

Houston and New Orleans along the Gulf of Mexico are beset by similar problems.

As freshwater is depleted and the water table is lowered, water is drawn in from adjacent regions. This can lead to groundwater contamination, either from marine water held in permeable sediments in coastal environments, or from contaminated groundwater drawn in from adjacent geological formations. An invasion by marine water occurred in the New York aquifer in the 1960s. Because salt water is denser than freshwater, it normally sinks below any freshwater in an aquifer. However, as freshwater is withdrawn and cones of depression form, the water pressure is reduced, which allows the saline water to migrate upward. This phenomenon is known as a **cone of ascension** (Fig. 14.21).

(a)

(b)

(c)

Figure 14.20   Groundwater Withdrawal

Groundwater withdrawal effects are visible in many populated communities including: (a) subsidence in the San Joaquin Valley; (b) the Leaning Tower of Pisa; and (c) invasion of sea water, Venice, Italy.

salt water. Beneath this layer there are important zones of high resistivity (blue colors) at distances between 300 and 600 meters (330 and 660 feet) and between 750 and 1000 meters (2500 to 3300 feet). These are interpreted as zones of good quality water. At depths between 100 and 200 meters (330 to 660 feet), the quality of the groundwater is highly variable. Beneath the 100- to 400-meter (330- to 1320-feet) distance interval, for example, the survey identified regions with conductive water. These regions are thought to represent cones of ascension (see Figure 14.21b) that are characteristic of the overpumping of an aquifer and the ascent of salty water.

To counteract this type of problem, some ingenious feats of engineering have been attempted. For example, the aquifer beneath Chicago is being replenished by pumping in freshwater from Lake Superior. In the same way, groundwater used in nonpolluting industrial processes is recycled by pumping it back into the ground. Nonhazardous biodegradable liquids can also be sprayed on the land surface; they are purified as they percolate down toward the water table. These examples of groundwater recycling may become more common as humans continue to mine our most precious resource.

The study of Salinas and the subsidence of the San Joaquin Valley are clear examples of the dilemma of groundwater exploitation. The San Joaquin Valley produces more than $2 billion worth of agricultural produce annually and a vast water supply is needed to support it. Yet overpumping of the aquifer threatens the very industry that it supports.

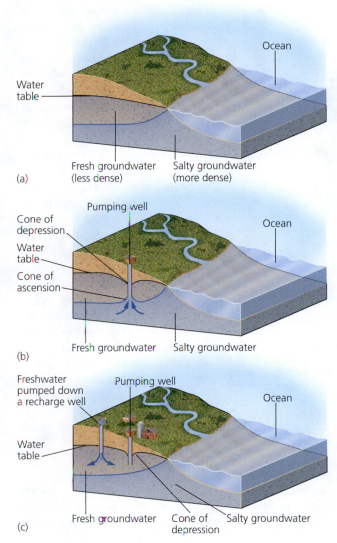

Figure 14.21   Cone of Ascension
These figures illustrate the invasion of an aquifer by marine water in a coastal region. (a) Freshwater is less dense than salt water and lies above it in the aquifer. (b) As groundwater is withdrawn, cones of depression form (see Figure 14.19), and a cone of ascension may form that results in the saline water rising to contaminate the freshwater aquifer. (c) The aquifer can be artificially recharged by pumping freshwater down a *recharge well*.

Electrical imaging of the subsurface readily detects contamination by marine water because salt water, with its abundant dissolved ions, conducts electricity far better than more resistant freshwater.

A study of marine contamination in Salinas, California, provides an example of how these images reveal the extent of the contamination (Fig. 14.22). Previous studies had indicated that seawater invasion had migrated more than 11 kilometers (7 miles) inland since 1980, but the seawater's influence on the aquifer was unclear. The electrical image in Figure 14.22, however, clearly defines a shallow (less than 50 meters, 165 feet) conductive portion of the aquifer, which identifies an area that has been extensively contaminated by

CHECK YOUR UNDERSTANDING
- How do cones of ascension form, and what is their relationship with cones of depression?
- Why does saltwater invade aquifers in coastal communities?
- How do electrical images of the subsurface reveal the location of groundwater resources?

HUMAN CONTAMINATION OF GROUNDWATER

Unfortunately, the unique qualities that make water a very special substance, also render it susceptible to human contamination. Water is an effective solvent, which means that many chemicals, benign and toxic, dissolve in water. Any soluble material, whether it occurs in the atmosphere, on the surface, or beneath the ground, can potentially contaminate groundwater. Some examples of this contamination are shown in Figure 14.23. In addition, rainwater scavenges soluble pollutants in the atmosphere, including sulfur (generating "acid rain") and toxic metals such as mercury or lead. This contaminated water may eventually infiltrate the ground to contaminate the aquifer.

CHECK YOUR UNDERSTANDING
- Why is groundwater so easily contaminated?

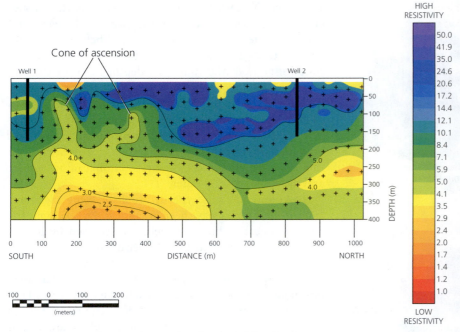

Figure 14.22　Saltwater Contamination in California

An electrical image 1 kilometer in length of the subsurface in Salinas, California, shows that the freshest water occurs in highly resistive zones (blue) at distances between 300 and 600 meters (1000 and 2000 feet) and 700 to 1000 meters (1330 to 2330 feet) along the survey. Note the cones of ascension marked by regions of highly conductive saltwater between 100 to 400 meters.

Landfill sites

Perhaps the most obvious example of human contamination occurs at landfill or municipal garbage sites, in which horrific cocktails of toxic substances are created from an array of discarded industrial and household goods. As water percolates through these sites, elements leached from the material they contain may contaminate nearby aquifers (Fig. 14.23a).

To protect against contamination, many landfills are lined with impermeable clay-like materials. However, leaks in the system inevitably occur, especially in old abandoned landfills that may have been built to less stringent standards. The identification and cleanup of such sites are controversial subjects that rightly involve the entire community.

Hydrogeologic studies that establish regional groundwater flow patterns are essential in the selection of new landfill sites. They are also essential for monitoring both abandoned and existing landfill sites in order to determine susceptibility to contamination and identify leakage at the earliest opportunity. In general, new sites are built in regions with thick deposits of impermeable clay well above the water table. Ideally, this blocks the passage of toxic substances into any underlying aquifer. Regions with permeable materials above the water table, such as sands and gravels, or rocks that typically contain fractures such as basalt or limestone, should be avoided.

Agricultural sources

Contamination from landfill sites, as serious as it is, pales in comparison to that from agricultural sources (Fig. 14.23b). The percolation of water through fields laced with insecticides

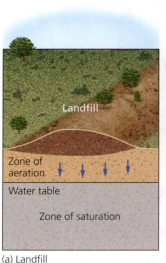

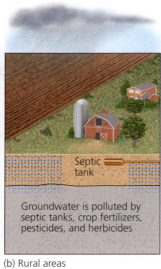

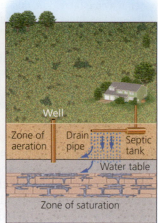

(a) Landfill　　(b) Rural areas　　(c) Urban areas　　(d) Septic tank placement problems

Figure 14.23　Human Sources of Groundwater Contamination

Humans contaminate groundwater in (a) landfill sites where pollutants can be carried into the zone of saturation; (b) rural areas in the vicinity of septic tanks and in general by fertilizers, pesticides, and herbicides; and (c) urban areas due to leaking gasoline tanks or sewage pipes. The identification of the source of contamination is possible by matching the chemistry of the pollutant in the groundwater with its source. (d) Contamination problems are associated with poor placement of septic tanks.

and fertilizers, for example, can contaminate an entire aquifer. Nitrate from fertilizers is the most hazardous contaminant in groundwater and has been linked to methemoglobenemia ("blue baby" sickness) as well as stomach cancer.

Industrial waste and sewage

Industrial waste and sewage are other potential contaminants (Fig. 14.23c). In many areas, sewage is stored in septic tanks housed in permeable sediment. These tanks are designed to slowly release the sewage into the sediment where organic toxins are decomposed and metal contaminants adhere to clay minerals. Problems with septic tanks arise if they are improperly designed, develop leaks, or are improperly positioned with respect to the water table. If the contamination continues for an extended period of time, the capacity of the sediment to filter the toxins is eventually exceeded.

One of the most notorious and tragic cases of groundwater pollution in North America occurred at Love Canal, Niagara Falls, in New York State. In the 1940s a chemical company purchased part of the canal and used it to dispose of nearly 20,000 tons of chemical waste loaded into steel drums. In 1953, the company covered one of the disposal sites with dirt and sold it to the Board of Education of Niagara Falls who constructed a school and playground on the site.

When the school was built, it was attached to the municipal water and sewer systems and the school board believed that the polluted site would not pose a health risk. They were tragically wrong. First, permeable gravel, used as a backfill to the water and sewerage schemes became a pathway that allowed the toxic chemicals to spread and pollute groundwater beneath various parts of the city. Then, in the spring of 1977, torrential rains brought the toxic groundwater to the surface. School children and local residents suffered a variety of serious ailments as a result, including severe headaches, epilepsy, and liver diseases.

SHORT-TERM AND LONG-TERM SOLUTION TO CONTAMINATION

Unfortunately, the Love Canal episode is by no means unique. In 2004, the U.S. Environmental Protection Agency estimated that as many as 350,000 contaminated sites would require cleanup during the next 25 years, and that the cost could be as much as $250 billion. There are some strategies, however, that can reduce the impact of human activity on the environment.

Some strategies involve a basic understanding of the effect of human activity on groundwater. In Figure 14.23d, for example, a septic tank is located too close to the water table. As a result, the contaminated water leaks into the groundwater and may reach neighboring wells. A better placement would be to locate the septic tank well above the water table, preferably in a low to moderately permeable sediment, so that water escaping from the tank is cleansed before it reaches the water table.

In ideal circumstances, the water escaping from a septic tank can be purified within a few tens of meters of the tank.

The key is to locate the septic tank in sediment with low permeability so that the sewage water moves very slowly and has time to react with the sediment as it passes through. Those of us who live in rural communities may be familiar with the standard application of this principle in the "percolation test," which generally must be passed before a building permit is issued. The purpose of the test is to determine whether a septic tank and drain field can be constructed for a house without harming the local environment. The test simply involves digging a hole and filling it with water. If the water drains away too quickly, it is likely that contaminants would pass through and could contaminate the water table. The sediment is consequently deemed to be too permeable and the test is failed. The test is also failed if the water fails to drain away because the sediment is too impermeable. A successful outcome requires the water to drain away slowly.

Aquifers likewise have the ability to clean themselves. In principal, the self-cleansing potential of most aquifers should allow sufficient time for the source of the pollution to be identified and eliminated. In practice, however, the changing chemistry of the aquifer is rarely monitored, and the contamination is often identified only after the cleansing capacity of the aquifer has been exceeded. The most obvious long-term solution to the cleansing of aquifers is the elimination of the sources of pollution. Given that groundwater can travel considerable distances, the culprit may sometimes be difficult to identify. However, sophisticated laboratory techniques applied to groundwater samples now analyze a wide range of chemical elements and a variety of isotopes, including those of hydrogen, carbon, and sulfur. These isotopes often serve as a fingerprint that can be used to identify the pollution source.

But one country's environmental policing can have political and economic consequences for citizens of other countries, as well as for its own citizens. The new global economy has given corporations the option of moving to countries with less stringent environmental regulations, rather than assuming the significant costs of making factories less polluting in their own countries.

As we have seen, groundwater and surface water are interconnected, so that contamination of one can have adverse affects on the other. For example, contaminated groundwater may leak into springs, streams, and lakes where the water table intersects the surface. The draining of the Aral Sea in Asia (see Living on Earth: **The Aral Sea—An Environmental Disaster** in Chapter 13) is an example of the environmental catastrophe that can happen when surface water diverted into canals results in contamination of both groundwater and surface water.

It is clear that we have much to learn about the consequences of our interference with the natural cycle of freshwater. We depend on water

> ### CHECK YOUR UNDERSTANDING
>
> ● What factors should influence the location of (a) landfill sites (b) septic tanks? Explain.
>
> ● How does groundwater have the ability to cleanse itself?

for our survival, yet our methods of exploitation are often poorly conceived and the results are unforeseen. Perhaps the best legacy we can leave our children is a sustainable planet. But if we wish to achieve this goal, we will need to learn more about water and its complex behavior.

14.5 SUMMARY

- When water is withdrawn from an aquifer by wells faster than nature can replenish it, cones of depression and land subsidence may occur. In coastal regions, cones of ascension may contaminate the aquifer with saline water.

- Aquifers are highly susceptible to contamination from a variety of sources.

- Aquifers have the capacity to cleanse themselves.

14.6 Groundwater and Plate Tectonics

Plate tectonic processes affect the properties of groundwater both locally and regionally. We have already discussed how the temperature of groundwater increases with depth. This is especially true in localities that are volcanically active. In these localities, plate tectonic forces are also responsible for the faults and fractures that bring groundwater to the surface in oases, geysers, and hot springs.

On a regional scale, plate tectonics processes create topography (as discussed in Chapter 6), and it is topography that provides the slope that permits groundwater to flow. Plate tectonics can also have a more direct influence on groundwater. Aquifers are commonly recharged by precipitation falling in mountainous regions, the existence of which owes its origin to the forces of plate tectonics.

The eastern flank of the Rocky Mountains, for example, is the recharge area for the High Plains aquifer.

We have also learned (see Living on Earth: **The Historical Importance of Aquifers** on page 447) that much of southeastern England, including the city of London, is underlain by a chalk aquifer that played a historically important role in the city's development. This aquifer exists because the chalk layer takes the form of a broad syncline, known as the London Basin, between recharge areas in hills to the east and west (see Fig. 4C). The formation of this syncline is a direct consequence of the forces of plate tectonics; it owes its origin to crustal movements related to the formation of the Alps. The syncline is consequently a distant expression of the continental collision that raised the Alps far to the southeast.

On an even larger scale, it has been argued that plate tectonics has controlled ancient movement of warm, saline groundwater across vast areas of North America. This deep groundwater, which is thought to be responsible for creating important deposits of lead and zinc in the limestone of the Mississippi Valley, is believed to have originated in the Appalachians and Rocky Mountains, and to have migrated into the continental interior at the time these mountain belts formed. In this way, plate tectonic forces operating along the margins of North America may have caused deep, saline groundwater to migrate long distances through the sedimentary rocks of the continental interior.

> **CHECK YOUR UNDERSTANDING**
>
> ○ Cite an example of a groundwater feature created by plate tectonics.

14.6 SUMMARY

- Plate tectonic processes create topography that influences regional groundwater flow.

Key Terms

aquiclude 442
aquifer 442
aquitard 442
artesian well 446
column 453
cone of ascension 458
cone of depression 457
confined aquifer 443
disappearing stream 453

discharge 441
fumarole 447
geyser 449
groundwater 439
hard water 452
hot spring 447
hydraulic head 443
karst 452
karst topography 452

monolith 453
oasis 446
permeability 440
porosity 440
pressure surface 445
recharge 441
sinkhole 453
soda straw 453
spring 441

stalactite 453
stalagmite 453
subsidence 457
unconfined aquifer 443
water table 439
zone of aeration 439
zone of saturation 439

Key Concepts

14.1 GROUNDWATER: A VITAL RESOURCE

- In the United States and Canada, 1.6 trillion liters per day of freshwater is consumed.

- Groundwater constitutes about 22 percent of the world's freshwater and is stored in rocks and sediment below the ground.

- Groundwater is by far the largest reservoir of accessible freshwater.

14.2 THE WATER TABLE

- The water table is the upper surface of the zone of saturation. The zone of aeration occurs above it.

- Porosity controls how much water subsurface sediment or rock can hold, and permeability is a measure of how easily groundwater can move through the sediment or rock.

- Groundwater supply is a delicate balance between recharge and discharge.

- Groundwater becomes surface water where the water table intersects the surface and is discharged into streams, lakes, springs, or swamps.

14.3 AQUIFERS AND GROUNDWATER FLOW

- Groundwater becomes an aquifer if a porous and permeable rock or sediment can store and transmit water. The upper surface of unconfined aquifers coincides with the water table. Confined aquifers occur where the water-bearing rocks or sediments are sandwiched between impermeable layers.

- Groundwater generally flows "downhill" toward valleys at a few meters per year under the influence of gravity and may take thousands of years to become surface water.

- Groundwater can be found beneath desert regions. An oasis forms where a spring occurs in desert regions.

- Deep penetrating groundwater may absorb Earth's internal heat and so become a source of geothermal energy and hot springs.

14.4 THE DISSOLVING POWER OF GROUNDWATER

- Because water is a strong solvent, its composition is influenced by the extent of chemical weathering and the composition of the subsurface materials. Water that contains large quantities of dissolved ions is known as hard water.

- Karsts are produced when groundwater dissolves soluble bedrock. Gypsum and limestone are very soluble, leading to the formation of caves, associated columns known as stalactites and stalagmites, and sinkholes.

14.5 EXPLOITATION OF GROUNDWATER

- When water is withdrawn from an aquifer faster than nature can replenish it, cones of depression form around each well, and land settling or subsidence may occur.

- Saline water sinks below freshwater in the aquifer, but may rise in a cone of ascension as freshwater is withdrawn from the aquifer.

- Because water is an effective solvent, it is susceptible to contamination, especially adjacent to landfill sites. Entire aquifers can be contaminated by the percolation of water through fields laced with insecticides and fertilizers.

14.6 GROUNDWATER AND PLATE TECTONICS

- On a local scale, plate tectonic processes result in volcanic activity that warms groundwater and the fractures that bring groundwater to the surface in oases, geysers, and hot springs.

- On a regional scale, plate tectonics govern the location and overall shape of aquifers by creating the mountains that act as recharge areas for aquifers, and the topography and slope that permits groundwater to flow.

Study Questions

1. Four features that characterize overpumping of aquifers have been mentioned in this chapter. What are they, and explain how aquifers are sensitive to overproduction?

2. Why is the water table level variable within the same aquifer?

3. Explain the significance of the pressure surface to the supply of water in a region. Why is the pressure surface in a confined aquifer gently sloped away from the recharge area?

4. Explain the relationship between geothermal resources and plate tectonic activity.

5. What features preserved in caves might help distinguish between caves formed by downward-migrating water and those formed by upward-migrating water? Explain.

6. Do you think global warming would serve to increase or decrease our groundwater supply?

7. Explain how electrical images of the subsurface help locate cones of depression or cones of ascension.

8. How may rerouting of surface waters affect the composition of groundwater? Explain.

9. What are the main sources of human contamination of groundwater and what measures can be taken to remedy them?

10. In what ways does plate tectonics contribute to groundwater flow?

15 Glaciers and Glaciation

15.1 Glaciers and Glacier Ice

15.2 Glacial Erosion

15.3 Glacial Deposition

15.4 The Pleistocene Ice Age

15.5 Plate Tectonics and Glaciation

15.6 Ice on Other Worlds

▶ Key Terms

▶ Key Concepts

▶ Study Questions

SCIENCE REFRESHER: Air Temperature in the Lower Atmosphere 468

LIVING ON EARTH: The Origin of the Great Lakes 477

IN DEPTH: Lake Missoula and the Channeled Scablands 490

Horn

Fjord

After glaciation

Valley glacier

Lateral moraine

Medial moraine

Meltwater lake

Outwash plain

Terminal moraine

Peak glaciation

After peak glaciation

After glaciation

Valley
glacier

Zone of
accumulation

Cirque glacier

Continental
ice sheet

Arête

Crevasse

Zone of ablation

Raised
beaches

Outwash
plain

Proglacial lake

Moraine-dammed
lake

Terminal moraine

After glaciation

Learning Objectives

15.1 Explain the formation and movement of glacier ice.

15.2 Describe the process of glacial erosion and the landforms that glacial erosion produces.

15.3 Describe the processes of glacial deposition and the effects glaciers have on Earth's landscape.

15.4 Describe the influence of the Pleistocene ice age on the landscapes of the northern hemisphere.

15.5 Summarize the evidence that is used to deduce the existence and causes of ancient ice ages.

15.6 List other instances of glaciation from across our Solar System.

O f all Earth's surface processes, few have affected the landscape of the world more than glaciation. *Glaciers* are thick masses of ice produced from compacted snow that move, ever so slowly, under the influence of their own weight. Most glaciers are the familiar "rivers" of ice that snake down the valleys of high mountain ranges. But some glaciers also flow outward from centers of glaciation, eventually turning into ice sheets that reach across entire continents. Beginning some 2.6 million years ago, such continental ice sheets have repeatedly advanced and retreated across much of the Northern Hemisphere.

Today, Earth is in an interglacial period between advances but a mere 20,000 years ago a vast ice sheet stretched from the Canadian Arctic to the American Midwest. It also blanketed much of Europe and Siberia. Most of the continent north of the United States-Canadian border was buried beneath a thousand meters or more of ice, while to the south, debris-laden *meltwater,* produced by glacial melting, drained all the way to the Gulf of Mexico. Although this massive ice sheet has now largely vanished, the influence of glaciation on the landscape is unmistakable. This period of past glaciation, which we refer to as the Ice Age, also has important implications for global climate. Today glaciers cover about 10 percent of Earth's land surface. But 20,000 years ago ice covered up to three times that area. In this chapter we explore glaciers and learn how they form and how they affect Earth's surface and Earth's climate.

15.1 Glaciers and Glacier Ice

Much of the landscape around us owes its origin to the processes of glacial erosion and deposition, collectively known as **glaciation**. But how do glaciers form and what are the unmistakable influences of glaciation on our landscape? What processes are involved in glacial erosion and deposition, and what are the consequences of the advance and retreat of ice on a continental scale? To answer these questions we begin by examining glacier ice and how it forms.

WHAT IS A GLACIER?

A **glacier** is a mass of ice that forms on land and is sufficiently thick to flow under its own weight. Unlike ice from the freezer, glacier ice is not produced simply from the freezing of water. Instead, *glacier ice* forms from the slow compaction of snow. The transformation from snow to glacier ice is a time-consuming one that takes place in several stages (Fig. 15.1). The gentle fall of snow reflects the fact that snowflakes are delicate lace-like crystals of ice. More than three-quarters of the volume of a snowflake is air. As snow accumulates, the delicate points of the star-shaped snowflakes melt under the weight of the snowpack and refreeze nearer the center of the flake, eventually producing granular particles of snow about the size of a pinhead. Packed together, these particles form a granular variety of

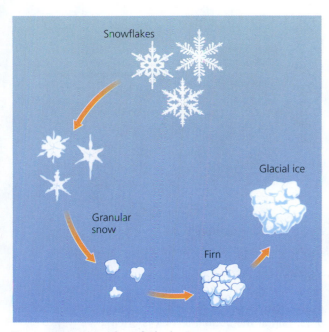

Figure 15.1   Formation of Glacier Ice
There are a number of stages in the conversion of snowflakes to glacier ice by way of granular snow and firn.

ice known as **firn**. With further burial, melting, and re-freezing, firn undergoes additional compaction and re-crystallization, a process that eventually fuses the grains together into solid glacial ice.

As more snow accumulates, more firn forms, and the glacial ice thickens beneath the snowpack. Once it reaches a thickness of about 60 meters (200 feet), the ice can no longer support its own weight and, like soft cheese or silly putty, the accumulating ice mass starts to slowly flow under the influence of gravity. Depending on the amount of snowfall, the time it takes to reach this point can range from several hundred to many thousands of years.

> **CHECK YOUR UNDERSTANDING**
> ● What is a glacier?
> ● How is snow converted to glacier ice?

HOW GLACIERS FORM

Glaciers form and grow in areas where the amount of snow that accumulates in the winter exceeds the amount of snow that is lost in the summer. In such areas, a permanent snow cover develops. The height above sea level at which this occurs defines the **snowline** (see Science Refresher: **Air Temperature in the Lower Atmosphere** on page 468). Above the snowline, snow remains throughout the year (Fig. 15.2).

Near the polar regions where summer temperatures rarely rise above the freezing point, glaciers form at or near sea level. In Greenland and Antarctica, huge thicknesses of ice have developed in this way (Fig. 15.3). Because this thick ice slowly spreads out in all directions and can

Figure 15.2   Snowline
The snowline separates the snowcapped mountains of the Austrian Alps from the warmer valleys between them.

Air Temperature in the Lower Atmosphere

As an airplane takes off and rises toward its cruising altitude of about 10,000 meters (33,000 feet), any window-seat passenger can observe that the atmosphere is not uniform. As the plane flies above the clouds into clear blue skies, the outside temperature plummets to a frigid −50°C (−58°F) and the air pressure outside drops so low that the cabin has to be pressurized to protect the passengers. As the flight attendants remind us, safety devices, such as oxygen masks, exist should the cabin pressure fall. We experience a pressure change whenever an aircraft rapidly changes its altitude, and our ears "pop" to adjust to it. The decrease in pressure with altitude occurs because the atmosphere gets thinner as we rise above Earth's surface and the weight that the atmosphere exerts on an airplane decreases as the plane climbs. But why does temperature decrease as well?

Air temperatures decrease with height because Earth's surface acts like a radiator in a room. Just as it is colder the farther one gets from a radiator, so the air temperature decreases with increasing height above Earth's surface. Have you ever sat on rocks exposed to sunlight on a hot day? The rocks are warm because they have absorbed some of the Sun's energy. In this way the Sun warms Earth's surface and some of the energy absorbed radiates back as heat into the atmosphere. As a result, air is warmest at Earth's surface and it cools with increasing altitude.

Air cools as it rises, because reduced atmospheric pressure allows it to expand. Cooling occurs because the air uses energy to push aside the surrounding air as it expands. This trend is

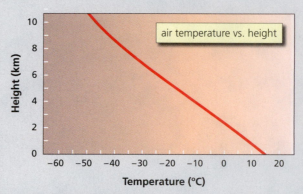

Figure 15A Temperature in the Atmosphere

The temperature of the air decreases with height from an average of about 15°C (60°F) at sea level to a chilly −50°C (−58°F) at 10,000 meters (33,000 feet).

especially obvious in mountainous regions, where, on wet days, rain at the bottom of valleys often gives way to snow before the precipitation reaches the mountain tops. As skiing enthusiasts will vouch, snow and glaciers persist on high mountain peaks because the air is colder at higher altitudes. In fact, the air temperature decreases by an average of 6.5°C per kilometer (or 3.6°F per 1000 feet) above Earth's surface and temperatures below −50°C (−58°F) occur at altitudes above 10,000 meters (33,000 feet) (Fig. 15A).

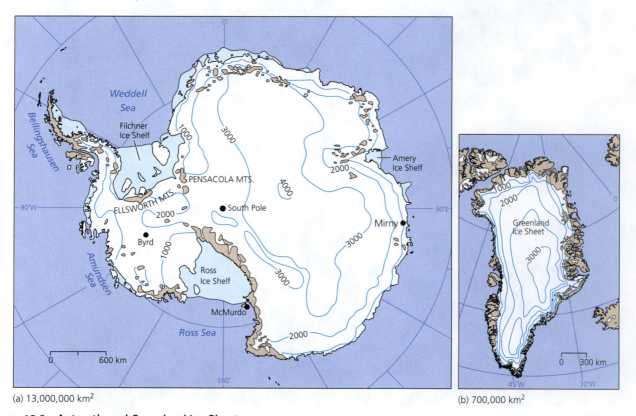

(a) 13,000,000 km²

(b) 700,000 km²

Figure 15.3 Antarctic and Greenland Ice Sheets

(a) The Antarctic ice sheet is up to 4.3 kilometers (2.7 miles) thick and covers almost 13 million square kilometers (5 million square miles). It is made up of two ice sheets, one covering West Antarctica and a larger one covering East Antarctica. Both ice sheets locally extend out into the sea to form floating *ice shelves*. (b) The smaller Greenland ice sheet covers more than 1.8 million square kilometers (700,000 square miles) to a depth of over 3.3 kilometers (2 miles).

eventually cover an entire landmass, the glaciers it forms are known as **continental glaciers**, or *ice sheets*.

At lower latitudes, the snowline generally lies at elevations that become progressively higher toward the equator. In regions where the snowline lies at high elevations, glaciers are localized in mountainous areas. These glaciers are known collectively as **alpine glaciers** and take several forms (Fig. 15.4). Those confined to mountainside hollows are called *cirque glaciers* because the amphitheater-shaped basins they carve out in these hollows are known as **cirques**.

Once cirque glaciers begin flowing downslope, they are called **valley glaciers** because they follow the route of a pre-existing stream valley. When a valley glacier is no longer confined by the valley, it can take one of three forms:

- A *tidewater glacier* forms where glaciated mountains lie close to the sea and valley glaciers extend beyond the coastline into tidal water.

- A *piedmont glacier* forms where valley glaciers emerge from a mountain front and spread out onto the lowlands.

- An *ice field* forms at the summit of a mountain range if enough ice accumulates so that individual valley glaciers coalesce to produce an upland (high elevation) ice mass.

(a) Cirque glacier

(b) Valley glacier

(c) Tidewater glacier

(d) Ice field

Figure 15.4   Types of Alpine Glaciers

There are several types of alpine glaciers. (a) Cirque glaciers are confined to amphitheater-shaped hollows on the mountainside called cirques. (b) Valley glaciers occupy former mountain stream valleys and may open out into piedmont glaciers at the mountain front. (c) Glaciers may reach the sea to form tidewater glaciers. (d) At the summit of a mountain range, valley glaciers may coalesce to produce an ice field.

CHECK YOUR UNDERSTANDING

◉ How do geologists classify glaciers?

HOW GLACIERS MOVE

As noted earlier, glaciers move when glacier ice becomes sufficiently thick (about 60 meters or 200 feet) to flow under its own weight. This movement resembles that of a very slow-moving fluid. But glacier ice is a solid material, so how does this movement take place? In fact, this behavior is not unusual. Recall from our discussions in Chapters 10 and 11, that solids are capable of *creep* under the right conditions, particularly when they are close to their melting points. For most solids this requires elevated temperatures and deep burial, but for glacier ice, which melts at low temperature, these conditions exist at Earth's surface.

Glacier ice moves in two main ways (Fig. 15.5). The glacier ice may change shape, a process known as *plastic flow*, or the glacier may slide along its base, a process known as *basal sliding*. During **plastic flow** the weight of the overlying ice causes crystals at the base of the ice pack to change shape (Fig. 15.5a). This movement is similar to that of a ball of silly putty, which flattens out under its own weight when left out overnight. Plastic flow is a slow process, occurring at rates of only a few meters per year, and is seen to occur only below about 60 meters (200 feet). Above 60 meters a glacier is brittle and cracks to form open fissures called **crevasses** as the ice beneath flows forward.

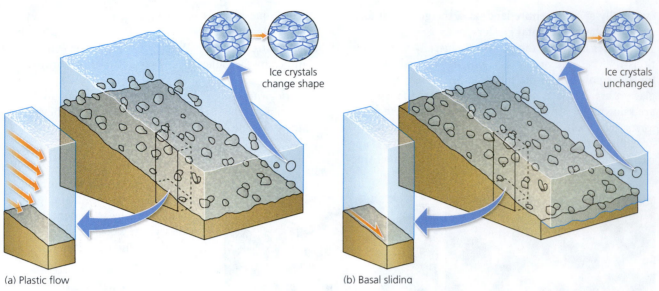

Ice crystals
change shape

Ice crystals
unchanged

(a) Plastic flow

(b) Basal sliding

Figure 15.5   Glacial Movement

(a) In plastic flow, the ice mass deforms internally and ice crystals change shape. (b) In basal sliding, the ice mass slides along its base without internal deformation.

In contrast to plastic flow, **basal sliding** involves mass movement of the ice sheet, with no change in crystal shape (Fig. 15.5b). It occurs when the ice at the base of a glacier melts to produce a film of lubricating water that allows the ice to slide over the material beneath it. This surface is called the **glacier bed**. Basal sliding is a faster process than plastic flow and is typical of glaciers in warmer climates. With it, glaciers move at rates of 500 meters (1650 feet) or more per year.

An abrupt increase in basal sliding is thought to be responsible for periodic episodes of rapid acceleration called **surges**. When a glacier surges, its flow rate may increase a hundredfold. In 1986, for example, the Hubbard Glacier in southeastern Alaska abruptly increased its rate of flow from the average rate of 100 meters (330 feet) per year it had maintained since the beginning of the twentieth century to 14 meters (46 feet) a day, or more than 5 kilometers (3 miles) per year. Surges, which usually last a few months to several years, apparently occur when sufficient water accumulates at the base of a glacier to raise it off its bed, thereby reducing friction and greatly facilitating basal sliding.

CHECK YOUR UNDERSTANDING

○ What are crevasses, and why are they restricted to the uppermost 60 meters of a glacier?

○ What two processes contribute to the movement of glacier ice and how do they operate?

HOW GLACIERS GROW AND SHRINK

A glacier grows by accumulating ice that is transported toward the front of the glacier, which is referred to as the **glacial terminus**. The water stored in this ice eventually returns to the hydrologic cycle as a result of melting and evaporation. Where a glacier terminates in a body of water,

as is the case with tidewater glaciers (see Fig. 15.4), ice may also be lost from the system by *calving*, the breaking off of blocks of ice from the front end of the glacier. The loss of ice from a glacial system by any means is **ablation**.

Glaciers move ice from an area of net gain to one of net loss. The upper, or inner, part of a glacier in which the net gain of ice occurs is the *zone of accumulation*. The lower, or outer, part of the system in which there is a net loss is the *zone of ablation*. At the boundary between these two zones, called the *equilibrium line,* the balance is approximately zero—accumulation equals ablation (Fig. 15.6). The position of this boundary generally coincides with the lower limit of snow cover on a glacier during the late summer. However, the equilibrium line is not fixed as the balance between accumulation and ablation may vary from one year to the next. This annual balance, described as the **glacial budget,** plays an important role in the behavior of glaciers.

Just as the balance on a bank account grows or shrinks as funds are deposited and withdrawn, glaciers expand and contract with accumulation and ablation. Providing the amount of ice that enters the system in the zone of accumulation matches that which is lost in the zone of ablation, a glacier remains constant in size and the position of its terminus remains stationary. But when accumulation exceeds ablation, the glacial budget becomes positive and the margins of the glacier advance. Conversely, when ablation exceeds accumulation, the budget is negative and the glacial margins retreat. This movement of the glacial terminus is entirely independent of movement due to glacial flow. Regardless of the budget, ice continually flows toward the terminus of a glacier from its

CHECK YOUR UNDERSTANDING

○ Define the term glacial budget.

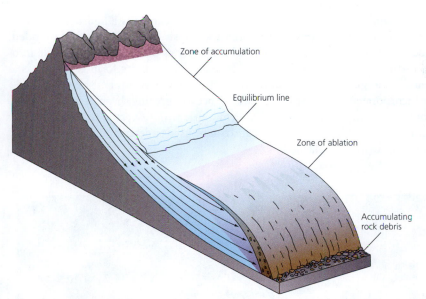

Figure 15.6   Glacial Budget
The glacial budget is the annual balance between ice gain in the zone of accumulation and ice loss in the zone of ablation. Arrows show the transport of snow and ice to the front of the glacier (the terminus) from the zone of accumulation to the zone of ablation. At the equilibrium line, accumulation equals ablation. The location of the equilibrium line may vary from one year to the next.

zone of accumulation. Glaciers do not retreat by moving backward. Instead, the ice at the margins of the glacier retreats by simply melting away, causing the glacier to lose mass. In the same way, individual hairs of a retreating hairline do not move backward, they just fall out.

15.1 SUMMARY

- Glaciers are moving masses of ice formed from the slow compaction of snow.

- There are two main types of glacier: vast ice sheets called continental glaciers and the familiar glaciers of mountainous regions called alpine glaciers.

- Glaciers form and grow if the amount of snow accumulation in winter exceeds that lost in the summer.

- Glaciers flow in one of two ways: a slow internal process of plastic flow or, in warmer regions, the faster external process of basal sliding.

- Glaciers move from an area of net gain to one of net loss. This annual balance is the glacial budget.

15.2 Glacial Erosion

Although its movement is imperceptibly slow, glacier ice serves as an agent of erosion and deposition. The effectiveness of glacier ice is almost unparalleled among surface processes. Glaciers act, on the one hand, like gigantic bulldozers capable of moving almost anything in their path, and on the other, like giant pieces of sandpaper because glacial ice carries with it huge quantities of rock debris that is capable of scraping, polishing, and loosening the bedrock it encounters. In this section of the chapter, we discuss these glacial processes and the landforms they produce.

ABRASION AND PLUCKING

In addition to simply shoving aside or bulldozing material in their paths, glaciers erode the landscape by scouring rock surfaces and by prying loose blocks of bedrock. These two processes of glacial erosion are respectively termed *abrasion* and *plucking*. **Abrasion** is the sandpaper-like wearing or grinding away of bedrock by rock particles transported by glacier ice. **Plucking** is the loosening and removal of rock fragments from the glacier bed.

Abrasion polishes, scratches, or gouges bedrock, depending on the size of the rock particles embedded in the glacier ice. Bedrock surfaces made shiny by this process are said to possess *glacial polish*. Scratch marks made in this fashion, termed glacial **striations**, record the direction of glacial movement (Fig. 15.7). Abrasion also grinds away at the tools of erosion. So while the bedrock surfaces are scoured, the transported rock particles may themselves become striated. Continued grinding acts like a mill, crushing and pulverizing rock particles and producing vast

Figure 15.7   Glacial Striations on Polished Bedrock
Glacial striations on exposed bedrock surfaces are produced by glacial abrasion and reveal the movement direction of the flowing ice. The striations shown here indicate that glacier ice has flowed over this glacially polished outcrop. Slight differences in the orientation of the striations (colors) reflect small changes in the ice flow direction over time.

quantities of finely powered rock known as *rock flour*. The characteristic milky color of glacial meltwater is due to the presence of suspended rock flour. Glacial plucking results when meltwater percolates into the fractured bedrock and freezes, expanding and prying the bedrock apart. The angular blocks of bedrock produced by plucking are incorporated into the moving ice and carried away as part of the glacier.

Glacial abrasion and plucking work together to shape rock masses on the glacier bed into distinctive forms.

When ice rides over and streamlines a bedrock obstacle, abrasion smooths the "upstream" slope into a rounded form, while plucking quarries the "downstream" side into a jagged cliff (Fig. 15.8). The resulting form is known as a **roche moutonnée**, from the French for "rock sheep," because of its resemblance to a grazing sheep. A typical roche moutonnée is asymmetric with respect to the glacier's line of movement and can be used to determine the direction of glacial flow.

Figure 15.8 Roche Moutonnée

A roche moutonnée is an asymmetric bedrock form produced by abrasion and plucking along the base of the ice sheet. Its shape can be used to determine the direction of glacial flow (in this case the flow is right to left).

Figure 15.9 Glacier National Park

The distant pyramidal peaks, broad U-shaped valleys, and ribbon lakes of Glacier National Park in Montana are typical of landscapes produced by alpine glacial erosion.

ALPINE GLACIAL LANDFORMS

Some of the world's most dramatic landscapes have been produced when alpine glaciers have eroded preexisting valleys. The breathtaking scenery of Yosemite and Glacier National Parks (Fig. 15.9), and Canada's Banff and Jasper National Parks, with their jagged mountains, precipitous cliffs, and frequent waterfalls and lakes, is typical of landscapes formed in this way. These landforms owe their origin to the enormous erosive power of alpine glaciation, and vividly attest to the former extent of ice in mountainous regions (Fig. 15.10).

Most alpine glaciers are born high on the sides of mountains in hollows or recesses above the snowline where ice first accumulates. These depressions become the cradle of glacial growth. Once sufficient ice has accumulated for glacial flow to commence, the depressions are rapidly hollowed out by glacial erosion into steep-walled, horseshoe-shaped bowls. These precipitous amphitheater-shaped basins are known

(a) Unglaciated topography

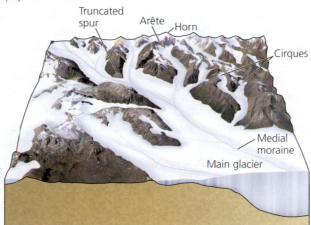

(b) Maximum glaciation

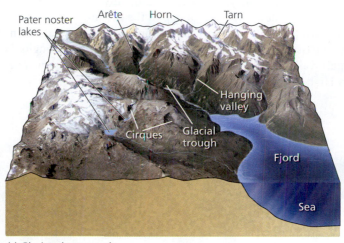

(c) Glaciated topography

Figure 15.10 Erosional Landforms of Alpine Glaciation

This series of figures illustrate the development of erosional landforms of alpine glaciation in a mountainous area (a) before, (b) during, and (c) after glaciation.

Figure 15.11   Cirque with Tarn

Melting of a cirque glacier (left) has produced a cirque basin containing a tarn (Upper Thornton Lake) in the North Cascades National Park in Washington State (right).

as cirques, from the French word for "arena." As shown in Figure 15.11, cirques form imposing features on the sides of formerly glaciated mountains. All cirque basins open in the downslope direction, and many have a ridge-like lip at their lower thresholds because their floors are bowl-shaped. For this reason, cirques are often occupied by small cirque lakes, which are called *tarns,* following deglaciation.

Sometimes several cirques start to form on the same mountain. Continuing erosion may then shape the mountain into three imposing glacial landforms: *arêtes, cols,* and *horns* (see Fig. 15.10). **Arêtes** are precipitous, knife-edged ridges that separate adjoining cirques. They form when glacial erosion expands the size of two adjacent cirques and narrows the divide between them. The two cirques may sit side by side, or they may converge by erosion from the opposite sides of the mountain. In either case, their precipitous walls gradually approach one another until the ridge separating the two is reduced to a narrow spine of rock (Fig. 15.12).

With sufficient erosion, cirques converging from the opposite sides of a mountain may largely remove the divide between them and create a low point or gap in the mountain range known as a **col**. Such low points form important mountain passes and are often used to cross from one high mountain valley to the next. Tioga Pass, over which California's Highway 120 crosses into Yosemite National Park, and Yellowhead Pass, over which Canada's Route 16 crosses into Jasper National Park, take advantage of cols in the Sierra Nevada and Canadian Rockies, respectively.

When three or more cirques develop on the sides of the same mountain, their expansion by glacial erosion can gradually sculpt the mountain into a pyramidal shape. The end result is one of the most impressive landforms of glacial erosion, the majestic pyramidal peak known as a **horn**. Glacial horns form the backdrop to the breathtaking scenery of the Grand Tetons of Wyoming and at Banff and Jasper in the Canadian Rockies. Best known of all, however, is the Matterhorn in the Swiss Alps (Fig. 15.13).

Recall from Section 15.1 that with sufficient accumulation, ice will flow out of a cirque and into the stream valley below to form a valley glacier. As it does so, it scours out the existing valley to produce a broad, steep-sided **glacial trough** that is wider, deeper, and straighter than the preglacial stream valley it replaces (see Fig. 15.10c). In this way, the V-shaped profile characteristic of stream valleys is replaced by a *U-shaped* profile that is one of the most distinctive features of valleys formerly occupied by ice. With its nearly vertical walls and broad, flat floor, Yosemite Valley in California (Fig. 15.14a) is typical of glacial troughs. Following glaciation, glacial troughs often become home to a series of small circular lakes linked by streams, rapids, and waterfalls. Known as *pater noster lakes,* these

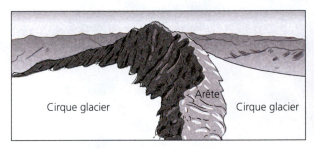

Figure 15.12   Arête

A panoramic image captures the arête known as Crib Goch in Snowdonia National Park, Wales, UK.

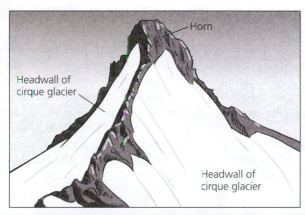

(a)

Figure 15.13   Matterhorn

The majestic pyramidal peak of the Matterhorn in the Swiss Alps is a glacial horn created by the erosion of cirques excavated on its flanks.

(b)

Figure 15.14   Yosemite National Park and Bridalveil Falls

(a) The spectacular scenery of Yosemite National Park, California, was sculpted by ice 10,000 years ago. The U-shaped valley of the Merced River (center), Bridalveil Falls plunging from a hanging valley (in shadow to right), and the precipitous truncated spurs of El Capitan (left) and Half Dome (center background) are characteristic features of alpine glaciation. (b) Bridalveil Falls cascades 190 meters (620 feet) from a hanging valley to the floor of the main glacial trough.

lakes fill basins in the unevenly scoured bedrock of the valley floors (see Fig. 15.10c).

In special circumstances, glacial troughs may be flooded by the sea, forming spectacular features called fjords. **Fjords** are glacial troughs that have been flooded by the sea. These deep, steep-sided sea inlets are found on high-latitude coastlines where glaciated mountains lie close to the sea (Fig. 15.15) and are characteristic of the coast of Norway, from whence they get their name.

As the valley glacier carves its U-shaped profile, the ridges that project into the preglacial valley as divides between tributary streams are beveled by the flowing ice as it widens and straightens the valley. These cut-off ridges are *truncated spurs* (See Fig. 15.10b). Following glaciation, truncated spurs form distinctive triangular-shaped cliffs on the valley sides. The precipitous walls of El Capitan and Half Dome (see Fig. 15.14a) in Yosemite National Park were formed in this way.

At the peak of glaciation, smaller alpine glaciers may occupy tributary stream valleys. Like the main glacier, which is also referred to as the *trunk glacier,* these tributary glaciers progressively deepen, widen, and straighten the tributary valleys. Because the erosive power of ice is controlled in part by a glacier's thickness, the floors of the smaller tributary valley glaciers are not deepened to the

Valley glacier carving
steep-walled trough

Figure 15.15   Fjord

Fjords, such as this one in Norway, are glacial troughs that were excavated well below sea level and flooded by the sea following glacial retreat.

same extent as that of the larger trunk glacier. Following glaciation, the mouths of the tributary valleys are consequently left high above the floor of the main glacial trough. Tributary valleys suspended in this way are known as **hanging valleys** and, like Yosemite's Bridalveil Falls, are commonly the sites of spectacular waterfalls (Fig. 15.14b).

CONTINENTAL GLACIAL LANDFORMS

Continental glaciers scour the land surface and leave behind an undulating landscape with countless lakes. This is because continental ice sheets are not confined to valleys, and so leave all features of the land surface over which they move smoothed and abraded. During the period of past glaciation we refer to as the Ice Age, continental glaciers eroded vast tracts of North America, particularly in Canada. Bare outcrops of striated and polished

CHECK YOUR UNDERSTANDING

○ What type of landscape do continental glaciers leave in their wake?

bedrock occur across the ancient rocks of the Canadian Shield, and countless lakes occupy depressions in the land surface. Some of the lake basins scoured by continental ice are very large. This is particularly true of the Great Lakes, which together occupy an area of 244,000 square kilometers (94,000 square miles) (see Living on Earth: **The Origin of the Great Lakes** on page 477).

15.2 SUMMARY

- Glaciers erode by abrasion and plucking; these processes can produce asymmetric rock forms known as roches moutonnées.

- Alpine glacial landforms include cirques, cols, arêtes, glacial troughs, hanging valleys, and fjords.

- Continental glaciers leave behind an undulating landscape of numerous lakes.

15.3 Glacial Deposition

In addition to its incredible power of erosion, the movement of glacier ice is an agent of transportation and deposition that is unequaled among surface processes. The movement of the ice carries huge quantities of rock debris produced by glacial erosion toward the glacial terminus, much of which is deposited beneath the ice on the glacier bed.

GLACIAL DRIFT

Glaciers carry material of all sizes, from the finest rock flour to the largest boulders. Glacial deposits, whether they are laid down by ice or meltwater, are known collectively as **drift**. Most of this glacial drift, particularly in alpine settings, is carried for little more than a few kilometers before being deposited. However, continental glaciers may transport debris, including boulders the size of houses, over great distances. When the ice finally melts, these boulders are dropped as the ice retreats (Fig. 15.16). Boulders of this sort, termed **erratics**, are typically of a rock type quite unlike that of the local bedrock. Indeed, as we learned in Chapter 8, erratics were mistakenly cited by Catastrophists as evidence of Noah's Flood. Erratics scattered across parts of the Great Lakes region, for example, comprise granites and high-grade metamorphic rocks quite unlike the sandstones, limestones, and shales that are typical of the bedrock in this region. From this, geologists can infer that glaciers transported the erratics over vast distances. In fact, rocks similar to the Great Lakes erratics are found in the Canadian Shield hundreds of kilometers to the north, and it

The Origin of the Great Lakes

The five Great Lakes—Superior, Michigan, Huron, Erie, and Ontario—cover an area of more than 244,000 square kilometers (94,000 square miles) to a maximum depth of over 400 meters (1300 feet). Together, they contain almost one-fifth of the world's supply of freshwater. The lakes provide a vital resource to the two nations, Canada and the United States, who share them, and for hundreds of years they have played a central role in the economic development of an entire continent. But why are they there and how did they form?

The Great Lakes were created along the margin of the North American (*Laurentide*) ice sheet. They owe their origin to several factors: the damming of northeasterly flowing rivers by the ice front; crustal subsidence caused by the weight of the ice sheet; glacial erosion and deposition; and processes associated with the glacial retreat.

The Great Lakes have a complex history. The main stages in their evolution are summarized in Fig. 15B. At the onset of glacial retreat, some 14,500 years ago, ice sheets extended well to the south of the Great Lakes region and no lakes existed. However, at that time erosion beneath the advancing glacial lobes was already producing the broad basins that the lakes would eventually occupy. The lakes themselves started forming as the ice sheet intermittently retreated and meltwater became dammed against the ice margin.

Some 14,000 years ago, ancestral portions of Lake Erie and southern Lake Michigan first appeared. They did so as the ice retreated and meltwater filled ice-dammed depressions along the glacial margin that eventually drained south. By 13,000 years ago, the Erie basin was occupied by glacial Lake Whittlesey, which drained westward through glacial Lake Chicago.

As the ice front retreated, the lakes began to assume their present outline. By 11,000 years ago, portions of all five lakes had developed. Ice-dammed along its northern margin, the former

Figure 15B *Stages in the Evolution of the Great Lakes*

These maps show six stages in the evolution of the Great Lakes with the retreat of the Laurentide ice sheet: (a) About 14,500 years ago, before the formation of the Great Lakes, the future positions of which are shown; (b) about 14,000 years ago; (c) about 13,000 years ago; (d) about 11,000 years ago, when drainage was established into the St. Lawrence Seaway; (e) about 9,500 years ago; and (f) about 6000 to 4000 years ago, following glacial retreat and uplift as a result of isostatic rebound.

(a) About 14,500 years ago.

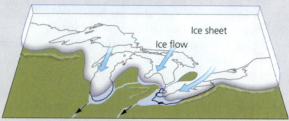

(b) About 14,000 years ago.

(c) About 13,000 years ago.

(d) About 11,000 years ago.

(e) About 9,500 years ago.

(f) About 6000 to 4000 years ago.

TIME

(Continued)

LIVING ON EARTH

The Origin of the Great Lakes (Continued)

Lake Algonquin occupied the future site of lakes Superior, Michigan, and Huron, and drained both to the south, into the Illinois River, and eastward through ancestral lakes Erie and Ontario. The development of this eastern outlet was responsible for the formation of the Niagara Falls as water from early Lake Erie spilled into early Lake Ontario over the resistant limestone beds of the Niagara Escarpment.

By 9500 years ago, the ice had retreated beyond the Great Lakes region and an ancestral Lake Superior had formed. However, the opening of an eastern outlet directly into the St. Lawrence River drained much of the former Lake Algonquin, severing all southward drainage and leaving

two relatively small lakes occupying the future sites of Lake Michigan and Lake Huron.

By about 6000 years ago, the ice had fully retreated and the land was progressively rising as a result of isostatic rebound following removal of the ice sheet. At this time, Lake Superior, Lake Michigan, and Lake Huron were filled beyond their present capacity. Drainage through lakes Erie and Ontario eventually lowered lake levels to their present elevations, thereby severing direct outlets from the other three lakes into the Mississippi and Ottawa rivers, and establishing the Great Lakes as we know them today.

Figure 15.16   Glacial Erratic
Large glacially transported boulders, known as glacial erratics, stand out from the glacial plains of the Great Lake states across which they are scattered.

is likely that these erratics were transported by glaciers from those regions. Beyond erratics, glacial drift is of two distinct types: drift from ice and drift from meltwater.

Drift from ice

Glacial drift deposited directly from ice takes the form of a sediment called **till**. Till comprises a chaotic mixture of

boulders, sand, and clay because glaciers neither sort nor stratify rock debris (Fig. 15.17). Thus, a till is typically a very poorly sorted clastic sedimentary deposit (see Chapter 6).

Till accumulates at the margins of glaciers in the form of mounds, ridges, and blankets. These features are known as **moraines** (Fig. 15.18). Geologists identify several types of moraines: end moraines accumulate in front of glaciers;

Figure 15.17   Till

This chaotic mixture of boulders (some grooved), sand and mud in the Pieria Mountains of northern Greece is typical of glacial till.

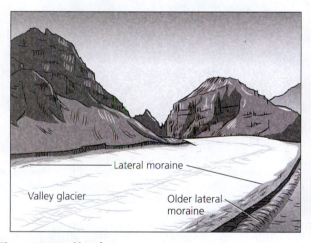

Figure 15.18   Moraine

Glacial moraines comprising low ridges of till were left behind by retreated glaciers above Lake Louise, Banff National Park, Canada.

ground moraines are deposited beneath glaciers; and lateral and medial moraines accumulate on top of alpine glaciers. Although usually deposited as a blanket, ground moraines may also form *drumlins,* distinctive egg-shaped landforms.

An *end moraine* is a bank of till that accumulates at a stationary glacier margin. End moraines that are created during the glacial advance are likely to be removed by the glacier as the ice moves forward. The outermost end moraine marks the position of the glacier's farthest advance and is known as a *terminal moraine* (Fig. 15.19). Scientists use terminal moraines to determine the farthest advance of ice during the Ice Age. During retreat, however, if a cold interval interrupts a warming trend, a retreating glacial margin may once again become stationary, creating a new end moraine—called a *recessional moraine*—behind, but roughly parallel to, the terminal moraine. A glacier may deposit a succession of recessional moraines during intermittent retreat.

A retreating glacier continuously liberates rock debris from the melting ice as the glacial margin retreats. This debris is deposited as a blanket of till known as *ground moraine.*

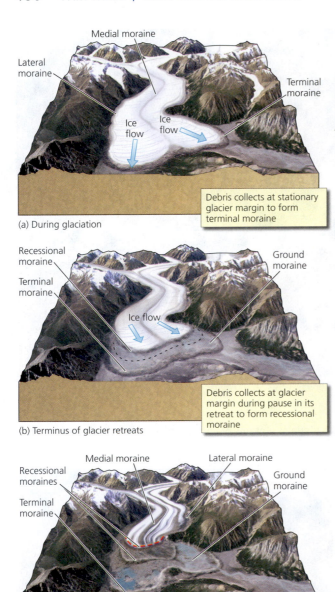

Debris collects at stationary glacier margin to form terminal moraine

(a) During glaciation

Debris collects at glacier margin during pause in its retreat to form recessional moraine

(b) Terminus of glacier retreats

Continued retreat leaves recessional moraine separated from terminal moraine by blanket of ground moraine.

(c) After further retreat of glacier margin.

Figure 15.19   Terminal and Recessional Moraines

(a) A glacier at its farthest advance builds a terminal moraine at its stationary glacial margin. (b) A glacier builds a recessional moraine during a pause in its retreat. (c) Continued retreat leaves both moraines behind and creates additional recessional moraines with each pause in the retreat. Till deposited beneath the glacier is left behind as ground moraine.

(a)

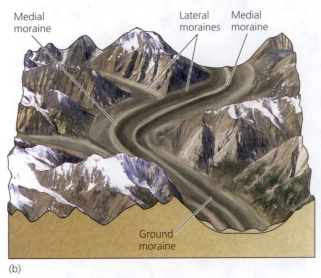

(b)

Figure 15.20   Lateral and Medial Moraines

Lateral and medial moraines accumulate on top of valley glaciers. (a) Where two valley glaciers join, lateral moraines made up of material transported along the glacier margins merge to form a medial moraine down the middle of the glacier. (b) Following retreat of the glacier, lateral moraines form prominent ridges down the side of the glacial valley and the medial moraine occupies the middle of the valley. Note also the ground moraine that is deposited beneath the glacier.

A blanket of till often tends to obscure any preexisting irregularities in the landscape so ground moraines are characterized by a gently rolling topography.

End moraines and ground moraines are depositional features of both alpine and continental glaciers. Alpine glaciers may additionally produce depositional ridge-forms known as lateral and medial moraines (Fig. 15.20). *Lateral moraines* are till ridges carried on or deposited near the *sides* of an alpine glacier, and *medial moraines* are till ridges carried near the *middle* of the glacier. Lateral moraines are made of material that is either loosened from the valley walls by glacial erosion, or falls onto the edge of the glacier from the bordering cliffs as a result of mass wasting processes such as frost wedging and rock slides (see Chapter 12). Medial moraines, on the other hand, form by the merging of lateral moraines below coalescing valley glaciers. When the glacier melts, these two types of ridges respectively occupy the sides and the middle of the glaciated valley. Because of their location on the valley floor, however, medial moraines are often strongly modified by meltwater during glacial retreat and so rarely survive as distinct landforms.

In some areas of past continental glaciation, the ground moraine is sculptured into streamlined mounds of till

Figure 15.21   Drumlin Field
Drumlin fields are swarms of elongate hills aligned in the direction of ice flow. Because of their shape, drumlins in large numbers create a landscape said to exhibit *basket-of-eggs* topography. The inset shows the characteristic shape of an individual drumlin.

called **drumlins**. Drumlins often occur in swarms of parallel, closely spaced individual hills. These *drumlin fields* may contain many hundreds or even thousands of drumlins aligned in the direction of glacial movement (Fig. 15.21). Drumlins are thought to form beneath a continental glacier as a result of the shaping of ground moraine by moving ice during glacial surges.

Drift from meltwater

In contrast to the unsorted drift deposited from ice, drift from glacial meltwater is stratified. While this **stratified drift** may be dominated by material derived from till, it is quite distinct from till because the role of water as a transporting agent makes it a better-sorted sediment with well-defined layering (Fig. 15.22). Deposits of stratified drift form several landforms, including *outwash plains, kettles, kames,* and *eskers.*

Although meltwater streams can flow above, within, and below a glacier, most of the sediment these streams transport is deposited downstream of the glacial terminus as *outwash*. Dominated by sands and gravels, these deposits

Figure 15.22   Stratified Drift
This stratified drift exposed in a gravel pit in Maine is better sorted and more clearly layered than till (compare with Fig. 15.17).

form a broad, gently sloping apron known as an **outwash plain** (Fig. 5.23). Finer material held in suspension by the meltwater streams may be transported far from the glacial terminus. Very fine grained glacial drift may even be picked up by the wind and eventually deposited far from the glacier as a homogeneous blanket of buff-colored silt known as *loess*. We will examine this type of deposit further in Chapter 16.

Outwash plains are often pock-marked by steep-sided basins or depressions called **kettles,** which commonly contain a small lake or swamp (Fig. 15.23). Kettles are produced when a retreating glacier leaves behind large blocks of stagnant ice. Partially or wholly buried by outwash, these blocks of ice melt and leave depressions in the glacial drift that can subsequently fill with water (Fig. 15.24).

Mound-like kames and sinuous eskers are meltwater landforms that are classified according to their shape. **Kames** are low conical hills or steep-sided mounds which,

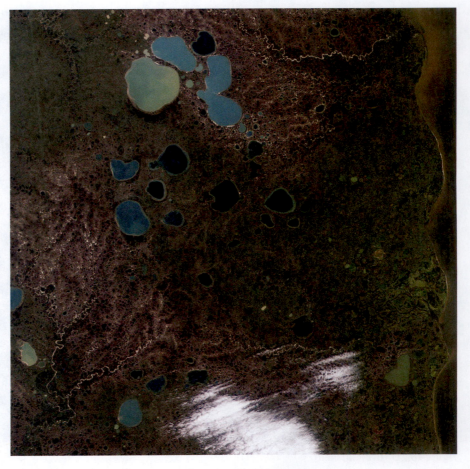

CHECK YOUR UNDERSTANDING

◗ Compare and contrast till and stratified drift.

◗ What are kettles, and how do they form?

◗ What is the difference between a kame and an esker?

unlike drumlins, usually occur in isolation and are made of stratified drift rather than unstratified poorly sorted till (Fig. 15.25). Most kames are the product of drift originally deposited in depressions on the glacier surface by meltwater streams flowing on top of the glacier. The deposits are left behind as isolated hummocks on the glaciated land surface when the glacier melts (see Fig. 15.24b).

Meltwater streams that flow in ice tunnels either within or beneath a glacier usually adopt a winding path so that

Figure 15.23   Kettle Lakes

This natural-color satellite photograph shows blue and green kettle lakes within an outwash plain in northern Siberia.

banks of stratified drift deposited within them are left behind as narrow, sinuous ridges when the ice melts. Ridges formed in this way are known as **eskers** (Fig. 15.26). Eskers are steep sided. They may branch and are often

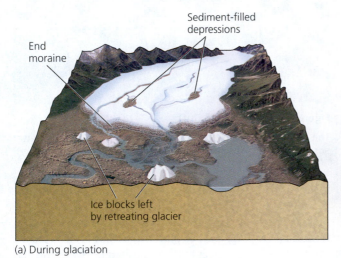

(a) During glaciation

(b) After glaciation

Figure 15.24   Origin of Meltwater Deposits

Stages in the origin of meltwater deposits are shown here (a) during glaciation and (b) after glaciation.

Figure 15.25   Kame

This kame in Wisconsin is Dundee Kame, a meltwater deposit shaped in the form of a conical hill.

Figure 15.26   Esker

This sinuous ridge snaking across an area of ground moraine in North Dakota is an esker, and was deposited by a meltwater steam running beneath the ice sheet.

Figure 15.27   Varve

Varves are thinly layered glacial lake sediments in which each pair of layers is thought to be an annual deposit. Note that the varve layering has been disrupted by a dropstone (center) that fell to the lake bed from floating ice.

discontinuous, and they are oriented at a high angle to the glacier margin (see Fig. 15.24b).

GLACIAL LAKE DEPOSITS

As glaciers retreat, lakes develop, particularly at glacial margins. The lake floors become reservoirs for the deposition of some of the finest sediment from glacial meltwater. Layered deposits consisting of very fine sediment deposited toward the center of glacial lakes are known as varves (Fig. 15.27). Each **varve** consists of two layers, a thicker, light layer of silt deposited during a thaw in the summer months, and a thinner, dark layer of clay that settled from suspension during the winter when the lake was frozen over. Because a single varve records a single year of deposition, varves establish the duration of glacial episodes and, in some cases, their age.

Because the clay layers of a varve are deposited beneath a cover of ice, the environment of deposition is very tranquil and these very thin layers have a remarkably planar geometry. However, another characteristic of these deposits is the presence of isolated rounded pebbles and boulders known as *dropstones*, which disrupt the otherwise planar geometry. Closer inspection shows that the layers underlying dropstones are depressed and the overlying layers are typically draped over the top. Dropstones acquire their name from the interpretation that they drop to the lake bed

from above, thereby disrupting and depressing the regular layering of the sediments on the lake floor. Subsequent deposits drape over the top of the dropstone. Most dropstones are the product of *ice rafting*, having been carried into the lake on floating ice and unceremoniously dropped when the ice melted.

> **CHECK YOUR UNDERSTANDING**
> ● What are varves, and how do they form?

15.3 SUMMARY

- All glacial deposits are known collectively as glacial drift; drift forms from either ice or meltwater.

- Drift from ice is a chaotic mixture of boulders, sand, and clay, called till, which is left behind in the form of mounds, ridges, or blankets called moraines, and elongate hills called drumlins.

- Drift from meltwater is stratified and its deposits form a broad, gently sloping apron called the outwash plain. Depositional landforms made of stratified drift include kettles, kames, and eskers.

- Varves form in glacial lakes and consist of alternating layers of silt (summer deposit) and clay (winter deposit). Varves are used to establish the age and duration of glacial episodes.

15.4 The Pleistocene Ice Age

Many current landscapes of the Northern Hemisphere owe their origin to the period of glaciation popularly known as the Ice Age. This period of glaciation started with the **Pleistocene Epoch**, which began some 2.6 million years ago at the beginning of the Quaternary Period (see Fig. 8.16). During this ice age, glaciers repeatedly advanced across much of North America, Europe, and Siberia (Fig. 15.28), influencing the landscape on a continental scale. At the height of each glaciation in North America, ice covered most of Canada, the mountains of Alaska, and the northeastern and central United States as far south as the Ohio and Missouri rivers. And the flow of meltwater during the advance and retreat of Pleistocene ice sheets dramatically influenced drainage patterns and landscapes far to the south of the glaciers. The cooler global climate at the time also meant that today's vast desert regions were more temperate and had well-developed drainage patterns (see Living on Earth: **Water in the Sahara Desert** in Chapter 14).

Louis Agassiz, the Swiss-born naturalist whose examination of erratics, moraines, and glacial striations convinced him that ice had once spread across much of central Europe, immigrated to the United States in the middle of the nineteenth century. There he noted remnants of many of the same glacial features he had seen in his fieldwork in the Alps. Agassiz became an immensely popular professor at Harvard University. His lectures on the evidence of continental glaciation in Europe and North America galvanized interest in the glacial geology of New England.

The studies of glacial deposits that were initiated in the wake of this interest revealed several episodes of glaciation alternating with warmer interglacial intervals. During these intervals soils developed on the deposits of the previous glacial advance. The interglacial warming brought about by the most recent glacial retreat marks the start of the Holocene Epoch, 10,000 years ago. If the Pleistocene history of glacial advances and retreats continues into the future (and there is no reason to suppose it will not), another period of glaciation should begin in about 10,000 years or so. When that happens, continental ice sheets will once again advance across much of the Northern Hemisphere.

GLACIAL ADVANCES

The effects of glaciation on the landscape did not end with the advent of the landforms of glacial erosion and deposition we explored in Section 15.3. Instead, the impact of glaciation ranged far beyond the ice sheet itself, influencing climate and sea level worldwide, disrupting and redirecting entire drainage systems, and creating a multitude of lakes.

Sea level

One of the most obvious global impacts of Pleistocene glaciation was the effect on sea level of each glacial advance and retreat. Throughout much of Earth history, the hydrologic cycle (see Chapter 1) has been balanced; that is, the volume of water being evaporated from the oceans and precipitated on land equals the volume of water returned to the oceans by continental drainage. Because of this balance, the change in sea level on an annual basis is negligible. During the Ice Age, however, some of the water evaporated from the oceans was stored on land as ice, and only returned to the oceans when the glaciers retreated. Not surprisingly, therefore, each glacial advance produced a worldwide fall in sea level, extending the coastline seaward by several hundred kilometers and exposing the continental shelves. Evidence such as stream channels, fossil forests, and the bones of land mammals on the continental shelves indicate that the continental shelves were once dry land.

Figure 15.28 Pleistocene Ice Age

The extent of continental glaciers in the Northern Hemisphere at the height of Pleistocene glaciation, some 20,000 years ago, is clear on this map. Note the dry "land bridges" that linked Alaska with Asia and Europe with the British Isles.

Land bridges

The fall in sea level associated with each glacial advance also linked many landmasses that are today separated by shallow seas. Some of these so-called *land bridges* played a vital role in the migration of land mammals, including our own ancestors. Forty thousand years ago, for example, giant deer came to Ireland on land bridges across the English Channel and the Irish Sea. In fact, the English Channel is thought to have formed as a result of the catastrophic failure of a land bridge some 8500 years ago.

Perhaps the most important land bridge in terms of human history linked Siberia and Alaska across the modern Bering Straits (see Fig. 15.28). Some archeologists hypothesize that it was across this land bridge that humans first migrated from Asia to the Americas. The land bridge also served as a crossing point for animals. Journeying with these possible early settlers were giant mammoths and mastodons, along with muskox and lemmings. Other animals, such as horses and camels, crossed in the opposite direction to populate Asia and Europe.

> **CHECK YOUR UNDERSTANDING**
>
> ○ What are land bridges and what role does glaciation play in their formation?

Pluvial lakes

The cooler temperatures that accompanied each glacial advance brought more temperate conditions and higher precipitation to regions ahead of the ice front. In many parts of the world, these temperate conditions resulted in the development of large lakes, particularly in the closed basins of otherwise arid regions that experienced increased runoff and lowered evaporation rates. Known as **pluvial lakes** (from the Latin word *pluvia* for rain), most of these bodies of water disappeared when glacial retreat brought about a return to arid conditions, leaving only lakebeds and abandoned shoreline features to mark their former extent. In North America, pluvial lakes were most widespread in Nevada, western Utah, and eastern California (Fig. 15.29). Today, the region's arid climate ensures that only dry lake beds and salt flats occupy the basin floors. But many of the basins show the telltale signs (such as shorelines) of having once been occupied by large freshwater lakes.

The largest of these pluvial lakes was Lake Bonneville (see Fig. 15.29), a vast body of water that once was larger than Lakes Erie and Ontario combined. When glacial retreat brought about a return to warmer, drier conditions, evaporation caused the lake to rapidly shrink and become salty. Today Lake Bonneville has all but dried up, leaving behind only the great salt deposits of the Bonneville Salt Flats and three present-day salty remnants: Great Salt Lake, Utah Lake, and Sevier Lake. On the sides of Antelope Island high above Great Salt Lake, the prominent wave-cut terraces produced by its shoreline are unmistakable evidence of a vast lake that would have dominated much of the Basin-and-Range province (Fig. 15.30).

GLACIAL RETREATS

Like each advance, each glacial retreat causes widespread changes as the ice retreats. During the last retreat, sea level rose again as the ice melted, submerging the continental shelf and land bridges. The continental crust, without its heavy weight of ice, rebounded much like an offloaded cargo ship stands higher in the water than one fully laden. And continental drainage systems, which had been severely disrupted by the previous glacial advance, were again completely redirected during its retreat, changing the network of rivers and creating many lakes.

Figure 15.29 Pluvial Lakes
When cooler, wetter climates prevailed during the last glacial advance, pluvial lakes (shown in blue) occupied many basins in the western United States.

Figure 15.30   Lake Bonneville

These wave-cut terraces on Antelope Island in Great Salt Lake were cut by the shoreline of Lake Bonneville, which was the largest of the pluvial lakes of the western United States.

Figure 15.31   Raised Beaches

Raised beaches on Lowther Island, which lies within the Canadian Arctic Archipelago, record progressive uplift of the shoreline in response to crustal rebound following the last glacial retreat.

(a) Before glaciation

(b) After glaciation

Figure 15.32 Glacial Modification of North American Drainage

(a) Before Pleistocene glaciation, most North American rivers between the Rockies and the Appalachians drained northeast into the Labrador Sea. (b) After the Pleistocene, a new system of drainage was established flowing east and south into the Mississippi or north and west into the Beaufort Sea. The larger lakes are proglacial lakes.

Sea level, crustal rebound, and raised beaches

As the ice sheets retreated and sea level rose once again, land bridges were submerged and continental shelves were flooded once more. In North America, the formation of barrier islands along the Atlantic and Gulf coasts and the development of drowned river valleys like those of Chesapeake Bay and the Gulf of St. Lawrence are testament to these rising sea levels. At high latitudes, the flooding of glacial troughs was similarly responsible for the formation of fjords. Yet despite the huge volume of ice that melted during the glacial retreat, the raising of some coastlines shows clear evidence of falling sea level during this period.

The existence of raised coastlines shows that, in certain areas, sea level has actually fallen since the last glacial retreat. So how can this be? As we discussed in Chapter 9, Earth's crust is in a state of floating balance called *isostasy*. Because of isostasy, the added weight of a continental ice sheet caused the crust beneath the sheet to subside into the upper mantle, just as a cargo ship sinks into the water as it is loaded with goods (see Fig. 9.7). When the ice sheet retreated, the crust rebounded and the upper mantle flowed back to reoccupy the position it previously vacated. Because flow in the mantle is sluggish, the process of crustal rebound was very slow, whereas changes in sea level as a result of glacial melting occurred almost immediately. As a result, crustal rebound lags well behind the rise in sea level following glacial retreat. Indeed, although the rise in sea level has long since taken place, many land areas once covered by ice sheets have yet to fully recover. Hudson Bay in the Canadian Arctic is just such an area. Sited over the former zone of accumulation where crustal subsidence was greatest, this area continues to rise at about 2 centimeters (1 inch) per year. Evidence of this rise is clear in many parts of the Canadian Arctic where uplift of the rebounding coastline has raised the shore above the high water mark. The result is a series of raised coastlines known as *raised beaches* (Fig. 15.31).

So, while the effect of the last glacial retreat on coastlines is generally one of sea level rise, continued uplift of the coast as a result of crustal rebound following the rise in sea level has locally caused sea level to fall in some areas. The crust beneath the center of Hudson Bay has already risen 300 meters (1000 feet) and must rise a further 80 meters (260 feet) before isostatic equilibrium is reestablished. Not until about 4000 years from now will that crust once again lie above sea level.

CHECK YOUR UNDERSTANDING

○ In what ways does continental glaciation affect sea level?

Continental drainage

The effect of Pleistocene glaciation on the drainage of North America was so profound that, of all the major drainage systems between the Rockies and the Appalachians, only the lower reaches of the Mississippi basin show a drainage pattern today that resembles the one they followed before glaciation. Before Pleistocene glaciation, much of the continental interior drained northeastward through Canada. For example, the Missouri River, along with Canada's Saskatchewan, Athabasca, Peace, and Liard rivers, drained across Hudson Bay and into the Labrador Sea (Fig. 15.32).

This preglaciation pattern of drainage was severely disrupted when ice advanced across the continent. Effectively dammed by a wall of ice, many of the northerly flowing rivers ponded along the front of the ice sheet to form vast lakes. Termed **proglacial lakes** because of their positions in front of the continental glacier, many of the Great Lakes evolved in this fashion (see Living on Earth: **The Origin of the Great Lakes** on page 477).

With a probable thickness of 3 kilometers (2 miles) in its zone of accumulation, the colossal weight of the ice sheet depressed the crust beneath it by several hundred meters. As a result, a regional slope was created immediately around the ice sheet that caused surface water to drain toward it. Spillwater from the proglacial lakes overflowed into a drainage system that followed the ice margin (see In Depth: **Lake Missoula and the Channeled Scablands** on page 490). At the same time, the preglacial drainage system beneath the ice sheet was largely obliterated. Following glacial retreat, the new pattern of drainage was established and Canada was left littered with lakes (see Fig. 15.32).

The largest of all the proglacial lakes once covered much of Manitoba and parts of central Canada, Minnesota, and North Dakota (Fig. 15.33). Lake Agassiz, named after the

Figure 15.33   Lake Agassiz

Lake Agassiz, a proglacial lake, drained when the ice sheet retreated, leaving Lakes Winnipeg, Manitoba, and Winnipegosis as remnants. Sediments of the former lake bed are responsible for some of the region's richest agricultural land.

IN DEPTH

Lake Missoula and the Channeled Scablands

One of the most dramatic events of the last glacial retreat left a scar across the eastern half of Washington State so massive in scale and so alien in origin that its likeness has been found only on the planet Mars. Cut into the basalt lava flows of the Columbia Plateau, the Channeled Scablands comprise a vast region of deeply scoured dry canyons, braided on a gigantic scale and floored by channel bars and ripple marks. These features are like those of normal river systems but of enormous size. Some areas of the Scablands are wrinkled into a thousand gravel ridges that are up to 10 meters (33 feet) high and 75 meters (250 feet) apart. So large are these features that only from the air are they recognizable as gigantic ripple marks (Fig. 15C).

What could have caused erosion and deposition on such a vast scale? The debate over the origin of this landscape developed into one of the great controversies of North American geology. The traditional interpretation was that the Scablands were carved slowly over a long period of time by normal rivers. But in the 1920s, American geologist J. Harlen Bretz challenged that assumption. He argued that the landscape was the work of a catastrophic flood of gigantic proportions. Bretz's hypothesis was dismissed by his fellow geologists, who were unconvinced by his arguments that only huge volumes of water traveling at tremendous speeds could have produced river features on such a gigantic scale. Bretz was later vindicated when both the cause of the flooding and the source of the flood water were found in glacial Lake Missoula.

While the retreating ice sheet (see Fig. 15.27) was setting the stage in North America for the formation of proglacial lakes (see Fig. 15.32b) from the St. Lawrence River to the Beaufort Sea, a lobe of the retreating Cordilleran ice sheet, a smaller ice mass centered on the Canadian Rockies, extended south into Washington and Idaho, damming the northwest-flowing Clark Fork River near the Idaho-Montana border (Fig. 15D). The ice dam created Glacial Lake Missoula, a proglacial lake that covered over 7800 square kilometers (3000 square miles) of western Montana, held over 2000 cubic kilometers (480 cubic miles) of water, and was about 600 meters (2600 feet) deep at the ice dam.

Like a grounded boat on an incoming tide, the ice dam became more buoyant as the water level rose in the lake.

Figure 15C   Giant Ripple Marks

It is only from a distance that gravel ridges like these in the Channeled Scablands of Camas Prairie, Montana, which resemble great waves in an ocean, can be recognized as giant ripple marks formed by a flood of gigantic proportions (note truck for scale).

Figure 15D   Glacial Lake Missoula

Failure of the ice dam at Glacial Lake Missoula in the Idaho panhandle produced the world's largest documented flood. Insets show the locations of some of the landforms the flood produced, including giant ripples, old shorelines, and dry waterfalls and channels. (Old shorelines photo by Don Hyndman, University of Montana.)

And just as a grounded boat eventually floats, so the base of the glacier finally lifted and water undercut the dam along a segment 180 kilometers (110 miles) long. The size of the resulting deluge is hard to imagine. Towering 600 meters (2000 feet) high at its source, a wall of water tore across eastern Washington at speeds estimated in excess of 100 kilometers per hour (60 miles per hour). With a volume 10 times greater than that of all the world's current rivers put together, the torrent ripped away 60 meters (200 feet) of soil and excavated a vast maze-like network of canyons, carved into the bedrock to a depth of over 70 meters (230 feet). Indeed, the power of the currents was so great that boulders measuring 10 meters (33 feet) across and weighing over

200 tons still lie scattered across the countryside like so much debris.

In all, over 200 cubic kilometers (50 cubic miles) of material was eroded. Much of it was redeposited in huge gravel bars and shaped into gigantic streambed ripples. Within a few days the torrent subsided and the great lake had drained. But in the wake of the flood, the dramatic dry canyons in the Columbia Plateau basalts, with their carved islands or "scabs" and abandoned waterfalls, were left as the Channeled Scablands (Fig. 15E).

Bretz originally believed that just a single flood had been responsible for the scouring of the Channeled Scablands. However, recent studies have shown that floods of this magnitude, each lasting just a few days, occurred here

(Continued)

IN DEPTH

Lake Missoula and the Channeled Scablands (Continued)

Figure 15E   Channeled Scablands
Like Niagara, but without the water, dry gorges and abandoned waterfalls of the Channeled Scablands at Dry Falls in eastern Washington State testify to the catastrophic flooding of glacial Lake Missoula.

more than 40 times between 15,000 and 12,800 years ago. While this may sound incredible, it is easy to see how it might happen, given the nature of the dam's collapse. The lake drained when the ice dam started to float. And as the lake level fell, the glacial wall would have once again formed, sealing off the lake basin. Once this had occurred,

the lake would have filled again. It is estimated that it would have taken just 50 years or so to fill. As the lake level rose, the ice dam would have become buoyant once more and once again let go catastrophically. In this way, the stage for catastrophic flooding could have been set and reset scores of times during the lifetime of the lake.

---

great Swiss proponent of the Ice Age, persisted until glacial retreat permitted it to drain, first into the Mississippi basin, then into the Great Lakes, and later into Hudson Bay. Water that failed to drain formed Lake Winnipeg, Lake Manitoba, and Lake Winnipegosis (see Fig. 15.32b). Although these are large lakes, they are but small remnants of the former Lake Agassiz, the abandoned shore-lines of which enclose an area of some 250,000 square kilometers (96,500 square miles). Thanks to the sediments deposited on the abandoned lake floor, much of the region today enjoys a fertile soil, explaining why this region of the prairies is one of the most agriculturally rich regions in North America.

15.4 SUMMARY

- Many current landscapes in the Northern Hemisphere owe their origin to the Pleistocene ice age, which influenced the scenery of North America on a continental scale.

- Besides creating glacial landforms, the Pleistocene ice age influenced climate and sea level worldwide. It also disrupted entire drainage systems and created a multitude of lakes.

15.5 Plate Tectonics and Glaciation

As we discussed in Chapters 12, 13, and 14, Earth's surface processes are profoundly influenced by topography, and are consequently linked to plate tectonic processes that have produced this topography. For example, there is direct and obvious connection between plate tectonics and the advance of the glaciers, which like mass wasting and running water, requires a slope. In this section of the chapter, we explore some of the less obvious but more profound connections between plate tectonic processes and longer-term cooling and warming intervals in Earth's geologic past.

Changes in Earth's climate over the long term—that is, in a time frame of many tens of millions of years—are an expected consequence of plate motion. As plates move, Earth's geography changes, altering the climate of the continents and changing the paths of ocean currents that transport heat from the equator to the poles. Moving plates also bring continents to the poles and so play a role in the formation of polar ice caps, just as Wegener argued in his case for continental drift (see Fig. 2.4).

More importantly, however, plate motion strongly influences the composition of the atmosphere, particularly with regard to the greenhouse gas, carbon dioxide. As we will discuss in more detail in Chapter 21, greenhouse gases are so-called because their presence allows Earth's atmosphere to trap heat, much like the panes of glass in a greenhouse. Hence, increasing levels of greenhouse gas (such as carbon dioxide) in the atmosphere lead to global warming. Carbon dioxide comes from volcanoes, which, as we have learned, are directly related to plate tectonics. Left unchecked, volcanoes steadily add this gas to the atmosphere, warming the planet in the process.

Plate tectonics also controls the arrangement of continents and the distribution of mountain belts, and these, in turn, influence the amount of weathering. Recall from Chapter 5 that weathering removes carbon dioxide from the atmosphere by reacting it with dissolved calcium to form carbonates. Increased weathering therefore causes climatic cooling by removing greenhouse gas from the atmosphere. Because chemical reactions occur faster at higher temperatures, the cooling effect of weathering serves to counteract any warming trend brought about by increased atmospheric carbon dioxide. Increased weathering also accompanies mountain building, which leads to rapid erosion. Indeed, some scientists argue that the onset of the cooling trend that resulted in the Pleistocene ice age was brought about by an increase in weathering associated with the rise of the Himalayas. In this scenario, calcium derived from the erosion of the Himalayas effectively lowered carbon dioxide levels in the atmosphere by promoting the deposition of carbonates in the Indian Ocean.

Other factors may have also played a role in the onset of this cooling trend since the trend broadly coincides not only with the rise of the Himalayas, but also with the evolution of grasses. Grasses may be climatically important because the type of photosynthesis they use to extract carbon dioxide from the atmosphere is significantly more efficient than that of other plants. Hence, the spread of grasses could have lowered atmospheric carbon dioxide

levels by increasing the amount of carbon dioxide stored in the soil.

The Pleistocene is not the only time interval in Earth history during which the planet has been plunged into an ice age. Evidence of other ice ages exists in the geologic past and plate tectonics likely played a key role in their development. Although widespread continental glaciation has occurred in the past, it is not commonplace in the geologic record. In fact, during Earth's entire 4-billion-year record of geologic history, there is evidence of such glaciation on no more than a dozen or so occasions (Fig. 15.34).

But what type of evidence is used to deduce the existence of ancient ice ages? The challenge is to recognize features similar to those found in modern glacial deposits that are preserved in the ancient rock record. This evidence takes the form of varves preserved in thinly bedded shales, other fine-grained deposits containing dropstones, and deposits of material called *tillite* (Fig. 15.35). **Tillite** is a chaotic, unstratified, and unsorted rock type made of lithified till. Sometimes containing striated boulders and occasionally resting on grooved bedrock, tillite deposits are diagnostic of past glacial conditions.

The oldest documented ice age occurred about 2.4 billion years ago, the evidence for which is preserved as tillites in Canada, the United States, South Africa, and Finland (Fig. 15.36). However, the most extensive period of glaciation in Earth history took place in the Late Precambrian, between about 900 and 580 million years ago. During this interval, at least six ice ages have been identified, four of which, between 750 and 580 million years ago, may have been so severe that Earth's entire surface, including its oceans, completely froze over.

The idea of an entirely frozen planetary surface, or *snowball Earth*, has been proposed to account for two enigmatic and seemingly contradictory characteristics of these late Precambrian glacial deposits. The first is the presence of such deposits near sea level in regions that were then in the tropics, and the second is the fact that the glacial deposits are directly overlain by warm-water limestone. How could such dramatic "freeze-fry" changes in temperature conditions occur so rapidly? Ice at the equator and a hothouse aftermath are precisely the conditions predicted for a snowball Earth.

Geologists propose that a number of steps led up to this climatic catastrophe (Fig. 15.37). First, an increase in weathering brought about by the drift of many small continents to the equator removed large quantities of carbon dioxide from the atmosphere, causing dramatic global cooling. Then the polar oceans froze, which amplified the cooling trend because the white surfaces of these frozen oceans reflected rather than absorbed the Sun's heat. Under normal circumstances the spread of ice across the continents would have curtailed their weathering, allowing temperatures to rise. Because the continents were at the equator, no continental ice sheets formed and weathering and cooling continued unchecked, plunging Earth into

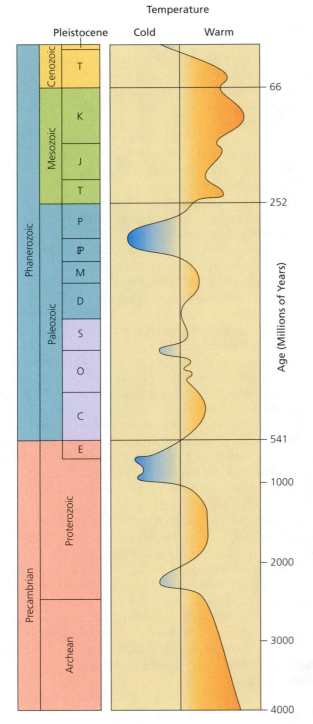

Figure 15.34   Earth's Climate Through Time

The Earth's long-term history of relative temperature changes with time shows that the planet has oscillated between warm (to the right) and cold (to the left) climates, and that ice ages (in blue) are unusual events that have occurred without regularity throughout much of Earth history. E though T are abbreviations for the geologic periods (See Fig. 8.16).

a deep freeze. With average temperatures as low as −50°C (−58°F), the entire planet would have been engulfed in ice, perhaps within as little as a thousand years. If the oceans were indeed frozen, they would have been unable to

Figure 15.35   Tillite

Characteristically unstratified and unsorted tillite of the Latest Precambrian Elatina Formation is located in the Flinders Ranges National Park, South Australia.

Figure 15.36   Earth's Oldest Glaciation

(a) Tillite and (b) varves with dropstones in 2.4 billion-year-old glacial deposits in Ontario, Canada, are evidence of Earth's oldest documented ice age.

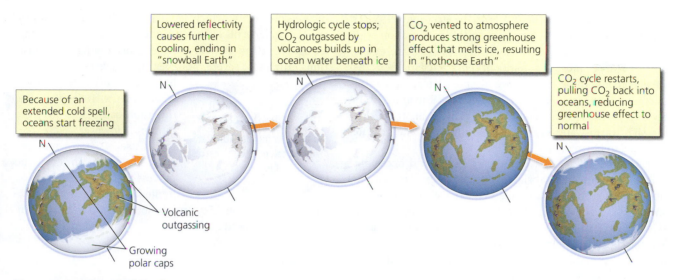

Because of an extended cold spell, oceans start freezing

Lowered reflectivity causes further cooling, ending in "snowball Earth"

Hydrologic cycle stops; CO_2 outgassed by volcanoes builds up in ocean water beneath ice

CO_2 vented to atmosphere produces strong greenhouse effect that melts ice, resulting in "hothouse Earth"

CO_2 cycle restarts, pulling CO_2 back into oceans, reducing greenhouse effect to normal

Volcanic outgassing

Growing polar caps

Figure 15.37   Snowball Earth

A specific sequence of events is thought to be associated with the "snowball Earth" scenario and the rapid "freeze-fry" changes in temperature conditions.

evaporate and the entire hydrologic cycle would have been switched off.

With the continents ice covered, weathering ceased and carbon dioxide vented from volcanoes accumulated in the atmosphere. As a result, atmospheric carbon dioxide concentrations increased dramatically, perhaps as much as a thousand-fold, over a period of some 10 million years. Global temperatures rose and the equatorial oceans thawed, seawater

Figure 15.38   Late Paleozoic Glacial Striations

This striated and polished bedrock surface in tropical southern Brazil was produced by the movement of continental ice sheets when India lay close to the South Pole in the late Paleozoic.

evaporated and added water vapor to the atmosphere. Water vapor is a strong greenhouse gas, so this evaporation triggered runaway heating that caused temperatures to soar to 50°C (122°F) or more in a matter of centuries. Finally, as the ice melted and the continents were exposed, torrential rainfall brought huge quantities of dissolved calcium and carbon dioxide to the oceans, cooling the planet and triggering massive warm-water carbonate deposition as the climate returned to normal.

Although ice ages have occurred on several occasions since these late Precambrian events, such extreme conditions are not thought to have occurred either before or since. The fact that the Sun has been getting progressively hotter since it first formed may have made Earth less susceptible to global freezing after the late Precambrian, but the main contributing factor is more likely to have been the unusual equatorial distribution of the continents at that time, a situation that directly reflects plate tectonic processes.

Other ice ages are thought to have occurred about 450 million years ago, and again about 360 million years ago. But the best-documented record of pre-Pleistocene glaciation is that of the late Paleozoic (between about 360 and 250 million years ago). The existence of this ice age, which coincided with the assembly of the supercontinent Pangea, has been known for more than a century. Indeed, as we learned in Chapter 2, it was the distribution of the glacial deposits of this ice age that Alfred Wegener used to show that much of the ancient continent of Gondwana was located near the South Pole in the Late Carboniferous. Often lying on striated bedrock (Fig. 15.38), glacial deposits of this period are widely distributed across parts of South America, Africa, Antarctica, Australia, and India (see Fig. 2.4).

From the geologic record, it is clear that ice ages are unusual events that have occasionally occurred throughout much of Earth history. Although there is no obvious regularity in their occurrence (see Fig. 15.34), the distribution of the continents, which is controlled by plate tectonics, appears to have played a central role in their timing. Most ice ages have initiated on continents occupying polar positions. The most severe, however, may have been triggered by an equatorial distribution of continents. In either case, the root cause lies in plate motion. However, plate tectonics itself is not a requirement for glaciation since the process of plate tectonics is limited to Earth, and, as we shall see in Section 15.6, ours is not the only celestial body to have experienced glaciation.

> **CHECK YOUR UNDERSTANDING**
>
> ○ What is tillite, and what does it tell us about Earth's history of ice ages?

15.5 SUMMARY

- Changes in Earth's climate over time are an expected consequence of plate motion.
- Evidence for a dozen or so ice ages occurs in the form of varved shales and tillite prior to the Pleistocene.
- The oldest well-documented ice age occurred about 2.4 billion years ago.
- On several occasions between 750 and 580 million years ago, Earth may have frozen over completely.
- The best-documented pre-Pleistocene glaciation occurred during the assembly of the Pangea.

15.6 Ice on Other Worlds

Human exploration of the Solar System by satellites has shown that ice is by no means restricted to Earth. Many of the moons of the outer planets, such as Jupiter's Callisto and Ganymede, Saturn's largest moon, Titan, and most of the moons of Uranus, are now thought to have extensive

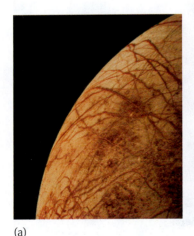

(a)   (b)

Figure 15.39   Europa

(a) The icy surface of Jupiter's moon Europa may rest on a layer of water as much as 100 kilometers (60 miles) thick. (b) In detail, broken portions of Europa's surface resemble a logjam of vast icebergs.

covers of ice. Ariel, one of the outer moons of Uranus, has broad, smooth-floored canyons that appear to have been smoothed by flowing ice, although it is unlikely to have been water ice. Other moons, such as Neptune's largest moon Triton, have large polar ice caps, in this case made of nitrogen ice.

Most dramatic of all is Jupiter's moon, Europa, which is about the size of our Moon and may hide an ocean of water beneath a thick layer of ice (Fig. 15.39). Resembling a cracked eggshell, Europa's surface is thought to consist of an icy shell some 10–30 kilometers (6–19 miles) thick, which covers an ocean perhaps 100 kilometers (60 miles) deep. Constantly kneaded by Jupiter's strong gravitational forces and bulged and broken by rising blobs of warmer ice, portions of Europa's surface have been broken into huge, iceberg-like blocks the size of small cities that appear to have initially separated in a sea of slush or water and then become frozen in place.

Polar ice caps have also been identified on Mars (Fig. 15.40), the presence of which testifies to the presence of water and so bears on the possible existence of life on the planet. Although the Martian ice caps are made up of frozen carbon dioxide (dry ice) as well as water, the ice caps expand and contract with the seasons just like those of our own poles. The Mars Global Surveyor spacecraft completed a survey of the planet's north pole, where the planet's larger ice cap is located, in 1999. This survey revealed an ice sheet about one and a half times the size of Texas, rising almost 3 kilometers (2 miles) above the surrounding terrain. Composed almost entirely of frozen water, the ice cap is estimated to contain some 2 million cubic kilometers (770,000 cubic miles) of ice. The survey also revealed smaller blocks of ice farther from the pole, suggesting that the ice cap was once larger. The same is true of the planet's smaller south polar ice cap around which astronomers have identified furrows and ridges that resemble glacial troughs and moraines. Whether the shrinking ice caps reflect some change in Martian climate is unknown, but as we will discuss in Chapter 21, such a shift would pale in comparison to some of the dramatic climatic upheavals the planet has experienced in its past.

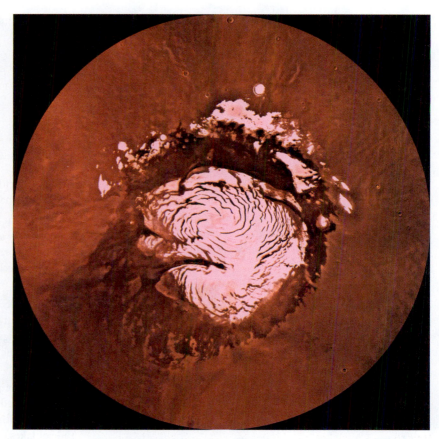

Figure 15.40   Martian Ice Cap
A computer-enhanced photomosaic shows the Martian north polar ice cap.

> **CHECK YOUR UNDERSTANDING**
>
> ○ Why do astronomers think that the polar ice caps of Mars may be shrinking?

15.6 SUMMARY

- Earth is not the only celestial body to have experienced glaciation, as evidenced by the existence of polar ice caps on Mars.

- The most dramatic example of extraterrestrial glaciation is Jupiter's moon Europa, which may hide an ocean of water beneath a thick layer of ice.

Key Terms

ablation   470
abrasion   471
alpine glacier   469
arête   474
basal sliding   470
cirque   469

col   474
continental glacier   469
crevasse   469
drift   476
drumlin   481
erratic   476

esker   482
firn   467
fjord   475
glacial budget   470
glacial terminus   470
glacial trough   474

glaciation   466
glacier   466
glacier bed   470
hanging valley   476
horn   474
kame   482

kettle   482
moraine   478
outwash plain   481
plastic flow   469
Pleistocene Epoch   485

plucking   471
pluvial lake   486
proglacial lake   489
roche moutonnée   472
snowline   467

stratified drift   481
striations   471
surge   470
till   478
tillite   493

valley glacier   469
varve   484

Key Concepts

15.1 GLACIERS AND GLACIER ICE

- Glaciers, or moving masses of ice, can have enormous impact on Earth's surface as agents of erosion and deposition. Today glaciers cover about 10 percent of Earth's surface. But much of Earth's surface was sculpted by glaciers that have since melted away.

- Glacier ice forms from the slow compaction of snow.

- There are two main types of glacier: vast ice sheets called continental glaciers and the glaciers of mountainous regions called alpine glaciers.

- Glaciers form and grow in areas where the amount of snow that accumulates in winter exceeds that which is lost in the summer. Glaciers flow in one of two ways: a slow internal process of plastic flow and, in warmer regions, the faster external process of basal sliding.

- Glaciers move from an area of net gain to one of net loss. This annual balance is known as the glacial budget.

15.2 GLACIAL EROSION

- Glaciers erode by abrasion and plucking. These two processes often shape rock masses into an asymmetric form known as a roche moutonnée.

- Typical alpine glacial landforms include cirques, arêtes, glacial troughs, hanging valleys, and fjords. Continental glaciers typically leave behind an undulating landscape of countless lakes.

15.3 GLACIAL DEPOSITION

- All glacial deposits are known collectively as glacial drift. Drift is of two types: drift from ice and drift from meltwater.

- Drift from ice is a chaotic mixture of boulders, sand, and clay, called till. This mixture is left behind in the form of mounds, ridges, or blankets of till, called moraines, and spoon-shaped hills called drumlins.

- Drift from meltwater is stratified. The deposits form a broad, gently sloping apron, the outwash plain. Depositional landforms made of stratified drift include kettles, kames, and eskers.

15.4 THE PLEISTOCENE ICE AGE

- Many current landscapes in the Northern Hemisphere owe their origin to the Pleistocene ice age. This glaciation influenced the scenery of North America on a continental scale.

- Besides the landforms of glaciation, the Pleistocene ice age influenced climate and sea level worldwide. It also disrupted entire drainage systems, and created a multitude of lakes.

15.5 PLATE TECTONICS AND GLACIATION

- Evidence of the dozen or so ice ages that occurred prior to the Pleistocene can be found as varve-like layers in shales and tillite.

- The oldest well-documented ice age occurred about 2.4 billion years ago. On several occasions between 750 and 580 million years ago, Earth may have frozen over completely.

- The best-documented pre-Pleistocene glaciation occurred in Gondwana during the assembly of the supercontinent Pangea.

15.6 ICE ON OTHER WORLDS

- Ice is not restricted to Earth. Polar ice caps have been identified on Mars and Jupiter's moon Europa may hide an ocean of water beneath a thick layer of ice.

Study Questions

1. How is it possible for ice to flow from the source of a glacier to its terminus while the glacier itself is retreating?

2. What landforms might you expect to see in a region formerly occupied by an alpine glacier?

3. What landforms might you expect to see in a region formerly occupied by a continental ice sheet?

4. Why don't the effects of continental glaciation stop at the glacial terminus?

5. Geologists believe that many of the dry basins of Nevada and western Utah were once occupied by large freshwater lakes. On what evidence is this claim based? What types of lakes were these? How did they form, and where did they go?

6. In what ways was the drainage of North America modified by the Pleistocene ice age?

7. How do moving plates influence global climate?

8. Why do some geologists think that between 750 and 580 million years ago a series of ice ages may have caused the entire Earth to freeze over?

9. Why might Earth be cooler if all the continents were located at the equator?

10. What makes scientists think the Martian polar ice caps were once larger than they are today?

16 Deserts and Winds

16.1 The Origins of Deserts

16.2 Weathering and Erosion in Deserts

16.3 Wind in Deserts

16.4 Desertification: Natural and Human Induced

16.5 Deserts and Plate Tectonics

16.6 Deserts on Mars

▶ Key Terms

▶ Key Concepts

▶ Study Questions

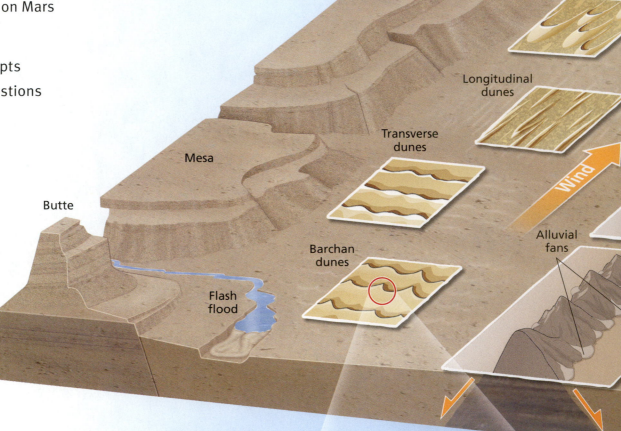

SCIENCE REFRESHER: Ocean Currents and the Coriolis Effect 506

IN DEPTH: Dust Devils 512

LIVING ON EARTH: The Dust Bowl 515

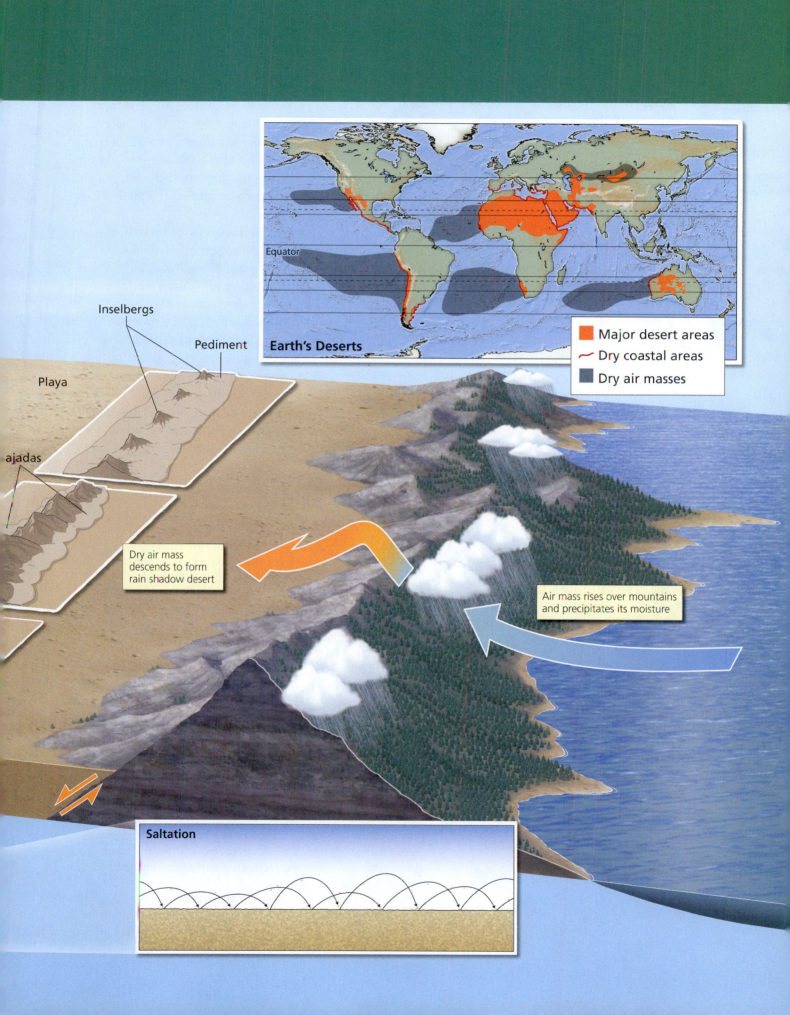

Inselbergs

Pediment

Playa

ajadas

Earth's Deserts

Equator

Major desert areas
Dry coastal areas
Dry air masses

Dry air mass
descends to form
rain shadow desert

Air mass rises over mountains
and precipitates its moisture

Saltation

16.1 Explain that deserts originate from a variety of processes and therefore form in a variety of settings.

16.2 Identify the surface processes that lead to weathering and erosion in deserts.

16.3 Describe the role wind plays in sculpting the desert landscape and the features it creates.

16.4 Explain how human activities such as cultivation and overgrazing pose a serious environmental threat in many parts of the world, resulting in the creation and expansion of deserts.

16.5 Show the influence of plate tectonics on the distribution of deserts using the evidence for ancient deserts in the rock record.

16.6 Use photographs of Mars to show that desert landforms like those on Earth are also present on other worlds.

Deserts, and the semiarid regions that surround them, form naturally whenever global atmospheric circulation, high mountain belts, cold ocean currents, or distance from the sea creates dry climates. These arid and semiarid conditions affect more land areas than any other climate, such that desert and semiarid regions account for almost a third of Earth's land surface.

In this chapter we pay special attention to the action of wind, because it is in deserts that the effects of wind are most apparent. Because desert soil is dry and dusty, it is readily picked up and transported by the wind. Armed with dust, the wind becomes a colossal sandblaster, etching rock faces and sculpting outcrops in a manner unlike that of any other surface process. As dust storms abate, this windblown material is deposited to form clastic sediments with unique features.

In addition to wind, water plays a role in the desert too. Although the role of wind in sculpting desert landscape is a more or less continuous process, running water remains the dominant agent of desert erosion and deposition, often taking the form of flash floods that are the result of intermittent storms.

16.1 The Origins of Deserts

Deserts are barren land areas that receive minimal rainfall. Although all deserts are dry regions, not all deserts are dry for the same reasons. In fact, five different types of desert are recognized, each with a dry climate of different origin. These five types are: (1) subtropical deserts; (2) rain-shadow deserts; (3) coastal deserts; (4) continental interior deserts; and (5) polar deserts. So what is the origin of each of these different types of desert and where and why do they form? In this section of the chapter, we will see that different processes are responsible for the formation of different desert types and that some deserts form from a combination of these processes.

> **CHECK YOUR UNDERSTANDING**
>
> ● How many different types of desert are recognized and what are they called?

SUBTROPICAL DESERTS

Although deserts occur on all of the world's major continents (Fig. 16.1), they are by no means randomly distributed. Most lie within one of two subtropical belts between 20 and 30 degrees north and 20 and 30 degrees south of the equator. These **subtropical deserts** result from a global pattern of atmospheric circulation that causes moisture to be removed from the air at the equator.

At the equator, warm moist air masses rise high into the atmosphere and spread out toward the poles, creating a band of low-pressure systems known as the *equatorial low* (Fig. 16.2). As the air masses rise, they cool, and this both increases the air's density (and, hence, the

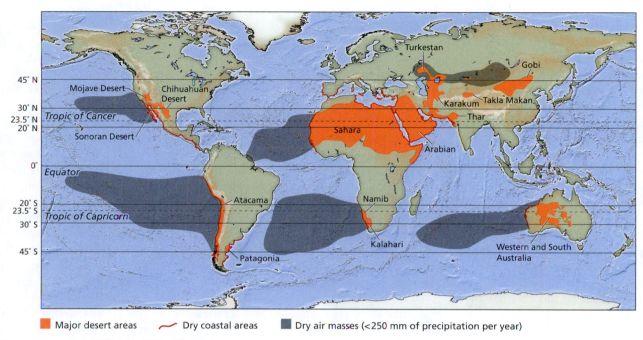

Figure 16.1 Global Distribution of Deserts

Note that most deserts and the semiarid regions that surround them lie between 20 and 30 degrees of latitude to the north and south of the equator.

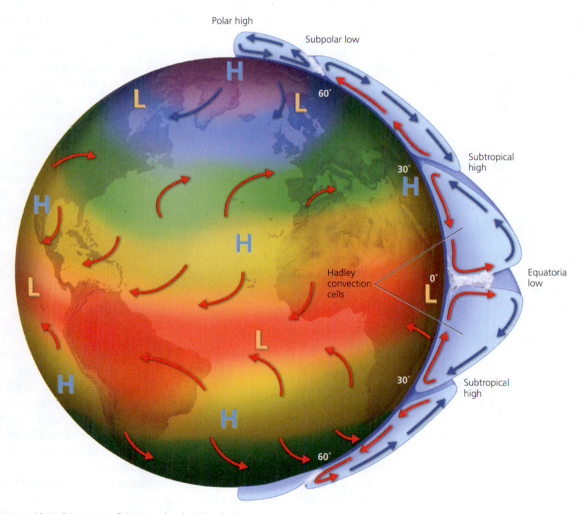

Figure 16.2 Global Pattern of Atmospheric Circulation

The pattern of atmospheric circulation between the equator and the poles involves a series of convection cells that give rise to areas of high (H) and low (L) atmospheric pressures. Subtropical deserts form beneath the subtropical highs between about 20 and 30 degrees of the equator. Color changes from red to blue track decreasing temperature from the equator to the poles. Red and blue arrows show circulation of warm and cool air, respectively.

atmospheric pressure) and decreases its capacity to hold water vapor. The released moisture falls as torrential equatorial rain, which produces the rain forests that dominate the equatorial regions of continents. Having lost their moisture as rainfall, the air masses spreading outward from the equator are relatively dry. These air masses are also compressed as they migrate poleward. This is because the air is moving on the surface of a spherical Earth whose circumference decreases from the equator to the poles. This convergence further increases the atmospheric pressure.

At about 30 degrees from the equator—in the vicinity of the tropics of Cancer and Capricorn—the increased atmospheric pressure causes the air to descend, creating a persistent band of high-pressure centers, known as the *subtropical high*, at these latitudes. This downwelling air is very dry since almost all of its moisture has already fallen as rain at the equator. Because the air becomes warmer as it descends, it is warm when it reaches the surface and its capacity to hold water vapor is very high. The downwelling air behaves like a huge blow dryer, rapidly evaporating any surface water on the land and creating a subtropical belt with a very dry climate. As it spreads out at the surface, part of the descending air moves back toward the equator, eventually rising once again to complete a loop of atmospheric circulation known as a *Hadley convection cell* (Fig. 16.2), named after the eighteenth-century amateur meteorologist George Hadley.

Because of this huge convection cell, areas beneath subtropical high-pressure zones experience arid climates, and many of the world's greatest deserts occur in these regions. In the Northern Hemisphere, four of these deserts straddle the Tropic of Cancer at 23.5 degrees north latitude: the great Sahara Desert of North Africa, the Arabian Desert of the Middle East, and the Sonoran and Chihuahuan deserts of the southwestern United States and Mexico. Likewise in the Southern Hemisphere, four deserts—the great Australian desert, the Namib and Kalahari deserts of southern Africa, and the Atacama Desert of Chile and Peru—straddle the Tropic of Capricorn at 23.5 degrees south.

> **CHECK YOUR UNDERSTANDING**
> ◐ Why do areas that lie beneath the subtropical high-pressure zones experience dry climates?

RAIN-SHADOW DESERTS

As we just learned, at the equator the air rises, cools, and loses its moisture as part of the global pattern of atmospheric circulation (Fig. 16.2). Moist air can also be forced upward if its path encounters a range of high mountains. Where this happens, the air is again cooled as it rises, its capacity to hold water decreases, and the moisture it contains falls as rain or snow. As a result, by the time the air reaches the far side of the mountain range, it has lost most of its moisture so that very little rain can fall. Because of this, mountain belts that block the path of the prevailing wind tend to be wet on the side facing the wind and dry on the sheltered, or leeward, side. The dry regions on the leeward side of such mountain ranges are called *rain shadows* and the deserts that they produce are known as **rain-shadow deserts**.

Most mountain ranges produce some form of rain shadow in which arid or semiarid conditions prevail, and sizeable rain-shadow deserts occur on the leeward side of high mountain ranges such as the Andes and Himalayas. But we need travel no farther than the mountainous regions of western North America to witness the profound influence of mountains on precipitation. Each of the various mountain ranges that make up the Rocky Mountains blocks the eastward passage of warm moist air from the Pacific, producing rain shadows on their eastern sides. In the mountainous region of the southwestern United States, for example, this moist Pacific air is forced upward by the Coast Range and the Sierra Nevada, so that precipitation occurs preferentially on the westward side of the mountains (Fig. 16.3). In this way, the air loses its moisture so that the regions east of these mountains receive little precipitation and tend to be dry. In fact, precipitation in the San Joaquin Valley around Los Banos, California (to the leeward of the Coast Ranges), and in the Great Basin to the east of Bishop (to the leeward of the Sierra Nevada) averages less than 25 centimeters (10 inches) a year. Because of their positions on the leeward side of mountain ranges, both of these dry regions qualify as rain-shadow deserts.

The desert regions of the southwestern United States are some of the hottest and driest places on Earth because they experience a combination of desert-forming processes. Not only do they lie in the rain shadow of the Rocky Mountains, but they also are located at 36°N, close to the downwelling dry air of the subtropical high-pressure zone.

> **CHECK YOUR UNDERSTANDING**
> ◐ What is a rain-shadow desert?

For example, Death Valley in California averages only 4.5 centimeters (2 inches) of rainfall annually and has endured some of the highest temperatures ever recorded, up to 57°C (134°F).

COASTAL DESERTS

In warm coastal areas next to cold ocean currents, conditions may be sufficiently dry to form **coastal deserts**. Coastal deserts form when the cool, dry offshore air becomes heated as it blows onshore and evaporates all surface moisture. Because of the rotation of Earth, the world's major ocean currents trace a clockwise path in the Northern Hemisphere and a counterclockwise path in the Southern Hemisphere, a phenomenon known as the *Coriolis effect* (see Science Refresher: **Ocean Currents and the Coriolis Effect** on page 506). In both hemispheres, warm currents originating near the equator consequently move poleward along the western sides of major ocean basins and return as cold currents along their eastern sides. Thus, coastal deserts are usually confined to the *west* coast of

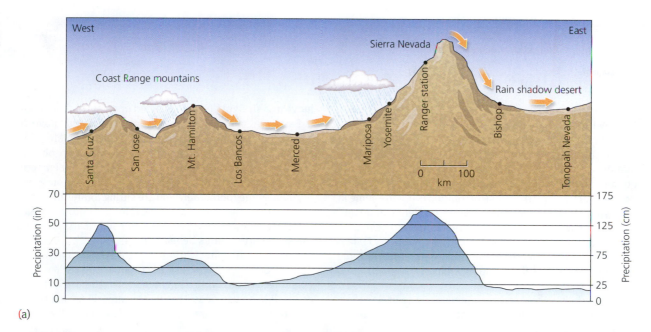

(a)

(b)

Figure 16.3   Rain Shadows and Deserts in the Pacific Southwest

(a) Moist air from the Pacific Ocean loses its moisture as it moves eastward across the mountains. (b) Warm air from the Pacific is cooled as it is forced aloft by the Sierra Nevada mountains, forming clouds that precipitate their moisture on the mountains' western (windward) side, leaving the eastern (leeward) side dry.

continental land masses and are best developed beneath subtropical high-pressure zones, where the climate is warm and dry. In the Northern Hemisphere, coastal deserts exist in Baja California and along the west coasts of Mexico and North Africa. In the Southern Hemisphere, they occur in coastal Peru and Chile and in coastal southwest Africa and western Australia.

The coastal Atacama Desert of northern Chile is one of the driest deserts in the world (Fig. 16.4); some parts of this desert have had no recorded rainfall in 500 years!

SCIENCE REFRESHER

Ocean Currents and the Coriolis Effect

Ocean currents, such as the Gulf Stream, are great rivers of relatively swift-moving ocean water. They occur in every ocean and are responsible for the circulation of surface water in the world's oceans (Fig. 16A). This motion is circular as warm water moves away from the equator toward the poles while cold water moves in the opposite direction. But the sense of rotation is not the same in all oceans; it is clockwise in the Northern Hemisphere and counterclockwise in the Southern Hemisphere. The reason for this, as demonstrated by the nineteenth-century French mathematician, Gaspard de Coriolis, is an apparent force that results from Earth's rotation, which is now known as the *Coriolis effect*.

The Coriolis effect is a consequence of one of Newton's laws of motion, which states that a body in motion will move in a straight line unless acted upon by another force. If Earth did not rotate, ocean currents would move in straight lines between the poles and the equator. But Earth does rotate. The effect of this rotation on objects pinned to Earth, such as people and plants, is not immediately obvious. For example, as you read this sentence, everything in the room where you are sitting rotates in unison with Earth so that the effect of the rotation is not apparent. But the oceans are not pinned to the solid Earth; instead the solid Earth rotates beneath them.

To mimic this effect, try drawing a straight line with a pencil from the edge of a slowly rotating object (like a circular piece of paper) to its center. Even if you draw the line as straight as you possibly can, the line will have a curvature because the paper is rotating beneath it (Fig. 16B).

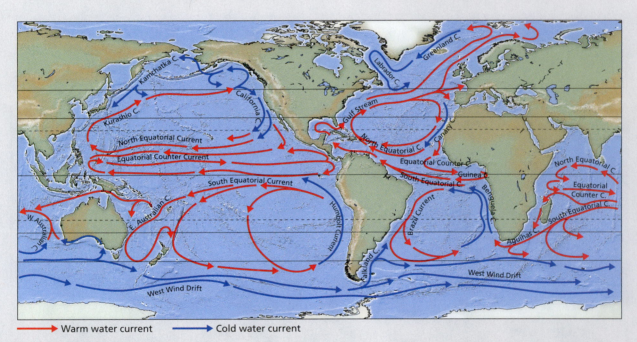

Warm water current   Cold water current

Figure 16A   Ocean Currents
The world's major ocean currents trace a clockwise path in the Northern Hemisphere and a counterclockwise path in the Southern Hemisphere.

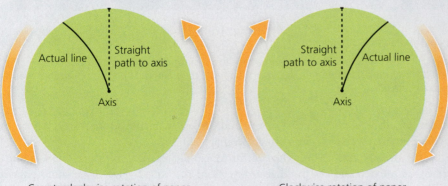

Counterclockwise rotation of paper   Clockwise rotation of paper

Figure 16B   Coriolis Effect
The Coriolis effect is illustrated by attempting to draw a straight line on a piece of paper rotating either counterclockwise (left) or clockwise (right). Although the hand guides the pencil in a straight line to the center, the result is a curved line because of the motion of the paper.

If you reverse the direction of spin, and try to draw the same line, the line will again be curved, but this time in the opposite sense. This is because the spinning object is moving in the opposite direction beneath the pencil.

If the circular piece of paper is rotating counterclockwise, the sense of deflection is such that the line traces a clockwise path (Fig. 16B, left). If the center is taken to represent the North Pole, and the edge of the paper the equator, the situation is similar to the influence of the Coriolis effect in the Northern Hemisphere. Since Earth has a counterclockwise sense of rotation as viewed from above the North Pole, currents that migrate from the equatorial regions toward the poles trace out a clockwise path, accounting for the clockwise rotation of ocean currents in the Northern Hemisphere (see Fig. 16A).

The reverse situation occurs in the Southern Hemisphere. From a viewpoint above the South Pole, Earth appears to rotate clockwise. Thus as ocean currents migrate from equatorial to polar regions, they rotate counterclockwise (Fig. 16B, right).

Figure 16.4   Atacama Desert
The coastal Atacama Desert, seen here in northern Chile, is one of the driest places on Earth with an average annual precipitation of only 1 millimeter (0.04 inch).

Here, cool dry air forms over the cold Humboldt Current (see Fig. 16A), which follows the coastline and is blown onshore. Once that air is over land, it warms and its capacity to absorb moisture dramatically increases. As a result, surface moisture is evaporated and desert conditions develop along the coast.

In reality, the extreme dryness of the Atacama Desert owes its origin to a combination of the three desert-forming processes we have examined in this chapter. Not only is the Atacama a coastal desert, but it also lies within the subtropical high-pressure zone of the Southern Hemisphere *and* in the rain shadow of the Andes Mountains. This combination of subtropical desert, rain-shadow desert, and coastal desert produces conditions of extreme aridity. Coastal ports such as

> **CHECK YOUR UNDERSTANDING**
> ● What circumstances produce dry conditions in coastal areas?

Iquique and Antofagasta, for example, receive rainfall only two or three times a century. Precipitation averages just one millimeter (0.04 inches) per year, and most of that is the result of coastal fog.

CONTINENTAL INTERIOR DESERTS

Continental interior deserts, such as those of central Asia, cannot be accounted for by any of the mechanisms we have so far examined: they are well removed from coastlines and major mountain belts, and they lie outside the subtropical high-pressure zone. Instead, these regions are dry because they are located so far from the sea that any rain they might otherwise have received falls before reaching them. As a result desert conditions can develop. The Turkestan and Gobi deserts are continental interior deserts, although the Gobi Desert also lies within the vast rain shadow produced by the high Himalayas.

POLAR DESERTS

Although we tend to associate deserts with heat, the extremely dry conditions that produce deserts do not necessarily go hand in hand with scorching temperatures. Indeed, some areas that receive virtually no precipitation are intensely cold. Among these are the **polar deserts** of northern Greenland, Arctic Canada, and Antarctica. Polar deserts, like subtropical deserts, owe their origin to global patterns of atmospheric circulation.

Figure 16.2 shows that the subtropical highs are not the only regions where convection causes downwelling of dry air. The same is also true of Earth's polar regions, which lie beneath another high-pressure zone known as the *polar high*. At these high latitudes, temperatures remain below freezing throughout the year and the air is intensely cold. And as was the case for the subtropical highs, the polar regions are also virtually devoid of moisture.

16.1 SUMMARY

- Deserts are areas of Earth's land surface that receive minimal rainfall. Deserts make up almost one-third of Earth's land surface.

- Deserts form where the climate is arid. There are five types of deserts: subtropical, rain-shadow, coastal, continental interior, and polar.

- Subtropical deserts form beneath subtropical high-pressure belts; rain-shadow deserts form in the dry regions on the leeward side of mountains; coastal deserts form where warm coastlines lie next to cold ocean currents; continental interior deserts form far from the sea in the middle of major landmasses; polar deserts form beneath polar high-pressure belts.

16.2 Weathering and Erosion in Deserts

Having already examined the glacial processes that operate in the cold deserts of the polar regions (see Chapter 15), we turn our attention in this chapter to the more familiar hot deserts of arid regions like the Sahara and the southwestern United States. As with landscapes in other climates, the austere splendor of the desert panorama with its vivid hues, rocky outcrops, steep-walled canyons, and sandy badlands is largely the result of the interaction between weathering and erosion by water and wind. Because the climate is dry and vegetation is sparse, however, the influence of these surface processes in desert regions differs from that in temperate regions and desert landforms are therefore distinctive. So how do these processes operate in the desert and how can water be involved if deserts have so little precipitation?

WEATHERING IN DESERTS

Weathering in deserts is largely the result of mechanical processes (see Chapter 5), which, as we discussed in Chapter 6, produce clastic sediments. In deserts, mechanical weathering, coupled with mass wasting, produces a debris of angular clasts of unaltered rock fragments instead of the soil that is produced in wetter, more lush environments. Because the desert air lacks moisture, it is unable to retain heat, and temperatures, while hot during the day, can be frigid at night. The repeated expansion and contraction caused by the extreme daily temperature fluctuations in deserts pries rocks apart and breaks them down into loose material. The growth of roots and salt crystals formed by evaporation plays a less important role in the breakdown.

Because of the lack of moisture and resulting paucity of vegetation, chemical weathering involving water and organic acids is greatly curtailed in desert climates. However, chemical weathering is not completely absent in deserts. This is because deserts, while dry, are not entirely without moisture.

During the night, for example, temperatures may fall sufficiently for dew to form, and in the winter months, many deserts receive some precipitation. Even in the driest deserts, this is sufficient to cause rusting of iron minerals in weathered rock debris and exposed rock surfaces by chemical reaction with oxygen. The vivid hues of orange, red, and purple that typify the desert landscape are the result of this process.

Another feature of many exposed rock surfaces in desert environments is the presence of a distinctive shiny coating known as **desert varnish**. Formed by a combination of chemical and physical weathering, this dark, often purplish veneer develops after long exposure. Native Americans produced rock paintings known as *petroglyphs* in the deserts of the Pacific Southwest (Fig. 16.5) by scratching or chipping away the dark varnish on exposed rock faces until fresh rock of lighter color was revealed beneath. The preservation of petroglyphs from prehistoric times shows that the rate at which desert varnish forms is extremely slow. The varnish is made up of a mixture of clay and iron and manganese oxides that is thought to be both concentrated on the rock surface by microorganisms (chemical weathering) and plastered to the surface by dusty desert winds (physical weathering).

> **CHECK YOUR UNDERSTANDING**
>
> ○ Why is weathering in deserts mostly mechanical and how does it operate?
>
> ○ What is the origin of the vivid colors that typify exposed rocks in desert climates?

WATER IN DESERTS

Surprisingly, the principal agent of erosion and deposition in desert environments is running water, just as it is in

Figure 16.5   Petroglyphs Carved in Desert Varnish

These native American petroglyphs exposed on a rock surface in Arches National Park, Utah, were carved by the Ute people hundreds of years ago. By removing the veneer of purple desert varnish that coats the outcrop, the lighter color of the fresh rock beneath was revealed. The petroglyphs have not been recoated with varnish since they were carved, which testifies to the extremely slow rate at which desert varnish forms.

Figure 16.6   Desert Landscape
Most panoramas of typical desert scenery, such as this view of the Painted Desert at Petrified Forest National Park in northern Arizona, reveal a landscape dominated by networks of streambeds and canyons. Although this landscape is dry, it could only have been carved by running water, which is the most important surface process in desert environments.

more humid regions. Most views of desert landscapes, such as the one shown in Figure 16.6, include numerous channels, valleys, and canyons that, although dry, could only have been carved by running water. In fact, the very dryness of the climate facilitates erosion by water. But if the climate is so dry, where does the water come from? The answer, as in more humid regions, is from rain.

Stream erosion and deposition
Rain in deserts often comes in the form of torrential cloudbursts that can set off flash floods. With little vegetation to slow it down, desert floodwater can flow rapidly. It easily picks up loose surface material, which lacks plant cover to bind it together. Armed with a sediment load, short-lived but fast-flowing streams become powerful agents of erosion, rapidly carving out steep-sided valleys, which for most of the year remain dry. They are termed **intermittent streams** because they flow only occasionally, but the dry valleys they carve go by various names in different parts of the world. In the southwestern United States and in arid parts of Mexico and Spain they are known by the Spanish word *arroyo*, in the deserts of North Africa and the Middle East by the Arabic word *wadi*, and in Pakistan and northern India by the Hindu word *nullah*.

The effectiveness of running water as an agent of desert erosion is vividly illustrated by the huge quantities of sediment found at the mouths of canyons in mountainous deserts like those in Nevada and Utah. As we discussed in Chapter 6, these deposits form triangular aprons of sediment known as **alluvial fans**, when intermittent, fast-flowing streams of debris-laden floodwater fan out and slow down as they discharge from narrow mountain canyons

onto the flatter desert floor (Fig. 16.7). With repeated flash floods, individual fans ultimately coalesce to form a continuous apron of sediment or **bajada** at the foot of the mountains.

Because the sediment is the product of erosion, the mountains slowly retreat as the bajada grows (Fig. 16.8). Eventually, the mountain range is worn back and the canyons slowly widen until only isolated erosional remnants of the mountains are left. These remnants are known as **inselbergs** and, until they eventually succumb to erosion, they rise conspicuously from a gently sloping, sediment-strewn bedrock erosion surface called **pediment** that is produced as the mountain front retreats.

Two additional types of erosional remnants (Fig. 16.9) form in areas of relatively flat-lying sedimentary rocks, such as those of the Colorado Plateau in the southwestern United States. Known as *mesas* and *buttes,* they are produced by the irregular retreat of vertical cliffs where fairly easily weathered rocks are capped by more resistant ones. **Mesas** are broad, flat-topped remnants bounded on all sides by steep cliffs. The ongoing retreat of these cliffs eventually leads to the formation of isolated pillar-like remnants known as **buttes**.

Torrential desert cloudbursts can produce so much runoff water that the ground cannot absorb it all. When this happens, water accumulates in low areas of the desert floor to form temporary lakes, called **playa lakes**. Playa lakes are usually shallow and dry up over a period of a few days or weeks, leaving behind a layer of mud that shrinks and develops mud cracks as the lake dries up. Because the water contains small amounts of dissolved salts, the lake becomes progressively saltier as it evaporates and may eventually leave a deposit of salt, or "salt pan" on the dry lake bed. This dry, salt-encrusted lake bed is called a **playa** (Fig. 16.10).

CHECK YOUR UNDERSTANDING
◐ How can running water be the principal agent of erosion and deposition in dry desert climates?
◐ What is the difference between a bajada and a pediment?
◐ What is the difference between a mesa and a butte?

Oases
Not all water in the desert results from rainfall. Some water comes from rare springs of groundwater. Springs of water in desert regions are called **oases**. As we learned in Chapter 14, springs are seeps of groundwater that form wherever the water table intersects the land surface or where groundwater is forced to the surface along fractures. Not surprisingly, oases are most common in valleys within mountainous regions or where deep faults penetrate the bedrock (see Fig. 14.10).

CHECK YOUR UNDERSTANDING
◐ What are oases and how do they form?

Figure 16.7   Alluvial Fans
An alluvial fan at the base of the Black Mountains at Badwater in Death Valley, California.

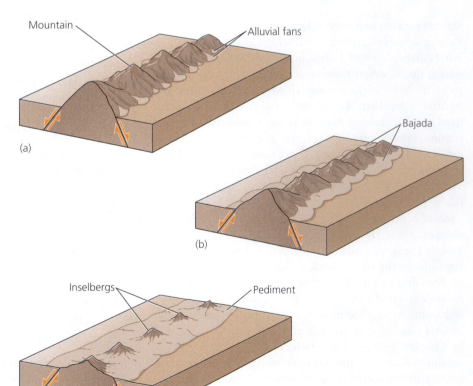

Figure 16.8   Inselbergs and Pediment

Progressive desert erosion of a steep-sided mountain range results in insel-bergs arising from pediment.

Figure 16.9   Monument Valley
The flat-topped mesas and chimney-like buttes in Monument Valley, Arizona, are erosional remnants produced by the irregular retreat of cliffs in flat-lying sedimentary rocks.

Figure 16.10   Badwater Basin Playa
Badwater Basin in Death Valley, California, occupies a depression in the desert floor. This depression occasionally becomes a temporary playa lake that leaves behind a playa as it evaporates.

16.2 SUMMARY

- Weathering in deserts is largely mechanical, resulting from rock expansion and contraction. Chemical weathering plays a less important role but accounts for the rusting of desert landscapes.

- The main agent of erosion and deposition in deserts is running water, which can take the form of flash floods and produces alluvial fans that coalesce to form a bajada at the foot of mountains.

- Inselbergs, pediment, mesas, and buttes are features of desert erosion.

- Flash floods may create temporary playa lakes that evaporate to form playas. Oases are desert springs.

16.3 Wind in Deserts

Although running water is the agent most responsible for shaping the desert landscape, it is by no means the only one. Imagine yourself in the desert looking out into the brunt of a strong desert wind. Your face can feel like it is being sand-blasted. In the distance you may see the iconic desert image of *dust devils*—those spiraling mini-tornadoes of sand and air that wax and wane in a matter of seconds (see In Depth: **Dust Devils** on page 512). You may also notice sand being whipped up from the windward side of dunes and being deposited on the leeward side. Wind is a relentless force of ongoing change in desert regions.

IN DEPTH

Dust Devils

One of the most evocative images of desert landscapes is the *dust devil,* that whirling column of dust, commonly several meters across and tens of meters tall, seen moving over the ground and spinning like a miniature tornado (Fig. 16C). Like tornadoes, dust devils comprise near-vertical rotating columns of air, but their origin is quite different. Unlike tornadoes, dust devils form on hot sunny days, not stormy ones, and they are rarely harmful.

Dust devils are small whirlwinds of air that are made visible by the dust they are carrying. They form when a small pocket of hot, buoyant air near the desert surface rises rapidly into the cooler, less buoyant air above. As the hot air rises, the cooler air rushes in to fill the void beneath the rising hot air and starts to spin, much like water as it drains from a bathtub. And just as an ice skater spins faster by drawing in his or her arms, so the inrushing air spins faster as it approaches the center of the developing dust devil. This increase in spin further reduces the air pressure at the center of the dust devil, which draws in yet more air and causes the air at the center, and the dust it has picked up, to rise in a slowly moving column.

Dust devils maintain their form providing they remain fueled by hot air at the desert surface. As a result, they develop best in flat, dusty terrain under clear sunny skies, when there is little or no wind; it is these conditions that best produce a blanket of hot surface air. Eventually, cooler air is sucked into the vortex, whereupon the dust devil rapidly dissipates.

Dust devils are not limited to the deserts of our own planet, but have also been observed on the surface of Mars. First photographed by the Viking orbiters in the 1970s, they have also been encountered by the Martian rovers Spirit and Opportunity, both of which benefited from the occasion by having their solar panels cleaned of accumulated dust. Martian dust devils are far larger than those on Earth and may pose a threat to future Martian landings. In 2012, for example, NASA's Mars Reconnaissance Orbiter photographed a Martian dust devil almost as wide as a football field and roughly 20 kilometers (12 miles) high (Fig. 16D)! Martian dust devils attain these extraordinary heights because the Martian atmosphere is so thin. In fact, the mass of an atmospheric column on Mars is less than 1 percent of that of one on Earth, which allows for much faster circulation and far higher penetration into the Martian atmosphere.

Figure 16C   Dust Devil
A dust devil whirls across the desert landscape of Arizona.

Figure 16D   Dust Devil on Mars
Twisted by high-altitude winds, this Martian dust devil was photographed wandering across the Amazonis Planitia region of Northern Mars in March 2012 by the High Resolution Imaging Science Experiment (HiRISE) camera on NASA's Mars Reconnaissance Orbiter. This impressive example is a little more than three quarters of a football field wide, and roughly 20 kilometers (12 miles) high!

Although, wind is simply air moving in response to differences in atmospheric pressure, it, like running water, is a moving fluid capable of picking up loose material and transporting it elsewhere. Of course, no climate is free of wind. But, as an agent of erosion and deposition, wind is particularly effective in desert regions because there is so little vegetation to bind the loose material together.

SEDIMENT TRANSPORT BY WIND

Like running water (see Chapter 13), wind transports material by either moving it along the ground as a *bed load,* or by carrying it in the air as a *suspended load*. Under normal conditions, however, the wind is unable to pick up material as coarse as that transported by water because wind has a much lower density than water. Even at high velocity, the wind is seldom able to carry material any coarser than sand.

Because, most sand grains are either too coarse or too heavy to be carried in suspension by the wind, the grains are either pushed or rolled along the ground surface, or skip along the ground in a series of bounces. Most windblown sediment transport occurs by way of this skipping motion (Fig. 16.11a), which, like that in streams, is known as **saltation**. During saltation, the wind lifts individual sand grains and carries them downwind a short distance before dropping them again. As the sand grains strike the surface, they hit other grains that bounce downwind to dislodge more sand in a sort of domino effect. In this way, significant quantities of sand move downwind. The collisions smooth and pit the rough edges of the sand grains so that windblown sand tends to be well rounded (Fig. 16.11b). Windblown sand is also typically well sorted. This is because the wind leaves behind grains too coarse to be carried and carries away the finer grains altogether.

In contrast to sand, finer particles of silt and clay can be blown high into the air and kept airborne for hours or even days. Although easily moved once airborne, fine material is not easily lifted by the wind. This is because, even on a windy day, a thin layer of more or less motionless air lies next to the ground. This thin boundary layer is created by the roughness of the ground surface, which prevents the air in contact with the ground from moving. For example, you may have noticed how dust is rarely raised from a dirt road

unless a passing vehicle or something else produces motion that disturbs this boundary layer. In the desert, saltating sand performs this function; it is the sand grains that generally move first, and, in so doing, disturb the finer particles.

Once airborne, fine particles form clouds of dust, and even choking dust storms (Fig. 16.12). Although the wind deposits most of the dust close to its source, it carries some of the finest material high into the atmosphere and transports it for thousands of kilometers. Dust from the Sahara Desert, for example, is routinely detected in the air over Florida and the West Indies. Similarly, during the historic Dust Bowl that occurred in the United States in the 1930s, dust from as far away as Texas fell on New England (see Living on Earth: **The Dust Bowl** on page 515).

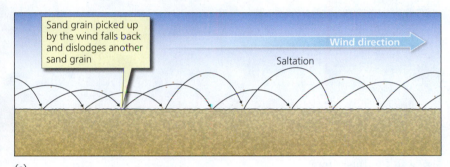

Sand grain picked up by the wind falls back and dislodges another sand grain

Wind direction

Saltation

(a)

Pits

(b)

Figure 16.11   Saltation

(a) Most windblown sand moves along the ground in a bouncing motion known as saltation. The wind picks up individual sand grains and carries them for a short distance; the grains fall back, whereupon they dislodge other grains, which bounce downwind in turn. (b) As a result of saltation, desert sand grains (seen here magnified about 40 times) are commonly well rounded, well sorted, and pitted.

Figure 16.12   Dust Storm
A dust storm sweeps across the floor of Death Valley, California.

FEATURES OF WIND EROSION

Wind is a very effective agent of desert erosion and produces a number of distinctive desert landforms. Wind erosion operates in two ways: *deflation* and abrasion. **Deflation** is a lowering of the ground surface due to wholesale removal of loose material by the wind. In the dry, poorly vegetated desert environment, wind gusts can lift the finer surface material and transport it elsewhere. Over time, the entire surface of an area can be lowered in this way. Typically the process of deflation is extremely slow, lowering the ground surface of an area by only a few centimeters per century. However, during the Dust Bowl of the 1930s, the surface was lowered over broad areas by as much as a meter in just a few years.

On a smaller scale, deflation can produce shallow depressions in the ground surface. These depressions, known as **blowouts**, are typically about 1 meter deep and tens of meters across (Fig. 16.13). But they can also be many kilometers across and several tens of meters deep. Blowouts occur throughout the semiarid Great Plains region of the United States, where they pepper the landscape from Texas north to the Dakotas. The depth of blowouts varies according to the level of the local water table. Deflation is inhibited once the water table is reached because damp grains adhere to each other.

Although deflation lifts silt- and sand-sized grains, larger and heavier material is left behind. As a result, coarser material becomes concentrated at the surface as the finer material is removed. In this way, a veneer of pebbles and boulders is eventually produced that is known as **desert pavement** (Fig. 16.14).

In addition to removing material from the ground surface by deflation, the wind blows sand that etches and polishes exposed rock surfaces. This form of natural sandblasting is known as *abrasion*. Automobiles stripped of their paint, glass so frosted that it is no longer transparent, and telephone posts felled at the base testify to the effectiveness of windblown sand as an agent of erosion. Saltating sand grains cause far more abrasion than suspended dust. However, this process only really occurs near ground level because saltation rarely lifts sand grains more than one meter above the surface.

Some of the most common products of wind erosion are faceted stones known as *ventifacts*. **Ventifacts** form when exposed pebbles and boulders

> ### CHECK YOUR UNDERSTANDING
> • How does deflation lead to the formation of blowouts and desert pavement?
>
> • What is abrasion and which desert landforms can be attributed to this process?

LIVING ON EARTH

The Dust Bowl

Desertification is the spread of desert conditions into formerly fertile regions. The human and ecological disaster that struck the southwestern Great Plains of the United States in the 1930s and came to be known as the "Dust Bowl" (Fig. 16E) is a vivid example of desertification brought about by a combination of poor land management and sustained drought. In the decades before the drought, thousands of settlers were lured to the grasslands of the southern plains by the promise of rich soils and prosperous farming. But the farming practices adopted by the new residents dramatically increased the region's vulnerability to drought. Periods of wet weather led the settlers to misjudge the region's climate and the settlers plowed millions of acres of former grassland using techniques that had worked well in the wetter climates where they had previously farmed. During these wet years, low crop prices and the high cost of farm machinery forced the settlers to plow ever more marginal farmland in order to make ends meet. This compounded the problems and depleted the soil of its nutrients and further increased the likelihood of crop failure.

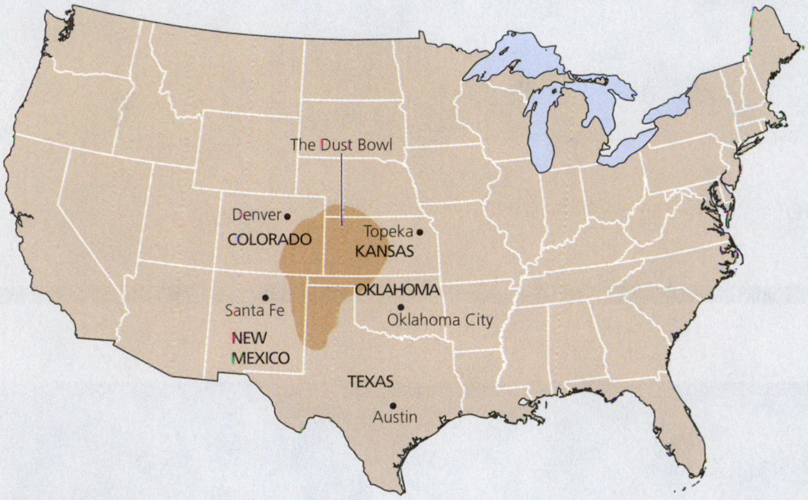

While the rains fell, wheat was plentiful. But in 1931, the midwestern and southern plains were hit by a period of severe drought that was to last for 8 years. As the crops died, the overplowed land was left without a protective cover of vegetation and the topsoil began to be blown away. Huge dust storms brought devastation to western Kansas, eastern Colorado, northeastern New Mexico, and the Oklahoma and Texas panhandles (Fig. 16F), carrying away hundreds of tons of topsoil from each square mile. Dense black clouds of choking dust darkened the sky as far east as New York City and hundreds of millions of acres of former farmland were rendered uninhabitable (Fig. 16G). In 1934, when the drought was at its most severe, an estimated 250 million acres of land was either rapidly losing its topsoil or had lost its topsoil entirely. Farm families in the hundreds of thousands fled west to California in a mass exodus that inspired John Steinbeck's epic novel *The Grapes of Wrath*.

By 1938, a variety of conservation methods that included planting trees to form "shelterbelts" and plowing the land into furrows had substantially reduced the amount of blowing soil. In 1939 the rains returned, bringing the drought to an end. Improved farming practices and the exploitation of the vast groundwater resource known as the High Plains aquifer, which underlies the High Plains from Nebraska south to the Texas Panhandle (see Chapter 14), have greatly reduced the effects of subsequent droughts.

Figure 16E   The Dust Bowl

The Dust Bowl was the name given to that portion of the southern plains most severely impacted by desertification in the 1930s.

Figure 16F   Dust Storm

An approaching dust storm darkens the sky during the Dust Bowl of the 1930s.

(Continued)

LIVING ON EARTH

The Dust Bowl (Continued)

Figure 16G   Abandoned Farm
Buried machinery on an abandoned farm testifies to the destruction brought about by dust storms of the Dust Bowl.

Figure 16.13   Blowout
This blowout with a floor of hardened mud is located in Death Valley, California.

Figure 16.14   Desert Pavement

The desert pavement seen here is in the Sturt Stony Desert of South Australia.

are subjected to sandblasting from a particular direction because of the prevailing wind. Such stones are abraded on only one side, becoming pitted and polished on the surfaces that face the wind. Only if the ventifact moves or there is a change in the prevailing wind direction do the stones become multifaceted (Fig. 16.15). On a larger scale, windblown sand may undercut cliff faces or preferentially abrade the base of rock columns to form distinctive mushroom-shaped features such as the one shown in Figure 16.16.

FEATURES OF WIND DEPOSITION

Although wind erodes the desert landscape less than running water, wind deposits result in several important landforms and sedimentary bodies. As a desert windstorm abates, airborne particles begin to fall. But they do not all fall at once. The saltating sand grains fall first, and accumulate in

Figure 16.15   Ventifacts

Multifaceted ventifacts are produced when pebbles and boulders exposed on the desert floor are subjected to windblasting by saltating sand from several prevailing wind directions.

Figure 16.16   Erosion by Sandblasting

In Death Valley, California, preferential erosion within a meter of the ground by sandblasting is responsible for the mushroom-like shape of this basalt outcrop.

mounds we call sand *dunes*. The finer silt is carried in suspension farther from its source and is eventually deposited over wide areas as a blanket of silt known as *loess*. In the material that follows we discuss dunes and loess deposits in more detail.

Dunes

Dunes are mound- or ridge-shaped deposits of windblown sand. They form wherever an obstacle in the wind's path slows the wind down, allowing windblown sand to accumulate. As the wind sweeps over and around such an obstruction, two areas of calmer air, called *wind shadows,* are produced—a larger one behind the obstacle and a smaller one in front. As the wind drops, saltating sand grains settle in these areas and accumulate to form mounds that may ultimately bury the obstruction. However, as the sand accumulates, the rising mound acts as an ever more imposing barrier to the wind, further slowing it down and forcing the deposition of more sand in the wind-shadow areas. In this way, the mound eventually grows into a dune as long as there is an adequate supply of sand and a wind of sufficient duration.

Most dunes are asymmetric in cross section. They slope relatively gently on the windward side and relatively steeply on the leeward (downwind) side (Fig. 16.17a). This characteristic shape begins to form when the wind deposits saltating sand grains in the wind shadow just beyond the dune crest. As sand accumulates, the slope of the leeward face increases until it becomes unstable and slides under the influence of gravity. As we learned in Chapter 13, the maximum slope that can be maintained by loose material is governed by its *angle of repose*. The leeward side, or **slip face**, of a dune maintains a fairly constant slope of about 34 degrees, which is the angle of repose for dry sand.

The process of dune growth involves the progressive removal of sand from the windward side of the dune and its addition to the leeward side, where it continually slides down the slip face. This process has the following three consequences:

- The continual transfer of sand from one side to the other causes the entire dune to migrate slowly downwind as it grows (Fig. 16.17b).

- Because the height to which sand grains can be moved up the leeward slope by saltation depends on the strength of the wind, the height of dunes is limited by the local wind conditions. For most dunes, this limit lies between to 10 and 25 meters (30–80 feet).

- The sand that accumulates on the slip face of a dune is deposited in inclined layers that dip downwind. Recall from Chapter 5 that these layers are known as *cross beds* and are easily distinguished from cross beds formed by moving water because of their much larger size; they are often on the order of meters rather than centimeters. Cross beds are often preserved in the rock record long after the shape of the dune has been destroyed. As a result, their presence in ancient sandstones can be used to identify fossil dune complexes and to determine the wind directions that produced them (Fig. 16.18).

Although all dunes look superficially similar, they do not all form in the same way. In fact, we can identify five different kinds of dune based on their shape and their orientation with respect to the prevailing wind (Fig. 16.19). Although intermediate forms occur between these shapes, the five main types are *barchan, transverse, longitudinal, parabolic,* and *star*.

Barchan dunes are solitary, crescent-shaped dunes with concave slip faces and crescent-shaped "horns" that point downwind (Fig. 16.19a). These dunes form on flat, barren areas of desert floor in regions where the wind is constant but the supply of sand is limited. Rarely more than 30 meters (100 feet) in height and 350 meters (1150 feet) from horn to horn, they migrate downwind at rates of up to 15 meters (50 feet) a year. Barchan dunes acquire their distinctive shape because the horns, which are lower in elevation than the dune crests, provide less of a barrier to the wind and therefore migrate downwind faster.

Transverse dunes are long, strongly asymmetric sand ridges (Fig. 16.19b). Like barchan dunes, transverse dunes

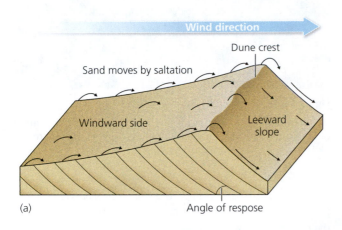

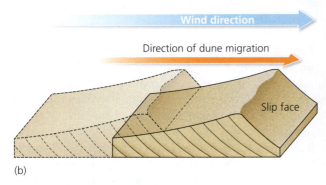

Figure 16.17   Profile of Dune

(a) This dune, seen in cross section, is asymmetric with a gently sloping windward side and a steeper leeward face, the slope of which is governed by the angle of repose for dry sand (about 34 degrees). (b) Movement of asymmetric dunes produces cross-bedding as sand grains that migrate up the leeward slope by saltation are deposited on the windward face.

Figure 16.18   Cross-Bedding

Beautifully exposed in Zion National Park, Utah, this dune bedding in the Navajo Sandstone demonstrates the existence of dune complexes (and therefore desert conditions) in this part of North America during the early Jurassic Period, some 200 million years ago. It also records the wind directions that produced them (from left to right).

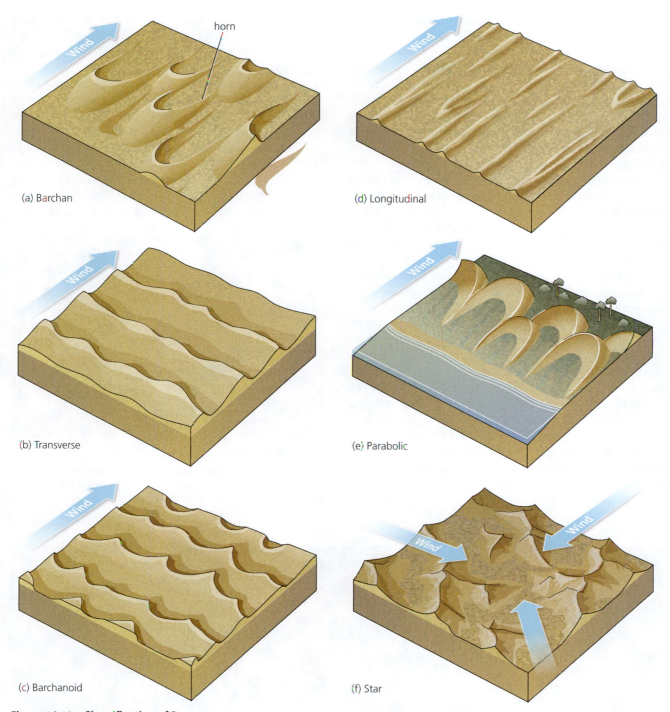

horn

(a) Barchan

(b) Transverse

(c) Barchanoid

(d) Longitudinal

(e) Parabolic

(f) Star

Figure 16.19   Classification of Dunes
Dunes are classified based on their shape and their orientation with respect to the prevailing wind. The five main types are barchan, transverse, longitudinal, parabolic, and star. Barchanoid dunes are intermediate between barchan and transverse.

form in flat, barren areas of desert floor where the wind is constant. But in contrast to barchan dunes, transverse dunes form where the supply of sand is plentiful. Aligned in parallel sets with crests perpendicular to the prevailing wind direction, transverse dunes resemble gigantic sand ripples with gentle windward slopes and steep leeward slip faces. They are perhaps most familiar to us as the common type of sand dune on coastlines and lakeshores, like those

of coastal Oregon and Lake Michigan. In parts of the Sahara, however, transverse dunes cover vast areas with crests in excess of 100 kilometers (60 miles) long and heights of up to 200 meters (650 feet).

Dune forms that are intermediate between barchan and transverse are known as *barchanoid* (Fig. 16.19c). These develop as the sand supply diminishes and the dune crests become increasingly sinuous and crescent shaped. At the edge of the dune field, barchanoid dunes may eventually separate into individual barchan dunes as the sand supply diminishes.

Longitudinal dunes are narrow, linear sand ridges with a broadly symmetrical profile that are aligned *parallel* to the prevailing wind direction (Fig. 16.19d). Smaller dunes of this shape, typically a few meters high and a few tens of meters long, form behind obstacles in areas of abundant sand and a strong and constant wind. But much larger longitudinal dunes form in areas of more moderate sand supply and a somewhat variable wind apparently as a result of wind eddies that blow toward the dune and so keep the intervening troughs swept clean. In the Arabian Desert (where they are called *seifs*), longitudinal dunes of this type reach lengths of 100 kilometers (60 miles) and heights of up to 100 meters (330 feet).

Parabolic dunes are U-shaped dunes that resemble barchans except that the horns of parabolic dunes point *upwind* (Fig. 16.19e) and reflect the role of vegetation on sand transport. Unlike other dunes, parabolic dunes typically develop in areas of partial vegetation and are common on coastlines with strong onshore winds and an abundant supply of sand. When partially vegetated transverse dunes, for example, are breached by strong winds, deflation can create blowouts in the areas where sand is exposed. The sand removed from each of the blowouts piles up in a curved rim downwind, and grows and moves downwind as the blowout enlarges. But the sides of the blowouts are covered in vegetation and so remain in place. As a result, the dunes develop an increasingly horseshoe or parabolic shape as they migrate downwind.

CHECK YOUR UNDERSTANDING

• How can dunes be aligned both parallel to and perpendicular to the prevailing wind direction?

• How do barchan dunes and parabolic dunes each acquire their distinctive shapes?

Star dunes are isolated hills of sand with three or four sharp-crested ridges radiating out from a central peak (Fig. 16.19f). Star dunes are the most complex type of dune and form in areas where the wind direction varies markedly. As a result, these dunes tend to build vertically so that the central peak, which can be as much as 100 meters (330 feet) high, may remain in place for centuries.

Figure 16.20   Loess
Loess deposits have been exposed on the Yukon River in northern Canada. (Photograph courtesy of the Illinois State Geological Survey. Photographer: Joel Dexter.)

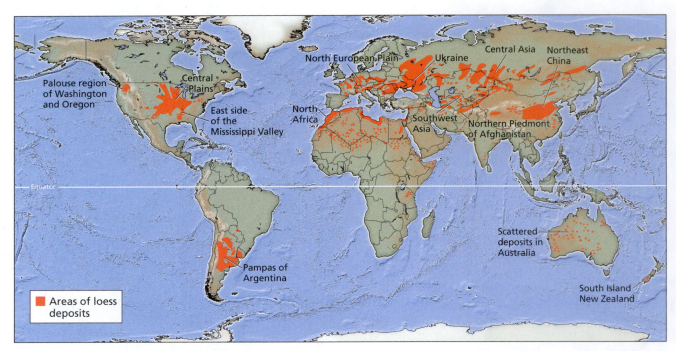

Figure 16.21   Global Distribution of Loess
Global distribution of loess deposits is shown in orange.

Loess deposits

In contrast to sand, finer grained silt and clay may be carried great distances by the wind as a suspension load. This load is deposited as a blanket over broad areas as the dust storm abates. Gradually thickening with each successive dust storm, such blanket deposits of windblown silt are called **loess**. Because loess is an airfall deposit, it is not confined to valleys, as river or lake deposits would be. Instead, it is distributed evenly over hill and vale. It also tends to be homogeneous and unlayered, and is typically buff (yellowish-brown) in color (Fig. 16.20). Loess forms a hospitable environment for local ecosystems and may contains fossil snails, the teeth and bones of mammals, and networks of vertical tubes left by the roots of grasses.

For several reasons, loess weathers to produce very fertile soils. Loess grains are made up of a variety of minerals and are the product of mechanical rather than chemical weathering, so they are commonly quite angular. As a result of the angularity, loess is also porous and so holds moisture well. Hence, soils produced from loess are fertile, being both mineral rich and moist.

Loess deposits are widespread, covering some 10 percent of the world's land surface (Fig. 16.21). The largest deposits occur in northern China, downwind of the huge Gobi Desert in central Asia. Covering a vast area and reaching a thickness of up to 300 meters (1000 feet), it is these deposits that give the Yellow River (Huang He) of China its name. Significant deposits of loess also occur in Europe and cover as much as 30 percent of the United States, where they form the rich farmlands of the Midwest and the Pacific Northwest. Loess can also be a by-product of glaciation and for the most part, the windblown silt of the North American deposits comes, not from deserts, but from the fine-grained glacial outwash sediments left behind following retreat of the ice sheets at the end of the last ice age.

> **CHECK YOUR UNDERSTANDING**
> ⟶ What is loess?

16.3 SUMMARY

- Wind transports material in the desert either on the ground as a bed load or in the air as a suspended load.

- Wind erosion removes material wholesale (deflation) or etches and polishes exposed bedrock (abrasion).

- Wind deposits form dunes (barchans, transverse, longitudinal, parabolic, and star) and blankets of windblown loess.

16.4 Desertification: Natural and Human Induced

Although there is a pattern in their worldwide distribution (see Fig. 16.1), deserts are not fixed features of Earth's surface but rather expand and contract over the course of time as conditions change. Major changes in the size and location of deserts have accompanied the slow movement of continents from one latitude to another over the course of geologic time. But the growth of deserts can also occur over a far more rapid time frame as can be seen by recent events along the southern margin of the Sahara.

Stretching from Mauritania and Senegal in the west to Ethiopia and Somalia in the east (Fig. 16.22), the *Sahel* is a semiarid region that separates the Sahara Desert from the more humid grasslands to the south. The dry lands that surround deserts are particularly sensitive to change and the dry lands in the Sahel region are no exception. Over the past four decades, this normally habitable region has experienced a significant decline in rainfall (Fig. 16.23) and portions have suffered severe drought with disastrous consequences for local inhabitants.

As we noted in Living on Earth: **The Dust Bowl**, the spread of desert conditions into formerly fertile regions is called *desertification*. But what causes desertification? Is it a natural phenomenon related to long-term climate change or is it caused by human activities? Certainly, drought can bring about desertification but natural processes tend to act gradually over the course of centuries or millennia. The change to desert-like conditions in the Sahel has been far more rapid than might be expected from natural processes, suggesting the involvement of human activities such as overpopulation and land mismanagement.

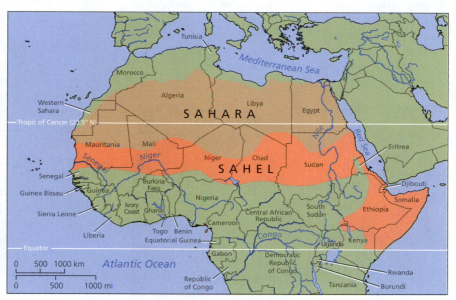

Figure 16.22   Map of the Sahel

Forming a transition zone between the Sahara Desert and regions of savannah and rain forest to the south, the Sahel is a semiarid region that has experienced rapid desertification over the past 30 years.

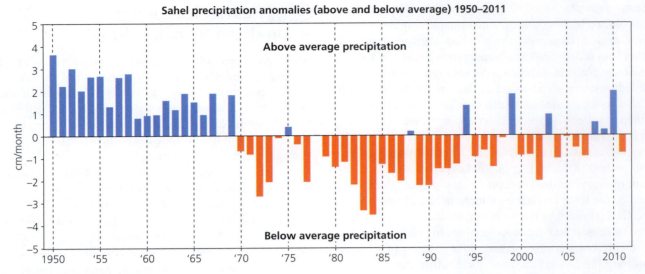

Figure 16.23   Sahel Rainfall

In the western Sahel 20 years of above-average rainfall has been followed by 40 years of drought (data from the National Oceanic and Atmospheric Administration).

CHANGES IN CLIMATE

Six thousand years ago, much of what is now North African desert was fertile grassland with many streams and lakes, abundant game and thriving Stone Age societies (Fig. 16.24). So what could have brought about such a dramatic change in climate? To a large extent, the Sahel is watered by seasonal rain from the tropical Atlantic, the northernmost reach of which marks the limit of vegetation. Six thousand years ago, this limit was 950 kilometers (600 miles) north of its present position. This change in reach was a consequence of subtle variations in Earth's orbit around the Sun—an astronomical cause of climate change that we will examine in more detail in Chapter 21. Six thousand years ago, Earth's shifting orbit brought the planet closer to the Sun during summers in the Northern Hemisphere, as a result of which the monsoon rains were dramatically stronger.

Figure 16.24   Saharan Petroglyphs

This 2500-year-old petroglyph from Niger's Ayr Mountains in the southern Sahara tells of a time when the land was wet and green enough to support teeming game and a sophisticated society.

With warmer summers and a warmer northern tropical Atlantic, moisture-laden air was able to penetrate farther into North Africa than it does today and rainfall was as much as 25 percent higher. As a result, grasslands flourished and the region enjoyed a more temperate climate than it now does. Since then, Earth's orbit during the Northern Hemisphere summer has become more distant from the Sun and the monsoon rains have weakened. Because of this, rainfall has decreased, the limit of vegetation has shifted south, and the grasslands have been replaced by semiarid desert (see Living on Earth: **Water in the Sahara Desert** in Chapter 14).

THE HUMAN FACTOR

Although the process of desertification in the Sahel may have been started by natural drought, the speed and severity of desertification in recent years is almost certainly linked to the destruction of vegetation by such human activities as overcultivation, overgrazing, and the felling of trees for fuel and shelter. These practices combine to deplete the soil of its nutrients and strip it of its protective vegetative cover. As a result, the soil's capacity to retain moisture is greatly reduced and it becomes highly susceptible to erosion. As the soil becomes less productive, crops start to fail and the unprotected topsoil is easily removed by the wind or washed away by occasional flash floods that rapidly evaporate.

Unfortunately, once started, the process of desertification tends to be amplified by a vicious circle of circumstances and human reactions to these circumstances. Drought conditions encourage local inhabitants to further degrade the land by yet more overplanting and overgrazing in an effort to sustain themselves in the face of failing crops. Although desertification may be temporarily halted during wetter years, this process is irreversible and the population is forced to migrate in search of food.

Although serious drought first struck the Sahel in the 1970s (see Fig. 16.23), the stage for the subsequent and disastrous desertification was set years earlier. In the 1950s and 1960s, a succession of anomalously wet years resulted in greatly increased productivity. The population of this marginally productive land doubled and, in order to provide food for this expanding population, the amount of cultivation, land clearing, and grazing by herds of domestic animals progressively increased. Because of this, the land was at its most vulnerable when the drought stuck in 1970. When crops failed, the scant grasses were stripped by foraging herds that were now too large for the land to support. As a result, the exposed topsoil was easily eroded by the wind and the ground was baked so hard that subsequent rainfall produced only runoff and still further soil erosion. Without forage, most of the animals died. The starving human population of 20 million migrated south, swelling cities beyond capacity with refugee camps. The drought lasted for almost 25 years.

Although the story of the Sahel is the most tragic example of desertification, it is by no means the only one. Desertification is a serious and growing problem in many of the world's semiarid regions, including the Middle East, central Asia, eastern Australia, Patagonia, northern Mexico, the western United States and the Canadian prairies. Indeed, it is estimated that worldwide, some 60,000 square kilometers (23,000 square miles) of new desert are being created annually. To reverse the process of desertification, only two remedies are readily available; either the surface must be revegetated or water must be delivered from elsewhere to supplement the meager rainfall of these semiarid regions.

overpopulation and land mismanagement. Once the soil is no longer protected from erosion, desertification is irreversible.

- Desertification is a serious problem in many of the world's semiarid regions.

16.5 Deserts and Plate Tectonics

By knowing the characteristic features of deposits associated with modern deserts, geologists can recognize similar features preserved in the rock record. Evidence from the rock record shows that deserts have been important features of Earth's surface for much of its history. Throughout this history, the controlling influence on the distribution of deserts has been plate tectonics. This is because deserts are best developed at particular latitudes, and it is plate tectonics that has controlled the movement of continents, and hence their latitude, throughout geologic time. Recall from Chapter 2, that the existence of ancient deserts in regions that now enjoy temperate climates was one of the enigmas that Wegener used in support of his hypothesis of continental drift.

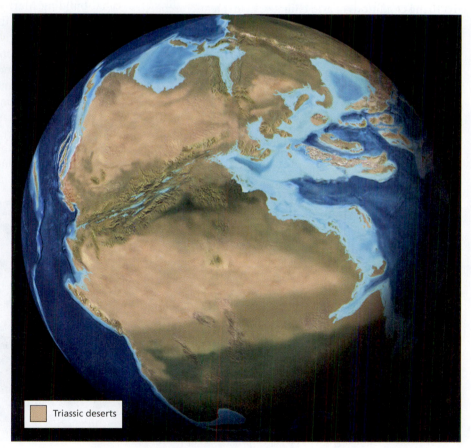

☐ Triassic deserts

Figure 16.25   Triassic World

During the Late Triassic Period, some 220 million years ago, widespread areas of subtropical and interior desert (buff-colored) covered much of North and South America, Europe, and Africa.

16.4 SUMMARY

- The spread of desert conditions is called desertification.
- Desertification can result from natural climatic processes and from human influences, such as

Figure 16.26   Triassic Cross-Bedding

Dune cross-bedding in red desert sandstones of the Triassic Period is visible on the shore of the Bay of Fundy, Canada.

The evidence of ancient deserts takes several forms. Although the shape of dunes is rarely preserved, the distinctive cross-bedding of wind-deposited sand can be readily identified in ancient sandstones (see Fig. 16.18). Such sandstones also tend to be orange or red in color, testifying to the rusting of iron minerals under the intensely oxidizing conditions of arid climates. Under a microscope, the individual sand grains are commonly well rounded, well sorted, and pitted as a result of their transportation by saltation (see Fig. 16.11b). In addition, the sandstones are often associated with deposits of rock salt and gypsum, which are produced by the evaporation of salt water in arid environments.

Best known for its desert climates is the Triassic Period, some 200 to 250 million years ago. During this period, the supercontinent Pangea stretched from pole to pole (Fig. 16.25). But much of North and South America, Europe and Africa lay beneath subtropical high-pressure zones and so experienced arid climates, and the shear size of the supercontinent fostered the development of continental interior deserts. In these regions, rocks of the Triassic Period are dominated by a variety of desert sediments that are loosely termed "redbeds" (Fig. 16.26). The breakup of Pangea, which began at the end of the Triassic, created vast shallow salt-water basins in what is now western Europe and the Gulf of Mexico as the continents rifted and separated. Exposed to the same arid climate of the subtropical highs, these basins experienced rapid rates of evaporation and so deposited thick sequences of evaporites, many of which are exploited today for rock salt and gypsum (see Chapter 6).

> **CHECK YOUR UNDERSTANDING**
>
> ◯ What evidence demonstrates the existence of deserts in the geologic past?
>
> ◯ How has the distribution of deserts in the geologic past been influenced by plate tectonics?

16.5 SUMMARY

- Deserts have existed for much of Earth's history as evidenced by the rock record.
- Plate tectonics plays a controlling influence on the distribution of deserts by controlling the movement of the continents.

16.6 Deserts on Mars

Deserts are by no means unique to our planet. And for evidence for the existence of deserts on other planets, we need only look at the dramatic pictures of the surface of Mars, the Red Planet, sent back by the Pathfinder mission in 1997, and by the Martian rover missions of 2003 and 2012. Despite its numbingly cold surface temperatures, Mars has long been considered a desert world. Indeed, in 1895, when the American astronomer Percival Lowell interpreted streaky marks on the planet's surface as a network of Martian canals, he proposed that the canals had been built by an intelligent race in a desperate effort to save the planet from desertification by bringing water from the poles to the equator. Although the space age laid any notion of canal-building Martians to rest, evidence of the planet's desert climate was dramatically demonstrated in 1971, when Mariner 9 arrived to find the Martian surface obscured by a dust storm that lasted 3 months.

Although the atmosphere of Mars is thin, under the planet's low gravity, hurricane-force winds blowing at over 200 kilometers (125 miles) per hour can cause dust to rise over 6 kilometers (4 miles) into the atmosphere. Some of these dust storms are so large that they change the very color of the planet's surface as vast areas of darker bedrock are blanketed by lighter dust only to be exposed again as

Figure 16.27   Martian Barchan Dunes

The Mars Reconnaissance Orbiter photographed these active barchan dunes in the Hellespontus region of Mars in 2008.

Figure 16.28   Martian Boulders

This image of the Martian surface taken by the Sojourer lander during the Pathfinder mission in 1998 shows wind-blown dust and sand-blasted and faceted boulders that closely resemble those of Earth's rocky deserts. The two hills in the distance, dubbed "Twin Peaks," are about 1–2 kilometers (1 mile) away.

Figure 16.29   Martian Ventifact

The Martian rover *Spirit* captured this image of a football-sized ventifact (dubbed Adirondack). Compare with Figure 16.16.

the dust moves on. Taken whenever the dust settles, satellite photographs of the Martian surface show many features of desert landscapes, including dune fields and loess-covered plains (Fig. 16.27). Land-based images from the Viking, Pathfinder, and rover missions also reveal windblown landscapes strikingly similar to those seen in rocky deserts on Earth (Fig. 16.28), including wind-faceted boulders or ventifacts (Fig. 16.29). Photographs from orbiting satellites have also revealed towering dust devils on the Martian surface (see In Depth: **Dust Devils**).

> **CHECK YOUR UNDERSTANDING**
>
> ◉ What is the evidence for the existence of deserts on Mars?

16.6 SUMMARY

- Mars is a desert world with dust storms and desert-like features.

Key Terms

alluvial fan 509	desert 502	loess 521	polar desert 507
bajada 509	desert pavement 514	longitudinal dune 520	rain-shadow desert 504
barchan dune 518	desert varnish 508	mesa 509	saltation 513
blowout 514	desertification 515	oases 509	slip face 518
butte 509	dune 517	parabolic dune 520	star dune 520
coastal desert 504	dust devil 512	pediment 509	subtropical desert 502
continental interior desert 507	inselberg 509	playa 509	transverse dune 518
deflation 514	intermittent stream 509	playa lake 509	ventifact 514

Key Concepts

16.1 THE ORIGINS OF DESERTS

- Deserts form whenever the climate is *arid*.

- Dry climates produce five types of deserts: subtropical, rain-shadow, coastal, continental interior, and polar.

- Subtropical deserts form in the subtropical high-pressure belts of the tropics of Cancer and Capricorn, where the air is warm and the atmospheric pressure is high. Rain-shadow deserts form in the dry regions on the sheltered, or leeward, side of mountains. Coastal deserts form where warm coastal areas lie next to cold ocean currents. Continental interior deserts develop in the interior of major landmasses because these areas are located so far from the sea that any rain falls before reaching them. Polar deserts, like subtropical deserts, lie beneath an atmospheric high-pressure zone. Polar deserts, however, are extremely cold.

16.2 WEATHERING AND EROSION IN DESERTS

- Weathering in deserts results primarily from mechanical processes when rocks break down as the temperature changes from day to night. Chemical weathering is limited in deserts but not completely absent, accounting for the rusting of desert landscapes.

- The principal agent of erosion and deposition in deserts is running water. Water flows in fast-flowing but short-lived streams, often in the form of flash floods.

- With repeated flash floods, debris-laden alluvial fans build out onto the flatter desert floor, eventually forming an apron of sediment, or bajada, at the foot of the mountains.

- Erosional features in deserts include inselbergs, pediment, mesas, and buttes.

- Flash floods may produce temporary playa lakes that evaporate to form playas.

16.3 WIND IN DESERTS

- The wind transports material in the desert either by moving it along the ground as a bed load, in the case of sand, or by carrying it in the air as a suspended load, in the case of finer particles such as silt and clay.

- Wind erosion operates in two ways: deflation and abrasion.

- Wind deposits materials to form several important landforms and sedimentary bodies. These include dunes—barchan, transverse, longitudinal, parabolic, and star—and blankets of windblown silt called loess.

16.4 DESERTIFICATION: NATURAL AND HUMAN INDUCED

- Over time, deserts expand and contract as conditions change. The spread of desert conditions into formerly fertile fringe areas is called desertification.

- Desertification can be brought about by natural climatic processes and by human activities, such as overpopulation and land mismanagement. Once the vegetation has been reduced to the point that the soil can no longer be protected from erosion, the process of desertification is irreversible.

- Desertification is a serious and growing problem in many of the world's semiarid regions.

16.5 DESERTS AND PLATE TECTONICS

- Deserts have been important features of Earth's surface for much of its history. We know this because features characteristic of modern deserts occur in the rock record.

- Because of the strong latitudinal control on the location of deserts, plate tectonics has had a controlling influence on the distribution of deserts throughout geologic time.

16.6 DESERTS ON MARS

- The planet Mars is a desert world, with huge dust storms and many desert-like deposits, including barchan dunes and ventifacts.

Study Questions

1. Under what circumstances do the dry climates responsible for the formation of deserts develop?

2. If subtropical deserts and polar deserts both owe their origin to the downwelling dry air of high-pressure belts, why are these two types of deserts so different?

3. Explain the combination of circumstances that make the Atacama Desert one of the driest places on Earth.

4. Which desert landforms owe their origin to running water and how do these features develop?

5. How does saltation lead to the formation of an internally cross-bedded dune?

6. What are the five main types of sand dune and how do they form?

7. What actions might you advise the political leader of a country that contains regions experiencing rapid desertification to take to attempt to reverse the process?

8. To what extent is the desertification of the African Sahel the result of human activities?

9. What features of modern desert deposits might you use to demonstrate the existence of ancient deserts in the geologic record?

10. How does plate tectonics account for the distribution of deserts in the Triassic Period?

11. Explain why Martian deposits resemble those of deserts on Earth despite the cold climate of the Red Planet.

17.1 Coastal Processes

17.2 Wind-Driven Waves

17.3 Tides

17.4 Coastal Erosion

17.5 Coastal Deposition

17.6 Coastlines

17.7 Coastal Management

17.8 Coastlines and Plate Tectonics

▶ Key Concepts

▶ Key Terms

▶ Study Questions

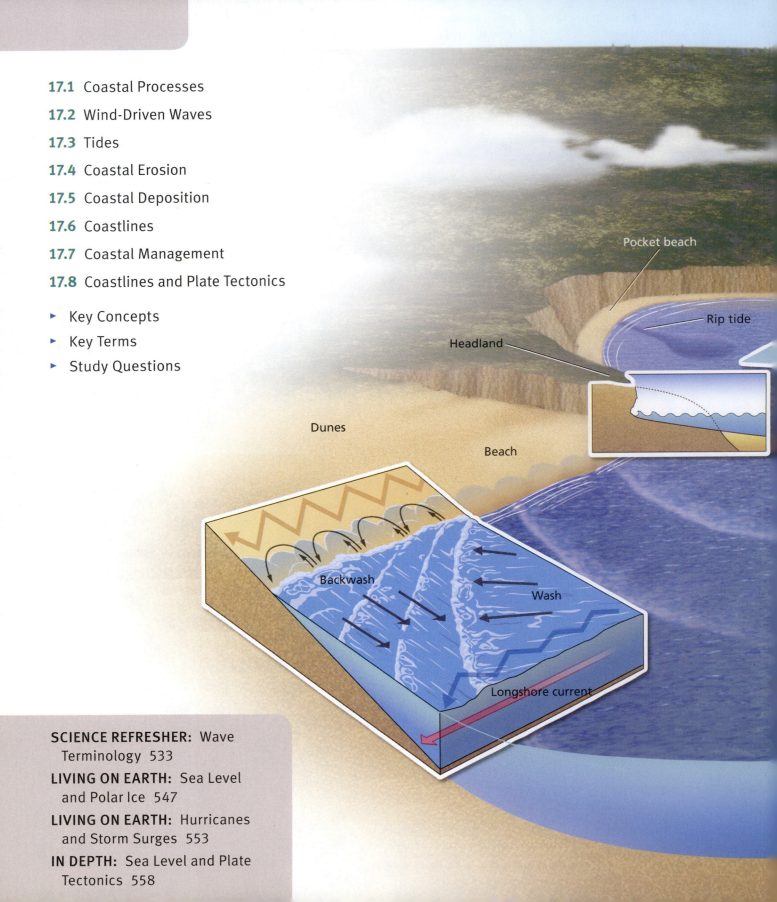

Pocket beach

Rip tide

Headland

Dunes

Beach

Backwash

Wash

Longshore current

SCIENCE REFRESHER: Wave Terminology 533

LIVING ON EARTH: Sea Level and Polar Ice 547

LIVING ON EARTH: Hurricanes and Storm Surges 553

IN DEPTH: Sea Level and Plate Tectonics 558

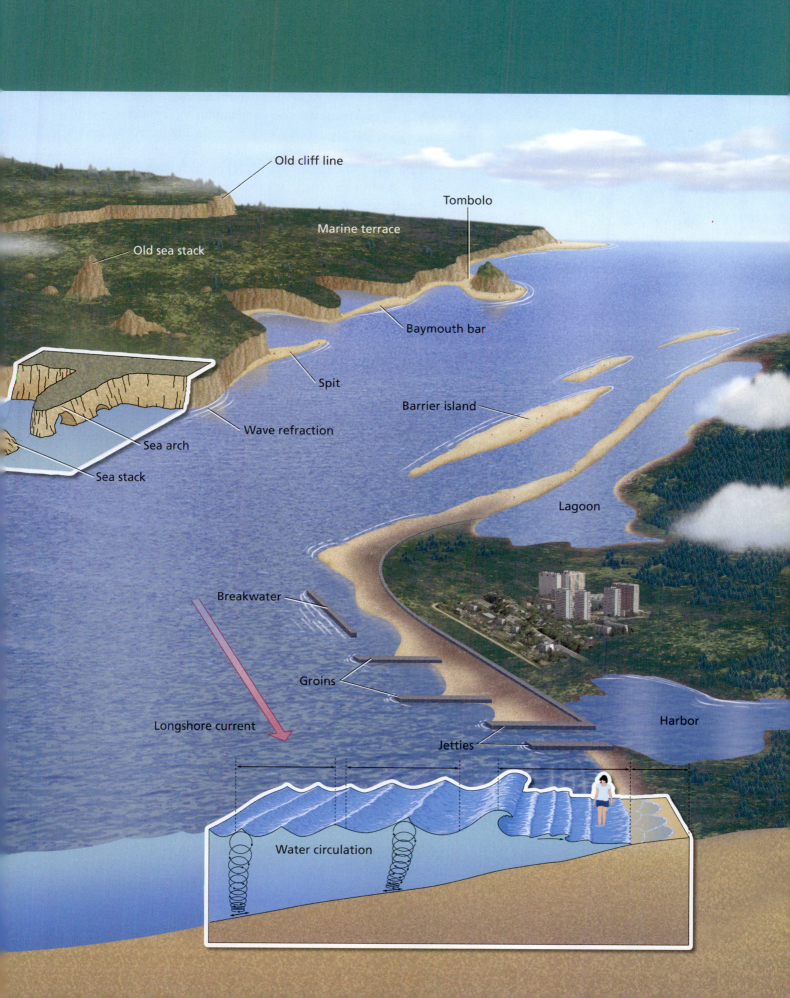

Old cliff line

Tombolo

Old sea stack

Marine terrace

Baymouth bar

Spit

Sea arch

Wave refraction

Barrier island

Sea stack

Lagoon

Breakwater

Groins

Longshore current

Harbor

Jetties

Water circulation

17.1 List the processes that shape coastlines.

17.2 Describe the nature of waves and compare and contrast wave behavior in the open ocean and near shore.

17.3 Identify the origin of tides and tidal patterns.

17.4 Explain the processes of coastal erosion and describe the landforms these processes produce.

17.5 Explain the processes of coastal deposition and describe the landforms these processes produce.

17.6 Summarize the evolution of coastal regions and the effects changing sea level have on coastlines.

17.7 Analyze the key issues in coastal management and the obstacles to sustainable coastal development.

17.8 Summarize the role that plate tectonics plays in governing the shape of coastlines.

The sea relentlessly pounds Earth's coastlines, creating spectacular scenery and producing some of the world's most sought-after vacation spots. Driven by wind, gravity, and heat from the Sun, the ceaseless motion of the world's oceans is manifest in coastal regions by the waves, tides, and currents that constantly buffet the shore. This battering continually modifies the shoreline, slowly eroding more resistant rocks to create dramatic cliffs and redistributing sand and gravel to form sweeping beaches. Although they are often considered prime sites for land development, coastlines are fragile regions ill-suited to this purpose. Coastlines are subject to violent storms and are home to delicate ecosystems. They are also short-lived features that move with the rise and fall of sea level. An understanding of the processes that shape our coastal regions is essential to the responsible management of coastal environments, where much of the world's population resides, and the sensitive coastal ecology upon which many of us depend.

Like streams, glaciers, and the wind, waves are important agents of erosion, transportation, and deposition. Pushed by the wind, waves expend their energy on the shore as breakers that batter the coastline with pressures that can reach 10 metric tons per square meter, equivalent to a head-on collision in an automobile at full speed. This relentless pounding both carves the coast into cliffs and builds up the shoreline with banks of sand and gravel in an endless cycle of erosion and deposition.

17.1 Coastal Processes

The ceaseless motion of the oceans provides some of the most dramatic evidence of the dynamic and unique nature of our planet. Nowhere is this more apparent than at the coastline, where the land meets the sea (Fig. 17.1). Here, *waves* pushed by the wind, *tides* pulled by the force of gravity, and *currents* driven by the Sun's heat, all expend their energy at the shoreline. This energy erodes the coastline and sweeps sediment produced by erosion out to sea and along the shore. The endless cycle of erosion and deposition creates the cliffs, beaches, and sandbars that are the familiar landforms of coastal regions and the environment that is home to a wide variety of coastal ecosystems. But what causes the waves, tides, and currents responsible for these processes? How do they operate as agents of erosion and deposition? And what is the impact of these processes on coastal development? We start to answer these questions in Section 17.2 by looking at waves and the currents they produce.

17.1 SUMMARY

- Coastlines are the result of erosion and deposition by waves, tides, and currents.

Figure 17.1 Cornish Coastline
Waves batter the cliffs at Botallack on the west coast of Cornwall, UK.

17.2 Wind-Driven Waves

An ocean **wave** is the familiar *swell* of the sea—an undulation in the water that moves across the sea's surface. In the context of the types of waves we discussed when examining earthquakes (see Chapter 11), an ocean wave is a form of *surface wave*. Ocean waves disturb the sea only near its surface, and their energy rapidly decreases with depth.

To those of us who have had the pleasure of strolling along the shore, waves are the most obvious example of the movement of ocean water. Waves are produced by the wind and occur when energy from the wind is absorbed by the water's surface. The relationship between waves and the wind is obvious to anyone on the shore that waves on a windy day are generally higher than they are on a calm day. However, as one travels out to sea, these same winds generate more gentle swells, which have smooth surfaces and are farther apart, in contrast to the *breakers* that form near shore.

The winds that blow over ocean waters are generated by variations in atmospheric pressure that exist above one part of an ocean and another. The strength of the wind depends on the atmospheric pressure gradient, much like the velocity of water flowing downhill depends on the hill's gradient or slope. Winds are highest where high and low pressures are close together, and gentlest where they are far apart. The effects of the wind depend on its velocity, duration, and the area over which it blows. For example, even gentle, short-lived winds of only one kilometer per hour or so can cause small ripples to move in the direction of the wind on a relatively smooth ocean surface. As the wind's velocity increases, so too does the height of the waves. This is because the presence of waves increases the surface area of the water, which results in increased frictional resistance as the wind blows across it. This, in turn, causes more energy to be transferred from the wind to the water so the waves become even larger (Fig. 17.2). As a result, even a gentle wind blowing over the ocean for a long enough period of time will produce increasingly larger waves.

Once a wave is initiated, it loses very little energy as it travels across an ocean. In fact, waves breaking on a shoreline could have been initiated anywhere within the ocean. As persistent, gentle winds can create large waves in the way we just described, it follows that major storms in the ocean are capable of generating very large waves that might not break until they reach coastlines thousands of kilometers away.

WAVE SHAPE AND WAVE MOTION

The terminology we use to describe the size and movement of ocean waves is the same terminology we use to describe any wave (see Science Refresher: **Wave Terminology** on page 533). The high and low points on a wave surface are, respectively, *crests* and *troughs*. To describe a wave's size,

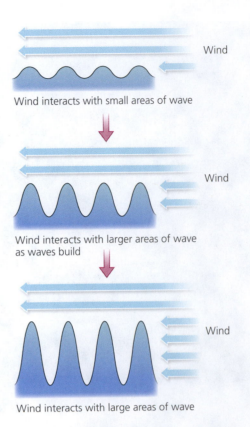

Wind interacts with small areas of wave

Wind interacts with larger areas of wave as waves build

Wind interacts with large areas of wave

Figure 17.2   Waves and the Wind
The wind interacts with more water as a wave builds, and as a result, more of the wind's energy is transferred to the water and the wave builds higher.

we use the terms **wave height**, which is the difference in height between the wave crest and trough, and **wavelength**, which is the horizontal distance from crest to crest or from trough to trough. The rate at which a wave moves is the **wave velocity** (Fig. 17.3).

The velocity of an ocean wave is influenced by the depth of the water. In deep water, subtle variations in water depth have little effect on wave velocity. But in shallow water, slight changes in water depth become important and wave velocity decreases dramatically as the water shoals.

WAVES IN THE OPEN OCEAN

The motion of waves is a form of energy that is rapidly lost when a wave breaks on the shore. When we view waves at sea, it appears that the water simply rises and falls, and does not move with the wave. A familiar example can illustrate this. Imagine a seagull floating on the ocean's surface. The actual movement of the water is approximated by

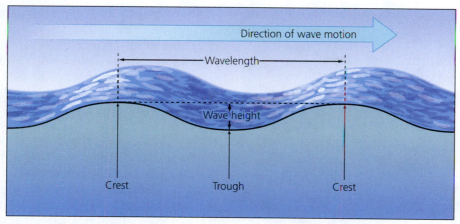

Figure 17.3   Wave Height and Wavelength
The height of an ocean wave is the difference in elevation between the crest of the wave and the trough. Wavelength is the horizontal distance from one crest to the next or from trough to trough.

SCIENCE REFRESHER

Wave Terminology

In addition to the terms *wave height*, *wavelength*, and *wave velocity*, three additional terms are used to describe the size and motion of ocean waves (Fig. 17A). The *amplitude* of an ocean wave is the vertical distance a wave moves above or below a level sea, or half the wave height. The *period* of a wave is the time required for successive wave crests to pass a particular point, and the *frequency* of a wave is the number of waves that pass a particular point each second.

As illustrated in Figure 17A, the period of a wave is the time required for the crest of a wave to travel one wavelength, that is, from points A to B in the figure, whereas the frequency is the number of wave crests that pass A (or B) every second. Thus, wave velocity can be measured using one of two equations:

$$V = \frac{L}{T} \text{ or } V = L \times F$$

where V is the wave velocity, L is the wavelength, T is the period, and F is the frequency. Put another way, the period of a wave is its wavelength divided by its velocity, whereas the frequency of a wave is its velocity divided by its wavelength.

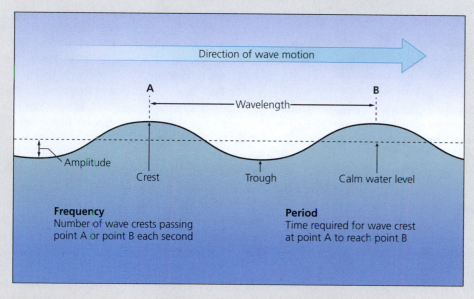

Figure 17A   Wave Terminology

Direction of wave motion

Figure 17.4   Circular Motion in a Wave

A floating seagull travels in a circular path (see dots) as the wave passes. The dots track the changing location of the center of the seagull; the location of the dots demonstrates that the water itself does not move with the wave.

the motion of the gull, which appears to bob up and down rather than moving forward with the water (Fig. 17.4). In fact, floating objects follow a circular path as the wave goes by. The seagull, for example, rises and is pushed forward as a wave crest goes past, only to fall backward into the trough that follows. Water behaves in a similar manner. It does not move horizontally with the waves but instead moves up and down in a more or less circular or orbital path. Consequently, there is no net movement of the water.

The circular motion of water in a wave progressively decreases with depth because the effect of the wind diminishes with depth. We can get an idea of the relationship between waves and the motion of water below the sea surface by tracing the movement of individual particles of water at varying depths. The motion changes with depth (Fig. 17.5). In the open ocean, where the water is deep, the individual water particles in a wave travel circular paths of progressively decreasing diameter with depth. Below a depth of about one-half the wavelength, known as the *wave base,* there is virtually no motion at all because, with increasing depth, the water is increasingly protected from surface winds. In this way, ocean waves are surface waves, the movement of which is focused at the sea surface. In coastal regions, where the water depth is relatively shallow, this pattern of movement is modified by the seabed. In shal-

> **CHECK YOUR UNDERSTANDING**
>
> ● Describe the motion of waves in the open ocean.

lower water, the circular pattern near the surface becomes oval shaped at depth as the "up and down" motion progressively diminishes. Therefore, in open ocean, waves generally reflect the interaction of surface sea water and the wind but, in shallower water, the seabed plays an increasing role in the wave's characteristics.

WAVES NEAR THE SHORELINE

Waves generated in mid-ocean migrate as an ocean swell virtually unhindered until the waves reach the shore. In certain circumstances—for example, during hurricanes—large waves generated at the storm center can travel thousands of miles to batter a coastline that may otherwise be experiencing tranquil weather conditions. This creates the rather spectacular sight of very high breakers on an otherwise calm day.

As waves approach a gently sloping shoreline, there is little loss of energy until the waves break on the shore. Recall that in the open ocean, the water itself does not actually move forward with the waves, it simply moves forward and backward (and up and down) in a circular path (see Fig. 17.4). But where the water depth is less than one-half of the swell's wavelength, the smooth surface of the ocean swell begins to transform into **breakers**, the familiar cresting and breaking waves of the coast. At this water depth, the restricting influence of shallow water on wave motion causes the waves to slow down. But the number of waves passing a particular point in any given time remains unchanged because the swell from the ocean continues to migrate shoreward unabated. This number is the wave *frequency* (see Science Refresher: **Wave Terminology** on page 533), which is equal to the wave velocity divided by its wavelength. For the frequency to remain the same, the wavelengths must also decrease as the waves slow down. As a result, the waves start to bunch together (Fig. 17.6), thereby concentrating the energy of the waves. The waves

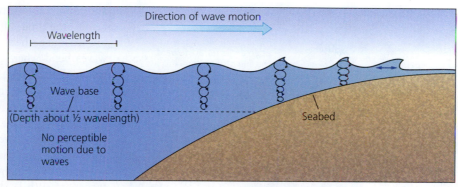

Figure 17.5   Motion of Wave below Surface

The circular motion of water particles in a wave progressively decreases with depth and ceases at the wave base, which lies at a depth approximately equal to half the wavelength.

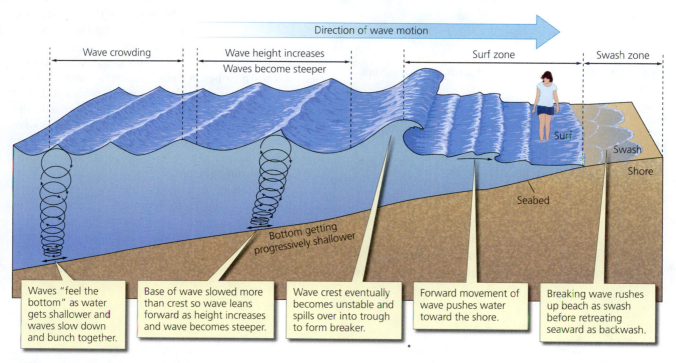

Direction of wave motion

Wave crowding

Wave height increases
Waves become steeper

Surf zone

Swash zone

Surf

Swash

Shore

Seabed

Bottom getting progressively shallower

Waves "feel the bottom" as water gets shallower and waves slow down and bunch together.

Base of wave slowed more than crest so wave leans forward as height increases and wave becomes steeper.

Wave crest eventually becomes unstable and spills over into trough to form breaker.

Forward movement of wave pushes water toward the shore.

Breaking wave rushes up beach as swash before retreating seaward as backwash.

Figure 17.6   Breakers and Surf Zone

As ocean waves approach the shore and the seabed becomes progressively shallower, wave crests heighten and wavelengths shorten as the wave's energy is concentrated into shallower water. As a result, the waves become peaked and more closely spaced, and start to lean forward, eventually forming breakers in the surf zone.

become higher and steeper but the base of the waves slows more than the crests, so the waves lean forward as their height increases.

A critical point of wave stability occurs when the wave attains a height equal to about one-seventh of the wavelength. When this happens, the water itself starts moving forward, and the wave crests become unstable and spill over into the troughs to form breakers (see Fig. 17.6). The area between the outermost breaker and the beach is known as the **surf zone**. At the beach, the breaking wave first rushes up the beach as **swash**, before retreating seaward as **backwash**. The region on the beach where this occurs is known as the **swash zone**.

WAVE REFRACTION AND LONGSHORE DRIFT

Waves rarely approach a shore head on; instead they approach at an angle. As waves near the shore, however, they bend so as to become more parallel to the shoreline (Fig. 17.7). The bending of waves is known as **wave refraction** (see Science Refresher: **Wave Refraction** in Chapter 11). Recall that wave refraction occurs because not all parts of a wave that

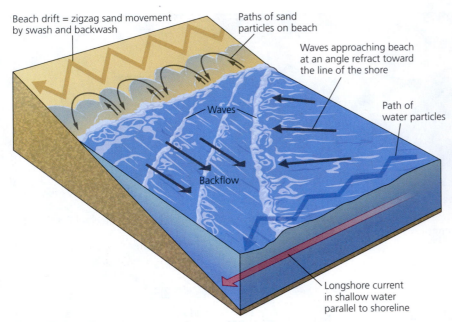

Figure 17.7   Wave Refraction, Beach Drift, and Longshore Currents

When waves are oblique to the shoreline, one end of the wave encounters shallow water before the other, causing the wave to bend or refract. However, the return flow is directly down the slope of the beach, so sand grains on the shore move down the beach in a zigzag movement called beach drift. Water in the surf zone also follows a zigzag path, giving rise to a longshore current parallel to the shoreline.

and is therefore traveling faster. As a result, the line of the wave bends toward that of the shore (just as seismic waves bend when crossing from one layer of Earth's interior to another).

Oblique waves have two components of movement. One, perpendicular to the shoreline, gives a measure of the erosive power of the crashing wave. The other, parallel to the shore, is responsible for significant sediment transport. While the swash of most waves hits the shore obliquely, the backwash always flows down the slope of the beach. Because of this, sediment particles tend to follow a zigzag path along the beach as they are carried in suspension, first by the advancing waves and then by the retreating backwash (see Fig. 17.7). This movement is known as **beach drift**. Incoming waves that approach the shore at an angle produce currents within the surf zone that move parallel to the shore. Because of the turbulence in the surf zone, these **longshore currents** are capable of carrying significant quantities of sediment in suspension, and moving it slowly along the shore, a process known as **longshore drift**.

The net effect of the two components of movement—parallel and perpendicular to the shore—is that beaches are constantly on the move, the sand moving in the direction of the longshore current. On many beaches along both the Atlantic and Pacific coasts, more than a million tons of sand are moved in this fashion each year. For this reason, it is

obliquely approaches a boundary (in this case the shoreline) reach it at the same time. Instead, the part of the wave nearest the shore encounters shallow water and so slows down, while the part farthest from it is still in deep water

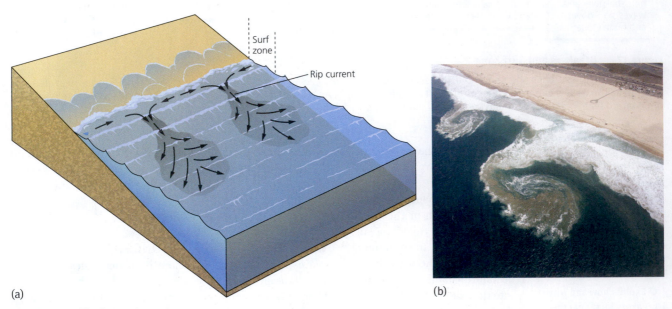

(a)   (b)

Figure 17.8   Rip Currents

(a) Fed by longshore currents, rip currents are narrow surface currents that flow straight out through the surf zone. (b) Rip currents carry suspended sediment offshore, as evidenced by the discoloration of the seawater.

essential to take into account the nature and strength of the currents in the surf zone when developing oceanfront property.

In areas where the waves are breaking strongly, the water pushed toward the shore is often forced sideways by the oncoming waves. This water streams along the shoreline as a longshore current until it can find a route back out to sea. Under certain conditions, the return flow can take the form of a **rip current** (Fig. 17.8a), a narrow surface current which runs straight out through the surf zone and dies out offshore and with depth. Rip currents generally occur in areas where the height of the incoming waves is lower than the height of the waves elsewhere, usually because the water is deeper as a result of a depression or channel on the bottom. As a result, the waves break less strongly and the shoreward current in the swash is much reduced.

Flowing at several kilometers per hour, rip currents are important agents of sediment transport, carrying fine-grained sediment offshore through the surf zone. Indeed, they are best recognized as areas of sea discolored by suspended sediment (Fig. 17.8b). Because of their speed, rip currents can also pose a threat to inexperienced swimmers who attempt to swim ashore against the flow of a rip current. However, because rip currents are narrow, a person can usually escape by swimming parallel to the shoreline for a short distance before turning toward the beach.

> ### CHECK YOUR UNDERSTANDING
>
> ○ Why do oblique waves bend as they approach the shore and what term describes this phenomenon?
>
> ○ What do the terms swash and backwash mean?
>
> ○ Describe the difference between beach drift and longshore drift.

17.2 SUMMARY

* Wind-driven waves have a wavelength, amplitude, velocity, frequency, and period.
* Ocean waves move water in a circle and die out rapidly with depth; as waves approach the shore, their height increases as they slow down and break forward onto the surf zone.
* Waves moving oblique to the shore are responsible for longshore drift and rip currents.

17.3 Tides

The daily rise and fall of ocean waters (Fig. 17.9) we call **tides** has been recognized for millennia, and the relationship of the tide cycle to the rising and setting of the Moon has been the source of many myths and legends. When the Moon is full or new, tides are amplified so that high tides are higher and low tides lower than usual. By contrast, the differences between high and low tides is more subdued during quarter moon phases (Fig. 17.10).

TIDES AND THE MOON

Earth's spin gives ocean water a tendency to move away from the poles (the spin axis) and toward the equator. This creates an equatorial bulge in the ocean surface that is distorted by the gravitational pull of the Moon and Sun. According to Newton's Laws of Universal Gravitation, *gravity* is the fundamental force of attraction between any two objects. It is proportional to the masses of the two objects and their distance apart—the greater the masses, the greater the attraction. However, the attraction diminishes very rapidly as the distance between the two objects increases. As a consequence, the Moon exerts a greater influence on Earth's tides than does the Sun. This is because the Moon, although it has far less mass than the Sun, is much closer to Earth.

Because the portion of Earth's surface facing the Moon is closer to the Moon than the side facing away, the gravitational pull from the Moon is stronger on that side. The side facing away from the Moon is farther away, so that the gravitational pull is weaker. The ocean water on the near

Figure 17.9 Bay of Fundy

High and low tides in the Bay of Fundy, Nova Scotia, Canada.

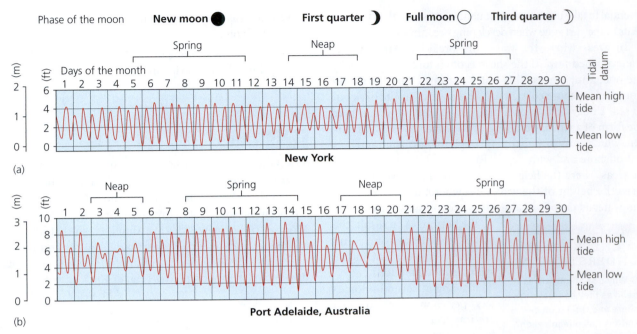

Figure 17.10   Tidal Record

Tidal records for a typical month in (a) New York and (b) Port Adelaide, Australia, illustrate that the biggest difference in high and low tide levels occurs at full moon and new moon; the smallest difference occurs during the quarter moon phases.

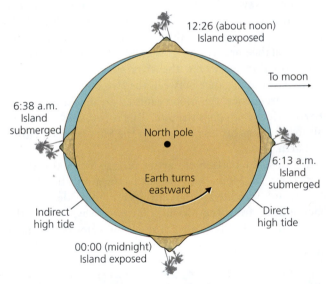

Figure 17.11   Tides on a Rotating Earth

On a rotating Earth, a bulge in the water surface (exaggerated for clarity) occurs on the portions of the surface that are aligned with the Moon. Note the island experiences two high tides and two low tides in 24 hours and 50 minutes as explained in Tidal Patterns.

less strongly than the solid earth. As a consequence, the solid earth shifts beneath the ocean water toward the Moon, literally leaving the ocean water behind. This creates a bulge in the ocean surface that points away from the Moon and is known as an *indirect high tide*.

Thus, at any given time, two bulges of ocean water occur on the Earth's surface, one where the surface is closest to the Moon, the other where it is farthest away. Furthermore, because the bulges form at the expense of other regions, depressions are produced in those areas farthest removed from the bulges. These are therefore regions of low tides. Since Earth rotates once every 24 hours about its axis, it moves beneath these tidal bulges and depressions, giving the appearance of migrating tides. As Earth completes one rotation, every location ideally experiences two high tides, one direct and the other indirect. Between the two high tides, two low tides occur (Fig. 17.11).

> ### CHECK YOUR UNDERSTANDING
>
> ○ The Sun has far more mass than the Moon, so why does the Moon exert a stronger influence on Earth's tides?
>
> ○ What is the difference between a direct high tide and an indirect high tide, and why do both occur?

side of Earth feels the pull toward the Moon more strongly than the solid Earth does because it is closer, which adds to the equatorial bulge in the ocean surface created by Earth's spin. The resulting bulge points toward the Moon and is known as a *direct high tide*.

On the far side of Earth, the equatorial bulge is farther away from the Moon and is therefore pulled toward the Moon

TIDAL PATTERNS

If the Moon were stationary, we would see high and low tides at identical times each day. The position of the bulges would not change and Earth would merely rotate beneath them. But the Moon is not stationary. It moves along its orbit as Earth completes each revolution. As a result, it takes Earth 50 minutes to "catch up" with the Moon, so

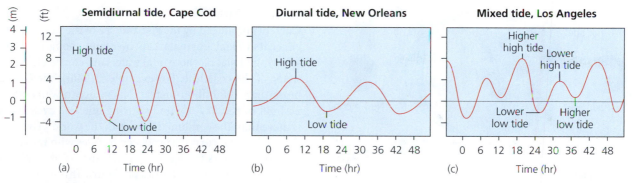

Figure 17.12 Tidal Patterns
Tidal curves are shown here for (a) semidiurnal, (b) diurnal, and (c) mixed tidal patterns.

that the two are once again in the same relative position. This time period—24 hours and 50 minutes—is commonly called a *lunar day*. It is during this time interval, rather than every 24 hours, that most coastal localities experience two high tides and two low tides.

Most coastal regions witness the rise and fall of the tides twice each lunar day, but not all coastlines experience this tidal pattern and significant differences can occur in the magnitude of successive tides in those regions that do. This is because factors other than Earth's rotation can influence the size and timing of successive tides. As a result, the following three major tidal patterns exist (Fig. 17.12):

- **Semidiurnal**, like those of the Atlantic, with two high tides each lunar day
- **Diurnal**, like those of the Gulf of Mexico, with only one high tide per lunar day
- **Mixed**, like those of the west coast of North America, with two high tides of quite different magnitudes each lunar day.

The explanation just discussed for the origin of tides accounts for the basic semidiurnal pattern in which the tides are just over 6 hours apart. This is the tidal pattern of the open ocean and many coastal regions, such as those of the Atlantic seaboard (Fig. 17.12a). However, our explanation assumes that Earth's oceans constitute a uniform layer of water and so does not take into account the complications associated with the positions of continental landmasses and any irregularities in the coastline or seafloor. Where the land partially encloses relatively small, confined bodies of water such as the Gulf of Mexico, for example, the sea behaves like water sloshing back and forth in a container. This motion can modify the basic semidiurnal pattern of tides, giving rise instead to a diurnal pattern in which the tides are just over 12 hours apart (Fig. 17.12b). Complications on the Pacific seaboard, on the other hand, give rise to mixed tidal patterns in which the tides are just over 6 hours apart, but one high tide is significantly higher than the other and one low tide is significantly lower (Fig. 17.12c).

In sum the gravitational attraction between Earth and Moon, together with the location and shape of the continents, explains the origin and timing of the three types of tidal pattern. But why do tidal ranges (that is, the difference in water height between high tide and low tide) at any one locality vary during the month (see Fig. 17.10)? And why do these variations match the phases of the Moon? We now turn to answering these questions.

> **CHECK YOUR UNDERSTANDING**
>
> ❍ What is a lunar day and how does it govern the interval between tides?

TIDES AND THE SUN

Variations in tidal range are also explained by gravitational attraction, but these variations are due to the influence of the Sun not the Moon (Fig. 17.13). Although the influence of the Sun on tides is less than that of the Moon, it is nonetheless significant. Depending on its position in the sky relative to the Moon, the Sun can either amplify or dampen daily tidal ranges. **Spring tides** (from the German "springen," to leap up) occur when the Sun and Moon are most closely aligned with Earth. During this time, the respective gravitational pulls of the Sun and the Moon work in concert, and the tidal range is amplified (Fig. 17.13a). **Neap tides** occur when the Sun and Moon are in maximum misalignment with Earth (Fig. 17.13b). In these positions, the respective gravitational pulls work against each other, and so dampen the tidal range.

But when do these alignments and misalignments occur? Like all objects in the Solar System, half of the Moon's surface is always illuminated by the Sun. The "phases of the Moon" occur because only a portion of the illuminated half is generally visible from Earth (Fig. 17.13). When the Moon is on the far side of Earth relative to the Sun, that is, in the opposite part of the sky, we see its entire illuminated half, which we refer to as a "full moon." When the Moon is aligned between the Sun and Earth, that is, in the same portion of the sky, we do not see the Moon at all because its illuminated half is facing away from us. We refer to this

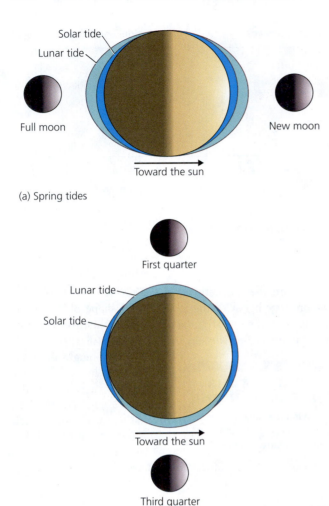

(a) Spring tides

(b) Neap tides

Figure 17.13   Spring and Neap Tides
The relationship between the Sun, Moon, and Earth governs the occurrence of (a) spring tides and (b) neap tides. At full moon and new moon, the tidal influences of the Sun and Moon act in concert, resulting in an amplified tidal range. At the first-quarter moon and third-quarter moon, the gravitational influences of the Sun and Moon act against each other, dampening the tidal range.

phase as the "new moon." Generally, the alignment during this phase is not perfect. (In the rare instances when it is, the Moon blots out the Sun and a solar eclipse occurs.) At both full moon and new moon phases, the Sun, Earth, and the Moon are most closely arranged in a line, and the additional gravitational attraction produces spring tides.

Quarter moons represent phases half way between these two alignments, when the Sun and Moon are most misaligned (that is, at right angles) with respect to Earth so that their respective gravitational pulls work against each other. The result is a dampening of the tidal range, which produces neap tides.

> **CHECK YOUR UNDERSTANDING**
>
> ◗ What are spring tides and neap tides, and why do they occur?

TIDAL CURRENTS

Currents produced by the rise and fall of the tides are **tidal currents**. Tidal currents can be swift enough to cause scouring in places where tidal waters are constricted by the shape of the coastline. For example, in Maritime Canada, more than 100 billion metric tons of ocean water flow in and out of the Bay of Fundy during one tide cycle, which is more than the combined flow of all the world's rivers (see Fig. 17.9). Tidal currents produced by tides with ranges that exceed 16 meters (50 feet) can reach speeds of 15 kilometers per hour (almost 10 miles per hour). This is faster than the average jogger can run, and every year tourists trapped against the cliffs by such currents are helicoptered to safety.

As the tide ebbs and flows, tidal currents periodically reverse their direction. **Flood currents** accompany a rising tide and flow inland, whereas **ebb currents** accompany a falling tide and flow seaward. The turn of the tide is marked by a period of *slack water* when little or no current flows.

Tidal currents can be agents of both erosion and deposition, scouring out tidal channels in areas of constricted coastline and depositing suspended particles of fine sediment on **tidal flats**, which are marshy areas that are alternately covered and uncovered by the tide (see Chapter 6). As we will discuss in Section 17.4, however, it is the erosive power of waves that is responsible for most features of coastal erosion.

> **CHECK YOUR UNDERSTANDING**
>
> ◗ What are tidal currents and in what way do they act as agents of erosion and deposition?

17.3 SUMMARY

- Tides are the result of the gravitational tug of the Moon and Sun on Earth's oceans.
- Tidal patterns can be semidiurnal, diurnal, or mixed.
- Spring tides occur when the Sun and Moon are aligned. Neap tides occur when the Sun and Moon are most misaligned.
- Tidal currents act as agents of erosion and deposition.

17.4 Coastal Erosion

Waves release their energy at the coastline, either by crashing on the beach or by pounding the cliffs that stand like immovable objects before them. Waves have enormous erosive power and are able to sculpt even the most resistant of rocks into precipitous sea cliffs (see Fig. 17.1).

WAVE EROSION

Although sea cliffs may seem like an impenetrable barrier, waves slowly but inevitably wear them down by erosion. They do so by dislodging rock fragments on impact and by compressing the air within fractures. They also erode through *abrasion*, the grinding action of water armed with sediment. The resistance of cliffs depends on the material of which they are made. The unconsolidated sediments on Cape Cod in Massachusetts, for example, are no match for ocean waves. Consequently, erosion is forcing these sediments to retreat at a hasty rate of about 1 meter per year (over 3 feet per year). On the other hand, hard crystalline rocks such as granite offer much sterner resistance and therefore retreat much more slowly. Many spectacular cliffs composed of hard crystalline rock occur around the coastlines of the Atlantic and Pacific oceans. An irregular indented shoreline is often related to a variable geology, with *headlands* composed of hard crystalline rocks and *bays* formed by the more effective erosion and retreat of less resistant rocks (Fig. 17.14).

But even the most resistant of rocks are no match for the erosive power of the sea. Under the relentless pounding of waves, the coastline is carved into a variety of erosional landforms, the origins of which we explore in this section of the chapter. The most common features of coastal erosion are *cliffs* and *wave-cut platforms*, and *headlands, sea arches,* and *sea stacks*.

CLIFFS AND WAVE-CUT PLATFORMS

No view of the erosive power of the sea is more spectacular than the sight of waves breaking against bedrock. The sheer impact of so many tons of water can be breathtakingly explosive, literally shaking the ground and sending sea spray rocketing skyward. The power is awesome. Breaking waves routinely exert pressures of 1000 kilograms (one metric ton) per square meter and, during major storms, can produce 10 times that pressure. The impact alone dislodges fragments of bedrock, but more important, as seawater is driven into every crack and crevice, the air within these openings is compressed so that the fractures are pressurized and pried open, and the material loosened. The impacting waves can also hurl boulders and, more effective

> ### CHECK YOUR UNDERSTANDING
> ⊙ What are the main ways waves erode the shoreline?

Figure 17.14   Headland and Bay

On Oregon's south coast, resistant bedrock in the background forms a headland, and weaker bedrock in the foreground forms a bay and a sandy beach. Ocean breakers define the surf zone (see Fig. 17.6).

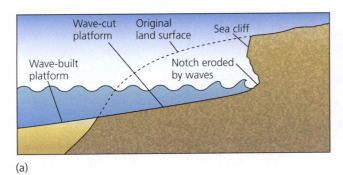

(a)

(b)

Figure 17.15   Wave Erosion

(a) Undercutting as a result of wave erosion creates a sea cliff, the retreat of which produces a gently sloping wave-cut platform. (b) This sea cliff and wave-cut platform are located along the coast of South Wales, UK.

still, abrade the bedrock with sediment held in suspension, relentlessly scouring surfaces like sandpaper. Soluble rocks such as limestone may even be dissolved. As a result, the breaking waves slowly but surely cut a horizontal path landward into the bedrock, producing in the process a wave-cut wall of rock known as a **sea cliff** (see Fig. 17.1).

Because the erosive power of the sea is focused in the surf zone, sea cliffs are progressively undercut at their base. As erosion proceeds, rocks above the overhanging notch cut into the base of the cliff eventually become unstable and fall into the surf. Thus sea cliffs retreat as a result of the combined processes of wave erosion and *mass wasting*—the collapse of material under the influence of gravity (see Chapter 12). The beveled surface left in the wake of this retreat is known as a

wave-cut platform. It forms a gently sloping bedrock bench extending seaward from the cliff base (Fig. 17.15).

The rate at which a cliff face retreats depends on a variety of factors, including the nature of the bedrock and the strength of the waves. However, in many areas the rate is noticeable even on a human timescale, so you can imagine the impact over millennia! Not surprisingly, the cliffs that recede the fastest are those composed of unconsolidated rocks, and most retreat takes place during storms, often with destructive consequences (Fig. 17.16). Given sufficient time,

CHECK YOUR UNDERSTANDING

● How do breaking waves create sea cliffs and wave-cut platforms?

Figure 17.16   Toppled Home, Nantucket Island

This home was a casualty of shoreline retreat on Nantucket Island off the coast of Massachusetts.

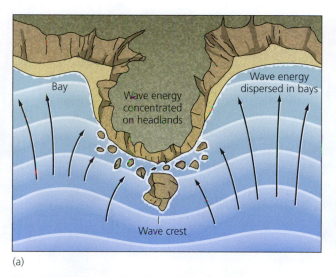

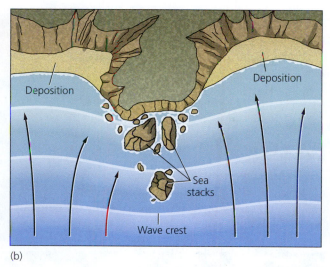

Figure 17.17   Wave Refraction at Headland

(a) Refraction of waves around a headland focuses their energy. (b) The headland is eroded from both sides as the waves bend around them.

however, even the most resistant cliffs suffer shoreward retreat, broadening the wave-cut platform as they do so.

HEADLANDS, SEA ARCHES, AND SEA STACKS

The retreat of a sea cliff does not occur uniformly because its resistance to erosion varies along its length. As a result, cliff lines tend to be irregular, with resistant promontories that jut out into the sea to form **headlands**, and less resistant segments that erode more rapidly to form pocket beaches and bays. The projecting headlands bear the brunt of the sea's attack because

wave refraction causes the waves to bend around them so that the wave energy is focused there (Fig. 17.17). As a result, headlands are eroded from both sides.

Where weaknesses exist in the side of a headland, the waves may exploit them to form *sea caves* at the foot of the cliff (Fig. 17.18a). Continued excavation of such caves on opposite sides of the headland may eventually unite the caves and form a bridge-like opening through the headland known as a **sea arch** (Fig. 17.18b). Ultimately, the span of the bridge may collapse to form a solitary pillar of rock known as a **sea stack** isolated from the headland by the sea (Fig. 17.18c). In this way, headlands are slowly eroded away until they are removed completely and the retreat of the cliff line continues.

> **CHECK YOUR UNDERSTANDING**
>
> ● Describe how the erosion of a headland can eventually result in the formation of a sea stack.

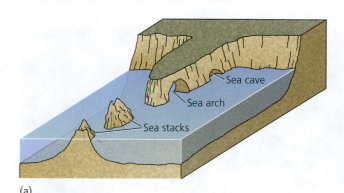

Figure 17.18   Sea Arches and Sea Stacks

(a) Erosion of a headland forms a sea cave, sea arch, and sea stack. (b) These sea arches on the coast of Victoria, Australia, shown in 1984, were called London Bridge until 1990, when the one on the left collapsed. (c) Sea stacks are a feature of Australia's south coast.

17.4 SUMMARY

- Waves erode by abrasion, creating an irregular coastline with headlands and bays.

- Erosional landforms include cliffs, wave-cut platforms, sea arches, and sea stacks.

17.5 Coastal Deposition

Much of the sediment produced by coastal erosion is swept out to sea by coastal currents and deposited offshore. However, a significant proportion is carried along the shoreline, where it is deposited to form a variety of landforms, including *beaches*, *bars*, and *barrier islands*.

BEACHES

Of all coastal landforms, none are more popular than beaches. **Beaches** are gently sloping bodies of unconsolidated sand, gravel, and shell fragments that extend landward from the low water line to the maximum reach of storm waves. Depending on the shape of the coastline, beaches may be small *pocket beaches* between headlands (Fig. 17.19a) or narrow *strands* continuing uninterrupted for long distances (Fig. 17.19b).

(a)

(b)

Figure 17.19   Pocket Beaches and Strands

(a) A pocket beach has developed between headlands on the coast of California at Big Sur. (b) Copacabana Beach in Rio de Janeiro, Brazil, is an iconic strand.

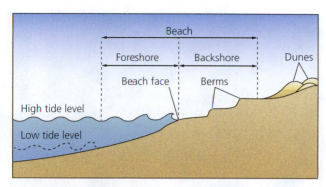

Figure 17.20 Beach Profile
A cross section of a typical beach shows the various components of a beach profile.

Typically, a beach is made up of two parts: a *foreshore* lying between the low-tide and high-tide lines; and *backshore* stretching landward from the high-tide line to the point where the topography changes and vegetation starts (Fig. 17.20). Beyond the backshore, windblown sand may accumulate to form *dunes*, as we learned in Chapter 16.

In profile, beaches typically have elements of varying slope. The *beach face*, at the head of the foreshore, is the steepest portion of the beach. It is routinely exposed to the uprush or swash of the breaking waves. The backshore, on the other hand, is exposed to wave action only during storms and may contain one or more horizontal or landward-sloping shelves, known as *berms,* formed from material thrown up by storm waves.

Although some of the sand responsible for the formation of beaches is derived from the erosion of headlands and cliffs, most is supplied from inland sources by coastal streams. Once deposited at the stream mouth, this sand is slowly transported along the shoreline by the action of waves. As we have learned, two processes contribute to this

> **CHECK YOUR UNDERSTANDING**
> ● Describe the features of a beach as seen in profile.

transport; *beach drift,* caused by the zigzag pattern of swash and backwash on the beach face and, more importantly, *longshore drift* caused by the movement of longshore currents. Beaches are therefore dynamic features that remain intact only if the rate of sediment supply equals that of sediment removal.

SPITS AND BAYMOUTH BARS

In addition to keeping beaches supplied with sediment, longshore drift is responsible for the development of a pair of related depositional landforms known as *spits* and *baymouth bars.* Since longshore currents are powered by the action of waves in the surf zone, their flow is abruptly slowed whenever they encounter deeper water, for example, when the line of the coast is interrupted by a bay or inlet. Where this occurs, the current deposits its sediment load to form a narrow tongue of sand that progressively builds out into the bay. Such fingerlike extensions of beaches

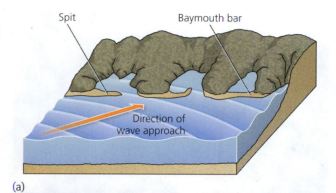

(a)

(b)

Figure 17.21 Spits and Baymouth Bars
(a) Spits form where longshore currents encounter deeper water at the mouth of a bay and deposit their load to form a sand bar that may eventually extend across the bay entrance as a baymouth bar. (b) The spit in this photo partly encircles Golden Bay at the northern tip of South Island, New Zealand.

that project part way across the entrance to a bay are known as **spits** and, with continued growth, may eventually extend completely across the mouth of the bay to form a **baymouth bar** (Fig. 17.21).

A less common form of spit links an offshore island or sea stack to the mainland. Known as a **tombolo** (from the Italian for sand dune), this type of spit is the result of wave refraction around the island (Fig. 17.22).

> **CHECK YOUR UNDERSTANDING**
> ● What is a tombolo, and how does it form?

As the waves bend shoreward, longshore currents are created, which converge and slow down on the landward side of the island. As a result, the sediment they are carrying is deposited to form a sand bar that eventually connects the island to the shore.

BARRIER ISLANDS

Barrier islands are low, offshore ridges of sand that parallel the coastline at some distance from the shore and are separated from the shore by a bay or by a body of quiet

(a)

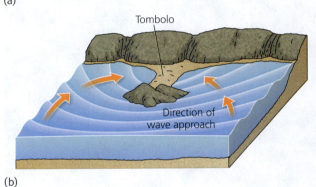

(b)

Figure 17.22   Tombolo

(a) A tombolo, like this one on Shōdo Island, Japan, is a spit connecting an island to the shore. (b) Tombolos are the result of wave refraction, which causes longshore currents to converge and deposit a sand bar on the landward side of the island.

(a)

(b)

Figure 17.23   Barrier Islands

(a) This satellite view is of the Outer Banks, a chain of barrier islands in North Carolina that includes Cape Hatteras (center). (b) On this barrier island off the coast of Florida near Pensacola, sand banks in the lagoon (right) were deposited by storm waves that broke across the island.

brackish (slightly salty) water known as a *lagoon*. Along the Atlantic and Gulf coasts of North America, barrier islands are the most common features of coastal deposition, often extending virtually uninterrupted for hundreds of kilometers (Fig. 17.23a). Barrier islands are typically less than 5 kilometers (3 miles) wide and rarely rise to more than 5 meters (17 feet) above sea level. Most comprise a narrow, linear beach bordered inland by a broader zone of dunes, a zone of vegetation, and an irregular swampy area bordering the quiet waters of the lagoon (Fig. 17.23b).

Barrier islands are characteristic of gently sloping coastlines with ample supplies of sand, relatively low wave energy, and limited tidal range. However, their origin has been the subject of considerable debate and it is likely that several processes contribute to their formation. Some may have originated as elongate spits that have become separated from the mainland by tidal currents, storm waves, and the rise in sea level that followed the last glacial retreat. Others may have simply been built up by storm waves breaking offshore above a gently sloping bottom. Yet others may be remnants of sand dune systems that developed along the shoreline during the last glacial period when sea level was significantly lower (see Living on Earth: **Sea Level and Polar Ice** on page 547). Later, when the ice retreated, sea level rose and flooded the dunes, isolating them from the mainland.

Whatever their origin, once formed, barrier islands are shaped by the combined processes of erosion and deposition. Because global sea levels are presently rising, most barrier islands are narrowing and are at risk of becoming submerged. Most are also migrating shoreward as storm waves break over them and transfer sand from the seaward side to the lagoon. Barrier islands along the Atlantic seaboard, for example, are moving at about a meter per year. However, this does not necessarily mean that the lagoons behind these islands are closing since the mainland shoreline is also retreating as sea level rises.

REEFS

Reefs are near-shore masses of limestone that result from the growth of corals, algae, and other colonial organisms.

LIVING ON EARTH

Sea Level and Polar Ice

One of the predicted consequences of climate change is the warming of Earth's polar regions. When these regions warm, sea level is expected to rise as the polar ice caps begin to melt. Indeed, evidence that this melting is already occurring can be seen in the decline in Arctic sea ice, both in extent and thickness, over the last several decades, and in the worldwide decline in the thickness of glaciers over the past half-century (Fig. 17B).

When polar ice caps begin to melt, the meltwater produced eventually joins the world's oceans, so that sea level will start to rise. How significant might this change in sea level be, and what effect could it have on our coastlines? In answer to this, it is instructive to examine what happened to sea level as the polar ice caps advanced and retreated during the last Ice Age.

If the amount of water draining into an ocean basin is less than that evaporating from it, sea level drops. During an ice age, evaporated water that becomes trapped in polar air masses falls as snow, rather than rain. If the snow does not melt in the spring, it eventually becomes compressed into glacial ice as it is buried beneath successive layers of snow. Since this water does not drain back to the ocean basin, sea level drops. The opposite process happens when glaciers melt. The water returns to the oceans and sea level rises.

During the last Ice Age, which peaked about 18,000 years ago, glacial ice up to 3 kilometers (almost 2 miles) deep covered many of the continents in the Northern Hemisphere. At that time, sea level is estimated to have been 100 meters (330 feet) lower than it is today. Thus, much of the present continental shelves was exposed. The North American Atlantic coastline, for example, was about 200 kilometers (125 miles) east of its present location (Fig. 17C), and Britain

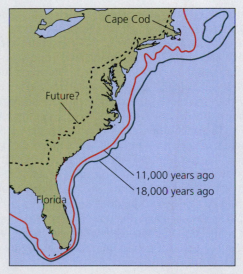

Figure 17C   Eastern North America Coastline

Approximately 18,000 years ago the coastline lay as much as 200 kilometers (125 miles) east of its present position, exposing the continental shelves. If all the polar ice were to melt, sea level would rise by about 60 meters (200 feet), submerging much of the eastern seaboard of North America.

and France were joined. Thus Britain became an island only after the ice began to melt and sea level rose. Similarly, a land bridge across the Bering Strait linked Alaska and Siberia, and many of the islands of Indonesia were connected to Southeast Asia. As anthropologists will vouch, this period of relatively low sea level had a profound

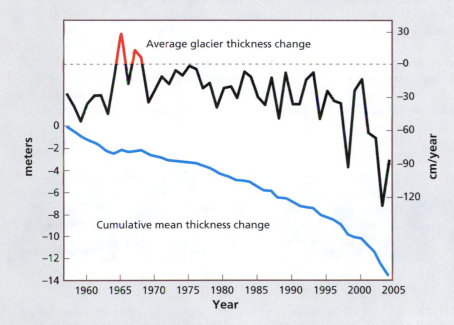

Figure 17B   Declining Thickness of the World's Glaciers

The pattern of average annual thickness change and accumulated thickness change in mountain glaciers around the world shows a steady decline over the past 50 years, with only 3 years (in red) witnessing a thickening.

(Continued)

LIVING ON EARTH

Sea Level and Polar Ice (Continued)

influence on the migration of animals, including humans. For example, humans migrated across the Bering Strait from Asia to North America during this glacial episode.

At the end of the last Ice Age, some 11,000 years ago, most of the polar ice caps melted and, as a consequence, sea level rose as the ice slowly retreated toward the poles. With rising sea levels, the coastlines likewise retreated and the continental shelves were, once again, flooded by the sea. This record indicates that if today's remaining glaciers start to melt as a climatic consequence of global warming, the sea will continue its invasion of coastal regions. Low-lying coastal states such as Florida, Louisiana, and the Carolinas will be threatened, and cities such as Boston, New York, Washington, Philadelphia, Orlando, Miami, New Orleans, and Houston will be submerged.

The realization that the sea level would change as the consequence of ice ages has spurred scientists to search for evidence of such events in the more ancient geologic record. In the mid-1970s, Peter Vail and a team of researchers from Exxon Corporation conducted a detailed analysis of drill core from sedimentary rocks deposited on the world's continental shelves. They proposed that similar changes in sea level revealed in these cores over the past 250 million years were related to the waxing and waning of ice sheets. According to their hypothesis, as the sea retreated during an ice age, portions of the continental shelf became exposed. As a result, they were eroded, creating a gap in the geologic record known as an *unconformity* (see Chapter 8). As all continental shelves appeared to possess unconformities of exactly the same age, the process responsible must have been global in scale. Vail and his team concluded that the only known process that was rapid enough to cause these unconformities was fluctuations in the sizes of major ice sheets.

A reef is a coastal deposition feature unlike bars and barrier islands. Reefs rise to within a meter of sea level along the coastlines and around the islands of warm, clear tropical seas. Because they are built of the skeletal remains of colonial organisms that need sunlight, only that portion of a reef that is within reach of the Sun's rays (a depth of about 75 meters or 250 feet) is active. But the most active part of a reef lies just a few meters below sea level, a position the reef maintains by growing upward if sea level rises or the coastline subsides. Indeed, because reefs are confined to tropical latitudes (between 30°N and 30°S) and have strict needs for warm, clear, shallow, and moderately salty water, ancient reefs preserved in the geologic record provide important clues to past climate and geography.

Reefs are chemical/biochemical deposits built of carbonate rocks and are of three types: *fringing reefs, barrier reefs,* and *atolls* (Fig. 17.24). **Fringing reefs** directly border the shore of a landmass or island, forming a table-like surface up to a kilometer across. Fringing reefs grow oceanward with time toward the food supply. **Barrier reefs** are narrow coral reefs roughly parallel to the shore but separated from it by a wide lagoon. As its name implies, a barrier reef protects the mainland shore by creating a barrier to the waves, which break against the reef. Some barrier reefs develop on the continental shelves of major landmasses, as is the case for the 2300-kilometer-long (1400 mile) Great Barrier Reef, which lies about 100 kilometers (60 miles) off the coast of northeastern Australia. Other barrier reefs enclose subsiding volcanic islands where they typically develop from the upward growth of fringing reefs (Fig. 17.24b). **Atolls** are roughly circular reefs surrounded by open sea that enclose a shallow lagoon in which there is no land. Atolls are the most common type of modern coral reef, especially in the Pacific, where their evolution from the upward growth of barrier reefs around volcanic islands that eventually subsided below sea level (Fig. 17.25) was first noted by Charles Darwin.

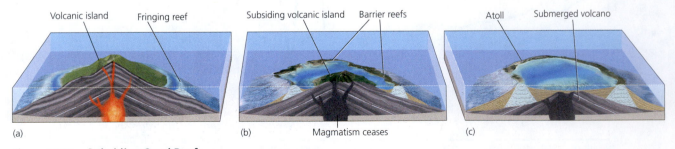

Figure 17.24   Subsiding Coral Reef

(a) A fringing reef develops around an active volcanic island. (b) As the inactive volcano subsides, the fringing reef grows upward to form a barrier reef separated from the shore by a lagoon. (c) Continued upward growth of the reef forms an atoll as the volcanic island subsides below sea level.

- Deposition by longshore currents produces spits, baymouth bars, and tombolos.
- Limestone reefs are built of the skeletal remains of colonial organisms and include fringing reefs, barrier reefs, and atolls.

17.6 Coastlines

Coastlines vary markedly from beautiful sandy beaches to daunting spectacular cliffs. Because of the many factors that influence their development, coastlines show a high degree of variability and do not lend themselves to classification. Nevertheless, depending on the dominant process operating on their shores and the direction of sea level change, we can broadly group coastlines into two pairs of categories: *erosional* or *depositional;* and *submergent* or *emergent.*

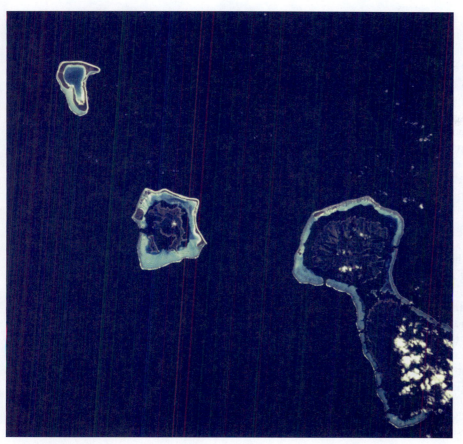

Figure 17.25   Society Islands

Coral reefs in the Society Islands in the South Pacific illustrate the evolution from fringing reef to atoll first recognized by Charles Darwin. The extinct volcanic islands of Ra'iatea (bottom right) and, beside it, Taha'a, are bordered by fringing and barrier reefs, respectively. The older and more heavily eroded volcanic island of Bora Bora (center) is bordered by a well-developed barrier reef. The oldest island, Tupai (top left), is an atoll, the original volcano around which the reef first formed having subsided below sea level.

The evolution from fringing reef to barrier reef to atoll reflects the fact that many Pacific islands are extinct hotspot volcanoes. As we learned in Chapter 9, such volcanoes subside and become progressively more eroded as plate motion carries them away from the hotspot (see Fig. 9.35), eventually forming a line of seamounts like that of the Hawaiian-Emperor chain. The progression in reef type occurs as reef growth attempts to keep pace with the subsidence in a race the reef builders are ultimately destined to loose.

> **CHECK YOUR UNDERSTANDING**
> ○ Under what conditions do reefs develop?

17.5 SUMMARY

- Coastal deposition produces pocket beaches and strands.

EROSIONAL AND DEPOSITIONAL COASTLINES

Although all coastlines are subject to both erosion and deposition, some are clearly dominated by either one process or the other. **Erosional coastlines** are dominated by features of coastal erosion, such as sea cliffs, headlands, and wave-cut platforms. As a result, they are usually rugged and irregular, with only narrow pocket beaches backed by steep cliffs. Much of the Pacific coastline (see Fig. 17.14 and Fig. 17.19a) and that of New England and Atlantic Canada fall into this broad category.

On the other hand, **depositional coastlines** like those of the Gulf Coast and much of the eastern seaboard of the United States (see Fig. 17.23) are dominated by features of coastal deposition such as beaches, bars, and barrier islands. Depositional coastlines typically lack prominent cliffs and instead show gently sloping shores with smooth outlines and well-developed beaches.

As we discussed in Chapter 9, one of the main factors that has contributed to the marked contrast between the Pacific coastline of North America and those of the Gulf Coast and much of the Atlantic seaboard is plate tectonics. The West Coast lies along the tectonically active leading edge of the North American plate and is subject to uplift and erosion. In contrast, the Gulf Coast and Atlantic seaboard lie far from any plate boundaries and so are far less tectonically active. Huge quantities of sediment have also been supplied to the Gulf and Atlantic coasts as a consequence of the advance and retreat of continental glaciers during the Ice Age.

SUBMERGENT AND EMERGENT COASTLINES

We can also group coastlines into two broad categories based on the influence of sea level change. Those that have subsided or have witnessed a rise in sea level are **submergent** (drowned) **coastlines**, whereas those that have been uplifted or have experienced sea level fall are **emergent coastlines**. Although sea levels have risen significantly since the last glacial retreat, not all coastlines are submergent because, in some places, uplift of the land as a consequence of glacial rebound (see Chapter 15) is occurring at a faster pace than sea level rise. Nevertheless, as a result of the melting of glacial ice, many coastlines have witnessed a rise in sea level since the Ice Age and show obvious features of submergence (see Living on Earth: **Sea Level and Polar Ice** on page 547).

Drowned coastlines commonly have highly irregular outlines as a result of the flooding of river valleys that formerly drained to a lower sea level. Such drowned river mouths, termed **estuaries**, are tidal areas where freshwater and seawater mix. Chesapeake and Delaware bays are prominent examples of estuaries on the drowned Atlantic Coast (Fig. 17.26). These two bays mark the courses of the Susquehanna and Delaware rivers at the height of the Ice Age when sea level was some 100 meters (330 feet) lower than it is today.

As we learned in Chapter 15, rising sea levels at high latitudes, where glaciated mountains lie close to the sea,

Figure 17.26   Chesapeake Bay

Chesapeake and Delaware bays are large estuaries on the Atlantic coast that formed when the lower reaches of the Susquehanna and Delaware rivers were drowned by the post-glacial rise in sea level.

Figure 17.27   Marine Terrace

This marine terrace in northern California represents an uplifted wave-cut platform and preserves the stumps of old sea stacks.

have resulted in the formation of *fjords,* steep-sided sea inlets produced by the flooding of glacial troughs (see Fig. 15.15). Fjords are common features of many high latitude coastlines in both the Northern and Southern hemispheres, such as the highly irregular coastline of Norway.

Other coastlines, such as those found in parts of California (see Fig. 17.19a), show clear evidence of emergence. On this coast, a relative fall in sea level has been brought about by tectonic uplift of the land, bringing sea cliffs and wave-cut platforms above the level of the waves. But on other coastlines, such as those of Hudson Bay and the Baltic Sea in Scandinavia, uplift is the result of crustal rebound following the retreat of continental ice as we discussed in Chapter 15.

Wave-cut platforms that have been elevated in this fashion are called *marine terraces* (Fig. 17.27), and may preserve ancient cliff lines and sea stacks, and uplifted stacks and old beach deposits known as *raised beaches.* With repeated uplift, a series of marine terraces can be produced, as is the case in the Palos Verdes Hills south of Los Angeles, where a succession of 13 terraces rises to 425 meters (1400 feet) above sea level.

Ancient marine terraces can be important to disciplines such as archeology. As we learned in Chapter 15, for example, the development of raised beaches in Arctic Canada records the progressive uplift of the shoreline in response to crustal rebound following the last glacial retreat (see Fig. 15.31). As a result of rebound, ancient coastal settlements are located beside the raised beaches that were the shoreline when the settlements were occupied. This knowledge not only permitted the settlements to be found but also allowed them to be dated, so that a detailed archeological record of settlement history could be established.

The sensitivity of coastline to erosion and deposition, to tectonic uplift and subsidence and the rise and fall of sea level, to the movement of beaches and barrier islands, to waves and storms and the varying range of the tides demonstrates the highly dynamic character of our coastal regions. As we shall discuss in Section 17.7, this sensitivity makes the development of the coastal environment for human purposes a difficult and risky undertaking and emphasizes the need for responsible coastal management.

> **CHECK YOUR UNDERSTANDING**
>
> ● What is the origin of submergent and emergent coastlines and which features characterize each?

17.6 SUMMARY

- Erosional coastlines feature cliffs, whereas depositional coastlines are dominated by beaches.

- Sea level rise produces submergent coastlines with estuaries, whereas sea level fall produces emergent coastlines with raised beaches and marine terraces.

17.7 Coastal Management

Coastlines are unique places. They are sites of remarkable beauty, they are home to a wide variety of ecosystems, and they are unrivaled areas of human recreation. As a result, they are in high demand for development and suffer some of the highest exploitation pressures in the world. Over the past century, coastal regions and shorelines have seen intense use for all kinds of activities, often to the detriment of the very resources that make them so attractive. Today, coastal regions are home to half of the world's population. As a result, they have witnessed unprecedented urban and industrial development, widespread residential and leisure use, and all of the problems of pollution and waste disposal that accompany human occupation.

Successful development requires a stable foundation on which to build and, as the content in this chapter makes clear, coastlines are anything but stable. Instead, they are dynamic features that are both easily damaged and short lived. Hence, the nature of coastlines is often incompatible with human desires and uses. As people develop the shoreline, build houses on barrier islands, and use inlets and coastal waterways for navigation, we operate on the assumption that beaches and shorelines are permanent features. But as anyone who has witnessed a hurricane can testify, this is not the case (see Living on Earth: **Hurricanes and Storm Surges** on page 553). Billions of dollars are spent annually in attempts to stabilize the shore and to repair the damage caused by natural forces beyond our control.

The need to combat the dynamic nature of coastal zones and prevent the natural migration of shorelines has led to a variety of human interventions into the processes that govern coastal evolution. Among these, the most important and most intrusive are those directed at combatting shoreline erosion.

PROBLEMS OF SHORELINE EROSION

Worldwide, sea level has risen 12 centimeters (4.7 inches) over the past century. As a result, shorelines are being eroded and barrier islands are on the move. To combat this process, several techniques have been employed in an attempt to stabilize the shoreline, some of which have had detrimental and unexpected consequences that are not easily rectified. Traditionally, there have been two approaches to shoreline protection, the use of protective structures and replenishment of the sand lost to beach erosion.

Protective structures

From the beginning of civilization, humans have enclosed harbors with piers and wharfs, and protected them with jetties. *Jetties* are structures, often built in pairs, that extend outward from the shore at the mouth of a harbor, river, or inlet (Fig. 17.28a). By confining currents and tides, jetties are designed to protect navigable passageways and

Figure 17.28   Jetties, Groins, and Breakwaters

(a) A pair of jetties shield the Tweed River on the New South Wales-Queensland state line, Australia. (b) Groins were built to curb the longshore drift of sand at Eastbourne Beach on the south coast of England. (c) Breakwaters were built along the Lake Erie shore to protect Presque Isle State Park at Eire, Pennsylvania. Note that sand is deposited in those areas protected from wave energy by human intervention. At Eastbourne Beach, longshore drift (from left to right) causes sand to pile up on the near side of the groins, while sand is removed from the far side. At Erie, deposition of sand shoreward of the breakwater has allowed the beach to build out.

LIVING ON EARTH

Hurricanes and Storm Surges

In coastal areas subjected to occasional severe low-pressure weather systems, such as hurricanes, damaging floods result from a sharp rise in sea level caused by the drop in atmospheric pressure and, more importantly, by the push of high winds on the ocean's surface. The abnormal rise of the sea above the usual tide level is known as a *storm surge*. The combined height of the tide and surge when the two occur together, constitutes the *storm tide* (Fig. 17D).

Powerful storms can produce storm surges of 5 meters (17 feet) or more, resulting in widespread coastal flooding that, in turn, has been historically responsible for most of the casualties associated with such storms. The devastating hurricane that struck Galveston, Texas, in 1900, for example, drove ashore a storm surge that took the lives of 6,000 to 12,000 people; to this day the deadliest natural disaster to strike the United States.

More recently, Hurricane Katrina, the costliest natural disaster in US history, inundated coastal Louisiana in 2005 with a storm surge of 8.5 meters (29 feet). Just 3 years later, in 2008, Hurricane Ike swept into coastal Texas with a 6-meter (20-foot) storm surge that moved inland nearly 50 kilometers (30 miles) (Fig. 17E).

Since a major hurricane makes landfall in the Gulf Coast region, on average, once every 2 years, storm surges are a major threat to coastal populations and coastal development. This is all the more troubling given the fact that the population density of the Gulf coastal counties increased by one-third between 1990 and 2008. According to the National Oceanic and Atmospheric Administration, a storm surge of 7 meters (23 feet) has the ability to inundate two-thirds of the interstates, almost half of the rail miles, 29 airports, and virtually all ports in the Gulf Coast area.

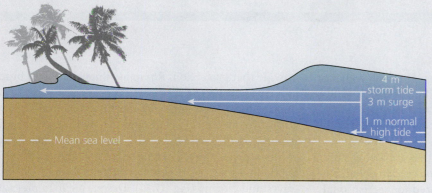

Figure 17D Storm Surge and Storm Tide

The storm surge is the rise of the sea above the usual tide level due to a storm, whereas the storm tide is the combination of tide and surge. In this case, a 3-meter storm surge on top of a tide that is 1 meter above mean sea level produces a storm tide of 4 meters.

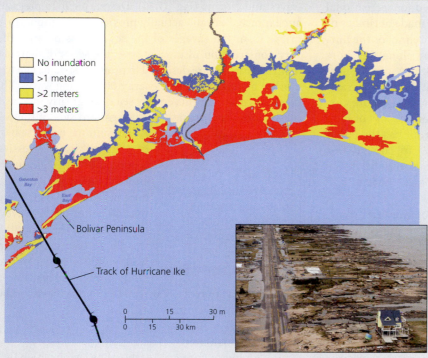

Figure 17E Hurricane Ike

This map shows the extensive inundation caused by Hurricane Ike in coastal southeastern Texas and southwestern Louisiana in 2008. The photo in the inset poignantly illustrates how little remained on the Bolivar Peninsula in Texas following the storm surge associated with Hurricane Ike.

(Continued)

LIVING ON EARTH

Hurricanes and Storm Surges (Continued)

Major hurricanes are less frequent in the northeastern United States, but the region's coastline is no less vulnerable to storm surges, as vividly illustrated by the impact of Hurricane Sandy on New York City in October 2012. Colliding with an Atlantic storm called a "nor'easter" before making landfall in New Jersey during a spring high tide (Fig. 17F), "Superstorm Sandy" caused the most devastating flooding in the city's history. Although the storm surge was just 2.8 meters (9 feet), the storm tide reached 4.2 meters (14 feet)—well above many of the city's sea walls. As a result, large sections of Lower Manhattan were inundated and seven subway tunnels beneath the East River were flooded. Nationwide, nearly 8.5 million people in 15 states lost power and the storm's damage amounted to $65 billion, making it the second costliest in US history.

Figure 17F   Hurricane Sandy

An aerial view shows the damage caused to the New Jersey coast by Hurricane Sandy.

---

prevent them from filling with sediment. Unfortunately, in areas of longshore drift, jetties block the passage of sand, causing it to build up against the upcurrent side of the jetty. The downcurrent side, on the other hand, is starved of sand and so is subject to beach erosion.

In an attempt to stabilize beaches that are losing sand to longshore drift, groins are sometimes constructed, especially in areas where the beach is used for recreational purposes. *Groins* are low walls, usually made of timber or concrete, built at right angles to the shoreline for the purpose of trapping sand that is migrating along the shore because of longshore drift (Fig. 17.28b). A groin effectively acts as a barrier to this migration so that sand builds up against it on the side facing the longshore current. As is the case with jetties, however, the loss of sand supply on the downcurrent side of a groin has the opposite effect, and fosters beach erosion. To combat this effect, a succession of groins, sometimes numbering in their hundreds, are built at intervals down the length of the beach (Fig. 17.29).

In areas prone to breaking waves, communities sometimes construct breakwaters for coastal protection. A *breakwater* is an offshore structure that, by breaking the force of the waves, provides shelter to a harbor, anchorage, or shoreline (Fig. 17.28c). In doing so, however, breakwaters encourage sand deposition in the quiet waters they produce on their leeward sides. As a result, beaches that are shoreward of a breakwater tend to build out and harbors protected by breakwaters may fill with silt.

Where the shoreline itself requires protection from breaking waves, riprap and seawalls are sometimes used to

provide that protection. *Riprap* is simply a heap of broken rock piled up along a shoreline to protect it from wave erosion. Although unsightly and quite quickly eroded, riprap is effective for short periods and inexpensive to replace. A *seawall,* on the other hand, is a constructed embankment of stone or reinforced concrete built along the shoreline for the purpose of protecting seafront development from waves and storms (Fig. 17.30). Despite their impregnable appearance, seawalls actually provide only temporary reprieve from the ravages of the sea. This is because their very construction ultimately sows the seeds of their own demise. Seawalls are designed to protect a specific site from erosion by fending off the energy of breaking waves. They do so by reflecting this energy seaward, greatly increasing the level of wave erosion on the seaward side of the wall.

As a result, the beach seaward of the wall is progressively eroded and, with time, may be entirely removed. And without a beach to sap their energy, the full brunt of the waves is expended on the seawall so that the destruction of the seawall is only a matter of time. Only through expensive repair or the construction of a new seawall can this inevitable outcome be delayed.

> ### CHECK YOUR UNDERSTANDING
>
> ○ Under what circumstances are (a) groins, (b) breakwaters, and (c) seawalls used to protect shorelines?

Beach replenishment

An alternative approach to shoreline stabilization is to artificially add new sand to the beach to replace sand lost to

Figure 17.29 Groins

Groins act as an artificial barrier to the migration of sand along the Jersey Shore. Note the accumulation of sand on the upcurrent (bottom) side of each groin.

Figure 17.30   Seawall

The seawall at Galveston, Texas, is a 5-meter-high (17-foot) structure built of granite and reinforced concrete designed to protect this barrier island community from waves and storm surges. Note that the beach that was seaward of the wall has been entirely eroded.

erosion. By widening the beach with new sand, both the beach and its protective function are improved. This approach has been adopted at Wallops Beach, Virginia, with dramatic results (Fig. 17.31). However, the process is expensive and the new sand is itself subject to erosion and must be periodically replaced. Additionally, because the new sand is rarely as clean as beach sand, *beach replenishment* (or *nourishment*) tends to cause the water offshore to become cloudy, often to the detriment of the coastal ecosystem. The original restoration of Miami Beach in 1976–1981 cost $64 million and periodic replenishment, amounting to more than 3 million cubic meters (4 million cubic yards) of sand, has been necessary in order to maintain the beach. As part of the second phase of the Miami-Dade County Shore Protection project, a further 190,000 cubic meters (250,000 cubic yards) of sand were added in 2012. Statewide, almost $1.5 billion has been spent since 1998 to restore and subsequently maintain over 350 kilometers (218 miles), or about 55 percent, of Florida's critically eroded beaches.

CHECK YOUR UNDERSTANDING

⊙ What is beach replenishment?

COASTAL MANAGEMENT AND PUBLIC POLICY

In an attempt to address the often conflicting needs of the various groups that utilize coastal regions, the Coastal Zone Management Act was passed in the 1970s. Virtually all coastal states and provinces now have coastal management programs, as do many counties and local communities. But while protective structures and beach replenishment are temporary solutions to some of the problems of shoreline stabilization, there is increasing recognition among coastal scientists that, ultimately, the only way to save a beach is to leave it alone. The dynamic nature of the shoreline is such that any human interference with shoreline processes can provide only short-term solutions. Protective structures are expensive and need frequent repair, and the sand used to replenish beaches must continually be replaced. In the long-term, these scientists argue, the only way to preserve our beaches is to allow them to remain in their natural state.

Given the inevitable shoreward movement of our barrier islands as sea level rises, and the ultimate futility of artificially maintaining beaches, many coastal planners are calling for a shift in policy toward much stricter controls on coastal development. North Carolina, for example, now prohibits the building of large structures any closer to the shoreline than 60 times the current annual amount of coastal erosion. Some coastal planners are now calling for a far more controversial *abandonment and relocation policy* such as the one that was adopted in the aftermath of the disastrous Mississippi Valley floods of 1993. Rather than endorsing the rebuilding of storm-damaged properties, this policy would have these structures relocated to higher and safer ground, and the beach is allowed to return to nature.

The adoption of such a policy is certain to meet with stiff resistance from coastal landowners, many of whom have huge sums invested in coastal properties. Yet some coastal planners are recognizing that this might be the best way forward, as they are becoming convinced that the implementation of current questionable shoreline protection practices has actually exacerbated the problem by encouraging further coastal development. Whatever the outcome, serious reevaluation of national and local coastal management policies will be necessary in the years to come if sea level continues to rise and storms intensify as a predicted climatic consequence of global warming (see Living on Earth: **Sea Level and Polar Ice** on page 547).

(a)

(b)

Figure 17.31 Beach Replenishment

Wallops Beach, Virginia, looks strikingly different (a) before and (b) after beach nourishment by the U.S. Army Corps of Engineers.

Sea Level and Plate Tectonics

As we discussed in Living on Earth: **Sea Level and Polar Ice** on page 547, the advance and retreat of Earth's polar ice caps can cause sea level to rise and fall by more than a hundred meters, with dramatic consequences to our coastlines.

However, the effects of plate tectonics on sea level, while slow to take effect, have been far more profound.

Plate tectonic activity slowly, but inexorably, changes the shape of ocean basins. Just as the same volume of

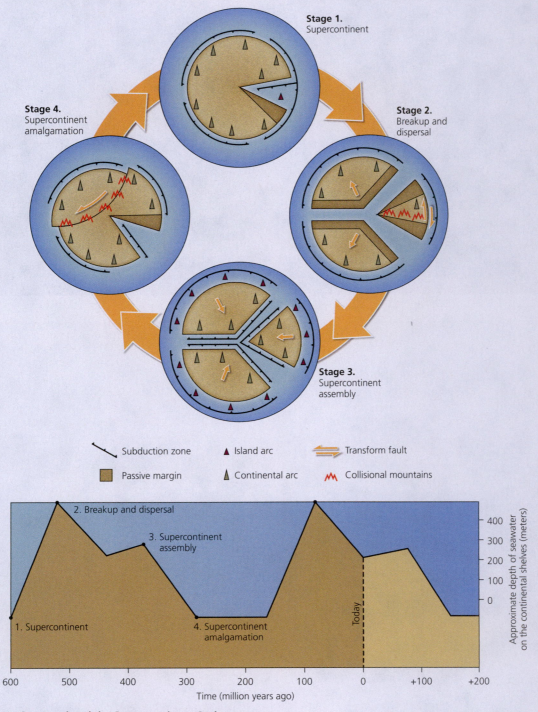

Figure 17G   Sea Level and the Supercontinent Cycle

In Stage 1, the supercontinent experiences uplift before continental breakup is initiated, resulting in a low sea level. Supercontinent breakup results in formation of new oceans. The development of mid-ocean ridges, increased volcanic activity, and the progressive subsidence of the crust of the dispersing continents contribute to a sharp rise in sea level during Stage 2. As the new oceans continue to widen, the average age of the ocean floor gets older, and sea level falls. Subduction of these oceans causes sea level to rise once more in Stage 3 only to fall again as amalgamation in Stage 4 traps mantle heat beneath the new supercontinent.

water will fill bathtubs of different size to different levels depending on the shape of the bathtub, so sea level changes with the tectonic evolution of both continental and oceanic crust.

Consider, for example, the effect on sea level of the supercontinent cycle we examined in Chapter 10, illustrated by the assembly and breakup of Pangea. Before breakup, the insulating properties of continental crust result in the trapping of mantle heat beneath the supercontinent (Fig. 17G; Stage 1). This causes the supercontinent to become elevated and, as a consequence, sea level drops. This situation changes rapidly when the supercontinent rifts, and the continental fragments drift apart.

As the new oceans form, mid-ocean ridges develop between each of the continental fragments. In addition, as the continental fragments drift apart, their edges or margins subside. The result is a dramatic rise in sea level amounting to many hundreds of meters (Fig. 17G; Stage 2). There are three reasons for this rise. First, the development of new mid-ocean ridges produces vast new underwater mountain ranges similar to the one at the center of the modern Atlantic Ocean. These mountains displace the water of the ocean just as placing a large object in a bathtub displaces the water upward. In this way, the new submarine mountains cause sea level to rise upward and outward across the continental margins. The result is the generation of shallow seas floored by continental crust.

Second, as drifting proceeds, the dispersed continental fragments cool and subside as they move away from the hot regions of mantle upwelling at the mid-ocean ridge crest. Just as lowering the sides of a bathtub filled to the brim causes the water to spill out, this subsidence allows sea water to flood over the continental margin.

Third, most geologists speculate that continental breakup results in global warming because greenhouse gases are introduced into the atmosphere by the volcanoes associated with the rifting process. Global warming would cause melting of any continental glaciers, resulting in a rise in sea level (see Fig. 17C). Unfortunately, it is difficult to obtain reliable estimates of the volume of ice that may have melted when Pangea broke up because the size of continental ice sheets during any stage of the supercontinent cycle is unknown.

As seafloor spreading becomes established, the edges of the newly formed oceanic crust become older. They also get farther away from the heat source at the mid-ocean ridge and so become cooler. As a result, the oceanic crust becomes denser and starts to sag, ultimately forming ocean trenches as subduction begins. Like a depression forming in the bottom of a bathtub, this sagging of the ocean floor causes sea level to slowly fall.

Subduction, however, begins the process of ocean destruction. As the width of the ocean shrinks, the oldest and heaviest oceanic crust is preferentially destroyed. As a consequence, the ocean floor becomes progressively younger and more buoyant. As the floor of the ocean rises, so too does sea level (Fig. 17G; Stage 3). If the ocean closes and a new supercontinent forms, sea level will begin to drop once more, as the continental crust is once again uplifted by trapped mantle heat and the cycle begins anew (Fig. 17G; Stage 4).

The geologic record corresponds remarkably to this long-term pattern of sea level change. For example, following the breakup of a supercontinent about 550 million years ago, extensive shallow marine continental shelf deposits of Early Cambrian age (about 540 million years old) formed at the edges of many of the dispersing continental fragments, consistent with rising sea level. Similarly, following the breakup of Pangea in the Triassic, continental shelf deposits of Jurassic and Cretaceous age (between 200 and 66 million years old) formed along the periphery of the Atlantic Ocean in response to rising sea level and the flooding of its continental edges. In both cases of supercontinent breakup, the newly flooded shelves formed an ideal habitat for flourishing ecosystems that are well preserved in the fossil record.

17.7 SUMMARY

- Coastlines are fragile regions in need of careful management.
- Stabilization of shorelines using jetties, groins, breakwaters, and beach replenishment is effective only in the short term.

17.8 Coastlines and Plate Tectonics

In 1912, Alfred Wegener used the remarkable jigsaw fit of the coastlines on either side of the Atlantic Ocean as his leading line of evidence for the existence of the supercontinent Pangea (Fig. 17.32). But coastlines, as we have learned, are temporary features that are highly sensitive to sea level change and can retreat at a rate apparent within a human lifetime, or be dramatically changed by a single storm. So how can today's coastlines have any bearing on the fit of continents more than a 100 million years ago? Wegener's critics raised this point and argued on this basis that the apparent fit of the coastlines must be entirely fortuitous. Today, we now know that the real fit lies, not with the coastlines, but with the edges of the continental shelves. In contrast to the coastlines, the edges of the continental shelves mark the edges of the continents and so are far more fundamental geologic features that can easily last 100 million years. But the fit of the coastlines exists because they crudely follow the edge of the continents. Hence, the shape of the coastlines around the Atlantic Ocean, as well as those around the Arctic, Southern, and much of the Indian

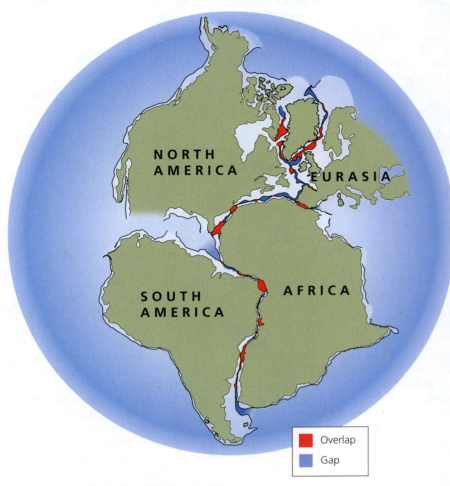

Figure 17.32   Jigsaw Fit of the Continents

This map shows the fit of the continents on either side of the Atlantic Ocean. Note that the fit between the continental edges (light blue), 130 meters (425 feet) below sea level, is far better than the fit between the coastlines.

oceans, are inherited from the breakup of Pangea, just as Wegener suggested.

Plate tectonics is responsible for all the coastlines of the world (see In Depth: **Sea Level and Plate Tectonics** on page 558). We can also apply plate tectonic theory to divide the margins of continents into three distinct types: *active* and *passive*, the coastal regions of which are very different, as well as *transform*, which can be active or passive. **Active continental margins** coincide with active plate boundaries and they are most frequently located at subduction zones (Fig. 17.33a). Active continental margins, such as the one that borders the Andes in South America, are typically associated with subduction zone features including an offshore deep-sea trench, a volcanically active mountain range that is parallel to the coast, and frequent earthquakes, some of which may launch destructive tsunamis.

Passive continental margins, on the other hand, are produced by continental rifting and the opening of new oceans. Recall from Chapter 9 that when continental rifting gives way to continental drift and a new ocean opens, the thinned margins of the separating continents cool and subside as they move farther from the mid-ocean ridge. As a result, these margins are

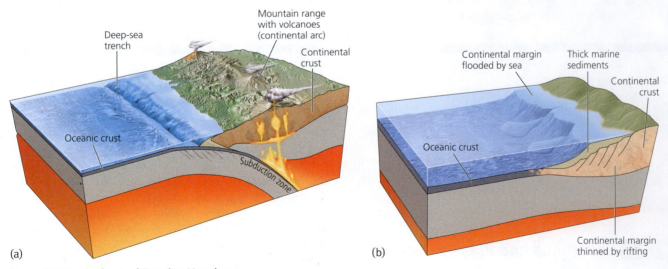

Figure 17.33   Active and Passive Margins

(a) Active continental margins exist along ocean–continent subduction zones such as the one that borders the Andes. (b) Passive continental margins, such as those that border the Atlantic Ocean, do not coincide with active plate boundaries.

progressively flooded by the sea and covered by thick sequences of marine sediments (Fig. 17.33b). As the ocean widens, passive continental margins come to lie progressively farther from any active plate boundary. So they are not bordered by deep-sea trenches nor by active mountain ranges. Earthquakes, if they occur at all, are very infrequent in these regions.

Plate tectonics also defines a third, less common, type of continental margin, the location and orientation of which is controlled by transform faults. **Transform continental margins** can be active or passive. The line of California's coast, for example, is controlled by offshore faults of the San Andreas fault system, and is consequently active because the margin coincides with an active transform plate boundary. Earthquakes are common along such transform margins, but unlike those associated with subduction zones, they rarely launch tsunami because the seabed along transform faults moves sideways, rather than up and down. (Recall from Chapter 9 that upward movement is characteristic of tsunami-launching earthquakes at subduction zones.).

Other transform continental margins lie far from plate boundaries and are passive as a result of their location. For example, the coastline of northern Brazil, and its counterpart in West Africa (see Fig. 17.32), are transform continental margins. Their orientation is inherited from major transform faults, which offset the Mid-Atlantic

Ridge north of Ascension Island. But as we discussed in Chapter 9, transform faults that offset mid-ocean ridges are only active between the offset ridge segments. Over the past 100 million years, Brazil and Africa have both drifted far away from these active segments.

Because transform faults move the crust sideways, rather than stretching it as continental rifting does, transform margins do not subside to the same degree as those produced at divergent plate boundaries. Therefore transform margins typically have narrower continental shelves than rifted margins, and a more abrupt transition from continental to ocean crust. But they are otherwise similar to passive continental margins.

> **CHECK YOUR UNDERSTANDING**
>
> ◉ Coastlines are temporary, so why do those around the Atlantic Ocean show such a remarkable jigsaw fit?
>
> ◉ What is the difference between active and passive continental margins?

17.8 SUMMARY

- The coastline of the Atlantic Ocean was inherited from the breakup of Pangea.

- Plate tectonics allows continental margins to be divided into three types: active, passive, and transform.

Key Terms

active continental margin 560
atoll 548
backwash 535
barrier island 545
barrier reef 548
baymouth bar 545
beach 544
beach drift 536
breaker 534
depositional coastlines 549
diurnal tidal pattern 539

ebb current 540
emergent coastline 550
erosional coastlines 549
estuary 550
flood current 540
fringing reef 548
headland 543
longshore current 536
longshore drift 536
mixed tidal pattern 539
neap tide 539
passive continental margin 560

reef 546
rip current 537
sea arch 543
sea cliff 542
sea stack 543
semidiurnal tidal pattern 539
spit 545
spring tide 539
submergent coastline 550
surf zone 535
swash 535
swash zone 535

tidal current 540
tidal flats 540
tide 537
tombolo 545
transform continental margin 561
wave 532
wave-cut platform 542
wave height 532
wave refraction 535
wave velocity 532
wavelength 532

Key Concepts

17.1 COASTAL PROCESSES

- Coastlines are dynamic features subject to constant erosion and deposition by waves, tides, and currents.

17.2 WIND-DRIVEN WAVES

- Waves are produced by the wind. Their shape is described by wavelength and wave height, their motion by velocity, frequency, and period.

- In open ocean, waves move the water in a circular fashion with no net forward movement of water. This circular motion dies out rapidly with depth.

- Near the shoreline, the seabed slows waves down, causing them to bunch together. This decreases their wavelengths and increases their wave heights until they break in the surf zone.

- Waves approaching the shoreline at an angle become refracted as they reach the shore. Oblique waves are responsible for the zigzag movement of sand particles on the beach known as beach drift, and the downshore transport of sand in the surf zone known as longshore drift.
- Rip currents are focused areas of return flow through the surf zone fed by longshore currents

17.3 TIDES
- Tides are caused by the gravitational pull of the Moon and Sun.
- The basic tide pattern is semidiurnal with two high tides per lunar day. Diurnal patterns have only one high tide per lunar day, and mixed patterns have two high tides, one of which is significantly higher.
- Spring tides occur when the Sun and Moon are aligned so that the tide is amplified. Neap tides occur at maximum misalignment.

17.4 COASTAL EROSION
- Waves erode by abrasion, creating headlands and bays, and erosional landforms such as cliffs, wave-cut platforms, sea arches, and sea stacks.

17.5 COASTAL DEPOSITION
- Coastal deposition produces pocket beaches and strands.
- Deposition of sand by longshore currents produces spits, baymouth bars, and tombolos.
- Barrier islands are offshore sand ridges that may have originated as spits, may have been piled up by storm waves, or may be remnants of dune systems formed during the last glacial period when sea level was lower.
- Reefs are near-shore masses of limestone built of the skeletal remains of colonial organisms. There are three types of reef: fringing reefs, barrier reefs, and atolls. Around subsiding islands of Pacific hotspots, fringing reefs evolve to barrier reefs and atolls as reef growth attempts to keep pace with subsidence.

17.6 COASTLINES
- Rugged erosional coastlines are dominated by features such as cliffs. Gently sloping depositional coastlines are dominated by features such as beaches.
- Submergent coastlines have experienced a relative rise in sea level as evidenced by estuaries. Emergent coastlines have experienced a relative sea level fall demonstrated by raised beaches and marine terraces.

17.7 COASTAL MANAGEMENT
- Coastlines are in high demand for exploitation and development. Yet they are dynamic, short-lived, and easily damaged and are therefore in need of careful management.
- Efforts to stabilize shorelines involve the use of protective structures such as jetties, groins, breakwaters, and seawalls, as well as beach replenishment. But such interventions provide only short-term solutions. In the long term, the only way to protect the shoreline may be to limit development.

17.8 COASTLINES AND PLATE TECTONICS
- The jigsaw fit of the coastlines around the Atlantic Ocean owes its origin to the breakup of Pangea.
- Plate tectonics allows continental margins to be divided into three types: active, passive, and transform. Active continental margins coincide with active plate boundaries, whereas passive continental margins do not.
- Transform continental margins coincide with transform plate boundaries and may be active or passive.

Study Questions

1. How and why do waves evolve as they approach the shore?
2. What processes govern the formation of longshore currents and rip currents?
3. If the tides are largely governed by gravitational attraction between Earth and the Moon, why doesn't the basic tidal pattern feature just one high tide every 24 hours?
4. What is meant by (a) semidiurnal, (b) diurnal, and (c) mixed tidal patterns and why do they occur?
5. Describe the process of coastal erosion and the most common landforms this process produces.
6. Compare and contrast the origin of spits and baymouth bars with that of barrier islands.

7. A progression of reef types is often associated with subsiding Pacific islands. What is this progression and what controls its evolution?

8. What is the difference between (a) erosional and depositional coastlines, and (b) submergent and emergent coastlines, and what are the main features of each?

9. Describe the ways in which protective structures have been used in an effort to stabilize shorelines.

10. What are the potential benefits and drawbacks to an "abandonment and relocation" policy for shoreline development in areas prone to coastal storms?

11. Wegener argued for the existence of Pangea based partly on the jigsaw fit of coastlines. His critics argued that coastlines were temporary features and were not relevant to discussions of the distant past. Who was right?

12. How can transform plate boundaries be both active and passive?

Chapter 18 | Mineral Deposits
            and Industrial
            Materials   566

Chapter 19 | Energy Resources   604

18 Mineral Deposits and Industrial Materials

18.1 Consumption of Minerals and Industrial Materials

18.2 Metals and Metallic Minerals

18.3 Nonmetallic Minerals and Industrial Materials

18.4 Formation of Mineral Deposits

18.5 Plate Tectonics and Mineral Deposits

18.6 Mineral Exploration

18.7 Future Demands on Mineral Deposits

▶ Key Terms

▶ Key Concepts

▶ Study Questions

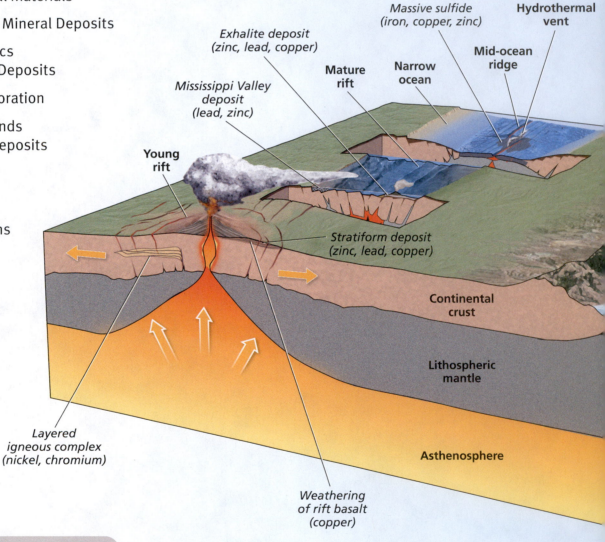

IN DEPTH: Mechanisms of Mineral Concentration 583

LIVING ON EARTH: The Environmental Impact of Mining 600

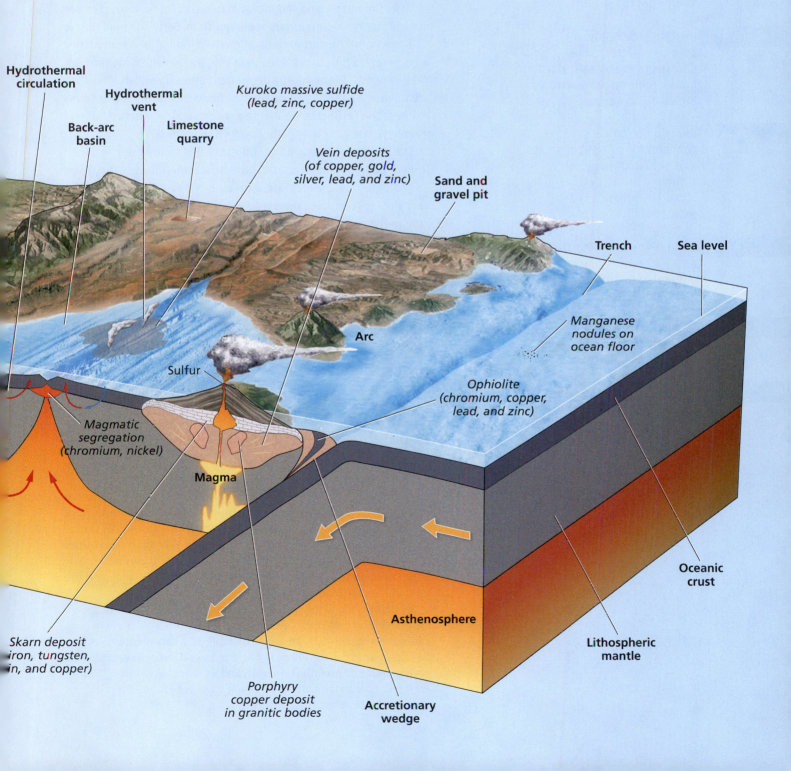

Hydrothermal circulation

Back-arc basin

Hydrothermal vent

Limestone quarry

Kuroko massive sulfide (lead, zinc, copper)

Vein deposits (of copper, gold, silver, lead, and zinc)

Sand and gravel pit

Trench

Sea level

Manganese nodules on ocean floor

Arc

Sulfur

Ophiolite (chromium, copper, lead, and zinc)

Magmatic segregation (chromium, nickel)

Magma

Asthenosphere

Oceanic crust

Skarn deposit iron, tungsten, in, and copper)

Porphyry copper deposit in granitic bodies

Accretionary wedge

Lithospheric mantle

Learning Objectives

18.1 Cite examples of the prevalent human use of mineral deposits and industrial materials.

18.2 Name the principal metals and metallic minerals and describe the relative abundance, human uses, and sources of these minerals.

18.3 Describe the relative abundance, human uses, and sources of the most important nonmetallic minerals and industrial materials.

18.4 Explain the processes that concentrate mineral deposits and industrial materials in sufficient quantities that they can be economically extracted.

18.5 Summarize and cite examples of the close link between mineral deposits and plate tectonics.

18.6 Describe the exploration methods by which mineral deposits are discovered.

18.7 Discuss what the future might hold as the demand for mineral deposits and industrial materials confronts a limited supply.

umans have been utilizing Earth's resources since the dawn of civilization. The very ages of human development are named for the resources that people first learned to use during each era of civilization. The earliest period was the Stone Age, in which people made tools from stone. Later, our ability to mix copper and tin ushered in the Bronze Age. The subsequent human mastery of iron marked the start of the Iron Age.

Modern society depends on the resources obtained from deposits of minerals and industrial materials. These resources are natural, inorganic materials removed from Earth for human use and they form the cornerstone of modern industrial society. They are so much a part of our daily lives that we take them for granted. Yet our way of life is wholly dependent upon them, whether they are the metals that make up our machines, computers, and smart phones, the fertilizers we use to grow food, or the industrial materials we use to construct buildings and roads.

The resources that are the subject of this chapter are obtained from deposits of minerals and industrial materials and they form at geologic rates, which are extremely slow. Although they are very slow to form, they are exploited at much faster rates to match human demand. Because they are finite commodities that are not replaced by nature as quickly as our society consumes them, they are termed *nonrenewable resources*.

Coal, oil, and natural gas are also nonrenewable resources, but these are traditionally considered separately from mineral deposits and industrial materials because they are organic in origin. Instead, they are classified as *energy resources* and are discussed in Chapter 19.

18.1 Consumption of Minerals and Industrial Materials

Mineral deposits is the term we use to collectively refer to concentrations of *metals* and *metallic* and *nonmetallic minerals* that are mined by humans for commercial purposes. However, some commercial resources are rocks rather than minerals. To include these resources, the term **industrial materials** is used to refer to all naturally occurring inorganic deposits of economic value, other than those from which metals are extracted. Industrial materials therefore include both nonmetallic mineral deposits and rock materials, such as those used in the construction industry.

By definition, all mineral deposits and industrial materials are *inorganic*, clearly setting them apart from fossil fuels such as coal, oil, and natural gas, which are *organic*. Table 18.1 lists the more important inorganic mineral deposits and industrial materials, including both metals and metallic minerals, and a variety of nonmetallic minerals and industrial materials that are used as building supplies, raw chemicals, abrasives, or gemstones.

TABLE 18.1   CLASSIFICATION OF MINERAL DEPOSITS AND INDUSTRIAL MATERIALS

Metallic Minerals		Nonmetallic Minerals		
Abundant Metals (greater than 0.1 percent of the crust)	Scarce Metals (less than 0.1 percent of the crust)	Building Materials	Raw Chemicals	Abrasives and Gemstones
Iron	Copper	Stone	Fertilizer	Diamond
Aluminum	Lead	Sand and Gravel	Nitrogen	Corundum
Magnesium	Zinc	Cement	Phosphorus	Garnet
Titanium	Chromium	Gypsum (Plaster)	Potash (Potassium)	Quartz
Manganese	Nickel		Halite	Emerald
	Gold		Sulfur	
	Silver			
	Platinum Group Elements			

Unlike some natural resources, such as agricultural crops and timber, which can be renewed with the passage of time, mineral deposits are **nonrenewable resources** because humans use them faster than nature can replenish them. Mineral deposits are vital to modern industrial society. *Mining* is the term we use to describe the for-profit extraction of all nonrenewable resources, including minerals and industrial materials, as well as fossil fuels (see Chapter 19) and, in some instances, even water (see Chapter 14).

Human consumption of mineral deposits and industrial materials is vividly illustrated by examining the role they play in our everyday lives. From the fluoride toothpaste we reach for in the morning to the tungsten-filament lightbulb we turn off at night, human use of Earth's inorganic resources is incessant. There is feldspar dust in our toothpaste and soda ash in our soap. We sprinkle talc in talcum powder and ingest dolomite in medicines. We drink beverages from cans of aluminum and eat food from cans of tin-plate. We bake with baking soda and season our meals with rock salt. We eat with stainless steel cutlery and cook with cast-iron skillets and copper-bottomed pans. There is quartz and lime in our glassware, and lead in our decorative crystal. We use powdered chert for scouring and chlorine bleach for whitening. There is china clay in the porcelain of our toilets, sinks, and bathtubs, and salt keeps our roads safer in winter. Rare earth elements (see Chapter 3) coat our television screens and run our cell phones, and mercury works our light switches.

We wear gold and silver jewelry, and trim our cars with chrome. In our cars, platinum can be found in the catalytic converters and fluorine in the air conditioners. Our pens are tipped with tungsten, our pencil "lead" is really graphite, and we still use chalk in the classroom. We use cobalt to form magnets and make paper using sulfur. Builders use limestone in cement, make concrete out of sand, employ gypsum for plastering, and make bricks and tiles from clay. There is manganese in siding and titanium in paint. We use copper in our wiring and add zinc to make it brass. We use potash, nitrate, and phosphorus if we fertilize our lawns, and we grace our public monuments with facades of polished stone.

18.1 SUMMARY

- Mineral deposits and industrial materials are nonrenewable metallic and nonmetallic minerals, metals, and other inorganic materials that are mined for human use.

- Mineral deposits and industrial materials form the cornerstone of modern industrial society and are a part of our daily lives.

18.2 Metals and Metallic Minerals

Mineral deposits are mined by humans in an effort to obtain **metals**, the common name we give to elements such as copper and iron, and mixtures of metals such as brass and bronze, that are shiny, hard, malleable, and good conductors of heat and electricity. Metals that occur commonly in the rocks of Earth's crust (Table 18.2) are termed *abundant metals*. Metals that are rare are called *scarce metals*. Few metals are mined in their pure or *native* state. Most metals occur as **metallic minerals** in which metals combine with elements such as oxygen, sulfur, or carbon to form mineral oxides, sulfides, or carbonates. Rocks containing commercially exploitable quantities of such minerals are **ores**.

ABUNDANT METALS

The abundant metals include iron, aluminum, magnesium, titanium, and manganese. These metals are among the most common elements in Earth's crust, forming an important component of many silicate minerals (recall from

> **CHECK YOUR UNDERSTANDING**
>
> ○ Citing examples, distinguish between renewable and nonrenewable resources.

TABLE 18.2   RELATIVE PERCENTAGES BY WEIGHT OF CHEMICAL ELEMENTS IN THE EARTH'S LITHOSPHERE

Element	% Weight
Oxygen (O)	46.40
Silicon (Si)	28.15
Aluminum (Al)	8.23
Iron (Fe)	5.63
Calcium (Ca)	4.15
Sodium (Na)	2.36
Magnesium (Mg)	2.33
Potassium (K)	2.09
Titanium (Ti)	0.48
Manganese (Mn)	0.10
All others	0.08
Total	100.00

Chapter 3 that most rock-forming minerals are silicates). But these metals are not easily extracted from silicate minerals because the metals are too tightly bonded within the silicate mineral structure to make this economic. Instead, these metals are more readily extracted from metallic minerals because the metal is more loosely bonded. The most familiar of these abundant metals are iron, aluminum, and titanium (Fig. 18.1).

Iron (Fe)

Of all Earth's mineral deposits, iron is the most important to humans. Current demand for iron ore in the United States alone is around 50 million metric tons per year, and worldwide, it is almost 3 billion metric tons per year and steadily increasing. Iron is the dominant metal used by industry (accounting for some 95 percent of all metal use in the world) and is the key ingredient of steel. When mixed (or *alloyed*) with other metals—such as nickel, chromium, tungsten, cobalt, or manganese—iron is strong, relatively inexpensive, and resistant to corrosion. It frames our buildings, shapes our vehicles, and is the metal of our machines.

Iron is mined either in its partially oxidized form *magnetite* (Fe_3O_4)—the mineral we encountered in our examination of paleomagnetism in Chapter 2—or in its fully oxidized form *hematite* (Fe_2O_3), a mineral more familiar to us as rust. The largest deposits of iron ore occur in Precambrian sedimentary rock known as *banded iron formation* (see In Depth: **The Role of Oxidation in Earth's History** in Chapter 5), which contain layers rich in iron that is thought to have been precipitated in seawater by oxygen-producing microorganisms (Fig. 18.2).

> **CHECK YOUR UNDERSTANDING**
> ◐ What is the most important abundant metal in terms of human use and how is it used?

Aluminum (Al)

Worldwide, the demand for aluminum is fast approaching 60 million metric tons per year, more than twice the amount that industries used just a decade ago. Aluminum is a lightweight, corrosion-resistant metal. Yet when alloyed with copper, manganese, or magnesium, it can be as strong as steel. It is widely used in the construction of airplanes, it is the metal of choice for packaging and window frames, and it is increasingly replacing iron in vehicle manufacturing.

> **CHECK YOUR UNDERSTANDING**
> ◐ What is bauxite and under what climatic conditions does it form?

The principal source of aluminum is the ore *bauxite* (Fig. 18.3). As we learned in Chapter 5, this complex mixture of aluminum oxides and hydroxides is a soil formed by the weathering of aluminum-rich rocks in tropical climates.

Titanium (Ti)

Titanium is almost as light as aluminum. But it is much stronger and resists corrosion even at high temperatures. It is therefore used by the aerospace and defense industries to make heat shields, such as those used to protect manned spacecraft during reentry into Earth's atmosphere. However, most of the demand for titanium—which presently exceeds 1.5 million metric tons per year in the United States alone—is for titanium oxide (TiO_2), which is used as a white pigment in paint.

Titanium is concentrated in several minor rock-forming minerals, such as *ilmenite* ($FeTiO_3$) and *rutile* (TiO_2), and most deposits of titanium ore occur in places where one of these two minerals, which are both resistant to weathering,

> **CHECK YOUR UNDERSTANDING**
> ◐ What are the principal uses of titanium?

is concentrated by sedimentary processes in beach sands (Fig. 18.4). However, deposits of ilmenite are also produced by igneous fractionation (see Chapter 4), during which early-formed ilmenite crystals separate in a cooling mafic magma and sink to form layers at the bottom of the magma chamber.

SCARCE METALS

Unlike abundant metals, which are plentiful enough that they need only be concentrated a few times above their average abundances to form minable deposits of ore, scarce metals must be concentrated by natural processes hundreds or even thousands of times above their average abundances in order to be profitably exploited. As a result, the discovery and management of these rare, nonrenewable resources are significant challenges. For abundant metals, concentrations are measured in percentages (or parts per hundred). For scarce metals, concentrations are given in parts per million (ppm), or even parts per billion (ppb). Scarce metals include nickel, molybdenum, tin,

Figure 18.1   Abundant Metals in North America
Abundant metal (iron, aluminum, manganese, and titanium) deposits occur throughout North America.

antimony, mercury, cadmium, lithium, and uranium, but the most important scarce metals, in terms of their significance to industry, are copper, zinc, lead, and chromium.

Some scarce metals, such as gold, silver, and platinum, are called *precious metals* because of their high monetary value. Economically exploitable deposits of these scarce minerals are rare and widely scattered.

Their distribution across North America is shown in Figure 18.5.

Copper (Cu)

Copper is the most important of the scarce metals. Although the average abundance of copper in Earth's crust (referred to as *crustal abundance*) is only 50 parts per million, it is the third most consumed metal behind iron and aluminum. Demand for copper in the United States alone is close to 2 million metric tons per year, and global

CHECK YOUR UNDERSTANDING

◉ What are the most important scarce metals?

Figure 18.2   Iron Ore
Banded iron formation (or BIF), a Precambrian sedimentary rock rich in hematite (dark layers), is the world's principal source of iron. The red layers are jasper (hematitic chert).

Figure 18.3   Bauxite
Bauxite, the principal ore of aluminum, is produced by the tropical weathering of aluminum-rich rocks. Bauxite commonly has a lumpy nodular appearance.

Figure 18.4   Mineral Sand
Dredging for titanium-rich sands on North Stradbroke Island takes place near Brisbane, Australia.

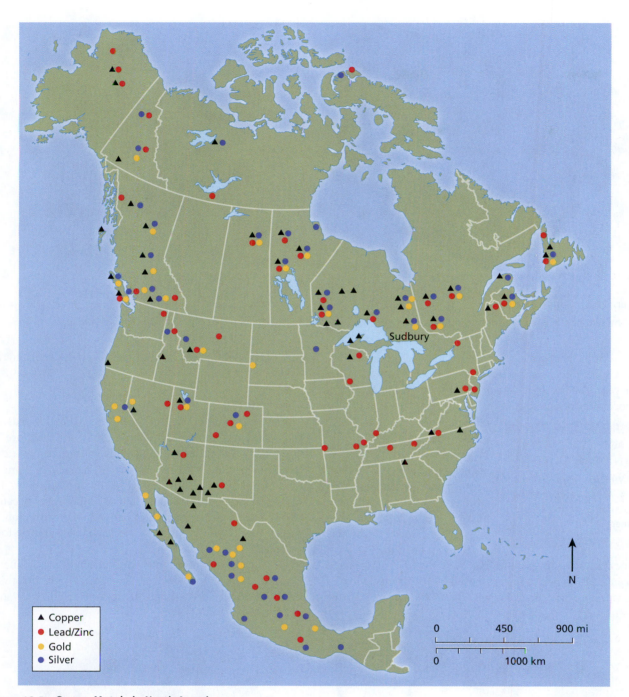

Figure 18.5   Scarce Metals in North America

This map shows the distribution of some important scarce metals (copper, lead, zinc, gold, and silver) in North America.

production currently exceeds 16 million metric tons. Copper can be shaped or drawn out into fine wire, and human use dates back at least 8000 years. Copper is an excellent electrical conductor, and it is this property that accounts for its primary use in electrical wiring. Copper is also used for plumbing and in coins, and is a key ingredient of brass (copper and zinc) and bronze (copper, tin, and zinc).

CHECK YOUR UNDERSTANDING

○ What is the most important scarce metal and how is it used?

The most important copper ore minerals are sulfides such as *chalcopyrite* ($CuFeS_2$) (Fig. 18.6a) and *chalcocite* (Cu_2S). However, copper is one of the few nonprecious metals to occur naturally in its metallic, or native, form (Fig. 18.6b).

Zinc (Zn) and lead (Pb)

Zinc and lead are described together here because they commonly accompany each other in natural settings. Although their average crustal abundance is low, zinc is the fourth most widely used metal after copper, and lead is the

(a)

(b)

Figure 18.6   Copper Ore
(a) The principal source of copper is the mineral chalcopyrite. (b) More rarely, copper occurs in its metallic, or native, form.

fifth. For the past decade, each metal has been consumed in the United States at a rate of 1–1.5 million metric tons per year. Zinc is used primarily for metal castings and as a corrosion-resistant coating on iron and steel. It is also alloyed with copper to make brass. Lead is used primarily in storage batteries. But it is also employed in construction because of its resistance to corrosion, and in ammunition and radiation shielding, because of its density. The use of lead in paint and as an antiknock additive in gasoline has been virtually eliminated because of the health hazard posed by its toxicity. Most of the world's production of zinc and lead come from just two minerals (Fig. 18.7), the zinc sulfide *sphalerite* (ZnS) and the lead sulfide *galena* (PbS).

> ### CHECK YOUR UNDERSTANDING
> ○ Which two types of mineral deposit are the principal producers of lead and zinc?

Chromium (Cr)

The average abundance of chromium in Earth's crust is only 100 parts per million. Chromium must therefore be concentrated hundreds of thousands of times before it can be mined profitably. Yet chromium is an essential component of stainless steel and a range of widely used alloys. It is of critical strategic importance to the United States because the country has no chromium deposits. Consumption in the United States is currently about 400,000 metric tons per year.

Most crustal chromium is concentrated in the mineral *chromite* ($FeCr_2O_4$), which is the metal's only ore mineral (Fig. 18.8). Chromium ores are restricted to iron- and

(a)

(b)

Figure 18.7   Ores of Zinc and Lead
The minerals (a) sphalerite and (b) galena are, respectively, the main sources of zinc and lead.

Figure 18.8   Chromite

The dark mineral in this sample from Mozambique is chromite, the only source of chromium.

Figure 18.9   Native Gold

Gold, shown here in a sample from Queensland, Australia, is one of the few metallic minerals found naturally as a native metal.

<table>
<tr><td>

CHECK YOUR UNDERSTANDING

⦾ Why is chromium considered a metal of critical strategic importance to the United States?

</td></tr>
</table>

magnesium-rich ultramafic igneous rocks, because, like olivine and pyroxene, chromite is concentrated by crystal setting to the base of the magma chamber (a form of fractionation, see Chapter 4) early in the cooling history of large bodies of mafic magma.

Gold (Au)

Gold, like silver and platinum, is a rare and precious metal. In addition to its use in jewelry, which accounts for more than 50 percent of demand, gold is an excellent electrical conductor and performs critical functions in electrical equipment and in the aerospace industry. Gold, which is mined as a native metal (Fig. 18.9), also has a unique status among commodities as a long-term store of value and is the monetary base for international trade. It is estimated that some 33,000 metric tons of gold, or about a fifth of all the gold ever mined, is stored in bank vaults and government treasuries, such as the one at Fort Knox, Kentucky.

<table>
<tr><td>

CHECK YOUR UNDERSTANDING

⦾ Why is almost a third of the gold that is mined stored rather than used?

</td></tr>
</table>

Silver (Ag)

Like gold, silver is an extremely rare metal. Although it is most familiar to us in the form of silverware and jewelry, its primary use is in the photographic and electronic industries. Silver occurs as a native metal but it is more common as a sulfide. Most silver is produced as a by-product of the extraction of copper, lead, and zinc from their ores. The lead ore *galena* (PbS) is the most common host to silver, which substitutes for some of the lead in the mineral's crystal structure.

18.2 SUMMARY

- Metallic minerals form the ores of abundant metals, which are among the most common elements in Earth's crust and include iron, aluminum, and titanium.

- Scarce metals, which include copper, zinc, lead, and chromium, are rare in Earth's crust and must be greatly concentrated by natural processes before they can be exploited by humans.

- The scarce metals gold and silver are also called precious metals because of their monetary value.

18.3 Nonmetallic Minerals and Industrial Materials

Mineral resources that are not used as a source of metals are **nonmetallic minerals**. Most nonmetallic minerals are very common and so are classified by their use rather than their abundance. Raw chemicals such as fertilizer form an important category of nonmetallic minerals, whereas abrasives and gemstones (some of which are very rare) form another. Often included in the nonmetallic minerals group, although they are not actually minerals, are the rock materials used in the construction industry. Recall from Section 18.1 that these rock resources are better termed industrial materials; industrial materials are the deposits used in the greatest volumes.

RAW CHEMICALS

A number of nonmetallic minerals, such as those that play a vital role in agriculture, are used to create blends of chemicals. For example, fertilizers contain mixtures of potassium and phosphorus that have been extracted from nonmetallic minerals. Other nonmetallic minerals, such as salt and sulphur, are used in their natural state. In the material that follows we discuss the uses for these unprocessed chemicals, which are referred to as "raw" chemicals.

Fertilizer

Current practices of intensive agriculture deplete the soil of many essential nutrients. To replace these nutrients, agriculture depends to a large degree on the use of *inorganic*

fertilizers, chemical substances added to the soil or land to increase its fertility. The three nutrients most essential to plant growth are potassium, phosphorus, and nitrogen. Materials containing these elements, combined with agricultural lime and gypsum, make fertilizer. Nitrogen (N) is extracted from the atmosphere for its use in plant growth–enhancing nitrates (NO_3^-), but the other components of fertilizer occur in rocks and minerals and must be mined (Fig. 18.10) from Earth's resources.

All economically exploitable potassium deposits are found in marine evaporates (Fig. 18.11). When ocean water evaporates, dissolved ions such as potassium are concentrated in solution until they become saturated and precipitate to form a variety of potassium-bearing minerals that

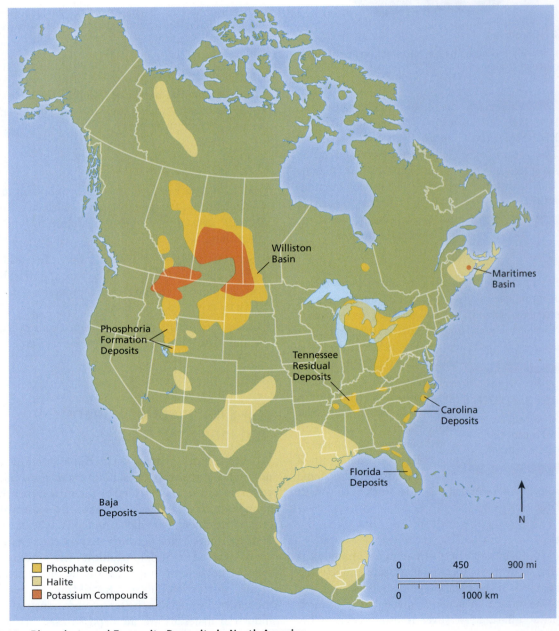

Figure 18.10   **Phosphate and Evaporite Deposits in North America**

Phosphate and marine evaporite (salt and potassium) deposits are located in various regions across North America.

Figure 18.11   Potash Mine

In the Mosaic K1 potash mine in Saskatchewan, Canada, 4700 kilometers (2900 miles) of tunnels have been bored to extract sylvite 1000 meters (3300 feet) below the surface.

are collectively marketed as *potash*. The most important of these is *sylvite* (KCl).

The principal source of phosphorus (P) is the mineral *apatite*, a calcium phosphate. Apatite is concentrated in marine sedimentary deposits called *phosphorites* and, less commonly, in phosphate-rich igneous rocks.

Salt

In addition to its well-known use as a food additive (common table salt) and road deicer, salt (NaCl), or *halite,* plays a far more important role in the industrial production of the chloralkalis used to produce the common plastic polyvinyl chloride (PVC), the chlorine and caustic soda used in pulp production for making paper, and the soda ash used in the manufacture of soaps and detergents.

The largest resource of salt is in the ocean, and salt is produced from the solar evaporation of *brine* (water saturated with salt) in many parts of the world. However, salt is also found on land. As we learned in Chapter 6, the evaporation of ancient seas resulted in the formation of bedded evaporite deposits and salt domes that are preserved in the rock record. These deposits form the main source of salt (Fig. 18.12).

Sulfur

Sulfur (S) is such an important element to modern industrial society that the level of its consumption is used as an index of a nation's industrial development. Sulfuric acid (H_2SO_4), which accounts for 90 percent of sulfur consumption, is a key industrial chemical. Approximately two-thirds of sulfuric acid is used in the production of fertilizer, but its powerful acidity also makes it vital to metal mining, paper and plastics production, oil refining, and the production of industrial chemicals. In fact, sulfuric acid is used at one stage or another in the manufacture of most industrial products, from detergents and drugs to lubricants and batteries.

Sulfur is the only nonmetallic element to occur in its native state (Fig. 18.13). It is

Figure 18.12   Salt Mine

The largest salt mines in Europe are in Slănic, Romania, an abandoned portion of which is now used as a spa, amusement center, and museum of the salt mining industry. The contorted layering visible in this photo reflects the glacier-like flow of salt within the salt dome as it moved upward under the weight of overlying sedimentary rocks.

Figure 18.13   Native Sulfur

Yellow crystals of sulfur, the only nonmetallic mineral to occur in native form.

found most commonly in volcanic vents and hot springs (Fig. 18.14), and at the top of salt domes where it forms from the bacterial breakdown of the evaporite mineral *anhydrite* ($CaSO_4$). The low melting point of sulfur allows it to be readily extracted from the top of salt domes. Hot water is pumped into the sulfur-bearing formation, and the resulting mixture of water and molten sulfur is then pumped out. The main source of sulfur, however, is from petroleum production, where it is a by-product of the refining of crude oil and the purification of natural gas.

ABRASIVES AND GEMSTONES

A variety of minerals are known for their natural hardness (7 and above on Mohs scale; see Chapter 3), and are used as abrasives for cleaning, polishing, cutting, grinding, and drilling. Some of these minerals have, in addition, great natural beauty and are also valued as gemstones.

Diamond

Diamond, the hardest of all minerals, is a crystalline form of carbon (C) created at very high pressure and found in rock only in intrusive bodies of *kimberlite,* known as *diamond pipes* (Fig. 18.15). Kimberlite is the rock product of violent, gas-rich volcanic explosions, which are capable of bringing up fragments from deep within the upper mantle. Diamonds that for reasons of color or impurity are not suitable for gemstones are valuable for drilling and cutting.

Figure 18.14 Volcanic Sulfur

Men gather sulfur from a vent in the crater of Kawah Ijen, a volcano in eastern Java.

Figure 18.15 Diamond Pipe

The Udachnaya diamond mine in eastern Siberia is Russia's largest, producing some 11 million carats of diamond annually from a kimberlite pipe. Active since 1971, the open-pit is now 600 meters (1000 feet) deep, making it the third deepest mine in the world (after the copper mines at Bingham Canyon in Utah and Chuquicamata in Chile).

Corundum

Corundum is aluminum oxide or alumina (Al_2O_3). It is the second-hardest mineral and occurs in metamorphic rocks of ancient continental shields (see Chapter 7). Although aluminum is a metal, corundum is considered nonmetallic since it is not a commercial source of aluminum. Instead, corundum is used industrially as an abrasive. It also occurs in gem quality, forming *ruby* and *sapphire* when its natural color is either red or blue respectively. These contrasting colors are caused by small amounts of impurities. Ruby's red color is caused by trace amounts of chromium in the crystal structure, whereas the deep blue of sapphire is the result of traces of iron and titanium. Natural mixtures of corundum and magnetite form *emery,* which is used in wear-resistant floors, polishing cloth, and nonslip surfaces.

> **CHECK YOUR UNDERSTANDING**
>
> • What minerals are most commonly used as abrasives?

INDUSTRIAL MATERIALS

Industrial materials include a wide variety of commodities used in the construction industry. As we noted in Section 18.1, industrial materials are often misleadingly categorized as mineral deposits, although most are actually rocks. Almost every rock type can be used for building. So it is not surprising that the natural resources used by humans in the greatest volume are building materials. The United States alone consumes more than $20 billion dollars worth of industrial materials every year. Of these materials, the most important are stone, sand, gravel, limestone, and gypsum.

Stone, sand, and gravel

A variety of rocks are quarried simply for their use as stone. These rocks include granite, limestone, sandstone, marble, and slate (Fig. 18.16). Rocks such as marble, granite, and slate form the polished floors of our banks and the facades and roofs of our public buildings and monuments. Most stone, however, is broken up. Some "crushed rock" is used in roads and buildings. The crushed rock used most widely is limestone, which is the primary ingredient of cement. Crushed limestone is also used in steelmaking, and finely powdered limestone is used in agriculture to treat soil or is burnt to produce the chemical *lime* used in concrete and plaster.

> **CHECK YOUR UNDERSTANDING**
>
> ● Which natural resources are used as construction materials?

As industrial materials, unconsolidated sand and gravel rank second in importance after crushed stone in terms of production tonnage. In the

Figure 18.16   Slate Quarry
The now-abandoned Penrhyn slate quarry in Bethesda, North Wales, UK, was once a major source of roofing slate.

Figure 18.17   Limestone Quarry

Limestone quarry on England's Isle of Portland, which gave its name to Portland cement.

Figure 18.18   Gypsum Deposit

Gypsum deposits of the Triassic Chugwater Formation near Medicine Bow, Wyoming.

United States, sand and gravel are readily accessible resources and are the only commodities produced in all 50 states. As we discussed in Chapter 6, sand and gravel deposits are sorted and deposited by flowing water. They are important components of concrete aggregate, road base, asphalt, and fill.

Cement and gypsum

Portland cement (Fig. 18.17), which hardens when mixed with water, is the chief binding agent for *concrete*. Cement is produced by heating the right proportions of crushed limestone, dolomite, quartz sand, and oxides in a high-temperature oven, and then adding the mixture to an aggregate of sand and gravel or crushed rock to make concrete.

The evaporite mineral *gypsum* ($CaSO_4 \cdot 2H_2O$), one of several minerals that form as the result of the evaporation of salt water (Fig. 18.18), is the natural source of building plaster and plasterboard. When the mineral is heated in low-temperature ovens, much of its water of crystallization is driven off, and *plaster of Paris* ($2CaSO_4 \cdot H_2O$) is formed. As those of us who have had the misfortune to break a bone can testify, this plaster rapidly sets upon the addition of water by recrystallizing back to gypsum.

18.3 SUMMARY

- Inorganic natural resources that are not used as a source of metals are known as nonmetallic minerals and industrial materials.

- Nonmetallic minerals are used as raw chemicals (fertilizer, salt, and sulfur) and as abrasives and gemstones (diamond, corundum, garnet, quartz, and emerald).

- Industrial materials, such as stone, sand and gravel, cement, and gypsum, are largely used for construction.

18.4 Formation of Mineral Deposits

Most elements in rocks occur in very low concentrations. If they are to form a mineral deposit, geologic processes must act to concentrate them. If the deposit can be exploited economically, it is known as an **ore deposit**. Whether a mineral deposit is an ore deposit or not depends on the market price of the ore in question, which fluctuates with economic and political factors. A mineral deposit is only an ore deposit if the price of the ore is high enough to make its extraction commercially viable. Because it may take several years to develop an ore deposit into a mine, mining companies must make investment decisions based on the projected price of the ore over the time interval during which the ore is extracted.

The elements sought in mineral deposits are usually rare or, when relatively abundant, are rarely concentrated in sufficient quantities to be exploited economically. All mineral deposits therefore require special circumstances during which natural processes act to concentrate the sought-after commodity. The degree of concentration needed for economic extraction is determined by the cost of recovering the commodity. This, in turn, depends on the average concentration of the element in Earth's crust. In this section of the chapter we discuss the composition of Earth's crust and

how Earth's minerals are concentrated in order to learn more about the viability of economic exploitation of the minerals we discussed in Sections 18.2 and 18.3.

There are 92 naturally occurring elements and Earth's crust contains all of them. Fully 99 percent of Earth's continental crust, however, is made up of just nine elements (Fig. 18.19a), and just two—oxygen and silicon—make up almost three-quarters of the continental crust. The same nine elements also dominate Earth's oceanic crust. It is not surprising, therefore, that most rock-forming minerals are silicates, and it is within these silicates that most of the common metallic elements occur.

The mineral composition of ocean water is quite different from that of Earth's crust; in seawater the relative abundance of dissolved elements reflects their respective solubilities. Excluding the water itself, more than 95 percent of the ocean's composition is made up of only six dissolved ions, and sodium and chlorine alone account for more than 80 percent (Fig. 18.19b). These statistics make it clear that massive concentration factors are required to produce economically exploitable ore deposits from the vast majority of elements.

The principal mechanisms of ore formation are physical and chemical and they depend upon physical and chemical differences between the ore mineral and its environment (see In Depth: **Mechanisms of Mineral Concentration** on page 583). The mechanism that acts to concentrate a particular mineral depends upon the ore in question and the geologic setting of the ore deposit. Ores that are produced by the segregation of metals as magma cools, for example, occur only in igneous rocks,

> **CHECK YOUR UNDERSTANDING**
> ◉ What are the two primary types of mechanisms that act to concentrate minerals?

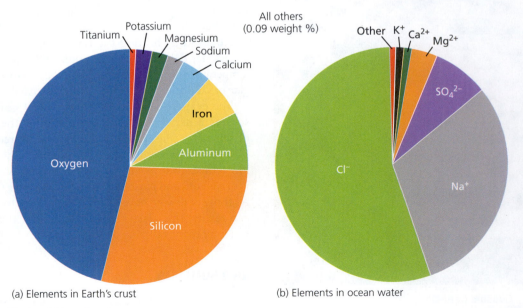

(a) Elements in Earth's crust

(b) Elements in ocean water

Figure 18.19   Element Abundance
These pie charts show the relative abundance of the elements in (a) Earth's crust, and (b) ocean waters.

IN DEPTH

Mechanisms of Mineral Concentration

Because most useful minerals are rare in Earth's crust, in order to form exploitable mineral deposits these minerals must be concentrated by natural processes. Natural mechanisms of mineral concentration fall into three groups: chemical, physical, and biological.

The most important mechanisms are either *chemical* or *physical*. These mechanisms depend upon differences in the physical or chemical properties of the ore mineral and those of its host. We can illustrate this concept with a simple example (Fig. 18A). Suppose you add a pinch of

finely ground pepper to half a cup of salt and stir the two together until they are thoroughly mixed. In this experiment, the pepper represents a valuable mineral resource but only when it is concentrated. Clearly, the mixture makes a hopeless substitute for pure pepper! How can you go about extracting or separating the pepper from the salt in order to concentrate the pepper?

Several options come to mind. If you happen to recall that salt dissolves in water and pepper does not, it might occur to you to pour the mixture into a glass of water and

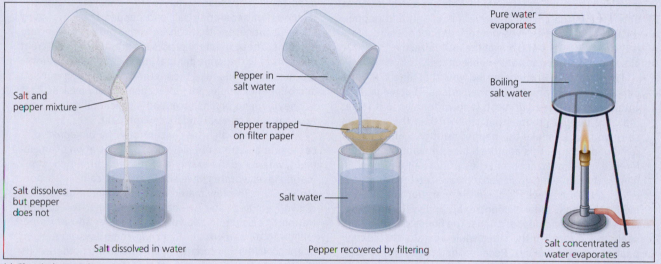

Salt and pepper mixture

Salt dissolves but pepper does not

Salt dissolved in water

Pepper in salt water

Pepper trapped on filter paper

Salt water

Pepper recovered by filtering

Pure water evaporates

Boiling salt water

Salt concentrated as water evaporates

(a) Chemical separation

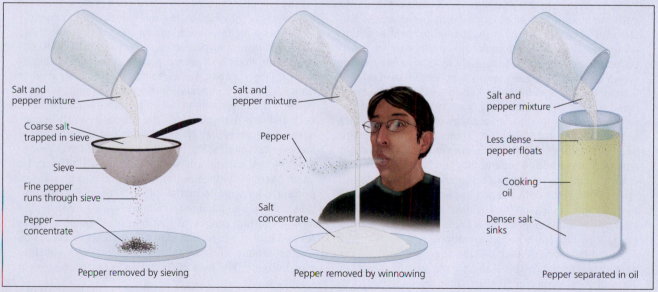

Salt and pepper mixture

Coarse salt trapped in sieve

Sieve

Fine pepper runs through sieve

Pepper concentrate

Pepper removed by sieving

Salt and pepper mixture

Pepper

Salt concentrate

Pepper removed by winnowing

Salt and pepper mixture

Less dense pepper floats

Cooking oil

Denser salt sinks

Pepper separated in oil

(b) Physical separation

Figure 18A   Methods of Concentration

Both chemical and physical methods can be employed to separate salt and pepper from a mixture of the two.

(Continued)

IN DEPTH

Mechanisms of Mineral Concentration *(Continued)*

stir the water until all the salt has dissolved. You could then pour the salt water off, leaving the pepper behind. This is similar to the natural weathering process that concentrates aluminum to form bauxite by the selective removal of other elements. If you employ this method, you leave the pepper behind and at the same time, the salt is concentrated in the water. That means that if the salt were the valuable commodity in our scenario (rather than the pepper), you could also use this method to isolate the salt. You could then recover the salt, like a natural evaporite deposit, by evaporating the salt water until the water was gone and the salt was left behind. This approach utilizes a difference in the chemical properties of salt and pepper, the fact that salt dissolves whereas pepper does not. This is the most obvious and one of the most effective methods of separation, but it is not the only one.

For example, it might strike you that finely ground pepper grains are smaller than salt grains. That means that with a sieve of the right mesh (one with holes in it larger than the pepper grounds but smaller than the grains of salt), you could shake the pepper through while the salt remained behind. This approach takes advantage of a difference in the physical properties of the salt and pepper, namely their different grain size. However, method number 2 is unlikely to achieve as clean a separation as the first method because there is bound to be some variation in the size of the salt and pepper grains. The finest salt grains are likely to fall through the sieve while some of the coarser pepper grounds will not. Recall from Chapter 4 that sedimentary processes take advantage of such physical differences to sort sediments according to their grain size.

Salt is also denser than pepper. So if you were to slowly pour the mixture out of its container, blowing across the stream of grains as you did so, the lighter pepper grains would be blown farther than the salt grains and you could collect the pepper grains separately. In a similar way, wind-blown sand in the desert tends to collect in sand dunes, leaving behind boulder fields known as desert pavement (see Chapter 16). Alternatively, you could spread out the salt and pepper mixture on a sheet of paper and gently shake until the lighter pepper grains separate from the heavier salt. These approaches are similar to the ancient agricultural method of separating grain from chaff by *winnowing* and take advantage of another difference in the physical properties of salt and pepper, their density. However, the mechanical processes involved in these physical methods of separation are fairly crude and are not likely to produce a particularly good separation.

Another approach that utilizes the difference in the density of salt and pepper is to pour the mixture into a tall glass of cooking oil. The heavier salt immediately sinks to the bottom of the glass and the lighter pepper settles much more slowly. As soon as all the salt has settled, you can pour the oil off into a second container, in which the pepper could be allowed to settle out of the oil separately. The oil could then be poured off, leaving the pepper behind. Done carefully, this method produces an excellent separation. However, it leaves you with the difficult "environmental" problem of separating the pepper from the oil.

Minerals are naturally concentrated to form mineral deposits by processes that are similar to these examples (although less simplistic). So are the industrial methods used to extract these minerals from their ores. Like the methods for separating salt and pepper, there are many different processes that can be employed to concentrate minerals. These different processes may apply to different minerals or to the same mineral under different conditions. But most methods, like the methods we used in our salt and pepper example, are either chemical or physical. That is, they depend on chemical and physical differences between the ore mineral and its environment.

Minerals can also be concentrated by *biological* processes. To illustrate the biological concentration of an element, we turn to the example of seashell formation. Many marine invertebrates have shells made of calcium carbonate ($CaCO_3$). The raw materials for calcium carbonate are drawn from seawater by shelled organisms. The concentration of calcium in seawater is just over 400 parts per million (0.04 percent), but calcium comprises 40 percent of the calcium carbonate shell. In order to produce its shell, a marine organism must therefore concentrate calcium biologically by a factor of about 10,000.

The role of organisms in the formation of ore deposits is not nearly as obvious as the role physical and chemical processes play. Yet, as marine animal shells demonstrate, living organisms can play an important part in mineral concentration. Plant and animal activity is also important in the formation of soils, and certain tropical soils weather to form bauxite, the principal source of aluminum. Similarly, bacteria that are common in hot, mineral-charged waters, such as those that spout from the submarine vents of mid-ocean ridges, may play an important role in the formation of some metal sulfides. The formation of sulfur from the evaporite mineral anhydrite is similarly facilitated by bacteria. Other bacteria synthesize iron oxides and may play an important role in the formation of certain sedimentary iron ores. In fact, banded iron formations (see Figure 18.2), the most important source of iron ore, are thought to have formed from the oxidation of iron in the world's oceans by photosynthetic microorganisms early in Earth's history (see In Depth: **The Role of Oxidation in Earth's History** in Chapter 5).

whereas those resulting from the circulation of hot, water-rich fluids occur in both igneous and metamorphic settings. Still other ores form as a result of deposition, evaporation, and weathering processes, and they are confined to sedimentary environments. To explore the different ways by which ore minerals are concentrated, we must therefore examine the various geologic settings in which ore deposits occur.

MINERAL CONCENTRATION IN IGNEOUS ROCKS

There are two primary mechanisms of mineral concentration in igneous settings:

- those that produce *magmatic deposits* during mineral separation in a cooling magma; and
- those that produce *hydrothermal deposits* during the circulation and cooling of magmatically heated waters.

In the following material, we discuss each of these processes in more detail.

Magmatic deposits

Ore deposits form in cooling magmas by way of **magmatic segregation**, a process that results in the physical separation of the magma's chemical components. Recall from Chapter 4 that as a body of magma cools, it undergoes crystal fractionation, a process that removes minerals from the magma under the influence of gravity (see Fig. 4.16). The minerals that first crystallize in the magma are either denser or less dense than the magma itself. Those that are denser tend to settle to the bottom of the magma chamber, whereas those that are less dense tend to float to the roof. The remaining magma consequently becomes depleted in the elements removed by the minerals at the same time it is enriched in the elements that are left behind. As a result, the magma physically segregates into components of different composition as it cools and crystallizes. A similar form of segregation is involved in the separation of cream from milk, or milk solids from clarified butter.

Depending on their chemical affinity, metallic elements join the separating crystals, or are left behind to become enriched in the remaining magma. In either case, they become concentrated. Metallic minerals with ore deposits that are the product of magmatic segregation include a variety of oxides and sulphides that are rich in chromium (Fig. 18.20), titanium, iron, platinum, and nickel.

Among the elements preferentially concentrated in the remaining magma during fractionation are those that are

Figure 18.20   Magmatic Segregation

Layers rich in chromite (black) are concentrated toward the base of the plutonic Bushveld complex in South Africa. The high density of the early-formed chromite crystals caused them to sink through the magma as the igneous body cooled. Chromite is a mineral rich in chromium.

not readily incorporated into the crystal structures of early-crystallizing minerals such as olivine, pyroxene, and plagioclase. Only elements with an ionic size or charge that can readily fit into these minerals are likely to be incorporated in their structures. Other so-called "incompatible" elements are excluded from these early crystallizing minerals, and therefore become extremely enriched in magma as crystal fractionation proceeds. At the same time, the concentration of water, which is also excluded from early crystallizing minerals, increases dramatically. When these liquids finally cool, they crystallize to form spectacularly coarse-grained bodies known as *pegmatites* (Fig. 18.21). The mineral grains in pegmatites are very coarse because the presence of water allows for efficient transport of ions to the sites where the crystals grow. Pegmatites can contain ore minerals enriched in beryllium and lithium and the rare earth elements, as well as gemstones such as emerald. Many of the mineral specimens displayed in museums are from pegmatites, which can feature spectacular crystals.

> **CHECK YOUR UNDERSTANDING**
>
> ● What is magmatic segregation?

Hydrothermal deposits

One of the most important ways metals become concentrated involves the circulation of magmatically heated water.

Figure 18.21   Pegmatite
Pegmatite from northern Nova Scotia, Canada, shows large quartz (white) and feldspar (orange) crystals with radiating needles of the semiprecious mineral tourmaline (black).

Mineral concentrations produced in this way are **hydrothermal deposits**. At high temperatures and pressures, water is a very efficient solvent and is typically enriched in metals. Heated water may be present in the host rocks around a cooling magma, or may be ejected from the magma itself. When the water can no longer hold the metals in solution, the metals precipitate out onto the walls of fractures to form mineral **veins** (Fig. 18.22). Metals may also permeate the host and precipitate within them in a more dispersed fashion as a **disseminated deposit**. Many important ore deposits, including those exploited for minerals rich in tin, copper, lead, zinc, tungsten, silver, and gold, are the product of hydrothermal activity.

MINERAL CONCENTRATION IN SEDIMENTARY ROCKS

Minerals are concentrated in sedimentary settings by weathering, erosion, deposition, evaporation, and solution. These processes work independently or in concert to produce rich deposits. The processes of weathering and erosion, for example, combine to reduce exposed rock to rubble or detritus. The detritus is picked up and transported by the various agents of erosion, such as wind, water, and ice, and may ultimately reach the sea. In the process, the detritus tends to become sorted according to its grain size (see Chapter 6). This is especially true if the transporting medium is one of flowing water. As a result, well-sorted deposits of gravel, sand, and clay are produced in environments such as streams, beaches, and lakes. These deposits can be of sufficient volume to exploit as construction materials. Many industrial materials owe their origin to the sorting process associated with sediment transport.

In addition, the transport of sedimentary detritus in flowing water provides the opportunity for minerals to dissolve, so that they may later be concentrated by evaporation (see In Depth: **Mechanisms of Mineral Concentration** on page 583). The effects of evaporation are well known to those familiar with arid environments. In desert climates, such as those of the southwestern United States and the Middle East, temporary lakes fed by flash floods or winter snows slowly grow saltier as they evaporate and finally form salt flats, or playas, as they dry up altogether (see Chapter 16).

When an entire sea dries up, as the Mediterranean is believed to have done some 6 million years ago, vast deposits of mineral salts are left behind on the seabed (see Living on Earth: **Salt of the Earth** on page 173). Similarly, thick deposits can form on the floor of any marine basin if it is subjected to very high levels of evaporation and is only weakly connected to an ocean. Protected by burial beneath younger marine sediments, these deposits of mineral salts become incorporated into the rock record as *evaporites*.

Evaporite deposits are sources of a wide variety of nonmetallic minerals. The most important of these—common rock salt (halite), gypsum, and potash—are produced by the evaporation of seawater. Others deposits, such as *borax*

Figure 18.22   Mineralized Vein

A gold vein in quartz. Sample is about 4 cm (1.6 inches) across.

(a mineral rich in sodium and boron), which is used as a cleaning agent, are produced by the evaporation of lakes. Because of the low density of salt and the ease with which it flows under pressure, beds of rock salt may become mobilized upon burial (see Fig. 18.12), rising up through the overlying strata in columns known as *salt domes*. These structures are particularly common in the Gulf Coast region of the southern United States and Mexico, where they play a major role in the trapping and accumulation of oil and natural gas (see Chapter 19), as well as being important sources of common salt and sulfur.

In addition to providing raw material for deposition or evaporation elsewhere, weathering may produce mineral concentrations in the material left behind. The material left behind forms **residual deposits** of less easily weathered material. These deposits can concentrate elements such as iron, manganese, and nickel, as well as clay minerals and bauxite, the principal ore of aluminum.

For minerals that do not dissolve, oxidize, or react with water, the transport of sedimentary detritus in streams provides another opportunity for concentration, that of **mechanical enrichment**. Stable metals and ore minerals such as gold and *cassiterite* (SnO_2, the principal ore of tin) are much denser than most sedimentary detritus and therefore they are not as easily moved by flowing water. As a result, if streams drain areas in which the bedrock is mineralized, eroded grains of metal and ore tend to be preferentially deposited in locations where the velocity of the stream slackens. Deposition can occur, for example, where the stream abruptly widens (Fig. 18.23) or in deltas where water slows down dramatically as it enters a lake or the sea. Concentrations of minerals formed in this way are known as **placer deposits**. Placer deposits are important sources of ore minerals containing platinum, titanium, and chromium and rare earth minerals (see Chapter 3), as well as diamonds and other gemstones.

Placer deposits are most important, however, as sources of tin and gold. Placers are the principal source of tin and are also responsible for the huge Witwatersrand gold field in South Africa. It was also placer gold that led to the California Gold Rush when it was discovered at the base of the Sierra Nevada Mountains in 1849. Interestingly, the search for gold by panning takes advantage of exactly the same principle that produced the placer deposits in the first place. The swirling action of the water in the pan effectively separates the heavier gold from the unwanted sedimentary detritus just as water concentrated the original gold placer deposits.

Placers are not restricted to streams. They are also found in the marine realm where they owe their origin to wave action and ocean currents. Gold-bearing placers, for example, occur off the coast of Alaska, chromite occurs

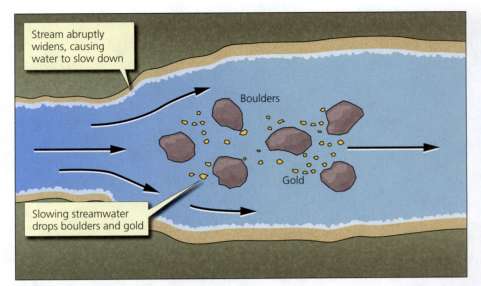

Stream abruptly widens, causing water to slow down

Boulders

Gold

Slowing streamwater drops boulders and gold

Figure 18.23   Placer Deposit Formation

Placer deposits form when running water slows down (for example, where a stream abruptly widens) and deposits both coarse sediment and dense minerals, such as gold.

off the coast of Oregon, and extensive titanium-mineral sands occur off the coasts of Florida, Georgia, and South Carolina. In each case, the ore minerals were derived by erosion of rock, transported by streams, and finally deposited in the sea.

Important marine sedimentary mineral deposits also occur where hot, mineral-charged hydrothermal fluids are discharged directly into the sea. These fluids cool rapidly causing dissolved chemicals to crystallize and form mineral-rich sedimentary strata known as *exhalites*. Significant deposits of iron, manganese, copper, and zinc have formed this way around submarine vents at divergent plate boundaries (see Fig. 18.24). Exhalite deposits on the deep ocean floor are likely to become one of the major environmental issues of the twenty-first century, as improving technology and rising metal prices make their exploitation both practical and financially attractive.

> **CHECK YOUR UNDERSTANDING**
>
> ○ What is mechanical enrichment, and what are mineral deposits produced by this process called?

MINERAL CONCENTRATION IN METAMORPHIC ROCKS

Hydrothermal fluids are also responsible for concentrating minerals in metamorphic environments, for example, in the contact aureole around an igneous body where the rocks have been baked adjacent to cooling magma (see Chapter 7). Under certain conditions, hydrothermal fluids emanating from magma may chemically react with the rocks with which they come in contact. As a result, the fluids remove certain elements from the rock and replace them with others. When hot fluids, rich in silica (SiO_2),

come into contact with limestone, for example, they react with the mineral calcite ($CaCO_3$, or calcium carbonate) by removing carbon dioxide and adding silicon, thereby replacing the calcium carbonate with calcium silicate ($CaSiO_3$). In the same way, the process of **replacement** may remove a rock-forming mineral and substitute an ore mineral in its place. Ore minerals concentrated by replacement are rich in elements such as iron, copper, zinc, lead, tin, tungsten, molybdenum, and manganese, and may contain graphite, as well as native gold and silver. The chemically altered metamorphic rocks produced by replacement are called *skarns*. Skarn formation is common in areas adjacent to granites that have been emplaced into limestones.

> **CHECK YOUR UNDERSTANDING**
>
> ○ What is replacement?

18.4 SUMMARY

- Because most elements are rare in Earth's crust, massive concentration factors are usually needed to form exploitable ore deposits.

- Mineral concentration occurs mainly as a result of chemical and physical processes that vary with the ore in question and the geologic setting of the ore deposit.

- Mineral deposits occur as magmatic and hydrothermal deposits in igneous rocks; as evaporites, residual deposits, placer deposits, and exhalites in sedimentary rocks; and as replacement deposits in metamorphic rocks.

18.5 Plate Tectonics and Mineral Deposits

As we discussed in Section 18.4, minerals concentrate in very specific geologic settings. These settings are commonly associated with activities along plate boundaries, which means that the distribution of ore deposits is often related to the plate tectonic history of the areas where they occur. An understanding of the relationship between plate tectonics and mineral deposition can be used to guide exploration strategy, and so provides a powerful tool for mineral exploration. Many of the processes of mineral concentration are either directly or indirectly linked to magmatism. Mineral deposits are therefore most frequently associated with the processes of rifting and spreading at divergent plate boundaries and with the process of subduction at convergent plate boundaries, because it is in these plate tectonic settings that magmatism is most common.

MINERAL DEPOSITS AT DIVERGENT PLATE BOUNDARIES

A wide variety of mineral deposits occur at divergent plate boundaries, where they are associated with the magmatism of continental rift and mid-ocean ridge environments (Fig. 18.24).

Continental rift environments

Major ore deposits in continental rift settings are produced during the early stages of rifting by magmatic concentration within huge igneous bodies of basaltic composition that cool and segregate to form what are known as *layered igneous complexes* (Fig. 18.24a). The best known layered igneous complex is the vast Bushveld complex in South Africa. This 2-billion-year-old igneous complex is the world's largest deposit of chromium, platinum, and vanadium-bearing magnetite; it is also mined for nickel. Here the ore minerals, which formed early in the crystallization process, were concentrated toward the base of the intrusion when their high density caused them to sink. Although they often occur in layers that are only a few inches thick (see Fig. 18.20), the layers extend over many thousands of square miles and contain huge reserves of ore.

The origin of the Bushveld complex is not fully understood, although some form of crustal rifting is certainly involved. Because the complex occupies an elliptical area and was intruded into a relatively isolated structural depression, some argue that it is the product of hotspot activity, while others surmise that it is the result of asteroid impact.

Continental rift environments are also associated with deposits of copper, zinc, and cobalt, such as those of the Zambian copper belt, the Kupferschiefer in central Europe, and the copper country of Upper Michigan's Keweenaw Peninsula. In the latter case, the chemical weathering of basaltic lavas produced by the process of rifting released the metal content held in the lavas, which was then transported in solution and deposited in sediments (see Fig 18.24a). Because these deposits occur in a specific part of the sedimentary succession, they are known as **stratiform deposits**.

Rifting in stable continental interiors is associated with carbonate-hosted stratiform lead-zinc deposits such as those of the Mississippi Valley. In these *Mississippi valley-type deposits,* ores are formed from lead and zinc released from the carbonates and precipitated as sulfides during low-grade metamorphism (Fig. 18.24b). Advanced stages of rifting are associated with siltstone- or shale-hosted stratiform deposits rich in zinc, lead, and copper, such as those located at Sullivan in British Columbia and Mount Isa

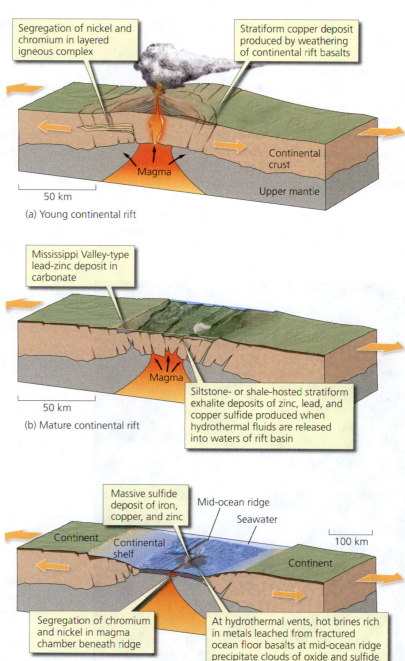

(a) Young continental rift

(b) Mature continental rift

(c) Mid-ocean ridge

Figure 18.24 Minerals Deposits at Divergent Plate Boundaries

Mineral deposits coupled to magmatism at divergent plate boundaries include those associated with: (a) young continental rifts; (b) mature rifts; and (c) mid-ocean ridges.

in Australia. These deposits formed as exhalites when hydrothermal fluids were released into the waters of the rift basin and precipitated their dissolved metals as sulfides. Advanced stages of rifting may also be associated with sandstone-hosted stratiform deposits of uranium. One example is found in Saskatchewan in Canada, where uranium removed in solution from sediments filling the rift basin was deposited along the unconformity at the base of the succession.

CHECK YOUR UNDERSTANDING

⊙ What metallic minerals might be found in a layered igneous complex?

Mid-ocean ridge environments

The most important mineral deposits formed at divergent plate boundaries are the product of the hydrothermal activity of seawater at the mid-ocean ridges (see Living on Earth: **The Bizzare World of Mid-Ocean Ridges** on page 267). Here, seawater seeps into the highly fractured rocks of the spreading center and is progressively heated as it approaches the magma below. At temperatures hot enough to melt lead, highly acidic brine leaches metals from the surrounding basalt and carries them in solution back to the ocean floor. On meeting the ice-cold, oxygenated seawater, the dissolved metals quickly precipitate as a sooty cloud of oxide and sulfide particles (Fig. 18.25). Belching from hydrothermal vents called *smokers*, the settling particles blanket the seafloor with *massive sulfide deposits* rich in iron, manganese, zinc, and copper (see Fig. 18.24c).

While the exploitation of these minerals on the seabed is presently not economically viable, fragments of ancient ocean floor that have subsequently been emplaced onto the margins of continents during plate convergence are important resources. Known as *ophiolites* (see Chapter 9), these oceanic fragments also provide a natural laboratory in which geologists can study igneous processes that occur in oceanic crust. On the island of Cyprus in the eastern Mediterranean, one such fragment has been mined since antiquity for massive sulfides of iron, copper, zinc, and cobalt formed by hydrothermal activity at an ancient mid-ocean ridge. In fact, the mines were so important in Roman and pre-Roman times that the name of the island is derived from the Latin *cuprum* and the Greek *kupros*, meaning copper. To date, deep-sea mining at an active spreading center is being investigated only in the narrow Red Sea where metal-rich sediments along the axis of the ridge lie close to the shores of Saudi Arabia and the Sudan.

Mineral deposits also form beneath mid-ocean ridges as a result of magmatic segregation. The ocean floor rocks of Cyprus, for example, have additionally been mined for their deposits of nickel sulfide and chromite. In contrast to the hydrothermal ores, however, these magmatic deposits were produced at depth in igneous bodies that cooled and crystallized beneath the ridge crest by the magmatic segregation of dense minerals rich in nickel and chromium.

MINERAL DEPOSITS AT CONVERGENT PLATE BOUNDARIES

A wide variety of mineral deposits occur at convergent plate boundaries, where they are associated with the magmatism of continental magmatic arcs and back-arc environments (Fig. 18.26).

Magmatic arc environments

The most important mineral deposits formed at convergent plate boundaries are those related to the subduction of oceanic lithosphere at continental margins. Metals are enriched in these settings in a two-stage process. First, the subducting ocean floor is enriched in metals as a result of the hydrothermal activity at mid-ocean ridges described above. In the second stage, these metals accompany the fluids that rise from the subducting oceanic plate and generate magmas in the overlying mantle wedge. The metals then become incorporated into the subduction zone magmatism.

Recall from Chapter 9 that subduction along continental margins produces magmatic arcs where linear belts of granitic magmas are emplaced into the continental crust of the overriding plate (see Fig. 9.21). Mineral concentrations resulting from hydrothermal activity are widespread in these granitic bodies as well as in the aureole of contact metamorphic rocks (see Chapter 7) around them, and include disseminated copper deposits and a wide variety of vein deposits and skarns.

Figure 18.25   Smoker

Dark, mineral-charged water spews from a "smoker," a hydrothermal vent on the floor of the Atlantic Ocean.

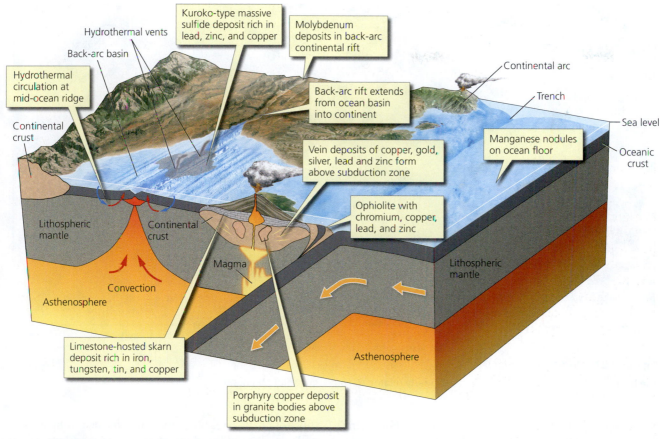

Figure 18.26 Mineral Deposits at Convergent Plate Boundaries
Mineral deposits are associated with magmatism in continental magmatic arc and back-arc settings at convergent plate boundaries.

Because of their immense volumes and wide distribution, some of these granitic bodies are attractive "high-tonnage" mining targets, the most important of which are the **porphyry copper deposits** (Fig. 18.27). Formed by hydrothermal activity within fractured granitic bodies, porphyry copper deposits often contain iron, molybdenum, gold, and silver, in addition to copper. The global distribution of these deposits, most of which are between 15 and 150 million years old, clearly shows their relationship to modern or geologically quite recent subduction zones (see Fig. 18.27). Most porphyry copper deposits, such as those of South America and the Philippines, occur adjacent to active subduction zones. Others, such as those of western North America, mark the sites of subduction in the recent geologic past.

Porphyry copper deposits actually contain quite low values of copper—between 0.2 and 2 percent—and are consequently classed as *low-grade* deposits. But they often cover large areas and so may contain as much as several million metric tons of copper in a single deposit. Because the extraction process is fairly simple, the high tonnage of these deposits means that they can be mined profitably despite their low copper values. In many cases, however, the economic viability of such deposits depends upon a secondary process of mineral enrichment. This process,

known as **supergene enrichment**, is made possible by weathering. Percolating groundwater leaches copper from near-surface parts of the deposit and precipitates it at depth. As a result, lower levels of a deposit that might otherwise have been uneconomic are sufficiently enriched in copper to form an ore deposit. However, huge volumes of ore must be processed to make these efforts worthwhile economically. The open-pit mines created, such as Chuquicamata in Chile and Bingham Canyon in Utah (see Fig. 18B in Living on Earth: **The Environmental Impact of Mining** on page 600), are the largest human-made excavations on Earth's surface. They are so large, they were the only excavations visible to the Apollo astronauts.

Skarn deposits are formed by the action of magmatic fluids on limestones adjacent to intrusive igneous bodies. Hydrothermal fluids, generated in response to arc magmatism, carry dissolved metals such as iron, tungsten, tin, and copper. These fluids react with the limestone and precipitate minerals rich in those metals. Hydrothermal fluids also produce vein deposits that often contain copper, gold, and silver. Closer to the surface,

> **CHECK YOUR UNDERSTANDING**
>
> • What are porphyry copper deposits and in what plate tectonic setting do they occur?

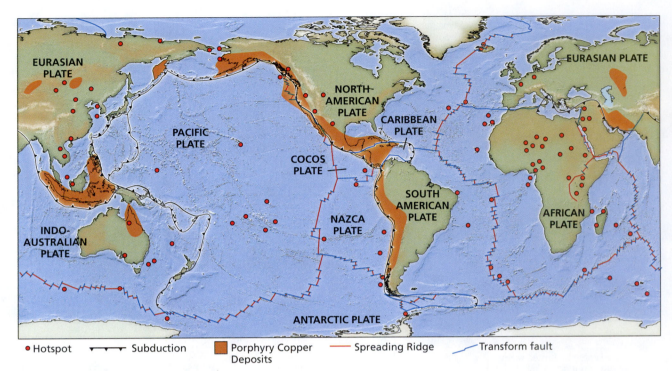

● Hotspot   ┬─┬─┬ Subduction   ▮ Porphyry Copper Deposits   ── Spreading Ridge   ── Transform fault

Figure 18.27   Porphyry Copper Deposits
The location of porphyry copper deposits is clearly related to convergent plate boundaries.

hydrothermal fluids may be responsible for hot-spring deposits rich in gold, mercury, antimony, and arsenic. Hydrothermal replacement and vein deposits on the inner (continent) side of the arc often contain zinc, lead, and silver, and the plutonic rocks in these locations may be rich in tin and tungsten.

Back-arc environments

Mineral deposits in convergent margin settings may also occur in the region behind the arc, where they are associated with crustal extension and the development of back-arc basins (see Fig. 18.26). Continental rift zones in areas of plate convergence, such as the Basin and Range Province and Rio Grande rift in the United States, have very high heat flow because the crust becomes heated as it is thinned by rifting and massive magmatic activity. Hydrothermal circulation fueled by this magmatism is responsible for the emplacement of large molybdenum deposits and smaller deposits of tungsten, tin, and uranium. The large molybdenum deposits associated with the Rio Grande rift, such as the Climax deposit of Colorado, are responsible for the disproportionately large reserves of this important steelmaking element in the United States.

Continental rifting behind the arc may lead to the opening of a back-arc basin floored by oceanic crust, such as the modern Japan Sea. Once established, these oceanic rift zones are characterized by hydrothermal mechanisms of mineral concentration at the spreading center that are similar to those at mid-ocean ridges. The most important of the resulting mineral deposits are the so-called *Kuroko-type* massive sulfides (see Fig. 18.26), named for

those of northern Japan, which were formed by hydrothermal activity on the floor of the Japan Sea some 12 to 15 million years ago. Kuroko-type massive sulfide deposits are less voluminous than porphyry copper deposits but they are significantly richer, forming *high-grade* deposits of zinc, lead, and copper.

18.5 SUMMARY

- At divergent plate boundaries, mineral deposits are associated with mid-ocean ridge (massive sulfides and chromite) and continental rift environments (stratiform copper and Mississippi Valley–type lead-zinc deposits).

- Mineral deposits at convergent plate boundaries include porphyry copper deposits, vein deposits, and skarns associated with magmatic arc environments, and molybdenum and Kuroko-type massive sulfide deposits associated with back-arc rift environments.

18.6 Mineral Exploration

Mineral exploration has always involved the search for mineral deposits and environments that might be favorable for their formation. In the days of the individual prospector with an eye for gold or ore, it was a skill learned from long experience on strenuous surveys. The prospector worked empirically, recognizing favorable patterns without necessarily understanding the processes involved.

Modern exploration benefits from our knowledge of the mechanisms of mineral enrichment and the plate tectonic settings in which they occur. We can select the broad geologic setting most likely to host the type of ore deposit being sought, and within this setting, identify the specific geologic environment most likely to be mineralized. Modern prospecting combines old discoveries and the wealth of accrued knowledge with sophisticated methods of exploration, including some that don't even involve visiting the area being explored.

The basic tool of exploration remains the geologic map. This is usually a simple topographic map that shows the distribution of rock types that make up the bedrock at or immediately below the surface. Coupled with satellite navigation, which provides precise locations using the Global Positioning System (GPS), such a map enables the geologist to identify areas where specific rock types occur and locate features, such as faults and fractures, that may have provided pathways for mineralizing fluids. However, although this approach allows potential target areas to be selected, it does not guarantee the existence of an ore deposit. Ore deposits, even when present, may not be exposed at the surface. To further narrow the target, therefore, geologists must usually conduct more sophisticated geophysical and geochemical surveys. Finally, all potential targets must be tested by drilling. When mineral exploration on the ground is not viable, remote sensing can play an important role in prospecting. In this section of the chapter, we discuss each of these methods of exploration.

GEOPHYSICAL METHODS

Geophysical methods of mineral exploration depend upon contrasts between the physical properties of ore bodies and those of their host rocks in four main areas: gravity, magnetism, electrical properties, and radioactivity.

Gravity and magnetism

Many ore bodies are made up of minerals of high density and so minutely increase the pull of Earth's gravity on the land surface above them. Although not detectable to human senses, minute changes in the local gravity field can be measured by sensitive instruments known as *gravimeters*. Because dense rocks in the subsurface cause gravity readings measured above them to be higher than average, their presence is revealed by regions in the local gravity field with anomalously high gravity. Such regions are said to have positive gravity anomalies (see In Depth: **Failed Rifts** on page 62).

Buried ore bodies can contain concentrations of weakly magnetic minerals. Just as dense minerals produce positive gravity anomalies, these minerals produce positive magnetic anomalies in the local magnetic field (Fig. 18.28). Although small, these anomalies can be detected by sensitive instruments known as *magnetometers*.

Electrical properties and radioactivity

Some mineral deposits contain metallic minerals, which have electrical properties that differ from those of silicate-rich

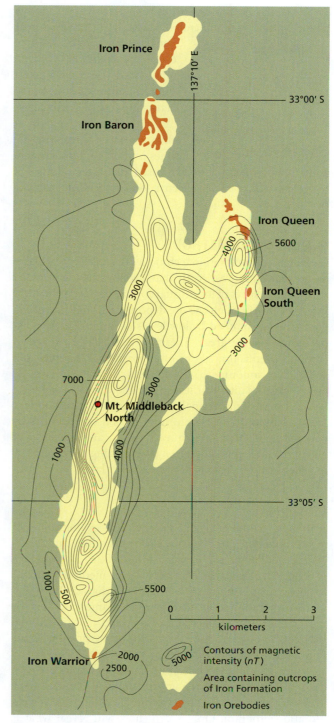

Figure 18.28   Ore Deposit Magnetic Anomaly

Contours of magnetic intensity over the Northern Middleback Range of South Australia, a major source of iron ore, show a striking correspondence with the area in which iron-bearing formations outcrop.

rocks. Prospectors have detected many massive sulfide deposits, for example, by using their conductive properties. When stimulated by an electrical current, such deposits generate weak magnetic fields that ground and airborne instruments can detect.

The search for uranium deposits makes use of yet another physical property of minerals—*radioactivity* (see Chapter 8).

CHECK YOUR UNDERSTANDING

○ Which geophysical properties of ore bodies are used in their exploration?

Radioactivity can be used to detect minerals that contain uranium, thorium, and potassium, the three most common naturally occurring radioactive elements. Because these three elements occur to varying degrees in most rocks, surveys of natural radioactivity can also be used to distinguish different rock types and so can be effective tools for geologic mapping purposes.

GEOCHEMICAL METHODS

Geochemical exploration techniques detect the presence of an ore deposit more directly than geophysical methods, by using some aspect of its unusual composition as a chemical fingerprint. Although the ore deposit itself may be buried, it may be surrounded by low levels of mineralization dispersed over wide areas. These minerals can be detected through chemical analyses of exposed rock samples and soils. Trace amounts of the mineralization may even be taken up by plants. In these cases, the deposit can be revealed by a chemical analysis of nearby plant materials. More commonly, the leaching of buried ore deposits by percolating groundwater affects the local surface water chemistry, and this is detected by analyzing the chemistry of water samples from nearby streams or lakes.

DRILLING

Although geophysical and geochemical exploration can locate potential ore deposits, they are indirect methods. Neither technique can prove whether a buried ore deposit is economically viable. The only direct method available for this purpose is to drill a hole with a diamond drill bit. Of all exploration tools, therefore, the diamond drill is the most important. But because of the high cost of drilling, indirect methods are commonly used before drilling begins in order to identify the best locations to drill.

Unless there are technical difficulties, samples are generally recovered along the entire length of the drill core. These samples provide the conclusive evidence in any exploration program. Only a drill hole will reveal what is actually present beneath the ground and only a series of carefully placed drill holes will allow an ore deposit to be outlined. Until recently, drill holes were essentially straight, but today geologists can use a joystick to change the direction of drilling as the core is examined, greatly increasing the versatility of the technique.

MINERAL EXPLORATION FROM A DISTANCE— REMOTE SENSING

Aerial photographs have long been used for geologic mapping and mineral exploration. Where topographic base maps are unavailable or of poor quality, aerial photographs are an essential element of any exploration program. Photos not only locate surface features but they can also be combined to allow the features to be perceived in three dimensions. With the advent of satellite technology, methods of mineral exploration have become available that do not require the area being examined to be visited at all. This form of "armchair" exploration or investigation from a distance is known as *remote sensing*. The dramatic discoveries we have made in our exploration of the Solar System through satellite images of the planets and their moons is a familiar example of knowledge gained through remote sensing. Similarly, the spectacular photographs of Earth's surface taken from the Space Shuttle are examples of the same technique applied to our own planet.

The application of remote sensing to exploration geology makes use of a variety of images of Earth's surface taken by satellites. Since the first Earth-resource satellite Landsat was launched in 1972, satellites have provided continuous surveillance of Earth's surface at ever greater resolution. The images the satellites produce, however, are not the same as photographs. Cameras take photographs using visible light, that is, they are sensitive to all wavelengths of the visible spectrum. Satellites, on the other hand, use sensors known as *multispectral scanners* to make images from specific wavelengths, some of which may not be visible to the naked eye.

As the satellite orbits Earth, these wavelength bands are scanned along a strip of Earth's surface directly below the scanner. The images received are then transmitted in digital format back to Earth, where the data carried in the various bands are superimposed and the image is reprocessed in the form of a picture. The colors used for the various spectral bands that are scanned by the satellite are often arbitrarily chosen to highlight particular features. The color red, for example, is often used to denote the infrared light reflected from lush vegetation because it provides better contrast than green. Areas of diseased crops or regions of clear-cut in a forest are consequently highlighted. Arbitrary colors must also be chosen for other portions of the spectrum which we cannot see. Satellite pictures, which superimpose each of the arbitrary colors assigned to the bands, are therefore referred to as *false color images* (Fig. 18.29). Using false color images, features that might be of particular significance for mineral exploration, such as differences in vegetation, rock type, or soil variety, can therefore be emphasized.

One limitation of multispectral scanning of reflected light is that the technique's success depends on clear weather. Luckily, some forms of remote sensing are not affected by poor visibility. The spectacular images of the surface of Venus (see Chapter 21), for example, could not have been produced with multispectral scanning of reflected light because of the planet's thick cloud cover. These images were instead produced by another form of remote sensing— radar. *Radar* is microwave radiation that can penetrate clouds. Radar images of Earth's surface, such as those produced by *SLAR (side-looking airborne radar)* and *SAR*

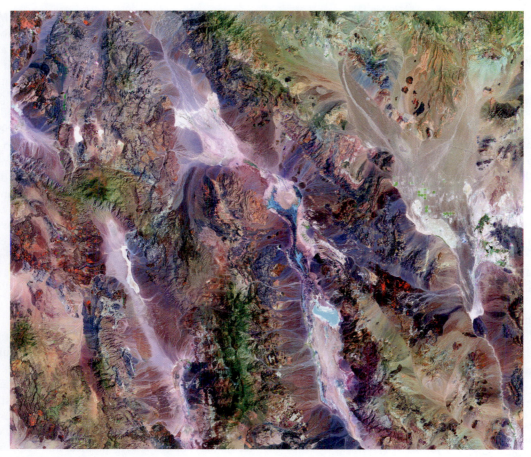

Figure 18.29   False Color Image

False color satellite image of the Death Valley area in eastern California acquired by Landsat 7 in 2000 highlights the region's varied geology. At 86 meters (282 feet) below sea level, Badwater saltpan (center in light blue) is the lowest place in North America.

(synthetic aperture radar), are sensitive to variations in the shape and roughness of Earth's surface. Like aerial photographs, radar images can be combined to create three-dimensional images (Fig. 18.30). Because the shapes of

surface features are often determined by the structure and composition of the underlying bedrock, radar images are also useful for exploration. By highlighting linear alignments of surface features known as *lineaments,* for

Figure 18.30   Radar Image

A shaded relief satellite radar image of part of Nova Scotia, Canada, highlights linear features of the landscape, or lineaments, many of which correspond to major faults. Color variations highlight differences in elevation.

example, radar images can reveal faults and fractures that may be the sites of mineralization. Also useful in locating faults and other geologic structures is *LiDAR (light detection and ranging)*, a remote sensing technique that resolves subtle variations in surface topography using laser light, which is able to probe through surface vegetation.

Remote sensing techniques have consequently provided the economic geologist with a range of new tools for mineral exploration. Unlike more conventional exploration methods on the ground, remote sensing can be applied worldwide. It is not limited by political frontiers and it is often far less expensive than ground-based expeditions and exploration. The resolution of satellite-based remote sensing techniques is continually improving. The multi-spectral scanner on the first Landsat, for example, was unable to resolve objects much smaller than 200 meters across. Today, the resolution of commercial satellite scanners, such as those of GeoEye launched in 2008, is down to less than 0.5 meters (20 inches)—a four hundred-fold improvement, and even higher resolutions are achieved by spy satellites. Airborne LiDAR can resolve differences in topography that are as little as 10 centimeters (4 inches).

> **CHECK YOUR UNDERSTANDING**
>
> ○ List two examples of remote sensing techniques that are used for mineral exploration purposes.

As with ground-based geophysical and geochemical exploration methods, however, remote sensing techniques can only focus the search for ore deposits. They cannot determine whether a commercial ore deposit is present, nor can they delineate the body if one is found. The only methods of exploration that can provide these answers remain basic geologic mapping followed by drilling.

18.6 SUMMARY

- Mineral exploration involves the search for mineral deposits and environments favorable for their formation.
- Mineral exploration involves geologic mapping, geophysical and geochemical surveying, remote sensing, and, ultimately, drilling.

18.7 Future Demands on Mineral Deposits

Mineral deposits are nonrenewable and cannot be replaced. Yet the rate at which humans consume them is only likely to increase as global populations rise and less-developed countries become increasingly industrial. Eventually, the demand for mineral deposits must confront the limited supply. So what are the future patterns of supply and demand likely to be, and what can we do to meet our future needs?

FUTURE PATTERNS OF SUPPLY AND DEMAND

As any stockbroker will tell you, attempting to determine the future availability of, and demand for, a particular mineral commodity is a very risky business. It is virtually impossible to predict new discoveries of a mineral or to predict what effects rapidly changing technology will have on mineral demand. One might have predicted, for example, that the "information age" with its dependence on electronic communications would have led to a spiraling increase in the demand for copper wiring. In reality, however, the opposite has occurred because the demand has been met technologically with the introduction of wireless systems and fiber-optic cables that contain no copper.

Future patterns of supply and demand serve to highlight the distinction between mineral resources and mineral deposits, and will no doubt be influenced by inequities in the global distribution of ore deposits and by future consumption trends.

Resources and reserves

As we discussed in Section 18.1, a resource, whether of minerals or industrial commodities, is a natural concentration of inorganic material used by humans. Whether a resource is exploitable, however, depends on the economics of its extraction. For this reason, geologists distinguish between *resources* and *reserves*. A **reserve** is a deposit of known size and quality that could be worked profitably with today's technology and in today's political and economic climate. Mineral reserves are usually defined on the basis of drilling and subsequent tests of the drill core. Because of factors such as changing technology, the world's reserves of a given mineral commodity can rise and fall significantly without actually changing the amount of the mineral known to be in the ground. A **resource**, in contrast, is a deposit that is, or may become, one of economic interest. All reserves are resources, but not all resources are reserves. Hence, resources go beyond reserves to include *potential* sources of a commodity—either from known but currently uneconomic deposits or from hypothetical ones. So whereas a mineral reserve could be mined today, there is no guarantee that a mineral resource will ever be mined.

> **CHECK YOUR UNDERSTANDING**
>
> ○ What is the distinction between mineral reserves and mineral resources?

Inequities in the global distribution of mineral deposits

Because the global allocation of mineral reserves is unevenly distributed (Fig. 18.31), future demands for a given mineral commodity are likely to vary dramatically from one nation to another. Consider the metal chromium, for example, which is a vital component of stainless steel and automotive chrome plate. Globally, there are adequate reserves of chromium ore to meet world demand for the immediate future. However, this is of little comfort to the United States, which has virtually no chromium reserves. To support a vast manufacturing industry, the country is

Metal Reserves of the World

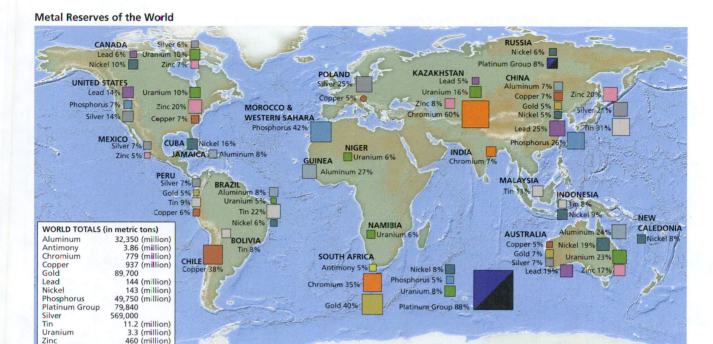

Figure 18.31   Metal Reserves of the World

A simplified map shows the global distribution of rare and precious metal reserves.

wholly dependent on foreign sources for this strategically important metal, the largest reserves of which are in South Africa. Future chromium supply to the United States is consequently of great concern despite the metal's plentiful reserves.

Mineral exploration techniques are now widely believed to be sufficiently sophisticated and global in scope that most of the world's major high-grade mineral deposits on the continents have already been found. This means that future mining activity, even for relatively abundant mineral commodities such as the common metals, will eventually have to focus increasingly on the recovery of lower-grade ores on the continents or high-grade ores on the ocean floors. Both of these activities are more expensive than current mining efforts. Reserves of some of the scarce metals, on the other hand, are sufficient for less than 50 years, at present rates of consumption (Fig. 18.32). Future shortages in these metals are therefore likely, with silver, tin, and lead probably leading the way. These trends predict a steady rise in future metal prices as their ore reserves become increasingly scarce and expensive to extract.

Consumption trends for mineral deposits

The future consumption of mineral deposits by any given country will most likely follow current models for the consumption of any limited resource. The history of metal consumption, for example, may follow the model illustrated in Figure 18.33. For any given country, the number of mines extracting a given metal initially increases as new deposits are found, and the amount of the metal produced increases in unison. As the deposits become depleted,

however, the number of working mines declines, which leads to a decline in production. Continued demand must therefore be met by importing the metal from foreign sources at ever increasing levels.

Within this model, individual countries currently occupy a wide spectrum of positions. Some, such as China and Russia, have huge mineral reserves and expanding patterns of exploitation. Others, like the United States, have passed peak productivity and are becoming increasingly dependent on imported metals. Still others, such as the United Kingdom and Japan, have almost exhausted their mineral deposits and are now heavily dependent on foreign sources.

Just as there is inequity in the global distribution of mineral deposits, so too is there inequity in their consumption. The developed industrial nations use the lion's share, such that some 90 percent of the world's mineral production is consumed by less than 10 percent of its population. These nations, however, have also produced the lion's share of mineral commodities. It was the rich mineral and energy resources of Europe, for example, that fueled the industrial revolution, and even today, it is the developed nations of China, Australia, Brazil, Russia, and South Africa that dominate world mineral production.

But this situation is rapidly changing. As the industrial nations exhaust their own

> **CHECK YOUR UNDERSTANDING**
>
> ● Describe the typical trend for the consumption of mineral deposits illustrated in Figure 18.33.

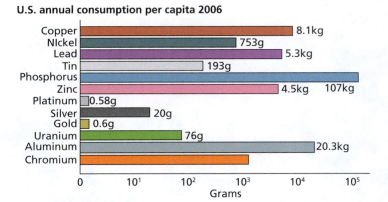

U.S. annual consumption per capita 2006

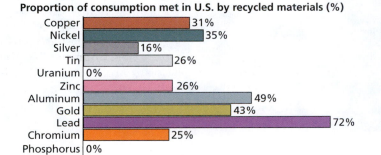

Proportion of consumption met in U.S. by recycled materials (%)

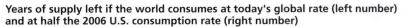

Years of supply left if the world consumes at today's global rate (left number)
and at half the 2006 U.S. consumption rate (right number)

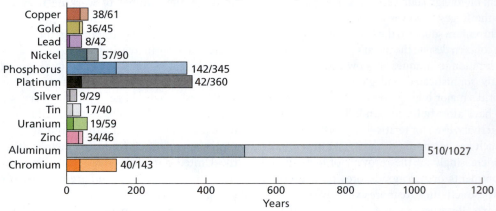

Figure 18.32 Expectations for Current Reserves

This graphic illustrates US consumption of some of the world's key metal resources and the proportion of this consumption met by recycled materials. It also shows the estimated length of time these resources will last if the world continues to consume them at today's rates or if world consumption increases to half the US consumption rate in 2006. For example, the world has 61 years of copper supply left at current consumption rates but only 38 years of supply if world consumption increased to just half that of the United States in 2006.

mineral reserves, they are becoming increasingly dependent upon minerals imported from less-developed countries to supply the raw materials used in manufacturing. This is viewed by some as the exploitation of these countries by the developed nations. Others, however, claim that it is the capital gained from the exportation of their raw materials that is likely to fuel the future development of the less-developed nations.

MEETING FUTURE NEEDS

Because mineral deposits are nonrenewable, the continued demand for mineral commodities must eventually result in shortages. Before this occurs, however, increases in ore prices are likely to convert many mineral deposits that are currently not economically viable into mineral reserves. Meeting the needs of the more immediate future is likely to involve technological developments aimed at reducing the

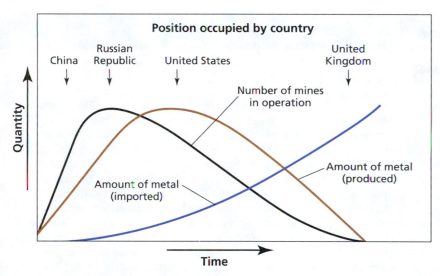

Figure 18.33   Consumption of Limited Resources
Models for the consumption of a limited resource illustrate the inevitable decline in supply over time.

cost of extracting and processing existing deposits, mineral exploration in frontier areas such as the deep sea and polar regions, and conservation and substitution.

Technological developments

For the exploitation of a mineral deposit to be profitable, the value of the product must outweigh the cost of its production. Production costs are those of extraction and processing, and future trends may see these rise or fall. Future technological developments resulting in increased recovery, for example, may reduce production costs by improving the techniques of extraction and processing. It is conceivable that old mine rock waste could be profitably reworked in this fashion. As the richest ore deposits become exhausted, however, the mining of progressively lower-grade and less easily processed ores may increase these costs. In addition, environmental concerns and political situations may prevent the working of an otherwise profitable deposit (see Living on Earth: **The Environmental Impact of Mining** on page 600). Given these uncertainties, it is both difficult and probably unwise to attempt to forecast what the future may hold for the mineral industry.

Mineral exploration

Exploration for new mineral deposits is certain to continue, and one area that is likely to receive considerable attention is the ocean floor. As we have discussed, rich deposits of minerals associated with submarine vents at mid-ocean ridges exist on the seabed and some of these, like those of the Red Sea and Bismarck Sea, are relatively accessible. However, these are not the only deposits known to occur on the

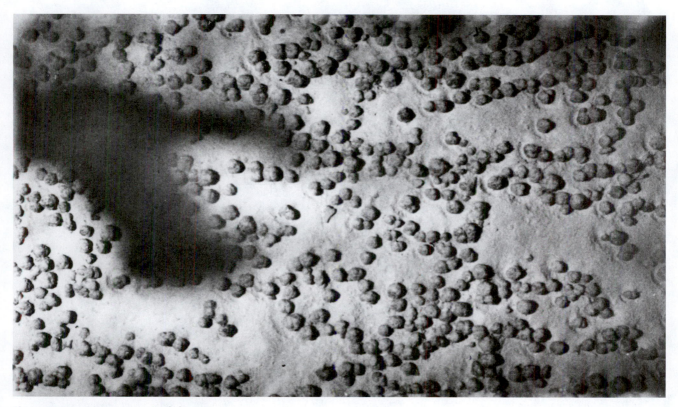

Figure 18.34   Manganese Nodules
Manganese nodules measuring 5 to 10 centimeters (2 to 4 inches) across are found on the deep ocean floor of the Pacific Ocean.

LIVING ON EARTH

The Environmental Impact of Mining

Exploitation of the world's mineral deposits has defined human development and lies at the very heart of modern industrial society. But the mining of our mineral deposits has serious environmental consequences. The quest for mineral riches has damaged vast tracts of otherwise unspoiled land, and today threatens national parks and other ecologically sensitive areas. The waste produced by mining far exceeds that from all other sources, and even outstrips the natural erosion of the world's streams. The extraction and processing of ore consume huge amounts of energy and so are major partners in the problem of climate change caused by the burning of fossil fuels. The refining of ore releases huge quantities of toxic gases into the atmosphere, and is a major contributor to acid rain. As a result, the environmental impact of our exploitation of mineral deposits can be severe.

The amount of rock waste produced by mining activities is truly staggering, particularly when low-grade ores are

recovered by surface or "open pit" methods. The Bingham Canyon copper mine in Utah, for example, has removed an entire mountain and excavated a pit over 3 kilometers (2 miles) across and 1 kilometer (3300 feet) deep (Fig. 18B). This mine alone has removed 3.3 billion metric tons of material, an amount seven times greater than was excavated in the building of the Panama Canal. At the height of its production, the mining operation removed 400,000 metric tons of material every day, three-quarters of which was rock waste. At only 0.3 percent copper, the remaining 100,000 metric tons of ore that the mine processed daily produced a further 97,000 metric tons of waste. Worldwide, close to 30 billion metric tons of material a year, or almost twice the estimated amount of sediment carried by the world's streams, is moved by mining activities.

Moving such vast quantities of material and processing the extracted ore requires huge amounts of energy. Aluminum production and steelmaking, for example, are particularly

Figure 18B   Bingham Canyon Mine
The Bingham Canyon copper mine in Utah is one of the largest in the world, over 3 billion metric tons of rock having been excavated. At peak production 400,000 metric tons of material were moved daily to recover just 3000 metric tons of copper.

energy-intensive processes. Worldwide, the excavation and refining of ore uses up to one-tenth of all the energy produced each year—enough to supply 80 percent of the world's annual electricity use! Because most of this energy is supplied by the burning of fossil fuels, mining is a major contributor to the environmental problems caused by energy use, such as air pollution, acid rain, and the climate changes linked to increased atmospheric carbon dioxide levels.

Serious contamination problems are also associated with leaching of the waste material produced by mining and with the refining process known as *smelting*, which extracts metal from its ore. Although most rock waste is chemically inert, the finely ground tailings left behind after ore has been concentrated can be highly reactive, producing acid mine drainage and contaminating surface waters with toxic metals. Up to 16,000 kilometers (10,000 miles) of streams are estimated to have been affected in this way in the United States alone. When mining efforts necessitate deforestation of vast areas, silting of lakes and streams can occur as a result of the consequent increased erosion.

The smelting of ore releases huge quantities of pollutants into the atmosphere. The extraction of metals from metal sulfide ores, for example, is estimated to be responsible for the annual emission of 6 million metric tons of sulfur dioxide, the primary cause of acid rain. Smelting also emits arsenic, lead, and other toxic chemicals, and when unchecked, can produce environmental disaster areas in which little or no vegetation will grow. A disaster area around the old nickel smelters in Sudbury, Ontario,

for example, once occupied over 100 square kilometers (almost 40 square miles).

The damage caused by the exploitation of mineral deposits is significant and poses one of the major dilemmas facing modern society. Mining is an inherently destructive industry that does not readily lend itself to environmental mitigation by pollution control. Put another way, mining is a very dirty business! Yet mining is an economic activity upon which civilization depends. The very prosperity of industrial nations like the United States and Canada, and the high standard of living that their inhabitants enjoy, depend on the use of mineral products.

If the devastating environmental impact of mineral extraction is to be lessened, it cannot be done simply by demanding cleaner mines. Instead, more mineral-efficient economies are needed that emphasize the recycling, repairing, and remanufacturing of mineral products. Also needed are substitutes for mineral products that can be produced with less damage to the environment. A particularly successful example of such a substitute is the replacement of copper wire by fiber optics made from glass.

Environmental issues associated with resource exploitation have led many citizens and groups to call for fundamental changes to our consumer-driven society. For example, although residents in the United States now recycle 65 percent of the aluminum they use, they still throw away 16 million metric tons of metal each year. Only by learning to reuse and recycle metals more efficiently while lessening our dependence on mineral products can we significantly reduce the environmental impact of their production.

---

seabed. On the Pacific Ocean floor, for example, huge deposits of manganese exist in the form of concretions known as *manganese nodules* (Fig. 18.34). Averaging 5 to 10 centimeters (2 to 4 inches) across, these nodules form as chemical precipitates in areas of slow deposition, and contain iron, copper, nickel, and cobalt in addition to manganese.

Estimates of the size of this resource, at least in terms of the metals manganese and copper, suggest it may be larger than those on land, and the technological difficulty of recovering the nodules by dredging or vacuuming the seabed is not insurmountable. However, the environmental impact of such an operation is entirely unknown and, because many such areas occur in international waters, there is presently no protocol for establishing national ownership and the right to mine, or for determining the responsibility for monitoring such mining activity. Some fundamental and politically difficult legal questions concerning mineral rights and policing responsibilities in international waters will need to be addressed and agreed upon before any such mining operations can be undertaken.

There are also significant tracts of land that for political, geographic, or technological reasons have seen little in the way of mineral exploration and development. Chief among these are the polar regions. Technological developments have now made mining possible in the high Arctic. Canada, for example, had an operating lead-zinc mine at Polaris on Little Cornwallis Island (75°N) until 2002, and has an intermittently operating gold mine at Lupin in Nunavut Territory (66°N). However, operating conditions are extreme, shipment of ore is difficult, and questions regarding the environmental impact of such mining in an ecologically sensitive area are still fiercely debated.

At the other pole, the Antarctic Treaty, which sets aside Antarctica for scientific research, has so far shielded the continent from mineral exploration and development. This protection, however, is not guaranteed indefinitely, and as future demands for mineral commodities rise, pressures to relax the conditions of the treaty are certain to increase.

This raises another issue in mineral exploitation that is likely to become an increasingly controversial topic in the future. As we have learned, mineral deposits occupy a minute fraction of Earth's surface, and occur only where the vagaries of mineral-forming processes have placed them. Consequently, if society continues to depend on mineral deposits for its well-being, future generations will be forced to carefully weigh the benefits that might be gained from mineral exploitation against those that are realized from setting aside the same land for uses (such as parks) that exclude mining activities.

In the long term, there is sure to be a debate about the necessity to extract minerals from ocean water, or even to expand the search for mineral deposits beyond Earth. The Moon, for example, is believed to be rich in iron and nickel, and the development of its mineral deposits might be possible if, in the future, a permanent lunar base is established and maintained.

Conservation and substitution

In the immediate future, the continuing supply of mineral commodities is likely to be maintained by an increasing effort to conserve, reuse, and recycle existing materials and, where possible, to develop cheaper and more abundant substitutes for scarce minerals. The **recycling** of resources is an increasingly important part of our daily lives. It is already worth our while to recycle aluminum because of the high energy costs associated with its production. In the future, the recycling of other metals is likely to become as familiar to us as that of aluminum. We can also expect to see the scarcer metals increasingly replaced by other less costly substitutes in the same way that fiber-optic cables are today replacing the use of copper in electrical wiring.

18.7 SUMMARY

- Mineral reserves are known mineral deposits that can be profitably exploited in today's marketplace, whereas a mineral resource is a potential source of a mineral.

- Mineral deposits are scarce and their global distribution is uneven.

- Meeting future needs for increasingly limited mineral deposits is likely to be achieved through improved extraction technology, further mineral exploration, the more widespread use of less expensive substitutes, and improved efforts at conservation and recycling.

Key Terms

disseminated deposit 586	metal 569	ore deposit 582	residual deposit 587
hydrothermal deposit 586	metallic mineral 569	placer deposit 587	resource 596
industrial material 568	mineral deposit 568	porphyry copper	stratiform deposit 589
magmatic	nonmetallic mineral 575	deposit 591	supergene
segregation 585	nonrenewable	recycling 602	enrichment 591
mechanical	resource 569	replacement 588	vein 586
enrichment 587	ore 569	reserve 596	

Key Concepts

18.1 CONSUMPTION OF MINERALS AND INDUSTRIAL MATERIALS

- Mineral deposits are nonrenewable sources of metallic and nonmetallic minerals, as well as industrial materials.

18.2 METALS AND METALLIC MINERALS

- Metals are considered abundant or scarce, the latter including the precious metals.

18.3 NONMETALLIC MINERALS AND INDUSTRIAL MATERIALS

- Nonmetallic minerals and industrial materials are used as raw chemicals, abrasives and gemstones, and building materials.

18.4 FORMATION OF MINERAL DEPOSITS

- Most resource minerals must be massively concentrated to form ore deposits that can be exploited in a way that makes economic sense.

- Mineral concentration is usually the result of chemical and physical processes but sometimes biological processes play a role as well.

- Magmatic and hydrothermal deposits occur in igneous rocks; evaporites, residual deposits, placer deposits, and exhalites occur in sedimentary rocks; and replacement deposits occur in metamorphic rocks.

18.5 PLATE TECTONICS AND MINERAL DEPOSITS

- Mineral deposits at divergent plate boundaries are associated with continental rift and mid-ocean ridge environments.

- Mineral deposits at convergent plate boundaries are associated with magmatic arc and back-arc rift environments.

18.6 MINERAL EXPLORATION

- Mineral exploration involves mapping, surveying, remote sensing, and drilling

18.7 FUTURE DEMANDS ON MINERAL DEPOSITS

- Mineral deposits are scarce and their global distribution is uneven.

- Future needs for mineral deposits are likely to involve improved extraction methods, the use of substitutes, and improved conservation and recycling, in addition to mineral exploration.

Study Questions

1. List as many examples as possible (other than those used to introduce Section 18.2), to illustrate the role mineral deposits play in your daily life.

2. Given that aluminum is the most abundant metal and the third most abundant element in Earth's crust, why do we make such an effort to recycle aluminum cans?

3. Describe how magmatic processes can result in hydro-thermal deposits.

4. Explain how processes at mid-ocean ridge spreading centers generate the submarine vents we call smokers, and describe the mineralization with which they are associated.

5. How do sedimentary processes result in the concentration of minerals?

6. What are false color images and how are they used for mineral exploration purposes?

7. Why are metal reserves not equally distributed world-wide and how has this influenced the distribution of mineral wealth?

8. Illustrate how models for the consumption of a limited resource can be used to predict future trends in the mining industry.

9. Why has there been relatively little mining in the polar regions? Is this predicted to change? If so, why?

10. What is the difference between substitution and conservation of mineral commodities and how will this ensure a maintained supply of metals in the immediate future?

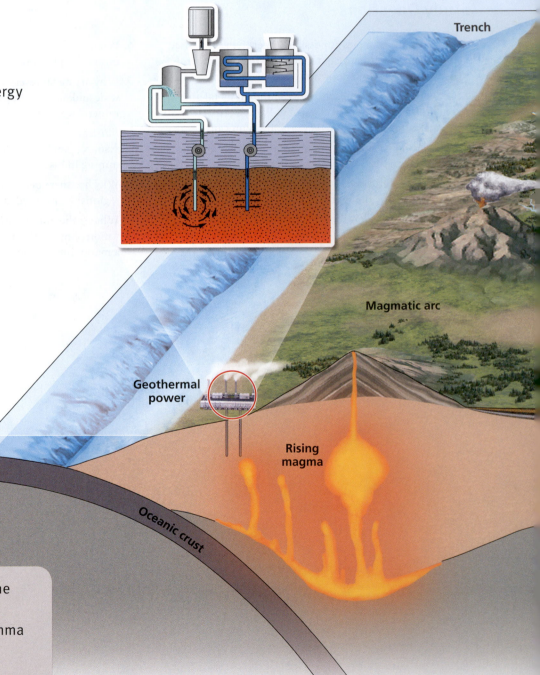

19 Energy Resources

19.1 Energy Use

19.2 Petroleum—Oil and Natural Gas

19.3 Coal

19.4 Nuclear Power

19.5 Renewable Energy

19.6 Plate Tectonics and Energy Resources

▶ Key Terms

▶ Key Concepts

▶ Study Questions

IN DEPTH: Correlation and the Search for Oil and Gas 614

LIVING ON EARTH: The Dilemma of Fracking 618

IN DEPTH: Is Ice That Burns the Fuel of the Future? 621

Trench

Magmatic arc

Geothermal power

Rising magma

Oceanic crust

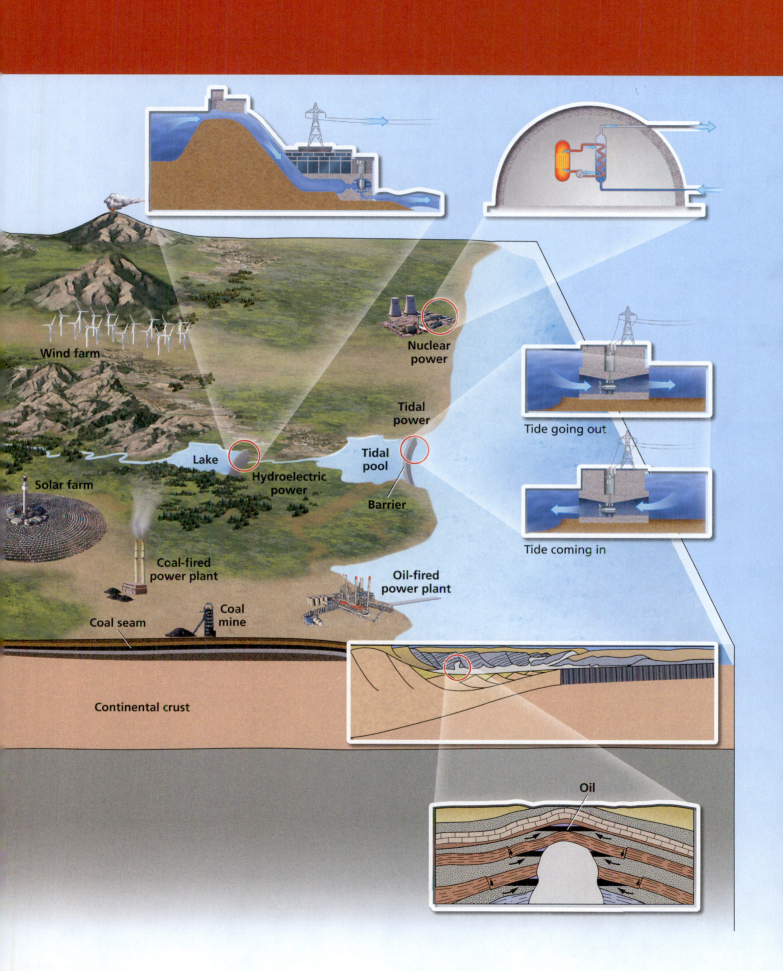

Wind farm

Nuclear power

Tidal power

Tidal pool

Tide going out

Lake

Hydroelectric power

Barrier

Tide coming in

Solar farm

Coal-fired power plant

Oil-fired power plant

Coal seam

Coal mine

Continental crust

Oil

Learning Objectives

19.1 Show that most of the energy that supports industrial economies comes from the nonrenewable fossil fuels—oil, natural gas, and coal.

19.2 Describe the origin of petroleum, the formation and distribution of petroleum reservoirs, and the means by which they are exploited.

19.3 Describe the formation of coal and the distribution of coal deposits.

19.4 Explain how nuclear power is generated and the issues of safety and waste disposal associated with this power source.

19.5 Discuss the principal sources of renewable energy—geothermal, water, wind, solar, and biomass—and explain why, as yet, none of these energy sources is an adequate replacement for fossil fuels.

19.6 Show how the distribution of our energy sources—both nonrenewable and renewable—is linked to the processes of plate tectonics.

Energy is the lifeblood of modern society, which consumes it at a staggering rate. Countries with industrial economies cannot function without huge supplies of energy, and the relatively cheap cost of energy is responsible for the high standard of living enjoyed by industrial nations. Energy powers the manufacturing, transportation, communication, and construction sectors of the world's economy. It fuels our vehicles, and with this our freedom to move. It also lights our buildings at night and controls the temperature inside them. Indeed, without energy, many of us would literally freeze in the dark.

Although there are many sources of energy, human society meets the vast majority of its energy needs by burning oil, natural gas, and coal. These energy sources are *fossil fuels* and, like the minerals deposits we examined in Chapter 18, they are *nonrenewable* resources because society uses them far faster than nature can replace them. These resources form in unusual geologic conditions, so an understanding of the natural processes that lead to their production is essential to the business of fossil fuel exploration. Unlike inorganic mineral resources, however, fossil fuels are organic in origin, and represent energy stored for millions of years in the fossilized tissue of plants and microorganisms.

Although we are completely dependent on fossil fuels, the total supply of this energy resource is finite. The future of these fuels is therefore likely to be one of diminishing production, such that alternative energy sources will have to be developed to take their place. Many of these alternative energy sources are already in use. Nuclear power, driven by the fission of naturally occurring radioactive uranium, is a proven source of energy. But like the fossil fuels, it is nonrenewable, being dependent in this case on a mineral resource. Geothermal and solar energy, hydroelectric power, wind and tidal power, and energy obtained from biomass have all been developed as alternative energy sources. Unlike fossil fuels and nuclear power, these are *renewable* energy sources because they do not depend on an exhaustible natural resource. However, only in certain geographic areas can power be produced from such sources, and, to date, the scale of these operations has been small. For the immediate future, therefore, we are likely to see little reduction in our dependency on fossil fuels.

19.1 Energy Use

Almost all of our present-day energy needs are met by the consumption of **nonrenewable resources**, so-called because we consume these resources faster than nature replenishes them as we discussed in Chapter 18. Each of these commodities is replaced naturally only on a geologic time scale and most cannot be recycled. **Renewable resources**,

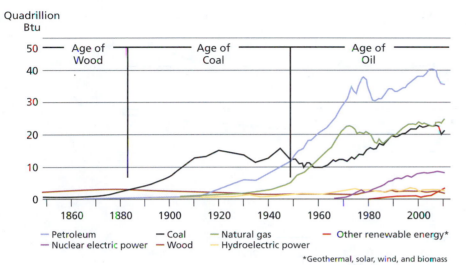

Figure 19.1   US Energy Consumption

A graph of energy consumption in the United States between 1850 and 2011 shows the change in principal energy source from wood to coal to oil. A BTU is the quantity of heat required to raise the temperature of one pound of water by 1oF. One million BTU equal approximately 1000 cubic feet of natural gas.

on the other hand, are those that can be used without exhausting the energy source. For much of human history, wood (a renewable resource) was the dominant fuel, and power was measured by the strength of individuals or the strength of their beasts of burden. Following the Industrial Revolution, the use of coal as a commercial fuel increased dramatically, and by the late nineteenth century, this nonrenewable resource had replaced wood as the dominant source of energy. By the middle of the twentieth century, another nonrenewable resource, oil, had surpassed coal as the principal energy source. Oil has remained the dominant source of energy to this day (Fig. 19.1).

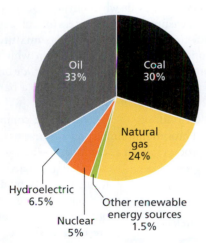

Figure 19.2   World Energy Consumption by Source

The world consumes energy from a variety of resources, the most important of which are the fossil fuels, oil, natural gas, and coal. Data from BP Statistical Review of World Energy 2012.

Just over 87 percent of the world's energy needs (Fig. 19.2) are currently met by the three **fossil fuels**—*oil, natural gas,* and *coal*—so-called because of their fossilized organic origin. A further 5 percent is met by nuclear fuel, a nonrenewable resource obtained from uranium ore. Only 8 percent of the world's energy is provided by renewable resources, the most important of which is **hydroelectric power**, which is the generation of electricity from running water.

A disproportionate share of this energy supply is consumed by the more industrialized nations (Fig. 19.3). Less industrialized countries still depend more on human and animal power, and wood as sources of energy. In contrast, the more industrialized nations expend huge quantities of generated energy powering their industries, transporting people and materials, and controlling the temperature of their buildings.

The United States, for example, consumes almost 20 percent of the world's total energy, yet accounts for less than 5 percent of the world's population. The less industrialized nations, on the other hand, use just over half of the global energy supply, yet make up about 88 percent of the world's population. But as these countries continue to advance technologically, their impact on energy demand will be enormous. The total consumption of energy by less industrialized countries exceeded that of the industrialized countries in 2007, and is projected to double in the next 20 years, largely as a result of rising living standards in China, India, and the Middle East (Fig. 19.3).

Perhaps more important than the simple fact that fossil fuels are destined to eventually run out is the impact of their use on the environment. Over the past few decades, it has become increasingly clear that our dependence on fossil fuels as an energy source is having a serious and detrimental effect on the climate of the entire planet. This dilemma, which has been dubbed by the media as *climate change* and *global warming,* stems from the fact that the burning of all fossil fuels releases carbon dioxide into the atmosphere. Carbon dioxide is known as a *greenhouse gas* because its presence allows the atmosphere to trap heat, much like the panes of glass in a greenhouse. The human use of fossil fuels has been accompanied by a sharp rise in atmospheric carbon dioxide

CHECK YOUR UNDERSTANDING

◗ What is the difference between renewable and nonrenewable energy sources?

◗ What are the fossil fuels and why are they so called?

◗ What percentage of global energy needs is met by each of the major energy sources?

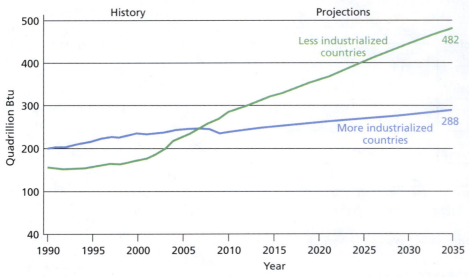

Figure 19.3 World Energy Consumption per Capita 1990–2035

Energy consumption by the less industrialized nations now exceeds that of the more industrialized nations. Per capita, however, developed nations consume a disproportionate share of the total energy supply. Global energy demand is predicted to rise by over 50 percent in the period 2008–2035, underpinned by rising living standards in China, India, and the Middle East.

levels and a lockstep rise in average global temperature. For this reason, the fossil fuels are not considered "*clean*" energy sources, unlike energy sources such as nuclear power, which release no carbon dioxide to the atmosphere and so do not contribute to climate change. The dilemma of climate change, which poses a potentially serious threat to modern society, is a topic we will examine in detail in Chapter 20.

19.1 SUMMARY

- Energy is consumed in huge quantities to power industry, control the environment, and transport the people and commodities of modern industrial society.

- Over 90 percent of the world's energy, a disproportionate share of which is consumed by the industrialized nations, is obtained from nonrenewable resources.

- Fossil fuels are the source of most of the world's energy, but their consumption releases carbon dioxide to the atmosphere and is responsible for global warming and climate change.

19.2 Petroleum—Oil and Natural Gas

The three fossil fuels—*oil, natural gas,* and *coal*—owe their origins to buried organic matter, as we noted in Section 19.1. In effect, the burning of fossil fuels reverses the process of *photosynthesis* by which plants grow. When fossil fuels are burned, the carbon dioxide extracted from the atmosphere by plant growth hundreds of millions of years ago is released back to the atmosphere. The type of fossil fuel that forms depends on the nature of the organic material

buried, the environment of deposition, and the changes it undergoes as a result of burial.

Both oil and natural gas are forms of **petroleum** (from the Greek for "rock oil"), which is derived largely from the remains of microscopic organisms. The name petroleum applies to all naturally occurring organic substances that consist mainly of *hydrocarbons* (compounds of hydrogen and carbon). Hydrocarbons occur in a wide variety of combinations that can be solid (bitumen or asphalt), semi-solid (paraffin wax), liquid (gasoline, kerosene) or gaseous (methane, propane). **Natural gas** is gaseous petroleum consisting predominantly of methane, and is used primarily for heating, cooking, and generating electricity. It is the "cleanest" of the fossil fuels because it releases the least amount of carbon dioxide when burned. Quantities of natural gas are measured in cubic meters (or cubic feet). **Oil** is liquid petroleum, comprising a complex mixture of liquid hydrocarbons. In the untreated form in which it is extracted from the ground, it is known as *crude oil*. Quantities of oil are measured in *barrels* (bbl), a barrel being equal to 42 gallons (158 liters).

During oil refining, the heavier, more complex molecules are first broken into simpler ones by means of heat and pressure, a process known as *cracking.* The lighter molecules are then separated according to their density by distillation, and removed (Fig. 19.4). Many of the products of oil refining are used as sources of energy for transportation (such as diesel oil, gasoline, and aviation fuel) and heating (such as heating oil and fuel oil). Others are used as lubricants (such as motor oil and grease), and yet others form vital raw materials for the chemical industry, where they are used in the

> **CHECK YOUR UNDERSTANDING**
>
> ● What is petroleum?

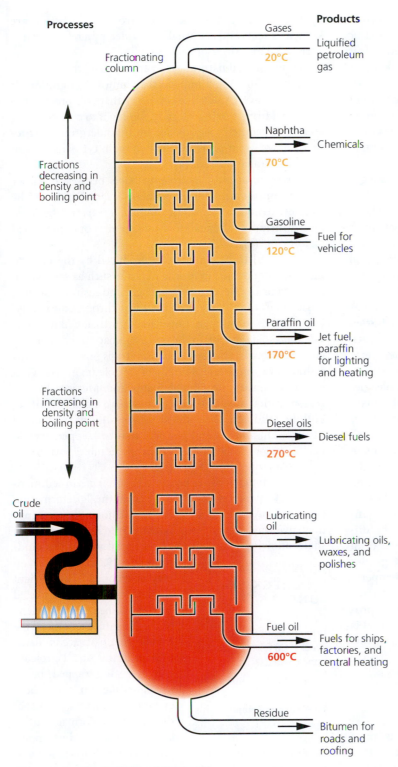

Processes

Fractionating column

Fractions decreasing in density and boiling point

Fractions increasing in density and boiling point

Crude oil

Products

Gases — Liquified petroleum gas — 20°C

Naphtha — Chemicals — 70°C

Gasoline — Fuel for vehicles — 120°C

Paraffin oil — Jet fuel, paraffin for lighting and heating — 170°C

Diesel oils — Diesel fuels — 270°C

Lubricating oil — Lubricating oils, waxes, and polishes

Fuel oil — Fuels for ships, factories, and central heating — 600°C

Residue — Bitumen for roads and roofing

Figure 19.4   The Refining of Crude Oil

During the refining of crude oil, rising temperatures break, or crack, complex hydrocarbons into simpler ones, which are then separated according to their density by distillation to produce a range of products from heavy bitumen to light gases. A similar process converts organic matter to oil and natural gas at depth in Earth.

production of such commodities as paints, fertilizers, insecticides, synthetic fibers, plastics, soaps, and pharmaceuticals.

PETROLEUM FORMATION

Petroleum occurs in rocks of all ages, but most petroleum reserves are geologically quite young, having formed during the Mesozoic Era following the rise in global temperatures and the proliferation of marine phytoplankton (microscopic plants from which most oil is derived) that accompanied the breakup of the supercontinent Pangea. The vast oil fields of the Middle East, for example, owe their origin to the warm waters of the Tethys Ocean, which separated Eurasia from Gondwana throughout the Mesozoic and closed in the early Tertiary to form the vast Alpine-Himalayan mountain belt (see Fig. 10.34).

For petroleum to form and accumulate in sufficient quantities to be commercially exploitable, four conditions must be met:

- A *source rock* that contains organic matter capable of producing oil or natural gas must be available and must be buried to the depth needed to convert its organic matter to petroleum.

- A *reservoir rock* is needed in which the petroleum, once formed, can be stored.

- An impervious *seal* is necessary to provide a barrier to the petroleum's upward movement.

- Some form of geologic structure or *oil trap* is required to pond the petroleum underground.

Even when all these conditions are met, however, the petroleum must still be found. Because oil and gas are difficult to detect underground, most exploration techniques focus on identifying potential traps in sedimentary basins (see Chapter 6) that are likely to have generated petroleum.

Petroleum is derived principally from the remains of microscopic plankton and bacteria incorporated in clays and buried to form organic-rich shale. As we noted at the beginning of this section, during their life cycle these organisms were *photosynthetic* and they converted solar energy into chemical energy. A simplified version of the photosynthesis reaction is:

$$6CO_2\,(gas) + 12\,H_2O\,(liquid) + solar\ energy \rightarrow$$

carbon dioxide   water   light

$$C_6H_{12}O_6\,(aqueous) + 6O_2\,(gas) + 6H_2O\,(liquid)$$

glucose   oxygen   water

Under normal circumstances, oxygen present in the environment ensures that when an organism dies its organic matter decays long before this matter can be buried. If this happens the energy stored in the organisms is lost because the decay reverses the process of photosynthesis and oxidizes the organic matter to water and carbon dioxide. For oil and natural gas to form, burial must take place in an oxygen-deficient or *anaerobic* environment. Such environments develop where the burial of organic matter is rapid or occurs under stagnant conditions. In stagnant conditions, there is little water circulation to replace the oxygen used in the decay process. Rapid burial is common in deltas, and restricted water circulation sometimes develops in lakes and marine environments. If these conditions are met, much of the organic matter (and the energy it retains) is preserved, and the slow transformation from rotting organisms to oil and natural gas can begin. Sedimentary rocks in which this transformation has occurred are **source rocks**.

The conversion of organic matter to petroleum is a complex process. As the source rock is progressively buried deeper and deeper beneath a cover of younger sediments, it is subjected to ever increasing pressures and temperatures. When this happens, a series of chemical reactions takes place that breaks down the heavy organic molecules into simpler hydrocarbons including *methane,* the main component of natural gas. The nature of this process is similar to the cracking procedure used to refine crude oil (see Fig. 19.4) and depends on the existing conditions of pressure and temperature, and the amount of time. The process also involves varying degrees of bacterial action. As a result, the composition of the resulting crude oil varies from one source rock to another, accounting for the differences found in the oils from different deposits. At greater depths and higher temperatures, crude oil breaks down and only natural gas is produced. The formation of oil is therefore restricted to a narrow range of temperature and pressure, which is appropriately termed the *oil window.*

> ### CHECK YOUR UNDERSTANDING
>
> ○ What four conditions must be met for petroleum to accumulate in commercial quantities?

PETROLEUM MIGRATION

Because oil and natural gas are low-density fluids, once they form, they tend to migrate upward toward the surface through the interconnecting cracks and spaces that exist in most rocks. Indeed, permeable sedimentary rocks such as sandstone and limestone often contain sufficient interconnecting spaces, or pores, that they hold oil and gas in much the same way a sponge holds water. Rocks that store petroleum in this way are called **reservoir rocks**.

If the oil and natural gas reach the surface, they disperse and their economic potential is lost. In order for a reservoir to form, oil and gas must be prevented from dispersing by an impenetrable rock layer that stops their upward movement. Barriers to petroleum migration are known as **seals** and are usually provided by a relatively impermeable mudstone or shale.

But for oil and natural gas to accumulate, the seal must do more than simply impede their upward migration. It must also have a vault-like geometry that will prevent the fossil fuel from moving laterally, forcing it to collect in one area. This geometry is provided by an underground structure known as an **oil trap**. The two main types of oil trap are structural and stratigraphic.

In *structural traps,* petroleum pools because of either bending or fracturing of sedimentary layers (Fig. 19.5). The simplest structural trap is the *anticlinal trap,* in which the layers (including the seal) fold into an arch or dome called an *anticline* (Fig. 19.5a).

Structural traps can also be produced by the upward movement of buried evaporite deposits such as salt, which is impermeable. Because they are light and easily mobilized, buried beds of salt may push their way through the overlying strata to form huge, finger-like protrusions known as *salt domes* (Fig. 19.5b). In the process, adjacent sedimentary layers are forced upward. Structural traps produced in this fashion are important sources of petroleum in the Gulf Coast region of the United States and Mexico. In this region, more than 500 salt domes have been identified, both on land and below the shallow waters of the Gulf of Mexico. Oil and natural gas may also be trapped against *faults,* which sometimes form impenetrable barriers to petroleum migration (Fig. 19.5c).

Stratigraphic traps are produced by changes in sedimentary rock type rather than by structural deformation (Fig. 19.6). A permeable layer of sandstone, for example, may give way, or *pinch out,* laterally into an impermeable shale, allowing petroleum to be trapped within the sandstone (Fig 19.6a). Similarly, oil may be trapped in an ancient coral reef that is later blanketed with impermeable mud. Many of the huge oil deposits of the Middle East have accumulated in traps of this sort. Petroleum may also be trapped below *unconformities* in tilted beds of permeable rock (Fig 19.6b) that are unconformably overlain and sealed against impermeable mudstones or shales (see In Depth: **Correlation and the Search for Oil and Gas** on page 614).

> ### CHECK YOUR UNDERSTANDING
>
> ○ What types of rock best function as (a) source rocks, (b) reservoir rocks, and (c) seals?
>
> ○ What are the three main types of (a) structural traps and (b) stratigraphic traps?

FINDING AND EXTRACTING PETROLEUM

Once petroleum has accumulated in a trap, it must still be found if it is to be exploited as an energy resource. Direct detection of oil and natural gas in the subsurface is very difficult. So most techniques are designed to detect potential petroleum traps instead. Even when such traps are found, however, most will not contain petroleum.

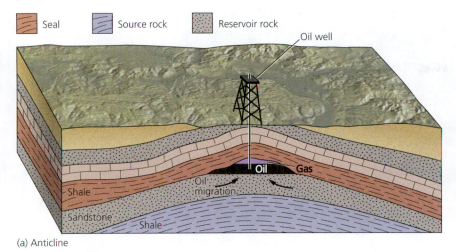

Seal   Source rock   Reservoir rock

(a) Anticline

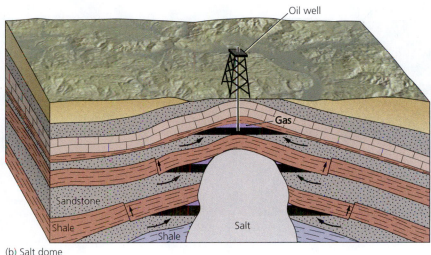

(b) Salt dome

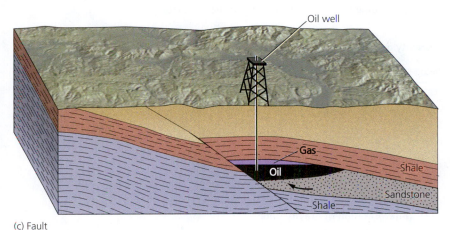

(c) Fault

Figure 19.5   Structural Petroleum Traps

Oil can be trapped by geologic structures, including (a) anticlines, (b) salt domes, and (c) faults. Natural gas, if present, would be located on top of the oil.

In addition to detailed geologic mapping, petroleum geologists use a wide variety of geophysical methods to search for potential petroleum traps in the subsurface. The most important of these exploration techniques employ gravity surveys, seismic reflection, and, ultimately, drilling.

Gravity surveys

Gravity surveys, which measure subtle variations in Earth's gravity due to density variations in the subsurface, often reveal the presence of buried structures. In sedimentary basins with the potential for petroleum production, these structures might act as oil traps. Buried salt domes (see Fig. 19.5b), for example, are readily identified by this method because salt is less dense than the rocks around the dome (Fig. 19.7). As a result, gravity measurements above buried salt bodies are anomalously low.

Seismic reflection

The most widely used geophysical search method is *seismic reflection.* This technique uses seismic waves to image the subsurface in much the same way that echo sounding is used to determine the depth of a body of water. Using small explosions or vibrating equipment, scientists shake the ground surface, generating shock waves that travel into the ground. These waves are reflected by rock layers underground so that part of the energy of the waves returns to the surface, where its arrival is monitored. Computer analysis of the time lapse between the explosion and the arrival of this reflected energy reconstructs a detailed picture of the underground structure of the rock layers, allowing potential exploration targets to be identified (Fig. 19.8).

CHECK YOUR UNDERSTANDING

◗ What is the most widely used geophysical method of petroleum exploration and how does it work?

Drilling

Ultimately, the presence of petroleum within a potential trap must be confirmed by drilling. If petroleum is discovered, further test wells are drilled to determine the size of the reservoir. If petroleum is found in sufficient quantity, the field may go into production. Any oil trapped in a reservoir rock at depth is usually under very high pressure. This is because buoyancy causes the water beneath the oil to push upward, while the pressure of the natural gas pushes downward from above (Fig. 19.9). In fact, the pressure is sometimes so high that when the seal is penetrated by a drill hole, oil, gas, water, and even the drill pipe itself may shoot into the air as a *gusher.*

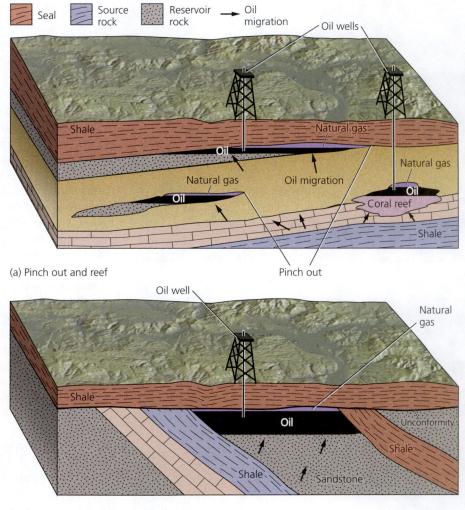

(a) Pinch out and reef

(b) Angular unconformity

Figure 19.6   Stratigraphic Petroleum Traps

Oil can be trapped by various stratigraphic features, including (a) pinch outs and reefs, and (b) unconformities.

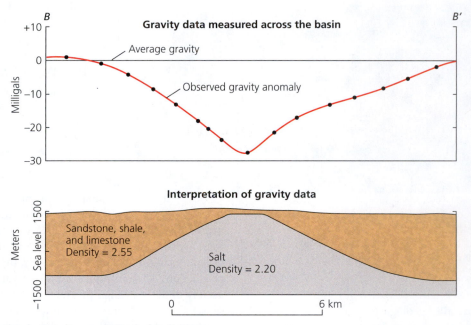

Figure 19.7   Variation in Gravity over a Buried Salt Dome

The force of gravity, expressed in milligals, shows a slight decrease (or negative anomaly) over a salt dome because the density of salt (2.2) is slightly less than the density of the surrounding sedimentary rocks (2.55). The variation in gravity is shown in the graph, and the salt dome is shown in the cross-section figure.

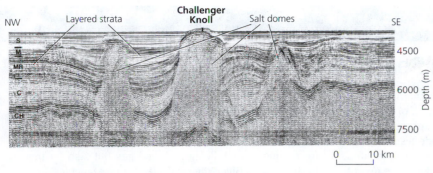

Figure 19.8   Seismic Image of a Salt Dome

This seismic profile reveals a series of salt domes centered on Challenger Knoll in the central Gulf of Mexico. Note the similarity with Fig. 19.5b. (Courtesy J. Lamar Worzel and C. A. Burk: *Geological and Geophysical Investigations of Continental Margins*. AAPG Memoir 29 © 1979. Reprinted by permission of the AAPG, whose permission is required for further use.)

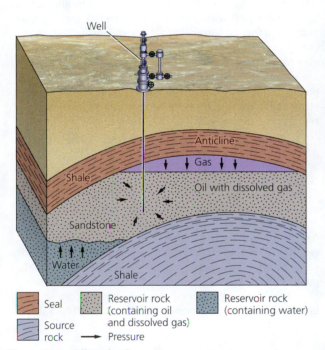

Figure 19.9   Oil Under Pressure

Oil in an anticline is driven upward by buoyant water pressure from below and meets with a downward pressure from the overlying natural gas.

In most wells, however, the pressure is not sufficient to bring the oil to the surface. Instead, the wells require pumping. Yet even with pumping and other recovery techniques, typically as much as 60 percent of the total oil is not recovered from the reservoir. The same problem exists for the recovery of natural gas. Thus, even the most successful exploration strategies leave much of the petroleum in the ground.

OIL SHALES, TAR SANDS, AND SHALE GAS

Several sources of petroleum exist that cannot be produced from a traditional oil or gas well because the petroleum does not flow naturally through the rock in which it is held

and so must be extracted by other methods. The oil and gas produced from such sources are collectively termed *unconventional* and they are usually considered separately from *conventional* petroleum produced from wells. The most important unconventional oil and gas deposits are *oil shales, tar sands,* and *shale gas*.

Oil shales are organic-rich mudstones that yield petroleum when they are heated and can even burn (Fig. 19.10a). The shales were originally deposited in swampy lakes, stagnant lagoons, or marine basins with restricted circulation, where oxygen-deficient (anaerobic) conditions allowed organic matter to be preserved. The process of oil extraction and refining from oil shales is expensive. As a result, only

(a)

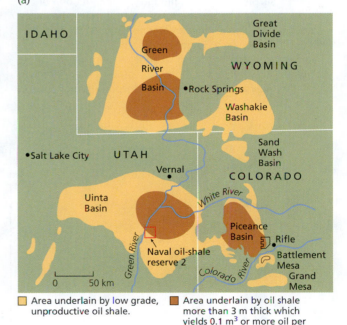

(b)

Figure 19.10   US Oil Shale

(a) This oil shale from Colorado contains sufficient heavy oil to burn. (b) In the United States, oil shales are widely distributed in the Green River Formation of Colorado, Utah, and Wyoming.

IN DEPTH

Correlation and the Search for Oil and Gas

Since the mid-twentieth century, North America has depended on oil and gas for more than half of its energy needs, and today that figure is fast approaching two-thirds. Oil and gas are the lifeblood of modern society, and as the larger, more easily located oil fields are exhausted, the search for new reservoirs becomes ever more urgent. At the heart of this search is the technique of correlation, a principle that has changed little since it was first proposed by William Smith more than two centuries ago. Recall from Chapter 8 that correlation is the ability to trace individual rock layers from one place to another.

As we have discussed in this chapter, for oil and gas to form a reservoir, an impermeable rock layer that stops their upward migration and causes them to pond below must exist. Energy companies extract the oil and gas by drilling a hole through this impermeable layer. The search for oil and gas involves the detection of areas where ponding has occurred. For a viable deposit to form several conditions must be met. Several steps are taken to find and exploit these deposits. First, it is necessary to identify sedimentary rocks that might provide a source for oil and gas. Then a route must be found that the oil and gas could use to migrate upward. Finally, other rocks must be identified that prevent the oil and gas from reaching the surface and dissipating.

Subsurface geology involves acquisition and interpretation of data on geologic features that occur below Earth's surface. In the oil industry, drilling is one of the main sources of such information. Drill holes, however, are just a few inches in diameter and are merely pinpricks into the subsurface. To obtain a larger picture of the subsurface geology, the data acquired from one drill hole must be correlated with data from others. Therefore, correlation plays a vital role in the interpretation of subsurface data.

During drilling, rock core or rock fragments are commonly recovered from the drill hole. These provide a wealth of information on the subsurface geology and are examined not only to determine their rock type and its potential for producing or collecting oil and gas but also for their fossil content. Given the small size of these samples, large fossils are rarely encountered, so fossils of microscopic organisms are examined instead. These *microfossils* follow the principle of faunal succession in exactly the same way as the larger fossils studied by Smith. So they can be used to determine the relative age of a sample and establish its correlation with samples from other drill holes. In this way, rock formations penetrated by one drill hole can be correlated in the subsurface with those penetrated by another.

A more sophisticated method of characterizing rock formations in the subsurface is that of *well logging* (Fig. 19A). In this procedure, an instrument package known as a logging tool is lowered into the drill hole, and measurements are made of a variety of physical properties in the rocks encountered. Rock properties such as fluid content, density, radioactivity, resistivity (resistance to electrical currents), and permeability (ability to transmit fluids) are recorded in this way. The data is printed out in the form of a wavy trace. The pattern of the trace is then compared with measurements made in other drill holes.

For example, in Figure 19B, well logs from three drill holes have been correlated to reveal the presence of a subsurface unconformity. Unconformities are important targets in oil and gas exploration because there is potential for oil and gas to pond beneath the unconformity surface (see Fig. 19.6b). In Figure 19B, this occurred where oil and gas that migrated upward though a sandstone unit became confined beneath the unconformity by units of impenetrable shale. Only by applying the principle of correlation, however, are the unconformity and the reservoir revealed.

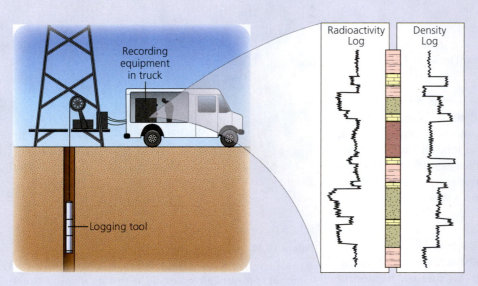

Figure 19A   Well Logging

Well logging is used to characterize rock formations in the subsurface. Measurements made by the logging tool as it is withdrawn from a drill hole are recorded at the surface and printed out as a log that shows how the physical properties of the rocks vary with depth.

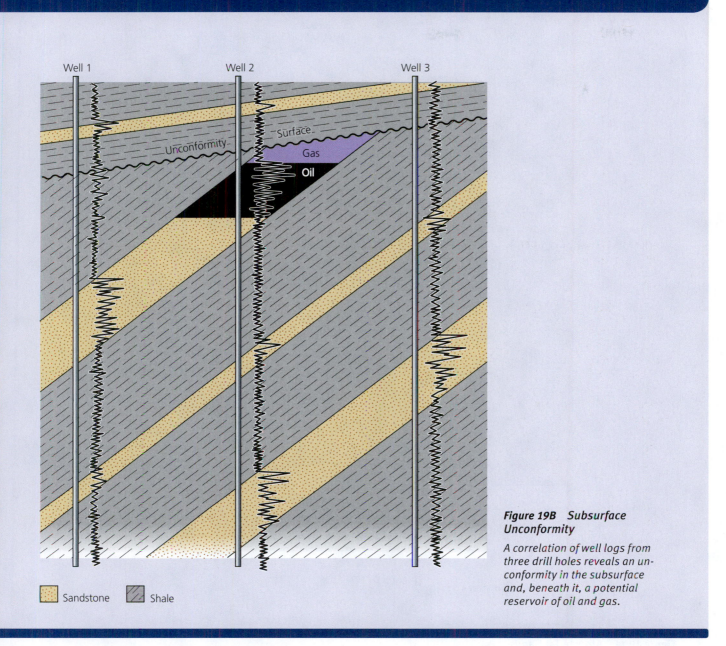

Figure 19B   Subsurface Unconformity

A correlation of well logs from three drill holes reveals an unconformity in the subsurface and, beneath it, a potential reservoir of oil and gas.

Sandstone   Shale

the richest deposits are commercially viable and, even then, significant environmental problems are associated with their exploitation. Nevertheless, when the price of crude oil is high, production of oil from this potentially vast resource booms, especially in North Dakota where production from the Bakken oil shale rose over 70 percent to 470,000 barrels a day in 2012.

The world's largest deposit of oil shale is the Green River Formation of Colorado, Utah, and Wyoming (Fig. 19.10b), which was deposited during the Eocene, some 50 million years ago, when the region was occupied by large stagnant freshwater lakes, in which organic matter was preserved. This huge oil shale deposit is estimated to contain up to

three trillion barrels of oil, which is three times the amount of oil consumed in all of human history! Even if only half of this oil is recoverable, it would equal the entire world's proven oil reserves.

Tar sands are sandstones that contain an asphalt-like hydrocarbon called *bitumen* that is too thick and viscous to be pumped out of the ground. To be recovered, bitumen must either be steam-heated prior to pumping, which encourages the tar to flow or, where the deposit is shallow, mined by open pit methods, treated, and refined.

Tar sands are thought to originate as oil but are later modified as a result of biodegradation and the escape of the more volatile components that leaves only the most viscous

Figure 19.11   Alberta Tar Sands
The Athabasca tar sands are the largest of three deposits in Alberta, Canada, that also include the Cold Lake and Peace River tar sands.

materials behind. The United States contains several small deposits, the largest of which is in Utah. Extensive deposits of tar sands occur in northern Canada, Russia, and Venezuela. The largest deposit in the world is the Athabasca tar sands of northern Alberta, Canada (Fig. 19.11), which is estimated to contain over 175 billion barrels of recoverable oil. Alberta's tar sands currently produce 1.5 million barrels of oil per day, a figure that is expected to increase to 3.3 million barrels per day by 2019.

Shale gas is natural gas produced from organic-rich (black) shales (Fig. 19.12a), which is becoming an increasingly

CHECK YOUR UNDERSTANDING

● What is the difference between oil shale and tar sands?

● What is shale gas and why is it becoming an increasingly important energy resource?

important resource, particularly in the United States, but also in Canada, Europe, Asia, and Australia. Long known as important natural gas source rocks, such shales have rarely been commercial sources of natural gas in the past because they lack the permeability needed to allow significant gas to flow to a drill hole. In

(a)

(b)

Figure 19.12   Shale Gas in the United States

(a) Organic-rich Marcellus shale is a rapidly emerging source of shale gas in western Pennsylvania. (b) Prospective shale gas fields (also known as *plays*) in the lower 48 states are shown on this map.

the 1990s, however, developments in *horizontal drilling* techniques and *hydraulic fracturing* ("*fracking*") methods for improving the flow of gas in such rocks (see Living on Earth: **The Dilemma of Fracking** on page 618) have dramatically increased the productivity of shales. In the United States, this has led to a natural gas boom, with shale gas production increasing eightfold in the first decade of this century, led by exploitation of the Bartlett

shale of Texas, the Marcellus shale of Pennsylvania, West Virginia, and New York, and the Ohio shale of Kentucky, the Virginias, and Ohio (Fig. 19.12b). Shale gas now accounts for 30 percent of US natural gas production—a figure that is predicted to increase to over 50 percent by 2040—and its development has increased the US proven natural gas reserve from 6 to 10 trillion cubic meters (210 to 350 trillion cubic feet).

LIVING ON EARTH

The Dilemma of Fracking

Hydraulic fracturing or **fracking** is a controversial technology used to extract shale gas. By effectively shattering the shale around a drill hole, fracking improves the rock's permeability so that the gas trapped within the rock can be released. Used in combination with another technique, known as *horizontal drilling,* the technology allows otherwise uneconomic reserves of natural gas to be commercially exploited. First used commercially on the Barnett

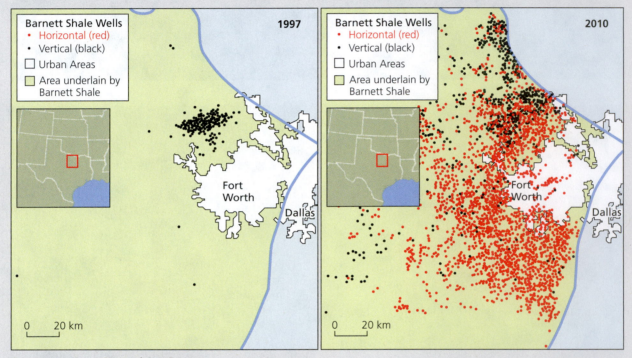

Figure 19C   Texas Natural Gas Boom

The dramatic increase in the number of wells in the Barnett Shale, Texas, between 1997 and 2010, was made possible by fracking and horizontal drilling (red dots).

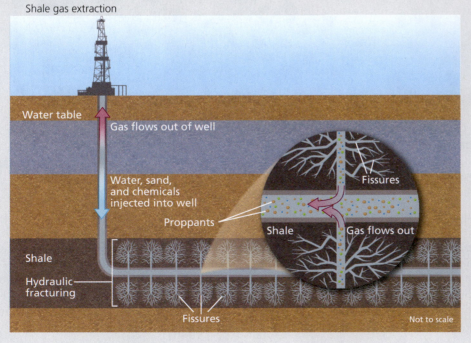

Figure 19D   Fracking

By shattering the shale around a horizontal drill hole and then keeping the fractures open by pumping in proppants, such as sand, fracking improves the rock's permeability. In this way, the gas trapped within the shale can be released.

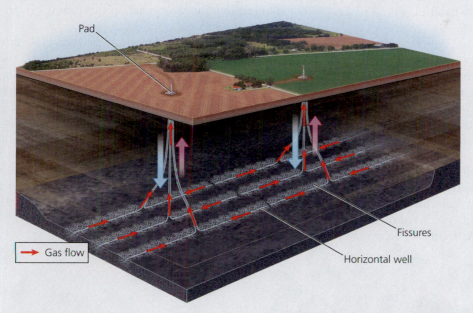

Pad

Fissures

Horizontal well

→ Gas flow

Figure 19E   Pad Drilling

In the technique known as pad drilling, up to 10 wells radiate horizontally for distances of up to 10 kilometers from a single site, or pad.

Shale in Texas in 1998 (Fig. 19C), these technologies have led to an economic boom in natural gas exploration in the United States and elsewhere.

Hydraulic fracturing is the cracking of a rock by a pressurized fluid, and occurs naturally in the formation of some rock fractures. When the process does not occur naturally, it is often referred to as *fracking*. When energy companies participate in fracking, they artificially create fractures around a borehole by injecting a fluid, known as "slickwater," into the well at very high pressure. This fracking fluid is largely water, but also contains a range of chemical additives to facilitate the process, and "*proppants*," such as sand, to keep the fractures open once they have formed (Fig. 19D).

Horizontal drilling is the technology that allows a vertical borehole to be continued horizontally once the target shale has been reached. Drilling can then be continued in any direction and a pattern of holes can be drilled that radiate horizontally from a single wellhead (Fig. 19E). The single wellhead is referred to as a *pad* and the process is called *pad drilling*.

The advantage of fracking is an economic one. It allows energy companies to extract vast quantities of natural gas from otherwise inaccessible reservoirs. Some 1.7 million jobs have been created by shale gas development with a total of 3.5 million projected by 2035, and, without fracking, it is estimated that the United States would lose almost 50 percent of its domestic natural gas production.

Fracking is not without significant environmental consequences, the most worrying of which is the potential for groundwater contamination by fracking fluid. There is also

Figure 19F   Jonah Field, Wyoming

The heavy development of the Jonah natural gas field near Pinedale, Wyoming, illustrates the impact that gas shale development can have on the landscape.

a risk of surface contamination and the millions of gallons of fracking fluid used in every well must be subsequently treated or disposed of safely. The exploitation of shale gas can also have a major impact on the landscape (Fig. 19F). As a result of these concerns, fracking has become a highly contentious issue, pitting industry against opposition in an emotionally charged public debate. This has led some local governments in North America to introduce new legislation, while some European countries have banned the process altogether.

PETROLEUM DISTRIBUTION AND RESERVES

Although sedimentary basins containing major petroleum fields occur on all continents, conventional oil and natural gas remain rare commodities and their distribution is erratic. More than three-quarters of the world's petroleum, for example, comes from just a few known petroleum deposits, and most comes from just a handful of huge fields, called *supergiants*.

The world's proven crude oil reserve is estimated to be about 1.5 trillion barrels, more than half of which lies in the Middle East, and over 70 percent of which is located in countries that are members of *OPEC (Organization of the Petroleum Exporting Countries)*—a cartel of oil-producing countries in the Middle East, Africa, and South America. North America possesses less than 1.5 percent of this reserve, the bulk of which lies in Canada (Fig. 19.13). Nevertheless, it is industrialized countries that produce the bulk of the world's oil. Because these nations are such rampant consumers but hold such a limited share, oil reserves in the United States and Canada are only predicted to last into the second half of this century at current production rates.

The world's natural gas reserve is estimated at just over 300 trillion cubic meters (1 quadrillion cubic feet), more than half of which lies in Russia, Turkmenistan, and the Middle East. North America possesses just 4 percent of this reserve, three-quarters of which lies in the United States. But as with oil, these reserves are only predicted to last into the second half of this century at current rates of production.

However, these estimates for oil and gas do not take into account "unconventional" petroleum resources, such as oil shales, tar sands, and shale gas, which could add very significantly to the lifespan of these commodities (see In Depth: **Is Ice That Burns the Fuel of the Future?** on page 621). The recent development of shale gas, for example, has increased the US natural gas reserve by 50 percent to 9 trillion cubic meters—enough to last more than a century at current rates of consumption. Similarly, the vast oil reserves locked in the oil shales of Colorado, Utah, and Wyoming (see Fig. 19.10b), and the tar sands of Alberta (see Fig. 19.11), could potentially exceed the world's entire proven reserve. This resource would meet North America's oil demand

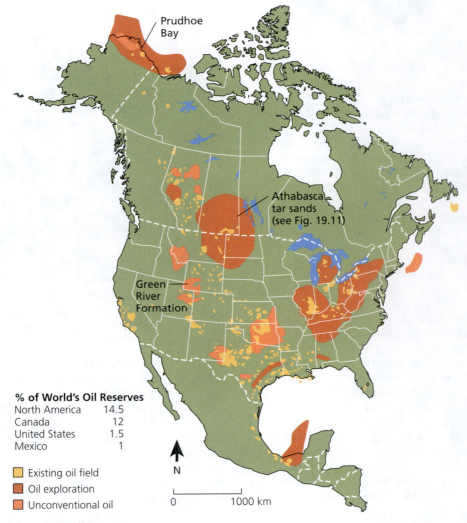

% of World's Oil Reserves

North America	14.5
Canada	12
United States	1.5
Mexico	1

- Existing oil field
- Oil exploration
- Unconventional oil

0   1000 km

Figure 19.13   North American Oil Reserves

Oil reserves in North America, estimated in 2011, are limited. (Unconventional oil includes reserves in oil shales and tar sands.)

IN DEPTH

Is Ice That Burns the Fuel of the Future?

At the bottom of the world's oceans lies a potential source of energy that could provide society with fuel well into the next century. These deposits are the product of ancient seafloor bacteria and they store twice as much carbon as all the world's fossil fuel reserves combined. This potential fossil fuel is not oil or coal, or even natural gas as we know it, but methane gas locked in crystals of ice. Resembling effervescent snow, this peculiar ice, known as *methane hydrate*, burns if ignited, leaving behind a pool of water (Fig. 19G). Methane ice forms under pressure and expands to more than 160 times its original volume when it is brought to the surface, much as propane does when it leaves the cylinder of a gas barbeque.

First discovered in 1970, methane hydrate has been found within just a few hundred kilometers of almost every coastline, as well as in permafrost regions on land, such as those of Alaska (Fig. 19H). And it has been found in enormous quantities. Off the coast of the Carolinas, for example, the US Geological Survey has discovered two deposits of methane hydrate, each about the size of Rhode Island. Together they are estimated to contain over 1300 trillion cubic feet of methane gas, or more than 50 times the amount of natural gas consumed in the United States in 2012. Is methane hydrate the fuel of the future? The sheer volume and richness of methane hydrate deposits make them a strong candidate for development as an energy resource. However, the challenges are enormous.

Figure 19G   Burning Ice

Burning methane hydrate is fueled by methane gas trapped within crystals of ice.

Although methane hydrate forms in layers that could potentially be mined, gaining access to the deposits beneath the seabed poses huge technical problems. In addition to the problems of mining it, the hydrate is stable only at the icy temperatures and crushing pressures of the deep ocean, and starts to break down long before it reaches the

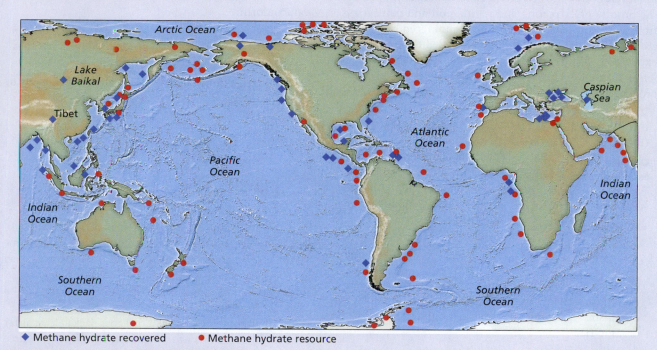

◆ Methane hydrate recovered   ● Methane hydrate resource

Figure 19H   World Methane Hydrate Deposits

Deposits of methane hydrate occur worldwide and contain more organic carbon that all other global reservoirs combined.

(Continued)

IN DEPTH

Is Ice That Burns the Fuel of the Future? *(Continued)*

surface, thus preventing the gas from escaping as the hydrate is recovered adds to the challenge of recovery.

Like conventional natural gas, the layers of methane hydrate could also be tapped by drilling. Several oil companies and government agencies are actively researching the feasibility of such an endeavor. Because the deposits are far deeper than most underwater oil and gas fields, special deep water drilling vessels would have to be constructed. Nevertheless, drilling is at least a feasible option, although the methane ice is under so much pressure, the challenge is akin to bursting a balloon and trying to capture all the escaping gas. One potential solution would be to expel the methane by pumping hot water or steam into the deposit through one drill hole and extracting the expelled methane through another. But once recovered, the methane would still have to be brought ashore and this would pose an additional challenge.

Methane hydrate would provide a "cleaner" source of energy than oil and coal because methane releases less than half the amount of carbon dioxide to the atmosphere when it is burned, and so contributes less significantly to climate change. However, its exploitation is not without potentially serious environmental consequences. It is estimated that the methane trapped in methane hydrate amounts to more than 3000 times the volume of methane in the atmosphere where it forms a potent greenhouse gas that is 10 times more effective at trapping heat than carbon dioxide. The global warming effect of methane accidentally released to the atmosphere is consequently ten times more serious than that of carbon dioxide. Breakdown of methane hydrate at the base of hydrate layers is also known to have triggered massive landslides on the ocean floor, adding a further hazard to the production of this resource.

Although the technological challenges facing the successful exploitation of methane hydrate are great, they are not insurmountable. Indeed, both Japan and China have projects aimed at commercial-scale extraction, and the oil industry is field testing the feasibility of extracting methane. We may therefore live to witness the exploitation of this resource as other reserves of fossil fuel dwindle and methane hydrate becomes an economically attractive alternative.

---

for several centuries at current rates of consumption. But the process of extracting and refining oil and natural gas from these sources is expensive and, until recently, the cost and difficulty of production have precluded their economic exploitation. With the rising price of oil, however, this is no longer the case, as illustrated by the oil boom in North Dakota brought about by the exploitation of the Bakken oil shale.

Indeed, in a dramatic endorsement of the potential of oil shale as an emerging resource, the International Energy Agency has predicted that, by 2020, the United States will have surpassed Saudi Arabia as the world's largest oil producer.

CHECK YOUR UNDERSTANDING

○ What are the world's petroleum reserves estimated to be and where do most of them occur?

19.2 SUMMARY

- Oil and natural gas are nonrenewable energy sources derived principally from the remains of microscopic organisms incorporated in clay and buried to form organic-rich shale.

- Four things—a source rock buried to an appropriate depth, a reservoir rock, a seal, and an oil trap—must exist for petroleum deposits to form.

- Oil traps, subsurface structures that allow petroleum to accumulate, can be structural (anticline, salt dome,

and fault) or stratigraphic (pinch out, reef, and unconformity).

- Oil shales yield petroleum when heated, tar sands produce petroleum from bitumen, and shale gas is natural gas produced from organic-rich shales.

- Seismic reflection, which uses shock wave reflections to penetrate the subsurface, is the geophysical technique most widely used for petroleum exploration.

19.3 Coal

The black, inflammable sedimentary rock called **coal** is the most abundant of the fossil fuels. Coal is found in rocks formed since plant life first became abundant some 390 million years ago. The vast deposits of coal in eastern North America and western Europe formed between 359 and 299 million years ago during the Carboniferous, a period named for its abundant coals.

Although the use of coal as a fuel goes back at least as far as the Middle Ages, its widespread use as an energy source came about with the introduction of the steam engine during the Industrial Revolution of the eighteenth century (see Fig. 19.1). It was the use of coal as an energy resource that made the fundamental transformation toward modern industrial society possible.

Because of its abundance, accessibility, and relative ease of extraction, coal is a relatively inexpensive fossil fuel. It is therefore expected to be an important part of the world's energy supply in the future, particularly as diminishing petroleum reserves force the price of oil and natural gas higher. Following a rise in coal consumption as a result of embargo-driven instabilities in the supply and cost of petroleum in the 1970s, world production of coal remained relatively stable at just over 5 billion metric tons per year until the turn of the century. Over the past decade, however, this figure has increased rapidly to 8 billion metric tons. At this level of production, the world's total recoverable coal reserves, which are presently estimated to be 860 billion metric tons, would be enough to last just over 100 years.

Of the coal consumed, most is used to generate electric power from steam turbines supplied from coal-fired boilers. However, coal is also converted into *coke,* a hard, porous and practically smokeless fuel that is produced when coal is baked and its volatile matter is driven off as coal gas. Coke plays a critical role in the smelting of iron and the manufacture of steel. Coal also provides the raw material for many organic chemicals as well as for nylon and many other plastics.

The use of coal as an energy source is not without severe environmental problems. *Strip mining* and the more controversial *mountaintop removal mining* for coal require elaborate land restoration, while underground mining is not only a hazardous operation but also can lead to surface subsidence and the outflow of acidic water from abandoned mines known as *acid mine drainage.* Methane gas released by underground coal mining operations is explosive; in addition, breathing methane gas in sufficient amounts is fatal. Indeed, historically the risk of explosion or asphyxiation was so severe that coal miners resorted to carrying caged canaries into the mine with them. Being less tolerant to the effects of methane, the canaries would succumb to the methane fumes before the miners: a clear warning that methane levels had become dangerously high. Furthermore, water used in the treatment of coal is too contaminated to be released to the environment. Coal dust is highly combustible and, if inhaled, presents a serious health risk in the form of the miner's disease known as "black lung."

Sulfur dioxide emissions associated with the burning of coal adversely affect ground level air quality and have been linked to the problem of acid rain. In addition, the burning of coal adds more carbon dioxide to the atmosphere than the combustion of any other fossil fuel. Carbon dioxide is viewed as the most important of the greenhouse gases associated with global warming and climate change. Despite these adverse environmental impacts, however, the International Energy Agency predicts a continued rise in global coal consumption. By 2017, coal is predicted to be close to surpassing oil as the world's chief energy source (see Fig. 19.2).

COAL FORMATION

Coal is a sedimentary rock consisting almost entirely of organic carbon. Coal forms as a result of the burial and compaction of accumulated plant remains in fresh water and brackish swamps (Fig. 19.14). In contrast to petroleum, which is largely the product of marine microorganisms, coal forms from decaying terrestrial plants. For the plant matter to be preserved, however, it must not decompose. This is best achieved under oxygen-starved conditions such as those produced in the stagnant waters of freshwater or brackish-water swamps.

To form coal, the accumulated plant remains must first form **peat,** a water-saturated, organic-rich humus produced by bacterial decay that, when dried, is itself a low-grade fuel. Peat often forms in warm, humid swamplands that border marine basins. The Okeefenokee Swamp in

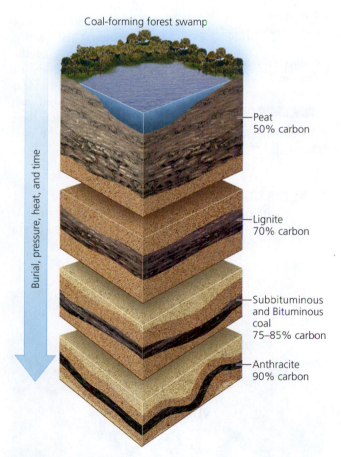

Coal-forming forest swamp

Burial, pressure, heat, and time

Peat
50% carbon

Lignite
70% carbon

Subbituminous
and Bituminous
coal
75–85% carbon

Anthracite
90% carbon

Figure 19.14   Formation of Coal

Compaction of swamp vegetation produces peat. Buried beneath increasing amounts of sediment, peat is progressively transformed into lignite (brown coal), subbituminous and bituminous coal, and anthracite as temperatures and pressures rise. With each transformation, the carbon content of the coal increases.

CHECK YOUR UNDERSTANDING

● Why are coal deposits unlikely to be more than 390 million years old?

● At present production rates, how long is the world's supply of coal expected to last?

Georgia and the Everglades of Florida, for example, contain large deposits of peat. In settings such as these, land subsidence allows occasional marine incursions to occur, which result in the deposition of sediments and the rapid burial of the peat deposit.

As peat is buried beneath overlying sediment, it becomes progressively compacted and heated, which squeezes out the water and drives off the organic gases. As a result, the carbon content of the peat becomes concentrated, eventually transforming it into woody-looking brown coal, or *lignite*. Continued burial raises temperatures and pressures, and concentrates carbon still further. The lignite undergoes additional changes, first becoming *subbituminous coal* and then *bituminous coal*, the two most common forms of coal in North America. This change in structure and composition is a form of very low grade metamorphism. Further metamorphism of bituminous coal, usually accompanied by structural deformation in regions of plate convergence, produces the jet-black coal called *anthracite*.

> ### CHECK YOUR UNDERSTANDING
>
> ● What are the steps that plant remains undergo to form coal?
>
> ● What is the difference between coal rank and coal grade?

This progression from lignite to anthracite coincides with an increase in the *rank* of the coal, which is based on its carbon content and heat value on combustion. Coal quality, referred to as its *grade*, on the other hand, is based on its purity. Typical contaminants include the sulfur and ash content, both of which have detrimental effects on the environment when the coal is burned.

COAL DISTRIBUTION AND RESERVES

The formation and periodic burial of peat in shallow swamps on broad coastal plains ultimately produces a series of relatively thin coal seams interlayered with other sedimentary strata as part of a normal stratigraphic succession. Indeed, most coal deposits extend over large areas in series of subhorizontal seams that range from several centimeters to a few meters in thickness (Fig. 19.15). As a result, coal seams are quite easily located and there is usually little effort involved in coal exploration. In fact, it is likely that most of the world's coal deposits have already been identified.

Coal deposits are widely distributed, but are far more abundant in the northern hemisphere than they are in the south. As we shall discuss in Section 19.6 (see Fig. 19.34), this bias is largely a function of the geography of Pangea because the majority of coal was formed just as Pangea was assembling in the Late Carboniferous. Today, almost 60 percent of the world's total recoverable coal occurs in the United States, Russia, and China (Fig. 19.16). China and the United States are both the leading producers and the leading consumers of coal.

The United States produces almost one billion metric tons (bmt) of coal annually, almost all of which is consumed in the production of electricity. Some of the largest reserves of high ranking bituminous coal and anthracite occur in the upper Midwest and Appalachia, and huge reserves of low-grade lignite lie beneath the northern Plain States (Fig. 19.17). However, with the introduction of the federal Clean Air Act in the United States, which requires reductions in sulfur emissions from coal combustion, coal production has shifted to the western states, led by Wyoming. This is because the coals in these regions are largely of Cretaceous age (see Fig. 19.15) and their sulfur content is low. At the same time, production of Carboniferous coals in Appalachia and the Midwest, led by Kentucky and West Virginia, has declined because of their high sulfur content. The need to either treat these high-sulfur coals to remove sulfur prior to combustion or "scrub" their emissions in order to reduce sulfur dioxide air pollution has increased the cost of these coals and made them less economic as an energy source for generating thermal electric power.

Figure 19.15   Coal Seam

A coal seam interlayered with Cretaceous sandstone near Helper, Utah.

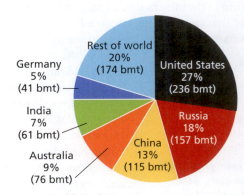

Figure 19.16   World Coal Reserves

The global share and total reserves (in bmt–billion metric tons) of recoverable coal are listed here by country.

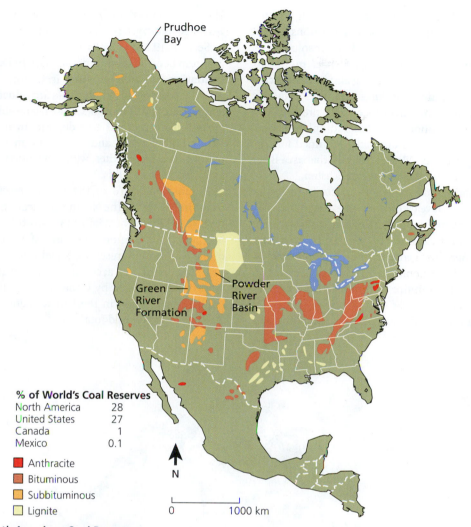

Figure 19.17   North American Coal Reserves
Coal reserves in North America and their share of the world reserves as estimated in 2008.

19.3 SUMMARY

- Coal is the end product of the progressive burial of swampy plant material that yields peat, lignite (brown coal), subbituminous and bituminous coal, and ultimately, anthracite.

- Coal rank is based on carbon content, whereas coal grade is based on purity.

19.4 | Nuclear Power

In addition to the fossil fuels, the solid Earth provides energy in the form of another nonrenewable resource: naturally occurring radioactive minerals. Nuclear power depends on the heat released when atoms of the naturally occurring radioactive element uranium (U) are split apart during radioactive decay. In a nuclear reactor, the heat produced by the decay of uranium is used to make steam, which is then used to generate electricity. Nuclear power consequently depends on the exploitation of a mineral resource and the utilization of its radioactive properties under controlled conditions.

URANIUM DEPOSITS AND THEIR EXPLOITATION

Uranium, with an atomic number of 92, is the heaviest of the naturally occurring elements and has an average crustal abundance of about 2 parts per million. It is found in minor quantities in a wide variety of rocks but is chiefly exploited as an oxide in the ore minerals *pitchblende* and *uraninite*. Economic deposits of these minerals are most common in permeable sedimentary rocks where uranium, dissolved in groundwater, has been concentrated as a result of chemical reduction.

Uranium exists in both oxidized (U^{6+}) and reduced (U^{4+}) forms (recall from Chapter 5 that iron likewise exists in both oxidized and reduced forms). Oxidized uranium is highly soluble in groundwater, whereas reduced uranium is highly insoluble. Uranium released into solution by the weathering of rocks under oxidizing conditions will therefore precipitate if percolating uranium-bearing groundwater encounters reducing (oxygen-poor) conditions. Where this occurs, the soluble U^{6+} is converted into insoluble U^{4+}, which is incorporated into minerals. Hence, motion of groundwater from oxidizing to reducing environments converts soluble uranium into insoluble uranium minerals. Over a long period of time, this process can produce very high uranium concentrations, most commonly in sandstones. Such sandstone-hosted deposits are mined for uranium in Wyoming and New Mexico, in the Athabasca Basin of Canada, in Central Europe, and in Kazakhstan.

Some of the richest uranium deposits occur in rocks that are more than 2.2 billion years old, and these deposits tell us much about the evolution of Earth's early atmosphere. During the Archean, which lasted from 4.0 to 2.5 billion years ago, atmospheric oxygen levels were extremely low—too low, in fact, to allow weathered uranium minerals to dissolve. As a result, uranium accumulated, along with other heavy minerals, in placer deposits (see Chapter 18), often in coarse grained sediments such as conglomerates.

The formation of such deposits ceased some 2.2 billion years ago as oxygen released by photosynthesis (see equation in Section 19.2) gradually built up in the atmosphere and uranium became soluble in water (see In Depth: **The Role of Oxidation in Earth's History** in Chapter 5). Archean and early Proterozoic placer deposits are mined for uranium in South Africa and Ontario, Canada. Uranium also forms important hydrothermal vein deposits in Australia, Canada, Africa, and France, and can occur in late-stage igneous rocks such as pegmatites, whose evolution has been affected by hydrothermal fluids.

Almost one-third of the world's reserves of uranium occur in Australia. The North American share of the world's reserve is 13 percent (Fig. 19.18), two-thirds of which occur in Canada. Wyoming and New Mexico host two-thirds of the US uranium reserves. Kazakhstan leads the world in uranium extraction, accounting for almost one-third of the world total, followed by Canada, Australia, and Niger. By contrast, extraction in the United States amounts to only 3 percent of the world total.

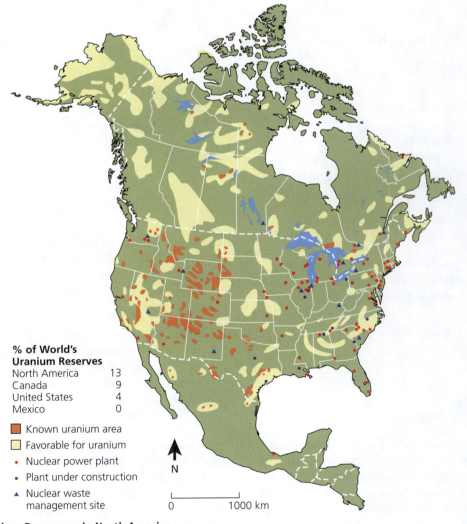

% of World's Uranium Reserves

North America	13
Canada	9
United States	4
Mexico	0

■ Known uranium area
▢ Favorable for uranium
• Nuclear power plant
• Plant under construction
▲ Nuclear waste management site

N

0   1000 km

Figure 19.18   Nuclear Resources in North America

This map highlights North American uranium reserves and the distribution of North American nuclear power plants and nuclear waste management sites.

Exploration for uranium ore and the evaluation of potential uranium deposits makes use of the element's radioactivity, which can easily be detected at the surface and measured in boreholes using instruments such as *geiger counters.* Uranium ore is often mined by open pit or underground methods. The uranium is extracted from the ore by chemical leaching following crushing and grinding. But where the host-rock is permeable, as is the case in sandstone-hosted deposits, the uranium is more economically removed by in-place leaching instead of underground mining. In-place recovery leaves the ore in the ground and removes the uranium by dissolving it by introducing an oxidizing or acidic agent, and pumping the resulting solution to the surface where the uranium is recovered. Almost half of the world's uranium in mined in this way, notably in Kazakhstan, as well as in the United States, where it is considered the most environmentally acceptable method of uranium mining.

NUCLEAR ENERGY

Although all radioactive isotopes are unstable and spontaneously break down (see Science Refresher: **Types of Radioactive Decay** in Chapter 8), some are so unstable that their atomic nuclei can be split by neutron bombardment, a process known as **nuclear fission**. When this occurs, the nucleus breaks into two fragments of roughly equal size, and heat energy is released along with a number of neutrons (Fig. 19.19). If these neutrons then split the nuclei of neighboring atoms, a *chain reaction* is initiated as more and more neutrons are released to split still further nuclei. Uncontrolled chain reactions lead to nuclear explosions, but controlled reactions are the source of heat that is used to generate electricity in nuclear power plants. In nuclear reactors, fission is controlled by inserting moderators that absorb some of the neutrons produced by each nuclear reaction. Failure to carefully monitor this process, however, can lead to accidents such as the reactor meltdowns that occurred at Chernobyl in the Ukraine in 1986 and at the Fukushima 1 reactor in Japan following the disastrous Tōhoku earthquake and tsunami of 2011.

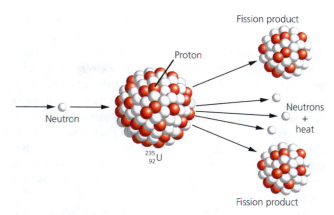

Figure 19.19   Nuclear Fission

During nuclear fission, a neutron strikes the nucleus of an atom of ^{235}U, producing fission products, neutrons, and heat. The free neutrons participate in additional fission reactions.

Almost all naturally occurring uranium is in the form of one of two isotopes, ^{238}U and ^{235}U, only one of which (^{235}U) is fissionable. The fissionable isotope makes up only 0.7 percent of natural uranium, the remaining 99.3 percent consisting of the heavier isotope ^{238}U. Some reactors, like those of the CANDU variety used in Canada, are designed to use natural uranium. In most commercial reactors, however, uranium containing about 3.5 percent of the fissionable isotope is needed to produce a chain reaction. So natural uranium must be processed to concentrate or enrich this isotope before it can be used. The enriched uranium is made into pellets and assembled into fuel rods that are inserted into the reactor core where the chain reaction takes place.

Most commercial reactors in use in North America are *light-water reactors,* in which ordinary water is used both as the moderator that controls and sustains the nuclear reaction, and as the medium that extracts the heat produced by nuclear fission (Fig. 19.20). The reactor core holding the fuel rods is housed within a containment structure and enclosed in a stainless steel reactor vessel through which water is circulated. Fission in the fuel rods starts the chain reaction by generating heat and releasing neutrons that

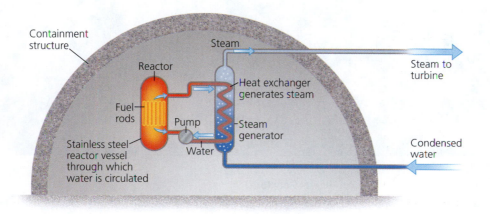

Figure 19.20   Light-Water Reactor

promote further fission, which generates further heat. The circulating water consequently becomes hot and is pumped through a heat exchanger that generates steam. The steam is then used to drive turbines that turn electrical generators. In this way, the water circulating through the reactor core is separated from that which supplies the turbines with steam. The chain reaction itself is regulated by control rods, which are made of material that absorbs excess neutrons and so prevents them from bombarding other nuclei. The chain reaction is consequently slowed when these are inserted into the core, and speeds up when they are removed.

Although conventional nuclear reactors account for about 20 percent of electric power generation in the United States and about 15 percent in Canada, only a small fraction of the uranium in their fuel rods is fissionable ^{235}U. As a result, the bulk of the uranium, which is nonfissionable ^{238}U, goes to waste. Under neutron bombardment, however, ^{238}U is converted into the fissionable isotope of plutonium, ^{239}Pu, by neutron capture. Thus the potential exists for converting part of the uranium waste into fissionable fuel by using some of the neutrons released during nuclear fission. This phenomenon is exploited in *breeder reactors,* so called because they produce more nuclear fuel than they use. In a breeder reactor, a fuel core of fissionable material (^{235}U or ^{239}Pu) is surrounded by a blanket of ^{238}U. Fission of the core material is used to generate electricity in the usual way. However, since each fission event releases several neutrons, yet is only required to produce one fission event of its own in order to sustain a chain reaction, the excess neutrons can be used to convert waste ^{238}U into fissionable ^{239}Pu fuel.

The most economical feature of breeder technology is its ability to use ^{238}U as a fuel. As a result, the amount of uranium needed to power a breeder reactor is only about one-hundredth of that used in a light-water reactor of equivalent capacity. Indeed, the amount of ^{238}U now stored in waste stockpiles could be used to meet breeder reactor needs for well over a century. Breeder reactor technology has been most widely developed in France, which uses nuclear power to generate three-quarters of its electrical energy. In North America, however, public concern over plutonium, which is not only highly toxic but also can be used to make nuclear weapons, has halted development of a breeder reactor program.

> **CHECK YOUR UNDERSTANDING**
> ● What is nuclear fission and under what conditions can it cause a chain reaction?

Despite this concern, the United States, with over 100 operating reactors (Fig. 19.21), leads the world in nuclear electric power generation, followed by France and Japan. Together, these three countries generated more than half of the world's nuclear electric power.

NUCLEAR SAFETY AND WASTE DISPOSAL

Nuclear power is "clean" energy because it does not release carbon dioxide to the atmosphere and so does not contribute to climate change. However, public confidence in the safety of nuclear power plants has been low in the United States ever since the nuclear accident at Three Mile Island in Pennsylvania in 1979. Although nuclear reactors are incapable of producing a nuclear explosion, core meltdown

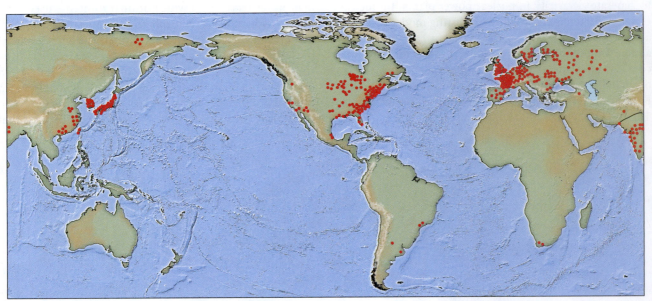

• Nuclear power plant

Figure 19.21   Global Distribution of Nuclear Power Reactors

In 2012, nuclear power accounted for 13.5 percent of the electricity generated worldwide with 435 commercial nuclear power reactors operating in 31 countries. An additional 65 nuclear units are under construction and 150 are planned, mostly in China and Russia.

resulting from a runaway chain reaction can destroy the containment structure and spread lethal radioactive contamination over wide areas. The 1986 Chernobyl disaster in the former Soviet Union and the 2011 Fukushima 1 disaster in Japan have further heightened public safety concerns. The latter resulted in an almost complete shutdown of Japan's nuclear reactors in 2012 with a loss of 30 percent of the country's generating capacity, and led to a decision by the German government to phase out all their nuclear power capacity by 2022. In the United States, these concerns, coupled with the high cost of constructing new nuclear power stations, have brought the nuclear industry to a virtual standstill. No new power plants have been constructed since the Three Mile Island accident and only five new units have been approved since then, all in the past decade and all at existing plants.

Nevertheless, many consider the safety record of the nuclear industry to be excellent and argue that coal-fired generating stations pose a far greater hazard to the environment than those using nuclear power. Uranium is also cheap in comparison to fossil fuels. One ton of nuclear fuel produces as much energy as 20,000 tons of coal at less than half the price. But these figures do not consider the hidden costs of nuclear energy.

The construction of nuclear power stations, for example, has proven to be expensive, and concerns persist over nuclear safety, acts of terrorism, and the security of the plutonium. There is also the unresolved problem of nuclear waste disposal. Much of the waste material from nuclear power stations will remain dangerously radioactive for thousands of years. How best to dispose of this material without contaminating the environment is a highly charged issue and one that has greatly deterred the development of the nuclear industry. Critics argue that waste disposal and construction costs must be considered as part of the fuel cost, in which case nuclear power is less economically attractive.

At present, most used nuclear fuel rods and other high-level radioactive wastes are stored temporarily in water-filled pools at individual reactor sites. These materials are ultimately supposed to be buried in permanent underground disposal facilities, but at present only one facility exists in the United States (the Waste Isolation Pilot Plant, or WIPP, near Carlsbad, New Mexico). The planned larger Yucca Mountain repository, that was to be built adjacent to the Nevada test site, was defunded in 2010. Worldwide, over 50 countries currently store their spent fuel in temporary locations, awaiting reprocessing or permanent disposal. But permanent disposal sites do not yet exist, although feasibility studies have been carried out in Canada, the United Kingdom, Germany, Sweden, and Japan; collaborative initiatives on regional repositories are being developed in Europe, the Middle East, and Southeast Asia; and potential sites have been proposed in Siberia, central Australia, southern Africa, Argentina, and western China.

Most waste repository studies have focused on sites in salt deposits (such as WIPP), because they are virtually impervious to groundwater, and on sites in stable homogeneous igneous rocks like those at Yucca Mountain. However, salt is capable of flow and no igneous rock is free of fractures. It is therefore difficult to guarantee that any storage site, even in the most ideal of geologic settings, would remain failsafe for the long period over which dangerous levels of radioactivity would persist.

19.4 SUMMARY

- Nuclear power is generated by the fission of naturally occurring radioactive uranium. The fissionable isotope (^{235}U) of uranium must usually be enriched in order to produce a chain reaction.

- In breeder reactors, fission of ^{235}U is used to convert waste ^{238}U into fissionable plutonium (^{239}Pu).

- Adoption of nuclear power has been slowed by public fear of accidents and concerns about how and where to dispose of nuclear waste.

19.5 Renewable Energy

In addition to nuclear energy and the fossil fuels, the supplies of which are nonrenewable, energy is obtained from a variety of renewable resources. An energy resource is renewable if it can be used without exhausting the energy source. Among these renewable resources are *geothermal energy* obtained from the solid Earth; *tidal, wave,* and *hydroelectric energy* derived from water power; *solar* and *wind energy* obtained from the atmosphere; and *biomass energy* generated by burning organic matter. Because most renewable energy is used to generate electricity, the resource is usually measured in *watts*, the familiar electrical power unit of lightbulbs, either with the prefix *mega* for a million, *giga* for a billion, or *tera* for a trillion. Although presently responsible for only a small fraction of the world's energy needs (see Fig. 19.2), renewable resources provide "clean" energy that does not contribute to global warming. Because of this, they are considered by many to be the energy sources of the future and their utilization is rapidly expanding.

GEOTHERMAL ENERGY

A source of energy within the solid Earth is provided by natural heat from Earth's interior. As we have learned, Earth becomes progressively warmer with depth, as any deep miner can testify. This heat is a reflection of the planet's accretion, the formation of its internal layers, and the ongoing decay of radioactive elements in its interior (see Chapter 1). The flow of heat from Earth's interior to its surface can be measured everywhere. But some near-surface regions are anomalously warm and provide a potential source of energy. The potential for such **geothermal energy** exists wherever water heated by shallow-level hot rocks or by deep circulation is close enough to Earth's surface that the water can be reached by borehole. To be economically viable, the water must be within 3 kilometers (about 10,000 feet) of the surface. This is most common in areas of recent volcanism (at active plate boundaries or near hotspots), where molten bodies of magma lie close to the surface. Alternatively, some near-surface rocks, such as granites, contain sufficient quantities of disseminated uranium minerals that they remain anomalously hot by virtue of radioactive decay. In this case, the energy source is the natural analogue of a nuclear power station.

In either case, hot rocks close to the surface are used to heat water, either to produce steam for generating electricity or to provide heat for buildings. Drilling into the hot rocks is usually necessary so that steam and hot water trapped in fractures can be brought to the surface to run a power plant (Fig. 19.22). Once used, the condensed steam and waste water are pumped underground again in

an injection well and reheated. Because the heat energy is replenished far faster than it is consumed, geothermal energy is classified as a renewable resource.

Today, the United States, the Philippines, Indonesia, Mexico, Italy, New Zealand, Iceland, and Japan lead the way in producing electricity from geothermal energy in regions close to active plate boundaries and hotspots where recent volcanism has occurred (see Fig. 19.36). In Iceland, geothermal power supplies 30 percent of the nation's electricity and provides almost 90 percent of its population with inexpensive central heating. The world's total geothermal output of electricity, however, is only 0.3 percent of global production (see Fig. 19.2).

The United States first developed geothermal energy in 1960 around The Geysers, a huge geothermal resource near Napa Valley in northern California (Fig. 19.23). Today, 77 geothermal plants operate on 19 fields, which lie in areas of anomalously high heat flow, in California, Nevada, Hawaii, and Utah. The Geysers remains the largest US facility and accounts for almost half the country's geothermal power. It provides over 1500 megawatts of electricity, enough to supply a city of over a million people. Nationwide, however, geothermal energy only provides about 2.5 percent of renewable energy consumption and 0.4 percent of the country's energy needs (see Fig. 19.1).

The world's total geothermal resource is vast, and geothermal energy is relatively clean. But it is not without environmental consequences. Geothermal waters leach minerals from the rocks through which they pass and so are often toxic brines that are highly corrosive to machinery. Subsidence and small tremors due to the extraction and injection of geothermal waters have also been recorded at The Geysers. It has also been estimated that only about 1 percent of the energy of a geothermal reservoir is recoverable, so that even with improved technology, it is unlikely that geothermal power will ever satisfy more than a fraction of human energy

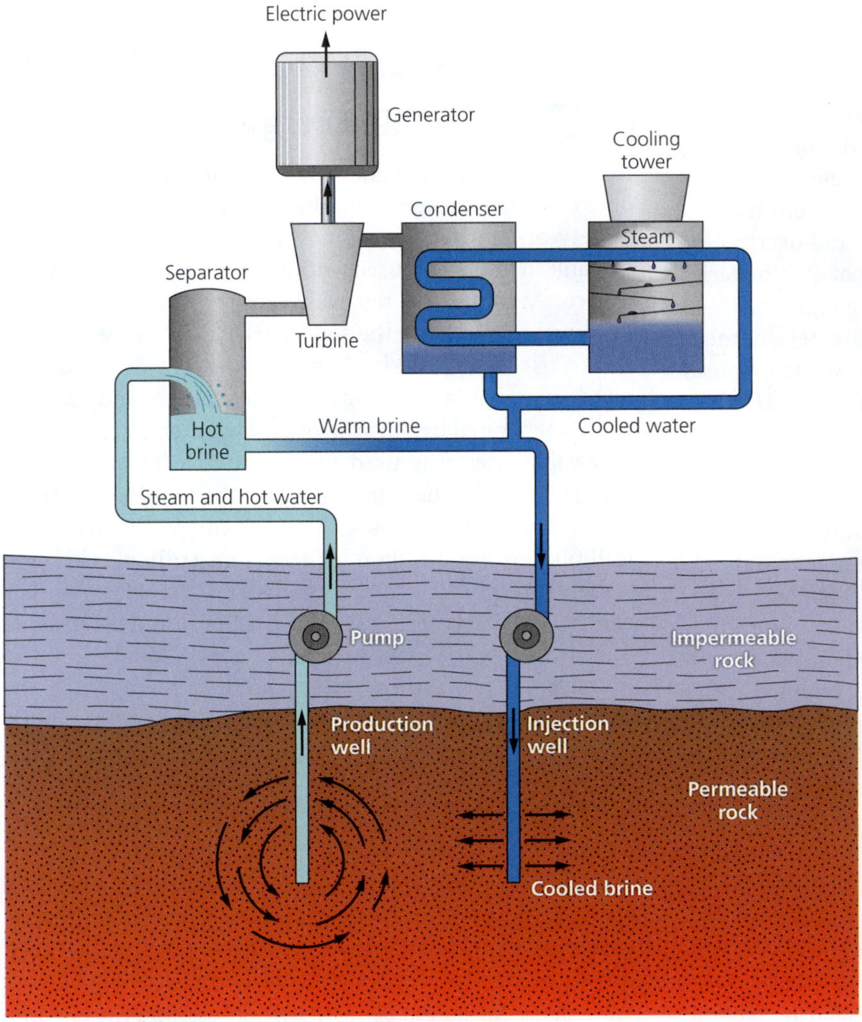

Figure 19.22 Electricity from Geothermal Energy
Geothermal energy is used to generate electricity from the steam produced when hot rocks near Earth's surface heat water.

CHECK YOUR UNDERSTANDING

● What conditions must be met for geothermal energy to be viable?

● Where are the most favorable conditions for geothermal energy likely to be found?

Figure 19.23   The Geysers

One of 18 active geothermal power plants (Sonoma Calpine 3) at The Geysers field in northern California, the largest complex of geothermal power plants in the world.

needs. Nevertheless, the supply of geothermal energy in the United States is expected to increase to about 1 percent of the country's energy needs when projects currently in development come on line.

ENERGY FROM WATER POWER

The energy of moving water is enormous, as anyone who has experienced whitewater rafting or a heavy coastal surf can testify. Indeed, water power has been used as a source of energy since the earliest of times. Prior to the Industrial Revolution, waterwheels powered grist mills, sawmills, cotton mills, and mining operations throughout western Europe and North America. Tidal mills have also been used since medieval times. More recently, water power has been applied to hydroelectric turbines, and efforts have been made to harness the energy in ocean waves and currents.

Tidal power

The ebb and flow of the tides, primarily under the influence of the Moon's gravitational attraction (see Chapter 17), have been used as a source of **tidal power** for more than 1000 years. Modern tidal power stations generate electricity by building a dam across a tidal basin. The basin is allowed to fill as the tide rises but is sealed at high tide, trapping the water as the tide begins to ebb. Once a sufficient head of water has developed, the basin is emptied through hydroelectric turbines, which turn and generate electricity from the motion of the emptying water (Fig. 19.24a). Alternatively, the focused flow of water

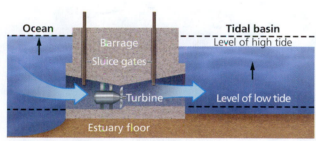

1. Basin fills as tide rises

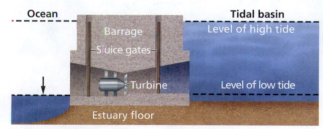

2. Sluice gates close until low tide

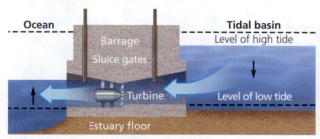

3. Sluice gates open at low tide and tidal basin drains
(a)

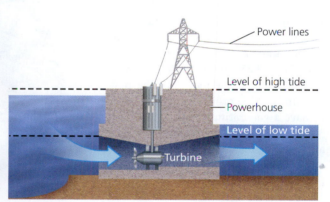

1. Tide coming in

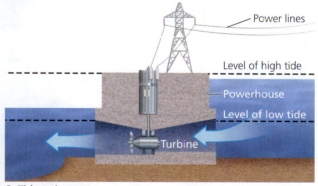

2. Tide going out
(b)

Figure 19.24   Tidal Power

Tidal power stations generate electricity in two ways: (a) by allowing a tidal basin to fill as the tide rises (step 1), trapping the water at high tide (step 2) and releasing it through hydroelectric turbines at low tide (step 3), or (b) by using the flow of water to power turbines as the tide comes in (step 1) and goes out (step 2).

associated with tides can be used to power turbines in the same way that wind is used to power windmills (Fig. 19.24b).

The sites of tidal power plants are consequently restricted to coastal areas where the *tidal range* (the difference in height between high and low tide) is at least 5 meters (16.5 feet) and a tidal basin of sufficient capacity exists. Worldwide, such sites number about a hundred. Several small plants have been built in Canada, Norway, China, Northern Ireland, and Russia, and others are being built in Scotland, India, and South Korea. The first tidal power station was built across the estuary of the River Rance in northwest France, where the tidal range is 8 to 13.5 meters (26 to 44 feet). Operating since 1967, the plant is capable of generating 240 megawatts of electricity from both the incoming and outgoing tides. The 254-megawatt Sihwa Lake facility in South Korea, completed in 2011, will remain the world's largest tidal power plant until a planned 1320-megawatt barrage is built in northern South Korea in 2017.

In the United States, tidal ranges of over 5 meters (16.5 feet) occur only in Alaska and Maine, where a pilot system near Eastport came on line in 2012. However, the world's greatest tidal range, of 16.3 meters (53.5 feet), occurs in Canada's Bay of Fundy. A tidal power project that would take advantage of this has been under consideration by Canada and the United States for more than 50 years. The proposed project, known as the Passamaquoddy Site (Fig. 19.25), involves damming a series of bays at the mouth of the Bay of Fundy. The project is anticipated to have a generating capacity of up to 345 megawatts. Feasibility studies have also been performed to explore the tidal power potential of the Severn Estuary in Britain.

Because of the limitations of tidal range and coastline shape, however, tidal power is unlikely to become anything more than a local energy option. Tidal power stations are also expensive to build, and while the energy source is pollution-free, the necessary interference with tidal flow is likely to be detrimental to the local marine life.

Wave energy

One only need witness a coastal storm to recognize the enormous amounts of energy contained in waves. But harnessing **wave energy** has proven difficult, and, to date, no large wave-powered electrical generating facility has been built, although an experimental wave farm opened in Portugal in 2008. Small wave-powered generators have also been used to power buoys and lighthouses, and a variety of wave power devices are being developed and tested in Europe, Australia, and the United States. Such systems are, however, susceptible to storm damage and corrosion by salt water.

Hydroelectric power

Hydroelectric power, which harnesses the potential energy of stream water as it flows downstream, is the world's primary source of renewable energy, accounting for 35 percent of total renewable energy in the United States and over 70 percent worldwide (see Fig. 19.1 and 19.2). To harness the energy effectively, however, a stream must first be dammed so that water from the top of the dam can be channeled through turbines at its base (Fig. 19.26).

Hydroelectric power, which was first used about 100 years ago in Appleton, Wisconsin, generates as much electricity today as 1300 medium-sized coal-burning power plants, and provides about 16 percent of the world's electricity needs. Indeed, Paraguay generates all of its electricity this way and Norway almost all, while Canada generates more hydroelectricity than it needs, exporting the surplus to the United States (see In Depth: **Dams and the Human Exploitation of Surface Water** in Chapter 13). China is the largest producer

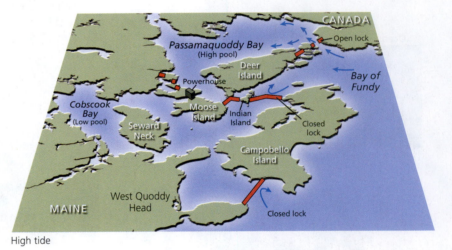

High tide

Low tide

Figure 19.25   Passamaquoddy Site

The proposed Passamaquoddy Site tidal power project, would be situated at the mouth of the Bay of Fundy.

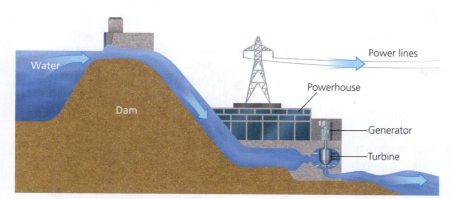

Figure 19.26   Hydroelectric Power Station

hydroelectric installation to as little as 20 years. There is also the potential threat of dam failure. For this reason, hydroelectric reservoirs should be installed in tectonically inactive regions.

Hydroelectric power currently supplies more than 60 countries with over half their electricity, and 14 large-scale projects are due to be completed in the next 5–10 years in China, India, Venezuela, and Myanmar, 10 of which are in China.

However, predictions of the future contribution of hydroelectric power to the world's energy consumption mix are complicated by the uncertain impact of climate change on worldwide stream flow, and it has been argued that the future of hydroelectricity is in small-scale generators on local rivers and streams rather than in massive new dams. In the United States, the potential for further development is likely to be limited since hydroelectric power is well established, and environmental and economic concerns are

of hydroelectricity, followed by Canada, Brazil, the United States, and Russia, the combined output of these countries accounting for more than half of the world's total supply. In the United States, the proportion of electricity generated hydroelectrically stands at 8 percent.

Hydroelectric power is commonly viewed as one of the simplest and most cost-effective forms of alternative energy. It is also easy to turn on and off, and so provides a useful source of energy during times of peak demand. In fact, many hydroelectric installations have built water storage facilities for just that reason. These function by using excess power during off-peak periods to pump water to a high reservoir so that it can be channeled back down through electrical turbines at times of peak demand (Fig. 19.27).

Although hydroelectricity is a clean energy source, it is not without disadvantages. Large hydroelectric installations are very expensive to build and may be difficult to justify economically unless oil prices are high. Hydroelectric reservoirs also drown vast tracts of land with significant ecological, social, and potential health consequences. In Egypt and Sudan, for example, 100,000 people were displaced when Lake Nasser was created behind the Aswan High Dam in 1970. The world's largest hydroelectric facility, the controversial Three Gorges Dam on the Yangtze River in China (completed in 2006), displaced a staggering 1.4 million people and flooded 13 cities, 140 towns, and 1350 villages (see Fig. 10C). Wildlife habitats were similarly impacted in the flooded area, and the presence of a reservoir and dam may irreversibly alter the ecological balance both upstream and down.

Because hydroelectric reservoirs are artificial basins, they are apt to become filled with sediment. Indeed, high sedimentation rates can limit the effective lifetime of a

> **CHECK YOUR UNDERSTANDING**
> ● How is hydroelectric power generated?

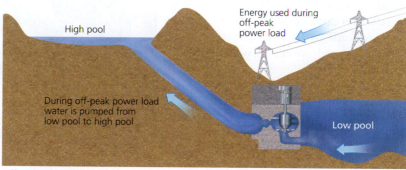

Pumping cycle

Energy used during off-peak power load

High pool

During off-peak power load water is pumped from low pool to high pool

Low pool

Off-peak power load

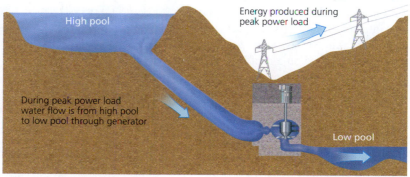

Generating cycle

Energy produced during peak power load

High pool

During peak power load water flow is from high pool to low pool through generator

Low pool

Peak power load

Figure 19.27   Hydroelectric Energy Storage

likely to preclude further dam building. Canada, on the other hand, has numerous plans to expand its hydroelectric resources.

ENERGY FROM THE ATMOSPHERE

Energy is obtained from the atmosphere either directly or indirectly from the Sun's radiation. Direct solar energy is derived from the Sun's heat. However, uneven heating of atmospheric gases by the Sun also causes the air to move, allowing us to obtain energy from the wind.

Solar energy

Direct use of **solar energy** from the Sun is an ancient technology and one that causes least harm to the environment. We witness the effect of passive solar radiation whenever we enter a warm, sunny room. Indeed, in the Northern Hemisphere many houses have windows facing south and east to take full advantage of this "greenhouse effect." Solar energy is also abundantly available, easily modulated, and readily accessible in remote areas. The process of converting solar energy into a usable form, however, is currently inefficient and expensive. Current solar energy technologies are of two types, thermal and photovoltaic. *Solar thermal devices* make use of the Sun's heat, and *photovoltaic devices* convert the Sun's energy directly into electricity.

Solar heating systems in homes typically involve thermal collection devices called *flat plate collectors*. These are black-backed, glass-fronted panels (Fig. 19.28a). Inside these panels fluid circulated through pipes is heated by the Sun. The heat is transferred to a storage tank and pumped to radiators (Fig. 19.28b). Usually mounted on the roof, flat plate collectors are used in many parts of the world to heat interior spaces, water supplies, and swimming pools.

The basic photovoltaic device is the *solar cell*, which is composed of semiconducting materials that produce electricity when sunlight is absorbed. To provide power, numerous interconnected cells are packaged into assemblies known as *solar panels*. Solar panels, either fixed on a roof (Fig. 19.29) or mounted to track the Sun, are used the world over to provide domestic electricity, and are particularly useful in remote regions.

Solar energy power plants operate on both the solar thermal and photovoltaic principles. Solar thermal plants make use of *concentrated solar power*. A field of parabolic mirrors, or *heliostats,* designed to track the Sun, focuses the sunlight onto a central power tower filled with a fluid, such as water

or molten salt, having the ability to store heat (Fig. 19.30a). Alternatively, a system of *parabolic troughs* is used, in which the parabolic mirrors are trough shaped and focus the sunlight on a fluid-filled glass absorber tube running the length of the trough (Fig. 19.30b). In both cases, the heat generated by the focused sunlight, which can reach 550°C (1020°F), is used to turn water into steam that drives a turbine to produce electricity.

Spain, with 25 solar thermal power stations, leads the world in utilizing concentrated solar power, and overtook the United States as the world's biggest producer of solar power in 2010. The world's largest solar power station, however, is the Ivanpah solar power facility, a heliostat project in the Mohave Desert southwest of Las Vegas. Opened in 2014,

(a)

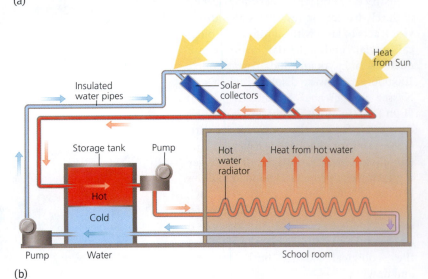

(b)

Figure 19.28   Solar Heating

(a) A flat plate collector is located on the roof of a hotel in Santorini, Greece. (b) Operation of a solar hot water heater using roof-mounted solar thermal collectors.

Figure 19.29   Solar Panels

Photovoltaic solar panels on the roof of a house near Boston, Massachusetts.

Figure 19.31   Photovoltaic Solar Park

Solar panels of the 250-megawatt California Valley Solar Ranch, the first phase of which was energized in 2012, under construction on the Carrizo Plain about 250 kilometers northwest of Los Angeles.

(a)

(b)

Figure 19.30   Concentrated Solar Power

(a) Solar thermal power plants such as the 11-megawatt PS10 (foreground) and 20-megawatt PS20 (background) generating stations near Seville in southern Spain, use parabolic mirrors (heliostats) to focus sunlight onto a central power tower. (b) The 354-megawatt Solar Energy Generating Systems (SEGS) installation in the Mohave Desert, California, a small portion of which is seen here, uses fields of parabolic troughs, each of which focuses sunlight onto a central tube.

the facility uses nearly 350,000 computer-controlled mirrors to focus sunlight on one of three 140-meter (459-foot) towers with a total generating capacity of 392 megawatts.

Photovoltaic power plants utilize large fields of Sun-tracking solar panels (Fig. 19.31). In recent years, such plants have been built in many parts of the world, especially Germany, where they currently produce 5 percent of the nation's electricity with plans to double this capacity by 2020. Indeed, solar photovoltaic power now ranks as the world's third most important renewable energy source after hydroelectricity and wind power. The largest plant is presently the 247-megawatt Agua Caliente project in Arizona, although plants of similar size have been built in India and China, and the Desert Sunlight and Topaz solar farms under construction in California are both 550-megawatt installations.

Solar energy is the most rapidly growing of all energy resources, the industry in the United States being one of the fastest growing sectors of the economy. The nation's total solar electric capacity grew from 4.4 to 7.6 gigawatts in 2011–2012, and is predicted to reach 26 megawatts by 2020. Indeed, according to a 2011 projection by the International Energy Agency, solar power generators could produce most of the world's demand for electricity, and half of all energy needs, by 2060, and dramatically reduce emissions of harmful greenhouse gases. But the cost would be high—the current price of solar energy per megawatt hour is several times that of coal and natural gas.

> **CHECK YOUR UNDERSTANDING**
>
> ◓ What are the two types of solar energy technologies and how do they work?

Wind power

Windmills have harnessed the power of the wind for grinding and pumping since antiquity, but their use for generating electricity on a commercial scale dates only to the 1980s. The potential of such **wind power** is enormous.

Figure 19.32   Wind Farm

The 619-megawatt San Gorgonio Pass wind farm, which straddles Interstate 10 near Palm Springs, California, is situated in one of the windiest places on Earth.

In the United States, for example, the electrical generating capacity of the wind has been estimated to be twice the country's existing generating capacity, with the steady winds in the wide-open states of Texas, North and South Dakota, Kansas, and Montana having the greatest potential power output. The wind is also a clean energy source. However, winds are subject to changing weather conditions and are strongly influenced by the local terrain, so the usable portion of this resource is quite small. Nevertheless, the United States currently has a wind generating capacity of over 60 gigawatts, second only to China at 80 gigawatts. In 2012, for example, wind-generated electricity in the United States amounted to 137 terawatt-hours, or enough to power 11 million homes. Wind power is also a rapidly growing energy resource, increasing at an annual rate of 15–20 percent. The worldwide wind-generating capacity, which amounted to almost 250 gigawatts in 2013, is expected to top 1000 gigawatts by 2025, and could account for more than 15 percent of global electricity usage by 2030.

Commercially, wind energy is harnessed by *wind farms* (Fig. 19.32). Wind farms are rows of *wind turbines* that are used to generate electricity. Sites are chosen on the basis of *wind power density,* a factor that takes into account the average wind speed, its variability, and the average air density. To date, the vast majority of sites utilized on this basis have been on land, but in Europe, wind farms are being increasingly installed offshore.

In the United States, wind farms exist in all but 11 states, with Texas leading the way with an installed capacity of 11 gigawatts, followed by California and Iowa, each with 4.5 gigawatts. Canada's wind power, which is expected to exceed 16 gigawatts by 2016, comes mostly from wind farms in Ontario, Quebec, and Alberta. America's largest wind farm is the Alta Wind Energy Center in California, with a capacity of just over 1 gigawatt. The Gansu Wind Farm in China has a current capacity of 6 gigawatts and is expected to grow to 20 gigawatts by 2020.

ENERGY FROM BIOMASS

The oldest energy source of all is the burning of organic matter, its use dating back to the discovery of fire. Today, this resource is termed **biomass**. Biomass is a renewable energy source (because it can be replenished); it includes wood, municipal solid wastes, and biomass-derived liquid fuels such as ethanol. Biomass is currently the most important renewable energy source overall, and is second only to hydropower in the generation of electricity. In the United States, biomass provides over 4 quadrillion Btu of energy, or almost 50 percent of total renewable energy consumption (see Fig. 19.1), and almost 60 billion kilowatt hours (or about 1.5 percent) of its electricity. Worldwide, biomass is a widely utilized source of energy, accounting for as much as 35 percent of the total energy supply in developing countries, where it is used mostly for cooking and heating (see Fig. 19.2).

The most familiar form of biomass energy is firewood, most of which is used for residential heating. Most biomass energy, however, is derived from three main sources: wood waste, such as the waste produced by the pulp and paper industry, *biofuels* produced from crops, and biomass derived from municipal solid waste, manufacturing waste, and landfill gas.

Like firewood, most biomass is simply burnt, and the heat produced is used to generate electricity, or steam for commercial heating. In the United States, hundreds of power plants and about 90 *waste-to-energy* installations (Fig. 19.33) utilize biomass in this fashion. Worldwide, more than 2000 biomass power plants (over 1000 in Europe alone) currently provide about 1.5 percent of global electricity demands—a figure that has been predicted to increase to as much as 10 percent by 2030.

But biomass can also be converted to other forms of usable energy. Crops such as corn and sugar cane can be fermented to produce the ethanol used in the gasoline

> **CHECK YOUR UNDERSTANDING**
> • What is meant by wind power density and how does it apply to the use of wind energy?

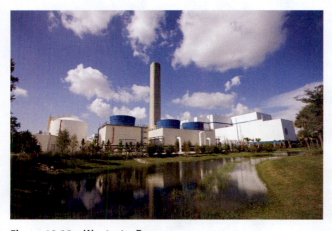

Figure 19.33 Waste-to-Energy

Lee County waste-to-energy facility, is a 53-megawatt biomass power plant near Fort Myers, Florida.

additive "gasohol," while waste vegetable oils and animal fats are used to produce *biodiesel,* another transportation fuel. The United States and Brazil currently lead the world in ethanol production, while the European Union is the world's largest biodiesel producer. Together these two biofuels currently make up nearly 3 percent of the world's road transportation fuel, and have the potential to make up more than a quarter by 2050.

Although burning biomass releases carbon dioxide to the atmosphere just as fossil fuels do, biomass energy is cleaner because the carbon dioxide released is simply that which the plants removed from the atmosphere while they were growing. Hence, biomass is carbon neutral since it contributes no "new carbon" to the atmosphere. Fossil fuels, on the other hand, do, because the carbon dioxide they release was removed from the atmosphere many millions of years ago.

Not surprisingly, the use of energy from biomass is very dependent upon proximity to a source of supply. In the United States, most energy obtained from wood is consumed in the forested East, Southeast, and Northwest; ethanol production is greatest amid the cornfields of the Midwest; and consumption of energy generated from waste is highest in population centers. Worldwide, energy from biomass accounts for almost 15 percent of the total energy supply, and is rapidly growing. Predictions by the International Energy Agency suggest that biomass energy will be producing about 10 percent of the world's electricity by 2030, and that consumption of biomass power and biofuels will have grown fourfold by 2035.

> **CHECK YOUR UNDERSTANDING**
>
> ◗ What are three main sources of biomass energy?

19.5 SUMMARY

- Renewable energy sources include geothermal energy (which uses natural heat from Earth's interior); tidal, wave, and hydroelectric energy (derived from water power); solar and wind energy (obtained from the atmosphere); and biomass energy (from burning organic matter).

- Renewable energy sources provide only 8 percent of the world's energy needs, but do not contribute to global warming and are considered by many to be the energy sources of the future.

19.6 Plate Tectonics and Energy Resources

Many of the energy resources we have discussed in this chapter occur in specific geologic settings created by plate tectonics. This is particularly true of the fossils fuels and uranium. The distribution of these resources is a consequence of plate tectonic history. An even closer link to plate tectonics exists for geothermal power, and even hydroelectricity and wind power require specific geographic settings that are directly or indirectly the result of plate tectonics.

PETROLEUM AND PLATE TECTONICS

Oil and natural gas have been recovered from sedimentary basins associated with all three types of plate boundaries: divergent, convergent, and transform (Fig 19.34). In the early stages of plate divergence, continental rifts and narrow ocean basins are often sites of organic-rich shale

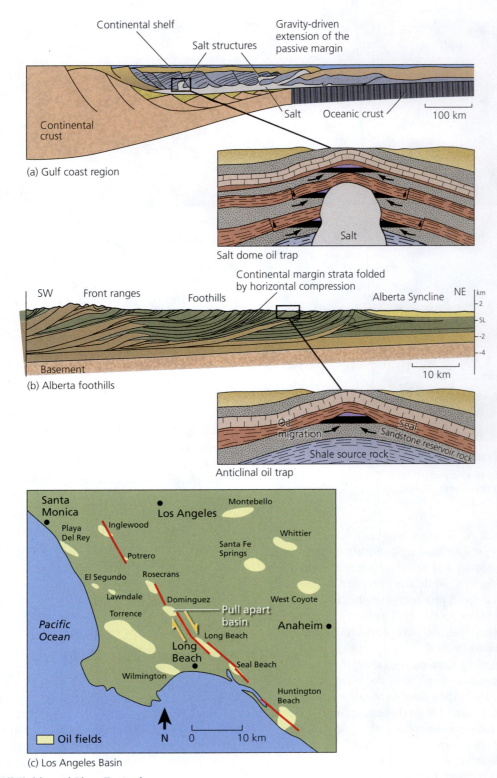

Figure 19.34   Oil Fields and Plate Tectonics

All types of plate boundaries influence the location of oil fields. (a) At divergent plate boundaries complex faulting and salt movement create traps for oil in the source and reservoir rocks of passive continental margins. The oil found in the Gulf coast region illustrates this series of events. (b) At convergent plate boundaries, compression of continental margin sediments traps oil in fold-thrust belts. This is the situations in the Alberta Foothills region of the Canadian Rockies. (c) The linear distribution of oil fields in the Los Angeles Basin of southern California reflects their close association with faults of the San Andreas transform plate boundary.

and salt deposition. This is because marine conditions are restricted and circulation is poor. These conditions foster low levels of dissolved oxygen as well as evaporation where climates are warm. Later, as the ocean opens, sediments eroded from the continent, such as sandstones, are deposited on the developing continental shelf. The continental shelf environment may therefore contain all the elements necessary for the formation and storage of petroleum deposits. Once buried, organic-rich shale provides a potential source rock and sandstones may form reservoirs. Salt provides an excellent seal and can form traps in which petroleum can accumulate. Just such a combination is responsible for the development of oil and natural gas fields along the coast of the Gulf of Mexico (Fig 19.34a).

At convergent plate boundaries, continental margin strata folded by horizontal compression produce anticlinal structural traps. Such is the case, for example, in some of the oil and natural gas fields in the mountain states of the western United States and the foothills of the Canadian Rockies (Fig 19.34b).

Transform plate boundaries are also a setting of sedimentary basins with important petroleum potential. In California, movement on faults associated with the San Andreas transform has created pull-apart basins (see Chapter 10) in which both source rocks and reservoir rocks have accumulated (Fig 19.34c). Fault movement has also deformed the sedimentary basin fill, resulting in the development of structural traps.

> **CHECK YOUR UNDERSTANDING**
>
> ● How do divergent, convergent, and transform boundaries foster the formation of petroleum deposits?

tropical conditions ideal for coal formation (Fig. 19.35). Sediments shed from the rising Appalachian Mountains (and their equivalent in Europe) were continually burying the developing coal swamps. Recall that the continuity of the coal deposits and mountain ranges from North America to Europe played an important role in Wegener's argument for the existence of the supercontinent Pangea. During this period, many of the southern continents of Gondwana lay in the south polar region and were experiencing the very ice age that Wegener also used in support of his hypothesis of Continental Drift. Although these conditions were unfavorable for coal formation in the southern continents, the burial of plant material near the Carboniferous equator was encouraged by the sea level changes associated with the advance and retreat of the polar ice sheet in the Southern Hemisphere. Hence, the formation of the world's Carboniferous coal deposits was strongly influenced by plate tectonics and the resulting geography of Pangea.

In the western United States, similar climatic conditions to those of the Late Carboniferous developed during the Cretaceous and Tertiary. A huge inland seaway extended from the Gulf Coast to Arctic Canada east of the developing Rocky Mountains as a result of the rise in sea level associated with the breakup of Pangea (see In Depth: **Sea Level and Plate Tectonics** in Chapter 17).

> **CHECK YOUR UNDERSTANDING**
>
> ● How are coal deposits and plate tectonics linked?

It was during the Cretaceous and Tertiary that much of the coal of western North America was deposited.

COAL DEPOSITS AND PLATE TECTONICS

The distribution of coal deposits has not only been controlled by plate movement but also was instrumental in Alfred Wegener's case for continental drift. Recall that coal deposits are far more abundant in the Northern Hemisphere than they are in the Southern Hemisphere. Because coal is a swamp deposit formed from plant material buried beneath sediment, this bias is largely a function of the world's geography at the time of the coal's formation.

Most of the world's coal deposits formed during the Late Carboniferous period, soon after the first extensive land colonization of plants in the Devonian and just as the supercontinent Pangea was assembling. At that time, much of what is now eastern North America and western Europe (which both host vast coal deposits) occupied positions near the equator and were experiencing warm, wet,

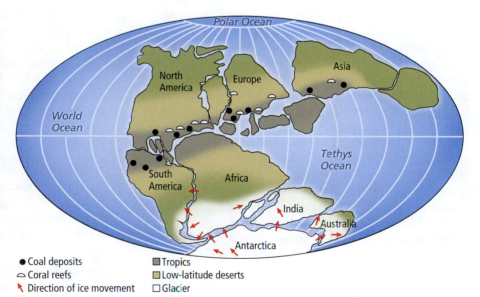

● Coal deposits
⌒ Coral reefs
↖ Direction of ice movement

■ Tropics
■ Low-latitude deserts
□ Glacier

Figure 19.35   Carboniferous Coal

The vast coal reserves of eastern North America and western Europe, together with those in southern Russia and China, owe their origin to the geography of Pangea and the existence of tropical conditions ideal for coal formation during the Carboniferous period.

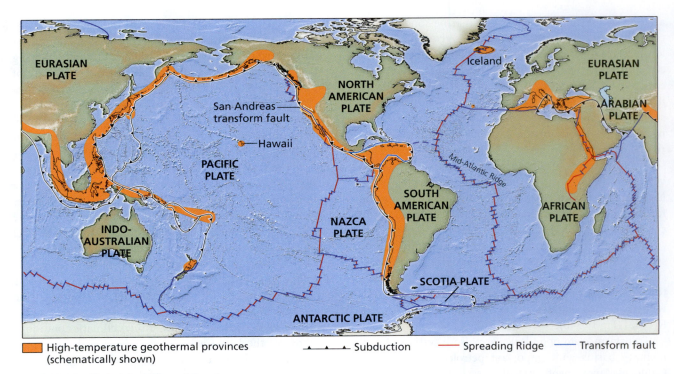

| High-temperature geothermal provinces (schematically shown) | ⊥—⊥—⊥ Subduction | — Spreading Ridge | — Transform fault |

Figure 19.36 Global Geothermal Provinces

Most geothermal fields, such as those of the Pacific Ring of Fire and Iceland, lie close to volcanically active plate boundaries or over volcanic hotspots, such as Hawaii, where hot rocks are close to the surface.

NUCLEAR ENERGY AND PLATE TECTONICS

Recall from Chapter 18 that the distribution of mineral deposits is closely related to the plate tectonic history of the areas in which they occur. This is why the relationship between the two can be used to guide exploration. Uranium mineralization is no exception and the formation of uranium ore that powers the nuclear industry is strongly linked to plate tectonics.

Although most economic deposits of uranium occur in sedimentary rocks, either as a result of precipitation from groundwater or through accumulation in placer deposits, uranium minerals are also common in vein deposits associated with the late stages of granite emplacement. This is because uranium is not readily incorporated into early crystallizing minerals in a magma and so becomes concentrated in the fluid left over as crystallization proceeds (see Chapter 18). These vein deposits of uranium minerals are consequently linked to the formation of granites, most of which occur in the magmatic arcs and collisional mountains of convergent plate boundaries, or are associated with the early stages of continental rifting at divergent plate boundaries.

RENEWABLE ENERGY AND PLATE TECTONICS

Of all energy sources, geothermal power is the most closely linked to plate tectonics. Geothermal power requires the existence of very hot rocks close to the surface, and therefore is generally harnessed in areas of volcanic activity. Most geothermal plants are located either at convergent plate boundaries, such as the Pacific Ring of Fire, or at divergent plate boundaries, such as Iceland, because these are the plate tectonic settings where volcanoes occur. Hotspots, such as Hawaii, also provide geothermal power (Fig. 19.36).

The link between plate tectonics and the other renewable energy sources—water, wind, solar, and biomass—is less obvious. However, hydroelectric power and wind energy do require specific drainage patterns and landscapes, while the availability of sunshine and the production of biomass depend on climate and topography. Hence, each of these energy sources is best developed in specific geographic settings, and these settings, like all of Earth's geography, are ultimately the product of the motion of Earth's plates.

19.6 SUMMARY

- In providing a mechanism for the development of petroleum-producing sedimentary basins and the formation of stratigraphic and structural traps at plate boundaries, plate tectonics plays an important role in the distribution and development of petroleum deposits

- Most coal was formed as Pangea was assembling in the Late Carboniferous and the global distribution of coal is strongly influenced by the plate movements responsible for the supercontinent's geography.

- Geothermal power depends on the existence of very hot rocks close to Earth's surface; it is generally harnessed in areas of volcanic activity, most of which lie along plate boundaries.

Key Terms

biomass 636	natural gas 608	peat 623	solar energy 634
coal 622	nonrenewable	petroleum 608	source rock 610
fossil fuels 607	resource 606	renewable resource 606	tar sands 615
geothermal energy 630	nuclear fission 627	reservoir rock 610	tidal power 631
hydraulic fracturing	oil 608	seal 610	wave energy 632
(fracking) 618	oil shale 613	shale gas 616	wind power 635
hydroelectric power 607	oil trap 610		

Key Concepts

19.1 ENERGY USE

- Energy is used in huge amounts to power industry, control the environment within buildings, and transport people and goods.

- Almost all of the world's energy is obtained from nonrenewable resources.

- Fossil fuels are nonrenewable energy sources of organic origin. They provide most of the world's energy but their consumption is responsible for climate change.

19.2 PETROLEUM—OIL AND NATURAL GAS

- Petroleum deposits require a source rock, a reservoir rock, a seal, and a trap.

- Oil traps can be structural or stratigraphic.

- Oil shales, tar sands, and shale gas are unconventional sources of petroleum.

- Petroleum exploration makes use of geophysical techniques and drilling.

19.3 COAL

- Coal forms when plant material in swamps is progressively buried.

- Coal is relatively inexpensive but mining and burning coal has environmental consequences.

19.4 NUCLEAR POWER

- Nuclear power is driven by fission and utilizes the radioactivity of naturally occurring uranium.

19.5 RENEWABLE ENERGY

- Renewable energy sources include geothermal energy, hydroelectric power, solar and wind energy, and biomass energy.

- Renewable energy sources do not contribute to global warming and so may become the energy sources of the future.

19.6 PLATE TECTONICS AND ENERGY RESOURCES

- In providing settings for the formation and accumulation of petroleum, the environments for coal formation, uranium mineralization and geothermal activity, and the geographic conditions for wind, solar, and biomass energy, plate tectonics plays an important role in all energy sources.

Study Questions

1. How has humanity's use of fuel evolved through time?

2. How does the energy use of industrial nations differ from that of less industrialized societies?

3. Which factors determine whether buried organic matter is converted to petroleum or coal?

4. At current levels of consumption, what would be the longevity of the world's supply of oil and natural gas, and why is it difficult to be certain of this estimate?

5. Describe the process by which coal is formed by the progressive burial of plant matter.

6. What is the difference between conventional reactors, such as those of the light-water variety, and breeder reactors?

7. Why is the safe disposal of nuclear waste difficult to achieve?

8. Compare and contrast the relative merits of the renewable sources of energy based on water power and those obtained from the atmosphere.

9. Which factors favored the formation of coal deposits in northeastern North America and western Europe during the Carboniferous period?

10. The US Energy Information Administration foresees no immediate reduction in our dependency on fossil fuels. Why do you think this is so?

PART **V**
Earth and Beyond

Chapter 20 | Physical Geology
and Climate
Change   644

Chapter 21 | The Planets   678

20

Physical Geology and Climate Change

20.1 Taking Earth's Pulse

20.2 The Greenhouse Effect: A Matter of Balance

20.3 The Carbon Cycle

20.4 Climate Change: A Geologic Perspective

20.5 Earth-Sun Geometry

20.6 Feedback Systems

20.7 Plate Tectonics and Climate Change

20.8 Putting It All Together: The Greenhouse Effect and Global Warmth

▶ Key Terms

▶ Key Concepts

▶ Study Questions

LIVING ON EARTH: Camels in the Arctic! 648

SCIENCE REFRESHER: The Physics of Global Warming 652

IN DEPTH: Melting of the Arctic Ice—An Omen of Global Warming? 666

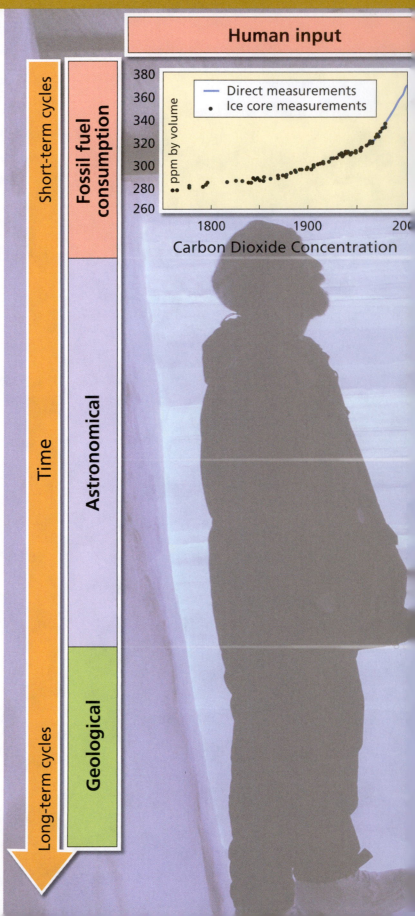

Human input

Short-term cycles

Fossil fuel consumption

Time

Astronomical

Long-term cycles

Geological

— Direct measurements
• Ice core measurements

ppm by volume

Carbon Dioxide Concentration

Physical phenomena	Measurements

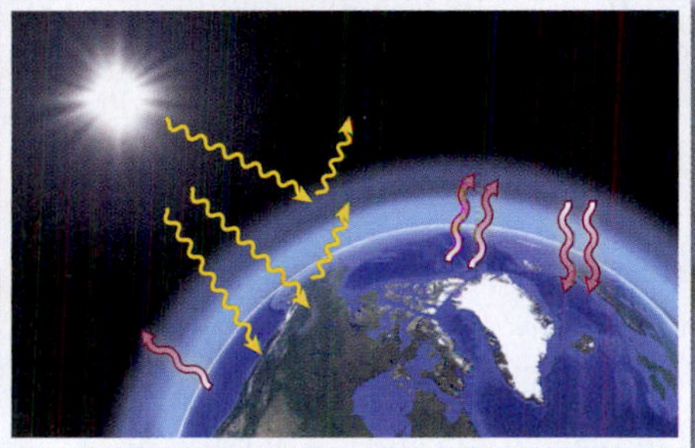

Greenhouse effect

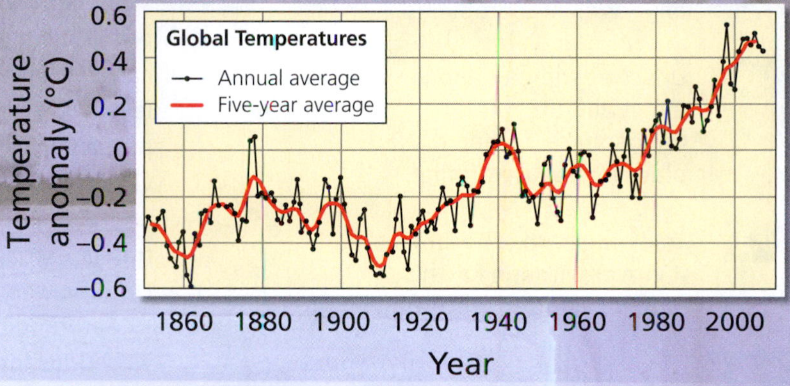

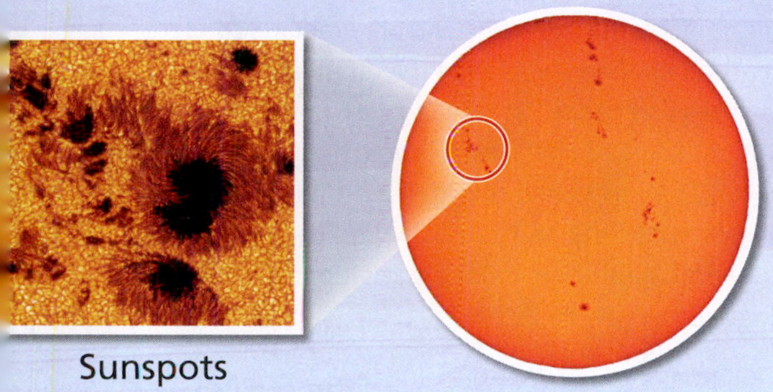

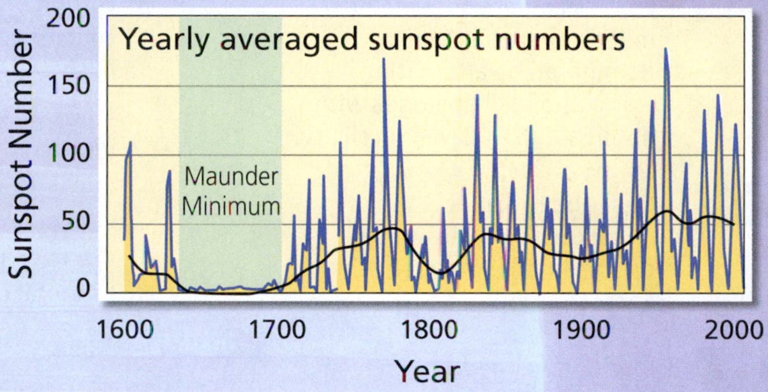

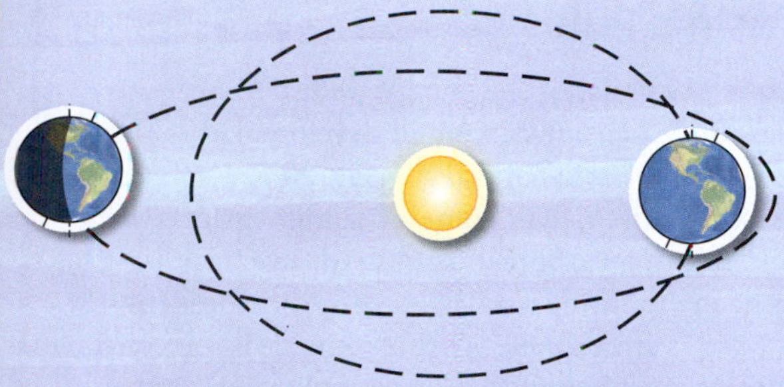

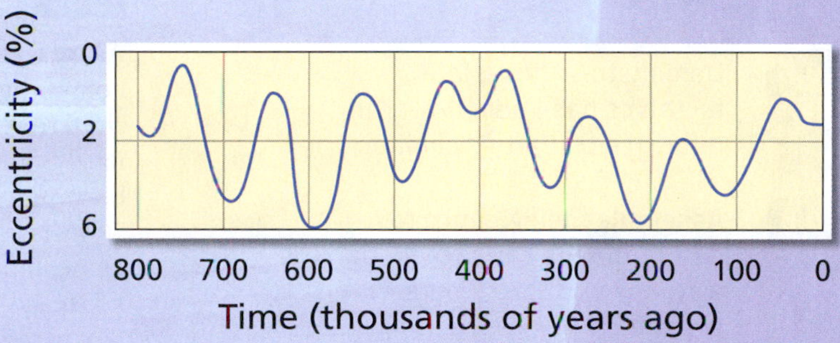

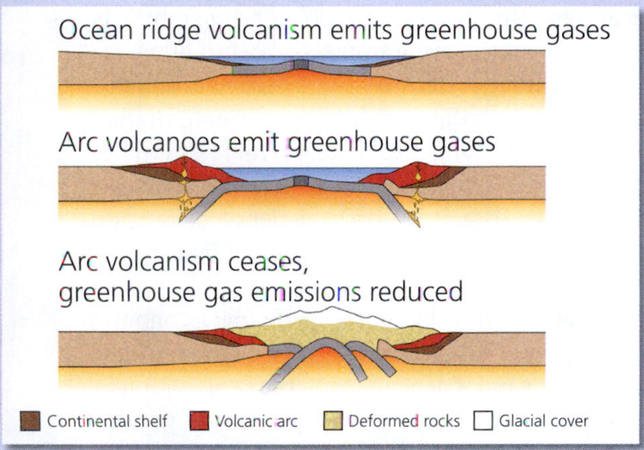

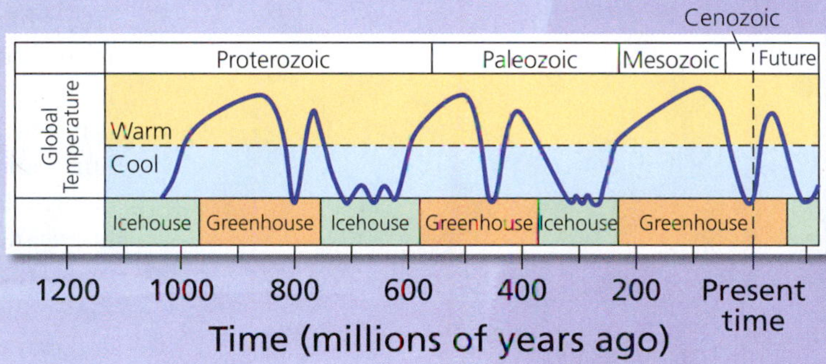

Global Temperatures
Annual average
Five-year average
Temperature anomaly (°C)
0.6
0.4
0.2
0
−0.2
−0.4
−0.6
1860
1880
1900
1920
1940
1960
1980
2000
Year

Sunspots

Yearly averaged sunspot numbers
Maunder Minimum
Sunspot Number
200
150
100
50
0
1600
1700
1800
1900
2000
Year

Eccentricity of Earth's orbit

Eccentricity (%)
0
2
6
800
700
600
500
400
300
200
100
0
Time (thousands of years ago)

Ocean ridge volcanism emits greenhouse gases
Arc volcanoes emit greenhouse gases
Arc volcanism ceases, greenhouse gas emissions reduced
Continental shelf
Volcanic arc
Deformed rocks
Glacial cover
Plate tectonics

Cenozoic
Proterozoic
Paleozoic
Mesozoic
Future
Global Temperature
Warm
Cool
Icehouse
Greenhouse
Icehouse
Greenhouse
Icehouse
Greenhouse
1200
1000
800
600
400
200
Present time
Time (millions of years ago)

20.1 Explain how understanding geologic processes and the geologic record provides a unique insight into modern environmental problems.

20.2 Describe the greenhouse effect and the role atmospheric carbon dioxide plays in global warming.

20.3 Track the path of carbon as it moves through Earth's surface environments.

20.4 Summarize the evidence that Earth's climate over the past million years has been dominated by alternating warm and cool episodes with about a 100,000-year cyclicity.

20.5 Describe the effects the Milankovitch Cycles have on the solar energy reaching Earth.

20.6 Identify the feedback systems that play a role in climate change.

20.7 Demonstrate how plate tectonics has had a powerful influence on Earth's climate.

20.8 Assemble the evidence to determine the cause of global warming and the climate change Earth is witnessing today.

There is no question that Earth's climate is changing. The weather we now experience is not what your parent's generation remembers, nor is it the weather your children will experience. We know what is causing this change. The culprit is rising concentrations of the so-called *greenhouse gas*, carbon dioxide, in our atmosphere. Atmospheric carbon dioxide levels and global temperature are tightly coupled by the laws of physics. If carbon dioxide levels go up, so does temperature, and vice versa. We also know that atmospheric carbon dioxide levels and global temperature have been rising in lockstep ever since the Industrial Revolution gathered momentum in the nineteenth century, when the burning of fossil fuels started in earnest.

Carbon dioxide gas has been vented to the atmosphere from volcanoes throughout Earth history and variations in atmospheric carbon dioxide concentrations are nothing new. Indeed, over the course of geologic time, there have been far more dramatic changes in atmospheric carbon dioxide levels than the relatively subtle increases we are witnessing today. So how do we know that these increases are not just part of some natural cycle that we are powerless to influence? The answer to this question has enormous implications for global society and the way we live. And as we discover in this chapter, geology is uniquely positioned to answer this vital question.

20.1 Taking Earth's Pulse

As we learned in Chapter 19, most of the energy we use to heat our homes, light our cities, and fuel our economic development is derived from fossil fuels—coal, oil, and natural gas (Fig. 20.1). Unfortunately, the use of these resources has come at a high price. The burning of fossil fuels releases carbon dioxide and, over the past 30 years, evidence has been mounting that this carbon dioxide is fundamentally changing Earth's atmosphere. Carbon dioxide is a **greenhouse gas**. The presence of greenhouse gases in the atmosphere traps heat in the same way that glass traps heat inside a greenhouse (Fig. 20.2). Rising atmospheric carbon dioxide levels therefore are linked to rising global temperatures. This increase in global temperature is called **global warming**. Global warming triggers a host of other related changes in Earth's weather patterns—such as stronger storms, more frequent hurricanes, record heat and drought. **Climate change** refers to significant and lasting changes to these weather patterns. It is important to note that a single significant weather event, such as a hurricane of record strength, is not by itself evidence of climate change. However, a pattern of increasing frequency and intensity of hurricanes over several decades is.

At the same time that Earth's climate is changing, economic forces dictate more energy consumption, and many communities and nations grow increasingly more dependent on the exploitation of fossil fuels and the employment it creates. The question of climate

Figure 20.1 North America at Night

Most of the energy North Americans use to light cities comes from fossil fuels—coal, oil, and natural gas.

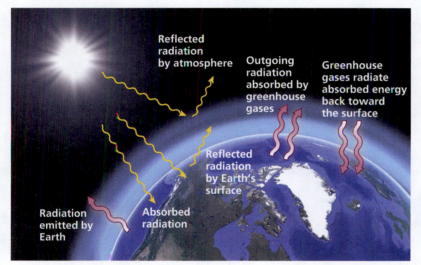

Figure 20.2 Greenhouse Gases

About 50 percent of solar radiation passes through the atmosphere and is reflected or absorbed by Earth's surface. Reflected radiation has the same short wavelength as incoming solar radiation and escapes back into space. But absorbed radiation is emitted from Earth's surface as heat, which has a longer wavelength and is partially absorbed by greenhouse gases in the atmosphere. As a result, the greenhouse gas molecules become excited and reradiate some of this heat energy back toward Earth's surface, which warms the planet.

change and our use of fossil fuels is one of the most serious issues facing modern society. We may have to change the way we live and fundamentally restructure our traditional economic base, or our climate may be irreversibly altered in ways that will be detrimental to our existence.

In 2014, the Intergovernmental Panel on Climate Change (IPCC), an international council of scientists and government representatives set up by the United Nations at the request of the United States, published a detailed report that includes the following findings:

- Warming of the climate is unequivocal and many of the changes that have been observed in Earth's climate

since the 1950s are unprecedented over decades to millennia.

- Atmospheric concentrations of greenhouse gases, such as carbon dioxide, methane, and nitrous oxide, have increased to levels unprecedented in at least the last 800,000 years.

- Human influence on the climate system is clear. It is 95–100 percent probable that human influence has been the dominant cause of global warming since 1951.

- Continued emissions of greenhouse gases will cause further global warming and changes in all components of the climate system.

- Limiting climate change will require substantial and sustained reductions of greenhouse gas emissions.

- Most aspects of climate change will persist for many centuries even if emissions of carbon dioxide are stopped.

In previous chapters in this book, we have discussed how the geologic record provides the perspective of deep time. This record tells us that Earth's climate has oscillated between periods of extreme warmth, referred to as **greenhouse conditions**, and extreme cold, referred to as **icehouse conditions**. Clearly, our modern world lies tolerably between these two extremes. So the geologic record presents us with a challenge. If nature can generate such extreme conditions, how can the Intergovernmental Panel on Climate Change be so certain that the subtle changes in climate we are witnessing today are human induced and not just part of some natural process (see Living on Earth: **Camels in the Arctic!** on page 648)?

In this chapter, we will discuss how nature works in cycles, and that these cycles have rhythms, just like the beat of a song. If a familiar song is cut off before it ends, we can usually continue to hum the tune because we recognize its rhythmic beat. In a similar way, if we are to interpret climate change in today's world, we must first recognize the climatic rhythms of Earth's geologic past. The accuracy of our predictions depends on how well we have learned these past rhythms. It may seem that 200 years of industrialization is a trivially short interval in a song lasting 4.54 billion years. Yet we can still seek distortions in this song, and even brief distortions are instantly recognizable just as they would be in a recording of familiar music. So, if human activity represents a distortion in the evolution of our planet, it should be readily identified. However, this is possible if, and only if, we know enough about Earth's history to recognize its natural pulse. Geology offers a unique perspective on modern environmental issues such as climate change because it has the potential to distinguish natural from human-induced global change.

Due to its relevance to modern society, the science of climate change is the subject of vigorous research. In this

Camels in the Arctic!

Targeted and careful field research combined with rigorous laboratory analyses of samples can provide powerful insights into ancient climates. This type of research is essential if we are to distinguish climate change induced by human activities from that induced by natural processes. The field expeditions led by Dr. Natalia Rybczynski of the Canadian Museum of Natural History to Ellesmere Island, high in the Canadian Arctic, provide an excellent example of this approach.

Because of modern global warming, strata previously hidden beneath glacial ice have been recently exposed. These new exposures are readily identified on satellite images. Working in strata about 3.5 million years old (i.e., mid-Pliocene in age), Dr. Rybczynski's research team uncovered a bone fragment less than the length of a human thumb. Laboratory analysis has revealed that naturally occurring proteins preserved in the bone are a perfect match for a camel. Further investigations yielded about 30 bone fragments that together comprise part of the limb bone of a camel (Fig. 20A).

Fossils in nearby layers indicate that local vegetation would have resembled a northern (boreal) forest (Fig. 20B), in contrast to its stark modern landscape. The mid-Pliocene has been established as a warm interval, with average global temperatures estimated to be about 2–3°C (4–5°F) warmer than today. But these camel bones and plant fossils indicate that the average temperatures in the high Arctic at the time were about 14 to 22°C warmer than today.

This find is about 1200 kilometers farther north than any previous camel fossil discovery. The camels in the high Arctic must have endured months of continuous darkness during the winter. The researchers speculate that the camel evolved several features, including its hump, which stored fat, to cope with the darkness.

This research also highlights the sensitivity of the Arctic to climate change in the past and may well be a harbinger for the changes in the Arctic that we will experience this century.

Figure 20A *Camel Bones in the Arctic*
These fossils comprise part of a limb bone of a camel.

Figure 20B *Ellesmere Island in the Pliocene*
This portrayal of the habitat in Ellesmere Island in the high Arctic some 3.5 million years ago is based on fossils in nearby strata.

chapter, we demonstrate that a knowledge and understanding of geologic processes and the geologic record provides a unique and essential insight into our modern environmental problems. In the process, we hope to make it clear why the Intergovernmental Panel on Climate Change speaks with such certainty in its 2014 report about the role humans play in modern climate change.

CHECK YOUR UNDERSTANDING

○ What is global warming?

○ What is climate change?

○ What unique perspective does geology offer to the study of climate change?

20.1 SUMMARY

- Rising atmospheric carbon dioxide levels are linked to rising global temperatures.

- The analysis of current climate change depends on our ability to distinguish between natural changes and changes caused by human activity.

20.2 The Greenhouse Effect: A Matter of Balance

Solar radiation that is absorbed and emitted from Earth's surface has a longer (infrared) wavelength than incoming solar radiation. Greenhouse gases, which we introduced in section 20.1, are gases that allow solar radiation to enter the atmosphere, but block the escape of the infrared radiation emitted from Earth's surface. In doing so, the gases become excited and reradiate some of this heat energy back toward Earth's surface, thereby trapping heat, much like the panes of glass in a greenhouse. The term **greenhouse effect** (see Science Refresher: **The Physics of Global Warming** on page 652) emphasizes that sunlight's influence on our atmosphere is similar to the Sun's effect on a greenhouse (see Fig. 20.2). On a sunny day, it is warmer inside a greenhouse than outside because the glass panes let the sunshine in but do not allow all of the warmth the sunshine creates to escape.

The most abundant greenhouse gases are water vapor (H_2O), carbon dioxide (CO_2), methane (CH_4), nitrous oxide (N_2O), and ozone (O_3); together they comprise a little under 1 percent of our atmosphere. The other approximately 99 percent of the gases in our atmosphere (nitrogen, 78%, and oxygen, 21%) do not act as greenhouse gases because they readily permit both the entry

of solar radiation from space and the exit of the infrared radiation that is emitted from Earth's surface.

It is commonly assumed that the presence of greenhouse gases in our atmosphere is bad. This is not true. Without the greenhouse effect Earth would be, on average, 35°C (63°F) degrees colder: an uninhabitable, icy planet. On the other hand, too much greenhouse gas in the atmosphere and the oceans would evaporate. For planet Earth to remain hospitable, a delicate balance is needed in the atmospheric content of greenhouse gases.

The Sun continuously bathes Earth with waves of radiation. The Sun, with a surface temperature of about 5700°C (10,300°F), emits solar radiation (known as the *electromagnetic spectrum*) with wavelengths that range from 10^{-12} meters to 10^4 meters (Fig. 20.3). Our eyes can only detect a very narrow range of these wavelengths, which is known as the *visible spectrum*. Within this range, our eyes resolve the wavelengths into various colors, which we see as the colors of the rainbow. Not all wavelengths of solar radiation are of equal intensity. The vast majority of the radiation has wavelengths between 10^{-8} meters and 10^{-6} meters (that is, from infrared to ultraviolet) and incorporates all the colors of the visible spectrum. The dominant wavelength is 5.0×10^{-7} meters, which also falls within the visible portion of the spectrum.

Although humans cannot see wavelengths outside of the visible spectrum, we know that radiation of other wavelengths exists because we can detect its influence. For example, we feel invisible infrared (IR) radiation as heat, and excessive exposure to invisible ultraviolet (UV) solar radiation can lead to sunburn. The characteristic wavelengths of IR and UV radiation are 10^{-6} meters and 10^{-8} meters, respectively, both of which lie outside the range of our eyes' sensitivity.

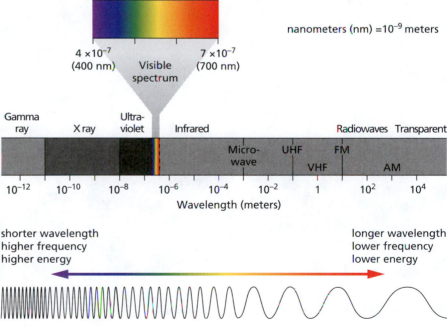

Figure 20.3   The Electromagnetic Spectrum

Solar energy encompasses a range of wavelengths, known as the electromagnetic spectrum, which vary from very short wavelength gamma rays (10^{-12} m) to very long wavelength radio waves (10^4 m).

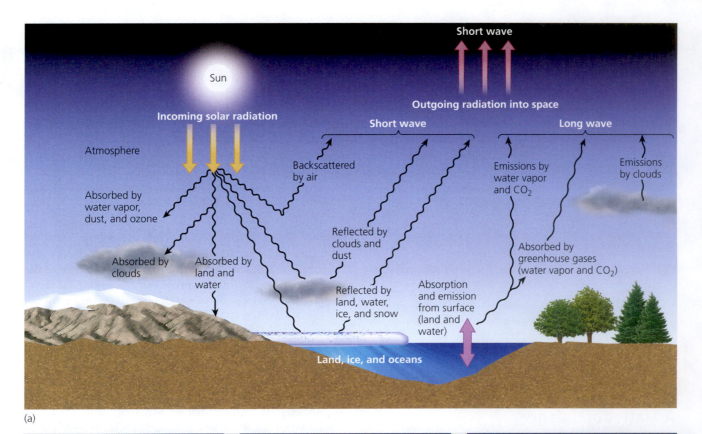

(a)

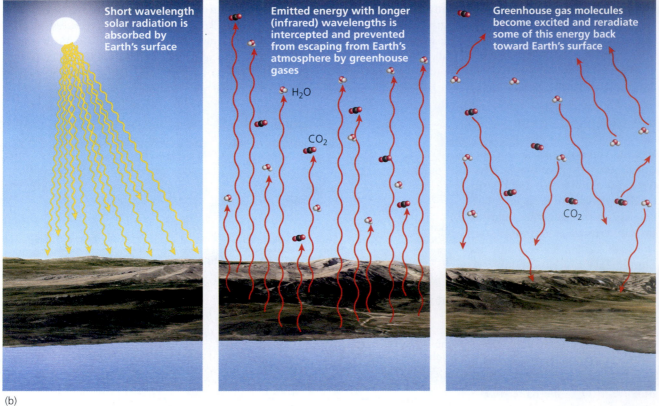

(b)

Figure 20.4   Solar Radiation and the Greenhouse Effect

(a) A good deal of solar radiation does not reach Earth's surface; it is either absorbed in the atmosphere or reflected back into space. Reflected energy has the same wavelength as incoming solar radiation and escapes into space. But absorbed energy emitted from the surface has a longer wavelength and the escape of this outgoing radiation is partially blocked by greenhouse gases. (b) This detailed view of the greenhouse effect illustrates how short wavelength solar radiation is absorbed by Earth's surface. This absorbed energy is emitted from the surface as heat, with a longer (infrared) wavelength that greenhouse gases intercept and block from escaping the atmosphere. The greenhouse gas molecules become excited and reradiate some of this heat energy back toward Earth's surface.

Solar radiation encounters numerous obstacles as it attempts to penetrate our atmosphere (Fig. 20.4a). Radiation is scattered upon entry, it is reflected by clouds and dust, and it is absorbed by gases such as water vapor and ozone. It is also both reflected and absorbed by Earth's surface. Energy reflected by Earth's surface passes through the atmosphere and escapes back into space because it has the same short wavelength as incoming solar radiation. Absorbed energy, on the other hand, heats the rocks and soils of Earth's surface and causes moisture to evaporate from the oceans.

Radiation absorbed by Earth's surface is consequently emitted as heat, with a longer (infrared) wavelength than the incoming solar radiation. This longer wavelength radiation is absorbed by greenhouse gases in the atmosphere and prevented from escaping (see Fig. 20.4b). The higher the concentration of greenhouse gases in the atmosphere, the more heat is trapped. This trapped heat energy, in turn, is radiated back toward Earth's surface, greatly influencing global temperatures, atmospheric circulation, evaporation from oceans, motion of ocean currents, and weather patterns.

The current debate on climate change has focused in particular on the atmospheric concentration of carbon dioxide. Given that there are several greenhouse gases, why has carbon dioxide been singled out? Water vapor is a very effective greenhouse gas (Table 20.1). This explains why daytime temperatures typically drop more dramatically on a clear night than they do on a cloudy one—because clear air has less water vapor in it. However, water vapor cycles through the atmosphere in a matter of days, returning to Earth's surface as precipitation. Consequently, water vapor is not considered to be the main driver of climate change. In contrast, carbon dioxide molecules can stay in the atmosphere for centuries. Methane molecules actually trap about 25 times as much heat as carbon dioxide molecules, but because methane is much less abundant in the atmosphere, its total effect on global warming is much smaller. Because of its relative abundance and longevity in

TABLE 20.1   GREENHOUSE GASES AND ESTIMATED CONTRIBUTION (IN PERCENTAGES) TO THE GREENHOUSE EFFECT

Water 36–72

Carbon Dioxide 9–25

Methane 4–9

Nitrous Oxide 1–3

Ozone 3–7

CFCs* ~1

*CFCs are greenhouse gases collectively known as chlorofluorocarbons that contain chlorine, fluorine, and carbon.

the atmosphere, the concentration of atmospheric carbon dioxide is thought to be the prime driver climate change.

The abundance of carbon dioxide in the atmosphere is so small that it is measured in parts per million (ppm). Yet its abundance has increased by about 40 percent since the start of the Industrial Revolution, from about 280 ppm in 1780 to 400 ppm in 2015. This correlation has focused attention on the burning of fossils fuels as a main reason for the increase in atmospheric carbon dioxide and global temperature increases.

Moreover, as carbon dioxide and global temperatures rise, the evaporation of surface water from oceans and lakes increases, which amplifies the warming effect still further. This is an example of a *positive feedback,* a vicious circle of change in which one event causes another that then reinforces the first (Fig. 20.5). (We discuss feedbacks in more detail in Section 20.6.) In this case, water vapor is not the prime driver of climate change, but the increase of water vapor in the atmosphere that results from the increasing evaporation caused by rising carbon dioxide levels amplifies the warming trend.

For Earth to remain hospitable to life as we know it, the greenhouse gas content of the atmosphere must be

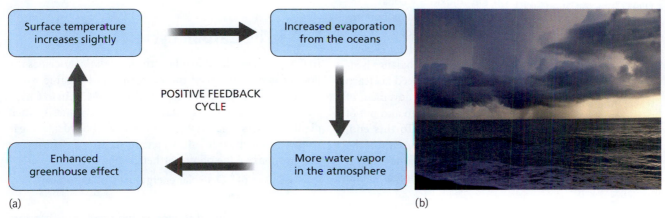

(a)   (b)

Figure 20.5   Positive Feedback Loop

(a) Increasing carbon dioxide concentrations in the atmosphere lead to global warming. This leads to the development of a positive feedback loop in which a warming atmosphere results in increased evaporation of water vapor, also a greenhouse gas, which further warms the atmosphere, leading to more evaporation, and so on. (b) Evaporating ocean water increases the greenhouse gas content of the atmosphere, which enhances the global warming effect. (Image courtesy of Barbara Zarrella)

SCIENCE REFRESHER

The Physics of Global Warming

Greenhouse gases trap heat because they allow solar radiation to enter the atmosphere, but partially block their escape (see Fig. 20.4). But how does this work? The correlation between the increase in greenhouse gases and the increase in global temperature is often taken as evidence of a cause-and-effect relationship between them. But correlations between two variables can have other explanations. A correlation could be caused by a third variable, or it could be fortuitous.

In science, it is important to look beyond a mere correlation, as suggestive as the relationship might be, and instead try to understand the processes involved. As it turns out, both scientific theory and experimentation provide a very elegant and robust explanation for why an increase in greenhouse gas concentrations causes an increase in average global temperatures.

The scientific explanation is quite straightforward. Temperature is the *average kinetic energy* of a substance (see Science Refresher: **Heat, Temperature, and Magma** in Chapter 4). Average kinetic energy is a measurement of the motion of the molecules in a substance. When infrared radiation attempts to pass through greenhouse gases, the gases react by rotating, bending, and vibrating. By so doing, they effectively intercept and block the passage of the radiation energy (see Fig. 20.4b). When a molecule of greenhouse gas absorbs the energy, the gas molecule vibrates and rotates at a faster speed and then collides with a neighboring molecule in the atmosphere, and transfers its energy to it. The net effect of this repeated excitation and collision is that molecules in the atmosphere are traveling faster, and therefore have higher average kinetic energy. By definition, therefore, the molecules have higher temperature. In this way, increasing greenhouse gas content in the atmosphere translates into an increasing average kinetic energy and an increasing temperature.

The comparison between the atmospheric temperatures of Mars and Venus is a very instructive example of how this process works. The atmospheres of both planets are overwhelmingly dominated by CO_2 (Mars, 95%; Venus 98%). The Martian atmosphere is very thin (its atmospheric pressure is about 100 times less than Earth). In the atmosphere of Mars, an excited molecule loses its energy before it can collide with another molecule and the greenhouse effect on Mars is very small. In contrast, the Venusian atmosphere is approximately 90 times thicker than Earth's atmosphere. As a result, excited molecules frequently collide with one another in Venus's atmosphere and in so doing they trap infrared radiation. The contrasting greenhouse effects on Mars and Venus are reflected in their surface temperatures. The average surface temperature on Mars is about −30°C (−22°F). In contrast, Venus is a nightmarish greenhouse with a surface temperature of about 470°C (880°F).

---

sufficient to prevent the planet from freezing but not so high as to cause the planet to become unbearably hot. As we have just established, the main factor in this delicate balance is carbon dioxide. Global temperatures rise and fall in response to the abundance of this gas in the atmosphere. To understand how this balance is attained, we first need to learn how to track the movement of carbon dioxide into and out of the atmosphere. To this end, in Section 20.3, we examine the *carbon cycle,* the name given to the ways in which carbon moves between the organic and inorganic world.

CHECK YOUR UNDERSTANDING

- What are the most common greenhouse gases in Earth's atmosphere?
- Why is it important that the greenhouse gas content of the atmosphere stay within a certain range?

20.2 SUMMARY

- Greenhouse gases, such as water vapor, carbon dioxide, and methane, permit solar radiation to pass through Earth's atmosphere but partially block the infrared radiation emitted from its surface.

- The presence of greenhouse gases in the atmosphere is vital to life on Earth, because without them Earth's atmosphere would be 35°C (63°F) colder.

20.3 The Carbon Cycle

In Chapter 5 we discussed how the *food chain* describes the flow of energy between the inorganic and organic world (see Fig. 5C and Living on Earth: **The Food Chain** in Chapter 5). This flow of energy starts with *photosynthesis,* when plants harness solar energy to create carbohydrate sugar from carbon dioxide and water. In its simplest form, we can summarize the photosynthetic reaction as:

$$6CO_2 + 6H_2O + \text{solar energy} \rightarrow C_6H_{12}O_6 + 6O_2$$
$$\text{Carbon dioxide} \quad \text{water} \qquad\qquad\qquad \text{sugar} \quad \text{oxygen}$$

The sugar produced is energy, and plants are therefore *energy producers* (see Fig. 5B). This energy is then transmitted down the food chain *by energy consumers.* Energy consumers are animals such as herbivores (which eat plants), primary carnivores (which eat herbivores) and secondary carnivores (which eat carnivores) (Fig. 20.6a). The energy

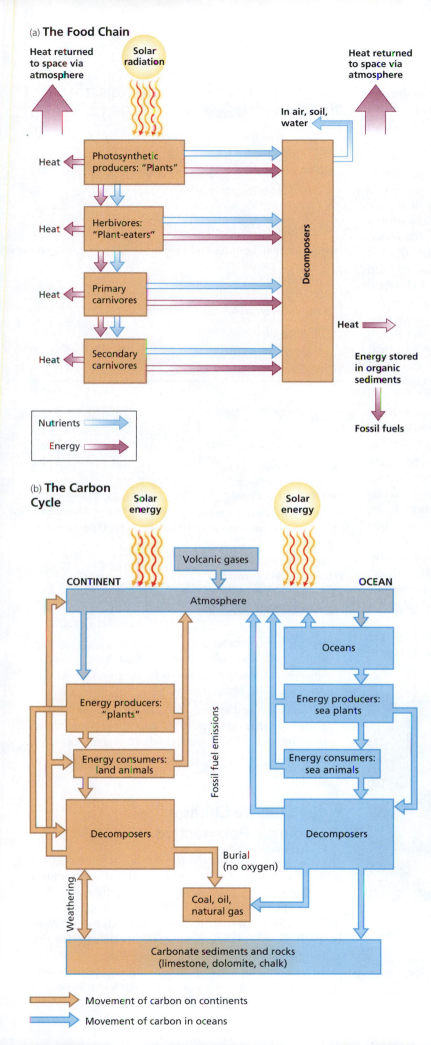

that plants and animals use while they are alive is not transmitted down the food chain (arrows labeled "Heat" in Fig. 20.6a). So only a tiny fraction of the energy initially produced by plants is transmitted to the secondary carnivores.

The **carbon cycle** is an adaptation of the food chain, which focuses on how carbon moves between the inorganic and organic worlds (Fig. 20.6b). Carbon is initially vented from volcanoes into the atmosphere in the form of carbon dioxide. The diagram shows the continental portion of the cycle on the left-hand side (in brown) and the oceanic portion on the right (in blue). Note that each portion is similar to the food chain, with carbon flowing from energy producers to energy consumers. At any one time, there are countless carbon atoms moving like cars along these congested routes. The boxes in Figure 20.6b are *reservoirs*, places where carbon is stored temporarily before it continues its travels.

On continents, (see Fig. 20.6b) carbon flows from energy-producing plants to energy-consuming animals (herbivores and carnivores). Upon the death and decomposition of the plant or animal, its unspent energy is stored as organic matter and the carbon it contains is transferred to the "*decomposers*" reservoir. What happens next depends on the availability of oxygen and on how rapidly the organic material is buried by sediment. If abundant oxygen is available, the decomposition converts the carbon into carbon dioxide, which may be temporarily stored in soil but eventually returns to the atmosphere. In oxygen-starved regions, or in regions where the rate of sedimentation is high, burial of organic matter and the carbon it contains produces fossil fuels

Figure 20.6   The Food Chain and the Carbon Cycle

(a) In the food chain, solar radiation stimulates photosynthesis in plants, which produces energy that is transmitted down the food chain from energy producers (plants) to energy consumers (herbivores, followed by primary carnivores and secondary carnivores). (b) The continental portion of the carbon cycle is on the left (in pale brown), and the oceanic portion is on the right (in blue). Carbon dioxide is introduced into the atmosphere by volcanoes. In both continental and oceanic environments, energy input from the Sun is harnessed by photosynthetic plants and is transmitted down the food chain by energy consumers (herbivores, carnivores, sea animals). Organic material decomposes when energy producers die. Depending on a variety of factors, decomposition may recycle the carbon to the atmosphere or the carbon may become buried in the form of fossil fuels or carbonate sediment.

(see Chapter 19). In this way, fossil fuels can be thought of as energy banks, storing the tiny fractions of the Sun's radiant energy that have been transmitted down the food chain.

In the oceans, the main energy producers are *phytoplankton,* which are microscopic sea plants. All energy consumers, from *zooplankton* (floating marine animals) to mammals, are grouped together and referred to as "sea animals" in Figure 20.6b. Ocean water has a strong affinity for carbon dioxide, and it extracts CO_2 from the atmosphere, forming carbonate ions. But ocean water also has a strong affinity for dissolved calcium ions, delivered from continents to the oceans by a combination of chemical weathering and erosion (see Chapter 5). Although complex in detail, sea animals absorb both calcium and carbonate ions during their lives and combine them to form calcium carbonate. As we learned in Chapter 6, when these animals die and decompose, the carbon they contain resides in carbonate sediments, which eventually becomes limestone (calcium carbonate, Fig. 20.7).

In detail, the carbon cycle is highly complex and the pathway of carbon is like the game Chutes (or Snakes) and Ladders. A mobile carbon atom may land on a "chute" or "snake" that brings it back toward the start. However, each atom eventually overcomes all barriers and comes to reside at its final destination either in fossil fuel deposits or in carbonate sediments. It is no accident that carbonate rocks are overwhelmingly the primary modern reservoir of carbon (Table 20.2). Carbonates are generally well preserved in the geologic record; only a small portion of the limestone produced throughout Earth history has been subsequently destroyed by weathering. When a carbon atom finally enters carbonate rock, it has little chance of being recycled and is effectively removed from circulation.

The carbon cycle explains how carbon moves in nature between the organic and inorganic world. The storage of

TABLE 20.2   AMOUNT OF CARBON (IN GIGATONS) IN EARTH'S SURFACE RESERVOIRS

Atmosphere	710
Continents	
Biomass	590
Litter (dead plant material)	60
Soil	1670
Fossil fuels	5000
Carbonate sediments and rocks	20,000,000
Oceans	
Biomass	4
Surface ocean water	680
Intermediate ocean water	8,200
Deep ocean water	26,000
Sediments	4900

CHECK YOUR UNDERSTANDING

◐ Why do fossil fuels represent stored energy within the food chain?

◐ What is a *reservoir* in the context of the carbon cycle?

◐ Describe the reservoirs that carbon may travel through in (a) an oceanic setting and (b) on land.

carbon in carbonate rock and fossil fuel deposits also demonstrates the important role that geological processes play in the carbon cycle. This understanding paves the way for looking at climate change from a geological perspective.

20.3 SUMMARY

• The carbon cycle describes the circulation of carbon at or near Earth's surface.

• During photosynthesis, plants extract carbon from the atmosphere. The carbon is transmitted down the food chain when plants are eaten by herbivores, and herbivores are eaten by carnivores.

• When plants and animals die, the carbon from their bodies is either returned to the atmosphere or incorporated into fossil fuel deposits and carbonate rock.

Figure 20.7   The Great Barrier Reef

The Great Barrier Reef in Australia is an impressive modern example of limestone deposition. Limestone is composed of calcium carbonate, which forms when the calcium and carbon dioxide dissolved in ocean water combine. The supply of calcium is critical to limestone formation.

20.4 Climate Change: A Geologic Perspective

The discussion of climate change centers on the relationship between average global temperature and the greenhouse gas content of the atmosphere. Most scientists, including those on the Intergovernmental Panel on Climate Change (IPCC), believe that the burning of fossil fuels has increased atmospheric greenhouse gas concentrations and that this increase is the prime cause of global warming.

To examine the case of global warming in a systematic fashion, in this section of the chapter we will trace the history of the greenhouse effect on our planet, peeling back the layers of time to look at progressively more and more of the geologic record. As we go farther back in geologic time, we learn more about the natural cycles of climate change and the processes responsible for them. We can then combine this understanding with our knowledge of the carbon cycle (section 20.3) to distinguish between the natural and human-induced contributions to climate change.

The greenhouse gas content of the atmosphere and average global temperatures are well documented for the past 100 years. However, as we go farther back in time, we cannot measure either of these quantities directly, and must turn instead to *climate proxies*. **Climate proxies** are the preserved characteristics of the past that allow scientists to reconstruct the climatic conditions for much of the Earth's history. For example, we can use oxygen isotope ratios as a proxy for measuring past temperatures because these isotope ratios are sensitive to temperature changes.

> **CHECK YOUR UNDERSTANDING**
>
> ◗ Why are climate proxies necessary in the study of long-term climate change?

(a)

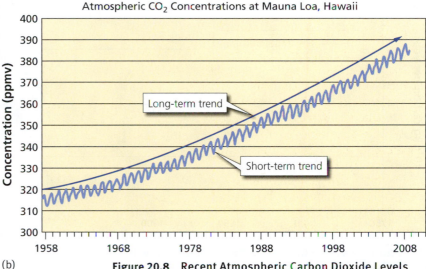

(b)

Figure 20.8   Recent Atmospheric Carbon Dioxide Levels

(a) Mauna Loa Observatory is located in Hawaii near the top of a volcano at an elevation of about 3500 m (11,000 feet). (b) Data from the Mauna Loa Observatory show the concentration of carbon dioxide in the air over the last 55 years, given in parts per million by volume (ppmv). 350 ppmv = 0.035 percent. The long-term trend shows an overall increase in carbon dioxide; short-term trends reflect annual fluctuations due to varying productivity in the biosphere.

CLIMATE CHANGE OVER THE LAST HALF-CENTURY

Since 1958, scientists have been acquiring accurate data on the composition of our atmosphere. The Mauna Loa Observatory on the island of Hawaii is the source of some of this recent atmospheric data collection. The observatory (Fig. 20.8a) is positioned near the top of a volcano that is suitable for monitoring Earth's changing atmospheric compositions because of its altitude and remoteness. The site is free from the effects of local pollution and the air at the altitude of Mauna Loa (3500 meters or 11,000 feet) is part of a pattern that circulates all around the globe at approximately the same latitude. Changes in atmospheric concentrations at this location are therefore indicative of global-scale change.

A graph of the carbon dioxide data collected at Mauna Loa over the last 55 years (Fig. 20.8b) shows two trends. The long-term trend shows a systematic increase in atmospheric carbon dioxide. Today's concentration, for example, is about 20 percent higher than it was about 50 years ago. Closer inspection suggests that the increase is not linear but is accelerating; the relationship between time and

CO_2 levels is *exponential*. In other words, atmospheric carbon dioxide levels are rising at an ever-increasing rate.

In Figure 20.8b we can also see less pronounced annual highs and lows, which represent short-term trends that we can attribute to patterns of plant growth within the biosphere in the Northern Hemisphere. During the spring and summer, plants are relatively active and absorb more carbon dioxide by photosynthesis than they do in the relatively dormant fall and winter months. Despite these annual fluctuations however, the data show that there has clearly been an overall increase in carbon dioxide concentrations in the atmosphere over the past 55 years.

During this same time period, there has been a parallel increase in the rate at which carbon dioxide has been added

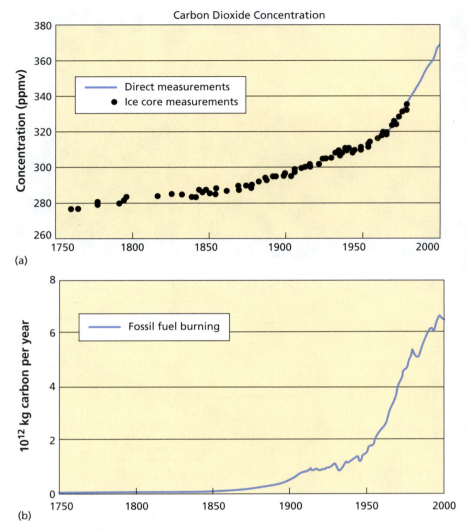

Figure 20.9 **Historic Atmospheric Carbon Dioxide Levels**

(a) Atmospheric concentrations of carbon dioxide have increased dramatically since the late eighteenth century. Direct measurements began in the 1950s. Atmospheric CO_2 concentrations for the years before the 1950s have been derived from ice cores, in which portions of the ancient atmosphere are trapped in bubbles in polar ice sheets. (b) Atmospheric concentrations of carbon dioxide began to increase at about the same time that humans began to burn fossil fuels as an energy source.

to the atmosphere through the burning of fossil fuels, a trend that as we have noted, can be traced back to the end of the eighteenth century (Fig. 20.9). Earth's ecosystems have the ability to absorb about half this amount, but the remaining half is not absorbed and causes carbon dioxide concentrations in the atmosphere to rise. In 2014, 32 billion metric tons of carbon dioxide were released into the atmosphere from fossil fuel emissions, and atmospheric CO_2 reached 398 ppm.

As we noted earlier, prior to the 1950s, the information needed to calculate average surface temperatures was sparse. Since that time, accurate temperature records have become available from many land stations. Most of these stations are located in the Northern Hemisphere. Each year, these records are manipulated in an identical fashion

in order to yield an average global temperature (Fig. 20.10). The data indicate that the average global temperature is nearly 1°C (1.8°F) higher than it was about 100 years ago, and that most of this increase has taken place since the 1920s. Computer models that reproduce temperature changes closely matching those recorded since 1860 project a further 1°C (1.8°F) increase over the next 25 years.

Although there is now almost universal agreement that the concentrations of greenhouse gases in the atmosphere have increased over the past 50 years, and that average global temperatures have risen over the same time period, how do we know whether a causal relationship is involved? Modern global temperatures are still comfortably within the bounds of natural temperature fluctuations as evidenced by the geologic record, as we noted in Section 20.1. There have been other periods in the past, millions of years before the present industrial era, when Earth was a hot global greenhouse and yet other times when it was a frozen global icehouse.

For example, during part of the Cretaceous Period, about 100 million years ago, average global temperatures were about 8°C (14°F) higher than they are today, and atmospheric carbon dioxide levels were probably more than 10 times higher. In contrast, about 70,000 to 11,000 years ago, average global temperatures were much lower and ice sheets covered more than half of continental North America, advancing well beyond the Canadian-US border. These natural swings are more dramatic and far more intense than anything witnessed over the past 50 years. Our present climate is within the bounds of proven natural variability and therefore a large component of natural variation in modern global change cannot be dismissed. So how can we gain the better understanding that we need?

CHECK YOUR UNDERSTANDING

○ Why is Mauna Loa a suitable place to monitor global atmospheric carbon dioxide concentrations?

In the material that follows, we delve into the past to look at preindustrial natural variations in global temperatures. We can use the geologic data we discover as a foundation to compare modern global change with global change in the geologic past. The geologic record offers a unique perspective because we know for certain that ancient temperature changes are entirely due to natural

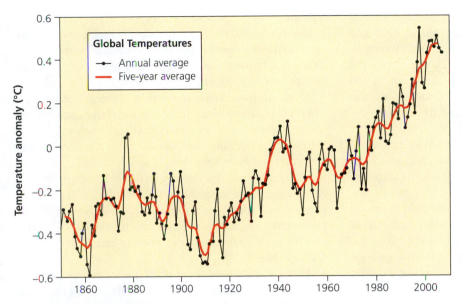

Figure 20.10   Earth's Surface Temperature

This graph charts the relative change in the Earth's average surface temperature since 1850 according to annual and 5-year averages. Current models suggest that the pace of global warming will accelerate during the first half of this century.

processes because they precede human dependence on fossil fuels. Fortunately, the geologic record of temperature variation is excellent over the last million years, and is reasonably good for up to a billion years.

CLIMATE CHANGE OVER SEVERAL CENTURIES: VARIATIONS IN SOLAR OUTPUT

In the preceding section, we made the assumption that changes in Earth's average surface temperatures are the result of earthly processes. By implication then, we assumed that the amount of solar radiation entering Earth's atmosphere is constant. Over a time interval of 200 years or so, this may well be a valid assumption. But the Sun is a star, and stars vary over time in their output of radiation. As the greenhouse effect is the result of trapped solar radiation near Earth's surface, variations in solar activity and the energy output of the Sun may be important variables in global warmth. When we examine climate change over a longer time interval, we find that variable solar output has a profound effect on global temperatures.

The Sun predominantly consists of hydrogen (72%) and helium (27%), the first two chemical elements in the periodic table. Energy radiated from the Sun is attributed to nuclear fusion reactions occurring within its core that convert hydrogen into helium (see Chapter 3). So far during the Sun's 4.57-billion-year life span, less than 5 percent of the

available hydrogen on the Sun has been converted to helium. The Sun has a diameter of about 1.4 million kilometers (900 million miles) (more than four times the distance from Earth to the Moon) and is approximately 150 million kilometers (93 million miles) away from us. It takes solar radiation about 8.3 minutes to make the journey to Earth.

The Sun's surface commonly contains small dark regions, which are up to 160,000 kilometers or 100,000 miles across, called **sunspots**. These areas are dark because their temperature is about 1500°C (2700°F) cooler than the approximately 5700°C (10,300°F) temperature of the surrounding regions (Fig. 20.11a b). Sunspots are cooler than the rest of the Sun's surface because their strong magnetism suppresses the escape of heat from the Sun's interior. The number of sunspots at any given time can be recorded with the aid of a telescope by projecting the Sun's image onto a piece of paper and counting the number of dark spots.

The number of sunspots varies considerably and, over the course of a year, fluctuates within an average range from 180 to nearly zero (Fig. 20.11c). Sunspot activity varies in a

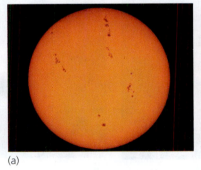
(a)

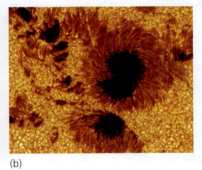
(b)

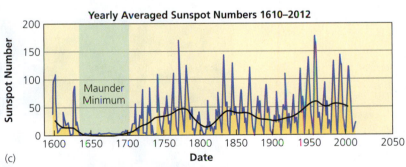
(c)

Figure 20.11   Sunspots

(a) Sunspots are dark regions of cooler, depressed, and anomalously magnetic areas on the Sun's surface. (b) This photo shows a close up view of a sunspot. (c) The annual mean number of sunspots has been recorded since their discovery in 1610. From about 1720 to the present day, sunspot numbers have varied on an 11.1-year period. Throughout much of the seventeenth century, however, few sunspots were recorded. This time period is known as the Maunder Minimum and overlaps the Little Ice Age.

cycle, known as the **sunspot cycle**; peak periods of activity are 11.1 years apart. Recent peaks occurred in 2002 and 2013. Minimum activity also occurs at 11.1- year intervals. Recent dips occurred in 1997 and 2008. The reason for the cycle is not well understood (and is not of direct concern to us here), but most scientists attribute this cycle to the interaction between the Sun's magnetism and its complex, dynamic internal structure.

Sunspots accelerate the **solar wind**, which is made up of tiny particles ejected from the solar atmosphere, to speeds of 500 kilometers per second (over 1 million miles an hour). This is important because the solar wind controls the abundance of charged particles that radiate away from the Sun. The effect of the solar wind is most visible in the tail of a comet (Fig. 20.12a), where the wind literally blasts the comet's surface and streams around it to form a gaseous tail that points directly away from the Sun (Fig. 20.12b). The effect of the solar wind on the Earth is most obvious at high latitudes where it bombards the Earth's magnetic field, resulting in the spectacular natural light display known as the **aurora** (Fig. 20.12c). But the solar wind also profoundly

influences the temperature of the upper atmosphere. During peak sunspot activity the solar wind is much more powerful, and the temperature of the uppermost atmosphere rises dramatically from the 225°C (400°F) typical of minimum sunspot activity to about 1225°C (2200°F).

Does this remarkable variation in the temperature of the uppermost atmosphere affect Earth's surface temperatures? There is some empirical (statistical) evidence that this is the case. For the most part, the sunspot cycle over the past 300 years has been very regular and predictable. However, during the seventeenth century relatively few sunspots were recorded and, for reasons that are not fully understood, the sunspot cycle appears to have been suppressed (see Fig. 20.11c). Historic records document episodes of prolonged cold during this time, which forms part of a period known as the **Little Ice Age** (Fig. 20.13).

During this extended period of limited sunspot activity, the solar wind was presumably less intense and the upper atmosphere was relatively cool. Could a decrease in the intensity of the solar wind over a long period result in low temperature conditions in the upper atmosphere, which

(a)

(b)

(c)

Figure 20.12   Effects of the Solar Wind

(a) The effect of the solar wind is most visible in the tail of a comet. (b) The solar wind bombards the comet's icy surface and forms a tail that points directly away from the Sun. (c) At high latitudes, the solar wind bombards Earth's magnetic field to create the aurora.

Figure 20.13   The Little Ice Age

A seventeenth-century contemporary artist's depiction of life on the Thames River, England, during the Little Ice Age, attests to rivers freezing deeply enough to support iceskaters in the winter. An annual "frost fair" was held on the River Thames between 1607 and 1814.

could, in turn, induce periods of cooling at Earth's surface? If so, the Little Ice Age could be related to decreased sunspot activity. The drop in average global surface temperature does not need to be substantial to initiate an ice age; average global temperatures during the Little Ice Age were probably less than 2°C (4°F) colder than they are today.

Thus, as we begin to peel back the layers of time, we find evidence that the output of solar energy varies in response to the development of sunspots and the strength of the solar wind. From a 300- to 400-year perspective, it seems that these phenomena may have had an important influence on our climate and could potentially cause global change.

> **CHECK YOUR UNDERSTANDING**
>
> ● Describe the relationship between sunspot activity and the Little Ice Age.

CLIMATE CHANGE OVER THE LAST MILLION YEARS: EVIDENCE FROM THE GEOLOGIC RECORD

Variation in solar activity is not the only way that the Sun influences global temperatures on Earth. As we go farther back in time, other factors come into play. Many scientists believe that variations in the relative positioning of Earth to the Sun have an even more profound effect on Earth's climate over time spans of up to one million years. But in attempting to assess the potential effect of these variations on climate we are immediately confronted with problems. Anecdotal evidence from historic records, like that used in the previous section to correlate sunspot activity with the Little Ice Age, is lacking, and we do not have a time machine at our disposal with which to directly measure global temperature!

As we discussed at the beginning of this section, in situations like this, because we cannot use direct temperature measurements, we turn instead to climate proxies, which allow us to determine ancient temperatures indirectly. To find a suitable temperature proxy, we must identify some measurable property that changes with temperature and is preserved in the geologic record. Furthermore, the property must be one that measures global rather than local temperature changes. In a sense, an ordinary thermometer is an example of a modern proxy method of measuring local temperature. What we are actually measuring with a thermometer is not the temperature (defined as the average kinetic energy of a body; see Science Refresher **The Physics of Global Warming** on page 652), but the height of a fluid (usually mercury) in a narrow tube. Using the numbers on the thermometer, we convert this height into temperature. This method is successful because we know from repeated experiments how mercury expands and rises up the thermometer tube as the temperature increases.

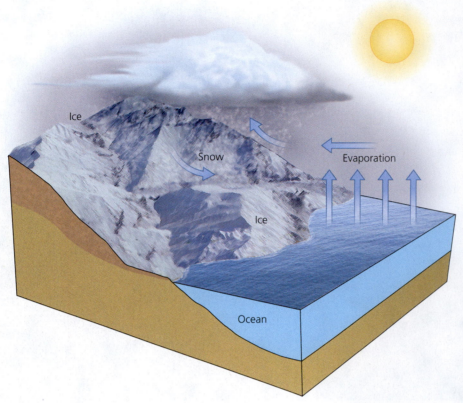

Figure 20.14 Ice and the Hydrologic Cycle

This version of the hydrologic cycle has been adapted to emphasize the storage of evaporated water in polar ice sheets. Evaporated water that circulates over polar regions during the winter falls as snow that does not melt during the following summer. Eventually each winter's snow compresses to form a layer of ice as the ice sheet grows.

The best place to look for suitable proxies of global temperature change is the oceans. Ocean water covers more than 70 percent of Earth's surface and is very sensitive to changes in global temperature. This sensitivity is reflected in variations in surface ocean water temperature, and in the amount of moisture that evaporates into the atmosphere. But is this evaporation history preserved in the geologic record? The answer is yes. Recall from Chapter 13 that most of the moisture evaporated from the oceans returns to the sea by way of the hydrologic cycle (Fig. 20.14). Moisture that evaporates from the oceans falls as rain on land, which feeds into valleys that drain back to the sea. Winter snow temporarily preserves the record of evaporation from the oceans, but in most regions this snow melts in the spring. In polar regions, however, the winter snow does not completely melt. In these regions, the snow that falls each year is compressed into a thin layer by the snowfall in subsequent years. With continued compaction, each year's snowfall is ultimately preserved as a thin layer of ice (Fig. 20.15). Each ice layer preserves a record of the evaporation for that year, accumulating in sequential order like the pages in a book. In accordance with the principle of superposition (see Chapter 8), we know that the topmost layers preserve a record of relatively recent evaporation from the oceans, whereas the lower layers preserve ancient records. In this way, the polar ice

> **CHECK YOUR UNDERSTANDING**
>
> ◔ How is evaporation history preserved in the geologic record?

sheets preserve up to 200,000 years of evaporation history. As we shall see in the material that follows, there are measurable features in this evaporation record that can be used as a proxy for ancient temperatures.

Bubbles of ancient air

As snow falls and is compressed, it traps bubbles of air. In this way, ancient layers of ice trap samples of the atmosphere at the time the snow fell (Fig. 20.16). Scientists have drilled cores into the ice sheets of both polar regions (see Fig. 20.15a). By carefully analyzing samples of the ancient atmosphere trapped in individual layers, they have obtained a record of changing atmospheric greenhouse gas composition. If we assume that the present relationship between greenhouse gas concentrations and global temperatures is typical of the past (that is, if we apply the principle of uniformitarianism; see Chapter 8), we can use these concentrations to deduce changes in global temperatures.

Gas bubbles analyzed in individual layers of these ice cores show systematic variations in the concentrations of carbon dioxide, methane, and other greenhouse gases from one layer to the next. Data from the Vostok station in

53–54 meters

1836–1837 meters

3050–3051 meters

(a)

(b)

Figure 20.15 Layers of Ice

The record of evaporation from the oceans is preserved in polar regions. The snow that falls each year is compressed and eventually forms a thin layer of ice, which preserves a record of the evaporation for that year, accumulating in accordance with the principle of superposition. (a) These cores of ice have been retrieved from the Greenland ice sheet. In shallow cores, such as the example shown here, each layer represents annual snow accumulation. In deeper cores, this relationship is more complex because the pressure of the overlying ice affects the integrity of the layer. Researchers calibrate the age of the ice in a variety of ways, most commonly by dating volcanic ash interbedded with the ice layers and relating these ages to known volcanic eruptions. (b) A researcher stands in a snow pit next to ice layers in an ice sheet in West Antarctica.

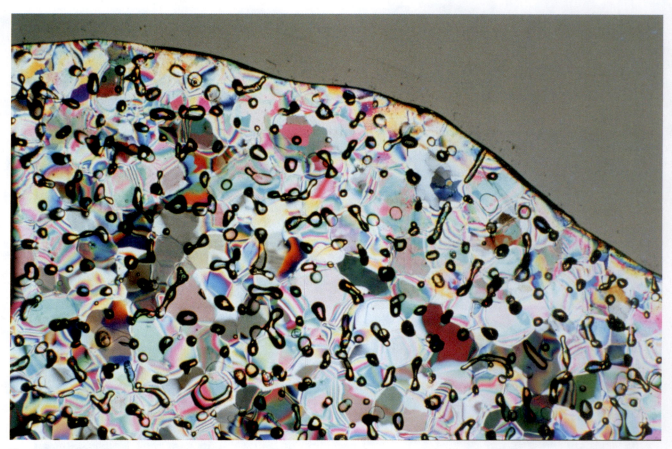

Figure 20.16   Ice Crystals

Polar ice crystals seen under the microscope feature small dark circular and rod-shaped areas, which are bubbles trapped within and between ice crystals. The bubbles preserve samples of the atmosphere at the time the ice formed. The gas in the bubbles may be extracted and its composition determined.

Antarctica (Fig. 20.17), for example, show significant fluctuations in both carbon dioxide and methane. Elevated values (to the right in each graph) suggest warm interglacial periods, whereas low values are interpreted as periods of glaciation.

As there are no direct measurements available, we do not know the exact temperature for each time period with any certainty. However, the ice core data provide evidence for an interval of high atmospheric greenhouse gas concentrations between about 150,000 and 120,000 years ago. This interval corresponds with a known interglacial period, implying that times of high atmospheric greenhouse gas concentrations correspond to known warm periods. Conversely, times of relatively low atmospheric greenhouse gas concentrations correspond with known cycles of glacier formation and melting between 110,000 and 11,000 years ago. This provides evidence of colder but fluctuating temperatures. The most recent interglacial rise in greenhouse gas

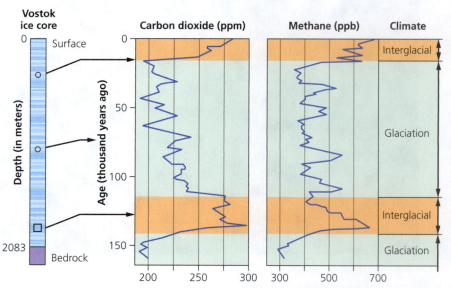

Figure 20.17   Analysis of Air Bubbles in Ice

A layer-by-layer analysis of air bubbles in ice cores from the Vostok Station in Antarctica show changing concentration of greenhouse gases in the atmosphere over the past 150,000 years. The trends in carbon dioxide (measured in parts per million, ppm) and methane (measured in parts per billion, ppb) data match the geologic record (right) of known glacial and interglacial stages.

concentrations began between 11,000 and 10,000 years ago and corresponds with a dramatic period of global warming. At one point within this interval it is estimated that the average global temperature climbed an astonishing 7°C (13°F) in only 40 years!

Therefore, we can see that existing ice sheets record variations in atmospheric composition over the past 150,000 years, implying that the relationship between greenhouse gas composition and temperature has existed over a considerable period of time, not merely during the industrial era. In the spirit of the scientific method, how do we test this hypothesis, and how do we know that data from narrow cores of ice are representative of global, rather than local phenomena?

First, ice cores from both the northern and southern polar ice sheets show very similar patterns. If they did not, scientists would be looking for local rather than global explanations for the patterns they show. Although there must be a local component, it does not appear to override the major global-scale patterns.

Second, as we learned in Chapter 15, there is abundant geologic field evidence for episodes of widespread glaciation between 90,000 and 11,000 years ago, which was followed by melting and the deposition of glacial sediments. This history is closely mirrored by fluctuations in the greenhouse gas concentrations preserved in the ice cores. Their concentrations were anomalously low between 90,000 and 11,000 years ago, but rapidly increased thereafter.

> **CHECK YOUR UNDERSTANDING**
>
> ● How do we know that data from cores of ice are representative of global phenomena?

Third, as we discuss in the following material, other proxy measures of ocean water evaporation, preserved both in the ice cores and in deep sea sediments, show similar patterns of global temperature change.

Patterns in oxygen isotopes

Although ocean water is salty (containing about 3.4% NaCl), water evaporated from the oceans is almost entirely devoid of salt. This contrasting composition indicates that evaporation selects some chemical components in preference to others. This subtle form of discrimination provides additional supporting evidence for the global temperature fluctuations deduced from the ice cores. The evidence is based on differences in oxygen isotopes.

Although all water molecules are composed of two parts hydrogen and one part oxygen (H_2O), both hydrogen and oxygen have more than one stable isotope (see Chapter 8 for an introduction to isotopes). Molecules of water vary in weight, mainly due to differences in the weights of the two main oxygen isotopes—lighter Oxygen-16 and heavier Oxygen-18.

By definition, all oxygen atoms contain eight protons in their nuclei. The vast majority (about 99.8%) also contain eight neutrons forming the isotope oxygen-16. However, a small proportion (0.2%) contains 10 neutrons, forming the heavier isotope oxygen-18. Molecules of water vary in weight

depending on whether or not they contain the heavier oxygen isotope, that is, the one that has more neutrons in its nucleus.

No matter what the temperature, water molecules containing the heavier oxygen find it more difficult to evaporate than those containing light oxygen. This difficulty is even more pronounced at low temperatures because, at high temperatures, more energy is available to help the heavier water molecules evaporate. In an "evaporation race," the lighter water molecule always wins, but if the conditions of the race change (for example, if the temperature rises), the winning margin may be reduced. The "margin of victory" can be used to infer the race conditions which, in this case, are dictated by global temperature.

When this method is applied to the layers of ice in the ice cores, an historical profile of global warmth is obtained. In this profile, we can identify warming and cooling trends. Figure 20.18 shows the layer-by-layer variations in the ratio of heavy oxygen (^{18}O) to light oxygen (^{16}O). This variation is shown on the x-axis with higher (less negative) $^{18}O/^{16}O$ ratios

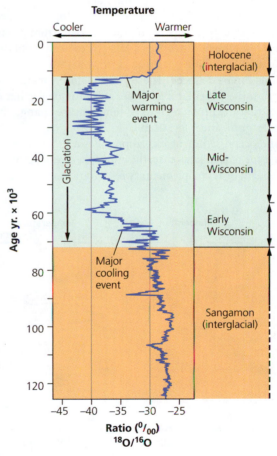

Figure 20.18   Oxygen Isotopes in Ice

Layer-by-layer variation in oxygen isotopic ratios ($^{18}O/^{16}O$) over the last 120,000 years can be obtained from the Greenland ice core. Higher (less negative) $^{18}O/^{16}O$ ratios (on the right) reflect warmer time intervals, whereas lower (more negative) $^{18}O/^{16}O$ ratios (on the left) reflect cooler time intervals. The data are in broad agreement with ancient temperatures inferred from the greenhouse gas content of trapped air bubbles in the Vostok ice core (see Fig. 20.17).

indicating increasing temperature to the right. The units for the $^{18}O/^{16}O$ ratio are in parts per thousand (‰) as opposed to part per hundred, or percent (%). Once again, we cannot be certain of the absolute temperature recorded by the $^{18}O/^{16}O$ ratio in each layer, but we can be confident of the warming and cooling trends that the ratios identify.

With some exceptions, the warm and cool periods predicted by oxygen isotopes are similar to those deduced from the analysis of the greenhouse gas content of the trapped atmospheric gases (see Fig. 20.17). Like the trapped gas record, the oxygen isotope data record a period of warmth about 120,000 years ago, and a cold period lasting from about 70,000 until 11,000 years ago, followed by a warming trend. Although there are some discrepancies around 90,000 years ago (compare Fig. 20.17 and Fig. 20.18), the general agreement between the two records gives added weight to the hypothesis that we are measuring global, rather than local, phenomena. Note that the global warming evident over the past 50 years or so (see Fig. 20.9) cannot be identified in this dataset because the recent change is dwarfed by effects of older natural fluctuations.

A further test of the use of oxygen isotopes as a proxy for ancient temperature is provided by data from glacially derived sediments in a Swiss lake. These sediments preserve about a 15,000-year record in which warm and cool periods can be deduced. These periods closely match those found in the Greenland ice sheet and provide further evidence that the proxy records of temperature changes have global significance (Fig. 20.19).

The ocean sediment record provides a different, yet related type of test. In a sense, the atmosphere and oceans can be thought of as complementary reservoirs. As temperatures drop, evaporation extracts moisture with an increasing abundance of light oxygen isotopes from the oceans. That means that during a period of glaciation, the oceans become increasingly depleted in light oxygen. Ice sheets preserve the evaporation record, but what preserves the record of what is left in the ocean? The answer lies in the fossils of marine organisms.

Floating microorganisms called *foraminifera* record important information about the average temperature of surface ocean waters in the composition of their shells. Oxygen in the shells of these marine organisms records the oxygen isotope ratio of the ocean water in which they live because the animals grow by exchanging water with their surroundings. As with evaporated moisture, this ratio is very sensitive to the temperature of the seawater. When the foraminifera die, they cease to chemically interact with the seawater and sink to the bottom of the ocean, where they are buried by subsequent layers of sediment. As more foraminifera die and sink to the seabed, they accumulate in progressively higher sediment layers.

Changes in the oxygen isotope ratio of foraminifera from one layer to the next consequently preserve an archive of the changes in ocean water temperature over time. When temperatures are cold and lighter ^{16}O-enriched moisture is preferentially evaporated to form the record preserved in polar ice, the oceans become relatively enriched in ^{18}O and high $^{18}O/^{16}O$ ratios are recorded in foraminifera shells. As the temperature rises, the polar ice sheets melt (see In Depth: **Melting of the Arctic Ice— An Omen of Global Warming?** on page 666), and ^{16}O-enriched meltwater is liberated into ocean water, leading to a decrease in the $^{18}O/^{16}O$ ratio in foraminifera shells.

These changes in foraminiferal $^{18}O/^{16}O$ ratios have been identified in sediment cores retrieved by ocean drilling surveys. Layer-by-layer analysis of the $^{18}O/^{16}O$ ratio in these fossil shells reveals alternating periods of warm and cold ocean surface water temperatures that match the alternation of glacial and interglacial stages (Fig. 20.20). Since ocean waters are well mixed, the $^{18}O/^{16}O$ ratios measured in fossil shells are thought to reflect past temperature variations on a global scale.

The proxy temperature records derived from ocean sediment and polar ice can be compared with the information from foraminifera shells. Figures 20.18–20.20 show relative temperatures increasing to the right. However, the oceanic oxygen isotope ratios on the horizontal axis in Figure 20.20 are opposite to those of the ice core records in Figures 20.18 and 20.19, they decrease *to the right*. This is consistent with their complementary relationship; the relative enrichment of an oxygen isotope in one reservoir necessitates its relative depletion in the other.

The oceanic record obtained from foraminifera in the Pacific Ocean (see Fig. 20.20) extends back one million years and confirms earlier findings. The periods of warming and cooling deduced in the upper part of the record (the last 120,000 years) compare very favorably with the ice core records shown in Figures 20.17 and 20.18. Each record

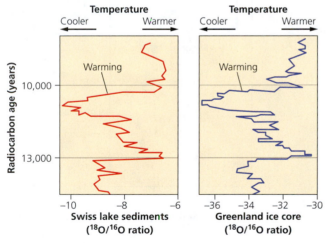

Figure 20.19   Comparing Oxygen Isotopes

Oxygen isotope data ($^{18}O/^{16}O$) from lake sediments in Switzerland is compared to the data from the Greenland ice core for the period between 9000 and 14,000 years ago that encompasses the end-Wisconsin warming trend (see Fig. 20.18). Note the broad agreement between these curves. For each location, higher $^{18}O/^{16}O$ ratios (to the right) reflect warmer time intervals, whereas lower $^{18}O/^{16}O$ ratios (to the left) reflect cooler time intervals.

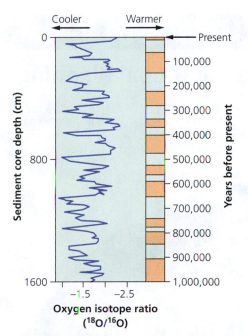

Figure 20.20   Oxygen Isotopes in Foraminifera Shells

Oxygen isotope data ($^{18}O/^{16}O$ ratio) from foraminifera shells in marine sediments extend over the past 1 million years. The samples were obtained from drill core extracted from the Pacific Ocean. The graph is arranged so that warmer time intervals or lower (more negative) $^{18}O/^{16}O$ ratios are on the right, whereas cooler time intervals, or higher (less negative) $^{18}O/^{16}O$ ratios are on the left. These match the alternation of interglacial (brown) and glacial (green) stages shown on the right. Note, however, that the oceanic oxygen isotope ratios on the horizontal axis have the opposite sense to those of the ice core records in Figures 20.17 and 20.18 because an increase in $^{18}O/^{16}O$ ratios in one reservoir necessitates depletion in the other.

indicates a major period of cooling about 80,000 years ago and a major warming trend starting about 11,000 years ago.

By looking at the oxygen isotope ratios preserved in polar ice and in marine sediments, we can deduce reasonably well-constrained patterns of global temperature change over the past 150,000 years and past million years, respectively. We know that these temperature changes show natural patterns of variation because they took place long before the modern industrial era.

The similarity between these results gives us confidence that each of the data sets reveals genuine periods of warming and cooling, allowing us to interpret warming and cooling trends over the past 1 million years. The temperature record over this time interval shows repeated oscillations between warm interglacial periods (such as the present one), and cold glacial climates (such as the Ice Age of 30,000 years ago). Variations of up to 10°C (18°F) are implied from geologic and biological evidence. Viewed in this context, our current global temperature is at the high end of the natural scale, but its variation is not inconsistent with the natural pulse of high and low temperature cycles that has dominated Earth's climate during the recent geologic past. By extrapolating this pulse into the future, some scientists claim that another ice age is inevitable and temperatures will begin to

decrease. If this is correct, then the effects of modern global warming may ultimately be entirely overridden.

In light of the data we have just presented, it could be argued that the current period of global warming began as a natural phenomenon 11,000 years ago. This interpretation assumes a long-standing relationship between greenhouse gases and global temperature. If this assumption is correct, then the questions we must address are what causes these temperature fluctuations, and do natural phenomena dwarf or override the effects of modern industrial pollution, or does the combination of the natural and industrial result in a multiplying effect?

It is difficult to evaluate the implications of these temperature variations on the global warming debate without understanding what causes the temperature oscillations in the first place. With the ocean sediment core data, we may finally be seeing enough of the record to identify some of Earth's natural rhythms. According to this record, there have been 10 warm periods in the last 1 million years (see Fig. 20.20), suggesting a cyclicity of about 100,000 years. It is also clear from this record that, as in a song, other rhythms are present as well.

> **CHECK YOUR UNDERSTANDING**
>
> ● How do oxygen isotopes serve as a climate proxy?

20.4 SUMMARY

- Recent and current global temperatures are well within the range of natural fluctuations. But over the last 40 years, the concentration of greenhouse gases in the atmosphere and the average global temperature have been rising in lockstep.

- Sunspot numbers shows a periodicity of about 11.1 years, with peak activity coinciding with anomalously warm temperatures in Earth's upper atmosphere.

- In the seventeenth century, a mysterious absence of sunspots coincided with part of the Little Ice Age, an anomalously cool period that lasted for most of that century.

- The evaporation record of oceans for the past 150,000 years is preserved as layers in the polar ice sheets, within which air bubbles trapped in pore spaces preserve remnants of the contemporary atmosphere. From the greenhouse gas content of these bubbles, Earth's surface temperatures can be deduced.

- Oxygen isotopes ($^{18}O/^{16}O$) also act as climate proxies. There is an impressive correlation between the oxygen isotope data for the past 150,000 years and the geologic record of global climate change; relatively low $^{18}O/^{16}O$ ratios coincided with times of major continental glaciation during this period.

- The shells of floating microorganisms, such as foraminifera, preserve the oxygen isotope ratio in ocean water. A layer-by-layer analysis of the microorganisms in ocean sediment consequently reveals changes in surface water temperatures with time.

IN DEPTH

Melting of the Arctic Ice—An Omen of Global Warming?

Because they are so finely tuned to temperature, Earth's polar ice caps are sensitive indicators of climate change, and there is no question that the Arctic ice is melting and melting fast. Since 1978, it is estimated that the ocean's ice cover has thinned by 60 percent and shrunk by 17 percent (Fig. 20C). And there are other signs of climate change.

During the same period, average temperatures in parts of Alaska, Canada, and Siberia have risen more than 3°C

Figure 20C   Shrinking Arctic Ice
Satellite-based images illustrate the decline of permanent Arctic sea ice (central white area) from 1980 to 2012. The grey area shows the full extent of winter sea ice.

Figure 20D   Melting Permafrost

Evidence of melting permafrost in the Arctic includes (top) "drunken" forests in Fairbanks, Alaska, and (bottom) "roller-coaster" roads along the Alaskan Highway.

(Continued)

Melting of the Arctic Ice—An Omen of Global Warming? *(Continued)*

(5°F), while average summer temperatures are the highest in 44,000 years. The beluga whale watch in Churchill, Manitoba, which normally starts in mid-June following the spring thaw, now occurs as early as May. Winter ice forms as much as two weeks later than it used to in Hudson Bay and polar bear populations are in sharp decline due to a lack of sea ice on which to hunt seals. Thawing of permanently frozen ground, or *permafrost*, has created new ponds and wetlands, caused houses to sink and coastal erosion to accelerate, and produced "drunken" forests and "rollercoaster" roads (Fig. 20D).

Although the effects of this warming trend are occurring far from any population centers, the consequences could be catastrophic for all of us. As the polar regions warm, the contrast in temperature between them and the equator lessens, and it is this contrast that drives both ocean circulation patterns and the world's climate system (see Chapter 16). While the outcome of any planetary circulation weakening is uncertain, it is likely to cause changes in ocean currents, prevailing wind directions, and rainfall patterns, the effects of which could be disastrous.

Not all the potential consequences of Arctic warming result in temperature increases. Disruption of the Gulf Stream, for example, could plunge much of the Northern Hemisphere into a mini ice age. By bringing warm water from the Gulf of Mexico to New England, Maritime Canada, and western Europe, this ocean current keeps the climate of these regions much milder than they otherwise would be. Any disruption in the current's path might therefore bring a deep freeze to some of the world's most populous areas.

How do we know this might happen? Because it has happened before. As we have seen in our examination of past global temperatures in Section 20.4, a major period of global warming occurred just 11,000 years ago at the end of the last glacial advance. The resulting influx of freshwater into the North Atlantic from the melting polar ice cap caused the Gulf Stream to stall, plunging Europe into a deep freeze, known as the Younger Dryas (after a tundra plant), which lasted 1300 years. A repetition of such an event could have a devastating impact, especially in western Europe.

20.5 Earth-Sun Geometry

Most scientists believe that the main cause of temperature fluctuations over the past million years relates to cyclic changes in the way in which the Earth orbits the Sun. In the early twentieth century, the Serbian astrophysicist Milutin Milankovitch recognized that the position of Earth's axis and its orbit around the Sun varies in predictable cycles—named **Milankovitch cycles** in honor of him—and that these variations can cause the amount of solar energy reaching Earth to vary by as much as 10 percent (Fig. 20.21).

Milankovitch described three types of changes. First, Earth's orbit around the Sun varies in a systematic manner over periods lasting 100,000 years. At times the orbit is highly *eccentric,* at other times it is almost circular. This means that the distance from Earth to the Sun, and hence the amount of the Sun's energy reaching Earth, varies systematically in a cycle of this duration (Fig. 20.21a). Second, the *tilt* of Earth's axis changes systematically from 21.50° to 24.5° in a motion that repeats itself every 41,000 years. A lower tilt means that the Northern Hemisphere gets less radiation in the summer and more in the winter. In other words, the contrast between summer and winter radiation becomes less pronounced (Fig. 20.21b). Third, the wobble of Earth's axis due to the varying gravitational pull of the Sun and the Moon on Earth's equatorial bulge causes

changes in the direction of Earth's tilt. This is a process known as *precession,* during which Earth's axis traces out the shape of a cone that, as Milankovitch showed, returns to an identical position every 26,000 years (Fig. 20.21c).

Milankovitch's calculations demonstrate the variations in the amount of solar radiation received by Earth as a result of each of these three different aspects of Earth-Sun geometry, as well as the variations in the total combined radiation produced by the summation of these effects. The result demonstrates a remarkable match between the combined signal of solar radiation received by Earth and the temperature variations derived from the oxygen isotope record (Fig. 20.22).

Recall from Section 20.4 that in the last million years, there have been 10 periods of global warming, each separated by a period of glaciation (see Fig. 20.20). Each cycle of warm or cold temperatures was about 100,000 years in duration, a period similar in length to the cycle of Earth's orbit around the Sun. The obvious correlations and intuitive appeal of relating global temperatures to the effects of changing Earth-Sun geometry on the amount of solar energy reaching Earth provides an elegant explanation of the temperature oscillations for the past few million years.

Although many scientists accept that Milankovitch cycles may initiate global warming and cooling events, they also maintain that these cycles alone cannot account for the extent of the observed fluctuations in temperature.

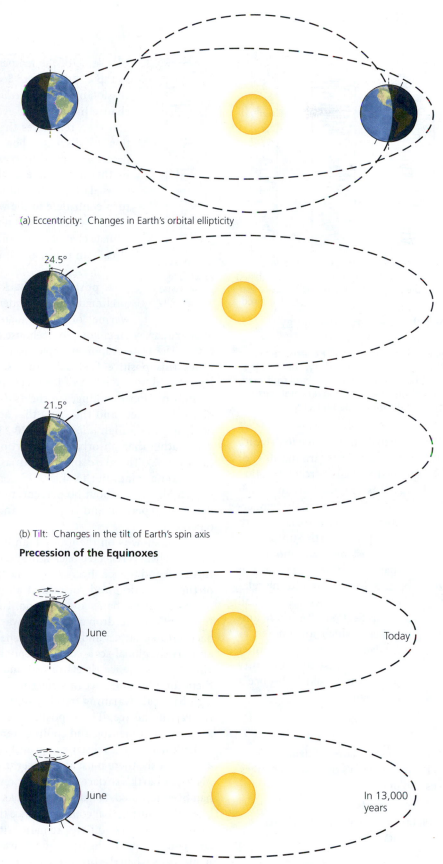

(a) Eccentricity: Changes in Earth's orbital ellipticity

(b) Tilt: Changes in the tilt of Earth's spin axis

Precession of the Equinoxes

(c) Precession: Changes in the direction of Earth's spin axis without changes in the tilt angle

Figure 20.21   Milankovitch Cycles

Milankovitch described three types of cyclical changes in Earth-Sun geometry. (a) Eccentricity: Earth has an elliptical orbit around the Sun. The shape of this orbit varies from being almost circular (low eccentricity) to more elliptical (high eccentricity). This change in shape occurs in a cycle lasting 100,000 years. (b) Tilt: The tilt of Earth's axis varies from 21.5° to 24.5°. Presently the tilt is 23.5°. The change in tilt angle occurs in a cycle lasting 41,000 years. (c) Precession: Earth wobbles on its axis like a spinning top, resulting in a change in the direction of tilt. The axis makes one revolution every 26,000 years.

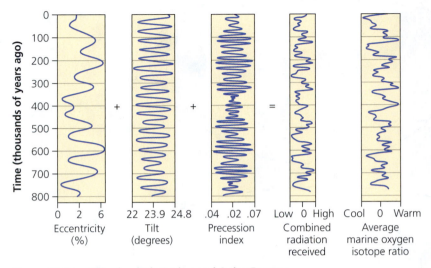

Figure 20.22   Milankovitch Cycles and Solar Energy Reaching Earth
Curves show variations in the amount of solar energy reaching the Earth as a result of variations in eccentricity, tilt, and precession as described by Milankovitch. The combined radiation, which represents the sum of these effects, closely matches the marine oxygen isotopic data (shown on the far right).

The calculated temperature fluctuations required to drive ice sheets back and forth across the continents (around 10°C or 18°F) are too great to be explained by the 10 percent variation in solar energy received by Earth as a result of Milankovitch cycles. Assuming these calculations are correct, the temperature swings associated with changes in Earth-Sun geometry must be made more dramatic by natural amplifiers in order for glacial and interglacial stages to occur. As we shall see in Section 20.6, these amplifiers, which are known as *feedbacks,* act in such a manner as to increase the intensity of a warm or cold cycle once it commences.

> **CHECK YOUR UNDERSTANDING**
>
> ○ Describe the three phenomena responsible for Milankovitch cycles.
>
> ○ Why can't Milankovitch cycles alone account for the observed fluctuations in temperature?

20.5 SUMMARY

- Milankovitch cycles describe climatic influences caused by the eccentricity of Earth's orbit, and changes in the tilt and precession of its axis.

- The existence of 10 warm periods over the past million years matches the 100,000 cyclicity caused by the eccentricity of Earth's orbit.

20.6 Feedback Systems

Milankovitch cycles elegantly explain the rhythmic oscillations between glacial and interglacial intervals over the past million years. However, to obtain a more complete understanding of the processes at work, we must also deduce why some of the temperature changes are so rapid. Recall from Section 20.2 our discussion of *positive feedback,* the pattern of change in which one event causes another that then reinforces the first. In that section, we discussed how rising global temperatures linked to rising carbon dioxide in the atmosphere can result in the increased evaporation of water, which can in turn contribute to the warming trend. Although water vapor is not a prime driver of climate change, it does amplify the changes already in place, and it is referred to as an *amplifier.*

A **positive feedback** (see Fig. 20.5) in a climate cycle renders warm climates warmer or cold climates colder, bringing about runaway greenhouse or icehouse conditions, respectively. The water vapor example from Section 20.2 illustrates this positive feedback process. Another example is provided by the effect of ice sheets on climate. During periods of global cooling, ice sheets expand in polar regions. Ice sheets and the snow that accumulates on top of them have a high *albedo* (Fig. 20.23); that is, they reflect, rather than absorb, a large portion of the Sun's heat (up to 90%). The glare and warmth associated with this reflection are familiar to skiers. In comparison, bare ice has an albedo of about 50 percent, most rocks and soils between 10 percent and 40 percent, and ocean water only 6 percent.

When snow is on the ground, an increased portion of the Sun's heat is reflected back into space. Surface temperatures fall and the ice sheets expand farther, covering more of the vegetation and rocks that absorb the Sun's heat more effectively than snow or ice do. The surface temperature continues to drop and in this way, the positive feedback creates larger and larger ice sheets which in turn amplify the global-scale cooling trend that produced them. Similarly, as the ice sheets retreat and the rocks beneath them (or additional ocean water) are exposed once again, a global-scale warming trend ensues, promoting further retreat of the ice. These positive feedback effects can amplify the warming and cooling trends initiated by the Milankovitch cycles discussed in Section 20.5.

Obviously, there must also be *negative feedback* mechanisms, or Earth's surface could never reverse a trend once it had been initiated. **Negative feedbacks** are self-regulators, like a thermostat, that counterbalance the effects of change to ensure that conditions remain within some reasonable constraints. If the door of a house is left open in the wintertime when the furnace is on, the heat loss is counterbalanced by the heating system of the house. Likewise, if a house gets too warm, the heating system shuts down until the house loses heat to the outdoors. In the water vapor example in Section 20.2, we discussed how a rise in global temperature causes an increase in the amount of evaporation, which results in a positive feedback loop. That same increase in evaporation, however, also causes the formation

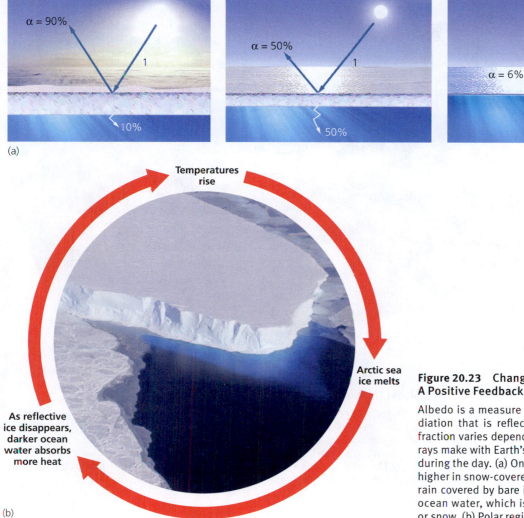

Ice with Snow Cover

α = 90%

1

10%

Bare Ice

α = 50%

1

50%

Open Ocean

1

α = 6%

94%

(a)

Temperatures
rise

Arctic sea
ice melts

As reflective
ice disappears,
darker ocean
water absorbs
more heat

(b)

Figure 20.23   Changing Albedo: A Positive Feedback

Albedo is a measure of the fraction of solar radiation that is reflected back into space. The fraction varies depending on the angle the Sun's rays make with Earth's surface and thus changes during the day. (a) On average, the albedo (α) is higher in snow-covered terrain, compared to terrain covered by bare ice. The albedo is low over ocean water, which is typically darker than ice or snow. (b) Polar regions have high albedo.

of more clouds, which reflect more sunlight back into space, a negative feedback (Fig. 20.24).

Positive and negative feedbacks have been battling for ascendancy throughout geologic time. For short time intervals, positive feedbacks amplify warming trends or cooling trends, and drive Earth into greenhouse or icehouse conditions. However, over the long term, the geologic record demonstrates that neither a runaway icehouse nor a runaway greenhouse has ever occurred in the past. This indicates that Earth must have been dominated by negative feedbacks throughout much of its history. These self-regulating mechanisms cooled warming trends and warmed cooling trends. To better understand Earth's *self-regulation*, in this section we examine the key role an element, such as calcium, plays in the feedback cycle (Fig. 20.25).

Suppose an interval of global warming begins with an increase in the carbon dioxide concentration in the atmosphere. We know from our discussions in Chapter 5, that carbon dioxide is highly soluble in water. When atmospheric carbon dioxide levels are high, rain water that falls

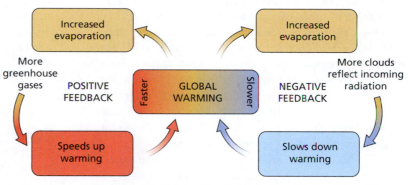

Increased
evaporation

Increased
evaporation

More
greenhouse
gases

POSITIVE
FEEDBACK

Faster

GLOBAL
WARMING

Slower

NEGATIVE
FEEDBACK

More clouds
reflect incoming
radiation

Speeds up
warming

Slows down
warming

Figure 20.24   Interaction between Positive and Negative Feedbacks

A rise in global temperature results in a positive feedback loop (see Fig. 20.5a). However, increased evaporation also causes the formation of more clouds, which reflect more solar radiation back into space and act as a negative feedback.

on land is warmer and more acidic, and is therefore much more effective at attacking the minerals it contacts. The degree of chemical weathering is thus greatly increased by the water's acidity. Recall from Chapter 5, that chemical

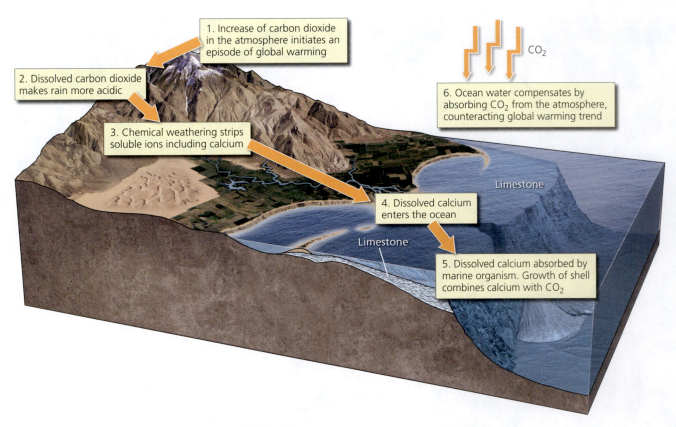

1. Increase of carbon dioxide in the atmosphere initiates an episode of global warming

2. Dissolved carbon dioxide makes rain more acidic

3. Chemical weathering strips soluble ions including calcium

CO_2

6. Ocean water compensates by absorbing CO_2 from the atmosphere, counteracting global warming trend

Limestone

4. Dissolved calcium enters the ocean

Limestone

5. Dissolved calcium absorbed by marine organism. Growth of shell combines calcium with CO_2

Figure 20.25   Carbon Dioxide, Calcium, and Global Warming

This schematic illustrates an example of a negative feedback. 1. A period of global warming begins because of increased carbon dioxide concentrations in the atmosphere. 2. As a result, rain falling on the continents becomes warmer and more acidic. 3. The more acidic rain enhances chemical weathering and strips soluble ions including calcium from minerals such as feldspar. 4. Calcium is transported in solution to the ocean. 5. Calcium is absorbed by marine organisms that combine it with carbon dioxide dissolved in ocean water to form hard calcium carbonate shells. 6. The ocean responds to the loss of dissolved carbon dioxide by extracting more carbon dioxide from the atmosphere, thereby lowering the greenhouse gas content and promoting global cooling.

weathering strips common minerals like feldspar of soluble ions such as sodium and calcium. These ions are carried in solution to the ocean (see Fig. 5.10). When the calcium content of the oceans increases, more calcium can combine with the carbon dioxide dissolved in ocean water to form calcium carbonate. This process removes carbon dioxide from ocean water. When this happens, the ocean water attempts to replenish its carbon dioxide by extracting it from the atmosphere. This, in turn, reduces the amount of carbon dioxide in the atmosphere, which results in global cooling, thereby reversing the warming trend.

The extraction of carbon dioxide from ocean water to form carbonate sediments such as limestone ($CaCO_3$) has been an important process throughout geologic history. The efficiency of this process throughout geologic time is evident in Table 20.2, which shows that 99.99 percent of carbon near Earth's surface resides in carbonate sediments. For these sediments to form, a vast supply of calcium is needed because the principal chemical component of carbonate sediment is calcium carbonate.

So far in this chapter, in our quest to understand the relationship between global warming, greenhouse gases, and industrialization, we have examined various factors that influence global temperatures and cause warming and cooling trends. We have noted cyclic oscillations in global temperature that may be initiated and terminated by the changing geometric relationships between the Sun and Earth known as Milankovitch cycles. And as we have just discussed, the effect of these cycles are then either amplified or dampened by a complex array of positive and negative feedback systems.

Taken at face value, this could lead us to a surprising conclusion. The natural temperature oscillations documented over the past million years show more dramatic swings and have had far more powerful effects than those associated with the current industrial era. If these cyclic temperature variations are projected into the future, they suggest that glaciers will once again advance across continental North America following today's interglacial climate.

CHECK YOUR UNDERSTANDING

- Describe the feedbacks that amplify the effects of Milankovitch cycles.

- What is the difference between a positive feedback and a negative feedback?

- How does calcium affect the greenhouse gas content of the atmosphere?

Does this mean that humanity is a relatively minor player when it comes to the environment?

In answering this question, it is important to keep in mind that we have still only viewed a tiny fraction of Earth's 4.54-billion-year history. Can we be sure, then, that we have heard enough of Earth's song to identify all of its rhythms? The answer is emphatically no, as we shall see in Section 20.7, when we look further back into the geologic record. When we examine time periods of many millions of years, we find that plate tectonic processes play a major role in controlling global temperature.

20.6 SUMMARY

- Although Milankovitch cycles may initiate global warming and cooling events, they must be amplified or dampened by feedback systems in order to produce the climatic fluctuations documented in the proxy temperature record.

- A positive feedback mechanism amplifies an existing trend, whereas a negative one dampens the trend.

- The roles of calcium and carbon dioxide in climate change provide important examples of feedback systems.

- As the calcium content of the oceans is increased, more carbon dioxide can be extracted from ocean water to form calcium carbonate (or limestone). The ocean then replenishes its carbon dioxide content by extracting it from the atmosphere. This reduces the greenhouse gas content in the atmosphere, causing global cooling.

20.7 Plate Tectonics and Climate Change

Carbon dioxide is primarily introduced into our atmosphere by volcanic activity. The vast majority of this carbon dioxide is preferentially taken up by the oceans because ocean waters can absorb 62 times more carbon dioxide than the atmosphere. But the oceans are not an infinite reservoir. The geologic record shows that carbon dioxide has been efficiently extracted from the oceans by organisms for at least 3.5 billion years to form carbonate sediments such as limestone. As we discussed in Section 20.6, the oceans have responded by extracting carbon dioxide from the atmosphere, causing global temperatures to fall.

The highway linking volcanic activity, the atmosphere, and carbonate rock is paved by life and death in the oceans (see Fig. 20.6). The linkage has been so successful that 99.99 percent of the world's carbon has traveled down this highway and, as we discussed in Section 20.3, is now stored in carbonate sediments (see Table 20.2). At the same time, the carbon dioxide content of the atmosphere has been reduced from about 98 percent early in Earth's history to its present value of 0.04 percent. If Earth's atmosphere had maintained its original level of carbon dioxide, the greenhouse effect would have run amok and surface temperatures now would be close to 300°C (570°F), which is far too hot to sustain life.

By providing the oceans with calcium, continental erosion indirectly affects the carbon dioxide content of the

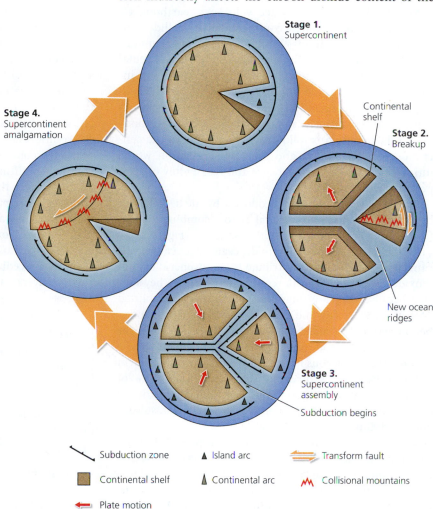

Figure 20.26   The Supercontinent Cycle

The supercontinent (stage 1) traps mantle heat beneath it, resulting in breakup (stage 2). Breakup results in the formation of new oceans and mid-ocean ridges. These oceans are flanked by continental shelves. Eventually, the aging oceans begin to subduct (stage 3), causing the continents to converge. This results in new subduction zones, abundant volcanism, the formation of continental arcs, and the destruction of the continental shelves. Ultimately, the oceans are destroyed by subduction and a new supercontinent assembles (stage 4). This supercontinent traps mantle heat once again and the cycle starts anew.

atmosphere, and hence the planet's average surface temperatures (see Fig. 20.25). Not surprisingly, the generation of calcium that results from the formation and erosion of mountain ranges has a particularly profound effect. A cyclic pattern of surface temperature variations that is related to the formation and erosion of mountain belts is evident in the geologic record over the last billion years.

Recall from Chapters 2 and 9 that continents drift at the rate of a few centimeters a year as they float on top of a denser, yielding layer beneath Earth's lithosphere. The lithosphere is a thin, rigid layer that resembles a cracked eggshell. According to plate tectonic theory, supercontinents form and are destroyed in a process referred to as *the supercontinent cycle* (see Chapter 10). The fractured segments join together to form a series of floating plates that constantly interact to produce volcanoes, earthquakes, and mountain belts. Where continental plates collide, orogenic (mountain-building) events result in major mountain belts. As we discussed in Chapter 10, continental collisions are not randomly distributed in time but, instead, are concentrated into narrow portions of the geologic record. These relatively short intervals of time reflect times of supercontinent amalgamation. Each supercontinent amalgamation is followed by breakup and the generation of new oceans, and then by subduction which ultimately leads to ocean closure and reassembly of a supercontinent (Fig. 20.26).

When a supercontinent produced by such collisions breaks up and disperses, several factors combine to raise the level of atmospheric greenhouse gases (Fig. 20.27a). First, the development of new mid-ocean ridges causes volcanism to increase and contribute to atmospheric carbon dioxide levels. Second, as the continents disperse, they subside because they are migrating away from the uplifted sites of active rifting and thermal upwelling. Once continental margins subside below sea level, they can no longer be eroded, and so they no longer contribute calcium to the oceans. The Cretaceous Period, with its anomalously warm temperatures and submerged continents (only 15% of Earth's surface was exposed), provides an example of the combined climatic effects of enhanced volcanism, subsidence, and diminished erosion rates.

Convergence begins when continental plates approach each other as the crust of intervening ocean is destroyed by subduction. This causes widespread volcanic activity, which pumps carbon dioxide into the atmosphere, resulting in global greenhouse warming (Fig. 20.27b). When the continents finally collide, however, several factors combine to turn the global greenhouse into an icehouse (Fig. 20.27c). First, subduction ceases and so the volcanic activity that was responsible for producing the greenhouse gas diminishes. Second, ice caps that develop and grow on the mountains produced by collision reflect solar radiation rather than absorb it due to their high albedo. Third, erosion of mountain belts rapidly contributes calcium and other nutrients to the oceans. In the feedback cycle described in

Section 20.6 and shown in Figure 20.25, the nutrients stimulate increased organic productivity and photosynthesis, and the calcium eventually combines with carbon dioxide in the ocean water to produce deposits of limestone. The ocean water responds by extracting carbon dioxide from the atmosphere, thereby diminishing the greenhouse effect.

An examination of the geologic record shows an impressive match between the history of ancient greenhouse and icehouse climates, and that of mountain building, limestone deposition, and spurts in organic productivity and biological diversity (Fig. 20.28). This correlation supports the current view of many geoscientists that plate tectonic processes such as mountain building and erosion dominate the natural controls of global warmth. Figure 20.28 shows the three most recent mountain building events: a period about 1100 million years ago and a period 750 to 550 million years ago, both thought to be associated with the amalgamation of supercontinents; and a period 470 to 300 million years ago, associated with the amalgamation of Pangea. The formation of each of these supercontinents coincided with the advent of icehouse conditions.

Basaltic dike swarms occur about 100 million years after supercontinent amalgamation and reflect the onset of continental rifting, which, as we discussed in Chapters 9 and 10, heralds the breakup of the supercontinent. The important evolutionary events cited in Figure 20.28 are times of biological innovation and occur when new habitats (continental shelves) form when ocean water is laden with nutrients from the erosion of the supercontinent as it breaks up.

In the preceding sections of this chapter, we have peeled back the layers of time. We started with the relatively modern record and, using temperature proxies, subsequently examined an ever increasing portion of the geologic record. By organizing our discussion in this way, we have been able to examine a variety of influences that operate with different intensities and with cycles of different duration. We are now in a position to put all these observations together. We now know how to recognize Earth's beats and rhythms, and it is time to identify its song in Section 20.8.

> ### CHECK YOUR UNDERSTANDING
> - How is continental erosion related to the supercontinent cycle?
> - How does the supercontinent cycle affect global climate?

20.7 SUMMARY

- Continental erosion is tied to plate tectonic activity and the supercontinent cycle, and is most pronounced during periods of mountain building.

- The extent of global change related to plate tectonic activity exceeds that which can be caused by changing Earth-Sun geometry.

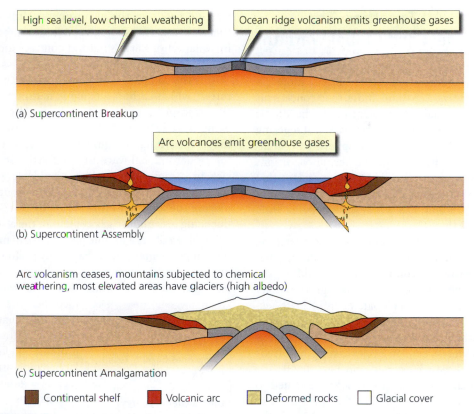

Figure 20.27   The Supercontinent Cycle and the Greenhouse Effect

(a) When a supercontinent breaks up, new oceans form and mid-ocean ridge volcanism produces abundant carbon dioxide, while the new continental margins subside. Sea level consequently rises, causing a decrease in continental weathering. The combination of these effects increases the abundance of carbon dioxide in the atmosphere and produces greenhouse conditions. (b) Convergence and subduction causes widespread volcanic activity, which pumps carbon dioxide into the atmosphere, resulting in global warming. (c) When a new supercontinent forms, subduction between the converging continents ceases and volcanic-derived greenhouse gas emissions are diminished. Ice caps on the mountains reflect solar radiation. Weathering and erosion of mountain belts transport calcium to the ocean water. The calcium eventually combines with dissolved carbon dioxide to produce deposits of limestone. The ocean water responds by extracting carbon dioxide from the atmosphere, thereby initiating icehouse conditions.

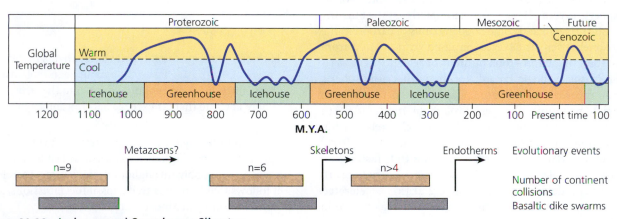

Figure 20.28   Icehouse and Greenhouse Climates

In this graph, we can see the relationship between global temperatures (blue curve), icehouse (when ice sheets form) and greenhouse climates, and the development of supercontinents based on episodes of continental collisions (mountain building associated with supercontinent amalgamation) for the past 1 billion years. Note that orogenic peaks culminate in cold icehouse climates. Supercontinents formed about 1100, 650, and 300 million years ago when a significant number of continental collisions occurred. When this occurs, subduction between the converging continents ceases and volcanic-derived greenhouse gas emissions are diminished. Hence, each supercontinent is associated with cold icehouse climates. Basaltic dike swarms reflect the rifting of the supercontinent, eventually resulting in supercontinent breakup and the development of new oceans flanked by continental shelves, conditions that stimulate major biological evolutionary events. Increasing volcanism associated with supercontinent breakup causes volcanic-derived greenhouse gas emissions to increase, ushering in greenhouse conditions.

20.8 Putting It All Together: The Greenhouse Effect and Global Warmth

From the perspective of the geologic record, the present global warming trend seems almost trivial. Far more extreme temperature variations than those that concern us today have occurred throughout geologic time. In the Middle Cretaceous, average global temperatures were a staggering 8°C (14°F) warmer than today, and atmospheric carbon dioxide levels were much higher. These extreme conditions lasted for millions of years. Their duration indicates that the shorter-term Milankovitch or sunspot cycles we discussed in Section 20.4 and 20.5 were clearly insufficiently powerful to override them.

The major beat in Earth's natural rhythms is, therefore, a tectonic one. As we saw in Section 20.7, the interplay between continental erosional rates, mountain building, the availability of continental shelves and biological activity exerts the primary control on global warmth. Plate tectonic processes, which involve the opening and closing of ocean basins and the fragmentation, dispersal, and amalgamation of continents, occur in cycles lasting hundreds of millions of years (see Chapter 10 and Fig. 20.28). The shorter Milankovitch and sunspot cycles are relegated to the role of amplifiers.

But what about the burning of fossil fuels? Since the amount of carbon stored in fossil fuels is dwarfed by that in carbonate sediments (see Table 20.2), the effects of burning fossil fuels might at first appear to be negligible. This is misleading, however. Carbon residing in carbonate sediments is effectively out of natural circulation because it is stored in bedrock. It is the amount of carbon in circulation that concerns us and, relative to only the carbon in circulation, the burning of fossil fuels is a very significant source of atmospheric carbon dioxide. It is estimated that the burning of fossil fuels will add some 5000 gigatons of carbon to the atmosphere over the next few hundred years, all of it from fossil fuels that nature took millions of years to form.

From the geologic record, we know that atmospheric carbon dioxide concentrations have been correlated with global temperature changes throughout much of Earth history. So it is reasonable to argue for such a relationship in the modern world. However, we are consuming fossil fuels many, many, many orders of magnitude faster than nature can replenish them. Ocean waters cannot absorb carbon dioxide from the atmosphere at this accelerated rate, so fossil fuel dependence directly increases carbon dioxide levels in the atmosphere. Human activity has therefore caused a short circuit in the carbon cycle; we are in violation of Earth's natural contract for the transfer of energy and nutrients.

It is the complete mismatch between the relatively slow rate at which carbon dioxide is naturally extracted from the atmosphere compared to the rapid rate at which it is being introduced into the atmosphere by human activity that is the strongest argument for a human-induced origin for modern global warming. In Earth's song, the modern rise in atmospheric carbon dioxide levels is a sudden screech! But we can only arrive at this conclusion after exploring the long-term role played by plate tectonics in influencing greenhouse gas content in the atmosphere.

Geology offers a unique perspective to modern environmental problems. In this chapter, by peeling back the layers of time, we have identified Earth's natural rhythms. At each step, we have gained additional insights. We have identified the main plate tectonic beat that lasts for hundreds of millions of years and profoundly affects global surface temperatures. We have seen how other shorter beats, such as those related to the geometry of Earth-Sun system and those related to solar variability, are less pronounced. We have also learned that only in identifying the primary rhythms can any distortions related to the human consumption of fossil fuels be recognized. Like a flaw in a well-known song, we can now instantly recognize the distortion of the current era.

> **CHECK YOUR UNDERSTANDING**
> • How does burning fossil fuels at current rates violate Earth's contract for the transfer of energy?
> • Is climate change fact or fiction?

20.8 SUMMARY

* Humans consume fossil fuels at a rate that is many orders of magnitude faster than nature can replenish them, introducing greenhouse gases into the atmosphere at a rate that far exceeds nature's ability to absorb them.

* This greenhouse gas content has correlated with average surface temperatures for at least the last billion years and probably throughout geologic time, therefore it follows that human activity, fossil fuel emissions, and global warming are inextricably linked.

Key Terms

aurora 658
carbon cycle 653
climate change 646
climate proxy 655

global warming 646
greenhouse conditions 647
greenhouse effect 649
greenhouse gas 646

icehouse conditions 647
Little Ice Age 658
Milankovitch cycles 668
negative feedback 670

positive feedback 670
solar wind 658
sunspot 657
sunspot cycle 658

Key Concepts

20.1 TAKING EARTH'S PULSE

- One of the most difficult challenges of environmental analysis is distinguishing between natural and human-induced global change.

20.2 THE GREENHOUSE EFFECT: A MATTER OF BALANCE

- Without greenhouse gases Earth would be uninhabitable, with an average surface temperature 35°C (63°F) cooler than today.

- Greenhouse gases allow solar radiation to penetrate Earth's atmosphere, but block some of the radiation emitted from Earth's surface.

20.3 THE CARBON CYCLE

- The carbon cycle is an adaptation of the food chain that shows how carbon moves between the inorganic and organic worlds.

- Carbon flows from energy producers to energy consumers to decomposers via the food chain.

- Decomposition in the presence of oxygen recycles carbon into the atmosphere.

- Decomposition in the absence of oxygen or during rapid burial can form fossil fuel deposits.

- Most carbon on Earth has ended up in carbonate sediments and sedimentary rocks.

20.4 CLIMATE CHANGE: A GEOLOGIC PERSPECTIVE

- Sunspot cycles, which are typically 11.1 years in duration, affect temperatures in the upper atmosphere. The mysterious absence of sunspots in the seventeenth century may be associated with a cooling event known as the Little Ice Age.

- To decipher the geologic record, average surface temperatures are deduced indirectly by climate proxies, such as oxygen isotope ratios ($^{18}O/^{16}O$).

- There is an impressive correlation between the geologic record of climate change and oxygen isotope data for both ice sheets and foraminifera. The record shows that Earth's climate over the past million years has been dominated by alternating warm and cool episodes with about a 100,000-year cyclicity.

20.5 EARTH-SUN GEOMETRY

- Milankovich cycles record variations in the eccentricity of Earth's orbit, and in changes in the tilt and precession of Earth's axis.

- The cycles, especially the 100,000-year cycle of eccentricity, provide an explanation for the initiation of cooling (glacial) and warming (interglacial) trends identified by oxygen isotope data.

20.6 FEEDBACK SYSTEMS

- Cooling and warming trends are amplified and dampened by positive and negative feedbacks.

- Continental erosion, which provides calcium to the oceans, is an important control on atmospheric carbon dioxide. Calcium combines with dissolved carbon dioxide in the oceans to form calcium carbonate or limestone. The oceans respond by extracting carbon dioxide from the atmosphere, resulting in global cooling.

20.7 PLATE TECTONICS AND CLIMATE CHANGE

- Continental erosion, and therefore calcium input into the oceans, is closely linked to the supercontinent cycle of mountain building, which lasts hundreds of millions of years.

20.8 PUTTING IT ALL TOGETHER: THE GREENHOUSE EFFECT AND GLOBAL WARMTH

- Humans consume fossil fuels much faster than nature can replenish them and introduce excess carbon dioxide into the atmosphere much faster than nature can remove it.

Study Questions

1. Why is the greenhouse gas content in the atmosphere "a matter of balance"?

2. Explain how increases in the concentration of greenhouse gases in the atmosphere cause global warming.

3. Why does 99.99 percent of carbon near Earth's surface reside in carbonate sediments (see Table 20.2)?

4. Explain the relationship between the food chain and the carbon cycle.

5. Explain how oxygen isotope data from (a) ice sheets and (b) foraminifera in marine sediments are used as a climate proxy.

6. Explain the relationship between Milankovitch cycles and the oxygen isotope data in foraminifera from sediments.

7. List an example of a positive feedback and a negative feedback and explain how each system works.

8. Why is the supply of calcium to the oceans an important factor to consider in the global warming debate?

9. Explain how plate tectonic cycles have been the predominant influence on global surface temperatures throughout most of geologic time.

10. Debate among yourselves the problems of fossil fuel consumption and its relationship to the greenhouse effect. Bring information from other sources to bear on the issue. The approach in this chapter was one of peeling back the layers of time. Can you suggest other approaches?

21 The Planets

21.1 Earth and Its Moon

21.2 The Terrestrial Planets and Their Moons

21.3 Plate Tectonics: A Planetary Perspective

21.4 The Jovian Planets, Their Moons, and the Kuiper Belt

21.5 Asteroids, Meteors, Meteorites, and Comets

21.6 Exoplanets and Other Solar Systems

▶ Key Terms

▶ Key Concepts

▶ Study Questions

SCIENCE REFRESHER: Center of Mass 685

IN DEPTH: Martian Life on Earth? 695

LIVING ON EARTH: Visits from the Celestial Junkyard 707

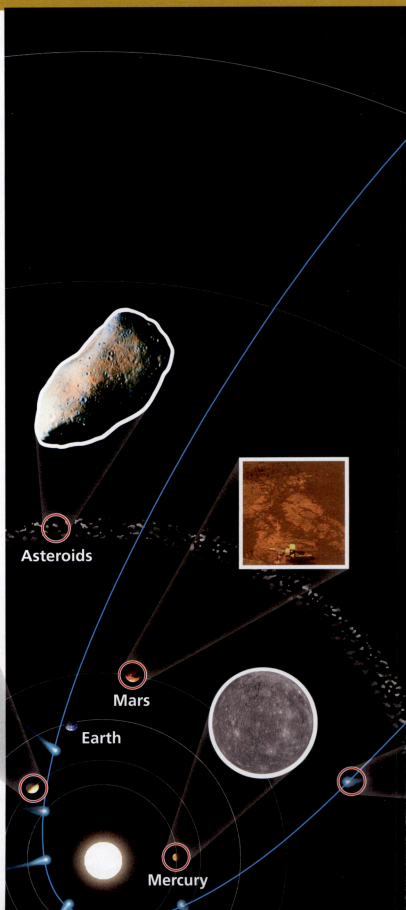

Asteroids

Mars

Earth

Venus

Mercury

Neptune

Jupiter

Saturn

Uranus

21.1 Compare and contrast Earth and its Moon and explain the origin of the Earth-Moon system.

21.2 Describe the terrestrial planets and show that each has its own unique set of characteristics while also sharing common traits.

21.3 Explain how Earth's unique ability to generate continental and oceanic crust was fundamental to the evolution of plate tectonics and the development of life on this planet.

21.4 Describe the Jovian planets and their moons.

21.5 Discuss the nature of asteroids, meteors, meteorites, and comets and what they can tell us about development of the Solar System.

21.6 Describe the search for solar systems like our own across the universe.

Over the centuries, humans have come to realize that Earth belongs to a Solar System made up of eight planets that orbit a single star we call the Sun. Throughout our history, humans have devised ingenious ways to observe phenomena in our Solar System and the universe. Many key observations were made centuries before the advent of optical telescopes. More recently, we have launched telescopes into space to search the heavens in ways never before possible. Humans have walked on the Moon, robot geologists now roam the Martian surface, and space probes have visited all of the eight planets in our Solar System. These sophisticated tools have dramatically increased our knowledge of the Solar System and the universe beyond.

In Chapter 1 we learned how the Solar System formed. In this chapter we will see that each planet in the Solar System has its own unique characteristics, and that Earth seems to be the only planet that can sustain life on its surface. This ability is related to a number of variables, including Earth's size, composition, active tectonics, and distance from the Sun. Together, these allow for a life-sustaining atmosphere and liquid water on our planet's surface. Plate tectonic activity on planet Earth has given rise to crust and lithosphere, and contributed to its oceans and atmosphere. But is plate tectonics unique to Earth? What do we know of our planetary neighbors and the composition of their surfaces and their interiors? Do they have atmospheres and, if so, what factors control their composition?

21.1 Earth and Its Moon

In Chapter 1, we discussed the solar nebula theory, which asserts that Earth and the other planets formed when a rotating cloud of dust and gas flattened into a disk as it contracted under the influence of gravity. In this chapter we will tour our Solar System. About 4.57 billion years ago, the Sun ignited to become a star and the center of the Solar System. As the inner nebula cooled, *planetesimals* grew in size by the process of accretion until they formed *protoplanets*, which evolved to become planets about 4.54 billion years ago. The innermost (*terrestrial*) planets (Mercury, Venus, Earth, and Mars) are rocky, whereas the outer (*Jovian*) planets (Jupiter, Saturn, Uranus, and Neptune) are gaseous. Many of these planets are tiny and have satellite bodies, which we call *moons*, that orbit around them.

Other important objects in the Solar System include *asteroids, meteoroids, meteorites,* and *comets*. An **asteroid** is a rocky celestial body that varies from 1000 kilometers (620 miles) to about 10 meters (33 feet) in diameter. A meteoroid is a smaller rocky body than an asteroid, and one that enters Earth's atmosphere is known as a **meteor**. Most meteors rapidly burn up in the atmosphere, but a meteor that survives and collides with the Earth's surface becomes a **meteorite**. Over time, the term meteorite has expanded to include all rocky celestial bodies

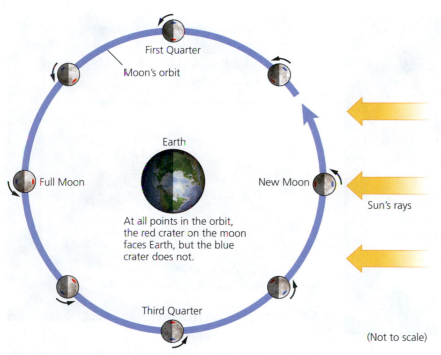

Figure 21.1 Synchronous Rotation of the Moon

The Moon rotates counterclockwise on its own axis as it revolves about Earth such that the same side always faces observers on Earth. In this figure, this relationship is schematically represented by a red crater that always faces Earth. If the Moon did not rotate as it revolved, the position of the red crater in each instance would be identical and different parts of the Moon's surface would face Earth.

that have collided with planets and moons. Asteroids, meteors, and meteorites are thought to represent the remnants of the early Solar System, those rocky objects that were not incorporated into planets (see Chapter 1). A **comet** is a mixture of rock, ice, and dust, some of which is carbon-rich. Most comets occur in the outermost parts of our Solar System, but some are thought to originate outside the Solar System.

During the course of our tour of the Solar System in this chapter, we will find that each planet has unique characteristics. By comparing the evolution of the various planets, we are able to place Earth's restless geology in a celestial context. In particular we will examine the effects that planet size and distance from the Sun have on planetary evolution. From the inner (*terrestrial*) planets we will gain a perspective on how Earth has the appropriate combination of size and distance from the Sun to harbor life, and how its plate tectonic activity has set it on a course that is fundamentally different from its closest celestial neighbors. From the outer (*Jovian*) planets,

and the mysterious icy bodies that inhabit the edges of the Solar System, we see the chilling effects that distance from the Sun and the waning influence of solar activity can have. We begin our tour of the Solar System with our closest celestial neighbor, Earth's Moon.

THE MOON

The Moon is much smaller than Earth, with a diameter of only 3476 kilometers (2155 miles) compared with Earth's 12,740 kilometers (7900 miles). It is our nearest neighbor, at an average distance of only 384,400 kilometers (238,330 miles). Our moon is held in its orbit by the balance between its motion and its gravitational attraction to Earth (see Science Refresher: **Center of Mass** on page 685). In 1680, Giovanni Cassini discovered that the Moon rotates about its own axis at the same rate that Earth rotates. Because of this, the same side of the Moon always faces us. This phenomenon is known as *synchronous rotation*. Although the Moon rotates as it orbits Earth, it does so only once per orbit. The illuminated surface we see at *full moon* is consequently the surface that always faces us (Fig. 21.1). Unless we venture into space, we cannot see the far side of the Moon directly.

In the late eighteenth century, mathematicians Joseph Lagrange and Pierre Laplace speculated that the synchronous rotation of the Moon was not a coincidence. Instead, it could be explained by Newton's law of gravity if the Moon were slightly elongated (like an egg) rather than perfectly spherical. The elongation was thought to be caused by the gravitational pull of Earth on the Moon. The Moon's shape, they speculated, should be deformed by this force and thus be elongated along a line pointing toward Earth. The gravitational attraction between Earth and the Moon would then act to keep the Moon's long axis pointing toward Earth. This hypothesis has since been confirmed by accurate measurements from space probes; one axis of the Moon is indeed about 2 to 3 kilometers (1 to 2 miles) longer than the other and points toward Earth.

Prior to 1969, when humans first landed on the Moon, the speculations of scientists concerning the composition and origin of the Moon were essentially untested. Viewed from Earth, the light and dark areas on the lunar surface are its most obvious visible features (Fig. 21.2). With the advent of telescopes, the light-colored areas were (correctly) interpreted to be **lunar highlands**, standing above the dark plains. Early astronomers mistakenly thought that the dark areas resembled Earth's ocean basins, and so called them **mare** (pronounced ma-ray), which is the Latin word for sea. At one time it was even thought that these basins might hold water, and that the Moon might therefore harbor life. However, even though the term *mare* (plural *maria*) is firmly entrenched in the literature, we now know that these basins are waterless depressions.

The space race of the 1960s culminated with the lunar landings of the Apollo space program, and between 1969 and 1972, 12 humans walked on the Moon's surface. In

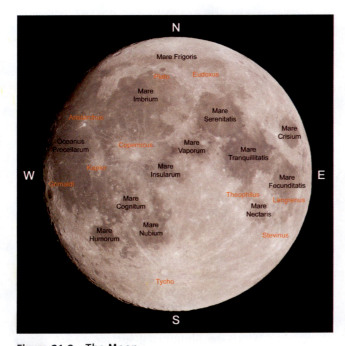

Figure 21.2   The Moon

Lunar maria are dark smooth areas; the lunar highlands are lighter and more heavily cratered.

Figure 21.3   Lunar Basalt

Basalt retrieved from the floor of a lunar mare reveals the presence of spherical cavities that are remnants of air bubbles. These cavities indicate that volcanic eruptions on the Moon once vented gas.

addition, several kilograms of lunar rocks were brought back to Earth, which have proven invaluable in unraveling the Moon's history. The Moon's surface is composed of igneous rocks that are broadly similar to the rocks on Earth. The most abundant rocks in the lunar maria are dark basalts (Fig. 21.3). These rocks are relatively rich in magnesium and iron, like the basaltic rocks that dominate oceanic crust on Earth (see Chapter 4). Basalts are known to be the product of volcanic activity, so their presence on the Moon suggests the existence of similar source materials beneath the surfaces of the Moon and Earth. The brightness of the lunar highlands reflects the presence of feldspar, which is one of the most common minerals in Earth's crust (see Chapter 3).

But here the similarities between the Moon and Earth end. The final Apollo mission, in December of 1972, specifically targeted a sampling area on the Moon near the Taurus Mountains that was thought to be a candidate for recent volcanism. But the age of the basalts in these samples, as determined by radioactive dating, proved to be in excess of 4 billion years! Further investigations revealed that volcanic activity on the Moon virtually ceased 3 billion years ago. Most rocks in the lunar maria were found to range in age from 3.0 to 3.8 billion years. Those of the lunar highlands prove to be even older, ranging up to 4.5 billion years in age.

The antiquity of these ages is significant. The age of the planets in our Solar System is thought to be about 4.54 billion years and the 4.5 billion-year ages from the Moon provided hard data in support of this since the Solar System must be at least as old as any of the objects (such as the Moon) it contains. The dates from the Moon therefore provide a minimum estimate for the age of the Solar System.

Because Earth is a geologically active planet, rocks in excess of 3 billion years old are very rare on its surface. We can therefore learn much about the early history of Earth by studying the evidence of the ancient processes that occurred on the Moon. The lunar landscape has abundant craters and has obviously been affected by the impact of innumerable meteorites. With its vanishingly thin atmosphere, the Moon has little protection from such bombardment. In contrast, Earth's atmosphere causes most celestial debris to burn up before it reaches the surface. The incineration of the debris causes the phenomenon misleadingly referred to as *shooting stars*. It has been calculated that impacts on the Moon had velocities close to 2.4 kilometer/second, or 5360 mph. Such impacts released sufficient heat energy to melt adjacent rocks and eject vast volumes of material to form craters (Fig. 21.4). It is estimated that the impact of a meteorite only 3 meters (10 feet) in diameter could result in a crater 150 meters (500 feet) wide. On the Moon, fallout from meteorite bombardment has produced a thin veneer of lunar "soil" that is 3 to 5 meters (10 to 15 feet) thick.

Craters are typically more abundant in the older lunar highlands than they are in the younger maria (Fig. 21.5), suggesting that meteorite bombardment was much more common early in the Moon's history and tapered off significantly about 3.1 billion years ago. Given that these meteorites traveled throughout the entire Solar System, it seems certain that Earth's early fragile crust was also affected by this activity. However, the evidence has since been obscured on Earth because of its restless geology.

It is widely believed that massive meteorite impacts are also responsible for the widespread volcanic activity that

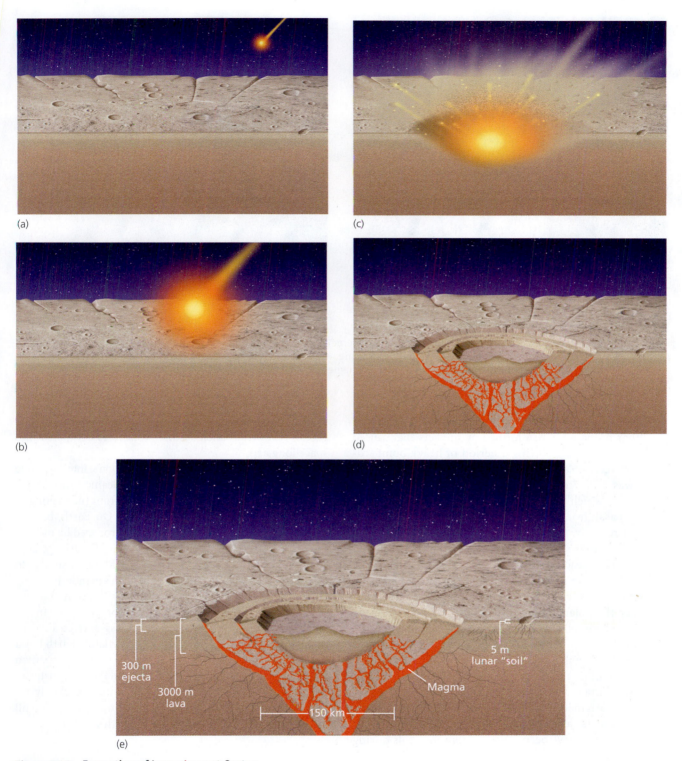

(a)

(b)

(c)

(d)

(e)

300 m
ejecta

3000 m
lava

150 km

Magma

5 m
lunar "soil"

Figure 21.4   Formation of Lunar Impact Crater

The meteorite (a) strikes the lunar surface, (b) penetrates the surface, and (c) vaporizes. Shock waves fracture and eject the rock to form a circular crater. (d) Rebound can result in a raised central portion of the crater. (e) The lunar maria were formed by major impacts that broke the lunar crust and produced multiringed basins, which were later flooded by basalt lava that rose up the cracks.

characterizes the Moon's early evolution. But what about the origin of the lunar maria? On Earth, plate tectonic processes provide an excellent explanation for the origin of the topographically depressed regions of the ocean basins because dense oceanic lithosphere rests on a plastic substrate

we call the asthenosphere. But there is no evidence of plate tectonic processes on the Moon, so we must search for a different mechanism.

The lower frequency of impact structures in the lunar maria implies that their basaltic floors formed *after* the

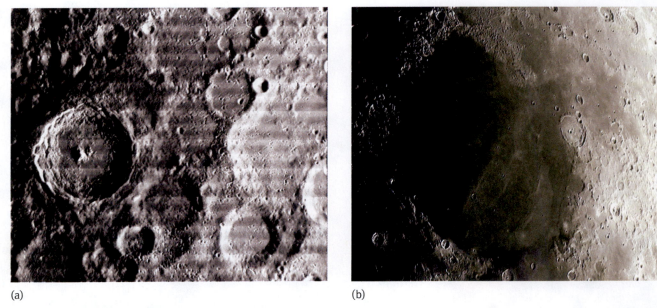

(a)   (b)

Figure 21.5   Lunar Highlands and Maria

Two photographs show the contrast between (a) the heavily cratered lunar highlands that represent the oldest portions of the Moon's surface, and (b) the comparatively smooth lunar lowlands, or maria, which are filled with basalt and contain few craters. Craters in the lunar highlands range up to about 100 kilometers (60 miles) in diameter; maria range up to 2500 kilometers (1550 miles) in diameter.

main bombardments. However, most scientists believe that the lunar maria were also formed by impacts that caused large craters to form during a period of heavy bombardment between 4.5 and 3.9 billion years ago. This bombardment was followed by widespread melting in the lunar subsurface, which flooded the crater floors with successive flows of basalt from 3.8 to 3.0 billion years ago (Fig. 21.4e).

Until recently, there was no evidence of water on the Moon. The Moon's surface is too cold for liquid water, and water vapor is readily burned off by solar radiation. However, in 1998 data from NASA's Lunar Prospector mission revealed that ice crystals make up 1 percent of the lunar soil near the polar regions and also occur in some deep trenches where solar radiation does not reach. In late 2009 and early 2010, Indian and NASA space probes detected water within lunar minerals and as crystalline water-ice. About 600 million metric tons of water-ice are thought to occur in 40 craters near the Moon's north pole. The water may have originated from the bombardment of water-bearing comets, asteroids, and meteoroids over billions of years, or may have been released from minerals during impact events.

The Moon's surface is characterized by the absence of oxygen. For example, rocks containing metallic iron are found on the Moon, whereas on Earth, metallic iron rapidly oxidizes in the presence of oxygen (that is, it absorbs oxygen from the atmosphere) to form iron oxide, or rust. The features of the Moon indicate that its surface is essentially ancient, static, and lifeless. The absence of modern volcanic activity may account for its lack of an atmosphere because no new gases are being vented from active volcanoes as they are on Earth. Because of the Moon's small size

(and, hence, its low gravitational attraction for atmospheric gases), gases from ancient lunar volcanoes dissipated into space long ago.

Clues to the composition of the Moon's interior come from the study of its ancient basaltic volcanic rocks, which originated from the melting of its interior. On Earth, basaltic magma is produced by melting of the mantle. The average density of the Moon is similar to that of Earth's mantle, implying a broadly similar composition.

The Moon preserves an ancient, lifeless surface in contrast to a dynamic Earth but it is nevertheless our companion in space and the early histories of both bodies are closely related. Because of this, studies of the Moon can be used to fill important gaps in our knowledge of Earth's early history.

> **CHECK YOUR UNDERSTANDING**
>
> ○ Why does one side of the Moon always face Earth?
>
> ○ What are the lunar maria and how did they form?
>
> ○ How does studying the surface of the Moon help us learn about Earth's early history?

ORIGIN OF THE EARTH-MOON SYSTEM

Recall from Chapter 1 that Earth's Moon formed in the aftermath of the impact of a Mars-sized object on Earth's surface (see Fig. 1.15). This impact occurred about 60 million years after Earth accreted. A splash of hot debris was probably ejected from the impact site, some of which cooled and accreted in orbit around Earth to form the Moon. The Moon's high angular momentum (about 10 kilometers per second or 6 miles per second), and possibly even Earth's tilt, are attributed to the impact. If this is

SCIENCE REFRESHER

Center of Mass

If you try to balance your plastic student ID card on your finger, you will find that there is a unique point about which this balance can be achieved. You have located the **center of mass** of the card, and it is balanced because its mass distribution is uniform around this point. The center of mass is an important concept in astronomy. It is the unique point around which the masses of two celestial bodies are balanced. For example, our Moon orbits Earth around a point where their respective masses balance. This point lies on an invisible line connecting the centers of Earth and the Moon, about 1710 kilometers (1062 miles) below Earth's surface.

For a single object of uniform density, the center of mass is simply the center of the object (Fig. 21A, top). For a pair of identical objects of uniform density, the center of mass lies midway along the line joining their centers (Fig. 21A, middle), much like the pivot point of a seesaw. For a pair of identical objects of unequal size, and hence unequal mass, the center of mass, once again, lies along the line joining their centers, but occupies a position closer to the object of larger mass, much like the fulcrum of a lever (Fig. 21A, bottom).

Accordingly, the center of mass in the Earth-Moon system is the point between them where the two celestial bodies balance each other (Fig. 21B). The masses of the Moon and Earth differ, much like the mass of a parent and a child. Imagine a parent and a young child on opposite ends of a seesaw. If the pivot of the seesaw is placed in the center, the system is hopelessly imbalanced, the adult is stuck on the ground while the child is hoisted into the air. In order for the system to be balanced, the pivot must be located much closer to the adult (see Fig. 21A bottom). Earth's mass (5.98×10^{24} kilograms or 1.3×10^{25} pounds) is nearly eighty times that of the Moon (7.35×10^{22} kilograms or 1.62×10^{23} pounds). Their respective masses balance on the line joining the center of Earth to the center of the Moon at a point about 1710 kilometers (1060 miles) *below* Earth's surface. The Moon orbits Earth around this point, and Earth and Moon orbit around this point as they travel around the Sun.

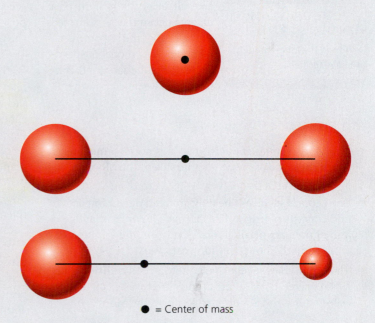

● = Center of mass

Figure 21A   Center of Mass

The center of mass of an object or group of objects is the point about which their masses are balanced; it varies according to the number of objects, their individual masses, and their relative separation.

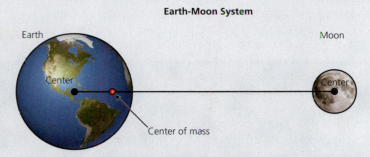

Earth-Moon System

Figure 21B   Center of Mass of the Earth-Moon System

The center of mass of the Earth-Moon system lies along the line joining their centers, but the large difference in mass between the two bodies places the center of mass at a depth of about 1700 kilometers (1060 miles) below Earth's surface.

correct, then the reason for the changing of the seasons may be traced back to the earliest history of our planet (see Living on Earth: **Earth's Seasons—A 4.5-Billion-Year Legacy?** in Chapter 1).

Earlier models suggested that the impact resulted in the expulsion of some of Earth's mantle, portions of which accreted to form the Moon. However, closer inspection shows that the abundances of iron and magnesium in lunar rocks differ significantly from their abundances in Earth's mantle. Also, lunar rocks and minerals appear to be completely devoid of water. This evidence indicates that the Moon is unlikely to have been carved out of Earth. Instead, it appears to have been formed from the mantle of the impacting object. Computer simulations of the impact event show that all volatiles, including water, were expelled by the heat generated from the impact event. The simulations also indicate that the metallic core of the impacting object sank into Earth's core, which explains the lower density of the Moon relative to Earth. The model suggests the feldspar-rich lunar highlands formed as a result of crystallization of a magma ocean that was formed by the heat of the impact event. Although this model is speculative, it fits the facts and constraints as we now know them.

> **CHECK YOUR UNDERSTANDING**
> • How is the Moon thought to have formed?

21.1 SUMMARY

- The Moon rotates about its own axis at the same rate as Earth. Because of this, the same side of the Moon always faces Earth.

- The Moon's surface is heavily cratered due to impacts and consists of bright, feldspar-rich lunar highlands and dark lunar maria floored by basalt, the eruption of which ceased about 3 billion years ago.

- The Moon may have formed after Earth's collision with a Mars-sized object. Part of the object was ejected and accreted in orbit about Earth to form the Moon.

21.2 The Terrestrial Planets and Their Moons

Figure 21.6 shows the arrangement of the major celestial bodies in the Solar System from the innermost planet, Mercury, to Neptune, the most distant. Pluto, which is no longer considered to be a planet, is included in Table 21.1 as a representative of the icy bodies at the outer reaches of the Solar System. Table 21.1 compares some of the planets' important properties. Every planet is unique, but as we will see as we continue our tour through the Solar System, important common themes emerge that provide insights into the processes that influenced Earth's evolution.

A general idea of each planet's chemical composition is given by its mean density (Table 21.1). The innermost planets, from Mercury to Mars, have relatively high densities, suggesting that, like Earth, they are essentially rocky spheres. These planets are collectively called the **terrestrial** planets, a name which underlines their similarity to Earth; *terra* is the Latin word for Earth. In contrast, the outer planets (from Jupiter to Neptune) have significantly lower densities and are essentially giant, gaseous bodies with poorly defined surfaces. The outer planets are collectively referred to as the *gas-giant* or **Jovian** (Jupiter-like) planets because they share defining features with Jupiter. The planets vary in important characteristics primarily because of their size and distance from the Sun. Some planets and their moons preserve ancient surfaces that have been undisturbed for billions of years. Others have very young surfaces, suggesting that they are still tectonically active. In this section of the chapter, we describe the terrestrial planets in some detail. In Section 21.4 we address the Jovian planets.

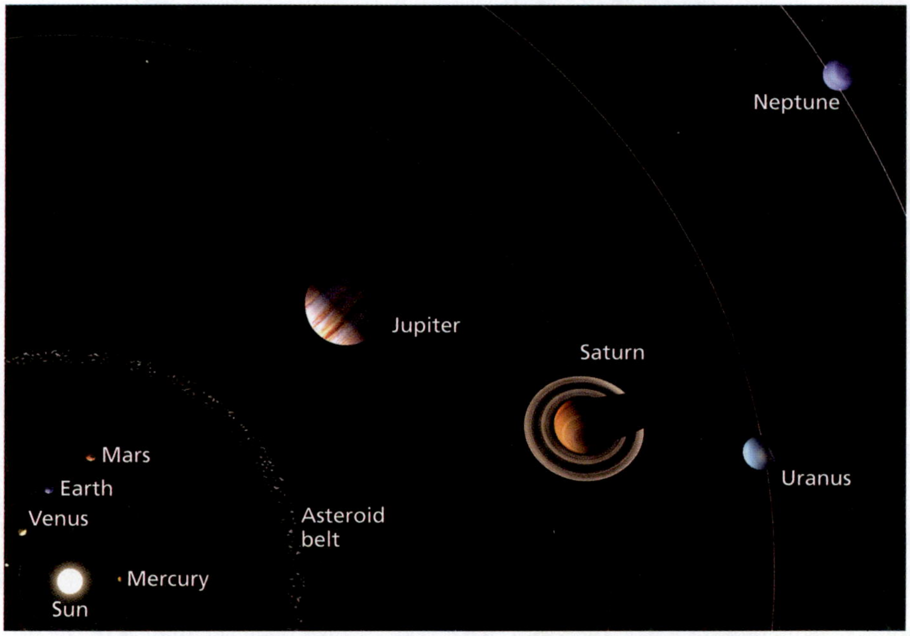

SIZE

Size is believed to play an important factor in planetary evolution. The innermost terrestrial planets are much smaller than the gaseous Jovian ones (see Table 21.1). For two terrestrial planets of similar composition, size gives an indication of the planets' cooling rates. Earth and Venus are the largest of the terrestrial planets and it may be no coincidence that these planets are still volcanically active. Smaller Mercury and Mars, on the other hand, cooled more rapidly and are essentially geologically dead.

> **CHECK YOUR UNDERSTANDING**
> • In what ways has size influenced the evolution of planets?

The presence of atmospheres on Earth and Venus and the virtual absence of atmospheres on Mercury and Mars may also be due to factors that ultimately depend on size. Volcanic activity, which plays an important role in the generation of atmospheres, ceases earlier on smaller

Figure 21.6   Solar System
There are eight planets in the Solar System. The asteroid belt is also shown.

TABLE 21.1   THE PLANETS

	Distance from Sun	Diameter	Density	Escape Velocity	Rotation	Revolution	Moons
Mercury	0.39	4.9	5.4	4.3	59 days	88 days	0
Venus	0.72	12.1	5.25	10.3	243 days	225 days	0
Earth	1.00	12.7	5.52	11.2	24 hours	365 days	1
Mars	1.52	6.8	3.93	5.0	24.5 hours	687 days	2
Asteroid Belt							
Jupiter	5.2	140	1.33	60	10 hours	11.9 years	67+
Saturn	9.54	121	0.71	36	10 hours	29.5 years	62+
Uranus	19.1	52	1.27	21	6 hours	84 years	27+
Neptune	30.1	49	1.7	24	6 hours	165 years	14+
Pluto*	39.4	2.3	1.99	3.2	6 days	248 years	1

*The data for Pluto is given for comparison purposes. Pluto is no longer considered a planet.

Note: The distance of a planet from the Sun is an average, and is compared to the distance from the Sun to the Earth (which is given a value of 1.0). The diameter of a planet is given in thousands of kilometers. The density of the terrestrial planets (given in grams/centimeter³) is significantly greater than the density of the Jovian planets. Escape velocities are given in kilometers per second. Rotation is the amount of time (in Earth days and hours) that the planet takes to complete one rotation around its own axis (e.g., Earth = 24 hours). Revolution is the amount of time (in Earth days and years) it takes a planet to complete one full revolution around the Sun. This time increases with distance from the Sun. The number of moons around the Jovian planets is a minimum as it is probable that more have yet to be discovered.

bodies than on larger ones. Furthermore, gases that may once have been expelled from ancient volcanoes are more loosely held by gravity on smaller planets and may escape into space. This is indicated by the lower escape velocity on Mercury and Mars relative to that of Earth and Venus (see Table 21.1). The escape velocity is the speed a molecule of gas needs in order to escape the gravitational attraction of a planet and dissipate into space.

DISTANCE FROM THE SUN

Our Solar System is just a very tiny portion of the universe; its size is dictated by the extent of the gravitational influence of the Sun on its nearest celestial neighbors. This gravitational attraction wanes rapidly with distance from the Sun. Pluto, for example, has a very poorly constrained orbit, consistent with this diminishing influence.

> **CHECK YOUR UNDERSTANDING**
>
> ◉ In what ways has distance from the Sun influenced the evolution of the planets in our Solar System?

A planet's distance from the Sun also influences the amount of solar radiation it receives. This distance is usually expressed relative to the distance between Earth and the Sun, which is defined as 1 astronomical unit (1 AU). For example, the distance of Mercury to the Sun (denoted 0.39 AU) is 39 percent of that of Earth. As a consequence, Mercury receives approximately 10 times more solar radiation than Earth. A planet's distance from the Sun also influences its orbital period, that is, the amount of time (in Earth years) it takes a planet to complete one revolution around the Sun. The more distant the planet, the longer it takes to complete a single orbit.

MERCURY

The Romans named the planet Mercury, noted for its rapid movements, after the messenger of the gods. Most of our information about Mercury comes from two spacecraft missions: Mariner 10 in 1974–1975, which mapped about 45 percent of its surface, and Mercury Messenger, which began its survey in 2008 and has mapped the remainder of the planet. Mercury is the smallest planet in the Solar System and, with a diameter of 4878 kilometers (3024 miles), is about 40 percent bigger than the Moon.

It takes Mercury only 88 Earth days to complete one orbit around the Sun. A Mercury "year" is consequently short, due to its proximity to the Sun. However, it takes Mercury 59 Earth days to complete one rotation around its own axis. That means a Mercury "day" (59 Earth days) is approximately two-thirds the length of a Mercury "year" (88 Earth days). The slowness of Mercury's rotation is probably due to the gravitational attraction of the Sun, which acts as a brake.

The Mariner 10 mission showed that the abundance of craters on Mercury's surface is similar to their abundance on the Moon (Fig. 21.7a), and scientists believe that Mercury's surface, like that of the Moon, must be similarly ancient.

By implication, scientists have also concluded that Mercury, although fractured (Fig. 21.7b), is essentially a dead planet with no recent volcanism. This lack of recent volcanism is consistent with its small size and, consequently, its relatively rapid rate of cooling. If ancient volcanic activity provided gases for a primitive atmosphere, those gas molecules have long since escaped into space. So Mercury has virtually no atmosphere and no chance of ever getting one. Only minute traces of sodium, hydrogen, helium, and other inert gases have been detected. Any gases that do occur on the surface are loosely held by Mercury's low gravity and

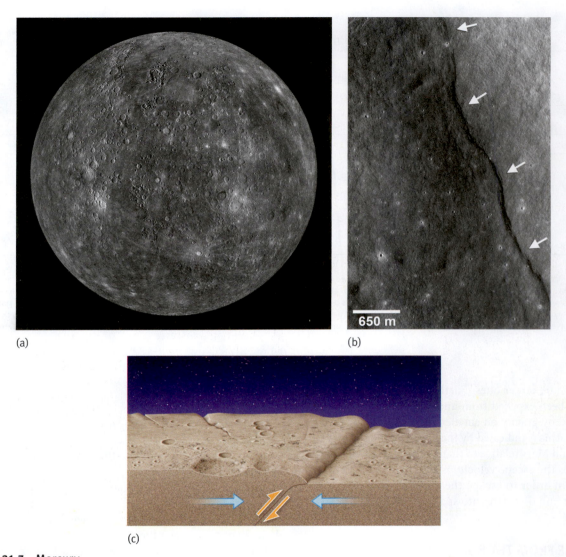

(a)

(b)

(c)

Figure 21.7   Mercury

(a) A photomosaic of Mercury made by the Mariner 10 spacecraft reveals craters that are typically between 100–150 kilometers (60–90 miles) across. (b) A fault scarp (arrows) on Mercury transects the floor of a crater, producing a cliff. (c) A schematic diagram illustrates the proposed geometry of these faults, which are thought to be related to contraction of the planet as it cooled.

readily acquire the necessary energy from the intense solar radiation to reach their escape velocity and leak into space.

Messenger documented the extraordinarily jumbled nature of Mercury's surface. The planet's surface is more heterogeneous than that of either Mars or the Moon, which both contain significant stretches of similar geology, such as maria and plateaus. The largest known crater on Mercury (Caloris Basin; 1550 kilometer, 980 mile diameter) was caused by a powerful impact which resulted in major ancient lava eruptions. On the opposite side of the planet, a jumbled terrain known as the "Weird Terrain" is attributed to shock waves that traveled around the planet's surface and converged at the antipode.

The surface temperatures on Mercury vary from 430°C (800°F) at "midday" to −180°C (−290°F) at "midnight." Because Mercury rotates very slowly, the side that faces the

Sun's intense radiation does so for a much longer period of time than the Sun-facing side of Earth does. It therefore becomes very hot. However, as Mercury has virtually no atmosphere, there is no "greenhouse effect" to store this solar heat. Once the daylight side rotates into night, Mercury loses heat rapidly and becomes very cold. In 1991, a team of scientists from the California Institute of Technology proposed that radar images of the planet's polar regions (an area shaded from solar radiation) look more like ice than rock. They suggested that ice may have been trapped there when vapor burned off impacting comets. Alternatively, it was proposed that ice could be present as a result of outgassing of water from the interior. In late 2012, images from Messenger were interpreted to indicate the presence of water-ice and perhaps organic carbon near the north pole. Frozen water may also be abundant in deep craters that are never exposed to direct sunlight. At present, the preferred interpretation is that the ice and organic carbon on Mercury were delivered to the planet by impacting comets.

The mean density of Mercury is 5.4 grams/centimeter³, which is comparable to that of Earth (average density 5.5 grams/centimeter³). However, calculations show that if Earth were the same size as Mercury, its density would be only 4.2 grams/centimeter³. This suggests that a larger proportion of Mercury consists of a metallic core than Earth (Fig. 21.8). A weak magnetic field (about 1% of Earth's magnetic field) has also been detected, consistent with the presence of an iron-rich core. Mercury's core probably occupies

---

CHECK YOUR UNDERSTANDING

• Why is Mercury considered to be a geologically dead planet?

---

about 42 percent of its volume, compared to Earth's core, which occupies just 17 percent of our planet. Scientists are still debating whether Mercury's core is predominantly solid or liquid.

VENUS

When Venus was named after the Roman goddess of love and beauty, scholars were unaware of the nightmarish conditions that exist at its surface. Instead, they saw it only as the brightest planet in the night sky. In the seventeenth century, Italian astronomer Galileo Galilei observed that Venus had "phases" much like those of the Moon. Like the Moon, half of the surface of Venus is always illuminated by the Sun at any given time, but when viewed from Earth different portions of the illuminated disk are visible.

Until the advent of the space age, Venus was viewed as Earth's planetary twin, perhaps even harboring life such as tree ferns and jungle creatures. This view was common because the two planets have approximately the same diameters (12,100 kilometers or 7500 miles for Venus, 12,740 kilometers or 7900 miles for Earth) and are neighbors in the Solar System. Observations also suggested that Venus had an atmosphere, further supporting the idea of life on its surface (Fig. 21.9). In reality, Venus's dense atmosphere is mostly carbon dioxide and camouflages a most inhospitable surface. Any theories that

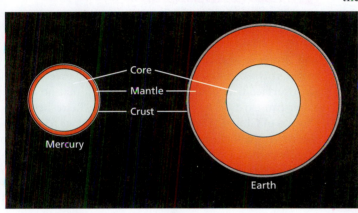

Figure 21.8   Mercury's Interior
When the internal layering of Mercury is compared to that of Earth, the thickness of the core of Mercury relative to its mantle, which is responsible for its relatively high density, is obvious.

Figure 21.9   Venus's Atmosphere
The swirling atmosphere of Venus is seen in two photographs taken 5 hours apart. To see the contrasting features, the photographs were taken in ultraviolet light.

Venus may have harbored life were laid to rest when intense microwaves emanating from the planet revealed a surface temperature of about 470°C (880°F).

Paradoxically, before the modern space age we knew more about the interior of Venus than its surface. Although Venus's mean density (5.25 grams/centimeter³) is lower than that of Earth (5.52 grams/centimeter³), calculations show that if Earth had the same diameter, it would have the same density as Venus. This suggests that the mean composition of the two planets is similar, and that Venus has both a silicate-rich mantle and an iron-rich core. Its magnetic field is much weaker than Earth's, and appears to be related to the interaction between Venus's outer atmosphere and the solar wind. Unlike Earth, Venus has no internally generated magnetic field, implying that its core lacks a dynamo.

In 1962, the deep cloud cover on Venus was penetrated using radar signals that bounced off its surface. This was the first of many probes that have used radar to study Venus, and the images gained from these surveys have greatly improved our understanding of the planet's surface. Radar has short wavelengths and can be concentrated into narrow beams that penetrate cloud cover (Fig. 21.10). By bouncing radar off Venus's surface, the time it takes for the planet to complete a single rotation about its own axis was determined. Prior to the 1962 survey, the thick pervasive cloud cover meant that this important information had eluded scientists. The results were astonishing. Venus proved to be the only terrestrial planet that is rotating clockwise rather than counterclockwise as viewed from above. On Venus, therefore, the Sun rises in the "west" and sets in the "east," the opposite pattern to that on Earth. Venus also

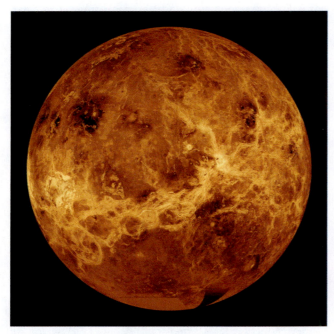

Figure 21.10   Surface of Venus
A mosaic image of the surface of Venus as revealed by radar, which is able to penetrate the planet's thick cloud cover, shows both impact craters and volcanoes (dark zones).

rotates rather slowly with a period of rotation (a Venus "day") that is 243 Earth days long. The reasons for the reverse spin and the slow period of rotation are uncertain but may account for the absence of a magnetic field on Venus.

A year on Venus is longer than a year on Mercury, but shorter than an Earth year. It takes 225 Earth days for Venus to complete one revolution around the Sun. This is consistent with the laws governing the orbital periods of planets, which requires orbits to increase as distance from the Sun increases. Note that a Venus year (225 Earth days) is actually shorter than a Venus day (243 Earth days)!

Recent space probes that have orbited Venus and landed on its surface reveal that its atmosphere is almost entirely carbon dioxide (96.5%) with a minor amount of nitrogen (nearly 3.5%). There are trace amounts of many gases, including sulfur dioxide and water vapor. Its abundant atmospheric carbon dioxide, immense atmospheric pressure, and closeness to the Sun make Venus a most inhospitable world. It is a celestial example of a "runaway greenhouse" with surface temperatures that are so high that no liquid water can exist on its surface (see Science Refresher: **The Physics of Global Warming** in Chapter 20, page 652). Furthermore, the surface pressure on Venus below its predominantly carbon dioxide atmosphere is believed to be about 95 times that of Earth. It is equivalent to the pressure that exists nearly one kilometer below Earth's surface. There is consequently little prospect of life on the planet. The first probe to land on its surface, the Soviet-built Venera 7, lasted only 23 minutes before being destroyed.

Atmospheric circulation on Venus is also very different from Earth's. The cloud layers of Venus occur at a height of about 55 to 60 kilometers (34 to 37 miles) above its surface, where the temperature is close to 0°C. This is more than 50 kilometers (31 miles) higher than typical cloud elevations above Earth. There is precipitation on Venus, but the drops of sulfuric acid that fall from the clouds do not reach the surface. Instead, the moisture heats rapidly as it falls, and rises again as it evaporates.

Between 1990 and 1994, the NASA-launched Magellan spacecraft used the reflection of radar signals from the surface of Venus to provide images of its topography (see Fig. 21.10). Like an echo sounder measuring water depth in a lake or ocean, the time elapsed between the transmission of the radar signal and the reception of an "echo" gives an indication of surface elevation. The surface consists of elevated mountainous regions (15%), comparable in height with the Himalayas, less elevated plateaus (25%), and relatively low-lying plains (60%). In comparison to either Mercury or the Moon, Venus has relatively few impact craters, implying that its surface is relatively young and perhaps still volcanically active. At least 1600 volcanic vent structures were identified; most of these are similar to shield volcanoes, but are broader and wider than those on Earth. Lava flows, and radial fracture patterns similar to those on Earth were also recognized. Shield volcanoes on Earth produce vast outpourings of basalt, and the dominant rock on the Venusian surface is also thought to be basalt.

Although no active volcanism has been detected directly, some ash layers thought to have been deposited from an active volcano have been recently identified. In addition, sensors have detected subtle changes in the atmospheric concentration of some key gases, most notably sulfur dioxide, that are similar to those accompanying volcanism on Earth (see Chapter 4).

If Venus is volcanically active, is some form of plate tectonics operating on its surface? Venus does have some surface features that resemble Earth's oceanic trenches and associated subduction zones. However, long linear mountain belts, like the Himalayas and Alps (which represent continent-continent collisions) or the Andes (which represent ocean-continent collisions), are absent on Venus. These differences may be due to the high surface temperatures on Venus that make its crust less rigid and more pliable than the crust of Earth. This would inhibit Earth-like subduction, which is greatly facilitated by the presence of old, cold, and dense oceanic crust.

Venus may resemble a very primitive Earth, with thin, fragile crustal plates or wafers. In contrast to Earth, which loses more than 70 percent of its heat at mid-oceanic ridges, Venus expels most of its heat from large currents of hot magma that rise beneath the crust. It is speculated that cooler magma may sink into the interior and that the combination may cause minor horizontal movement of the crust. If so, these processes may give us a glimpse of the very primitive style of plate tectonic activity that occurred on Earth during the Archean.

> **CHECK YOUR UNDERSTANDING**
>
> ◔ Why does Venus have such a high surface temperature?
>
> ◔ What evidence indicates that Venus is still volcanically active?

There is an intriguing possibility that, as Venus cools, it could follow an evolutionary path similar to Earth's. Venus may well develop some form of plate motion in the future as its cooling surface conditions may generate a rocky exterior that would begin to resemble that of Earth. Since life on Earth initially evolved beneath an atmosphere similar to that of present-day Venus, we cannot exclude the possibility that some form of life may eventually evolve on Venus as well.

MARS

Mars was named by the Romans in honor of their god of war. If Venus is a model for the primitive Earth, Mars may shed light on Earth's future. Of all the planets, Mars was the one people have historically thought of as most likely to support life and the one whose surface has been most intensely studied by telescope. In 1895, American astronomer Percival Lowell, drawing on the ideas of Father Angelo Secchi in Rome, interpreted streaky marks on the surface of Mars as canals built for the purpose of transporting water from the cold polar areas to the hot equatorial regions. These interpretations inspired H. G. Wells to write

The War of The Worlds. In 1938, the famous Orson Wells radio production of the book convinced many Americans that they were being invaded by Martians.

The space age put all fantasies of a race of Martians to rest. The first close-up pictures of Mars were obtained in 1965 from the Mariner 4 spacecraft, which was followed by Soviet (1971–1974) and American (1976, 1997) space probes. From these probes, scientists concluded that there is little possibility of life on the Martian surface today. However, it remains possible that life might exist in warmer environments below the surface. There also remains a tantalizing possibility that primitive life forms may have existed either at or near the Martian surface in the distant past (see Living on Earth: **Martian Life on Earth?** on page 695). The red-orange color of Mars (Fig. 21.11) is due to the presence of oxidized iron, or rust, on its surface. On Earth, rust is formed by absorption of oxygen from the atmosphere, the origin of which is attributed to photosynthesis. If a similar process of oxygen absorption is responsible for the coloration on Mars, then photosynthetic organisms may have once produced free oxygen in the Martian atmosphere.

Mars has two tiny, potato-shaped moons, Phobos (28 kilometers long by 20 kilometers wide, or 17 miles by 12 miles) and Deimos (16 by 12 kilometers, or 10 miles by 7 miles), named after the sons of Ares, the Greek god of war. These small moons are thought to be asteroids that were captured into orbit around Mars as they traversed the Solar System.

Mars is farther from the Sun (at about 1.5 AU) than Earth is, and takes the equivalent of 687 Earth days to complete a single revolution about the Sun. But a Martian day is similar in length to an Earth day because Mars takes 24.5 hours to complete one rotation about its axis. At 3.93 grams/centimeter^3, the average density of Mars is the lowest of all the terrestrial planets, suggesting that its iron-rich core may be poorly developed, an interpretation that is also consistent with the planet's lack of an internally-driven magnetic field. The iron-rich Martian surface is also atypical of the other terrestrial planets, and it is speculated that, in contrast to the iron on Earth, iron never sank to the core of Mars. This, in turn, would imply that Mars never experienced a meltdown such as the one that is believed to have occurred on Earth during its early history.

The diameter of Mars is just over half that of Earth, so it probably cooled significantly faster than Earth did. In the southern hemisphere, much of the surface of Mars is densely cratered terrain similar to that of the Moon and Mercury. This suggests that the surfaces of these three bodies are of similar age. However, the northern hemisphere of Mars is less cratered, and suggests a period of renewed volcanism about one billion years ago. There is only limited evidence of more recent volcanic activity. These observations again emphasize the importance of size in the evolution of a planet. It appears that Mars, which is between Mercury and Venus in size, became geologically dead after smaller Mercury but before larger Venus.

Mars has only a very thin atmosphere with a surface atmospheric pressure only 0.7 percent of the pressure at sea level on Earth. The relative proportions of gases in the

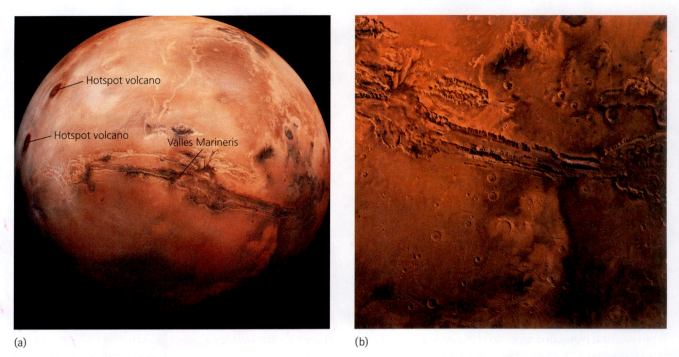

(a)

(b)

Figure 21.11   Surface of Mars

Photographs of the Martian surface from the Viking spacecraft show (a) a dark streak across the middle of the planet, which is the 4000-kilometer (2500-mile) Valles Marineris, the longest and deepest canyon in the Solar System. Two examples of Martian volcanoes are visible along the left horizon. (b) A close-up of the western end of the Valles Marineris exposes greater detail of the planet's surface. The red color of Mars indicates the presence of iron oxides in its soil.

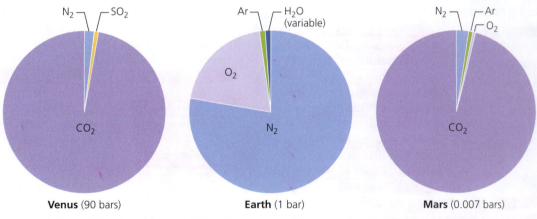

Figure 21.12   Comparison of the Composition of the Atmospheres of Venus, Earth, and Mars

The atmospheres of Venus and Mars are both dominated by carbon dioxide (CO_2), with minor amounts of nitrogen (N_2). In contrast, Earth's atmosphere is dominated by nitrogen and oxygen (O_2). The atmospheric pressure at Venus's surface is 90 times that of Earth (i.e., 90 bars), whereas the atmospheric pressure on Mars is about 0.7 percent that of Earth (i.e., 0.007 bars).

atmosphere are very similar to those of Venus (Fig. 21.12). Carbon dioxide (95%) dominates and water vapor and oxygen are only present in trace amounts. However, the total amount of atmospheric gas on Mars is so small (as indicated by the very low atmospheric pressure it exerts) that the planet experiences virtually no greenhouse effect. The average atmospheric pressure and temperature are also too low for any long-term presence of liquid water.

Surface temperatures on Mars are highly variable. The average is about −30°C (−22°F). This is consistent with the trend of decreasing average surface temperatures with increasing distance from the Sun. Like Earth, surface temperatures vary considerably from pole to equator. At the equator, the Martian daytime temperature may rise to about 25°C (77°F), but at night the temperature might drop below −70°C (−94°F). At the Martian poles, intensely cold temperatures of −120°C (−184°F) have been recorded by the Viking space probes. Despite the thin atmosphere, the contrast between polar and equatorial temperatures is sufficient to cause vigorous atmospheric circulation and is

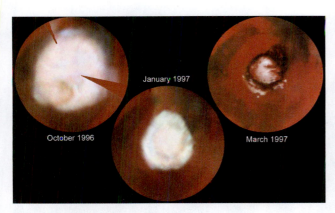

Figure 21.13   Martian Ice Caps

Images from the Hubble Space Telescope show the waxing and waning of the polar ice caps on Mars.

thought to be responsible for dust storms (see In Depth: **Dust Devils** in Chapter 16, page 512.) and the generation of Earth-like weather frontal systems.

Because Mars is tilted on its axis at a similar angle and in a similar direction to Earth, it too has four seasons. It also has a day similar in length to an Earth day. As is the case on Earth, each Martian season is about one-quarter of its year. However, as Mars has an orbital period of 687 days, each season is just over 170 days in length, or close to six Earth months. Evidence of Martian seasons is confirmed by photographs taken above the poles which reveal the presence of ice caps believed to be composed of frozen carbon dioxide and water. These ice caps wax and wane with the seasons (Fig. 21.13)—they expand in the winter and contract in the summer, suggesting that they are formed by freezing of the Martian atmosphere.

Mars has a highly exaggerated topography, with the deepest valleys and highest mountains known in the Solar System. Valles Marineris (see Fig. 21.11) is a system of canyons that straddles the Martian equator and is ten times longer, seven times wider and seven times deeper than the Grand Canyon of Arizona. The volcano Olympus Mons stands 25 kilometers (16 miles) high (Fig. 21.14), dwarfing the very highest mountains on Earth, which are only 8.8 kilometers (5.5 miles) high, and Earth's largest volcano, Mauna Kea in Hawaii, that rises a mere 10 kilometers (6 miles) above the floor of the Pacific Ocean. But the Martian surface has none of the features associated with plate tectonics, such as earthquake zones defining plate boundaries. In the absence of plate tectonic processes, most planetary geologists believe that the Martian volcanoes were produced by hotspot-type volcanism (see Chapters 4 and 9), possibly initiated by meteorite impacts. This raises the possibility that similar features on Earth, such as the Hawaiian chain, could have been initiated in a similar fashion. The exaggerated height of the volcanoes on Mars may reflect the greater extent of melting.

Alternatively, since there are no moving tectonic plates on Mars, the products of hotspot volcanism may have accumulated in one place instead of producing the linear island chains characteristic of oceanic hotspots on Earth (Fig. 21.14). The height of the volcanoes on Mars implies that its crust is thick enough to support the load. On Earth, elevation is influenced by isostasy because the lithospheric plates float buoyantly on the asthenosphere (see Chapter 9). It is therefore unlikely that Mars has an asthenosphere beneath its crust. The absence of an asthenosphere implies no detachment of the Martian crust from its deep interior and, hence, no possibility of plate motion.

A series of space probes over the last 40 years have provided strong evidence for the presence of flowing water across the Martian surface in the past. The Valles Marineris (see Fig. 21.11) is just one of many deep valleys that are currently deep waterless natural canyons. However, they may not always have been waterless. In detail, their patterns resemble drainage systems on Earth, but on a much grander scale. Some estimates suggest that the water that carved these canyons may have flowed at rates up to 10,000 times greater than that of the Mississippi River.

In 2003, NASA began a robotic space mission to explore the Martian surface and its geology. Since that time two robotic geologists, *Spirit* and *Opportunity*, have been providing high quality images of the Martian landscape (Fig. 21.15a),

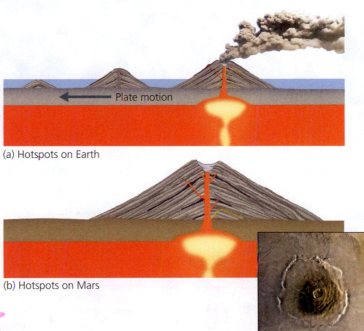

(a) Hotspots on Earth

(b) Hotspots on Mars

Figure 21.14   Hotspots on Earth and Mars

(a) As Earth's lithospheric plates move, hotspot volcanoes cool and subside as they are transported away from the stationary hotspot forming a volcanic island chain. (b) A Martian volcano also develops above a stationary hotspot, but on a crust that is not moving. As a result, Martian volcanoes can build to great size. Inset: Olympus Mons is 600 kilometers (370 miles) wide and its peak is 25 kilometers (16 miles) high—more than three times the height of Mount Everest.

(a)

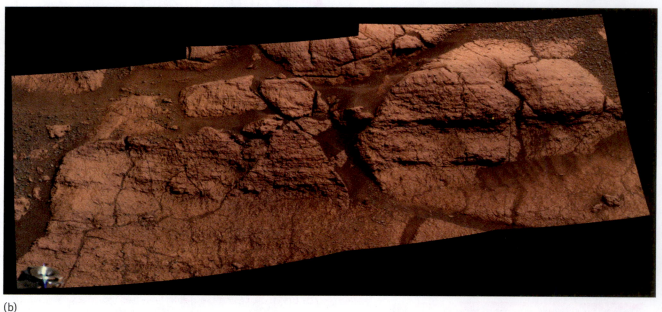

(b)

Figure 21.15   Martian Landscapes

(a) Matijevic Hill is captured in a photo taken by *Opportunity*'s panoramic camera. (b) *Opportunity* took this photo of a rock outcrop named El Capitan. The very fine layers are typical of those found in sediments deposited in water.

and have found a great deal of evidence for past water on the Martian surface. This evidence includes the presence of very fine layers typical of sediments deposited in water (Fig. 21.15b). Analysis of this sediment identified a rare mineral called *jarosite* that contains hydroxide ions in its crystal structure (see Chapter 3), implying the presence of water when the mineral formed. NASA scientists have concluded that this region once was covered by a shallow salty sea that later evaporated. In 2012, *Spirit* and *Opportunity* were joined by *Curiosity,* which also found evidence of ancient stream-beds and has detected water molecules, chlorine, and sulfur in Martian soils.

In 2005, NASA's Mars Reconnaissance Orbiter was launched. From its orbit above the Martian surface, this spacecraft provides detailed photographs of the Martian surface, monitors daily weather, and analyzes the composition of its atmosphere as well as its surface minerals. In addition,

it sends out radar signals that penetrate the Martian surface to search for evidence of subsurface water. In 2013, scientists interpreted images from this survey as providing the first evidence of ancient channels in the subsurface. These channels are thought to have been carved by massive floods, and were subsequently buried by basaltic lava that vented from ancient volcanoes. These channels provide additional evidence of the important role water played in the planet's history.

Collectively, these data and interpretations have enormous significance because they strongly suggest that water once flowed across the Martian surface. But if so, where did all the water go? There are three possibilities. The very low atmospheric pressure on Mars would have promoted rapid evaporation. The moisture may have therefore escaped the gravitational clutches of the planet, in which case it is lost forever. Alternatively, the water may exist as ice in the polar

IN DEPTH

Martian Life on Earth?

There have been very few discoveries that have set the scientific world abuzz like the suggestion that life may have existed on Mars. If validated, it would have huge implications for the origin of life on Earth and our place in the cosmos. Two independent research investigations by teams of American, Canadian, and British scientists have provided new and controversial evidence that supports such a claim. These scientists admit that their evidence is circumstantial, but in the true spirit of scientific investigation they have drawn attention to the possibility so that other scientists may find ways to test their hypotheses. By either confirming or refuting their findings, science will move forward.

The material that these teams studied was not recovered from Mars at all; it was recovered from the ice sheet of Antarctica (Fig. 21C). It has been accepted for several years that some meteorites on Earth have a Martian origin. These meteorites possess the chemical and isotopic signature characteristic of Martian crust, and contain trapped gases whose composition is identical to that of the Martian atmosphere. To have come from Mars, these meteorites must have been blasted off the Martian surface by an asteroid collision that expelled vast chunks of rock into space. Eventually some of these chunks became caught in Earth's gravitational field and fell to its surface as

meteorites. Dark meteorites are relatively easy to find in the unspoiled icy terrain of Antarctica. They have been recovered both from the surface itself, and from the front of glaciers where the heaving of ice along fractures churns up buried meteorites.

The first sample analyzed by the American and Canadian team was dated using radiometric techniques (see Chapter 8). It was found to be 4.5 billion years old, and to have been fractured 4 billion years ago. Within it, carbonate material believed to be organic and thought to have been formed by solutions percolating along the fractures, was dated at 3.6 billion years. Mars was probably much warmer and wetter 3.6 billion years ago than it is today, so that conditions may have been favorable for life. Age dating also shows that this sample exploded off the Martian surface 16 million years ago and landed in Antarctica 13,000 years ago. The British team reanalyzed this rock and confirmed these results. The British team also analyzed a second sample that was found to be only 175 million years old, which was blasted off the Martian surface 600,000 years ago.

Both teams reported the presence within the meteorites of minuscule rods similar in shape to fossilized bacteria. The British team also found the telltale signature of "microbially-produced methane" similar to that produced by common bacteria on Earth. The isotopic character (see Chapter 8) of this material matches that of the most primitive life forms on Earth.

In addition to the obvious implications for Martian life, the analysis of the first sample raises the possibility that life on Earth may have been "seeded" by Mars. If life did exist on Mars 3.6 billion years ago, it would correspond in age to the earliest indications of life found on Earth. Given the abundance of asteroid impacts on planetary surfaces in the early history of the Solar System, chunks of Martian crust could easily have made the journey to Earth. If so, these early meteorites may well have provided the raw materials for life to commence on Earth.

The second, younger meteorite sample is also significant because it suggests the presence of Martian life in the relatively recent past, sometime between 175 million years ago (the age of the meteorite) and 600,000 years ago (the time of asteroid impact). If so, life on Mars must have been relatively robust to have survived the enormous changes in the Martian climate. Furthermore, as the climatic conditions of Mars have not changed substantially over the last 200 million years, the sample suggests that life may still exist there today.

Figure 21C   Martian Meteorite

This Martian meteorite recovered from the ice sheet of Antarctica contains miniscule rod-shaped material, similar in shape to fossilized bacteria.

regions of the planet (see Fig. 21.13). The polar ice caps on Mars may store vast quantities of liquid that if unleashed by melting would have tremendous powers of erosion. However, the percent of this ice that is dry ice, consisting of carbon dioxide rather than water, is uncertain. A final possibility is that the water exists as permafrost (see Chapter 15) in the Martian soils. On melting, it too could fuel catastrophic flooding and erosion.

The number of craters on Mars indicates that the huge floods that carved the canyons may have occurred between 1 and 3 billion years ago. But if water flowed across the Martian surface in the past, how did it happen? One possibility is that meteorite impacts triggered the release of water from the interior, like a needle piercing a water balloon. Another possibility is that the Martian climate was warmer in the past, as indicated by past periods of intense volcanic activity. Ancient volcanic activity would certainly have vented greenhouse gases from the Martian interior to its atmosphere. During periods of enhanced volcanism, atmospheric pressures on Mars would have been temporarily greater, so the greenhouse effect would have been more intense and surface temperatures would have been higher. This would have facilitated the melting of polar ice and/or permafrost and may have triggered catastrophic flooding.

Other more subtle influences may affect Martian surface temperatures just as they affect temperatures on Earth. For example, recall from Chapter 20 that over the past million years, Earth's climate has been driven in and out of ice ages by Milankovitch cycles—cyclic changes in Earth-Sun geometry produced by variations in Earth's tilt, by the eccentricity of its orbit, and by its precession. There is nothing inherent in these cycles that would restrict their occurrence to Earth. Changes in Mars-Sun geometry may also have triggered the melting or growth of ice sheets on Mars.

Whether water was present on the Martian surface long enough to provide a habitat for life remains unknown. If Mars was warmer in its past, then it may have experienced similar conditions to those that existed when life was initiated on Earth (see In Depth: **Martian Life on Earth?** on page 695). Discoveries made by the rover *Curiosity* in 2013 suggest that this was indeed the case. Analysis of a clay-rich mudstone in an ancient streambed identified minerals likely to have formed in relatively fresh water, and revealed the presence of sulfur, nitrogen, hydrogen, oxygen, phosphorus, and carbon, all of which are critical building blocks for living organisms. The presence of iron oxide–rich rocks at the surface implies an ancient atmosphere that must have been far richer in oxygen than the present one. However, "the window of opportunity" for life on the Martian surface may have been quite short. Mars cooled faster than Earth, which means that any primitive life forms, if they had existed, would not have evolved to any significant degree. The planet's inability to replenish its atmosphere once volcanic activity ceased would have doomed whatever fragile surface life had gained a foothold. The very thin carbon dioxide–dominated atmosphere and the very low average temperature on Mars suggest that it is very unlikely that life could exist at its surface today. However, life may be possible in the warmer subsurface.

According to recent calculations, several billion years from now Earth will have cooled to the point where plate tectonics will cease and it too will become unable to replenish its atmosphere or its hydrosphere. Hence, Mars may illustrate the ultimate fate of our own planet.

21.2 SUMMARY

- The size of a terrestrial planet influences its rate of cooling, the longevity of its volcanic activity, and its atmospheric composition, whereas its distance from the Sun influences the amount of solar radiation it receives.

- The age of the planetary surfaces can be estimated by the abundance of impact craters.

- Mercury, the smallest planet, has an ancient heavily cratered surface resembling that of the Moon. It is geologically dead with no recent volcanism, it has essentially no atmosphere, so surface temperatures fluctuate between 430°C (800°F) and −180°C (−290°F), and it has a large metallic core.

- Venus is similar to Earth in size and has phases like the Moon. It has a dense atmosphere dominated by carbon dioxide and a surface temperature of 470°C (880°F), and has relatively few craters, suggesting its surface is quite young and possibly volcanically active.

- Mars has exaggerated relief, polar ice caps, a very thin atmosphere, and very cold average surface temperatures. It may once have harbored life, and might still do so in its subsurface. Recent evidence implies the former existence of flowing water on the Martian surface.

21.3 Plate Tectonics: A Planetary Perspective

As we discussed in Section 21.2, Earth is apparently the only planet that can harbor life as we know it. Only Earth is the right distance from the Sun, and has the winning combination of atmospheric composition, active plate tectonics, and a biosphere capable of photosynthesis. In the simplest sense, Earth is like the proverbial porridge in the fairy tale about Goldilocks and the three bears. Venus is too hot. Mars is too cold. But Earth is just right!

But recall that planet size is also important, and it is clear that Earth's evolutionary path rapidly diverged from that of Mars because of its larger size. Why Earth's fate diverged from that of Venus is less obvious. Many of the processes ascribed to the early Earth apply equally to Venus. Admittedly, Venus is closer to the Sun and so may have had a hotter atmosphere. But Earth's early atmosphere was also dominated by greenhouse gases. Nevertheless, the geologic record suggests that there was a fundamental parting of

---

CHECK YOUR UNDERSTANDING

🔵 Why does Mars have the lowest density of the terrestrial planets?

🔵 Why is Mars considered geologically dead and what are the implications of this for its atmosphere?

🔵 Summarize evidence for the role of water in the evolution of the Martian landscape.

the evolutionary paths of Earth and Venus early in their respective histories.

Perhaps the key to this divergence is the contrasting tectonics on the two planets; thanks to these differences Earth developed continental crust, whereas Venus did not. The earliest records of life on Earth are found on ancient continental platforms, which provided relatively hospitable environments. Buoyant continental crust was first generated early in Earth's history (certainly by 4.0 billion years ago) by a process known as *differentiation,* which results in the separation of light continental crust from the relatively dense interior. Initially this crust would have been rather thin, fragile, and surrounded by molten magma. Much of this early crust may have been destroyed by meteorite bombardment (Fig. 21.16). Continued meteorite bombardment may also have ultimately enhanced crustal differentiation, in the same way that shaking a soda generates bubbles that rise to the surface. Meteorite bombardment, therefore, may have destroyed the fragile crustal surface, but at the same time may have sped up the exhumation of light continental crust from the deep interior.

At some stage, a critical thickness of continental crust must have been attained that could survive meteorite bombardment. When this happened, Earth's evolutionary path irrevocably diverged from that of Venus. Lighter continental crust is more buoyant than denser oceanic crust, so that, in contrast to Venus, Earth developed elevated regions and vast ocean basins where water could collect. In short, exhumation of the continental crust began Earth's hydrologic cycle (see Chapter 13) and drainage of water from elevated continental crust to depressed ocean basins commenced. As carbon dioxide is soluble in water, the establishment of

a hydrologic cycle would have significantly reduced the greenhouse gas content of the atmosphere, further amplifying the contrast between Earth and Venus.

Continental crust is also a much better insulator than oceanic crust and blocks the escape of heat from Earth's interior to a far greater degree than oceanic crust. Seventy percent of Earth's heat loss today occurs along mid-ocean ridges. This contrast indicates a profound asymmetry in the distribution of heat loss from Earth's surface, and may have provided a basis for the initiation of a primitive form of plate tectonics. Plate tectonics, in turn, provided the environment for the origin of life. It is conceivable, therefore, that for life to evolve, a critical thickness of continental crust must first be attained.

Earth's early evolution occurred at a hectic pace when compared to the more pedestrian changes of the recent geologic past. Earth became layered according to density very early in its history, from a dense iron-nickel rich core, through a silicate mantle, to a buoyant crust, above which lay water. The lightest molecules became part of the atmosphere. The development of this layering was primarily an internally driven process in which plate tectonics played little or no role. Plate tectonic activity similar to today may not have developed until the end of the Archean (about 2.5 billion years ago). However, a primitive form of plate tectonics may have commenced in the Early Archean with the exhumation of continental crust and the development of asymmetric heat loss at Earth's surface.

It is generally believed that plate tectonics would have been a more rapid process in its early primitive form than it is today. It therefore makes sense to ask if there may be a long-term trend of decreasing rates of plate motion as Earth

Figure 21.16 Early Earth

Primitive crust (dark) on early Earth was surrounded by pools of molten magma and was subject to frequent meteorite impacts. By agitating the magma, these impacts may have encouraged the formation of buoyant continental crust.

cools. If so, will plate motions eventually cease on Earth? This leads us to ask the question: as plate tectonics is fundamental to the replenishment of the hydrosphere, atmosphere, and biosphere, will our planet end up as geologically dead as Mars?

21.3 SUMMARY

- Earth's evolution may have diverged from that of Venus because Earth developed buoyant continental crust that allowed for the formation of oceans, the evolution of life, and the eventual development of plate tectonics.

21.4 The Jovian Planets, Their Moons, and the Kuiper Belt

The Jovian planets occupy the outer reaches of the Solar System (see Fig. 21.6). The one nearest the Sun, Jupiter, is more than five times farther from the Sun than the distance between the Sun and Earth (see Table 21.1). The most distant planet, Neptune, is on average more than 30 times this distance. The great distances from the Sun are reflected in progressively longer periods of revolution. Jupiter, for example, takes 11.9 Earth years to travel its 5 billion-kilometer (3 billion-mile) orbit around the Sun.

Several features of the Jovian planets yield important information about the origin of the Solar System. Their mean densities (ranging from 1.70 to 0.71 grams/centimeter³) show that these bodies are fundamentally different in composition from the inner terrestrial planets. The Jovian planets are largely composed of hydrogen and helium, the lightest elements in the periodic table. The fact that these planets are dominated by lighter elements in the outer reaches of the Solar System is central to models for their origin. Many astronomers believe that when the raw materials of the Solar System cooled, the Jovian planets grew rapidly enough to capture gas directly out of the solar nebula. The internal structures of both Jupiter and Saturn are dominated by liquid molecular hydrogen (liquid H₂), which with increasing depth (and pressure) gives way to metallic hydrogen (H) around relatively small cores of rock and ice (Fig. 21.17). Less is known of Neptune and Uranus, but it is generally assumed that they have internal structures that are similar to those of Jupiter and Saturn.

The enormous size and mass of the Jovian planets relative to the terrestrial planets implies a greater gravitational attraction for their atmospheres. In addition, the Jovian planets tend to have a larger number of moons. Given the gaseous nature of the surfaces of these planets, there is no way that these moons could have been jettisoned from the planets by a gigantic impact like the one that is believed to have created the Earth-Moon system. The 1994 collision of the Shoemaker-Levy 9 comet with Jupiter demonstrated that impacting bodies merely sink into

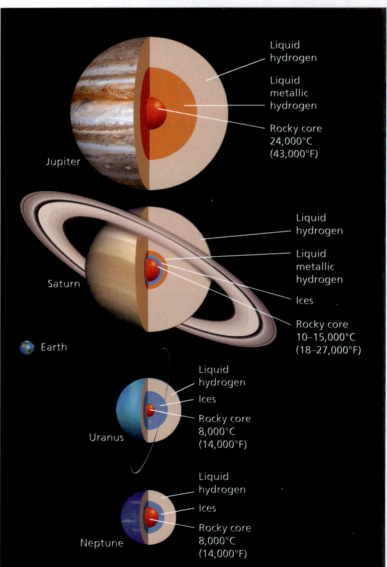

Figure 21.17   **Interiors of the Jovian Planets**

The Jovian planets—Jupiter, Saturn, Uranus, and Neptune—are thought to be largely composed of hydrogen and helium. Uranus and Neptune do not have liquid metallic hydrogen near their cores because their smaller masses provide insufficient interior pressures. Planet Earth is shown for scale.

Comet on collision course

(a)

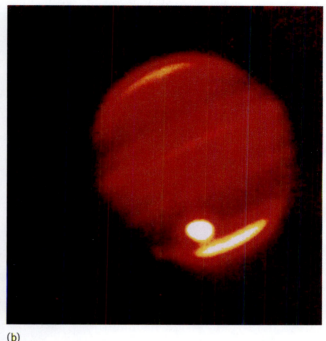

(b)

Figure 21.18   Comet Shoemaker-Levy 9

Images from the Hubble Space Telescope capture the Shoemaker-Levy 9 comet, which collided with the surface of Jupiter in July 1994. (a) Stages of the advancing comet are shown in time elapsed images. (b) The impact created a "hotspot" on Jupiter.

a gaseous interior (Fig. 21.18). Instead, the Jovian moons are likely to have different origins than the moons that orbit the terrestrial planets. Some are believed to be bodies of rock and ice that were captured and held in orbit by the strong gravitational attraction associated with these large planetary bodies. Others may have formed at the same time as their adjacent planet and were held in by gravitational attraction.

JUPITER

Jupiter, the largest planet, was named after the most important of the Roman gods. This giant planet contains 71 percent of all planetary mass in the Solar System. Its mass is 318 times that of Earth, and its diameter is about 11 times larger. It is remarkably similar in composition to the Sun and is made up mainly of hydrogen and helium with minor amounts of methane, ammonia, and water.

Although Jupiter's composition resembles that of a star, it did not attain the size necessary to ignite nuclear reactions. A critical size is needed to ignite a star's nuclear furnace because the heat at the core of a star is largely generated by gravitational collapse. Because of its comparatively small size, the temperature at Jupiter's core is only 24,000°C (43,000°F), which is much too low to initiate a nuclear furnace. Jupiter would have to be 30 times larger before this could happen.

Jupiter's atmosphere appears to be layered with an uppermost cloud layer dominated by ammonia and hydrogen

sulfide, and a lowermost layer dominated by water vapor. The temperature of the highest clouds is estimated at about −123°C (−189°F). Jupiter's most recognizable features are the color bands in the atmosphere (Fig. 21.19a) that encircle the planet and are in constant and rapid motion. These color bands are caused by variations in temperature and chemical composition (particularly elements such as carbon, sulfur, and phosphorous). These variations result in different degrees of reflectivity of sunlight. In general, dark orange bands appear to represent cold descending clouds, whereas bright blue bands are hot and rising (Fig. 21.20). In between the bands are complex vortices that move in the opposite direction.

Observation of the motion of Jupiter's most famous feature, the **Great Red Spot** (Fig. 21.21) indicates that Jupiter has a 10-hour period of rotation. This implies an incredible rotational speed at the equator of almost 45,000 kilometer per hour (28,000 mph)! By way of comparison, Earth's equatorial speed of rotation is a mere 1670 kilometer per hour (1040 mph).

The Great Red Spot is the largest of numerous storm systems in Jupiter's atmosphere. Winds at its outer edge move at speeds greater than 400 kilometer per hour (240 miles per hour). It is an anticyclone (circulation about a region of unusually high atmospheric pressure) located in its southern hemisphere and has a diameter close to 38,000 kilometers (23,560 miles), which is three times the diameter of Earth. This storm has lasted for at least 400 years. It is

(a)

(b)

(c)

(d)

Figure 21.19   Jovian Planets

The Jovian planets, (a) Jupiter, (b) Saturn, (c) Uranus, and (d) Neptune, are seen here in photographs taken by the Voyager spacecraft. In (a), Jupiter's color bands and the Great Red Spot (lower center) are clearly visible. In (b), Saturn's rings and relatively poorly developed color bands can also be seen, as can the moons Rhea and Dione. In (c), note the unusual orientation of the color bands of Uranus. In (d), the Great Dark Spot of Neptune is visible in the center of the image.

unclear when it started. The storm is driven by heat from the planet's interior, in contrast to storms on Earth, which are driven by solar radiation. Jupiter and Saturn emit more heat from their surfaces than they absorb from the Sun. This excess heat may be responsible for the enormous scale and longevity of Jupiter's storms. However, most recent observations suggest that the Great Red Spot may be shrinking in size.

CHECK YOUR UNDERSTANDING

○ Why doesn't Jupiter have a well-defined surface?

Because Jupiter is a gaseous planet, the distinction between the planetary surface and its atmosphere is rather vague. However it is generally considered that a major compositional change does take place. Jupiter's core is thought

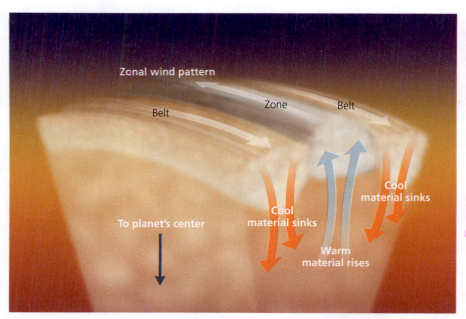

Figure 21.20   Jupiter's Color Bands

The color bands of Jupiter are interpreted as ascending (light blue) and descending (orange) clouds high in Jupiter's atmosphere.

Figure 21.21   Jupiter's Great Red Spot

A photograph of the Great Red Spot of Jupiter, taken by the Voyager 1 spacecraft in 1979, depicts details of its internal turbulence and rotation.

to be made up of a rock and ice mixture enclosed by a thick layer of liquid metallic hydrogen (see Fig. 21.17).

JUPITER'S MOONS

Jupiter has 67 confirmed moons, the highest number in the Solar System. Of these, 16 moons are *regular satellites,* that is, they have regular elliptical orbits and are held in place by strong gravitational attraction. The others have only recently been identified, and are classified as *irregular satellites* because of their highly eccentric orbits. It is believed that they were captured by Jupiter's gravity field and are more distant from Jupiter than the regular satellites.

The four moons (Io, Europa, Ganymede, and Callisto), first observed in the early seventeenth century by Italian astronomer Galileo Galilei, are the innermost of Jupiter's 16 regular satellites. These so-called Galilean moons revolve in nearly circular orbits in the planet's equatorial plane. This fact, together with the decrease in the average density of these satellites with distance from Jupiter, suggests that these moons may have formed along with Jupiter.

Images from the Voyager space probe taken in the 1970s and more recently by the Galilean spacecraft shows that the four Galilean moons are very different from one another (Fig. 21.22). Io is comparable in size and density to our Moon, with a diameter of 3632 kilometers (2252 miles) and a density of 3.55 grams/centimeter³, indicating that it is dominated by silicate rock and an iron-rich core. It has a smooth, crater-free surface, which suggests that its surface is extremely young and geologically active. Its yellow-to-red coloration indicates an abundance of sulfur-rich compounds. Astonishingly, the Voyager images captured an erupting volcano (Fig. 21.23), making Io the only other body in the Solar System on which active volcanism has so far been observed. We now know that Io has at least 400 active volcanoes, and that some lava flows can be traced for more than 500 kilometers (300 miles). It is thought that the volcanism is caused by the immense gravitational pull of Jupiter and its nearest satellite neighbor, Europa.

Europa is the next closest moon to Jupiter and is a mixture of rock and ice (see Fig. 21.22). It has a diameter of 3126 kilometers (1938 miles) and a density of about 3.0 grams/centimeter³, suggesting that it too is dominated by silicate rock and an iron-rich core. Europa's icy surface was once thought to be very smooth, like a billiard ball. However, pictures taken by the Galileo spacecraft show that the surface is intricately cracked. Ice flows and ice volcanoes were probably produced by water that erupted between the cracks. There are speculations on the presence of a subterranean ocean on Europa and the possibility of life. It has also been speculated that Europa's atmosphere may contain oxygen. The fragments of ice move relative to each other, and some have suggested that the movement of the ice is analogous to plate movements on Earth.

Ganymede, with a diameter of 5262 kilometers (3262 miles), is bigger than Mercury. It is the largest moon in the Solar System. With a density of 1.9 grams/centimeter³, Ganymede is thought to be primarily composed of approximately equal amounts of surface ice overlying a rocky core. Measurements taken by the Galileo spacecraft in 1996 indicate that Ganymede has a strong magnetic field, a clearly differentiated internal structure with a metallic core surrounded by a

Figure 21.22   Galilean Moons

These photographs are of the Galilean moons of Jupiter. Earth's Moon (which has a diameter of 3474 kilometers or 2159 miles) is pictured in the center for scale. Io (upper left) has a sulfur-rich surface, Europa (upper right) displays an icy crust, and Callisto (bottom right) and Ganymede (bottom left) have icy surfaces with cores of rock.

Figure 21.23 Io

A computer-enhanced image taken by the Voyager spacecraft shows a volcano on Io (left) venting a plume of sulfur-rich ash.

silicate mantle, and an ice shell up to 800 kilometers (500 miles) thick. Ganymede's surface also displays evidence of ice tectonics. Water appears to have oozed through fractures in the ice and crystallized on its surface. It is therefore speculated that on Ganymede, as on Europa, a layer of liquid water may exist below the icy surface.

Callisto is the outermost Galilean moon, and the third largest moon in the Solar System. It is virtually the same size as Mercury, with a diameter of 4800 kilometers (3000 miles), but only has about one-third of Mercury's mass. Given its density of 1.8 grams/centimeter³, Callisto is thought to be composed of roughly equal proportions of rock and ice like Ganymede. Analysis has identified water ice, carbon dioxide, organic compound, and silicates on its surface. Its crater density is similar to both the Moon and Mercury, suggesting that it preserves an ancient 4.5 billion-year-old surface, and so is likely to be the least active of all the Galilean moons.

CHECK YOUR UNDERSTANDING

● Compare and contrast the defining characteristics of the four Galilean moons.

SATURN

Saturn, the second largest planet in the Solar System, has a diameter of 120,660 kilometers (75,400 miles). Named after the father of Jupiter, Saturn is the lightest of all the planets with a density of only 0.71 grams/centimeter³, which is even less than that of ice (see Table 21.1). Saturn is thought to have an internal structure and composition that is dominated by hydrogen and helium and similar to that of Jupiter. However, Saturn's smaller size implies that its internal layers are under less pressure than the equivalent layers on Jupiter (see Fig. 21.17). The planet's pale yellow coloration is due to trace amounts of ammonia crystals in the upper atmosphere. The outermost gaseous layer surrounds a layer of liquid hydrogen and liquid helium, followed by a deeper layer of metallic hydrogen, which is responsible for a magnetic field similar in intensity to that of Earth. Saturn also has color bands in its atmosphere, although the colors are more subdued than those of Jupiter (see Figure 21.19b).

Saturn is most famous for the spectacular flat rings that encircle the planet above its equator (Fig. 21.24a). These extend for 170,000 kilometers (105,400 miles) but vary in width, and can be as narrow as 2 kilometers (1.2 miles). It has long been known that Saturn's rings are not solid, because distant stars, and indeed the planet's surface, can be seen through them. However, their composition remained a mystery until the early 1970s when observations suggested that they consisted of chunks of ice, rock debris, and dust varying in size from 5 centimeters (2 inches) to 100 meters (330 feet) across. This conclusion was later confirmed by images from the Voyager space probe.

At last count, 62 moons have been discovered orbiting Saturn, the largest of which is the ice body, Titan. Titan is the only moon in the Solar System to have a relatively thick atmosphere. Its atmosphere is dominated by nitrogen, with minor amounts of methane and ethane, and until 2004, this thick atmosphere prevented imaging of the moon's surface. Since then images from the Cassini-Huygens mission greatly increased our understanding of this moon. It has a surface temperature of −180°C (−290°F), so cold that methane flows as a liquid. Indeed, one of the most intriguing discoveries is the detection of a hydrocarbon lake near its north pole that is the size of the Caspian Sea (Fig. 21.24b). The moon's surface has few impact craters, suggesting it is geologically young. Mountains and possible volcanoes have also been identified, and images reveal surface drainage features like Earth, including streams, lakes, deltas, and seas. Instead of water, however, the draining liquid is probably methane or ethane, and many scientists maintain that Titan has a methane cycle that is analogous to the hydrologic cycle on Earth. If this is the case, Titan may harbor microbial life.

In 2005, NASA's Cassini spacecraft produced startling images that allowed detailed mapping of Enceladus, the sixth-largest of Saturn's moons (about 500 kilometers, or 310 miles, in diameter). Enceladus's northern hemisphere is dominated by craters, but photographs taken of its south pole reveal an icy surface fractured by fissures (Fig. 21.24c) and evidence of water-rich geysers erupting along these fissures. Recently, researchers have interpreted these images as indicating the presence of relatively warm ocean water

(a)

(b)

(c)

Figure 21.24   Saturn's Rings and Moons

(a) Saturn's rings are made up of ice particles and are transparent. The rings' transparency is apparent where they cross in front of the planet's surface (left). (b) This view of Saturn's moon, Titan, is partially obscured by the moon's hazy atmosphere. Dark patches are lakes of hydrocarbons. (c) A composite image of Saturn's moon, Enceladus, shows the northern hemisphere dominated by craters and the southern hemisphere dominated by a fractured icy surface.

> **CHECK YOUR UNDERSTANDING**
> • How was the nature of Saturn's rings determined?

beneath the moon's icy surface. Many researchers maintain that Enceladus is now the most likely place to look for the existence of extraterrestrial life in our Solar System.

URANUS AND NEPTUNE

In comparison with Jupiter and Saturn, little is known about the remaining outer planets, Uranus and Neptune. These are often called "ice giants," because in addition to hydrogen and helium, their atmospheres contain a relatively high amount of "ices" mainly made up of water, methane, and ammonia.

Uranus was named for the Roman god of the heavens, the grandfather of Jupiter and the father of Saturn. Uranus is unique among the planets in that its axis of rotation is almost on its side, rather than upright like the other planets. As a consequence, it spins in the plane of its orbit, like a football, in contrast to the other planets, which spin perpendicular to their orbits like spinning tops. This unusual orientation is thought to be related to a massive celestial impact that occurred when the planet was forming. Uranus has nine narrow rings, and a clear outer atmosphere consisting of hydrogen and helium as well as ices such as ammonia, water, and methane (see Fig. 21.17). Unlike Jupiter and Saturn, the surface of Uranus is almost featureless.

Uranus's density of 1.27 grams/centimeter³ suggests that its internal structure probably consists of a rocky core, an icy mantle, and a gaseous hydrogen/helium-dominated atmosphere. Uranus's internal heat is much lower than the other Jovian planets, and, unlike these planets, it does not radiate more heat than it receives from the Sun. The reason for this low heat flow is not known.

Neptune was named for the Roman god of the sea because of its blue color (see Fig. 21.19d). This color is attributed to the abundance of methane in its atmosphere. Images

from Voyager portray Neptune as a stormy planet with a large storm center called the Great Dark Spot that is similar to Jupiter's Great Red Spot. Images from Voyager 2 show that Neptune has active and vigorous weather patterns, with recorded wind speeds of up to 2100 kilometers per hour (1300 miles per hour), Neptune also has the strongest sustained winds measured on any planet in the Solar System.

> **CHECK YOUR UNDERSTANDING**
> • What is thought to have caused Uranus to spin like a football rather than a spinning top?

PLUTO AND THE KUIPER BELT

Until recently, Pluto was considered to be the most distant planet of the Solar System. But astronomers had long realized that some of Pluto's characteristics did not fit with its classification as a planet. With the discovery of similar objects in the outer Solar System, the classification began to be seriously questioned and, in 2006, Pluto was formally stripped of its full planetary status and reclassified as a dwarf planet.

Pluto, named for the Roman god of the underworld, has a density of 1.99 grams/centimeter³ (which is higher than any of the Jovian planets), and its composition is thought to consist of rock (80%) and ice (20%). Pluto is only 2300 kilometers (1426 miles) in diameter and is smaller than Earth's Moon. Its surface, which ranges in temperature from −233°C to −223°C (−387°F to −369°F), appears to be dominated by nitrogen ice (97%) and traces of methane ice. Recent observations that stars do not abruptly disappear when they pass behind Pluto, but rather "wink out," suggests that Pluto has a gaseous atmosphere.

Pluto orbits the Sun, albeit in a highly irregular manner. This irregular orbit is probably due to the waning gravitational attraction felt by Pluto because of its small size and its position at the outer edge of the Solar System. From January 1979 to March 1999 Pluto was closer to the Sun than Neptune was.

Figure 21.25   Pluto and Charon

An image from the Hubble Space Telescope shows Pluto (lower left) with its moon Charon (upper right).

In 1978, images of Pluto also revealed a satellite body, 1300 kilometers (800 miles) across, named Charon. In 1990, the Hubble Space Telescope took the first clear photo images of both bodies (Fig. 21.25). The presence of a satellite body was used by some astronomers as evidence of Pluto's status as a planet. However, more recent observations from the Hubble telescope indicate that Pluto is probably just a large representative of a vast number of icy bodies, at least 70,000 in number, that orbit the Sun at the outermost fringes of the Solar System. These bodies are thought to represent remnants left over from the Solar System's formation and are collectively known as the **Kuiper Belt**.

> **CHECK YOUR UNDERSTANDING**
>
> • Why is Pluto's orbit so eccentric?

21.4 SUMMARY

- Jupiter is dominated by hydrogen and helium. Its color bands reflect subtle changes in the composition and temperature of its atmosphere. Its massive storms, such as the Great Red Spot, are driven by energy released from its interior.

- Jupiter's innermost moons include Io (which has volcanism), Europa (which has a cracked icy surface that may overlie a subterranean ocean), Ganymede (which shows evidence of ice tectonics), and Callisto (with an ancient, heavily cratered surface).

- Saturn shares many similarities with Jupiter. It has subdued color bands and is most famous for its flat rings, which consist of chunks of ice and dust.

- Uranus and Neptune are the so-called ice giants. Although these two planets are dominated by hydrogen and helium, their atmospheres contain important ices such as water, ammonia, and methane.

- Pluto is no longer considered a planet, and instead is a large representative of ice-rock bodies that comprise the Kuiper Belt, located at the extremity of the Solar System.

21.5 Asteroids, Meteors, Meteorites, and Comets

As we discussed in Chapter 1, when the Sun and planets formed, a vast amount of debris was left over. This debris is preserved as an assortment of celestial bodies that have been virtually unmodified since that time, and therefore provide vital clues as to the age and origin of our Solar System. Asteroids, which we introduced in Section 21.1, vary from 1000 kilometers (620 miles) to about 10 meters (33 feet) in diameter and are composed of rocky, metallic, and carbonaceous material (Fig. 21.26). The collision of a large asteroid (about 10 kilometer in diameter) with Earth is believed to have wiped out the dinosaurs and a large number of other species about 66 million years ago, bringing the Mesozoic Era to a close.

Most asteroids are located in a belt between Mars and Jupiter (2 to 3 AU from the Sun) known as the **asteroid belt** (see Fig. 21.6). This belt was first discovered in 1801 and although the exact number of asteroids it contains is unknown, there are at least one million with diameters of more than 1 kilometer (0.6 miles). About half the total mass of the asteroid belt is contained in its four largest asteroids. Many of these asteroids have impact craters on their surface (see Fig. 21.26), indicating that they too were bombarded early in the history of the Solar System. Indeed, many of the smaller asteroids are thought to be the splintered fragments of such collisions. The failure of the belt to accrete into a planet is attributed to the gravity field of Jupiter, which is thought to have supplied the material within the belt with too much energy to accrete.

Once an asteroid or a smaller celestial object, known as a meteoroid, enters Earth's atmosphere it is called a meteor. A meteor is commonly called "a shooting star," although meteors are not stars at all. Meteors burn up in the atmosphere, producing a bright trail as they do so. The light we

Figure 21.26   Asteroid

A photograph taken by the Galileo spacecraft of an asteroid reveals its pitted surface, which is interpreted to record collisions with small celestial objects. Note the irregular shape of the asteroid, a common characteristic of such bodies.

Figure 21.27 Meteor
This photograph captures the path of the meteor that exploded over the Ural Mountains in central Russia in February of 2013.

see is not due to nuclear fusion (the essential process that lights the stars), instead it merely records the vaporization of meteors as they travel through Earth's atmosphere (Fig. 21.27). Meteor showers occur when clusters of meteors simultaneously enter the atmosphere. The most famous, the *Perseid shower,* occurs every year near the middle of August and many people enjoy watching the display as dozens of meteors per hour travel through the sky. These showers may be related to the dust scattered from comets. For example, the Perseid shower occurs when Earth's orbit around the Sun takes us through the orbit of a comet. Most meteors never reach Earth's surface; they disintegrate in

the upper atmosphere. Nevertheless, their disintegration adds some 1000 tons of mass to Earth each year. The late Luis Alvarez, a Nobel laureate physicist from the University of California, has likened this to someone with a pepper shaker perpetually sprinkling meteoritic dust over the globe.

Meteorites are meteors that collide with any of the planets or their moons. It is thought that about 500 meteorites collide with Earth each year, but only 1 percent of these are found. They typically impact at speeds between 40,000 and 200,000 kilometers per hour (25,000 and 125,000 miles per hour). Major craters resulting from impacts have been found on all continents (Fig. 21.28), but they record only a tiny fraction of the impact events that have occurred over geologic time.

Of course, the chances of any given person's life being influenced by a meteorite impact are very remote, far less, for example, than that person's chances of winning a million dollar lottery. However, Earth has had some close calls with large asteroids (see Living on Earth: **Visits from the Celestial Junkyard** on page 707). The asteroid Toutatis, which is 5 kilometers (3 miles) long and 2.5 kilometers (1.5 miles) wide, has been involved with several close encounters. In 1992, it came within 3.3 million kilometers (2 million miles) of Earth. It returned in November 1996, missing Earth by 5.3 million kilometers (3.3 million miles), and again in September of 2004, missing Earth by less than 1.6 million kilometers (1 million miles, just four times the distance from Earth to the Moon). In April 2029, Earth will have another close call, this time with the asteroid Apophis, which is predicted to pass inside the orbits of our communication satellites.

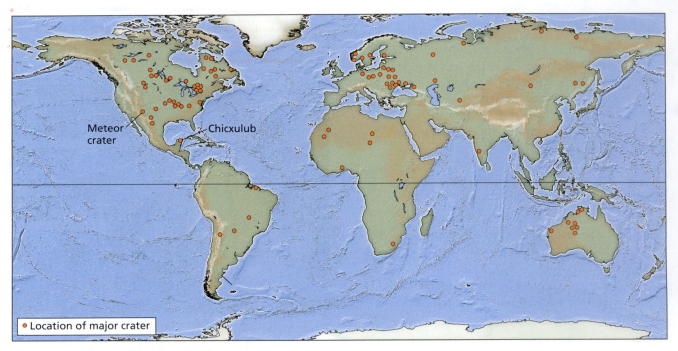

Figure 21.28 Impact Craters on Earth
The dots show the sites of major craters interpreted to be the result of meteorite impacts. The locations of Meteor Crater (see Fig. 21.29) and Chicxulub Crater (thought to be related to the mass extinction event about 66 million years ago) are also identified.

Visits from the Celestial Junkyard

On February 15, 2013, astronomers predicted the close approach of an asteroid 30–40 miles in diameter. Asteroid 2012DA14 had been discovered in 2012 and was predicted to pass by Earth at a distance of 27,700 kilometers, which is closer than many communication satellites. The asteroid had been identified well in advance, its trajectory had been calculated, and there was never any danger to any part of Earth. Yet its approach was anticipated with a great degree of excitement. The asteroid passed by exactly as predicted; the best views were in Europe, Asia, and Australia which were closest to its approach. On a galactic scale, this is classified as a close miss—the closest for an asteroid of its size in over a century.

Oddly enough, about 16 hours earlier, 2012DA14 was upstaged by a completely unknown rogue asteroid that did not miss, but rather entered Earth's atmosphere at an estimated speed of 67,000 kilometers/hour (41,000 mph). At the point of entry, the asteroid was about 20 meters in diameter with a mass of 11,000 metric tons—the largest asteroid to enter Earth's atmosphere since 1930. It became a meteor as it blazed across the sky (see Fig. 21.27), and its spectacular path was witnessed by millions. The light emitted from it was intense enough to cast shadows.

The meteor exploded into a fireball about 23 kilometers (14 miles) above the town of Chelyabinsk, Russia, releasing energy the equivalent of 440 kilotons of TNT, or 20 to 30 times more energy than that released by the two atomic bombs detonated in Japan at the end of World War II. About 1500 people were injured, and more than 7000 buildings in three cities were damaged by effects of the shockwave. The intense heat that the meteor emitted was felt by people on the ground below.

Asteroids are difficult to detect because, in contrast to comets, they are not very reflective in sunlight. Remarkably, the two asteroids that arrived in February 2013 had widely different trajectories and so are certain to have been unrelated phenomena. But the event provided a timely reminder that many asteroids have not yet been detected.

The geologic record reminds us of the potential significance of a large impact event. There is a general acceptance that a major impact event was responsible for the mass extinction at the end of the Cretaceous period, some 66 million years ago. The Chicxulub crater associated with the impact event lies off the Yucatan peninsula in the Gulf of Mexico (see Fig. 21.28). The crater is buried beneath more recently deposited sediments, but it is estimated to have been about 180 kilometers (112 miles) in diameter. Scientists debate whether this crater was caused by the impact of an asteroid or a comet. Until recently, the general consensus was that the crater was made by an asteroid, estimated to have been about 10 kilometers (6.2 miles) in diameter, which released energy equivalent to 100 trillion tons of TNT. But this interpretation has recently been challenged. Recent calculations suggest that there was less debris than previously supposed, which would imply that a smaller body hit Earth, possibly on the order of 5 kilometers across. But for a 5-kilometer object to make a 180-kilometer crater, the object would need to have been traveling more rapidly than previously thought. Asteroids do not travel at such high velocities, suggesting that the culprit may have been a long-period comet.

The consequences of extraterrestrial impacts on Earth have altered the course of its evolutionary history. Although most scientists subscribe to the principles of a slow progressive evolution of life on Earth, they are equally aware that catastrophic events, such as major impacts, have had a dramatic influence on the course of evolution, and in some instances may have reset the evolutionary clock. The mass extinction about 66 million years ago wiped out the dinosaurs and paved the way for mammals, including humans, to become the dominant animals on Earth.

In North America, one of the best examples of a crater thought to have been produced by the impact of large meteorites is Barringer Crater, also known as Meteor Crater, near Flagstaff, Arizona (Fig. 21.29). This impact occurred 50,000 years ago and produced a crater 1.2 kilometer (0.75 miles) across and 200 meters (660 feet) deep. It was formed by a meteorite thought to be little more than 90 meters (295 feet) across.

Meteorites are quite variable in composition and probably have no single mode of origin. As we discussed in Section 21.2, some meteorites appear to have originated on Mars, however, most appear to be remnants of asteroids that formed early in the history of the Solar System.

Meteorites are classified according to their composition and texture. Most are *stony meteorites,* which are made up of silicate minerals and classified as either *chondrites* or *achondrites*. Chondrites have peculiar textures that have been simulated in the laboratory by rapid heating of meteoritic material followed by rapid cooling. This is consistent with melting and rapid crystallization associated with the collisional events at the birth of the Solar System. Many scientists believe the composition of these meteorites approximates the average composition of the primitive Earth, before its iron- and nickel-rich core formed. This interpretation is supported by hundreds of age dates on meteorite samples, all of which suggest that they formed between 4.50 and 4.57 billion years ago. Achondrites lack the peculiar texture of chondrites and are similar in composition to Earth's mantle, suggesting that they were originally part of a larger body with internal layering similar to that of Earth.

Figure 21.29 Meteor Crater

Meteor Crater is located near Flagstaff, Arizona (see Fig. 21.28). The crater is 1.2 kilometers (0.75 miles) in diameter and 200 meters (660 feet) deep. It formed 50,000 years ago when a meteorite 90 meters (295 feet) in diameter blazed through Earth's atmosphere. Although it splintered in flight, the majority of it impacted near Flagstaff and produced the impressive crater seen here.

A few meteorites are metallic. These *iron meteorites* may represent the splintered cores of asteroids and, as such, are pieces of the primitive Solar System. This interpretation is, once again, supported by their ages of 4.50 to 4.57 billion years. The composition of iron meteorites is similar to Earth's core, suggesting that iron may collapse to the core in many celestial bodies.

Comets are mixtures of ice and dust, some of which is carbon-rich, with a small, rocky core. Although popularly known as "dirty celestial snowballs," recent evidence suggests that most of the ice occurs below the surface in comets, which is very dark. Comets have a central nucleus ranging in diameter from 100 meters to 50 kilometers (0.06 to 31 miles). This icy nucleus vaporizes as a comet approaches the Sun, releasing dust and gases to form the head—the *coma*—and the tail of the comet. The coma may be up to 100,000 kilometers (62,500 miles) in diameter and is the most visible part of the comet, whereas the tail may be millions of kilometers in length. About 4200 comets have been documented, but most astronomers maintain that this is a tiny fraction of the total number.

Recall from Chapter 20 that comets are famous for their tails, which always point directly away from the Sun (Fig. 21.30). The tails form as comets are blasted by the *solar wind* (see Chapter 20). Analysis of comets reveals that they contain many kinds of carbon-, nitrogen-, and oxygen-bearing molecules. This implies that the primitive components of the Solar System contained the raw materials needed to initiate life (Table 21.2).

The most recent comets to visit Earth are the comet *Hyakutake* (pronounced hiyakoo-ta'-kee) in 1996, comet Hale-Bopp in 1997, comet C/2011 L4 Pan-Starrs in 2013 (Fig. 21.31) and Comet Lovejoy in 2015. Comet Hyukatake was the first comet in which the organic compounds methane (CH_4) and ethane (C_2H_6) were detected.

The origin of comets was a mystery until the 1950s, when Jan Oort, a Dutch astronomer, showed that the overwhelming majority of comets come from the remotest parts of the Solar System, between 20,000 and 150,000 AU from the Sun. This region is now known as the **Oort cloud**. It is thought that comets plunge toward the Sun when their orbits become disturbed by the gravity fields of the massive outer planets of the Solar System, or by passing stars. These comets may take a few hundred to several million years to orbit the Sun. Another source of comets is believed to be located beyond Pluto in the Kuiper belt, and some comets, known as **exocomets**, may originate in regions beyond the Solar System.

In contrast to other celestial bodies, comets have highly elliptical orbits around the Sun (see Fig. 21.30). Although these orbits may be highly eccentric, some are very regular. In 1704, Edmond Halley deduced that the comets sighted in 1456, 1531, 1607, and 1682 had identical orbits so that the sightings were likely the result of the return of the same comet every 75 years. His prediction of its return in 1758 was confirmed when a comet, later named Halley's Comet, was sighted on Christmas Day of that year. The most recent sojourn of Halley's Comet into the inner Solar System occurred in 1986; it will reappear in 2061. Comets with very long-term reappearances may have profoundly affected the evolution of life on Earth. Although hotly debated, some studies of mass extinctions suggest that extinctions occur

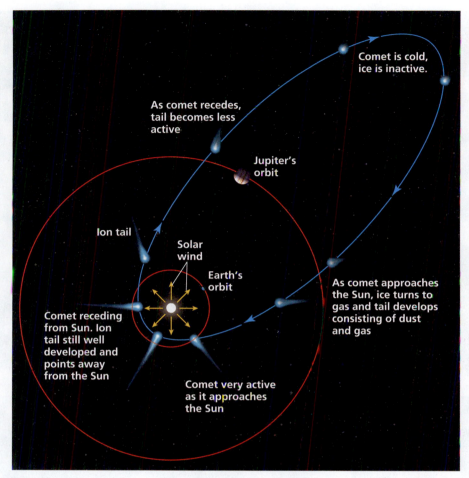

Figure 21.30   Orbit of Comet

Schematic representation (not to scale) of a comet's eccentric orbit compared to that of Earth and Jupiter. Note that the comet's tail, which develops as volatile ices become gaseous, always points away from the Sun and is produced by the interaction of the comet's icy surface with the solar wind.

TABLE 21.2   COMPOUNDS AND IONS DETECTED IN COMETS

Coma	Gas Tail
H, OH, O	CO, CO_2
C, C_2, C_3, CH	H_2O, OH
CN, CO, CS, S	CH, CN, N_2
NH, NH_2, HCN, CH_3CN	C, Ca
Na, Fe, K, Ca, V	
Cr, Mn, Co, Ni, Cu	

every 26 to 28 million years on Earth, possibly as the result of cometary impacts.

One of the best documented examples of the impact of a comet was on Jupiter's surface. In the summer of 1994, fragments of Comet Shoemaker-Levy 9 plunged toward Jupiter's surface. The path of the comet had previously been determined and so the effects of this impact were extensively documented (see Fig. 21.18). Traveling at 100 kilometers per second (62 miles per second), the energy released at impact from one fragment of the comet was an order of magnitude higher than the world's entire nuclear arsenal would have been able to produce at the height of the Cold War.

The effects of such impacts on Jupiter are very different from what they would be on Earth. Jupiter's larger mass means that it has a much stronger gravitational attraction, and many astronomers believe that the Shoemaker-Levy 9 comet came close enough to Jupiter to be snared by its gravity. After that, the comet fragmented, and its fate was sealed. Although the probability of comet impacts is far greater on Jupiter than it is on Earth, the long-term effects of these impacts are less obvious. Recall from Section 21.5 that Jupiter has no well-defined surface where the planet ends and its atmosphere begins. Because of this, astronomers debated whether the effect of the impact of the Shoemaker-Levy 9 comet on Jupiter would be any more dramatic than throwing a stone into a pool of water. When the comet did

CHECK YOUR UNDERSTANDING

○ What are the two main types of meteorites and how did each originate?

○ What important information do meteorites yield about the early history of the Solar System?

○ What are comets and where do they come from?

(a)

(b)

(c)

Figure 21.31   Comets

(a) A photograph of Comet Hyakutake was taken in 1996, around the time of its closest approach to Earth. Note how the tail points away from the setting sun. (b) Comet Hale-Bopp reached its most favorable position for viewing in the Spring of 1997. (c) Comet C/2011 L4 Pan-Starrs was visible over Australia in February 2013.

hit, huge plumes extending more than 3000 kilometers (1860 miles) above the surface were observed. Bright spots the size of Earth were caused by fragments burning in the atmosphere (see Fig. 21.18). But the visible evidence of impact lingered only for weeks, as dark spots that were gradually stretched and blended to form bands.

21.5 SUMMARY

- Asteroids and comets represent the remaining debris left over when the Solar System formed. Meteorites are rocky celestial bodes that collide with planetary surfaces. Their ages and chemistry provide vital information about the early history of the Solar System.

- Most, but not all, of the asteroids occur in a confined belt between the orbits of Mars and Jupiter.

- Comets are mixtures of dust and ice and have a rocky core. Many come from the remotest parts of the Solar System and have highly eccentric orbits. Their famous tails always point directly away from the Sun, revealing the presence of the solar wind.

21.6 Exoplanets and Other Solar Systems

The countless numbers of stars in the universe make the likelihood of other solar systems almost a certainty. Astronomers have long hypothesized about the existence of other solar systems, in which a collection of planets revolves around a central star. Only recently, however, has this been confirmed. There are several reasons why confirmation of orbiting bodies has proven difficult. Planets do not shine by themselves; they only reflect the light of the star around which they orbit, and so may be too dim to be observed directly. In addition, the brightness and size of a star makes it very difficult to detect the more subdued outlines of nearby orbiting bodies. If our Solar System is any indication, planets will be much smaller than the adjacent star and their size, coupled with their distance from Earth, may prevent their detection.

But important hints of other solar systems were discovered in the mid-1980s. In 1984, astronomers at an observatory in Chile found evidence of a disk-shaped cloud of debris surrounding a star in the constellation Pictor (Fig. 21.32). This debris was interpreted as a solar system in the process of formation.

In late 1995 and early 1996, the breakthrough came in separate observations from two different observatories. First, two Swiss astronomers reported the existence of a planet orbiting a star called 51 Pegasi, demonstrating for the first time the existence of other solar systems. Second, Geoffrey Marcy at San Francisco State University found evidence of two planets, one orbiting the star 70 Virginis in the constellation Virgo, and the other orbiting 47 Ursae Major, a star within the Big Dipper.

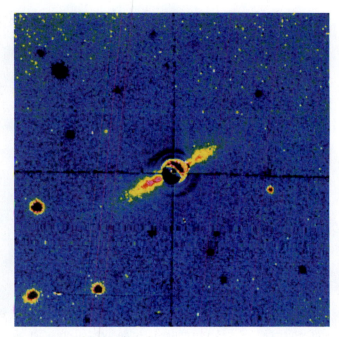

Figure 21.32   Beta Pictoris

Beta Pictoris, a star in the constellation Pictor, is surrounded by a disk-shaped cloud of matter that may contain planet-sized objects. This cloud is interpreted to be a solar system in the process of formation.

Both discoveries were subsequently confirmed by other observatories and caused considerable excitement among the scientific community. The search was now on in earnest, and soon more planets had been identified outside the Solar System than exist within it. As of 2015, some 2000 of these **exoplanets** have been identified, and 20,000 more candidates are being evaluated, including more than 250 that may have conditions hospitable to life. According to some estimates there are likely to be 100–400 billion exoplanets within our own galaxy, the Milky Way, alone, and of these, 17 billion are likely to be larger than Earth.

Because the planets themselves are invisible, their existence must be determined indirectly. Consider the following analogy. During a hammer-throwing event, the thrower wobbles as the hammer revolves as a consequence of gravitational interaction (Fig. 21.33). So even if the spectator's view of the hammer is obstructed, the viewer can still deduce the hammer's existence by the rhythmic wobble of the thrower. Although the relationship between a star and its planets is considerably more complicated, planets revolve around a star as if tied taut to it by an invisible string. As they do so, their gravitational pull results in rhythmic oscillations in the star, which can be observed. So while a planet may be hidden from our view, the rhythmic oscillations of the star can be used to betray its presence.

Most astronomers believe that there is nothing particularly special about our Solar System and that there must be billions of similar systems in the universe. If so, the odds that life exists elsewhere in the universe are very good indeed. If the relationship between Earth and the Sun is

Figure 21.33   Detecting Exoplanets Using Newton's Laws
A traditional Scottish sport, hammer-throwing, illustrates Newton's Laws of motion and gravity. The rhythmic wobble of the thrower is analogous to the way in which planets revolve around a star. Their gravitational pull results in rhythmic oscillations in the star, which can be observed and used to deduce the presence of a planet.

anything to go by, then the inhabited planet would have to lie at an appropriate distance from its sun and would need to be volcanically active so as to have an atmosphere and surface water.

But if there is life, what form would it take? Again, our own experience on Earth tells us that life evolves such that the fittest, most adaptable species survive. However, evolution is not as simple as that because it also contains an element of luck. For example, it is most unlikely that we humans would have become the dominant life form if the dinosaurs had not become extinct about 66 million years ago. If theories that this extinction was the result of an extraterrestrial impact are correct, then our own existence may be largely the result of chance; we are potentially the fortuitous benefactors of a cataclysmic event.

If life forms exist on the planets of other solar systems, it is likely fruitless to speculate what form they would take. Nevertheless, the discovery of exoplanets has added fuel to a scientific and philosophical question that refuses to go away: Is there anybody out there?

> **CHECK YOUR UNDERSTANDING**
> ○ How is the existence of exoplanets determined?

21.6 SUMMARY

- Although difficult to detect, over some 2000 exoplanets have been found, more than 250 of which may support conditions hospitable to life.

- Exoplanets are invisible but can be detected from the wobble their gravitational attraction causes in the star they orbit.

Key Terms

asteroid 680	exoplanets 711	lunar highlands 681	meteoroid 680
asteroid belt 705	Great Red Spot 699	mare (maria) 681	Oort cloud 708
comet 681	Jovian planet 686	meteor 680	terrestrial planet 686
exocomets 708	Kuiper Belt 705	meteorite 680	

Key Concepts

21.1 EARTH AND ITS MOON
- The Moon's surface is heavily cratered and consists of bright lunar highlands and dark lunar maria.

- Volcanic activity on the Moon ceased about 3 billion years ago.

- The Moon formed after a Mars-sized object collided with Earth.

21.2 THE TERRESTRIAL PLANETS AND THEIR MOONS
- The innermost planets (Mercury, Venus, Earth, and Mars) are terrestrial (rocky) planets.

- Planets vary in character primarily because of differences in their size and distance from the Sun.

- Mercury has a heavily cratered surface. It is a geologically dead planet with no recent volcanism and a vanishingly thin atmosphere.

- Venus has a dense atmosphere dominated by carbon dioxide and a hot surface temperature. Modern volcanic activity is suspected.

- Mars has polar ice caps, a very thin atmosphere, and is very cold. More than a billion years ago it had volcanoes and flowing water on its surface.

21.3 PLATE TECTONICS: A PLANETARY PERSPECTIVE
- Earth is unique in having active tectonics, an oxygen-rich atmosphere, and abundant surface water and life.

21.4 THE JOVIAN PLANETS, THEIR MOONS AND THE KUIPER BELT

- The outer planets (Jupiter, Saturn, Uranus, and Neptune) are Jovian (gaseous) planets.

- Jupiter is mostly hydrogen and helium. Its atmosphere is characterized by color bands and massive storms.

- Saturn is similar in many ways to Jupiter. Its flat rings are chunks of ice.

- Uranus and Neptune are gas giants like Jupiter and Saturn but their atmospheres contain water, ammonia, and methane ice.

- Pluto is no longer considered a planet; instead it is a dwarf planet and a prominent member of a group of icy bodies that occupy the Kuiper Belt.

21.5 ASTEROIDS, METEORS, METEORITES, AND COMETS

- Asteroids, comets, meteoroids, and meteorites are various forms of celestial debris that can tell us important information about the Solar System and its origins.

21.6 EXOPLANETS AND OTHER SOLAR SYSTEMS

- Some 2000 exoplanets have been discovered orbiting suns beyond the Solar System.

Study Questions

1. The early history of Earth is thought to have been profoundly influenced by meteorite impacts. However, there is no direct evidence for this on Earth. Where does the evidence come from and why has it been applied to Earth?

2. Why do Mercury and the Moon have virtually no atmosphere?

3. Why is Earth's atmosphere dominated by nitrogen (N_2) and oxygen (O_2), whereas the atmospheres on Venus and Mars are dominated by carbon dioxide (CO_2)?

4. Compare and contrast tectonics on Venus and Earth.

5. Why is Earth the only terrestrial planet in our Solar System with a water-rich surface?

6. What evidence is there that iron-rich meteorites are similar in chemical composition to that of Earth's core?

7. Where do planetary atmospheres come from and when do they form?

8. What evidence points to the age of the Solar System as 4.57 billion years?

9. Do you think that there was once life on Mars? If so, what are your main arguments in favor of the idea? If not, what are your main objections?

10. Why is Pluto no longer considered a planet?

Appendix A

This appendix provides a table of units and their conversion from older units to Standard International (SI) units.

Length

Metric Measure

1 kilometer (km)	= 1000 meters (m)
1 meter (m)	= 100 centimeters (cm)
1 centimeter (cm)	= 10 millimeters (mm)

Nonmetric Measure

1 mile (mi)	= 280 feet (ft)
	= 1760 yards (yd)
1 yard (yd)	= 3 feet (ft)
1 foot (ft)	= 12 inches (in)
1 fathom (fath)	= 6 feet (ft)

Conversions

1 kilometer (km)	= 0.6214 mile (mi)
1 meter (m)	= 3.281 feet (ft)
	= 1.094 yards (yd)
1 centimeter (cm)	= 0.3937 inch (in)
1 millimeter (mm)	= 0.0394 inch (in)
1 mile (mi)	= 1.609 kilometers (km)
1 foot (ft)	= 0.3048 meter (m)
1 inch (in)	= 2.54 centimeters (cm)
	= 25.4 millimeters (mm)

Area

Metric Measure

1 square kilometer (km^2)	= 1,000,000 square meters (m^2)
	= 100 hectares (ha)
1 square meter (m^2)	= 10,000 square centimeters (cm^2)
1 hectare (ha)	= 10,000 square meters (m^2)

Nonmetric Measure

1 square mile (mi^2)	= 640 acres (ac)
1 acre (ac)	= 4840 square yards (yd^2)
1 square foot (ft^2)	= 144 square inches (in^2)

Conversions

1 square kilometer (km^2)	= 0.386 square mile (mi^2)
1 hectare (ha)	= 2.471 acres (ac)
1 square meter (m^2)	= 10.764 square feet (ft^2)
	= 1.196 square yards (yd^2)
1 square centimeter (cm^2)	= 0.155 square inch (in^2)
1 square mile (mi^2)	= 2.59 square kilometers (km^2)
1 acre (ac)	= 0.4047 hectare (ha)
1 square foot (ft^2)	= 0.0929 square meter (m^2)
1 square inch (in^2)	= 6.4516 square centimeters (cm^2)

Volume

Metric Measure

1 cubic meter (m^3)	= 1,000,000 cubic centimeters (cm^3)
1 liter (l)	= 1000 milliliters (ml)
	= 0.001 cubic meter (m^3)
1 milliliter (ml)	= 1 cubic centimeter (cm^3)

Nonmetric Measure

1 cubic foot (ft^3)	= 1728 cubic inches (in^3)
1 cubic yard (yd^3)	= 27 cubic feet (ft^3)

Conversions

1 cubic meter (m^3)	= 264.2 gallons (US) (gal)
	= 35.314 cubic feet (ft^3)
1 liter (l)	= 1.057 quarts (US) (qt)
	= 33.815 fluid ounces (US) (fl oz)
1 cubic centimeter (cm^3)	= 0.0610 cubic inch (in^3)
1 cubic mile (mi^3)	= 4.168 cubic kilometers (km^3)
1 cubic foot (ft^3)	= 0.0283 cubic meter (m^3)
1 cubic inch (in^3)	= 16.39 cubic centimeters (cm^3)
1 gallon (gal)	= 3.784 liters (l)

Mass

Metric Measure

1000 kilograms (kg)	= 1 metric ton (t)
1 kilogram (kg)	= 1000 grams (g)

Nonmetric Measure

1 short ton (ton)	= 2000 pounds (lb)
1 long ton	= 2240 pounds (lb)
1 pound (lb)	= 16 ounces (oz)

Conversions

1 metric ton (t)	= 2205 pounds (lb)
1 kilogram (kg)	= 2.205 pounds (lb)
1 gram (g)	= 0.03527 ounce (oz)
1 pound (lb)	= 0.4536 kilogram (kg)
1 ounce (oz)	= 28.35 grams (g)

Pressure

standard sea-level air pressure	= 1013.25 millibars (mb)
	= 14.7 lb/in^2

Temperature

To change from Fahrenheit (F) to Celsius (C)

$$°C = °F − 32/1.8$$

To change from Celsius (C) to Fahrenheit (F)

$$°F = °C \times 1.8 + 32$$

Energy and Power

1 calorie (cal)	= the amount of heat that will raise the temperature of 1 g of water 1°C (1.8°F)
1 joule (J)	= 0.239 calorie (cal)
1 watt (W)	= 1 joule per second (J/s)
	= 14.34 calories per minute (cal/min)

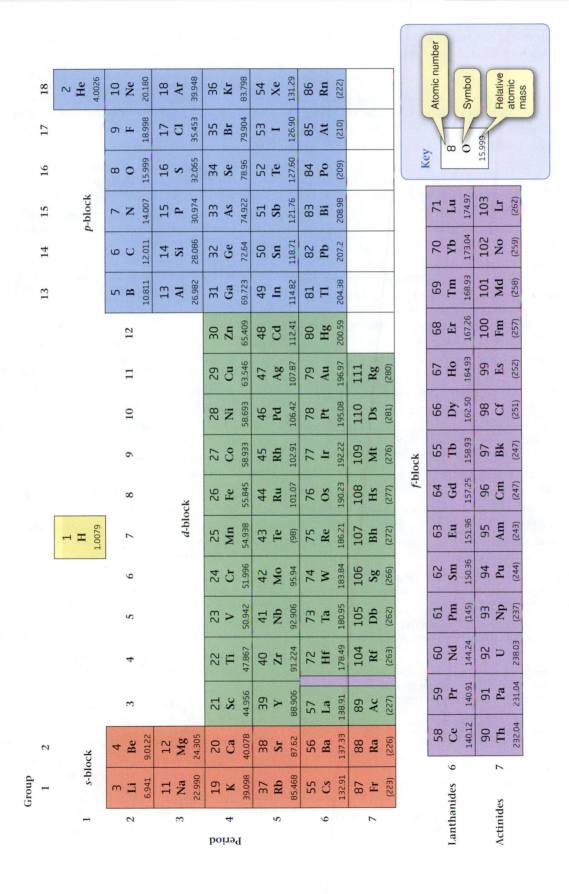

Glossary

Note: The number at the end of each definition indicates the chapter in which the term first appears.

A horizon   The layer of soil in which dead plant material is broken down, resulting in the formation of organic-rich humus; the second layer from the top in the soil profile. (5)

Ablation   The loss of ice and snow from a glacial system by any means. (15)

Abrasion   The sandpaper-like grinding away of bedrock by rock particles transported by water, wind, or glacier ice; one of two main types of glacial erosion (*plucking* is the other). (15)

Absolute age   The amount of time (in years) that has elapsed since a rock or mineral formed. (8)

Accreted terrane   A crustal block plastered to the edge of a continent along zones of ocean-continent convergence (*see* **Terrane**). (10)

Accretionary wedge   A wedge-shaped package of highly deformed rocks scraped off a subducting slab at a deep ocean trench and plastered onto the leading edge of the overriding plate at a subduction zone. (9)

Acid   A chemical substance that can release a proton, or hydrogen ion (H^+). (5)

Acid rain   Rainwater containing high concentrations of hydrogen ions (H^+). (5)

Active margin   Continental margin adjacent to a deep ocean trench where subduction occurs and a magmatic arc develops; associated with earthquakes, volcanoes, mountain building. (2)

Aerosols   A mixture of gases vented into the atmosphere during explosive eruptions. (4)

Agglomerate   Rock that forms from the pyroclastic deposition of blocks or bombs from a volcano. (4)

Alfred Wegener   The German meteorologist who proposed the hypothesis of continental drift. (2)

Alluvial fan   Triangular-shaped apron of sediment that forms at the mouth of a mountain canyon in desert regions due to an abrupt decrease in a stream's gradient. (13)

Alpine glacier   Any of a variety of glaciers that form in regions where the snowline lies at high elevations, so that glaciers are localized in mountainous areas. (15)

Amphibole group   A complex set of double-chain silicate minerals with highly variable chemical compositions and water in their crystal structure; hornblende is the most common. (3)

Amphibolite   A medium-grade, nonfoliated metamorphic rock for which the parent rock is mafic.

Amphibolite facies   The metamorphic facies at which hornblende becomes stable; characterized by coarse-grained schists and amphibolites formed under conditions of medium temperature and pressure. (7)

Andesite   A fine-grained intermediate igneous rock; commonly contains phenocrysts of plagioclase and/or hornblende in a fine-grained groundmass. (4)

Angle of repose   The maximum stable slope developed by loose material; the steepness of this slope varies with the type of material. (12)

Angular unconformity   An angular discordance between an upper and lower set of sedimentary strata; the result of uplift, tilting, and erosion of the lower set, followed by deposition of the upper set. (10)

Anion   A negatively charged ion. (3)

Antecedent stream   A stream that cuts through a ridge because its course was established before the ridge was uplifted and was maintained because downcutting kept pace with uplift of the ridge. (13)

Anticline   A type of fold that forms when a sequence of rock strata in normal succession arches upward so that the oldest rocks occur in the center of the fold. (10)

Apparent polar wander   The explanation for the polar-wander curves in terms of movement of the continents relative to the poles rather than movement of the poles themselves. (2)

Aquiclude   A rock body or sediment layer that is impermeable to water. (14)

Aquifer   A rock body or layer of unconsolidated sediment that is porous and permeable and can store and transmit water below Earth's surface; the most productive source of groundwater. (14)

Aquitard   A rock body or sediment layer that does not readily transmit water but instead slows its motion. (14)

Archean Eon   The second of the four eons of geologic time between 4000 and 2500 million years ago. (8)

Arête   An alpine glacial landform; precipitous, knife-edged ridge that separates adjoining cirques. (15)

Arkose   A sandstone in which feldspar accounts for more than 25 percent of the grains. (6)

Artesian well   A well located on ground that is below the pressure surface of the confined aquifer into which it is drilled, so that water gushes out. (14)

Assimilation   Processes that occur as magma reacts chemically and physically with the wall rock it comes into contact with; these processes change the magma's chemical composition. (4)

Asteroid   A rocky celestial body that varies from 1000 kilometers (620 miles) to about 10 meters (33 feet) in diameter. (21)

Asteroid belt   The area of the Solar System between the orbits of Mars and Jupiter in which most asteroids are located. (21)

Asthenosphere   The weak portion of Earth's upper mantle above which the rigid lithosphere "floats"; allows the lithospheric plates to move. (2)

Atmosphere   Earth's gaseous envelope and the air we breathe. (1)

Atoll   A roughly circular reef surrounded by open sea that encloses a shallow lagoon in which there is no land. (17)

Atom   The smallest unit of an element that displays all the physical and chemical properties of that element. (3)

Atomic number   The number of protons in an atom of an element. (3)

Aulacogen   The failed arm of a three-armed continental rift that fails to open into an ocean and slowly fills with sediment; often channels major rivers into the new ocean formed as the other two arms open. (9)

Aureole   The halo of contact metamorphic rocks around an igneous body produced by heating adjacent to a body of cooling magma. (7)

Aurora   The spectacular natural light display that occurs when the solar wind bombards Earth's atmosphere at high latitudes. (20)

Axial plane   An imaginary surface that divides a fold into two halves and includes the hinge of each folded layer; it is perpendicular to the direction of shortening. (10)

B horizon   Soil layer in which material leached from the A and O horizons accumulates; the third layer in the soil profile, below the A horizon. (5)

Back-arc basin   Basin floored by oceanic crust that opens behind a volcanic arc in response to subduction zone rollback. (9)

Back-arc spreading   A localized form of seafloor spreading responsible for the opening of a back-arc basin. (9)

Backwash   Occurs when the water from a breaking wave retreats away from the beach and toward the sea. (17)

Bajada   Forms when repeated flash floods cause individual alluvial fans from adjacent valleys to coalesce into a continuous apron of sediment at the foot of a mountain. (16)

Baked contact   Where the heat from a lava, dike, or sill transforms a few centimeters of the adjacent rock into a metamorphic rock. (4)

Barchan dune   A solitary, crescent-shaped dune with a concave slip face and crescent-shaped "horns" that point downwind; found on flat, barren areas of desert floor in regions where the wind is constant but the supply of sand is limited. (16)

Barrier island   A low, offshore ridge of sand that parallels the coastline at some distance from the shore and is separated from the shore by a bay or by a body of quiet brackish water. (17)

Barrier reef   Narrow coral reef roughly parallel to the shore but separated from it by a lagoon. (17)

Basal sliding   One of two ways in which glaciers move (*plastic flow* is the other), in which lubricating meltwater accumulates at the base of a glacier allowing the ice to slide over the material beneath it; involves mass movement of the ice sheet with no change in crystal shape; faster than plastic flow. (15)

Basalt   A fine-grained volcanic rock; one of the most common mafic igneous rocks. (4)

Base level   The lowest elevation to which a stream can erode; dictated by the elevation where a stream enters an ocean, a lake, or another stream, or meets a waterfall or rapids. (13)

Basin   A type of fold structure without a hinge in which rocks dip inward to the center to form a bowl-shaped depression. (10)

Batholith   A pluton with a surface outcrop area greater than 100 square kilometers (38 square miles). (4)

Baymouth bar   A spit that grows until it extends completely across the mouth of a bay; formed by deposition from longshore currents. (17)

Beach   Gently sloping body of unconsolidated sand, gravel, and shell fragments that extends landward from the low water line to the maximum reach of storm waves. (17)

Beach drift   The zigzag path along a beach followed by sediment particles suspended in an advancing, then retreating, wave. (17)

Bed   A layer of sediment with distinctive characteristics, such as the composition, shape, size, and sorting of its particles. (6)

Bed load   Products of erosion that are too large or too heavy to be lifted off a stream bed and carried in suspension and are moved, instead, by traction (sliding and rolling) or saltation (skipping). (13)

Bedding   Layering developed in sedimentary rocks. (6)

Bedding plane   The surface that separates one bed from another in sediment and in sedimentary rock; reflects the end of one depositional event and the beginning of another. (6)

Bedrock   Consolidated rock underlying, or exposed at, Earth's surface; usually covered in soil. (1)

Biochemical sedimentary rock   Sedimentary rock that is the end product of chemical weathering and whose formation is aided by the action of organisms. (6)

Biomass   Total mass of organic matter within an area or ecosystem. Also used as fuel or to produce fuel; a renewable energy source. (19)

Biosphere   The realm of life on Earth including plants and animals that are either living or in the process of decay. (1)

Block   A solid angular fragment ejected from a volcano; a constituent of pyroclastic deposits. (4)

Blocky lava   Describes the solid surface of a lava flow that has broken up into jagged blocks even as the molten interior continues to advance. (4)

Blowout   A depression in the ground surface produced by a turbulent vortex of wind; its depth depends on the level of the local groundwater table. (16)

Blueschist facies   The metamorphic facies in which schists containing a blue amphibole become stable; formed under conditions of low temperature and high pressure. (7)

Body wave   A seismic wave that travels through Earth's interior. (11)

Bomb   (*see* **volcanic bomb**)

Bottom structures   Sedimentary structures that occur at the base of a bed. (6)

Boulder   A particle in detrital sedimentary rocks with a diameter greater than 256 millimeters (10 inches). (6)

Bowen's continuous reaction series   A process that occurs in the formation of igneous rock; a continuous cycle of crystallization and reaction in cooling mafic magma in which early forming calcium-rich plagioclase is continuously replaced by more sodium-rich plagioclase. (4)

Bowen's discontinuous reaction series   A process that occurs in the formation of igneous rock in which a sequence of minerals is produced in cooling mafic magma as one mineral becomes converted to another at a specific temperature in the cooling history. (4)

Braided stream   A stream that develops wherever there is more sediment than the stream can carry; results in continuous midstream deposition and an elaborate network of shallow, interlacing channels. (13)

Breaker   A wave of water that crests (reaches a maximum height), becomes unstable, then spills over into the trough. (17)

Breccia   Detrital sedimentary rock containing angular fragments greater than 2 millimeters (0.08 inches). (6)

Brittle   Physical characteristic of a rock that breaks, cracks, or fractures when placed under stress. (10)

Brittle failure   The deformation produced when brittle rocks break when stress is applied. (10)

Buoyancy   A characteristic of Earth's crust that describes the way in which the crust and the rigid mantle beneath it "float" on the more pliable and denser asthenosphere; the degree of buoyancy depends on the thickness of the crust and the density contrast with the asthenosphere. (9)

Burial metamorphism   Metamorphism of rocks caused by burial beneath a thick succession of overlying rock layers. (7)

Butte   An isolated, pillar-like mountain remnant produced in an arid region by the irregular erosion of vertical cliffs where easily weathered rocks are capped by more resistant ones. (16)

C horizon   The bottommost layer in the soil profile (beneath the B horizon); consists mostly of fragmented and weathered bedrock. (5)

Caldera   Large steep-walled, roughly circular depression produced by the collapse of the center of a volcano following an eruption; common at the summit of a shield volcano. (4)

Capacity   The maximum amount of sediment that a stream can carry at any one time; one of two measures (competence is the other) that describes a stream's ability to carry solid particles. (13)

Carbon cycle   Describes the circulation of carbon at or near Earth's surface: how it moves between the inorganic and organic worlds and flows from energy producers to energy consumers. (20)

Carbonate mineral   A mineral in which common cations such as calcium and magnesium are bonded to a combination of carbon and oxygen to form minerals such as calcite and dolomite; the most common nonsilicate. (3)

Cation   A positively charged ion. (3)

Cenozoic Era   The most recent era of the Phanerozoic Eon, from 66 million years ago to the present, characterized by essentially modern animals and plants; separated from the preceding Mesozoic Era by a major mass extinction. (8)

Center of mass   The unique point around which the masses of two bodies are balanced. (21)

Chalk   A soft, porous biochemical sedimentary rock made up of the skeletal remains of microorganisms , (6)

Channel   A ribbon-like depression that carries surface water as it drains from land. (13)

Channel bar   Forms when sand and gravel accumulate midstream; characteristic of sediment-laden braided streams with interlacing channels. (13)

Chemical formula   A description of a mineral that lists its component ions according to the proportion in which they occur in the mineral. (3)

Chemical sedimentary rock   A sedimentary rock formed from the inorganic precipitation of minerals from solution; one of the two main types of sedimentary rock (along with *detrital sedimentary rock*). (6)

Chemical weathering   A process in which chemical reactions break down or decompose the unstable minerals in a rock and convert them into stable products. (4)

Chert   A sedimentary rock made up of microcrystalline quartz (SiO_2) that can be either chemical or biochemical in origin. (6)

Chilled margin   Where magma in a lava, dike, or sill is cooled rapidly as it flows through the cold rocks with which it is in contact so that the igneous rock adjacent to the contact is fine grained or even glassy. (4)

Cinder cone   A subaerial volcanic edifice in which cinders ejected from the volcano rapidly build a relatively steep-sided cone-shaped edifice around the vent. (4)

Cirque   An amphitheater-shaped basin carved by a glacier located in a mountainside hollow. (15)

Clast   A fragment of a preexisting mineral or rock in a detrital sedimentary rock. (6)

Clastic texture   The characteristic appearance of all detrital sedimentary rocks due to the presence of clasts. (6)

Clay   The finest particles of sediment, less than 0.004 millimeters across. (6)

Clay mineral   A rock-forming mineral with a sheet-like silicate structure; commonly formed by the breakdown of feldspar during weathering. (3)

Cleavage   The tendency of a mineral to break (or "cleave") along preferred directions; reflects systematic weaknesses in the internal structure of a mineral in the directions of their weakest bonds. (3)

Climate change   Significant and lasting changes to Earth's weather patterns. (Ch. 20)

Climate proxy   A preserved characteristic of the past that allows scientists to reconstruct the climatic conditions at that time. (20)

Coal   A biochemical sedimentary rock composed of more than 50% carbon and formed from highly compressed and altered plant remains. (6, 19)

Coarse-grained   A characteristic of the minerals in plutonic igneous rocks; large minerals (visible to the naked eye) are formed when magma deep within Earth's interior cools slowly. (4)

Coastal desert   A desert that forms in warm coastal areas next to cold ocean currents when the cool, dry offshore air becomes heated as it blows onshore and evaporates all surface moisture. (16)

Cobble   A particle in detrital sedimentary rock with a diameter between 64 millimeters (2.5 inches) and 256 millimeters (10 inches). (6)

Cohesion (rock)   A rock's resistance to sliding or fracturing. (12)

Col   An alpine glacial landform produced when cirques converge from the opposite sides of a mountain and create a low point, gap, or pass in the mountain range. (15)

Color   A visual characteristic of a mineral that is often related to its overall chemical composition. (3)

Column   A cave formation created when a downward growing stalactite connects with an upward growing stalagmite; an important structure that supports the roof of the cave. (14)

Columnar joint(s)   A system of joints that divide fine-grained igneous rock like basalt into long polygonal columns; produced by contraction on cooling. (10)

Competence   The maximum size of the material a given stream can carry; determined by a stream's velocity; one of two measures (*capacity* is the other) that describes a stream's ability to carry solid particles. (13)

Composite volcano   A volcanic edifice that consists of alternating layers of pyroclastic deposits from explosive eruptions and lava flows that erupted from either the central vent or along the flanks of the volcano; Earth's most picturesque volcanoes. (4)

Compositional convection   The process by which Earth's outer core circulates; light material rises while heavy material sinks. This circulation is thought to be driven by the buoyancy of the less dense material left when iron solidifies and attaches itself to the surface of the inner core. (11)

Compound   A substance that contains two or more different chemical elements that are joined by bonds and can be separated into simpler substances by chemical reactions; a molecule is the smallest particle of a compound. (3)

Compressional stress   Force that pushes rocks together, causing them to strain by bending or becoming shorter and thicker. (10)

Cone of ascension   Occurs when saline water migrates upward after freshwater is withdrawn and cones of depression form; cones of ascension may contaminate an aquifer with saline water. (14)

Cone of depression   A cone-shaped depression in the water table that forms around a well when water is withdrawn from the ground at a rate that is faster than nature is able to replenish it. (14)

Confined aquifer   An aquifer in which the water-bearing rocks or sediments are sandwiched between impermeable layers of rock; water in confined aquifers is supplied by precipitation in an adjacent mountainous region. (14)

Confining pressure   One of two types of pressure (along with directed pressure) that affect crustal rocks; produced by the weight of the overlying rocks, it pushes equally in all directions causing a rock to change its volume, but not its shape. (7)

Conglomerate   Detrital sedimentary rock containing rounded fragments (such as gravel, pebbles, and boulders) greater than 2 millimeters (0.08 inches) in diameter. (6)

Contact   The boundary between adjacent rocks. (4)

Contact metamorphism   Metamorphism of rocks around an igneous body produced as a result of heating by an adjacent body of magma. (7)

Continental crust   The least dense of Earth's solid layers and the one on which we stand; varies widely in thickness and is predominantly made up of granitic rocks. (9)

Continental drift   The hypothesis that all the continents have moved slowly across Earth's surface. (2)

Continental glacier   (also called an *ice sheet*) A glacier that forms in a continental interior, spreads slowly in all directions, and eventually covers a large area. (15)

Continental interior desert   A desert region that is dry because it is located so far from the sea that any rain it might otherwise receive falls before reaching it. (16)

Continental rifting   The extension and splitting of a continent as the lithosphere is stretched and thinned. (2)

Continental shelf   The broad submerged region bordering a continent where the sea is quite shallow; forms when thinned margins of a rifted continent slowly subside and become flooded and covered by marine sediments. (6)

Continental slope   Structure that marks the edge of a continental shelf and separates areas of land and shallow seas from the deep ocean basins. (2)

Convection   The dominant mechanism of heat flow within Earth in which material adjacent to a heat source is warmed, becomes less dense and rises, and cooler, denser material sinks to take its place. (1)

Convergent plate boundary   (also called *convergent margins*) The boundary along which two plates collide and either oceanic crust is destroyed as one plate is subducted beneath another or two continents collide. (2)

Coral limestone   A type of biochemical sedimentary rock that consists of abundant fossils cemented by crystalline calcite. (6)

Coral reef   A network of ridges and islands built by corals in areas of clear, tropical seawater. (6)

Core   Earth's innermost layer consisting of a liquid outer core and solid inner core. (1)

Correlation   The process by which individual rock layers can be traced from one place to another on the basis of their distinctive characteristics, such as the fossils they contain. (8)

Covalent bond   A bond created when two or more atoms share electrons. (3)

Creep   (also called *solid-state flow*) The extremely slow movement by which solid materials can flow (11); the continuous downslope movement of soil and weathered bedrock under the force of gravity; the slowest but the most important type of mass wasting because of the huge volume of material the process moves. (12)

Crevasse   A fissure that occurs above a depth of 60 meters in a glacier, where the glacier is brittle and cracks, showing little in the way of plastic deformation before breaking, cracking, or fracturing. (15)

Cross bed   A sedimentary structure that occurs within a bed; a layer of sediment is deposited at an angle to the main bedding by a current of moving air or water. (6)

Crust   The rigid, rocky, outermost layer of Earth. (1)

Crystal   The perfect, regular geometric shape of a mineral. (3)

Crystal face   The outer surface of a crystal with straight edges and sharp corners. (3)

Crystal form   The geometric arrangement of crystal faces; an external expression of a mineral's internal bonding. (3)

Crystal settling   The process by which minerals sink and accumulate as layers on the floor of the magma chamber because they are denser than the magma from which they are crystallizing. (4)

Crystal structure   The stable, orderly arrangement of ions within a crystal. (3)

Daughter isotope   The final, stable, nonradioactve isotope that remains after a parent isotope breaks down through radioactive decay. (8)

Debris flow   The downslope movement of a slurry of poorly sorted rock fragments, soil, and mud in which more than half the material is sand sized or coarser. (12)

Deep ocean trench   Narrow, curved and elongate depression in the ocean floor that marks the downward bending of oceanic lithosphere into a subduction zone. (2)

Deflation   A lowering of the ground surface due to wholesale removal of loose material by the wind; one of two types of wind erosion (*abrasion* is the other). (16)

Deformation   A general term to describe the process by which rocks change their original position, shape, and/or volume as a result of the application of stress. (10)

Delta   A triangular tract of land that forms at the mouth of a stream where it enters the sea or lake; the stream branches into many smaller streams and abruptly dumps most of its sediment load as its velocity suddenly falls. (13)

Dendritic drainage   Pattern of drainage in which a trunk stream and its tributaries within a single drainage basin resemble the irregular branching of a tree. (13)

Density   The mass of an object divided by its volume; a measure of how tightly an object's constituent atoms are packed. (3)

Depositional coastline   Coastline dominated by features of coastal deposition such as beaches, bars, and barrier islands; characterized by smooth shores and lack of prominent cliffs. (17)

Depositional environment   The physical setting in which sediments are deposited. (6)

Desert   Barren land area that receives minimal precipitation; deserts make up almost one-third of Earth's land surface. (16)

Desert pavement   A layer of large, heavy pebbles and boulders left behind after deflation has removed silt- and sand-sized grains in a desert. (16)

Desertification   The spread of desert conditions into formerly fertile regions. (16)

Detachment   A low-angle fault associated with large-scale extension; typically occurs when normal faults merge downward. (10)

Detrital sedimentary rock   Sedimentary rock formed from the transported and deposited detritus produced by weathering; one of the two main types of sedimentary rock (along with chemical/biochemical sedimentary rock). (6)

Diagenesis   A variety of biological, physical, and chemical processes that sediment undergoes after deposition; responsible for lithification. (6)

Dike   A sheet-like body of intrusive igneous rock that cuts across existing layers in the surrounding rock at an angle; important in transporting magma to Earth's surface. (4)

Diorite   A coarse-grained intermediate plutonic igneous rock generally dominated by plagioclase and hornblende. (4)

Dip   The measurement of the maximum slope of a planar geologic structure; the angle the line of maximum slope makes with the horizontal; one of two parameters (strike is the other) that describes the orientation of all planar surfaces. (10)

Dip-slip fault   A fault in which the direction of slip (movement) is generally parallel to the dip of the fault plane; the movement on such a fault is either down dip (normal fault) or up dip (reverse fault). (10)

Directed pressure   One of two types of pressure (along with confining pressure) that affects metamorphic rocks; directed pressure pushes on a rock with more vigor in one direction than in others, causing a change in a rock's volume *and* its shape. (7)

Disappearing stream   A stream that flows into a sinkhole, leaving a dry valley downstream; may eventually return to the surface as a spring. (14)

Discharge   Groundwater that reaches Earth's surface to form a spring. This term is also used as a measure of the total volume of water in a stream that passes a given location within a given period of time. (13)

Disconformity   An unconformity in which rock strata above and below the unconformity surface are parallel, but evidence of erosion of the underlying strata indicates a time gap between the deposition of the two sequences. (10)

Disseminated deposit   Ore deposit in which metals have permeated the host rock and precipitated within them in a dispersed fashion. (18)

Dissolution   A process by which minerals with weak ionic bonds dissolve in water and break down into their constituent ions. (5)

Dissolved load   The portion of a stream's sediment load that is carried in solution; consists of the soluble products of chemical weathering. (13)

Distributaries   Smaller channels into which a stream branches at its mouth. (13)

Diurnal tidal pattern   One high tide per lunar day. (17)

Divergent plate boundary   The boundary along which two plates move apart and separate and new oceanic lithosphere is created by seafloor spreading; typically initiated by continental rifting. (2)

Divide   Narrow tract of higher ground that separates drainage basins. (13)

Dolostone   A chemical/biochemical sedimentary rock that contains a large percentage of a calcium-magnesium carbonate mineral called dolomite. (6)

Dome   A fold structure in rocks that dips outward from the center in all directions to form a broadly circular bulge. (10)

Drainage basin   The total land area that contributes water to a stream system. (13)

Drift   Material of all sizes that is carried by a glacier and laid down by ice or meltwater. (15)

Drumlin   Elongate, parallel mounds of boulders, sand, and clay formed by the interaction of glaciers with glacial till. (15)

Ductile   Bending, flowing, and smearing of rock when subjected to stress. (10)

Dune   Mound- or ridge-shaped deposits of windblown sand. (16)

Dust devil   Small whirlwinds of air made visible by the dust they carry. (16)

Dynamic metamorphism   Metamorphism that occurs along fractures in Earth's crust where significant movement has taken place; involves crushing or smearing of rocks adjacent to fault zones. (7)

E horizon   A soil layer between the A and B horizons of the soil profile that has undergone substantial leaching of its mineral and organic content; occurs only in older, well-developed soils. (5)

Earthflow   The slow, downslope movement of a tongue-shaped, fine-grained mass of clay-rich soil and weathered bedrock along a surface more or less parallel to the ground. (12)

Earthquake focus   The point in Earth's interior at which an earthquake is generated. (2)

Ebb current   Accompanies a falling tide and flows seaward. (17)

Eclogite facies   The metamorphic facies at which rocks containing garnet and omphacite become stable; rocks in this facies are usually unfoliated and form under conditions of extremely high pressure and moderate temperature; usually associated with subduction zones. (7)

Elastic deformation   Occurs when the change in a rock's shape and volume in response to a stress is temporary so that the rock returns to its original form once the stress is removed; promoted by low temperatures and pressures and high strain rates. (10)

Elastic rebound theory An explanation for the origin of earthquakes: earthquakes occur when stress builds up on locked faults, straining the rocks on either side until they abruptly rupture and snap back to their original positions, sending out seismic waves. (11)

Electron A negatively-charged subatomic particle with almost no mass; electrons orbit the nucleus of an atom. (3)

Element A substance that cannot be broken down using chemical methods. All atoms of an element have the same number of protons. (3)

Emergent coastline Coastline that has been uplifted or has experienced a fall in the sea level. (17)

Eon The longest subdivision of the geologic time scale; some have lengths of billions of years. (1, 8)

Epicenter The point on Earth's surface directly above the focus of an earthquake. (11)

Epoch An interval of geologic time that is a subdivision of a period. (8)

Era Time interval in the geologic time scale with a duration of hundreds of millions of years that is a subdivision of an eon; separated by major extinction events. (8)

Erosion The process that loosens, dissolves, and transports rocks and sediment by mobile agents, such as wind, water, and ice. (6)

Erosional coastline Coastline dominated by features of coastal erosion, such as sea cliffs, headlands, and wave-cut platforms; rugged and irregular, with narrow pocket beaches backed by steep cliffs. (17)

Erratic A boulder transported and deposited by glaciers that differs from the local bedrock. (15)

Esker Steep-sided, sinuous ridge made of stratified drift deposited by meltwater streams that flow in ice tunnels within or beneath a glacier. (15)

Estuary Partly enclosed tidal area at the mouth of a stream where freshwater and seawater mix. (17)

Evaporite mineral A nonsilicate rock-forming mineral produced by the evaporation of saltwater; often the product of the evaporation of ancient seas. (3)

Evaporites A family of chemical sedimentary rocks that form by the precipitation of salts left behind by the evaporation of seawater. (6)

Exfoliation joint A gently curved expansion fracture that is broadly parallel to the land surface; common in granites and formed by expansion due to reduced pressure as a result of uplift and erosion of overlying rocks (*see* **joint**). (5, 10)

Exocomet A comet that originates in regions beyond the Solar System. (21)

Exoplanet A planet that is located outside the Solar System. (21)

Fabric Describes the geometric arrangement of grains in metamorphic rocks; used to describe their texture. (7)

Failed rift Occurs when a rift valley does not successfully open into a new ocean because initial spreading is not sustained or because it shifts to another location. (2)

Fall A mechanism of mass wasting that occurs when material detaches from a precipitous slope and free-falls to the ground, either directly or in a series of bounds. (12)

Fault A fracture in rock along which significant movement has occurred. (10)

Feldspar group The most abundant non-ferromagnesian silicate minerals in Earth's crust. (3)

Felsic Category of igneous rock comprising more than 65% silica, which is low in iron and magnesium. (4)

Ferromagnesian silicate Category of silicate minerals that contains significant amounts of iron (*ferrum* in Latin) and magnesium. (3)

Field relationship Describes how igneous rocks relate to one another and to their neighbors. (4)

Fine-grained A characteristic of the minerals in igneous rock; very small minerals form when lava is cooled rapidly. (4)

Firn A granular variety of ice. (15)

Fissure eruption Occurs when the magma in dikes reaches the surface and eruption occurs along a fracture or sets of fractures. (4)

Fjord A glacial valley that has been flooded by the sea. (15)

Flexural basin A type of sedimentary basin that forms at convergent plate margins where one continental plate is thrust over another and the additional weight causes the adjacent crust to flex downward. (6)

Flood basalt Basalt that erupts on land in huge volumes as a result of voluminous melting in the underlying mantle. (4)

Flood current Accompanies a rising tide and flows inland. (17)

Floodplain The land area inundated whenever a stream overflows its banks during flooding. (13)

Flow Mass wasting that occurs when unconsolidated and usually water-saturated materials move downslope in a fluid-like fashion, often spreading out in toe-like fans. (12)

Flow texture A pattern that occurs when phenocrysts are extruded with molten lava at Earth's surface and become aligned like logs floating in a stream. (4)

Flute cast Sand-filled, spoon-shaped hollow in a mud layer scooped out by a turbidity current; a sedimentary bottom structure visible at the base of the overlying bed. (6)

Focus The source of an earthquake; occurs in brittle rocks below Earth's surface on a locked fault that suddenly lets go. (11)

Fold A bend in layered rock formed deep in Earth's crust in response to compression; the product of plastic deformation that shortens the rock layer. (10)

Fold hinge The line joining the points of greatest curvature in a folded layer; one of two measurements (the other is the *axial plane*) that defines a fold's geometry. (10)

Foliated metamorphic rock A metamorphic rock in which the mineral grains have a preferred, layered orientation as the result of directed pressure; one of two groups of metamorphic rock (nonfoliated is the other). (7)

Foliation The planar fabric produced in metamorphic rock by the alignment and segregation of minerals as they grow under conditions of directed pressure. (7)

Food chain A concept that describes the movement of energy and nutrients through the biosphere from energy producers to energy consumers (i.e., from plants to herbivores and carnivores). (5)

Fossil The petrified remains of ancient life or an impression of an ancient life form. (1)

Fossil fuels Fuels (such as oil, natural gas, and coal) consisting of hydrocarbons; formed from fossilized organic material. (19)

Fractional crystallization   A process in the formation of igneous rock during which early-crystallizing minerals are separated and prevented from reacting with the magma, thus changing the magma's chemical composition. (4)

Fractionation   A variety of processes that occur during the formation of igneous rock whereby the composition of magma changes as it cools and crystallizes. (4)

Fracture   A surface along which rupture occurs in a rock near Earth's surface; a mineral possesses a fracture when it breaks along irregular surfaces rather than cleavage planes. (3)

Fringing reef   A reef that directly borders the shore of a landmass, forming a table-like surface; grows oceanward toward the food supply. (17)

Frost wedging   An expansion caused when water located in the cavities, joints, and cracks in rocks freezes to form ice and expands in volume; causes a rock to mechanically disintegrate. (5)

Fumarole   A steam vent formed when heated groundwater is converted to steam and escapes to the surface through fissures. (14)

Gabbro   A coarse-grained plutonic rock dominated by plagioclase and pyroxene; one of the most common mafic igneous rocks. (4)

Garnet group   A group of minerals with an isolated silicate structure like that of the olivine group, but with different compositions; occurs most commonly in metamorphic rocks. (3)

Gems   Crystals that have intrinsic value because of their beauty, durability, rarity, or size. (3)

Geologic time scale   A calendar that divides Earth history into blocks of time (eons, eras, periods and epochs); places natural events in the sequence in which they occurred and seeks to establish the exact time when these events took place. (1)

Geology   The study of planet Earth: the materials of which it is made, the processes that act on these materials, and the history of the planet and its lifeforms. (1)

Geosphere   The solid Earth; consists of the rocky outer layer (the lithosphere) and its hot interior. (1)

Geothermal energy   A renewable source of energy generated when hot rocks close to the surface are used to heat water, either to produce steam for generating electricity or to provide heat for buildings; once used, the condensed steam and waste water are pumped underground and reheated. (19)

Geyser   Cyclic and explosive emissions of superheated groundwater from a hot spring. (14)

Gigapascal   A unit of pressure that equals a billion pascals. (7)

Glacial budget   The annual balance between accumulation (gain in ice) and ablation (loss of ice) in any type of glacier. (15)

Glacial terminus   The front of a glacier. (15)

Glacial trough   A broad, steep-sided valley created when the ice flowing out of a cirque and into a V-shaped stream valley scours out the existing valley and makes it wider, deeper, and straighter (U-shaped). (15)

Glaciation   The processes of glacial erosion and deposition. (15)

Glacier   A moving mass of ice that forms on land from the slow compaction and recrystallization of snow. (15)

Glacier bed   The material beneath a glacier. (15)

Global warming   An increase in the average temperature of Earth's climate system. (20)

Gneiss   A coarse-grained metamorphic rock with alternating layers of different mineral composition. One of four types of foliated metamorphic rocks (along with slate, phyllite, and schist). (7)

Gneissic foliation   A type of foliation in metamorphic rock in which minerals are separated into distinct light and dark layers; characteristic of gneisses. (7)

Gondwana   A supercontinent consisting of the united southern continents (South America, Africa, Antarctica, Australia and India). (2)

Graben   A crustal block in a rift that has been displaced downward between two inward dipping normal faults. (10)

Graded bedding   A sedimentary structure that occurs within a bed; characterized by a systematic decrease in particle size from the base to the top. (6)

Graded stream   A stream that has reached equilibrium flow such that sediment input is balanced by sediment removal. (13)

Granite   A coarse-grained felsic rock dominated by quartz and feldspar; the most familiar igneous rock. (4)

Granulite facies   The metamorphic facies in which gneisses become common; rocks in this facies are usually foliated and consist of minerals that form under conditions of high temperature and pressure. (7)

Gravel   Somewhat rounded particles in detrital sedimentary rocks with diameters greater than 2 millimeters (0.08 inches). (6)

Gravity anomaly   Variation in Earth's gravity from its predicted value. (2)

Graywacke   A sandstone in which more than 15 percent of the rock is composed of a matrix dominated by clay minerals. (6)

Great Red Spot   The largest of numerous storm systems in Jupiter; the planet's most famous feature has a diameter of almost 38,000 kilometers (23,560 miles), has been raging for at least 400 years, and is driven by heat from the planet's interior. (21)

Greenhouse conditions   A period of extreme warmth in Earth's climate. (20)

Greenhouse effect   Occurs when greenhouse gases allow shorter-wave solar radiation to enter Earth's atmosphere, but partially block the escape of the longer-wave solar radiation reflected off Earth's surface, thus causing the atmosphere to trap heat. (20)

Greenhouse gas   Gases, such as water vapor, carbon dioxide and methane, that permit shorter-wave solar radiation to enter Earth's atmosphere but partially block the escape of the longer-wave radiation reflected off Earth's surface. (20)

Greenschist facies   The most common metamorphic facies characterized by a variety of green minerals; rocks in this facies are usually foliated and form under conditions of low temperatures and pressures. (7)

Groundmass   The fine-grained material found in igneous rock or volcanic glass; consists of interlocking crystals or glass. (4)

Groundwater   Water stored in rocks and soils below the ground; the largest reservoir of *accessible* fresh water on Earth and a nonrenewable resource. (14)

Habit   A mineral's characteristic outward appearance. (3)

Hadean Eon   The division of geologic time that covers the interval from the formation of planet Earth to the appearance of its oldest rocks. (8)

Half-graben   A down-dropped block with an asymmetric shape; down-faulted on one side. (10)

Half-life   The fixed length of time it takes for the number of parent atoms of a radioactive isotope to decay to exactly half that number; a measure of an isotope's rate of decay. (8)

Hanging valley   Produced when tributary valleys previously occupied by relatively small alpine glaciers join the main, deeper glacial trough; commonly the sites of spectacular waterfalls. (15)

Hard water   Water that contains large quantities of bedrock-derived dissolved ions such as calcium, magnesium, and iron. (14)

Headland   An erosion-resistant promontory that juts out into the sea due to the uneven retreat of a sea cliff. (17)

Headward erosion   Process by which a stream increases its length by eroding upslope at the stream head; occurs because runoff becomes focused into a channel at the head of a stream, which greatly increases its erosive power. (13)

Horn   A pyramidal peak formed when three or more cirques develop on the sides of the same mountain and expand by headward erosion; an alpine glacial landform. (15)

Hornfels   A fine-grained, nonfoliated metamorphic rock produced by recrystallization adjacent to a cooling igneous rock. (7)

Horst   A high-standing crustal block bound by outward-dipping normal faults along which the adjacent blocks have been down-dropped. (10)

Hot spring   Occurs when heated groundwater rises rapidly and naturally to the surface where its temperature is significantly higher than that of other surface water. (14)

Hotspot   Small, isolated area of Earth's crust below which a mantle plume impinges on the base of the lithosphere and gives rise to localized areas of higher than average heat flow and volcanism; may be located far from plate boundaries. (9)

Humus   Organic-rich soil immediately below the surface that is rich in nutrients released by the decomposition of plant roots, bacteria, fungi, and animals. (5)

Hydration   Occurs when a mineral absorbs water into its crystal structure which then expands. (5)

Hydraulic fracturing (fracking)   a controversial technology used to extract shale gas that involves shattering the shale around a drill hole to improve the rock's permeability so that the gas trapped within the rock can be released; allows otherwise uneconomic reserves of natural gas to be commercially exploited. (19)

Hydraulic head   A measure of water pressure created by differences in the elevation of the water table. (14)

Hydroelectric power   Electricity generated by running water, a renewable energy resource. (19)

Hydrologic cycle   The motion of water in Earth's atmosphere and on its surface. Water evaporates from oceans and enters the atmosphere as moisture that is carried by the wind until it condenses to form clouds and precipitation. Most precipitation falls into the oceans, but some falls on land and, if it is not absorbed by the ground, eventually returns to the oceans. (1)

Hydrolysis   A reaction in which the hydrogen ions present in water extract loosely bound ions from a mineral's crystal structure and in the process form a new mineral. (5)

Hydrosphere   Earth's water, found in oceans, lakes, or rivers, frozen in ice and snow, and trapped underground in soils and rock fractures. (1)

Hydrothermal deposit   A concentration of metallic minerals that form from the circulation and cooling of magmatically heated water. (18)

Hydrothermal metamorphism   Metamorphism that occurs when rocks react with adjacent hot circulating fluids. (7)

Hypothesis   A step in the scientific method; a tentative explanation of data gathered by scientists who are investigating a process or phenomenon. (1)

Icehouse conditions   A period of extreme cold in Earth's climate. (20)

Igneous rock   One of the three major rock groups (along with sedimentary and metamorphic); forms from the cooling and crystallization of molten rock (magma). (1, 4)

Incised meander   A meander that retains its shape as it is deepened by downcutting, so that the stream comes to lie within a deep meandering valley cut into bedrock and with no floodplain; characteristic of rejuvenated streams. (13)

Inclusion   A piece of one rock unit contained in another, for example, a remnant of wall rock in an igneous rock that was unmelted by the magma, or a clast contained in a detrital sedimentary rock. (4)

Index mineral   The specific mineral that is characteristic of a given metamorphic zone. (7)

Industrial material   All naturally occurring inorganic deposits of economic value, other than those from which metals are extracted; includes nonmetallic mineral deposits and raw materials for manufacturing products such as chemicals, concrete, and wallboard. (18)

Infiltration   The process by which water seeps into sediment. (12)

Inner core   The solid part of the Earth's center or core. (1)

Inorganic   Compounds that lack carbon derived from living matter. (3)

Inselberg   Isolated mountain remnant produced by desert erosion that rises conspicuously from the gently sloping pediment that surrounds it. (16)

Intermediate   A category of igneous rock comprising 52% to 65% silica. (4)

Intermittent stream   Short-lived, fast-flowing stream that flows occasionally in a desert after torrential rain. (16)

Intrusive rock   Coarse-grained igneous rock that forms below Earth's surface as magma slowly cools and solidifies; cuts across preexisting rock layers as it rises through the crust. (4)

Ion   An atom that has an electric charge because of an imbalance in the number of protons and electrons. (3)

Ionic bond   A chemical bond that results from the transfer of an electron from one atom to another. (3)

Island arc   A curved chain of volcanic islands adjacent to a deep ocean trench; the product of magmatism produced above a subduction zone. (9)

Isograd   A line that represents the location on Earth's surface of the metamorphic reaction that produces each index mineral. (7)

Isostasy   The balance reached by Earth's crust and the lithospheric mantle as they float upon the denser, more pliable asthenosphere. According to this principle, the elevation of the crust depends upon its thickness and the density contrast with the asthenosphere. (9)

Isotopes   Atoms of the same element that have the same number of protons, but different masses due to different numbers of neutrons in the nucleus. (8)

Joint   A regularly spaced fracture in a rock outcrop along which there has been no appreciable movement; one of the most common geologic structures and one of two main types of fractures (the other is a *fault*). (10)

Joint set   Broadly parallel, evenly spaced groups of joints. (10)

Jovian planet   A planet in the outer part of the Solar System (from Jupiter to Neptune) with a significantly lower density than the terrestrial planets; essentially a giant, gaseous body with a poorly defined surface. (21)

Kame   Low, conical, isolated hill or steep-sided mound of stratified drift originally deposited by a meltwater stream in a depression on the surface of a glacier and left behind when the glacier melts; a depositional landform. (15)

Karst   Surface and below-ground features produced when significant amounts of bedrock are dissolved. (14)

Karst topography   The surface expression of a karst, characterized by a very irregular landscape that includes broad plains with tall monoliths, circular depressions (*sinkholes*), and disappearing streams; caves often develop beneath a karst topography. (14)

Kettle   Steep-sided depression produced when a receding glacier leaves behind a large block of stagnant ice that melts and leaves a space in the glacial drift that can subsequently fill with water. (15)

Kilobar   A unit used to measure the enormous pressures that exist below Earth's surface; equals 1000 bars (10 gigapascals). (7)

Komatiite   An ultramafic lava; not erupted by modern volcanoes but important early in Earth history. (4)

Kuiper Belt   Location of a vast number of ice-rock bodies that orbit the Sun at the outermost fringes of the Solar System. (21)

Lahar   A special type of fast-moving mudflow that occurs on volcanoes that are capped by snow and ice or exist in wet climates; composed of hot, unconsolidated, fine-grained pyroclastic deposits that are mobilized by meltwater or rainfall. (12)

Laminar flow   In a stream, the lines along which the water moves are parallel to each other and to the sides of the stream channel so the flow is smooth; typical of slow-moving streams with smooth channels. (13)

Lapilli   Fine-grained particles in pyroclastic deposits. (4)

Laterite   A rust-colored soil rich in iron and aluminum, common in wet, tropical climates. (5)

Lava   Magma that reaches Earth's surface and flows from volcanoes. (4)

Lava fountain   A jet of incandescent lava that shoots out from a volcanic vent or fissure; results from rapid expansion and escape of gas bubbles trapped in fluid magma as the magma reaches Earth's surface. (4)

Lava lake   Occurs when basaltic lava pools within a volcanic caldera to form a molten lake. (4)

Lava tubes   The complex labyrinth of passages that form in the interior of a cooling lava flow as the lava moves through its congealing surroundings; when eruption ceases, the lava drains from the tubes, leaving them hollow. (4)

Law   An expression or statement that accounts for a natural process or processes from which there is no known deviation. (1)

Layered mantle convection   A model for mantle convection in which more vigorous circulation in the upper mantle is thought to occur independently of slower convection in the lower mantle so that the convective systems are decoupled across the transition zone. (11)

Left-lateral fault   A strike-slip fault in which the fault block on the far side of the fault line has moved to the left relative to the block on the near side. (10)

Limestone   A common sedimentary rock primarily composed of the mineral calcite ($CaCO_3$) that can be either chemical or biochemical in origin. (6)

Liquefaction   A phenomenon that occurs when quick clay or unconsolidated, water-saturated soil and sediment are transformed into materials that act like fluids when disturbed by vibrations usually generated by an earthquake; can cause serious damage. (12)

Lithification   The process of compaction and cementation through which loose, unconsolidated sediment binds together to form solid, coherent sedimentary rock. (6)

Lithosphere   Earth's rigid outermost layer, which is broken into huge moving slabs, or tectonic plates; consists of the crust and the uppermost mantle and floats on the asthenosphere. (9)

Lithospheric mantle   That part of the upper mantle immediately below the crust that is cool enough to be rigid and so forms part of a tectonic plate. (9)

Little Ice Age   An anomalously cool period between AD 1500 and 1800 that was most pronounced in the seventeenth century. (20)

Load cast   A sedimentary bottom structure commonly formed after the passage of a turbidity current; occurs when the coarser sediment first deposited by the turbidity current sinks into the underlying mud to form bulbous protrusions, while the mud rises into sand in relatively narrow ridges. (6)

Loess   A thick, homogeneous, unlayered blanket of fine-grained silt and clay that has been carried great distances by the wind and deposited successively after dust storms. (16)

Longitudinal dune   A narrow, linear sand ridge with a symmetrical profile that is aligned *parallel* to the prevailing wind direction. (16)

Longitudinal profile   The downslope path of a stream. (13)

Longshore current   Current produced within the surf zone by incoming waves approaching the shore at an angle; carries significant quantities of sediment in suspension. (17)

Longshore drift   The slow, lateral movement of significant quantities of sediment suspended in the turbulent water in the surf zone. (17)

Lower mantle   The lower part of the middle layer of Earth's interior where solid rocks are hotter and denser than those of the upper mantle and can flow, but are not molten; extends from about 700 kilometers to 2900 kilometers in depth. (11)

Low-velocity zone   A discontinuous region within the asthenosphere that is associated with decreased seismic wave velocities likely caused by the presence of small amounts of melt; occurs between depths of 100 and 250 kilometers. (11)

Lunar highlands   Higher-elevation, light-colored areas of the Moon's surface. (21)

Luster   Refers to the way in which a mineral reflects light. (3)

Mafic   A category of igneous rock that is 45% to 52% silica and is high in iron and magnesium. (4)

Magma Molten or partially molten rock found beneath Earth's surface. (1, 2, 4)

Magma chamber A large cavern deep underneath Earth's surface filled with magma. (4)

Magma mixing The mixing of magmas of different compositions. (4)

Magmatic segregation The process by which ore deposits form in cooling magmas; the physical separation of the magma's chemical components. (18)

Magnetic anomaly The difference between measured and expected strength of Earth's magnetic field at any location. (2)

Magnetic inclination The angle, relative to the horizontal, of the lines of Earth's magnetic field. (2)

Magnetic polarity reversal A global event during which Earth's magnetic field reverses itself. (2)

Mantle The middle of Earth's three layers; the hot but solid layer beneath the crust that separates the crust from its core. (1)

Mantle drag One of three types of force (along with ridge push and slab pull) involved in the movement of plates; occurs when a plate is carried by convective circulation in the asthenosphere like a raft in a moving stream. (9)

Mantle plume Giant column of hot, solid, and buoyant material that rises from deep within Earth's interior; induces melting in the mantle that produces magma; gives rise to hotspots. (2, 4, 9)

Mantle wedge The wedge-shaped area of mantle above a subduction zone where magma is produced. (9)

Marble A nonfoliated rock produced from the metamorphism of limestone and dolostone. (7)

Mare (pl.: *maria*) Dark plains on the Moon's surface that are waterless depressions and floored by basalts. (21)

Marine regression Occurs when the retreat of the sea due to falling sea level causes the relative distribution of sand, mud, and lime to migrate offshore so that the rock record shows sandstone on top of shale and shale on top of limestone. (6)

Marine transgression Occurs when the landward advance of the sea due to rising sea level causes the relative distribution of sand, mud, and lime to migrate onshore so that the rock record shows limestone on top of shale, and shale on top of sandstone. (6)

Mass wasting The downslope movement of materials under the direct influence of gravity; classified according to how the material moves, what it consists of, and how fast it travels. (12)

Matrix The finer-grained material between larger grains in a rock. (6)

Meandering stream A stream for which the gradient of the channel is gentle, so that it develops sinuous loops (meanders) that carve out a broad, flat floodplain on the valley floor. (13)

Mechanical enrichment Concentration of stable metals or heavy minerals that are not easily moved by flowing water and so are preferentially deposited in locations where the water velocity slackens. (18)

Mechanical weathering The physical process by which a rock disintegrates by breaking into smaller fragments; does not change a rock's chemical composition. (5)

Mesa A broad, flat-topped mountain remnant bounded on all sides by steep cliffs and produced in arid regions by the irregular retreat of vertical cliffs where relatively easily weathered rocks are capped by more resistant ones. (16)

Mesozoic Era Meaning "middle life," the middle era of the Phanerozoic Eon, from 252 to 66 million years ago, characterized as the age of the dinosaurs; separated the preceding (Paleozoic) and following (Cenozoic) eras by major mass extinctions. (8)

Metal A substance composed of one or more metallic elements that is shiny, hard and malleable if solid, and a good conductor of heat and electricity. (18)

Metallic bond When atoms combine to form a piece of metal, each atom gives up an outer electron. Collectively, the atoms form a layer of positive ions, with a freely moving, common cloud of shared electrons swarming around them; the strength of a metal is directly related to the strength of its metallic bonds. (3)

Metallic mineral A mineral in which metals combine with themselves or with elements such as oxygen, sulfur, or carbon to form mineral oxides, sulfides, or carbonates. (18)

Metamorphic facies An association of metamorphic rocks that were metamorphosed under similar conditions of temperature and pressure; all the minerals in a facies are stable within a certain range of pressure and temperature. (7)

Metamorphic grade A measure of the intensity of the temperatures and pressures under which metamorphic rocks are formed. (7)

Metamorphic rock One of the three basic rock groups (along with sedimentary and igneous rocks); a preexisting rock that has been subjected to heat, pressure, and fluids, so that its mineral composition and texture are changed significantly from its original state; this transformation occurs in the solid state below Earth's surface. (1, 7)

Metamorphic zone Systematic variation in the mineral content of metamorphic rocks caused by differences in the temperature and pressure conditions; each zone is characterized by the appearance of a new metamorphic mineral. (7)

Metamorphism The process of change in the texture and mineral composition of a rock subjected to heat, pressure, and/or chemically active fluids; occurs while the rock is still solid. (7)

Meteor Small piece of extraterrestrial material that burns up upon entering Earth's atmosphere. (21)

Meteorite Extraterrestrial material made up of rock or metal alloy that does not completely burn up on entry into Earth's atmosphere and falls to Earth's surface. (21)

Mica group A set of minerals made up of sheets of silicate tetrahedra connected to one another by cations. (3)

Microcontinent A small continental block that may reside in an ocean and may be accreted to an active continental margin. (10)

Mid-ocean ridge A submarine mountain chain buoyed by heat that marks the site where oceanic lithosphere is being formed at a divergent plate boundary; the elevated ridges rise about 3 kilometers (1.9 miles) from the deep ocean floors and girdle the Earth. (2, 9)

Migmatite A metamorphic rock that forms when gneiss is heated to the point of partial melting; indicative of very high metamorphic grade. (7)

Milankovitch cycle Describes how the position of Earth's axis and its orbit around the Sun vary in predictable cycles over periods lasting 100,000 years; results in climate cycles like the advance and retreat of continental ice sheets over the past 2 million years. (20)

Mineral The building block of rocks; a naturally occurring, inorganic crystalline solid. The internal structure is an orderly arrangement of atoms, ions, or molecules. (1, 3)

Mineral deposit Concentration of metals and metallic and nonmetallic minerals that are mined for economic benefit. (18)

Mineraloid A solid material that differs from a mineral because it lacks a regular atomic structure and is therefore not crystalline. (3)

Mixed tidal pattern Two high tides and two low tides each lunar day, but each of significantly different heights. (17)

Moho (also called the Mohorovičić Discontinuity) The sharp boundary between crust and mantle characterized by significant change in seismic velocity. (9)

Molecule A stable combination of two or more atoms linked together by chemical bonds; the smallest part of a compound that retains the properties of the compound and can participate in a chemical reaction. (3)

Monolith A landform composed of hard, relatively insoluble rock that stands out above a broad plain. (14)

Moraine A sediment mound deposited by glaciers. (15)

Mudcrack A sedimentary structure that occurs on the surface of a mud-rich layer; forms when mud is exposed to air and shrinks as it dries out to produce a polygonal pattern of cracks. (6)

Mudflow A type of mass wasting featuring downslope movement of water-saturated silt or clay-sized sediment. (12)

Mudstone A very fine-grained detrital sedimentary rock dominated by clay-sized particles. (6)

Muscovite A non-ferromagnesian sheet silicate that incorporates aluminum and water into its crystal structure; the sheet-like arrangement of silicate tetrahedra results in a strong cleavage. (3)

Mylonite A fine-grained foliated metamorphic rock produced by shearing at elevated temperature during dynamic metamorphism. (7)

Native metal A type of ore mineral made up of a single metallic element. (3)

Natural gas Gaseous petroleum consisting predominantly of methane; the "cleanest" of the fossil fuels. (19)

Natural levee A ridge of coarse-grained sediment adjacent to a stream channel deposited during flood events. (13)

Neap tide Diminished tidal range; occurs when the Sun and Moon are in maximum misalignment with Earth (first- and third-quarter phases of the Moon). (17)

Negative feedback Natural process that counterbalances an existing trend by dampening it; slows a process down or causes a reversal in trend (*see* **Positive feedback**). (20)

Neutron One of the particles in an atom's nucleus (a proton is the other); a neutron carries no charge and has approximately the same mass as a proton. (3)

Nonclastic texture Describes the texture of chemical and biochemical sedimentary rocks that are made up of interlocking crystals and not of broken fragments (clasts). (6)

Nonconformity Occurs when a plutonic igneous rock or a metamorphic rock is exhumed by uplift and erosion prior to being directly overlain by younger sedimentary strata. (10)

Non-ferromagnesian silicate A type of silicate mineral that does not contain iron or magnesium. (3)

Nonfoliated metamorphic rock One of two main types of metamorphic rock (the other is foliated); has a random fabric because its mineral grains have no preferred orientation; forms in the absence of directed pressure or because the rock lacks aligned minerals. (7)

Nonmetallic mineral Minerals, like gypsum, sulfur, and halite, that may be commercially exploited but not as a source of metals. (18)

Nonrenewable resource Resource (such as mineral deposits) that is consumed faster than nature can replenish it. (18, 19)

Nonsilicate mineral A mineral that does not contain silicon, such as calcite. (3)

Normal fault A type of dip-slip fault in which the rocks above the fault surface (hanging wall) move *down* relative to the rocks below the fault surface (footwall); commonly the result of extensional forces (*see* **Reverse fault, Thrust fault**). (10)

Normal magnetic polarity Earth's present magnetic field; the north ends of all compass needles point toward the north magnetic pole. (2)

Nuclear fission Occurs when the nucleus of a heavy radioactive isotope, such as uranium, is split by neutron bombardment, releasing a great amount of energy. (19)

Nucleus An atom's dense central unit, which contains protons and (with the exception of hydrogen) neutrons and is orbited by electrons. (3)

O horizon The uppermost layer of the soil profile; rich in loose or partly decayed organic matter. (5)

Oasis A vegetation-rich region surrounded by desert, occurring where natural springs provide surface water. (14)

Ocean wave An undulation in ocean water that moves across the ocean's surface; produced when energy from the wind is absorbed by the water's surface. (17)

Oceanic crust Formed by seafloor spreading, consists overwhelmingly of basaltic rocks with an average thickness of 7 km; overlain by a thin blanket of marine sediment. (9)

Oil A liquid petroleum formed from the remains of animal and plant matter by heat and pressure within a buried, oxygen-deficient environment. (19)

Oil shale Organic-rich, fissile, mud-rich rock that yields petroleum when heated. (19)

Oil trap A geologic structure that allows oil or natural gas in an underground reservoir rock to collect in one area. (19)

Olivine group Ferromagnesian silicate minerals characterized by isolated tetrahedra that are linked by iron and/or magnesium ions. (3)

Ooid A spherical grain consisting of concentric layers of a chemically precipitated mineral (typically calcite) around a central fragment, such as a shell particle. (6)

Oolitic limestone A chemical sedimentary rock formed from cementation of ooids. (6)

Ophiolite A fragment of oceanic lithosphere (oceanic crust) that has escaped subduction and has been uplifted or thrust onto the continent. (9)

Ore Rocks containing economically exploitable quantities of metal-rich minerals. (18)

Ore deposit A mineral deposit that can be exploited economically; a mineral deposit is only an ore deposit if the price of the ore is high enough to make its extraction commercially viable. (18)

Ore mineral A mineral from which a metal can be economically extracted; typically a metal sulfide or oxide. (3)

Organic A substance that was part of a living organism; contains carbon-hydrogen bonds. (3)

Original horizontality One of three fundamental principles for determining a rock's relative age (along with superposition and lateral continuity); holds that sedimentary strata were deposited as flat-lying layers. Hence, inclined layers were once horizontal and were tilted *after* they were deposited. (8)

Original lateral continuity One of three fundamental rules for determining a rock's relative age (along with original horizontality and superposition); holds that flat-lying sedimentary strata would have initially extended laterally in all directions. Hence, identical strata now separated by an erosional feature, such as a valley, were originally continuous. (8)

Orogeny The process of mountain building by tectonic forces, resulting in folding, faulting, and associated igneous and metamorphic activity. (10)

Outcrop A body of rock exposed at Earth's surface. (1)

Outer core The region beneath Earth's mantle composed mostly of liquid iron alloy, enveloping a solid inner core. (1)

Outwash plain A broad, gently sloping apron of stream-deposited sands and gravels fed by glacier meltwater. (15)

Oxidation The process by which an element loses electrons during a chemical reaction. (5)

Paleocurrent The direction of flowing water in the geologic past; typically recorded by sedimentary sructures. (6)

Paleomagnetism The study and measurement of Earth's magnetic field in the geologic past, as preserved in minerals and rocks. (2)

Paleowind The direction of the wind in the geologic past; typically recorded by sedimentary structures. (6)

Paleozoic Era Meaning "ancient life." The oldest era in the Phanerozoic Eon, from 541 to 252 million years ago, characterized by abundant marine, shelly fossils; separated from the following Mesozoic Era by a major mass extinction. (8)

Pangea Meaning "all lands." A single landmass or supercontinent that amalgamated about 300 million years ago and began to break up about 200 million years ago. (2)

Parabolic dune A U-shaped dune with a concave slip face and crescent-shaped "horns" that point upwind; common on coastlines with partial vegetation, strong onshore winds, and an abundant supply of sand. (16)

Paradigm A unifying principle that forms the framework of a scientific discipline. (2)

Parent isotope The original (unstable) radioactive isotope that breaks down through radioactive decay. (8)

Partial melting A process of incomplete melting in a rock, whereby those minerals with the lowest melting temperatures melt, whereas those with higher melting temperatures do not; occurs when a decrease in pressure or an increase in temperature destabilizes, expands, and breaks the chemical bonds of some of the minerals in a source rock. (4)

Passive continental margin A continental shelf that forms when thinned margins of a continent slowly subside, settle, flood, and become covered by marine sediments following continental rifting; these margins come to lie far from any plate boundary because the ocean that opened when the margins first formed widens. (2, 9)

Peat A water-saturated, organic-rich humus produced by bacterial decay of plant remains that is a precursor to the formation of coal; often forms in warm, humid swamplands that border marine basins. (19)

Pebble A gravel particle that is between 2 and 64 millimeters (0.075–2.5 inches) across. (6)

Pedalfer A type of soil that forms in temperate climates; it is typically rich in clay minerals and is red due to its high iron content. (5)

Pediment A gently sloping, sediment-strewn bedrock erosion surface produced in desert environments at the base of a retreating cliff. (16)

Pedocal A type of soil that forms in arid climates; it contains little organic matter and is saturated with salts, especially calcite. (5)

Peridotite A coarse-grained ultramafic rock with more than 40% olivine. (4)

Period Time interval in the geologic time scale that is a subdivision of an era; the beginning and end of a period is determined, respectively, by the advent and demise of fossil life forms; typically has a duration of tens of millions of years. (1, 8)

Permafrost Permanently frozen ground that underlies large areas at high latitudes. (12)

Permeability A measure of how well a fluid flows through the network of pores and cracks in sedimentary material. (12, 14)

Petroleum Naturally occurring organic gaseous, liquid, or semisolid substance derived from the remains of microscopic organisms and consisting of hydrocarbons (compounds of hydrogen and carbon). (19)

Phanerozoic Eon The youngest of the four eons of the geologic time scale, the other three being Hadean, Archean, and Proterozoic; started 541 million years ago with the first appearance of shelly fossils. (8)

Phase change Changes in the internal structure of a mineral but not in its chemical composition; occurs within Earth's interior where changes in temperature and/or pressure cause minerals to adopt new, more stable crystal structures. (11)

Phenocrysts Relatively large minerals surrounded by much-finer-grained material (groundmass) in an igneous rock. (4)

Photosynthesis The process by which organisms in the oceans and vegetation on land convert the radiant energy from sunlight into energy, extracting carbon dioxide from, and introducing oxygen into, the atmosphere as they do so. (4, 5)

Phyllite One of four types of foliated metamorphic rocks (along with slate, schist, and gneiss). Slightly coarser than a slate and characterized by a sheen; produced at low metamorphic grade. (7)

Physical geology   The study of geology that emphasizes the processes that interact at Earth's surface and have shaped its evolution throughout geologic time. (1)

Pillow lava   Forms when basaltic eruptions occur underwater; lava extruded onto the ocean floor rapidly freezes against the cold seawater, producing pillow-like bulges. (4)

Placer deposit   Mechanical concentration of stable metals or heavy minerals (such as gold or diamond) in a sediment that were deposited where the water carrying these minerals abruptly slowed down, for example, where a stream abruptly widens or enters a lake or the sea. (18)

Plastic deformation   Occurs when mineral grains in a rock behave like plastic and change shape and volume in response to a stress and do not return to their original shape when the stress is removed, so the changes become permanent; rocks that deform plastically smear or fold; promoted by high temperatures and pressures and low strain rates. (10)

Plastic flow   A slow, internal process during which the weight of the overlying ice in a glacier causes crystals at the base of the ice pack to change shape, so that the glacier flattens out under its own weight; one of two ways in which glaciers move (*basal sliding* is the other). (15)

Plate   Large slabs of the Earth's rigid outer shell (lithosphere) that are in constant motion and that may be thousands of kilometers across but are only 50 to 150 kilometers (30 to 100 miles) thick. (2)

Plate boundary   The margins where two plates meet; lines along which plates diverge, converge, or slide past each other. (2)

Plate tectonics   A unifying theory that states that Earth's outer shell (lithosphere) consists of moving slabs or plates that interact along their margins; provides an explanation for continental drift and the origin of most of the world's largest geologic features. (2)

Playa   A dry, salt-encrusted lake bed formed when all the lake water evaporates; found on a desert floor. (16)

Playa lake   A temporary lake that forms in low areas of a desert floor when torrential desert cloudbursts produce so much runoff water that the ground cannot absorb it all; eventually dries up to form a playa. (16)

Pleistocene Epoch   The geologic time interval that starts with the beginning of the Ice Age (the period of glaciation across North America, Europe, and Siberia) about 2.6 million years ago and ends with the last glacial retreat about 12,000 years ago. (15)

Plucking   The loosening and removal of rock fragments from a glacier bed; one of two types of glacial erosion (*abrasion* is the other). (15)

Plume   (*see* **Mantle plume**)

Plume (mantle convection)   Possible explanation for mantle convection proposing that plumes rise from a thermally unstable layer at the core–mantle boundary through the entire mantle until they reach the base of the lithosphere. There they spread outward, carrying plates away from the plume center. The return flow in the convection process is thought to be focused at sites of subduction, where cold, dense oceanic lithosphere descends into the mantle. (11)

Pluton   A body of intrusive igneous rock, ranging in size from tens of meters to hundreds of kilometers across, that forms when the contents of a large magma chamber cool and crystallize. (4)

Plutonic   Describes coarse-grained igneous rocks that form from cooling magma deep within Earth's interior. (4)

Pluvial lake   A large lake formed in regions ahead of an ice sheet due to the cooler temperatures and higher precipitation accompanying a glacial advance, particularly in otherwise arid regions. (15)

Point bar   A curved ridge of sand and gravel that forms on the inside curve of a meander when a stream's flow rate drops and sediment is deposited. (13)

Polar desert   An area that is intensely cold and receives virtually no precipitation due to global patterns of atmospheric circulation. (16)

Polar wander   The name given to the apparently changing magnetic pole positions with time derived from basalt samples of various ages; correctly explained by movement of the continents (*see* **Apparent polar wander**). (2)

Polymorphs   Minerals with the same chemical composition but different crystal structures. (3)

Poorly sorted   Refers to deposits of detritus in which the grains vary widely in size. (6)

Pore(s)   The small spaces between particles in sediment or sedimentary rock that are filled with air or water. (6, 12)

Porosity   The volume of spaces (pores) relative to the volume of particles in a sediment or a sedimentary rock. (6, 12)

Porphyritic texture   The texture of an igneous rock in which larger minerals (phenocrysts) are enclosed in fine-grained material (groundmass). (4)

Porphyry copper deposit   Deposits of copper (often with iron, molybdenum, gold, and silver) that are formed by hydrothermal activity within fractured granitic bodies; form above active subduction zones. (18)

Positive feedback   Natural process that amplifies the effects of an existing trend. (20)

Pothole   A distinctive circular depression produced by abrasion in the bedrock of the floor of a stream's channel. (13)

Precambrian   The vast interval (88%) of geologic time, encompassing the Hadean, Archean, and Proterozoic Eons; precedes the Phanerozoic Eon. (8)

Pressure surface   The imaginary surface that defines the level to which water rises in each well when a large number of wells are drilled into a confined aquifer. (13)

Primary (P) wave   A seismic body wave generated by sudden compression or extension of the ground at the site of an earthquake; the first seismic wave to arrive at a seismic station. (11)

Principle of faunal succession   Builds on the principle of original lateral continuity in establishing a rock's relative age and holds that individual rock layers can be identified on the basis of their fossil content; based on the observation that the fossils in sedimentary strata succeed each other vertically in a specific order that can be identified over wide areas. (8)

Proglacial lake   A vast lake that forms in front of a continental glacier when the preglaciation pattern of drainage is disrupted and rivers are ponded along the front of the ice sheet. (15)

Proterozoic Eon   The second eon of the geologic time scale (the other three being the Hadean, Archean, and Phanerozoic); from 2500 to 541 million years ago. (8)

Proton   The positively-charged particle in an atom's nucleus; the number of protons determines the atomic number of an element. (3)

Pull-apart basin   A long, narrow depression formed where a bend in a stick-slip fault causes the crust to stretch and subside; may collect sediment or become filled with water. (9)

Pumice   A rock produced when the frothy magma ejected during a pyroclastic eruption cools; it consists of glass, rock fragments, and stretched vesicles, and initially floats on water. (4)

Pyroclastic deposit   Forms from an explosive volcanic eruption when fragments of hot magma and surrounding rock that are ejected into the air from a volcano eventually fall to earth. (4)

Pyroclastic flow   A ground-hugging avalanche of hot volcanic ash, red-hot rock fragments, molten lava, and explosive gases; the most devastating of all violent eruptions. (4)

Pyroxene group   A group of calcium-magnesium-iron silicates that are usually green, brown, or black in color. (3)

Quartz   A silicate mineral made up almost entirely of silicon and oxygen. (3)

Quartz sandstone   A sandstone in which more than 90 percent of the mineral grains are made up of quartz. (6)

Quartzite   A very hard nonfoliated metamorphic rock formed from quartz-rich sandstone. (7)

Quaternary   Most recent period of geologic time scale from 2.6 million years ago to the present; characterized by intervals of continental glaciation. (8)

Quick clay   A type of sediment that is rich in clay and flows like water when shaken, typically by an earthquake. (12)

Quicksand   Forms when loose, water-saturated sand is suddenly agitated and undergoes liquefaction because the water is unable to escape; such highly fluid sand cannot support a load. (12)

Radial drainage   Pattern of drainage characterized by streams that radiate outward from a central high area like the spokes on a wheel. (13)

Radioactivity   The spontaneous and predictable breakdown of an atom with an unstable nucleus (a parent isotope) until it becomes a stable daughter isotope. (8)

Radiometric age   The age of a rock determined by measuring the radioactive decay of certain isotopes present in a mineral in the rock. (8)

Rain-shadow desert   A desert that forms on the leeward side of a mountain range because the height of the mountains effectively prevents most moisture in the atmosphere from reaching it. (16)

Ray path   The line of travel of an earthquake wave. (11)

Recharge   Replenishment of a groundwater aquifer by downward percolation of precipitation. (14)

Rectangular drainage   Pattern of drainage in which both the trunk stream and its tributaries show frequent right-angle bends and often join each other at right angles due to perpendicular fractures in the bedrock. (13)

Recycling   Process by which resources are saved after use so that they can be reused. (18)

Reduction   Occurs when a compound gains electrons during a chemical reaction (*see* **Oxidation**). (5)

Reef   Near-shore mass of limestone that results from the growth of corals, algae, and other colonial organisms. (17)

Refraction   The change of direction waves experience when they pass from one material into another. (11)

Regional metamorphism   The metamorphic process that occurs when large regions of crustal rocks are subjected to elevated temperatures and pressures; takes place during mountain building. (7)

Regolith   A loose layer of rock and mineral fragments produced by weathering; overlies bedrock and covers many parts of the land surface. (5)

Rejuvenation   The process that occurs when sea level falls and the gradient and energy of a stream increases; starts a new cycle of erosion, transport, and deposition as the stream endeavors to attain new equilibrium. (13)

Relative age   How geologists describe the order in which geologic events occur. (8)

Renewable resource   A resource (such as wood) that can be used without exhausting the supply. (19)

Replacement   The ore mineral concentration process that removes a rock-forming mineral and substitutes an ore mineral in its place; occurs in metamorphic rock. (18)

Reserve   A deposit of minerals or industrial commodities that is of known size and quality and can be extracted profitably with today's technology and in today's political and economic climate. (18)

Reservoir rock   Permeable sedimentary rocks such as sandstone and limestone that contain sufficient interconnecting spaces (pores) that they can hold oil and gas as a sponge holds water. (19)

Residual deposit   A mineral concentration formed as a result of being left behind during the weathering process; occurs in sedimentary rock. (18)

Residual soil   Soil that develops from the weathering of the bedrock directly beneath it, so has not been subjected to significant transport. (5)

Residue   The most stable part of an igneous source rock; remains after other parts of the source rock have melted. (4)

Resource   A natural concentration of inorganic material (minerals or industrial commodities) used by humans that *is,* or may become, of economic interest. (18)

Reverse fault   The type of dip-slip fault in which the rocks above the fault surface (hanging wall) move *up* relative to the rocks below the fault surface (footwall); typically the result of compressional forces (*see* **Dip-slip fault, Normal fault**). (10)

Reversed magnetic polarity   The opposite of Earth's present magnetic field; the north ends of all compass needles point toward the south magnetic pole. (2)

Rhyolite   A fine-grained felsic igneous rock. (4)

Richter magnitude scale   A measurement scale used to determine the size or *magnitude* of earthquakes; based on the amplitude of the largest ground motion recorded on a seismogram. (11)

Ridge push   One of three types of force (along with slab pull and mantle drag) involved in the movement of plates; occurs when a plate slides downslope from a ridge crest under the influence of gravity. (9)

Rift basin   A sedimentary basin that forms when continents start to break apart. (6)

Rift valley   A steep-walled linear trough that forms when large blocks of Earth's crust break and subside along faults during rifting. (2, 9)

Right-lateral fault   A strike-slip fault in which the fault block on the far side of the fault line has moved to the right relative to the block on the near side. (10)

Rip current   Zone of focused return flow in areas where the waves are breaking strongly; a narrow fast-moving surface current that runs straight out through the surf zone and dies out offshore. (17)

Ripple mark   A type of sedimentary structure that occurs on the surface of a bed; a small curving ridge that forms at right angles to the direction of current flow when wave motion or wind or water currents agitate the surface of a sediment layer (usually sand). (6)

Roche moutonnée   An asymmetric mass of bedrock smoothed by glacial abrasion in front and broken up by glacial plucking at the rear. (15)

Rock   An aggregate of one or more minerals either cemented together or grown together in an interlocking mosaic; does not have a specific chemical composition. (1)

Rock avalanche   A type of rockslide in which the rock fragments ride on a cushion of compressed air; a particularly destructive form of slide. (12)

Rock cycle   A concept that illustrates the origin of the three basic rock groups (sedimentary, metamorphic, and igneous) and the relationship between them involving the processes of creation and destruction. (1, 4)

Rockfall   Occurs when blocks of rock fall freely from a cliff face. (12)

Rock-forming mineral   Minerals that are common in rocks. (3)

Rockslide   The rapid downslope sliding and tumbling of masses of angular rock fragments that occurs along planes of weakness in the bedrock such as bedding planes, foliation surfaces, or fractures. (12)

Ropy lava   A lava flow with irregular wrinkles on its otherwise smooth surface; formed when the solid, exposed surface of the flow is deformed by the still-molten lava moving beneath it. (4)

Rounding   The process by which angular fragments produced by mechanical weathering are made less angular during transport. (6)

Rubble   Somewhat angular particles in detrital sedimentary rocks with diameters greater than 2 millimeters (0.08 inches). (6)

Rule of cross-cutting relationships   A principle of relative dating of rocks which holds that any feature that cuts across a rock must be younger than the rock it cuts. (8)

Rule of inclusions   A principle of relative dating of rocks which holds that any rock must be younger than the fragments it contains. (8)

Saltation   (1) *In water:* Occurs when small grains of sand and gravel are lifted briefly off the channel floor by a stream's turbulent flow and move downstream in a series of skips or hops; one of two methods (traction is the other) by which particles move along a stream bed (13). (2) *With wind:* Occurs when the wind lifts individual sand grains and carries them downwind a short distance before dropping them again, so that the grains skip along the ground in a series of bounces; most windblown sediment transport occurs by saltation. (16)

Sand   A sediment particle that ranges in diameter from 0.067 millimeters (0.003 inches) to 2 millimeters. (6)

Sandstone   A type of detrital sedimentary rock made up of sand-sized particles with diameters between 0.067 millimeters and 2 millimeters (0.003 inches to 0.08 inches). (6)

Saturation   (1) *In groundwater:* The condition that exists when the pores between grains of unconsolidated sediment become filled with water. (12) (2) *In chemical sedimentary rocks:* The maximum threshold of ion concentration in water, above which minerals precipitate. (6)

Schist   One of four types of foliated metamorphic rocks (along with slate phyllite, and gneiss); has visible minerals and a schistosity and occurs in regions that experience moderate to high temperatures and pressures. (7)

Schistosity   A type of foliation in which the minerals defining the foliation planes are visible to the naked eye; characteristic of schists. (7)

Scientific method   A systematic approach to investigation that scientists use based on observation, measurement, and experiment, and the formulation, testing, and modification of hypotheses. (1)

Sea arch   A bridge-like opening through a headland that is created when sea caves unite. (17)

Sea cliff   A wave-cut wall of rock formed when breaking waves cut a horizontal path landward into the bedrock. (17)

Sea stack   A solitary pillar of rock isolated from a headland by the seas; forms when the narrow bridge of a sea arch collapses. (17)

Seafloor spreading   The process by which new oceanic lithosphere is produced at a divergent plate boundary. (2, 9)

Seal   An impenetrable rock layer (usually mudstone or shale) that stops the upward and lateral movement of oil and gas and forces it to collect. (19)

Secondary (S) wave   A seismic body wave generated by the sudden shearing or sliding motion at an earthquake site; the second wave to arrive at a seismic station. (11)

Sediment   Accumulation of loose fragments of bedrock, minerals, shells, and/or crystals that precipitate directly from water; the product of physical and chemical weathering. (6)

Sedimentary basin   A region where Earth's crust has subsided and been filled with thick deposits of sediments and sedimentary rocks. (6)

Sedimentary facies   An association of sedimentary beds with differing characteristics, all of which were deposited at the same time within a particular sedimentary environment. (6)

Sedimentary rock   One of the three major rock groups (along with igneous and metamorphic); formed at or near Earth's surface by the cementation of loose material or by precipitation of minerals from water. (1, 6)

Sedimentary structure   Physical feature contained in detrital sediments and sedimentary rocks that reflect the conditions under which the sediment was deposited; these structures can occur within individual layers, at the top of a layer, or at the bottom of a layer. (6)

Seismic discontinuity   Abrupt changes in the speed of seismic waves at a particular depth within Earth's interior; indicates the presence of a boundary between rocks of contrasting composition or physical properties. (11)

Seismic ray   An alternate way to describe a seismic wave; a line of seismic activity that originates at an earthquake's focus and "radiates" outward. (11)

Seismic tomography   A computer imaging technique that gives a 3-dimensional "X-ray" picture of Earth's interior; suggests that convective flow in the mantle is complex. (11)

Seismic wave   A shock wave emitted from the focus of an earthquake when the rocks on either side of a fault jerk violently past each other, and abruptly release energy; seismic waves spread outward in all directions and travel faster through denser, rigid materials. (11)

Seismogram   The wiggly line readout produced by a seismometer whenever movement of the ground shakes its frame; used to record earthquakes. (11)

Seismology   The study of earthquake waves. (11)

Seismometer   (*seismograph*) A device that detects seismic waves; can detect horizontal or vertical motion. (11)

Semidiurnal tidal pattern   Two high tides of similar magnitudes each lunar day. (17)

Settling velocity   (terminal velocity) The velocity of a clast before it settles at the base of a body of water. (6)

Shadow zone   A belt encircling the globe in which direct seismic waves emanating from an earthquake focus fail to arrive because of refraction at the seismic discontinuity that marks the core-mantle boundary; lies between 103 degrees and 143 degrees from an earthquake's epicenter. (11)

Shale   A type of mudstone that is thinly layered and tends to split along those layers. (6)

Shale gas   Natural gas produced from organic-rich (black) shales. (19)

Shear   A metamorphic process that occurs in fault zones where stresses cause rock masses to slide past one another; this process occurs on a large number of closely spaced planes in a parent rock and it grinds and pulverizes the original grains so that their size is reduced. (7)

Shear strength   The natural strength of Earth materials that resists sliding. (12)

Shear stress   Stresses on a fault plane that act parallel to one another but in the opposite directions, causing rocks to strain by slipping past each other sideways; results in a change in a rock's location and shape. (10)

Shield volcano   A gently-sloped subaerial volcano that spreads out over a wide area and has its vent at the summit; dominated by basalts. (4)

Shock metamorphism   The metamorphism produced by the high-velocity impact of an extraterrestrial object such as a meteorite or an asteroid on Earth's surface. (7)

Silicate minerals   Minerals in which a variety of cations are bonded with silicon and oxygen. (3)

Silicate tetrahedron   A building block of a silicate mineral consisting of a central silicon cation bonded with four negatively charged oxygen anions positioned at the corners of the block. (3)

Sill   An intrusive igneous rock that occurs when magma cools into a sheet-like body that intrudes into surrounding wall rock along the boundary between layers so that its contact is parallel to the layering in the wall rock. (4)

Silt   A sediment particle that ranges from 0.004 millimeters to 0.067 millimeters (0.0002 inches to 0.003 inches). (6)

Siltstone   A very fine-grained variety of detrital sedimentary rock dominated by silt-sized particles. (6)

Sinkhole   A circular depression formed either directly when groundwater dissolves soluble bedrock at or near Earth's surface, or indirectly by collapse of the roof of a cave under the weight of the overlying rocks. (14)

Slab pull   One of three types of force (along with ridge push and mantle drag) involved in the movement of plates; occurs when the weight of a subducting slab pulls the rest of a plate behind it toward a trench and down a subduction zone. (9)

Slate   One of the four main types of foliated metamorphic rock (along with phyllite, schist, and gneiss); a very fine-grained rock with very closely spaced foliation planes; formed from the metamorphism of shale at low metamorphic grade. (7)

Slaty cleavage   A type of foliation in which the foliation planes are very closely spaced and the minerals defining the planes are visible only with the aid of a microscope; produced by directed pressure during the low-temperature, low-pressure metamorphism of clay-rich sedimentary rocks. (7)

Slide   The mechanism of mass wasting that occurs when material moves downslope in a relatively coherent mass along a well-defined surface; there are two types: *rockslides* and *slumps*. (12)

Slip   The movement along a fault. (10)

Slip face   The leeward side of a dune. (16)

Slump   A slide in which blocks of material move downslope along a curved surface, rotating as they slide. (12)

Snowline   The height above sea level at which a permanent, year-round snow cover develops. (15)

Soda straw   The hollow tube in the center of a stalactite. (14)

Soil   The portion of loose material covering bedrock that supports plants and their root systems. (5)

Soil profile   The layered structure of mature soil (soil in which soil-forming processes have been operating for a long time). (5)

Solar energy   Energy from the sun. (19)

Solar wind   Consists of subatomic particles ejected from the solar atmosphere; controls the abundance of charged particles that radiate away from the Sun. (20)

Solifluction   An extremely slow (a few centimeters per year), downslope movement of waterlogged surface material that occurs on slopes underlain by frozen ground; common in permafrost regions. (12)

Sorting   The process by which particles of the same size, shape, or density are naturally selected and separated from dissimilar particles. (6)

Source area   A geographic area from which the various clasts found in detrital sedimentary rock could have been derived. (6)

Source rock   (1) *In igneous rocks:* Rock that partially melts to form magma when sufficient heat is available to break the bonds of some of its minerals (4). (2) *In sedimentary rocks:* Source of clasts in a detrital sedimentary rock (3). *In Petroleum geology:* Organic-rich sedimentary rocks from which the petroleum in a reservoir was derived. (18)

Specific gravity The ratio of the mass of an object relative to the mass of an equal volume of water at 4°C. (3)

Spit Fingerlike extension of a beach that projects part way across the entrance to a bay; formed from deposition by longshore currents. (17)

Spring A natural outlet of groundwater on Earth's surface. (14)

Spring tide Occurs when the Sun and Moon are most closely aligned with Earth (new and full Moon); amplifies the tidal range. (17)

Stalactite A column formed by layers of calcite projecting downward from the roof of a cave. (14)

Stalagmite A stubby mound of layers of calcite that grows upward from the floor of a cave. (14)

Star dune An isolated hill of sand with three or four sharp-crested ridges radiating out from a central peak; forms where wind directions are variable. (16)

Stock A pluton with a surface outcrop area of less than 100 square kilometers (38 square miles). (4)

Strain The product of deformation; a measure of the change in shape or volume that a rock body undergoes as a result of an applied stress. (10)

Strain rate The amount of strain per unit time, or the change in shape or volume of a rock divided by the time it takes to achieve it; a measure of the intensity of deformation such as crushing and smearing along a fracture zone. (7)

Stratified drift Glacial deposit that has been redistributed by glacial meltwater; because water is the transporting agent, the deposit is sorted with well-defined layering. (15)

Stratiform deposit Deposit of ore minerals transported in solution and deposited within specific sedimentary strata. (18)

Stratigraphic column A representation that arranges sedimentary strata into a sequence of younger-upon-older layers. (8)

Streak The color of a mineral in powdered form. (3)

Stream Running water that flows within a channel; part of the hydrologic cycle that delivers water from land back to the oceans; flow of water in streams plays a major role in sculpting the landscape. (13)

Stream piracy (also called *stream capture*) Occurs when the headwater erosion of one stream breaches the land barrier between it and an adjacent stream so that the flow of the adjacent stream is diverted into the headwaters of the other stream. (13)

Stream terrace One of a series of elevated broad, flat benches that mark the former levels of a floodplain and extend from the modern floodplain toward the sides of a valley in a series of steps; characteristic of rejuvenated streams. (13)

Stress The cause of the deformation of a rock body measured as force per unit area; the push, pull, or shear that occurs when rocks are subjected to pressure and undergo a change in volume and/or shape. (7, 10)

Striations Linear scratch marks made by abrasion on the bedrock beneath a glacier that record the direction of glacial movement. (15)

Strike The orientation of an imaginary horizontal line on a plane of a geologic structure, measured with a compass as an angle relative to north; one of two parameters (dip is the other) that describes the orientation of all planar surfaces. (10)

Strike-slip fault A fault in which the slip caused by shear stresses is parallel to the strike of the fault plane; the fault blocks on either side of a strike-slip fault move sideways. (10)

Subduction Process at convergent plate boundaries where two plates come together and one (the denser plate) angles down beneath the other (the more buoyant plate) and sinks into Earth's heated interior. (2, 9)

Subduction zone The inclined zone along which subduction occurs. (2, 9)

Submergent coastline Coastline along which the land is subsiding relative to sea level. (17)

Subsidence (1) *groundwater:* Happens when land settles after groundwater is extracted from unconsolidated subsurface sediment. (14) (2) *tectonic:* Vertical sinking of the crust. (14)

Subtropical desert Results from a global pattern of atmospheric circulation that causes moisture to be removed from the air as it migrates from the equator. (16)

Sunspot Small region (up to 160,000 kilometers or 100,000 miles across) on the Sun's surface that is dark because its temperature is lower than the temperature of the surrounding regions; sunspots are cooler because their strong magnetism suppresses the escape of heat from the Sun's interior. (20)

Sunspot cycle Regular variation in sunspot activity; peak periods of activity are 11.1 years apart and coincide with anomalously warm temperatures in Earth's upper atmosphere. (20)

Supercontinent cycle The episodic amalgamation and breakup of supercontinents through Earth history; when a supercontinent breaks up, the continental fragments may later come together to form another supercontinent; explains the apparently cyclic pattern of mountain building, sea level change, and climate in the geologic record. (10)

Supergene enrichment A secondary process of mineral enrichment in which percolating groundwater leaches metals from near-surface sulfide minerals and precipitates them at depth so that lower levels of a deposit become sufficiently enriched in metals to form an ore deposit. (18)

Superposed stream A stream that cuts down into and across a geologic structure from a path established above. (13)

Superposition One of three fundamental rules for determining a rock's relative age (along with original horizontality and lateral continuity); holds that if sedimentary strata are deposited in sequence, one on top of the other, then the oldest layers will occur at the base. (8)

Surf zone The area between the outermost breaker and the shoreline. (17)

Surface tension The tendency of liquids to acquire the least surface area possible. Where water is in contact with air, surface tension results from the greater attraction of water molecules to each other than to the molecules in the air; this attractive force exists because water molecules are polarized (have opposite electrical charges at either end). (12)

Surface water A collective term for liquid or seasonally frozen water that occurs at Earth's surface; includes oceans, streams, lakes, and swamps. (14)

Surface wave   A seismic wave that travels along Earth's surface; the main cause of destructive earthquake damage. (11)

Surge   An abrupt increase in basal sliding during which a glacier's flow rate may increase a hundredfold; occurs when water accumulates at the base of a glacier and raises it off its bed, reducing friction. (15)

Suspended load   The products of erosion that are small enough to be lifted off a stream bed and held in suspension by the turbulent flow of the stream; almost three-quarters of the sediment carried by the world's streams is transported in suspension. (13)

Suture   The boundary between formerly separate tectonic plates or geologic terranes. (10)

Swash   Occurs when the water from a breaking wave rushes up the beach. (17)

Swash zone   The region on a beach where a wave breaks, rushing up the beach, then retreating back toward the sea. (17)

Syncline   A type of fold that forms when a sequence of rock strata in normal succession flexes downward; as a result, the youngest rocks occupy the fold core. (10)

Talus   An accumulation or apron of loose rock debris at the base of a cliff. (12)

Tar sands   Sandstone reservoir that contains a sticky asphalt-like hydrocarbon called *bitumen* and must be steam-heated prior to extraction by pumping (to encourage flow); can be mined by open-pit methods if the deposit is shallow. (19)

Tectonic joint   A fracture along which movement is minimal, produced by tectonic stresses that cause rocks in Earth's crust to break in a systematic fashion (*see* **Joint**). (10)

Tensional stress   Stress that pulls rocks apart, causing them to strain by stretching, so they become longer and thinner. (10)

Terrane   A fault-bounded fragment of crustal material with a stratigraphic, structural, and geologic history that is different than the tectonic plate to which it is currently affixed. (10)

Terrestrial planet   One of the inner planets of the Solar System (Mercury, Venus, Earth, and Mars); essentially rocky spheres with relatively high densities. (21)

Texture   The overall appearance of a rock; describes the size, shape, and arrangement of the particles (minerals, clasts, and matrix) in a rock. (4)

Theory   A step in the scientific method; a hypothesis that passes many tests that rival hypotheses have failed. A theory is not a fact and is subject to further testing. (1)

Thermal convection   The process that drives the circulation of Earth's mantle; hot, less dense material from Earth's interior rises toward the surface, while cool, denser material sinks; in this way the heat produced in Earth's interior travels to the surface where it is radiated away into space. (11)

Thermal subsidence basin   A type of sedimentary basin formed during rifting of continental crust; as the crustal fragments cool, their margins subside to form a basin in which sediments progressively accumulate. (6)

Thrust fault   A low-angle reverse fault along which the hanging wall moves up the slope of the fault plane relative to the footwall; common in mountain belts where they may accommodate significant movement. (10)

Tidal current   Alternating current produced by the rise and fall of the tides; agent of erosion and deposition. (17)

Tidal flat   Marshy area that is alternately covered and uncovered by the rise and fall of the tide. (17)

Tidal power   Energy generated by the flow of water associated with tides. (19)

Tide   Daily rise and fall of ocean waters caused by the gravitational tug of the Moon and Sun. (17)

Till   An unconsolidated mixture of boulders, sand, and clay deposited directly from the ice of a glacier; it is neither sorted nor layered because glaciers do not sort or stratify rock debris. (15)

Tillite   A chaotic, unstratified, unsorted rock type made of till that has turned to rock (lithified). (15)

Tombolo   A form of spit that links an offshore island or sea stack to the mainland; formed as a result of deposition by longshore currents. (17)

Topography   Changes in elevation and general arrangement of the physical features of an area of Earth's surface. (5)

Topsoil   The A and O horizons of the soil profile; typically dark and nutrient-rich. (5)

Traction   Occurs when boulders and cobbles hug the channel floor and move downstream in an intermittent sliding, dragging, or rolling motion; one of two ways in which particles move along a stream bed (saltation is another). (13)

Transform continental margin   Type of continental margin, the location and orientation of which is controlled by transform faults. (17)

Transform fault   A major fracture in the lithosphere along which one plate slides by another; found at transform plate boundaries where plates slide past each other. (2, 9)

Transform plate boundary   A type of plate boundary where one lithospheric plate slides past another, creating great crustal fractures; lithosphere is neither created nor destroyed, instead one plate boundary is linked to another by a transform fault. (2, 9)

Transition zone   Region between about 400 and 700 km depth that separates the upper mantle from the lower mantle in which seismic velocities gradually increase; part of the asthenosphere. (11)

Transported soil   Soil that develops from mineral and rock material that has been transported from elsewhere to its current location by mobile agents such as running water, ice, or wind. (5)

Transverse dune   A long, strongly asymmetric sand ridge that buries the ground surface and forms in flat, barren areas of desert floor where the wind is constant and the supply of sand is plentiful. (16)

Transverse ridge   A mountain range formed across the line of a strike-slip fault where a bend in the fault causes the crust to be compressed and uplifted. (9)

Travertine   A special type of chemical sedimentary rock composed of calcium carbonate that forms from hot springs or in caves and is in great demand as a decorative stone. (6)

Trellis drainage   Pattern of drainage developed in a landscape of parallel ridges and valleys, in which broadly parallel trunk streams are more or less perpendicular to their much shorter tributaries. (13)

Tributary   A smaller stream that feeds into the main trunk of a larger stream; a tributary may have tributaries of its own. (13)

Triple point A Y-shaped plate junction between three lithospheric plates. (9)

Tsunami A massive seismic sea wave that, in shallow water, can reach a height of over 40 meters (130 feet); generated when the seabed is heaved up or down during an earthquake and a vast amount of the energy released is transferred to the ocean water. (9)

Tuff A rock produced by an explosive volcanic eruption consisting of volcanic debris (ash, pumice, and other fragments) jettisoned from the volcano; the debris may be dispersed hundreds of kilometers from its original source volcano. (4)

Turbidity current Downslope movement of a dense current of sediment-rich water; typically developed when sand and mud on the continental shelf or slope become unstable. (6)

Turbulent flow Occurs when the flow lines of a fast-flowing stream crisscross in a chaotic fashion, continually mixing the water. Characterized by swirling water and eddies, the degree of turbulence depends on the velocity of the water and the irregularity of the surface over which it flows; turbulence keeps a stream's sediment load in suspension. (13)

Ultramafic A type of igneous rock that contains less than 45% silica and is high in iron and in magnesium. (4)

Unconfined aquifer An aquifer whose upper surface coincides with the water table. (14)

Unconformity A boundary between two rock bodies that represents a major time gap in the geologic record during which no deposition is preserved (*see* **Angular unconformity, Disconformity, Nonconformity**). (8)

Upper mantle A chemical division of the mantle that extends to a depth of about 400 km and consists predominantly of dense ultramafic rock. It is divided into two zones based on contrasting physical characteristics: (i) the rigid lithospheric mantle that is part of a tectonic plate and extends from the base of the crust to a depth of 50–150 km, and (ii) the weaker upper mantle below that is part of the asthenosphere and capable of flow. (9)

Valley glacier Formed when a cirque glacier begins flowing downslope, typically following the route of a preexisting stream valley. (15)

Varve Deposit found toward the center of a glacial lake consisting of a pair of thin, alternating layers: a thicker, light-colored layer of silt deposited in the summer months, and a thinner, dark layer of clay that settled from suspension during the winter when the lake was frozen over. Because a single varve records a single year of deposition, varves are used to establish the age and duration of glacial episodes. (15)

Vein A thin, sheet-like seam of minerals formed when ions in solution precipitate onto the walls of fractures. (18)

Vent An opening in a volcano from which lava, ash, steam, and hot gases flow or are ejected during eruptions. (4)

Vent eruption Occurs when magma is extruded at the central vent of a volcano. (4)

Ventifact A faceted stone formed in deserts when exposed pebbles and boulders are abraded by sandblasting. (16)

Vesicle A small, open cavity in lava that forms from the preservation of bubbles in frothy magma. (4)

Viscosity A measure of a liquid's stickiness or resistance to flow; the less viscous a magma, the more easily it flows. (4)

Volatiles Elements or compounds such as water, carbon dioxide, and nitrogen that exist as gases at Earth's surface. (4)

Volcanic arc A curved chain of volcanoes that develop above a subduction zone at a convergent plate margin (*see* **Continental arc, Island arc**). (4, 9)

Volcanic bombs Blobs that are liquid when ejected from a volcano, become rounded during their flight, then fall to earth as a solid; a constituent of pyroclastic deposits. (4)

Volcanic breccia Pyroclastic rock consisting of broken fragments of material (blocks, bombs, ash, rock fragments, minerals) ejected from a volcano, cemented together by a fine-grained matrix dominated by volcanic ash. (4)

Volcanic glass Forms when lava cools so rapidly that minerals do not get time to form. (4)

Volcanic island arc (*see* **Volcanic arc**)

Volcano A mountain formed by the accumulation of volcanic rocks vented from a magma chamber below the surface. (4)

Wall rock The rock that surrounds magma as it rises to Earth's surface. (4)

Water table The upper surface of the zone of saturation; overlain by the zone of aeration. (14)

Wave A disturbance that transmits energy through a substance with no net movement of the substance (*see* **Ocean wave, Seismic wave**). (17)

Wave energy Energy harnessed from ocean waves and used to generate electricity. (19)

Wave height The difference in height between the wave crest (the highest point) and trough (the lowest point). (17)

Wave refraction The bending of a wave that obliquely approaches a boundary (such as a shoreline or a boundary between Earth's internal layers) because not all parts of the wave reach the boundary at the same time. (17)

Wave velocity The speed and direction of wave movement; influenced by water depth. (2)

Wave-cut platform A gently sloping, beveled surface of bedrock left at the low-tide line when erosion causes the base of a sea cliff to retreat; extends from a cliff to the sea. (17)

Wavelength Measures the horizontal distance between two waves, either from crest (highest point) to the nearest crest, or from trough (lowest point) to the nearest trough. (17)

Weathering The physical, chemical, and biological processes by which rocks and minerals exposed at Earth's surface slowly break down as a result of their interaction with water, ice, and air; the first step in the formation of sedimentary rock. (1, 5)

Welded tuff Rock formed when a highly mobile mixture of liquid, solid, and gas produced by a pyroclastic eruption comes to rest and retains enough heat that glass shards and rock fragments meld together. (4)

Well sorted Refers to deposits of detritus in which the components of the detritus are very similar in size. (6)

Whole mantle convection A model of mantle convection that proposes that large convection cells extend through the entire mantle from the core–mantle boundary to the base of the lithosphere; warm, buoyant mantle material rises beneath the mid-ocean ridges, then spreads laterally beneath the lithosphere, which it carries in a conveyor-belt fashion. Eventually, the warm buoyant material cools, sinks back into the mantle at subduction sites, and is reheated. (11)

Wilson cycle   A cycle of tectonic activity that commences with rifting and the development of an ocean, followed by subduction and the formation of volcanic arcs, and finally continent-continent collision. (10)

Wind power   Energy harnessed from blowing winds and used to power various processes and to generate electricity. (19)

Xenocryst   An inclusion in an igneous rock that is a crystal fragment derived from a different rock. (4)

Xenolith   An inclusion in an igneous rock that is a fragment of a different rock. (4)

Zeolite facies   The metamorphic facies in which zeolite minerals are stable; rocks in this facies are usually formed at low pressures and low temperatures; most of the original features of the parent rock are preserved. (7)

Zone of aeration   The area above the water table and the zone of saturation in which the pore spaces of the sediment are filled with air as well as water. (14)

Zone of saturation   The zone above impermeable bedrock and below the water table in which water fills all the available spaces in fractures and pores in soil, sediment, or rock; contains groundwater. (14)

Credits

Chapter 7   Figure 7.1a: © Wyckoff, Jerome/Animals Animals. Figures 7.1b, 7.16a: E. R. Degginger/Science Source. Figures 7.1c, 7.16c: Michael C. Rygel via Wikimedia Commons. Figures 7.1d, 7.16d: © Kent, Breck P./Animals Animals. Figures 7.8a, 7.23: © Geology.com. Figure 7.8b: © Ken Lucas/Visuals Unlimited/Corbis. Figure 7.9: Dr. Theodore C. Labotka. Figure 7.11a: Howard Donohoe. Figure 7.11b: Woudloper via Wikimedia Commons. Figure 7.12a: © Kent, Breck P./Animals Animals. Figure 7.12b: Planet Observer/Getty Images. Figure 7.13: Dirk Wiersma/Science Source. Figure 7.16b: Chadmull via Wikimedia Commons. Figure 7.18: © Zbynek Burival/Dreamstime.com. Figure 7.19: © Geology.com. Figure 7.20: Dr S. A. Wellings. Figure 7.22: © Geology.com. Figure 7.26: Cin-Ty Lee.

Part II Opener   © Corbis.

Chapter 8   Figure 8.1: Damian Nance. Figure 8.2a: Ramsay, Andrew. *Physical Geology and Geography of Great Britain*. Figure 8.2b: NoelWalley via Wikimedia Commons. Figure 8.3: G. R. Roberts/Science Source. Figure 8.4: Marcel Nijhuis/Shutterstock. Figure 8.5a: King's College London/Science Source. Figure 8.5b: Dea L. Romano/Getty Images. Figure 8.9b: Dr. Marli Miller. Figure 8.13: CP14/058 British Geological Survey © NERC. All rights reserved. Figure 8.15 (Grand Canyon): © Brownold, Thomas J./Animals Animals. Figure 8.15 (Zion): bjul/Shutterstock. Figure 8.15 (Bryce Canyon): Damian Nance. Figure 8.17: Photo by Denis Finnin, © American Museum of Natural History.

Chapter 9   Figure 9.5: National Geophysical Data Center. Figure 9.22: David Harlow/Getty Images. Figure 9.24: © Logan Abassi/UN Handout/UN/MINUSTAH/Corbis. Figure 9.31 (main photo): Kevin Schafer/Getty Images. Figure 9.31 (inset photo): © Corbis. Figure 9A (left): Dr Ken Macdonald/Science Photo Library. Figure 9A (right): Science Source. Figure 9C: NOAA. Figure 9D: © Philip A. McDaniel/CNP/Corbis. Figure 9E: Kyodo/AP Images.

Chapter 10   Figures 10.1, 10.11a, 10.24: Brendan Murphy. Figures 10.5, 10.6, 10.7, 10.8a, 10.9b, 10.15b, 10.16a, 10.16b, 10.16c, 10.18b, 10.23, 10.25a: Damian Nance. Figure 10.8b: Scott T. Marshall. Figure 10.8c: Michael Collier. Figure 10.9c: Garry Hayes/GeoTripper Images. Figure 10.11b: Norris W. Jones. Figure 10.13b: Andrew Birrell. Figure 10.15a: Dr. Marli Miller. Figure 10.21b: © NASA/Corbis. Figure 10.25b: Bruce Railsback. Figure 10.26: phdpsx/iStock. Figure 10.30 (upper left): Bonita Murchey/USGS. Figure 10.30 (upper right): Cin-Ty Lee. Figure 10.30 (lower right): William Elder. Figure 10.33: © Bartosz Hadyniak/iStock. Figure 10B: NASA. Figure 10C: Le Grand Portage via Wikimedia Commons.

Chapter 11   Figure 11.1e: © Corbis. Figure 11.2: © edfuentesg/iStock. Figure 11.8: Scott Dickerson/TandemStock.com. Figure 11A: USGS.

Part III Opener   © Andria Hautamaki/Dreamstime.com

Chapter 12   Figures 12.2, 12.3, 12.4, 12.24d: Damian Nance. Figure 12.7: USGS, Lynn Highland. Figure 12.8: USGS. Figure 12.11: David F. Walter. Figure 12.12: Marek Ślusarczyk via Wikimedia Commons. Figure 12.13: Dr. George Plafker, USGS. Figure 12.15: USGS. Figure 12.16: Associated Press. Figure 12.17: Montreal Gazette: The Gazette photo archives © 1971. Figure 12.18: Same as Fig 4.25b. Figure 12.19: Bob Schuster, USGS. Figure 12.20: Terrence Spencer/Getty Images. Figures 12.21, 12.22, 12.24b: B. Bradley,

University of Colorado/NOAA. Figure 12.24b: Figure 12.24c: Douglas MacDonald. Figure 12.26b: California Department of Transportation. Figure 12.28: Japanese Photographer (20th century)/Private Collection/Peter Newark Pictures/Bridgeman Images. Figure 12F: Lunar and Planetary Institute. Figure 12G, 12H: NASA. Figures 12I, 12J (right): Ted S. Warren/Associated Press. Figure 12J (left): Google Earth.

Chapter 13   Figure 13.6a: © Image Source/Corbis. Figure 13.6b: Michael Hall via Wikimedia Commons. Figures 13.9, 13.14: Damian Nance. Figure 13.10: Sirpa Fagerholm. Figure 13.11: © W. Perry Conway/Corbis. Figure 13.12: Dr. Marli Miller. Figure 13.16b: Pete Mcbride/National Geographic Creative. Figure 13.23: Leruswing via Wikimedia Commons. Figure 13.25: © Kent, Breck P./Animals Animals. Figure 13.28: Phil Degginger. Figure 13.29: SuzanneKn via Wikimedia Commons. Figure 13.30e: katorisi via Wikimedia Commons. Figure 13.33: © NASA/Corbis. Figure 13.36: © Andrews, Michael/Animals Animals. Figure 13.37a: Georg Gerster/Science Source. Figure 13.37b: Jim Edds/Science Source. Figure 13.38: © Andrew Holbrooke/Corbis. Figure 13.40: Mike Caplinger, Malin Space Systems. Figure 13.41a: © visualphotos.com. Figure 13.41b: NASA/JPL-Caltech/Cornell/USGS. Figure 13A: NASA.

Chapter 14   Figure 14.4: © Alan Towse/Ecoscene/Corbis. Figure 14.5b: © David Hall/LOOP IMAGES/Loop Images/Corbis. Figure 14.10a: © Henglein and Steets/cultura/Corbis. Figure 14.10b: James Steinberg/Science Source. Figure 14.11c: Gary S. Chapman/Getty Images. Figure 14.12: Zuki/iStock. Figure 14.14: © Martin Puddy/Corbis. Figure 14.15a: G. R. 'Dick' Roberts/NSIL/Getty Images. Figure 14.16a: Associated Press. Figure 14.16b: © Dave Bunnell. Figure 14.18a: © Gibson, Mickey/Animals Animals. Figure 14.18b: NCKRI. Figure 14.20a: Richard Ireland, USGS. Figure 14.20b: © Liz Leyden/iStock. Figure 14.20c: © Michael-Blackburn/iStock. Figure 14D: NASA.

Chapter 15   Figure 15.2: Chmee2 via Wikimedia Commons. Figure 15.4a: Sonny Bou. Figure 15.4b: USGS. Figure 15.4c: Ron Sanford/Science Source. Figure 15.4d: Rufus Hawthorne via Wikimedia Commons. Figure 15.7: From Dowdeswell, Julian & Hambrey, Michael: Islands of the Arctic 2002, Cambridge University Press. Figure 15.8: Mat Fascione. Figure 15.9: SunChan/iStock. Figure 15.11: Walter Siegmund via Wikimedia Commons. Figure 15.12: Diliff via Wikimedia Commons. Figure 15.13: Camptocamp.org via Wikimedia Commons. Figure 15.14: chensiyuan via Wikimedia Commons. Figure 15.15: © Tomassino/iStock. Figure 15.16: Bryn Scrivner. Figure 15.17: Damian Nance. Figures 15.18, 15.30: Mark A. Wilson via Wikimedia Commons. Figure 15.21 (inset): Boschfoto via Wikimedia Commons. Figure 15.21 (main photo): Michael Gibbons, WalkingIreland.com. Figure 15.22: Geoffrey Smith. Figure 15.23: Jesse Allen, Earth Observatory, using data obtained courtesy of the University of Maryland's Global Land Cover Facility. Figure 15.25: Tom Bean Photography. Figure 15.26: © Tom Bean/Corbis. Figure 15.27: Dr. Marli Miller. Figure 15.31: Trevor Bell. Figure 15.35: Bahudhara via Wikimedia Commons. Figure 15.36: Damian Nance. Figure 15.38: SamirNosteb via Wikimedia Commons. Figure 15.39a: Science Source. Figure 15.39b: NASA/JPL/University of Arizona. Figure 15.40: NASA/JPL/USGS. Figures 15C, 15D (upper right): Tom Foster/HugeFloods.com. Figure 15D (lower right): Don Hyndman photo, University of Montana. Figures 15D (bottom), 15E: Ikiwaner via Wikimedia Commons.

Chapter 16   Figure 16.3: © Ingram, Stephen/Animals Animals. Figures 16.4, 16.10: Beth Kordesch. Figures 16.5, 16.6, 16.14, 16.18, 16.26: Damian Nance. Figures 16.7, 16.12: Dr. Marli Miller. Figure 16.9: Frans Lanting/MINT Images/Science Source. Figure 16.11b: David Scharf/Science Source. Figure 16.13: © Dr. Marli Miller/Visuals Unlimited/Corbis. Figure 16.15: © Dennis, David M./Animals Animals. Figure 16.16: © Thompson, Michael/Animals Animals. Figure 16.20: http://www.isgs.uiuc .edu/outreach/geology-resources/loess. © 2014 University of Illinois Board of Trustees. All rights reserved. Photograph courtesy of the Illinois State Geological Survey. Photographer: Joel Dexter. Figure 16.24: Albert Backer via Wikimedia Commons. Figure 16.25: © Ron Blakey @ Colorado Plateau Geosystems, Inc. Figure 16.27: NASA/JPL/University of Arizona. Figure 16.28: Dr. Timothy Parker, JPL/NASA. Figure 16.29: NASA/JPL. Figure 16C: NASA. Figure 16D: NASA/JPL-Caltech/UA. Figure 16F: NASA/ Goddard Institute for Space Studies. Figure 16G: NOAA.

Chapter 17   Figure 17.1: Damian Nance. Figure 17.8b: Tom Cozad. Figure 17.9: © Degginger, Phil/Animals Animals. Figure 17.14: Tracy Knauer/Science Source. Figure 17.15b: Yummifruitbat via Wikimedia Commons. Figure 17.16: © Stephen Rose—Rainbow/ Science Faction/Corbis. Figure 17.18b: © Mago-World-Image/ Animals Animals. Figure 17.18c: Phillip Hayson/Science Source. Figure 17.19a: Wordydave via Wikimedia Commons. Figure 17.19b: marche11074/Shutterstock. Figure 17.21b: NASA. Figure 17.22b: 663highland via Wikimedia Commons. Figure 17.23a: Science Source. Figure 17.23b: Dr. Frank Hanna/Getty Images. Figure 17.25: NASA. Figure 17.26: NASA GSFC Landsat/LDCM EPO Team. Figure 17.27: Andrew Alden. Figure 17.28a: © Mark A. Johnson/ Corbis. Figure 17.28b: © Susan Robinson/Dreamstime.com. Figure 17.28c: US Army Corps of Engineers. Figure 17.29: Jim Wark/Air Photo. Figure 17.30: © M. F. Fitzpatrick. Figure 17.31: NASA. Figure 17E (inset): ©Houston Chronicle/Smiley N. Pool. Used with permission.

Part IV Opener   Associated Press.

Chapter 18   Figure 18.2: André Karwath via Wikimedia Commons. Figure 18.3: siphon via Wikimedia Commons. Figure 18.4: Friends of Stradbroke Island Inc. Figure 18.6a: © Vladimir Blinov/Dreamstime.com. Figure 18.6b: Jonathan Zander via Wikimedia Commons. Figure 18.7a: Daniel Schwen. Figure 18.7b: R. Klopfer & A. Koplov/James Madison University Mineral Museum. Figure 18.8: Dirk Wiersma/Science Source. Figure 18.9: Rob Lavinsky/iRocks.com. Figure 18.11: Martin Mraz. Figure 18.12: Romaniadacia.wordpress.com. Figure 18.13: © Gibson, Mickey/ Animals Animals. Figure 18.14: Claude GRANDPEY. Figure 18.15: Stapanov Alexander via Wikimedia Commons. Figures 18.16, 18.30: Damian Nance. Figure 18.17: Mark A. Wilson via Wikimedia Commons. Figure 18.18: Dr. Charles E. Jones. Figure 18.20: Jackie Gauntlett. Figure 18.21: Brendan Murphy. Figure 18.22: © Dr. Marli Miller/Visuals Unlimited/Corbis. Figure 18.25: OAR/ National Undersea Research Program (NURP); NOAA. Figure 18.29: NASA. Figure 18.34: Woods Hole Oceanographic Institute. Figure 18B: Spencer Musick via Wikimedia Commons.

Chapter 19   Figure 19.10a: © Ted Wood/Aurora Photos/Corbis. Figure 19.12a: Lvklock via Wikimedia Commons. Figure 19.15: Bart

Kowallis. Figure 19.23: Stepheng3 via Wikimedia Commons. Figure 19.28a: Stan Zurek via Wikimedia Commons. Figure 19.29: Gray Watson via Wikimedia Commons. Figure 19.30a: Koza1983 via Wikimedia Commons. Figure 19.30b: Alan Radeck via Wikimedia Commons. Figure 19.31: © Steve Proehl/Proehl Studios/ Corbis. Figure 19.32: Matthew Field via Wikimedia Commons. Figure 19.33: Photo courtesy of HDR (http://www.hdrinc.com/). Photography by © Keith Philpott (http://www.keithphilpott.com/). Figure 19F: Jane Pargiter/EcoFlight. Figure 19G: USGS.

Part V Opener   NASA/JPL-Caltech/Cornell.

Chapter 20   Figure 20.1: NASA Earth Observatory image by Robert Simmon, using Suomi NPP VIIRS data provided courtesy of Chris Elvidge (NOAA National Geophysical Data Center). Figure 20.5a: Barbara Zarrella. Figure 20.7: Science Source. Figure 20.8a: Alan L via Wikimedia Commons. Figure 20. 11a: SCIENCE SOURCE/SCIENCE PHOTO LIBRARY. Figure 20.11b: Scharmer et al, Royal Swedish Academy of Sciences/Science Source. Figure 20.12a: © 2007 Miloslav Druckmüller. Figure 20.12c: NASA. Figure 20.13: © Heritage Images/Corbis. Figure 20.15a: National Ice Core Lab. Figure 20.15b: Kendrick Taylor. Figure 20.16: CSIRO Science Image. Figure 20.23b: NASA. Figure 20A: © Martin Lipman/Lipman Still Pictures. Figure 20B: © Julius T. Csotonyi. Figure 20C: NASA/Goddard Scientific Visualization Studio. Figure 20D: Joe Moore/USDA-NRCS.

Chapter 21   Figure 21.2: © Natursports/Shutterstock. Figure 21.3: NASA/Science Photo Library. Figure 21.5a: NASA/ Lunar and Planetary Institute. Figure 21.5b: Jeff Barton via Flickr. Figure 21.7a: NASA. Figure 21.7b: NASA/Goddard/Arizona State University/Smithsonian. Figures 21.9, 21.10: NASA/Goddard Space Flight Center. Figure 21.11a: © NASA/Roger Ressmeyer/ CORBIS. Figure 21.11b: NASA/JPL/USGS. Figure 21.13: Phil James (Univ. Toledo), Todd Clancy (Space Science Inst., Boulder, CO), Steve Lee (Univ. Colorado), and NASA/ESA. Figure 21.15a: NASA/JPL-Caltech/Cornell/Arizona State U. Figure 21.15b: NASA/JPL. Figure 21.18a: NASA, ESA, H. Weaver and E. Smith (STScI) and J. Trauger and R. Evans (Jet Propulsion Laboratory). Figure 21.18b: Science Source. Figure 21.19a: MarcelClemens/ Shutterstock. Figure 21.19b: NASA/JPL. Figure 21.19c: Science Source. Figure 21.19d: NASA. Figure 21.21: NASA, JPL; Digital processing: Björn Jónsson (IAAA). Figure 21.22 (Io): NASA/JPL/ University of Arizona. Figure 21.22 (Europa): NASA/JPL/DLR. Figure 21.22 (Moon): verbaska/MorgueFile. Figure 21.22 (Ganymede): NASA. Figure 21.22 (Callisto): Science Source. Figure 21.23: NASA/JPL/University of Arizona. Figure 21.24a: Björn Jónsson. Figure 21.24b: NASA/JPL-Caltech/Space Science Institute. Figure 21.24c: © NASA/JPL-Caltech/Corbis. Figure 21.25: Dr. R. Albrecht, ESA/ESO Space Telescope European Coordinating Facility; NASA. Figure 21.26: Science Source. Figure 21.27: Associated Press. Figure 21.29: Shane Torgerson via Wikimedia Commons. Figure 21.31a: Gordon Garradd/Science Source . Figures 21.31b, 21.31c: Chris Samuel via Flickr. Figure 21.32: Science Source. Figure 21.33: Associated Press. Figure 21C: Science Source.

Index

aa (blocky) lava, 121–122
Aberfan (South Wales), 389–392
ablation, 470
abrasion, 410, 471, 514, 541
abrasives, 578–581
absolute age, 227
absolute dating
    geologic time scale and, 248–250
    half-life and, 245–246
    radioactivity and radioactive decay, 241–245
    radiometric dating, 246–248
absolute time, 12
abyssal plains, 259
acceleration, 17
accreted terrane, 319–321
accretion, planetary, 21
accretionary wedges, 271, 319
accumulation, zone of, 155
achondrites, 707
acid mine drainage, 623
acid rain, 143–144, 459
acid rock drainage, 147
acids, 144
active continental margins, 276, 560
active margins, 58
aeration, zone of, 439–440
aerosols, 31, 123
Agassiz, Lake, 489, 492
Agassiz, Louis, 485
agents of erosion, 16, 165, 409, 508
agents of metamorphism, 198–201
agglomerates, 124
agricultural contamination of groundwater, 460–461
A horizon, 155
air bubbles in polar ice, 660–663
air temperature in the lower atmosphere, 468
albedo, 670
Aleutian Islands, 56–57, 61
alluvial fans, 187, 427–428, 509
alluvium, 423
alpha particle emission, 244
alpine glacial landforms, 473–476
alpine glaciers, 469, 473

aluminum (Al), 570
Amasia, 331
Amazon River, 10, 406, 432
amphibole group, 88
amphibolite, 209, 214
amphibolite facies, 217
amplifiers, 670
amplitude of a wave, 533
Amu Darya River, 419
anaerobic environments, 610
andalusite, 95
Andes, 57, 61, 132, 318–319
andesite, 116
angle of repose, 373–375, 518
angular unconformity, 233, 313–314
anhydrite, 176, 578
anions, 73, 76
Antarctic Treaty, 601
antecedent streams, 414
anthracite coal, 177, 624
anticlinal traps, 610
anticlines, 311
apatite, 577
Appalachian-Caledonian mountain belt, 327–330
apparent pole wander, 47–48
aquicludes, 442
aquifers
    confined, 445–446
    High Plains (Ogallala) aquifer, 444–445
    historical importance of, 447
    Nubian aquifer, Sahara Desert, 450
    plate tectonics and, 462
    self-cleansing potential of, 461
    terminology, 442
    unconfined, 443–445
    See also groundwater
aquitards, 442
Aral Sea, 419–420
Archean Eon, 240
arc-trench gap, 275–276
area, 16
arêtes, 474
arkose, 175–176
arroyos, 509
artesian wells, 446
Ascension Island, 53–54
ash, 124–125
assimilation, 120
asteroids and asteroid belts, 680, 705, 707

asthenosphere, 26, 256–257, 354–355
astronomical units (AUs), 687
Aswan High Dam (Egypt), 431
asymmetric folds, 311
Atacama Desert, 505, 507
Athabasca tar sands (Canada), 616
Atlantic realm, 329
atmosphere
    air temperature in the lower atmosphere, 468
    atmospheric temperature of Mars and Venus, 652
    circulation pattern, global, 503
    definition of, 23, 32–33
    on other planets, 686–687, 689–690, 691–692, 699
    volcanoes and composition of, 108–109
atolls, 286, 548–549
atomic mass, 76
atomic nucleus, 76
atomic number, 71, 76, 241
atoms, 16, 70–71, 76
aulacogens, 290, 432
aureoles, 204
aurora, 658
average gradient, 405, 430
average kinetic energy, 104, 652
axial plane, 310, 311

back-arc basins, 274, 322, 592
back-arc spreading, 274
backshore, 545
backswamps, 425
backwash, 535
Badwater Basin, Death Valley, 511
Baikal, Lake (Siberia), 326–327
bajadas, 509
baked contacts, 110
banded iron formation (BIF), 148, 570, 584
banded marble, 213
Banff National Park (Canada), 473
barchan dunes, 518

barchanoid dunes, 519–520
barrier islands, 188, 545–546
barrier reefs, 286, 548–549
Barringer Crater (Arizona), 707
Barrow, George, 214
basal sliding, 470
basalt, 14, 116, 682, 690
base level of a stream, 406–407
basins, 312
batholiths, 107, 112
bauxite, 149, 570
baymouth bars, 545
bays, 541
beach deposits, 188
beach drift, 536, 545
beaches
    definition of, 544
    as depositional landforms, 544–545
    pocket, 544
    raised, 489, 551
    replenishment of, 554–556
    strands, 544
beach face, 545
bedding, 180–182
bedding planes, 180
bed load, 420–421
bedrock, 12, 153
bedrock slopes, 374–375
beds, 180–181
beheaded streams, 411
benching, 397
Benue Trough, 432
berms, 545
beta particle emission, 244
B horizon, 155
Big Bang theory, 20
Bingham Canyon copper mine, 600
biochemical sedimentary rocks, 166, 170–174, 176–179
bioclastic limestone, 172
bioclastic textures, 172
biodiesel, 637
biomass, 636–637
biosphere, 33
biotite, 114–115
bitumen, 615
bituminous coal, 177, 624
black smokers, 268
blocks, 123–124
blocky (*aa*) lava, 121–122

blowouts, 514, 520
blueschist facies, 217
body waves, 339. *See also* seismic waves
bombs, 123–124
bonding, chemical, 71–74
bonds, 71, 76
Bonneville, Lake, 173, 486
borax, 586–587
boreholes, 456
bottomset beds, 426
bottom structures, 182, 185
boulders, 175
Bowen, Norman, 117, 118
Bowen's Continuous Reaction Series, 117–118
Bowen's Discontinuous Reaction Series, 117–118, 120–121
braided streams, 424
breakers, 534
breakwaters, 554
breccias, 175
breeder reactors, 628
Bretz, Harlen, 490
brittle failure, 301
brittle materials, 300–301
Bryce Canyon National Park, 238, 240
buoyancy, 258–259
burial metamorphism, 207
Bushveld complex (South Africa), 589
buttes, 509

C/2011 L4 Pan-Starrs comet, 708
calcite, 91, 144–145, 166
calcium carbonate, 584, 672
calderas, 121, 129, 291
caliche, 157
Callisto, 703
camel-bone fossils in the Arctic, 648
canyons, 410
capacity, 422
capillary fringe, 439
carbonate minerals, 91
carbon cycle, 652–654
carbon dioxide
  atmospheric, 651, 655–656
  biomass and, 637
  calcium carbonate and, 672
  coal and, 623
  glaciation and, 493, 495–496
  global warming and, 646
  weathering and, 158–159
  See also climate change

carbon isotopes, 241, 243
Carlsbad Caverns (New Mexico), 453
Cascade Range, 128
Cassini, Giovanni, 681
cassiterite, 587
catastrophism, principle of, 227
cations, 73, 76
CAT (computerized axial tomography) scans, 361
caves, limestone, 453–455
cement, 581
cementation, 166
Cenozoic Era, 238, 240
center of mass, 685
chain reactions, 627
chalk, 166, 176
channel bars, 423–424
Channeled Scablands, 490–492
channels, 403, 409–410, 426–427
Charon, 705
chemical bonding, 71–74
chemical formula of a mineral, 79
chemical reactions, 76
chemicals, raw, 576–578
chemical sedimentary rocks, 166, 170–174, 176–179
chemical weathering
  acids and acid rain, 143–144
  climate, 149
  definition of, 138–139
  dissolution, 144–145
  hydration, 149
  hydrolysis, 145–146
  mechanical weathering vs., 143
  oxidation, 147–149
  time, 149
chemistry, basic concepts in, 16–18
chemosynthesis, 267
chert, 178
Chesapeake Bay, 550
chilled margins, 110
chlorine (Cl), 72–73
chondrites, 707
C horizon, 155
Christchurch, New Zealand, 398
chromite, 574
chromium (Cr), 574–575, 596–597
chrysotile (white asbestos), 93
cinder cones, 121, 127
cirques, 469, 474
clast composition, 167–169
clastic textures, 167, 174

clasts, 105, 166
clay minerals, 90–91
clay particles, 175, 423
Clean Air Act, 624
cleavage, 85–86
cleavage planes, 85
Cleopatra's Needle, 141
cliffs, 541–543
climate
  chemical weathering and, 149
  desertification and, 522
  mass wasting and, 376
  over geological time, 493
  plate tectonics and, 43–44
  soil development and, 153–154
climate change
  Arctic camel-bone fossils and, 648
  carbon cycle and, 652–654
  definition of, 646
  Earth's natural pulse and IPCC findings, 646–647
  Earth-Sun geometry and Milankovitch cycles, 668–670
  feedback systems and, 670–673
  fossil fuels and, 607–608
  global warmth and the greenhouse effect, 676
  greenhouse gases, 178, 493, 646 (*See also* carbon dioxide; methane)
  ice cores and proxies of global temperature change, 659–665
  melting of Arctic ice and, 666–668
  over last half-century, 654–657
  physics of global warming, 652
  plate tectonics and, 158–159, 673–675
  solar output variations and, 657–659
  solar radiation and greenhouse effect, 649–652
climate proxies, 655, 659–660
closed drainage basins, 416, 419
coal
  about, 622–623
  as biochemical sedimentary rock, 170, 176–177

distribution and reserves of, 624
  formation of, 623–624
  historical importance of, 178
  plate tectonics and, 639
  as rock, not mineral, 93
coarse-grained igneous rocks, 107
coastal deserts, 504–507
coastal environments
  barrier islands, 545–546
  beaches, 544–545, 554–556
  coastlines, erosional and depositional, 549
  coastlines, submergent and emergent, 550–551
  deposition in, 188, 544–549
  erosion in, 384
  headlands, sea arches, and sea stacks, 543–544
  hurricanes and storm surges, 553–554
  jetties, groins, and breakwaters, 551–554
  management of, 551–557
  plate tectonics and, 558–561
  polar ice, sea level, and coastlines, 547–548
  processes (overview), 530
  reefs, 546–549
  sea level change and, 188–190
  spits and baymouth bars, 545
  tides, 537–540
  wave erosion, 540–543
  wind-driven waves, 532–537
Coastal Zone Management Act, 556
cobbles, 175
cohesion, 374
coke, 623
collapse sinkholes, 453
collisional zones
  metamorphism and, 218–219
  mountain building by continental collision, 321–327
  terrane accretion, 319–321
color of a mineral, 82
cols, 474
columnar joints, 140, 304
columns, 453
comets, 681, 708–711
compaction, 166
competence, 422

composite batholiths, 112
composite volcanoes, 121, 127
compositional convection, 359
compounds, 70, 76
compression, 200, 261
compressional stresses, 300
concentrated solar power, 634–635
concentrations of minerals and ore deposits, 582–588
conchoidal fracture, 86
concrete, 581
conduction, 28–29
cones of depression, 457
confined aquifers, 445–446
confining pressure, 200, 300
conglomerates, 175
conservative plate boundaries, 61
constructive plate boundaries, 60
contact metamorphism, 113, 204
contact relationships, 110–112
continental arcs
    formation of, 58, 61, 276, 278
    island arcs vs., 132
    metamorphism and, 218
continental crust
    about, 353–354
    definition of, 25
    on Earth vs. Venus, 697
    as layer of lithosphere, 257
    partial melting and, 317–318
continental divides, 416
continental drift, 40–46, 639. See also plate tectonics and plates
continental environments, deposition in, 187–188
continental glaciers, 469, 476
continental hotspots, 289–292
continental interior deserts, 507
continental margins, types of, 60, 276, 560–561
continental platforms, 259
continental red beds, 148
continental rifting, 59–60, 263–264, 589–590, 592
continental rift valleys, 263
continental shelves
    edges of, 559
    plate boundaries and, 60
    polar ice, sea level, and, 548
    rifting and, 263
    turbidity currents and, 176
continental slopes, 42

continental transform faults, 281–286
continent-continent collision, 270
convection, 27, 28–29, 359–364
convergent plate boundaries
    about, 61, 262, 269
    characteristics of subduction zones, 270–271
    continent-continent collision, 270, 321–327
    metamorphism and, 217–218
    mineral deposits at, 590–592
    ocean-continent convergence, 276–278
    ocean-ocean convergence, 271, 274–276
    petroleum and, 639
    subduction and growth of continents, 278–280
    subduction and subduction zones, 269–270
    See also subduction
copper (Cu), 571, 573, 591
coral limestone, 171–172
coral reefs, 171–172
core
    definition of, 26
    inner and outer, 351–352, 356, 359
    iron-rich, 356
    magnetism and, 357–358
Coriolis effect, 504, 506–507
correlation of rock strata, 234–237, 614–615
correlation vs. causation, 652
corundum, 580
covalent bonds, 73
cracking, 608
Crate Lake (Oregon), 129
craters, impact, 682–684, 706, 707
creep, 360, 392–394, 469
Cree people, 431
crest of a flood, 428
crevasses, 469
cross beds and cross-bedding, 180–181, 518, 522
cross-cutting relationships, rule of, 231–233
crude oil, 608–609
crushed rock, 580
crushing, 204
crust
    about, 353–354
    definition and types of, 25, 257

on Earth vs. Venus, 697
    partial melting and, 317–318
    rebound and sea level changes, 489
crustal abundance, 573
crystal faces, 78
crystal form, 84
crystal growth, 201
crystalline solids, 77, 78–79
crystallization, 103–104, 118–119, 176
crystals, about, 78
crystal settling, 118
crystal structure, 73
crystal wedging, 141
Cynognathus, 44
Cyprus, 590

dams, 430–431, 632–634
Darcy, Henri, 444
Darwin, Charles, 118, 235, 548
dating
    absolute, 241–250
    radiometric, 246–248
    relative, 237–241
daughter isotopes, 245, 246
Dead Sea Transform, 282, 283–284
debris flows, 388–392
decomposers reservoir, 653
deep ocean trenches, 54–55, 271, 319
deformation
    about, 298–299
    elastic brittle vs. plastic ductile, 300–301
    factors influencing, 301–302
    faults and, 304–309
    folds, 302, 309–313
    joints, 304
    orientation of geologic structures and, 302–304
    stress and strain, 299–300
    unconformities and, 313–315
Deimos, 691
Delaware Bay, 550
Delaware River, 414
deltas, 180, 181, 426–427
dendritic drainage, 417
density
    concept of, 16–17, 344
    Earth's interior and, 336–337, 356
    of other planets, 687, 689, 690, 703, 704
    plates and, 258–259
    specific gravity and, 86

depleted mantle, 318
deposition
    coastal, 188, 544–549
    in continental environments, 187–188
    glacial, 476–484
    loess, 521
    by streams, 187, 423–428, 509
    by wind, 517–521
depositional coastlines, 549
depositional environments, 164, 185–190
Descartes, René, 19
desertification, 515, 521–523
desert pavement, 514
deserts
    coastal, 504–507
    continental interior, 507
    definition of, 502
    dunes, 517–521
    global distribution map, 503
    on Mars, 524–525
    oases, 446, 450, 509
    plate tectonics and, 523–524
    polar, 507
    rain-shadow, 504
    subtropical, 502–504
    weathering and erosion in, 508–511
    wind in, 511–521
desert varnish, 508
destructive plate boundaries, 61
detachments, 307
detrital sedimentary rocks, 166, 167–169, 175–176
detritus, 165, 586
Devonian Period, 240
diagenesis, 166
diamond, 73, 77–78, 86, 95, 578
diamond pipes, 578
differential stresses, 300
differentiation, 697
dikes, 109–110, 112, 113, 279
diorite, 116
dip, 302–304
dipoles, 378
dip-slip faults, 306–307
directed pressure, 200–201
direct high tide, 538
disappearing streams, 453
discharge, 406, 441–442
disconformities, 234, 314
disseminated deposits, 586
dissolution, 144–145, 451–452
dissolved load, 421–422
distributaries, 415
diurnal tidal pattern, 539

divergent plate boundaries
  about, 59–60, 262
  continental rifting and
    ocean formation,
    263–264
  mid-ocean ridges,
    development of,
    266–269
  mid-ocean ridges and
    ocean opening,
    264–265
  mineral deposits at,
    589–590
  petroleum and, 637–639
divides, 416
dolomite, 91, 172
dolostone, 174
domes, 312
downcutting, 410
dowsing, 456
drainage basins, 415–417
drainage systems, 414–418
drift, 476–484
drilling, 594, 611–613,
    618–619, 622
dropstones, 484
drowned coastlines, 550
drumlin fields, 481
drumlins, 479, 481
ductile behavior, 301
dunes, 517–520
Dust Bowl, 514, 515–516
dust devils, 511, 512
dust storms, 513
dynamic metamorphism,
    204–205

Earth
  age of, 10, 227–228, 249
  axis tilt and wobble, 668
  Earth-Sun geometry, 659,
    668–670
  evolution of (*See* time,
    geologic)
  internal heat and heat
    flow, 27–30
  internal layering of, 23–26
  magnetic field of, 46–47,
    357–358
  mass and volume of, 356
  rotation of, 538
  seasons and tilt of, 22–23
  self-regulation of, 671
  size of, 686
  snowball Earth, 493–495
  as system, 30–34
  Venus compared to,
    696–697
  See also Earth's interior
earthflows, 392
Earth-Moon system, origin
    of, 21, 684–686

earthquake foci, 54–55, 337
earthquakes
  Christchurch, New
    Zealand, 398
  Haiti (2010), 279
  liquefaction and mass
    wasting, 376, 397–398
  Loma Prieta (1989),
    340–341
  at mid-ocean ridges, 270
  Northridge (1994),
    282–283, 376
  prediction of, 340–342
  Richter magnitude scale,
    344–346
  San Francisco (1906),
    284, 340
  subduction and, 54,
    270–271
  See also seismic waves
Earth's interior
  circulation and
    convection in, 359–365
  core, 351–352, 356–359
  crust layer, 353–354
  earthquakes and elastic
    rebound theory, 337
  general characteristics of
    layers, 352–353
  heat and density as clues
    to, 336–337
  magnetism and, 357–358
  mantle layer, 354–355
  the Moho and the shadow
    zone, 350–351
  plate tectonics and fate of
    subducted slabs,
    365–366
  seismology and seismic
    rays, 347–350
  temperatures within, 361
  See also seismic waves
East Pacific Rise, 265, 266
eclogite facies, 217
economic geologists, 34
E horizon, 155
Einstein, Albert, 20, 357
elastic deformation, 300–301
elastic rebound theory, 337
electrical properties, mineral
    exploration with, 593
electromagnetic
    spectrum, 649
electron capture, 244
electrons, 70–74, 76
elements
  atoms and, 70–71
  chemical bonding and,
    71–74
  definition of, 16, 70, 76
  origin of chemical
    elements, 74–75

Ellesmere Island
    (Canada), 648
El Niño events, 384, 385, 395
embayments, 325
emergent coastlines, 550–551
emery, 580
Enceladus, 703–704
end moraines, 479
energy
  definition of, 17
  kinetic, 17–18, 104, 652
  thermal, 18
  types of, 17–18, 104
energy producers and
    consumers, 652–654
energy resources
  coal, 178, 622–625
  methane hydrate, 621–622
  nonrenewable vs.
    renewable resources,
    606–607
  nuclear, 625–629
  oil and natural gas (*See*
    petroleum)
  plate tectonics and,
    637–640
  renewable (*See* renewable
    energy)
  use and consumption of,
    607–608
engineering geologists, 34
environmental geologists, 34
Environmental Protection
    Agency, US, 461
eons, 12, 240
epicenter, 337
epochs, 241, 485
equatorial low, 502
equilibrium line, 470
equilibrium profiles, 407
eras, 12, 238, 240
erosion
  agents of, 16, 165, 409, 508
  coastal and wave,
    540–543
  coastal management of,
    551–556
  definition of, 16
  in deserts, 508–511
  El Niño events and coastal
    erosion, 384
  glacial, 471–476
  isostacy and, 260–261
  mass wasting and, 373
  mineral concentration
    and, 586
  sediment formation
    and, 165
  by streams, 409–418, 509
  subduction erosion, 319
  by wind, 514–517
erosional coastlines, 549

erosional unloading, 304
erratics, 227, 476, 478
eruptions. *See* volcanic
    eruptions
eskers, 482–483
Eskola, Pentti, 216
estuaries, 550
Europa, 701
evaporation of ocean
    water, 660
evaporite minerals, 91, 170,
    176, 586–587
Everest, Mt., 325
evolution of the earth. *See*
    time, geologic
exfoliation joints,
    140–141, 304
exhalites, 588
exocomets, 708
exoplanets, 36, 711–712
exotic terranes, 321
exponential decay, 246
exponential relationships, 247
extension, 200, 261
exterior oceans, 332
extinctions, mass
    (Cretaceous), 707,
    708–709
extrusive rocks, 107
Eyjafjallajökull volcano
    (Iceland), 107

fabric, 207
failed rifts, 62, 290
fall, 380–382
false color images, 594
fault breccia, 204, 304
fault creep, 285
fault gouge, 304
fault plane, 304
faults
  about, 304–306
  Cabot Fault, 286
  continental drift and
    continuity of, 43
  continental rifting
    and, 263
  dip-slip, 306–307
  dynamic metamorphism
    and, 204
  elastic rebound theory
    and, 337
  Great Glen Fault, 286
  normal, 306–307
  as petroleum traps, 610
  reverse, 306, 307
  San Andreas Fault,
    282–283, 284–285
  strike-slip, 307–309, 326
  transform, 60–61, 309
fault scarps, 306
fault zones, 304

faunal succession, principle of, 234–237
feedback, 651, 670–671
feldspar group, 89
feldspars, 146, 169
felsic igneous rocks, 113–115
ferromagnesian silicates, 82, 87–88
fertilizers, 576
field relationships, 110–113
fine-grained igneous rocks, 104
fire-flood sequences, 388
firns, 467
fission, nuclear, 627–628
fissure eruptions, 121
fjords, 475, 551
flat plate collectors, 634
flexural basins, 191
flood basalts, 121
flooding, 187, 428–432
floodplains
    deposits and natural levees, 187, 425–426
    meanders and, 411
    sediment formation and, 165
    soil development on, 153
    stream terraces and, 413
flow
    creep, 392–394
    debris flows, 388–392
    definition of, 384
    earthflows, 392
    mudflows, 384–388
    solifluction, 392
flow textures, 105
fluids and metamorphism, 201–202
flute casts, 182
fold hinges, 311–312
folds, 302, 309–313
foliated metamorphic rock, 208–212
foliation, 302, 312–313
food chain, 151, 152–153, 652–653
footwalls, 306
foraminifera, 664–665
force, 17
foreset beds, 426
foreshore, 545
fossil fuels (overview), 607. See also climate change; energy; petroleum
fossil record, 44, 52
fossils
    absolute dating and, 248
    Appalachian-Caledonian fossil assemblages, 327, 329

camel-bone fossils in the Arctic, 648
definition of, 13
index, 241
microfossils, 614
sedimentary rocks and interpretation of, 179–180
fracking (hydraulic fracturing), 617, 618–619
fractional crystallization, 118–119
fractionation, 118
fractures in minerals, 86
fractures in rock, 302, 304–309. See also faults
frequency of a wave, 533, 534
fringing reefs, 286, 548–549
frost wedging, 141, 380
fumaroles, 447
Fundy, Bay of, 540, 632

gabbros, 116, 279
Galapagos Rift, 267
galena (PbS), 575
Galileo Galilei, 689, 701
Ganges-Brahmaputra River, 427
Ganymede, 701–703
garnet group, 87
garnets, 203
gaseous phase, 77
geiger counters, 627
gemologists, 82
gems, 78
gemstones, 578–581
geochemical methods of mineral exploration, 594
geochemists, 34
GeoEye, 596
geologic time. See time, geologic
geologic time scale, 12, 238–241, 248–250
geologists, 7, 34–36
geology
    birth of, 9–10
    definition of, 6
    scientific method and, 7–9
    subdisciplines and specialties, 34
geophysical methods, 337, 593–594
geophysicists, 34, 44–45
geosphere, 32
geothermal energy, 449–451, 629–631, 640
geothermal gradient, 198
The Geyers (California), 630

geysers, 449
gigapascals (GPas), 200
Gilbert, William, 357
glacial budget, 470
glacial lake deposits, 484
glacial terminus, 470
glacial troughs, 474
glacier beds, 470
Glacier National Park, 473
glaciers and glaciation
    abrasion and plucking, 471–472
    alpine landforms, 473–476
    atmospheric air temperatures and, 468
    continental landforms, 476
    definitions, 466–467
    deposits from glacial lakes, 484
    drift and deposition, 476–484
    formation of glaciers, 467–469
    Great Lakes, origin of, 477–478
    growth and shrinking of glaciers, 470–471
    Lake Missoula and Channeled Scablands, 490–492
    movement of glaciers, 469–470
    on other worlds, 496–497
    plate tectonics and, 492–496
    Pleistocene glacial advances, 485–486
    Pleistocene glacial retreats, 486–492
    soil development on glacial deposits, 153
    thickness of glaciers, change in, 547
glaucophane, 217
global warming, 646. See also climate change
Glomar Challenger, 52
Glossopteris flora, 44
gneiss, 212
gneissic foliation, 208
gold (Au), 575, 587
Gondwana, 43, 63, 496, 639
grabens, 307
graded bedding, 181–182
graded streams, 407
gradients, 405, 430
grading, 397
gradualism, principle of, 228–229
grain shape, 169

grain size, 169
Grand Canyon, 180, 238, 240, 374
granite, 15, 93, 114–115, 179
granulite facies, 217
graphite, 74, 77–78
gravel, 175, 423, 580–581
gravimeters, 593
gravitational fields, 17
gravity
    concept of, 17
    law of, 8
    mineral exploration and, 593
    specific gravity of a mineral, 86
    tides and, 537–540
    weight and, 76
gravity anomalies, 62
gravity surveys, 611
graywackes, 176
Great Barrier Reef, 548
Great Dark Spot, 704
Great Glen Fault, 286
Great Lakes, origins of, 477–478
Great Red Spot, 699–700
Great Rift Valley, 264, 265
Great Slave Lake (Canada), 249
greenhouse conditions, 647, 674
greenhouse effect, 649–652, 676, 690
greenhouse gases, 178, 493, 646. See also carbon dioxide; climate change; methane
Greenland, 467
Green River Formation, 615
greenschist facies, 216–217
green turtles, 53–54
groins, 554
groundmass, 104, 110, 116
ground moraines, 479–480
groundwater
    aquifers, aquitards, and aquicludes, 442
    aquifers, historical importance of, 447
    confined and unconfined aquifers, 443–446
    dissolution by, 451–452
    exploring for, 456–457
    factors in flow of, 442–443
    geothermal energy, 449–451
    hot springs, fumaroles, and geysers, 446–449
    human contamination of, 459–462

groundwater (*Continued*)
  infiltration and, 404
  karst topography,
    452–453
  limestone caves and,
    453–455
  plate tectonics and, 462
  porosity and permeability,
    440–441
  recharge-discharge
    balance, 441–442
  as resource, 438–439
  Sahara Desert and, 450
  as solvent, 166
  springs and oases, 446
  water pressure at
    depth, 444
  water table, zone of
    saturation, and zone of
    aeration, 439–440
  withdrawal, and seawater
    invasion, 457–459
gushers, 611
Gutenberg, Beno, 350–351
gypsum, 176, 581

habit in minerals, 84–85
Hadeon Eon, 249
Hadley convection cells, 504
Hadrian's Wall, 9, 229
Hale-Bopp comet, 708
half-grabens, 307
half-life, 245–246
halite (NaCl; table salt)
  chemical formula of, 79
  concentration of, 583–584
  as crystal structure, 73
  dissolution of, 144
  as nonmetallic
    mineral, 577
  salt domes, 174, 577, 578,
    587, 610
  salt flats and rock salt,
    173–174
  shape of, 78
Halley, Edmund, 249, 708
Halley's Comet, 708
hanging valleys, 476
hanging walls, 306
hardness of minerals, 85
hard water, 452
Haughton, Samuel, 249
Hawaiian-Emperor
    Seamount Chain,
    287–289
Hawaiian islands, 64–65,
    133, 287–289, 655
headlands, 543
headwall scarps, 383
headward erosion, 410
headwaters, 405
heat, 18, 104, 198–199, 337

heat flow, 27–30, 337
heliostats, 634–635
helium, 74
hematite, 570
high grade metamorphic
    rocks, 209
High Plains (Ogallala)
    aquifer, 444–445
high tide, direct and
    indirect, 538
Himalayan Mountains, 325,
    327, 328–329, 493
homogeneous model of
    planetary accretion, 21
horizontal drilling, 617,
    618–619
horizontal folds, 312
hornblende, 88, 114–115
hornfels, 209, 213
horns, 474
horsts, 307
hotspots
  about, 64, 286–287
  composition of, 287
  continental, 289–292
  geothermal energy
    and, 640
  magma composition
    and, 133–134
  on Mars, 693
  mountain building
    and, 316
  oceanic, 287–289
  reefs and, 549
hot-spring deposits, 592
hot springs, 447
Huang He (Yellow River),
    328, 420, 521
Hubbard Glacier, 470
Hubble, Edwin, 20
Hubble Space Telescope, 705
Hudson Bay, 489
human activities
  climate change and,
    647–649, 673, 676
  groundwater
    contamination and,
    459–462
  mass wasting and,
    379, 389
  mechanical weathering
    and, 143
  See also energy
humus, 153
hurricanes, 553–554
Hutton, James, 9–10, 138,
    228–229, 313
hydration, 149
hydraulic fracturing
    (fracking), 617,
    618–619
hydraulic heads, 443–444

hydrocarbons, 77, 608
hydroelectric power, 607,
    632–634
hydrogen bonds, 144, 378
hydrogeologists, 34, 456
hydrologic cycle, 33,
    402–404
hydrologists, 34
hydrolysis, 145–146
hydrosphere, 32, 33–34, 438.
    See also streams
hydrothermal deposits,
    586, 590
hydrothermal
    metamorphism, 206
hypothesis, 7–8
hypsometric curve, 259
Hyutake comet, 708

ice ages
  climate over time and, 493
  Little Ice Age, 658–659
  Pleistocene, 485–492,
    547–548
ice caps. *See* polar ice caps
ice cores, 660–663
ice fields, 469
ice giants, 704
icehouse conditions, 647, 674
Iceland, 449, 467
ice rafting, 484
ice sheets, 416–417, 469
igneous complexes,
    layered, 589
igneous rocks
  appearance and texture
    of, 104–107
  chemical evolution of
    magma, 117–120
  definition of, 13
  extrusive and intrusive,
    107–110
  felsic, intermediate, mafic,
    and ultramafic
    classifications, 113–116
  field relationships,
    observation of,
    110–113
  fissure eruptions vs. vent
    eruptions, 121
  intermediate and felsic
    magma eruptions,
    123–129
  mafic magma eruptions,
    121–123
  magma formation,
    transport, and cooling,
    101–103
  mineral concentration in,
    585–586
  physical evolution of
    magma, 120–121

plate tectonics and,
    131–134
  rock cycle and, 14–16,
    100–101
impact craters, 682–684,
    706, 707
incised meanders, 414
inclusions, 113, 231
index fossils, 241
index minerals, 214
indirect high tide, 538
industrial materials,
    568–569, 580–581.
    See also mineral
    deposits
industrial waste and
    sewage, 461
infiltration, 376, 404
infrared (IR) radiation, 649
inhomogeneous model of
    planetary accretion, 21
inner core, 26, 351, 356, 359
inorganic compounds, 77
inorganic fertilizers, 576
inselbergs, 509
Intergovernmental Panel on
    Climate Change
    (IPCC), 647, 654
interior oceans, 332
intermediate igneous rocks,
    113, 116
intermittent streams, 509
intrusive rocks, 107–110
Io, 701
ionic bonds, 71–73
ionic transport, 201
ions, 73, 76
iron (Fe), 570
iron meteorites, 356, 708
iron oxides, 584
iron-rich core, 356
irregular satellites, 701
island arcs, 57, 61, 132, 271,
    274–275
isostacy, 259–261, 489
isostatic equilibrium, 260
isostatic rebound, 260
isotopes, 241, 243–246

**James Bay Project (Quebec,
    Canada), 431**
Japanese earthquakes, 271
jarosite, 694
Jasper National Park
    (Canada), 473, 474
Jeffreys, Harold, 45
jetties, 551–554
jigsaw fit of continents,
    42–44
joints, 139–141, 157, 304
joint sets, 304
Joly, John, 249

Jovian planets, about, 686, 698–699. *See also* Solar System and planets
Jupiter, 698, 699–701
Jupiter's moons, 701–703
Jurassic Period, 240

kames, 482
Kant, Immanuel, 19
kaolinite, 211
karst topography, 452–453
Kazakhstan, 419
Kelvin, William Thomson, Lord, 27, 249
Kepler, Johannes, 8
kettles, 482
kilobar (kbars), 200
kimberlite, 578
kinetic energy, 17–18, 104, 652
komatiites, 116
Kuiper Belt, 705
Kuroko-type massive sulfide deposits, 592
kyanite, 95

lagoons, 188, 545
La Grande River, 431
Lagrange, Joseph, 681
lahars, 125, 387–388
laminar flow, 404
land bridges, 485–486
landfill sites, 460
Landsat, 594, 596
landslides. *See* mass wasting
lapilli, 124
Laplace, Pierre-Simon, 19, 681
lateral accretion, 425
lateral moraines, 480
laterite soils, 157
lava, 31, 100. *See also* volcanic eruptions
lava fountains, 122–123
lava lakes, 121
lava tubes, 122
laws and the scientific method, 8
layered igneous complexes, 589
layered mantle convection model, 360–361
leaching, zone of, 155
lead (Pb), 573–574
Leaning Tower of Pisa, 457
left-lateral strike-slip faults, 309
length, 16
levees, natural, 187, 425, 428
LiDAR (light detection and ranging), 596
light-water reactors, 627–628
lignite, 177, 624

lime, 580
limestone
    about, 93, 171–172, 178–179
    caves, 453–455
    crushed, 580
    karst topography and, 452
    Portland cement, 581
    reefs, 171–172, 546–549
limonite, 147
lineaments, 595–596
linear relationships, 247
liquefaction, 387, 397–398
liquid phase, 77
lithification, 16, 166
lithosphere, 26, 256–258
lithospheric mantle, 258
Little Ice Age, 658–659
load casts, 182, 185
local base levels, 406
Loch Ness, 285
loess, 481, 521
London Basin, 447, 462
longitudinal dunes, 520
longitudinal profile, 405
longshore currents, 536
longshore drift, 536–537, 545
Love Canal, Niagara Falls, 461
Lowell, Percival, 524, 691
lower mantle, 355
lower plate, 269
low-grade deposits, 591
low grade metamorphic rocks, 209
low-velocity zone, 355
lunar days, 539
lunar highlands, 681
luster, 84
Lyell, Charles, 229–230
Lystrosaurus, 44

mafic igneous rocks, 113
magma
    cooling of, 103, 104
    definition of, 100
    igneous rocks and, 13
    intrusion mechanisms, 103
    rift valleys and, 59–60
    seafloor spreading and, 51
    source rock and formation of, 101–102
    subduction and, 55–58
    transport of, 102–103
    See also subduction; volcanic eruptions
magma chambers, 101
magma mixing, 119–120
magmatic arc environments, 590–592
magmatic deposits, 585–586
magmatic segregation, 585, 590
magnetic anomalies, 50

magnetic field, 357–358
magnetic inclination, 46–47
magnetic polarity reversal, 49–52, 358
magnetic poles, 357–358
magnetism of Earth, 46–47, 357–358
magnetite, 570
magnetometers, 593
magnetotellurics, 456–457
Mammoth Cave (Kentucky), 453
manganese nodules, 601
mantle, 26, 102, 257–258, 354–355
mantle convection, 360–364
mantle drag, 292, 293
mantle plumes, 64, 133–134, 286–287, 316, 361, 365
mantle wedges, 275
marble, 209, 213
Marcy, Geoffrey, 711
maria (mare), lunar, 681, 682
Mariana Trench, 271
marine environments, deposition in, 188. *See also* coastal environments
marine regression, 189–190
marine terraces, 551
marine transgression, 189–190
Mars
    atmosphere on, 686
    atmospheric temperature of, 652
    deserts and wind features on, 512, 524–525
    life on, 695, 696
    mass wasting on, 386–387
    polar ice caps on, 497
    running water, evidence of, 433–434
    size of, 686
Mars Exploration Rovers, 8
mass, 16, 76
mass, center of, 685
mass extinctions (Cretaceous), 707, 708–709
massive sulfide deposits, 590, 592
mass number, 241
mass wasting
    classification of, 380
    climate, vegetation, and, 376
    definition of, 372–373
    earthquakes and other triggers, 376, 379
    fall, 380–382
    flows, 384–394

identifying, predicting, and preventing slope failure, 395–397
    on the Moon and Mars, 386–387
    plate tectonics and, 397–398
    sea cliffs and, 542
    sedimentation and, 165
    slides, 382–384, 390–391
    slope stability and angle of repose, 373–375
    surface tension and sandcastles, 378–379
    water content and, 375–376, 396–397
matrix, 167, 175–176
matter, 16, 76
Matterhorn, 474
Matthews, Drummond, 50, 51–52
Mauna Loa Observatory (Hawaii), 655
meandering streams, 411, 414, 424
mechanical energy, 18
mechanical enrichment, 587
mechanical weathering, 138, 139–143
medial moraines, 480
meltwater, drift from, 481–484
Mercury, 686, 687–689
mesas, 509
Mesosaurus, 44
Mesozoic Era, 238, 240, 248
metallic bonds, 73–74
metals and metallic minerals
    abundant, 569–570
    definition of, 570
    scarce, 570–575
metamorphic facies, 216–217
metamorphic grade, 199
metamorphic rock
    classification of, 209–214
    definition of, 13
    foliated, 208–212
    mineral concentration in, 588
    nonfoliated, 209, 212–214
    rock cycle and, 16–19, 196–198
    texture and, 197, 207–209
metamorphic zones, 214–216
metamorphism
    definition of, 196
    deformation and, 312–313
    factors controlling, 198–203
    plate tectonics and, 217–219
    rock composition and, 202

metamorphism (*Continued*)
rock cycle and, 196–198, 215
types of, 204–207
zones and facies, 214–217
Meteor Crater (Arizona), 707
meteorite bombardment, 682–684, 697
meteorites, 356, 680–681, 706–708
meteors, 680, 697, 705–706, 707
meteor showers, 706
methane, 610, 622, 651
methane hydrate, 621–622
mica, 86
mica group, 88
microcontinents, 319
microfossils, 614
microprobe analysis, 203
Mid-Atlantic Ridge
Ascension Island and, 53–54
exploratory dives at, 267
hotspots along, 290
Iceland and, 449
magnetic anomalies and, 50
seafloor age and, 52
spreading rates along, 265
topography of, 266
mid-ocean ridges
bizarre world of, 267–268
development of, 266–269
earthquakes and, 270
hotspots near, 290
igneous rocks and, 131
magnetic reversals and, 50
metamorphism and, 206, 217–218
mineral deposits at, 590
mountain building and, 315–316
ocean opening and, 264–265
transform offsets in, 281
volcanism and, 278
Midway Island, 289
migmatite, 212
Milankovitch, Milutin, 668
Milankovitch cycles, 668–670, 676, 696
Milky Way Galaxy, 20
mineral assemblages and metamorphic facies, 216
mineral deposits
classification of, 569
definition of, 568
environmental impact of mining, 601–602

exploration and prospecting, 592–596, 599–602
formation and concentrations of, 582–588
future demands on, 596–602
global distribution of, 596–597
human consumption of, 569, 597–598
industrial materials, 568, 580–581
metals and metallic minerals, 569–575
nonmetallic minerals, 575–580
plate tectonics and, 588–592
mineralogists, 34, 82
mineralogy, 82
mineraloids, 78
minerals
chemical bonding and, 71–74
chemical composition and formulas, 79
cleavage and fracture, 85–86
color and streak, 82–83
crystal form and habit, 84–85
as crystalline solids, 78–79
definition of, 13, 77
elements as building blocks of, 70–77
ferromagnesian silicates, 82, 87–88
hardness, 85
luster, 83
as naturally occurring and inorganic, 77–78
non-ferromagnesian silicates, 82, 88–91
nonsilicate, 91–92
people and, 93–95
physical properties of, 82–87
plate tectonics and, 95
polymorphs, 79–81, 95
rock-forming (silicate and nonsilicate), 81–82
rocks and, 93
silicate tetrahedron, 87
specific gravity and density, 86
mining, 569, 623. *See also* mineral deposits
Mississippi River, 167–168, 188, 406, 415, 416, 427, 429

Mississippi valley-type deposits, 589
Missoula, Lake, 490–492
mixed tidal pattern, 539
Moho (Mohorovičić Discontinuity), 258, 350, 354
Mohorovičić, Andrija, 350
Mohs, Friedrich, 85
Mohs scale of hardness, 85
Mojave Desert, 446
molecules, 71, 76
molybdenum deposits, 592
monoliths, 453
Monument Valley, 511
Moon
about, 681–684
age of, 249
fractional crystallization and, 119
mass wasting on, 386
origin of, 21
phases of, 539–540
tides and, 537–540
moons
definition of, 680
of Jupiter, 701–703
of Saturn, 703–704
moraines, 478–480
Morgan, Jason, 286, 289
Morley, Lawrence, 50, 51–52
mountain belts, ancient, continuity of, 42–43
mountain building
Amazon River and, 10
climate change and, 674
compression and, 200
by continental collision, 321–327
hotspots and, 316
at mid-ocean ridges, 315–316
by subduction, 316–319
supercontinent cycle and, 330–332
terrane accretion, 319–321
transform plate margins and, 316
Wilson cycle, 327–330
mountaintop removal mining, 623
mudcracks, 182, 187
mudflows, 384–388
mudstones, 176
multispectral scanners, 594
muscovite, 90, 114–115
mylonite, 204, 208

native minerals, 92
natural gas, 608. *See also* petroleum
natural levees, 187, 425, 428

Navajo Sandstone, Zion National Park, 180, 181, 238, 240
neap tides, 539
nebula, 21
negative feedback, 670–671
Neptune, 698, 704
neptunists, 228
neutrons, 71, 76, 241
Nevado Huascarán rockfall (Peru), 380–381, 382–383, 397
Newton, Isaac, 8, 537
Newton's First Law of Motion, 17
Niagara Falls, 407
Nile River, 427, 432
nitrogen (N), 576
nonclastic sedimentary rocks, 170
nonclastic textures, 174
nonconformities, 234, 314
non-ferromagnesian silicates, 82, 88–91
nonfoliated metamorphic rock, 209, 212–214
nonmetallic minerals, 575–580
nonrenewable resources, 569, 606–607. *See also* energy; mineral deposits; petroleum
nonsilicate minerals, 81, 91–92
normal faults, 306–307
normal magnetic polarity, 49
Nova Scotia unconformity (Canada), 313–314
Nubian sandstone and aquifer, 450
nuclear fission, 627–628
nuclear power, 625–629, 640
nucleus, 70, 76
nullahs, 509

oases, 446, 450, 509
ocean-continent convergence, 276–278
ocean currents and Coriolis effect, 504, 506–507
oceanic crust, 25, 257, 354
oceanic energy producers, 654
oceanic hotspots, 287–289
oceanic plateaux, 319
oceanic transform faults, 281
ocean-ocean convergence, 271, 274–276
octet rule, 71
Ogallala (High Plains) aquifer, 444–445
O horizon, 155
oil. *See* petroleum

oil shales, 613–615
oil traps, 609, 610
Old Faithful geyser, Yellowstone, 449
olivine group, 87
Olympus Mons, 693
ooids, 171
oolitic limestone, 171
Oort, Jan, 708
Oort cloud, 708
OPEC (Organization of the Petroleum Exporting Countries), 620
ophiolite complexes, 206, 218
ophiolites, 279, 319, 590
Ordovician Period, 329
ore deposits, 582. *See also* mineral deposits
ore minerals, 91–92
ores, 569. *See also* mineral deposits
organic activity and soil development, 154–155
organic chemicals, 77
organic processes in weathering, 143
orientation of geologic structures, 302–304
original horizontality, 231, 302
original lateral continuity, 231
orogeny, 315. *See also* mountain building
Oso landslide (Washington State), 390–391
outcrops as forensic arena, 12
outer core, 26, 46, 351, 356
outwash plains, 481
oversteepened slopes, 374, 379, 383
oxbow lakes, 424
oxidation, 147–149
oxidation-reduction (redox) reactions, 147
oxides, 92
oxygen isotopes, 663–665

Pacific realm, 329
Pacific Ring of Fire, 56–57
pad drilling, 619
pahoehoe (ropy lava), 121
Painted Desert, Petrified Forest National Park, 509
paired metamorphic belts, 218
paleocurrents, 181
paleomagnetism, 46–49
paleowinds, 180
Paleozoic Era, 238, 240, 248
Palm Springs, California, 446

Pangea
    Atlantic Ocean and, 329
    coal deposits and, 178
    mountain building and, 159
    running water and, 432
    supercontinent cycle and, 331, 559
    Wegener's hypothesis, 40, 63
Pangea Proxima, 331
parabolic dunes, 520
parabolic troughs, 634
paradigms, 63
parent isotopes, 245, 246
parent rock, 196
partial melting, 102, 317–318
particle size, 175–176
Passamaquoddy Site (Bay of Fundy), 632
passive continental margins, 60, 560–561
passive margins, 263–264
pater noster lakes, 474–475
peat, 177, 623–624
pebbles, 175
pedalfer soils, 156–157
pediment, 509
pedocal soils, 157
pegmatites, 586
Pelée, Mt. (Martinique), 125, 130
percolation test, 461
peridotite, 116, 257–258, 279, 318, 354
period of a wave, 533
periods, 12, 240
permafrost, 387, 392
permeability, 375, 440–441
Perseid shower, 706
petroglyphs, 508, 522
petroleum
    conventional vs. unconventional, 613, 620
    correlation and search for, 614–615
    definitions, 608
    distribution and reserves of, 620–622
    finding and extracting, 610–613
    formation of, 609–610
    hydraulic fracturing (fracking), 617–619
    migration of, 610
    oil refining, 608–609
    oil shales, 613–615
    plate tectonics and, 637–639
    shale gas, 616–617
    tar sands, 615–616

petroleum geologists, 34
petrologists, 34
Phanerozoic Eon, 240
phase change, 355
phenocrysts, 104–105, 110, 116
Phobos, 691
phosphorites, 577
phosphorus (P), 577
photosynthesis
    atmospheric oxygen and, 109
    ecosystem not based on, 267
    food chain and, 152
    oxidation and, 148
    petroleum and, 608, 609
    uranium and, 626
photovoltaics, 634–635
phyllite, 212
physical geology, definition of, 6
physics, basic concepts in, 16–18
phytoplankton, 654
piedmont glaciers, 469
pillow lavas, 123, 267, 279
Pinatubo, Mount (Philippines), 31, 107, 108, 119, 277–278
pinch out, 610
pitchblende, 625
placer deposits, 587–588
plagioclase, 114
plagioclase feldspars, 89, 146
planes, 302
planetary accretion, 21
planetary geologists, 34
planetesimals, 21, 23, 680
planets. *See* Earth; Mars; Solar System and planets; Venus
plaster of Paris, 581
plastic deformation, 300–301
plastic flow, 469
plate boundaries
    about, 261–262
    convergent, 61, 262, 269–280, 590–592, 639
    divergent, 59–60, 262, 263–269, 589–590, 637–639
    failed rifts, 62
    igneous rocks and, 131–133
    transform, 60–61, 262, 280–286, 639
plate tectonics and plates
    about, 58–59
    buoyancy, 258–259
    climate change and, 673–675
    coastlines and, 559–561

deserts and, 523–524
energy resources and, 637–640
geologic time and, 251
glaciation and, 492–496
groundwater and, 462
hotspots, 64–65, 286–292
igneous rocks and, 131–134
isostacy, 259–261
lithosphere, asthenosphere, and crust, 25–26, 256–258
metamorphism and, 217–219
mineral deposits and, 588–592
minerals and, 95
mountain building and, 315–327
moving plates and plate boundaries, 58–64 (*See also* plate boundaries)
paleomagnetism and, 46–49
planetary perspective on, 696–698
plate-driving mechanisms, 292–293
running water and, 432
seafloor spreading and, 50–54
sea level and, 558–559
sedimentary basins and, 190–191
subducted slabs in Earth's interior and, 365–366
subduction, 54–58 (*See also* subduction)
supercontinent cycle, 330–332, 559, 674
as theory, 8, 40
Venus and, 691
weathering, soils, and, 157–159
Wegener's continental drift hypothesis, 40–46
Wilson cycle, 327–330
platy minerals, 208, 211
playa lakes, 509
playas, 509
Pleistocene Epoch, 485
Pleistocene Ice Age
    about, 485
    glacial advances, 485–486
    glacial retreats, 486–492
    Lake Missoula and Channeled Scablands, 490–492
    sea level and, 547–548
plucking, 471–472

plume push, 293
plunge pools, 407
plunging folds, 312
Pluto, 687, 704–705
plutonic rocks, 107, 110
plutonists, 228
plutonium (Pu), 628
plutons, 107, 111–112
pluvial lakes, 486
pocket beaches, 544
point bars, 424
polar deserts, 507
polar high, 507
polar ice caps
    cores of, 660–663
    global warming and
        melting of, 666–668
    on Mars, 497
    Martian, 693
    sea level and, 547–548
polar mineral exploration
    and development, 601
polar molecules, 144, 378
pole wander, 47–48
polflucht ("flight from the
    poles"), 45
polymorphs, 79–81, 95
poorly sorted deposits, 169
pores, 166, 375
pore spaces, 150
pore water pressure, 201
porosity, 166, 375, 440–441
porphyritic texture, 104, 110
porphyry copper
    deposits, 591
Portland cement, 581
positive feedback, 651,
    670–671
potash, 577
potassium (K), 359,
    576–577
potassium-argon dating, 248
potassium feldspars, 89, 146
potential energy, 18
potentiometric (pressure)
    surfaces, 445
potholes, 410
the Precambrian, 240
precession, 668
precious metals, 571
precipitates, 76
precursors, 341
prediction, 7
preferred orientation, 207
pressure
    concept of, 17
    confining, 200, 300
    deformation and, 301
    directed, 200–201
    metamorphism and,
        200–201
    units of, 200

volume change with, 16–17
water pressure at depth, 444
pressure (potentiometric)
    surfaces, 445
pressure-temperature
    gradient, 216
primary (P) waves, 339–344,
    355, 356. *See also*
    seismic waves
principle of catastrophism, 227
principle of faunal
    succession, 234–237
principle of gradualism,
    228–229
principle of original
    horizontality, 231, 302
principle of original lateral
    continuity, 231
principle of superposition,
    230–231, 311
principle of
    uniformitarianism,
    229–230
proglacial lakes, 489
prograde metamorphism, 211
promontories, 325
proppants, 619
Proterozoic Eon, 240
protons, 71, 74, 76, 241
protoplanets, 21, 23, 680
pull-apart basins, 282
pumice, 126
pyrite, 70, 78, 147
pyroclastic deposits, 105–107,
    123–126
pyroclastic flows, 125–126
pyroxene group, 87–88

quartz, 78, 79, 86, 89–90
quartzite, 209, 214
quartz sandstone, 176
quick clays, 385–387
quicksand, 387

radar, 450, 594–596, 690
radial drainage, 417
radiant energy, 18
radiation, 28–29
radiation, solar, 649–651
radioactive decay, 227
radioactivity, 594, 627.
    See also uranium (U)
radiocarbon dating, 243
radiometric ages, 246
radiometric dating, 246–248
radon, 242–243
rainfall-triggered debris
    flows, 388
rain-shadow deserts, 504
raised beaches, 489, 551
rampart craters, 387
random fabric, 207

rapids, 407
rare earth minerals, 93
raw chemicals, 576–578
ray paths, 347–350
recessional moraines, 479
recharge, 441–442
rectangular drainage, 417
recycling, 602
redbeds, 524
redox (oxidation-reduction)
    reactions, 147
Red River, 429
reduction, 147
reefs, 286, 546–549
refraction, 349
regional metamorphism, 204
regoliths, 149
regular satellites, 701
rejuvenation, 407, 413–414
relative age, 227
relative dating, 237–241
relative time, 12
releasing bends, 282
remote sensing, 594–596
renewable energy
    about, 629
    biomass, 636–637
    geothermal, 449–451,
        629–631, 640
    hydroelectric, 607,
        632–634
    plate tectonics and, 640
    solar, 634–635
    tidal power, 631–632
    wave energy, 632
    wind, 635–636
renewable resources,
    606–607
replacement, 588
reserves
    of coal, 624
    of mineral deposits,
        596–598
    of petroleum, 620–622
    of uranium, 626
reservoir rocks, 609, 610
reservoirs, 653
residual deposits, 587
residual soil, 149–150
residue, 102
resources
    geologists and, 7
    groundwater, 438–439
    renewable and
        nonrenewable, 569,
        606–607
    reserves vs., 596
    See also energy; mineral
        deposits
restraining beds, 282, 316
retrograde metamorphism, 215
reverse faults, 306, 307

reverse magnetic polarity,
    49–52
rhyolite, 114, 115
Richter, Charles, 344
Richter magnitude scale,
    344–346
ride-dominated deltas, 427
ridge push, 292, 293
rift basins, 190–191
rifting, 59–60, 263–264,
    289–290, 589–590, 592
rift valleys, 59–60, 62, 191,
    263, 266
right-lateral strike-slip
    faults, 309
rip currents, 537
ripple marks, 182
riprap, 554
road cuts, 374
roche moutonnée, 472
rock avalanches, 382
rock bolts, 396
rock cycle, 14–19, 100–101,
    196–198
rockfalls, 380–382
rock flour, 472
rock-forming minerals,
    81–82. *See also*
    minerals
rocks
    crushed, 580
    definition of, 13
    minerals and, 93
    quarried, 580
    types of, 13
rock salt, 173–174
rockslides, 382–383
Rocky Mountains, 321
ropy lava (*pahoehoe*), 121
rounding, 169
rubble, 175
ruby, 580
rule of cross-cutting
    relationships, 231–233
rule of inclusions, 231
runoff, 404
Rybczynski, Natalia, 648

Sahara Desert, 450
Sahel, 521–523
Saint-Jean-Vianney (Quebec,
    Canada), 387
Salinas, California, 459
salt. *See* halite
saltation, 187, 420–421, 513
salt domes, 174, 577, 578,
    587, 610
San Andreas Fault, 282–283,
    284–285, 340
sand, 175, 580–581
sandstone facies, 186
sandstones, 175–176

San Francisco Bay and Bay Area, 320–321, 340–341
mudflows and, 385
sapphire, 580
SAR (synthetic aperture radar), 594–595
satellite images, 594–596
satellites, regular and irregular, 701
saturation, 170, 376
saturation, zone of, 439
Saturn, 698, 703–704
schist, 212
schistosity, 208
scientific method, 7–9
sea arches, 543
sea caves, 543
sea cliffs, 542–543
seafloor spreading, 50–54, 251, 319
sea level changes
coastal environments and, 188–190
crustal rebound and, 489
plate tectonics and, 558–559
Pleistocene glaciation and, 485, 489
polar ice and, 547–548
stream base levels and, 407
submergent and emergent coastlines and, 550–551
seals, 609, 610
seamounts, 286, 287–289
seashell formation, 584
seasons, 22–23
sea stacks, 543
seawalls, 554
seawater, invasion of, 457–459
Secchi, Angelo, 691
secondary (S) waves, 339–344, 355, 356.
See also seismic waves
sediment and sedimentation
definition of sediment, 164
depositional environments and sedimentary facies, 185–190
formation of sediment, 165
plate tectonics and, 190–191
"the present is the key to the past" and, 10, 185
soil development and, 153
transport of sediment, 167–169, 418–423, 513, 586
See also erosion
sedimentary basins, 190–191
sedimentary facies, 185–190

sedimentary rock
bedding and other structures, 180–185
chemical and biochemical, 166, 170–174, 176–179
classification of, 174–179
definition of, 13
detrital, 166, 167–169, 175–176
formation of, 166
fossil content and identification, 179–180
interpretation of, 179–185
mineral concentration in, 586–588
plate tectonics and, 190–191
in rock cycle, 16
source area, 179
See also dating
sedimentary structures, 180–185
seismic discontinuities, 350
seismic rays, 347–350, 363
seismic reflection, 611
seismic tomography, 361
seismic velocity, 344, 363
seismic waves
elastic rebound theory and, 337
measurement of, 344–346
refraction of, 349
transmission of, 339, 342–344
types of, 337–339
seismometers, 344
semidiurnal tidal pattern, 539
septic tanks, 461
serpentine, 93
settling velocities, 182, 420
shadow zone, 350–351
shale gas, 616–619
shales, 176, 211
shear, 200–201, 261
shear strength, 373
shear stresses, 300
sheeted dikes, 279
shield volcanoes, 121
shock metamorphism, 205–206
Shoemaker-Levy 9 comet, 709–711
shooting stars, 682, 705–706
shorelines. See coastal environments
Siccar Point (Scotland), 233, 313
Sierra Nevada batholith, 112
siesmograms, 344
siesmology, 347
silicate minerals, 81, 87–91, 146

silicate tetrahedron, 87, 146
sillimanite, 95
sills, 109–110, 112, 113
silt, 175
siltstones, 176
silver (Ag), 575
sinkholes, 145, 453
skarns, 588, 591
slab pull, 292, 293
SLAR (side-looking airborne radar), 594–595
slate, 211–212
slaty cleavage, 208
slides, 382–384
slip, 306
slip face, 518
slope stability, 373–375, 395–397
slope-stability maps, 395
slumps, 383–384, 386–387, 390–391
slump toes, 383
smearing, 204–205, 208
smelting, 601
Smith, William, 614
smokers, 590
snowball Earth, 493–495
snowline, 467
soda straw, 453
sodium (Na), 72–73
soil profiles, 155–156
soils
composition of, 150
controls on soil development, 151–155
definition of, 149
lava and chemical enrichment of, 31
nutrients in, 150
plate tectonics and, 157–159
processes and chemistry of, 150–151, 155–156
profiles and horizons, 155–156
residual vs. transported, 149–150
types of, 156–157
solar energy, 634–635
solar nebula theory, 19–21, 680
solar radiation, 649–651
Solar System and planets
asteroids and asteroid belts, 680, 705, 707
comets, 681, 708–711
definitions, 680–681
deserts on, 524–525
Earth-Moon system, origin of, 21, 684–686
exoplanets and other solar systems, 36, 711–712

formation of, 21
glaciers and glaciation, 496–497
Jovian planets, about, 686, 698–699
Jupiter, 698, 699–701
Jupiter's moons, 701–703
Kuiper Belt, 705
Mars, about, 691–696
Mercury, 687–689
meteorites, 356, 680–681, 697, 706–708
meteors, 680, 705–706, 707
Moon, about, 681–684
Neptune, 698, 704
plate tectonics, planetary perspective on, 696–698
Pluto, 704–705
running water on, 433–434
Saturn, 698, 703–704
terrestrial planets, about, 686–688
theories on origins of, 19–26
Uranus, 698, 704
Venus, about, 689–691
See also Earth; Mars; Moon; Sun; Venus
solar thermal devices, 634
solar wind, 21, 658–659, 708
solid phase, 77
solifluction, 392
solution, 76
solution sinkholes, 453
sorting, 169, 423, 586
source area, 167, 179
source rock, 101–102, 165
source rocks for petroleum, 609, 610
specific gravity, 86
spillways, 428
spits, 545
springs, 441–442, 446, 509
spring tides, 539
stalactites, 453
stalagmites, 453
star dunes, 520
Steno, Nicholas, 78, 230
Steno's rules, 230–231
St. Helens, Mount (Washington State), 107, 125, 379, 388, 398
stick-slip motion, 280, 337
stishovite, 206
stocks, 107, 111–112
Stokes' Law, 182
stony meteorites, 356, 707
storm surges, 553–554
storm tides, 553
strain, 299

strain rate, 205, 301
Strait of Gibraltar, 170
strands, 544
stratified drift, 481
stratiform deposits, 589
stratigraphic columns, 235, 238
stratigraphic traps, 610
stratigraphy, 238, 456
streak in minerals, 82–84
stream deposition
    about, 423
    alluvial fans, 427–428
    channel deposits, 423–425
    deltas, 426–427
    in deserts, 509
    floodplain deposits, 425
stream discharge, 406
stream-dominated deltas, 427
stream erosion
    antecedent and super-posed streams, 414
    Aral Sea disaster, 419–420
    channels, 409–410
    in deserts, 509
    drainage basins, systems, and patterns, 414–418
    incised meanders, 414
    stream piracy, 411
    terraces, 411–414
    valley profiles, 411
streamflow, 404–407
stream piracy (capture), 411
stream profiles, 430
streams (running water)
    about, 402
    cross beds in, 180–181
    dams and human exploitation, 430–431
    definition of, 403
    deposit sorting in, 169
    flooding, 187, 428–432
    hydrologic cycle and movement of water, 402–404
    intermittent, 509
    on other worlds, 433–434
    plate tectonics and, 432
    saltation and deposition in, 187
    trunk streams, tributaries, and distributaries, 415
stream transport
    bed load, traction, and saltation, 420–421
    capacity and competence, 422
    dissolved load, 421–422
    mineral concentration and, 586
    suspended load, 418–420

stress, 299–300
striations, 169, 471
strike, 302–304
strike-slip faults, 307–309, 326
strip mining, 623
structural traps, 610
subbituminous coal, 624
subduction
    continental growth and, 278–279
    convergent plate boundaries and, 269–271
    deep ocean trenches and, 54–55
    definition of, 55
    fate of subducted slabs, 365–366
    mineral deposits and, 590
    mountain building by, 316–319
    polarity of, 276
    sea level and, 559
    tsunamis and, 272–274
    See also convergent plate boundaries
subduction erosion, 319
subduction zones, 55–58, 218, 269–280
submarine fans, 182
submergent coastlines, 550–551
subsidence, 457
substance, 76
subsurface geology, 614
subtropical deserts, 502–504
subtropical high, 504
suites, 118
sulfides, 92, 590, 592
sulfur (S), 577–578
sulfur dioxide emissions, 623
sulfuric acid, 577
Sun
    characteristics of, 657
    Earth-Sun geometry, 659, 668–670
    solar nebula theory, 19–21
    tides and, 539–540
    variations in solar output, 657–659
    See also entries at solar
sunspot cycles, 658, 676
sunspots, 657–658
supercontinent cycle, 330–332, 559, 674
supergene enrichment, 591
supergiants, 620
supernova, 21
supernova explosions, 74–75
superposed streams, 414
superposition, 230–231, 311
surface tension, 376, 378–379

surface water, 404
surface waves, 339, 532
surf zone, 535
surges, 470
suspect terranes, 321
suspended load, 418–420
Susquehanna River, 414
suture, 325
swarms, 112
swash, 535
swash zone, 535
swells, 532
sylvite, 577
symmetric folds, 311
synchronous rotation, 681
synclines, 311, 462
Syr Darya River, 419

talus, 380
tarns, 474
tar sands, 615–616
tectonic escape, 325
tectonic joints, 304
temperature
    air temperature in the lower atmosphere, 468
    crystallization of magma and, 104
    definition of, 18
    deformation and, 301
    in Earth's interior, 361
    metamorphism and, 199
    on other planets, 652, 688, 692, 696, 699, 704
    pressure-temperature gradient, 216
    thermometers, 659
    of water at depth, 444
Temple of Jupiter Serapis, 229
tensional stress, 300
tephra, 123
terminal moraines, 479
terminal velocity, 182
terraces, 411–413
terrane accretion, 319–321
terrestrial planets, about, 686–687. See also Solar System and planets
Tethys Ocean, 322, 325, 327, 432
texture, 166–169, 174–175, 197, 207–209
theory and hypothesis, 8
Theory of Earth, The (Hutton), 9, 228
thermal convection, 359
thermal effects, 112–113
thermal energy, 18
thermal expansion and contraction, 141–143
thermal subsidence basins, 191
thermometers, 659

thermonuclear fusion, 74–75
thorium, 359
Three Gorges Dam (China), 329, 431, 633
thrusts, 307
Tibetan Plateau, 325, 327, 328
tidal currents, 540
tidal flats, 188
tidal power, 631–632
tidal range, 632
tides, 537–540
tidewater glaciers, 469
till, 478
tillite, 493
tilt of Earth, 22–23
time
    chemical weathering and, 149
    deformation and, 301
    metamorphism and, 203
    soil development and, 153
time, geologic
    about, 10
    absolute dating, 241–250
    age of Earth, quest for, 249
    correlation and the principle of faunal succession, 234–237
    geologic time scale, 238–241, 248–250
    half-life and, 245–246
    Hutton and principle of gradualism, 228–229
    ice ages and climate over time, 493
    Lyell and principle of uniformitarianism, 229–230
    neptunists vs. plutonists, 228
    plate tectonics and, 251
    "the present is the key to the past," 10, 33, 185, 230
    radioactivity and radioactive decay, 227, 241–245
    radiometric dating, 246–248
    reasons for studying, 11
    reconstructing geologic history, 237–238
    relative dating, 237–241
    relative vs. absolute age, 226–227
    relative vs. absolute time, 12
    rule of cross-cutting relationships, 231–233
    rule of inclusions, 231
    sense of time, 250–251

Steno's rules (superposition, original horizontality, and original lateral continuity), 230–231
stratigraphy and the stratigraphic column, 238
two-pronged approach to, 11–12
unconformity, 233–234
Ussher and the principle of catastrophism, 227
tin (Sn), 587
titanium (Ti), 570
tombolos, 545
topography and soil development, 154, 158
topset beds, 426
topsoil, 155
traction, 420–421
transform boundaries
about, 60–61, 262, 280–281
continental transforms, 281–286
mountain building along, 316
oceanic transforms, 281
petroleum and, 639
transform continental margins, 561
transform faults, 60–61, 309
transition zone, 355
transpiration, 404
transportation
definition of, 16
of magma, 102–103
of sediment, 167–169, 418–423, 586
stream transport, 418–423, 586
by wind, 513
transported soil, 149–150
transverse dunes, 518–519
transverse ridges, 282–283, 316
travertine, 170
trellis drainage, 417
Triassic Period, 524
tributaries, 415
triple points, 264, 289–290
truncated spurs, 475
trunk glaciers, 475
trunk streams, 415
tsunamis, 272–274
tuffs, 124, 126
turbidity currents, 176
turbulent flow, 404–405
Turtle Mountain rockslide (Alberta, Canada), 382
turtles, green, 53–54

ultimate base level, 406
ultramafic igneous rocks, 113
ultraviolet (UV) solar radiation, 649
unconfined aquifers, 443–445
unconformities, 233–234, 313–315, 548, 610
unconsolidated slopes, 374
uniformitarianism, principle of, 229–230
universe, origin of, 20
upper mantle, 257–258, 355
upper plate, 269
uraninite, 625
uranium (U), 242–243, 245–246, 359, 625–629
uranium-lead dating, 246–248
Uranus, 698, 704
U-shaped valleys, 474
Ussher, James, 227, 249
Uzbekistan, 419

Vail, Peter, 548
Valles Marineris, 693
valley glaciers, 469
valleys
hanging, 476
stream valleys, 410–414
U-shaped, 474
V-shaped, 373, 409, 411
van der Waals force, 74
varves, 484
vegetation and mass wasting, 376
veins, 586
velocity
concept of, 17
seismic, 344, 363
settling velocities, 182, 420
of streamflow, 405–406, 420
Venice, Italy, 457
vent eruptions, 121
ventifacts, 514, 517, 525
vents, 105
Venus
about, 689–691
atmospheric temperature of, 652
plate tectonics and Earth vs., 696–697
size of, 686
water on, 433
vesicles, 126
Vesuvius, Mt. (Italy), 128–129, 130, 449
Vine, Fred, 50, 51–52
Vine-Matthews-Morley hypothesis, 50
viscosity, 77, 120–121, 420

visible spectrum, 649
volatile compounds, 108
volatiles, 102
volcanic arcs, 132–133, 271
volcanic ash, 124–125, 376, 379
volcanic breccias, 124
volcanic eruptions
Earth as system and, 30–31
Eyjafjallajökull (Iceland), 107
fissure vs. vent eruptions, 121
of intermediate and felsic magma, 123–129
of mafic magma, 121–123
mass wasting and, 376, 379, 398
Mount Pinatubo (Philippines), 31, 107, 108, 119, 130, 277–278
Mount St. Helens (Washington State), 107, 125, 379, 388, 398
Mt. Pelée (Martinique), 125, 130
Mt. Vesuvius (Italy), 128–129, 130
prediction of explosive eruptions, 130
pyroclastic deposits, 105–107
volcanic glass, 104
volcanism
atmospheres and, 686–687
on Mars, 691, 696
mid-ocean ridges and, 267, 278
on Moon, 682–683
Pacific Ring of Fire, 56–57
plate boundaries and, 61
soils and, 157–158
subduction and, 56–58
in subduction zones, 277–278
on Venus, 691
See also hotspots
volcanoes
cinder cones, 121, 127
composite, 121, 127
definition of, 101
shield, 121
volcanologists, 34
volume, 16–17, 344
V-shaped valleys, 373, 409, 411

wadis, 509
Wallops Beach (Virginia), 556
wall rock, 102

Wasatch Mountain (Utah), 388
waste-to-energy installations, 636
water, running. See streams
waterfalls, 407
water in deserts, 508–511
water on planets or Moon, 433–434, 684, 688–689, 694–696
water pressure at depth, 444
water table, 439–442
wave base, 534
wave-cut platforms, 542–543
wave-dominated deltas, 427
wave energy, 632
wave height, 532
wavelength, 532
wave refraction, 349, 535–536
waves
erosion by, 540–543
longshore drift and rip currents, 536–537
in open ocean, 532, 534
shape and motion of, 532, 533
shorelines and, 534–535
surface waves, 532
wind-driven, 532–537
See also seismic waves
wave velocity, 532
weathering
chemical, 138–139, 143–149
crystal wedging, 141
definition of, 15–16, 138
in deserts, 508
dissolution, 144–145
frost wedging, 141
human activities, 143
hydration, climate, and time, 149
hydrolysis, 145–146
joints, 139–141, 157
mechanical, 138, 139–143
mineral concentration and, 586
minerals and differential resistance to, 451–452
organic processes, 143
oxidation, 147–149
plate tectonics and, 157–159
sediment formation and, 165
stream channels and, 410
thermal expansion and contraction, 141–143
Wegener, Alfred, 40–46, 48, 63, 364, 493, 496, 559–560, 639
weight, 16, 76

welded tuff, 126
well logging, 614
Wells, H. G., 691
well sorted deposits, 169
white asbestos (chrysotile), 93
white smokers, 268
whole mantle convection model, 360
Wilson, J. Tuzo, 43, 286, 327
Wilson cycle, 327–330
windblown deposits, 169, 187
wind gaps, 411

wind in deserts
    deposition features and dunes, 517–521
    Dust Bowl, 514, 515–516
    dust devils, 511, 512
    erosion features, 514–517
    sediment transport by, 513
wind power, 635–636
wind shadows, 517
Wrangellia, 321

xenocrysts, 113
xenoliths, 113
X-rays, 361, 456

Yangtze River, 328, 329, 431
Yellowstone hotspot, 291–292
Yellowstone National Park, 291–292, 449
Yosemite National Park, 381–382, 473, 474, 475

zeolite facies, 216
zigzag outcrop patterns, 312
zinc (Zn), 573–574
Zion National Park, 180, 181, 238, 240
zircon, 246–248
zone of ablation, 470
zone of accumulation, 155, 470
zone of aeration, 439–440
zone of leaching, 155
zone of saturation, 439
zooplankton, 654